Recommended Dietary Allowances (RDA) and Adequate Intakes (AI) for Vitamins

Age (yr)	Thiamin (mg/day) RDA	Riboflavin (mg/day) RDA	Niacin (mg/day)[a] RDA	Biotin (μg/day) AI	Pantothenic acid (mg/day) AI	Vitamin B6 (mg/day) RDA	Folate (μg/day)[b] RDA	Vitamin B12 (μg/day) RDA	Choline (mg/day) AI	Vitamin C (mg/day) RDA	Vitamin A (μg/day)[c] RDA	Vitamin D (IU/day)[d] RDA	Vitamin E (mg/day)[e] RDA	Vitamin K (μg/day) AI
Infants														
0–0.5	0.2	0.3	2	5	1.7	0.1	65	0.4	125	40	400	400 (10 μg)	4	2.0
0.5–1	0.3	0.4	4	6	1.8	0.3	80	0.5	150	50	500	400 (10 μg)	5	2.5
Children														
1–3	0.5	0.5	6	8	2	0.5	150	0.9	200	15	300	600 (15 μg)	6	30
4–8	0.6	0.6	8	12	3	0.6	200	1.2	250	25	400	600 (15 μg)	7	55
Males														
9–13	0.9	0.9	12	20	4	1.0	300	1.8	375	45	600	600 (15 μg)	11	60
14–18	1.2	1.3	16	25	5	1.3	400	2.4	550	75	900	600 (15 μg)	15	75
19–30	1.2	1.3	16	30	5	1.3	400	2.4	550	90	900	600 (15 μg)	15	120
31–50	1.2	1.3	16	30	5	1.3	400	2.4	550	90	900	600 (15 μg)	15	120
51–70	1.2	1.3	16	30	5	1.7	400	2.4	550	90	900	800 (20 μg)	15	120
>70	1.2	1.3	16	30	5	1.7	400	2.4	550	90	900	800 (20 μg)	15	120
Females														
9–13	0.9	0.9	12	20	4	1.0	300	1.8	375	45	600	600 (15 μg)	11	60
14–18	1.0	1.0	14	25	5	1.2	400	2.4	400	65	700	600 (15 μg)	15	75
19–30	1.1	1.1	14	30	5	1.3	400	2.4	425	75	700	600 (15 μg)	15	90
31–50	1.1	1.1	14	30	5	1.3	400	2.4	425	75	700	600 (15 μg)	15	90
51–70	1.1	1.1	14	30	5	1.5	400	2.4	425	75	700	800 (20 μg)	15	90
>70	1.1	1.1	14	30	5	1.5	400	2.4	425	75	700	800 (20 μg)	15	90
Pregnancy														
≤18	1.4	1.4	18	30	6	1.9	600	2.6	450	80	750	600 (15 μg)	15	75
19–30	1.4	1.4	18	30	6	1.9	600	2.6	450	85	770	600 (15 μg)	15	90
31–50	1.4	1.4	18	30	6	1.9	600	2.6	450	85	770	600 (15 μg)	15	90
Lactation														
≤18	1.4	1.6	17	35	7	2.0	500	2.8	550	115	1200	600 (15 μg)	19	75
19–30	1.4	1.6	17	35	7	2.0	500	2.8	550	120	1300	600 (15 μg)	19	90
31–50	1.4	1.6	17	35	7	2.0	500	2.8	550	120	1300	600 (15 μg)	19	90

NOTE: For all nutrients, values for infants are AI. The glossary on the inside back cover defines units of nutrient measure.

[a]Niacin recommendations are expressed as niacin equivalents (NE), except for recommendations for infants younger than 6 months, which are expressed as preformed niacin.

[b]Folate recommendations are expressed as dietary folate equivalents (DFE).

[c]Vitamin A recommendations are expressed as retinol activity equivalents (RAE).

[d]Vitamin D recommendations are expressed as cholecalciferol and assume an absence of adequate exposure to sunlight.

[e]Vitamin E recommendations are expressed as α-tocopherol.

Recommended Dietary Allowances (RDA) and Adequate Intakes (AI) for Minerals

Age (yr)	Sodium (mg/day) AI	Chloride (mg/day) AI	Potassium (mg/day) AI	Calcium (mg/day) RDA	Phosphorus (mg/day) RDA	Magnesium (mg/day) RDA	Iron (mg/day) RDA	Zinc (mg/day) RDA	Iodine (μg/day) RDA	Selenium (μg/day) RDA	Copper (μg/day) RDA	Manganese (mg/day) AI	Fluoride (mg/day) AI	Chromium (μg/day) AI	Molybdenum (μg/day) RDA
Infants															
0–0.5	120	180	400	200	100	30	0.27	2	110	15	200	0.003	0.01	0.2	2
0.5–1	370	570	700	260	275	75	11	3	130	20	220	0.6	0.5	5.5	3
Children															
1–3	1000	1500	3000	700	460	80	7	3	90	20	340	1.2	0.7	11	17
4–8	1200	1900	3800	1000	500	130	10	5	90	30	440	1.5	1.0	15	22
Males															
9–13	1500	2300	4500	1300	1250	240	8	8	120	40	700	1.9	2	25	34
14–18	1500	2300	4700	1300	1250	410	11	11	150	55	890	2.2	3	35	43
19–30	1500	2300	4700	1000	700	400	8	11	150	55	900	2.3	4	35	45
31–50	1500	2300	4700	1000	700	420	8	11	150	55	900	2.3	4	35	45
51–70	1300	2000	4700	1000	700	420	8	11	150	55	900	2.3	4	30	45
>70	1200	1800	4700	1200	700	420	8	11	150	55	900	2.3	4	30	45
Females															
9–13	1500	2300	4500	1300	1250	240	8	8	120	40	700	1.6	2	21	34
14–18	1500	2300	4700	1300	1250	360	15	9	150	55	890	1.6	3	24	43
19–30	1500	2300	4700	1000	700	310	18	8	150	55	900	1.8	3	25	45
31–50	1500	2300	4700	1000	700	320	18	8	150	55	900	1.8	3	25	45
51–70	1300	2000	4700	1200	700	320	8	8	150	55	900	1.8	3	20	45
>70	1200	1800	4700	1200	700	320	8	8	150	55	900	1.8	3	20	45
Pregnancy															
≤18	1500	2300	4700	1300	1250	400	27	12	220	60	1000	2.0	3	29	50
19–30	1500	2300	4700	1000	700	350	27	11	220	60	1000	2.0	3	30	50
31–50	1500	2300	4700	1000	700	360	27	11	220	60	1000	2.0	3	30	50
Lactation															
≤18	1500	2300	5100	1300	1250	360	10	13	290	70	1300	2.6	3	44	50
19–30	1500	2300	5100	1000	700	310	9	12	290	70	1300	2.6	3	45	50
31–50	1500	2300	5100	1000	700	320	9	12	290	70	1300	2.6	3	45	50

NOTE: For all nutrients, values for infants are AI. The glossary on the inside back cover defines units of nutrient measure.

Tolerable Upper Intake Levels (UL) for Vitamins

Age (yr)	Niacin (mg/day)[a]	Vitamin B6 (mg/day)	Folate (μg/day)[a]	Choline (mg/day)	Vitamin C (mg/day)	Vitamin A (IU/day)[b]	Vitamin D (IU/day)	Vitamin E (mg/day)[c]
Infants								
0–0.5	—	—	—	—	—	600	1000 (25 μg)	—
0.5–1	—	—	—	—	—	600	1500 (38 μg)	—
Children								
1–3	10	30	300	1000	400	600	2500 (63 μg)	200
4–8	15	40	400	1000	650	900	3000 (75 μg)	300
9–13	20	60	600	2000	1200	1700	4000 (100 μg)	600
Adolescents								
14–18	30	80	800	3000	1800	2800	4000 (100 μg)	800
Adults								
19–70	35	100	1000	3500	2000	3000	4000 (100 μg)	1000
>70	35	100	1000	3500	2000	3000	4000 (100 μg)	1000
Pregnancy								
≤18	30	80	800	3000	1800	2800	4000 (100 μg)	800
19–50	35	100	1000	3500	2000	3000	4000 (100 μg)	1000
Lactation								
≤18	30	80	800	3000	1800	2800	4000 (100 μg)	800
19–50	35	100	1000	3500	2000	3000	4000 (100 μg)	1000

[a] The UL for niacin and folate apply to synthetic forms obtained from supplements, fortified foods, or a combination of the two.

[b] The UL for vitamin A applies to the preformed vitamin only.

[c] The UL for vitamin E applies to any form of supplemental α-tocopherol, fortified foods, or a combination of the two.

Tolerable Upper Intake Levels (UL) for Minerals

Age (yr)	Sodium (mg/day)	Chloride (mg/day)	Calcium (mg/day)	Phosphorus (mg/day)	Magnesium (mg/day)[d]	Iron (mg/day)	Zinc (mg/day)	Iodine (μg/day)	Selenium (μg/day)	Copper (μg/day)	Manganese (mg/day)	Fluoride (mg/day)	Molybdenum (μg/day)	Boron (mg/day)	Nickel (mg/day)	Vanadium (mg/day)
Infants																
0–0.5	—	—	1000	—	—	40	4	—	45	—	—	0.7	—	—	—	—
0.5–1	—	—	1500	—	—	40	5	—	60	—	—	0.9	—	—	—	—
Children																
1–3	1500	2300	2500	3000	65	40	7	200	90	1000	2	1.3	300	3	0.2	—
4–8	1900	2900	2500	3000	110	40	12	300	150	3000	3	2.2	600	6	0.3	—
9–13	2200	3400	3000	4000	350	40	23	600	280	5000	6	10	1100	11	0.6	—
Adolescents																
14–18	2300	3600	3000	4000	350	45	34	900	400	8000	9	10	1700	17	1.0	—
Adults																
19–50	2300	3600	2500	4000	350	45	40	1100	400	10,000	11	10	2000	20	1.0	1.8
51–70	2300	3600	2000	4000	350	45	40	1100	400	10,000	11	10	2000	20	1.0	1.8
>70	2300	3600	2000	3000	350	45	40	1100	400	10,000	11	10	2000	20	1.0	1.8
Pregnancy																
≤18	2300	3600	3000	3500	350	45	34	900	400	8000	9	10	1700	17	1.0	—
19–50	2300	3600	2500	3500	350	45	40	1100	400	10,000	11	10	2000	20	1.0	—
Lactation																
≤18	2300	3600	3000	4000	350	45	34	900	400	8000	9	10	1700	17	1.0	—
19–50	2300	3600	2500	4000	350	45	40	1100	400	10,000	11	10	2000	20	1.0	—

[d] The UL for magnesium applies to synthetic forms obtained from supplements or drugs only.

NOTE: An upper Limit was not established for vitamins and minerals not listed and for those age groups listed with a dash (—) because of a lack of data, not because these nutrients are safe to consume at any level of intake. All nutrients can have adverse effects when intakes are excessive.

SOURCE: Adapted with permission from the *Dietary Reference Intakes series*, National Academies Press. Copyright 1997, 1998, 2000, 2001, 2002, 2005, 2011 by the National Academies of Sciences.

Nutrition Now

About the Author

JUDITH E. BROWN is Professor Emerita of Nutrition at the School of Public Health, and of the Department of Obstetrics and Gynecology at the University of Minnesota. She received her Ph.D. in human nutrition from Florida State University and her M.P.H. in public health nutrition from the University of Michigan. Dr. Brown has received competitively funded research grants from the National Institutes of Health, the Centers for Disease Control and Prevention, and the Maternal and Child Health Bureau and has over 100 publications in the scientific literature including the New England Journal of Medicine, the Journal of the American Medical Association, and the Journal of the American Dietetic Association. A recipient of the Agnes Higgins Award in Maternal Nutrition from the March of Dimes, Dr. Brown is a registered dietitian and the successful author of Everywoman's Guide to Nutrition, Nutrition for Your Pregnancy, and What to Eat Before, During, and After Pregnancy.

Courtesy of the author

SEVENTH EDITION

Nutrition Now

Judith E. Brown

University of Minnesota

WADSWORTH
CENGAGE Learning·

Australia • Brazil • Japan • Korea • Mexico • Singapore • Spain • United Kingdom • United States

WADSWORTH
CENGAGE Learning®

Nutrition Now, **Seventh Edition**
Judith E. Brown

Publisher: Yolanda Cossio

Acquisitions Editor: Peggy Williams

Developmental Editor: Nedah Rose

Assistant Editor: Elesha Feldman

Editorial Assistant: Sean Cronin

Media Editor: Miriam Myers

Market Development Manager: Tom Ziolkowski

Content Project Manager: Carol Samet

Art Director: John Walker

Manufacturing Planner: Karen Hunt

Rights Acquisitions Specialist: Tom McDonough

Production Service and Compositor: Lachina Publishing Services

Photo Researcher: Q2A/Bill Smith

Text Researcher: Pablo D'Stair

Text and Cover Designer: Riezebos Holzbaur/ Andrei Pasternak

Cover Image: © Jon Larson / photodisc / Getty Images

For product information and technology assistance, contact us at
Cengage Learning Customer & Sales Support, 1-800-354-9706.

For permission to use material from this text or product, submit all requests online at **www.cengage.com/permissions**. Further permissions questions can be e-mailed to **permissionrequest@cengage.com**.

Library of Congress Control Number: 2012943422

ISBN-13: 978-1-133-93653-4

ISBN-10: 1-133-93653-9

Wadsworth
20 Davis Drive
Belmont, CA 94002-3098
USA

Cengage Learning is a leading provider of customized learning solutions with office locations around the globe, including Singapore, the United Kingdom, Australia, Mexico, Brazil, and Japan. Locate your local office at **www.cengage.com/global**.

Cengage Learning products are represented in Canada by Nelson Education, Ltd.

To learn more about Wadsworth, visit **www.cengage.com/Wadsworth**

Purchase any of our products at your local college store or at our preferred online store **www.CengageBrain.com**.

Printed in United States of America
1 2 3 4 5 6 7 16 15 14 13 12

Contents in Brief

Contents

nutrition timeline

1621
First Thanksgiving feast at Plymouth colony

1702
First coffeehouse in America opens in Philadelphia

1734
Scurvy recognized

1744
First record of ice cream in America at Maryland colony

Alistair Berg/Getty Images

Lind publishes "Treatise on Scurvy," citrus identified as cure

PhotoDisc

Ojibway and Sioux war over control of wild rice stands

Sandwich invented by the Earl of Sandwich

PhotoDisc

Potato heralded as famine food

Americans drink more coffee in protest over Britain's tea tax

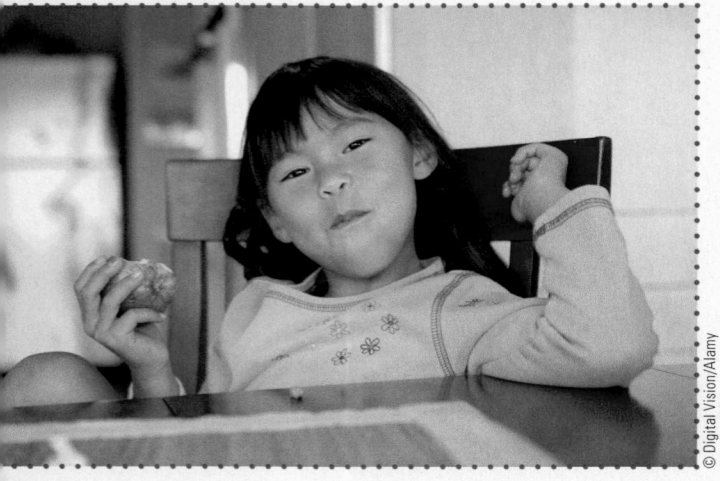
© Digital Vision/Alamy

nutrition timeline

1775
Lavoisier ("the father of the science of nutrition") discovers the energy- producing property of food

Stefano Bianchetti/CORBIS

1816
Protein and amino acids identified, followed by carbohydrates and fats in the mid-1800s

1833
Beaumont's experiments on a wounded man's stomach greatly expand knowledge about digestion

© Bettmann/Corbis

1862
U.S. Department of Agriculture founded by authorization of President Lincoln

1871
Proteins, carbohydrates, and fats determined to be insufficient to support life; there are other "essential" components

© Scott Goodwin Photography

First milk station providing children
with un-contaminated milk opens in
New York City
Bettmann/CORBIS

Atwater publishes Proximate
Composition of Food Materials

Pure Food and Drug Act passed
by President Theodore Roosevelt
to protect consumers against
contaminated foods
Bettmann/CORBIS

Pasteurized milk introduced
PhotoDisc

Funk suggests scurvy, beriberi, and
pellagra caused by deficiency of
"vitamines" in the diet

Elena Schweitzer/Shutterstock.com

nutrition timeline

1913
First vitamin discovered (vitamin A)

PhotoDisc

1914
Goldberger identifies the cause of pellagra (niacin deficiency) in poor children to be a missing component of the diet rather than a germ as others believed

1916
First dietary guidance material produced for the public released; title is "Food for Young Children"

1917
First food groups published the Five Food Groups: Milk and Meat; Vegetables and Fruits; Cereals; Fats and Fat Foods; Sugars and Sugary Foods

1921
First fortified food produced: iodized salt, needed to prevent widespread iodine-deficiency goiter in many parts of the United States

© Scott Goodwin Photography

1928
American Society for Nutritional Sciences and the Journal of Nutrition founded

1929
Essential fatty acids identified

PhotoDisc

1930s
Vitamin C identified in 1932, followed by pantothenic acid and riboflavin in 1933 and vitamin K in 1934

PhotoDisc

1937
Pellagra found to be due to a deficiency of niacin

1938
Health Canada issues nutrient intake standards

1941
First refined grain enrichment standards developed

Creatas/Fotosearch

nutrition timeline

1941
First Recommended Dietary Allowances (RDAs) announced by President Franklin Roosevelt on radio

Franklin D. Roosevelt Presidential Library and Museum

1946
National School Lunch Act passed

PhotoDisc

1947
Vitamin B$_{12}$ identified

1953
Double helix structure of DNA discovered

PhotoDisc

1956
Basic Four Food Groups released by the U.S. Department of Agriculture

photoshut/Shutterstock.com

1965
Food Stamp Act passed, Food Stamp program established

1966
Child Nutrition Act adds school breakfast to the National School Lunch Program

1968
First national nutrition survey in U.S. launched (The Ten State Nutrition Survey)

1970
First Canadian national nutrition survey launched (Nutrition Canada National Survey)

1972
Special Supplemental Food and Nutrition Program for Women, Infants, and Children (WIC) established

PhotoDisc

© Digital Vision/Getty Images

nutrition timeline

1977
Dietary Goals for the U.S. issued

1978
First Health Objectives for the Nation released

1989
First national scientific consensus report on diet and chronic disease published

1992
The Food Guide pyramid is released by the USDA

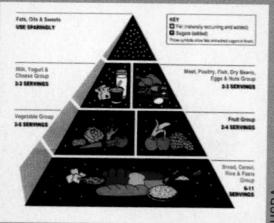
USDA

1997
RDAs expanded to Dietary Reference Intakes (DRIs)

living the dream/Flickr/Getty Images

1998
Folic acid fortification of refined
grain products begins

Richard Anderson

2003
Sequencing of DNA in the human
genome completed; marks
beginning of new era of research
in nutrient–gene interactions

2009
Obesity and diabetes become
global epidemics

Jose Luis Pelaez Inc/Getty Images/Blend Images

PhotoDisc

Preface

"Everything should be made as simple as possible. But not simpler."
—ALBERT EINSTEIN

Welcome to the new *Nutrition Now*. It is reborn in the seventh edition. Take a look. In this edition of *Nutrition Now* you will see a crisp design, instructive photos, tables, and illustrations—and uncluttered margins. You will find content that is rewritten and reorganized around learning objectives and additional interactive learning opportunities. After Unit 1, if you read the first section of a unit you will find that specific, key nutrition concepts that apply to the unit's content are listed. You will notice that ChooseMyPlate food guidance materials are featured in this edition, as well as the 2010 Dietary Guidelines for Americans. The recently released DRIs for vitamin D and calcium appear in this edition, as well as the 2020 Health Objectives for the Nation. Multiple-choice questions based on real-life case scenarios are added to the Review Question sections. Additional opportunities for interactive learning, and activities that focus on real-life situations that require decision-making skills, are expanded in this edition. Revised, interactive, and real-life-based problem-solving activities are added to the seventh edition's *Instructor's Activity Manual*.

The principles of the science of nutrition as presented in this text have not changed, but much else has. Knowledge gains in nutrition continue to advance, and the implications of the advances to human health and well-being are impressive. Growth in our understanding of the role of plant foods on health has changed dietary guidance. New information about dietary fats and health is changing recommendations for fat intake. Advances in knowledge about the health effects of nutrient-gene interactions are changing fundamental concepts about the origins and prevention of disease. The epidemics of obesity and type 2 diabetes are reaching into younger age groups, and much more public attention is being paid to the roles of the food environment, diet, and physical activity in their prevention and management. The seventh edition covers these and other emerging topics directly related to the state of the science of nutrition now. It attempts to introduce these topics to students in a straightforward and clear way that keeps coverage as simple as possible, but no simpler than that.

Nutrition Now continues to be oriented toward enhancing instructor's teaching experiences and helping students build a firm foundation of scientific knowledge about nutrition that will serve them well throughout their careers and life.

Pedagogical Features

Units now begin with learning objectives. Unit content and Review Questions at the end of units are organized by learning objectives.

Also new to this edition are the Activity Cards at the back of the book. Each card contains an activity and its worksheet. Activities can be assigned individually or as group or special projects. They are designed to involve students in interactive application of topics to new situations. Activities include taste testing to identify genetically determined sensitivity to bitterness, developing a dietary behavioral change plan, anthropometry lab, designing fraudulent nutrition products, a physical activity assessment, and an assessment of three days of dietary intake.

Features widely utilized in past editions are updated and expanded in this one:

- Nutrition Scoreboard, a feature that pre-tests student knowledge and understanding of specific nutrition-related topics, has new questions that correspond to updated content in the Units. Answers for the Nutrition Scoreboard are moved to the end of units.

- The Reality Check feature is now included in every unit after Unit 1.

- Take Action, a feature that encourages students to consider options for diet or physical activity improvements, is included in most Units.
- Health Action, a feature that directs student's attention to diet, physical activity, and health recommendations, is updated and revised throughout.
- Nutrition Up Close activities have been revised in a number of units to correspond to updated and revised nutrition content. Answers to the Nutrition Up Close activities are now included in Appendix G.
- Review Questions are now organized around learning objectives and include multiple choice questions based on real-life scenarios. Answers to the Review Questions now appear in Appendix G.
- The Glossary is expanded and includes 46 mainly new definitions that correspond to the revised content of the units.

Several unit titles have been changed to reflect updated content. The term *genetically modified foods* is deleted from the title of Unit 21 and its contents. Unit 24 now emphasizes dietary supplements while content on functional foods is abbreviated. Prebiotics and probiotics are still covered in this unit. Primary coverage of lactose intolerance is now located in the unit on carbohydrates.

Changes to the Seventh Edition

To keep the list of changes appearing in the seventh edition to a reasonable length, I have attempted to select and list the substantial changes.

Unit 1: Key Nutrition Concepts and Terms

- The latest DRI report, released in 2010 with updated calcium and vitamin D intake recommendations, is presented
- Dietary Reference Intakes (DRIs) section is rewritten and reorganized
- New student activities encourage students to use the DRI tables
- ChooseMyPlate.gov content replaces MyPyramid
- A number of the illustrations have been modified or updated

Unit 2: The Inside Story about Nutrition and Health

- Tables and figures are updated and revised
- Ethnic-group specific data added
- ChooseMyPlate content replaces MyPyramid materials
- Nutrition Up Close feature modified from MyPyramid to ChooseMyPlate

Unit 3: Ways of Knowing about Nutrition

- New illustrations replace many from the sixth edition
- The FDA's Food Safety Modernization Act of 2011 Information section added
- Reality Check scenario modified

Unit 4: Understanding Food and Nutrition Labels

- Replaced a number of Illustrations with updated ones
- Added content on new calorie labeling requirements for chain restaurants
- Added discussion of new food labeling standards proposals

Unit 5: Nutrition, Attitudes, and Behaviors

- Revised illustrations
- Updated content on trends in food consumption

Unit 6: Healthy Diets, Dietary Guidelines, ChooseMyPlate and More

- Updated illustrations and tables
- Replaced content on the 2005 Dietary Guidelines with information from the 2010 Dietary Guidelines
- Replaced content on MyPyramid with ChooseMyPlate materials
- Updated and condensed information on dietary guidelines from other countries
- Modified the Health Action feature to focus on the 2010 Dietary Guidelines

Unit 7: How the Body Uses Food: Digestion and Absorption

- Changed Nutrition Scoreboard questions and answers
- Expanded content on the development and practical meaning of individual food preferences
- Expanded content on absorption, including the absorption of alcohol
- Modified content of digestive processes tables
- Expanded content on the lymph and circulatory systems
- Added sections titled "Other functions of the gastrointestinal tract," "Functions of taste," and "Gut bacteria"
- Revised content or replaced tables
- Revised content on ulcers, heartburn, constipation, and irritable bowel syndrome
- Added a Reality Check
- Moved content on lactose intolerance to the unit on carbohydrates
- Modified the Nutrition Up Close activity

Unit 8: Calories! Food, Energy, and Energy Balance

- Added a section on basal metabolism
- Added photo of indirect calorimeter plus added brief discussion on indirect calorimetry

Unit 9: Obesity to Underweight: The Highs and Lows of Weight Status

- Modified presentation of culturally-defined ideal body size for women and men
- Updated definitions of overweight and obesity in children and adolescents and added new example of CDC's BMI charts for age and sex that provide an example of interpretation of growth chart results in practice
- Updated data on obesity and overweight incidence and maps
- Added content on "environmental triggers" and genetic susceptibility to obesity
- Replaced the Reality Check
- Added discussion of replacement terms for "obese" and "fat" and on some health care provider's bias against obese people

Unit 10: Weight Control: The Myths and Realities

- Updated or replaced illustrations and table content
- Reorganized presentation of content
- Revised section on diet pills
- Rewrote section on popular diets
- Modified content on dietary supplements and internet products for weight loss
- Updated content on effectiveness of organized weight-loss programs
- Added section on successful weight control methods and programs, including lifestyle programs
- Revised content on bariatric surgery
- Added a separate section on physical activity and weight control

Unit 11: Disordered Eating: Anorexia Nervosa, Bulimia, and Pica

- Reorganized section on female athlete triad
- Expanded content on prevention of eating disorders
- Updated information about Health at Every Size
- Modified and updated presentation on plumbism and proposed eating disorders

Unit 12: Useful Facts about Sugars, Starches, and Fiber

- Modified Nutrition Scoreboard questions and answers
- Updated content on health effects of high sugar intake and particular effects in Latinos.
- Moved content on lactose maldigestion and intolerance to this unit
- Revised content on types of fiber
- Modified the Reality Check

Unit 13: Diabetes Now

- Updated content on prediabetes
- Updated and expanded content on fatty liver disease
- Added content on hemoglobin A1c
- Updated presentation on glycemic index and glycemic load
- Revised table on the glycemic index of foods
- Added Australia's GI food label
- Added content on vitamin D and diabetes
- Added an illustration showing projected increases in rates of type 2 diabetes
- Revised content on hypoglycemia

Unit 14: Alcohol: The Positives and Negatives

- Expanded presentation on fatty liver disease and steatohepatitis
- Modified content on fetal alcohol spectrum disorder
- Increased information on alcohol absorption and metabolism

Unit 15: Proteins and Amino Acids

- Modified illustrations and tables
- Expanded coverage on the functions of protein
- Added illustration and presentation of the basic structure of an amino acid
- Added section on protein structural types (e.g. primary, secondary, tertiary, and quaternary structures of proteins)
- Added section on nitrogen balance, presentation on how to calculate it
- Updated content on amino acids supplements, protein powders, and muscle mass and strength
- Added section on protein deficiency
- Deleted content on tryptophan supplements

Unit 16: Vegetarian Diets

- Updated and expanded content on vegetarian diets and health
- Added table on fortified foods that provides key nutrients needed by some vegetarians

Unit 17: Food Allergies and Intolerances

- Modified titles on headers
- Added section on the prevalence of food allergies
- Added sections on food allergy prevention and food allergy treatment
- Increased presentation of celiac disease and gluten-free regulations and products
- Moved content on lactose intolerance to Unit 12

Unit 18: Fats and Cholesterol in Health

- Expanded coverage of functions of fats
- Added content on the structure of fats
- Updated the table on food sources of *trans* fat
- Updated content on fat intake and health

Unit 19: Nutrition and Heart Disease

- Updated and expanded discussion of saturated fat and cholesterol intake, genetic traits, and heart disease risk
- Updated content on dietary and lifestyle changes and plasma lipid levels
- Updated and expanded presentation of oxidation and inflammation and heart disease
- Updated content on sex differences in heart disease risk and heart attack symptoms
- Heavily modified the Health Action
- Revised the Reality Check
- Added a section on dietary interventions for the prevention and treatment of heart disease
- Added content on added sugars, fatty liver, insulin resistance, metabolic syndrome, and heart disease risk
- Extensively modified content of tables
- Changed the Nutrition Up Close activity

Unit 20: Vitamins and Your Health

- Updated and revised table on vitamin functions, deficiency, toxicity, and food sources
- Extensively updated and modified table on food sources of vitamins
- Expanded content on vitamin deficiency disease prevention
- Updated content on folate and neural tube defects
- Updated content on recommendations for vitamin D intake and vitamin D and health relationships
- Reorganized and updated content on the antioxidant vitamins

Unit 21: Phytochemicals

- Reorganized content on phytochemicals and health, and phytochemical functions
- Deleted or reorganized content on phytochemical work in groups, beta-carotene supplements, vegetable extracts, and the table on examples of phytochemicals, their food sources, and potential mechanisms of action in disease prevention
- Revised content of tables
- Added a section on caffeine
- Added a section on food sources of phytochemicals
- Revised the Reality Check
- Deleted content on genetically modified organisms
- Modified the Nutrition Up Close activity

Unit 22: Diet and Cancer

- Revised Nutrition Scoreboard questions and answers
- Revised and updated section on how cancer develops
- Expanded presentation on how DNA becomes damaged
- Extensively revised sections on diet and cancer prevention, including tables and illustrations
- Added section on obesity and cancer
- Modified the Nutrition Up Close activity

Unit 23: Good Things to Know about Minerals

- Updated content on calcium, and on calcium, vitamin D, and osteoporosis
- Updated and revised table on mineral functions, deficiency, toxicity, and food sources
- Extensively updated and modified table on food sources of minerals
- Deleted the Health Action and other redundant content on osteoporosis
- Updated content on prehypertension and hypertension
- Added a Reality Check
- Updated DASH eating plan guidance
- Added content on potassium and health
- Updated content of tables
- Modified the Take Action feature to focus on potassium intake
- Changed the Nutrition Up Close activity to focus on the sodium content of processed foods

Unit 24: Dietary Supplements

- Added content on herbs versus drugs and herbs with drugs
- Extensively modified content of tables and some illustrations, added new tables and deleted others
- Added a presentation on dietary supplement realities

Unit 25: Water Is an Essential Nutrient

- Changed Nutrition Scoreboard questions and answers
- Expanded content on functions of water
- Updated illustrations
- Addressed new threats to safe and sufficient water supplies worldwide

Unit 26: Nutrient-Gene Interactions in Health and Disease

- Extensively modified and updated section on nutrient-gene interactions
- Extensively modified and updated tables
- Added content on gene variants, nutrients, and disease risk; and on epigenetics

- Added content on galactosemia, deleted content on celiac disease and lactose intolerance
- Expanded content on the effects of nutrient-gene interactions on the development of cancer, hypertension, and obesity
- Added section on the genetics of taste
- Revised section on the future of personalized diets for disease prevention

Unit 27: Nutrition and Physical Fitness for Everyone

- Updated photos and table content
- Expanded coverage of cardiorespiratory fitness, flexibility, and delayed onset muscle soreness
- Added a new focus on achieving physical fitness
- Revised the Reality Check
- Added a Take Action feature on aerobic exercise options
- Added new information on physical fitness in children

Unit 28: Nutrition and Physical Performance

- Added a table on eight common myths about nutrition and physical performance
- Revised illustrations and tables
- Revised the Reality Check
- Revised and updated content on amino acids, protein powders, muscle mass, and muscle strength
- Deleted table on ergogenic aids: claims and evidence. Added a table on the effects of dietary supplements and ergogenic aids on strength, endurance, or body composition
- Modified the Nutrition Up Close

Unit 29: Good Nutrition for Life: Pregnancy, Breast-Feeding, and Infancy

- Updated information in tables and illustrations
- Replaced MyPyramid information with ChooseMyPlate content
- Simplified illustration of the interior of a breast
- Replaced CDC growth chart with WHO chart
- Updated content on nutrition recommendations for pregnant and breast-feeding women, and for infants
- Modified the Nutrition Up Close to correspond with ChooseMyPlate guidance

Unit 30: Nutrition for the Growing Years: Childhood through Adolescence

- Added a discussion on and a table summarizing the RDAs for selected nutrients for children (4-8 years) and adolescents (14-16 years)

- Added tables on average nutrient and total sugar intake of children and adolescents compared to recommended levels of intake
- Updated presentation on diets on children and adolescents; added a discussion and comparison of usual intake versus recommendations
- Added sections on nutrient needs of children and adolescents, and on diet and health status of children and adolescents
- Added an example of a one-day diet for a child from ChooseMyPlate

Unit 31: Nutrition and Health Maintenance for Adults of All Ages

- Modified organization of content on section headers
- Added a section on factors that influence life expectancy, and on genes and life expectancy
- Updated tables and illustrations
- Replaced MyPyramid content with ChooseMyPlate dietary guidance content
- Added a table on RDAs for selected nutrients for adults 19-50 and over 70 years old

Unit 32: The Multiple Dimensions of Food Safety

- Modified content of tables, illustrations
- Reorganized sections of content
- Added a table on the top 10 foods related to foodborne illness outbreaks
- Added content on bisphenol A, and on antibiotics and pesticides in foods

Unit 33: Aspects of Global Nutrition

- Updated content in tables and illustrations
- Reorganized presentation of content
- Expanded section on the nutrition transition
- Added content and an illustration on increasing rates of chronic diseases in some developing nations

Appendices

- Updated food composition tables
- Inserted Health Canada's new Food Guide illustrations and information
- Inserted updated Food Exchange lists
- Modified the Glossary to reflect additions and revisions made in the seventh edition of *Nutrition Now*

Although a number of elements have changed in this new edition, many of the basic tenets of the text's approach have stayed the same. The text remains focused on meeting the needs of instructors offering introductory nutrition courses to students from a variety of majors. The 33 units in the text stand alone and can be covered in the order of the instructor's choosing. Instructors may choose to customize their selection of units to be included in the text. The text remains heavily illustrated, concise, and interactive.

Resources for the Instructor

Instructor's Manual with Test Bank. This manual features chapter outlines, assignment worksheets, suggested classroom activities, Web resources, critical thinking questions, and chapter-by-chapter test questions that are also available for the ExamView Computerized Testing program.

ExamView ® Computerized Testing. An easy-to-use assessment and tutorial system facilitates creation, delivery, and customizing of tests and study guides, both print and online.

Power Lecture DVD-ROM. This one-stop course preparation and presentation resource makes it easy for you to assemble, edit, publish, and present custom lectures for your course, using PowerPoint®. The PowerLecture includes PowerPoint® lecture slides, animations, BBC video clips, the Instructor's Manual, the test bank, "clicker" content, and ExamView computerized testing.

Transparency Acetates. This set of full-color transparencies features key illustrations in the text.

CourseMate. This feature brings course concepts to life with interactive learning, study, and exam preparation tools that support the printed textbook, or the included eBook. With CourseMate, professors can use the included Engagement Tracker to assess student preparation and engagement. Use the tracking tools to see progress for the class as a whole or for individual students. Students can access an interactive eBook, chapter-specific interactive learning tools, including flashcards, quizzes, Pop-up Tutors, Nutrition Tutorials, and BBC video clips, and more in their Nutrition CourseMate.

WebTutor ™. Provides customizable, text-specific content that allows instructors to edit, reorganize, or delete content to meet their course needs, It offers quizzing, videos, animations, Pop-up Tutors, and testbank materials along with direct access to Diet Analysis Plus, Global Nutrition Watch and an interactive eBook.

Diet Analysis Plus™. *Diet Analysis Plus* enables you to track and assess your diet and physical activity online! You can create a personal profile based on height, weight, age, sex, and activity level, and use this tool to easily analyze the nutritional value of the food you eat, adjust your diet to meet your personal health goals, and gain a better understanding of how nutrition relates to your life. Diet Analysis Plus includes a 20,000+ food database, 10 reports for analysis, a food recipe feature, the latest Dietary References, and goals and actual percentages of essential nutrients, vitamins, and minerals. Diet Analysis Plus is a valuable tool that you can use in your nutrition course and then continue to use after the course is over.

Global Nutrition Watch. Bring currency to the classroom with Global Nutrition Watch from Cengage Learning! This user-friendly website provides convenient access to thousands of trusted sources, including academic journals, newspapers, videos, and podcasts, for you to use for research projects or classroom discussion. Global Nutrition Watch is updated daily to offer the most current news about topics related to nutrition.

Acknowledgments

Peggy Williams, Senior Acquisitions Editor, and Nedah Rose, Senior Developmental Editor, helped me plan and implement a major revision of a textbook. Both have invested themselves fully in making the seventh edition of *Nutrition Now* and its pedagogical aids effective teaching and learning tools. The results of their efforts are reflected in the new design, photographs, and features presented in this edition of *Nutrition Now*. Elesha Feldman has been with *Nutrition Now* since the early editions and I love the work she does on the Instructor's Manual with Test Bank. Her help and advice continues to be a major and reliable asset. The very helpful feedback provided by Dr. Steven Nizielski and Dr. Eugene Fenster during their review of the seventh edition for instructional materials development is much appreciated. I think I made Dana Richards from Lachina Publishing Services work overtime on the seventh edition. Thanks for your spirit of collaboration and the perfectly clear instructions you write for authors related to production.

It is said that instructors adopt a specific textbook but that students play a major role in instructors' decision to keep it. I am honored that you chose to adopt *Nutrition Now* and deeply pleased with the thought that students are helping you decide to keep it.

Reviewers' feedback is the lifeline of text writing, and the reviewers for the sixth edition conveyed very useful advice that became incorporated into the seventh edition. The advice led me to some very interesting places on specific topics that changed my thinking and writing. May you remain students of *Nutrition Now* the textbook and keep your comments coming.

Reviewers

Janet Colson, PhD, RD
Middle Tennessee State University

Shannon Gleave, RD
Central Arizona College

Margaret Gunther
Palomar Community College

Teresa Johnson, DCN, RD
Troy University

Joe LaVilla, PhD
The Art Institute of Phoenix

Lane Miller, MBA
Medical Careers Institute

Kara Montgomery
University of South Carolina

Steven L. Silverman, BS, DC
Palmer College of Chiropractic

Jennifer Wegmann
Binghamton University

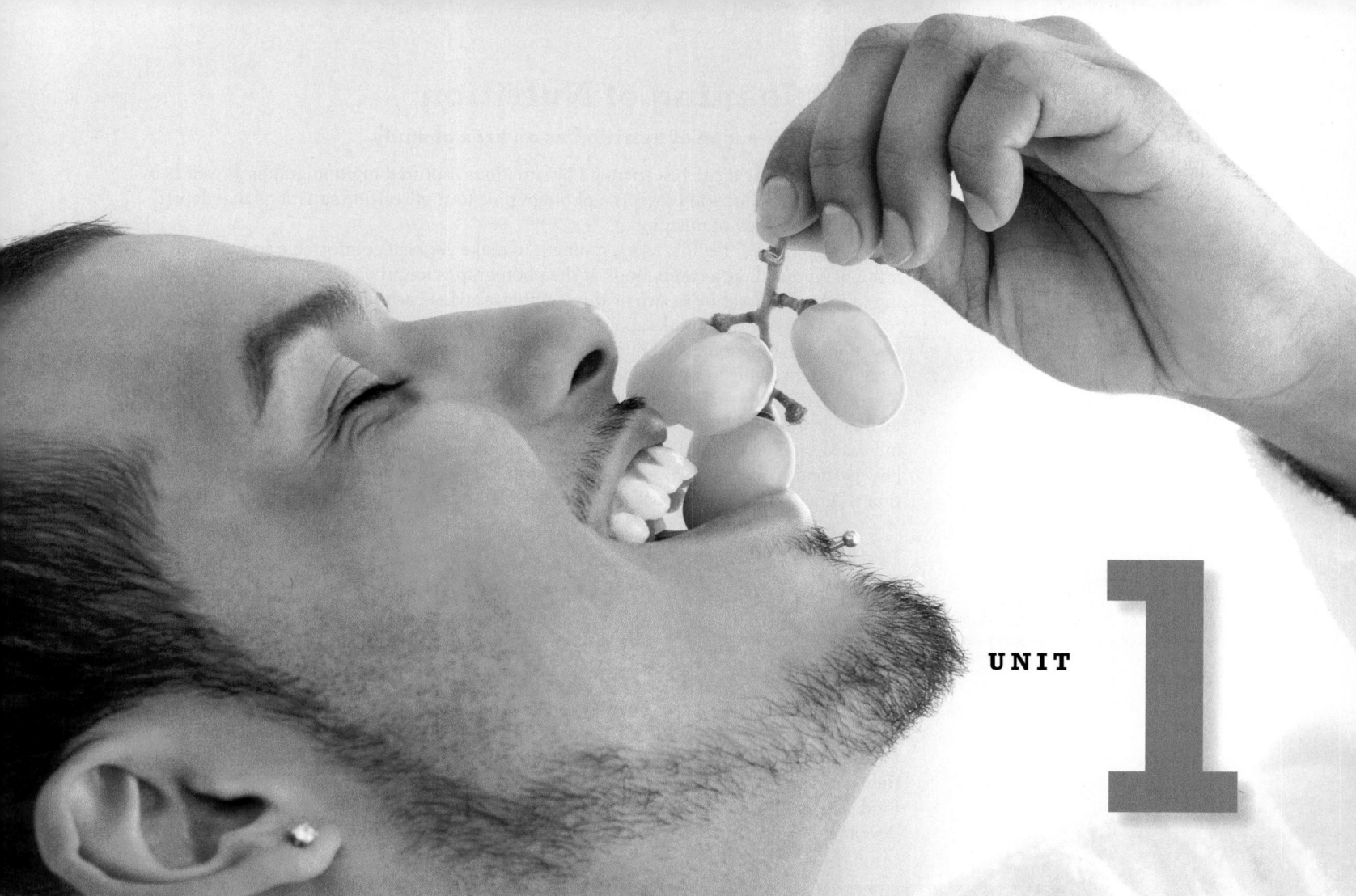

^{UNIT} **1**

Key Nutrition Concepts and Terms

NUTRITION SCOREBOARD

1 Calories are a component of food.   **True/False**

2 Nutrients are substances in food that are used by the body for growth and health.   **True/False**

3 Inadequate intakes of vitamins and minerals can harm health, but high intakes do not.   **True/False**

4 "Dietary Reference Intakes" (DRIs) provide science-based standards for nutrient intake.   **True/False**

Answers can be found at the end of the unit.

After completing Unit 1 and its interactive learning features, you will be able to:

- Explain the scope of nutrition as an area of study.
- Demonstrate a working knowledge of the meaning of the 10 nutrition concepts.

The Meaning of Nutrition

• Explain the scope of nutrition as an area of study.

What is nutrition? It can be explained by situations captured in photographs as well as by words. This introduction presents a photographic tour of real-life situations that depict aspects of the study of nutrition.

Before the tour begins, take a moment to make yourself comfortable and clear your mind of clutter. Take a careful look at the photographs found on pages 1-3 and 1-4, pausing to mentally describe in two or three sentences what each photograph shows.

Not everyone who thinks about the photographs will describe them in the same way. Reactions will vary somewhat due to personal experiences, interests, attitudes, and beliefs. An individual trying to gain weight will probably react differently to the photograph of the person on the scale than someone who is trying to lose weight. If you grew up in a family that farmed for a living, the picture of pesticides being sprayed on a crop may mean increased food production to you; another person may see the photograph and think about how pesticide residues on foods may be harmful to health. Although knowledge about nutrition is generated by impersonal and objective methods, it can be a very personal subject.

Nutrition Defined

nutrition The study of foods, their nutrients and other chemical constituents, and the effects that food constituents have on health.

In a nutshell, **nutrition** is the study of foods and health. It is a science that focuses on foods, their nutrient and other chemical constituents, and the effects of food constituents on body processes and health. The scope of nutrition extends from food choices to the effects of diet and specific food components on biological processes and health.

Nutrition Is a "Melting Pot" Science The broad scope of nutrition makes it an interdisciplinary science. Knowledge provided by the behavioral and social sciences, for example, is needed in studies that examine how food preferences develop and how they may be changed. Information generated by the biological, chemical, physical, and food sciences is required to propose and explain diet and disease relationships. Methods developed by quantitative scientists such as mathematicians and statisticians are needed

© Gary Conner/PhotoEdit

© Royalty free/Corbis

Jupiterimages/FoodPix/Getty Images

© Phil Schermeister/Corbis

Photo Disc

Uschi Gerschner/Newscom

Photo Disc

© Pat Shearman/Alamy

Photodisc

AP Images

to guide decision making about the significance of results produced by nutrition research. The study of nutrition will bring you into contact with information from a variety of disciplines.

Nutrition Knowledge Is Applicable   As you study the science of nutrition, you will discover answers to a number of questions about your own diet, health, and eating behaviors. Is obesity primarily due to eating habits, physical inactivity, or your genes? How do you know whether new information you hear about nutrition is true? Can sugar harm more than your teeth? Can the right diet or supplement give you a competitive edge? What is a healthful diet and how do you know if you have one? If improvements seem warranted, what's the best way to go about changing your diet for the better? These are just a few of the questions that will be addressed during the course of your study of nutrition. You will take from this learning experience not only knowledge about nutrition and health, but skills that will keep the information and insights working to your advantage for a long time to come.

Foundation Knowledge for Thinking about Nutrition

- **Demonstrate a working knowledge of the meaning of the 10 nutrition concepts.**

You don't have to be a bona fide nutritionist to think like one. What you need is a grasp of the language and basic concepts of the science. The purpose of this unit is to give you that background. The essential topics covered here are explored in greater depth in units to come, and they build on this foundation of knowledge. With a working knowledge of nutrition terms and concepts, you will have an uncommonly good sense of nutrition.

• •

NUTRITION CONCEPT #1

Food is a basic need of humans.

Humans need enough food to live, and they need the right assortment of foods for optimal health. In the best of all worlds, the need for food is combined with the condition of **food security.** People who experience food security have access at all times to a sufficient supply of the safe, nutritious foods that are needed for an active and healthy life. They are able to acquire acceptable foods in socially acceptable ways; they do not have to scavenge or steal food in order to survive or to feed their families. **Food insecurity** exists whenever the availability of safe, nutritious foods—or the ability to acquire them in socially acceptable ways—is limited or uncertain (Illustration 1.1).[1]

food security   Access at all times to a sufficient supply of safe, nutritious foods.

food insecurity   Limited or uncertain availability of safe, nutritious foods—or the ability to acquire them in socially acceptable ways.

Illustration 1.1   "It is possible to go an entire lifetime without knowing about people's experiences with hunger." —MEGHAN LECATES, CAPITOL AREA FOOD BANK

Adults who live in food-insecure households are more likely to have poor-quality diets, to be overweight, and to have heart disease or diabetes than adults who are poor but food secure.[2] Poor nutritional quality of food consumed, the absence of local supermarkets, lack of exercise, and episodic food shortages may be partly responsible for the higher rates of overweight and chronic disease among food-insecure adults.[3] Although many children living in food-insecure households are nourished adequately, successful in school, and develop high levels of social skills, as a group they are at higher risk of poor school performance and social and behavioral problems.[4]

Food insecurity exists in 14.5% of U.S. and 7.7% of Canadian households.[3,5] It is most likely to occur in poor, female-headed households with young children living in inner-city areas. The rate in the United States is over twice as high as the target of 6% established as a national health goal.[6]

• •

Food Terrorism The term *food security* now means more than it used to. Food is a potential weapon of bioterrorism. Although public health concerns related to bioterrorism center on disease threats such as smallpox and anthrax, food and water could also be used to intentionally spread illness. Simply recalling the latest announcement about a food-borne illness outbreak makes it easy to imagine the panic and health consequences that could result from intentional contamination of food or water supplies.

Toxic substances that could be introduced into food and water supplies include botulism toxin, ricin, radioactive particles, and microorganisms such as *Salmonella, E. coli 0517:H7,* and *Shigella.* Botulism toxin is the single most poisonous substance known. Most bacteria are killed when heated above 160°F or when municipal water supplies are treated with chlorine; however, it takes about 15 minutes of boiling to destroy botulism toxin.[7,8]

• •

NUTRITION CONCEPT #2

Foods provide energy (calories), nutrients, and other substances needed for growth and health.

People eat foods for many different reasons. The most compelling reason is that we need the calories, nutrients, and other substances supplied by foods for growth and health.

Calories

A **calorie** is a unit of measure of the amount of energy in a food—and of how much energy will be transferred to the person who eats it. Although we often refer to the number of calories in this food or that one, calories are not a substance present in food. And, because calories are a unit of measure, they do not qualify as a nutrient.

Nutrients

Nutrients are chemical substances present in food that are used by the body to sustain growth and health (Illustration 1.2). Essentially everything that's in our body was once a component of the food we consumed.

There are six categories of nutrients (Table 1.1). Each category (except water) consists of a number of different substances that are used by the body for growth and health. The carbohydrate category includes simple sugars and complex carbohydrates (starches and dietary fiber). The protein category includes 20 amino acids, the chemical units that serve as the "building blocks" for protein. Several different types of fat are included in the fat category. Of primary concern are the saturated fats, unsaturated fats, essential fatty acids, *trans* fats, and cholesterol. The vitamin category consists of 14 vitamins, and the mineral category includes 15 minerals. Water makes up a nutrient category by itself.

Carbohydrates, proteins, and fats supply calories and are called the "energy nutrients." Although each of these three types of nutrients performs a variety of functions, they share the property of being the body's only sources of fuel. Vitamins, minerals, and water are chemicals that the

calorie A unit of measure of the amount of energy supplied by food. (Also known as a kilocalorie, abbreviated kcal, or the "large Calorie" with a capital *C.*)

nutrients Chemical substances in food that are used by the body for growth and health. The six categories of nutrients are carbohydrates, proteins, fats, vitamins, minerals, and water.

Illustration 1.2 Foods provide nutrients. "Please pass the complex carbohydrates, thiamin, and niacin . . . I mean, the bread!"

© Brand X Pictures/Getty Images

Photo Disc

Illustration 1.3 Examples of good food sources of phytochemicals.

Table 1.1 **The six categories of nutrients**
1. Carbohydrate
2. Protein
3. Fat
4. Vitamins
5. Minerals
6. Water

body needs for converting carbohydrates, proteins, and fats into energy and for building and maintaining muscles, blood components, bones, and other tissues in the body.

Other Substances in Food

Food also contains many other substances, some of which are biologically active in the body. One major type of these substances are the **phytochemicals**. There are thousands of them in plants. Illustration 1.3 presents examples of plant foods that are particularly rich sources of phytochemicals. Phytochemicals provide plants with color, give them flavor, foster their growth, and protect them from insects and diseases. In humans, consumption of certain phytochemicals in diets is strongly related to a reduced risk of developing certain types of cancer, heart disease, infections, and other disorders.[9]

Specific phytochemicals have names that may be hard to pronounce and difficult to remember. Nevertheless, here are a few examples. Plant pigments, such as lycopene (like-o-peen), which help make tomatoes red, anthocyanins (an-tho-sigh-an-ins), which give blueberries their characteristic blue color, and beta-carotene (bay-tah-kar-o-teen), which imparts an orange color to carrots, are phytochemicals that act as **antioxidants**.

They protect plant cells—and in some cases, human cells, too—from damage that can make them susceptible to disease. Various types of sulfur-containing phytochemicals are present in cabbage, broccoli, cauliflower, brussels sprouts, and other vegetables of the same family. These substances may help prevent a number of different types of cancer.[10]

Some Nutrients Must Be Provided by the Diet Many nutrients are required for growth and health. The body can manufacture some of these from raw materials supplied by food, but others must come assembled. Nutrients that the body cannot generally produce, or produce in sufficient quantity, are referred to as **essential nutrients**. Here "essential" means "required in the diet." Vitamin A, iron, and calcium are examples of essential nutrients. Table 1.2 lists all the known essential nutrients.

Nutrients used for growth and health that can be manufactured by the body from components of food in our diet are considered nonessential. Cholesterol, creatine, and glucose are examples of **nonessential nutrients**. Nonessential nutrients are present in food and used by the body, but they are not required parts of our diet because we can produce them ourselves.

Both essential and nonessential nutrients are required for growth and health. The difference between them is whether or not we need to obtain the nutrient from a dietary source. A dietary deficiency of an essential nutrient will cause a specific deficiency disease, but a dietary lack of a nonessential nutrient will not. People develop scurvy (the vitamin C–deficiency disease), for example, if they do not consume enough vitamin C. But you could have zero cholesterol in your diet and not become "cholesterol deficient," because your liver produces cholesterol.

Our Requirements for Essential Nutrients The amount of essential nutrients humans need each day varies a great deal, from amounts measured in cups to micrograms. (See Table 1.3 to get a notion of the amount represented by a gram, milligram, and other measures.) Generally speaking, adults need 11 to 15 cups of water from fluids and

phytochemicals (phyto = plant) Chemical substances in plants. Some phytochemicals perform important functions in the human body. They give plants color and flavor, participate in processes that enable plants to grow, and protect plants against insects and diseases. Also called phytonutrients.

antioxidants Chemical substances that prevent or repair damage to cells caused by exposure to oxidizing agents such as environmental pollutants, smoke, ozone, and oxygen. Oxidation reactions are a normal part of cellular processes.

essential nutrients Nutrients required for normal growth and health that the body can generally not produce, or produce in sufficient amounts. Essential nutrients must be obtained in the diet.

nonessential nutrients Nutrients required for normal growth and health that the body can manufacture in sufficient quantities from other components of the diet. We do not require a dietary source of nonessential nutrients.

Table 1.2 Essential nutrients for humans: A reference table

Energy Nutrients	Vitamins	Minerals	Water
Carbohydrates	Biotin	Calcium	Water
Fats[a]	Folate	Chloride	
Proteins[b]	Niacin (B_3)	Chromium	
	Pantothenic acid	Copper	
	Riboflavin (B_2)	Fluoride	
	Thiamin (B_1)	Iodine	
	Vitamin A	Iron	
	Vitamin B_6 (pyroxidine)	Magnesium	
	Vitamin B_{12}	Manganese	
	Vitamin C (ascorbic acid)	Molybdenum	
	Vitamin D	Phosphorus	
	Vitamin E	Potassium	
	Vitamin K	Selenium	
	Choline[c]	Sodium	
		Zinc	

[a]Fats supply the essential nutrients linoleic and alpha-linolenic acid.
[b]Proteins are the source of nine "essential amino acids": histidine, isoleucine, leucine, lysine, methionine, phenylalanine, threonine, tryptophan, and valine. The other 11 amino acids are not a required part of our diet; they are considered "nonessential."
[c]A dietary source of choline may not be required during all stages of the life cycle.

Table 1.3 Units of measure commonly employed in nutrition

Measure	Abbreviation	Equivalents
Kilogram	kg	1 kg = 2.2 lb = 1000 grams
Pound	lb	1 lb = 16 oz = 454 grams = 2 cups (liquid)
Ounce	oz	1 oz = 28 grams = 2 tablespoons (liquid)
Gram	g	1 g = 1/28 oz = 1000 milligrams
Milligram	mg	1 mg = 1/28,000 oz = 1000 micrograms
Microgram	mcg, μg	1 mcg = 1/28,000,000 oz

1 egg = 50 grams or 1³/₄ oz; 212 milligrams (0.2 grams) of cholesterol in yolk

1 slice of bread = 1 oz = 28 grams

1 nickel = 5 grams

Photo Disc

1 teaspoon of sugar = 4 grams, 1 grain of sugar = 200 micrograms

foods, 9 tablespoons of protein, one-fourth teaspoon of calcium, and only one-thousandth teaspoon (a 30-microgram speck) of vitamin B_{12} each day.

We all need the same nutrients, but not always in the same amounts. The amounts needed vary among people based on:

- Age

- Sex

- Growth status

- Body size

- Genetic traits

and the presence of conditions such as:

- Pregnancy
- Breastfeeding
- Illnesses
- Drug/medication use
- Exposure to environmental contaminants

Each of these factors, and others, can influence nutrient requirements. General diet recommendations usually make allowances for major factors that influence the level of nutrient need, but they cannot allow for all of the factors.

Nutrient Intake Standards Recommendations for daily levels of nutrient intake were first developed in the United States in 1943 and have been updated periodically since then. Called the Recommended Dietary Allowances (RDAs), these standards were established in response to the high rejection rate of World War II recruits, many of whom were underweight and had nutrient deficiencies. The recommended levels of nutrient intake provided are based on age, gender, and condition (pregnant or breastfeeding). Because the science underlying nutrient intake and health advances with time, these standards are periodically revised and expanded (Illustration 1.4).

Dietary Reference Intakes Nutrient intake standards now in place are referred to as "Dietary Reference Intakes" (DRIs) and include categories of nutrient intake in addition to the RDAs. The current RDAs referenced in the DRIs reflect nutrient intake levels that protect most all healthy individuals from deficiency disease development and that also reduce the risk of common chronic diseases. Table 1.4 provides examples of endpoints aimed at chronic disease prevention used by the DRI Committee to determine the RDAs. A category labeled "Adequate Intakes" (AIs) has been added to indicate "tentative RDAs" for a few nutrients such as vitamin K and fluoride where too few, reliable scientific studies have been done to establish an RDA.

The DRI standards include a category called "Estimated Average Requirement," or EAR. This category represents nutrient intake levels that are estimated to meet the nutrient intake requirements of 50% of individuals within an age, sex, and condition (pregnant or breastfeeding) group.

DRI standards consider the effects of excessively high intake of nutrients, primarily from supplements and fortified foods, on health. These standards are labeled "Tolerable Upper Intake Levels," and are abbreviated "ULs" for "upper levels." Table 1.5 graphically displays the relationships between nutrient intake level and the various categories of the DRI standards now in use, and presents definitions of the DRI nutrient intake categories.

Developed by nutrition scientists from the United States and Canada, the RDAs apply to 97 to 98% of all healthy people in both countries. The fundamental premise that nutrient intake should come primarily from foods stated when the first RDAs were developed is maintained in the current nutrient intake standards.

DRI tables are shown on the inside front and back covers of this text. Check out these tables. They can be use to identify recommended, daily levels of essential nutrient intake and levels of intake that should not normally be exceeded.

· ·

NUTRITION CONCEPT #3

Health problems related to nutrition originate within cells.

Cells are the main employers of nutrients (Illustration 1.5). All body processes required for growth and health take place within cells and the fluid that surrounds them. The human body contains more than one hundred trillion (100,000,000,000,000) cells. (The most common cell in the body is the red blood cell. There are more than 20 billion of them.) Functions of each cell are

Illustration 1.4 The latest DRI report on calcium and vitamin D, 2010.

Table 1.4 Examples of primary endpoints used to estimate DRIs[11]

Carbohydrate: Amount needed to supply optimal levels of energy to the brain.

Total Fiber: Amount shown to provide the greatest protection against heart disease.

Folate: Amount that maintains normal red blood cell folate concentration.

Iodine: Amount that corresponds to optimal functioning of the thyroid gland.

Selenium: Amount that maximizes its function in protecting cells from damage.

Illustration 1.5 Schematic representation of the structure of a human cell.

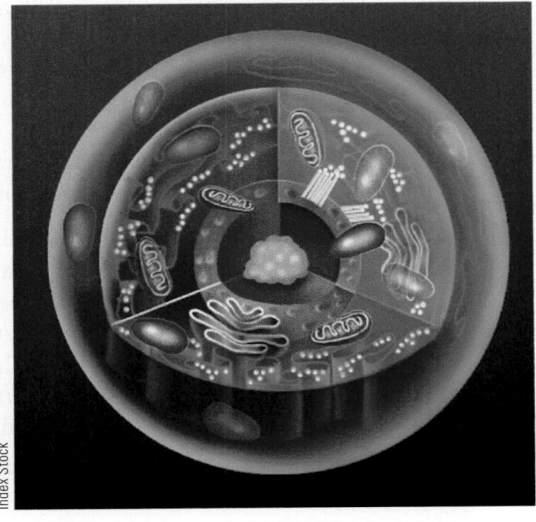

Index Stock

Table 1.5 Terms and abbreviations used in the DRIs and a graphic representation of their meaning[11]

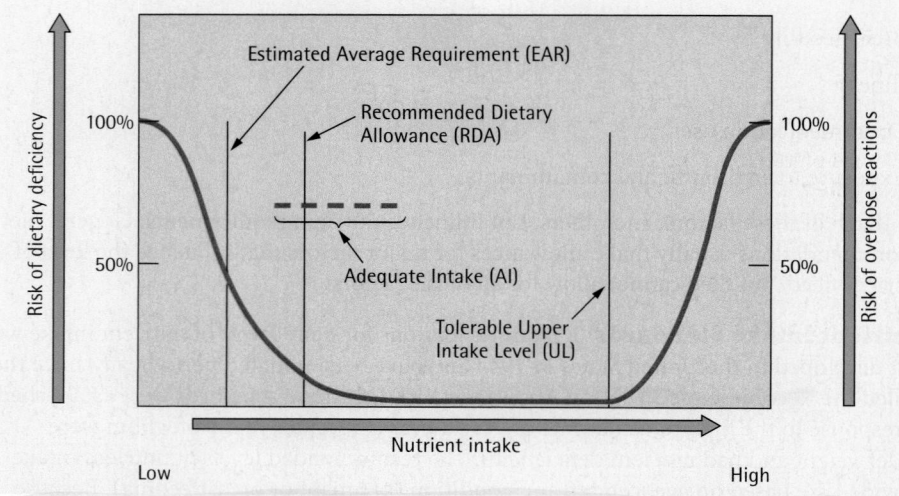

- **Dietary Reference Intakes (DRIs).** This is the general term used for nutrient intake standards for healthy people.

- **Recommended Dietary Allowances (RDAs).** These are levels of essential nutrient intake judged to be adequate to meet the known nutrient needs of practically all healthy persons while decreasing the risk of certain chronic diseases.

- **Adequate Intakes (AIs).** These are "tentative" RDAs. AIs are based on less conclusive scientific information than are the RDAs.

- **Estimated Average Requirements (EARs).** These are nutrient intake values that are estimated to meet the requirements of half the healthy individuals in a group. The EARs are used to assess adequacy of intakes of population groups.

- **Tolerable Upper Intake Levels (ULs).** These are upper limits of nutrient intake compatible with health. The ULs do not reflect desired levels of intake. Rather, they represent total, daily levels of nutrient intake from food, fortified foods, and supplements that should not normally be exceeded.

maintained by the nutrients it receives. Problems arise when a cell's need for nutrients differs from the available supply.

· ·

Nutrient Functions at the Cellular Level Cells are the building blocks of tissues (such as muscles and bones), organs (such as the kidneys, heart, and liver), and systems (such as the respiratory, reproductive, circulatory, and nervous systems). Normal cell health and functioning are maintained when a nutritional and environmental utopia exists within and around the cells. Such circumstances allow **metabolism**—the chemical changes that take place within and outside of cells—to proceed flawlessly. Disruptions in the availability of nutrients—or the presence of harmful substances in the cell's environment—initiate diseases and disorders that eventually affect tissues, organs, and systems. Here are two examples of how cell functions can be disrupted by the presence of low or high concentrations of nutrients:

1. Folate, a B vitamin, is required for protein synthesis within cells. When too little folate is available, cells produce proteins with abnormal shapes and functions. Abnormalities in the shape of red blood cell proteins, for example, lead to functional changes that produce loss of appetite, weakness, and irritability.[12]
2. When too much iron is present in cells, the excess reacts with and damages cell components. If cellular levels of iron remain high, the damage spreads, impairing the functions of organs such as the liver, pancreas, and heart.[13]

Health problems in general begin with disruptions in the normal activity of cells. Humans are as healthy as their cells.

metabolism The chemical changes that take place in the body. The formation of energy from carbohydrates is an example of a metabolic process.

NUTRITION CONCEPT #4

Poor nutrition can result from both inadequate and excessive levels of nutrient intake.

For each nutrient, every individual has a range of optimal intake that produces the best level for cell and body functions. On either side of the optimal range are levels of intake associated with impaired body functions.[14] This concept is presented in Illustration 1.6. Inadequate essential nutrient intake, if prolonged, results in obvious deficiency diseases. Marginally deficient nutrient intakes generally produce subtle changes in behavior or physical condition. If the optimal intake range is exceeded, mild to severe changes in mental and physical functions occur, depending on the amount of the excess and the nutrient. Severe zinc deficiency, for example, is related to diarrhea, respiratory infection, and stunted growth. Mild zinc deficiency causes disturbances in the sense of taste and smell, and reduces appetite and food intake. Excessive intake of zinc is associated with vomiting and a decline in the body's ability to fight infections.[15] Nearly all cases of vitamin and mineral overdose result from the excessive use of supplements or errors made in the level of nutrient fortification of food products. They are almost never caused by foods. For nutrients, "enough is as good as a feast."

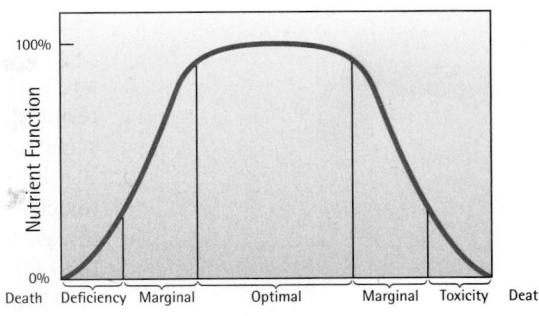

Increasing concentration or intake of nutrient ➡️

Illustration 1.6 For every nutrient, there is a range of optimal intake that corresponds to the optimal functioning of that nutrient in the body.

Steps in the Development of Nutrient Deficiencies and Toxicities Poor nutrition due to inadequate diet generally develops in the stages outlined in Illustration 1.7. To help explain the stages, this illustration includes an example of how vitamin A deficiency develops.

After a period of deficient intake of an essential nutrient, the body's tissue reserves of the nutrient become depleted. Blood levels of the nutrient then decrease because there are no reserves left to replenish the blood supply. Without an adequate supply of the nutrient in the blood, cells get shortchanged. They no longer have the supply of nutrients needed to

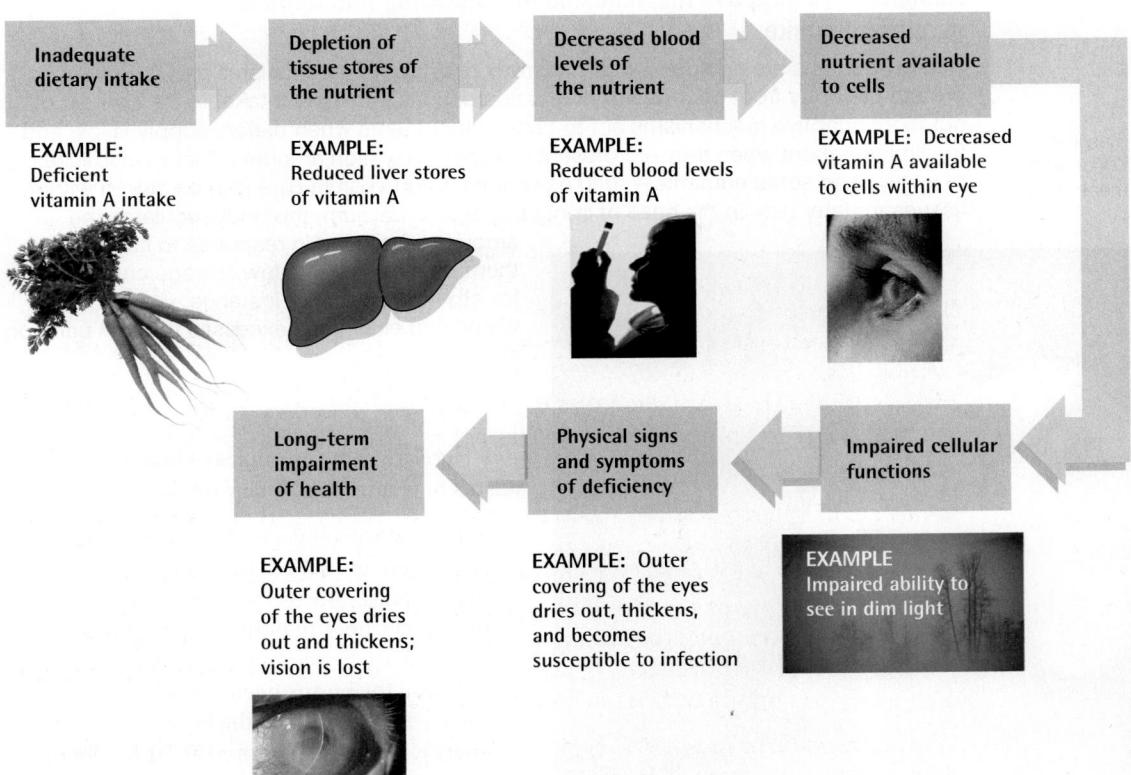

Illustration 1.7 The usual sequence of events in the development of a nutrient deficiency and an example of how vitamin A deficiency develops.[16,17]

Top row: Photo Disc; Bottom: ICI Pharmaceuticals

maintain normal function. If the dietary deficiency is prolonged, the malfunctioning cells cause sufficient impairment to produce physically obvious signs of a deficiency disease. Eventually, some of the problems produced by the deficiency may no longer be repairable, and permanent changes in health and function may occur. In most cases, the problems resulting from the deficiency can be reversed if the nutrient is supplied before this final stage occurs.

Excessively high intakes of many nutrients such as vitamin A and selenium produce toxicity diseases. The vitamin A toxicity disease is called "hypervitaminosis A," and the disease for selenium toxicity is called "selenosis." Signs of the toxicity disease stem from increased levels of the nutrient in the blood and the subsequent oversupply of the nutrient to the cells. The high nutrient load upsets the balance needed for normal cell function. The changes in cell functions lead to the signs and symptoms of the toxicity disease.

For both deficiency and toxicity diseases, the best time to correct the problem is usually at the level of dietary intake, before tissue stores are adversely affected. In that case, no harmful effects on health and cell function occur—they are prevented.[18]

Nutrient Deficiencies Are Often Multiple Most foods contain many nutrients, so poor diets will affect the intake level of more than one nutrient (Illustration 1.8). Inadequate diets generally produce a spectrum of signs and symptoms related to multiple nutrient deficiencies. For example, protein, vitamin B_{12}, iron, and zinc are packaged together in many high-protein foods. The protein-deficient, starving children you may see in news reports are rarely deficient just in protein. They may also be deficient in iron, zinc, and vitamin B_{12}.

The "Ripple Effect" Dietary changes affect the level of intake of many nutrients. Switching from a high-fat to a low-fat diet, for instance, generally results in a lower intake of cholesterol and vitamin E and increased intake of vitamin A, vitamin C, and iron.[19] So, dietary changes introduced for the purpose of improving the intake level of a particular nutrient produce a ripple effect on the intake of other nutrients.

• •

NUTRITION CONCEPT #5

Humans have adaptive mechanisms for managing fluctuations in nutrient intake.

Healthy humans are equipped with a number of adaptive mechanisms that partially protect the body from poor health due to fluctuations in dietary intake. In the context of nutrition, adaptive mechanisms act to conserve nutrients when dietary supply is low and to eliminate them when they are present in excessively high amounts. Dietary surpluses of energy and some nutrients—such as vitamin A and vitamin B_{12}—can be stored within tissues for later use. In the case of iron, copper, and calcium, the body regulates the amounts absorbed in response to its need for them. The body has a low storage capacity for other nutrients, for instance, vitamin C and water, and eliminates excesses through urine or stools.

• •

Here are some examples of how the body adapts to changes in dietary intake:

• When calorie intake is reduced by fasting, starvation, or dieting, the body adapts to the decreased supply by lowering energy expenditure. Declines in body temperature and the capacity to do physical work also act to decrease the body's need for calories. When caloric intake exceeds the body's need for energy, the excess is stored as fat for energy needs in the future.

Illustration 1.8 This woman has iron deficiency anemia. Chances are good that she has poor status of other nutrients in addition to iron.

© Chad Johnston/Masterfile

- The ability of the gastrointestinal tract to absorb dietary iron increases when the body's stores of iron are low. To help protect the body from iron overdose, the mechanisms that facilitate iron absorption in times of need shut down when enough iron has been stored.

- The body can protect itself from excessively high levels of intake of vitamin C from supplements by excreting the excess in the urine.

Although these built-in mechanisms do not protect humans from all the consequences of poor diets, they do provide an important buffer against the development of nutrient-related health problems.

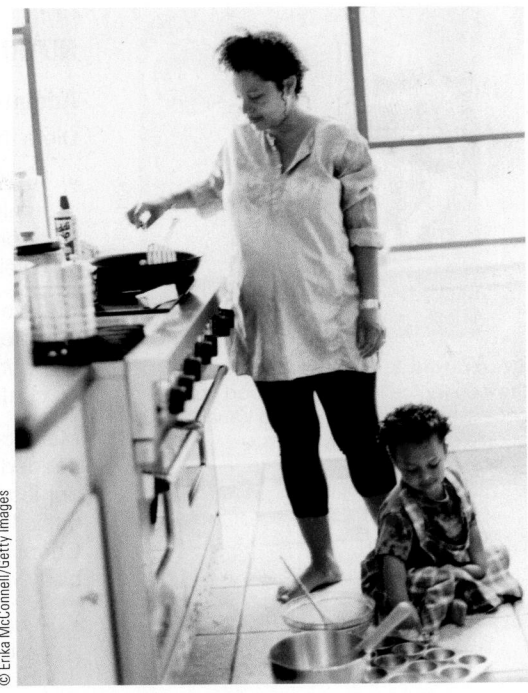
© Erika McConnell/Getty Images

NUTRITION CONCEPT #6

Malnutrition can result from poor diets and from disease states, genetic factors, or combinations of these factors.

Malnutrition means "poor" nutrition and results from both inadequate and excessive availability of calories and nutrients in the body. Vitamin A toxicity, obesity, vitamin C deficiency (scurvy), and underweight are examples of malnutrition.

Malnutrition can result from poor diets and also from diseases that interfere with the body's ability to use the nutrients consumed. Diarrhea, alcoholism, cancer, bleeding ulcers, and HIV/AIDS, for example, may be primarily responsible for the development of malnutrition in people with these disorders.

In addition, a percentage of the population is susceptible to malnutrition and increased disease risk due to genetic factors. For example, people may be born with a genetic tendency to produce excessive amounts of cholesterol, absorb high levels of iron, or use folate poorly. Some cases of obesity, diabetes, heart disease, and cancer are related to a combination of genetic and dietary factors.[20]

Illustration 1.9 Women who are pregnant or breastfeeding and infants are among the people who are at a higher risk of becoming inadequately nourished.

malnutrition Poor nutrition resulting from an excess or lack of calories or nutrients.

NUTRITION CONCEPT #7

Some groups of people are at higher risk of becoming inadequately nourished than others.

Women who are pregnant or breastfeeding, infants, growing children, the frail elderly, the ill, and those recovering from illness have a greater need for nutrients than other people. As a result, they are at higher risk of becoming inadequately nourished than other people (Illustration 1.9). In cases of widespread food shortages, such as those induced by natural disasters or war, the health of these nutritionally vulnerable groups is compromised the soonest and the most.

Within the nutritionally vulnerable groups, certain people and families are at particularly high risk of malnutrition. These are people and families who are poor and least able to secure food, shelter, and high-quality medical services. The risk of malnutrition is not shared equally among all persons within a population.

NUTRITION CONCEPT #8

Poor nutrition can influence the development of certain chronic diseases.

Poor nutrition does not result only in nutrient deficiency or toxicity diseases. Faulty diets play important roles in the development of heart disease, hypertension, cancer, osteoporosis, and other **chronic diseases**. Diets high in animal fat, for example, are related to the development of heart disease; those low in vegetables and fruit to cancer; low-calcium diets and poor vitamin D status to osteoporosis; and high-sugar diets to tooth decay. The harmful effects of poor dietary practices on chronic disease development generally accumulate over the course of years.

chronic diseases Slow-developing, long-lasting diseases that are not contagious (e.g., heart disease, diabetes, and cancer). They can be treated but not always cured.

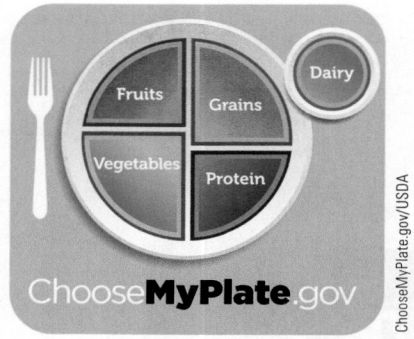

Illustration 1.10 ChooseMyPlate is the icon for the USDA's new food guidance system. It is intended to visually help people choose healthy meals.

NUTRITION CONCEPT #9

Adequacy, variety, and balance are key characteristics of healthful diets.

Diets that promote growth, development, and the maintenance of health provide:

- An adequate amount of essential nutrients from foods while delivering a level of calorie intake that corresponds to a healthy weight. Adequate amounts of essential nutrients approximate the RDA (Recommended Dietary Allowance) or the AI (Adequate Intake) levels cited in the DRI (Dietary Reference Intake) tables.

- A variety of foods from each of the basic food groups. A variety of foods is needed to obtain the wide assortment of nutrients and beneficial phytochemicals we need for the optimal functioning of our bodies. No one food (except breast milk for young infants) contains them all, and most single foods don't come close.

Many combinations of foods can supply the variety of nutrients and phytochemicals needed for a healthful diet, so no specific foods are required. Healthy diets could include apples, snails, sea cucumbers, ants, and burdock root. Diets that include the variety of basic foods recommended in the United States Department of Agriculture's (USDA's) ChooseMyPlate food guidance system (Illustration 1.10) are most likely to supply the body with adequate amounts of nutrients and other beneficial substances.[21]

- A balanced selection of food types and amounts from the ChooseMyPlate basic food groups. Regular consumption of fried foods, high fat meats, and sweets; and infrequent intake of whole grains and colorful vegetables, for example, can knock a diet out of balance.

To bring diets into balance, Americans are being urged to consume more vegetables, fruits, whole grains, dried beans, fish, low-fat meats, and dairy products; and to consume less sugar, animal fat, and high-salt foods than is the case now.[23]

energy-dense foods Foods that provide relatively high levels of calories per unit weight of the food. Fried chicken; cheeseburgers; a biscuit, egg, and sausage sandwich; and potato chips are energy-dense foods.

empty-calorie foods Foods that provide an excess of energy or calories in relation to nutrients. Soft drinks, candy, sugar, alcohol, and animal fats are considered empty-calorie foods.

nutrient-dense foods Foods that contain relatively high amounts of nutrients compared to their calorie value. Broccoli, collards, bread, cantaloupe, and lean meats are examples of nutrient-dense foods.

Energy and Nutrient Density Most Americans consume more calories than needed, become overweight as a result, *and* consume inadequate diets. This situation is partly due to over-consumption of **energy-dense foods** such as processed and high-fat meats, chips, candy, many desserts, and full-fat dairy products. Energy-dense foods have relatively high-calorie values per unit weight of the food. Intake of energy-dense diets is related to the consumption of excess calories and to the development of overweight and diabetes.[23]

Many energy-dense foods are nutrient poor, or contain low levels of nutrients given their caloric value. These foods are sometimes referred to as **empty-calorie foods** and include products such as soft drinks, sherbet, hard candy, alcohol, and cheese twists. Excess intake of energy-dense and empty-calorie foods increases the likelihood that calorie needs will be met or exceeded before nutrients needs are met.[22] Diets most likely to meet nutrient requirements without exceeding calorie need contain primarily **nutrient-dense foods**, or foods with high levels of nutrients and relatively low calorie value. Nutrient-dense foods such as non-fat milk and yogurt, lean meat, dried beans, vegetables, and fruits provide relatively high amounts of nutrients compared to their calorie value.[23] Illustration 1.11 shows a comparison of the calorie and nutrient content of an empty-calorie and a nutrient-dense food.

NUTRITION CONCEPT #10

There are no "good" or "bad" foods.

People tend to classify foods as being "good" or "bad," but such opinions over-simplify each food's potential contribution to a diet.[24] Typically hot dogs, ice cream, candy, bacon, and french fries are judged to be bad, whereas vegetables, fruits, and whole grain products are given the "good" stamp. Unless we're talking about spoiled stew, poison mushrooms, or something similar, however, no food can be firmly labeled as "good" or "bad." Ice cream can be a "good" food for physically active, normal-weight individuals with a

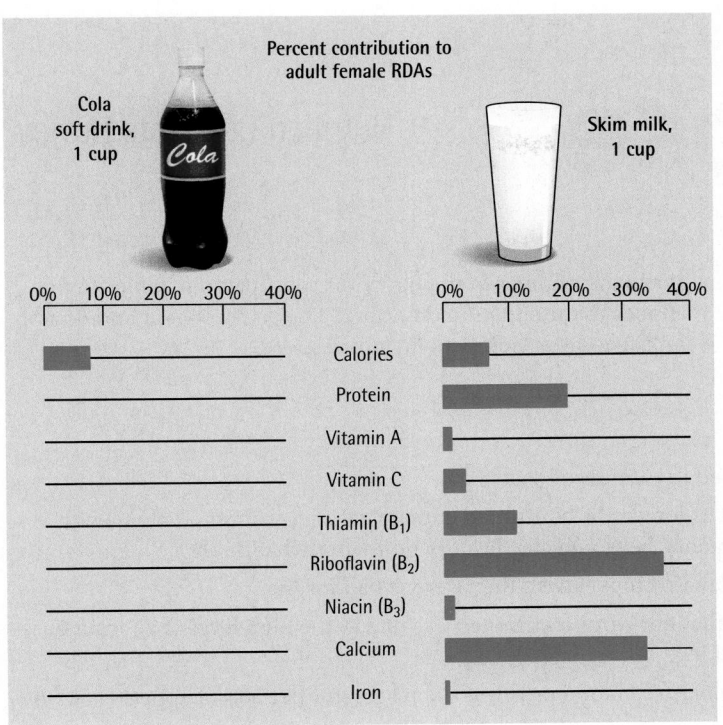

Illustration 1.11 Calorie and nutrient content of an empty-calorie and a nutrient-dense food. Percentages given represent percent contributions to adult female RDAs.

high-calorie need who have otherwise met their nutrient requirements by consuming nutrient-dense foods. Some people who eat only what they consider to be "good" foods such as broccoli, berries, brown rice, and tofu may still miss the healthful diet mark due to inadequate consumption of essential fatty acids and certain vitamins and minerals.

All foods can fit into a healthful diet as long as nutrient needs are met at calorie intake levels that maintain a healthy body weight.[24] If nutrient needs are not being met and calorie intake levels are too high, then the diet likely includes too many energy-dense or empty-calorie foods. Substituting nutrient-dense for energy-dense foods would help bring the diet back into balance.[25]

• •

The basic nutrition concepts presented here are listed in Table 1.6. It may help you to remember the concepts and to start "thinking like a bona fide nutritionist," if you go back over each concept and give several examples related to it. If you understand these concepts, you will have gained a good deal of insight into nutrition.

Table 1.6 Nutrition concepts

1. Food is a basic need of humans.
2. Foods provide energy (calories), nutrients, and other substances needed for growth and health.
3. Health problems related to nutrition originate within cells.
4. Poor nutrition can result from both inadequate and excessive levels of nutrient intake.
5. Humans have adaptive mechanisms for managing fluctuations in nutrient intake.
6. Malnutrition can result from poor diets and from disease states, genetic factors, or combinations of these factors.
7. Some groups of people are at higher risk of becoming inadequately nourished than others.
8. Poor nutrition can influence the development of certain chronic diseases.
9. Adequacy, variety, and balance are key characteristics of healthful diets.
10. There are no "good" or "bad" foods.

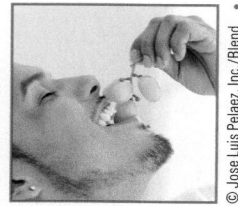

© Jose Luis Pelaez, Inc./Blend Images/Corbis

NUTRITION
up close

Nutrition Concepts Review

Focal Point: Nutrition concepts apply to diet and health relationships.

Write the number of the nutrition concept from Table 1.6 that applies to the situation described. Use each concept and do not repeat concept numbers in your responses.

Nutrition concept number	Situation
1. _____	The Irish potato famine caused thousands of deaths.
2. _____ and _____	Otis mistakenly thought that as long as he consumed enough calories from food along with vitamin and mineral supplements, he would stay healthy no matter what he ate.
3. _____	I feel guilty every time I eat potato chips. I wish they weren't bad for me.
4. _____	Phyllis was relieved to learn that her chronic diarrhea was due to the high level of vitamin C supplements she had been taking.
5. _____	A low amount of iron in Tawana's red blood cells was the reason for her loss of appetite and low energy level.
6. _____	Far more young children than soldiers died as a result of the 10-year civil war in Sudan.
7. _____	For the past 20 years, Don's idea of dinner was a big steak and potatoes. His recent heart attack changed his view of what's for dinner.
8. _____	During the two weeks they were backpacking in the Netherlands, Tomás and Ozzie ate very few vegetables and fruits. Their health remained robust, however.
9. _____	Zhang wasn't aware that he had the inherited condition hemochromatosis until he began taking iron supplements and developed iron overload symptoms.

Feedback to the Nutrition Up Close is located in Appendix G.

REVIEW QUESTIONS

- **Explain the scope of nutrition as an area of study.**

1. Nutrition is defined as "the study of foods, their nutrients and other chemical constituents, and the effects that food constituents have on health."   **True/False**

2. ____ Cassandra is on her way to her first nutrition class and is thinking about what the course will cover. Listed below are her ideas. Three of the ideas correspond to the scope of the study of nutrition. Which idea is *not* a component of the scope of the study of nutrition?

   a. diet and disease relationships
   b. components of healthful diets
   c. magical powers of super foods for weight loss
   d. nutrient composition of foods

- **Demonstrate a working knowledge of the meaning of the 10 nutrition concepts.**

3. The word *nonessential* as in *nonessential nutrient* means that the nutrient is *not* required for growth and health. **True/False**

4. Food insecurity is a problem in developing countries, but it is *not* a problem in the United States or Canada. **True/False**

5. Nutrients are classified into five basic groups: carbohydrates, protein, fats, vitamins, and water.   **True/False**

6. The development of standards for nutrient intake levels was prompted in part by the high rejection rate of World War II recruits due to underweight and nutrient deficiencies.   **True/False**

7. Tissue stores of nutrients decline after blood levels of the nutrients decline.   **True/False**

8. To maintain health, all essential nutrients must be consumed at the recommended level daily.   **True/False**

9. An individual's genetic traits play a role in how nutrient intake affects disease risk.   **True/False**

10. ____The Recommended Dietary Allowance for protein for a 7-month-old infant in grams/day (g/d) is:

   a. 52
   b. 9
   c. 71
   d. 11

11. ____The Tolerable Upper Intake Level for iron for a 65-year-old male in milligrams per day (mg/d) is:

   a. 1,100
   b. 45
   c. 350
   d. 3.5

12. Certain phytochemicals, or_____, such as lycopene and anthocyanins act as _____.

13. Groups of people at higher risk than others of becoming inadequately nourished include _____ and _____.

Match the term in column A with its definition in column B.

Column A	Column B
____14. Essential nutrients	a. Chemical substances in food that are used by the body for growth and health.
____15. Nutrients	b. Chemical substances that prevent or repair damage to cells caused by exposure to oxidizing agents such as environmental pollutants, smoke, ozone, and oxygen.
____16. Food insecurity	c. Foods that contain relatively high amounts of nutrients in relation to their calorie value.
____17. Antioxidants	d. The chemical changes that take place in the body.
____18. Calorie	e. Limited or uncertain availability of safe, nutritious foods—or the ability to acquire them in socially acceptable ways.
____19. Metabolism	f. A unit of measure of the amount of energy supplied by food.
____20. Malnutrition	g. Nutrients required for normal growth and health that the body can generally not produce, or produce in sufficient amounts.
____21. Nutrient dense	h. Poor nutrition resulting from an excess or lack of calories or foods nutrients.

Answers to these questions can be found in Appendix G.

NUTRITION SCOREBOARD ANSWERS

1. Calories are a measure of the amount of energy supplied by food. They're a property of food, not a substance present in food.   **False**

2. That's the definition of nutrients.   **True**

3. Excessive as well as inadequate intake levels of vitamins and minerals can be harmful to health.   **False**

4. The DRIs are shown on the inside front and back covers of this text.   **True**

The Inside Story about Nutrition and Health

NUTRITION SCOREBOARD

1 How long people live and how healthy they are primarily depends on four factors: lifestyle behaviors, the environments to which people are exposed, genetic factors, and access to quality health care. **True/False**

2 Diet is related to the top two causes of death in the United States. **True/False**

3 Biological processes of modern humans were designed over 40,000 years ago. **True/False**

Answers can be found at the end of the unit.

After completing Unit 2 and its interactive learning features, you will be able to:

- Identify characteristics of diets related to the development of specific diseases.

- Explain how differences in diets of early versus modern humans may promote the development of certain diseases.

- List the types of food that are core components of healthful diets.

Nutrition in the Context of Overall Health

- **Identify characteristics of diets related to the development of specific diseases.**

Think of your body as a machine. How well this machine performs depends on a number of related factors: the quality of its design and construction, the appropriateness of the materials used to produce it, and how well it is maintained.

A machine designed to produce 10,000 copies a day will break down sooner if it is used to make 20,000 copies a day. The repair call will, in all probability, come earlier if the machine is overused *and* poorly maintained or if it has a part that doesn't work well. On the other hand, chances are good the copy machine will function at full capacity if it is free from design flaws, skillfully constructed from appropriate materials, properly used, and kept in good shape through regular maintenance.

Although much more complex and sophisticated, the human body is like a machine in some important ways. How well the body works and how long it lasts depend on a variety of interrelated factors. The health and fitness of the human machine depend on genetic traits (the design part of the machine), the quality of the materials used in its construction (your diet), and regular maintenance (your diet, other lifestyle factors, and health care).

Lifestyles exert the strongest overall influence on health and longevity (Illustration 2.1).[2] Behaviors that constitute our lifestyle—such as diet, smoking habits, illicit drug use or excessive drinking, level of physical activity or psychological stress, and the amount of sleep we get—largely determine whether we are promoting health or disease. Of the lifestyle factors that affect health, our diet is one of the most important.[3] In a sense, it is fortunate that diet is related to disease development and prevention. Unlike age, gender, and genetic makeup, our diets are within our control.

People have an intimate relationship with food—each year we put over a thousand pounds of it into our bodies! Food supplies the raw materials the body needs for growth and health; these, in turn, are affected by the types of food we usually eat. The diet we feed the human machine can hasten, delay, or prevent the onset of an impressive group of today's most common health problems.

Key Nutrition Concepts

Unit 1 presents 10 key nutrition concepts that are fundamental to the science of nutrition. Content on diet and health covered in this unit directly relates to three of them:

1. Health problems related to nutrition originate within cells.

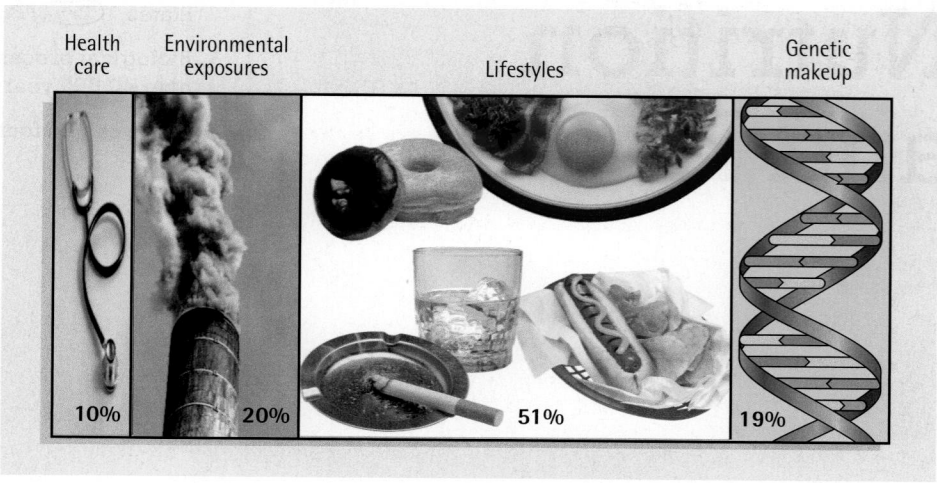

Illustration 2.1 Conditions that contribute to death among adults under the age of 75 years in the United States. Health care refers to access to quality care; environmental exposures include the safety of one's surroundings and the presence of toxins and disease-causing organisms in the environment; lifestyle factors include diet, exercise, obesity, smoking, genetic traits, and alcohol and drug use.[4,5]

Health care   Environmental exposures   Lifestyles   Genetic makeup

10%   20%   51%   19%

Photo Disc (1–3); © Cengage Learning (4)

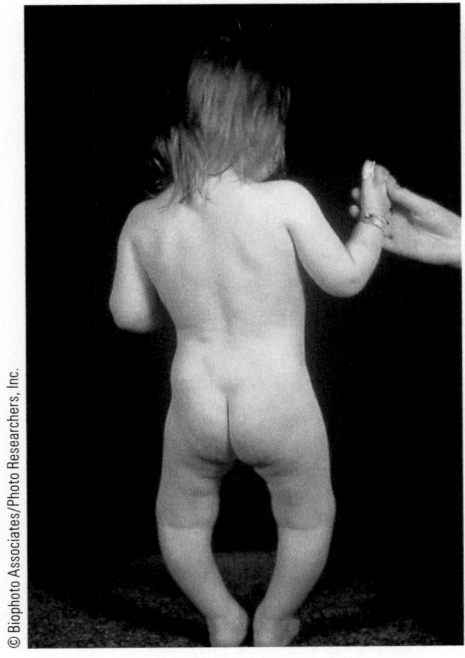
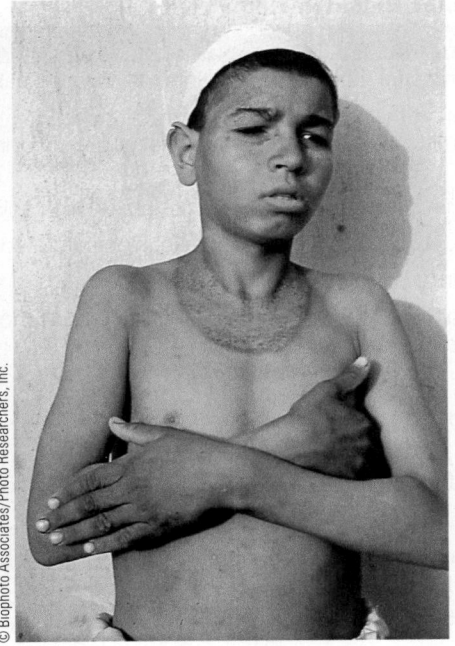

Illustration 2.2 Vitamin D deficiency (rickets shown on the left) and niacin deficiency (pellagra pictured on the right) were leading causes of hospitalization of children in the United States in the 1930s.

2. Malnutrition can result from poor diets and from disease states, genetic factors, or combinations of these factors.

3. Poor nutrition can influence the development of certain chronic diseases.

The Nutritional State of the Nation

Since early in the twentieth century, researchers have known that what we eat is related to the development of vitamin and mineral deficiency diseases, to compromised growth and impaired mental development in children, and to the body's ability to fight infectious diseases. Seventy years ago in the United States, widespread vitamin deficiency diseases filled children's hospital wards and contributed to serious illness and death in adults (Illustration 2.2). Now, however, dietary excesses are filling hospital beds and reducing the quality of life for millions of Americans.

Today, the major causes of death among Americans are slow developing, lifestyle-related **chronic diseases**. Based on government survey data, 44% of Americans have a chronic condition such as **diabetes**, heart disease, cancer, **hypertension**, or high cholesterol levels; 13% have three or more of these conditions.[7]

The leading causes of death among Americans are heart disease and cancer (see Illustration 2.3). Together they account for 48% of all deaths. Western-type diets high in saturated and *trans* fats, and low in vegetables, fruits, and whole grain products are linked to the development of heart disease.[9] Six types of cancer—including colon, pancreatic, and breast cancer—are related to obesity, habitually low intakes of vegetables and fruits, and high intake of processed meats.[8]

Diet is related to many other diseases and disorders, including stroke, osteoporosis, and obesity. Examples of such relationships are overviewed in Table 2.1.

Shared Dietary Risk Factors A number of the diseases and disorders listed in Table 2.1 share the common risk fac-

chronic diseases Slow-developing, long-lasting diseases that are not contagious (e.g., heart disease, cancer, diabetes). They can be treated but not always cured.

diabetes A disease characterized by abnormal utilization of glucose by the body and elevated blood glucose levels. There are three main types of diabetes: type 1, type 2, and gestational diabetes. The word *diabetes* in this text refers to type 2 diabetes, by far the most common. Diabetes is short for the term *diabetes mellitus*.

hypertension High blood pressure. It is defined as blood pressure exerted inside of blood vessel walls that typically exceeds 140/90 mm Hg (or, millimeters of mercury).

Illustration 2.3 Percentage of total deaths for the top 10 leading causes of death in the United States, 2009.[6]

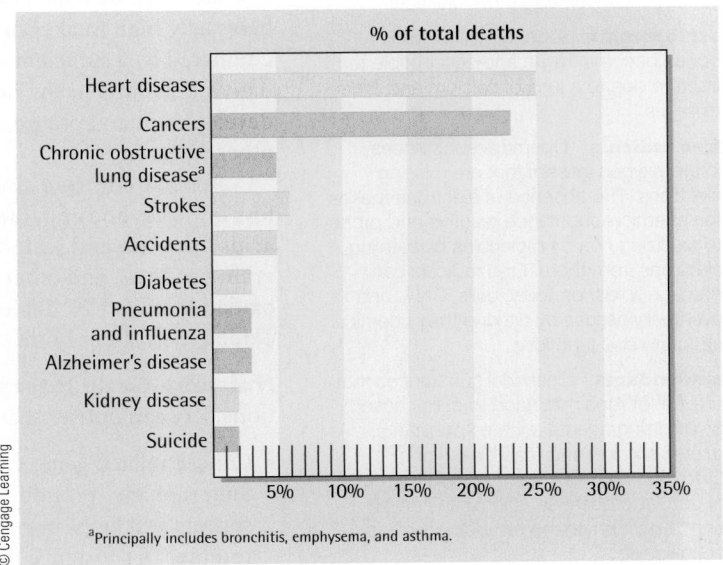

% of total deaths

Heart diseases
Cancers
Chronic obstructive lung disease[a]
Strokes
Accidents
Diabetes
Pneumonia and influenza
Alzheimer's disease
Kidney disease
Suicide

5% 10% 15% 20% 25% 30% 35%

[a]Principally includes bronchitis, emphysema, and asthma.

© Cengage Learning

Table 2.1 Examples of diseases and disorders linked to diet[9–13]

Disease or disorder	Dietary connections
Heart disease	High saturated and *trans* fat, and cholesterol intakes; low vegetable, fruit, and whole grain intakes; excessive body fat
Cancer	Low vegetable and fruit intakes; excessive body fat and alcohol intake, regular consumption of processed meats
Stroke	Low vegetable and fruit intake; excessive alcohol intake, high animal-fat diets
Diabetes (type 2)	Excessive body fat; low vegetable and fruit intake; high saturated fat and energy-dense food intake
Cirrhosis of the liver	Excessive alcohol consumption; poor overall diet
Hypertension	Excessive sodium (salt) and low potassium intake, excess alcohol intake; low vegetable and fruit intake; excessive levels of body fat
Iron-deficiency anemia	Low iron intake
Tooth decay and gum disease	Excessive and frequent sugar consumption; inadequate fluoride intake
Osteoporosis	Inadequate calcium and vitamin D, low intakes of vegetables and fruits
Obesity	Excessive calorie intake, overconsumption of energy-dense, nutrient-poor foods.
Chronic inflammation and oxidative stress	Excessive calorie intake, excessive body fat, high animal-fat diets, low intake of whole grains, vegetables, fruit, and fish, poor vitamin D status
Alzheimer's disease	Regular intake of high animal-fat products; low intake of olive oil, vegetables, fruits, fish, wine, and whole grains

chronic inflammation Low-grade inflammation that lasts weeks, months, or years. Inflammation is the first response of the body's immune system to infectious agents, toxins, or irritants. It triggers the release of biologically active substances that promote oxidation and other reactions to counteract the infection, toxin, or irritant. A side effect of chronic inflammation is that it also damages lipids, cells, and tissues.

oxidative stress A condition that occurs when cells are exposed to more oxidizing molecules (such as free radicals) than to antioxidant molecules that neutralize them. Over time, oxidative stress causes damage to lipids, DNA, cells and tissues. It increases the risk of heart disease, type 2 diabetes, cancer, and other diseases.

osteoporosis A condition in which bones become fragile and susceptible to fracture due to a loss of calcium and other minerals.

free radicals Chemical substances (often oxygen-based) that are missing electrons. The absence of electrons makes the chemical substance reactive and prone to oxidizing nearby molecules by stealing electrons from them. Free radicals can damage lipids, proteins, cells, DNA, and eventually tissues by altering their chemical structure and functions.

antioxidants Chemical substances that prevent or repair damage to cells caused by oxidizing agents such as pollutants, ozone, smoke, and reactive oxygen. Oxidation reactions are a normal part of cellular processes. Vitamins C and E and certain phytochemicals function as antioxidants.

tors of low intakes of vegetables, fruits, and whole grains; excess calorie intake and body fat, and high-animal fat intake. These risk factors are associated with the development of **chronic inflammation** and **oxidative stress**, conditions that are strongly related to the development of heart disease, diabetes, **osteoporosis**, Alzheimer's disease, cancer, and other chronic diseases.[14]

Chronic Inflammation and Oxidative Stress Inflammation is a normal response of the body to the presence of infectious agents, toxins, or irritants. It neutralizes these threats, in part, by triggering the release of oxidizing agents such as **free radicals** that destroy the offending substances. In the process, however, free radicals also oxidize fats (lipids), cell membranes, and DNA inside of cells. In the short term, damage induced by oxidation reactions can generally be repaired by **antioxidants** produced by the body and consumed in vegetables, fruits, whole grain products, and other plant foods.[14,15]

Inflammation is related to chronic disease development when it is present at a low level for a long time, or is "chronic." Chronic inflammation and the resulting oxidative stress are sustained by irritants continually present in the body. Excess body fat and habitually high intakes of saturated and *trans* fats are examples of such irritants. If not countered by a sufficient supply of antioxidants, chronic inflammatory processes and oxidative stress impair the normal functioning of cells and tissues in ways that promote the development and progression of diabetes, heart disease, Alzheimer's disease, cancer, and other diseases.[16,17]

Adverse effects of chronic inflammation and oxidative stress can be diminished by loss of excess body fat, diets low in saturated and *trans* fats, and intake of omega-3 fatty acids from fish and seafood. Adequate intake of antioxidant-rich vegetables, fruits, whole grain products, and other plant foods can reduce the damage done to cells and tissues by oxidative stress.[9,11,15] Table 2.2 lists types of foods that decrease and increase inflammation, oxidative stress, or both.

Nutrient–Gene Interactions and Health Some diseases are promoted by interactions between nutrients and genes. Here are a few examples:

- Cancer-related genes can be activated or deactivated by certain components of food. Sulforaphane (pronounced sul-four-ah-phane) found in cabbage, broccoli, and brussels sprouts and helps prevent several types of cancer. Sulforaphane inactivates a gene that produces a substance that encourages cancer development.[19]

- About half of the U.S. population is genetically susceptible to cholesterol in the diet. For them, blood cholesterol levels increase with high cholesterol intake. People who lack the trait are considered "cholesterol nonresponders" because their blood cholesterol levels do not increase in response to a high cholesterol intake.[20]

- Fish and seafood rich in omega-3 fatty acids improve mental functioning in adults who are genetically susceptible to dementia. Initial evidence indicates that omega-3 fatty acids may lower the risk of Alzheimer's disease in genetically susceptible adults.[21]

Knowledge of nutrient–gene interactions in health and disease is expanding rapidly and is greatly enhancing our understanding of the relationship between diet and health.

The Importance of Food Choices

People are not born with an internal compass that directs them to select a healthy diet— and it shows. If given access to a food supply like that available in the United States, people show a marked tendency to choose a diet that is high in energy-dense, nutrient-poor foods[22] (Illustration 2.4). Such a diet tends to include processed foods high in saturated fat, salt, or sugar and low in whole grains, vegetables, fruits, and other basic foods. This type of diet poses the greatest risks to the health of Americans.[23]

Diet and Diseases of Western Civilization

- **Explain how differences in diets of early versus modern humans may promote the development of certain diseases.**

Why is the U.S. diet—a "Western" style of eating—hazardous to our health in so many ways? What is it about a diet that is high in animal fat, salt, and sugar and low in vegetables, fruits, and whole grains that promotes certain chronic diseases? A good deal of evidence indicates that the chronic diseases now prevalent in the United States and other Westernized countries have roots in dietary changes that have taken place over centuries.

Table 2.2 Types of food associated with decreased or increased inflammation, oxidative stress, or both[9–14,18]

Decreased
Colorful fruits and vegetables
Dried beans
Whole grains
Fish and seafood, fish oils
Red wines
Dark chocolates
Olive oil
Nuts
Coffee

Increased
Processed and high-fat meats
High-fat dairy products
Baked products, snack foods with *trans* fats
Soft drinks, other high-sugar beverages

Illustration 2.4 Lopsided, all-American food choices.

Photo Disc

Our Bodies Haven't Changed

The biological processes that control what the human body does with food were developed more than 40,000 years ago. These hardwired, evolution-driven processes exist today because they are firmly linked to the genetic makeup of humans, and genes change very little over great spans of time.[24]

Then . . . For the first 200 centuries of their existence, humans survived by hunting and gathering (Illustration 2.5). They were constantly on the move, pursuing wild game or following the seasonal maturation of fruits and vegetables. Meat, berries, and many other plant products obtained from successful hunting and gathering journeys spoiled quickly, so they had to be consumed in a short time. Feasts would be followed by famines that lasted until the next successful hunt or harvest.[24]

. . . and Now The bodies of modern humans, adapted to exist on a diet of wild game, fish, fruits, nuts, seeds, roots, vegetables, and grubs; to survive periods of famine; and to sustain a physically demanding lifestyle are now exposed to a different set of circumstances. The foods we eat bear little resemblance to the foods available to our early ancestors (Illustration 2.6). Sugar, salt, alcohol, food additives, oils, margarine, dairy products, refined grain products, and processed foods were not a part of their diets. These ingredients and foods came with Western civilization.[24] Furthermore, we do not have to engage in strenuous physical activity to obtain food, and our feasts are no longer followed by famines.

The human body developed other survival mechanisms that are not the assets they used to be. Mechanisms that stimulate hunger in the presence of excess body fat stores, conserve the body's supply of sodium, and confer an innate preference for sweet-tasting foods—as well as a digestive system that functions best on a high-fiber diet—were advantages for early humans. They are not advantageous for modern humans, however, because our diets and lifestyles are now vastly different.

Although the human body has a remarkable ability to adapt to changes in diet, health problems of modern civilization such as heart disease, cancer, hypertension, and diabetes are thought to result, in part, from diets that are greatly different from those of our early ancestors. The human body was built to function best on a diet that is low in sugar and sodium, contains lean sources of protein, and is high in fiber, vegetables, and fruits.[25] Strong evidence for this conclusion is provided by studies that track how disease rates change as people adopt a Western style of eating.

© Marjorie Shostak/Anthro-Photo

© Marjorie Shostak/Anthro-Photo

Illustration 2.5 Hunter-gatherers still exist in the world, but their numbers are diminishing. It is estimated that hunter-gatherers consume approximately 3,000 calories daily due to their physically demanding way of life.

Illustration 2.6 The disconnect between high animal fat, high salt, high sugar, and processed foods in Western-type diets (right) and wild plants and animal foods consumed by our early ancestors (left). Foods consumed by hunter-gatherers shown in the photograph include bird's eggs, wild cucumbers, roots, nuts, and berries. Not shown are grubs and other insects, which might be consumed as quickly as they are discovered.

Changing Diets, Changing Disease Rates

Many countries are adopting the Western diet and the pattern of disease that accompanies it. People in Japan, for example, live longer than anyone else in the world—until they move to the United States (Table 2.3). In Japan, the traditional diet consists mainly of rice, vegetables, fish, shellfish, and meat (Illustration 2.7).

When Japanese people move to the United States, their diets change to include, on average, more fat, sugar, and calories; and less fish and vegetables.[27] Japanese living in the United States are much more likely to develop diabetes (Illustration 2.8), heart disease, breast cancer, and colon cancer than people who remain in Japan.[27, 28]

Dietary habits in Japan are rapidly becoming similar to those in the United States. Hamburgers, fries, steak, ice cream, and other high-fat foods are gaining in popularity. Rates of diabetes, heart disease, and cancer of the breast and colon are on the rise in Japan.[27] Similarly, the "diseases of Western civilization" are occurring at increasing rates in Russia, Greece, Israel, and other countries adopting the Western diet.[29] Obesity, diabetes, and heart disease rates tend to increase among some population groups after they immigrate to the United States. Latinos who move to the United States, for example, tend

Table 2.3 Life expectancy at birth for countries with high life expectancies, 2007[26]

Country	Life expectancy (years)	Country	Life expectancy (years)
Japan	82.6	Germany	80.0
Switzerland	82.0	Belgium	79.9
Australia	81.4	Ireland	79.8
Spain	81.1	Finland	79.6
France	81.0	Greece	79.5
Italy	80.8	United Kingdom	79.4
Canada	80.7	Portugal	79.1
Norway	80.6	Denmark	78.4
Netherlands	80.2	United States	77.9
New Zealand	80.2	Mexico	75.6
Austria	80.1		

Illustration 2.7 Traditional Japanese foods. Compare these to the "all-American" food choices shown in Illustration 2.4.

Photo Disc

to consume poorer quality diets, are more likely to become obese, and are more likely to develop diabetes than people in their home countries.[30]

Today's food supply makes it a bit challenging to eat more like our early ancestors did. What types of foods would you choose if you wanted to shape a diet that is closer to that of hunter-gathers? Join Beth and Shandra in making these dietary decisions in the Reality Check.

The Power of Prevention

Heart disease, cancer, and other chronic diseases are not the inevitable consequence of Westernization. The types of diets that promote chronic disease can be avoided or changed. Although heart disease is still the leading cause of death in the United States, its rate has declined by over 50% in the last 30 years. About half of this decline is related to improvements in risk factors such as smoking, hypertension, and elevated blood cholesterol levels, the other half is primarily related to medical interventions for people with heart disease.[31] The American Heart Association concludes that future gains in heart health among Americans will primarily stem from improved dietary intakes, declines in rates of overweight and obesity, increased physical activity, and decreased smoking.[32]

Illustration 2.8 An increased rate of diabetes in Japanese men immigrating to Seattle corresponds to dietary changes. Source: Tsunehara CH, et al. Diet of second-generation Japanese-American men with and without non-insulin-dependent diabetes. *Am J Clin Nutr.* 1990;52:731–38.

Table 2.4 Examples of the *Healthy People 2020* nutrition objectives for the nation[33]

Healthier food access

- Increase the proportion of schools that offer nutritious foods and beverages outside of school meals.
- Increase the proportion of Americans who have access to a food retail outlet that sells a variety of foods that are encouraged by the Dietary Guidelines.

Weight status

- Increase the proportion of adults who are at a healthy weight.
- Reduce the proportion of adults who are obese.
- Reduce the proportion of children and adolescents who are considered obese.
- Prevent inappropriate weight gain in youths and adults.

Food insecurity

- Eliminate very low food security among children.
- Reduce household food insecurity and in doing so reduce hunger.

Food and nutrient consumption

- Increase the variety and contribution of vegetables to diets.
- Increase the contribution of whole grains to diets.
- Reduce the consumption of calories from solid fats and added sugars.
- Reduce the consumption of saturated fat.
- Reduce the consumption of sodium.
- Increase the consumption of calcium.
- Reduce the consumption of sodium.

Iron deficiency

- Reduce iron deficiency among young children and females of childbearing age.
- Reduce iron deficiency among pregnant females.

Health care and worksite settings

- Increase the proportion of physician office visits that include counseling or education related to nutrition and weight.

Improving the American Diet

• List the types of food that are core components of healthful diets.

Many efforts are under way to improve the diet and health status of Americans. Like other countries with high rates of "Western" diseases, the United States sets national health goals and implements programs aimed at improving health. Goals and objectives for changes in health status in the United States are presented in the report *Healthy People 2020*. Examples of the *Healthy People 2020* objectives for nutrition are shown in Table 2.4. These objectives highlight the national emphasis on improving weight status and dietary intake of the population by the year 2020.

REALITY CHECK
Getting Back to the Basics—But How?

Beth and Shandra have been roommates for a year and usually shop for groceries together. For the next trip to the grocery store, they decide that each of them will make up a shopping list that includes foods that resemble those their early ancestors might have eaten. Here are the results.

Which list do you think comes closest to matching the basic foods consumed by our early ancestors? Answers on page 2-10

Beth: rice, yogurt, pork, honey, olive oil

Shandra: carrots, nuts, asparagus, fish, blueberries

Shandra's list contains unprocessed plant foods and fish, which would have been available to our early ancestors.

Beth:

Shandra:

What Should We Eat?

Evidence-based conclusions about nutrition and health are translated into food choices by the United States Department of Agriculture's (USDA's) food guidance materials. Intended for public consumption, the materials specify food choices that build healthy diets and contribute to the prevention of common diseases such as diabetes, hypertension, and heart disease.

USDA's advice on healthy food choices changes as knowledge about diet and health relationships advances, and so do the names it uses to label the new food guidance materials. In the past nine decades, USDA's food choice guidance has changed from the "Basic Five Food Groups," to the "Basic Four Food Groups," to "MyPyramid." Beginning in 2011, food guidance materials became labeled "MyPlate" and "ChooseMyPlate" and reflect current concerns about food choices, nutrition, and heath (Table 2.5).

Current food intake recommendations focus on basic, nutrient-dense foods such as whole grain products, vegetables, fruits, lean meats, low-fat dairy products, dried beans, and fish. They call for reduced consumption of soft drinks and other sweetened beverages, high-fat meats and dairy products, and foods high in salt. Sweets, desserts, and packaged snacks are not excluded from the recommendations. They can be included in a healthy diet as long as food group recommendations are met and overall calorie needs are not exceeded. ChooseMyPlate interactive tools for planning and evaluating dietary intake are available at ChooseMyPlate.gov.

Nutrition Surveys: Tracking the American Diet

The food choices people make and the quality of the American diet and food supply are regularly evaluated by national surveys (Table 2.6). The first survey began in 1936. It was conducted in conjunction with the original national program aimed at reducing hunger, poor growth in children, and vitamin and mineral deficiency diseases. Results of nutrition surveys are used to identify problem areas within the food supply, characteristics of diets consumed by the public, and the prevalence of nutrition-related health disorders. The surveys provide information ranging from the amount of lead and pesticides in certain foods to the adequacy of diets of low-income families. Together with the results of studies conducted by university researchers and others, they provide the information needed to give direction to food and nutrition programs, and policies aimed at improving the availability and quality of the food supply.

Table 2.5 ChooseMyPlate.gov food guidance

Balancing calories
- Enjoy your food, but eat less.
- Avoid oversized portions.

Foods to increase
- Make half your plate fruits and vegetables.
- Make at least half your grains whole grains.
- Switch to fat-free or low-fat (1%) milk.

Foods to reduce
- Compare sodium in foods like soup, bread, and frozen meals—and choose the foods with lower numbers.
- Drink water instead of sugary drinks.

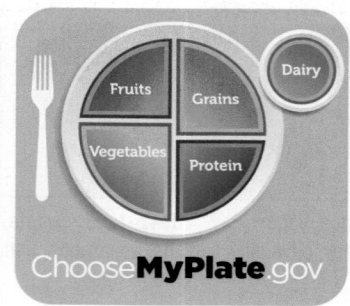

Source: www.ChooseMyPlate.gov

Table 2.6 Periodic, national surveys of food, diet, and health in the United States

Survey	Purpose
1. National Health and Nutrition Examination Survey (NHANES)	Assesses dietary intake, health, and nutritional status in a sample of adults and children in the United States on a continual basis
2. Nationwide Food Consumption Survey (NFCS)	Performs regular surveys of food and nutrient intake and understanding of diet and health relationships among a national sample of individuals in the United States
3. Total Diet Study (sometimes called Market Basket Study)	Ongoing studies that determine the levels of various pesticide residues, contaminants, and nutrients in foods and diets

NUTRITION
up close

Food Types for Healthful Diets

Focal Point: Separating the types of foods that fit into the ChooseMyPlate basic food groups from those that characterize Western-type diets.

The ChooseMyPlate food groups consist of foods that make up a healthful, disease-preventing diet. The typical Western-type diet, on the other hand, is related to the development of a number of diseases and includes many types of food that are not part of the ChooseMyPlate basic food groups. To which dietary pattern do the following types of food most appropriately belong?

Food types	Dietary pattern	
	ChooseMyPlate	Western
Mixed vegetables	mixed vegetables	cold cuts
Cold cuts (ham, bologna, salami)	fish + seafood	fruit jams + jellies
Fish and seafood	whole grain bread	potato, tortilla + snack chips
Whole grain breads	dried beans	gravy
Fruit jams and jellies	poultry	ice cream
Potato, tortilla, and other snack chips	fruit	soft drink
Dried beans	vegetables	salad dressing
Poultry (chicken, turkey)	skim milk	cake
Fruit		
Vegetables		
Gravy		
Ice cream		
Soft drinks		
Salad dressing, mayonnaise		
Skim milk		
Cake		

Feedback to the Nutrition Up Close is located in Appendix G.

REVIEW QUESTIONS

- **Identify characteristics of diets related to the development of specific diseases.**

1. Vitamin deficiency diseases remain a major health problem of children in the United States. **True/False**

2. Diets are related to the top two causes of death. **True/False**

3. Low intake of vegetables and fruits is related to the development of heart disease, cancer, hypertension, and osteoporosis. **True/False**

4. The incidence of chronic diseases such as diabetes, cancer, and heart disease increases as countries adopt a Western style of eating. **True/False**

- **Explain how differences in diets of early versus modern humans may promote the development of certain diseases.**

5. Our genetic makeup changes as our diets change. **True/False**

6. List two major characteristics of the lifestyles of early humans cited in this Unit that are rare in modern humans and don't involve types of foods available.
   a. _____
   b. _____

7. ____ Chronic inflammation and oxidative stress are sustained by irritants continually present in the body. Two

examples of irritants that promote the presence of these conditions in the body are:

a. physical inactivity and meal skipping
b. weight loss and lack of sleep
c. excess body fat and continually high intake of saturated fat
d. high intake of water and excess dietary fiber

8. ____ Adverse effects of chronic inflammation and oxidative stress can be diminished by:

a. loss of excess body fat
b. increased water consumption
c. decreased intake of high-salt foods
d. daily consumption of potatoes

9. Although individuals cannot change the genes they inherit at birth, they can change their risk for chronic disease development by making healthful eating and lifestyle changes.   **True/False**

- **List the types of food that are core components of healthful diets.**

10. National surveys in the United States assess food intake, nutritional health of the population, and the safety of the food supply.   **True/False**

11. The *Basic Four Food Groups* is the most recently published guide to selection of a healthful diet.   **True/False**

12. ____ Which of the following statements about food choices is true?

a. Given the availability of a wide assortment of foods, people tend to select and consume a healthful diet.
b. People are *not* born with an internal compass that directs them to select and consume a healthy diet.
c. Food choices usually change very little as people move from one country to another.
d. Recommendations for healthy diets include a narrow range of food choices.

Questions 13–15 refer to the following case scenario:

Assume you are attending a family picnic and are staring at a table full of food. Your choices from the table are: fried chicken, cold-cut platter (bologna, salami), whole grain rolls, shrimp, jell-o, spinach salad, fruit salad, baked beans, potato chips, zucchini squash, cake, and low-fat milk.

13. ____ From the food options available, which three foods would you put on your plate if you wanted to consume the basic foods?

a. fried chicken, fruit salad, and baked beans
b. zucchini squash, jell-o, and baked beans
c. spinach salad, cold cuts, and shrimp
d. spinach salad, zucchini squash, and shrimp

14. ____ Which of the following sets of foods would be considered creations of modern humans?

a. potato chips and cold cuts
b. cake and spinach salad
c. jell-o and fruit salad
d. fried chicken and zucchini squash

15. ____ You decide to have a glass of milk along with shrimp, fruit salad, baked beans, and spinach salad. Which basic food group is missing from your plate?

a. vegetables
b. fruits
c. grains
d. protein foods

Answers to these questions can be found in Appendix G.

NUTRITION SCOREBOARD ANSWERS

1. There are no secrets to a long, healthy life.   True

2. Diet is associated with the development of heart disease and cancer, which cause about half of all deaths in the United States.[1]   True

3. Hairstyles may be different, but our bodies are the same as they were 40,000 years ago.   True

Ways of Knowing about Nutrition

UNIT

3

NUTRITION SCOREBOARD

1 It is illegal to convey false or misleading information about nutrition in magazine and newspaper articles and on television.   **True/False**

2 Knowledge about nutrition is gained by scientific studies.   **True/False**

3 "Double-blind" studies are used to diminish the "placebo effect."   **True/False**

Answers can be found at the end of the unit.

- Evaluate the reliability of advertisements and other information about nutrition and nutrition-related products and services.

- Identify sources of reliable nutrition information.

- Explain each component of the scientific methods used to identify reliable information about nutrition.

How Do I Know if What I Read or Hear about Nutrition Is True?

- **Evaluate the reliability of advertisements and other information you hear or read about nutrition and nutrition-related products and services.**

Maria, a preschool teacher: *"You really ought to try this new Herbal Melt Down Diet I found on the Internet. I used the herbs for a week and lost 5 pounds!"*

Newspaper headline: *"Eating Cauliflower Daily Prevents Breast Cancer"*

Web advertisement: *"Lose fat, gain muscle and energy! Eat Complete Cereal Pro bars every day!"*

How do you know if what you read or hear about nutrition is true? With only the information given in these examples, you couldn't know. In reality, however, this is how much of the information we receive about nutrition comes to us—in bits and pieces. The nutrition information offered to the public is a mix of truths, half-truths, and gossip. The information does not have to meet any standard of truth before it can be represented as true in books, magazines, newspapers, TV and radio reports and interviews, pamphlets, the Internet, and speeches.

Opinions expressed about nutrition are protected by the freedom of speech provisions of the U.S. Constitution. Although it is misleading and fraudulent for a tabloid article or an Internet site to announce "19 foods have negative calories," it is not illegal. The promotion of nutritional remedies that are not known to work, such as amino acid supplements for hair growth and tea extract for the treatment of cancer, is likewise protected by freedom of speech. It is illegal, however, to put false or misleading information about a product or service on a product label, in a product insert, or in an advertisement. In addition, the U.S. and Canadian mail systems cannot be used to send or to receive payments for products that are fraudulent.

With so much misinformation available, it is difficult to know what to do when we hear or read something about nutrition that may benefit us personally or perhaps help a friend or relative. Why does such a mix of nutrition sense and nonsense exist? How can you separate the sound information from the highly questionable? Where does nutrition information you can trust come from? These questions are answered in this unit.

Key Nutrition Concepts

Content in Unit 3 on how trustworthy information about nutrition is generated and how to identify it relates to these key nutrition concepts:

- Health problems related to nutrition originate within cells.

- Malnutrition can result from poor diets and from disease states, genetic factors, or combinations of these factors.

Profits, Prophets, and Proof

Consumers are bombarded with nutrition information, much of it about products and services that may be ineffective or scientifically untested. Likely reasons for this situation can be grouped into two major categories: profit, and personal beliefs and convictions. The first and most important reason is the profit motive.

Motivation for Nutrition Misinformation #1: Profit

". . . [join] a special company with 22nd century breakthrough nutritional products and unequaled compensation plan!"
—USA TODAY *BUSINESS ADVERTISEMENT*

*"Uncover new ways to beautify your bottom line—
use strategic nutrition
to develop great-tasting foods and drinks
that offer age-defying benefits."*
—NUTRITION MARKETING COMPANY ON THE WEB

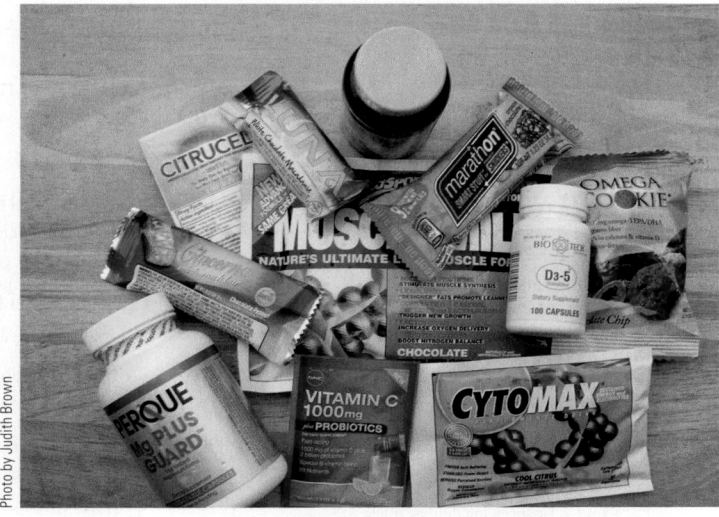

Illustration 3.1 Some examples of the many nutrition products available.

As long as consumers seek quick and easy ways to lose weight, build muscle, slow aging, and reduce stress—goals that cannot be achieved quickly or easily—there will continue to be a huge market for nutrition products and services that offer assistance (Illustration 3.1).

Not everyone who is in the business of nutrition has the goal of maintaining or improving people's health. Many seek to make money from people who are willing to believe their advertisements and buy products or services because they sound like they will work.

People may believe highly questionable claims about nutrition-related products, and be willing to buy and try the products for many reasons:

- Many people believe what they want to hear.

- People tend to believe what they see in print.

- Promotional materials sound scientific and true.

- Individuals may lack the knowledge and skills needed to evaluate product claims.

- The products offer solutions to important problems that have few or no solutions in orthodox health care.

- Promotional materials may appeal to people who are dissatisfied with traditional medical care, fear the costs or side effects of medications, or want a "natural" remedy.

- Many people believe that vitamins, minerals, proteins, amino acids, herbs, and similar products are healthful and harmless.

Instead of evidence and facts, profit-oriented companies may use testimonials ("It worked for me, it will work for you!"), "medical experts," and Hollywood stars and sports heroes to promote their products. Their advertisements promote ideas like "a wonder to science" or "miraculous" that appeal to some people's inclinations toward the mysterious or the divine. Real remedies aren't advertised in such terms. Nor are they portrayed as being effective because they come from Europe, the Ecuadoran highlands, the ancient Orient, or organic algae ponds.

Illustration 3.2 offers a formula for developing and marketing a fraudulent nutrition product. Try following the steps and make up your own miraculous nutritional cure. Once you've devised your own product, you'll find it easier to detect fraudulent marketing.

Controlling Profit-Motivated Nutrition Frauds The Federal Trade Commission (FTC) has the authority to remove advertisements that make false claims from the airwaves and the Internet. In the past, the FTC has exerted its authority by removing blatantly false and misleading advertisements, including those for the "European Weight Loss Patch," colon detoxifiers, juices with herbal extracts, coral calcium, male enhancement dietary supplements, and bee pollen. Nevertheless, misleading and inaccurate advertisements still appear because enforcement efforts are weak and are concentrated on very dangerous products. The FTC and other federal agencies do respond when several consumers register complaints about a nutritional product or advertisement. You can notify the FTC of a complaint from the website ftccomplaintassistant.gov.

Science for Sale Here's a true story:

Friday, 2:00 p.m. A call came into the Nutrition Department from a man who wanted to talk to a nutrition expert. He had heard on last night's news that zinc lozenges were good for treating

A Step-by-Step Guide

1. Identify a common problem people really want fixed that cannot easily or quickly be fixed another way.

   EXAMPLES: Obesity, low energy, weak muscles.

2. Make up a nutritional remedy and connect it to a biological process in the body. Try to think of a remedy that probably won't harm anyone. While you are at it, create a catchy name for you product.

   EXAMPLES: An herb that burns fat by speeding up metabolism, vitamins that boost the body's supply of energy, protein supplements that go directly to your muscles.

3. Develop a scientifically-sound ding explanation for the effect the product has on the body. Refer to the results of scientific studies.

   EXAMPLES: "Research has shown that the combination of herbs in this product stimulates the production of chemicals that trigger the energy-production cycle in the body." "As nutritional scientists have known for a century, the body requires B vitamins to form energy. The more you consumer of this special formulation of B vitamins, the more energy you can produce and the more energy you will have!" "The unique combination of amino acids in this product is the same as that found in the jaw muscles of African lions. It is well known that an African lion can lift a 600-pound animal in its teeth."

4. Dream up testimonials from bogus previous users of the product, before-and-after photos, or "expert" opinions to quote. Use terms like "magical," miraculous," suppressed by traditional medicine," "secret," or "natural" as much as possible.

   EXAMPLES: Jane Fondu, Indianapolis: "I didn't think this product was going to work. The other products I tried didn't. My doctor couldn't help me lose weight. Thank God for [insert name of product]! I miraculously lost 20 pounds a week while eating everything I wanted." Dr. J. R. Whatsit, DM, NtD, Director of Nutritional Research: "We discovered this secret herb in our laboratories after years of looking for the substance in plants that keeps them from getting fat. Voila! We found it. This discovery may win us a Nobel Prize." Or, show photos of a skinny person and a beefed-up person. (The photos don't have to be of the same person; the people only need to look similar.

5. Offer customer a money-back guarantee.

Illustration 3.2 Create your own fraudulent nutrition product.

cold symptoms. Because he had a cold and didn't think the zinc would hurt, he purchased and consumed a whole roll of lozenges. Now he had a horrible, metallic taste in his mouth that he couldn't get rid of, no matter how often he brushed his teeth or gargled with mouthwash. He was worried that the taste was going to stay in his mouth forever. The nutrition expert assured him that it wouldn't. He had overdosed on zinc, and the taste would go away slowly.

This man was not the only one who heard the newscast about zinc and consumed too many lozenges the next day. Sales of zinc lozenges skyrocketed after the newscast, and as it happened, the author of the research study profited handsomely from the increased sales. After completing the study, the author bought shares of stock in the company that made the lozenges.[1] Such a financial tie should have been reported in the article. It is possible that a financial incentive could have influenced how the author presented the study's results.

Some researchers have a vested interest in their research results (Illustration 3.3). A study of 1,000 Massachusetts scientists who had published research articles found that one-third held a patent for the product tested, were paid industry consultants, or had another form of financial stake in the research.[1] Another study found that authors of research supportive of a new artificial fat were four times more likely to have financial affiliations with the company producing the product than were authors with no such company ties.[2]

Who Is Conducting the Research? Nutrition research is sometimes conducted by people or companies that have a stake in the results. Although this doesn't mean the research results are invalid, it does mean that the study design should be carefully scrutinized

before it is published in a scientific journal and broadcast on the news. People and companies with vested interests in the results of studies may not report findings that reflect negatively on a product.[3] Also, their promotional materials may neglect to mention studies that produced different results. For these reasons, it is important to consider whether the results of the study and other claims may be tainted by purely financial motives.

A Checklist for Identifying Nutrition Misinformation   Illustration 3.4 lists some common features of fraudulent information for nutrition products and services. If you find any of these characteristics in articles, advertisements, websites, or pamphlets, or hear them on infomercials or TV and radio interviews, beware of the product or service. People offering legitimate information are cautious and don't exaggerate nutritional benefits to health. They usually aren't selling anything but rather are trying to inform people about new findings so they can make better decisions about optimal nutrition.

The Business of Nutrition News   Newspapers, TV stations, magazines, books, and other sources of information make money when the number of viewers or readers is high. To increase viewers or readers, the media attempts to present information that will pique people's interest. Nutrition and health news tends to do that. The media have access to a large number of nutrition studies; over 500 nutrition-related research articles are published weekly (Illustration 3.5).

Studies reporting new results related to obesity, cancer, diabetes, heart disease, vitamins, and food safety are particularly hot topics. To keep people interested, the media may sensationalize and oversimplify nutrition-related stories. They may report on one study one day and describe another study with opposite results the next. A classic example of headlines about nutrition that confused the public is shown in Illustration 3.6. Such study-by-study coverage of nutrition news leaves consumers not knowing what to believe or what to do. A Nutrition Trends Survey by the American Dietetic Association uncovered a strong consumer preference (81% of those sampled) for learning about the latest nutrition breakthroughs only *after* they had been generally accepted by nutrition and health professionals.

Nutrition studies are complex, and the results of one study are almost never enough to prove a point. Decisions about personal

A checklist for identifying nutrition misinformation

1. Is something being sold? 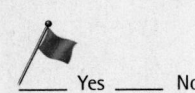 _____ Yes _____ No

2. Does the product or service offer a new remedy for problems that are not easily or simply solved (for example, obesity, cellulite, arthritis, poor immunity, weak muscles, low energy, hair loss, wrinkles, aging, stress)?  _____ Yes _____ No

3. Are such terms as "miraculous," "magical," "secret," "detoxify," "energy restoring," "suppressed by organized medicine," "immune boosting," or "studies prove" used? 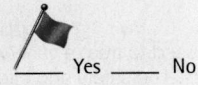 _____ Yes _____ No

4. Are testimonials, before-and-after photos, or expert endorsements used?  _____ Yes _____ No

5. Does the information sound too good to be true? 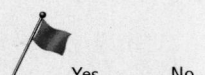 _____ Yes _____ No

6. Is a money-back guarantee offered?  _____ Yes _____ No

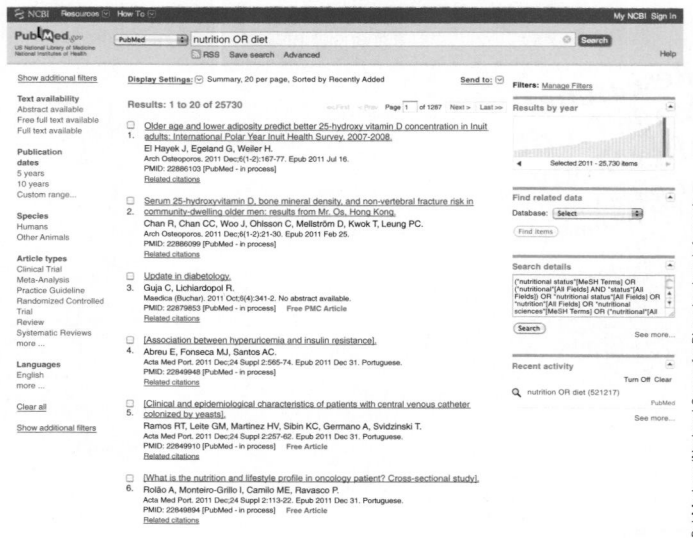

PubMed.gov/National Center for Biotechnology Information, U.S. National Library of Medicine

Illustration 3.5 Searching the term "nutrition OR diet" on the National Library of Medicine's PubMed site indicates that almost 26,000 nutrition research papers were published in 2011.

nutrition should be based on accumulated evidence that is broadly supported by nutrition and other scientists.

Motivation for Nutrition Misinformation #2: Personal Beliefs and Convictions

Numerous alternative health practitioners (such as nutripaths, irridologists, electrotherapists, scientologists, and faith healers) use nutrition remedies. Although their approaches to and philosophies about health may vary, these practitioners often have strong beliefs in the benefits of what they do.

There's no way to say whether most of the remedies used by alternative nutrition practitioners work. (Some of the unproven remedies employed are shown in Illustration 3.7.) The nutritional cures offered are often not based on evidence provided by scientific studies. Until they are evaluated scientifically, the logical conclusion is that they cannot be considered effective or safe.

Professionals with Embedded Beliefs Professionals who work in health care and research are not immune to the pull of deeply rooted convictions about diet and health relationships. They sometimes remain wedded to theories and beliefs even after they have been disproved. The nurse or doctor who holds onto the belief that salt intake should be restricted in pregnancy and the university professor who is convinced that pesticides on foods pose no risk to health are two examples of professional sources of misinformation. The failure of such professionals to give up erroneous convictions about diet and health adds to the flow of nutrition misinformation as well as to consumer confusion and increased health risk.

Illustration 3.6 How to confuse the public: a lesson delivered by the headlines about vitamin E and heart disease.

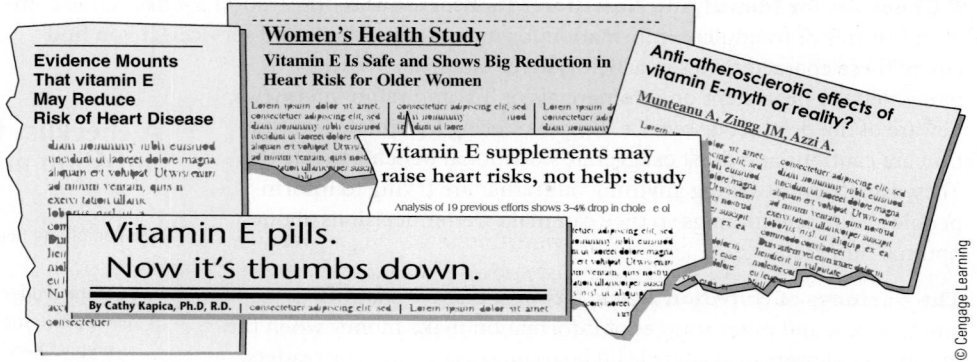

Illustration 3.7 Supplements that may be used to treat a variety of health problems. The safety and effectiveness of many of these products are yet to be proved.

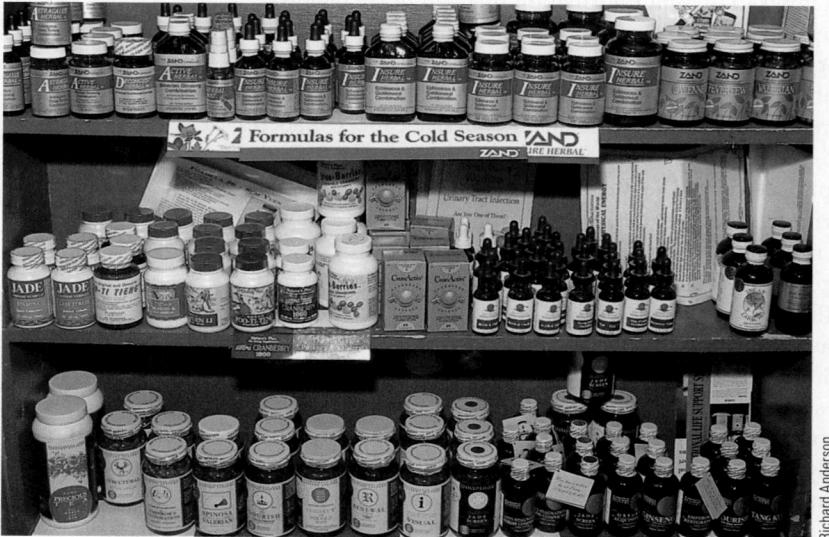

FDA FOOD SAFETY MODERNIZATION ACT

Food Safety Reform Updates

Get the latest information at www.fda.gov/fsma

The one-stop reference for the most current information in all key areas of food safety reform. Find out how to participate in public meetings and comment on proposed rules.

NEW PREVENTION STANDARDS

NEW IMPORTER ACCOUNTABILITY REQUIREMENTS

NEW COMPLIANCE AND INSPECTION PRACTICES

NEW INTEGRATED FEDERAL, STATE AND LOCAL EFFORTS

U.S. Food and Drug Administration

Illustration 3.8 The FDA's Food Safety Modernization Act of 2011 requires that science-based standards for the safe production and harvesting of fruits and vegetables be implemented by the produce industry.

How to Identify Nutrition Truths

- **Identify sources of reliable nutrition information.**

Where does accurate nutrition information come from? There is only one way to identify sound nutrition information: put it to the test and see if it survives the dispassionate, systematic examination dictated by science.

Science is a unique and powerful explanatory system. It produces information based on facts and evidence. Our understanding of nutrition is based on scientifically determined facts and evidence obtained from laboratory, animal, and human studies. These studies provide information that qualifies for use when developing public policies about nutrition and health (Illustration 3.8) and clinical nutrition practices. Science delivers information eligible for inclusion in textbooks about nutrition.

Sources of Reliable Nutrition Information

Reliable sources of information about nutrition meet the standards of proof required by science. They report decisions about nutrition and health relationships that are based on multiple studies and achieve scientific consensus. These decisions represent the majority opinion of scientists who are knowledgeable about a particular nutrition topic.

Nutrition recommendations made to the public, such as the Dietary Guidelines for Americans and the Dietary Reference Intakes, are based on the consensus of scientific opinion. The information is made available to the public not to sell a product or to pass on an ideology, but to inform consumers honestly about nutrition and to help people use the

---

REALITY CHECK

Microwaves and Plastic

You've got mail: DO NOT microwave foods in plastic! Hot plastic releases highly toxic dioxins that can cause cancer . . .

No source for the facts stated is given in the e-mail. Cyndi's and Scott's responses:

Who gets the thumbs up? Answers on page 3-8

Photo Disc

Cyndi: I'll check it out at the WebMD .com.

Photo Disc

Scott: I was afraid of that. Now I know microwaving food in plastic containers is dangerous.

ANSWERS TO **REALITY** CHECK
Microwaves and Plastic

Good thinking, Cyndi. Information gained from the fda.gov site indicates that plastic containers may contain bisphenol A, a softening compound that makes plastic flexible. Heat increases its release from plastic. Preliminary evidence indicates that high levels of bisphenol A may be related to changes in organ development in animals. Canada and a few other countries have banned use of the substance in baby bottles and food containers intended for young children, but the FDA is awaiting the arrival of stronger evidence. The FDA recommends that people concerned about bisphenol A use glass or ceramic containers in the microwave rather than plastic containers or plastic wrap, and don't use plastic baby bottles.[4,5]

Cyndi:

Scott:

Table 3.1 **Reliable sources of nutrition information**[a]

Source of nutrition information	Examples
Nonprofit, professional health organizations	American Heart Association American Cancer Society Academy of Nutrition and Dietetics (formerly known as the American Dietetic Association) American Diabetes Association
Scientific organizations	National Academy of Sciences American Society for Clinical Nutrition American Society for Nutrition
Government publications: nutrition, diet, and health reports	National Institutes of Health Surgeon General Food and Drug Administration Centers for Disease Control U.S. Department of Agriculture
Registered dietitians	Hospitals Public health departments Extension service Universities
Nutrition textbooks	College and university nutrition courses Nutrition faculty of accredited universities

[a]See Appendix B for more details about reliable sources of nutrition information.

information to maintain or improve their health. Organizations and individuals offering reliable nutrition information are listed in Table 3.1.

Nutrition Information on the Web Web-based search engines are transforming the way we get information about foods, diets, supplements, and health. Eight in ten Internet users, including scientists and consumers, look for health and nutrition information online.[6] As is the case for many other media sources, the accuracy of nutrition information made available on the web varies considerably. In general, the most reliable sites are those developed by government health agencies (web addresses that end in or include .gov), educational institutions (.edu), and nonprofit professional health and science organizations (.org).[7]

Who Are Qualified Nutrition Professionals? Many people refer to themselves as nutritionists, but only some of them are qualified based on education and experience. These individuals are registered, licensed, or certified dietitians or nutritionists who meet qualifications established by national and state regulations. They have demonstrated a mastery of knowledge about the science of nutrition and appropriate clinical practices. Table 3.2 describes the qualifications of those who legitimately use the title *dietitian* or *nutritionist*. The specific titles vary somewhat depending on state regulations and laws governing the practice of nutrition and dietetics.

Table 3.2 Who's who in nutrition and dietetics: Laws and regulations governing practice[8,9]

Title	Qualifications
Registered dietitian	Individual who has acquired the knowledge and skills necessary to pass a national registration examination and participates in continuing professional education. Qualification is conferred by the Commission of Dietetic Registration.
Licensed dietitian/nutritionist	Individual (usually a registered dietitian) who is qualified to practice nutrition counseling based on education and experience. Nonlicensed individuals who practice nutrition counseling can be prosecuted for practicing without a license. Licenses are conferred by state law in 48 states.
Certified dietitian/nutritionist	Individual who meets certain educational and experience qualifications. Unqualified individuals may practice nutrition counseling but have to call themselves something else (such as "nutritional counselor"). Certification is established by state regulation.

It is difficult to make sound decisions about every nutrition question or issue that arises. When trying to make sense out of the information you read or hear, don't hesitate to get help. Visit reliable websites, check out the index of this book for the relevant topic, or call your local health department or area Food and Drug Administration office. The more solid your information, the better your decisions will be.

The Methods of Science

- **Explain each component of the scientific methods used to identify reliable information about nutrition.**

Illustration 3.9 The scientific method is not mysterious at all but a carefully planned process for answering a specific question.

Does the term *scientific method* conjure up an image of beady-eyed, white-coated scientists working diligently in windowless basement laboratories? Although we tend to assume that scientists are busy advancing knowledge about nutrition and many other fields, for most people the methods of science are a mystery. They should not be (Illustration 3.9). Consumers use a lot of nutrition information. To make sound judgments about nutrition and health, people need to know how to distinguish results produced by scientific studies from those generated by personal opinion, product promotions, and bogus studies.

There are many established methods for conducting scientific studies. The specific methods employed vary from study to study depending on the type of research conducted. Nevertheless, all types of scientific studies have one feature in common: They are painstakingly planned. Planning is the most important, and often the most time-consuming, part of the entire research process.

Developing the Plan

The first part of the planning process entails clearly stating the question to be addressed—and, one hopes, answered—by the research. For the purposes of illustration, let's say our research will address the effects of vitamin X supplements on hair loss (Table 3.3 and Illustration 3.10). The idea for the research came from a study that hinted that users of vitamin X supplements had increased hair loss. In that study, adults were given 25,000 international units (IU) of vitamin X for three months to test its safety. Although the study found that vitamin X had no adverse effects on health and offered some benefits, several subjects who had received the supplements complained of hair loss. Accordingly, our study poses the question, Do vitamin X supplements increase hair loss?

The Hypothesis: Making the Question Testable

The question is then transformed into an explicit hypothesis that can be proved or disproved by the research. (**Hypothesis** and other research terms used in this unit are defined in Table 3.4.) The results of the research must provide a true or false response to the hypothesis and not an explanatory sentence or paragraph. So, in this example, "vitamin X increases hair loss" wouldn't do as a hypothesis because it leaves too many questions about the relationship unanswered. The effect of vitamin X on hair loss may depend on the amount of vitamin X given; how long it is taken; or on the research subjects' health,

Table 3.3 Overview of a human nutrition research study: Vitamin *X* supplements and hair loss[a]

A. **Pose a clear question:** "Do vitamin *X* supplements increase hair loss?"

B. **State the hypothesis to be tested:** "Vitamin *X* supplements of 25,000 IU per day taken for three months increase hair loss in healthy adults."

C. **Design the research:**

  1. What type of research design should be used? "In this study, a clinical wtrial will be used. The supplement and placebo will be allocated by a double-blind procedure."

  2. Who should the research subjects be? "The study will exclude subjects who may be losing or gaining hair due to balding, hair treatments, medications, or the current use of vitamin *X* supplements."

  3. How many subjects are needed in the study? "The required sample size is calculated to be 20 experimental and 20 control subjects."

  4. What information needs to be collected? "Information on hair loss, conditions occurring that might affect hair loss, and the use of supplements and placebos will be collected."

  5. What method for measuring hair loss should be used? "The study will use the reliable "60-second hair count" technique.[12]

  6. What statistical tests should be used to analyze the results? "Appropriate tests identified."

D. **Obtain approval for the study from the committee on the use of humans in research:** "Approval obtained."

E. **Implement the study design:** "Implemented."

F. **Evaluate the findings:** "Subjects receiving the 25,000 IU of vitamin *X* for three months lost significantly more hair than subjects receiving the placebo. Hypothesized relationship found to be true."

G. **Submit paper on the research for publication in a scientific journal or other document.**

[a]This is a fictitious study used for illustrative purposes only.

Illustration 3.10 Do high doses of vitamin *X* increase hair loss?

Photo Disc

age, and other characteristics. The hypothesis "vitamin *X* supplements of 25,000 IU per day taken for three months by healthy adults increase hair loss" is concrete enough to be addressed by research.

The Research Design: Gathering the Right Information

Poor research design, or the lack of a solid plan on how the research will be conducted, is often a weak link that renders studies useless. It's the "oops, we forgot to get this critical piece of information!" or the "how did all the measurements come out wrong?" at the end of a study that can ruin months or even years of work. Each step in the research process must be thoroughly planned. In research, there are no miracles. If something can go wrong because of incomplete planning, it probably will.

Research designs are often based on the answers to the following questions:

- What type of research design should be used?

- Who should the research subjects be?

- How many subjects are needed in the study?

- What information needs to be collected?

- What are accurate ways to collect the needed information?

- What statistical tests should be used to analyze the findings?

What Type of Research Design Should Be Used? Several different research designs can be used to test hypotheses. We could use an **epidemiological study** design to determine if hair loss is more common among people who take vitamin *X* supplements than among people who do not. Researchers commonly use this type of study to identify conditions that are related to specific health events in humans. To provide preliminary evidence, we could use animal studies that determine hair loss in supplemented and unsupplemented animals. Or we could use another design, such as a clinical trial. Since this design would work well for the proposed hypothesis about vitamin *X* supplements,

Table 3.4 A short glossary of research terms

Association	The finding that one condition is correlated with, or related to another condition, such as a disease or disorder. For example, diets low in vegetables are associated with breast cancer. Associations do not prove that one condition (such as a diet low in vegetables) causes an event (such as breast cancer). They indicate that a statistically significant relationship between a condition and an event exists.
Cause and effect	A finding that demonstrates that a condition causes a particular event. For example, vitamin C deficiency causes the deficiency disease scurvy.
Clinical trial	A study design in which one group of randomly assigned subjects (or subjects selected by the "luck of the draw") receives an active treatment and another group receives an inactive treatment, or "sugar pill," called the placebo.
Control group	Subjects in a study who do not receive the active treatment or who do not have the condition under investigation. Control periods, or times when subjects are not receiving the treatment, are sometimes used instead of a control group.
Double blind	A study in which neither the subjects participating in the research nor the scientists performing the research know which subjects are receiving the treatment and which are getting the placebo. Both subjects and investigators are "blind" to the treatment administered.
Epidemiological study	Research that seeks to identify conditions related to particular events within a population. This type of research does not identify cause-and-effect relationships. For example, much of the information known about diet and cancer is based on epidemiological studies that have found that diets low in vegetables and fruits are associated with the development of heart disease.
Experimental group	Subjects in a study who receive the treatment being tested or have the condition that is being investigated.
Hypothesis	A statement made prior to initiating a study of the relationship sought to be tested by the research.
Meta-analysis	An analysis of data from multiple studies. Results are based on larger samples than the individual studies and are therefore more reliable. Differences in methods and subjects among the studies may bias the results of meta-analyses.
Peer review	Evaluation of the scientific merit of research or scientific reports by experts in the area under review. Studies published in scientific journals have gone through peer review prior to being accepted for publication.
Placebo	A "sugar pill," an imitation treatment given to subjects in research.
Placebo effect	Changes in health or perceived health that result from expectations that a "treatment" will produce an effect on health.
Statistically significant	Research findings that likely represent a true or actual result and not one due to chance.

we'll follow the rules for conducting a **clinical trial** in this example. (You can see a map of the research design used for this hypothetical study in Illustration 3.11.)

The purpose of clinical trials is to test the effects of a treatment or intervention on a specific biological event or other measureable outcome. In our example, we will test the effect of vitamin X supplements on hair loss.

The Experimental and Control Groups Clinical trials, as well as other research studies that address questions about nutrition, require an **experimental group** (the group of subjects who receive vitamin X in this example) and a **control group** (the comparison

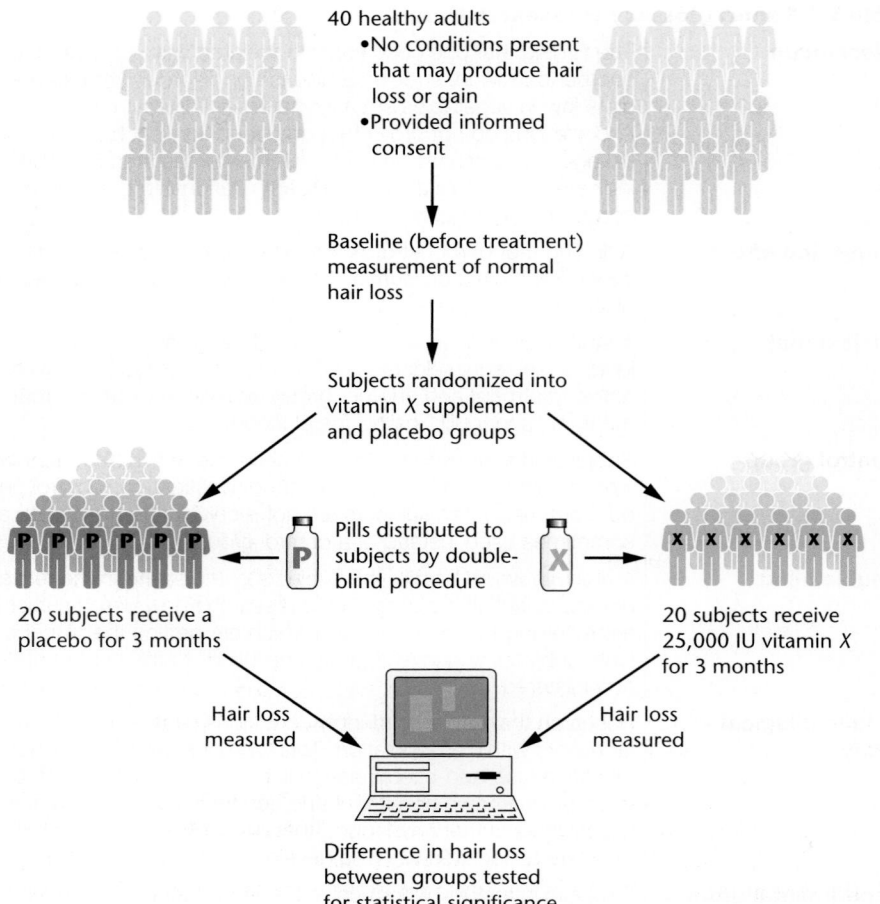

Illustration 3.11 Map of the design of the vitamin *X* supplement study.

40 healthy adults
- No conditions present that may produce hair loss or gain
- Provided informed consent

Baseline (before treatment) measurement of normal hair loss

Subjects randomized into vitamin *X* supplement and placebo groups

Pills distributed to subjects by double-blind procedure

20 subjects receive a placebo for 3 months

20 subjects receive 25,000 IU vitamin *X* for 3 months

Hair loss measured

Hair loss measured

Difference in hair loss between groups tested for statistical significance

© Cengage Learning

Stan Maddock

Illustration 3.12 Which is the vitamin *X* supplement? Neither the subject nor the investigator should know which is vitamin *X* and which is the placebo.

group that receives a **placebo** and not vitamin *X*). Never trust the results of a study that didn't employ both. Here's why. How do we know the effect of a certain treatment isn't due to something *other* than the treatment? What if we gave a group of adults vitamin *X* supplements and they lost hair? Does this mean the vitamin *X* caused the hair loss? Could it be that the subjects lost no more hair during the treatment period than they would have lost without the vitamin *X*? We can't know whether vitamin *X* supplements produce hair loss if we don't know how much hair is lost without vitamin *X*.

After measuring their usual hair loss, we will randomly assign (by the "luck of the draw") people in the study to serve in either the experimental or the control group. Individuals assigned to the control group will be given pills that look, taste, and feel like the vitamin *X* supplements but have no effect on hair loss or gain. Neither the research staff nor the people in the study will know who is getting vitamin *X* and who is getting the placebo (Illustration 3.12). Scientists use this **double-blind** procedure because knowing which group is which may affect people's expectations and change the results.

The "Placebo Effect" The placebo effect can cause a good deal of confusion in research. That's because people tend to have expectations about what a treatment will do, and those expectations can influence what happens.

A good example of the **placebo effect** occurred in a study that tested the effectiveness of a medication meant to reduce binge eating among people with bulimia.[10] After the usual number of binge-eating episodes was determined, 22 women with bulimia were given either the medication or a placebo in a double-blind fashion. The number of binge-eating episodes was then reassessed. At the end of the study period, a whopping 78% reduction in binge-eating episodes was found among women taking the medication. But binge-eating episodes dropped by 70% in the control group. Was the medication effective? No. Was the expectation that a medication would help effective? Yes, at least for the

Illustration 3.13 How many subjects are needed in the study? It's important to get the number right.

time covered by the study. Due to the placebo effect, a reduction of greater than 90% in episodes of binge eating would have been needed to conclude that the medication had a real effect on binge eating.

Who Should the Research Subjects Be?
Although expedient, it is not enough to say, "Well, let's use 10 faculty members in the study." The type and number of subjects employed by the research are important considerations.

The type of subjects involved in research is important because we need to exclude people who have conditions that might produce the problem the research is examining. For example, hair loss may result from periodic balding, a bad perm, the use of mousse or hair spray, or illnesses or medications. If we include people with these conditions and circumstances in the study, it will be difficult to determine whether hair loss is due to vitamin X or something else. In addition, it is important to exclude from the study people who already take vitamin X supplements or are bald. An inappropriate or biased selection of subjects could make the results useless or, in the worst case, make the study's results come out in a predetermined way.

How Many Subjects Are Needed in the Study?
The number of subjects needed for a study is a mathematical question, and we won't go into the formulas here. The number of subjects is based on the number needed to separate true differences between the experimental and control groups from differences that are due to chance (Illustration 3.13). Daily hair loss normally ranges from 50 to 150 strands per day.[11] We'll need to include enough subjects in the study to make sure differences in hair loss between the groups are not due to normal fluctuations in hair loss. This point is important because studies employing too few subjects lead to inconclusive results and a great deal of controversy about nutrition. One or a few subjects are never enough to prove a point about nutrition.

Let's assume that the mathematical formulas applied to the vitamin X study show that 20 people are needed in both the experimental group and the control group.

What Information Needs to Be Collected?
Information recorded by researchers must represent an evenhanded approach to discovering the facts. We must identify not only the findings that may support the hypothesis, but *those that may refute it*. In order to know if vitamin X supplements increase hair loss, we need to know how much hair is lost, whether the subjects faithfully took the supplement, and whether something came up during the study (such as a bad perm) that might alter hair loss. The presence of such conditions should be determined by methods known to provide accurate results.

What Are Accurate Ways to Collect the Needed Information? "Garbage in—garbage out." If the information obtained on and from subjects is inaccurate, then the study is worthless. To avoid this problem, good research employs methods of collecting information known to produce accurate results.

Let's consider the problem of measuring hair loss. How can we do that accurately? First, we look for a method that has already been demonstrated by research to be accurate. We find one called "the 60 second hair count"[12] and employ it.

What Statistical Tests Should Be Used to Analyze the Findings? Before collecting the first piece of information, we'll select appropriate tests for identifying statistically significant results of the research. Statistical tests are used to identify significant differences between the findings from the experimental and control groups. Without such tests, it is often very hard to decide what the findings mean. What if the study shows the group that took vitamin X lost an average of 5% more hair than the control group? Is that 5% difference due to the vitamin X or to something else, such as chance fluctuations in normal hair loss? Well-chosen statistical tests will tell us whether the differences between groups are in all probability real or due to chance occurrences or coincidence.

Obtaining Approval to Study Human Subjects

An important step in the planning process is applying for approval to conduct the proposed research on human subjects. Universities and other institutions that conduct research have formal committees, called institutional review boards, that scrutinize plans to make sure proposed studies follow the rules governing research on human subjects. For this study, we will have to show to the committee's satisfaction that the level of vitamin X employed is safe. As a part of this process, the committee usually requires that human subjects consent, in writing, to participate in the study.

Implementing the Study

With the design in place and the appropriate approvals obtained, it is time to implement the study. Assume that, due to the study's solid design and importance, we have been awarded a grant to fund the research. We can now recruit subjects for the study and enroll them if they are eligible and consent to participate. Measurements of hair loss are performed according to schedule, use of the pills by subjects is monitored, results are checked for errors and entered into a computerized data file, and statistical tests are applied. This process usually ends with computer printouts that exhibit the findings.

Making Sense of the Results

Let's say subjects who received the supplement lost 30% more hair than the control subjects did. Having applied the appropriate statistical test, we find that 30% is a highly significant difference. Does that mean that vitamin X supplements *cause* hair loss? *Cause* is a strong word in research terms. It implies that a **cause-and-effect** relationship (that the vitamin X supplements caused the hair loss) exists. But many factors can contribute to hair loss, and since many causes may be unknown or not measured by the study, it is difficult to conclude with absolute certainty that vitamin X by itself caused the hair loss. Assume 4 of the 20 subjects in the experimental group who took the vitamin X supplement lost less hair than usual. The vitamin X supplements *didn't* cause them to lose hair.

We could conclude from this research (if, as a reminder, there was such a thing as vitamin X) that vitamin X supplements, given to healthy adults at a dose of 25,000 IU per day for three months, are strongly *associated* with hair loss. The term *associated* as used here means that the vitamin X supplements were strongly *related* to hair loss but may not have *caused* the hair loss. The hypothesized relationship between supplemental vitamin X and hair loss would be found to be true. Although the research strongly indicates a cause-and-effect relationship, additional studies would be needed to prove the relationship exists in other groups of people and at different doses of vitamin X.

After determining the results, a paper is written describing the research and its conclusions and submitted for publication in a scientific journal. If judged to be acceptable after peer review (review by qualified experts), the paper is published.

Complete coverage of how nutrition questions are addressed through research would take a three-course sequence. Nevertheless, it is hoped that the information presented here makes it easier to judge the likely accuracy of reports about nutrition studies in the popular media. There is a lot more to identifying nutrition truths than meets the eye.

Science and Personal Decisions about Nutrition Science is based on facts and evidence. The grounding ethic of scientists is that facts and evidence are more sacred than any other consideration. These characteristics of science and scientists are strong assets for the job of identifying truths.

Although imperfect because they are undertaken by humans in an environment of multiple constraints, only scientific studies produce information about nutrition you can count on. Evidence is the single best ingredient for decision making about nutrition and your health.

NUTRITION
up close

Checking Out a Fat-Loss Product

Focal Point: How to make an informed decision about the truthfulness of an advertisement for a fat-loss product.

Read the accompanying advertisement for a fat-loss product. Then, check out the information using the checklist for identifying nutrition misinformation.

Obesity Research Ce
P.O. Box 281752,
Miami, FL

Lose Over 2 Inches In 3 Weeks

A Scientifically Proven Fat Reduction Cream That Actually Works!

A scientific, double-blind, placebo-controlled research study of both sexes demonstrated that LIPID MELT fat reduction cream reduces inches from the thigh area. Subjects lost an average of 2 inches from each thigh in only 3 weeks!

Subjects from a preliminary pilot study reported that LIPID MELT cream reduced inches from their abdomen. Both studies reported no skin rashes, discomfort or sensitivity. Formulated with patented liposome technology, LIPID MELT contains no drugs, only natural active ingredients. Initially sold only in salons and spas.

Six-week supply: $25 + $4.95 s/h. We guarantee that you will lose inches or your money will be refunded.

Obesity Research

1. Is something being sold?
   Yes   No

2. Is a new remedy for problems that are not easily or simply solved being offered?
   Yes   No

3. Are terms such as *miraculous, magical, secret, detoxify, energy restoring, suppressed by organized medicine, immune boosting*, or *studies prove* used?
   Yes   No

4. Are testimonials, before-and-after photos, or expert endorsements used?
   Yes   No

5. Does the information sound too good to be true?
   Yes   No

6. Is a money-back guarantee offered?
   Yes   No

Optional: Repeat the activity using a nutrition-related advertisement from the Internet.

Feedback to the Nutrition Up Close is located in Appendix G.

REVIEW QUESTIONS

- **Evaluate the reliability of advertisements and other information about nutrition and nutrition-related products and services.**

1. By law, claims about nutrition that are presented on product labels and packaging must be scientifically accurate.   **True/False**

2. The leading motivation underlying the presentation of nutrition misinformation is the profit motive.
   True/False

3. The Food and Drug Administration (FDA) has the authority to remove advertisements making false claims about nutrition from the airwaves and the Internet.
   True/False

4. Individuals can help combat the proliferation of bogus nutrition advertisements and products by notifying the Federal Trade Commission (FTC) of their existence.   **True/False**

5. Nutrition products and services that offer a "money-back guarantee" are most likely to work as advertised. **True/False**

6. Nutrition information offered by trained sales staff at nutrition product stores can be counted on as being accurate. **True/False**

• **Identify sources of reliable nutrition information.**

7. The term *registered dietitian* means that individuals with that title have acquired the knowledge and skills necessary to pass a national examination on nutrition and participate in continuing education. **True/False**

8. _____ Recently, a talk show–induced scare about the safety of apple juice due to its arsenic content spread across the Internet. The news was hard to believe, but it might be true so you decide to look it up on the Web. From the options listed below, select the most reliable source of information on the safety of apple juice.
   a. www.doctoroz
   b. www.cbsnews.com
   c. www.fda.gov
   d. www.usapple.org

• **Explain each component of the scientific methods used to identify reliable information about nutrition.**

9. _____ While talking with your brother on the phone your brother complains that he "feels a cold coming on." You tell him to take a vitamin D supplement because you have not gotten a cold since you started taking vitamin D. Your brother takes the pill and the next day he feels better. Which of the following statements is *least* likely to be true regarding the relationship between the vitamin D pill and prevention of the brother's cold?
   a. The brother's experience proves vitamin D prevents colds.
   b. Although it is possible vitamin D helps prevent colds, the brother may have felt better the next day even if he hadn't taken the vitamin D.
   c. Reliable information about the effects of vitamin D on colds cannot be based solely on one person's experience.
   d. Believing that vitamin D prevents colds is not the same as scientifically demonstrating that it does.

10. _____ You hear on *The Nightly News* that scientists have discovered an association between high levels of soft drink consumption and violent behavior in teens. Which of the following statements represents the correct interpretation of this study's results?
   a. Violent behavior in teens is triggered by soft drink consumption.
   b. Teens who engage in violent behavior tend to consume more soft drinks than teens who do not engage in violent behavior.
   c. Violent behavior is only exhibited by teens who consume high amounts of soft drinks.
   d. Eliminating soft drink consumption would end violent behavior among teens.

11. An advantage of clinical trials over other study designs is that they consistently identify cause and effect relationships. **True/False**

12. Research results may be unreliable if too few subjects are used in studies. **True/False**

13. Research articles published in scientific journals undergo peer review, a process by which experts in the areas under investigation evaluate articles prior to their acceptance for publication. **True/False**

Match the term in Column A with its abbreviated definition shown in Column B.

Column A	Column B
_____ 14. Meta-analysis	a. "Sugar pill" or imitation treatment used in clinical trials.
_____ 15. Double blind	b. Subjects in a study who do not receive the active treatment or who not have the condition under investigation.
_____ 16. Placebo	c. An analysis of data from multiple studies.
_____ 17. Statistically significant	d. Research finding that is likely true and not due to chance.
_____ 18. Control group	e. A study in which both research subjects and investigators do not know who is receiving an active or placebo treatment.

Answers to these questions can be found in Appendix G.

· ·

NUTRITION SCOREBOARD ANSWERS

1. Freedom of speech applies to information about nutrition in articles, speeches, pamphlets, and broadcasts. However, it is illegal to make false or misleading claims about nutrition in advertisements or on product labels and packaging. **False**

2. The "way of knowing" for the field of nutrition is science. **True**

3. In a double-blind study, neither the research staff nor the research subjects know which subjects are getting the real treatment and which are receiving the fake treatment. This reduces the placebo ("sugar pill") effect, or changes in health that are due to the expectation that a specific treatment will have a particular impact on health. **True**

Understanding Food and Nutrition Labels

NUTRITION SCOREBOARD

1 Nutrition labels are required on all foods and dietary supplements sold in the United States. **True/False**

2 Nutrition labeling rules allow health claims to be made on the packages of certain food products. **True/False**

3 Nutrition labels contain all of the information people need to make healthy decisions about what to eat. **True/False**

Answers can be found at the end of the unit.

Nutrition Labeling

• Apply knowledge about the four key elements of nutrition labeling to decisions about the nutritional value of foods.

Misleading messages, hazy health claims, and the slippery serving sizes that characterized food labels in the past led to a revolution in nutrition labeling. Consumers—especially those responsible for buying food for the family; people with weight concerns, food allergies, or diabetes; and the health-conscious—made it clear they wanted to end the mystery about what's in many foods. Passage of the 1990 Nutrition Labeling and Education Act by Congress indicated that their concerns had been heard. In 1993 the Food and Drug Administration (FDA) published rules for nutrition labeling,[1] and implementation and revisions of the new standards have been ongoing since then.

Key Nutrition Concepts

Content in this unit on nutrition labeling directly relates to three key nutrition concepts:

1. Foods provide energy (calories), nutrients, and other substances needed for growth and health.

2. Poor nutrition can result from both inadequate and excessive levels of nutrient intake.

3. Malnutrition can result from poor diets and from disease states, genetic factors, or combinations of these factors.

Key Elements of Nutrition Labeling Standards

Nutrition labeling regulations cover four major areas:

- The Nutrition Facts panel
- Nutrient content claims
- Health claims
- Structure/function claims

The Nutrition Facts panel is required on most foods sold in grocery stores. Rules established for the other three areas must be followed when a claim about the product is made on the packaging.

The Nutrition Facts Panel With the exception of foods sold in very small packages or in stores like local bakeries, foods containing more than one ingredient must display a Nutrition Facts panel. Single fresh foods, like pears, a head of cabbage, or fresh shrimp, do not have to be labeled, but grocery stores are encouraged to present nutrition information on posters (Illustration 4.1).

Nutrition Facts panels provide specific information about a food's serving size, calorie value, nutrient content, and ingredients. Illustration 4.2 shows a Nutrition Facts panel and provides explanations of what various components of the panel mean. The panel highlights the fat, saturated fat, *trans* fat, cholesterol, sodium, carbohydrate, dietary fiber, sugars, vitamins A and C, and calcium and iron content of a serving of food. The serving size listed in the Nutrition Facts panel must be based on a standard serving size as defined by the FDA.

All labeled foods must provide the information shown on the Nutrition Facts panel in Illustration 4.2. Additional information on specific nutrients can be added to the panel on a voluntary basis

---

After completing Unit 4 and its interactive learning features, you will be able to:

- Apply knowledge about the four key elements of nutrition labeling to decisions about the nutritional value of foods.

- Evaluate nutrient content and health claims made on dietary supplement labels.

- Compare the characteristics of organically and conventionally produced food products.

- Identify strengths and weaknesses of various nutrition labeling systems on food packaging and calorie listings for food items.

Illustration 4.1 Nutrition information for produce, fish, and seafoods can be presented on posters.

Seafood Nutrition Facts

Cooked (by moist or dry heat with no added ingredients), edible weight portion. Percent Daily Values (%DV) are based on a 2,000 calorie diet.

Seafood Serving Size (84 g/3 oz)	Calories	Calories from Fat	Total Fat (g)	%DV	Saturated Fat (g)	%DV	Cholesterol (mg)	%DV	Sodium (mg)	%DV	Potassium (mg)	%DV	Total Carbohydrate (g)	Protein (g)	Vitamin A %DV	Vitamin C %DV	Calcium %DV	Iron %DV
Blue Crab	100	10	1	2	0	0	95	32	330	14	300	9	0	20g	0%	4%	10%	4%
Catfish	130	60	6	9	2	10	50	17	40	2	230	6	0	17g	0%	0%	0%	0%
Clams, about 12 small	110	15	1.5	2	0	0	80	27	95	4	470	13	6	17g	10%	0%	8%	30%
Cod	90	5	1	2	0	0	50	17	65	3	460	13	0	20g	0%	2%	2%	2%
Flounder/Sole	100	15	1.5	2	0	0	55	18	100	4	390	11	0	19g	0%	0%	2%	0%
Haddock	100	10	1	2	0	0	70	23	85	4	340	10	0	21g	2%	0%	2%	6%
Halibut	120	15	2	3	0	0	40	13	60	3	500	14	0	23g	4%	0%	2%	6%
Lobster	80	0	0.5	1	0	0	60	20	320	13	300	9	1	17g	2%	0%	6%	2%
Ocean Perch	110	20	2	3	0.5	3	45	15	95	4	290	8	0	21g	0%	2%	10%	4%
Orange Roughy	80	5	1	2	0	0	20	7	70	3	340	10	0	16g	2%	0%	4%	2%
Oysters, about 12 medium	100	35	4	6	1	5	80	27	300	13	220	6	6	10g	0%	6%	6%	45%
Pollock	90	10	1	2	0	0	80	27	110	4	370	11	0	20g	2%	0%	0%	2%
Rainbow Trout	140	50	6	9	2	10	55	18	35	1	370	11	0	20g	4%	4%	8%	2%
Rockfish	110	15	2	3	0	0	40	13	70	3	440	13	0	21g	4%	0%	2%	2%
Salmon, Atlantic/Coho/Sockeye/Chinook	200	90	10	15	2	10	70	23	55	2	430	12	0	24g	4%	4%	2%	2%
Salmon, Chum/Pink	130	40	4	6	1	5	70	23	65	3	420	12	0	22g	2%	0%	2%	4%
Scallops, about 6 large or 14 small	140	10	1	2	0	0	65	23	310	13	430	12	5	27g	2%	0%	4%	14%
Shrimp	100	10	1.5	2	0	0	170	57	240	10	220	6	0	21g	4%	4%	6%	10%
Swordfish	120	50	6	9	1.5	8	40	13	100	4	310	9	0	16g	2%	0%	0%	6%
Tilapia	110	20	2.5	4	1	5	75	25	30	1	360	10	0	22g	0%	0%	0%	2%
Tuna	130	15	1.5	2	0	0	50	17	40	2	480	14	0	26g	2%	2%	2%	4%

Seafood provides negligible amounts of *trans* fat, dietary fiber, and sugars.

U.S. Food and Drug Administration (January 1, 2008)

© Cengage Learning

(Table 4.1). However, if the package makes a claim about the food's content of a particular nutrient that is not on the "mandatory" list, then information about that nutrient must be added to the Nutrition Facts panel. Nutrition Facts panels also contain a column headed **% Daily Value (%DV)**. Figures given in this column are intended to help consumers answer such questions as "Does a serving of this macaroni and cheese contain more fat than the other brand?" and "How much fiber does this cereal provide compared to my daily need for it?"

Daily Values Daily Values (DVs) are standard amounts of nutrients developed specifically for use on nutrition labels (see Table 4.1). They are based on an earlier edition of the Recommended Dietary Allowances.[3] The %DV figures listed on labels represent the percentages of the standard nutrient amounts obtained in one serving of the food product. Standard values for total fat, saturated fat, and carbohydrates are based on a daily intake of 2,000 calories. The %DV for total fat intake is based on 30% of total calories from fat (65 grams), the saturated fat standard on 10% of total calories (20 grams), and the standard for carbohydrate intake on 60% of total calories (300 grams). If, for example, a serving of a food product contains 10 grams of saturated fat, the %DV listed on the nutrition label would be 50% because the standard for saturated fat intake is 20 grams. In general,

% Daily Value (%DV) Daily Values are scientifically agreed-upon standards of daily intake of nutrients from the diet developed for use on nutrition labels. The "% Daily Values" listed in nutrition labels represent the percentages of the standards obtained from one serving of the food product.

Illustration 4.2 Inside the Nutrition Facts panel.

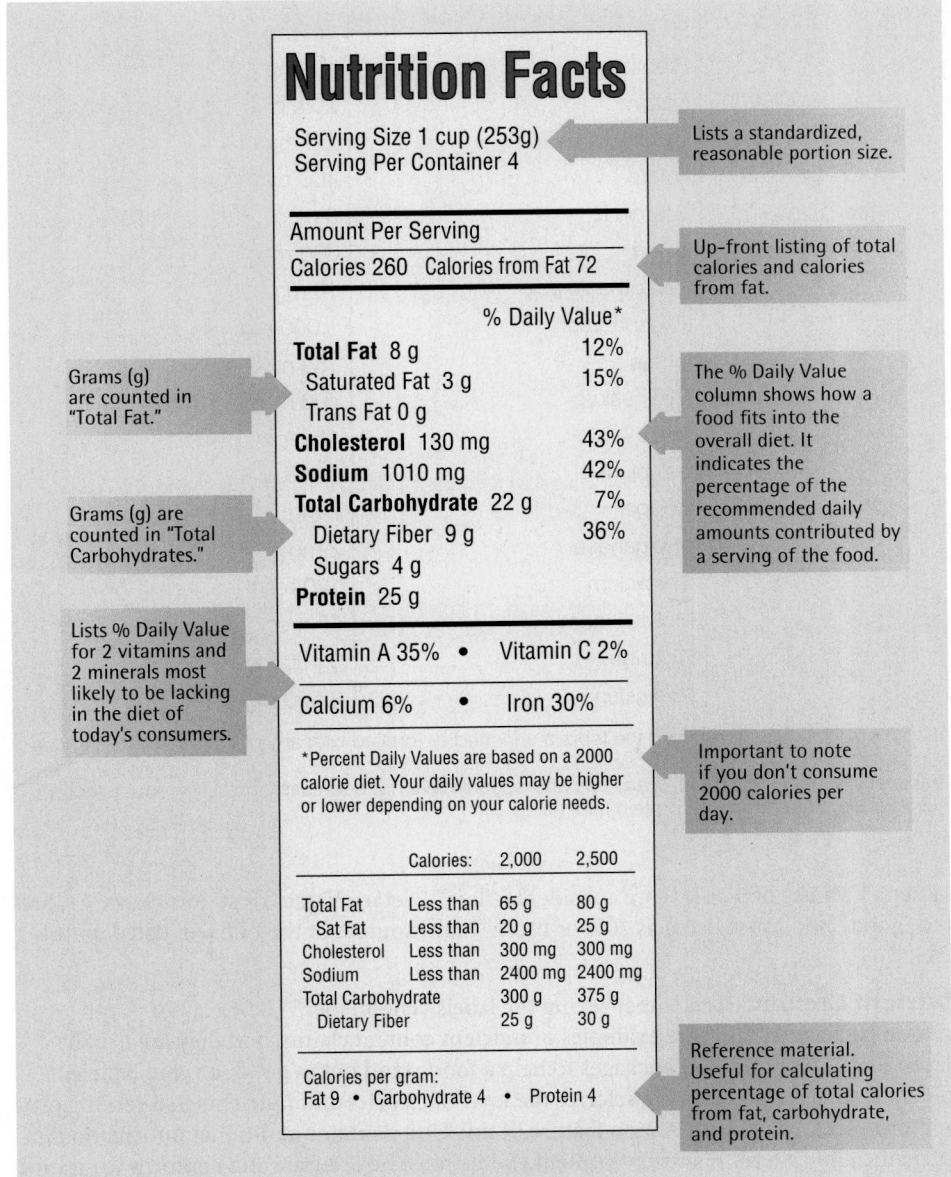

Table 4.1 Mandatory and voluntary components of the Nutrition Facts panel and assigned Daily Values (DVs)

Components are listed in the order in which they must appear on the nutrition panel; unapproved components may not be listed.[a]

Mandatory	DV	Voluntary	DV
Total calories	—	Calories from saturated fat	
Calories from fat	30%	Polyunsaturated fat	
Total fat	65 g	Monounsaturated fat	
Saturated fat	20 g	Stearic acid	
Trans fat	—[b]	Insoluble fiber	
Cholesterol	300 mg	Other carbohydrates	
Sodium	2,400 mg	Soluble, Insoluble fiber	
Total carbohydrate	300 g	Sugar alcohols (xylitol,	
Dietary fiber	25 g	mannitol, sorbitol)	
Sugars	—[b]	Vitamin D	400 IU
Protein	—[b]	Vitamin E	30 IU
Vitamin A	5,000 IU	Vitamin K	80 mcg
Vitamin C	60 mg	Thiamin	1.5 mg
Calcium	1,000 mg	Riboflavin	1.7 mg
Iron	18 mg	Niacin	20 mg
		Vitamin B_6	2.0 mg
		Folate	400 mcg
		Vitamin B_{12}	6 mcg
		Biotin	300 mcg
		Pantothenic acid	10 mg
		Phosphorus	1,000 mg
		Iodine	150 mcg
		Magnesium	400 mg
		Zinc	15 mg
		Selenium	70 mcg
		Copper	2 mg
		Manganese	2 mg
		Chromium	120 mcg
		Molybdenum	75 mcg
		Chloride	3,400 mg
		Potassium	3,500 mg

[a]If the food package makes a claim about any of the voluntary components, or if the food is enriched or fortified with any of them, they become a mandatory part of the nutrition panel.
[b]DVs are not shown on Nutrition Facts panels for *trans* fats, sugars, or protein. *Trans* fats and sugars have not been assigned DVs, and, although there is a DV for protein, it is not listed because most people have adequate protein intakes.

% Dietary Values of 5 or less are considered "low," Dietary Values of 10 to 19% are considered "good," and those listed as 20% or more "high" sources of the nutrient listed on the label.[2]

Nutrient Content Remember seeing the labels "High Fiber," "Low Fat," or "Lean" on food packages? These are examples of nutrient content claims, and they are usually placed on the front of food packages to help a food stand out as good for you. Nutrient content claims are used to characterize the level of calories and nutrients in a serving of a food product. (The Health Action feature in this Unit contains additional information on nutrient claim criteria.) Nutrient content claims must be accurate and conform to specific criteria developed by the FDA. Foods labeled "low fat," for example, must contain 3 grams

health action Some Examples of What "Front of the Package" Nutrient-Content Claims Must Mean

Term	Examples	Means That a Serving of the Product Contains:
Fresh	Fresh spinach	Foods are raw, not frozen or heated, and contain no preservatives.
Healthy	Healthy burritos, canned vegetables	No more than 60 milligrams of cholesterol, 3 grams of fat, and 1 gram of saturated fat; more than 10% of the Daily Value of vitamin A, vitamin C, iron, calcium, protein, or fiber. "Healthy" foods must also contain 360 milligrams or less of sodium.
Extra lean	Extra-lean pork, extra-lean hamburger	Fewer than 5 grams of fat, fewer than 2 grams of saturated fat and *trans* fat combined, *and* fewer than 95 milligrams of cholesterol (applies to meats only).
Lean	Lean beef, lean turkey	Fewer than 10 grams of fat, fewer than 4.5 grams of saturated fat and *trans* fat combined, *and* fewer than 95 milligrams of cholesterol (applies to meats only).
Free	Fat-free, *trans* fat-free, sugar-free, sodium-free foods	No—or negligible amounts of—fat, sugars, *trans* fat, or sodium.
Good source	Good source of fiber, good source of calcium or antioxidants	From 10 to 19% of the Daily Value for a particular nutrient or for vitamin A (beta- carotene), vitamin C, or vitamin E in the case of antioxidants.
High	High iron, high fiber foods	20% or more of the Daily Value for a particular nutrient.
Low cholesterol	Low-cholesterol egg product	20 milligrams or less cholesterol (applies to animal products only).
Low fat	Low-fat cheese, low-fat ice cream	3 grams or less of fat.
Low saturated fat	Low-saturated-fat pancake mix, low-saturated-fat eggnog	1 gram or less saturated fat and 0.5 grams or less *trans* fat.
Low sodium	Low-sodium soup, low-sodium hot dogs	140 milligrams or less sodium.

of fat or less per serving. Low-fat foods can be labeled with a "percent fat free" label, such as "98% fat-free" turkey. This label means that the product contains approximately 2% fat on a weight basis. Meat products labeled "lean" must contain fewer than 10 grams of fat, 4.5 grams of saturated fat and *trans* fat combined, and 95 milligrams of cholesterol per serving. Some labels promote meat as lean based on the percentage of the meat's weight that consists of fat. So, a meatloaf mix that is 16% fat on a weight basis might be labeled as "lean." It would not be lean according to nutrition labeling standards. Illustration 4.3 provides an example of an appropriately and an inappropriately labeled meat.

Other claims, such as "natural," "hormone-free," and "pure" are sometimes placed on food labels for sales appeal (Table 4.2). These terms have not been defined by the FDA, may not represent a health benefit, and may mislead consumers about the health value of products. The word *natural*, for example, can be found on labels for foods such as potato chips, butter-flavored microwave popcorn, and ice cream (see Illustration 4.4). The term is so broadly and inconsistently used that consumers tend to mistrust it.[5]

Health Claims In 1984 the Kellogg Company launched an ad campaign for All-Bran cereal that announced, "eating the right foods may reduce your risk of some kinds of cancer." Sales of the high-fiber cereal increased 37% in one year, but then Kellogg had to withdraw the ads. The FDA ruled the All-Bran statement was equivalent to a claim for a drug. The campaign, however, started the nutrition and health claims revolution.[7]

Table 4.2 Examples of claims not approved by the FDA for use on food labels

Natural
All natural
Pure
Antibiotic-free
Raised without antibiotics
Additive-free
Pesticide-free
Hormone-free
Nutritionally improved
No cholesterol (on plant foods)
Free-range
Eco-friendly
Pasture-fed

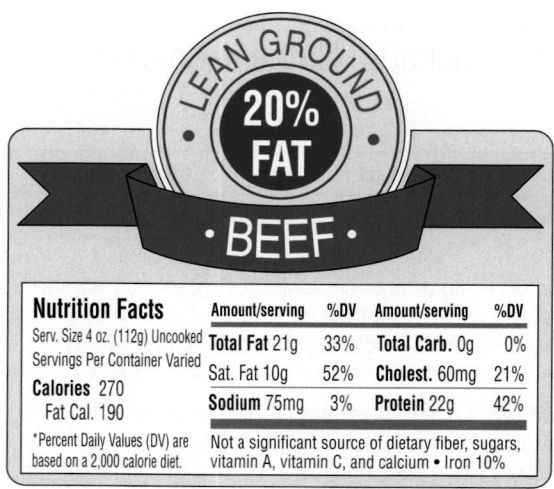

Nutrition Facts	Amount/serving	%DV	Amount/serving	%DV
Serv. Size 4 oz. (112g) Uncooked	**Total Fat** 21g	33%	**Total Carb.** 0g	0%
Servings Per Container Varied	Sat. Fat 10g	52%	**Cholest.** 60mg	21%
Calories 270	**Sodium** 75mg	3%	**Protein** 22g	42%
Fat Cal. 190				
*Percent Daily Values (DV) are based on a 2,000 calorie diet.	Not a significant source of dietary fiber, sugars, vitamin A, vitamin C, and calcium • Iron 10%			

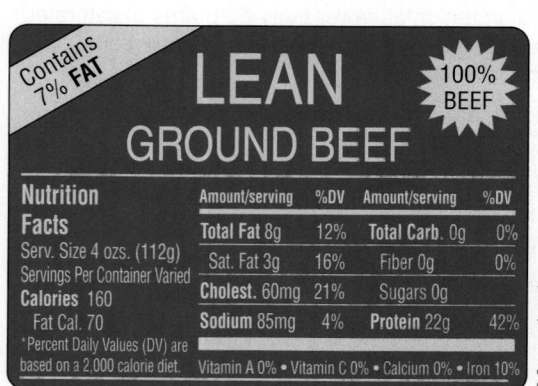

Nutrition Facts	Amount/serving	%DV	Amount/serving	%DV
Serv. Size 4 ozs. (112g)	**Total Fat** 8g	12%	**Total Carb.** 0g	0%
Servings Per Container Varied	Sat. Fat 3g	16%	Fiber 0g	0%
Calories 160	**Cholest.** 60mg	21%	Sugars 0g	
Fat Cal. 70	**Sodium** 85mg	4%	**Protein** 22g	42%
*Percent Daily Values (DV) are based on a 2,000 calorie diet.	Vitamin A 0% • Vitamin C 0% • Calcium 0% • Iron 10%			

© Cengage Learning

Illustration 4.3   Two examples of labels claiming "lean" ground beef. Only the 7% fat beef is actually lean. "20% fat" ground beef is far from lean, actually providing 21 grams of fat and 70% of total calories from fat per serving.

Illustration 4.4   Here's the ingredient list for the ice cream labeled "Made with natural ingredients": skim milk, cream, sugar, corn syrup, whey, molasses, acacia gum, guar gum, ground vanilla beans, vanilla extract, carob bean gum, carrageenan, xanthan gum.

Judith Brown

On approval by the FDA, foods or food components with scientifically agreed-upon benefits to disease prevention can be labeled with a health claim (Table 4.3). However, the health claim must be based on the FDA's "model claim" statements. For instance, scientific consensus holds that diets high in fruits and vegetables may lower the risk of cancer, so a health claim to this effect is allowed. The FDA's model claim for labeling fruits and vegetables is "Low-fat diets rich in fruits and vegetables may reduce the risk of some types of cancer, a disease associated with many factors." The FDA approves health claims only for food products that are not high in fat, saturated fat, cholesterol, or sodium. Model health claims approved by the FDA are shown in Table 4.4.

Labeling Foods as Enriched or Fortified   The vitamin and mineral content of foods can be increased by **enrichment** and **fortification**. Definitions for these terms were established more than 50 years ago. Enrichment pertains only to refined grain products, which lose vitamins and minerals when the germ and bran are removed during processing. Enrichment replaces the thiamin, riboflavin, niacin, and iron lost in the germ and bran. By law, producers of bread, cornmeal, pasta, crackers, white rice, and other products made from

Table 4.3 FDA-approved health claims[7]

1. Calcium, vitamin D, and osteoporosis
2. Dietary lipids (fats) and cancer
3. Dietary saturated fat and cholesterol and risk of coronary heart disease
4. Dietary non-cariogenic carbohydrate sweeteners and dental caries
5. Fiber-containing grain products, fruits, vegetables, and cancer
6. Folic acid and neural tube defects
7. Fruits, vegetables, and cancer
8. Fruits, vegetables, and grain products that contain fiber, particularly soluble fiber, and risk of coronary hearth disease
9. Sodium and hypertension
10. Soluble fiber from certain foods and risk of coronary heart disease
11. Soy protein and risk of coronary heart disease
12. Stanols/sterols and risk of coronary heart disease

© Scott Goodwin Photography

Table 4.4 Examples of model health claims approved by the FDA for labels of foods that qualify based on nutrient content; model claims are often abbreviated on food labels

Food and related health issues	Model health claim
Whole grain foods and heart disease, certain cancers	Diets rich in whole grain foods and other plant foods and low in fat, saturated fat, and cholesterol may reduce the risk of heart disease and certain cancers.
Sugar alcohols and tooth decay	Frequent between-meal consumption of food high in sugars and starches promotes tooth decay. Xylitol, the sugar alcohol in this food, may reduce the risk of tooth decay.
Saturated fat, cholesterol, and heart disease	Development of heart disease depends on many factors. Eating a diet low in saturated fat and cholesterol and high in fruits, vegetables, and grain products that contain fiber may lower blood cholesterol levels and reduce your risk of heart disease.
Calcium, vitamin D, and osteoporosis	Regular exercise and a healthy diet with enough calcium and vitamin D help maintain good bone health and may reduce the risk of osteoporosis later in life.
Fruits and vegetables and cancer	Low-fat diets rich in fruits and vegetables may reduce the risk of some types of cancer, a disease associated with many factors.
Folate and neural tube defects	Women who consume adequate amounts of folate daily throughout their childbearing years may reduce their risk of having a child with a brain or spinal cord defect.

refined grains must use enriched flours. Beginning in 1998, federal regulations mandated that folate (a B vitamin) in the form of folic acid be added to refined grain products. This new regulation was put into effect to help reduce the risk of a particular type of inborn, structural problem in children (neural tube defects) related to low blood levels of folate early in pregnancy.

Any food product can be fortified with vitamins and minerals—and many are. One of the few regulations governing the fortification of foods is that the amount of vitamins and minerals added must be listed in the Nutrition Facts panel. Illustration 4.5 shows some examples of enriched and fortified foods.

Food enrichment and fortification began in the 1930s to help prevent deficiency diseases such as rickets (vitamin D deficiency), goiter (from iodine deficiency), pellagra (niacin deficiency), and iron-deficiency anemia. Today, foods are increasingly being fortified for the purpose of reducing the risk of chronic diseases such as osteoporosis, cancer, and heart disease. Other foods, such as snack bars, fruit drinks, and sweetened ready-to-eat cereals, are being fortified with an array of vitamins and minerals. Regular consumption of fortified foods increases the risk that people will exceed Tolerable Upper Intake

enrichment The replacement of thiamin, riboflavin, niacin, and iron lost when grains are refined.

fortification The addition of one or more vitamins and/or minerals to a food product.

Illustration 4.5 Examples of foods that are enriched (*left*) and fortified (*right*).

Table 4.5 All foods with more than one ingredient must have an ingredient label. Here are the rules.

1. Ingredients must be listed in order of their contribution to the weight of the food, from highest to lowest.

2. Beverages that contain juice must list the percentage of juice on the ingredient label.

3. The terms *colors* and *color added* cannot be used. The name of the specific color ingredients must be given (for example, caramel color, turmeric).

4. Milk, eggs, fish, and five other foods to which some people are allergic must be listed on the label.

Levels (ULs) of nutrients designated in the Recommended Dietary Allowances (RDAs). Regular use of multiple vitamin and mineral supplements, along with liberal intake of fortified foods, enhances the likelihood that excessive amounts of some nutrients will be consumed.[24]

The Ingredient Label Still more useful information about the composition of food products is listed on ingredient labels. The label of any food that contains more than one ingredient must list the ingredients in order of their contribution to the weight of the food (Table 4.5).

The FDA now requires that ingredient labels include the presence of common food allergens in products. Potential food allergens consist of milk, eggs, fish, shellfish, tree nuts, wheat, peanuts, and soybeans. These foods, sometimes called the "Big Eight," account for 90% of food allergies. Precautionary statements, such as "may contain," or "processed in a facility with" are allowed.[8]

Food Additives on the Label Specific information about **food additives** must be listed on the ingredient label. Nearly 3,000 chemical substances may be added to food to enhance its flavor, color, texture, cooking properties, shelf life, or nutrient content. Food additives on the FDA's GRAS (*Generally Recognized As Safe*) list can be used in food without preapproval. New additives must be approved by the FDA prior to use. Table 4.6 provides examples of functions of some additives used in food. The most common food additives are sugar and salt, but trace amounts of polysorbate, potassium benzoate, and many other additives that are not so familiar are also included in foods. Although the new labeling regulations are more comprehensive than the old ones, they don't help consumers understand what some additives do. Appendix D lists many of the most common additives and indicates their function in foods.

Trace amounts of substances such as pesticides, hormones and antibiotics given to livestock, fragments of packaging materials such as plastic, wax, aluminum, or tin, very small fragments of bone, and insects may end up in foods. These are considered "uninten-

food additives Any substances added to food that become part of the food or affect the characteristics of the food. The term applies to substances added both intentionally and unintentionally to food.

Table 4.6 Solving the mystery of ingredient label terms[a]

Cake mix ingredient label		Additive	Function
	Ingredients: Sugar, enriched flour bleached (wheat flour, niacin), iron thiamin Mononitrait, (vitamin B1), riboflavin (vitamin B2), vegetable shortening (contains partially hydrogenated soybean cottonseed oil), **sodium aluminum phosphate**, dextrose, leavening, (baking soda, monocalcium phosphate, diacalcium phosphate, aluminum sulphate), wheat starch, **propylene glycol monoesters**, modified corn starch, salt, egg white, vanilla, dried corn syrup, polysorbate 60, nonfat milk, **mono and diglycerides**, sodium caseetrate, **xanthan gum**, soy lecithin.	Sodium aluminum phosphate	Gives baked products a light texture
		Propylene glycol monoesters	Helps blend ingredients uniformly, enhances moisture content and texture
		Mono- and diglycerides	Maintains product softness after baking
		Xanthan gum	Thickening agent, helps hold product together after baking

Note: All foods with more than one ingredient must have an ingredient label.
[a]See Appendix D for more information on the definitions of ingredient terms.

tional additives" and do not have to be included on the label.

Irradiated Foods Food irradiation is an odd example of a food additive. It is actually a process that doesn't add anything to foods. Irradiation uses X-rays, gamma rays, or electron beams to kill insects, bacteria, molds, and other microorganisms in food. Food irradiation enhances the shelf life of food products and decreases the risk of foodborne illness.[10] Irradiation must be performed according to specific federal rules.

Irradiated foods retain no radioactive particles. The process is like having your luggage X-rayed at the airport. Your luggage doesn't become radioactive, nor has anything in it changed. The process leaves no evidence of having occurred. Actually, this lack of change creates a challenge because inspection agencies can't determine whether a food has been irradiated or not. All irradiated foods—except spices that are added to processed foods—are required to display the international "radura" symbol and to indicate that the food has been irradiated (Illustration 4.6). Irradiation is approved for use on chicken, turkey, pork, beef, eggs, grains, fresh fruits and vegetables, and other foods in the United States.[9]

Irradiated foods have a long history but have not yet won the trust of some American consumers.[11] The desire to limit radioactive processes of all sorts, as well as radioactive waste products, is widespread; many people fear anything that involves radioactivity, especially when it comes to food.

Illustration 4.6 The "radura," as the symbol is called, must be displayed on irradiated foods. The words "treated by irradiation, do not irradiate again" or "treated with radiation, do not irradiate again" must accompany the symbol.

Dietary Supplement Labeling

- **Evaluate nutrient content and health claims made on dietary supplement labels.**

A **dietary supplement** is a product taken by mouth that contains a "dietary ingredient" intended to supplement the diet. The dietary ingredients in these products include vitamins, minerals, proteins, enzymes, herbs, hormones, and organ tissues. Nutrition labeling regulations place dietary supplements in a special category under the general umbrella of "foods," not drugs. Dietary supplements differ from drugs in that they do not have to undergo vigorous testing and obtain FDA approval before they are sold. In return, dietary supplement labels cannot claim that the products treat, cure, or prevent disease.[12]

According to FDA regulations, dietary supplements must be labeled as such and include a "Supplement Facts" panel that lists serving size, ingredients, and percent Daily Value (%DV) of essential nutrient ingredients (Illustration 4.7). Like foods, qualifying dietary supplements can be labeled with nutrient claims (e.g., "high in calcium,"), and health claims (e.g., "diets low in saturated fat and cholesterol that include sufficient soluble fiber may reduce the risk of heart disease"). Dietary supplement labels can also include **structure/function claims**.

Structure or Function Claims

Dietary supplements can be labeled with statements that describe effects the supplement may have on body structures or functions (shown in Illustration 4.8). Under this regulation, for example, certain supplements can be labeled with the following statements: "Promotes healthy heart and circulatory function," "Helps maintain mental health," or "Supports the immune system." If a structure/function claim is made on the label or package inserts, the label or insert must acknowledge that the FDA does not support the claim: *"This statement has not been evaluated by the FDA. This product is not intended to diagnose, treat, cure, or*

dietary supplement Any product intended to supplement the diet, including vitamins, minerals, proteins, enzymes, herbs, hormones, and organ tissues. Such products must be labeled "Dietary Supplement."

structure/function claim Statement appearing primarily on dietary supplement labels that describes the effect a supplement may have on the structure or function of the body. Such statements cannot claim to diagnose, cure, mitigate, treat, or prevent disease.

Nutrition Facts
Per 125mL (87g)

Amount		% Daily Value
Calories 80		
Fat 0.5 g		1%
Saturated 0 g + Trans 0 g		0%
Cholesterol 0 mg		
Sodium 0 mg		0%
Carbohydrate 18 g		6%
Fibre 2 g		8%
Sugars 2 g		
Protein 3 g		
Vitamin A 2%	Vitamin C	10%
Calcium 0%	Iron	2%

Scott Goodwin Photography

Illustration 4.7 Dietary supplement label.

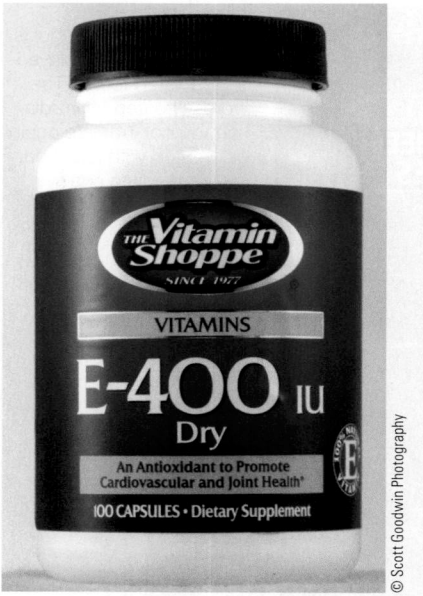

Illustration 4.8 Example of a structure, function claim.

prevent any disease." Dietary supplements labeled with misleading or untruthful information, and those that are not safe, can be taken off the market and the manufacturers fined.

Structure and function claims do not have to be approved before they can appear on product labels. The FDA and FTC have taken action against companies responsible for fraudulent claims.[13] The FDA has issued new guidelines that define the quality of scientific evidence needed to substantiate structure/function claims.[12]

The COOL Rule

USDA requires retailers to display a country-of-origin label (COOL) on certain products (Illustration 4.9).[14] The rule is meant to expand informed consumer choices and to help track down food-borne illness outbreaks. The rule applies to meats, fish, seafoods, fruits, vegetables, many nuts, and some herbs.

Organic Foods

- **Compare the characteristics of organically and conventionally produced food products.**

For many years, consumers and producers of organic foods urged Congress to set criteria for the use of the term *organic* on food labels. Consumers wanted to be assured that foods were really organic, and producers wanted to keep the business honest. Standards were needed because consumers cannot distinguish organically produced foods from others by looking at them, tasting them, or reading their nutrient values. The U.S. Department of Agriculture developed and is implementing standards for organic foods.[15]

Labeling Organic Foods

Rules that qualify foods as organic are shown in Illustration 4.10. If organic growers and processors qualify according to USDA-approved certifying organizations, they can place the green-and-white USDA Organic seal on product labels (Illustration 4.11). The Canadian organic label, also shown in Illustration 4.11, is equivalent to the USDA Organic label. The USDA and Canadian authorities can impose financial penalties on companies that use the seal inappropriately. Organically grown and produced foods can be labeled in four other ways:

1. "100% Organic" if they contain entirely organically produced ingredients.
2. "Organic" if they contain at least 95% organic ingredients.
3. "Made with organic ingredients" if they contain at least 70% organic ingredients.
4. "Some Organic Ingredients" if the product contains less than 70% organic ingredients.

Illustration 4.9 Country-of-origin labels are now required on certain foods.

Illustration 4.10 USDA rules for qualifying foods as organic.

1. Plants
   - Must be grown in soils not treated with synthetic fertilizers, pesticides, and herbicides for at least three years
   - Cannot be fertilized with sewer sludge
   - Cannot be treated by irradiation
   - Cannote be grown from genetically modified seeds or contain genetically modified ingredients

2. Animals
   - Cannot be raised in "factory-like" confinement conditions
   - Cannot be given antibiotics or hormones to prevent disease or promote growth
   - Must be given feed products that are 100% organic

Some people choose organic foods because they perceive them to be better for the environment and for health than conventionally grown foods.[16] Most organic foods, however, are not totally free of some of these ingredients. Due to "pesticide drift" from sprayed crops, past use of pesticides on farmland, and the leaching of chemicals used on crops into groundwater, organically grown plants may contain traces of pesticides. There are positive and negative characteristics of both organic and conventional foods, and so far neither type has been shown to be superior overall in nutrient content, safety, or effects on health.[16,17]

The New Wave of Nutrition Labels

- **Identify strengths and weaknesses of various nutrition labeling systems on food packaging and calorie listings for food items.**

If you are a nutrition label reader you have probably noticed that a variety of food company–initiated labeling systems appear on the front of food packages. Three of these, "Nutrition at Glance," "Smart Choices Made Easy," and the "Smart Choices Program," are shown in Illustration 4.12. The Smart Choices Program features a symbol that identifies more nutritious choices within specific food product categories, and lists the calories per serving and number of servings per package. This system has been selected for use by a number of food companies in Europe and the United States.[18] Other nutrition labeling systems, including those from the American Heart Association and various grocery store chains, are being used on food product packaging.

Industry-initiated labeling systems are voluntary and generally employ science-based criteria for judging the nutrient qualities of food products.[20] The labels supplement the Nutrition Facts panel of the food products. The primary weaknesses of the food industry-backed labeling systems are that they may cause confusion among consumers, criteria

Canada Organic Regime (COR), Canadian Food Inspection Agency (CFIA)

Illustration 4.11 Foods certified as organic by the USDA can display the "USDA Organic" seal on food packages. The approved seal for Canada is below it.

All © Scott Goodwin Photograph

Illustration 4.12 A few examples of food industry–initiated nutrition labeling systems used on food packages.

ANSWERS TO REALITY CHECK
Nutrition Labeling

Ronald got it. "Fat free" does not equal "calorie free." Fat-free Caesar salad dressing, for example, provides 40 calories in a 2-tablespoon serving.

Jared:

Ronald:

Illustration 4.13 An example of the required calorie labeling of chain restaurant menu items.

used to label foods vary, they may favor the company's products, and the labels only appear on selected food products.[18,19]

Labeling systems in place may not adequately emphasize information consumers need to know to help curb high-calorie intakes and obesity. This concern is being addressed through increased emphasis on the labeling of calories in portions of food sold in chain restaurants.

Calories on Display

Within the past few years several large cities and states have implemented legislation that requires chain restaurants with more than 20 locations to post the calorie value of a serving of each standardized menu item offered (Illustration 4.13). Details regarding the item's content of 11 other nutrients must be available on request. In 2010, this calorie labeling system for chain restaurants was adopted by the federal government for nationwide implementation in 2012. The labeling requirement also extends to food offered in vending machines for establishments with 20 or more locations nationwide.[21]

Preliminary evidence indicates that calorie labeling of restaurant menu items is associated with no change or somewhat lower calorie food choices among diners, and increased availability of lower calorie menu options in some restaurants.[21,22]

Beyond Nutrition Labels: We Still Need to Think

Understanding and applying the information on nutrition labels calls for more nutrition knowledge on the public's part. As highlighted in the nearby Reality Check, people need to know about nutrition before they can understand nutrition labels and incorporate labeled foods appropriately into an overall diet. Good diets include more than foods with nutrition labels. The ice cream cone from the stand in the mall; the orange, potato, or fish we buy in the store; and the pizza delivered to the dorm are unlabeled parts of many diets. We need to know enough about the composition of unlabeled foods to fit them into a healthy diet. Nutrition labels often list a few vitamins and minerals, but many more are required for health. It's particularly important for people to know if their diet is varied enough to supply needed vitamins and minerals. In addition, nutrient and health claims made on food packaging do not address potentially negative aspects of food products, such as "low fiber," "high saturated fat," or "may promote tooth decay." Use of the MyPlate food group guide should go hand in hand with label reading for diet planning.

Not every food we eat has to have the "right" label profile. Serving mostly low-fat or low-calorie foods to children, for example, might have unintended, unhealthy effects. Young children need calories and fat for growth and development. If diets are severely restricted, growth and development will be impaired.

Nutrition labels are an important tool for helping people make informed food-purchasing decisions. About 6 in 10 adults use them to guide their food purchasing decisions, and those who do use them tend to consume healthier diets than those who don't.[23] However, labels do not now—nor will they ever—provide all the information needed to make wise decisions about food. Only people who are well informed about nutrition can do that.

NUTRITION
up close

Comparison Shopping

Focal Point: Comparing label information can help you choose the most nutritious product for your money.

Whole-Grain Cereal

Nutrition Facts

Serving Size **1 cup (50g)**
Serving Per Container 10

Amount Per Serving

Calories 170   Calories from Fat 5

	% Daily Value*
Total Fat 0.5 g	1%
Saturated Fat 0 g	0%
Trans Fat 0 g	
Cholesterol 0 mg	0%
Sodium 0 mg	0%
Total Carbohydrate 41 g	14%
Dietary Fiber 5 g	21%
Sugars 0 g	
Protein 5 g	

Vitamin A 0%	Vitamin C 0%
Calcium 2%	Iron 8%

*Percent Daily Values are based on a 2000 calorie diet. Your daily values may be higher or lower depending on your calorie needs:
**Intake should be as low as possible.

	Calories:	2,000	2,500
Total Fat	Less than	65 g	80 g
Sat Fat	Less than	20 g	25 g
Cholesterol	Less than	300 mg	300 mg
Sodium	Less than	2400 mg	2400 mg
Total Carbohydrate		300 g	375 g
Dietary Fiber		25 g	30 g

Health Granola Cereal

Nutrition Facts

Serving Size **1 cup (50g)**
Serving Per Container 10

Amount Per Serving

Calories 210   Calories from Fat 30

	% Daily Value*
Total Fat 3 g	5%
Saturated Fat 0 g	0%
Trans Fat 0 g	
Cholesterol 0 mg	0%
Sodium 120 mg	5%
Total Carbohydrate 43 g	14%
Dietary Fiber 3 g	12%
Sugars 16 g	
Protein 5 g	

Vitamin A 2%	Vitamin C 0%
Calcium 2%	Iron 10%

*Percent Daily Values are based on a 2000 calorie diet. Your daily values may be higher or lower depending on your calorie needs:
**Intake should be as low as possible.

	Calories:	2,000	2,500
Total Fat	Less than	65 g	80 g
Sat Fat	Less than	20 g	25 g
Cholesterol	Less than	300 mg	300 mg
Sodium	Less than	2400 mg	2400 mg
Total Carbohydrate		300 g	375 g
Dietary Fiber		25 g	30 g

© Cengage Learning

Assume both cereals cost the same. Rate the nutrition value of these two cereals by completing the second paragraph below using information from the Nutrition Facts panels. The first paragraph has been done for you.

The **Whole-Grain Cereal**, providing **170 calories** per serving, contributes **5 calories from fat** (or **1**% of the % Daily Value). Each serving also provides **0** mg of **sodium** (or **0**% of the % Daily Value), **5** g of **dietary fiber** (or **21**% of the % Daily Value), and **0** g of **sugars**.

The **Health Granola Cereal**, providing _____ **calories** per serving, contributes _____ **calories from fat** (or _____ % of the % Daily Value). Each serving also provides _____ mg of **sodium** (or **5**% of the % Daily Value), _____ g of **dietary fiber** (or _____ % of the % Daily Value), and _____ g of **sugars**.

Which cereal provides the better nutritional value?

Feedback to the Nutrition Up Close is located in Appendix G.

REVIEW QUESTIONS

- **Apply knowledge about the four key elements of nutrition labeling to decisions about the nutritional value of foods.**

1. Almost all multiple-ingredient foods must be labeled with nutrition information.   **True/False**

2. Food manufacturers can list any serving size they want on Nutrition Facts panels.   **True/False**

3. In general, %DV listed for nutrients in Nutrition Facts panels of 10% or more are considered "low," and those listed as 50% or more are considered "high."   **True/False**

4. An overriding principle of nutrition labeling regulations is that nutrient content and health claims made about a food on the packaging must be truthful.   **True/False**

5. The term *enriched* on a food label means that extra vitamins and minerals have been added to the food to bolster its nutritional value.   **True/False**

6. The ingredient that makes up the greatest portion of a food product's weight must be listed first on ingredients labels.   **True/False**

7. Food labeling regulations do *not* require food manufacturers to list the presence of major food allergens on ingredient labels.   **True/False**

8. _____ A tea manufacturer labels a green tea product as "high in the antioxidant vitamin C." After brewing, a serving of the green tea was found to contain 10 mg vitamin C. Which of the following statements about the green tea nutrient content claim is true?
   a. The green tea qualifies for the "high in" nutrient content claim made.
   b. The green tea qualifies for a "natural" content claim.
   c. The green tea qualifies for a "healthy" nutrient content claim.
   d. The green tea qualifies for a "good source of" nutrient content claim.

9. _____ You see a yogurt product at the grocery store labeled "fat free." This nutrient content claim means that a standard serving of the yogurt contains:
   a. No, or negligible amounts of fat.
   b. No, or negligible amounts of fat, *trans* fat, and sodium.
   c. Fewer than 10 grams of fat and 4.5 grams of saturated fat.
   d. Three grams of fat or less.

10. _____ John eats three cookies for a bedtime snack while reading the Nutrition Facts panel on the cookie package. He notices that a three-cookie serving provides 2 grams of saturated fat and 0 grams *trans* fat. The cookie qualifies for a nutrient content claim of:
   a. Low saturated fat.   c.   Extra lean.
   b. *Trans* fat free.   d.   None of the above.

Questions 11 through 14 refer to the following case scenario.

You have just been informed by your health care provider that your cholesterol level is "borderline-high" and that you need to lose 10 pounds. The health care provider recommends that you reduce your calorie intake by 400 calories a day, cut back on saturated fats, and come back to the clinic in three months.

   Here's your plan on how to decease calories and saturated fat: You will cut back on your intake of snack food. You have been eating three 1-ounce servings of potato chips, and two 1-ounce servings of cheese crackers nightly. You notice from the Nutrition Facts panels on the chips and crackers that each ounce serving of potato chips (15 chips) provides 160 calories, and each ounce serving of cheese crackers (25 crackers) 140.

11. How many calories were you consuming from the chips and crackers daily? _____ calories

12. If you reduced your intake of both potato chips and cheese crackers to one serving per day you would reduce your intake from these snack foods by _____ calories.

13. One serving of potato chips provides 1 gram (5% Daily Value) saturated fat, and one serving of cheese crackers has 1.5 grams (8% Daily Value) for saturated fat. The amount of saturated fat in terms of grams and % Daily Value provided by three servings of potato chips and two servings of cheese crackers is _____ grams and _____ % Daily Value.

14. If you decrease your intake of potato chips and cheese crackers to one ounce of each a day, you will reduce your saturated fat intake from these snacks by _____ grams and by _____ % Daily Value.

• **Evaluate nutrient content and health claims made on dietary supplement labels.**

15. Dietary supplements, such as those for herbs and vitamins, are considered drugs and are not regulated by FDA's Nutrition Labeling rules.   **True/False**

16. Foods, but *not* dietary supplements, can be labeled with nutrient content and health claims.   **True/False**

17. _____ Assume a dietary supplement made from pomegranate is labeled with the claims "lowers plaque formation in the arteries" and "improves blood flow." It turns out that the FDA becomes aware of the claims and subsequently sues the maker of the dietary supplement. What would be the most likely reason for the FDA suit?
   a. There are no approved nutrient claims for "lowers plaque formation" and "improves blood flow."
   b. Dietary supplement labels cannot claim that a product treats a disease.
   c. There was insufficient research to support the claims.
   d. The FDA did not preapprove the claim before it was made for the juice.

• **Compare the characteristics of organically and conventionally produced food products.**

18. Animals providing meats labeled "organic" cannot be given antibiotics or hormones.   **True/False**

19. Foods bearing the USDA Organic seal are certified as organic by the U.S. Department of Agriculture.   **True/False**

• **Identify strengths and weaknesses of various nutrition labeling systems on food packaging and calorie listings for food items.**

20. Nutrition labels provide all the nutrition information we need to make healthful food choices.   **True/False**

Answers to these questions can be found in Appendix G.

NUTRITION SCOREBOARD ANSWERS

1. Labeling is required for almost all processed foods and for dietary supplements, but remains largely voluntary for fresh fruits, vegetables, and fish.[1,2]   **False**

2. Health claims for food products are allowed on many food packages. The claims must be truthful and adhere to FDA standards.   **True**

3. Nutrition labels are necessarily short and can't tell the whole story about food and health. They help people make several key decisions about a food's composition.   **False**

Nutrition, Attitudes, and Behavior

NUTRITION SCOREBOARD

1 Food preferences are genetically determined.
   True/False

2 Food habits never change.   **True/False**

3 Skipping breakfast does not affect school performance
   in well-nourished children.   **True/False**

Answers can be found at the end of the unit.

- Identify factors that influence an individual's food choices and preferences.

- Apply the process for making healthful changes in food choices to a specific change in food intake.

- Differentiate between scientifically supported and unsupported conclusions about diet and behavior relationships.

Origins of Food Choices

- **Identify factors that influence an individual's food choices and preferences.**

Horse meat is a favorite food in a large area of north-central Asia. Pork, which is widely consumed in North and South America, Europe, and other areas, is rigidly avoided by many people in Islamic countries. Bone-marrow soup and sautéed snails are delicacies in France, while kidney pie is traditional in England. Dog is a popular food in Borneo, New Guinea, the Philippines, and other countries, whereas snake is a delicacy in China. In some countries, people enjoy insects (Illustration 5.1). Some people consider corn and soybeans fit only for animal feed. And then there are steamed clams and raw oysters—food passions for some, but absolutely disgusting to others.[3]

When did you first think "yecck!?" The food choices just described would elicit that response among people from a variety of cultures, but they would not necessarily be responding to the same foods.

Why do people eat what they do? People learn from their culture and the society in which they live what animals and plants are considered food and which are not.[4] Once items are identified as food, they develop a legacy of strong symbolic, emotional, and cultural meanings. Comfort foods, health foods, junk foods, fun foods, soul foods, fattening foods, mood foods, and pig-out foods, for example, have been identified in the United States. All cultures have their "super food": in Russia and Ireland, it's potatoes; in Central America, it's corn and yucca (a starchy root, also called manioc); in Somalia, it's rice. The designation refers to the cultural significance of the food and not to its nutritional value.[5]

In countries such as the United States where a wide variety of foods are available and people have the luxury of selecting which foods they will eat, food choices are influenced by a range of factors (Illustration 5.2). Of these factors, food preference has the largest impact. Food preferences vary a good deal among individuals and lead to a wide array of specific food choices. Rather than being inborn, food preferences are primarily learned.[7]

Illustration 5.1 Grasshoppers are Mexican delicacies, served at Girasoles Restaurant in Mexico City.

Gavriel Jecan/Photodisc/Getty Images

Key Nutrition Concepts

Our examination of the factors involved in food choices relates to three of the key nutrition concepts introduced in Unit 1:

1. Food is a basic need of humans.

2. Some groups of people are at higher risk of becoming inadequately nourished than others.

3. Poor nutrition can influence the development of certain chronic diseases.

We Don't Instinctively Know What to Eat

The food choices people make are not driven by a need for nutrients or guided by food selection genes. People deficient in iron, for example, do not seek out iron-rich foods. If we're overweight, no inner voice tells us to reject high-calorie foods. Women who are pregnant don't instinctively know what to eat to nourish their growing fetuses. No evidence indicates that young children, if offered a wide variety of foods, would select and ingest a well-balanced diet.[8]

Humans are born with mechanisms that help them decide when and how much to eat, however.[7] An inborn attraction to sweet-tasting foods, a dislike for bitter foods, and the response of thirst when water is needed all influence food and fluid intake to an extent (Illustration 5.3).[10] There is evidence to suggest that people deficient in sodium experience an increased preference for salty foods.[7]

Illustration 5.2 Factors influencing food selection and dietary quality. Each of these sets of factors interacts with the others.[6] *Why do we prefer the foods we do?*

Culture
- Acceptable foods
- Customs
- Food symbolism
- Religious beliefs

Nutrition Knowledge and Beliefs
- Health concerns
- Nutritional value of foods
- Attitudes and values
- Education
- Experience

Food Preferences
- Food taste, smell, color, texture, and temperature
- Heredity
- Familiarity

FOOD SELECTION
Dietary quality

Practical considerations
- Food cost
- Convenience
- Level of hunger
- Food availability
- Health status

Photo Disc

Most people have a strong aversion to foods they think smell bad, which are likely spoiled. Whether this response is inborn or learned is not clear.

Food Preferences

The strong symbolic, emotional, and cultural meanings of food come to life in the form of food preferences. We choose foods that, based on our cultural background and other learning experiences, give us pleasure.[6] Foods give us pleasure when they relieve our hunger pains, delight our taste buds, provide the right feeling of texture in our mouth, or give us comfort and a sense of security. We find foods pleasurable when they outwardly demonstrate our superior intelligence, our commitment to total fitness, or our pride in our ethnic heritage. We reject foods that bring us discomfort, guilt, and unpleasant memories, and those that run contrary to our values and beliefs.[6]

The Symbolic Meaning of Food Food symbolism, cultural influences, and emotional reasons for food choices are broad concepts that may become clearer with concrete examples. Here are a few examples to consider.

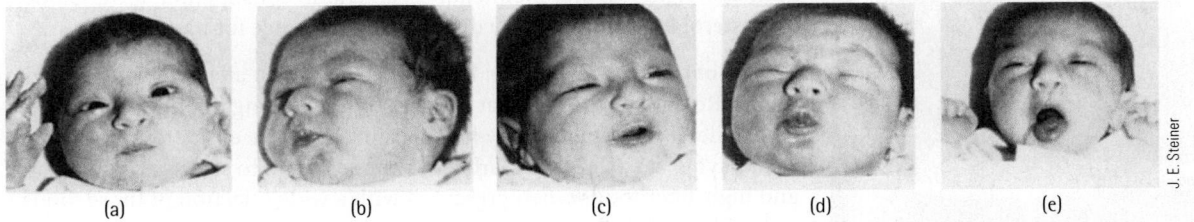

(a)　　(b)　　(c)　　(d)　　(e)

J. E. Steiner

Illustration 5.3 Newborn infants respond to different tastes: (a) Baby at rest, (b) tasting distilled water, (c) tasting sugar, (d) tasting something sour, and (e) tasting something bitter.
Source: Taste-induced facial expressions of neonate infants from the classic studies of J. E. Steiner, in *Taste and Development: The Genesis of Sweet Preference*, ed. J. M. Weiffenbach, HHS Publication no. NIH 77-1068 (Bethesda, MD: U.S. Department of Health and Human Services, 1977), pp. 173–89, with permission of the author.

Status Foods Vance Packard, in his book *The Status Seekers*, provides a memorable example of the symbolic value of food:

> *As a lad, this man had grown up in a poor family of Italian origin. He was raised on blood sausages, pizza, spaghetti, and red wine. After completing high school, he went to Minnesota and began working in logging camps, where—anxious to be accepted—he soon learned to prefer beef, beer, and beans, and he shunned "Italian" food. Later, he went to a Detroit industrial plant, and eventually became a promising young executive. . . . In his executive role he found himself cultivating the favorite foods and beverages of other executives: steak, whiskey, and seafood. Ultimately, he gained acceptance in the city's upper class. Now he began winning admiration from people in his elite social set by going back to his knowledge of Italian cooking, and serving them, with the aid of his manservant, authentic Italian treats such as blood sausage, spaghetti, and red wine![11]*

Comfort Foods Ice cream, apple pie, chicken noodle soup, boxed chocolates, meat loaf and mashed potatoes: these are popular comfort foods in the United States. The feelings of security and love that came along with the tea and honey or chicken soup that your mother or father gave you when you had a cold, or with the ice cream and popsicles lovingly given to soothe a sore throat, are renewed with comfort foods. Some comfort foods can bring pleasure and reduce anxiety just by their image (Illustration 5.4).[12]

Once the symbolic value of a food is established, its nutritional value will remain secondary.[13] Food status is a strong determinant of food choices; and after all, as a noted nutritionist once said, "Life needs a little bit of cheesecake."[14]

"Discomfort Foods" Memories of bad experiences with food, and expectations that certain foods will harm us in some way, each contribute to our learning about food and affect our food preferences. Eating a piece of blueberry pie right before an attack of the flu or overdosing on sweet pickles or olives, for example, may take these foods off your preferred list for a long time.

Cultural Values Surrounding Food A team of scientists observed that the diet of certain groups in the Chin States of Upper Burma was seriously deficient in animal protein. After considerable study a way was found to improve the situation by cross-breeding the small, local black pigs raised by the farmers with an improved strain to obtain progeny, giving a greater yield of meat. The entire operation, however, completely failed to benefit the nutrition of the population because of one fact, which had been viewed as irrelevant. The cross-bred pigs were spotted. And it was firmly believed—as firmly as we believe that to eat, say, mice, would be disgusting—that spotted pigs were unfit to eat.[3]

Dietary change introduced into a culture for the purpose of improving health can be successful only if it is accepted by the culture. Cultural norms are not easily modified.[4]

Other Factors Influencing Food Choices and Preferences Food preferences and selections are also affected by the desire to consume foods that are considered healthy. Reducing fat intake, eating more fruits and vegetables, and cutting down on sweets bring rewards and pleasures such as weight loss and maintenance, an end to constipation, lower blood cholesterol level, and a newly discovered preference for basic foods.

Food Cost and Availability Food choices are also affected by the cost of food.[15] Researchers found that college students eating in dining halls have better diets when they prepay for their meals for the entire term rather than paying at each meal.[16] Grocery shoppers tend to select more low-fat and high-fiber foods when presented with a wide selection of those foods rather than just a few.[17]

Genetic Influences Genetics plays an important role in food preference development. Some people are born with a strong sensitivity to bitterness and may reject foods such as brussels sprouts and broccoli (especially if

Illustration 5.4 Do you find comfort in this photo?

Jon Edwards Photography/Healthy Food Images/age fotostock

Illustration 5.5   Changes in Americans' food choices from 2006 to 2010–2011.[22,23]

they are over-cooked!), bitter-tasting teas and wines, and tonic water.[18,19] Some people with these genetic traits will eat strong-flavored vegetables if the bitterness is disguised with seasonings or sauces. Differences in how food tastes can strongly influence individual food likes and dislikes.[19]

Food Choices Do Change

- **Apply the process for making healthful changes in food choices to a specific change in food intake.**

Who says old dogs can't learn new tricks? Most Americans aren't eating the way they used to.[20] A survey conducted by the American Dietetic Association (now called the Academy of Nutrition and Dietetics) in 2011 found that 48% of American adults reported consuming more berries, 43% more low-fat foods, and 31–34% said they are consuming more low-sodium and low-sugar foods than they did five years ago. They reported decreased consumption of beef (39%), pork (35%), and dairy products (22%).[21] Examples of these changes in food consumption since 2006 are shown in Illustration 5.5.

Food choices are largely learned and do change as we learn more about foods. Perhaps your food choices have changed over time. How do the food choices you make now compare with the choices you made five years ago?

How Do Food Choices Change?

What are the ingredients for change in food choices? Why do some people succeed in improving their food choices while other people find that very hard to do? Nutrition knowledge, attitudes, and values have a lot to do with changing food choices for the better.

Nutrition Knowledge and Food Choices Sound knowledge about good nutrition necessarily precedes the selection of a healthful diet. But is knowledge enough to ensure that healthy changes in diet will be made? The answer is "yes" for some people and "no" for others. Higher levels of knowledge about diet and health have been related to healthier eating practices among students at James Madison University, the University of Texas at Austin, and the University of Vermont (Illustration 5.6).[47–49] For some Americans, news that animal fats raise blood cholesterol levels was enough to convince them to cut down on beef and whole milk. Knowledge that diets containing an adequate amount of fiber reduce the risk of certain types of cancer led many Americans to switch to high-fiber breakfast cereals.[22] Knowledge alone is often not enough, however.

When Knowledge Isn't Enough Many people know far more about the components of a good diet than they put into practice. The problem is that between knowledge and practice lie multiple beliefs and experiences that act as barriers to change (Illustration 5.7). Change of any type is most likely to succeed when the benefits of making the change outweigh the disadvantages. This makes changes in food choices a very individual decision.

Daily Tribune

Nutrition Course Improves Diets of College Students

The diets of college students improved after taking a basic nutrition course.

Results indicated that students decreased the amount of total fat from 82 to 68 grams and made other changes putting their diets more in line with the Dietary Guidelines.

Illustration 5.6   A true story.

"I feel guilty about eating the foods I like."

"Eating right is too expensive."

"I tried eating better, but I didn't stick with it."

"I don't have the time to eat right."

"The vegetables I like aren't available."

"I'm healthy now... Why should I worry about my diet?"

Photo Disc

Illustration 5.7 Why knowledge about a good diet may not be enough to improve food choices.

Individuals decide whether a change is in their best interests. But what sorts of circumstances, in addition to increased knowledge, make changes in food choices worthwhile for individuals—and even highly desired?

Nutrition Attitudes, Beliefs, and Values The value individuals place on diet and health is reflected in the food choices they make. A survey of restaurant patrons found that food choices varied according to the consumer's perceptions of the importance of diet to health:[24]

- "Unconcerned" consumers—people who are unconcerned about the connection between diet and health and who tend to describe themselves as "meat and potato eaters"—select foods for reasons other than health.

- "Committed" consumers believe that a good diet plays a role in the prevention of illness. They tend to consume a diet consistent with their commitment to good nutrition.

- "Vacillating" consumers—people who describe themselves as concerned about diet and health but who do not consistently base food choices on this concern—tend to vary their food choices depending on the occasion. These consumers are likely to abandon diet and health concerns when eating out or on special occasions, but they generally adhere to a healthy diet.

Avoiding illness and curing or diminishing current health problems are likewise strong incentives for changing food choices (Table 5.1).[25] One study found that nurses and dietitians based dietary changes primarily on the benefits of good diets to future health.[26] In almost all instances, the key to lasting improvements in diet is to make changes you can live with. Changes that bring you more pleasure than inconvenience or discomfort have staying power, while those that require a lot of willpower to maintain rarely last.[27]

Successful Changes in Food Choices

The primary reason efforts to improve food choices fail is that the changes attempted are too drastic. Improvements that last tend to be the smallest acceptable changes needed to do the job.[28,29]

The Process of Changing Food Choices Assume you need to lose weight and want to keep it off by modifying your food choices. A promising plan to accomplish this goal would begin by identifying food choices you would like to change and lower calorie food options you would be willing and able to eat (Table 5.2). Then you could make the plan more specific by identifying the changes that would be easiest to implement. For example, assume the low-calorie foods you like include frozen nonfat yogurt and oranges. You might decide to eat yogurt or an orange in place of your usual bedtime snack of ice cream. A specific dietary change such as this is much easier to implement than a broad notion, such as "eat less." Although weight loss will take a while, such a small acceptable change has a much better chance of working than a drastic change in diet.

Table 5.1 Factors that enhance food changes

- Attitude that nutrition is important
- Belief that diet affects health
- Perceived susceptibility to diet-related health problems
- Perception that benefits of change outweigh barriers to change

Table 5.2 Changing food choices for the better[28],[29]

The process	An example
1. Identify a healthful change in your diet you'd like to make.	1. I'd like to lower my fat intake.
2. Identify two food choices you make that should change because they contribute to the need for the healthful change you identified.	2. I eat at fast-food restaurants three times a week. I usually have a large order of fries, and once a week I eat fried chicken.
3. Identify two or more specific, acceptable options for more healthful food choices than the ones identified under number 2.	3. Options identified: • Order tossed salad with low-calorie dressing instead of fries. • Eat a grilled chicken sandwich every other week instead of fried chicken. • Eat Mexican fast food more often.
4. Decide which option is easiest to accomplish and requires the smallest change to get the job done.	4. I love Mexican food. It would be easy to eat tacos instead of french fries or fried chicken.
5. Plan how to incorporate the change into your diet.	5. Mondays and Fridays, I'll eat tacos.
6. Implement the change. Be prepared for midcourse corrections.	6. Midcourse correction: On Fridays, when I'm with my friends, it's easier to eat at the restaurant they like. I'll order the grilled chicken sandwich and coleslaw.

Planning for Relapses When making a change in your diet, be prepared for relapses. Relapses happen for a number of reasons, and they don't mean the attempt has failed. People often return to old habits because the change they attempted was too drastic or because they tried to make too many changes at once. If the change undertaken doesn't work out, rethink your options and make a midcourse correction.

Does Diet Affect Behavior?

- **Differentiate between scientifically supported and unsupported conclusions about diet and behavior relationships.**

Food affects behavior in some rather striking ways. Irritable, hungry, crying infants rapidly change into cooing, sleepy angels after they are fed. Low-on-sleep employees perk up after their morning coffee. A high-calorie lunch makes many people feel calm and sleepy.[30] Not only do our behaviors affect our diet, but our diets can affect our behaviors. Examples of associations between dietary characteristics and behavior are listed in Table 5.3. One common belief about food and behavior is the subject of this unit's "Reality Check."

REALITY CHECK
Can Food Be a Love Potion?

Toward the end of a friend's birthday party, Glenda and Cassell got into a heated debate about the existence of food aphrodisiacs. Part of the conversation went like this:

Is it all in Glenda's head?

Answers on page 5-8

Glenda: Chocolate and vanilla are natural love potions, there's no doubt about it. If I eat a chocolate truffle and spritz on vanilla flavoring like perfume before a date, the guy always goes nuts for me!

Cassell: Are you kidding? No way! I've heard about oysters and this tree bark that are supposed to work miracles, too. It's all in your head.

The idea that food can act as an aphrodisiac has been around since ancient times. Although many have looked and others have tried, no one has found a food that acts like a love potion.[38]

Glenda:

Cassell:

Table 5.3 Examples of ways in which diet may affect behavior

Dietary characteristic	Behavioral outcomes
Malnutrition, growth stunting in early childhood	Lower intellectual functioning and school performance, increased antisocial behavior in childhood[31]
Nutritional supplementation of malnourished young children	Improved growth and intellectual functioning in adulthood[32]
Very low carbohydrate intake (less than 20 grams per day)	Reduced short-term memory, slower reaction times, increased attention span[33]
Ingestion of certain color additives and a preservative food additive	Moderate increases in hyperactivity, and inattentive behaviors in children[34]
Lead consumption	Higher risk of violent and aggressive behavior, hyperactivity, and mental and behavioral problems[35]
Iron deficiency in young children	Long-term deficits in learning ability and social skills[36]

Malnutrition and Mental Performance

Like growth and health, mental development and intellectual capacity can be affected by diet. The effects range from mild and short term to serious and lasting, depending on when the malnutrition occurs, how long it lasts, and how severe it is. The effects are most severe when malnutrition occurs while the brain is growing and developing.

Severe deficiency of protein, calories, or both early in life leads to growth retardation, low intelligence, poor memory, short attention span, and social passivity. When the nutritional insult is early and severe, some or all of these effects may be lasting (Illustration 5.8).[31] In Barbados, for example, children who experienced protein-calorie malnutrition in the first year of life did not fully recover even with nutritional rehabilitation. Growth improved but academic performance did not. Compared to well-nourished children, those experiencing protein and calorie deficits during infancy were more likely to drop out of school and were four times more likely to have symptoms of attention deficit hyperactivity disorder (ADHD).[37]

Protein-calorie malnutrition that occurs later in childhood, after the brain has developed, produces behavioral effects that can be corrected with nutritional rehabilitation. Correction of other deficits that often accompany malnutrition, such as the lack of educational and emotional stimulation and harsh living conditions, hastens and enhances recovery.[31]

Protein-calorie malnutrition severe enough to cause permanent delays in mental development rarely occurs in the United States. When it does, malnutrition is usually due to neglect or inadequate care giving. More common dietary events that impair learning in U.S. children are breakfast skipping, fetal exposure to alcohol, iron deficiency, and lead toxicity.

Does Breakfast Help You Think Better? Short-term fasting, such as skipping breakfast, reduces the late-morning problem-solving performance of children and adolescents.[21] Only a third of Americans and three out of four young children eat breakfast regularly.[39] (No one has yet studied the effects of skipping breakfast on college students' midterm exam scores. . . .)

Early Exposure to Alcohol Affects Mental Performance Mental development can be permanently delayed by exposure to alcohol during fetal growth. Although growth is also retarded, the most serious effects of fetal exposure to alcohol are permanent delays in mental development and behavioral problems associated with them. Women are advised not to drink if they are pregnant or may become pregnant.[40]

Iron Deficiency Impairs Learning Most cases of iron-deficiency anemia in children result from inadequate intake of dietary iron. Iron-deficiency anemia in children is a widespread problem in developed and developing countries and likely is the most common single nutrient deficiency.[41] The potential impact of iron-deficiency anemia on the functional capacity of humans represents staggering possibilities.

Illustration 5.8 Malnutrition in early childhood has long-lasting effects. Some children never fully recover.

Until recently, it was thought that the effects of iron-deficiency anemia on intellectual performance were short term and could be corrected by treating the anemia. It now appears that some of the effects may be lasting. Five-year-old children in Costa Rica treated for iron-deficiency anemia during infancy scored lower on hand–eye coordination and other motor skill tests than similar children without a history of anemia. Studies in the United States have detected shortened attention span and reduced problem-solving ability in iron-deficient children.[36]

Overexposure to Lead Our concern about exposure to lead has recently increased, due in large measure to information indicating that exposure to low levels of lead has long-term behavioral effects.[42] There are many opportunities for overexposure to lead. Approximately 84% of U.S. houses built before 1980 contain some lead-based paint.[43] Children living in or near these houses may eat the paint flakes (they taste sweet), or the old paint may contaminate the soil near the houses (Illustration 5.9). Lead also ends up in soil from industrial and agricultural chemicals, in water from lead-based pipes and solder, and in the air from the days when leaded gas was used.

Although the use of lead in cans, pipes, and gasoline has decreased dramatically, lead remains in the environment for long periods. Lead also stays in the body, stored principally in the bones, for a long time—20 years or more. It takes over a year of treatment to reduce blood lead levels. The effects of excessive exposure to lead include increased absenteeism from school, impaired reading skills, higher dropout rates, and increased aggressive behavior.[35] Blood lead levels in children have dropped substantially in recent decades. Despite this drop, however, 250,000 young children in the United States still have elevated blood lead levels.[44]

Food Additives, Sugar, and Hyperactivity The notion that certain food additives are related to hyperactivity in children has been popular since the related Feingold Hypothesis was announced in the mid-1970s. Research has failed to clarify this claim . . . until recently. In a study involving a large group of healthy three-, eight-, and nine-year-old children, intake of a beverage containing four food colorants (types of yellow, orange, and red color additives) and a preservative (sodium benzoate) was related to the development of hyperactivity. Signs of hyperactivity detected in the children consuming the additive-containing beverage included over-activity, impulsiveness, and short attention span. Not all children consuming the beverage demonstrated hyperactive behaviors,

Illustration 5.9 There are many opportunities for overexposure to lead. Young children are especially vulnerable.

iStockphoto.com/Paz Ruiz Luque

indicating that some children may be vulnerable to the effects of these food additives while others are not.[34]

Studies examining the effects of sugar intake on hyperactivity in children have not demonstrated that such a relationship exits.[45,46] The excitement that often accompanies high–sugar eating occasions such as Halloween and birthday parties—or the expectation that sugar causes hyperactivity—may be responsible for the reported effect. (See Illustration 5.10.)

The Future of Diet and Behavior Research

Identifying the effects of nutrition on behavior is a tricky business. Many factors in addition to diet influence behavior, making it difficult to separate influences from social, economic, educational, and genetic influences. We still have much to learn, and many assumptions about diet and behavior must await confirmation through research.

iStockphoto.com/Don Bayley

NUTRITION
up close

Focal Point: Developing a plan for healthier eating.

Identify a change in your diet that you would like to make. Then develop a plan for making the change by thinking through and responding to each element of the dietary change process listed. (Refer to Table 5.2 for examples of responses.)

Dietary Change Process

1. Identify a healthful change in your diet you'd like to make.
2. Identify two food choices you make that should change because they contribute to the need for the healthful change you identified.
3. Identify two specific, acceptable options for food choices more healthful than the ones identified under number 2.
4. Decide which option is easiest to accomplish and requires the smallest change to get the job done.
5. Plan how to incorporate the change into your diet.

Your Response

1. _____
2. _____

3. _____

4. _____

5. _____

Feedback to the Nutrition Up Close is located in Appendix G.

REVIEW QUESTIONS

- **Identify factors that influence an individual's food choices and preferences.**

1. Food preferences are universal—everyone likes the same foods.   **True/False**

2. Foods will be rejected by a population, no matter how nutritious they may be, if the foods don't fit into a culture's definition of what foods are appropriate to eat.   **True/False**

3. Food choices are driven largely by a person's need for nutrients.   **True/False**

4. Potatoes, rice, corn, and yucca are examples of "super foods" in specific countries due to their cultural significance.   **True/False**

5. Nutrition knowledge is an important prerequisite for making healthful food choices.   **True/False**

6. _____Emanuel is at a dinner party and the hostess asks, "Would you like to try the brussels sprouts? They're fresh from the garden and delicious!" Emanuel immediately responds, "No, thank you. I've tried brussels sprouts at least 10 times before and they just don't taste good to me." Chances are good that Emanuel

   a. is being a picky eater.
   b. dislikes all vegetables.
   c. is upset about something.
   d. really does not like the taste of brussels sprouts.

7. _____ Last week Yu asked her neighbor where she should take her visiting parents to dinner. The neighbor recommends Al's Restaurant, saying it has "the best food in town." Later that week Yu spoke with her neighbor and the neighbor asked how her parents liked Al's Restaurant. You respond, "Well, actually, we thought it was awful. They used way too much garlic in everything." The neighbor is shocked. What's the most likely reason for this reaction?

   a. The neighbor assumed that because she likes the food at Al's, everyone will.
   b. Yu's family doesn't recognize good food.
   c. Yu's family members are genetically sensitive to the taste of garlic.
   d. The neighbor is unable to detect the presence of garlic in food.

- **Apply the process for making healthful changes in food choices to a specific change in food intake.**

8. Broad dietary changes, such as a decision to simply "eat less," are more likely to produce lasting behavioral changes than are small changes in diets, such as snacking on favorite fruits rather than candy.   **True/False**

9. Changes in dietary intake that are acceptable to an individual and easy to implement are the types of changes that are most likely to last. **True/False**

10. Changing food choices for the better takes planning and includes individual decisions on specifically how the change in food choices will be implemented. **True/False**

11. Individuals need to plan for modifying their approach to improving food choices because even the best planned changes in food choices sometimes fail. **True/False**

The next two questions refer to this scenario:

Assume your wife decides to increase her vegetable intake. She travels from place to place for work and eats out a lot. She loves the premade mixed greens salads and the single-serve yogurt dressing packet you can get at some gas station stores. She decides she would happy to have that for lunch twice a week instead of her usual burger or pizza slices.

12. _____ Does this plan have a chance of working?
   a. Yes, because she seems to be a person with great will power.
   b. Yes, she is planning a specific and acceptable change.
   c. No, the plan isn't likely to work because the change is too drastic.
   d. No, the plan won't work because she will grow tired of eating salads.

13. _____ The salad-for-lunch plan worked so well that your wife decides she will lose 10 pound in the next two months by only eating salads. Will this plan likely work?
   a. It has a good chance of working. She really loves salads.
   b. Yes, she won't be consuming very many calories.
   c. It's unlikely to work because it is too drastic a change in her diet.
   d. It probably won't work because she will start craving burgers and pizza.

• **Differentiate between scientifically supported and unsupported conclusions about diet and behavior relationships.**

14. Examples of ways in which dietary intake affects behavior include the relationship between sugar intake and hyperactivity in children. **True/False**

15. Results of recent research studies make it clear that the consumption of breakfast makes little difference to children's academic performance or weight status. **True/False**

Answers to these questions can be found in Appendix G.

NUTRITION SCOREBOARD ANSWERS

1. Although genetics plays a role in food preferences, the predominant influences are environmentally determined.[1] **False**

2. The idea that food habits don't change is a myth. **False**

3. Well-nourished children who don't eat breakfast tend to score lower on problem-solving tests than children who eat breakfast.[2] (If you got this one wrong, answer one more question: Did you skip breakfast this morning?) **False**

Alistair Berg/Digital
Vision/Getty Images

Healthy Diets, Dietary Guidelines, MyPlate, and More

UNIT

6

NUTRITION SCOREBOARD

1 Over half of U.S. adults fail to consume three or more servings of vegetables and two or more servings of fruits daily. **True/False**

2 The basic food groups include a "healthy snack" food group. **True/False**

3 "Supersized" fast-food meals may provide two to three times more calories than regular-sized fast-food meals. **True/False**

Answers can be found at the end of the unit.

After completing Unit 6 and its interactive learning features, you will be able to:

- Apply the characteristics of healthful diets to the design of one.

- Identify characteristics of diets that promote health and those that do not.

- Utilize ChooseMyPlate.gov guidance materials and interactive tools for dietary planning and evaluation.

Healthy Eating: Achieving the Balance between Good Taste and Good for You

- **Apply the characteristics of healthful diets to the design of one.**

May I have your attention please? For a moment, think about the foods in Illustration 6.1. If your mouth is watering and you're ready to go out and buy some ripe peaches, you have found the balance between good taste and good for you. Who said foods that taste good aren't good for you?

Key Nutrition Concepts

This unit explores what constitutes a healthy diet and ways of evaluating and assessing diets. The discussion directly relates to four key nutrition concepts:

1. Foods provide energy (calories), nutrients, and other substances needed for growth and health.

2. Poor nutrition can result from both inadequate and excessive levels of nutrient intake.

3. Poor nutrition can influence the development of certain chronic diseases.

4. Adequacy, variety, and balance are key characteristics of healthful diets.

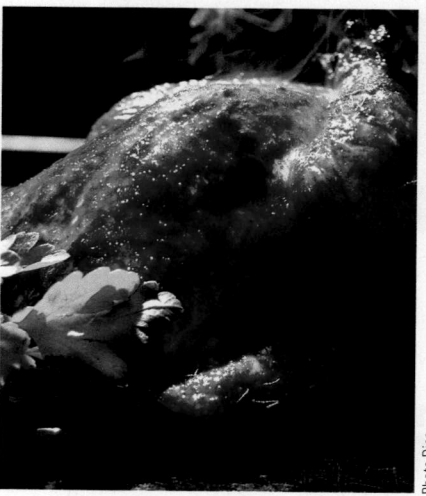

Illustration 6.1 Can you smell it? Can you taste it? A plump, golden peach. It's so ripe that juice spurts from it and drips down your chin when you take a bite. . . . A golden brown turkey just taken out of the oven. The wonderful smell fills the kitchen. A steaming loaf of homemade bread just set out to cool. A perfect ripe tomato just picked from the garden. It melts in your mouth.

Characteristics of Healthful Diets

Healthful diets come in a variety of forms. They may be based on bread, olives, nuts, fruits, beans, vegetables, lamb, and chicken (as in Greece); rice, vegetables, and small amounts of fish and other meats (as in China); or black beans, rice, meat, and tropical fruits (as in Cuba and Costa Rica). Illustration 6.2 gives examples of the diverse foods that can be part of a healthy diet.

Although the types of foods that go into them can vary substantially, healthful diets all share three basic characteristics: adequacy, variety, and balance.

Adequate diets include a wide variety of foods that together provide sufficient levels of calories and **essential nutrients**. What's sufficient? For calories, it's the number that maintains a healthy body weight. For essential nutrients, sufficiency corresponds to intakes that are in line with recommended intake levels represented by the *Recommended Dietary Allowances (RDAs)* and *Adequate Intakes (AIs)*. (Tables showing these levels appear on the inside covers of this book.) Recommended amounts of essential nutrients should be obtained from foods to reap the benefits offered by the variety of naturally occurring substances in foods that promote health.

Variety is a core characteristic of healthy diets because the nutrient content of food differs. Consumption of an assortment of foods from each of the basic food groups increases the probability that the diet will provide enough of them all. You could, for example, eat three servings of vegetables a day by eating only potatoes. But you would consume a much broader variety of vitamins, minerals, and beneficial substances in plants such as antioxidants if you consumed spinach and tomatoes along with the potatoes.

adequate diet A diet consisting of foods that together supply sufficient protein, vitamins, and minerals and enough calories to meet a person's need for energy.

essential nutrients Substances the body requires for normal growth and health but cannot manufacture in sufficient amounts; they must be obtained in the diet.

variety A diet consisting of many different foods from all of the food groups.

Illustration 6.2 Foods that contribute to healthy diets in different countries. a. Dinner in Italy might be linguine primavera (pasta with vegetables). b. Pad Thai, rice noodles and vegetables, is a favorite dish in Thailand. c. Tamales are a celebration food in many Latin cultures. d. Dal, curry dishes, vegetables, and chicken are popular parts of the cuisine of India.

a.
Photo Disc

b.
Photo Disc

c.
Photo Disc

d.
Photo Disc

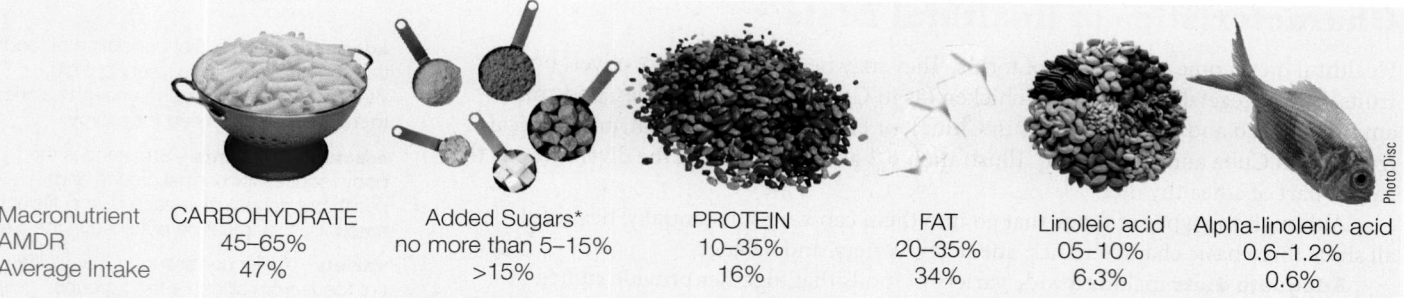

Macronutrient	CARBOHYDRATE	Added Sugars*	PROTEIN	FAT	Linoleic acid	Alpha-linolenic acid
AMDR	45–65%	no more than 5–15%	10–35%	20–35%	05–10%	0.6–1.2%
Average Intake	47%	>15%	16%	34%	6.3%	0.6%

*The Dietary Guidelines 2010 recommends that "discretionary calorie" intake, which includes added sugars and fats, be kept within the range of 5–15% of calories.[2]

Illustration 6.3 The match between Acceptable Macronutrient Distribution Ranges (AMDRs) and average intake by adults in the United States.[3,37]

balanced diet A diet that provides neither too much nor too little of nutrients and other components of food such as fat and fiber.

macronutrients The group name for carbohydrate, protein, fat, and water. They are called macronutrients because we need relatively large amounts of them in our daily diet.

saturated fats The type of fat that tends to raise blood cholesterol levels and the risk for heart disease. They are solid at room temperature and are found primarily in animal products such as meat, butter, and cheese.

trans fats A type of unsaturated fat present in hydrogenated oils, margarine, shortening, pastries, and some cooking oils that increases the risk of heart disease.

added sugars Sugars, syrups, and other caloric sweeteners that are added to foods during processing, preparation, or consumed separately. Added sugars do not include naturally occurring sugars such as those in milk or fruits.

whole grains Cereal grains that consist of the intact, ground, cracked, or flaked kernel, which includes the bran, the germ, and the inner most part of the kernel (the endosperm).

refined grains Whole grains that have been processed to remove the bran covering the grain kernel and germ that stores nutrients for seedling growth. The endosperm, the portion of the grain kernel that stores complex carbohydrate and other nutrients, remains.

A **balanced diet** provides calories, nutrients, and other components of food in the right proportion—neither too much nor too little. Diets that contain too much sodium or too little fiber, or are high in fat or sugar, for example, are out of balance. Diets that provide more calories than needed to maintain a healthy body weight are also out of balance.

Current dietary intake standards include guidelines to help individuals balance their intake of carbohydrate, protein, and fat; and to consume appropriate amounts of water. Because we need relatively large amounts of carbohydrates, proteins, fats, and water, these nutrients are collectively classified as **macronutrients**. Guidelines indicating the percentages of total caloric intake that should consist of carbohydrate, protein, and fat are listed in a DRI table labeled "Acceptable Macronutrient Distribution Ranges," or AMDRs. AMDRs have also been set for linoleic acid and alpha-linolenic acid (the two essential fatty acids). Illustration 6.3 shows the acceptable ranges for carbohydrate, protein, fat, added sugars, and essential fatty acid intake, as well as the average intake levels of U.S. adults. It is recommended that diets be kept as low in **saturated fat** and **trans fat** as possible. The AMRDs apply to individuals over the age of four years.[2]

How Balanced Is the American Diet?

Comparisons of the recommended ranges of macronutrient intake with actual levels of intake by adults show that the U.S. diet is out of balance in several ways. The average intake of fat and added sugars by U.S. adults averages near the top of the acceptable intake ranges, indicating many adults consume more fat and sugars than recommended. Foods high in fat, like sausages and potato chips for example, or **added sugars**, such as desserts and regular soft drinks, tend to be high in calories and low in nutrients. Excess consumption of these foods can knock a diet out of balance. Although the average intake of fat is on the high side, intakes of the two essential fatty acids (linoleic and alpha-linolenic acid) average near the bottom of the acceptable ranges. Adults tend to consume a lot of fat, but too much of it is in the form of animal and milk fats that contain relatively low amounts of the essential fatty acids.[2]

Diets in the United States are out of balance in other respects. Only 14% of adults consume three or more servings of vegetables and two or more servings of fruits daily. About 10% include dark-green or colorful vegetables that are high in antioxidants. The vegetable most commonly consumed in the United States is french fries.[4] We tend to consume too few **whole grain** products, opting instead for breads and cereals made with **refined grains**. Caloric intake is becoming out of balance with body weight. Average calorie intake in the United States has increased by approximately 505 calories since 1970, and the incidence of obesity has increased along with it.[5] This dietary profile prompted one leading nutrition researcher to announce, "We're becoming an over-fed but undernourished population."[6]

National Guides for Healthful Diets

Due to the impact of food choices on the health of individuals and population groups, many countries have established recommendations for dietary intake. Such recommendations are periodically updated as new discoveries about diet and health emerge. Many of the guidelines include recommendations for physical activity as well.

National recommendations for diet and physical activity usually apply to children over the age of two years and aren't appropriate for every individual in a population. General guidelines may not completely match the needs of individuals who, for example, are strict vegetarians, disabled, or have disorders such as hypertension or diabetes. A basic premise of national dietary guidelines is that nutrient needs should be met primarily through food consumption.

Benefits of population-based dietary and physical activity recommendations are multiple. The information is science-based, free, and made widely available to the public on the web. Adherence to the information can help people stay healthy and lower their risk of developing disorders such as cancer, osteoporosis, heart disease, and obesity.

Some of the national guidelines also give credit to the cultural and social importance of food. "Enjoy your food!" is listed as Ireland's first dietary guideline, and the guidelines for Japanese people include "Happy eating makes for happy family life; sit down and eat together and talk; treasure family taste and home cooking."[7] Table 6.1 provides examples of dietary guidelines established for a variety of countries.

National guidance on healthful diets for some countries is accompanied by extensive information on how to select a healthful diet and achieve recommended levels of physical activity. In the United States, the national guidelines for diet are called the "Dietary Guidelines for Americans," and the major how-to guide is represented by "MyPlate."

Dietary Guidelines for Americans

- **Identify characteristics of diets that promote health and those that do not.**

The Dietary Guidelines for Americans provide science-based recommendations to promote health and to reduce the risk for major chronic diseases through diet and physical activity. Due to their credibility and focus on health promotion and disease prevention for the public, the Dietary Guidelines form the basis of federal food and nutrition programs and policies.

By law, the Dietary Guidelines for Americans are updated every five years. The first edition of the Dietary Guidelines was published in 1980. The 2010 Dietary Guidelines highlight the importance of weight control and reductions in the rates of common diseases associated with overweight, obesity, and unhealthful diets.

The Dietary Guideline's Overarching Concepts

Overall, the Dietary Guidelines encompass two broad concepts:

1. Maintain calorie balance over time to achieve and sustain a healthy weight.
2. Focus on consuming **nutrient-dense foods** and beverages.

The first concept paves the path toward weight control for people who weigh too much and for others who want to maintain a healthy weight. Adults who are most successful in achieving and maintaining a healthy weight do so by limiting calorie intake to levels needed to maintain a healthy weight and by being physically active.

nutrient-dense foods Foods that contain relatively high amounts of nutrients compared to their calorie value.

Table 6.1 **Examples of dietary guidelines from around the world**[7,8]

Country	Example of Dietary Guidelines
Japan	Eat 30 or more different kinds of foods daily.
China	Eat clean and safe food.
Norway	FOOD + JOY = HEALTH.
United Kingdom	Encourage and support the production of lower saturated fat foods.
Mexico	Eat more dried beans and less food of animal origin.
South Africa	Enjoy a variety of foods. Be active.
Cuba	Fish and chicken are the healthiest meats.

Table 6.2 Examples of types of foods Americans are being urged to eat less or more of by the Dietary Guidelines

Eat less:	
High-salt (sodium) foods:	salami, bologna, hot dogs, some canned soups, salted snack foods
Saturated and *trans* fats:	sausage, beef and other types of animal fat, butter, cream, pastries, ice cream, full-fat dairy products
Foods and beverages with added sugar:	desserts, candy, soft drinks, fruit drinks, sweet teas and coffees, sweetened breakfast cereals
Refined grain products:	white bread, flour tortillas, white rice, breakfast cereals, cornmeal, and breads not labeled "whole grain"
Eat more:	
Vegetables:	carrots, collard greens, beets, sweet potatoes, squash, tomatoes, spinach, cabbage, broccoli
Fruits:	oranges, apples, berries, grapefruit, grapes, bananas, cherries, pineapple, papaya, mangos, avocados, star fruit, kiwis
Fat-free or low-fat milk and milk products:	low-fat cottage cheese, cheese, yogurt, and milks
Fish and seafood:	salmon, white fish, bass, anchovies, herring, scallops, clams[a]
Lean meats:	chicken, turkey, lean beef, and pork
Beans and peas (legumes):	black beans, navy beans, pinto beans, chick peas, peas, lentil beans
Nuts and seeds:	almonds, cashews, walnuts, pecans, hazel nuts
Oils:	olive oil, vegetable oils, grape seed oil, walnut oil
Whole grains:	whole grain bread, crackers, pasta, cereals, and rolls; corn, millet, wheat berries, quinoa, brown rice, corn tortillas, popcorn, oats

[a]The Dietary Guidelines report notes that the benefits of consuming 8 ounces per week of seafood far outweigh the risks, even for pregnant women.

eating pattern The combination of the types of foods and beverages normally consumed over time.

The second concept addresses the need to improve the quality of our diets through adoption of a healthful **eating pattern** rich in nutrient-dense foods and beverages, and in plant foods. Regular consumption of vegetables, fruits, whole grains, fat-free or low-fat milk and milk products, seafood, lean meats and poultry, eggs, beans and peas, and nuts and seeds is encouraged. These food should replace high sodium processed foods, sources of saturated and trans fats; foods and beverage with added sugar, and refined grain products. Table 6.2 lists examples of types of foods that correspond to these recommendations. The basic premises underlying this concept are that nutrient needs are primarily met through consuming foods, and that health is more closely related to overall diet than to any specific foods.

Key Recommendations The 2010 Dietary Guidelines present four categories of key recommendations for healthful eating and physical activity, and lists special advice for three specific population groups (women capable of becoming pregnant, pregnant and breastfeeding women, and individuals over the age of 50). The overarching concept's key recommendation areas, and examples of advice given under each key recommendation area, are summarized in the nearby Health Action.

Most Americans do not currently consume diets that match the recommendations stated in the Dietary Guidelines.[2] There are many reasons for this, including access to affordable and nutritious foods, food preferences, opportunities for physical activity, and fast-paced lifestyles. This edition of the Dietary Guidelines includes practical suggestions for eating better. Such suggestions include "Think ahead before attending parties." "Eat a small, healthy snack before heading out," and "Fruits and vegetables should fill half your plates."

Application of the Dietary Guidelines to Public Programs Due to their credibility and focus on health promotion and disease prevention, the Dietary Guidelines for Americans form the basis of federal food and nutrition programs and policies (Table 6.3). Food options served in School Lunch Programs, nutrition labeling standards, USDA's food guidance materials, and foods served to military personnel, for example, are based on the Dietary Guideline recommendations.

health action The 2010 Dietary Guidelines for Americans: Overarching Concepts

Balancing Calories to Manage Weight

- Prevent and/or reduce overweight and obesity through improved eating and physical activity behaviors.
- Increase physical activity and reduce time spent in sedentary behaviors.
- Control total calorie intake to manage body weight.

Foods and Food Components to Reduce

- Reduce daily sodium intake to less than 2,300 milligrams (mg) and further reduce intake to 1,500 mg among persons who are 51 and older and those of any age who are African American, or have hypertension, diabetes, or chronic kidney disease.
- Consume less than 10% of calories from saturated fatty acids by replacing them with monounsaturated and polyunsaturated fatty acids.
- Keep *trans* fatty acid consumption as low as possible.
- Reduce intake of calories from solid fats and added sugars.
  (Note: Solid fats are often high in saturated and/or *trans* fatty acids.)
- If alcohol is consumed, it should be consumed in moderation—up to one drink per day for women and two drinks per day for men.

U.S Department of Agriculture/U.S. Department of Health and Human Services

Foods and Nutrients to Increase

- Increase vegetable and fruit intake.
- Eat a variety of vegetables, especially dark-green and red and orange vegetables, beans, and peas.
- Consume at least half of all grains as whole grains.
- Increase intake of fat-free or low-fat milk and milk products.
- Choose a variety or protein foods, which include seafood, lean meat and poultry, eggs, beans and peas, soy products, and unsalted nuts and seeds.
- Increase the amount and variety of seafood consumed by choosing seafood in place of some meat and poultry.
- Choose foods that provide good amounts of potassium, dietary fiber, calcium, and vitamin D—nutrients of concern in American diets.

Building Healthy Eating Patterns

- Select an eating pattern that meets nutrient needs over time at an appropriate calorie level.
- Follow food safety recommendations when preparing and eating foods to reduce the risk of food-borne illnesses.

Interactive dietary assessment and planning tools for individuals and families based on the Dietary Guideline recommendations are becoming increasingly available. The most important and useful materials on how-to implement the Dietary Guidelines are found under "MyPlate.gov," USDA's icon for the "ChooseMyPlate" food intake guidance materials.

Table 6.3 Examples of federal food and nutrition programs that utilize the U.S. Dietary Guidelines recommendations[9]

- Supplemental Food and Nutrition Assistance Program (SNAP, formerly known as Food Stamps)
- Head Start
- WIC (Special, Supplemental Nutrition Program for Women, Infants and Children)
- National School Lunch Program
- Military food allowance program
- Nutrition labeling
- MyPlate educational materials
- Indian Health Service
- Healthy People 2020 (national objectives for improvements in weight status and diet)
- Older Americans Nutrition Program

MyPlate

- **Utilize ChooseMyPlate.gov guidance materials and interactive tools for dietary planning and evaluation.**

Food group guides from the USDA have been available in the United States since 1916. Known by such names as the Basic Four Food Groups, and the Food Guide Pyramid, these guides are periodically updated to reflect advances in knowledge about foods, diets, and health. The latest revision of USDA's food guidance materials is called MyPlate and is represented by the MyPlate icon. Illustration 6.4 shows this icon and a plate of foods set up to match the MyPlate icon's messages. The MyPlate icon is intended to give consumers a visual reminder of the types and proportions of food that make up healthy meals. The icon shows a plate with four sections in different colors representing the proportion of vegetables, fruits, grains, and protein foods that should be represented on your plate. Next to the plate is a circle that represents a dairy product such as low-fat milk or other low-fat dairy product. The types of foods included on the plate are basic and nutrient dense.

ChooseMyPlate materials are replacing the previous USDA guidance materials called MyPyramid. Most all of the information offered by ChooseMyPlate is available in English and Spanish from the main web address "MyPlate.gov."

Illustration 6.4 Foods shown on the plate consist of stewed turkey, barley, Swiss chard, mandarin oranges, and 1% milk.

ChooseMyPlate.gov Healthy Eating Messages

ChooseMyPlate.gov supports the healthy eating messages in the Dietary Guidelines through offering the following key pieces of advice:

- Make at least half your plate fruits and vegetables.

- Enjoy your food but eat less.

- Make half your grains whole grains.

- Eat fewer foods that are high in saturated fat, *trans* fats, added sugar, and sodium.

- Avoid oversized portions.

- Switch to fat-free or low-fat (1%) milk.

- Drink water instead of sugary drinks.

- Compare sodium in foods like soup, bread, and frozen meals?and choose the foods with lower numbers.

The importance of consuming enough calories for growth and health while not eating extra calories and gaining weight is also stressed. Regular physical activity (60 minutes per day for children and adolescents and $2^{1}/_{2}$ hours or more per week of moderate activity for adults weekly) is stressed because it contributes to weight control and provides many other health benefits.

The USDA's Food Groups

The USDA's long tradition of offering consumers a list of basic food groups is upheld in the current recommendations (Illustration 6.5). Grains, vegetables, fruits, dairy, and protein foods are the designated groups. Interactive, educational material provided by ChooseMyPlate includes details about the types of foods that belong to each group.

Fruits	Vegetables	Grains	Protein Foods	Dairy
Focus on fruits.	**Vary your veggies.**	**Make at least half your grains whole.**	**Go lean with protein.**	**Get your calcium-rich foods.**
>> See Fruit Group	>> See Vegetable Group	>> See Grains Group	>> See Protein Foods Group	>> See Dairy Group

Illustration 6.5 USDA's basic food groups and priority messages related to each group.

Portion Sizes and Food Measure Equivalents Unlike previous food group guides, the current version does not recommend serving sizes or numbers of servings individuals in general should consume from the food groups. This information is provided if requested using a personalized interactive tool on menu planning such as the Daily Food Plan. The personalized information generated shows amounts of each food group to consume and food portion sizes that correspond to that amount.

Information on amounts of basic foods contained in a cup or an ounce is provided separately in the ChooseMyPlate materials. Table 6.4 highlights examples of this information. You can get additional examples of serving size equivalents by highlighting a food group shown on the Daily Food Plan output and clicking on "What counts as an ounce?" or "What counts as a cup?"

Example Menus What does an eating pattern based on USDA's food group guidelines look like? Seven days of menus based on a 2,000-calorie diet that meet USDA's food group recommendations and nutrient needs are available from the "Sample Menus and Recipes" link on the ChooseMyPlate.gov homepage. Three days of the menus are shown in Illustration 6.6. The menus are intended to give consumers specific and general ideas about the types of foods to include in meals on a daily basis.

USDA's Interactive Diet Planning Tools Do you want to lose weight? Want to learn about healthful foods for your preschooler or when you are pregnant? Do you need to look up information on the calorie value of different foods or keep track of your food intake? You can access this information and more from the ChooseMyPlate.gov site. To help lose weight, for example, access the Daily Food Plan interactive tool. This feature can be used to identify the amount of food from each group you should consume daily based on your age, sex, weight, height, and physical activity level (Illustration 6.7). An allowance for teaspoons of oil, and an "empty calorie" allowance for extra fats and sugars are included in the Daily Food Plan results.

Access SuperTracker from the ChooseMyPlate home page and you will see six interactive tools for diet, weight management, and physical activity planning, assessment, and analysis (Illustration 6.8).

- Food-A-Pedia enables you to identify and compare the nutrient content of foods.

- Food Tracker can be used to identify the calorie, nutrient, and food group content of a diet. Its interactive features allow you to modify food selections and portion sizes. The

Table 6.4 How much food counts as a cup or an ounce?[36]

Vegetables: 1 cup = 1 cup raw or cooked vegetables or vegetable juice or 2 cups leafy salad greens
Fruits: 1 cup = 1 cup raw or cooked fruit or 100% fruit juice or $1/2$ cup dried fruit
Dairy: 1 cup = 1 cup milk, yogurt, or fortified soy milk, or $1^1/_2$ ounces natural or 2 ounces processed cheese
Grains: 1 ounce = 1 slice of bread, $1/2$ cup cooked rice, cereal, or pasta; or 1 ounce ready-to-eat cereal
Protein: 1 ounce = 1 ounce lean meat, poultry, or fish; 1 egg; 1 Tbsp peanut butter; $1/2$ ounce nuts or seeds; $1/4$ cup cooked dried beans or peas

Illustration 6.6 Sample menus for a 2,000-calorie food pattern

Sample Menus for a 2000 Calorie Food Pattern

DAY 1	DAY 2	DAY 3
BREAKFAST	**BREAKFAST**	**BREAKFAST**
Creamy oatmeal (cooked in milk):	Breakfast burrito:	Cold cereal:
½ cup uncooked oatmeal	1 flour tortilla (8" diameter)	1 cup ready-to-eat oat cereal
1 cup fat-free milk	1 scrambled egg	1 medium banana
2 Tbsp raisins	⅓ cup black beans*	½ cup fat-free milk
2 tsp brown sugar	2 Tbsp salsa	1 slice whole wheat toast
Beverage: 1 cup orange juice	½ large grapefruit	1 tsp tub margarine
	Beverage:	Beverage: 1 cup prune juice
LUNCH	1 cup water, coffee, or tea**	
Taco salad:		**LUNCH**
2 ounces tortilla chips	**LUNCH**	Tuna salad sandwich:
2 ounces cooked ground turkey	Roast beef sandwich:	2 slices rye bread
2 tsp corn/canola oil (to cook turkey)	1 small whole grain hoagie bun	2 ounces tuna
¼ cup kidney beans*	2 ounces lean roast beef	1 Tbsp mayonnaise
½ ounce low-fat cheddar cheese	1 slice part-skim mozzarella cheese	1 Tbsp chopped celery
½ cup chopped lettuce	2 slices tomato	½ cup shredded lettuce
½ cup avocado	¼ cup mushrooms	1 medium peach
1 tsp lime juice (on avocado)	1 tsp corn/canola oil (to cook mushrooms)	Beverage: 1 cup fat-free milk
2 Tbsp salsa	1 tsp mustard	
Beverage:	Baked potato wedges:	**DINNER**
1 cup water, coffee, or tea**	1 cup potato wedges	Roasted chicken:
	1 tsp corn/canola oil (to cook potato)	3 ounces cooked chicken breast
DINNER	1 Tbsp ketchup	1 large sweet potato, roasted
Spinach lasagna roll-ups:	Beverage: 1 cup fat-free milk	½ cup succotash (limas & corn)
1 cup lasagna noodles(2 oz dry)		1 tsp tub margarine
½ cup cooked spinach	**DINNER**	1 ounce whole wheat roll
½ cup ricotta cheese	Baked salmon on beet greens:	1 tsp tub margarine
1 ounce part-skim mozzarella cheese	4 ounce salmon filet	Beverage:
½ cup tomato sauce*	1 tsp olive oil	1 cup water, coffee, or tea**
1 ounce whole wheat roll	2 tsp lemon juice	
1 tsp tub margarine	½ cup cooked beet greens (sauteed in 2 tsp corn/canola oil)	**SNACKS**
Beverage: 1 cup fat-free milk	Quinoa with almonds:	¼ cup dried apricots
	½ cup quinoa	1 cup flavored yogurt (chocolate)
SNACKS	⅓ ounce slivered almonds	
2 Tbsp raisins	Beverage: 1 cup fat-free milk	
1 ounce unsalted almonds		
	SNACKS	
	1 cup cantaloupe balls	

ChooseMyPlate.gov/USDA

program provides instant feedback on how the changes would affect calorie, nutrient, and food group intake.

- Physical Activity Tracker can be used to compare your level of physical activities to the Physical Activity Guidelines for Americans. It can also be used to plan your activities and evaluate progress in meeting your physical activity goals.

- My Weight Manager provides tips for weight loss and helps you track progress in meeting weight loss goals.

- My Top 5 Goals presents 19 options for goals related to weight management, physical activity, calorie intake, food group intake, and nutrient intake. You can select and track progress on meeting up to five of these goals. For example, you can set a weight goal and receive a calorie intake plan that will help you reach the goal. Graphs on changes in weight over time and tips for weight loss can be generated by this tool.

- My Reports can be used to track changes in food group, calorie, and nutrient intake; and changes in physical activity level. Over 40 different reports can be generated. The reports automatically compare your results to the appropriate recommendations.

Stay tuned to MyPlate.gov. Additional useful tools are periodically added to this site.

Limitations of MyPlate Materials available from MyPlate are almost entirely made available on the web, making the information inaccessible to people who do not use computers or have access to the Internet. MyPlate does not provide specific recommendations for infants, individuals on therapeutic diets, or strict vegetarians.

Menus suggested by MyPlate interactive tools may not correspond to individual food preferences and contain relatively few ethnic foods. As with past food guides, planning and evaluating how mixed dishes (such as stews, soups, salads, and various types of pizza) fit into the foods groups still can be perplexing. In addition, it can be challenging to convert the calorie allotment for "extra fats and sugars" into food portions.

My Daily Food Plan

Based on the information you provided, this is your daily recommended amount for each food group.

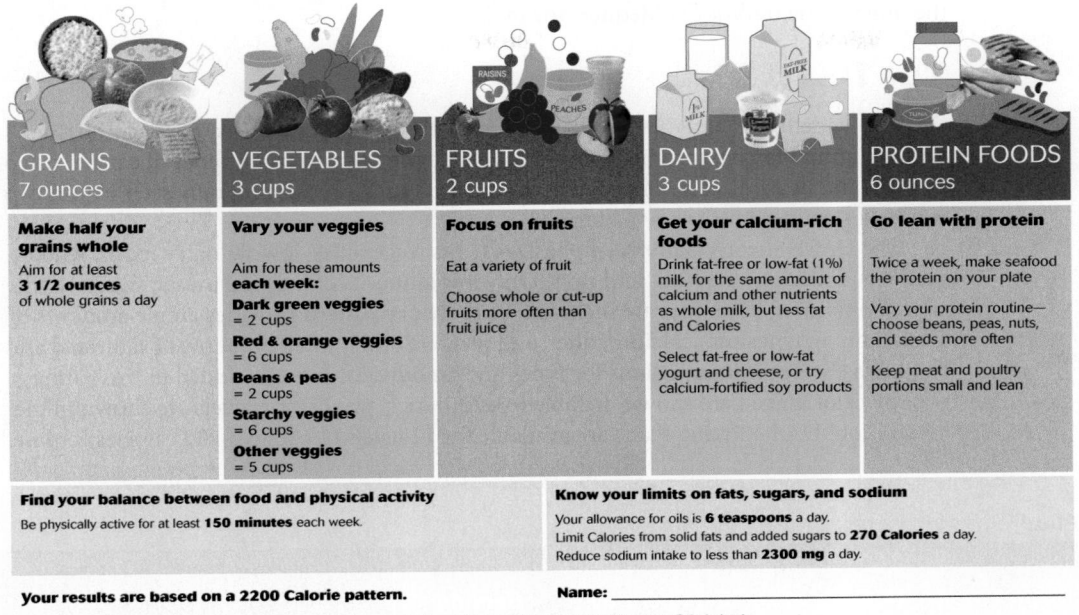

GRAINS 7 ounces	VEGETABLES 3 cups	FRUITS 2 cups	DAIRY 3 cups	PROTEIN FOODS 6 ounces
Make half your grains whole Aim for at least **3 1/2 ounces** of whole grains a day	**Vary your veggies** Aim for these amounts each week: **Dark green veggies** = 2 cups **Red & orange veggies** = 6 cups **Beans & peas** = 2 cups **Starchy veggies** = 6 cups **Other veggies** = 5 cups	**Focus on fruits** Eat a variety of fruit Choose whole or cut-up fruits more often than fruit juice	**Get your calcium-rich foods** Drink fat-free or low-fat (1%) milk, for the same amount of calcium and other nutrients as whole milk, but less fat and Calories Select fat-free or low-fat yogurt and cheese, or try calcium-fortified soy products	**Go lean with protein** Twice a week, make seafood the protein on your plate Vary your protein routine—choose beans, peas, nuts, and seeds more often Keep meat and poultry portions small and lean

Find your balance between food and physical activity

Be physically active for at least **150 minutes** each week.

Know your limits on fats, sugars, and sodium

Your allowance for oils is **6 teaspoons** a day.
Limit Calories from solid fats and added sugars to **270 Calories** a day.
Reduce sodium intake to less than **2300 mg** a day.

Your results are based on a 2200 Calorie pattern.

Name: _____

This Calorie level is only an estimate of your needs. Monitor your body weight to see if you need to adjust your Calorie intake.

ChooseMyPlate.gov/USDA

Illustration 6.7 Daily food plan results for a person with a 2,200 calorie need. Source: ChooseMyPlate.gov. Food group recommendations presented are for a 28-year-old female, 5'6" tall weighing 135 pounds and exercising 30–60 minutes a day.

Illustration 6.8 Interactive tools available from SuperTracker.

Daily Food Plan
Get a personalized plan just for you

Daily Food Plans for Preschoolers
Get your child's Plan today

Daily Food Plans for Moms
Start out right as a new mom or mom-to-be

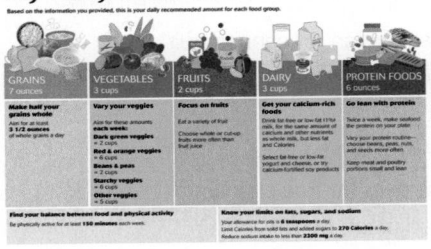

SuperTracker
Helps you plan, analyze, and track your diet and physical activity.

Other Healthful Eating Patterns Two major types of eating patterns have been shown to promote health and prevent disease in addition to the USDA's food guide plan. Both of these eating patterns are consistent with the Dietary Guideline's recommendations for dietary intake.[2] One pattern is the Dietary Approaches to Stop Hypertension (DASH) and the other is the traditional Mediterranean diet.

The DASH Diet

Originally identified as a diet that helps control mild and moderate **hypertension**, the DASH Eating Plan has also been found to reduce the risk of certain types of cancer, osteoporosis, and heart disease. Improvements in blood pressure are generally seen within two weeks of starting this dietary pattern.[12,13]

The DASH dietary pattern emphasizes fruits, vegetables, low-fat dairy foods, whole-grain products, poultry, fish, and nuts. Only low amounts of fats, red meats, sweets, and sugar-containing beverages are included. This dietary pattern provides ample amounts of potassium, magnesium, calcium, fiber, and protein; and limited amounts of saturated and trans fats.[10,11] Recommendations for types and amounts of foods included in this eating plan by calorie need are shown in Table 6.5. Although two calorie levels are shown in the illustration, DASH Eating Plans are available for 12 levels (1,600 to 3,200 calories) online.[10]

hypertension High blood pressure. It is defined as blood pressure exerted inside blood vessel walls that typically exceeds 140/90 millimeters of mercury.

Table 6.5 The DASH Eating Plan[10,11]

Food group	2,000 calories	2,600 calories	Serving size
	Number of servings per day		
Grains[a]	6–8	10–11	1 slice bread
			1 oz dry cereal
			$1/2$ cup cooked rice, pasta, cereal
Vegetables	4–5	5–6	1 cup raw leafy vegetable
			$1/2$ cup raw or cooked vegetables
			$1/2$ cup vegetable juice
Fruits	4–5	5–6	1 medium fruit
			$1/4$ cup dried fruit
			$1/2$ cup fresh, frozen, or canned fruit
			$1/2$ cup fruit juice
Fat-free or low-fat milk and milk products	2–3	3	1 cup milk
			1 cup low-fat or fat-free yogurt
			$1 1/2$ oz low-fat or fat-free cheese
Lean meats, poultry, fish	2 or fewer	2	3 oz cooked meat, poultry, or fish
Nuts, seeds, legumes	4–5/week	1/day	$1/3$ cup or $1 1/2$ oz nuts
			2 Tbsp peanut butter
			2 Tbsp or $1/2$ oz seeds
			$1/2$ cup cooked dry beans or peas
Fats and oils	2–3	3	1 tsp soft margarine
			1 Tbsp low-fat mayonnaise
			2 Tbsp light salad dressing
			1 tsp vegetable oil
Sweets	5/week	2 or fewer/day	1 Tbsp sugar, jelly, jam
			$1/2$ cup sorbet and ices
			1 cup lemonade

[a]Whole grain products primarily.
Source: Dietary Guidelines for Americans, Appendix A-1: The DASH Eating Plan at 1,600-, 2,000-, and 3,100-Caloric Levels. Available at www.health.gov/dietaryguidelines/dga2005/document/html/appendixA.htm.

The Mediterranean Diet

The traditional Mediterranean diet ranks with the USDA's Food Guide and the DASH Eating Plan when it comes to health promotion and chronic disease prevention.[14] The Mediterranean diet was originally based on foods consumed by people in Greece, Crete, southern Italy, and other Mediterranean areas where rates of chronic disease were low and life expectancy long.[15]

The Mediterranean dietary pattern, represented by the Mediterranean Diet Pyramid shown in Illustration 6.9, emphasizes that diets should be primarily based on plant foods, such as fruits, vegetables, grains (mostly whole), beans, and nuts. Fish and seafood should be consumed at least twice a week, poultry and eggs twice weekly or less, and cheese and yogurt one to seven times a week. Meats and sweets form the smallest part of the pyramid and should be consumed infrequently. Wine in moderation is a traditional part of the Mediterranean diet, and water intake is encouraged. A number of studies have shown that this dietary pattern is associated with lower risks of heart disease, stroke, diabetes, several forms of cancer, and overall mortality.[2,16,17]

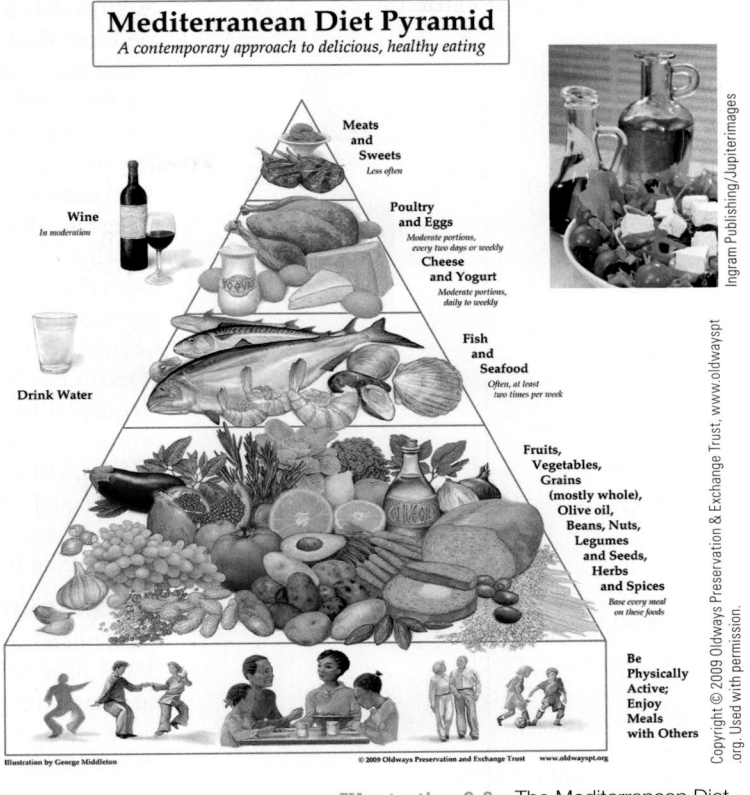

Illustration 6.9 The Mediterranean Diet Pyramid.

Portion Distortion

Do you know how much food you ate yesterday? If you had meat, what was the size of the serving of meat you consumed? Did you drink something with that? How much of that beverage did you consume?

Few people are aware of how much food they eat. Some people think a serving of food is the same as the portion of food they are served or eat. Portion sizes or "servings" of food today tend to exceed standard serving sizes developed by the USDA for use in planning healthful diets. In this era of growing portion and people sizes it's particularly important to become aware of how much food is eaten.[18]

Individual ideas of normal serving amounts are based on past experiences at family meals, the size of portions provided by restaurants, and packaged food and beverage sizes. Supersized meals at fast-food restaurants, large portions served by other restaurants, large bakery products, and larger cups of soft drinks are contributing to the problem of portion distortion. Table 6.6 provides examples of how food portion sizes are expanding. Table 6.7 provides a guide for estimating food portion sizes.

Table 6.6 **Typical portion sizes and calorie content of foods in the marketplace versus calorie content and portion sizes 20 years ago**[18,19]

Food	Portion size Calories 20 years ago	Marketplace portion size Calories now
Bagel	3-inch diameter	6-inch diameter
	140 calories	350 calories
Cheeseburger	4.3 ounces	7.1 ounces
	343 calories	535 calories
French fries	2.4 ounces	6.9 ounces
	210 calories	610 calories
Soft drink	6.5 ounces	20 ounces
	85 calories	820 calories
Muffin	1.5 ounce	6.5 ounce
	167 calories	724 calories

Table 6.7 Portion size estimators

1 cup = baseball	skizer/Shutterstock.com
½ cup = tennis ball	vlad09/Shutterstock.com
¼ cup = golf ball or extra large egg	Cameramann/Shutterstock.com
2 Tablespoons = ping-pong ball	© Tomas1111/Dreamstime.com
1 teaspoon = fingertip	© RunPhoto/Stockbyte/Jupiterimages
3 ounces of meat = deck of cards, palm of hand	Dedyukhin Dmitry/Shutterstock.com

For single-serve foods, check the weight given on the food package label

Is Supersizing Leading to Supersized Americans? Supersizing fast foods can double or triple the caloric content of the foods compared to their regular-sized counterparts. A single, supersized meal of a cheeseburger, large fries, and thick shake provides more calories (about 2,200) than many people need in a day. Larger portions don't cost restaurants much more than smaller portions, they increase sales volume, and they encourage people to eat more.[21] Many Americans are eating a good deal more food than needed, and it is appears that rising rates of obesity are partly related to increased portion sizes.[22-24]

Children and adults tend to eat more when offered larger portions of foods than smaller portions. Children and adolescents have been reported to consume 5–12% more soft drink, pizza, french fries, and salty snacks when offered large portions.[33] Among adults, a 50% increase in portion sizes of meals has been found to increase daily energy intake by 423 calories.[34] Frequent dining at fast food restaurants (three or more times per week) that primarily serve burgers and french fries is associated with higher intakes of calories, soft drinks, and fat; and to a higher risk of overweight and obesity than frequent use of full-service restaurants.[35] On a positive note, some restaurants have recently begun to offer smaller portion sizes and healthier menu options than in the past.[26]

Can You Still Eat Right When Eating Out? The question about what to eat often boils down to choosing the right restaurant. According to recent USDA data, 41% of American adults eat out at least weekly.[25] In general, foods eaten away from home have lower nutrient content and are higher in fat than foods eaten at home (see Illustration 6.10).[26] In addition, children and teenagers who eat dinner with their families most days tend to have more healthful diets and eating patterns than others who never or occasionally eat dinner with their family.[28]

Staying on Track While Eating Out You'll find it easier to stick to a healthy diet if you decide what to eat before you enter a restaurant and look over the menu (Illustration 6.11). You could make the decision to order soup and a salad, broiled meat, a half-portion of the entrée (or to split an entrée with someone else), or no dessert *before* entering the restaurant. "Impulse ordering" is a hazard that can throw diets out of balance. If you're going to a party or a business event where food will be served, decide before you go what types of food you will eat and what you will drink. If only high-calorie foods are offered, plan on taking a small portion and stopping there. Some people find it helps to have a healthy snack before going to a party or an event where high-calorie food will be served to avoid being really hungry when they get there.

Illustration 6.10 Hamburgers, french fries, and pizza are the top-selling food items in U.S. restaurants. [27]

© Ocean/Corbis

©Envision/Corbis

Can Fast Foods Be Part of a Healthy Diet? As the information in Table 6.8 demonstrates, many of the foods served in fast-food restaurants deserve their reputation as being high in calories because they are often high in fat. Many of the foods offered do not fit into a healthy diet if eaten often. In recognition of this fact, and in response to legislative requirements that chain restaurants label the caloric value of menu items, lower calorie and more nutrient-dense foods are being added to menus. Some of these foods are listed in Table 6.9 along with their content of calories, saturated fat, fiber, and the contribution of the serving to vegetable or fruit intake. Several higher calorie foods representing traditional offerings at fast-food restaurants are shown in this table for comparative purposes.

The addition of low-fat milk, apple slices, a variety of salads with low-fat dressings, and baked potatoes to fast-food menus makes it easier to eat nutritiously when eating out. However, there is room for improvement. Only 3% of kids' meals served at fast-food restaurants meet the National School Lunch Program's criteria for nutritional quality.[29]

The Slow Food Movement An interesting trend in food preparation and consumption is making its way across the globe. The trend is away from fast and processed foods, and toward sustainable, eco-friendly agricultural practices and locally grown foods. The Slow Food movement represents some aspects of this trend. Part of an international group, Slow Food USA is an educational organization that supports ecologically sound food production; the revival of the kitchen and the table as centers of pleasure, culture, and community; and living a slower and more harmonious rhythm of life.[30,31] The trend is

Illustration 6.11 "No, no thank you. Extra cheese isn't part of what I planned to eat."

Table 6.8 Calorie and fat content of some fast foods[a]

Food	Calories	Percentage of calories from fat	Food	Calories	Percentage of calories from fat
Sausage Breakfast Croissanwich	538	69	Hamburger	350	41
Bacon cheeseburger	610	58	Beef tostada	239	41
Fried chicken breast	276	55	Beef and cheddar sandwich	490	39
Fried chicken drumstick	147	55	Beef burrito	357	38
Quarter Pounder with Cheese	525	55	Roast beef sandwich	353	38
Big Mac	570	55	Tostada	179	30
Sausage McMuffin	427	55	Grilled chicken breast on bun	320	28
Whopper with cheese	711	54	Pizza with pepperoni (2 slices)	380	28
French fries, regular	227	52	Bean burrito	357	25
Chicken McNuggets® (6)	323	50	Pizza with cheese (2 slices)	376	24
Cheeseburger	318	45	Vanilla shake	280	19
Burrito Supreme	457	43	Chocolate shake	447	16
Fillet-o-fish	373	43	Mashed potatoes, without gravy	62	15
Egg McMuffin	340	42	Baked potato, plain	250	7

[a]Calories and fat content may vary somewhat depending on the fast-food restaurant or chain.

REALITY CHECK
Portion Distortion

Mohammad and Kevin decided to eat at an Italian restaurant after soccer practice. Just for the fun of it, they agreed to guess how many cups of spaghetti they would consume if they ate the portion served to them.

Who has the better guess?

Answers on page 6-16

Mohammad: Let me see . . . I'd guess it would be 3 cups.

Kevin: I bet it's about a cup. It looks like a normal serving to me.

The average portion size of pasta served by restaurants is nearly 3 cups.[18] For people on a 2,000 calorie per day diet, that's a day's allotment for the foods from the grain group.

Mohammad:

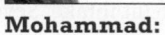

Kevin:

Table 6.9 Improvements in the calorie and nutrient profiles of fast foods[a]

Newer items	Calories	Saturated fat	Fiber	Vegetable/fruit
Side garden salad with low-fat dressing	135	1.5 g	1 g	1.0 c
Sliced apples	25	0 g	1 g	1.0 c
Raisins	130	0 g	2 g	0.5 c
Low-fat turkey sub, 6"	280	1.5 g	4 g	1.0 c
Fruit and walnut salad	210	1.5 g	2 g	1.0 c
Traditional items				
Quarter-pound hamburger on bun	410	7 g	2 g	0 c
Crispy chicken sandwich	530	3.5 g	3 g	0.25 c
Bacon ranch salad with crispy chicken and ranch dressing	540	8.5 g	4 g	2.5 c
Large french fries	550	3.5 g	6 g	2.0 c

[a]Calorie and nutrient content given per portion served by a fast-food restaurant offering the item. Content may vary depending on the restaurant.

placing the topic of healthy eating in a new light for many individuals and communities and may help bring people closer to family, friends, and the environment.

What If You Don't Know How to Cook? With so many convenience foods available and time at a premium, there is growing concern that we're becoming a nation of cooking illiterates. Cooking at home gives you control over what you eat and how it's prepared. Some people immensely enjoy cooking and get a thrill out of making their specialties for friends and family. It's becoming a popular leisure-time activity: 43% of U.S. adults have taken it up for enjoyment.[32]

If you don't know how to cook, there are several ways to learn. You could start on your own by using the recipes on food packages like pasta, tomato sauce, or dried beans. You could search for recipes online or buy a basic cookbook and make simple dishes like salads, tacos, shish kebab, and lentil soup. You could even take a community education course. Illustration 6.12 shows some examples of good starter cookbooks. Read the sections on basic cooking skills and learn what types of equipment and utensils you need to prepare basic dishes. Get the foods and other supplies you need. Select the recipes that look doable and be sure they pass your taste and nutritional standards tests. Voilà! You're cooking.

Bon Appétit!

Dietary guidelines from some other countries contain one other rule of healthy eating that would serve Americans well: "Enjoy your meals." Eating a healthy diet should be enjoyable. If it's too much of a struggle, the healthy diet won't last. The best diets are those that keep us healthy and enhance our sense of enjoyment and well-being. The trick is to remember the broad array of nutritious foods we like that give us good taste and enjoyment when we eat them. And remember not to feel guilty when you occasionally eat a hamburger or some ice cream. Enjoy them to the utmost—as much as ripe oranges, papaya, homemade soups, roast turkey, hummus, and countless other nutritious delicacies.

Illustration 6.12 Some good starter cookbooks and recipe sources.

Alistair Berg/Digital Vision/
Getty Images

NUTRITION
up close

Dietary Guidelines for Americans: What are healthful types of food choices?

Focal Point: Food types to emphasize and de-emphasize in the diets of Americans.

The Dietary Guidelines for Americans make specific recommendations for the types of foods Americans tend to eat too often and those they do not eat often enough. From the options listed below, check the column that reflects whether the types of foods listed correspond to the recommendation to decease or to increase consumption.

Types of foods	Increase consumption	Decrease consumption
Fish and seafood		
Cheese		
Fruits		
High sodium		
Beans and peas		
Nuts and seeds		
Oils		
Whole grains		
High sugar		
Low-fat milk		
Refined grain products		

Feedback to the Nutrition Up Close is located in Appendix G.

REVIEW QUESTIONS

- **Apply the characteristics of healthful diets to the design of one.**

1. Adequate diets are defined as those that provide sufficient calories to relieve hunger and maintain a person's body weight.   **True/False**

2. The diets of Americans tend to be out of balance in a number of ways.   **True/False**

3. _____ "Macronutrients" consist of:
   a. fiber, vitamins, carbohydrates, and minerals
   b. water, calories, fat, and fiber
   c. protein, minerals, vitamins, and fats
   d. carbohydrates, proteins, fat, and water

4. ___ Andre, a normal weight man with little body fat, takes a multivitamin and mineral supplement each morning to make sure he gets all the vitamins and minerals needed daily. The rest of the day he eats whatever he wants, such as candy, donuts, pizza, burgers, and soft drinks. Is Andre consuming a healthful diet?
   a. Most likely yes because he is getting all the vitamins and minerals he needs.

   b. Yes because he is normal weight.
   c. Probably not because his diet lacks balance and variety.
   d. It depends on the composition of the multivitamin and mineral supplement

- **Apply the characteristics of healthful diets to the design of one.**

5. The Mediterranean food guide is an example of a vegetarian diet plan that promotes health and decreases chronic disease risk.   **True/False**

6. The 2010 Dietary Guidelines for Americans emphasize the importance of selecting nutrient-dense foods and balancing calorie intake with output to achieve and maintain a healthy weight.   **True/False**

7. The DASH Eating Plan represents a dietary pattern that is consistent with the 2010 Dietary Guidelines recommendations for dietary intake.   **True/False**

8. Reduction in saturated fat intake is no longer recommended by the Dietary Guidelines.   **True/False**

9. Key Recommendations in the 2010 Dietary Guidelines include a recommendation related to food safety. **True/False**

10. _____ As the head of food service for Lincoln Elementary School you have been charged with making sure the new cafeteria lunch menus conform to the Dietary Guidelines recommendations. Which of the following sets of menu items would you limit offering in the lunch menu?

   a. tossed salads and salad dressing
   b. corn and hot dogs
   c. hamburgers and fries
   d. fruit salad and corn bread

11. _____ The DASH Eating Plan and traditional Mediterranean diet are similar in that they both emphasize:

   a. whole grains and fish
   b. wine and low-fat milk
   c. soy and nuts
   d. cheese and vegetable oil

- **Utilize ChooseMyPlate.gov guidance materials and interactive tools for dietary planning and evaluation.**

12. ChooseMyPlate provides the major tools used in the United States to help people implement the Dietary Guidelines. **True/False**

13. _____ Jack wants to get into shape for the baseball season and decides a healthful diet may help him do that. Which interactive tool from ChooseMyPlate should he use to best help him plan a healthful diet given his size and current physical activity level?

   a. Food-A-Pedia
   b. Food Tracker
   c. Athlete's Food Guide
   d. Daily Food Plan

14. _____ Marta has recently been diagnosed as having "high cholesterol" and decides to assess the amount of saturated fat in her usual diet. Which of the following ChooseMyPlate tools will best help her get this information?

   a. Food-A-Pedia
   b. Food Tracker
   c. MyCholesterol Count
   d. Daily Food Plan

15. _____ Lindsay lost 10 pounds and would like to keep the weight off. Tonight she's headed to a party where lots of delicious pastries will be served and she doesn't want to eat too many of them. What can she do to help with that?

   a. Try to convince herself that she really doesn't like pastries.
   b. Decide before she goes to the party to enjoy one pastry.
   c. Skip the party. It would be impossible to resist the pastries.
   d. Go to the party late because the pastries may be gone.

16. _____ Which of the following statements about large portion size is true?

   a. Adults and children tend to eat more when offered large portions of foods rather than small portions.
   b. People eat until they feel full and then stop eating, regardless of portion size.
   c. Large portions tend to make adults and children eat less than if given smaller portions.
   d. Children tend to eat more when offered more food but adults do not.

Answers to these questions can be found in Appendix G.

NUTRITION SCOREBOARD ANSWERS

1. 86% of U.S. adults fail to meet this recommendation.[1] **True**

2. The basic food groups do not include a "healthy snack" food group. **False**

3. Supersizing fast-food meals piles on calories. **True**

© Digital Vision/Alamy

UNIT

How the Body Uses Food: Digestion and Absorption

NUTRITION SCOREBOARD

1   Almost all of the carbohydrate and fat you consume in
    foods is absorbed by the body, but only about half of the
    protein is absorbed.   **True/False**

2   Disorders of the digestive system are a leading cause of
    hospitalizations in the United States and Canada.
    True/False

3   Most stomach ulcers are caused by overeating spicy
    foods.   **True/False**

Answers can be found at the end of the unit.

After completing Unit 7 and its interactive learning features, you will be able to:

- Identify specific mechanical and chemical processes involved in the digestion of carbohydrates, proteins, and fats.

- Describe the ways in which diet is related to common types of digestive disorders.

My Body, My Food

- **Identify specific mechanical and chemical processes involved in the digestion of carbohydrates, proteins, and fats.**

You are not the same person you were a month ago. Although your body looks the same and you don't notice the change, the substances that make up the organs and tissues of your body are constantly changing. Tissues we generally think of as solid and permanent, such as bones, the heart, blood vessels, and nerves, are continually renewing themselves. The raw materials used in the body's renewal processes are the nutrients you consume in foods.

Each day, about 5% of our body weight is replaced by new tissue. Existing components of cells are renewed, the substances in our blood are replaced, and body fluids are recycled. Taste cells, for example, are replaced about every seven days, and cells lining the intestinal tract are replaced every one to three days. All of the cells of the skin are replaced every month. Red blood cells turn over every 120 days.[5] If you thought it was hard to maintain a car, an apartment, or a house, just imagine the difficulty of maintaining a body! Maintenance is just one of the body's ongoing functions that require nutrients as raw material.

Key Nutrition Concepts

Two key nutrition concepts directly relate to the content on digestion, absorption, and digestive disorders covered in this unit:

1. Foods provide energy (calories), nutrients, and other substances needed for growth and health.

2. Humans have adaptive mechanisms for managing fluctuations in nutrient intake.

How Do Nutrients in Food Become Available for the Body's Use?

The components of food used to form and maintain body tissues are nutrients. Through the processes of **digestion** and **absorption**, nutrients are made available for use by every cell in the body.

The Internal Travels of Food: An Overview The "food processor" of the body is the digestive system, shown in Illustration 7.1. It consists of a 25- to 30-foot-long muscular tube and organs such as the liver and pancreas that secrete digestive juices. The digestive juices break foods down into their molecular components that can be absorbed and utilized by the body.

Much of the work of digestion is accomplished by **enzymes** manufactured by components of the digestive system such as the salivary glands, stomach, and pancreas. Enzymes are complex protein substances that speed up the reactions that break down food. A remarkable feature of enzymes is that they are not changed by the chemical reactions they affect. This makes them reusable.

Carbohydrates, fats, and proteins each have their own set of digestive enzymes. All together, over a hundred different enzymes participate in the digestion of carbohydrates, fat, and proteins. Table 7.1 presents information on some of the enzymes involved in digestion and highlights their specific roles. In Table 7.2 you will see how these enzymes are involved in the digestion of carbohydrate, fat, and protein.

Digestive Processes Digestive processes actually begin before the first bite of food enters the mouth. All a person needs to do to get digestive juices flowing is to think about food, smell food, or see it.[6] You can put this information to a test by clearing your mind of all thoughts, turning the page, and then concentrating only on the photo shown in Illustration 7.2.

digestion The mechanical and chemical processes whereby ingested food is converted into substances that can be absorbed by the intestinal tract and utilized by the body.

absorption The process by which nutrients and other substances are transferred from the digestive system into body fluids for transport throughout the body.

enzymes Protein substances that speed up chemical reactions. Enzymes are found throughout the body but are present in particularly large amounts in the digestive system.

Illustration 7.1   The digestive system.

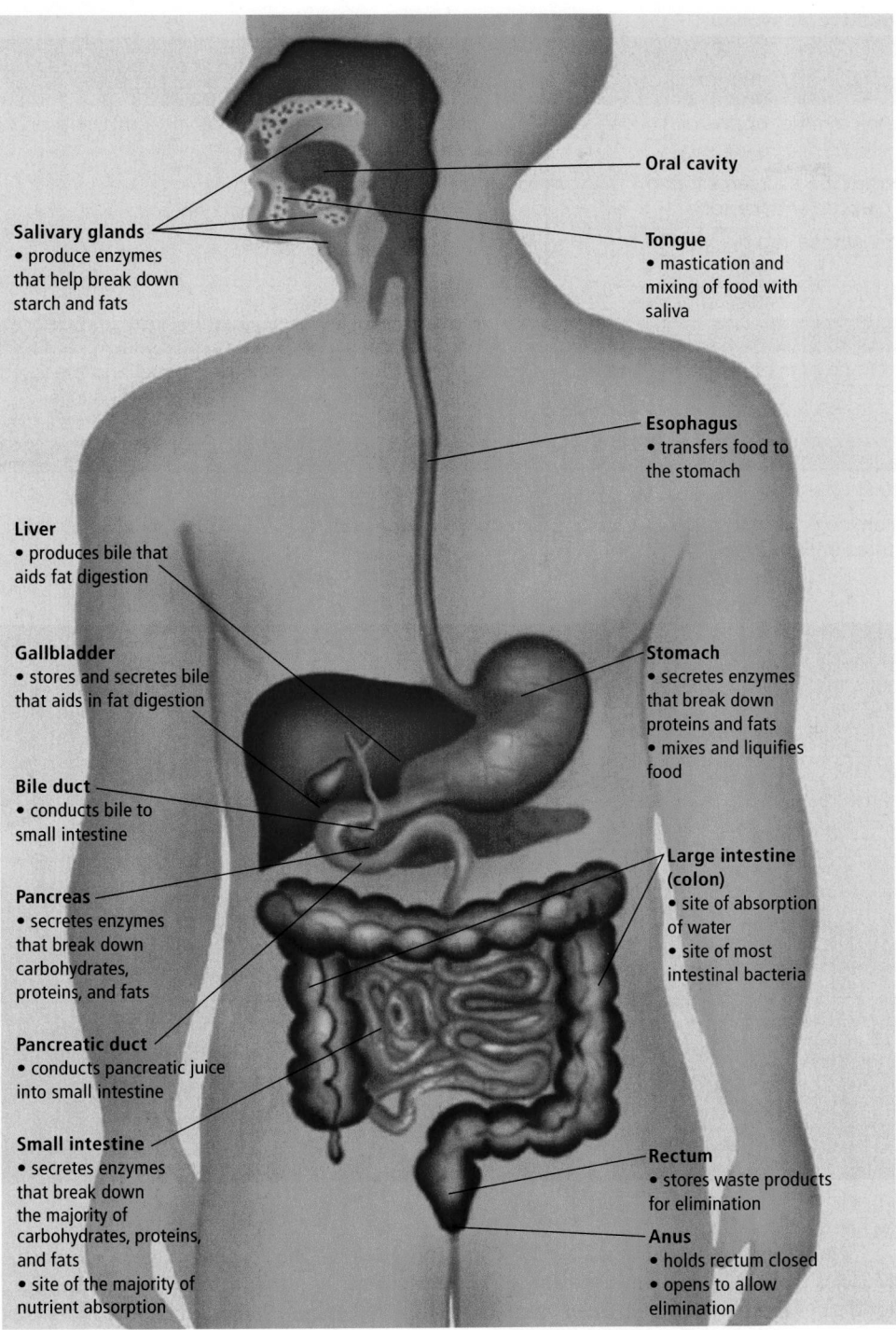

Oral cavity

Salivary glands
• produce enzymes that help break down starch and fats

Tongue
• mastication and mixing of food with saliva

Esophagus
• transfers food to the stomach

Liver
• produces bile that aids fat digestion

Gallbladder
• stores and secretes bile that aids in fat digestion

Stomach
• secretes enzymes that break down proteins and fats
• mixes and liquifies food

Bile duct
• conducts bile to small intestine

Pancreas
• secretes enzymes that break down carbohydrates, proteins, and fats

Large intestine (colon)
• site of absorption of water
• site of most intestinal bacteria

Pancreatic duct
• conducts pancreatic juice into small intestine

Small intestine
• secretes enzymes that break down the majority of carbohydrates, proteins, and fats
• site of the majority of nutrient absorption

Rectum
• stores waste products for elimination

Anus
• holds rectum closed
• opens to allow elimination

© Cengage Learning

As you chew, glands under the tongue release saliva that lubricates the food so that it can be swallowed and easily passed along the intestinal tract. Saliva gets food digestion started. It contains salivary amylase and lingual lipase that begin to break down carbohydrates (amylase) and fats (lipase).[7]

The amount of salivary amylase produced by individuals varies a good deal based on genetic traits. Differences in salivary amylase production can alter the "feel," or texture, of food in the mouth and food preferences. People who produce high levels of amylase break down and liquefy starch in foods to a greater extent than do people who produce low amounts. This gives the foods a soft feel and enhances their flavor. The same foods consumed by low amylase producers may feel firm and have a less desirable flavor.[8,9]

Table 7.1 Primary function of some digestive enzymes

Enzyme	Enzyme function	Enzyme source
A. Carbohydrate digestion		
Amylase	Breaks down **starch** into smaller chains of glucose molecules	Produced in the salivary glands (salivary amylase) and the pancreas (pancreatic amylase)
Sucrase	Separates the **disaccharide** sucrose into the **monosaccharides** glucose and fructose	Produced in the small intestine
Lactase	Splits the disaccharide lactose into glucose and galactose	Produced in the small intestine
Maltase	Separates maltose into two molecules of glucose	Produced in the small intestine
B. Fat digestion		
Lipase	Breaks down fats into fragments of fatty acids and glycerol	Produced in salivary glands (lingual lipase) and the pancreas (pancreatic lipase). The action of lipase is enhanced by **bile**.
C. Protein digestion		
Pepsin	Separates protein into shorter chains of amino acids	Produced by the stomach
Trypsin	Splits short chains of amino acids into molecules containing, one, two, or three amino acids	Produced by the pancreas

Illustration 7.2 Testing, testing. This is a test of your salivary secretions. Did the lemon speak directly to your salivary glands? If you want to turn the digestive processes off, quit thinking about food.

©Michael Newman/PhotoEdit

starch Complex carbohydrates made up of complex chains of glucose molecules. Starch is the primary storage form of carbohydrate in plants. The vast majority of carbohydrate in our diet consists of starch, monosaccharides, and disaccharides.

disaccharide Simple sugars consisting of two sugar molecules. Sucrose (table sugar) consists of a glucose and a fructose molecule, lactose (milk sugar) consists of glucose and galactose, and maltose (malt sugar) consists of two glucose molecules.

monosaccharides (*mono* = one, *saccharide* = sugar) Simple sugars consisting of one sugar molecule. Glucose, fructose, and galactose are monosaccharides.

bile A yellowish-brown or green fluid produced by the liver, stored in the gall-bladder, and secreted into the small intestine. It acts like a detergent, breaking down globs of fat entering the small intestine to droplets, making the fats more accessible to the action of lipase.

After food is chewed, it is swallowed and passed down the esophagus to the stomach. Muscles that act as one-way valves at the entrance and exit of the stomach ensure that the food stays there until it's liquefied, mixed with digestive juices, and ready for the digestive processes of the small intestine. Solid foods tend to stay in the stomach for two to four hours, whereas most liquids pass through it in about 20 minutes.[10]

When the stomach has finished its work, it ejects 1 to 2 teaspoons of its liquefied contents into the small intestine through the muscular valve at its end. Stomach contents continue to be ejected in this fashion until they are totally released into the small intestine. These small pulses of liquefied food stimulate muscles in the intestinal walls to contract and relax; these movements churn and mix the food as it is digested by enzymes. When the diet contains sufficient fiber, the bulge of digesting food in the intestine tends to be larger. Larger food bulges stimulate a higher level of intestinal muscle activity than do smaller food bulges. Thus, high-fiber meals pass through the digestive system somewhat faster than low-fiber meals.

Table 7.2 Summary of the digestion of carbohydrates, fats, and proteins

	Mouth	Stomach	Small intestine, pancreas, liver, and gallbladder	Large intestine (colon)
Carbohydrates (excluding fiber)	The salivary glands secrete saliva to moisten and lubricate food; chewing crushes and mixes the food with salivary amylase that initiates starch digestion.	Digestion of starch continues while food remains in the stomach. Some alcohol is absorbed here. Acid produced in the stomach aids digestion and destroys many bacteria in food.	Pancreatic amylase continues starch digestion. Sucrase, lactase, and maltase break down disaccharides into monosaccharides that are absorbed. Some alcohol is absorbed here.	Undigested carbohydrates reach the colon and can be partly broken down by intestinal bacteria.
Fiber	The teeth crush fiber and mix it with saliva to moisten it for swallowing.	No action.	Fiber binds cholesterol and some minerals.	Most fiber is excreted with feces; some fiber is digested by bacteria in the colon.
Fat	Fat-rich foods are mixed with saliva. Small amounts of lingual lipase accomplish some fat breakdown.	Fat tends to separate from the watery stomach fluid and foods and float on top of the mixture. About 10-30% of fat is broken down by lingual lipase. Fat is last to leave the stomach.	Bile readies fat for the action of lipase from the pancreas. Lipase splits fats into fatty acids and glycerol that are absorbed.	A small amount of fatty materials escapes absorption and is carried out of the body with other wastes.
Protein	In the mouth, chewing crushes and softens protein-rich foods and mixes them with saliva.	Stomach acid works to uncoil protein strands and to activate the stomach's protein-digesting enzyme. Pepsin breaks the protein strands into smaller chains of amino acids.	Trypsin splits protein into molecules containing one, two, or three amino acids. These amino acids are absorbed.	The large intestine concentrates and carries undigested fiber and other residue out of the body.

Digestion, as well as the absorption of nutrients, is greatly enhanced by the structure of the intestines (Illustration 7.3). Fingerlike projections called "villi" line the inside of the intestinal wall and increase its surface area tremendously. If laid flat, the surface area of the small intestine would be about the size of a baseball infield, or approximately 675 square feet. This large mass of tissue requires a high level of nutrients for maintenance. Much of this need (50% in the small intestine and 80% in the large intestine) is met by foods that are being digested.[11]

Absorption Digestion is complete when carbohydrates, fats, and proteins are reduced to substances that can be absorbed, and when vitamins and minerals are released from food. The end products of the digestion of approximately 99% of the carbohydrate, 92%

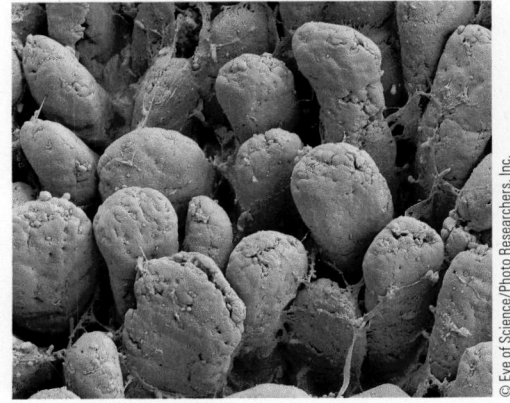
© Eye of Science/Photo Researchers, Inc.

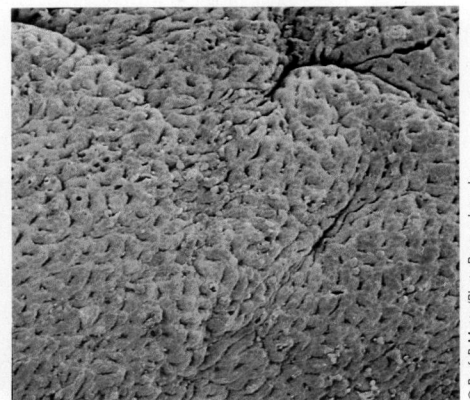
© Prof. P. Motta/Photo Researchers, Inc.

Illustration 7.3 Scanning electron micrographs of cross-sections of the small intestine (*left*) and the large intestine (*right*). Note the high density of villi in the small intestine and the relative flatness of the lining of the large intestine.

of the protein, and 95% of the fat in our diets are absorbed. The primary end products of carbohydrate, fat, and protein digestion that are absorbed are

- Carbohydrate: glucose
- Fat: fatty acids and glycerol
- Protein: amino acids

Some of the end products of digestion are absorbed in the stomach and large intestine, however, nutrient absorption primarily occurs in the small intestine. About 30% of alcohol consumed with meals is absorbed in the stomach, the remainder is absorbed in the small intestine.[12] Water, sodium, and chloride are mainly absorbed by the large intestine. Substances in food that cannot be digested or absorbed, along with bacteria and fragments of cells from the intestinal lining that are being discarded, are concentrated by the large intestine and excreted as stools. The Reality Check feature located near-by addresses the digestion of raw vegetables and the question some people have about whether our bodies can absorb nutrients from them at all.

The Lymphatic and Circulatory Systems The end products of digestion are taken up by the **lymphatic system** (Illustration 7.4) and the **circulatory system** (Illustration 7.5) for eventual distribution to all cells in the body. Lymph vessels and blood

Illustration 7.4 (*left*) The lymphatic system.
Illustration 7.5 (*right*) The circulatory system includes the heart and blood vessels. This system serves as the nutrient transportation system of the body.

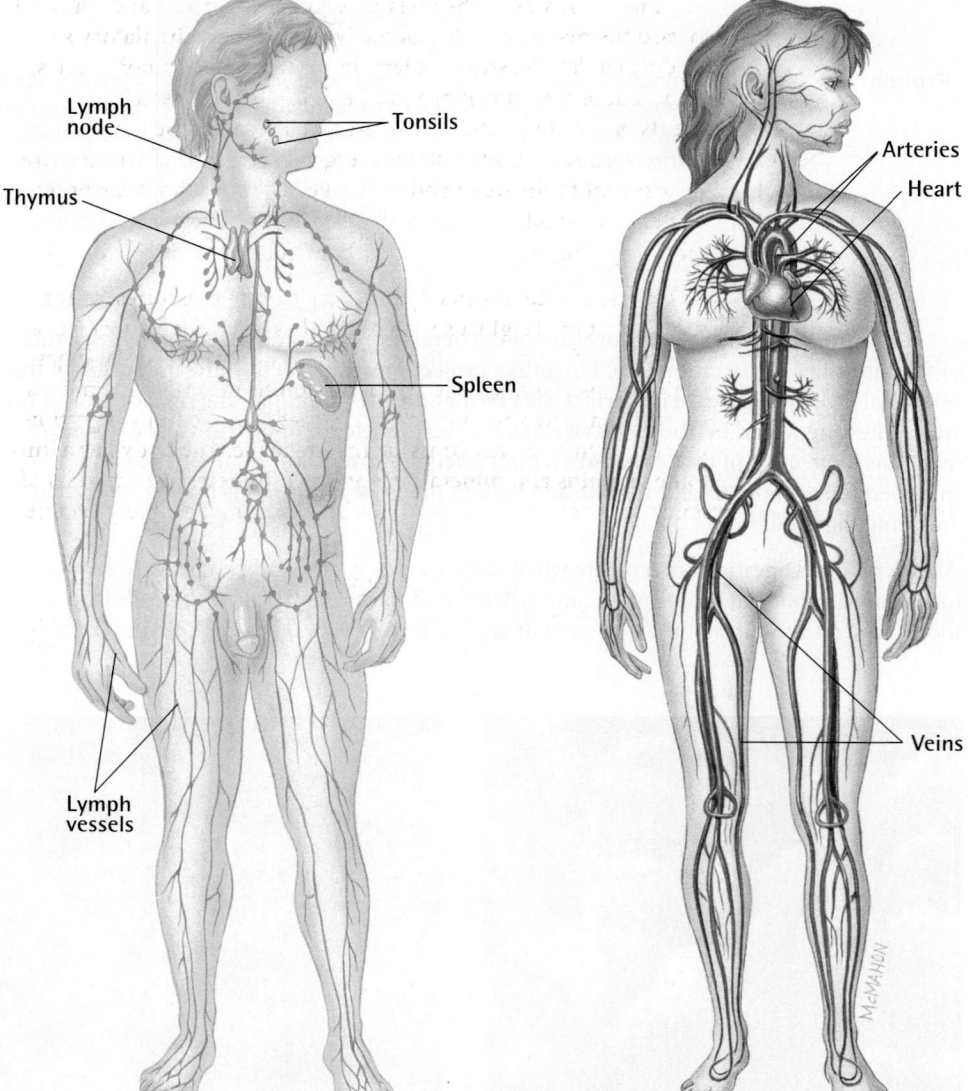

Lymph node
Tonsils
Thymus
Spleen
Lymph vessels
Arteries
Heart
Veins

McMAHON

© Cengage Learning

lymphatic system A network of vessels that absorb some of the products of digestion and transport them to the heart, where they are mixed with the substances contained in blood.

circulatory system The heart, arteries, capillaries, and veins responsible for circulating blood throughout the body.

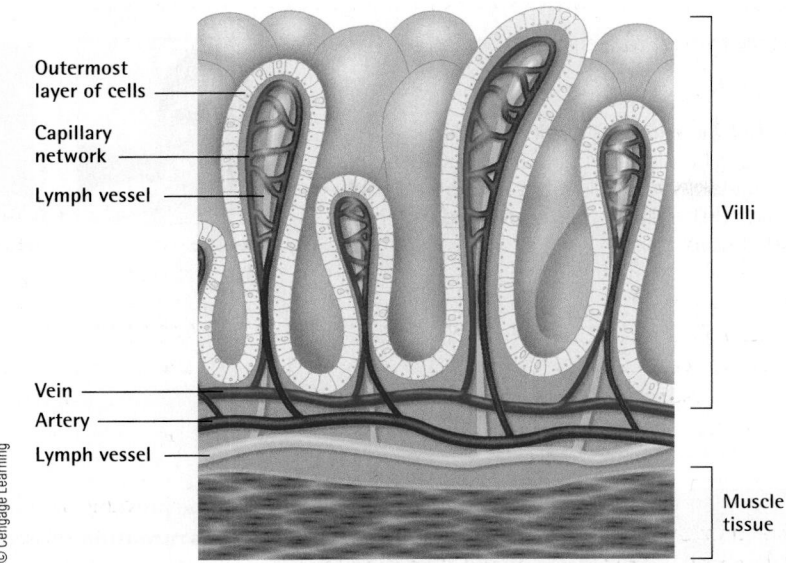

Outermost layer of cells

Capillary network

Lymph vessel

Villi

Vein

Artery

Lymph vessel

Muscle tissue

© Cengage Learning

Illustration 7.6 Structure of villi, showing blood and lymph vessels.

vessels infiltrate the villi that line the inside of the intestines (Illustration 7.6) and transport absorbed nutrients toward the major branches of the lymphatic and circulatory systems. The broken down products of fat digestion are largely absorbed into lymph vessels, whereas carbohydrate and protein broken down products enter the blood vessels.

The nutrient-rich contents of the lymphatic system are transferred to the bloodstream at a site near the heart where vessels from both systems merge into one vessel. From there the lymph and blood mixture is sent to the heart and subsequently throughout the body by way of the circulatory system. Nutrients delivered by the circulatory system reach every organ and tissue in the body, supplying cells with the nutrients obtained from food.

Beyond Absorption Cells can use nutrients directly for energy, body structures, or the regulation of body processes. For example, glucose can be used as is for energy formation or converted to glycogen and stored for later use. Fatty acids, an end product of fat digestion, can be incorporated into cell membranes, used in the synthesis of certain hormones, or used as fuel for energy formation. Vitamins and minerals freed from food by digestion can be used by cells to regulate enzyme activity or tissue maintenance. The body has a limited storage capacity for some vitamins and minerals. Consequently, excessive amounts of certain vitamins and minerals such as vitamin C, thiamin, and sodium are largely excreted in urine.

Other Functions of the Gastrointestinal Tract The gastrointestinal tract is involved in body processes beyond digestion and absorption. These functions include regulation of digestion and absorption processes by taste sensors and the roles of gut bacteria

REALITY CHECK
Can You Digest Raw Vegetables?

Owen and Raja had just finished eating a shredded cabbage salad when Owen suddenly comes up with a statement about something he recently read on the Internet: "You can't digest raw cabbage. It's like trying to digest raw carrots, broccoli, or corn. They just go right through you."

Do they?

Answers on page 7-8

AISPIX by Image Source/ Shutterstock.com

iStockphoto.com/Juanmonino

Owen: Our digestive juices can't break down raw vegetables so we don't get any nutrients from them.

Raja: Really? I eat raw vegetables and think they're nutritious.

ANSWERS TO **REALITY** CHECK

Can You Digest Raw Vegetables?

Nutrients are absorbed from raw vegetables but the extent of nutrient availability largely depends on cell wall break down. Plant cell walls resist digestion because they are fibrous, and nutrients contained in vegetable cells will not be made available for absorption if cell walls stay intact. Disrupting cell walls by thoroughly chewing raw vegetables increases nutrient availability and absorption. [45-47]

Owen: **Raja:** 👍

Illustration 7.7
The taste of food activates many digestive tract processes.

© Doug Menuez/Photodisc/Jupiterimages

inflammation Reactions of the body to the presence of infectious agents, toxins, or irritants. Inflammation triggers the release of biologically active substances that promote oxidation and other reactions that counteract the infectious agent, toxin, or irritant.

umami (u-mam-e) A Japanese word meaning "pleasant savory taste." The taste is described as brothy or meaty and is recognized as the fifth basic taste. Foods containing glutamate, an amino acid derivative, elicit the umami taste and include fish, meats, mushrooms, ripe tomatoes, and aged cheese.

microbes Microscopic organisms, including bacteria and fungi. Some microbes are beneficial, some pathogenic (harmful), and some have little effect on the body. Also called "microorganisms."

in the prevention of infection and **inflammation** related disorders, fiber digestion, and vitamin production.

Functions of Taste Sensors Our gastrointestinal tract plays an important role in identifying the five basic tastes of food (salty, sweet, bitter, sour, and **umami**). The taste of food helps humans identify which foods are safe and beneficial to eat and which may not be. Sweet, savory, and salty foods generally provide pleasurable sensations (Illustration 7.7) whereas the taste of bitter and sour may be disliked and identified as potentially toxic.[13,14]

Taste and other sensors that identify the composition of foods consumed are located throughout the gastrointestinal tract, and not just in the mouth. Messages about how a food tastes and its composition are relayed from these sensors to tissues involved in digestion and absorption processes, and in the utilization of nutrients by the body. The sweet taste message, for example, begins processes that prepare the body to absorb and utilize glucose. The detection of bitter substances can lead to mechanisms that promote the development of diarrhea that will flush potential toxins out of the body. Taste signals also activate processes that decrease appetite and food intake over the course of a meal.[13,14]

Gut Bacteria The large intestine is home to thousands of different species of **microbes**, most prominently bacteria, introduced by foods, fluids, and the environment (Illustration 7.8). Rather than causing disease, a healthy balance of bacteria in the gut promotes health by initiating signals that prompt the body to produce specific, infection-fighting substances such as white blood cells and chemicals that destroy harmful bacteria and other microbes.[15-17]

Bacteria in the large intestines digest a portion of the fiber in our diet. Bacteria excrete fatty acids and gases as the end products of fiber digestion. The fatty acids are absorbed and utilized by the body and the gases are excreted.[18]

Some bacteria in the large intestine produce vitamins that are absorbed and utilized by the body. The two primary examples of vitamins that are produced by bacteria are vitamin K and biotin. Although vitamin production by bacteria is insufficient to meet requirements, the production by bacteria does contribute to vitamin K and biotin adequacy.[19,20]

Digestive Disorders

- **Describe the ways in which diet is related to common types of digestive disorders.**

Digestive disorders such as **heartburn**, **hemorrhoids**, **irritable bowel syndrome**, and **duodenal and stomach ulcers** are the leading cause of hospitalization among U.S. adults aged 45 to 64 and Canadian adults aged 35 to 54 years. Digestive disorders account for more than 104 million medical visits each year in the United States alone.[2,21] Diet and weight status are among the important factors that influence the development and treatment of many of the common disorders of the digestive tract.

Constipation

Constipation is medically defined to exist when an individual has fewer than three bowel movements per week.[22] Some people think they are constipated if they do not have a bowel movement every day. However, normal stool elimination may be three times a day or three times a week, depending on the person. Constipation is characterized by difficulty passing stools because they are hard and dry. People with constipation feel "blocked up" and lousy overall. They may see small amounts of bright red blood on the stool or toilet paper caused by bleeding hemorrhoids or a slight tearing of the anus. This generally disappears after constipation is controlled. Constipation is a symptom, not a disease. Almost everyone experiences constipation at some point in life, and a poor diet typically is the cause.[23]

There are a number of other causes of constipation: the presence of a disorder or disease in the intestinal tract, immobility, medication use, the habitual use of laxatives, or a slow transit time of food in the digestive tract. Constipation caused by a slow transit time of food through the digestive tract is often due to the consumption of too little fiber. This is a common cause of constipation and can generally be relieved, and subsequently prevented, by including 25–30 grams of dietary fiber daily. Good sources of fiber are shown in Illustration 7.9. People with severe constipation may not benefit from high fiber intake, however.[23]

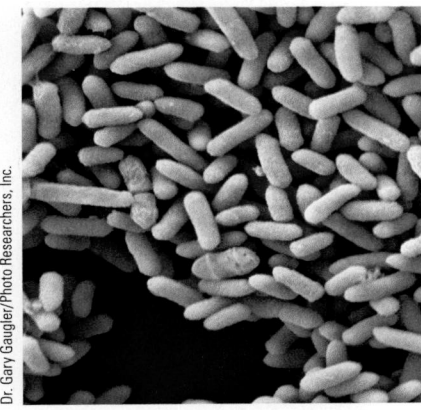

Illustration 7.8 An example of a health-promoting species of bacteria (*B. infantis*) found in the large intestine.

Illustration 7.9 Food sources of dietary fiber. Together, the foods shown provide 29 grams of dietary fiber, an amount that helps prevent constipation.

heartburn A condition that results when acidic stomach contents are released into the esophagus, usually causing a burning sensation.

hemorrhoids (hem-or-oids) Swelling of veins in the anus or rectum.

irritable bowel syndrome (IBS) A disorder of bowel function characterized by chronic or episodic gas; abdominal pain; diarrhea, constipation, or both.

duodenal (do-odd-en-all) and stomach ulcers Open sores in the lining of the duodenum (the uppermost part of the small intestine) or the stomach.

Table 7.3 Myths about constipation [25,38,44]

1. Poisonous substances are absorbed from stools and cause "autointoxication" diseases.
2. Colon cleansing "detoxifies" the body.
3. Extra-long colons cause constipation.
4. All cases of constipation are caused by inadequate fiber intake.
5. You can treat constipation by drinking plenty of fluids.
6. You can lose weight and stay healthy if you take laxatives regularly.
7. If you do not have a bowel movement every day there is something wrong with you.

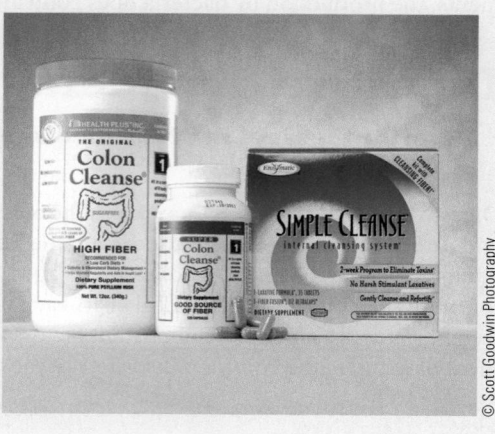

© Scott Goodwin Photography

Fiber increases a person's fluid need so people who increase their fiber intake should also increase their fluid intake. Fiber can increase the production of gas by bacteria that digest fiber. It should be introduced gradually to allow the bacteria in the colon to adjust to the new levels of fiber.[24]

Myths Related to Constipation Some common beliefs about constipation do not hold up to scientific scrutiny (see Table 7.3). For example, there is no evidence that stools contain toxins that can be absorbed and cause harm to the body, or that you can improve your health by periodically "cleansing" or "detoxifying" the colon.[25] Unless dehydration is a problem, consuming plenty of fluids will not treat constipation.[22] The presence of soft stools that are easily excreted is a strong sign that constipation does not exist.[26]

Ulcers

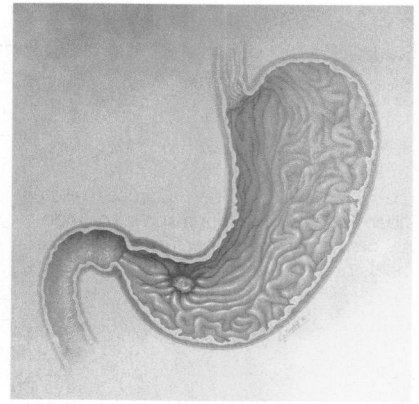

Illustration 7.10   A stomach ulcer.

Photo Researchers/Getty Images

Ulcers are sores that occur primarily in the lining of the stomach and duodenum (Illustration 7.10). They occur in millions of Americans each year.[27] Ulcers are caused by *H. pylori* bacterial infection or the overuse of aspirin, ibuprofen, and similar medications. *H. pylori* is acquired by the ingestion of foods and other substances contaminated with saliva, vomit, or feces from a person harboring the bacteria in the stomach. Rates of *H. pylori* infection are highest in countries with poor sanitary conditions.[28]

H. pylori infection and over use of some types of pain medications break down the protective mucus layer that coats the inside of the stomach and small intestine. When this happens, digestive juices and stomach acid are allowed to erode the stomach and intestinal lining. Symptoms of ulcers include abdominal pain, reduced appetite, weight loss, and feelings of being bloated or nauseated after eating.[29,30]

The development of ulcers has not been found to be related to the intake of specific foods or to high levels of stress. Ulcers can be treated in almost all cases by antibiotics that destroy *H. pylori* bacteria or drugs that reduce stomach acid.[4,30]

Heartburn

Heartburn occurs at least weekly in 20% of American adults.[27] The most common symptom of heartburn is a painful, burning feeling in the chest or throat near the area where the heart is located. It occurs when the valve at the top of the stomach that normally keeps digesting food in the stomach, relaxes. The weakened valve allows the acidic contents of the stomach to back up into the esophagus.[31]

The causes of heartburn are not completely understood, but a variety of factors such as obesity, overeating, pregnancy, and the use of certain medication appear to be related to its development. Certain dietary factors, such as those listed in Table 7.4, have been reported to worsen the symptoms of heartburn in some people. Lifestyle changes, such as weight loss, dietary changes, elevation of the head during sleep; and medications that reduce stomach acid are commonly used for the treatment heartburn.[31,32]

Irritable Bowel Syndrome

Irritable bowel syndrome (IBS), has a name that matches its primary symptom: an irritated bowel. People with irritable bowel syndrome have trouble moving food along the intestines. Food tends to go though too quickly (that produces diarrhea), or too slowly (which leads to constipation). They experience pain and cramping. Overgrowth of bacteria in the large intestine and sensitivity to certain foods and stress appear to be parts of the problem in IBS. The diagnosis of IBS is made when people experience continuous or recurring symptoms of abdominal pain, bloating, and diarrhea or constipation for three months or longer. As many as one in five adult Americans have symptoms of IBS. Although IBS is not is not associated with serious disease development, it has important, negative impacts on quality of life.[36]

Most people with IBS can control their symptoms with diet, stress management, and medications. Avoidance of large meals, adding **soluble fiber** to the diet, and the use of specific **probiotics** have been found to ease the symptoms of IBS.[36,37] It is recommended that people with IBS keep track of what they eat, note foods that seem to promote distress, and speak with a health care provider, such as a dietitian, about the results.[36]

Diarrhea

Diarrhea is a common problem in the United States, occurring an average of four times a year in adults.[38] It is a leading public health problem in many developing countries. Most cases of diarrhea are due to bacterial- or viral-contaminated food or water, lack of immunizations against infectious diseases, and interactions between malnutrition and infection. Children are particularly susceptible to diarrhea. It can deplete the body of fluids and nutrients and produce malnutrition. If it lasts more than two weeks or is severe, diarrhea can lead to dehydration, heart and kidney malfunction, and death. An estimated 3.5 million deaths from diarrhea diseases occur each year to the world's population of children five years of age or under.[39]

The vast majority of cases of diarrhea can be prevented through food and water sanitation programs, immunizations, and adequate diets. The early use of oral rehydration fluids (e.g., the formula provided by the World Health Organization and commercial formulas such as Pedialyte and Rehydralyte) shortens the duration of diarrhea. Rehydration generally takes four to six hours after the fluids are begun.[40]

Rather than "resting the gut" during diarrhea (as used to be recommended), children and adults, once rehydrated, should eat solid foods. Foods such as yogurt, lactose-free or regular milk, chicken, potatoes and other vegetables, dried beans, and rice and other cereals are generally well tolerated and provide nutrients needed for the repair of the intestinal tract. It is best to avoid sugary fluids such as soft drinks. High-sugar beverages tend to draw fluid into the intestinal tract rather than increase the absorption of fluid.[41]

Flatulence

Everyone experiences **flatulence**—it's normal. Gas can occur in the esophagus, stomach, small intestine, and large intestine due to swallowed air or bacterial breakdown of food in the large intestine. Air may be swallowed along with food and beverages or while chewing gum. Eating and drinking while in a rush generally increases air ingestion. Bacterial production of gas in the large intestine can be related to the ingestion of dried beans, broccoli, cauliflower, brussels sprouts, onions, corn, and other vegetables containing "resistant starch" that bacteria digest. Fructose, which is used to sweeten a variety of food products and beverages, and sorbitol (used in some types of candy and gum) may lead to gas formation by bacteria that produce gas as a waste product of carbohydrate digestion. Heartburn and other gastrointestinal tract disorders and medications such as antibiotics are also associated with gas production.[38]

Table 7.4 Diet-related factors and foods that may worsen the symptoms of heartburn.[31,33-35]

- Obesity
- Overeating
- High-fat foods
- Carbonated beverages
- Alcohol
- Spicy foods
- Tomato sauce
- Inadequate fiber intake
- Coffee
- Mint flavorings

soluble fiber Types of dietary fiber that dissolves in water, forming a gel. Good sources of soluble fiber include oats, oatmeal, oat bran, apples, pears, dried beans, carrots, and psyllium.

probiotics Nonharmful bacteria and some yeasts that help colonize the intestinal tract with beneficial microorganisms and that sometimes replace colonies of harmful microorganisms. The most common probiotic strains are *Lactobacilli* and *Bifidobacteria*.

diarrhea The presence of three or more liquid stools in a 24-hour period.

flatulence (flat-u-lens) Presence of excess gas in the stomach and intestines.

People often think they produce too much gas, even when they don't. The amount of gas swallowed and produced by gut bacteria varies a good deal among individuals and within the same individual. Gas production changes depending on what foods are eaten, the types of bacteria populating the large intestine, the medications used, and the presence of gastrointestinal tract disorders. Severe and painful symptoms related to gas production may signal the presence of a digestive disorder.[38,42]

Stomach Growling Gas in the stomach can make your stomach growl. When your stomach growls, you know that gas and food or fluids are mixing in your stomach. The growling tends to be louder when your stomach is empty, when there's no food to muffle the noise. The thought, sight, or smell of food can also trigger stomach growling.[43]

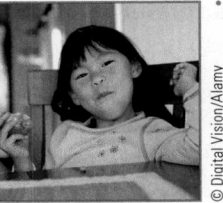
© Digital Vision/Alamy

NUTRITION
up close

Carbohydrate, Fat, and Protein Digestion

Focal Point: Digestion makes carbohydrates, proteins, and fats available for absorption and utilization by the body.

Nikki just finished lunch. She had a chicken sandwich with mayonnaise and a glass of skim milk. She's wondering what is happening to the food in her digestive tract. What is happening?

List the primary enzymes involved in the digestion of the carbohydrate, fat, and protein in Nikki's meal. Also list the primary end products of carbohydrate, fat, and protein digestion that are absorbed.

	Primary enzymes	Primary end products of digestion
Carbohydrate	1.	1.
	2.	2.
	3.	3.
	4.	
Fat	1.	1.
		2.
Protein	1.	1.
		2.

Feedback to the Nutrition Up Close is located in Appendix G.

REVIEW QUESTIONS

- **Identify specific mechanical and chemical processes involved in the digestion of carbohydrates, proteins, and fats.**

1. Cells lining the intestinal tract are replaced every 30 days.   **True/False**

2. Enzymes are protein substances that speed up chemical reactions and are reusable.   **True/False**

3. The end products of digestion are taken up by the lymphatic and circulatory systems.   **True/False**

4. The nutrient-rich contents of the lymphatic system do not mix with blood. These nutrients are distributed throughout the body by a separate system of lymph vessels.   **True/False**

5. The body has a limited capacity to store some vitamins and minerals.   **True/False**

6. About 75% of the protein, carbohydrate, and fat we consume in food is digested and absorbed.   **True/False**

- **Identify specific mechanical and chemical processes involved in the digestion of carbohydrates, proteins, and fats.**

7. The amount of salivary lipase produced by individuals affects the "feel," or texture, of food in the mouth and food preferences.   **True/False**

8. The digestion of fats in the small intestine is largely accomplished by the action of pancreatic lipase.   **True/False**

9. Sweet, sour, bitter, salty, and umami represent the five basic tastes.   **True/False**

10. Taste sensors are located only in the mouth.   **True/False**

11. Bacteria present in our large intestine digest a portion of the fiber in our diets. This is a function human digestive enzymes *cannot* perform.   **True/False**

12. Bacteria in the large intestine produce several vitamins but they are not in a form humans can absorb.   **True/False**

13. _____ Chris underwent surgery for weight loss that greatly reduced the amount of the small intestine that could be used for food digestion and nutrient absorption. Digestion and absorption of which of the following nutrients would be most adversely affected by the surgery?

   a. fat
   b. protein
   c. carbohydrate
   d. alcohol

The next three questions refer to the following situation:

Assume a pharmaceutical company makes a drug that must be swallowed whole. It comes in a pill coated with protein.

14. _____ In which part of the gastrointestinal tract would the drug release begin?

   a. mouth
   b. stomach
   c. small intestine
   d. large intestine

15. _____ This same company makes another drug that comes in a pill coated with starch that must be swallowed whole. In which part of the gastrointestinal tract would this drug start to be released?

   a. mouth
   b. stomach
   c. small intestine
   d. large intestine

16. _____ The company also manufacturers a fiber supplement. In which part of the gastrointestinal tract would the fiber be digested to some extent?

   a. mouth
   b. stomach
   c. small intestine
   d. large intestine

- **Describe the ways in which diet is related to common types of digestive disorders.**

17. Avoidance of high protein meals, adding calcium to the diet, and the use of specific probiotics are recommended for the treatment of constipation.　**True/False**

18. Overeating, and consumption of high fat foods and carbonated beverages worsen heartburn symptoms in some people.　**True/False**

19. Foods containing resistant starch, such as dried beans and corn, may increase gas production by bacteria in the large intestine.　**True/False**

20. Constipation is medically defined to exist when an individual has fewer than three bowel movements per week.　**True/False**

21. Colon cleansing has been shown to prevent disease by removing toxic products from the large intestine.　**True/False**

22. It is recommended that people with diarrhea who are not dehydrated consume solid foods.　**True/False**

Answers to these questions can be found in Appendix G.

NUTRITION SCOREBOARD ANSWERS

1. Over 90% of all the carbohydrates, fats, *and* proteins consumed in food are absorbed and become part of the body.　**False**

2. Digestive disorders are the leading cause of hospitalizations among 45- to 64-year-olds in the United States and 35- to 54-year-olds in Canada.[1,2]　**True**

3. Stomach ulcers have not been shown to be caused by overeating spicy for other types of foods.[3,4]　**False**

Calories! Food, Energy, and Energy Balance

NUTRITION SCOREBOARD

1 Carbohydrates provide more calories than do fats. **True/False**

2 A teaspoon of butter has a higher calorie value than a teaspoon of margarine. **True/False**

3 Energy can be neither created nor destroyed. It can, however, change from one form to another. **True/False**

Answers can be found at the end of the unit.

After completing Unit 8 and its interactive learning features, you will be able to:

- Calculate total calorie need using the formulas presented.

- Calculate the calorie value of a food from its content of carbohydrate, protein, fat, and alcohol (if present).

calorie (calor = heat)   A unit of measure used to express the amount of energy produced by foods in the form of heat. The calorie used in nutrition is the large calorie, or the kilocalorie (kcal). It equals the amount of energy needed to raise the temperature of 1 kilogram of water (about 4 cups) from 15 to 16°C (59 to 61°F). The term *kilocalorie*, or *calorie* as used in this text, is gradually being replaced by the kilojoule (kJ) in the United States; 1 kcal = 4.2 kJ.

basal metabolism   Energy used to support body processes such as growth, health, tissue repair and maintenance, and other functions. Assessed while at rest, basal metabolism includes energy the body expends for breathing, the pumping of the heart, the maintenance of body temperature, and other life-sustaining, ongoing functions.

dietary thermogenesis   Thermogenesis means the production of heat. Dietary thermogenesis is the energy expended during the digestion of food and the absorption, utilization, storage, and transport of nutrients. Some of the energy escapes as heat. It accounts for approximately 10% of the body's total energy need. Also called diet-induced thermogenesis and thermic effect of foods or feeding.

Energy!

• Calculate total calorie need using the formulas presented.

When you think of calories, do you think of energy? That is the "scientifically correct" way to think about them. Energy is what calories are all about.

The **calorie** is like a centimeter or pound in that it is a unit of measure. Rather than serving as a measure of length or weight, the calorie is used as a measure of energy. Specifically, a calorie is the amount of energy needed to raise the temperature of 1 kilogram of water (about 4 cups) from 15°C to 16°C, or 59°F to 61°F (Illustration 8.1). This amount of energy is used as a standard for assigning caloric values to foods. Because calories are a unit of measure, they are not a component of food like vitamins or minerals. When we talk about the caloric content of a food, we're really talking about the caloric value of the food's energy content.

The caloric value of food is determined in a "bomb calorimeter" by burning it completely in a container surrounded by a specific amount of water (Illustration 8.2). The energy released by the food in the form of heat raises the temperature of the surrounding water. The rise in temperature indicates how many calories were released from the portion of food. Although the body doesn't literally burn food, the amount of heat released by food while burning is approximately the same as the amount of energy it supplies to the body.

Key Nutrition Concepts

The following key nutrition concepts relate to this unit's content on calories, energy, and energy balance:

- Foods provide energy (calories), nutrients, and other substances needed for growth and health.

- Poor nutrition can result from both inadequate and excessive levels of nutrient intake.

- There are no good or bad foods.

The Body's Need for Energy

The body uses energy from foods to fuel muscular activity, growth, and tissue repair and maintenance; to chemically process nutrients; and to maintain body temperature (to name a few examples). These needs for energy are subdivided into three categories: **basal metabolism**, physical activity, and **dietary thermogenesis** (Table 8.1 and Illustration 8.3).

The largest single contributor to energy need is basal metabolism or "resting metabolism," as it is also called. It accounts for 60 to 75% of the total need for calories in the vast majority of people.[1] Energy-requiring processes of basal metabolism include breathing, the beating of the heart, maintenance of body temperature, renewal of muscle and bone tissue, and other ongoing activities that sustain life and health. Growth is considered a component of basal metabolism. The proportion of total calories needed for basal metabolism is particularly high during the growing years.

Some organs and tissues in the body are more metabolically active, or require more energy to sustain their functions, than others. Body fat is less metabolically active than lean tissues and accounts for under 20% of basal metabolic calorie expenditure in most people. The brain, liver, kidneys, and muscle are metabolically active; together, they account for 80% or more of the energy used for basal metabolism.[2]

Only a small percentage of people have an unusually low or high **basal metabolic rate (BMR)**. In healthy individuals, slight differences in BMR rarely account for the ease with which weight is gained, maintained, or lost. Uncommon metabolic disorders, such as underactive thyroid and Cushing's syndrome, can modify basal metabolic processes and can lead to changes in weight.[4] The Reality Check for this unit addresses the topic of slow metabolism and weight loss.

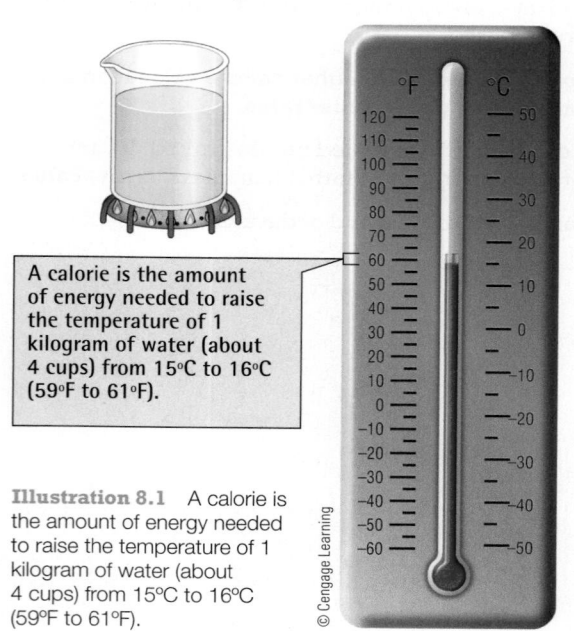

A calorie is the amount of energy needed to raise the temperature of 1 kilogram of water (about 4 cups) from 15°C to 16°C (59°F to 61°F).

Illustration 8.1   A calorie is the amount of energy needed to raise the temperature of 1 kilogram of water (about 4 cups) from 15°C to 16°C (59°F to 61°F).

© Cengage Learning

Energy-using activities of basal metabolic processes require no conscious effort on our part; they are continuous activities that the body must perform to sustain life. The energy needed to carry out basal metabolic functions is assessed when the body is in a state of complete physical and emotional rest.

Basal metabolic rate, or resting energy expenditue as it is also called, is determined clinically and in research studies by **indirect calorimetry**. Illustration 8.4 shows an example of an indirect calorimeter and its use in a clinical setting. This method determines energy expenditure indirectly through a determination of the body's oxygen utilization for energy production in the body.[21]

Indirect calorimetry is a fairly expensive and time-consuming process that is utilized to accurately determine calories required for basal metabolism among people who are seriously ill.[4] Equations for calculating basal metabolic rate are used when estimates suffice.

How Much Energy Do I Expend for Basal Metabolism? You can quickly estimate the calories needed for basal metabolic processes as follows:

- For men: Multiply body weight in pounds by 11.

- For women: Multiply body weight in pounds by 10.

Thus, a man who weighs 170 pounds needs approximately 170 × 11, or 1,870 calories per day for basal metabolic processes. A 135-pound woman needs 135 × 10, or 1,350 calories.

This formula gives an estimate of calories used for basal metabolism in adults based on sex and weight. Physical activity level, muscle mass, height, health status, and genetic

Table 8.1 The three energy-requiring processes of the body

1. **Basal metabolism**
   Energy required to maintain normal body functions while at rest

2. **Physical activity**
   Energy needed for muscular work

3. **Dietary thermogenesis**
   Energy use related to food ingestion. (The process gives off heat.)

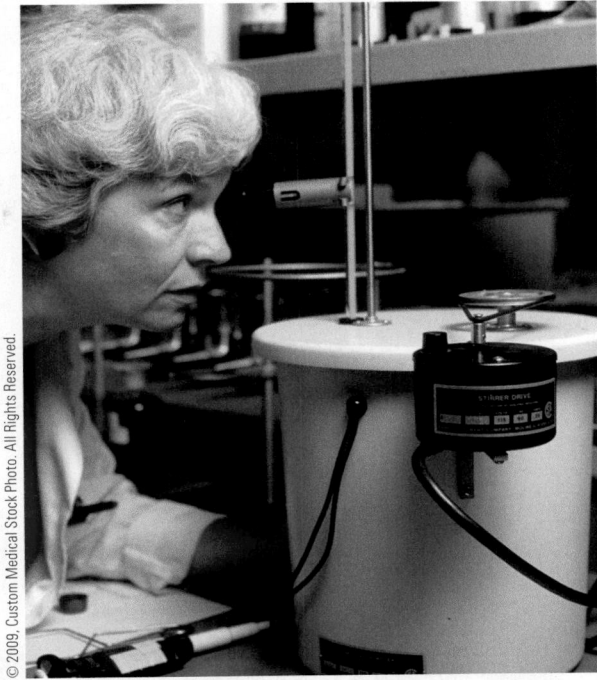

Illustration 8.2 A bomb calorimeter used to measure the calorie value of foods. A food's calorie value is determined by the amount of heat released and transferred to water when the food is completely burned.

basal metabolic rate (BMR) The rate at which energy is used by the body when it is at complete rest. BMR is expressed as calories used per unit of time, such as an hour, per unit of body weight in kilograms. Also commonly called *resting metabolic rate* (*RMR*).

indirect calorimetry A method of measuring energy expenditure from determination of the amount of oxygen utilized by the body during a specific unit of time.

basal metabolism

physical activity

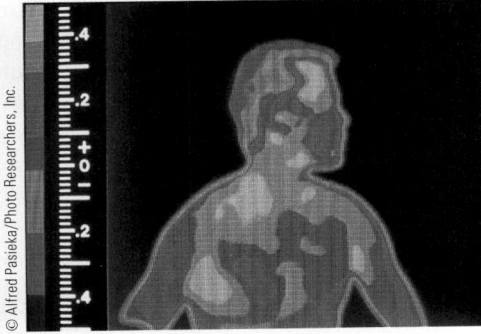

dietary thermogenesis

Illustration 8.3 Examples of the three types of energy-requiring processes in the body.

REALITY CHECK
Is There Any Such Thing as a Slow Metabolism?
Answers on page 8-5

Mel: I've got a slow metabolism. It doesn't matter how little I eat, I still can't lose weight.

Juanita: Some of my friends who have trouble losing weight say that, too, but it can't be right. "Slow metabolism" is a myth.

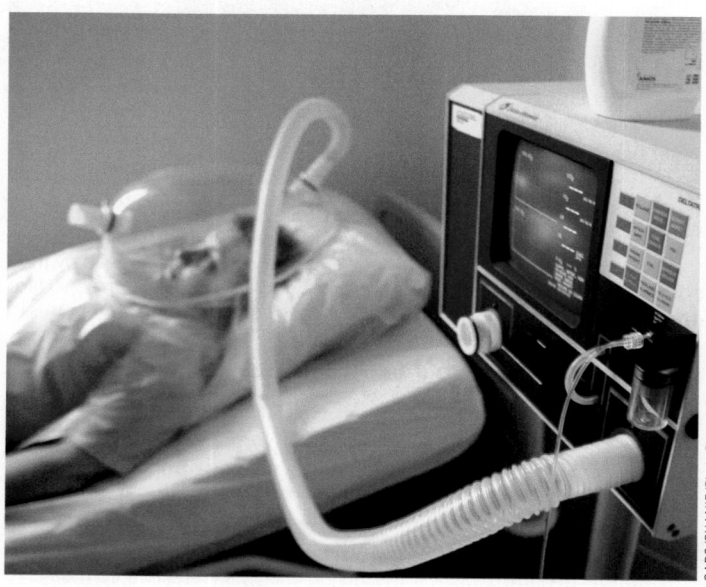

Illustration 8.4 An example of an indirect calorimeter used to assess energy expenditure in clinical and research settings.

traits also influence calorie expenditure for basal metabolism to some extent. Consequently, results obtained using this quick formula may be 10 to 20% lower or higher than the true number of calories required for basal metabolism.[6,7]

How Much Energy Do I Expend in Physical Activity? The caloric level needed for physical activity can vary a lot, depending on how active a person is. It usually accounts for the second highest amount of calories we expend. The energy cost of supporting a physically inactive lifestyle (Table 8.2) is about 30% of the number of calories needed for basal metabolism. An "average" activity level requires roughly 50% of the calories needed for basal metabolism, and an "active" level requires approximately 75%.[9] A physically inactive person needing 1,500 calories a day for basal metabolism, for example, would require about 450 calories (1,500 calories × 0.30 = 450 calories) for physical activity.

People have a tendency to overestimate time spent in physical activity. This in turn tends to overestimate calories needed for physical activity, and depending on the amount of overestimation, can lead to highly inaccurate estimates of total calorie need.[10] Physical activity level should be based on time spent actually engaged in physical activity and not include time spent getting ready for the activity, standing around, or between activities.

How Many Calories Does Dietary Thermogenesis Take? A portion of the body's energy expenditure is used for digesting foods, absorbing, utilizing, and storing nutrients, and transporting nutrients into cells. Some of the energy involved in such activities escapes as heat. These processes are referred to as dietary thermogenesis. Calories expended for dietary thermogenesis are estimated as 10% of the sum of basal metabolic and usual physical activity calories. For instance, say a person's basal metabolic need is 1,500 calories and 450 calories are required for usual activity: 1,500 calories + 450 calories = 1,950

Table 8.2 Energy expenditure by usual level of activity[7,8]

Activity level	Percentage of basal metabolism calories
Inactive. Sitting most of the day; less than two hours of moving about slowly or standing	30
Average. Sitting most of the day; walking or standing two to four hours, but no strenuous activity	50
Active. Physically active four or more hours each day; little sitting or standing; some physically strenuous activities	75

Table 8.3 Summary of calculations for estimating total calorie need of a 130-pound, inactive woman

	Calories
1. Basal metabolism Multiply body weight in pounds by 10. (For men the figure is 11.)	$130 \times 10 = 1{,}300$
2. Physical activity Multiply basal metabolism calories by 0.30 (for 30%) based on the usual energy expenditure level of "inactive" (see Table 8.2).	$1{,}300 \times 0.30 = 390$
3. Dietary thermogenesis Add calories needed for basal metabolism and physical activity together: $1{,}300 + 390 = 1{,}690$ Multiply the result by 0.10 (for 10%).	$1{,}690 \times 0.10 = 169$
4. Total calorie need Add calories needed for basal metabolism, physical activity, and dietary thermogenesis together. $1{,}300 + 390 + 169 = 1{,}859$	**Total calorie need = 1,859**

calories. Calories expended for dietary thermogenesis would equal approximately 10% of the 1,950 calories, or 195 calories.

Adding It All Up   Your estimated total daily need for calories is the sum of calories used for basal metabolism, physical activity, and dietary thermogenesis. In the preceding example, total calories needed would be 2,145 calories (1,500 + 450 + 195). A complete example of calculations involved in estimating a person's total calorie need is provided in Table 8.3. Although the caloric level calculated won't be exactly right, it should provide a reasonable estimate of your total caloric need.

Where's the Energy in Foods?

- **Calculate the calorie value of a food from its content of carbohydrate, protein, fat, and alcohol (if present).**

Any food that contains carbohydrates, proteins, or fats (the "energy nutrients") supplies the body with energy. (That *includes* the foods listed in Table 8.4!) Carbohydrates and proteins supply the body with four calories per gram, and fat provides nine calories per gram. Alcohol also serves as a source of energy. There are seven calories in each gram of alcohol (Table 8.5).

If you enjoy grilling foods on an outdoor barbecue, you have probably observed firsthand the high level of stored energy in fats (Illustration 8.5). Unlike the drippings of low-fat foods such as shrimp or vegetables, drips from high-fat foods cause bursts of flames to shoot up from the grill. The high-energy content of alcohol can be seen in the alcohol-fueled flames that adorn cherries jubilee and bananas Foster.

Table 8.4 Foods that *do* have calories

1. A candy bar eaten with a diet soda
2. Celery and grapefruit
3. Hot chocolate, cheesecake, or soft drinks consumed to make you feel better
4. Cookie pieces
5. Foods "taste-tested" during cooking
6. Foods you eat while on the run
7. Foods you eat straight from their original containers (like ice cream from the carton, milk from a jug, or chips from a large bag)

Table 8.5 Calorie values of the energy nutrients of alcohol

	Cals/gm
Carbohydrate	4
Protein	4
Fat	9
Alcohol	7

ANSWERS TO REALITY CHECK
Is There Any Such Thing as a Slow Metabolism?

Mel's basal metabolism may be lower than average BMR for his weight, but his metabolism wouldn't be "slow." Calories needed for basal metabolism can be somewhat lower in people with relatively high amounts of body fat relative to muscle mass.[3]

Mel:

Juanita:

Illustration 8.5 If you have observed the flames produced by fat dripping from a steak or hamburger on a grill, you have seen the powerhouse of energy stored in fat. The carbohydrate and protein contents of grilled foods don't burn with nearly the same intensity. They have less energy to give.

© Felicia Martinez/PhotoEdit

If you know the carbohydrate, protein, fat, and (if present) alcohol content of a food or beverage, you can calculate how many calories it contains. For example, say a cup of soup contains 15 grams of carbohydrate, 10 grams of protein, and 5 grams of fat. To calculate the caloric value of the soup, multiply the number of grams of carbohydrate and protein by 4 and the number of grams of fat by 9. Then add the results together:

$$
\begin{aligned}
15 \text{ grams carbohydrate} \times 4 \text{ calories/gram} &= 60 \text{ calories} \\
10 \text{ grams protein} \times 4 \text{ calories/gram} &= 40 \text{ calories} \\
5 \text{ grams fat} \times 9 \text{ calories/gram} &= \underline{45 \text{ calories}} \\
& \quad 145 \text{ calories}
\end{aligned}
$$

You can calculate the percentage of total calories from carbohydrate, protein, and fat in the soup by dividing the number of calories supplied by each energy nutrient by the total number of calories in the soup, and then multiplying the results by 100:

- Carbohydrate: $\dfrac{60 \text{ calories}}{145 \text{ calories}} = 0.41 \times 100 = 41\%$

- Protein: $\dfrac{40 \text{ calories}}{145 \text{ calories}} = 0.28 \times 100 = 28\%$

- Fat: $\dfrac{45 \text{ calories}}{145 \text{ calories}} = 0.31 \times 100 = \underline{31\%}$
$$ 100\%$$

Given this information about the caloric value of carbohydrates, proteins, and fats, which of the items in Illustration 8.6 would you expect to be highest in calories?

College students have been found to miss this question 47% of the time.[22] It's the margarine! Margarine contains the most fat; it is made primarily from oil. *High-fat foods*

Illustration 8.6 Which contains the most calories—a tablespoon of margarine, sugar, or pork? (Calorie values are shown on page 8-7.)

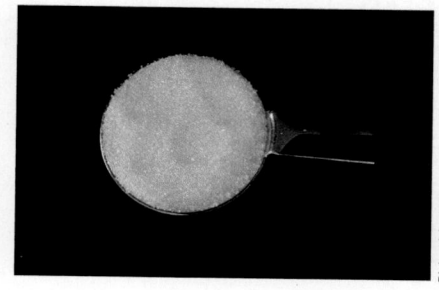

Richard Anderson

provide more calories ounce for ounce than foods that contain primarily carbohydrate or protein. That means bread and potatoes (rich in carbohydrates), catsup (rich in water and a low-calorie vegetable), lean meats, and other low-fat or no-fat foods provide fewer calories than equal amounts of foods that contain primarily fat.

Most Foods Are a Mixture

Some foods (such as oil and table sugar) consist almost exclusively of one energy nutrient, but most foods contain carbohydrates, proteins, and fats in varying amounts.

Bread is high in complex carbohydrates ("starch"), but it also contains protein and a small amount of fat. Likewise, steak is not all protein. Although protein constitutes about 32% of the total weight of lean sirloin steak, fats make up about 8%, and the largest single ingredient is water—60% of the weight. Yet we don't think of steak as a "high-water food"; instead, it's often thought of as pure protein.

So, even though some foods provide relatively more carbohydrate, protein, or fat than other foods, most foods contain a mixture of energy nutrients.

Resources for Estimating the Caloric Value of Food Some people say you can estimate the caloric value of food by how it tastes or by how appetizing it looks. Actually, it's not that simple. How many calories do you think are contained in a half cup of peanuts; a boiled, medium-sized potato; and a cup of rice (Illustration 8.7)?

Taste, appearance, and reputation do not make good criteria for determining the caloric value of foods. Here's an example:

> *I used to order the fish sandwich at fast-food restaurants because I was trying to avoid all the calories in hamburgers. Then I found out the fish sandwich had about the same number of calories as the quarter-pound hamburger! Where did I get the idea that fried fish loaded with tartar sauce has fewer calories than a hamburger?*

This doesn't have to happen to you! By referring to Appendix A, you could determine that a typical fast-food fish sandwich and a quarter-pound hamburger weigh in at about 400 calories each. If you're interested in the caloric value of foods you frequently eat, look them up in this appendix.

Energy Density

The obesity epidemic in the United States is strongly related to increased calorie intake over recent decades.[11] Why are so many people consuming more calories than they need and gaining weight? Part of the answer appears to be related to the **energy density** (or calorie-density) of foods that have become a regular part of the U.S. diet.[12,13]

Energy density represents number of calories per gram of a food item. Twenty grams of potato chips (about 10 chips), for example, has 107 calories. Its energy density equals 107 calories divided by 20 grams, or 5.4. A 202-gram baked potato that provides 212 calories, on the other hand, has an energy density of 1.0 (212 calories divided by 202 grams).

Answer to Illustration 8.6
Margarine.
There are about 101 calories in one tablespoon of margarine, 46 in a tablespoon of sugar, and 40 in a tablespoon of relatively lean pork.

energy density The number of calories per gram of food. It is calculated by dividing the number of calories in a portion of food by the food's weight in grams. May also be calculated as calories per 100 grams of food.

Illustration 8.7 What's the caloric value of these foods? One contains 420 calories, another 205, and the third 118 calories. Try matching the caloric values with the foods, and then check your answers on page 8-8.

Richard Anderson

Richard Anderson

Richard Anderson

Answer to Illustration 8.7

	Calories
$\frac{1}{2}$ cup of peanuts	420
1 medium boiled potato	118
1 cup of white rice	205

Diets that regularly provide energy-dense foods are associated with overeating, excess calorie intake, weight gain, obesity, and type 2 diabetes.[13-15] Diets high in energy-dense foods may interfere with normal food-intake regulation mechanisms by delaying the onset of satiety.[16] In addition, foods high in energy density tend to be nutrient poor and those of low energy density are often nutrient rich.[17] Because they are not energy dense, nutrient-rich foods can be consumed in higher quantities while keeping calorie intake in check. Regular consumption of nutrient-rich foods that are low in energy density is associated with favorable nutrient intakes and reduced weight gain.[13] Vegetables, fruits, whole grain products, and lean meats have lower energy densities than foods such as processed meats, fried foods, and high-fat, sweet desserts.

Small Differences in Energy Density Make a Big Difference Rather small differences in the energy density of foods can make a sizeable difference to calorie intake. On an equal weight basis, there is 38% more calories in macaroni and cheese (energy density = 1.2 calories/gram) than in Ramen noodles (energy density = 0.7 calories/gram). More examples of the energy density of foods are given in the Health Action feature.

How Is Caloric Intake Regulated by the Body?

Because energy is critical to survival, the body has a number of mechanisms that encourage regular caloric intake. It has less effective means of discouraging an excessive intake of calories.[18]

Mechanisms that encourage food intake don't depend on weight status. Rather, they are keyed to encouraging eating on a regular basis so that food, if available, will be consumed and carry the body through times when food isn't around. Humans developed on a schedule of "feast and famine." Those who could store enough fat to see them through the times when food was scarce had an advantage. So, no matter how thin or fat a person is, she or he experiences **hunger** if food is not consumed several times throughout the day. The "hungry" signal is thought to be sent by a series of complex mechanisms when cells run low on energy nutrients supplied by the last meal or snack.[19]

When we eat, we reach a point when we feel full and are no longer interested in eating. The signal is due to hormones and internal sensors in the brain, stomach, intestines, liver, and fat cells that indicate **satiety**—the feeling that we've had enough to eat.[19]

For some people, hunger and satiety mechanisms adjust energy intake to match the body's need for energy. However, for other people, internal signals that urge us to eat or to stop eating are overridden some of the time. People can resist eating, no matter how strong the hunger pains. On the other hand, even after the "I'm full" siren has sounded, people can go on eating. Sometimes people eat because they have an **appetite** for specific foods and the pleasure foods can bring.[20]

Appetite may or may not be related to being hungry. It can be triggered when we smell or see a tasty food right after a meal or when we're really hungry. Have you ever seen those TV commercials for a juicy burger and fries that air around 11:00 p.m.? How many people who jump into their cars and head to the carry-out window are actually hungry? Or have you noticed how appealing the idea of eating is and how good food tastes when you're very hungry? That's appetite at work.

hunger Unpleasant physical and psychological sensations (weakness, stomach pains, irritability) that lead people to acquire and ingest food.

satiety A feeling of fullness or of having had enough to eat.

appetite The desire to eat; a pleasant sensation that is aroused by thoughts of the taste and enjoyment of food.

The Question of Energy Balance Unless you are currently losing or gaining weight, the number of calories you need is the number you usually consume in your diet. Adults who maintain their weight are in a state of energy balance (Illustration 8.8). Because they are not losing weight (using fat and other energy stores) or gaining it (storing energy), their body's expenditure of energy and its intake of energy are balanced.

When energy intake is less than the amount of energy expended, people are in negative energy balance. In this case, energy stores are used and people lose weight. When a positive energy balance exists, weight and fat stores are gained because more energy is available from foods than is needed by the body. Small increases in calorie intake above calorie need may not amount to noticeable gains in body weight on a day-to-day basis but can add up to significant weight gains over the course of months or years.

health action   Moving Toward Foods Lower in Energy Density

EAT MORE. CONSUME FEWER CALORIES!

No, this isn't an ad for a weight loss product. It's a public service announcement about the benefits of lower energy-dense foods. You *can* improve your nutrient intake while eating more food and reducing calorie intake. The key is to select foods that are relatively low in energy density often. The list below compares high energy dense foods (left column) with their lower energy-dense alternatives (right column).

Higher energy-dense food	Calories/g	Lower energy-dense food	Calories/g
Taco shell	4.7	Corn tortilla	2.2
Bologna	3.1	Sliced turkey breast	0.9
Fried chicken	2.8	Grilled chicken	1.7
Fried pork chop	2.8	Broiled pork chop	2.0
Cheeseburger	2.7	Bean burrito	1.9
Hash brown potatoes	2.2	Boiled potato	0.9
Fried fish	2.2	Broiled fish	1.2
Fried rice	1.6	Rice	1.3
Potato salad	1.4	Tossed salad with salad dressing	1.1
Frozen, sweetened strawberries	1.1	Fresh strawberries	0.3

Sometimes, a positive energy balance is a healthy and normal circumstance. For example, a positive energy balance is normal when growth is occurring, as in childhood or pregnancy, or when a person is regaining weight lost during an illness.

Keep Calories in Perspective

Calories is not a word that means fattening or bad for you. Calories are a life and health-sustaining property of food. Diets compatible with health contain a mixture of foods providing various amounts of calories. It's not the caloric content of individual foods that makes them good or not. It's the sum of calories and nutrients in foods that make up our total diets.

Illustration 8.8   The body's energy status.

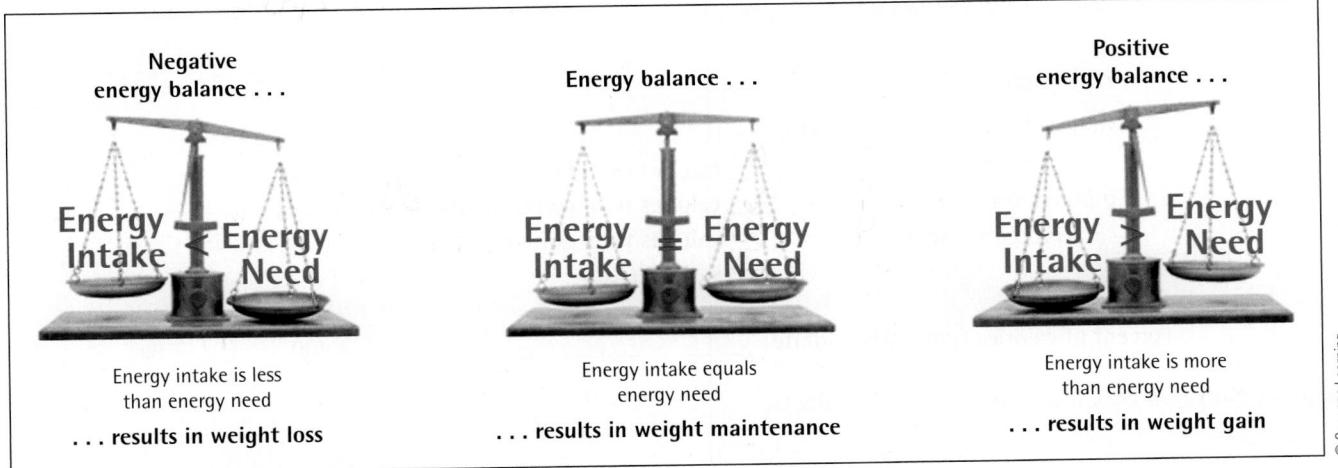

Negative energy balance . . .
Energy Intake   Energy Need
Energy intake is less than energy need
. . . results in weight loss

Energy balance . . .
Energy Intake   Energy Need
Energy intake equals energy need
. . . results in weight maintenance

Positive energy balance . . .
Energy Intake   Energy Need
Energy intake is more than energy need
. . . results in weight gain

© Cengage Learning

© Howard Kingsnorth/Stone/
Getty Images

NUTRITION
up close

Food as a Source of Calories

Focal Point: Examine the distribution of calories in the foods you eat.

What are the sources of calories in food? Determine the caloric contribution of the fat, carbohydrate, and protein content of the following snack foods using the composition data and calorie conversion factors listed. Then answer the two questions that follow.

Which snack is lower in total calories?

_____

Which snack is lower in fat calories?

_____

Photo Disc   Photo Disc

Calorie conversion factors

1g fat	= 9 calories
1g carbohydrate	= 4 calories
1g protein	= 4 calories

Potato Chips

Serving size: 1 oz (about 20 chips)

Fat: 10g × 9 = 90 calories from fat $\frac{90}{158} = 60$ 6

Carbohydrate: 15g × 4 = 60 calories from carbohydrate   69 58 37   3 7   5

Protein: 2g × 4 = 8 calories from protein   958   58 .05

158 Total calories

Percent of calories from fat: _____ ÷ _____ = _____ × 100 = 48 %

Mini Pretzels

Serving size: 1 oz (about 17 pieces)

Fat: 0g × 9 = 0 calories from fat   0

Carbohydrate: 24g × 4 = 96 calories from carbohydrate   88

Protein: 3g × 4 = 12 calories from protein   (1

108 Total calories

Percent of calories from carbohydrate: _____ ÷ _____ = _____ × 100 = 99 %

Feedback to the Nutrition Up Close is located in Appendix G.

REVIEW QUESTIONS

- **Calculate total calorie need using the formulas presented.**

1. A person's calorie need is based on his or her energy expenditure from basal metabolism, physical activity, and dietary thermogenesis.   **True/False**

2. Basal metabolic rate varies substantially among people due to differences in genetic traits and disease history.   **True/False**

3. George, a city bus driver, weighs 220 pounds. The number of calories he expends for basal metabolism daily would equal about 2,420 calories.   **True/False**

4. A common source of error in estimations of physical activity level is underestimation of the amount of time spent in physical activities.   **True/False**

5. Dietary thermogenesis accounts for approximately 30% of a person's total calorie need.   **True/False**

6. Appetite is related to the desire for food, whereas hunger refers to a physiological need for food.   **True/False**

7. A person who is gaining weight is in positive energy balance.   **True/False**

The next three questions refer to this scenario:

Janelle lost 10 pounds in the past few months and now weighs 162.

8. _____ What is the approximate difference in Janelle's calorie need for basal metabolism from before to after the 10-pound weight loss?
   a. 50 calories
   b. 100 calories
   c. 150 calories
   d. 200 calories

9. _____ Janelle continues to lose weight gradually until she reaches her target weight of 140 pounds. Her new calorie need for basal metabolism will be _____ calories lower than before her weight loss.
   a. 120   c. 320
   b. 220   d. 420

10. _____ Janelle starts strength training and increases her muscle mass by 10% while she maintains her weight. What happens to her calorie need for basal metabolism now?
   a. It will decrease then gradually increase.
   b. It will stay the same.
   c. It will increase somewhat.
   d. It will decrease somewhat.

11. _____ You notice a customer buying a 20-ounce bottle of ginger ale in a store and wonder how many calories that represents. Use the information in Appendix A to determine the calorie value of 20 ounces of ginger ale. Your answer is
   a. 248 calories   c. 124 calories
   b. 186 calories   d. 207 calories

12. _____ Use Appendix A to identify how many grams of white granulated sugar (table sugar) are in a teaspoon. Your answer is
   a. 2 grams   c. 6 grams
   b. 4 grams   d. 8 grams

13. _____ If all of the calories provided by 20 ounces of ginger ale is in the form of table sugar, how many teaspoons of table sugar would be in the ginger ale?
   a. 32.1   c. 12.9
   b. 10.0   d. 51.5

- **Calculate the calorie value of a food from its content of carbohydrate, protein, fat, and alcohol (if present).**

14. A calorie is also referred to as a kilocalorie (kcal).   **True/False**

15. Calories are a component of food that provides the body with energy.   **True/False**

16. Say you ate a medium serving of french fries that contained 3 grams of protein, 38 grams of carbohydrate, and 14 grams of fat. The calorie value of the french fries would be 290.   **True/False**

17. The serving of french fries referred to in question 16 would provide 60% of calories from fat.   **True/False**

18. The composition of lean steak is nearly 100% protein.   **True/False**

The next two questions refer to this scenario:

You are in your favorite coffee shop and notice the calorie value of the foods offered is listed right below the foods. You focus on the whole wheat English muffin, the oat bran muffin, and the mini cupcake because they all look delicious. The table below shows the calories listed, as well as the serving size and fat content of each food.

	Weight	Calories	Fat, grams
whole wheat English muffin	2 oz (57 grams)	127	1
oat bran muffin	3 oz (84 grams)	231	6
mini cupcake	2 oz (57 grams)	180	10

19. _____ Which of these menu items has the lowest energy density?
   a. whole wheat English muffin
   b. oat bran muffin
   c. mini cupcake

20. _____ Which menu item has the highest percentage of total calories from fat?
   a. whole wheat English muffin
   b. oat bran muffin
   c. mini cupcake

Answers to these questions can be found in Appendix G.

NUTRITION SCOREBOARD ANSWERS

1. Carbohydrates provide four calories per gram. Fat, on the other hand, provides nine calories per gram. **False**

2. Butter and margarine contain the same amount of fat, so they provide the same number of calories (35 per teaspoon). **False**

3. This is the conservation of energy principle, the first law of thermodynamics. **True**

UNIT 9

Obesity to Underweight: The Highs and Lows of Weight Status

NUTRITION SCOREBOARD

1 There is usually little difference between ideal body weights defined by cultural norms and those defined by science.   **True/False**

2 Fat stores located in the stomach area present a greater hazard to health than do fat stores located around the hips and thighs.   **True/False**

3 The basic cause of obesity is calorie intake that exceeds calorie needs.   **True/False**

4 Healthy underweight people often find it nearly impossible to gain weight.   **True/False**

Answers can be found at the end of the unit.

After completing Unit 9 and its interactive learning features, you will be able to:

- Identify weight status based on weight, height, age (where appropriate), and sex using appropriate formulas and growth charts.

- Delineate the potential causes and health consequences of overweight and obesity.

- Delineate the potential causes and health consequences of underweight.

- Demonstrate an awareness of the roles of weight bias on psychological and other aspects of the health status of children and adults.

Variations in Body Weight

- **Identify weight status based on weight, height, age (where appropriate), and sex using appropriate formulas and growth charts.**

For humans, one size does not fit all. Human bodies come in a range of sizes that vary from heavy and tall to thin and short. Over recent decades, however, the distribution of body sizes has become lopsided. The proportion of people in the United States and other countries classified as overweight or obese is increasing, and rates of underweight are dropping. What lies underneath these trends? Why are humans so susceptible to gaining weight?

Obesity appears to represent a weak link in the biological evolution of humans. Body processes that regulate food intake developed more than 40,000 years ago when "feast and famine" cycles were common and all foods were basic. Being underweight was a distinct disadvantage. The more body fat people could store after a feast, the better their chances of surviving the subsequent famines. Consequently, multiple mechanisms that favored food intake and body fat storage developed.[2] These body mechanisms continue to encourage food intake even when food is constantly available and obesity poses a greater threat to survival than does famine. In the words of Theodore Van Itallie, a noted obesity researcher, in environments with an abundant supply of food and no requirement for vigorous physical activity, "perhaps thin people are the ones who are abnormal."

Circumstances today are very different than they were 40,000 years ago. These differences may go a long way to explaining why obesity rates are increasing in many countries.[3]

Key Nutrition Concepts

Evidence addressed in this unit on weight status is related to the key nutrition concepts of:

- Health problems related to nutrition originate within cells.

- Poor nutrition can result from both inadequate and excessive levels of nutrient intake.

- Poor nutrition can influence the development of certain chronic diseases.

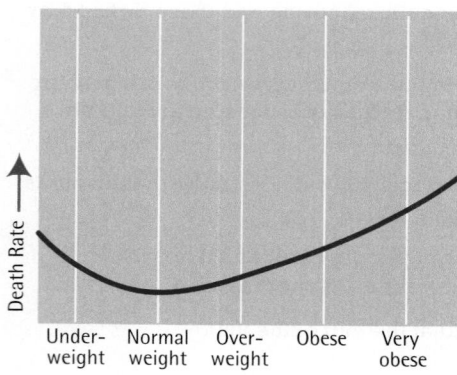

Illustration 9.1 The relationship between body weight status and deaths from all causes for adults.

body mass index (BMI) An indicator of the status of a person's weight for his or her height. It is calculated by dividing weight in kilograms by height in meters squared. It can also be calculated using inches and pounds as shown in Table 9.2.

How Is Weight Status Defined?

Culture and science define the appropriateness of body size. Culturally-defined body weight and shape preferences change with time and in ways that have little to do with health. In the 15th century, culturally-defined body shape standards in Europe favored thin, muscular men and plump women.[4,5] In America in the early 1900s, moon-faced and pear-shaped women and muscular men were the rage. It was not until the latter part of the 20th century that ideal body shape took a turn in Europe and America toward thinness, and at times, to unhealthy levels of extreme thinness. In contrast, the culturally defined, ideal weight status in some poor nations is overweight or obesity. People with ample fat stores are considered wealthier, higher class, and more able to produce children than thin people who tend to be perceived as sickly and poor.[5]

Science defines standards for body weight for adults primarily based on the risk of death from all causes. Death rates are highest among adults who have very high body weights for height; the next highest rates are for underweight adults; and the lowest rates are among adults who are normal weight for height (Illustration 9.1).[6] You can get an idea of what a normal weight for height is by following the steps detailed in the Health Action.

In the past, standards used to identify healthy body weights for adults were developed by the insurance industry. Height and weight tables were created to estimate life and death expectancy and the risk of death for adults. These tables have been replaced by standards that employ **body mass index (BMI)**.

There is a quick way to estimate within ±10% what is a normal, or healthy, weight for height. It is called the Hamwi[61] method.

Women

Begin with 5 feet equals 100 pounds, and then add 5 pounds for each additional inch of height. Here's an example of how you would estimate a healthy weight for a woman who is 5 feet, 7 inches tall:

5 feet = 100 pounds
7 inches x 5 pounds = 35 pounds
100 pounds + 35 pounds = 135 pounds

Men

For men, 5 feet equals 106 pounds, and each additional inch of height adds on 6 pounds. So, for example, to estimate a healthy weight for a man who is 5 feet 10 inches tall the following way:

5 feet = 106 pounds
10 inches x 6 pounds = 60 pounds
100 pounds + 66 pounds = 166 pounds

Body Mass Index

Most commonly referred to as BMI, body mass index is a measure of weight for height that provides a fairly good estimate of body fat content.[7] Ranges of BMI are used to define weights for height that correspond to underweight, normal weight, overweight, and obesity in adults (Table 9.1). BMI has the advantage of being calculated the same way for adult females and males. Calculating BMI involves dividing weight in pounds by height in inches, then dividing this result by height in inches again, and then multiplying that result by 703. An example of BMI calculation is given in Table 9.2. You can use the chart shown in Illustration 9.2 to identify weight status based on BMI.

Assessing Weight Status in Children and Adolescents Standards used to assess weight status in children and adolescents employ BMI percentile ranges for girls and boys (Illustration 9.3). Percentiles of BMI are based on the proportion of children and adolescents who have different levels of BMI at given ages. For example, if a child's BMI is at the 50th percentile for his or her age, then half of children will have BMIs below and half will have values above the 50th percentile. Table 9.3 shows BMI percentile ranges by age that correspond to underweight, healthy weight, overweight, and obesity in children and adolescents.

BMI percentile ranges for children and adolescents do not provide information on growth progress in terms of weight or height for age. Consequently, growth in height cannot be assessed using BMI. Other standards for assessment of weight and height for age in children and adolescents are available at cdc.gov/growthcharts.

Overweight and Obesity

- **Delineate the potential causes and health consequences of overweight and obesity.**

Being overweight or obese is the norm in the United States. The combined incidence of overweight and obesity among adults has increased to 68% (Illustration 9.4).[9]

Over one in six children and adolescents in the United States are overweight, and this percentage is trending upward. Overweight and obesity are becoming America's number-one health problems. Obesity, which represents the largest risks to health, varies considerably among U.S. states (Illustration 9.5). In 2010, for example, rates of obesity were highest in Mississippi and West Virginia, and lowest in Colorado and the District of Columbia.[10]

The trend toward higher rates of overweight and obesity in the United States is shared by many other countries of the world.[3] A sampling of overweight plus obesity rates by country is shown in Table 9.4. The high and growing incidence of overweight

Table 9.1 Classifying adult weight status by body mass index[8]

	Body mass index
Underweight	under 18.5 kg/m²
Normal weight	18.5–24.9 kg/m²
Overweight	25–29.9 kg/m²
Obese	30 kg/m² or higher[a]

[a]Obesity is subdivided for some purposes into the BMI groups of moderate obesity (30–34.9 kg/m²), severe obesity (35–39.9 kg/m²), and very severe obesity (40+ kg/m²).

Table 9.2 Calculating BMI: An example

Say Chris weighs 140 pounds and is 5 feet, 3 inches tall. To calculate BMI:

Figure out how many inches tall Chris is:

5 feet x 12 inches per foot = 60 inches
60 inches + 3 inches = 63 inches

Divide Chris's weight by his height in inches:

$$\frac{140 \text{ pounds}}{63 \text{ inches}} = 2.22$$

Divide the result (2.22) by Chris's height in inches again:

$$\frac{2.22}{63 \text{ inches}} = 0.035$$

Multiply this result by 703:

0.035 × 703 = 24.8 This is Chris's BMI.

Illustration 9.2 BMI chart.

Body Mass Index (BMI)

Height	18	19	20	21	22	23	24	25	26	27	28	29	30	31	32	33	34	35	36	37	38	39	40
										Body Weight (pounds)													
4'10"	86	91	96	100	105	110	115	119	124	129	134	138	143	148	153	158	162	167	172	177	181	186	191
4'11"	89	94	99	104	109	114	119	124	128	133	138	143	148	153	158	163	168	173	178	183	188	193	198
5'0"	92	97	102	107	112	118	123	128	133	138	143	148	153	158	163	168	174	179	184	189	194	199	204
5'1"	95	100	106	111	116	122	127	132	137	143	148	153	158	164	169	174	180	185	190	195	201	206	211
5'2"	98	104	109	115	120	126	131	136	142	147	153	158	164	169	175	180	186	191	196	202	207	213	218
5'3"	102	107	113	118	124	130	135	141	146	152	158	163	169	175	180	186	191	197	203	208	214	220	225
5'4"	105	110	116	122	128	134	140	145	151	157	163	169	174	180	186	192	197	204	209	215	221	227	232
5'5"	108	114	120	126	132	138	144	150	156	162	168	174	180	186	192	198	204	210	216	222	228	234	240
5'6"	112	118	124	130	136	142	148	155	161	167	173	179	186	192	198	204	210	216	223	229	235	241	247
5'7"	115	121	127	134	140	146	153	159	166	172	178	185	191	198	204	211	217	223	230	236	242	249	255
5'8"	118	125	131	138	144	151	158	164	171	177	184	190	197	203	210	216	223	230	236	243	249	256	262
5'9"	122	128	135	142	149	155	162	169	176	182	189	196	203	209	216	223	230	236	243	250	257	263	270
5'10"	126	132	139	146	153	160	167	174	181	188	195	202	209	216	222	229	236	243	250	257	264	271	278
5'11"	129	136	143	150	157	165	172	179	186	193	200	208	215	222	229	236	243	250	257	265	272	279	286
6'0"	132	140	147	154	162	169	177	184	191	199	206	213	221	228	235	242	250	258	265	272	279	287	294
6'1"	136	144	151	159	166	174	182	189	197	204	212	219	227	235	242	250	257	265	272	280	288	295	302
6'2"	141	148	155	163	171	179	186	194	202	210	218	225	233	241	249	256	264	272	280	287	295	303	311
6'3"	144	152	160	168	176	184	192	200	208	216	224	232	240	248	256	264	272	279	287	295	303	311	319
6'4"	148	156	164	172	180	189	197	205	213	221	230	238	246	254	263	271	279	287	295	304	312	320	328
6'5"	151	160	168	176	185	193	202	210	218	227	235	244	252	261	269	277	286	294	303	311	319	328	336
6'6"	155	164	172	181	190	198	207	216	224	233	241	250	259	267	276	284	293	302	310	319	328	336	345

Underweight (<18.5) | Healthy Weight (18.5-24.9) | Overweight (25-29.9) | Obese (30)

Find your height along the left-hand column and look across the row until you find the number that is closest to your weight. The number at the top of that column identifies your BMI. The area shaded in green represents healthy weight ranges.

and obesity in the United States and around the world represents a time bomb for a future explosion in obesity-related disease rates.

The Influence of Obesity on Health People who are obese are more likely to experience diabetes, heart disease, certain types of cancer, hypertension, and other disorders (listed in Table 9.5) than are people of normal weight.[12]

The increased risk of disease appears to be primarily due to a higher prevalence of **metabolic** abnormalities in many obese people.[11] (Metabolic disorders are discussed further in the pages ahead). Approximately 70% of obese persons have two or more metabolic abnormalities such as:

- Hypertension

- Elevated triglycerides, glucose, and/or insulin

- Low HDL-cholesterol (the "good cholesterol")

- High **C-reactive protein** (a key marker of inflammation)

About 23% of normal weight adults, and 50% of overweight individuals have two or more metabolic abnormalities that increase disease risk.[13]

Weight loss of 10–15% of initial body weight, paired with exercise that improves physical fitness level, reduces metabolic abnormalities and the risk of disease.[14,15]

metabolism The chemical changes that take place in the body. The conversion of glucose to energy or to body fat is an example of a metabolic process.

C-reactive protein (CRP) A key inflammatory factor produced in the liver in response to infection or inflammation. Elevated concentrations of CRP are associated with heart disease, obesity, diabetes, inactivity, infection, smoking, and inadequate antioxidant intake.

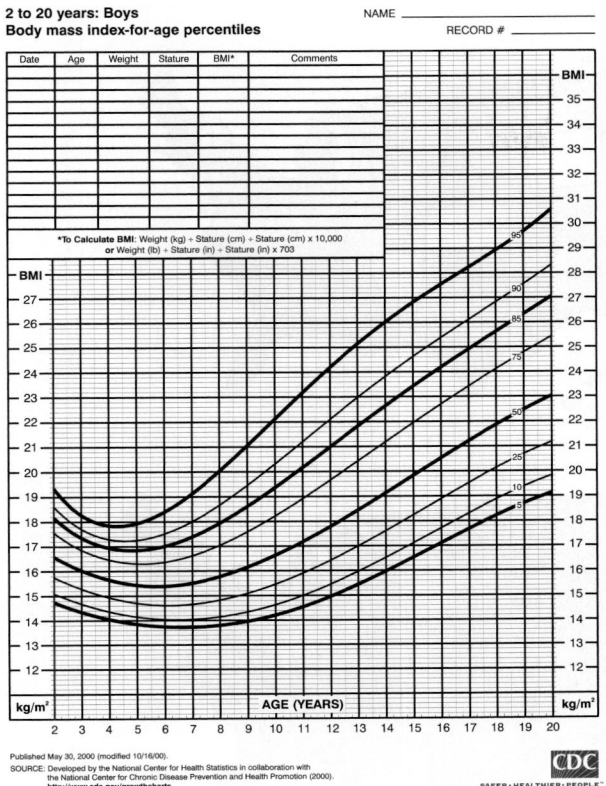

2 to 20 years: Boys
Body mass index-for-age percentiles

NAME _____

RECORD # _____

*To Calculate BMI: Weight (kg) ÷ Stature (cm) ÷ Stature (cm) x 10,000
or Weight (lb) ÷ Stature (in) ÷ Stature (in) x 703

AGE (YEARS)

Published May 30, 2000 (modified 10/16/00).
SOURCE: Developed by the National Center for Health Statistics in collaboration with
the National Center for Chronic Disease Prevention and Health Promotion (2000).
http://www.cdc.gov/growthcharts

CDC
SAFER · HEALTHIER · PEOPLE™

Illustration 9.3 An example of the CDC's BMI charts by age and sex, including examples of their interpretation for a 10-year-old boy. *Copies of all growth charts can be obtained from www.cdc.gov/growthcharts.*

Table 9.3 Weight status standards for 2- to 19-year-olds based on BMI-for-age and sex

	Percentile range(s)
Underweight	< 5th
Healthy weight	5th–85th
Overweight	85th–95th
Obese	95th or higher

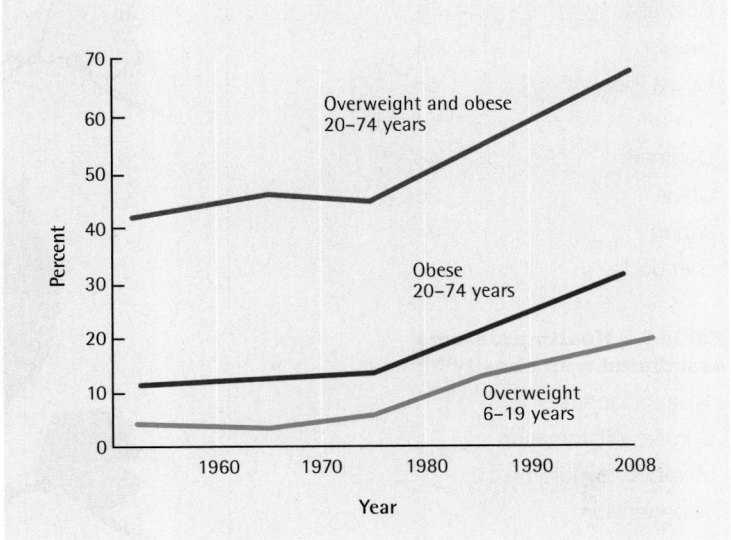

Illustration 9.4 Overweight and obesity by age in the United States, 1960–2008.[9]

Obesity and Psychological Well-Being An important aspect of obesity is the prejudice to which obese people may be subjected. Children who are obese are more likely to suffer unfair or indifferent treatment from teachers than other children. They experience more isolation, rejection, and feelings of inferiority than other children. Obese adults are likely to be discriminated against in hiring and promotion decisions and to be thought of as lazy or lacking in self-control, sometimes even by the health professionals.[17,18] Society's prejudice against people who don't conform to the cultural ideal of body size may be one of the most injurious aspects of being obese. It is also a preventable part of reduced health and well-being experienced by many obese people.[5]

Body Fat and Health: Location, Location, Location

It is becoming increasingly clear that many of the health problems associated with obesity are directly related to where excess fat is stored. Humans store fat in two major locations: under the skin over the hips, upper arms, and thighs; and in the abdomen. Fat stored under the skin is called **subcutaneous fat** and that stored in the abdomen under the skin and a layer of muscle **visceral fat** (see Illustration 9.6). People who store fat primarily in their hips, upper arms, and thighs are said to have a "pear shape" and those who store fat principally in the abdomen an "apple shape." These body shapes are shown in Illustration 9.7. You may be better off health-wise if you're a "pear" rather than an "apple."

Visceral fat is much more metabolically active, and more strongly related to disease risk than subcutaneous fat.[20] Metabolic processes initiated by visceral fat produce **chronic inflammation** and oxidation reactions that disrupt normal body functions. These disruptions promote the development of **insulin resistance**, **metabolic syndrome**, elevated blood glucose and triglyceride concentrations, high blood pressure, and hardening of the

subcutaneous fat (sub-q-tain-e-ous)
Fat located under the skin.

visceral fat (vis-sir-el) Fat located under the skin and muscle of the abdomen.

chronic inflammation Low-grade inflammation that lasts weeks, months, or years. Inflammation is the first response of the body's immune system to infection or irritation. Inflammation triggers the release of biologically active substances that promote oxidation and other potentially harmful reactions in the body.

insulin resistance A condition in which cells "resist" the action of insulin in facilitating the passage of glucose into cells.

metabolic syndrome A constellation of metabolic abnormalities generally characterized by insulin resistance, abdominal obesity, high blood pressure and triglyceride levels, low levels of HDL cholesterol, and impaired glucose tolerance. Metabolic syndrome predisposes people to the development of type 2 diabetes, heart disease, hypertension, and other disorders. It is common: One in five U.S. adults has metabolic syndrome.[26]

Table 9.4 Prevalence of adult overweight and obesity in a sampling of countries[16]

Country	Percent overweight and obese
United States	68
Czech Republic	66
Canada	63
Australia	61
Israel	61
Costa Rica	58
France	51
Denmark	48
China	25
Japan	24
Cambodia	12

Table 9.5 Health problems associated with obesity[11,12]

Type 2 diabetes

Chronic inflammation

Metabolic syndrome

Hypertension

Stroke

Elevated cholesterol level

Low HDL-cholesterol level

Heart disease

Certain types of cancer

Gallbladder disease

Fatty liver disease

Shortened life expectancy

Discrimination

Depression

Accidents

Skin disorders

Sleep disorders

type 2 diabetes A disease characterized by high blood glucose levels due to the body's inability to use insulin normally or to produce enough insulin (previously called adult-onset diabetes).

fatty liver disease A reversible condition characterized by fat infiltration of the liver (10% or more by weight). If not corrected, fatty liver disease can produce liver damage and other disorders. The condition is associated primarily with obesity, diabetes, and excess alcohol consumption. The disease is called steatohepatitis when accompanied by inflammation.

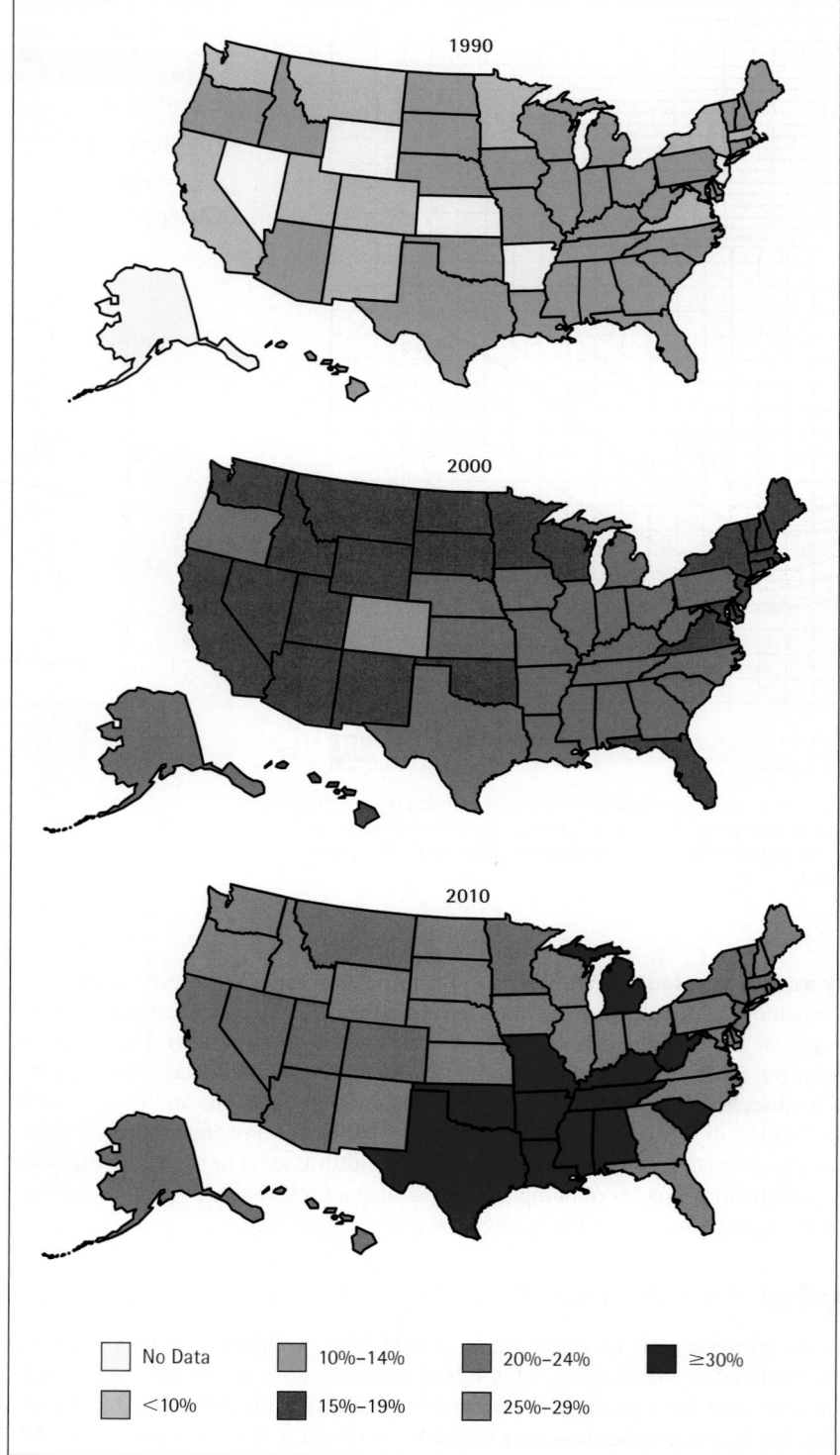

Illustration 9.5 Percent of obese (BMI ≥ 30 kg/m²) adults by state in 1990, 2000, and 2010.[10]

arteries. These changes, in turn, can lead to the development of heart disease, some types of cancer, **type 2 diabetes**, hypertension, **fatty liver disease**, and other disorders.[11,21] Normal weight and overweight individuals with excessive visceral fat deposits are also at increased risk of metabolic abnormalities and diseases associated with them.[22]

Metabolic abnormalities and disease risk associated with visceral fat can be reduced by regular exercise and improved physical fitness level. Higher levels of health benefits are

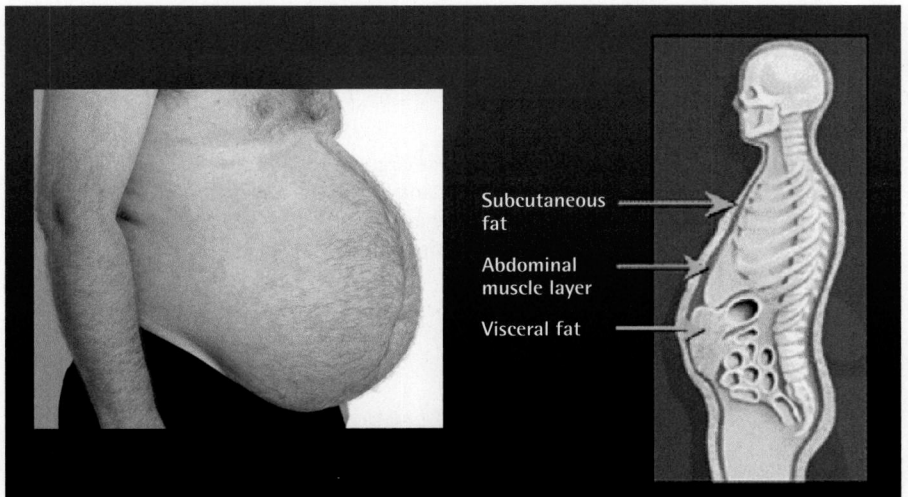

Illustration 9.6 A diagram of the location of subcutaneous and visceral fat.
Source: Photo (l): Victor de Schwanberg/Photo Researchers, Inc.; Caballero E. Dyslipidemia and vascular function: The rationale for early and aggressive intervention. Available at: *www.medscape.com*. Accessed April 18, 2006.

Subcutaneous fat

Abdominal muscle layer

Visceral fat

PhotoDisc

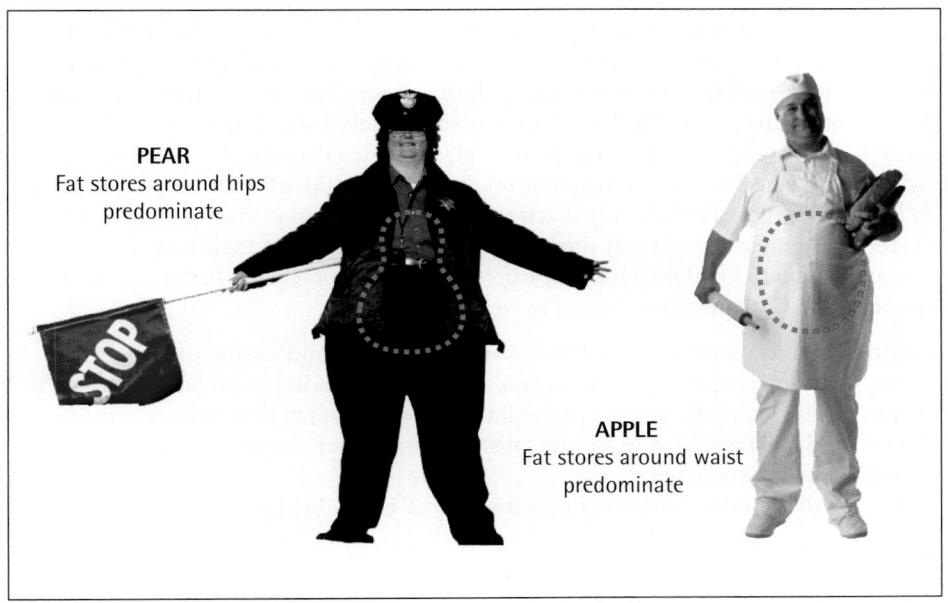

PEAR
Fat stores around hips predominate

APPLE
Fat stores around waist predominate

Illustration 9.7 Basic body shapes.
The pear normally has narrow shoulders, a small chest, and an average-size waist. Fat is concentrated in the hips, upper arms, and thighs.
Apples are round in the middle. The apple shape is riskier than the pear shape due to the presence of visceral fat.

PhotoDisc

achieved if both aerobic (walking, jogging, swimming, gardening) and resistance (strength training) exercises are part of the program.[23] Weight loss combined with exercise is the bonus pack for large reductions in risks.[27]

Visceral Fat and Waist Circumference The size of visceral fat deposits can be closely estimated by measuring waist circumference (Illustration 9.8).[24] In men, waist circumferences over 40 inches (102 cm), and in women, over 35 inches (88 cm) are related to excess visceral fat.[25] Waist circumference may not accurately estimate visceral fat content in large, muscular individuals, however.[24]

Because waist circumference is such a strong indicator of disease risk, its measurement in clinical practice is being encouraged.[28] This is particularly true in Japan due to the exaggerated effects of waist circumference on health risk in Asians.[70] Japan is attempting to prevent and control the country's quickly growing rates of obesity and health care expenditures by populationwide screening of central body fat. The country has instituted a policy that requires all citizens aged 50–74 years (that's 56 million people) to have their waist circumference measured yearly (Illustration 9.9). Individuals with high circumferences and weight-related health problems are given dieting guidance.[26]

Illustration 9.8 Determining your waist circumference.

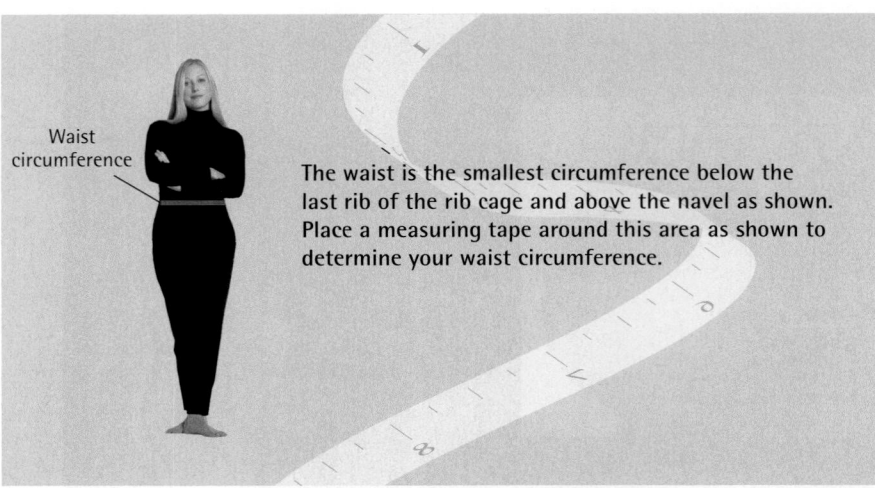

Waist circumference

The waist is the smallest circumference below the last rib of the rib cage and above the navel as shown. Place a measuring tape around this area as shown to determine your waist circumference.

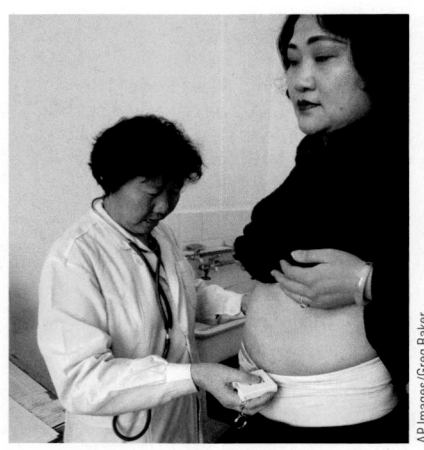

AP Images/Greg Baker

Illustration 9.9 Waist circumference measurements are required annually in Japan for adults ages 50–74 years.

Assessment of Body Fat Content

Body mass index is commonly used to approximate body fat content because the two measures correspond closely in groups of people. This is not always the case for individuals. Take a 165-pound, 30-year-old woman who is 5 feet, 6 inches tall and weight trains heavily. Her body weight for height would indicate obesity, but she may have a normal body fat content and be healthy. If BMI were used to assess body fat levels in professional football players, many of them would be wrongly categorized as obese.[24] Sometimes people who are classified as normal weight or underweight by BMI standards have too much body fat because they are physically inactive. Certain medications make people retain fluid. Their weight-for-height may qualify them as overweight, but their body fat content may actually be very low. Obviously, measures of body fat content are better estimators of health status than are measures related to weight for height.

Methods for Assessing Body Fat Content Accurate and inexpensive tools for assessing body fat content are available, and their use is spreading. Standards for classifying percent body fat, or the percent of weight that consists of fat, have been developed (Table 9.6) and will be refined as additional studies on the relationships between body fat and health risk are conducted.

Here are the most common methods for determining body fat content:

- Skinfold thickness measures

- Bioelectrical impedance analysis (BIA)

- Underwater weighing

- Magnetic resonance imaging (MRI)

- Dual-energy X-ray absorptiometry (DEXA)

- Whole body air displacement

The theories underlying each of these methods and its advantages and limitations are presented in Table 9.7. The tests are most likely to provide accurate results when they are performed by skilled, experienced technicians using proper, well-maintained equipment, and when the measurements are converted into percent body fat by the appropriate formulas.

Table 9.6 Percentages of body weight as fat considered low, average, and high[19]

	Percent body fat		
	Low	Average	High
Women	less than 12	32	35 or more
Men	less than 5	22	25 or more

Table 9.7 Commonly used methods for assessing body fat content

Skinfold measurement

Tom Pantages

Body fat content can be estimated by measuring the thickness of fat folds that lie underneath the skin. Calipers are used to measure the thickness of fat folds, preferably over several sites on the body. Body fat content is estimated by "plugging" the thicknesses into the appropriate formula.

Advantages: Calipers are relatively inexpensive (they cost about $250–$350). The procedure is painless if done correctly and can yield a good estimate of percent body fat.

Limitations: This measurement is often performed by untrained people. Skinfolds may be difficult to isolate in some individuals.

Bioelectrical impedance analysis (BIA)

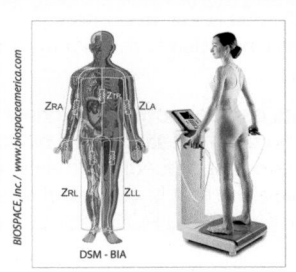

BIOSPACE, Inc./ www.biospaceamerica.com

Because fat is a poor conductor of electricity and water and muscles are good conductors, body fat content can be estimated by determining how quickly electrical current passes from the ankle to the wrist.

Advantages: The equipment required is portable, and the test is easy to do and painless. The results are fairly accurate for people who are not at the extremes of weight-for-height.

Limitations: Equipment may be expensive; inferior equipment produces poor results. Hydration status and meal ingestion may affect electrical conductivity and produce inaccurate results, as may inaccurate formulas used to calculate body fat content from test results (The InBody device shown to the left, however, is advanced and the technology employed overcomes many of the limitations of other BIA devices *through Direct Segmental Multi-Frequency BIA*). May cost $20 at fitness centers.

Underwater weighing

© Yoav Levy/Phototake

The subject is first weighed on dry land; next he or she is submerged in water and exhales completely; then his or her weight is measured. The less the person weighs under water compared to the weight on dry land, the higher the percent body fat. (Fat, but not muscle or bone, floats in water.)

Advantages: If undertaken correctly and if appropriate formulas are used in calculations, this technique gives an accurate value of percent body fat.

Limitations: The equipment required is expensive and not easily moved. The test doesn't work well for people who don't swim or who are ill or disabled in some way. A body fat assessment costs $50 or more.

Magnetic resonance imaging (MRI)

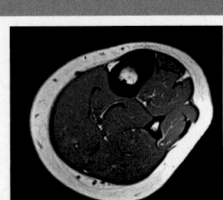

Needell M.D./Custom Medical Stock Photo, Inc.

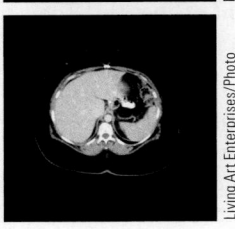

Living Art Enterprises/Photo Researchers, Inc.

MRI and CT or CAT scans provide similar results. Using this technology, a person's body fat and muscle mass can be photographed from cross-sectional images obtained when the body, or parts of it, is exposed to a magnetic field (or, in the case of CAT scans, to radiation). Based on the volume of fat and muscle observed, total fat and muscle content can be determined. Liver fat content is shown on the MRI scan to the left.

Advantages: Provides highly accurate assessment of fat and muscle mass.

Limitations: The test is expensive (around $1,000 per assessment) and largely used for clinical and research purposes.

Continued on next page

Table 9.7   Continued

Dual-energy X-ray absorptiometry (DEXA)		
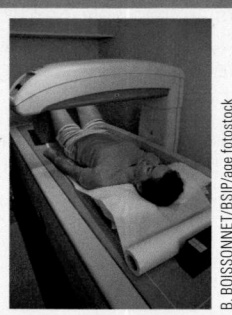	DEXA (or DXA) is based on the principle that various body tissues can be differentiated by the level of X-ray absorption. The measure is made by scanning the body with a small dose of X-rays (similar to the level of exposure from a transcontinental flight) and then calculating body fat content based on the level of X-ray absorption.	*Advantages*: Provides highly accurate results when measurements are undertaken correctly. DEXA is safe and user friendly for people being measured and can also be used to assess bone mineral content and lean tissue mass. *Limitations*: DEXA machine is expensive, the cost of individual assessments is about $500 per person. Machine must be operated by a trained and certified radiation technologist in many states.
Whole body air displacement		
	This is an established method that has become practical for broader use due to development of the BOD POD. The method is similar to that of underwater weighing but uses air displacement for determining percent body fat. Individuals sit in an enclosed "cabin" for about five minutes while wearing a tight-fitting swim suit and cap. Computerized sensors determine body weight and the amount of air that is displaced by the body. The PEA POD is used for assessing body fat in infants.	*Advantages*: This method provides a quick, comfortable, automated, and reasonably accurate way to assess body fat content. It is suitable for disabled individuals, the elderly, and children. *Limitations*: Results can be modified by drinking, eating, or exercising before testing; a full bladder or failure to adhere to test procedures may lead to erroneous results. The test is costly but less expensive than DEXA. A BOD POD assessment costs around $50.

B. BOISSONNET/BSIP/age fotostock

Courtesy, Live Measurement, Inc., www.bodpod.com

Everybody Needs Some Body Fat A certain amount of body fat—3 to 5% for men and 10 to 12% for women—is needed for survival. Body fat serves essential roles in the manufacture of hormones; it's a required component of every cell in the body, and it provides a cushion for internal organs. Fat that serves these purposes is not available for energy formation no matter how low energy reserves become. Low body fat levels are associated with delayed physical maturation during adolescence, infertility, accelerated bone loss, and problems that accompany starvation.[29]

What Causes Obesity?

Simply stated, obesity results when the intake of calories exceeds caloric expenditure. But the cause of obesity is not that simple. Whether people accumulate excess body fat or not is due to complex and interacting factors that include:

- Diet

- Physical activity

- Environmental exposures

- Genetic traits[30]

Some medications such as antipsychotics, antidepressants, insulin, and beta-blockers used for hypertension are also associated with the development of obesity.[31] Their contribution to the overall incidence of obesity is small (although meaningful to the affected people).

Are Some People Born to Be Obese? With rare exception, people are not genetically destined to become obese. The current epidemic of obesity is primarily driven by environmental factors and not by our genes.[32] However, genetic traits inherited at birth and changes in gene function acquired early in life can influence a person's susceptibility to becoming obese.

People are born with a specific set of genes and that does not change throughout life. However, the function of certain genes can change based on early life exposures, or "**environmental triggers**." For example, small size at birth related to maternal malnutrition is associated with alterations in gene function that affects insulin production and increases the risk of obesity in the offspring.[67] Exposure to high saturated fat diets is related to gene functions that increase food intake and some people's weight.[68] For other people, changes in gene function accompany reduction in physical activity level and trigger overeating and weight gain. When food availability is not limited, gene functions in some individuals lead to changes in appetite hormone production, a delayed sense of satiety, and overeating.[32,69]

environmental trigger An environmental factor, such as inactivity, a high-fat diet, or a high sodium intake, that causes a genetic tendency toward a disorder to be expressed.

Do Obese Children Become Obese Adults? The link between weight status very early in life and adult obesity appears to be weak or nonexistent. The relationships of early weight status to later obesity becomes meaningful among older children and adolescents, especially if one or both parents are obese.[33–35] Only 8% of obese children who are heavy at one to two years of age and who do not have an obese parent are obese as adults. However, nearly 80% of children who are obese between the ages of 10 and 14 and have at least one obese parent have been found to be obese as adults.[62]

The Role of Diet in the Development of Obesity

Americans are consuming more calories from food than in previous decades and the increase accounts for the rise in rates of obesity.[36] Fruits, vegetables, and fiber are often missing from diets that provide too many high-calorie, energy-dense foods.[37] The generous availability of inexpensive, energy-dense foods, eating out regularly at fast-food restaurants and all-you-can-eat buffets, and large portions of foods tend to increase calorie intake.[38] Adults served large portions of food consume an average of 30–50% more food than when presented with small food portions.[39]

Low Levels of Physical Activity Promote Obesity Low levels of physical activity are related to the development of obesity among Americans.[41] To a large extent, physical activity has become voluntary. Many farmers now plow, sow, and reap in air-conditioned tractors with power steering. Lawn mowers propel themselves and sometimes their operators, too. Instead of walking or biking two and a half miles to the store and back, we drive. An average-size adult driving 2.5 miles burns about 17 calories. Biking that distance would use seven times more calories (122). But walking 2.5 miles would burn around 210 calories, more than 12 times the calories it takes to drive! We have traded physical activity for convenience, time, and, perhaps, personal fat stores. Too much television watching, in particular, has been cited as encouraging obesity development in children.[42]

Obesity: The Future Lies in Its Prevention

Whether obesity is related to genetic predisposition, environmental factors, or a combination of both, certain steps can be taken to help prevent it.

Preventing Obesity in Children For children, the prevention of obesity includes the early development of healthy eating and activity habits. Parents should offer a nutritious selection of food, but children themselves should be allowed to decide how much they eat.[43] Physical activities that are fun for every child—not just those who show athletic promise—should be routine in schools and summer programs.

Interactions between parents and children around eating and body weight can set the stage for the prevention or the promotion of overweight in children. Parents who overreact to a child's weight by focusing on it, restricting food access, and making negative comments to the child may increase the likelihood that eating and weight problems will develop or endure. Lifestyle changes for the whole family—such as incorporating fun physical activities into daily schedules; making a wide assortment of nutritious foods available in the home; and decreasing a focus on eating, foods, and weight—are some of the positive changes families can make to promote healthy eating and exercise habits, and normal weight in children.[43]

Preventing Obesity in Adults Action needs to be taken to prevent weight gain during the adult years. Many adults gain weight at a slow pace (about a pound per year) as they age, whereas others gain substantial amounts of weight over short periods of time.[44] Data from a national nutrition and health survey indicate that major gains in weight are most likely to occur in adults between the ages of 25 and 34 years.[45] Regular exercise may prevent or lessen the amount of weight gain that occurs with age,[46] as may decreased portion sizes at home and in restaurants.[47] Paying attention to the "I'm hungry" and "I'm full" signals can help moderate food intake.[40] For some people, regularly getting eight hours of sleep at night appears to reduce weight gain.[48]

Changing the Environment Environmental changes in food portion sizes, the increased availability of energy-dense, inexpensive foods, and sedentary lifestyles are among the key factors that underlie the current obesity epidemic.[32] From a public health point of view, it is being reasoned that if environmental changes got us into the obesity epidemic, then environmental changes will help get us out of it.[49]

Many communities are taking action to promote healthy environments. These actions include limiting access to "junk" foods in schools, building exercise parks, and developing community gardens. Sidewalks, bicycle and walking paths, and nature trails are being planned or added in urban residential areas.[32] Fast-food and other restaurants are gradually making changes toward selling smaller portions of energy-dense foods and offering more nutrient-dense items. Some restaurants are adding a wider selection of not-fried entrées and half portions to their menus. The general trend in weight- and health-friendly environmental change is toward providing individuals and families with widespread opportunities for making healthful choices.[51]

Illustration 9.10 Some underweight people are genetically thin and are healthy.

Jane Holmes/Shutterstock.com

Underweight

- **Delineate the potential causes and health consequences of underweight.**

Illustration 9.11 The appearance of one man who follows a calorie-restricted diet.

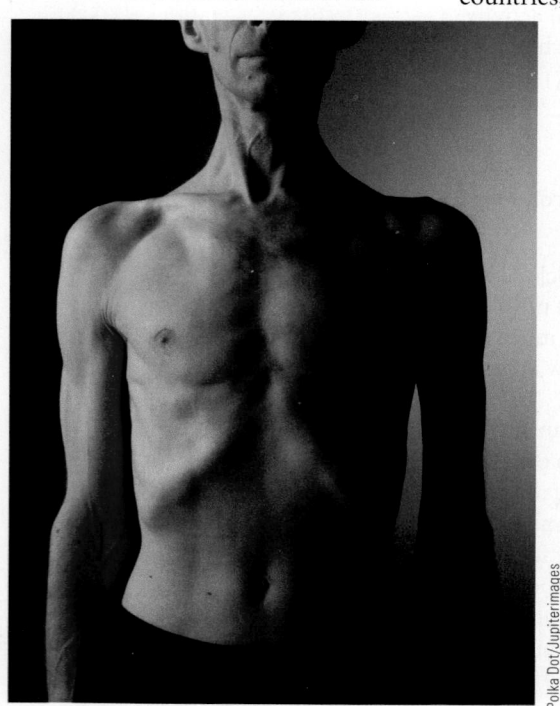

Polka Dot/Jupiterimages

In contrast to the desperate situations faced by many people in economically emerging countries, underweight in developed nations largely results from illnesses such as HIV/AIDS, pneumonia, cancer, or an eating disorder (anorexia nervosa). An important cause of underweight in some poor nations is poverty.[2,3]

Underweight Defined

People who are underweight have too little body fat, or less than 12% body fat in adult females and 5% body fat in males.[52] They have BMIs below 18.5 kg/m² as indicated in Table 9.1. A portion of the 2% of U.S. adults classified as underweight by BMI will not actually be underweight, just as a subset of people categorized as obese by BMI are not really obese.

Some people assessed as underweight for height are healthy and have a normal body composition. Like the person in Illustration 9.10, they are probably genetically thin. People who are naturally thin often have as much difficulty gaining weight as obese people have losing it.[2,53] People who are thin and unhealthy are more likely than others to experience apathy, fatigue, and illnesses frequently, and they take longer to recover from illness. They tend to have reduced bone mineral density and more bone fractures, be intolerant of cold temperatures, and have impaired concentration.[54]

Underweight and Longevity in Adults

Longevity can be extended in adult mice and monkeys by feeding them a nutritious, calorie-restricted diet that produces underweight.[55] Although it is not known whether caloric restriction and underweight would serve as a fountain of youth for humans,[55] there are groups of adults who believe that it will. Devotees of calorie restriction for longer life tightly control their food intake and activity level to maintain a lean body. Calorie restrictors tend to consume 1,600 to 1,700 calories per day, emphasize nutrient-rich foods, and exercise regularly.[63] In general, they appear thin but are healthy (see Illustration 9.11).

Adults following calorie-restricted diets tend not to have chronic inflammation nor the health problems associated with it. They are at risk of iron deficiency, osteoporosis, infertility, infections, and of becoming irritable. The theory that life expectancy would be increased by calorie restriction is challenged by data on humans that show increased disease and death rates among underweight compared to normal weight people.[6] Several long-term studies of health outcomes related to calorie restriction in men are underway.

Toward a Realistic View of Body Weight

- **Demonstrate an awareness of the roles of weight bias on psychological and other aspects of the health status of children and adults.**

A widespread belief among Americans is that individuals can achieve any body weight or shape they desire if they just diet and exercise enough. It's a myth. People naturally come in different weights and shapes, and these can only be modified so much (Illustration 9.12). Half of all women in the United States wear sizes 14 to 26, yet many clothing models are very underweight. Many men, no matter how hard they work out, will never have a washboard stomach or fit into slim jeans.

Illustration 9.12 People come in many different sizes and shapes.

Size Acceptance

"The war is on obesity, not on the obese."[64] The U.S. obsession with body weight and shape is spreading to other industrialized countries as part of popular culture. Ironically, strong societal bias against certain body sizes may contribute broadly to weight and health problems. Intolerance of overweight and obese children and adults tends to increase discrimination against them. This type of societal bias lowers the individual's feeling of self-worth and may promote eating disorders, including the consumption of too much food.[5] Females are hardest hit by negative attitudes about body size. Although the incidence of overweight and obesity in females and males is similar,[56] obesity in females carries with it many more negative stereotypes.[5]

Acceptance of people of different sizes and a more realistic view of obtainable body weights and shapes may be two of the most important things society can do to ameliorate harmful effects of obesity. Some changes are underway. Unproductive attitudes and harmful biases against heavier people by clinical practitioners is being address by various institutions. Health professionals are being urged to examine their own biases against individuals of various sizes and shapes by increasing their awareness of their personal biases. They are being asked to think about whether they do any of the following:

- Judge a person's competence or worth by their size.

- Blame people for their size even though many factors are involved and there are no easy solutions.

- Attempt to change their attitudes to become sensitive to the needs of obese people.[17,65]

The nearby Reality Check feature addresses the issue of weight bias.

REALITY CHECK
A Weighty Topic

On their way through the park Carlos and Sandra see two kids playing. One is pulling a wagon while the other is riding in it. The child pulling the cart is thin and the child riding it is plump. Here's the thought that flashed through Carlos's and Sandra's heads when they saw the kids:

Answers are below.

Carlos: The heavier kid should be pulling the wagon.

Sandra: I wonder what they're doing. They're probably taking turns pulling the wagon. We used to do that.

The Health at Every Size Program A kinder and more effective approach to health improvement in obese people has been developed. Called "Health at Every Size," the program is gaining acceptance among consumers and health professionals in the United States, Canada, and other countries.[5,57,58] The program emphasizes the following: eating when hungry and stopping when full; engaging in enjoyable, life-enhancing physical activities; accepting body size; using healthy eating behaviors; having a peaceful relationship with food; showing self-esteem; and using social support networks. The program reduces a number of health problems related to obesity even though it does not lead to weight loss in the short term. Health at Every Size program participation is associated with reduced blood pressure and LDL-cholesterol, increased HDL-cholesterol levels, improved self-esteem, decreased disordered eating behaviors, and improved body image.[59,60] The Health at Every Size philosophy and program components could help future generations of children, men, and women achieve healthful eating patterns and high levels of well-being and quality of life, as well as improved long-term health.[60]

ANSWERS TO REALITY CHECK
A Weighty Topic

You can't know why one child is pulling a wagon while another child rides in it by looking. There are many possible reasons for this besides the children's weight statuses.

Carlos:

Sandra:

NUTRITION
up close

Are You an Apple?

Focal Point: Determining your waist circumference.

Follow the directions given in Illustration 9.9 for determining your waist circumference. If you don't have a measuring tape, use a string. Mark the string where the two ends intersect when placed around your waist. Then use a ruler to measure the distance between the end and the mark on the string. Are you an apple?

Feedback to the Nutrition Up Close is located in Appendix G.

REVIEW QUESTIONS

- **Identify weight status based on weight, height, age (where appropriate), and sex using appropriate formulas and growth charts.**

1. Death rates are lowest among adults who remain underweight as they age.   **True/False**

2. According to the Hamwi method for estimating normal weight in adults, a 6 foot, 1 inch male should weigh approximately 204 pounds.   **True/False**

3. A child with a body mass index higher than the 95th percentile on the CDC growth charts is considered obese.   **True/False**

4. The prevalence of overweight and obesity in the United States has declined over the past decade.   **True/False**

5. Culturally defined standards for ideal body shape and size have not changed throughout the course of history.   **True/False**

6. Rates of obesity are substantially higher for females than for males in the United States.   **True/False**

7. _____ A 57-year-old person weighs 176 pounds and is 5 foot, 10 inches tall. The person's BMI would qualify as:

    a. underweight
    b. normal weight
    c. overweight
    d. obese

8. _____ A 6-year-old child's weight was measured and her BMI percentile was plotted on the appropriate growth chart. Her BMI is at the 25th percentile. She would be considered:

    a. underweight
    b. healthy weight
    c. overweight

- **Delineate the potential causes and health consequences of overweight and obesity.**

9. Excess visceral fat poses higher risks to health than excess subcutaneous fat.   **True/False**

10. Features of metabolic syndrome can include excess abdominal fat and insulin resistance.   **True/False**

11. Women but not men require a minimal amount of body fat stores for health.   **True/False**

12. If undertaken correctly, and if appropriate formulas are used, underwater weighing produces accurate values of a person's percent body fat.   **True/False**

13. The use of medications that cause weight gain is a leading reason for the obesity epidemic in the United States.   **True/False**

14. Vegetables, fruits, and fiber are often consumed in low amounts by people who regularly eat energy-dense, high-calorie foods.   **True/False**

15. Adults tend to eat 30–50% more food when given large portions of food than when given smaller portions.   **True/False**

16. _____ Which of the following statements about the potential causes of obesity is true?

    a. Obesity is caused primarily by genetic traits.
    b. Obesity is related primarily to lack of self-esteem and control.
    c. Low levels of physical activity are the main cause of obesity.
    d. The current epidemic of obesity is related primarily to environmental factors.

17. _____ Which of the following statements does *not* represent an effective approach to the prevention of weight gain in adults?

    a. Consume smaller portion sizes.
    b. Pay attention to satiety signals.
    c. Sleep at least nine hours daily.
    d. Get regular physical activity.

- **Delineate the potential causes and health consequences of underweight.**

18. Calorie-restricted diets have been shown to prolong life in mice, monkeys, and humans.   True/False

19. Unlike obesity, underweight has been related to specific genetic traits.   True/False

20. Underweight is associated with an increased risk of illness and delayed recovery from illnesses.   True/False

- **Demonstrate an awareness of the roles of weight bias on psychological and other aspects of the health status of children and adults.**

21. Health care professionals intentionally show bias against obese people because this behavior has been shown to encourage people to lose weight.   True/False

22. People who lack an ideal body shape and size lack the willpower to stay on a diet and exercise program.   True/False

23. A healthy diet and regular exercise helps to improve the health status of people of all sizes.   True/False

Answers to these questions can be found in Appendix G.

NUTRITION SCOREBOARD ANSWERS

1. Cultural norms of ideal body weights are often at odds with the "healthy" weights defined by science.   False

2. Health is affected by the location of body fat stores as well as by obesity.[1] Adults who store body fat in the stomach, or central area of the body, are at higher risk for a number of health problems than are adults who store fat primarily in their hips and thighs.   True

3. Many factors contribute to the development of obesity. However, the basic cause of obesity is calorie intakes that exceed calorie needs.   True

4. Gaining weight is as difficult for many underweight people as losing weight is for many overweight people.[2]   True

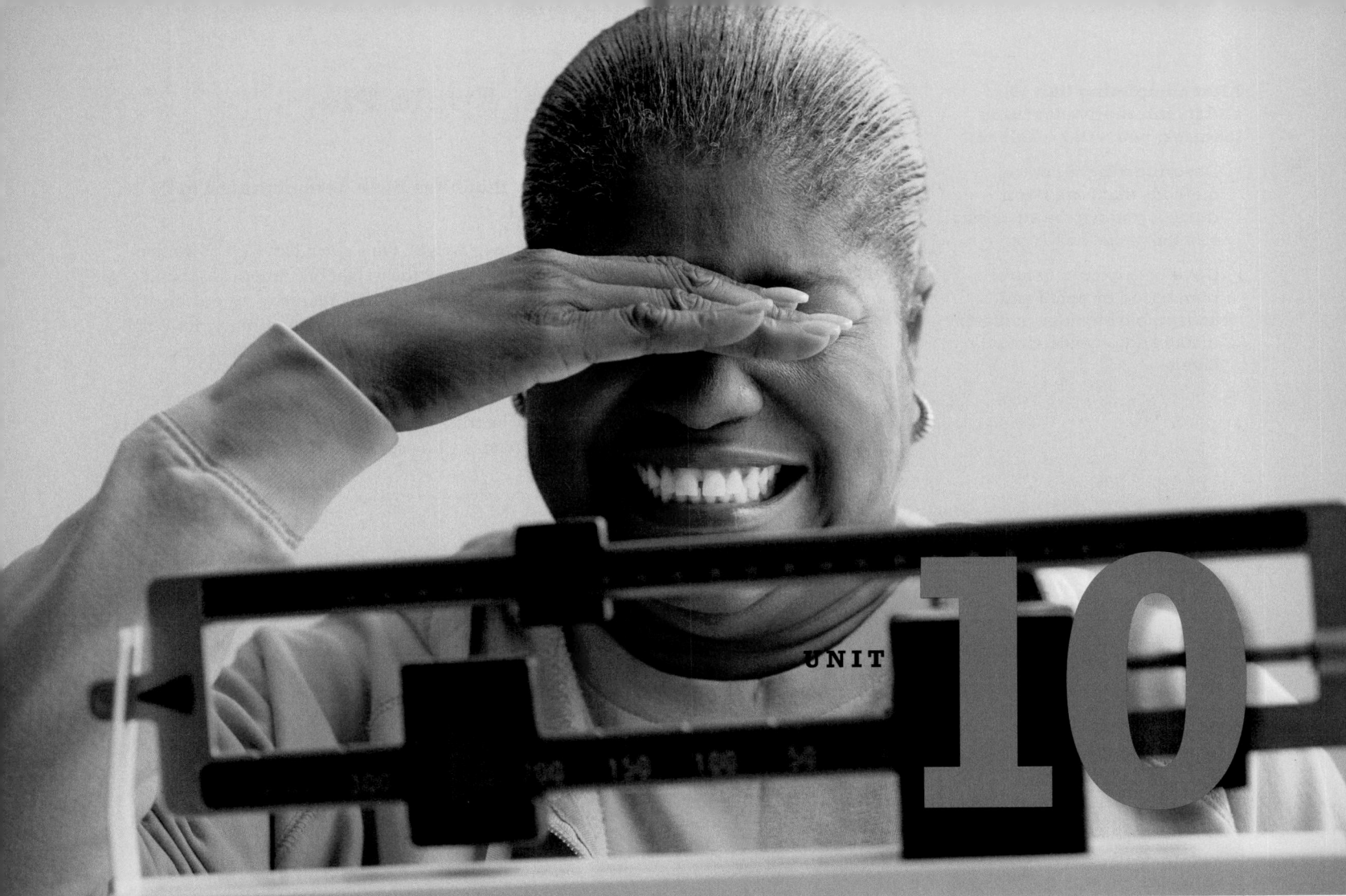

Weight Control: The Myths and Realities

NUTRITION SCOREBOARD

1 Anybody who really wants to can lose weight and keep it off. **True/False**

2 Weight loss is the cure for obesity. **True/False**

3 Weight loss can be accomplished using many types of popular diets. Such diets rarely help people maintain the weight loss in the long run. **True/False**

4 Weight-loss products and services must be shown to be safe and effective before they can be marketed. **True/False**

5 Small and acceptable improvements in eating and exercise behaviors are more likely to produce weight loss and weight-loss maintenance than are large and unpleasant changes in behaviors. **True/False**

Answers can be found at the end of the unit.

After completing Unit 10 and its interactive learning features, you will be able to:

- Describe weight-control methods that have been demonstrated to be successful and those that have not.

- Develop a weight-control plan based on small and acceptable changes in dietary intake and physical-activity level.

Baseball, Hot Dogs, Apple Pie, and Weight Control

- **Describe weight-control methods that have been demonstrated to be successful and those that have not.**

Americans in general are preoccupied with their weight. On a given day, 57% of women and 40% of men are trying to lose weight.[3] To help get the weight off, Americans spend over *$60 billion* annually on weight-loss products and services.[4] Yet many Americans are gaining weight faster than they are losing it. Consumers are paying handsomely for weight loss without experiencing the desired cosmetic changes or health benefits that come when weight loss is maintained.

Why do so many Americans fail at weight control? For some people, achieving permanent weight reduction on their own may be truly impossible. For others, the problem is the ineffective methods employed, not the people who use them.

Key Nutrition Concepts

Information presented in this unit on weight control closely relates to these three key nutrition concepts:

1. Foods provide energy (calories), nutrients, and other substances needed for growth and health.

2. Poor nutrition can influence the development of certain chronic diseases.

3. Adequacy, variety, and balance are key characteristics of healthful diets.

Illustration 10.1 Weight-loss books and products.

© Scott Goodwin Photography

Weight Loss Versus Weight Control

Dietary torture and quick weight loss are not the cure for overweight. If losing weight was all it took to achieve the cultural ideal of thinness, nobody would be overweight. Any popularized approach to quick weight loss (and some get pretty spectacular) that calls for a reduction in caloric intake can produce weight loss in the short run. These methods fail in the long run, however, because they become too unpleasant. Feelings of hunger and deprivation that tend to occur while on a rapid weight-loss diet can lead to a return to previous habits and weight regain.[5] Humans are creatures of pleasure and not pain. Any painful approach to weight control is bound to fail. Improved, enjoyable, and sustainable eating and exercise experiences are needed to keep excess weight off.[6] Quick weight-loss approaches don't promote sustainable lifestyle changes required for weight control for the long term.[7]

Unfortunately, many dieters think weight-control methods are successful if they lead to a rapid loss of weight.[8] When the lost weight is regained, dieters tend to blame themselves and not the faulty method. Blaming themselves for the failure, many people try other quick weight-loss methods. But they usually fail, too.[9]

The Business of Weight Loss Thousands of weight-loss products and services are available. Some of these are shown in Illustration 10.1. Most of them either don't work at all or don't prevent weight regain. "Quick-fix" weight-loss approaches that don't lead to long-term changes in behavior are the primary reasons approximately 90–95% of people who lose weight quickly gain some or all of it back.[10] The demand for such products and services is so great, however, that many are successful—financially.

One reason so many weight-loss products and services are available is that almost none of them work. If any widely advertised approach helped people lose weight and keep it off, manufacturers of bogus methods would go out of business. The weight-loss industry also thrives because of the social pressure to be thin. Many people try

Table 10.1 A brief history of discontinued weight-loss methods

Year	Method	Reasons for discontinuation
1940s–1960s	Amphetamines	Highly addictive, heart and blood pressure problems
	Vibrating machines	Did not work
1950s	Jejunoileal; bypass surgery	Often caused chronic diarrhea, vitamin and mineral deficiencies, kidney stones, liver failure, arthritis
1960s	Liquid protein diet	Poor-quality protein caused heart failure, deaths
1980s	Intestinal bypass surgery	Excessive risk of serious health problems
1990	Oprah Winfrey liquid diet success (she lost 67 pounds)	
1991	Oprah Winfrey gains 67 pounds and declares "No more diets!"	
1997	Phen-Fen (Redux)	Heart valve defects, hypertension in lung vessels
2004	Ephedra	Excessive risk of stroke, heart attack, and psychiatric illnesses
2010	Sibutramine (Merida)	Excessive risk of stroke and heart attack

new weight-loss methods even though they sound strange or too good to be true. Often people believe that a product or service must be effective, or it wouldn't be allowed on the market. Although reasonable, this belief is incorrect.

The Lack of Consumer Protection The truth is that general societal standards for consumer protection do not apply to the weight-loss industry.[11] No laws require products to be effective; in most cases, companies do not have to show that weight-loss products or services actually work before they can be sold. That's why products like herbal remedies, forks with stop and go lights, weight-loss skin patches, colored "weight-loss" glasses, electric cellulite dissolvers, mud and plastic wraps, and inflatable pressure pants are available for sale. These, like many other weight-loss products, do not work, and there is no reason why they should. Promotions for the products just have to make it *seem* as though they might work wonders on appetite or body fat.

Furthermore, weight-loss products and services are usually not tested for safety before they reach the market. The history of the weight-loss industry is littered with abject failures: fiber pills that can cause obstructions in the digestive tract, very low-calorie diets and intestinal bypass surgeries that may lead to nutrient-deficiency diseases and serious health problems, amphetamines that can produce physical addiction, diet pills that have caused heart valve problems, and liquid protein diets that have led to heart problems and death from heart failure.[12] A brief history of weight-loss method failures is chronicled in Table 10.1.

Pulling the Rug Out Fraudulent weight-loss products and services can be investigated and taken off the market. Currently, the Federal Trade Commission (FTC) monitors deceptive practices and weight-loss claims on a case-by-case basis. The FTC has filed suits against companies that make exaggerated claims and has identified the most common dubious claims made by the industry (Table 10.2).[13] The FTC performs most investigations in response to consumer complaints; the investigations are not proactive.[14] Illustration 10.2 shows three examples of products that were taken off the market and explains why.

In 2011, the FTC took action against 40 bogus weight-loss products, one of which was for hoodia, a supplement marketed as an appetite suppressant. It was ordered off the market because it does not work, and false and misleading claims were made for its effectiveness.[17] Another product, HCG, a hormone produced by the placenta during pregnancy, was widely advertised as promoting 20–30 pound weight loss in 30–40 days. Users were to follow a 500-calorie-per-day diet while using HCG. Both the HCG and the 500-calorie-a-day diet were deemed dangerous the FDA. The product was taken off the market in 2011. HCG does not promote weight loss and 500-calorie-a-day diets require close medical supervision due the risk of irregular heartbeat and death.[18]

Table 10.2 The top seven features of weight-loss ads that make false or misleading claims.[15,16]

1. Use testimonials, before-and-after photos.
2. Promise rapid weight loss.
3. Require no special diet or exercise.
4. Guarantee long-term weight loss.
5. Include a "clinically proven," "guaranteed," "scientific breakthrough," or "doctor-approved" statement.
6. Make a "safe," "natural," or "easy" claim.
7. Are marketed by mass e-mails.

Fat magnet pills were purported to break into thousands of magnetic particles once swallowed. When loaded with fat, the particles simply flushed themselves out of the body. The FTC found the product too hard to swallow. The company took the product off the market and made $750,000 available for customer refunds.

© Scott Goodwin Photography

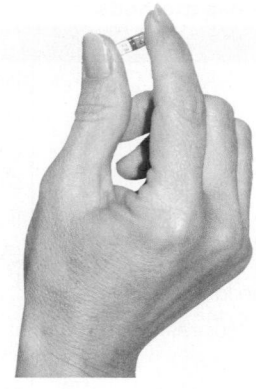

"Blast" away 49 pounds in less than a month with Slim Again, Absorbit-All, and Absorbit-AllPlus pills, claimed ads in magazines, newspapers, and on the Internet. The company got blasted by the FTC to the tune of $8 million for making fraudulent claims.

Stan Maddock

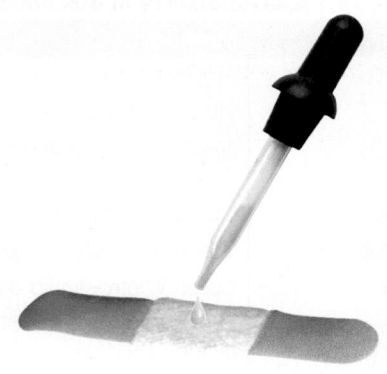

Take a few drops of herbal liquid, put them on a bandage patch, and voilà—a new weight-loss product! The herbal liquid was supposed to reach the appetite center of the brain and turn the appetite off. Federal marshals weren't impressed. They seized $22 million worth of patch kits and banned the sale of others.

© Scott Goodwin Photography

Illustration 10.2 Examples of bogus weight-loss products.

Requirements for truth in labeling keep many bogus products from including false or misleading information on their labels. These laws, however, do not keep outrageous claims from being made in infomercials, mass e-mails, on the Internet, or printed in newspapers, books, magazines, and advertisements.[19]

What if the Truth Had to Be Told? Suppose the weight-loss industry had to inform consumers about the results of scientific tests of the effectiveness and safety of weight-loss products and services. What if this information had to be routinely included on weight-loss product labels (Illustration 10.3)? What do you think would happen to consumer choices and the weight-loss industry? Increased federal enforcement of truth-in-advertising laws may help put an element of honesty into promotions for many weight-loss products. In addition, the FTC has proposed that claims about long-term weight loss "must be based on the experience of patients followed for at least two years after they complete the [weight-loss] program."[20] If this proposal ever becomes law, it will change the weight-loss industry in the United States.

Illustration 10.3 Just imagine what would happen if weight-loss approaches were required to divulge their effectiveness and risks.

Consumer Protection Label

Effectiveness: Ineffective for 95% of users

Risks: Bee pollen may cause adverse reactions in allergic people. Weight regain may exceed weight loss. If the diet fails, people may blame themselves, thereby precipitating anxiety and self-doubt.

The Amazing BEE POLLEN-GRAPEFRUIT DIET

Dr. Quickfix

MELTS AWAY POUNDS OF CELLULITE IN JUST 10 DAYS!

Weight Control

- **Develop a weight-control plan based on small and acceptable changes in dietary intake and physical-activity level.**

People usually begin weight-loss efforts with the goal of losing weight and maintaining the new, lower body weight forever. What programs help individuals do that?

The only way to know whether a weight-control method, drug, product, or surgery works is to test its safety and effectiveness in scientific studies. In general, weight-control methods found to be safe that lead to an average of 5% or greater loss of initial body weight among participants over a two-year period are identified as being successful.[22-24] Translated into pounds of weight loss, this means that a group starting a weight-loss program with an average weight of 200 pounds would need to lose an average of 10 pounds or more in two years to obtain a 5% or greater average weight loss. Weight-control approaches identified as successful do not guarantee that everyone who uses the approach will lose 5% or more of their body weight and keep it off for two years, but that some will.

Table 10.3 Examples of fad diets

1. *The Popcorn Diet.* Like popcorn? You'll get to eat all the unbuttered, unsalted popcorn on this diet you want. The diet also emphasizes fruits, vegetables, smaller portions, and exercise.

2. *The Grapefruit Diet.* This diet is based on the myth that, when combined with protein, grapefruit triggers fat burning and weight loss. The high grapefruit–high protein diet has been around since the 1930s.

3. *The Chocolate Diet.* A weight-loss diet for chocolate addicts, this nonsense diet categorizes chocolate lovers into types and provides a diet plan for each type. Includes a liquid chocolate diet shake.

4. *The Metabolism Diet.* As the name implies, this diet purports to produce weight loss by speeding up metabolism. The low-carbohydrate, low-calorie diet recommended can only be used for seven days.

5. *The Three-Day Diet.* If you follow the food prescriptions for what to eat, how much to eat, and when to eat, you'll supposedly be rewarded with ramped-up metabolism and a 10-pound weight loss in three days. But wait, there's more. The diet also provides internal cleansing, cholesterol reduction, and more energy.

6. *The Cabbage Soup Diet.* It's a seven-day weight-loss plan based on cabbage soup with other vegetables.

7. *Japanese Morning Banana Diet.* You get to eat all the bananas you want on this diet— and nothing else.

Loss of 5% or more of body weight in obese individuals is considered clinically significant. Losses of this magnitude are associated with lowered blood concentrations of glucose, insulin, triglycerides, cholesterol, and markers of inflammation. They are also associated with reduced blood pressure, risk of type 2 diabetes, fatty liver disease, and heart disease. The magnitude of risk reduction increases as body weight becomes closer to normal weight.[22–24]

Weight-Control Methods: The Evidence

Most weight-control products and services available to consumers have not been evaluated for safety or effectiveness. This situation is the case for most all fad diet books available in bookstores and on the Internet, and weight-loss products sold on store shelves and over the Internet. Table 10.3 provides examples of the types of weight-loss approaches offered by fad diets that have a very slim chance of working in the long run, primarily because they don't change behavior. Although many of the approaches that call for calorie restriction lead to weight loss in the short run,[9] that is not the measure of a successful weight-control method. Successful measures help people keep the weight off.[26]

Popular Diets Five popular diets that have undergone scrutiny by researchers are Dr. Atkins's New Diet Revolution, Weight Watchers Pure Points Program, the Slim-Fast Plan,

REALITY CHECK
Do Some Foods Have Negative Calories?

You're looking for a way to shed the 10 pounds you gained during your freshman year and are seriously thinking about adding grapefruit and vinegar to your diet. You have heard these foods have "negative calories" because they make the body burn fat.

Answers on page 10-6

Natalie: I know vinegar cleans the grease off windows. Maybe it will melt away my fat, too.

Chuck: My grandfather lives in Florida and eats a lot of grapefruit. He's as thin as a rail.

The negative calorie myth surrounding vinegar or grapefruit lives on because it strikes many people as reasonable. Neither vinegar nor grapefruit, nor any other food, has negative calories or causes the body to gear up metabolism and burn fat.[21]

Natalie:

Chuck:

the Zone Diet, Ornish diet, and Rosemary Conley's Eat-Yourself-Slim Diet. Studies have compared nutrient composition, weight loss, adherence to the diet, and/or health effects of these diets.[24,27,28]

Approaches to weight loss described in these books represent low-carbohydrate, high-protein (see Illustration 10.4), or low-fat diets. All of the diets were found to produce a reduction in calorie intake if followed and weight loss. Several of the diets were found to be low in certain vitamins and minerals.[28] Weight loss varied by how long a person stayed on the diet and averaged about seven pounds for five of the diets. As has been found in other studies, calorie restriction, rather than dietary composition of protein, carbohydrate, or fat influenced the amount of weight lost.[29] Most of the people in the studies were unable to stick to the diet for a year, primarily because it was too hard to follow.[24,27]

Internet Weight-Loss Products Weight-loss products sold over the Internet are particularly untrustworthy. Other than an FDA-approved weight-loss drug (orlistat, also known as Alli or Xenical) that can be purchased without a prescription, there is no dietary supplement or medication sold over the Internet that is currently approved by the FDA as safe or effective for weight loss.[30,31] Some of the products sold as orlistat over the Internet have been found to contain another, potentially dangerous drug.[31] A range of the fraudulent products advertised on the Internet are also available in stores.[16]

Efforts are underway to limit the availability of bogus weight-loss products. The Federal Trade Commission (FTC) issues red flag alerts for bogus weight-loss products advertised over the Internet and is requesting consumer help in identifying such products (Illustration 10.5).

Illustration 10.4 An example meal from a low-carbohydrate, high-protein diet.

Organized Weight-Loss Programs A number of for-profit weight-loss programs that provide diet and exercise guidance and support, or structured programs using behavioral counseling and prepackaged meals and snacks, are available to consumers. Two of the most popular, Weight Watchers and Jenny Craig, have been evaluated in research studies. Participants in the research were given free membership, lifestyle counseling, support group participation, or free meals and diet and exercise counseling for one to two years. Over a year's time, individuals participating in the Weight Watchers program lost an average of 14.6 pounds versus an average loss of 7.2 pounds by participants receiving standard weight-loss care in primary health care settings.[32] Participants in the Jenny Craig program lost an average of 16.3 pounds over two years (7.9% of initial body weight) compared to a 4.4 pounds loss (2.1% loss in initial body weight) among participants in standard weight-loss counseling.[33] It is not known whether the same results would be achieved if participants in real life had to pay for the services and meals.[7]

Successful Weight-Control Programs

Two approaches to weight control are related to weight loss and weight-loss maintenance. One approach is lifestyle-based programs and the other is weight-loss surgery. Diet drugs cannot be evaluated separately for weight control because it is required that they be used with diet, exercise, and behavioral counseling.[35]

Lifestyle Programs Weight-control programs based on individually tailored, sustainable lifestyle changes that focus on behavioral strategies for reducing calorie intake and increasing physical activity have been demonstrated to be successful.[36] Client-centered lifestyle programs generally offer positive and supportive counseling that encourage and motivate participants to adopt healthful eating and exercise lifestyles the participant will find rewarding. People participating in these programs who are highly motivated to lose weight and keep it off are most likely to succeed.[37] An example of such a program is one that offered monthly lifestyle coaching sessions with goal setting, behavior change strategy development, and follow-up sessions to evaluate and fine-tune personal approaches.[38]

Studies of client-centered lifestyle programs for weight control show that many lead to average weight loss of 5% or more in a third of participants followed for one or two years.[34,36,38] For many participants, weight loss achieved is sufficient to improve blood cholesterol and blood pressure levels, and to decrease inflammatory markers.[23]

Physical Activity and Weight Control Physical activity is a core component of lifestyle programs for weight control.[37,39] Regular physical activity promotes health by decreasing abdominal fat stores, improving blood cholesterol and glucose levels, and decreasing blood pressure. A regular program of strength building increases lean muscle mass and reduces fat mass without weight loss.[40] Regular exercise "fine-tunes" appetite regulation mechanisms in many people.[25]

Recommendations for physical activity in overweight and obese adults for the prevention of additional weight gain, weight loss, and maintenance of weight loss have been developed by the American College for Sports Medicine.[40] These recommendations are summarized in Table 10.4. The minutes of physical activity listed in the table refer to moderate intensity activities such as brisk walking, soccer, jogging, tennis, calisthenics, rope skipping, and aerobic dancing. Low-intensity physical activities are also related to improvements in health risks and reductions in weight gain, but to a lesser extent than are moderate-intensity activities.[34]

www.ftc.gov/bcp/edu/pubs/business/adv/bus60.pdf

Illustration 10.5 The Internet is home to hundreds of bogus weight-loss sites. File a complaint at www.ftc.gov/bcp/edu/microsites/redflag/beyond.html.

Table 10.4 American College of Sports Medicine's recommendation for moderate physical activity and weight management in overweight and obese adults[40]

Weight outcome for most people	Average number of minutes per day of moderate physical activity
Prevention of weight gain	21 minutes or more
Weight loss	32 to 60 minutes
Weight-loss maintenance	29 to 43 minutes

Daily calorie intake reductions of 100 calories or more below the number of calories needed to maintain weight combined with regular physical activity are related to higher levels of weight loss than exercise alone. The combination creates larger calorie deficits and greater use of fat stores than either one alone.[2] Although uncommon, some people have health problems requiring medical supervision of exercise programs. Individuals with health concerns should get an "all clear" from their health care provider before undertaking a higher level of physical activity than usual.

There are indications that consumers are growing tired of popular, money-making approaches to weight loss. A survey undertaken by the food industry in the United Kingdom and the United States found that 29% and 44% of the adults studied were aware that extreme diets *cannot* produce sustained weight loss. Instead of popular weight-loss diets, many consumers are deciding that the path to weight control is paved by acceptable and sustainable changes in diet and physical activity.[9,41]

Diet Pills Two types of diet pills are currently approved for use in the United States. One of them (Qsymia) was recently approved and has yet to be used in clinical practice. The other is Orlistat, also know by the names Alli and Xenical.[42]

Orlistat works by partially blocking fat absorption in the intestines. The primary side effect associated with the use of orlistat is oily stools, which can be particularly bothersome if high-fat meals are consumed. The malabsorption of fat caused by orlistat reduces absorption of a number of fat-soluble nutrients such as vitamin D, vitamin E, and beta-carotene.[43]

Orlistat is intended for use in conjunction with a reduced-calorie diet and exercise. Use of orlistat as part of a weight-reduction program is related to a loss of five pounds up to a total of about 10% of initial body weight.[44] Along with a reduced-calorie diet, the over-the-counter version of orlistat is associated with weight loss of up to 5% of initial body weight.[43] Weight regain after pill use stops is common.[45]

Many other diet drugs are under development or being tested. Much activity currently centers around medications that decrease appetite and increase satiety. Even though people using prescribed diet drugs are carefully screened and monitored by medical professionals, serious side effects can develop. No diet drug is absolutely safe, and none are known to cure obesity forever.[42,46]

Obesity Surgery

Weight-loss surgery may be needed by individuals whose disease risks do not improve enough with lifestyle interventions.[34] This type of surgery is called **bariatric** surgery. Candidates for the surgery must be assessed as to their understanding of the surgery and the lifestyle changes required after surgery. Bariatric surgery usually leads to weight losses in the range of 75 to 90 pounds, and most of the weight loss is maintained over the years. It is the most effective method for weight control available.[48] A beneficial side effect of weight-loss surgery is reduction in appetite due to surgery-induced changes in appetite mechanisms located in the stomach.[49]

A number of bariatric surgical techniques are used clinically, but the two most utilized techniques by far are gastric bypass surgery and adjustable gastric banding (lap-band).[51] Gastric bypass surgery is not reversible, but the lap-band procedure is.[50] The use of surgery for weight loss and weight-loss maintenance is expanding worldwide. In 2009, 220,000 bariatric surgeries were performed in the United States.[50]

Gastric Bypass Surgery Gastric bypass surgery is the most effective method for weight loss and weight maintenance available. On average, individuals undergoing this

bariatrics The field of medicine concerned with weight loss.

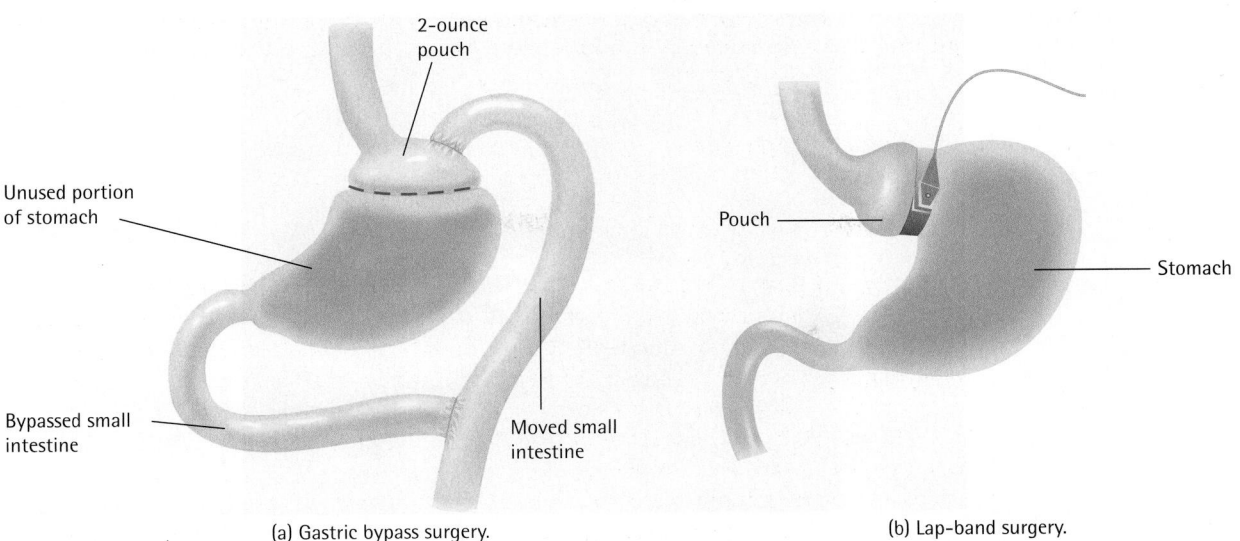

(a) Gastric bypass surgery.

(b) Lap-band surgery.

Illustration 10.6 Gastric bypass surgery (*left*), lap-band surgery (*right*).

surgery lose 60% of their excess body weight (the amount of weight they would need to lose to achieve normal weight), and they often maintain much of the loss over the long term.[51] Gastric bypass surgery is approved for use by the FDA for adults with BMIs over 40 kg/m^2 and for others with BMIs over 35 who have diabetes or other serious health problems related to obesity.[48] Although a higher risk procedure, gastric bypass surgery is associated with greater levels of weight loss than the less-risky lap-band procedure.[52] Health status generally improves dramatically as a result of the weight loss.[53] Resolution of type 2 diabetes, hypertension, sleep disorders, and elevated LDL cholesterol blood levels often follow gastric bypass surgery.[49,51] The surgery comes at a high cost ($14,000 to $26,000)[50] and may be accompanied by complications. There is a 0.1 to 0.5% chance of death from the operation, but that risk is considered worth it because obese people who receive the surgery live significantly longer than those who stay obese.[51] About half of the individuals who undergo bypass surgery struggle with emotional issues around food when old habits begin to come back. People who undergo this surgery have to be committed to long-term lifestyle changes and follow-up care.[49]

Lap-Band Surgery Adults with BMIs over 35, or with BMIs of 30–35 and an obesity-related health problem, can qualify for lap-band surgery.[47] Lap-band surgery is performed laparoscopically by inserting a tube through small incisions made in the abdomen. The surgery produces a small stomach pouch by constricting the upper part of the stomach with a band (Illustration 10.6). The band can be adjusted, or inflated by the injection of saline water to control the amount of food allowed to enter the stomach.[54] Individuals receiving this surgery tend to lose less weight (48% of excess body weight on average) than people having gastric bypass surgery.[51]

Concerns Related to Bariatric Surgery Some of the complications arising from bariatric surgery are related to nutrient deficiencies. The small stomach and changes in absorption that result from the surgery lead to malabsorption of a number of vitamins and minerals, most notably vitamin D, vitamin B$_{12}$, folate, calcium, and iron. Multivitamin and mineral supplements are a routine component of care after bypass surgery.[55] Other complications related to adverse consequences of anesthesia—such as infection, nausea, vomiting, dehydration, and gallstones—can also develop.[56]

   Some amount of lost weight is usually regained in the years following bariatric surgery and most individuals retain some amount of excess body weight.[57] Increasing food intake can stretch people's stomachs. The two-tablespoon pouch left after the surgery can be enlarged to hold half to two-thirds cup of food.[58] (A regular-size stomach has a capacity of about four cups.)

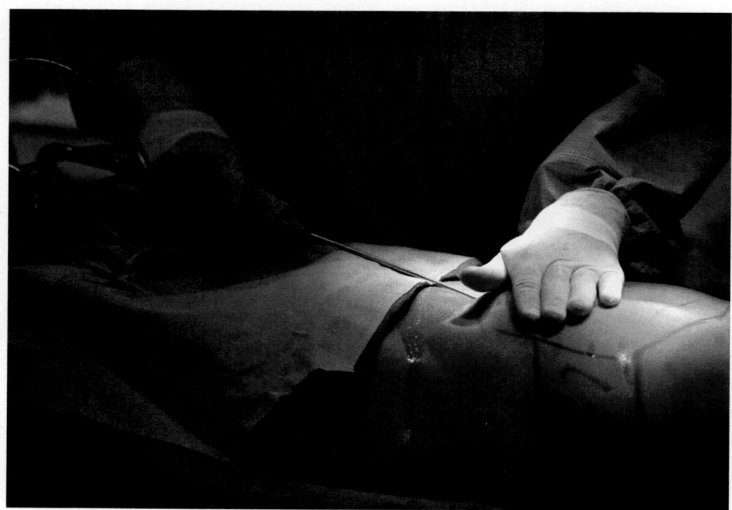

ZUMA Press/Newscom

Illustration 10.7 Liposuction surgery. In some cases, lasers are used during surgery to heat the fat, essentially turning the semi-soft fat in to a liquid to facilitate removal.

Bariatric surgery leaves some people with folds of excess skin where fat stores were lost. Skin tissue previously stretched by high levels of fat stores does not retract on its own after fat stores are lowered. The only known way of removing it is body-contouring surgery. The surgery may have to be repeated a number of times and can cost tens of thousands of dollars.

Liposuction At a cost of $2,000 to $8,000 per surgery (prices vary by fat deposit site, practice location, and surgeon), fat deposits in the thighs, hips, arms, back, or chin can be partially removed by liposuction (Illustration 10.7). The procedure is the most common type of cosmetic surgery performed in the United States.[59] Only fat present underneath the skin, and not fat the lies underneath muscles or that surrounds organs, is removed by liposuction.

Liposuction is intended for body shaping like that shown in Illustration 10.8. It is not recommended for weight loss. Surgical standards require that no more than eight pounds of fat be removed by liposuction.[60] Although considered a cosmetic procedure, liposuction that removes fat underneath abdominal skin may decrease plasma triglyceride levels in people who enter surgery with high triglyceride levels.[61] Weight and fat losses from liposuction will not be permanent if weight is gained. With weight gain, fat deposits will increase at the surgical sites and other locations in the body.[18] In addition, surgery always carries a risk of infection and other complications, so it cannot be taken lightly.

Weight Loss: Making It Last

People who lose weight and maintain their weight afterward tend to carefully watch portion sizes and to exercise regularly (Illustration 10.9). These are the major common-alities among people who maintain their weight after weight loss.[37] Other characteristics

Illustration 10.8 The body-contouring effects of liposuction. The photo on the left shows the "spare tire" of a man before liposuction, and the photo on the right shows the same man's waist after liposuction.

© Courtesy of the American Society for Aesthetic Plastic Surgery

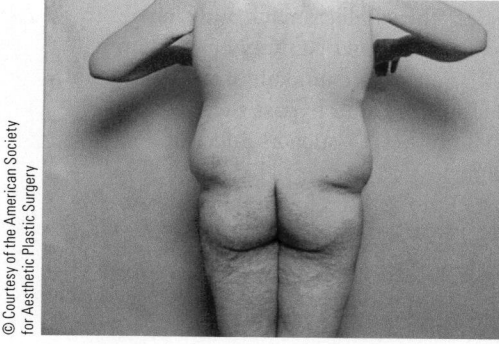
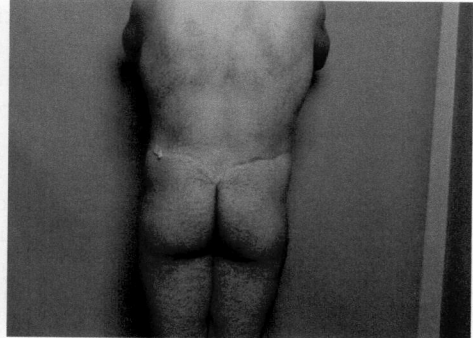

Illustration 10.9 Small, acceptable changes are the key to a successful weight-loss/weight-maintenance program. Try to find activities that you enjoy; get out there and play!

of people who lose weight and keep it off and characteristics of people who regain lost weight are listed in the Health Action below. Changes in diet and physical activity likely to be maintained tend to be small, easy to implement, and acceptable—even preferable—to existing behaviors.[26,62]

Small, Acceptable Changes For most people, excess body fat accumulates slowly over time. Consuming an extra 50 calories daily above the level of calories needed to maintain weight, for example, will lead to a gain of approximately five pounds in one year. In general, an excess of calorie intake over calorie need of 3,500 will produce a weight gain of about one pound. A 3,500-calorie deficit below calorie need will produce a loss of one pound in body weight.[67] (It's important to note that calorie intake has to be adjusted downward as weight is lost. The number of calories needed to maintain weight gradually decreases as body weight goes down.) By cutting back food intake by 100 calories daily below calorie need, a person could lose 10 pounds in a year. Table 10.5 provides examples of small changes in diet and physical activity level that, if all else about a person's diet and physical activity remain the same, would produce about a 100-calorie deficit a day.

Table 10.5 Small changes in diet and physical activity worth about 100 calories[a]

Diet	Physical activity
1. Consume 1 cup of fat-free yogurt with fruit instead of 1 cup of low-fat yogurt with fruit.	1. Walk an additional 20 minutes.
2. Drink 1½ cups of skim milk rather than 1½ cups of whole milk.	2. Lift arm weights for 15 minutes.
3. Eat a roasted or grilled chicken sandwich rather than a fried chicken sandwich.	3. Garden for 15 minutes.
4. Eat ½ cup rather than a cup of rice.	4. Play Frisbee for 30 minutes.
5. Order a regular rather than a large fish sandwich.	5. Clean house for 30 minutes.

[a]Physical activity calories are based on a 150-pound person.

health action Weight-Loss Maintainers Versus Weight Regainers[26,37,63,64]

Weight-Loss Maintainers

- Exercise regularly.
- Make small and comfortable changes in diet and exercise.
- Eat breakfast.
- Choose low-fat foods.
- Keep track of their weight, dietary intake, and physical activity level.

Weight Regainers

- Exercise little.
- Use popular diets.
- Make drastic and unpleasant changes in their diets and physical activity levels.
- Take diet pills.
- Cope with problems and stress by eating.

Identifying Small, Acceptable Changes To identify changes in eating and activity that have staying power, first list the weak points in your diet and activity. Dietary weak points might include the consumption of high-fat foods due to eating out often or relying on high-fat convenience foods that can be heated up in seconds. Another weak point might be skipping breakfast and overeating later in the day because of extreme hunger. Weak points in physical activity might include driving instead of walking, not engaging in sports, or spending too little time playing outside.

For each weak point, identify options that seem acceptable and enjoyable. A person who enjoys broiled chicken with barbecue sauce might not mind eating that at a restaurant instead of fried chicken. That's a change people can make if they plan ahead. A person who gets too full from a large serving of fries might be happier not eating as many and ordering a small serving. A breakfast skipper might find grabbing a piece of fruit and a slice of cheese for breakfast acceptable and doable. People who enjoy walking may not mind leaving the car or bus behind and letting their feet carry them to class, the grocery store, or a friend's house.

Many acceptable options for making small improvements in diet and activity may be available (see the "Take Action—Small Steps Can Make a Big Difference"). The easiest changes to accomplish are the ones that should be incorporated into the overall lifestyle improvement plan. Some people, for example, lose weight and keep it off simply by consciously cutting down on portion sizes. Others avoid eating too much at any meal and walk more. Simply adding breakfast helps some people lose weight and maintain the loss.[65] Increasingly, people are losing weight and keeping it off by eating more nutrient-dense foods such as vegetables and fruits, and fewer high-fat, high-sugar, energy-dense foods.[66] This change doesn't require that people eat less food but rather select and prepare foods that are nutrient-dense rather than calorie-dense. The more acceptable and easier the changes are to follow, the more likely they are to succeed.

Individualized plans that don't work out often include unacceptable or unenjoyable changes. The changes may be too large or too different from your usual activity. In these cases, go back to the drawing board and modify the plan to include small changes that are acceptable in the long run. Perhaps the original plan included jogging, but it turns out that you don't enjoy jogging. In that case, take jogging out of the plan! Replace it with another physical activity that would be enjoyed. Midcourse corrections should be expected. Some experimentation may be required to identify the small changes that will last.

take action Small Steps Can Make a Big Difference

Trying to lose weight or keep it off? Pick one eating and one physical activity option from the lists below that you find attractive. Give it a try for a day or two and see how it works out for you. Write down the changes you made and how you felt about making them. Could you see yourself making the changes for a week? The rewards offered by small steps may be greater than you expect.

Small Steps for Healthy Eating

1. Eat cereal for breakfast.
2. Eat a piece of fruit before dinner.
3. Eat larger portions of vegetables than meat.
4. Drink skim milk.
5. Switch from soft drink to flavored water. You can make your own by diluting fruit juice with water.
6. Stop eating as soon as you start to feel full.
7. Eat only when you feel hungry.
8. Dish up a smaller than usual portion of dessert.
9. If you're right-handed, eat using your spoon or fork in your left hand. Do it the other way if you're left-handed.
10. Devote your full attention to eating. Chew your food thoroughly.

Small Steps for Physical Activity

1. Walk an additional 2,000 steps (20 minutes).
2. Park farther from the store than usual.
3. Take a 10-minute walk in a park and look for flowers.
4. Lift a weight (or bottle of water, your textbook, a phone book) for 10 minutes while watching TV.
5. Do 10 sit-ups in the morning.
6. Use the stairs.
7. Take your neighbor's dog for a walk.
8. Jump rope for two minutes.
9. Dance for five minutes while no one is watching.

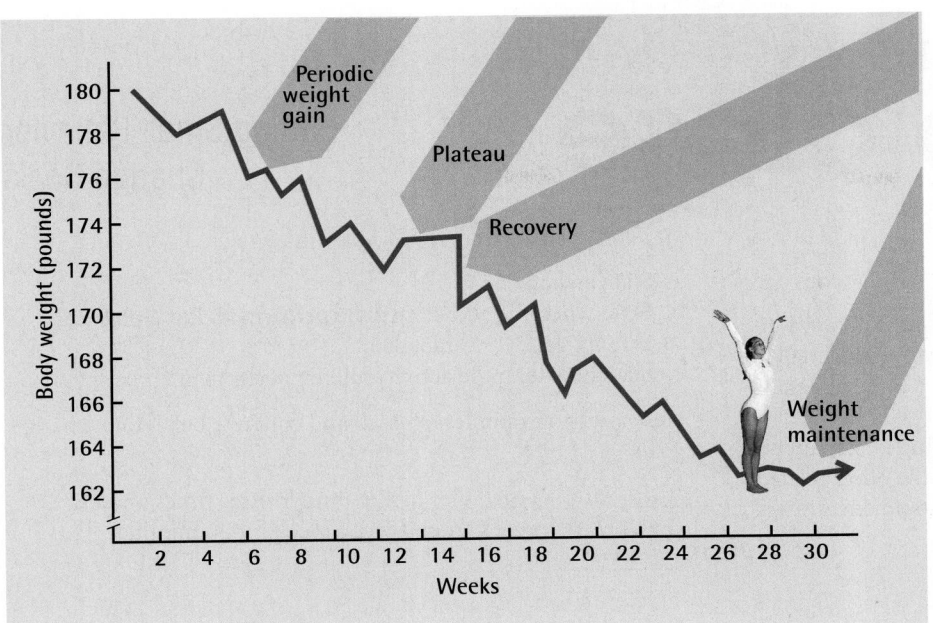

Illustration 10.10 People who lose weight gradually are more likely to keep it off than those who lose weight rapidly. The weight loss graphed here averages half a pound per week.

What to Expect for Weight Loss If you're happy with the changes and have found the right levels of calorie intake and physical activity, weight loss will be gradual but lasting. The pattern of loss should be somewhat like that graphed in Illustration 10.10, where the person lost 18 pounds over 30 weeks. The pattern will include peaks, valleys, and plateaus—not a straight downward curve.[67] Sometimes a bit of weight will be gained, and other times more weight than expected will be lost. It's more important to enjoy and continue improved eating and activity patterns than to concentrate on the number of pounds lost.[9]

If diet and exercise behaviors are improved in acceptable ways, there may be little need to become preoccupied with the number of calories consumed, the number of calories burned off in a bout of exercise, or the number of pounds lost last week. The goal is reached when the small changes become an enjoyable part of life on a day-to-day basis. Improved and rewarding eating and activity patterns offer many benefits. Weight control is one of them.

© Jose Luis Pelaez Inc./Getty Images/Blend Images

NUTRITION
up close

Focal Point: Small changes practice

Small changes in diet and exercise can return large benefits to health over time. What's the best way to get started on small and important behavioral changes? For many people, it begins with setting small but concrete goals.

This activity asks you to practice developing two small behavioral change goals, one related to diet and the other related to physical activity. Each small change goal should encompass the following:

a. State an activity.
b. State when the activity will be performed. For a physical activity goal, it should also:
c. State how long the activity will be performed.

Here are two examples of small and concrete behavioral change goals:

"I will eat my vegetables first at dinner three times a week."
"I will work in the garden twice a week for 20 minutes."

Your Small Change Goals

Goal 1—Diet change:
_____

Goal 2—Physical activity change:
_____

Checklist: Do your goals meet the criteria listed above? Do they represent small rather than large changes? If not, rework the goals so they do.

Feedback to the Nutrition Up Close is located in Appendix G.

REVIEW QUESTIONS

- **Describe weight-control methods that have been demonstrated to be successful and those that have not.**

1. Claims made for the effectiveness of weight-loss products in advertisements must, by federal regulation, be true.   True/False

2. Commercial weight-loss services and products must be tested for safety before they are made available to the public.   True/False

3. Some foods, including celery and grapefruit, have negative calories and help overweight people burn fat.   True/False

4. Many different types of quick weight-loss diets lead to weight loss if followed.   True/False

5. Liposuction is a recommended method for weight loss and weight-loss maintenance.   True/False

6. A person who weighs 250 pounds and has 95 pounds of excess body weight before gastric bypass surgery and loses the average amount of excess body weight would be expected to weigh 170 pounds two years after the procedure.   True/False

7. ____ Your friend Amoni developed her own weight-loss method. It's simple: Just eat fish, papaya, and broccoli. She lost 5 pounds in two weeks using her method and recommends it to everyone she hears talking about weight loss. This diet:
    a. Is nutrient-rich and likely meets people's need for nutrients.
    b. Would likely change people's eating behavior in the long run.
    c. Would be followed by weight regain in most people who try it.
    d. Would help people get started on losing weight forever.

8. ____ Your computer was on the receiving end of a mass e-mail describing a new weight-loss powder that guarantees users will safely lose 10 pounds in 10 days. Which term or word used in the above sentence should serve as the first clue to the fact that the weight-loss product is bogus?
    a. safely
    b. mass e-mail
    c. guarantees
    d. 10 pounds

- **Develop a weight-control plan based on small and acceptable changes in dietary intake and physical-activity level.**

9. It is recommended that people using a prescription diet pill for weight loss should also follow a reduced-calorie diet and exercise routine.   True/False

10. The major problem shared by quick weight-loss approaches is that they do not help people maintain weight in the long run.   True/False

11. A successful way to lose weight and keep it off is by making small and acceptable changes in diet and physical activity.   True/False

12. Lifestyle approaches to weight control include individually based and sustainable behavioral changes.   True/False

13. ____ Your cousin weighs 183 pounds and wants to lose 5% of his current weight. How many pounds would your cousin have to lose to meet the goal of a 5% weight loss?
    a. 5
    b. 9
    c. 15
    d. 18

The next two questions refer to the following situation.

A research study finds that participants in a meal-replacement weight-control program lost an average of 4.4 pounds over two years. The control group (similar people who did not receive the program) lost 2 pounds over the same period. The average weight of people in the meal replacement program prior to the onset of the study was 213 pounds, and the average weight of the control group was 210 pounds.

14. ____ Which of the following statements about the results of this study is accurate?
    a. Neither program can be considered successful.
    b. Both programs can be considered successful.
    c. Only the meal replacement program was successful.
    d. If participants in the control group had lost an average of three pounds more, the control group would have achieved a 5% or greater weight loss.

15. ____ Assume the meal replacement program was not successful. What is the most likely reason for its failure?
    a. Participants lacked the willpower to stay on the diet.
    b. The meal replacements provided too many calories.
    c. The program did not focus on increasing physical activity.
    d. The program did not achieve sustainable and acceptable lifestyle changes.

Answers to these questions can be found in Appendix G.

NUTRITION SCOREBOARD ANSWERS

1. If this claim were true, hardly anybody would be obese. **False**

2. Maintenance of weight loss is the cure for obesity. **False**

3. Many popular diets can lead to weight loss, but none successfully help people prevent weight regain.[1] **True**

4. Unfortunately, many weight-loss products and services on the market have not been shown to be safe or effective. Laws and regulations do not fully protect the consumer from the introduction of bogus products and services. **False**

5. Small and acceptable behavioral changes are easier to live with over time than drastic and disliked changes.[2] **True**

Wavebreakmedia Ltd/Jupiterimages

Disordered Eating: Anorexia and Bulimia Nervosa, Binge-Eating Disorder, and Pica

NUTRITION SCOREBOARD

1 The United States has one of the world's highest rates of anorexia nervosa. **True/False**

2 Eating disorders result from psychological, and not biological, causes. **True/False**

3 People in many different cultures may consume clay, dirt, and other nonfood substances. **True/False**

Answers can be found at the end of the unit.

After completing Unit 11 and its interactive learning features, you will be able to:

- Recognize the potential causes of eating disorders and their effects on nutritional status and health.

Eating Disorders

- **Recognize the potential causes of eating disorders and their effects on nutritional status and health.**

Three square meals a day, an occasional snack or missed meal, and caloric intakes that average out to match the body's need for calories—this set of practices is considered "orderly" eating. Self-imposed semistarvation, feast and famine cycles, binge eating, **purging**, and the regular consumption of nonfood substances such as paint chips and clay—these behaviors are symptoms of disordered eating.

Three specific types of disordered eating patterns are officially recognized as eating disorders and have been assigned diagnostic criteria. They are (1) anorexia nervosa, (2) bulimia nervosa, and (3) pica.[2] A fourth eating disorder, binge eating, is provisionally classified as an eating disorder. Other forms of eating disorders such as compulsive overeating and nighttime-eating syndrome have been proposed. They are classified as Eating Disorders Not Otherwise Specified (EDNOS). A variety of other forms of disordered eating are under consideration for designation as eating disorders. [2]

Key Nutrition Concepts

Fundamental knowledge of nutrition science supports this content on eating disorders through these three key nutrition concepts:

1. Food is a basic need of humans.

2. Malnutrition can result from poor diets and from disease states, genetic factors, or combinations of these factors.

3. Adequacy, variety, and balance are key characteristics of healthful diets.

Anorexia Nervosa

It's about 9:30 on a Tuesday night. You're at the grocery store picking up sandwich fixings and some milk. Although your grocery list contains only four items, you arrive at the checkout line with a half-filled cart. The woman in front of you has only five items: a bag with about 10 green beans, an apple, a bagel, a green pepper, and a 4-ounce carton of non-fat yogurt (Illustration 11.1). As she carefully places each item into her shopping bag, you notice that she is dreadfully thin.

The woman is Alison. She has just spent half an hour selecting the food she will eat tomorrow. Alison knows a lot about the caloric value of foods and makes only low-calorie choices. Otherwise, she will never get rid of her excess fat. To Alison, weight is

Illustration 11.1 A day's diet for Alison. The foods shown provide approximately 562 calories.

Richard Anderson

purging The use of self-induced vomiting, laxatives, or diuretics (water pills) to rid the body of food.

everything—she cannot see the skeleton-like appearance others see when they look at her.

Alison has an intense fear of gaining weight and of being considered fat by others. She is annoyed when her parents and friends express their concern about her weight. You didn't know this about Alison when you saw her. There is much more to anorexia nervosa than meets the eye.

Individuals with **anorexia nervosa** such as described in the real-life case example of Alison, starve themselves. They can never be too thin—no matter how emaciated they may be. As shown in Illustration 11.2, people with anorexia nervosa look extraordinarily thin from the neck down. The face and the rest of the head usually look normal because the head is the last part of the body to be affected by starvation.

Women with anorexia nervosa have relatively little body fat (15% of body weight compared to 27% in control women).[3] They often keep their body fat content low by consuming fewer than 1,200 calories a day. They become cold easily and have unusually low heart rates and sometimes an irregular heartbeat, dry skin, and low blood pressure. Women with anorexia nervosa experience absent or irregular menstrual cycles, infertility, and poor pregnancy outcomes (Table 11.1).[4] Men with anorexia nervosa tend to see themselves as being too fat and want less body fat and a more muscular body.[5,6] Low testosterone levels in males with this eating disorder produce diminished sexual drive and impaired fertility.[7]

Approximately 9 in 10 women with anorexia nervosa have significant bone loss, and 38% have osteoporosis. The extent of bone loss correlates strongly with undernutrition: the lower the body weight, the lower the bone density. Males with anorexia nervosa lose bone mass, too. Improving calcium and vitamin D intake is recommended along with nutritional rehabilitation experience.[4]

Illustration 11.2 Eating disorders occur in males as well as in females, but females make up approximately 90% of all cases.

The Female Athlete Triad Pediatricians, nutritionists, and coaches are beginning to be on the lookout for eating disorders, menstrual cycle dysfunction, and decreased bone mineral density in young female athletes. Calorie intakes below calorie need and underweight related to eating disorders can lower estrogen levels and disrupt menstrual cycles. The lack of estrogen decreases calcium deposition in bones and reduces bone density at a time when peak bone mass is accumulating.[9] A triad of adverse affects occur in approximately 20% of female athletes as well as 8% of male athletes[12] so the title "female athlete triad" is in need of change. It is most common in athletes engaged in gymnastics, distance running, figure skating, diving, and ballet.[10]

Irregular or absent menstrual cycles used to be thought of as no big deal. That attitude has changed, however, due to research results indicating that abnormal cycles in young females are related to delayed healing of bone and connective tissue injuries, and to bone fractures and osteoporosis later in life.[11]

Motivations Underlying Anorexia Nervosa The overwhelming desire to become and remain thin drives people with anorexia nervosa to refuse to eat, even when ravenously hungry, and to exercise intensely. Half of the people with anorexia turn to **binge eating** and purging—features of bulimia nervosa—in their efforts to lose weight.[13] Preoccupied with food, people with anorexia may prepare wonderful meals for others but eat very little of the food themselves. Family members and friends, distressed by their failure to persuade the person with anorexia to eat, report high levels of anxiety. Although adults often describe people with anorexia as "model students" or "ideal children," their personal lives are usually marred by low self-esteem, social isolation, and unhappiness.[14]

What Causes Anorexia Nervosa? The cause of anorexia nervosa isn't yet clear. It is likely that many different conditions, both psychological and biological, predispose an individual to become totally dedicated to extreme thinness. The value that Western societies place on thinness, the need to conform to society's expectations of acceptable body weight and shape, low self-esteem, and a need to control some aspect of one's life completely are commonly offered as potential causes for this disorder (Illustration 11.3).[4,15]

anorexia nervosa An eating disorder characterized by extreme weight loss, poor body image, and irrational fears of weight gain and obesity.

binge eating The consumption of a large amount of food in a small amount of time.

Table 11.1 Features of anorexia nervosa

A. Essential features

1. Refusal to maintain body weight at or above 85% of normal weight for age and height
2. Intense fear of gaining weight or becoming fat, despite being underweight
3. Disturbance in the way in which body weight or shape is experienced, undue influence of body weight or shape on self-evaluation, or denial of the seriousness of current low body weight
4. Lack of menstrual periods in teenage females and women (missing at least three consecutive periods)
5. Behavior falls into two types:
   a. *Restricting type*: Person does not regularly engage in binge-eating or purging behavior.
   b. *Binge-eating type*: Person regularly engages in binge-eating or purging behavior (self-induced vomiting; laxative, diuretic, or enema use).

B. Common features in females

1. Low-calorie diet, extensive exercise, low body fat
2. Soft, thick facial hair, thinning scalp hair
3. Loss of heart muscle; irregular, slow heartbeat
4. Low blood pressure
5. Increased susceptibility to infection
6. Anemia
7. Constipation
8. Low body temperature (hypothermia)
9. Dry skin
10. Depression
11. History of physical or sexual abuse
12. Low estrogen levels
13. Low bone density
14. Infertility, poor pregnancy outcome

C. Common features in males

1. Most of the common features in females
2. Substance abuse
3. Mood and other mental disorders, self-loathing
4. Decreased testosterone level, sex drive, and fertility

Source: Reprinted with permission from the *Diagnostic and Statistical Manual of Mental Disorders, Text Revision, Fourth Edition*. (Copyright 2000). American Psychiatric Association.

How Common Is Anorexia Nervosa? It is estimated that 1% of adolescent and young women in the Western world and less than 0.1% of young males have anorexia nervosa. The disorder has been reported in girls as young as five and in women through their forties,[16] however, it usually begins during adolescence. It is estimated that 1 in 10 females between the ages of 16 and 25 has "subclinical" anorexia nervosa, or exhibits some of the symptoms of the disorder.[5]

Certain groups of people are at higher risk of developing anorexia nervosa than others (Table 11.2). People at risk come from all segments of society but tend to be overly concerned about their weight and food and have attempted weight loss from an early age.[5]

Treatment There is no "magic bullet" treatment that cures anorexia nervosa quickly and completely. In all but the least severe cases, the disorder generally takes five to seven years and professional help to correct. Treating the disorder is often difficult because few people with anorexia believe their weight needs to be increased.[13]

Illustration 11.3 There is a need for the use of more realistic body shapes and sizes in the media.

Treatment programs for anorexia nervosa are multidisciplinary in approach. They generally focus on the prompt restoration of nutritional health and body weight, psychological counseling to improve self-esteem and attitudes about body weight and shape, medically supervised use of antidepressant or other medications, family therapy, and normalizing eating and exercise behaviors.[2] These programs are successful in 50% of people and partially successful in most other cases.[4] One-third of people who recover fully from anorexia nervosa will relapse within seven years or less. Eight years after diagnosis, 3% of people with anorexia nervosa will die from the disorder, and 33 years after diagnosis, 18% will die. Results of treatment are often excellent when the disorder is treated early.[17] Unfortunately, many people with the condition deny that problems exist and postpone treatment for years. Initiation of treatment is often prompted by a relative, coach, or friend.[18]

Table 11.2 Risk groups for anorexia nervosa[5]

- Dieters
- Ballet dancers
- Competitive athletes (gymnasts, figure skaters)
- Fitness instructors
- Dietetics majors
- People with type 1 (insulin dependent) diabetes

Bulimia Nervosa

Finally home alone, Lisa heads to the pantry and then to the freezer. She has carefully controlled her eating for the last day and a half and is ready to eat everything in sight.

It's a bittersweet time for her. Lisa knows the eating binge she is preparing will be pleasurable, but that she'll hate herself afterward. Her stomach will ache from the volume of food she'll consume, she'll feel enormous guilt from losing control, and she'll be horrified that she may gain weight and will have to starve herself all over again. Lisa is so preoccupied with her weight and body shape that she doesn't see the connection between her severe dieting and her bouts of uncontrolled eating. To get rid of all the food she is about to eat, she will do what she has done several times a week for the last year. Lisa avoids the horrible feelings that come after a binge by "tossing" everything she had eaten as soon as she can.

In just 10 minutes, Lisa devours 10 peanut butter cups (the regular size), a 12-ounce bag of chocolate chip cookies, and a quart of ice cream. Before five more minutes have passed Lisa will have emptied her stomach, taken a few deep breaths, thrown on her shorts, and started the five-mile route she jogs most days. As she jogs, she obsesses about getting her 138-pound, 5-foot 5-inch frame down to 115 pounds. She will fast tomorrow and see what news the bathroom scale brings.

Lisa is not alone. **Bulimia nervosa** occurs in 1 to 3% of young women and in about 0.5% of young males in the United States.[4] The disorder is characterized by regular episodes of dieting, binge eating (see Illustration 11.4), and attempts to prevent weight gain by purging or using laxatives, diuretics, excessive exercise, or enemas. In most cases,

bulimia nervosa An eating disorder characterized by recurrent episodes of rapid, uncontrolled eating of large amounts of food in a short period of time. Episodes of binge eating are followed by compensatory behaviors such as such as self-induced vomiting, dieting, excessive exercise, or misuse of laxatives, to prevent weight gain.

Illustration 11.4 Bulimia nervosa is characterized by the consumption of a large amount of food (such as shown here) followed by purging and dieting, or other behavior such as excessive exercise.

bulimia nervosa starts with voluntary dieting to lose weight. At some point, voluntary control over dieting is lost, and people may engage in binge eating and vomiting.[4] The behaviors become cyclic: Food binges are followed by guilt, purging, and dieting. Dieting leads to a feeling of deprivation and intense hunger, which leads to binge eating, and so on. Once a food binge starts, it is hard to stop.

Table 11.3 lists the features of bulimia nervosa. Approximately 86% of people with this condition vomit to prevent weight gain and to avoid post-binge anguish. A smaller proportion of people use laxatives, ipecac (a vomiting-inducing medication), diuretics (water pills), or enemas alone or in combination with vomiting.[2] These approaches do not prevent weight gain, however, and their regular use can be harmful. The habitual use of laxatives and enemas causes "laxative dependency"—these products become necessary for bowel movements. Long-term use of ipecac may damage heart muscle, and diuretics can cause illnesses by disturbing the body's fluid balance.[4]

The lives of people with bulimia nervosa are usually dominated by conflicts about eating and weight. Some affected individuals are so preoccupied with food that they spend days securing food, bingeing, and purging. Others experience only occasional episodes of binge eating, purging, and fasting.[13]

Unlike those with anorexia nervosa, people with bulimia usually are not underweight or emaciated. They tend to be normal weight or overweight. Like anorexia nervosa, bulimia nervosa is more common among athletes (including gymnasts, weight lifters, wrestlers, jockeys, figure skaters, physical trainers, and distance runners) and ballet dancers than in other groups.[4]

Bulimia nervosa leads to major changes in metabolism. The body must constantly adjust to feast and famine cycles and mineral and fluid losses. Salivary glands become enlarged, and teeth may erode due to frequent vomiting of highly acidic foods from the stomach.[13] There is evidence that satiety signals are decreased among people with this eating disorder and that may support overeating behaviors.[19]

Is the Cause of Bulimia Nervosa Known? The cause of bulimia nervosa is not known with certainty, but the scientific finger is pointing at depression, abnormal mechanisms for regulating food intake, and feast-and-famine cycles as possible causes. Fasts and **restrained eating** may prompt feelings of deprivation and hunger that may trigger binge eating.[4] The ideal thinness may become more and more difficult to achieve as the feast-and-famine cycles continue.

restrained eating The purposeful restriction of food intake below desired amounts in order to control body weight.

Table 11.3 Features of bulimia nervosa[13]

A. Essential features

1. Recurrent episodes of binge eating. An episode of binge eating is characterized by both of the following:
   a. Eating an amount of food within a two-hour period of time that is definitely larger than most people would eat in a similar amount of time and under similar circumstances.
   b. A sense of lack of control over eating during the episode; a feeling that one cannot stop eating or control what or how much one is eating.

2. Recurrent inappropriate compensatory behavior in order to prevent weight gain, such as self-induced vomiting; misuse of laxatives, diuretics, enemas, or other medications; fasting; or excessive exercise.

3. The binge eating and inappropriate compensatory behaviors both occur, on average, at least twice a week for three months.

4. Self-evaluation is unduly influenced by body weight and shape.

5. The disturbance does not occur exclusively during episodes of anorexia nervosa.

6. Behavior falls into two types:
   a. *Purging type*: The person regularly engages in self-induced vomiting or the misuse of laxatives, diuretics, or enemas.
   b. *Nonpurging type*: The person regularly engages in fasting or excessive exercise but does not regularly engage in self-induced vomiting or the misuse of laxatives, diuretics, or enemas.

B. Common features

1. Weakness, irritability
2. Abdominal pain, constipation, bloating
3. Dental decay, tooth erosion
4. Swollen cheeks and neck
5. Binging on high-calorie foods
6. Eating in secret
7. Normal weight or overweight
8. Guilt and depression
9. Substance abuse
10. Dehydration
11. Impaired fertility
12. History of sexual abuse

Source: Reprinted with permission from the *Diagnostic and Statistical Manual of Mental Disorders, Text Revision, Fourth Edition.* (Copyright 2000). American Psychiatric Association.

Treatment The goal of bulimia treatment is to break the feast-and-famine cycles via nutrition and psychological counseling. Replacing the disordered pattern of eating with regular meals and snacks often reduces the urge to binge and the need to purge. Psychological counseling aimed at improving self-esteem and attitudes toward body weight and shape goes hand in hand with nutrition counseling. In many cases, antidepressants are a useful component of treatment.[15] The full recovery of women with bulimia nervosa is higher than that of women with anorexia nervosa. Nearly all women with bulimia achieve partial recovery, but one-third will relapse into bingeing and purging within seven years.[20] Bulimia nervosa usually improves substantially during pregnancy; about 70% of pregnant women with the condition will improve their eating habits for the sake of their unborn baby.[21]

Binge-Eating Disorder

Features of **binge-eating disorder** are shown in Table 11.4. People with this condition tend to be overweight or obese, and it affects an equal number of males and females.[4] Like individuals with bulimia nervosa, people with binge-eating disorder eat several thousand

binge-eating disorder An eating disorder characterized by periodic binge eating, which normally is not followed by vomiting or the use of laxatives. People must experience eating binges twice a week on average over a period of six months to qualify for the diagnosis.

Table 11.4 Features of binge-eating disorder[13]

1. Rapid consumption of extremely large amounts of food (several thousand calories) in a short period of time

2. Two or more such episodes of binge eating per week over a period of six months

3. Binge eating by oneself

4. Lack of control over eating or an inability to stop eating during a binge

5. Post–binge-eating feelings of self-hatred, guilt, and depression or disgust

6. Purging, fasting, excessive exercise, or other compensation for high-calorie intakes not present

calories' worth of food within a short period of time during a solitary binge, feel a lack of control over the binge, and experience distress or depression after the binge occurs. People must experience eating binges twice a week on average over a period of six months to qualify for the diagnosis. Unlike individuals with bulimia nervosa, however, people with binge-eating disorder don't vomit, use laxatives, fast, or exercise excessively in an attempt to control weight gain.[4]

It is estimated that 9 to 30% of people in weight-control programs[22] and 1 to 3% of U.S. adults experience binge-eating disorder.[23] Stress, depression, anger, anxiety, and other negative emotions appear to prompt binge-eating episodes. Preliminary evidence indicates that binge-eating disorder aggregates in families and has both genetic and environmental orgins.[4]

Treatment The treatment of binge-eating disorder focuses on both the disordered eating and the underlying psychological issues. Persons with this condition will often be asked to record their food intake, indicate bingeing episodes, and note feelings, circumstances, and thoughts related to each eating event (Illustration 11.5). This information is used to identify circumstances that prompt binge eating and practical alternative behaviors that may prevent it. Individuals being treated for binge-eating disorder are usually given information about it, attend individual and group therapy sessions, and receive nutrition counseling on *mindful eating*. Components of the mindful eating approach include paying attention to hunger and satiety cues, slowing down the pace of eating, and identifying triggers to eating. Antidepressants may be part of the treatment.[25] Treatment is successful in 85% of women treated for binge-eating disorder.[4,26]

Resources for Eating Disorders

Information and services related to eating disorders are available from a variety of sources. Services are best delivered by health care teams that specialize in the treatment of eating disorders. Contact with a primary care physician, dietitian, or nurse practitioner is often a good start to the process of identifying qualified health care teams. Reliable sources of information about eating disorders can be located from the Academy of Eating Disorders website (www.aedweb.org).

One of the most important resources for people with an eating disorder may be a trusted friend or relative. This unit's Reality Check explores this resource in a very personal way—by putting you in the shoes of a person whose sister has bulimia.

Prevention of Anorexia and Bulimia Nervosa, and Binge-Eating Disorder

The pressure to conform to society's standard of beauty and acceptability is thought to be a primary force underlying the development of eating disorders.[27] Children acquire prevailing cultural values of beauty before adolescence. As early as age five, American children learn to associate negative characteristics with people who are overweight and positive characteristics with those who are thin.[16] Standards of beauty are defined by models and movie and television stars often include thinness, but the body shape portrayed as most

REALITY CHECK
Close to Home

Although she hides it, you are sure your sister has bulimia nervosa and that she is not getting help. You are deeply concerned for her health and well-being but don't know what to do about it.

Here's what Heather and Crystal say they would do if it was their sister. Who do you think has the better idea?

Answers on page 11-9

Heather: I'd talk with her about getting help.

Crystal: I'd spend more time with her to let her know I love her.

desirable is often unhealthfully thin and unattainable by many.[28,29] The disparity between this ideal and what people normally weigh can foster low self-esteem and body image dissatisfaction. Approximately 50% of normal-weight adult women are dissatisfied with their weight; many diet, binge, purge, or fast occasionally in an attempt to reach the standard of beauty set for them.[29] Men are more likely to express discontent with their body shape and fat content. They may strive obsessively for muscular bodies with little visible body fat.[30]

A movement toward acceptance of body size, fashionable attire for larger people, full-size models, and a more realistic view of individual differences in body shapes is emerging in the United States and Europe[30–32] (Illustration 11.6). Acceptance of a realistic standard of body weight and shape—one that corresponds to health and physical fitness—and respect for people of all body sizes may be the most effective measures that can be taken to prevent anorexia nervosa, bulimia nervosa, and binge-eating disorder.

Daily Food Record

Date _____

Time	Type and amount of food and beverage	Meal, Snack, Binge?	Eating triggers (feelings, situation)
7:30 am	coffee, 2 cups	M	Hunger!
	sugar 2 tsp		
	cornflakes, 2 cups		
	skim milk, 1 cup		
11:30 am	tuna sandwich	M	Bored, hungry
	ice tea, 2 cups		
7:30 pm	3 hamburgers	B	Stressed out, angry at my coach
	2 large fries		
	24 oreos		
	1/2 gallon ice cream		

Illustration 11.5 Example of a food diary of a person with binge-eating disorder.

Table 11.5 Helping a family member or friend with an eating disorder[2]

Whether at work, home, or play, many of us experience anxiety and a sense of helplessness when someone we love is living with an eating disorder. We may feel compelled to take action to help but aren't sure what to do or how to do it. Here are some tips on how to express your concerns to a friend or relative with an eating disorder:

1. Gather information about services for people with eating disorders to share with your friend or relative.

2. Talk with your friend or relative privately when there is enough time to discuss the issue fully. Tell them you are worried and that they may need to seek help.

3. Encourage your friend or relative to express his or her feelings and then listen intently. Be accepting about the feelings that are expressed. Be ready to talk to the friend or relative more about it in the future.

4. Do not argue with your friend or relative about whether she or he has an eating disorder. Let your friend or relative know you heard what was said but that you are concerned that he or she may not get better without treatment.

5. Seek emergency medical help in life-threatening situations.

6. Understand that affected individuals would rather not have an eating disorder and that parents are not to be blamed for them.

Only individuals with an eating disorder can make the decision to get help. Knowledge that people who love them will be around to support them and their decision to seek treatment may help encourage people with an eating disorder to take action.

ANSWERS TO REALITY CHECK
Close to Home

Both ideas are admirable and deserve a thumbs-up. Heather's idea is aimed directly at helping her sister consider treatment and may sometimes be the appropriate action to take. There are appropriate ways to talk to relatives or friends about your concerns for them. Learn more about it from the information presented in Table 11.5.

Heather:

Crystal:

Richard Anderson

Illustration 11.6   The trend toward size acceptance.

I am the shape of the future.

Richard Anderson

pica (pike-eh)   The regular consumption of nonfood substances such as clay or laundry starch.

geophagia (ge-oh-phag-ah)   Clay or dirt eating.

pagophagia (pa-go-phag-ah) Ice eating.

amylophagia (am-e-low-phag-ah) Laundry starch or cornstarch eating.

plumbism   Lead (primarily from old paint flakes) eating.

PhotoDisc

Pica

When did I start eating clay? I know it might sound strange to you, but I started craving clay in the summer of '58. It was a beautiful spring morning—it had just rained. I smelled something really sweet in the breeze coming in my bedroom window. I went outside and knew instantly where the sweet smell was coming from. It was the wet clay that lies all around my house. I scooped some up and tasted it. That's when and how I started my craving for that sweet-smelling clay. I keep some in the fridge now because it tastes even better cold.

A most intriguing type of eating disorder, **pica** has been observed in chimpanzees and in humans in many different cultures since ancient times.[33] The history and persistence of pica might suggest that the practice has its rewards. Nevertheless, important health risks are associated with eating many types of nonfood substances.

The characteristics of pica are summarized in Table 11.6. Young children and pregnant women are most likely to engage in the practice; for unknown reasons, it rarely occurs in men.[34] It most commonly takes the form of **geophagia** (clay or dirt eating), **pagophagia** (ice eating), **amylophagia** (laundry starch and cornstarch eating), or **plumbism** (lead eating). A potpourri of nonfood substances, listed in Table 11.7, may be consumed. It is not clear why pica exists, although several theories have been proposed.

Geophagia   Some people like to eat certain types of clay or dirt. Those who do often report that the clay or dirt tastes or smells good, quells a craving, or helps relieve nausea or an upset stomach. The belief that certain types of clay provide relief from stomach upsets may have some validity: A component of some types of clay is used in nausea and diarrhea medicines. There is no evidence that geophagia is motivated by a need for minerals found in clay or dirt, however.[35]

Although the reasons given for clay and dirt ingestion make the practice understandable and helps explain its acceptance in some cultures, the consequences to health outweigh the benefits. Clay and dirt consumption can block the intestinal tract and cause parasitic and bacterial infections.[13] The practice is also associated with iron deficiency and sickle-cell anemia in some individuals.[35]

Pagophagia   Have you ever known somebody who constantly crunches on ice? That person may have a 9-in-10 chance of being iron deficient. Regular ice eating, to the extent of one or more trays of ice cubes a day, is closely associated with an iron-deficient state. Ice eating usually stops completely when the iron deficiency is treated.[36]

Ice eating may be common during pregnancy. In one study of women from low-income households in Texas, 54% of pregnant women reported eating large amounts of ice regularly. Ice eaters had poorer iron status than other pregnant women who did not eat ice.[37]

Amylophagia   The sweet taste and crunchy texture of flaked laundry starch are attractive to a small number of women, especially during pregnancy. If the laundry starch pre-

Table 11.7 A partial list of nonfood substances reported to be consumed by individuals with pica

Animal droppings	Coffee grounds	Leaves	Plaster
Baking soda	Cornstarch	Mothballs	Sand
Burnt matches	Crayons	Nylon stockings	String
Cigarette butts	Dirt	Paint chips	Wool
Clay	Foam rubber	Paper	
Cloth	Hair	Paste	
Coal	Laundry starch	Pebbles	

PhotoDisc

ferred is not available, cornstarch is sometimes eaten in its place. Laundry starch is made from unrefined cornstarch. The taste for starch almost always disappears after pregnancy.[34]

Laundry starch and cornstarch have the same number of calories per gram as do other carbohydrates (4 calories per gram). Consequently, starch eating provides calories and may reduce the intake of nutrient-dense foods. In addition, starch may contain contaminants because it is not intended for consumption. Starch eaters' diets are generally inferior to the diets of pregnant women who don't consume starch, and their infants are more likely to be born in poor health.[34]

Plumbism The consumption of lead-containing paint chips poses a major threat to the health of children in the United States and many other countries (Illustration 11.7). Many older homes and buildings, especially those found in substandard housing areas, may be covered with lead-based paint and its dried-up flakes. Children may develop lead poisoning if they eat the sweet-tasting paint flakes or inhale lead from contaminated dust and soil near the buildings. Approximately 1.4% of young children in the United States have elevated blood lead levels.[38]

High levels of exposure to lead can cause profound mental retardation and death in young children. Low levels of exposure can lead to hearing problems, growth retardation, reduced intelligence, and poor classroom performance. Children with lead poisoning are more likely to fail or drop out of school than children not exposed to lead in their environment.[38,39] Cases of lead poisoning due to plumbism in adults has been observed among people who consume ground-up clay pottery, clay, lead-containing nontraditional medicines, and lead-containing cooking utensils.[39]

Proposed Eating Disorders Several other patterns of abnormal eating behaviors, referred to as EDNOS—eating disorders not otherwise specified, have been described and may become official diagnoses in the future.[3] These tentative eating disorders include:

- Nighttime eating syndrome—high food consumption during the night accompanied by sleep disturbances and psychological distress

- Compulsive overeating—the uncontrolled ingestion of large amounts of food, as found in binge-eating disorder

© Peter Essick/Aurora Photos

Illustration 11.7 The regular consumption of lead-based paint chips from old houses is a major cause of lead poisoning in young children.

- Purging disorder—frequent purging without binge eating

- Restrained eating—the consistent limitation of food intake to avoid weight gain

- Orthorexia nervosa (pronounced ortho-rex-e-ah)—an unhealthy fixation with the health value and purity of food

- Selective eating disorder—children and adults who are picky eaters; they consume a very limited variety of food

- Individuals displaying these eating characteristics have been identified, but the core features are not sufficiently understood to develop reliable diagnoses criteria and treatment approaches.

Wavebreakmedia Ltd/
Jupiterimages

NUTRITION
up close

Eating Attitudes Test

Focal Point: Assess whether your eating attitudes and behaviors are likely to be within a normal range.

Date _____ Age _____ Gender _____
Height _____ Present weight _____ How long at present weight? _____
Highest past weight _____ How long ago? _____
Lowest past weight _____ How long ago? _____

Answer the following questions using these responses:
A = always   S = sometimes   U = usually   R = rarely   O = often   N = never

_____ 1. I am terrified of being overweight.

_____ 2. I avoid eating when I am hungry.

_____ 3. I find myself preoccupied with food.

_____ 4. I have gone on eating binges where I feel that I may not be able to stop.

_____ 5. I cut my food into very small pieces.

_____ 6. I am aware of the calorie content of the foods I eat.

_____ 7. I particularly avoid foods with a high-carbohydrate content.

_____ 8. I feel that others would prefer that I ate more.

_____ 9. I vomit after I have eaten.

_____ 10. I feel extremely guilty after eating.

_____ 11. I am preoccupied with a desire to be thinner.

_____ 12. I think about burning up calories when I exercise.

_____ 13. Other people think I am too thin.

_____ 14. I am preoccupied with the thought of having fat on my body.

_____ 15. I take longer than other people to eat my meals.

_____ 16. I avoid foods with sugar in them.

_____ 17. I eat diet foods.

_____ 18. I feel that food controls my life.

_____ 19. I display self-control around food.

_____ 20. I feel that others pressure me to eat.

_____ 21. I give too much time and thought to food.

_____ 22. I feel uncomfortable after eating sweets.

_____ 23. I engage in dieting behavior.

_____ 24. I like my stomach to be empty.

_____ 25. I enjoy trying new rich foods.

_____ 26. I have the impulse to vomit after meals.

Source: From Garner, D.M., Olmsted, M.P., Bohr, Y., and Garfinkel, P.E. (1982). "The Eating Attitudes Test: Psychometric features and clinical correlates," *Psychological Medicine*, 12, 871-878. Copyright © 1982 Cambridge University Press. Reprinted by permission.

The questions are not intended to diagnose eating disorders but rather to screen for them. Feedback to the Nutrition Up Close is located in Appendix G.

REVIEW QUESTIONS

- **Recognize the potential causes of eating disorders and their effects on nutritional status and health.**

1. Critical aspects of anorexia nervosa include low body fat and low bone density.   **True/False**

2. Individuals with anorexia nervosa are at risk of anemia.   **True/False**

3. The adverse effects of the female athlete triad on health are primarily related to low calorie intakes and underweight.   **True/False**

4. More individuals have anorexia nervosa than any other recognized eating disorder.   **True/False**

5. Essential features of bulimia nervosa are binge eating and excessive sleeping.   **True/False**

6. Obese people are more likely to experience binge-eating disorder than are people who are not obese.   **True/False**

7. Treatment is effective for 85% of women treated for binge-eating disorder.   **True/False**

8. *Pica* refers to the regular consumption of raw whole grains, grasses, and other foods generally used to feed livestock.   **True/False**

9. People who consume lead, such as from lead-based paints, have a condition called *geophagia*.   **True/False**

10. Nutritional rehabilitation and eating behavior counseling are key components of the treatment of eating disorders.   **True/False**

Answers to these questions can be found in Appendix G.

NUTRITION SCOREBOARD ANSWERS

1. Anorexia nervosa is most common in the United States and other Westernized countries.[1]   **True**

2. The causes of eating disorders are not known with certainty. Both psychological and biological factors play a role.   **False**

3. Although not recommended for health reasons, people in many different cultures practice pica—the regular ingestion of nonfood items such as clay and dirt.   **True**

Useful Facts about Sugars, Starches, and Fiber

NUTRITION SCOREBOARD

1 Pasta, bread, and potatoes are good sources of complex carbohydrates. **True/False**

2 Ounce for ounce, presweetened breakfast cereals and unsweetened cereals provide about the same number of calories. **True/False**

3 A 12-ounce can of soft drink contains about 3 tablespoons (9 teaspoons) of sugar. **True/False**

4 All foods high in fiber are fibrous. **True/False**

5 Cooking vegetables destroys their fiber content. **True/False**

6 Sugar consumption is strongly related to the development of tooth decay. **True/False**

Answers can be found at the end of the unit.

After completing Unit 12 and its interactive learning features, you will be able to:

- Describe the food sources and functions of simple sugars, complex carbohydrates, fiber, and alcohol sugars.

- Explain relationships between carbohydrates and dental health.

- Identify foods and beverages that can be consumed by individuals with lactose maldigestion to assure adequate intake of calcium and vitamin D.

carbohydrates Chemical substances in foods that consist of a simple sugar molecule or multiples of them in various forms.

Carbohydrates

- **Describe the food sources and functions of simple sugars, complex carbohydrates, fiber, and alcohol sugars.**

Carbohydrates are the major source of energy for people throughout the world. They are the primary ingredient of staple foods such as pasta, rice, cassava, beans, and bread. On average, Americans consume fewer carbohydrates than people in much of the world: approximately half of total calories.[1] This level of intake is on the low end of the recommended range of carbohydrate intake of 45 to 65% of total calories.[2]

The carbohydrate family consists of three types of chemical substances:

1. Simple sugars
2. Complex carbohydrates ("starch")
3. Total fiber

Some food sources of these different types of carbohydrates are shown in Illustration 12.1. Carbohydrates consist of carbon, hydrogen, and oxygen. They perform a number of functions in the body, but their primary function is to serve as an energy source. Simple sugars and complex carbohydrates supply the body with four calories per gram. Dietary fiber, on average, supplies two calories per gram. Although humans cannot digest fiber, bacteria in the colon can digest some types of fiber. These bacteria excrete fatty acids as a waste product from fiber digestion. The fatty acids are absorbed and used as a source of energy.[3] The total contribution of fiber to our energy intake is modest (around 50 calories) and supplying energy is not a major function of fiber. Certain carbohydrates perform roles in the functioning of the immune system, reproductive system, and blood clotting. One simple sugar (ribose) is a key a component of the genetic material DNA.

Key Nutrition Concepts

The various forms of carbohydrates are important components of our diets and influence health in many ways. Knowledge about carbohydrates presented here relates to these three key nutrition concepts:

1. Foods provide energy (calories), nutrients, and other substances needed for growth and health.

2. Poor nutrition can result from both inadequate and excessive levels of nutrient intake.

3. Adequacy, variety, and balance are key characteristics of healthful diets.

Alcohol sugars and alcohol (ethanol) are similar in chemical structure to carbohydrates (Illustration 12.2) and are presented in this text. The alcohol sugars are presented in this unit and alcohol in Unit 14. Increasingly, carbohydrates and carbohydrate-containing foods are being classified by their glycemic index, or the extent to which they increase blood glucose levels. This topic is introduced in this unit and explored further in Unit 13 on diabetes.

Illustration 12.1 The carbohydrate family. Some food sources of simple sugars (left), and some food sources of starch and dietary fiber (right) are shown.

Illustration 12.2 The similar chemical structures of glucose, fructose, xylitol (an alcohol sugar), and ethanol (an alcohol).

Glucose   Fructose   Xylitol   Ethanol

Simple Sugar Facts

Simple sugars are considered "simple" because they are small molecules that require little or no digestion before they can be used by the body. They come in two types: **monosaccharides** and **disaccharides**. The monosaccharides consist of one molecule and include glucose (blood sugar or dextrose), fructose (fruit sugar), and galactose. Disaccharides consist of two monosaccharide molecules (see Table 12.1). The combination of a glucose molecule and a fructose molecule makes sucrose (or table sugar); maltose (malt sugar) is made from two glucose molecules; and lactose (milk sugar) consists of a glucose molecule plus a galactose molecule. Honey, by the way, is a disaccharide. It is composed of glucose and fructose just as sucrose is, but it's a liquid rather than a solid because of the way the two molecules of sugar are chemically linked together. Disaccharides are broken down into their monosaccharide components during digestion; only glucose, fructose, and galactose are absorbed into the bloodstream.

High-fructose corn syrup—a liquid sweetener used in many soft drinks, fruit drinks, breakfast cereals, and other products is also considered a simple sugar. It generally consists of 55% fructose and 45% glucose, compared to sucrose that contain 50% glucose and 50% fructose.[4] Most of the simple sugars have a distinctively sweet taste.

The simple sugars the body uses directly to form energy are glucose and fructose. Galactose is readily converted by the body to glucose. When the body has more glucose than it needs for energy formation, it converts the excess to fat and to **glycogen**, the body's storage form of glucose. Glycogen is a type of complex carbohydrate and storage of it is limited. It consists of chains of glucose units linked together in long strands. Glycogen is produced only by animals and is stored in the liver (Illustration 12.3) and muscles. When the body needs additional glucose, glycogen is broken down, making glucose available for energy formation. Glucose can also be derived from certain amino acids and the glycerol component of fats. Illustration 12.4 shows the various ways glucose becomes available to the body. A constant supply of glucose is needed because the brain, red blood cells, white blood cells, and specific cells in the kidneys require glucose as an energy source.[5]

Table 12.1 The monosaccharides and the disaccharides they form

Monosaccharides	Disaccharide formed
glucose + glucose	maltose
glucose + fructose	sucrose
glucose + galactose	lactose

simple sugars   Carbohydrates that consist of a glucose, fructose, or galactose molecule, or a combination of glucose and either fructose or galactose. High-fructose corn syrup and alcohol sugars are also considered simple sugars. Simple sugars are often referred to as sugars.

monosaccharides   (mono = one, saccharide = sugar): Simple sugars consisting of one sugar molecule. Glucose, fructose, and galactose are common examples of monosaccharides.

disaccharides   (di = two, saccharide = sugar): Simple sugars consisting of two molecules of monosaccharides linked together. Sucrose, maltose, and lactose are disaccharides.

glycogen   The body's storage form of glucose. Glycogen is stored in the liver and muscles.

Illustration 12.3   Glycogen in a liver cell. The black "rosettes" are aggregates of glycogen molecules. This cell was photographed under an electron microscope at a magnification of 65,000X.

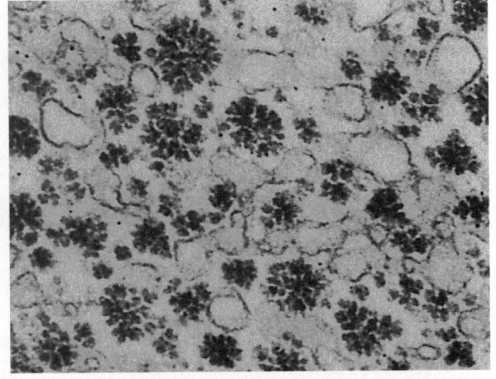

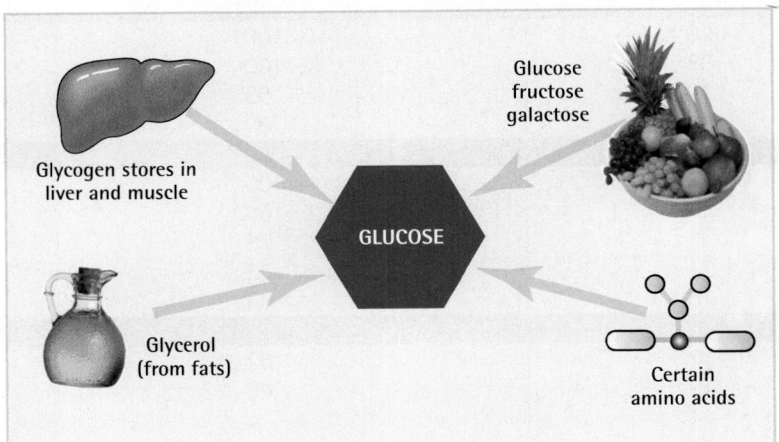
Glucose fructose galactose

Glycogen stores in liver and muscle

GLUCOSE

Glycerol (from fats)

Certain amino acids

Illustration 12.4   The body's sources of glucose.

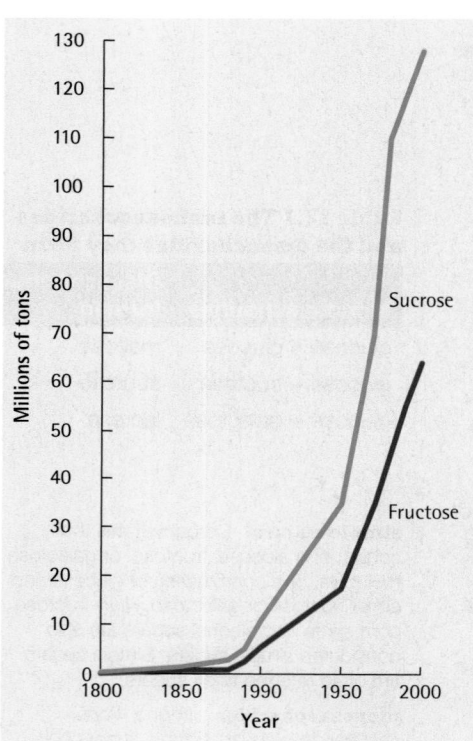

Simple Sugar Intake Most of the simple sugar in our diet comes from foods and beverages sweetened with sucrose and high-fructose corn syrup. As a matter of fact, simple sugars are the most commonly used food additive.[6] Per person consumption of added sucrose and fructose has increased dramatically across the globe over time (Illustration 12.5). Added sugars make up 15% of the total calorie intake of Americans.[1] That's a lot of sugar and far more than is good for health.[7]

Most all of the added sugar in diets comes from soft drinks, fruits drinks, energy and sports drinks, ready-to-eat cereals, candy, and desserts.[7] One 12-ounce serving of a soft drink contains about 9 teaspoons of simple sugar. Simple sugars are also present in fruits, and small amounts are found in some vegetables (Table 12.2). Fruits and vegetables, however, provide an array of vitamins and minerals, and beneficial phytochemicals that are absent from many foods high in added sugar.[2] Milk is the only animal product that contains significant amounts of a simple sugar (lactose).

Nutrition Labeling of Sugars Nutrition labels must list the total amount of mono- and diglycerides per serving of food under the heading "sugars" (Illustration 12.6). In addition, in the ingredient list, all simple sugars contained in the product must be listed in order of weight.[8] Labels contain information on total sugars per serving and do not distinguish between sugars naturally present in foods and added sugars.

Illustration 12.5 Worldwide trends in sucrose and fructose consumption between 1800 and 2000.
Source: Figure drawn from data presented in Bray GA.[21]

Table 12.2 The simple sugar content of some common foods

	Amount	Simple Sugars (grams)[a]	Percent of Total Calories from Simple Sugars
Sweeteners:			
Corn syrup	1 tsp	5	100%
Honey	1 tsp	6	100
Maple syrup	1 tsp	4	100
Table sugar	1 tsp	4	100
Fruits:			
Apple	1 medium	16	91
Peach	1 medium	8	91
Watermelon	1 wedge (4" × 8")	25	87
Orange	1 medium	14	86
Banana	1 medium	21	85
Vegetables:			
Broccoli	½ cup	2	40
Corn	½ cup	3	30
Potato	1 medium	1	4
Beverages:			
Fruit drinks	1 cup	29	100
Soft drinks	12 oz	38	100
Skim milk	1 cup	12	53
Whole milk	1 cup	11	28
Candy:			
Gumdrops	1 oz	25	100
Hard candy	1 oz	28	100
Caramels	1 oz	21	73
Fudge	1 oz	21	73
Milk chocolate	1 oz	16	44
Breakfast cereals:			
Apple Jacks	1 oz	13	52
Raisin Bran	1 oz	19	40
Cheerios	1 oz	14	4

[a]4 grams sucrose = 1 teaspoon.

	% Daily Value**	
Total Fat 1 g*	2%	2%
Saturated Fat 0 g	0%	0%
Monounsaturated Fat 0 g		
Polyunsaturated Fat 0.5 g		
Trans Fat 0 g		
Cholesterol 0 mg	0%	0%
Sodium 5 mg	0%	3%
Potassium 200 mg	6%	12%
Total Carbohydrate 48 g	16%	18%
Dietary Fiber 6 g	24%	24%
Sugars 12 g		
Other Carbohydrate 30 g		
Protein 6 g		
Vitamin A	0%	4%
Vitamin C	0%	0%
Calcium	0%	15%
Iron	90%	90%
Thiamin	25%	30%
Riboflavin	25%	35%
Niacin	25%	25%
Vitamin B$_6$	25%	25%
Folic Acid	25%	25%
Vitamin B$_{12}$	25%	35%
Phosphorus	15%	25%
Magnesium	15%	20%
Zinc	10%	15%
Copper	10%	10%

	Calories	2,000	2,500
Total Fat	Less than	65 g	80 g
Saturated Fat	Less than	20 g	25 g
Cholesterol	Less than	300 mg	300 mg
Sodium	Less than	2,400 mg	2,400 mg
Potassium		3,500 mg	3,500 mg
Total Carbohydrate		300 g	375 g
Dietary Fiber		25 g	30 g
Calories Per gram: Fat 9 • Carbohydrate 4 • Protein 4			

INGREDIENTS: WHOLE GRAIN WHEAT, SUGAR, HIGH FRUCTOSE CORN SYRUP, GELATIN, **VITAMINS AND MINERALS:** REDUCED IRON, NIACINAMIDE, ZINC OXIDE, PYRIDOXINE HYDRO-CHLORIDE (VITAMIN B$_6$), RIBOFLAVIN (VITAMIN B$_2$), THIAMIN HYDROCHLORIDE (VITAMIN B$_1$), FOLIC ACID AND VITAMIN B$_{12}$, TO MAINTAIN QUALITY, BHT HAS BEEN ADDED TO THE PACKAGING.

Illustration 12.6 Labeling the sugar content of breakfast cereal

Added Sugars and Health Foods to which simple sugars have been added are often not among the top sources of nutrients. By themselves, simple sugars are among the few foods that provide only calories. Many foods high in simple sugars, such as cake, sweet rolls, cookies, pies, and ice cream, are also high in fat. The likelihood that diets will provide insufficient amounts of vitamins and minerals increases along with sugar intake.[2] High levels of consumption of sucrose and fructose are related to increased blood levels of triglycerides.[11] Neither sucrose nor high fructose corn syrup has been found to cause obesity, type 2 diabetes, or attention deficit hyperactivity isorder.[9,10,56] However, high intakes of these sweeteners may contribute to the risk of developing types 2 diabetes and obesity.[9] Recent research indicates the possibility that regular intake of sugar-sweetened beverages in young children increases the risk of obesity versus the risk in young children who do not consume them.[12] High sugar intake appears to be related to fat accumulation in the liver, particularly in Hispanic children, due to a genetic susceptibility.[12] Furthermore, it is perfectly clear that the frequent consumption of sugary foods causes tooth decay.[2]

Interestingly, sucrose has the property of dampening pain in infants. A few drops of sucrose placed on an infant's tongue produces pain-relieving effects that last several minutes. Sucrose solution is being used in hospitals to control pain in infants undergoing heel pricks for blood collection and other minor procedures.[13]

Advice on Added Sugar Intake What's the bottom line on eating sugary foods? Enjoy them in limited amounts as part of a healthful diet. The American Heart Association recommends that women consume no more than 100 calories in added sugar (about 6 teaspoons a day) and men 150 calories of added sugar (about 9 teaspoons).[7]

If you are concerned about the amount of sugar in your diet, don't wait until it's time for a New Year's resolution. Get some suggestions for small changes that will have a positive impact in this unit's Take Action feature.

The Alcohol Sugars—What Are They?

Nonalcoholic in the beverage sense, the **alcohol sugars** (or polyols) are like simple sugars except that they include a chemical component of alcohol. Like simple sugars, the alcohol

alcohol sugars Simple sugars containing an alcohol group in their molecular structure. The most common are xylitol, mannitol, and sorbitol. They are a subgroup of chemical substances called polyols.

Illustration 12.7 Examples of products sweetened with xylitol, mannitol, and sorbitol.

sugars have a sweet taste. Xylitol is by far the sweetest alcohol sugar—it's much sweeter than the other two common alcohol sugars, mannitol and sorbitol.

Alcohol sugars are found naturally in very small amounts in some fruits. They are mostly used as sweetening agents in gums and candy (Illustration 12.7). Unlike the simple sugars, xylitol, mannitol, and sorbitol do not promote tooth decay because bacteria in the mouth that cause tooth decay cannot digest them.[6] Foods sweetened with alcohol sugars can use the health claim "Does not promote tooth decay" on labels.

Like dietary fiber, the alcohol sugars are slowly and incompletely broken down in the gastrointestinal tract and provide fewer calories per gram than other carbohydrates.[14] On average, alcohol sugars provide two calories per gram, so foods labeled "sugar free" will not be calorie free. High intake of alcohol sugars can, like fiber, cause diarrhea. This characteristic limits their use in foods. The "diarrhea dose" for alcohol sugars defined by the FDA equals 50 grams of sorbitol and 20 grams of mannitol. A food product's content of alcohol sugars per serving must be listed on the nutrition label as "Sugar Alcohols" or by the name of the sugar alcohol.[8]

Artificial Sweetener Facts

Unwanted calories in simple sugars, the connection of sucrose with tooth decay, the need for a sugar substitute for people with diabetes, and sugar shortages such as occurred during the two world wars have all provided incentives for developing sugar substitutes. Six artificial sweeteners are currently on the market in the United States (Table 12.3) and

take action Lower Your Sugar Intake

Concerned about your sugar intake? Want to consider some ways to lower it and to protect your teeth? Consider these actions and check the options you'd be willing to try.

_____ 1. When you want something sweet to eat, try:
    _____ sweet cherries
    _____ melon
    _____ a banana
    _____ a mango
    _____ unsweetened applesauce

_____ 2. Replace a serving of soft drink or fruit drink with water or a no-added sugar, 100% juice serving such as:
    _____ tomato juice
    _____ vegetable juice
    _____ dark grape juice
    _____ apple cider/juice
    _____ pineapple juice
    _____ cranberry juice
    _____ grapefruit tangerine juice
    _____ other juice

_____ 3. Taste-test beverages or gum sweetened with alcohol sugars or artificial sweeteners for acceptability. Try:
    _____ iced tea sweetened with aspartame or sucralose
    _____ soft drinks sweetened with Reb A or aspartame
    _____ gum sweetened with xylitol or other alcohol sugar

_____ 4. Keep sugar off your teeth. After you eat a food with added sugar:
    _____ rinse your mouth with water
    _____ brush your teeth
    _____ floss in-between your teeth

Table 12.3 Artificial sweeteners currently approved for use in the United States

Trade name	Product name	Calories/gram
Saccharin	Sweet and Low	0
Aspartame	NutraSweet, Equal Sugar Twin	4
Sucralose	Splenda	0
Acesulfame potassium	Acesulfame K, Sunnette, Sweet One	0
Neotame	–	0
Rebiana	Reb-A, Truvia, PureVia	0

more are being developed. None of the artificial sweeteners that are currently approved for use exactly mimic the taste and properties of sugar.[15]

Artificial sweeteners are also known as nonnutritive sweeteners because they are not a significant source of energy or nutrients.[15] They have chemical properties that invoke an intensively sweet taste on the tongue. Gram for gram, artificial sweeteners are 160 to 13,000 times sweeter than sucrose, and only small amounts are needed to sweeten food products. Of the artificial sweeteners, only aspartame provides calories (4 calories/gram).[6]

The artificial sweeteners currently on the market do not promote tooth decay because they are not utilized by bacteria in the mouth that cause decay.[6] They do not appear to promote weight loss without calorie restriction.[2] Artificial sweeteners are used to sweeten thousands of products, a few of which are shown in Illustration 12.8.

Saccharin Saccharin was the first artificial sweetener developed. Did you know that it was discovered in a laboratory in the late 1800s? That's right—saccharin is over 100 years old. The availability of this artificial sweetener, which is 300 times as sweet as sucrose, helped relieve the sugar shortages that occurred during World Wars I and II.

In 1977 saccharin was taken off the market after very high doses were found to cause cancer in laboratory animals. At that time, however, saccharin was the only no-calorie artificial sweetener available, and its removal sparked a public outcry. After many consumers complained to Congress, saccharin was returned to the market by congressional mandate. Saccharin was deemed safe in 2000 after scientists concluded there was no clear evidence that it causes cancer in humans.[16] It is used in some types of toothpastes, diet sodas, mouthwash, pill coatings, juices, and jellies.[17]

Aspartame Early in the 1980s, the artificial sweetener aspartame was approved for use in the United States and more than 90 other countries. Primarily known as NutraSweet, this artificial sweetener is about 200 times sweeter than sucrose.

Illustration 12.8 Some of the thousands of foods that contain artificial sweeteners.

© Scott Goodwin Photography

NEOTAME
SUCRALOSE
SACCHARIN
REB-A
ACESULFAME K
ASPARTAME
FRUCTOSE
SUCROSE
XYLITOL
GLUCOSE
SORBITOL
MANNITOL
GALACTOSE
MALTOSE
LACTOSE

Illustration 12.9 A ranking of various types of artificial sweeteners and naturally occurring sugars in order of sweetness.

Illustration 12.10 You would have to consume more than 20 cans of soft drinks sweetened with aspartame (NutraSweet) a day to exceed the safe limit set for this artificial sweetener.

phenylketonuria (feen-ol-key-tone-u-re-ah) (PKU) A rare genetic disorder related to the lack of the enzyme phenylalanine hydroxylase. Lack of this enzyme causes the essential amino acid phenylalanine to build up in blood.

Aspartame is made from two amino acids (phenylalanine and aspartame). Although both are found in nature, it took chemists to arrange their chemical partnership. Because aspartame is made from amino acids (the building blocks of protein), it supplies four calories per gram. Aspartame is so sweet, however, that very little is needed to sweeten products. Illustration 12.9 shows the relative sweetening power of various artificial sweeteners and naturally occurring sugars.

Aspartame is used in more than 6,000 products worldwide, including soft drinks, whipped toppings, jellies, cereals, puddings, and some medicines. Products containing aspartame must carry a label warning people with **phenylketonuria (PKU)** (an inherited disease) and others with certain liver conditions about the presence of phenylalanine. People with these disorders are unable to utilize the amino acid phenylalanine, causing it to build up in the blood. Because high temperatures tend to break down aspartame, it is not used in baked or heated products.[6]

Is Aspartame Safe? A safe level of aspartame intake is defined as 50 milligrams per kilogram of body weight per day in the United States and as 40 milligrams per kilogram of body weight in Canada.[18] In food terms, the limit in the United States is equivalent to approximately 20 aspartame-sweetened soft drinks or 55 desserts per day (Illustration 12.10). The average intake of aspartame in the United States, Canada, Germany, and Finland, for example, ranges from 2 to 10 milligrams per day, well below the level of intake considered safe.[18,19]

A small proportion of individuals, however, report that they are sensitive to aspartame and develop headaches, dizziness, or anxiety when they consume small amounts. Studies have failed to confirm these effects.[19] Aspartame has not been found to promote cancer, nerve disorders, or other health problems in humans.[20]

Sucralose This noncaloric, intense sweetener is made from sucrose, is safe, is very sweet (600 times sweeter than sucrose), and does not leave a bitter aftertaste. Known primarily as Splenda on product labels, it is used in both hot and cold food products, including soft drinks, baked goods, frosting, pudding, and chewing gum.

Acesulfame Potassium Also known as acesulfame K, Sunette, and Sweet One, acesulfame potassium was approved by the FDA in 1988. It is added to at least 4,000 foods and is used in food production in about 90 countries. It is 200 times as sweet as sucrose, provides zero calories, and does not break down when heated.

Neotame Neotame is derived from the same amino acids as aspartame and is extraordinarily sweet. Its sweetness potency is 7,000 to 13,000 times that of sucrose. Only minute

Richard Anderson

Illustration 12.11 The stevia plant is the source of the artificial sweetener rebiana.

amounts of neotame are absorbed when it is consumed, and it is not considered harmful to individuals with PKU. It is marketed as having a clean sweet taste with little bitter or metallic aftertaste.[6]

Rebiana In December 2008, the FDA approved Rebiana for use as an artificial sweetener. Called Reb-A, Truvia, and PureVia, this sweetener is derived from the herb stevia (Illustration 12.11). Stevia grows in subtropical and tropical areas and has been known in these areas as sugar leaf for centuries.

Stevia leaves contain the chemical rebaudioside that imparts a sweet taste and it is used in purified form in Reb-A. It is currently used to sweeten beverages (Illustration 12.12) and reportedly tastes best with citrus flavors.[15]

Illustration 12.12 Example of beverages sweetened with rebiana (Reb-A).

Complex Carbohydrate Facts

Starches, glycogen, and dietary fiber constitute the **complex carbohydrates**. They are known as **polysaccharides**. Only plant foods such as grains, potatoes, dried beans, and corn that contain starch and dietary fiber are considered dietary sources of complex carbohydrates (Table 12.4). Very little glycogen is available from animal products.

complex carbohydrates The form of carbohydrate found in starchy vegetables, grains, and dried beans and in many types of dietary fiber. The most common form of starch is made of long chains of interconnected glucose units.

polysaccharides (poly = many, saccharide = sugar): Carbohydrates containing many molecules of monosaccharides linked together. Starch, glycogen, and dietary fiber are the three major types of polysaccharides. Polysaccharides consisting of 3 to 10 monosaccharides may be referred to as oligosaccharides.

Which Foods Have Carbohydrates? Food sources of complex carbohydrates include whole grain breads, cereals, pastas, and crackers, as well as these same foods produced from refined grains. Whole grain products provide more fiber and beneficial substances naturally present in grains than do refined grain products. Regular intake of whole grain foods reduce the risk of heart disease and some types of cancer. [2]

Are They Fattening? Starchy foods are caloric bargains (Illustration 12.13). A medium baked potato weighs in at only 122 calories, a half cup of corn at 85 calories, and a slice of bread at 70 calories. You can expand the caloric value of complex carbohydrates quite easily by adding fat, sauces, and cheese. One cup of macaroni (about 200 calories) gains around 180 calories when it comes as macaroni and cheese. Adding a quarter cup of gravy to potatoes elevates calories by 150.

Illustration 12.13 Which has more calories?

Check your answers below.

One medium baked potato (four ounces) *or* three ounces of lean hamburger?

One slice of bread *or* a half cup of low-fat cottage cheese?

One cup of spaghetti noodles *or* 17 french fries (three ounces)?

Answers
Potato = 122 calories;
Lean hamburger = 239 calories.
Bread = 70 calories;
Cottage cheese = 102 calories.
Spaghetti = 197 calories;
French fries = 265 calories.

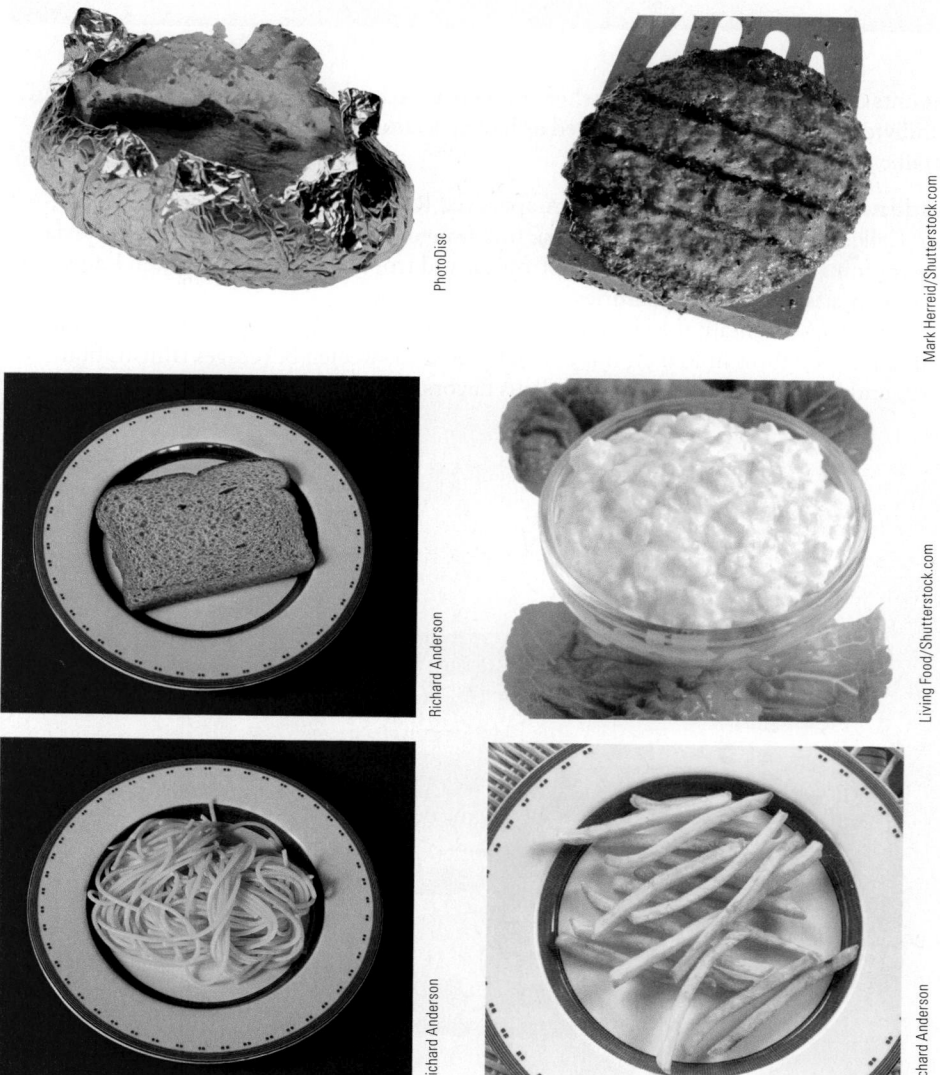

PhotoDisc

Mark Herreid/Shutterstock.com

Richard Anderson

Living Food/Shutterstock.com

Richard Anderson

Richard Anderson

Table 12.4 The complex carbohydrate content of some common foods

	Amount	Complex carbohydrate (grams)	Percent of total calories from complex carbohydrate
Grain and grain products:			
Rice (white), cooked	½ cup	21	83%
Pasta, cooked	½ cup	15	81
Cornflakes	1 cup	11	76
Oatmeal, cooked	½ cup	12	74
Cheerios	1 cup	11	68
Whole wheat bread	1 slice	7	60
Dried beans (cooked):			
Lima beans	½ cup	11	64
White beans	½ cup	13	63
Kidney beans	½ cup	12	59
Vegetables:			
Potato	1 medium	30	85
Corn	½ cup	10	67
Broccoli	½ cup	2	40

Fiber Intake and Health What is low in calories; prevents constipation; may lower the risk of heart disease, obesity, and diabetes; and is generally underconsumed by people in the United States? The answer is dietary fiber.[22–24]

Total fiber intake by U.S. children and adults (15 grams per day) is well below the amount recommended (28 grams for women and 35 grams for men).[25] People who consume the recommended amount of fiber tend to select whole grain breads, high-fiber cereal, and dried beans most days and eat at least five servings of vegetables and fruits daily.[26] Food sources of fiber are listed in Table 12.5. It doesn't matter whether the fiber foods are mashed, chopped, cooked, or raw. They retain their fiber value through it all.

In the past it was assumed that fiber had no calorie value because it is not broken down by human digestive enzymes. Recent studies, however, suggest that the calorie contribution of dietary fiber be reconsidered. Bacteria in the colon are able to break down many types of fiber to some extent. The bacteria excrete fatty acids as a waste product, and they are used as an energy source by the colon and the rest of the body. On average, fiber provides two calories per gram.[3] Although not yet required on U.S. or Canadian nutrition information labels, those in the European Union must include the calorie contribution of fiber in food products.[27]

REALITY CHECK
Sugar Addiction

Have you ever pined or whined for something sweet? Do you think you can get addicted to carbohydrates?

Who gets a thumbs-up?

Answers on page 12-12

Terry: Maybe you could really love to eat them, but addicted? I don't think so.

Wolfgang: The more I don't eat candy, the more I want to eat some. I swear, I'm addicted.

ANSWERS TO REALITY CHECK
Sugar Addiction

The theory that consumption of sweet foods may become addictive in some people is being intensely studied, especially among individuals with restrictive followed by binge-eating patterns. Sweet foods have *not* been found to cause addiction in healthy humans.[33,34]

Terry:

Wolfgang:

Table 12.5 **Examples of good sources of fiber**

	Amount	Fiber (grams)
Grain and grain products:		
Bran Buds	¹/₂ cup	12.0
All Bran	¹/₂ cup	11.0
Raisin Bran	1 cup	7.0
Granola (homemade)	¹/₂ cup	6.0
Bran Flakes	³/₄ cup	5.0
Oatmeal	1 cup	4.0
Spaghetti noodles	1 cup	4.0
Shredded Wheat	1 biscuit	2.7
Whole wheat bread	1 slice	2.0
Bran (dry; wheat, oat)	2 Tbsp	2.0
Fruits:		
Raspberries	1 cup	8.0
Avocado	¹/₂ medium	7.0
Mango	1 medium	4.0
Pear (with skin)	1 medium	4.0
Apple (with skin)	1 medium	3.3
Banana	6" long	3.1
Orange (no peel)	1 medium	3.0
Peach (with skin)	1 medium	2.3
Strawberries	10 medium	2.1
Vegetables:		
Lima beans	¹/₂ cup	6.6
Green peas	¹/₂ cup	4.4
Potato (with skin)	1 medium	3.5
Brussels sprouts	¹/₂ cup	3.0
Broccoli	¹/₂ cup	2.8
Carrots	¹/₂ cup	2.8
Green beans	¹/₂ cup	2.7
Collard greens	¹/₂ cup	2.7
Cauliflower	¹/₂ cup	2.5
Corn	¹/₂ cup	2.0

Table 12.5 continued

	Amount	Fiber (grams)
Nuts:		
Almonds	¼ cup	4.5
Peanuts	¼ cup	3.3
Peanut butter	2 Tbsp	2.3
Dried beans (cooked):		
Pinto beans	½ cup	10.0
Peas, split	½ cup	8.2
Black beans (turtle beans)	½ cup	8.0
Lentils	½ cup	7.8
Kidney or navy beans	½ cup	6.9
Black-eyed peas	½ cup	5.3
Fast foods:		
Big Mac	1	3
French fries	1 regular serving	3
Whopper	1	3
Cheeseburger	1	2
Taco	1	2
Chicken sandwich	2	1
Egg McMuffin	1	1
Fried chicken, drumstick	1	1

Types of Fiber A classification system for defining edible fibers based on type of fiber has been developed.[5] Fibers are classified as **functional fiber**, **dietary fiber**, and **total fiber**. All fibers share the property of not being digested by human digestive enzymes.

Functional fibers are non-digestible carbohydrates extracted from foods or produced commercially for use in fortifying foods with fiber, as well as in fiber supplements.[5] Psyllium, pectin, and seed and plant gums are classified as functional fibers.

Dietary fiber consists of nondigestible carbohydrates present in plant foods. They are found in the bran component of oats and wheat (Illustration 12.14), in cellulose (a rigid component of plant cell walls), in vegetables and fruits, and in the nondigestible starch components of dried beans. The recommended daily intake of fiber is based on total fiber, which is the sum of functional plus dietary fiber intake.[5]

There is a long history of studies showing health benefits of dietary fiber. Relatively few studies have examined the health effects of functional fiber.[28] Several functional fibers classified as soluble fiber have been approved for health claims related to decreased risk of heart disease and reduced blood cholesterol levels.[29]

Soluble and Insoluble Fiber Soluble fiber slows glucose absorption, thereby lowering peak blood levels of glucose, and reduces fat and cholesterol absorption.[24,30] This type of fiber is found in oats, barley, fruit pulp, dried beans, and psyllium. They are called *soluble* because this type of fiber combines chemically with water.[26] They are fibers that are not fibrous. Soluble fibers form thick liquids similar to gels when combined with water.[30]

Insoluble fibers, such as wheat bran and other brans, the fiber in legumes, the coating on seeds, and the skin on fruits and vegetables do not combine chemically with water. They are particularly beneficial for preventing constipation.[30]

Be Cautious When Adding More Fiber to Your Diet Newcomers to adequate fiber diets often experience diarrhea, bloating, and gas for the first week or so of increased fiber intake. These side effects can be avoided. They occur when too much fiber is added to the diet too quickly. Adding sources of dietary fiber to the diet gradually can prevent these side effects. In addition, dietary fiber can be constipating if consumed with too little

functional fiber Specific types of nondigestible carbohydrates that have beneficial effects on health. Two examples of functional fibers are psyllium and pectin.

dietary fiber Naturally occurring, intact forms of nondigestible carbohydrates in plants and "woody" plant cell walls. Oat and wheat bran, and raffinose in dried beans, are examples of this type of fiber.

total fiber The sum of functional and dietary fiber.

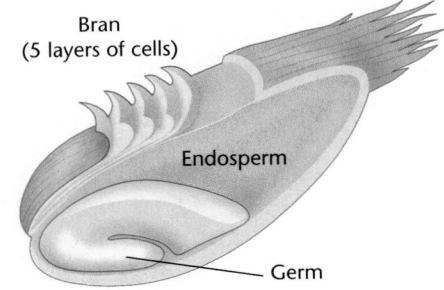

Illustration 12.14 Diagram of a grain of wheat showing the bran that is a rich source of dietary fiber. The germ contains protein, unsaturated fats, thiamin, niacin, riboflavin, iron, and other nutrients. (The bran and germ are removed in the refining process.) The endosperm primarily contains starch, the storage form of glucose in plants.

health action Putting the Fiber into Meals and Snacks

HIGH–FIBER OPTIONS FOR BREAKFAST

Whole grain toast		2 g per slice
Bran cereal:		
Bran flakes	1 cup	7 g
All Bran	$\frac{1}{2}$ cup	11 g
Raisin bran	1 cup	7 g
Oat bran	$\frac{1}{3}$ cup	5 g
Bran muffin, with fruit	1 small	3 g
Strawberries	10	2 g
Raspberries	$\frac{1}{2}$ cup	3 g
Bananas	1 medium	2 g

LUNCHES THAT INCLUDE FIBER

Whole grain bread		2 g per slice
Baked beans	$\frac{1}{2}$ cup	10 g
Carrot	1 medium	2 g
Raisins	$\frac{1}{4}$ cup	2 g
Peas	$\frac{1}{2}$ cup	4 g
Peanut butter	2 tablespoons	2 g

FIBER ON THE MENU FOR SUPPER

Brown rice	$\frac{1}{2}$ cup	2 g
Potato	1 medium	3 g
Dried cooked beans	$\frac{1}{2}$ cup	8 g
Broccoli	$\frac{1}{2}$ cup	3 g
Corn	$\frac{1}{2}$ cup	3 g
Tomato	1 medium	2 g
Green beans	$\frac{1}{2}$ cup	3 g

Photo Disc

FIBER–FILLED SNACKS

Peanuts	$\frac{1}{4}$ cup	3 g
Apple	1 medium	2 g
Pear	1 medium	4 g
Orange	1 medium	3 g
Prunes	3	2 g[a]
Sunflower seeds	$\frac{1}{4}$ cup	2 g
Popcorn	2 cups	2 g

[a]Prunes contain fiber, but their laxative effect is primarily due to a naturally occurring chemical substance that causes an uptake of fluid into the intestines and the contraction of muscles that line the intestines.

fluid. Your fluid intake should increase along with your intake of dietary fiber.[31] You know you've got the right amount of fiber in your diet when stools float and are soft and well formed.

The Health Action on this page suggests some food choices that will put fiber into your meals and snacks. Fiber should be added carefully to children's diets, because the large volume of some high-fiber diets can fill the children up before they consume enough calories.[32]

Glycemic Index of Carbohydrates

In the not-too-distant past it was assumed that "a carbohydrate is a carbohydrate is a carbohydrate." It was thought that all types of carbohydrates had the same effect on blood glucose levels and health, so it didn't matter what type was consumed. As is the case with many untested assumptions, this one fell by the wayside. It is now known that some types of simple and complex carbohydrates in foods elevate blood glucose levels more than do others. Such differences are particularly important to people with disorders such as **insulin resistance** and **type 2 diabetes**.[35,36]

Carbohydrates and carbohydrate-containing foods are now being classified by the extent to which they increase blood glucose levels. This classification system is called the **glycemic index**. Carbohydrates that are digested and absorbed quickly have a high glycemic index and raise blood glucose levels to a higher extent than do those with lower glycemic index values. Carbohydrates and carbohydrate-containing foods with high glycemic index values include glucose, white bread, baked potatoes, and jelly beans. Fructose, xylitol, hummus, apples, and all-bran cereal are examples of carbohydrates and carbohydrate-containing foods with low glycemic indexes. Fructose has a low glycemic index because it does not raise blood glucose levels. It is absorbed as fructose and does not require insulin for passage into cells.[37]

Carbohydrates and Dental Health

- **Explain relationships between carbohydrates and dental health.**

The relationship between sugar and **tooth decay** is very close, and the history of tooth decay closely parallels the availability of sugar. The incidence of tooth decay is estimated to have been very low (less than 5%) among hunter-gatherers who had minimal access to sugars.[38] Tooth decay did not become a widespread problem until the late 17th century, when great quantities of sucrose were exported from the New World to Europe and other parts of the world. When sugar shortages occurred in the United States and Europe during World War I and World War II, rates of tooth decay declined; they rebounded when sugar became available again.[38] Rates of tooth decay in children vary substantially among countries, but the highest rates are in countries where sugar is widely available in processed foods and beverages. Tooth decay is spreading rapidly in developing countries where sugar, candy, soft drinks, and fruit drinks are becoming widely available.[39]

Sweets are not the only culprit. Simple sugars that promote tooth decay can also come from starchy foods, especially pretzels, crackers, and breads that stick to your gums and teeth. Some of the starch is broken down to simple sugars by enzymes in the mouth.

To reduce the incidence of tooth decay, a number of countries have developed campaigns to help inform consumers about cavity-promoting foods. Switzerland and other countries label foods that are safe for the teeth with a "Tooth Friendly" symbol (Illustration 12.15) and encourage the use of alcohol sugars (which don't promote tooth decay) in gums and candy.[14,40] Other countries recommend that sweets be consumed with meals or that teeth be brushed after sweets are eaten.

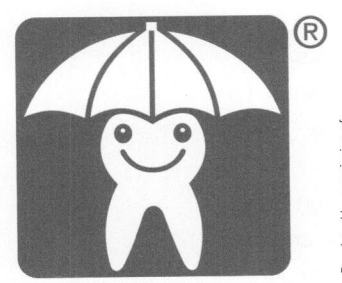

Illustration 12.15 Switzerland's "Tooth Friendly" symbol. It has become an internationally used symbol.

Reprinted by permission of Toothfriendly International

There's More to Tooth Decay Than Sugar Per Se

How frequently sugary and starchy foods are consumed and how long they stick to gums and teeth make a difference in their tooth-decay-promoting effects.[41] Marshmallows, caramels, and taffy, for example, are much more likely to promote tooth decay than are apples and milk chocolate. Nevertheless, all of the foods listed in Table 12.6 can promote tooth decay if allowed to remain in contact with the gums and teeth. Drinking coffee or tea with sugar throughout the day or consuming three or more regular soft drinks between meals hastens tooth decay (more than if these beverages are consumed with meals).[42] Candy, cookies, and crackers eaten between meals are much more likely to promote tooth decay than are the same foods consumed as part of a meal. Chewing as few as two sticks of sugar-containing gum a day also significantly increases tooth decay.[42,43]

Why Does Sugar Promote Tooth Decay? Sugar promotes tooth decay because it is the sole food for certain bacteria that live in the mouth and excrete acid that dissolves teeth. In the presence of sugar, bacteria in the mouth multiply rapidly and form a sticky, white material called **plaque**. Tooth areas covered by plaque are prime locations for tooth decay because they are dense in acid-producing bacteria. Acid production by bacteria increases within five minutes of exposure to sugar. It continues for 20 to 30 minutes after the bacteria ingest the sugar.[44] If teeth are frequently exposed to sugar, the acid produced

insulin resistance A condition is which cell membranes have a reduced sensitivity to insulin so that more insulin than normal is required to transport a given amount of glucose into cells.

type 2 diabetes A disease characterized by high blood glucose levels due to the body's inability to use insulin normally or to produce enough insulin.

glycemic index A measure of the extent to which blood glucose is raised by a 50 gram portion of a carbohydrate-containing food compared to 50 grams of glucose or white bread.

tooth decay The disintegration of teeth due to acids produced by bacteria in the mouth that feed on sugar. Also called dental caries or cavities.

plaque A soft, sticky, white material on teeth; formed by bacteria.

Table 12.6 The "stickiness" value of some foods
The stickier the food, the worse it is for your teeth.

Very sticky	Sticky	Somewhat sticky	Barely sticky
Caramels	Doughnuts	Bagels	Apples
Chewy cookies	Figs	Cake	Bananas
Crackers	Frosting	Cereal	Fruit drinks
Cream-filled cookies	Fudge	Dry cookies	Fruit juices
Granola bars	Hard candy	Milk chocolate	Ice cream
Marshmallows	Honey	Rolls	Oranges
Pretzels	Jelly beans	White bread	Peaches
Taffy	Pastries		Pears
	Raisins		
	Syrup		

Table 12.7 Foods that don't promote tooth decay[5,14,41]

Artificial sweeteners
Gum and candy sweetened with alcohol sugars
Peanut butter
Cheese
Tea
Coffee (no sugar)
Meats
Water
Eggs
Milk
Yogurt (plain)
Fats and oils
Nuts
Vegetables
Fresh fruit

by bacteria may erode the enamel, producing a cavity. If the erosion continues, the cavity can extend into the tooth and allow bacteria to enter the inside of the tooth. That can cause an infection and the loss of the tooth. It can be prevented if the plaque is removed before the acid erodes much of the enamel. Teeth are capable of replacing small amounts of minerals lost from enamel.[45] Table 12.7 lists foods that do not promote tooth decay.

Water Fluoridation In the early 1930s, lower rates of tooth decay were observed among children living in areas where water naturally contained fluoride. This provided the initial evidence that led to the fluoridation of many community water supplies. Fluoridated water reduces the incidences of tooth decay by 50% or more and is primarily responsible for declining rates of tooth decay and loss.[46]

Credit for declines in tooth decay and tooth loss in the United States is also shared by fluoride supplements, toothpastes, rinses and gels, protective sealants, and improved dental hygiene and care.[47] Further improvements in rates of dental caries will occur with reduced intake of sugars and sticky carbohydrates. Fluoridation is a safe, effective, and inexpensive method of controlling dental disease.[45] Despite the advantages of fluoride, 31% of Americans consume water from a less than optimally fluoridated water supply.[46] Few bottled waters contain fluoride.

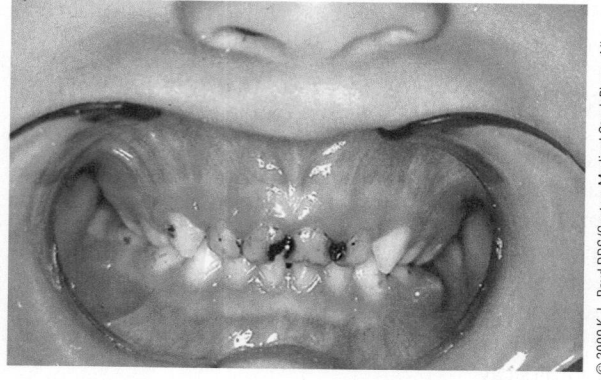

Baby Bottle Caries A startling example of the effect that frequent and prolonged exposure to sugary foods can have is "baby bottle caries" (Illustration 12.16). Infants and young children who routinely fall asleep while sucking a bottle of sugar water, fruit drink, milk, or formula—or while breastfeeding—may develop severe decay. After the child falls asleep, the fluid may continue to drip into the mouth. A pool of the fluid collects between the tongue and the front teeth, bathing the teeth in the sweet fluid for as long as the child sleeps. The upper front teeth become decayed first because the tongue protects the lower teeth. Baby bottle caries occur in 5 to 10% of infants and young children and can lead to the destruction of all baby teeth.[48]

Illustration 12.16   "Baby bottle caries" (also called "nursing bottle syndrome") occurs in infants who habitually receive sweet fluids or milk in bottles when they go to sleep. Cavities occur first in the upper front teeth because that's where fluid pools when babies sleep.

lactose maldigestion   A disorder characterized by reduced digestion of lactose due to the low availability of the enzyme lactase.

Lactose Maldigestion and Intolerance

- **Identify foods and beverages that can be consumed by individuals with lactose maldigestion to assure adequate intake of calcium and vitamin D.**

A very common digestive disorder is **lactose maldigestion**. The lactose found in milk and milk products presents a problem for most of the world's adults, who cannot digest it, either partially or completely (Table 12.8).[49] The condition occurs more commonly in population groups that have no historical links to dairy farming and milk drinking.[50] Early humans in central and northwestern Europe and the regions of Africa and China high-lighted in Illustration 12.17 tended to raise dairy animals and drink milk.

Lactose maldigestion is caused by a genetically determined low production of lactase, the enzyme that digests lactose.[51] People who lack this enzyme end up with free lactose in their large intestine after they consume milk or milk products. The presence of lactose in the large intestine produces the symptoms of **lactose intolerance**: a bloated feeling, diar-rhea, gas, and abdominal cramping.

Lactose maldigestion is rare in very young children and affects adults to various degrees. Some adults produce little or no lactase and develop symptoms of lactose intoler-ance when they consume only small amounts of milk or milk products. Others produce some lactase and can tolerate limited amounts of lactose-containing milk and milk prod-ucts, such as a cup of milk at a time or two cups of milk consumed with meals during the day.[51] Regular consumption of milk may improve lactose digestion due to enhanced bacte-rial breakdown of lactose in the gut.[52] Lactase tablets can help improve lactose digestion, but the benefit is limited due to the inactivation of lactase by digestive processes in the stomach. Once lost, the body's ability to produce sufficient lactase cannot be restored.[53]

Many people who are lactose maldigesters have no trouble eating yogurt and other fermented milk products such as cultured buttermilk, kefir, and aged cheese. The bacteria used to culture yogurt can digest half or more of the lactose. This reduction in lactose content is sufficient to prevent adverse effects in many people with lactose maldigestion.[53]

Milk solids, milk, and other lactose-containing components of milk may be added to foods you wouldn't expect. Consequently, it's best to examine food ingredient labels when in doubt. Milk, for instance, is a primary ingredient in some types of sherbet, and milk solids are added to many types of candy.

The single-most reliable indicator of lactose maldigestion is the occurrence of lactose intolerance within hours after consuming lactose.[54] If you consistently experience symp-toms of lactose maldigestion, visit your health care provider for a diagnosis. The symp-toms could be due to lactose, other substances in milk, or another problem.

How Is Lactose Maldigestion Managed?

Lactose maldigestion should *not* be managed by omitting milk and milk products from the diet! Doing so would exclude a food group that contributes a variety of nutrients that cannot easily be replaced by other foods. The omission of milk and milk products from

Table 12.8 Estimated incidence of lactose maldigestion among older children and adults in different population groups[49,50]

	Incidence of lactose maldigestion
Asian Americans	90%
Africans	70
African Americans	70
Asians	65 or more
American Indians	62 or more
Mexican Americans	53 or more
U.S. adults (overall)	25
Northern Europeans	20
American Caucasians	15

lactose intolerance The term for gas-trointestinal symptoms (flatulence, bloating, abdominal pain, diarrhea, and "rumbling in the bowel") resulting from the consumption of more lactose than can be digested with available lactase.

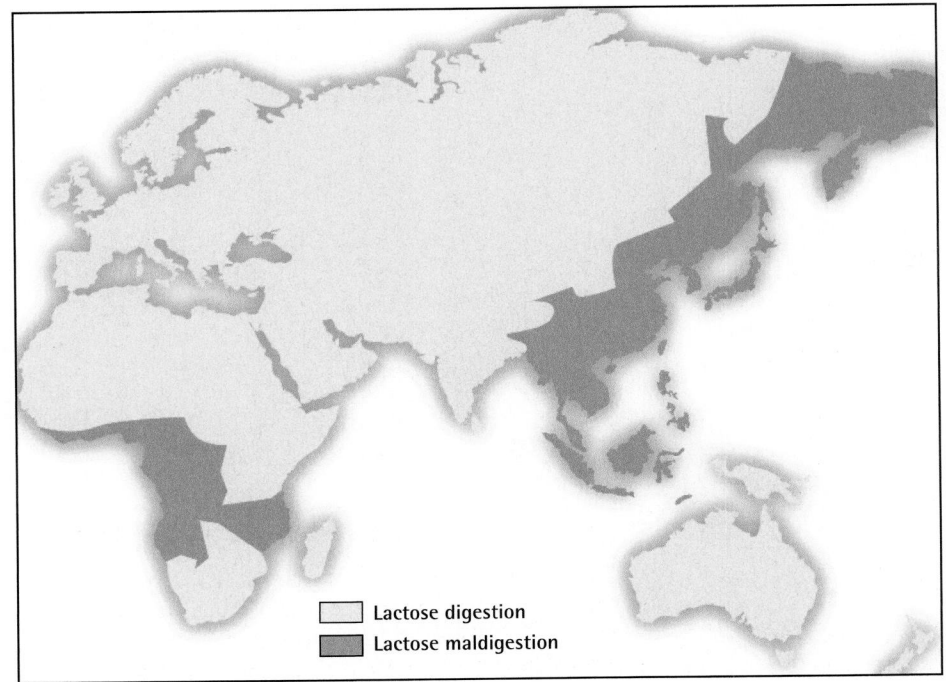

Legend:
Lactose digestion
Lactose maldigestion

Illustration 12.17 Lactose maldigestion is less common among descendants of people who consumed milk from domesti-cated animals during prehistoric times (light areas) than among people whose early ancestors did not drink milk (dark areas).[55]

Illustration 12.18 Dairy and fortified products generally well tolerated by individuals with lactose maldigestion.

Richard Anderson

the diet of people with lactose intolerance promotes the development of osteoporosis.[51] Rather, fortified soy or rice milk, low-lactose cow's milk, milk pretreated with lactase drops, yogurt, and other fermented milk products (if tolerated), and ready-to-eat cereals and fruit juices fortified with calcium and vitamin D should be consumed. Illustration 12.18 shows a variety of dairy products that are generally well tolerated by people with lactose maldigestion.

Elena Schweitzer/Shutterstock.com

NUTRITION
up close

Does Your Fiber Intake
Measure Up?

Focal Point: Approximate the amount of fiber your diet contains.

Are you meeting your fiber quota, or do you consume the typical low-fiber American diet? To determine if your fiber intake is adequate, award yourself the allotted number of points for each serving of the following foods that you eat in a typical day. For example, if you normally eat one slice of whole grain bread each day, give yourself two points. If you eat two slices daily, give yourself four points. After tallying your score, refer to the Feedback section for the results.

High-fiber food choices	Number of points
Fruits: 2 points for each serving	
1 whole fruit (e.g., apple, banana)	
$1/2$ cup cooked fruit	
$1/4$ cup dried fruit	
Grains: 2 points for each serving	
$1/2$ cup cooked brown rice	
1 whole grain slice of bread, roll, muffin, or tortilla	
$1/2$ cup hot whole grain cereal (e.g., oatmeal)	
$3/4$ cup cold whole grain cereal (e.g., Cheerios)	
2 cups popcorn	
Nuts and seeds: 2 points for each serving	
$1/4$ cup seeds (e.g., sunflower)	
2 tablespoons peanut butter	
Vegetables: 3 points for each serving	
1 whole vegetable (e.g., potato)	
$1/2$ cup cooked vegetable (e.g., green beans)	
Bran cereals: 7 points for each serving	
$1/2$ cup cooked oat bran cereal	
1 cup cold bran cereal	
Legumes: 8 points for each serving	
$1/2$ cup cooked beans (e.g., baked beans, pinto beans)	
Total score	

Special note: You can also calculate your fiber intake using the Diet Analysis Plus software. Input your food intake for one day. Choose the Reports Tab at the top, then select Source Analysis. Once you are on the Source Analysis page, choose Dietary Fiber from the drop-down menu. Record your results.

Feedback to the Nutrition Up Close can be found in Appendix G.

REVIEW QUESTIONS

- **Describe the food sources and functions of simple sugars, complex carbohydrates, fiber, and alcohol sugars.**

1. Excluding fiber, one gram of carbohydrate provides four calories.   True/False

2. Fructose is a monosaccharide, maltose is a diasaccharide, and glycogen is a trisaccharide.   True/False

3. When the body has more glucose than it needs for energy formation, the excess glucose is converted to glycogen and fat.   True/False

4. Sugars are the most commonly used food additive.   True/False

5. Excess sugar intake is the primary cause of obesity in developed countries.   True/False

6. Xylitol and sorbitol are examples of alcohol sugars.   True/False

7. Excess intake of aspartame (also called NutraSweet) causes headaches in a majority of children and adults.   True/False

8. Complex carbohydrates are also known as polysaccharides.   True/False

9. All types of fiber share the characteristic of providing four calories per gram.   True/False

10. Carbohydrates and carbohydrate-containing foods are classified by their glycemic index, or the extent to which they increase blood glucose levels.   True/False

11. _____ Which of the following statements represents a documented relationship between simple sugar intake and health?
    a. High simple sugar intakes are related to the risk of development of type 2 diabetes and obesity.
    b. Excessive simple sugar intake increases the onset of attention deficit hyperactivity disorder in children.
    c. High sugar intakes are related to increased blood lead levels.
    d. High intake of beverages high in simple sugars, and not solid foods, are related to the risk of type 2 diabetes development.

- **Explain relationships between carbohydrates and dental health.**

12. Declining rates of dental caries in the United States are primarily related to increased access to fluoridated water supplies.   True/False

13. Children who consume water at home from a fluoridated water supply are protected from developing dental caries if they consume candy and other high-sugar foods regularly.   True/False

14. _____ Which of the following measures is associated with the prevention of dental caries?
    a. Frequent intake of dried fruits.
    b. Regular consumption of sweetened beverages.
    c. Use of chewing gum sweetened with xylitol.
    d. Consumption of milk right before bedtime.

- **Identify foods and beverages that can be consumed by individuals with lactose maldigestion to assure adequate intake of calcium and vitamin D.**

15. Lactose intolerance refers to the disorder associated with low or absent lactase availability whereas lactose maldigestion refers to the symptoms of gas and bloating experienced by people who cannot digest lactose.   True/False

16. Individuals who lack the enzyme lactase are at increased risk of developing osteoporosis.   True/False

17. _____ When he was young, Hank remembers hearing his father say, "I can't drink milk. It tears me up inside." This means:
    a. His father has lactose maldigestion.
    b. Hank will develop lactose maldigestion, too.
    c. His father is lactose intolerant.
    d. Unless diagnosed, you cannot know why his father has a problem he associates with milk drinking.

18. _____ Individuals with lactose maldigestion can replace the calcium and vitamin D they do not get from milk with food such as:
    a. Ice cream, cream, and coffee creamer.
    b. Fortified soy and rice milk, low-lactose milk.
    c. Fortified margarine, fruit drink, and salad dressings.
    d. Raisins, cheese, and whole grain bread.

Answers to these questions can be found in Appendix G.

NUTRITION SCOREBOARD ANSWERS

1. If you get this right, you may be in the minority. In one study of college students, only 38% could identify good sources of complex carbohydrates. Rice, crackers, grits, dried beans, corn, peas, tortillas, biscuits, and oatmeal are also good sources of complex carbohydrates.   True

2. That's true. Both sweetened and unsweetened cereals consist primarily of carbohydrates. A gram of carbohydrate provides 4 calories, whether the source is sugar or flakes of corn.   True

3. That's a lot of sugar!   True

4. Not all foods high in fiber are fibrous, or crunchy when you eat them. Some form gels.   False

5. Cooking doesn't destroy dietary fiber.   False

6. Rates of tooth decay increase in populations as sugar intakes increases.   True

UNIT

13

Diabetes Now

NUTRITION SCOREBOARD

1 **Excess sugar consumption causes diabetes.**
   True/False

2 **Diabetes generally develops over the course of many
   years.** True/False

3 Glycemic index is a measure of the simple sugar content
   of foods. True/False

Answers can be found at the end of the unit.

Upon completing Unit 13 and its interactive learning activities, you will be able to:

- Describe risk factors for development of diabetes and approaches to the management and prevention of diabetes.

- Estimate your risk of developing type 2 diabetes, and identify lifestyle changes that would reduce the risk.

diabetes A disease characterized by abnormal utilization of carbohydrates by the body and elevated blood glucose levels. There are three main types of diabetes: type 1, type 2, and gestational diabetes. The word *diabetes* in this unit refers to type 2, which is by far the most common form of diabetes.

type 1 diabetes A disease characterized by high blood glucose levels resulting from destruction of the insulin-producing cells of the pancreas. This type of diabetes was called juvenile-onset diabetes and insulin-dependent diabetes in the past, and its official medical name is type 1 diabetes mellitus.

type 2 diabetes A disease characterized by high blood glucose levels due to the body's inability to use insulin normally, or to produce enough insulin. This type of diabetes was called adult-onset diabetes and non-insulin-dependent diabetes in the past, and its official medical name is type 2 diabetes mellitus.

Illustration 13.1 Diabetes headlines say it all.

The Diabetes Epidemic

- **Describe risk factors for development of diabetes and approaches to the management and prevention of diabetes.**

It's not the plague, yellow fever, or heart disease. The latest worldwide disease epidemic is **diabetes**, and the rising rates are directly related to the global increase in obesity (Illustration 13.1). Diabetes affects 6% of adults worldwide,[4] and 11% of U.S. adults.[5] Less than 1% of U.S. adults were diagnosed with type 2 diabetes in 1960. One-fourth of U.S. adults with diabetes have yet to be diagnosed and are not receiving health care services for the disorder.[7]

Key Nutrition Concepts

Two key nutrition concepts underlie this content on carbohydrates and other components of diet on the development of diabetes:

1. Health problems related to nutrition originate within cells.

2. Poor nutrition can influence the development of certain chronic diseases.

There are three major forms of diabetes: **type 1**, **type 2**, and **gestational diabetes**. Type 2 diabetes is the most common by far and is fueling the diabetes epidemic. Table 13.1 summarizes key features of type 1 and type 2 diabetes. Both types are diagnosed when fasting levels of blood glucose are 126 milligrams/deciliter (mg/dl) and higher; these types generally take years to develop.[2] In all cases of diabetes, the central defect is an elevated blood glucose level caused by an inadequate supply of insulin, an ineffective utilization of insulin, or both.[7]

Insulin is a hormone produced by the pancreas that performs many functions, one of which is to reduce blood glucose levels after meals. By facilitating the passage of glucose into cells, insulin keeps a steady supply of glucose going into cells. Glucose is needed by cells as a source of energy for thousands of chemical reactions that participate in the maintenance of ongoing body functions and health. If insulin is produced in insufficient amounts, or if cell membranes are not sensitive to the action of insulin, cells become starved for glucose. Functional levels of multiple tissues and organs in the body degrade as a result. High levels of blood glucose are related to adverse side effects in the body, too, such as elevated blood levels of triglycerides, **chronic inflammation**, increased blood pressure, and hardening of the arteries.[8]

Symptoms associated with high blood glucose levels vary somewhat among individuals. The most common, early signs of elevated glucose level include:

- Increased thirst
- Headaches
- Difficulty concentrating
- Blurred vision
- Frequent urination
- Fatigue (weak, tired feeling)
- Weight loss[9]

© Cengage Learning

HEALTH
The Global Epidemic of Diabesity: A Major Threat to Human Health
Associated Press
WASHINGTON

WORLD & NATION
Diabetes Threat on the Rise among U.S. Children, Specialists Say
Associated Press
WASHINGTON, D.C. — A laboratory

CDC's Forecast for Diabetes is Grim

By Maria Elena Baca
Star Tribune Staff Writer

In a recent government study, the Centers for Disease Control and Prevention (CDC) estimated that obesity is fast approaching tobacco as the top underlying preventable

percent of U.S. adults are considered overweight, and an additional 31 percent are obese.
Anyone with a body mass index (a ratio between your height and weight) of 25 or above -- that's someone, for example, who

considered overweight, according to the National Institutes of Health. Anyone with a body mass index of 30 or above -- such as someone who is 5-foot-6 and 186 pounds -- is considered obese. Check your body mass index here!

Health Consequences of Diabetes

Health effects of diabetes vary depending on how well blood glucose levels are controlled and on the presence of other health problems such as hypertension or heart disease. In

Table 13.1 Key characteristics of type 1 and type 2 diabetes[9]

Characteristic	Type 1 diabetes	Type 2 diabetes
Insulin deficiency?	Yes	Occurs in advanced stages of the disease
Proportion of cases	5–10%	90–95%
Risk factors	Viral infection early in life (or other triggers in genetically sensitive individuals) that destroys the part of the pancreas that produces insulin.	Obesity (especially abdominal fatness), sedentary lifestyle, insulin resistance, low weight at birth, certain ethnicities, family history, older age
Treatment	Insulin, individualized diet and exercise	Weight loss (in most cases), increased physical activity, individualized diet, sometimes oral medications, and/or insulin

gestational diabetes Diabetes first discovered during pregnancy.

chronic inflammation Low-grade inflammation that lasts weeks, months, or years. Inflammation is the first response of the body's immune system to infection or irritation. Inflammation triggers the release of biologically active substances that promote oxidation and other potentially harmful reactions in the body.

the short run, poorly controlled and untreated diabetes produces blurred vision, frequent urination, weight loss, increased susceptibility to infection, delayed wound healing, and extreme hunger and thirst. In the long run, diabetes contributes to heart disease, hypertension, nerve damage, blindness, kidney failure, stroke, and the loss of limbs due to poor circulation. The number-one cause of death among people with diabetes is heart disease.[10] Many of the side effects of diabetes can be diminished or their onset delayed if blood glucose levels are well controlled.[6]

Type 2 Diabetes

The development of this common type of diabetes is most likely to occur in overweight and obese, inactive people (Illustration 13.2). The term *diabesity* is being used to describe the close relationship between obesity and type 2 diabetes, and the current surge in worldwide rates of both is being called the "diabesity epidemic."[11] Approximately 60% of people who develop type 2 diabetes are obese, and 30% are overweight.[12] Illustration 13.3 shows the close relationship between rising rates of obesity and type 2 diabetes in the United States.

Although most often diagnosed in people over the age of 40, type 2 diabetes is becoming increasingly common in children and adolescents.[13] There are genetic components to this disease, as evidenced by the fact that it tracks in families and is more likely to occur in certain groups (Hispanic American, African American, Asian and Pacific Islanders, and Native Americans) than others.[14] Rather than inheriting genes that cause diabetes, certain individuals inherit or acquire very early in life multiple genetic traits that increase the likelihood that diabetes will develop. These genetic traits lead

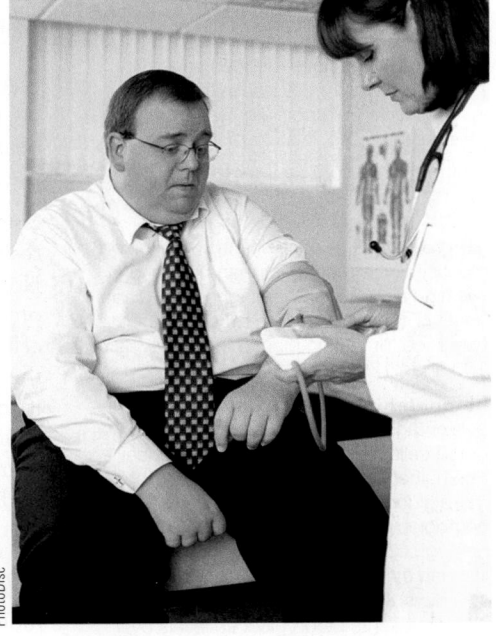

PhotoDisc

Illustration 13.2 Obesity characterized by central-body fat stores and physical inactivity is a strong risk factor for type 2 diabetes.

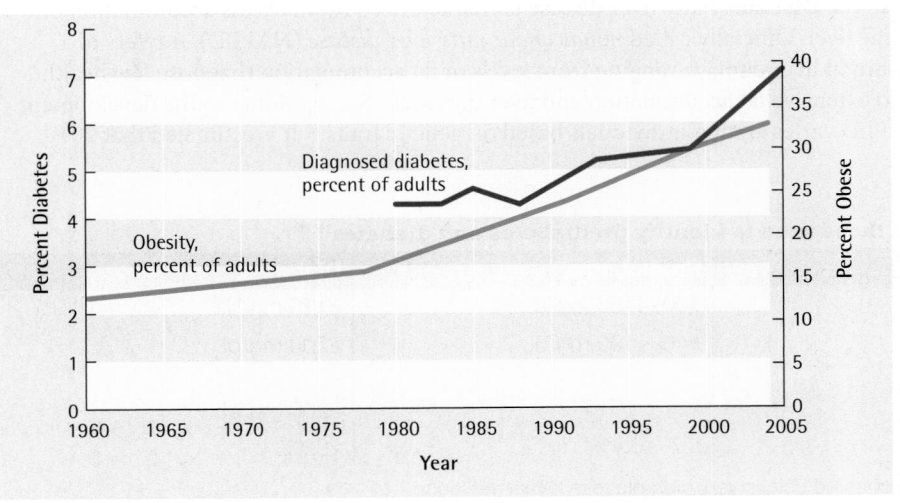

© Cengage Learning

Illustration 13.3 The incidence of type 2 diabetes increases as rates of obesity increase.[5,12]

to metabolic changes that promote the development of diabetes when a person is exposed to certain environmental triggers, such as a diet high in saturated fats or excess calories.[15]

The development of type 2 diabetes is characterized by a number of metabolic changes that can be identified before blood levels of glucose become high enough to qualify for the diagnosis. Individuals on their way to developing type 2 diabetes tend to have somewhat higher that normal blood glucose levels and a disorder called **insulin resistance**.

Prediabetes Individuals at risk of developing type 2 diabetes based on somewhat elevated measures of blood glucose are considered to have "**prediabetes**." According to the American Diabetes Association, three different tests can be used to identify prediabetes.[16] Blood glucose values considered normal, and those used as cut points for the identification of prediabetes as well as diabetes are shown in Table 13.2. Approximately 35% of U.S. adults are at risk of type 2 diabetes due to this condition.[5] The presence of prediabetes increases a person's odds of developing type 2 diabetes by about 10% per year.[17] Abdominal obesity, physical inactivity, and genetic predispositions are common risk factors for the development of prediabetes.[18]American Indians, Alaskan Natives, African Americans, and Hispanics appear to be more genetically susceptible to prediabetes than are Whites or Asians in general.[5]

The American Diabetes Association recommends that individuals with prediabetes receive individualized medical nutrition therapy aimed at reducing body weight about 7% (if needed), increasing physical activity level to 150 minutes per week, and instituting healthy eating practices. If accomplished, these measures help prevent the progression of prediabetes into type 2 diabetes.[19]

Insulin Resistance Most people with prediabetes or type 2 diabetes have insulin resistance. Insulin resistance is due to abnormalities in the way the body uses insulin. Normally, insulin is able to lower blood glucose levels after meals by binding to receptors on cell membranes. These receptors are activated by insulin and allow glucose to pass into cells. With insulin resistance, cell membranes "resist" the effects of insulin, and that lowers the amount of glucose transported into cells. The pancreas responds to the low availability of glucose within cells by producing additional insulin. Higher than normal levels of insulin are generally sufficient to keep blood glucose levels under control for a number of years. Cells in the pancreas, however, may become exhausted from years of overwork. In such cases, the production of insulin slows and may eventually stop, leading to elevated blood glucose levels.[9]

Many other adverse health effects of insulin resistance have been identified. The reduction in glucose availability to cells caused by insulin resistance forces the body to mobilize fat from fat stores and use it as the primary source of energy. High levels of fat mobilization from fat stores in people with insulin resistance contribute to the development of elevated blood levels of free fatty acids and triglycerides. These changes increase insulin resistance further and promote the development of **fatty liver disease**.[21]

Fatty Liver Disease Fatty liver disease is characterized by an excess accumulation of fat in the liver. Officially called *nonalcoholic fatty liver disease* (NAFLD), it refers to a spectrum of liver damage ranging from levels of fat accumulation that pose few health risks to extensive fat accumulation and liver damage.[21] Susceptibility to the development of NAFLD varies among individuals based on genetic traits.[22] It is estimated that 20%

---

insulin resistance A condition in which cell membranes have reduced sensitivity to insulin so that more insulin than normal is required to transport a given amount of glucose into cells. It is characterized by elevated levels of serum insulin, glucose, triglycerides, and increased blood pressure.

prediabetes A condition in which blood glucose levels are higher than normal but not high enough for the diagnosis of diabetes. It is characterized by impaired glucose tolerance, or fasting blood glucose levels of 100 and ≤125 mg/dL.

hemoglobin A1c A measure of the percentage of hemoglobin proteins in red blood cells that are attached to glucose. The higher the blood glucose levels, the more glucose will become attached to hemoglobin. Once glucose binds with hemoglobin, it will stay there for the lifespan of hemoglobin (normally about 120 days). Consequently, hemoglobin A1c represents blood glucose levels over several months.

fatty liver disease A reversible condition characterized by fat infiltration of the liver (10% or more by weight). If not corrected, advanced forms of fatty liver disease can produce liver damage and other disorders. The condition is primarily associated with obesity, diabetes, and excess alcohol consumption. Also referred to as nonalcholoic fatty liver disease (NAFLD).

Table 13.2 **Normal blood glucose levels and those used to identify prediabetes and diabetes**[16,20]

Blood glucose test[a]	Normal values	Prediabetes	Diabetes
fasting blood glucose	<100 mg/dL	100–≤125 mg/dL	≥126 mg/dL
blood glucose measured 2-hr post glucose load	<140 mg/dL	140–199 mg/dL	≥200 mg/dL
random blood glucose	—	—	≥200 mg/dL
hemoglobin A1c	4–6%	5.7–≤6.4%	≥6.5%

[a]It is recommended that blood glucose measures be repeated unless very high glucose levels are found.

or more of the U.S. population has fatty liver disease.[23] It occurs in people of all weight statuses, however, individuals with excess abdominal fat, people with diabetes, and those with metabolic syndrome are at higher risk for the disorder.[22]

Mechanisms related to insulin resistance that increase blood levels of free fatty acids, promote oxidation and chronic inflammation, and decrease the body's ability to use insulin are related to fat accumulation in the liver. Recent evidence suggests that fructose, a major caloric sweetener used in beverages and other foods, plays a significant role in its development. Unlike glucose, fructose can initiate oxidation reactions in the liver and be converted to fatty acids that can be deposited in the liver.[24]

The first line of treatment is lifestyle modification. NAFLD can be reversed over time by gradual weight reduction, consumption of a healthful diet, and increased physical activity. Diets based on whole grain foods, colorful vegetables and fruits, fish, seafood, poultry, and low-fat dairy products have been found to reverse NAFLD.[22,25]

Insulin resistance is also related to the development of a spectrum of metabolic abnormalities that have far-reaching effects. Collectively, the adverse effects of insulin resistance are included in a disorder called **metabolic syndrome**.

Metabolic Syndrome Physicians have known for decades that obese people with hypertension and type 2 diabetes are at high risk of heart disease. What they didn't know is why. Over time, research studies discovered that part of the answer was insulin resistance.[26] There is currently no simple and inexpensive test for insulin resistance. Clinically, it is assumed to be present in people with high waist circumferences.[27] Insulin resistance is related to a cluster of metabolic abnormalities included in the metabolic syndrome that increase the risk of type 2 diabetes as well as heart disease. Symptoms related to the metabolic syndrome include:

- Waist circumference 40"or over in males, 35" or over in females

- High blood pressure (130/85 mm Hg or higher)

- Elevated blood triglycerides levels (150 mg/dL or higher)

- Low levels of protective HDL cholesterol (less than 50 mg/dL in women and 40 mg/dL in men)

- Elevated fasting blood glucose levels (100 mg/dL or higher)

The diagnosis of metabolic syndrome is made when three or more abnormalities are identified.[28] Individuals with four or five metabolic abnormalities have a 3.7 times greater risk of heart disease, and a 25 times higher risk for diabetes than do people with no abnormalities.[9] It is estimated that 32% of men and women in the United States and 20% of Canadians have metabolic syndrome.[29] Most, but not all, people who develop the metabolic syndrome are overweight or obese and are physically inactive.[29] Weight loss, exercise, adequate vitamin D status, and the other factors listed in Illustration 13.4 are the key components of preventing and reversing metabolic syndrome.[23,31]

metabolic syndrome A constellation of metabolic abnormalities that increase the risk of heart disease and type 2 diabetes. Metabolic syndrome is characterized by insulin resistance, abdominal obesity, high blood pressure and triglycerides levels, low levels of HDL cholesterol, and impaired glucose tolerance. It is also called *Syndrome X* and *insulin* resistance syndrome.

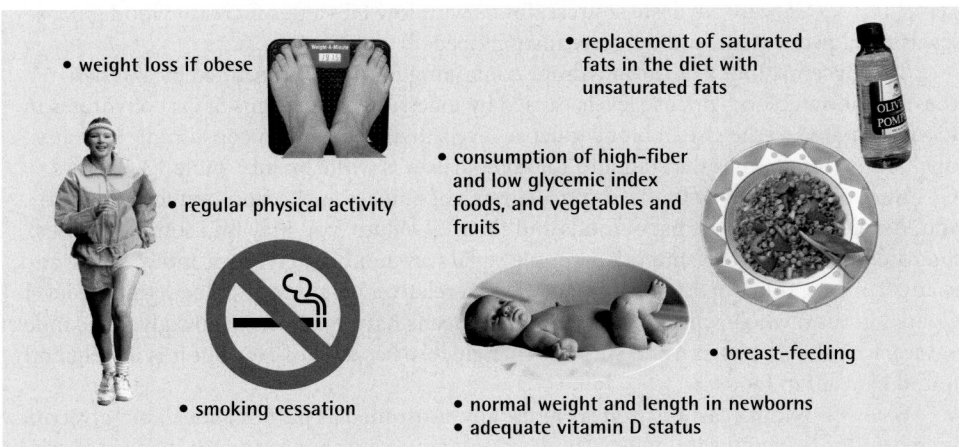

- weight loss if obese
- regular physical activity
- smoking cessation
- replacement of saturated fats in the diet with unsaturated fats
- consumption of high-fiber and low glycemic index foods, and vegetables and fruits
- normal weight and length in newborns
- adequate vitamin D status
- breast-feeding

Illustration 13.4 Key components of the prevention and management of metabolic syndrome.

PhotoDisc

Managing Type 2 Diabetes

As is the case for all forms of diabetes, diet is a cornerstone of the treatment of type 2 diabetes.[6] Modest weight loss alone (5–10% of body weight) has been repeatedly shown to significantly improve blood glucose control in overweight and obese people with type 2 diabetes.[32]

In general, diets developed for diabetes emphasize:

- Calorie reduction if overweight or obese.

- Low **glycemic index** carbohydrate foods, including whole grain breads and cereals, vegetables, fruits, low-fat milk, and lean meats and fish.

- Foods and beverages without added sugars.

- Unsaturated fats.

- Regular meals and snacks.[6,33]

Dietary management of diabetes should focus on reducing heart disease risk as well as controlling blood glucose.[34] Food sources of unsaturated fats, such as oils, nuts, and seeds, are recommended over foods high in saturated fats such as cheese and processed meats. Diet and weight-loss interventions may be supplemented by oral medications that decrease insulin resistance and blood lipids, and by insulin when needed.[32] In some cases, weight-loss surgery will be recommended for obese individuals with type 2 diabetes. This surgery successfully eliminates high blood glucose levels in many people, and lowers blood glucose levels in all cases.[35]

Regular physical activity is an important component of the management of type 2 diabetes. In addition to facilitating weight loss, physical activity reduces insulin resistance, decreases blood pressure and body fat content, and improves blood lipid and glucose levels.[36] Moderate intensity aerobic and strength-building activities, such as weight lifting, jogging, fast walking, aerobic dancing, and swimming are recommended. The target generally set for the duration of physical activity is 150 minutes or more per week, or an average of 21 minutes per day or more.[37]

Knowledge of the role of diet, exercise, insulin, and other factors in the management of diabetes is expanding so rapidly that the American Diabetes Association updates its management recommendations yearly. Dietary recommendations are currently not consistent across developed countries, indicating that scientific consensus is yet to be reached on a number of important issues related to diet and diabetes.

Glycemic Index Carbohydrates in the diet have the largest effect on blood glucose levels. Some carbohydrate-containing foods cause a rapid rise in blood glucose levels and others raise blood glucose levels more slowly. Foods that increase blood glucose to relatively high levels require more insulin to move glucose into cells than foods that produce lower levels of glucose. Many carbohydrate-containing foods have been tested for their effect on blood glucose levels and assigned a glycemic index value. Compared to high glycemic index (GI) carbohydrate sources, foods with low GI values increase blood glucose levels to a lower extent, and decrease insulin needs.[38]

The glycemic index of carbohydrate-containing foods is determined by assessing the elevation in blood glucose levels caused by ingestion of 50 grams of carbohydrates in food compared to the rise in blood glucose levels that results from consuming 50 grams of glucose. (Sometimes the standard for comparison is white bread.) Table 13.3 shows GI values of a number of foods, using 50 grams of glucose as the standard for comparison. As you read over the list of foods and their GI values, you may find some surprises. Sucrose, honey, fructose, and other simple sugars are not high glycemic index foods, and many fruits we think of as sweet do not cause a relatively high rise in blood glucose level. Coarse-ground whole wheat breads and dried beans have medium-to-low glycemic index values. It probably comes as no surprise that glucose has a GI of 100, but it is infrequently found by itself in foods.

Foods providing carbohydrates are usually consumed as part of a meal, but glycemic index is determined for individual foods. Blood glucose response to carbohydrate foods

glycemic index (GI) A measure of the extent to which blood glucose levels are raised by consumption of an amount of food that contains 50 grams of carbohydrate compared to 50 grams of glucose. A portion of white bread containing 50 grams of carbohydrate is sometimes used for comparison instead of 50 grams of glucose.

Table 13.3 Glycemic index (GI) of selected foods[3,71–73]

High GI	(70–100)	Medium GI	(56–69)	Low GI	(55 or lower)
glucose	100	breadfruit	69	honey	55
French bread	95	Fruit Loops	69	oatmeal	54
scone	92	orange soda	68	corn	53
sticky rice	87	pita bread	68	cracked wheat bread	53
broken rice	86	sucrose	68	orange juice	52
potato, baked	85	taco shells	68	banana	52
potato, instant mashed	85	croissant	67	mango	51
Special K, rice	84	angel food cake	67	potato, boiled	50
Corn Chex	83	fruit punch	67	corn tortilla	49
pretzel	83	cherries	66	green peas	48
Rice Krispies	82	Cream of Wheat	66	pasta	48
cornflakes	81	brown rice	66	carrots, raw	47
Corn Pops	80	couscous	65	lactose	46
Gatorade	78	Quaker Quick Oats	65	milk chocolate	43
jelly beans	78	raisins	64	All-Bran	42
Cocoa Pops	77	chapati	62	orange	42
doughnut, cake	76	French bread with butter and jam	62	peach	42
waffle, frozen	76	Raisin Bran	61	apple juice	40
doughnuts	75	sweet potato	61	apple	38
french fries	75	bran muffin	60	pear	38
Grape Nuts	75	Just Right cereal	60	tomato juice	38
Shredded Wheat	75	blueberry muffin	59	yam	37
white rice	75	Mini Wheats	59	yogurt	31
Cheerios	74	Coca-Cola	58	flour tortilla	30
popcorn	72	power bar	56	dried beans	25
watermelon	72	Special K	56	grapefruit	25
carrots, diced, cooked	70			milk	25
wheat bread	70			fructose	19
white bread	70			pinto beans	14
				hummus	6

may vary based on the extent of food processing, protein and fat content of a food or meal, whether a food is consumed cooked or raw, and other factors such as the ripeness of fruit.[39,40]

Some high-GI foods, such as baked potatoes and French bread, are good sources of a number of nutrients. Just because a food has a high glycemic index doesn't mean it should not be consumed as part of a balanced diet.[41] Adjusting food choices toward selection of mainly low-GI foods is most helpful for people attempting to prevent or control type 2 diabetes or to diminish the effects of insulin resistance.[38,42]

Some countries have developed policies that allow nutritious foods with low GI values to bear a glycemic index label. Australia was the first country to do that, and a "Glycemic Index Foundation Certified" label (Illustration 13.5) can be found on nutritious, low glycemic index food products in that country. Consumers report that the label takes some of the confusion out of identifying low GI foods and influences their food purchase decisions toward the selection of healthy low-GI foods.[43]

Glycemic Load Glycemic Index estimates the blood glucose–raising effect of 50 grams (1.75 ounces) of a carbohydrate-containing food on blood glucose level. That amount represents a $1\frac{1}{3}$ cup serving of flaked breakfast cereal or one-third of a medium boiled potato.

Illustration 13.5 Low-GI, nutritious foods have been labeled with this symbol in Australia since 2002.

REALITY CHECK
Will the Real Whole Grain Please Stand Up?

Which bread is made from whole grains?

Who gets a thumbs-up?

Answers on page 13-9

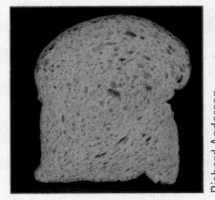

Wheat Bread

Whole Wheat Bread

glycemic load (GL) A measure of the extent to which blood glucose level is raised by a given amount of a carbohydrate-containing food. GL is calculated by multiplying a food's GI by its carbohydrate content.

But what would be the expected blood glucose response if you ate more or less cereal or potato than 50 grams? The **glycemic load (GL)** measure was developed to estimate blood glucose-response based on the amount carbohydrate-containing foods consumed. The GL is also used to estimate blood glucose response to snacks, meals, and diets.[44]

Glycemic load is calculated by multiplying the grams of carbohydrate in a specific amount of food times the food's glycemic index. This result is then divided by 100 to calculate glycemic load. For example, a raw carrot provides approximately 7 grams of carbohydrate and has a glycemic index of 47. Its glycemic load would be calculated as:

$$7 \times 47 = 329$$

$$329 \div 100 = 3.29$$

$$\text{Glycemic load} = 3.29$$

The blood glucose–raising effect of one carrot doesn't amount to much. If you consumed four slices (4 ounces) of French bread, the result would tell a different story. The glycemic load supplied by this amount of French bread is 49.4, a level that raises blood glucose and insulin levels far more than a carrot or a slice of French bread.

Consumption of low-GI foods and a low-GL diet is a recommended component of the dietary management of type 2 and gestational diabetes in a number of countries, and is becoming increasingly employed in the United States.[45] Consumption of low-GI foods is viewed as a useful part of the management of insulin for resistance and metabolic syndrome,[46] and as a secondary aid to blood glucose control for people with all forms of diabetes.[44,47,48]

Sugar Intake and Diabetes Does sugar intake cause diabetes? High sugar intakes have not been found to directly cause type 2 diabetes.[1] However, high sugar diets that provide excessive levels of calories could contribute to diabetes by promoting weight gain.[1] Recent research indicates the possibility that regular intake of sugar-sweetened beverages by young children increases the risk of obesity versus the risk in young children who do not consume them.[49] High sugar intake appears to be related to fat accumulation in the liver particularly in Hispanic children due to genetic susceptibility.[12] Regular consumption of high sugar or high glycemic index foods has been associated with the development of type 2 diabetes in certain population groups.[49,50]

Should people with diabetes exclude sugars from their diet? Sugar does not have to be eliminated from the diet of people with diabetes, but the amount consumed should be limited due to its effect on blood glucose level and insulin need.[45] Intake of total carbohydrates and the glycemic index of the carbohydrate-containing foods, rather than sugar intake specifically, are most strongly related to blood glucose levels.[51] High-sugar diets increase blood triglyceride levels in people with metabolic syndrome, and that may increase the risk of heart disease.[52]

Prevention of Type 2 Diabetes

The effects of weight loss and exercise in preventing type 2 diabetes can be quite dramatic. In one large study that took place over a three-year period, people with prediabetes reduced their risk of developing type 2 diabetes by over 50% by losing around 7% of body weight and exercising for 150 minutes a week.[19]

Diets rich in whole grain and high-fiber foods are protective against the development of type 2 diabetes and appear to aid weight loss (Illustration 13.6). Components of high-fiber, whole grain foods raise blood glucose levels marginally and appear to provide nutrients and other biologically active substances that lessen the risk of this disease.[53] Consumption of regular or decaffeinated coffee (1 to 4 cups daily) and moderate alcohol intake (one to two drinks per day if appropriate) also appear to increase insulin sensitivity and decrease the risk of developing type 2 diabetes.[54,55] Vitamin D supports insulin production by the pancreas. Adequate vitamin D status has been associated with a reduced risk of type 2 diabetes.[56]

Finally, type 2 diabetes may be prevented or postponed by not gaining weight during the young adult years. Weight gains of 10 to 15 pounds between the ages of 25 and 40 years have been found to increase the risk of type 2 diabetes sevenfold compared to not gaining weight or gaining it after the age of 40.[57]

It is anticipated that 800,000 new cases of type 2 diabetes may develop each year in the United States due to rising rates of obesity.[14] The additional burdens such an increase would place on individuals and health care costs clearly convey the message that prevention is urgently needed. Public health campaigns are now underway to encourage people to lose weight if overweight, to exercise regularly, and to select whole grain products and other high-fiber foods along with ample intake of vegetables and fruits.

Are you at risk of developing type 2 diabetes? Find out if you are, and what people at risk can do to help prevent it in the accompanying Health Action.

PhotoDisc

Illustration 13.6 Whole grain products, other high-fiber foods, and ample servings of vegetables and fruits can help prevent type 2 diabetes.

Type 1 Diabetes

Type 1 diabetes is an **autoimmune disease** that produces a deficiency of insulin. It accounts for 5–10% of cases of diabetes.[58] The onset of type 1 peaks around the ages of 11 to 12 years and usually occurs before age 40.[59] Although individuals may inherit a tendency to develop type 1 diabetes, most people who have the disease have no family history of it. Unlike type 2 diabetes, worldwide rates of type 1 diabetes are stable.[60]

The insulin deficiency that marks type 1 diabetes appears to develop when a person's own **immune system** destroys beta cells in the pancreas that produce insulin. The destruction is likely triggered by a viral infection (mumps, rubella, measles, the flu) in genetically susceptible people.[61] Although the specific cause or causes of type 1 diabetes are yet to be fully understood, breast-feeding infants for the first four months or more of life and adequate vitamin D status may confer a level of protection against the development of type 1 diabetes.[62,63]

Managing Type 1 Diabetes The main goals of the management of type I diabetes are blood glucose control and health maintenance.[32] Blood glucose levels are managed primarily by regular, healthful meals controlled in carbohydrate content that are consumed at planned times and in specific amounts. Diets are designed to match insulin dose so that

autoimmune disease A disease initiated by the destruction of the body's own cells by components of the immune system that mistakenly recognize the cells as harmful.

immune system Body tissues that provide protection against bacteria, viruses, foreign proteins, and other substances identified by cells as harmful.

ANSWERS TO REALITY CHECK
Will the Real Whole Grain Please Stand Up?

You can't tell by looking, but the slice of bread on the right contains more fiber (4 versus 2 grams). Want to find whole grain breads? Look for the term *whole grain* or *whole wheat* on the label. Bread products carrying that label contain 51% or more whole grains. The ingredient label on whole grain products will list one or more whole grains before listing enriched flour.

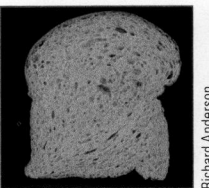
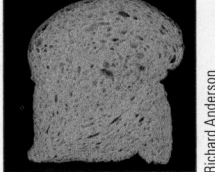
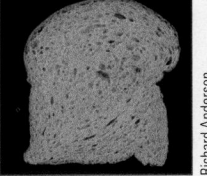
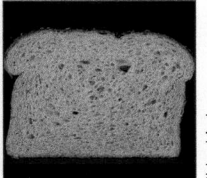

Richard Anderson

Wheat Bread

Whole Wheat Bread

health action Preventing Type 2 Diabetes

Are You at Risk?

Check each category that applies to you:

_____ I have a brother, sister, mother, or father with type 2 diabetes.

_____ My BMI is 25 or higher.

_____ My waist circumference is over 35 inches (if female) or 40 inches (if male).

_____ I have been diagnosed with prediabetes or insulin resistance.

_____ I am habitually inactive.

_____ I have been diagnosed with high blood pressure.

_____ I have been diagnosed as having high blood triglyceride levels.

If you checked one or more of the categories above, you may be at increased risk of developing type 2 diabetes.

What to Do Now?

1. Recognize that there are no quick fixes for the prevention of type 2 diabetes.
2. If you are overweight or obese, gradually cut back on your calorie intake by making small and acceptable changes in your diet. Aim for a reduction in caloric intake of 100 to 200 calories a day.
3. Gradually increase your physical activity level until you are moderately to vigorously active for at least 150 minutes per week (21 minutes per day). Select activities you enjoy and will keep doing in the long run.
4. Choose low glycemic index foods for your carbohydrate-containing food choices as often as possible. (You can use Table 13.3 to check out your options for low glycemic index foods).

5. Consume enough fiber (26 grams daily).
6. Eat your favorite vegetables and fruits often.
7. Choose low-fat dairy products.
8. Select and consume whole grain products rather than refined grain products.
9. Consume enough vitamin D.[30]
10. See your health care provider if you think you have a problem with your blood glucose levels.

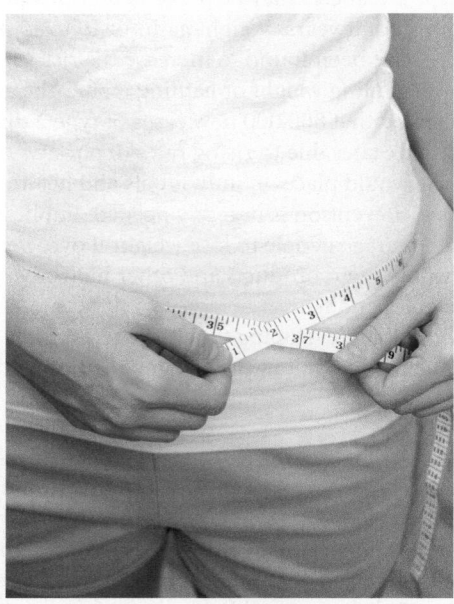

© Natasha Paterson/Design Pics/Corbis

---

blood glucose levels remain within normal ranges. They are controlled in carbohydrate content because carbohydrates raise blood glucose levels and increase insulin need to a greater extent than protein or fats.[22]

People with type 1 diabetes are urged to replace simple sugars with reasonable amounts of artificial sweeteners or with nonsugar added foods and beverages. Sugar and sweetened foods do not have to be eliminated from the diet but rather should be consumed in limited amounts. Foods low in glycemic index and high in fiber (especially soluble fiber such as oatmeal) are encouraged, as are brightly colored fruits and vegetables, low-fat meat and dairy products, fish, dried beans, and nuts and seeds.[22,45] Reduced calorie diet plans should be included as part of the care for individuals with type 1 diabetes who would benefit from weight loss.

Physical activity is generally part of a diabetes care plan because it improves blood glucose levels, physical fitness, and insulin utilization. Both strength and aerobic exercises are recommended.[27] Individualized meal and physical activity plans and follow-up care for individuals with type I diabetes should be provided by an experienced health care team that includes a registered dietitian.[33]

Insulin and New Technologies in the Management of Type 1 Diabetes People with type 1 diabetes require insulin to control blood glucose levels. The amount and type of insulin required depends on diet and physical activity level and the presence of conditions such as pregnancy, stress, illness, and physical activity. To get the insulin dose right, individuals with type 1 diabetes measure their blood glucose level a number of times daily. Blood glucose levels have been traditionally tested using a pinprick blood sample and a device that measures blood glucose level in a droplet of blood. The glucose measurement would be followed by the appropriate amount of insulin delivered in an injection. Although nearly painless if very small needles are used, many people hate the idea of getting pricked by a needle.

Technological advances are taking some of the perceived sting out of type 1 diabetes management. Insulin pumps that deliver programmed doses of insulin are now combined with continuous glucose monitors that assess the body's glucose level (Illustration 13.7). Continuous glucose monitors issue a warning signal if glucose levels are rising or dropping too much. The signal gives the individual enough advance warning to take appropriate action, such as eating or using insulin, before a problem develops. It must be used with self-monitoring of blood glucose levels.[64,65]

Many new types of insulin that better manage blood glucose levels are now available, and it is hoped that in the future a cure for type 1 diabetes will be identified. The cure may take the form of pancreatic islet cell transplants and stem cell therapy that increases the mass of cells in the pancreas that produce insulin. These techniques are being intensively studied now and preliminary results are promising.[67]

Gestational Diabetes

Approximately 5–6% of women develop gestational diabetes during pregnancy, but the incidence varies a good deal based on ethnicity. Native Americans, African Americans, and Asians are at higher risk for gestational diabetes than other groups. Women over the age of 35 years, obese women, and those with habitually low levels of physical activity are at higher risk than other women.[66] A number of dietary risk factors have been associated with the development of gestational diabetes and include diets high in red and processed meats, low-fiber intake, and diets high in gylcemic load.[66] Gestational diabetes is likely caused by the same basic environmental and genetic predisposition factors as is type 2 diabetes.[64] The prevalence of gestational diabetes is increasing in many countries and parallels increases in obesity and type 2 diabetes.[68] Infants born to women with poorly controlled diabetes may be excessively fat at birth, require cesarean delivery, and have blood glucose control problems after delivery. They are at greater risk for developing diabetes later in life.[74]

As is the case for type 2 diabetes, women with gestational diabetes are insulin resistant and can often control their blood glucose levels with an individualized diet and exercise plan. Some women require daily oral medication or insulin injections for blood glucose control.[64]

Gestational diabetes often disappears after delivery, but type 2 diabetes may appear after delivery or later in life. Exercise, maintenance of normal weight, and consumption of a healthy diet that includes good amounts of fiber, low-glycemic index carbohydrates, vegetables, fruits, and low-fat meats reduce the risk that diabetes will return.[66]

New criteria for the diagnosis of gestational diabetes have been proposed that would require only one abnormal glucose value for a diagnosis compared to the two abnormal values recommended now. It is estimated that implementation of the new criteria would increase the percent of women diagnosed with gestational diabetes to 18%.[5]

Hypoglycemia

Hypoglycemia, or low blood glucose level, is a fairly common side effect of diabetes treatment with certain oral medications and insulin. It is often mild and accompanied by a feeling of weakness, a negative mood, nervousness, hunger, shakiness, and confusion. Severe cases of hypoglycemia can lead to coma and death. Symptoms of hypoglycemia can usually be resolved by consuming 3 or 4 glucose tablets, eating 5 or 6 pieces of hard candy, eating a tablespoon of sugar or honey, or by drinking a cup of milk or a half cup of fruit juice.[69]

AP Images/Tom Strattman

Illustration 13.7 An insulin pump with a built-in glucose sensor can warn individuals about potentially harmful changes in blood glucose levels.

hypoglycemia A disorder resulting from abnormally low blood glucose levels. Symptoms of hypoglycemia include irritability, nervousness, weakness, sweating, and hunger.

Hypoglycemia can also occur in people not taking medications for diabetes. Other causes of hypoglycemia include the consumption of excess alcohol, insulin-producing tumors, prolonged fasting or starvation, and certain medications and illnesses that lower blood glucose levels.[69]

Diabetes in the Future

The anticipated surge in the worldwide incidence of type 2 diabetes shown in Illustration 13.8 is not inevitable. It could be lowered substantially by individuals making environmental and lifestyle changes that reduce the risk for, and incidence of, overweight and obesity. Increased awareness of the connection between diabetes and body weight may help. The desired future of diabetes is the one negating the dire forecasts by the experts.

Illustration 13.8 The red bars indicate millions of cases of type 2 diabetes in 2000 and the green bars the projections for 2030. The projected percentage increases in cases of type 2 diabetes are shown below the bars.[70]
Source: From Hossain et al., "Obesity and Diabetes in the Developing World — A Growing Challenge," *New England Journal of Medicine* 356 (3): 214. Copyright © 2007 Massachusetts Medical Society. All rights reserved. Reprinted by permission.

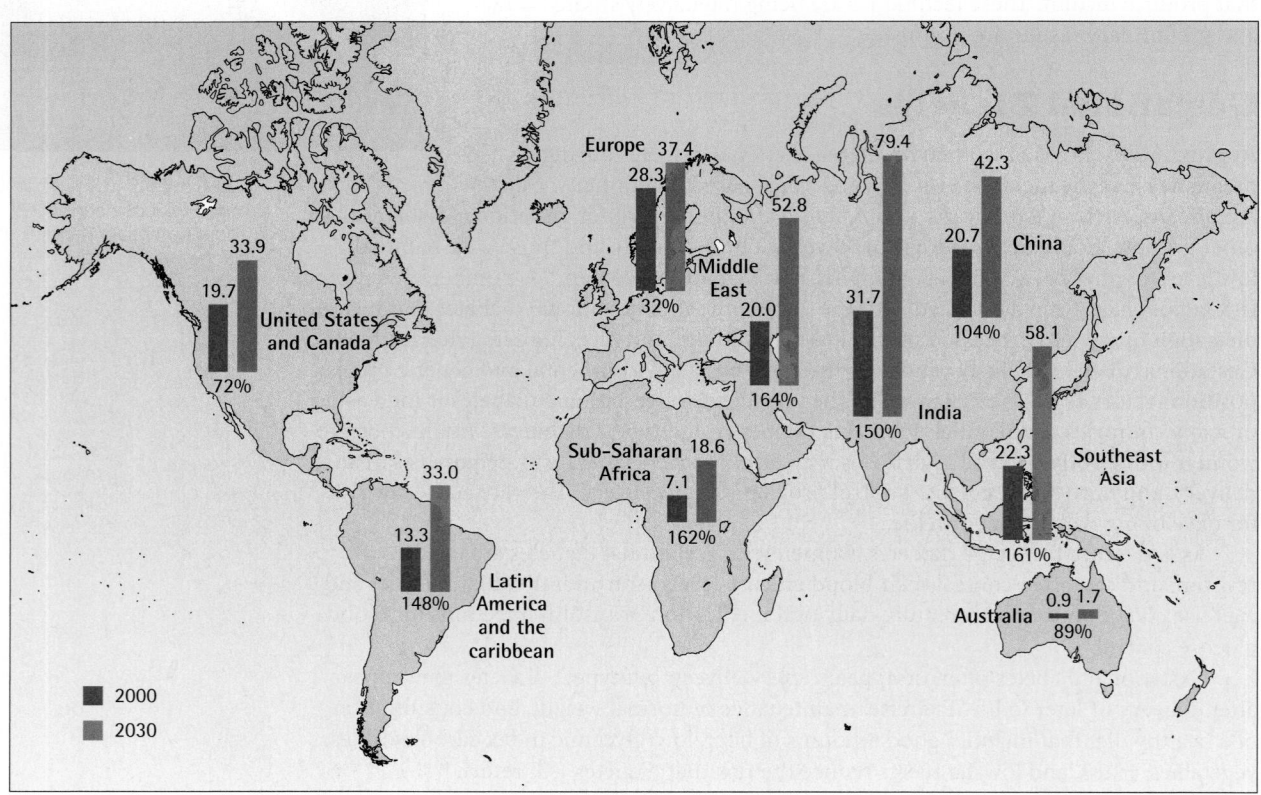

© Scott Goodwin Photography

NUTRITION
up close

Calculating Glycemic Load

Focal Point: To gain an appreciation of the effect of source and amount of carbohydrate consumed on blood glucose levels.

Glycemic index provides an estimate of the rise in blood glucose levels expected from consuming 50 grams of carbohydrate in a food. Glycemic load, on the other hand, is a measure of the expected rise in blood glucose related to the ingestion of other amounts of this food. Consequently, glycemic load estimates the blood-glucose-raising potential of the amount of carbohydrate-containing food actually consumed.

1. Use Appendix A to look up the carbohydrate content of each food for the serving size listed.
2. Use Table 13.3 to identify the glycemic index for each food.
3. Calculate glycemic load based on this formula:

Glycemic load = grams carbohydrate × glycemic index of the food, divided by 100

Here's an example for 2 tsp sucrose:

$$\text{grams carbohydrate in 2 tsp sucrose} = 8$$
$$\text{glycemic index of sucrose} = 68$$
$$8 \times 68 = 544$$
$$\text{glycemic load} = 544 \div 100 = 5.44$$

Food	Serving size	Grams carbohydrate	×	Glycemic index	÷	Glycemic load 100
cola beverage	36 oz (3 cans)					
potato, baked, no skin	1					
apple juice, bottled	1 cup					
milk chocolate, plain	1 oz					
hummus	½ cup					

Feedback to the Nutrition Up Close can be found in Appendix G.

REVIEW QUESTIONS

- **Describe risk factors for development of diabetes and approaches to the management and prevention of diabetes.**

1. The incidences of type 2 diabetes and of gestational diabetes increase as rates of overweight and obesity increase.   True/False

2. Insulin facilitates the passage of vitamins into cells.   True/False

3. Individuals with type 2 diabetes and gestational diabetes generally have the condition known as insulin resistance.   True/False

4. The most effective strategies for the prevention and control of type 2 diabetes are weight loss and exercise.   True/False

5. Prediabetes is defined as a condition in which fasting blood glucose levels are 100 to ≤125 mg/dL.   True/False

6. Gestational diabetes appears to be caused by the same environmental and genetic predisposition factors that apply to type 2 diabetes.   True/False

7. All foods high in complex carbohydrates have low glycemic indices.   True/False

8. Excess sugar intake from soft drinks is a direct cause of type 2 diabetes.   True/False

9. Hypoglycemia in people with diabetes is most often caused by an excess consumption of sugar.   True/False

The next four questions refer to this situation:

Professor Batista has been worried about her weight for a while. She is 40 pounds overweight and is physically inactive. For the past few months she has felt unusually tired, thirsty, and has had difficulty concentrating on her class preparation notes.

Professor Batista's favorite foods are white rice, pinto beans, corn, sweet potatoes, flour tortillas, and fruit punch. At her last physical exam, her nurse practitioner let Professor Batista know that her fasting blood glucose level was 136 mg/dL, her triglyceride level was elevated (190 mg/dL), and her HDL level was 32 mg/dL. Her care provider also suggested she should consider losing weight.

10. _____ Clinical findings indicate that Professor Batista most likely qualifies for a diagnosis of:

    a. prediabetes
    b. type 1 diabetes
    c. metabolic syndrome
    d. fatty liver disease

11. _____ Clinical findings, such as those for Professor Batista, may also likely indicate the presence of:

    a. hypoglycemia
    b. type 1 diabetes
    c. autoimmune disease
    d. type 2 diabetes

12. _____ Which of the following sets of foods preferred by Professor Batista would be expected to raise blood glucose levels the most:

    a. rice and sweet potato
    b. fruit punch and rice
    c. flour tortilla and corn
    d. fruit punch and sweet potato

13. _____ Over the following year, Professor Batista lost 20 pounds (12% of her initial body weight), switched to whole grain tortillas and brown rice, substituted unsweetened iced coffee for fruit punch, and walked vigorously for 30 minutes daily. Which of the following improvements in the clinical findings at her next checkup would *not* be expected?

    a. increased insulin sensitivity
    b. reduced blood glucose level
    c. improved blood lipid levels
    d. increased insulin resistance

- **Estimate your risk of developing type 2 diabetes, and identify lifestyle changes that would reduce the risk.**

14. A female with a waist circumference of 29 inches, high blood pressure, and low levels of LDL cholesterol would be diagnosed as having "metabolic syndrome."   **True/False**

15. Roger has a BMI of 35 mg/kg[2] and is clinically considered obese. His fasting blood glucose level is 97 mg/dL. Roger probably has diabetes.   **True/False**

NUTRITION SCOREBOARD ANSWERS

1. Simple sugar intake does not cause diabetes.[1]   **False**

2. Disease processes underlying the development of diabetes exist for years before the onset of diabetes in most cases.[2]   **True**

3. Glycemic index is a measure of the potential of carbohydrate-containing foods to raise blood glucose level. Only a few simple sugars have a high glycemic index.[3]   **False**

Image Source/Jupiterimages

Alcohol: The Positives and Negatives

NUTRITION SCOREBOARD

1 Alcohol is produced by the fermentation of carbohydrates. **True/False**

2 You can protect your body from the harmful effects of consuming excessive amounts of alcohol by eating a nutritious diet. **True/False**

3 Alcohol abuse plays a major role in injuries and deaths in the United States. **True/False**

4 Dark beer provides more calories than the same amount of regular beer. **True/False**

Answers can be found at the end of the unit.

After completing Unit 14 and its interactive learning activities, you will be able to:

- Objectively assess the effects of the quantity of alcohol intake on behavior, health, and disease.

Alcohol Facts

- **Objectively assess the effects of the quantity of alcohol intake on behavior, health, and disease.**

Alcohol is both a food and a drug. It's a food because alcohol is made from carbohydrates, and the body uses it as an energy source. Alcohol is a drug because it modifies various body functions.

The type of alcohol people consume in beverages is ethanol. (We refer to ethanol by the broader term *alcohol* in this unit.) Alcohol is produced from carbohydrates in grains, fruits, and other foods by the process of **fermentation**. Wines, brews made from grains, and other alcohol-containing beverages are a traditional part of the food supply of many cultural groups.[1] In high doses, however, alcohol is harmful to the body and can cause a wide variety of nutritional, social, and physical health problems.

Key Nutrition Concepts

Content and activities included in this unit on alcohol are related to the following key nutrition concepts:

- Health problems related to nutrition originate within cells.

- Malnutrition can result from poor diets and from disease states, genetic factors, or combinations of these factors.

The Positive

Whether alcohol has harmful effects on health depends on how much is consumed. The consumption of moderate amounts of alcohol by healthy adults who are not pregnant appears to cause no harm. In fact, moderate alcohol consumption versus abstinence is associated with a significant level of protection against heart disease, type 2 diabetes, hypertension, stroke, dementia (cognitive decline), and all cause mortality.[2,3] In the United States, a moderate level of alcohol consumption is considered to be one standard-sized drink per day for women and two drinks for men (Illustration 14.1).[4]

Alcohol at moderate doses increases HDL-cholesterol levels (the "good" cholesterol) and decreases **chronic inflammation**. Decreased inflammation helps prevent the formation of plaque in arteries and improves circulatory function and maintenance of normal cell health. Alcohol improves the body's utilization of insulin and glucose, lowers post-meal blood glucose levels somewhat, and improves cognitive function.[2,6–8]

Some of the beneficial effects of alcohol on health are related to the phytochemical content of the fruit, vegetable, or grain fermented to produce it. Red wine, and to a lesser extent beer and white wine, contain pigments and other phytochemicals that act as powerful antioxidants and decrease inflammation and artery plaque formation.[5,9] Purple grape juice and other purple and blue-colored fruit juices also provide antioxidants and have anti-inflammatory effects that benefit health, although to a lesser degree than red wine.[2,10]

People don't have to consume alcohol to reduce their risk of heart disease. Diets low in saturated and *trans* fat, liberal intakes of vegetables and fruits, ample physical activity, and not smoking also reduce the risk of heart disease.[11]

The Negative

Heavy drinking, often defined as the consumption of five or more drinks per day,[4] poses a number of threats

fermentation The process by which carbohydrates are converted to ethanol by the action of the enzymes in yeast.

chronic inflammation Inflammation that is low grade and lasts weeks, months, or years. Inflammation is the first response of the body's immune system to infection or irritation. It triggers the release of biologically active substances that promote oxidation and other potentially harmful reactions in the body.

Illustration 14.1 Standard serving sizes of alcohol-containing beverages. Standard servings shown each contain 0.6 oz (17g) of alcohol.

LIGHT BEER
12 oz.

MALT BEVERAGE
12 oz

BEER
12 oz.

80-PROOF LIQUOR
1 1/2 oz.

WINE COOLER
12 oz.

WINE
5 oz.

PhotoDisc

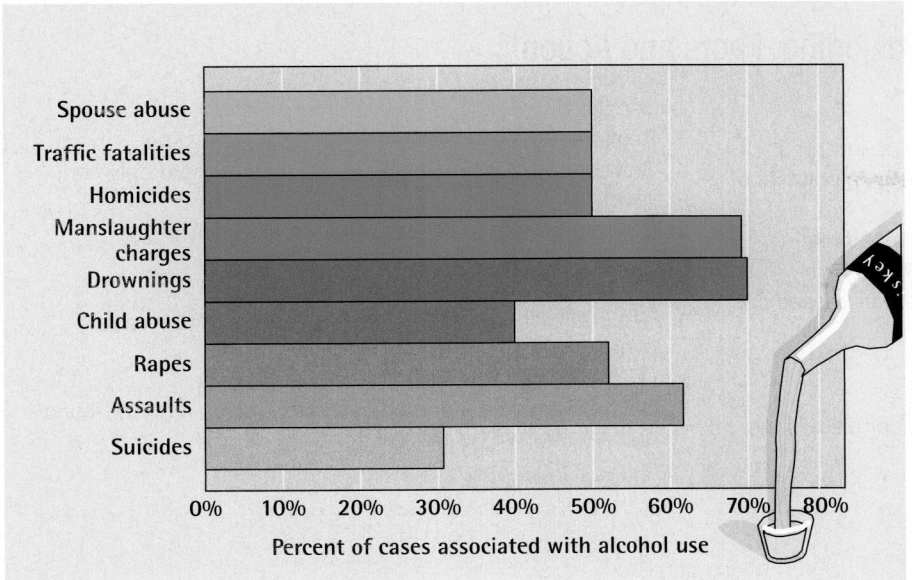

Illustration 14.2 Violence and injuries associated with alcohol.
Source: National Institute on Alcohol Abuse and Alcoholism, 2001, 2006.

Percent of cases associated with alcohol use

to the health of individual drinkers and often to other people as well. Although health can be damaged by the regular consumption of high amounts of alcohol, the ill effects of alcohol are most obvious in people with alcohol dependency, or **alcoholism**.

Habitually high alcohol intakes and alcoholism increase the risk of developing high blood pressure, stroke, and dementia; throat, stomach, and bladder cancer; central nervous system disorders; and vitamin and mineral deficiency diseases.[12] Alcohol abuse is associated with a high proportion of deaths from homicide, drowning, fires, traffic accidents, and suicide (Illustration 14.2). It is also involved in a large proportion of rapes and assaults and can devastate families.[13] **Alcohol poisoning** from the consumption of a large amount of alcohol in a short period of time can cause death—and does to about 1,700 college students each year.[14] You can read more about the important topic of alcohol poisoning in the accompanying Health Action.

Long-term, excessive alcohol intake is related to the development of **steatohepatitis**, a condition marked by the build-up of fat in the liver, inflammation, and eventually **cirrhosis** (Illustration 14.3). As a result of cirrhosis, scar tissue forms in the liver and decreases the liver's ability to function normally in synthesizing protein, combating infection, and processing the nutrients it receives from the bloodstream. Cirrhosis can lead to easy bruising, bleeding, or nosebleeds; swelling of the abdomen or legs; kidney failure; and other health problems.[15]

Alcohol consumption by pregnant women can cause lifetime physical and developmental problems in their offspring. Alcohol is readily transferred from mother to fetus. Tissue and organs undergoing development are particularly sensitive to the effects of alcohol.

alcoholism An illness characterized by a dependence on alcohol and by a level of alcohol intake that interferes with health, family and social relations, and job performance. Also referred to as alcohol dependency.

alcohol poisoning A condition characterized by mental confusion, vomiting, seizures, slow or irregular breathing, and low body temperature due to the effects of excess alcohol consumption. It is life-threatening and requires emergency medical help.

steatohepatitis Steatohepatitis (pronounced ste-at-oh-hep-ah-tie-tis) is a disease characterized by inflammation of and fat accumulation in the liver. It is associated with alcoholism and may occur in obesity and diabetes. Steatohepatitis may progress to cirrhosis.

cirrhosis Cirrhosis (pronounced sear-row-sis) is a disease of the liver characterized by widespread fibrous tissue buildup and disruption of normal liver structure and function. It can be caused by a number of chronic conditions that affect the liver.

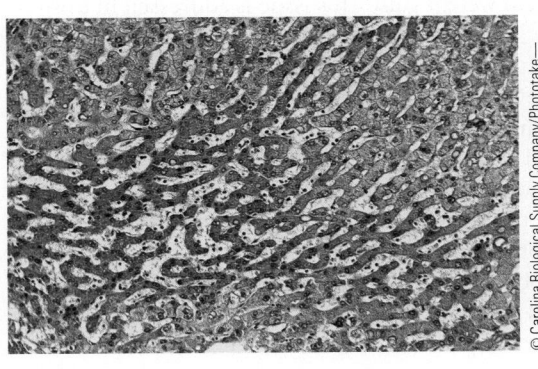

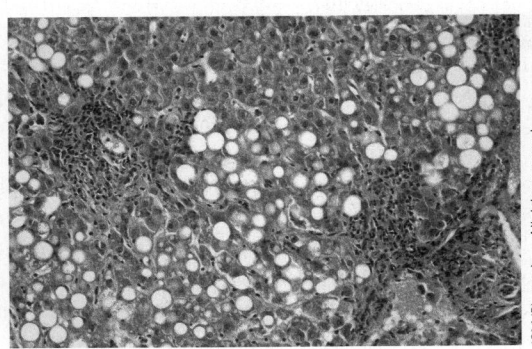

Illustration 14.3 Normal liver tissue is shown in the left-hand photo. The photo on the right shows a liver with inflammation and fatty tissue, or steatohepatitis.

health action Alcohol Poisoning: Facts and Action[18]

Facts

- Alcohol poisoning is a serious and sometimes deadly result of drinking excessive amounts of alcohol, usually in an episode of binge drinking.
- A high level of alcohol in the blood depresses breathing, promotes choking and vomiting, and slows heart rate.
- Blood alcohol levels continue to rise after a person passes out.

CRITICAL SIGNS of alcohol poisoning include the following:

1. Mental confusion and loss of consciousness; sometimes the person cannot be roused.
2. Vomiting
3. Seizures
4. Slow or irregular breathing (10 seconds or more between breaths)
5. Off-colored skin (paleness or bluish skin color)

ACTION

1. CALL 911 for help if there is any suspicion of an alcohol overdose.
2. DO NOT wait for all the symptoms to appear.
3. DO NOT assume a person can safely "sleep it off" or that a cold shower will help.

Bill Roth/Anchorage Daily News/MCT/Newscom

Illustration 14.4 Children with FASD experience physical and developmental impairments to varying degrees. This photograph shows facial characteristics associated with a severe form of FASD. Alcohol-containing beverages must show a warning statement on labels.

Fetal Alcohol Spectrum Disorder Once referred to as fetal alcohol syndrome, the term *fetal alcohol spectrum disorder* (FASD)[16] is now used to refer to the range of physical and developmental effects on offspring induced by alcohol consumption during pregnancy. Signs and symptoms present in individuals affected by various levels of severity of the FASD range from mild to severe. Some individuals affected by FASD may exhibit physical characteristics such as those shown in Illustration 14.4. These characteristics include a smooth ridge between the nose and upper lip, a small head size, and short height and low weight for height. Other problems observed range from poor coordination, hyperactive behavior, and learning difficulties to hearing, vision, heart, and kidney problems. Some individuals affected to some degree by FASD are physically normal, experience some level of interrupted and impaired development, and become accomplished adults.[16]

The severity of the effects of alcohol exposure during pregnancy depends on how much alcohol was consumed during pregnancy, whether the mother is genetically susceptible to adverse effects of alcohol, and if excessive intake occurred early or late in pregnancy. Because there is no known safe level of alcohol intake, it is recommended that women who are or may become pregnant not drink.[16,17]

Alcohol Intake, Diet Quality, and Nutrient Status

Alcohol provides seven calories per gram, making alcohol-containing beverages rather high in caloric content (Table 14.1). Because many alcohol-containing beverages provide calories and few or no nutrients, they are considered energy-dense, empty-calorie foods. On average, alcohol accounts for 3–9% of the caloric intake of U.S. adults who drink. The average goes up to around 50% among heavy drinkers.[18] Although beer, wine, and mixed drinks are known to contain alcohol and to provide calories, some are still confused about whether calories from alcohol contribute to weight gain. This issue is addressed in this unit's Reality Check.

Although diet quality tends to be better than average in moderate drinkers, as caloric consumption from alcohol-containing beverages increases, the quality of the diet generally decreases. Heavy drinkers are often malnourished, and their diets often provide too little thiamin, niacin, vitamins B_{12}, A, and C, and folate.[19] Deficiencies of nutrients, as well as direct, toxic effects of high levels of alcohol ingestion on liver function, nutrient absorption, and cellular utilization of nutrients, produce most of the physical health problems associated with alcoholism.[19,20] The lack of thiamin, for example, impairs the brain's utilization of glucose. When people with alcoholism initially withdraw from alcohol, thiamin deficiency may be expressed and result in "delirium tremens," a condition called the DTs by people who staff detoxification centers. People with delirium tremens experience

Table 14.1 Caloric value of common alcohol-containing beverages

	Serving	Calories
Beer, regular	12 oz	150
Beer, light	12 oz	110
Beer, dark	12 oz	168
Malt beverage	12 oz	225
80-proof liquor	1.5 oz	100
Wine, red	5 oz	105
Wine, white	5 oz	100
Wine cooler	12 oz	215

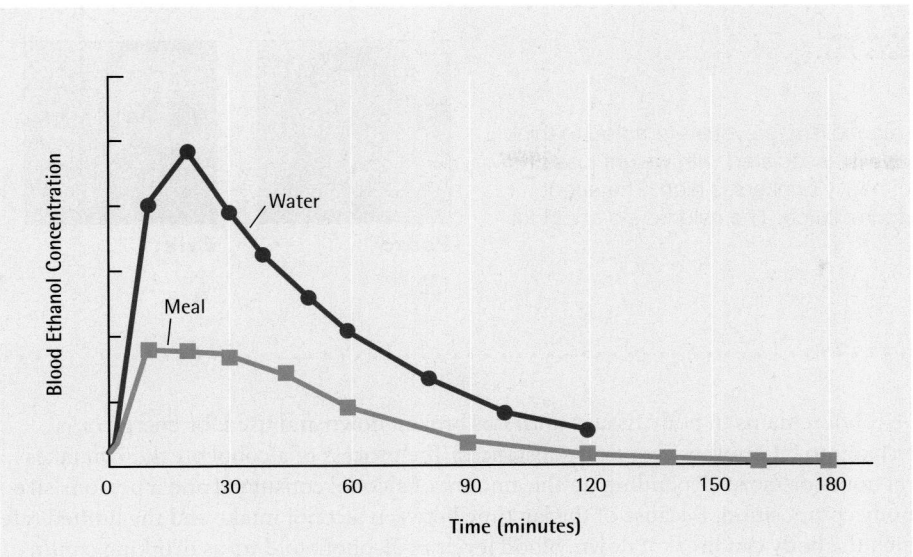

Illustration 14.5 Blood ethanol (alcohol) concentrations over time in response to ethanol administration with water or with a meal.
Source: Adapted from Figure 4, Levitt M D, et al. *Am J Physiol Gastrointest Liver Physiol.* 1997; 273:G951–7.

convulsions and hallucinations, and are severely confused. Thiamin injections are a key component of treatment for delirium tremens.[21] Because alcohol in excess is directly toxic to body tissues, consuming an adequate diet protects heavy drinkers from only some of the harmful effects of alcoholism.[14] Nutritional rehabilitation is a cornerstone of the treatment for people recovering from alcoholism.[19]

How the Body Handles Alcohol

Alcohol (ethanol) is easily and rapidly absorbed in the stomach and small intestine. Within minutes after it is consumed, alcohol enters the circulatory system where it increases blood alcohol concentration, and is on its way to the liver, brain, and other tissues throughout the body. The extent to which blood alcohol concentrations rise after alcohol consumption depends not only on the amount of alcohol consumed but also on whether the alcohol is consumed without food on an empty stomach or with food. Approximately 10% of alcohol consumed without food is absorbed in the stomach, whereas 30% of alcohol is absorbed in the stomach when consumed with food.[22] Alcohol consumed without food rapidly leaves the stomach and enters the small intestine where most of it is absorbed. The absorbed alcohol quickly enters the bloodstream and increases blood alcohol concentration. Alcohol consumed with food stays in the stomach and small intestine longer, slowing the rate of alcohol absorption. Consequently, blood alcohol concentrations increase less when consumed with food than without food.[22] Illustration 14.5 shows the effect of alcohol (ethanol) consumption with and without food on blood alcohol concentration over time.

REALITY CHECK
Do Alcohol Calories Count?

Perhaps you've heard the popular opinion that alcohol intake does not increase the risk of obesity. Is that the same as the not-so-popular opinion of scientists?

Do your thoughts side with Pedro or Erik?

Answers on page 14-6

Pedro: I started drinking a beer at night over the summer, and my weight never changed.

Erik: The six-pack around my abdomen is really a six-pack.

Maybe you've never seen an obese person with alcoholism, so you're tempted to think alcohol calories don't count. Chronic alcohol abuse is associated with weight loss and muscle wasting, even though the calorie intake of heavy drinkers is high. The effect appears to be due to an inhibition of fat tissue accumulation. The calories do count for light and moderate drinkers, however.[36]

Pedro:

Erik:

Alcohol remains in body tissues until it is broken down and used for energy or is converted into fat or other chemical substances. The process of alcohol breakdown takes several hours or more, depending on the amount of alcohol consumed and a person's size and body composition. Because of the lag time between alcohol intake and the limited rate at which the body can break it down, blood levels of alcohol build up as drinking continues (Table 14.2).[23]

Not all individuals break down alcohol at the same rate or to the same extent. Certain genetic traits exist that affect the liver's ability to break down alcohol. Most of the alcohol consumed is broken down in the liver by enzymes that convert it to products that can be used for energy formation, converted to fat, or used to form other chemical substances. The production of enzymes involved in the breakdown of alcohol is controlled by a number of genes. Some of the genes produce enzymes that quickly and thoroughly break down alcohol, whereas other forms of the enzymes break it down slowly and incompletely. Individuals who completely break down alcohol quickly are more tolerant of alcohol and more prone to excess alcohol intake than are individuals who break it down slowly. People who break down alcohol slowly experience nausea, a rapid heart rate, and develop facial flushing when they drink alcohol.[23] The experience is unpleasant so they tend to drink less. These and other differences in genetic traits account for approximately 50% of the heritable risk for alcohol dependency.[24]

Approximately 90% of the alcohol consumed is broken down by the body. The remaining 10% is lost in sweat, urine, or breath. The excretion of alcohol through a person's breath is the reason you can smell alcohol on the breath of someone who has been drinking recently, and why blood alcohol concentration can be estimated by a breathalyzer test.[25]

Alcohol Intake and Blood Alcohol Concentration The intoxicating effects of alcohol correspond to blood alcohol levels. It doesn't matter if the alcohol comes from beer, wine, or hard liquor; the intoxicating effects are the same. A drink or two in an hour raises blood levels of alcohol to approximately 0.03% in most people who weigh about 140 pounds. Blood alcohol levels of 0.03% correspond to mild intoxication. At this level,

Table 14.2 Alcohol doses and estimated percent blood alcohol levels

Number of drinks[a]	Percent blood alcohol by body weight					
	100 LB	120 LB	140 LB	160 LB	180 LB	200 LB
1	0.04	0.03	0.03	0.02	0.02	0.02
2	0.04	0.03	0.03	0.03	0.02	0.02
3	0.07	0.06	0.05	0.05	0.04	0.04
4	0.11	0.09	0.08	0.07	0.06	0.06
5	0.14	0.12	0.10	0.09	0.08	0.07
6	0.18	0.15	0.13	0.11	0.10	0.09

[a]Taken within an hour or so; each drink equal to 1/2 ounce pure ethanol. Effects may vary based on food intake, sex, and other factors.
Source: University of Oklahoma Police Dept., Blood Alcohol Calculator. Available at: www.ou.edu/oupd/bac.htm. Accessed August 2006.

people lose some control over muscle movement and have slowed reaction times and impaired thought processes. A person's ability to drive or operate equipment in a safe manner is decreased at this level of blood alcohol content (Illustration 14.6). Blood alcohol levels of around 0.06% are associated with an increased involvement in traffic accidents. The legal limit for intoxication according to all states' highway safety ordinances is 0.08%—beyond the point where driving is impaired. When blood alcohol content increases to 0.13% speech becomes slurred, "double vision" occurs, reflexes are dulled, and body movements become unsteady. If blood alcohol level continues to increase, drowsiness occurs and people may lose consciousness. Levels of blood alcohol above 0.35% can cause death.[26]

A given amount of alcohol intake among women produces higher blood levels of alcohol than it does for men of the same body weight. Pound for pound, women's bodies contain less water than men's bodies, so blood alcohol levels in women increase faster than in men.[37]

Over 150 medications, including sleeping pills, antidepressants, and painkillers interact harmfully with alcohol. Combining three or more drinks per day with aspirin or nonaspirin pain relievers (acetaminophen, ibuprofen) may cause stomach ulcers or liver damage.[27]

A Note about Alcohol Proof Alcohol proof is an old measure of how much alcohol is contained in alcohol-containing liquids. The term was derived hundreds of years ago by testing whether an alcohol containing liquid would ignite. If it did, that was "proof" of a high enough content of alcohol.[28] The proof of an alcohol-containing liquid is twice its alcohol content. So, a beer labeled 3.5% alcohol by volume would be 7 degrees proof, or a liquor labeled 35% alcohol by volume would be 70 degree proof. In the United States and in many other countries, it is required that alcohol-containing liquids be labeled with the percent alcohol by volume.[29]

PhotoDisc

Illustration 14.6 The legal limit for intoxication is 0.08%—beyond the point where driving is impaired.

How to Drink Safely if You Drink Many of the problems related to alcohol intake can be prevented by not drinking or by drinking responsibly. That means:

- Not drinking if you are or could become pregnant.

- Not drinking on an empty stomach (which can make you intoxicated surprisingly fast).

- Slowly sipping rather than gulping drinks.

- Limiting alcohol to an amount that doesn't make you lose control over your mind and body.

- Never driving a car or boat, hunting, or operating heavy equipment while under the influence of alcohol.

Caffeine does not counteract the effects of alcohol on impaired judgment, reaction time, or motor skills.[30]

What Causes Alcohol Dependency?

One in thirteen adults in the United States abuse alcohol or has alcohol dependency.[31] Alcoholism tends to run in families (about half of alcohol-dependent people have a family history of the disease), and there are documented genetic components to alcoholism.[24] Its development is also influenced by environmental factors. In general, the younger individuals are when they begin to drink, the greater likelihood that they will develop a drinking problem at some point in life.[33] Individuals who begin drinking before the age of 15, for example, are four times more likely to become alcohol dependent than people who do not drink before age 21 (Illustration 14.7). Close association with friends or peers who drink, high levels of stress, and availability of alcohol may also increase the risk of alcoholism. Television ads depicting youth-oriented parties, fun, and beer may increase underage drinking.[34]

Illustration 14.7 The younger a person is when drinking begins, the higher the probability that a drinking problem will develop.

Alcohol Use Among Adolescents Alcohol use among adolescents is increasing, and the age when teens begin drinking is going down. Underage drinking accounts for 11% of all the alcohol consumed in the United States. The average age when teens begin drinking is 14 years. These trends are particularly disturbing because they may lead to higher rates of alcohol dependency and alcohol-related problems.[35] Reduction in alcohol intake by adolescents is a major public health initiative of the Health Objectives for the Nation.

Help for Alcohol Dependence Alcohol dependency is a chronic disease that can be successfully managed but not always cured.[1] Many options for treatment are available. Treatments generally involve behavioral therapy, medications, or both. Behavioral therapy is successful in about one-third of people with alcoholism. Medications now available successfully treat alcohol dependency in certain individuals with a genetic predisposition toward developing the disease. Other medications that are under development would act by reducing the craving for and intake of alcohol among individuals without a genetic predisposition for the disease.[1]

Image Source/Jupiterimages

NUTRITION
up close

Effects of Alcohol Intake

Focal Point: Estimating blood alcohol levels and side effects.

Scenario: Ligia and Mark attend a wedding reception. Prior to the meal, they both drink a glass of champagne to toast the bride. Fifteen minutes later, they drink another glass to toast the groom. Ligia weighs 140 pounds and Mark weighs 180.

Questions: Using the information in Table 14.2 and the information given on how the body handles alcohol, answer the following questions:

A. After two glasses of champagne, what would be Ligia's estimated blood alcohol level? What is her percent blood alcohol? What would be Mark's estimated blood alcohol level? What is his estimated percent blood alcohol?

B. List three side effects of 0.03% blood alcohol content:

(1)   (2)   (3)

Feedback to the Nutrition Up Close can be found in Appendix G.

REVIEW QUESTIONS

- **Objectively assess the effects of the quantity of alcohol intake on behavior, health, and disease.**

1. Moderate consumption of alcohol-containing beverages decreases the risk of heart disease.   **True/False**

2. In the United States, a "moderate" intake of alcohol-containing beverages is considered to be two standard servings daily for men and one for women.   **True/False**

3. Alcohol is considered a food because it is a good source of a variety of vitamins and minerals.   **True/False**

4. Alcohol is used by the body for energy or is converted to glycogen.   **True/False**

5. Alcohol poisoning represents an emergency situation requiring medical care.   **True/False**

6. The idea that alcohol is absorbed more slowly if you drink after you eat or while eating is a myth.   **True/False**

7. Some people are genetically susceptible to developing alcoholism.   **True/False**

8. Caffeinated energy drinks counteract the affects of alcohol on the body.   **True/False**

9. Alcohol is converted to glucose before it is absorbed into the bloodstream.   **True/False**

10. About 90% of alcohol consumed is absorbed in the stomach.   **True/False**

11. _____ The alcohol content of 12 ounces of regular beer is equivalent to the alcohol content of:
    a. 1 ounce of liquor
    b. 16 ounces of "light" beer
    c. 3$\frac{1}{2}$ ounces of wine
    d. 1$\frac{1}{2}$ ounces of liquor

12. _____ Moderate alcohol intake is related to:
    a. increased ability of cells to utilize insulin
    b. decreased HDL cholesterol levels
    c. increased chronic inflammation
    d. increased risk of type 2 diabetes

13. _____ Heavy drinking is defined as the daily consumption of _____ drinks per day.
    a. 3
    b. 5
    c. 7
    d. 9

14. _____ Which of the following is *not* usually related to heavy drinking?
    a. obesity
    b. traffic accidents
    c. assaults
    d. high blood pressure

Answers to these questions can be found in Appendix G.

NUTRITION SCOREBOARD ANSWERS

1. Alcohol (actually ethanol) is produced by the fermentation of carbohydrates in grains, fruits, and other plant foods.   **True**

2. High intakes of alcohol are harmful to the body, regardless of the quality of the diet.   **False**

3. The statistics on alcohol abuse, injury, and death are startling. Alcohol abuse is a major personal, social, and public health problem in the United States.   **True**

4. Dark beers provide more calories than regular beer.   **True**

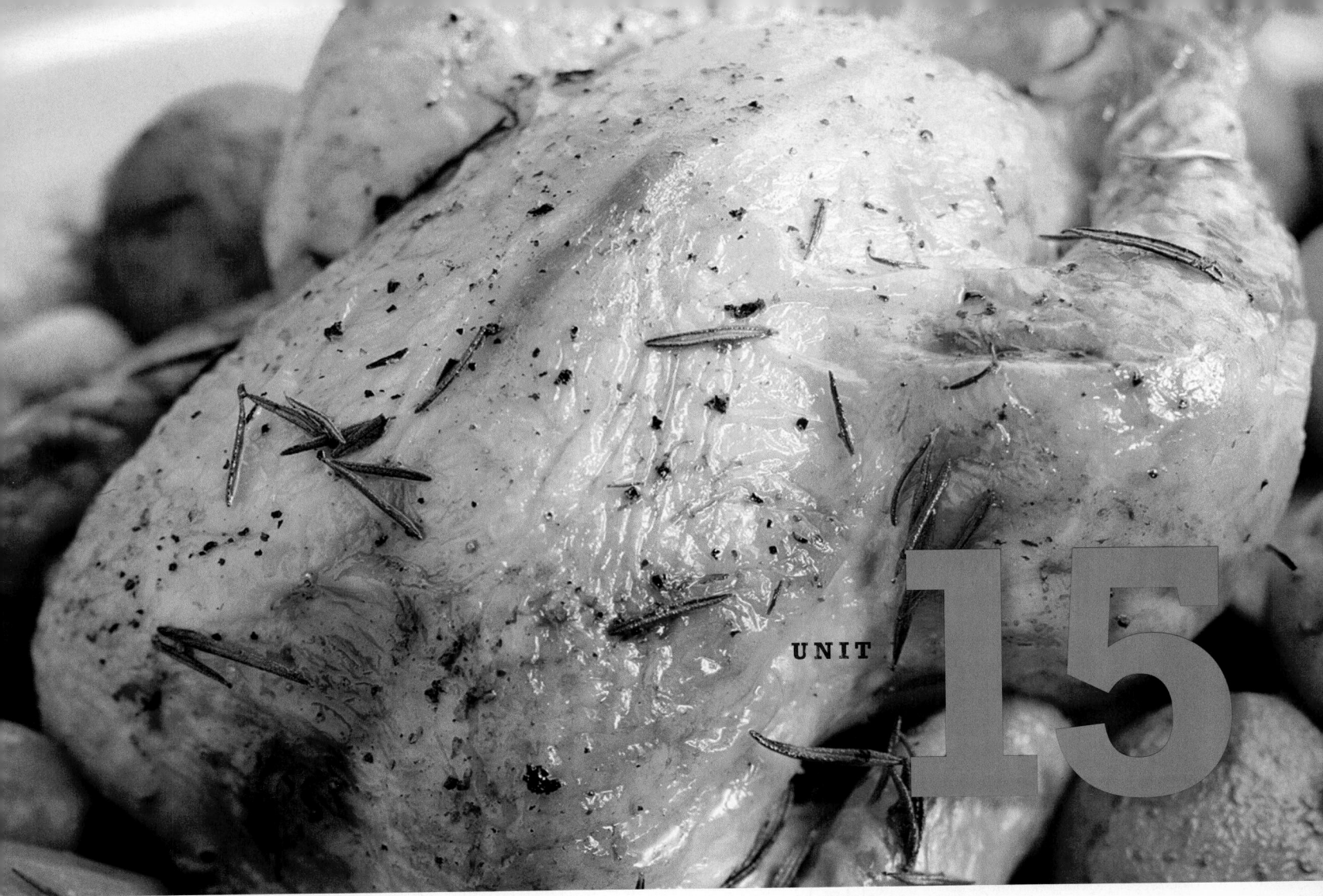

Proteins and Amino Acids

NUTRITION SCOREBOARD

UNIT **15**

1 The primary function of protein is to provide energy. **True/False**

2 "Nonessential amino acids" are not required for normal body processes. Only "essential amino acids" are. **True/False**

3 High-protein diets and amino acid supplements by themselves increase muscle mass and strength. **True/False**

Answers can be found at the end of the unit.

After completing Unit 15 and its interactive learning activities, you will be able to:

- Recall the basic functions, structures, and food sources of protein.

- Summarize the effects of amino acid supplements and protein on muscle mass and strength.

protein   Chemical substance in foods made up of chains of amino acids.

hormone   A substance, usually a protein or steroid (a cholesterol-derived chemical), produced by one tissue and conveyed by the bloodstream to another. Hormones affect the body's metabolic processes such as glucose utilization and fat deposition.

immunoproteins   Blood proteins such as antibodies that play a role in the functioning of the immune system (the body's disease defense system). Antibodies attack foreign proteins.

Illustration 15.1   The protein perception.

Protein

Other nutrients

PhotoDisc

Protein

- **Recall the basic functions, structures, and food sources of protein.**

The term **protein** is derived from the Greek word *protos*, meaning "first." The derivation indicates the importance ascribed to this substance when it was first recognized. Protein is an essential structural component of all living matter and is involved in virtually every biological process that occurs in cells. Protein has a very positive image (Illustration 15.1). The perception is so positive that you don't have to talk about the importance of protein—people are already convinced of it.

Key Nutrition Concepts

Content covered in Unit 15 on protein and amino acids relates to the following key nutrition concepts:

- Foods provide energy (calories), nutrients, and other substances needed for growth and health.

- Adequacy, variety, and balance are key characteristics of healthful diets.

Nearly all people in the United States get enough protein in their diets.[2] The average intake of protein by adults in the United States is 98 grams per day, approximately twice the recommended daily allowance (RDA) for men (56 grams) and for women of (46 grams). Approximately 16% of total calories in the average U.S. adult diet are supplied by protein.[3]

High-protein intakes are generally accompanied by high-fat and low-fiber intakes. That's because foods high in protein, such as hamburger, cheese, nuts, and eggs, tend to be high in fat and contain little or no fiber. Even lean meats provide a considerable proportion of their total calories as fat (Illustration 15.2).

Functions of Protein

Proteins perform structural and functional roles in the body (Table 15.1). They are an integral structural component of skeletal muscle, bone, connective tissues (skin, collagen, and cartilage), organs (such as the heart, liver, and kidneys), red blood cells and hemoglobin, hair, and fingernails. Proteins are the basic substance that make up thousands of enzymes in the human body; they are a major component of **hormones** such as insulin and growth hormone, and serve as other substances that perform important biological functions. Tissue maintenance and the repair of organs and tissues damaged from illness or injury are functions of different types of protein. Albumin, a protein made by the liver, is the blood's "tramp steamer." It attaches to and transports fatty acids, calcium, and other substances through the circulatory system to cells throughout the body.[4] Protein serves as an energy source at the level of four calories per gram.

The body of a 154-pound man contains approximately 24 pounds of protein. Nearly half of the protein is found in muscle; the rest is present in the skin, collagen, blood, enzymes, and **immunoproteins**; and in organs such as the heart, liver, and intestines; and other body parts. All protein in the body is continually being turned over, or broken down and rebuilt. This process helps maintain protein tissues in optimal condition so they continue to function normally. The process of protein turnover utilizes roughly nine ounces of protein each day. Yet, we consume only two to three ounces of protein daily. Most of the protein used for maintenance is recycled from muscle and other protein tissues being turned over. Proteins play key roles in the repair of body tissues by serving as substances—such as fibrin—that help blood clot (Illustration 15.3) and by replacing tissue proteins damaged by illness or injury.[5]

Protein serves as a source of energy in healthy people, but not nearly to the extent that carbohydrates and fats usually do. Protein is unlike carbohydrate and fat in that it contains nitrogen and does not have a storage form in the body. In order to use protein for energy, amino acids that make up proteins must first be stripped of their nitrogen. The free nitrogen can be used as a component of protein formation within the body, or, if pres-

(a) Hamburger (90% lean): 45%

(b) Tenderloin: 43%

(c) Sirloin: 33%

(d) Pork chop, lean: 48%

(e) Pork loin roast: 36%

(f) Pork tenderloin: 28%

(g) Chicken thigh, no skin: 47%

(h) Baked chicken breast, no skin: 19%

Illustration 15.2   The fat content of three-ounce portions of "lean" meats. The percentage of calories from fat is indicated for each portion. (A three-ounce portion of meat is about the size of a deck of cards.) Each portion of meat provides approximately 21 grams of protein.

ent in excess, it is largely excreted in urine. Excretion of nitrogen requires water, so high intake of protein increases water need. Amino acids missing their nitrogen component are converted to glucose or fat that then can be used to form energy. A small amount of protein (1%) can be obtained from the liver and blood and used to cover occasional deficits in protein intake.[6]

Nitrogen Balance   **Nitrogen balance** represents the balance between protein intake and protein utilization by the body. Protein intake is estimated as the nitrogen content of protein consumed. The nitrogen content of protein is estimated as 16% of the weight of protein consumed. Nitrogen excretion is assessed as the amount of nitrogen excreted in the form of **urea**. Nitrogen balance is measured as the difference between nitrogen intake and nitrogen excreted.[22]

- A person is considered to be in "nitrogen balance" when her or his intake of nitrogen equals nitrogen excreted. Example: A person consumes 10 grams of nitrogen and excretes 10 grams of nitrogen: 10 grams nitrogen – 10 grams nitrogen = 0 grams nitrogen. People who are in nitrogen balance are consuming as much protein as they are utilizing.

- A person who consumes less nitrogen than excreted is in "negative nitrogen balance." Example: A person consumes 10.2 grams nitrogen and excretes 14.0 grams: 10.2 grams

Table 15.1 **Examples of functions of protein**

Structural

- Serves as a structural material in muscles, connective tissue, organs, and hemoglobin
- Maintains and repairs protein-containing tissues

Functional

- Serves as the basic component of enzymes, hormones, and other biologically important chemicals
- Serves as an energy source
- Helps maintain body fluid balance
- Helps maintain acid-base balance in body fluids

nitrogen balance   The difference between nitrogen intake and excretion. It is assessed as the difference between nitrogen intake and nitrogen excreted.

urea   Nitrogen released from the breakdown of proteins for energy is largely excreted in the urine in the form of urea. It can be measured in urine, or in blood as blood urea nitrogen (BUN).

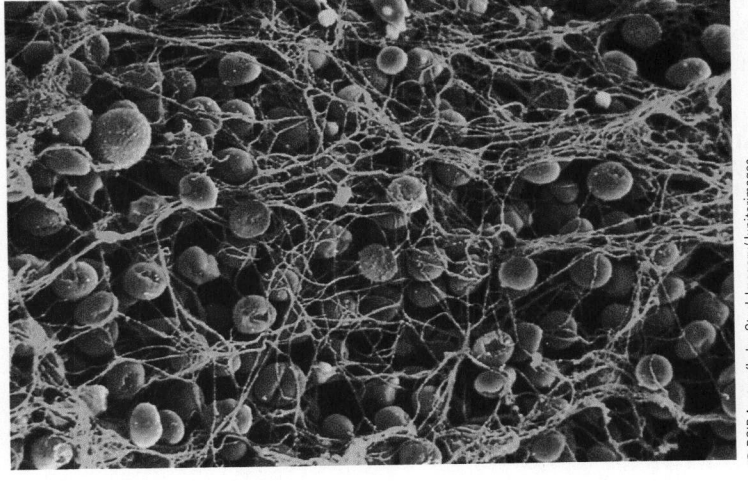
Illustration 15.3   Red blood cells enmeshed in fibrin in a color-enhanced microphotograph. Red blood cells and fibrin (which helps stop bleeding by causing blood to clot) are made primarily from protein.

nitrogen – 14.0 grams nitrogen = –3.8 grams nitrogen. A negative nitrogen balance means the person is consuming less protein than the body is utilizing. A negative nitrogen balance may occur with undernutrition, fasting, burn and other serious injuries, fever, and other illnesses.

- A person is considered to be in "positive nitrogen balance" when his or her intake of nitrogen is higher than nitrogen excretion. Example: A person consumes 18 grams of nitrogen and excretes 14.5 grams: 18 grams nitrogen – 14.5 grams nitrogen = +3.5 grams nitrogen. A positive nitrogen balance indicates that some of the protein consumed is being retained and used to build up body protein tissues. Positive nitrogen balance occurs during growth, pregnancy, breastfeeding, and recovery from illness or injury.

Results of nitrogen balance studies help determine protein need and are sometimes used clinically to adjust a person's protein intake to meet protein need.[22]

Amino Acids

essential amino acids Amino acids that cannot be synthesized in adequate amounts by humans and therefore must be obtained from the diet. They are sometimes referred to as "indispensable amino acids."

nonessential amino acids Amino acids that can be readily produced by humans from components of the diet. Also referred to as "dispensable amino acids."

DNA (deoxyribonucleic acid) Genetic material contained in cells that directs the production of proteins in the body.

The building blocks of protein are amino acids, which share the characteristic of containing nitrogen. Illustration 15.4 shows an example of the basic structure of an amino acid and its nitrogen component. Twenty common amino acids (Table 15.2) form proteins, and every protein in the body is composed of unique combinations of amino acids. Nine of the twenty common amino acids are considered **essential**, and 11 are **nonessential**. The essential amino acids are called *essential* because the body cannot produce them or produce enough of them, so they must be provided by the diet. Healthy individuals can produce the nonessential amino acids, so we don't require a dietary source of them. Despite the labels, all 20 amino acids are required to build and maintain protein tissues. Proteins in foods contain both essential and nonessential amino acids.

Amino Acids and Protein Structure The assembly of amino acids into proteins is directed by **DNA**, the genetic material within each cell. Some proteins are made of only a few amino acids, while other proteins contain over 25,000. The arrangement of amino acids determine whether a protein functions as an enzyme or a hormone, or it becomes a component of red blood cells, muscle, or other substance.

Proteins produced vary in size and complexity based on their role in cellular processes. They are classified by their properties as having primary, secondary, tertiary, or quaternary structures (Illustration 15.5). Proteins with primary structure consist of a linear arrangement of linked amino acids, whereas proteins with secondary structures have folded chains of amino acid. Some hormones and chemical messengers that initiate cellular processes have these structures. Tertiary structures represent the three-dimensional structure of more complex and larger proteins. These proteins have elaborate folding patterns such as found in insulin (Illustration 15.6). Proteins with quaternary structure are the largest and consist of multiple, linked chains of amino acids folded and formed in such a way that they can perform specific functions. Hemoglobin, a component of red blood cells that transports oxygen to cells and removes carbon dioxide from cells, is an example of this type of protein structure.[6]

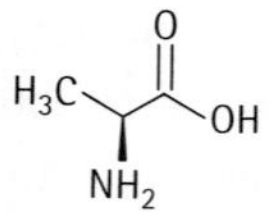

Illustration 15.4 The basic chemical structure of alanine, a nonessential amino acid.

Table 15.2 Essential and nonessential amino acids

Essential		Nonessential	
Histidine	Threonine	Alanine	Glutamine
Isoleucine	Tryptophan	Arginine	Glycine
Leucine	Valine	Asparagine	Proline
Lysine		Aspartic acid	Serine
Methionine		Cysteine	Tyrosine
Phenylalanine		Glutamic acid	

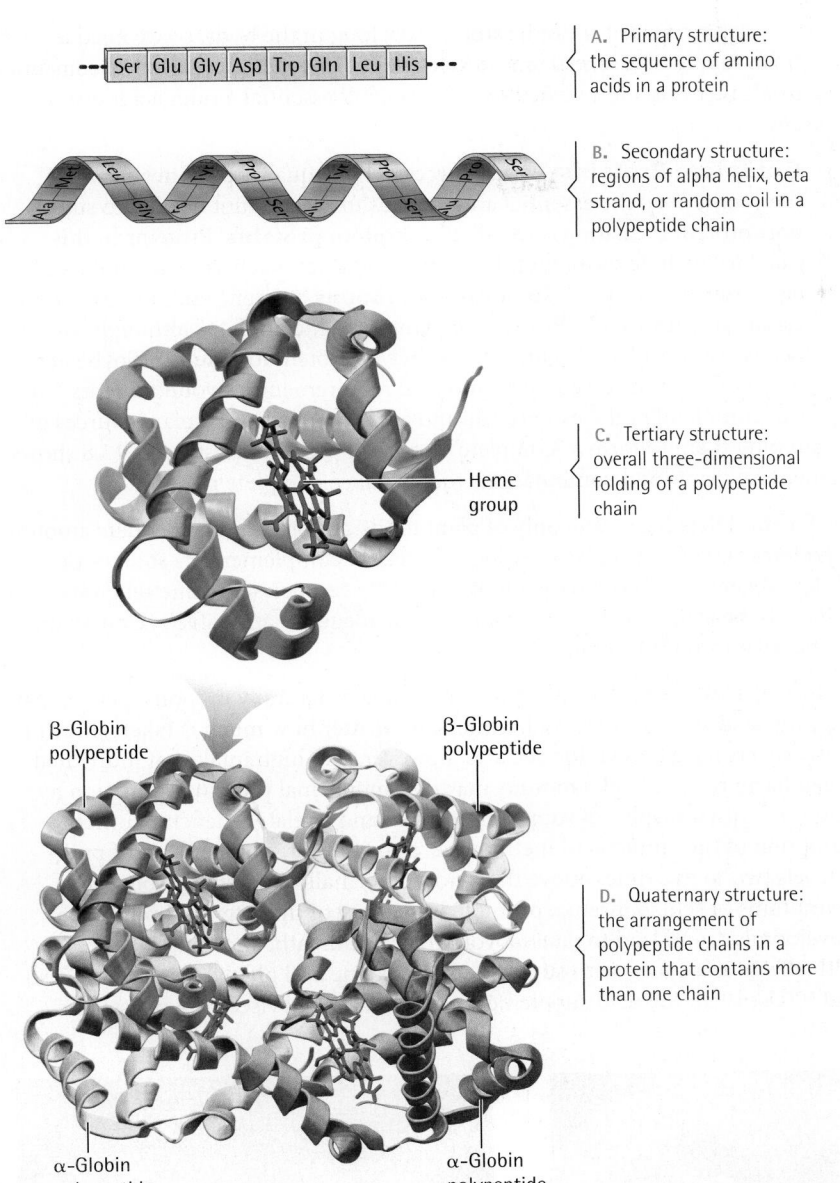

A. **Primary structure:** the sequence of amino acids in a protein

-- -- Ser Glu Gly Asp Trp Gln Leu His -- --

B. **Secondary structure:** regions of alpha helix, beta strand, or random coil in a polypeptide chain

C. **Tertiary structure:** overall three-dimensional folding of a polypeptide chain

Heme group

β-Globin polypeptide

β-Globin polypeptide

D. **Quaternary structure:** the arrangement of polypeptide chains in a protein that contains more than one chain

α-Globin polypeptide

α-Globin polypeptide

Illustration 15.5 The primary, secondary, tertiary, and quaternary structures of proteins.
Source: Russell, *Biology: The Dynamic Science,* 2nd ed., chap. 3, fig. 3.18, p. 58.

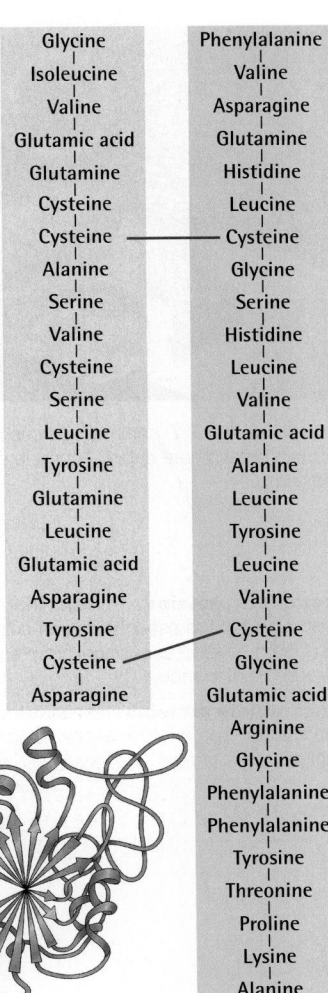

Glycine — Phenylalanine
Isoleucine — Valine
Valine — Asparagine
Glutamic acid — Glutamine
Glutamine — Histidine
Cysteine — Leucine
Cysteine —— Cysteine
Alanine — Glycine
Serine — Serine
Valine — Histidine
Cysteine — Leucine
Serine — Valine
Leucine — Glutamic acid
Tyrosine — Alanine
Glutamine — Leucine
Leucine — Tyrosine
Glutamic acid — Leucine
Asparagine — Valine
Tyrosine — Cysteine
Cysteine — Glycine
Asparagine — Glutamic acid
Arginine
Glycine
Phenylalanine
Phenylalanine
Tyrosine
Threonine
Proline
Lysine
Alanine

Illustration 15.6 Amino acid chains and folding of insulin, a tertiary protein.

Protein Quality

The ability of proteins to support protein tissue construction in the body varies depending on their content of essential amino acids. How well dietary proteins support protein tissue construction is captured by tests of the protein's "quality."

Proteins of high quality contain all the essential amino acids in the amounts needed to support protein tissue formation by the body. If any of the essential amino acids are missing in the diet, proteins are not formed—even those proteins that could be produced from available amino acids. Shutting off all protein formation for want of an amino acid or two may appear inefficient; but if the body did not cease all protein formation, cells would end up with an imbalanced assortment of proteins. This would seriously affect cell functions. When the required level of an essential amino acid is lacking, the remaining amino acids are primarily used for energy.

Illustration 15.7 Animal sources of protein shown here supply "complete proteins."

complete proteins Proteins that contain all of the essential amino acids in amounts needed to support growth and tissue maintenance.

incomplete proteins Proteins that are deficient in one or more essential amino acids.

Amino acids cannot be stored very long in the body, so we need a fresh supply of essential amino acids daily. This means we need to consume foods that provide a sufficient amount of all essential amino acids every day.

Complete Proteins Food sources of high-quality protein (meaning they contain all the essential amino acids in the amount needed to support protein formation) are called **complete proteins**. Proteins in this category include those found in animal products such as meat, milk, and eggs (Illustration 15.7). **Incomplete proteins** are deficient in one or more essential amino acids. Proteins in plants are "incomplete," although soybeans are considered a complete source of protein for adults.[7] (Soybeans may not meet the essential amino acid requirements of young infants.) You can complement the essential amino acid composition of plant sources of protein by combining them to form a "complete" source of protein. Illustration 15.8 shows a few complementary plant food combinations that produce complete proteins.

Vegetarian Diets Diets consisting only of plant foods can provide an adequate amount of complete proteins. A key to success is eating a variety of complementary sources of protein each day. Vegetarian diets have been practiced for centuries by some religious and cultural groups, bearing testimony to their general adequacy and safety. (Unit 16 on vegetarian diets expands on this topic.)

Amino Acid Supplements Because amino acids occur naturally in foods, people may assume that amino acid supplements are harmless, no matter how much is taken (Illustration 15.9). Researchers have known for decades, however, that high intakes of individual amino acids can harm health. High amounts may disrupt normal protein production by overwhelming cells with a surplus of some amino acids and a relative deficit of others.[8] Excess consumption of the amino acid methionine, in particular, causes a host of problems. Intake levels two to five times above the amount normally consumed from foods worsen the symptoms of schizophrenia, promote hardening of the arteries, impair fetal and infant development, and lead to nausea, vomiting, bad breath, and constipation.[8,9] Adverse health effects associated with supplemental cysteine and phenylalanine have also been reported.[8,10] Use of amino acid supplements should be supervised by a physician.

Illustration 15.8 Each of these combinations of plant foods provides complementary protein sources.

Safe doses of certain amino acids or their derivatives are being used to manage certain conditions. For example, melatonin (a derivative of the essential amino acid tryptophan), is used to promote sleep.[11] Melatonin supplements have been found to reset the sleep clock in shift workers, pilots, jetlagged travelers, and people with sleep disorders.

The biggest users of amino acids supplements are athletes who believe they will increase muscle mass and strength.[12] Additional information on the topic of physical performance aids is provided in Unit 28.

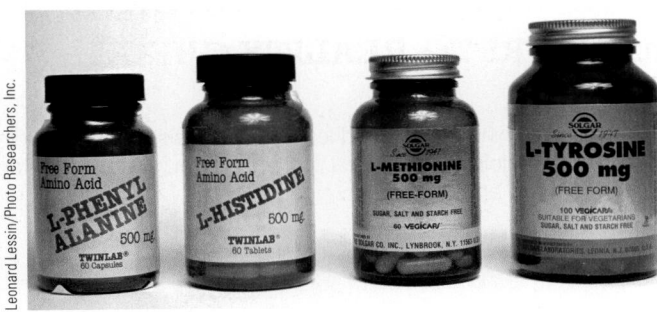

Illustration 15.9 A wide variety of amino acid supplements are available on the web or in stores.

Amino Acid Supplements, Protein Powders, Muscle Mass, and Strength You can't just consume amino acids or protein powders and watch your muscles grow—no matter how convincing the ads that sell such products are. If that happened, anyone who wanted a ripped stomach and bulging triceps could have them. Neither essential amino acids nor protein supplements by themselves increase muscle size and strength.[1] Muscle size and strength are built slowly from the raw ingredients of a healthy diet and resistance training (Illustration 15.10).[13] Current research indicates, however, that you may be able to enhance muscle protein synthesis and muscle mass by consuming 20 grams of high-quality protein or about 10 grams of certain amino acids within an hour following resistance exercise workouts.[14,15] Skim milk, lean meat, fish, eggs, and beans and rice are examples of high-quality protein foods that could be consumed. Protein intakes over 20 grams following exercise do not appear to offer additional benefit.[14-16]

As is the case for other dietary supplements, the purity, dose, and safety of supplements available on the web and in stores is not guaranteed. Even if the amino acid supplements are pure, safe, and the dose reported on the label correct, it is likely that the supplement does not work better than 20 grams of high-quality protein from food. Additional studies are needed on the safety of amino acid supplements.[17]

Illustration 15.10 Building muscles like these takes time, a good diet, and lots of resistance exercise.

Food as a Source of Protein

Approximately 70% of the protein consumed by Americans comes from meats, milk, and other animal products.[3] Dried beans and grains are not as well known for their protein, but are nevertheless good sources (Table 15.3). Plant sources of protein are generally low in fat, making them a wise choice for consumers who are trying to limit their intake of fat. Nearly all food sources of protein provide an assortment of vitamins and minerals as well. Beef and pork are particularly good sources of iron, a mineral often lacking in the diets of women (Table 15.4).

Protein Deficiency Protein deficiency has been found to occur in combination with a deficiency of calories and nutrients.[2] Because food sources of protein generally contain

REALITY CHECK
Pure Protein

You have a nutrition exam coming up, so you and your classmate Carole have gotten together to study. You get into a discussion about food sources of protein that goes like this:

Who gets the thumbs-up?

Answers on page 15-8

Carole: Lean meats are the best source. They're pure protein!

Lauren: Pure protein? Even the driest, toughest meats contain more than protein.

ANSWERS TO REALITY CHECK
Pure Protein

Some people think muscle and lean meat consist only of protein. They don't. By weight, lean cooked sirloin steak is 29% protein, 8% fat, and 62% water. Lean pork is 29% protein, 9% fat, and 61% water.

Carole:

Lauren:

Table 15.3 Food sources of protein

The adult RDA is 46 grams for women aged 19 to 30 years and 56 grams for men aged 19 to 30 years.

Food	Amount	Grams	Protein content percentage of total calories
Animal products			
Tuna (water packed)	3 oz	24	89%
Shrimp	3 oz	11	84
Cottage cheese (low-fat)	½ cup	14	69
Beef steak (lean)	3 oz	26	60
Chicken (no skin)	3 oz	24	60
Pork chop (lean)	3 oz	20	59
Beef roast (lean)	3 oz	23	45
Skim milk	1 cup	9	40
Fish (haddock)	3 oz	19	38
Leg of lamb	3 oz	22	37
Yogurt (low-fat)	1 cup	13	34
Hamburger (lean)	3 oz	24	34
Egg	1 medium	6	32
Swiss cheese	1 oz	8	30
Sausage (pork links)	3 oz	17	28
2% milk	1 cup	8	26
Cheddar cheese	1 oz	7	25
Whole milk	1 cup	8	23
Dried beans and nuts			
Tofu	½ cup	14	38
Soybeans (cooked)	½ cup	10	33
Split peas (cooked)	½ cup	5	31
Lima beans (cooked)	½ cup	6	27
Dried beans (cooked)	½ cup	8	26
Peanuts	½ cup	9	17
Peanut butter	1 tbs	4	17
Grains			
Corn	½ cup	3	29
Egg noodles	½ cup	4	25
Oatmeal (cooked)	½ cup	3	15
Whole wheat bread	1 slice	2	15
Macaroni (cooked)	½ cup	3	13
White bread	1 slice	2	13
White rice (cooked)	½ cup	2	11
Brown rice (cooked)	½ cup	2	10

essential nutrients such as iron, zinc, vitamin B_{12}, and niacin, diets that produce protein deficiency usually cause a variety of other deficiencies, too. Protein does not generally serve as an important source of energy, but body protein will be used as a major energy source during starvation. To meet the need for energy, the body will extract protein from the liver, intestines, heart, muscles, and other organs and tissues. Loss of more than about 30% of body protein results in reduced body strength for breathing, susceptibility to infection, abnormal organ functions, and death.[5] Inadequate protein intake is related to decreased growth in children and loss of muscle mass and strength in adults.[18,19]

In the past it was thought that **kwashiorkor**, a devastating disease that can affect severely undernourished children, was primarily due to a protein deficiency. Although kwashiorkor is related to protein-calorie malnutrition, the disease does not appear to be due only to a lack of protein.[2] This conclusion is supported by studies that show that improving protein intake in children with kwashiorkor does not correct the disease. It appears that some children with protein-calorie malnutrition develop kwashiorkor due to an inability to utilize protein and fat normally during starvation.[20,21]

How Much Protein Is Too Much? Adults can consume a substantial amount of protein—approximately 35% of total calories—for months at a time without ill effects. This observation is based on studies of the diets of Eskimos, explorers, trappers, and hunters in northern America.[2] The very high-protein diets would generally contain a good deal of fat in the form of whale blubber, lard, or fat added to dried meat. Consumption of 45% of total calories from protein is considered too high and is related to the development of symptoms such as nausea, weakness, and diarrhea. Diets very high in protein may result in death after several weeks. A complex disease resulting from excess protein intake was termed "rabbit fever" after it occurred in trappers attempting to exist exclusively on wild rabbit.[5]

High-protein diets have been implicated in the development of weak bones, kidney stones, cancer, heart disease, and obesity. The National Academy of Sciences has concluded that the risk of such disorders does not appear to be increased among individuals consuming 10–35% of total calories from protein, and on average adults consume 16%.[3,5]

Table 15.4 Iron content in a 3-ounce serving of various meats

The RDA for women aged 19 to 30 years is 15 milligrams. The RDA for men aged 19 to 30 years is 10 milligrams.

Meat	Iron content (mg)
Pork chop (lean)	3.4
Round steak (lean)	3.1
Hamburger (lean)	3.0
Shrimp	2.6
Tuna	1.6
Baked chicken (no skin)	1.4
Lamb (lean)	1.3

kwashiorkor A form of severe protein-energy malnutrition in young children. It is characterized by swelling, fatty liver, susceptibility to infection, profound apathy, and poor appetite.

© Envision/Corbis

NUTRITION
up close

My Protein Intake

Focal Point: Determine the amount of protein in your diet yesterday.

For *each serving* of a food item you ate yesterday, write the grams of protein the food contains in the corresponding blank. For example, a standard serving of meat is three ounces (about the size of the palm of your hand or a deck of cards). If you had *one* three-ounce pork chop yesterday, write *20 grams* in the cor-

responding blank. If you had *two* three-ounce pork chops, write *40 grams*. If a protein food you ate yesterday is not included, choose the item on the list closest to it. Then, total the grams of protein you ate yesterday from both plant and animal sources. Finally, compare your protein intake with the RDA of 46 grams for women or 56 grams for men.

Food	One serving	Protein in one serving (grams)	Protein you ate (grams)
Animal products			
Milk (whole)	1 c (8 oz)	8	
Yogurt	1 c (8 oz)	13	
Cottage cheese	1/2 c (4 oz)	14	
Hard cheese	1 oz	7	
Hamburger (lean)	3 oz	24	
Beef steak (lean)	3 oz	26	
Chicken (no skin)	3 oz	24	
Pork chop (lean)	3 oz	20	
Fish	3 oz	19	
Hot dog	1	6	
Sausage	3 oz	17	
		Subtotal from animal foods:	
Plant products			
Bread	1 slice	2	
Rice	1/2 c (4 oz)	2	
Pasta	1/2 c (4 oz)	3	
Cereals	1/2 c (4 oz)	3	
Vegetables	1/2 c (4 oz)	2	
Peanut butter	1 tbs	4	
Nuts	1/4 c (2 oz)	7	
Cooked beans (legumes)	1/2 c (4 oz)	8	
		Subtotal from plant foods:	
		Total grams of protein from plant and animal foods:	
		Amount above/below RDA:	

Special note: You can also calculate your protein intake using Cengage's Diet Analysis Plus software. To use this, input your food intake for one day. Then go to the Analyses/Reports section to view the total number of grams of protein in your diet.

Feedback to the Nutrition Up Close can be found in Appendix G.

REVIEW QUESTIONS

- **Recall the basic functions, structures, and food sources of protein.**

1. The only known function of protein is to serve as a structural component of muscle, bones, and other solid tissues.   **True/False**

2. Low protein intake is a fairly common problem among adults and children in the United States.   **True/False**

3. Immunoproteins are a part of the body immune system.   **True/False**

4. Fibrin is a protein that helps blood clot.   **True/False**

5. The nitrogen component of amino acids must be removed before they can be used for energy.   **True/False**

6. There are 22 amino acids, and half of them are "essential."   **True/False**

7. DNA provides the blueprint for the production of specific proteins by the body.   **True/False**

8. People need to eat animal sources of protein in order to consume enough high-quality "complete" proteins.   **True/False**

9. Individual amino acids supplements may be hazardous to health.   **True/False**

10. Kwashiorkor is the name of a protein deficiency disease.   **True/False**

11. Protein intake should constitute 10–35% of total calorie intake.   **True/False**

12. Proteins with primary structure consist of a linear arrangement of linked amino acids, whereas proteins with secondary structures have folded chains of amino acid.   **True/False**

- **Summarize the effects of amino acid supplements and proteins on muscle mass and strength.**

13. Amino acid supplements plus resistance exercise are the keys to building muscle mass and strength.   **True/False**

The next three questions refer to Bertie's resistance training and protein consumption.

14. _____ Bertie decides to consume 20 grams of high-quality protein after her resistance training workout. How many glasses of skim milk would Bertie need to consume to get about 20 grams of protein?

 a. 1
 b. 2
 c. 3
 d. 4

15. _____ Bertie consumes 20 grams of high-quality protein after each resistance training session. Over time, the protein would likely _____ Bertie's muscle mass.

 a. have no effect on
 b. decrease
 c. increase
 d. increases then decrease

16. _____ Bertie decides to consume 20 grams of high-quality protein daily as a supplement to her diet but to give up on resistance training. What do you expect would happen to her muscle mass?

 a. The extra protein would maintain her muscle mass.
 b. The extra protein would increase her muscle mass.
 c. Her muscle mass would decline.
 d. Her muscle mass would stay the same.

Answers to these questions can be found in Appendix G.

NUTRITION SCOREBOARD ANSWERS

1. Energy is a function of protein, but it's not the primary one.   **False**

2. "Nonessential amino acids" are required by the body, but they are not required components of our diet.   **False**

3. Muscles contain protein, but you can't increase muscle mass and strength by consuming a high-protein diet, amino acid supplements, or protein powders.[1]   **False**

Vegetarian Diets

NUTRITION SCOREBOARD

1 The human body developed to function best on a vegetarian diet. **True/False**

2 People who don't eat meat have more health problems than people who do. **True/False**

3 Macrobiotic diets cure some types of cancer. **True/False**

4 In order to consume enough high-quality protein, vegetarians need to consume combinations of plant foods that provide a complete source of protein at every meal. **True/False**

Answers can be found at the end of the unit.

After completing Unit 16
and its interactive learning
features, you will be able to:

• Understand the health
benefits and limitations of
various types of vegetarian
diets.

Perspectives on Vegetarian Diets

• **Understand the health benefits and limitations of various types of vegetarian diets.**

Vegetarianism in the United States, Canada, and other economically developed countries is moving from the realm of counterculture to the mainstream.[2] Yet even with the increased acceptance of vegetarian diets, people tend to be for vegetarianism or against it, often without knowing much about it. Few of those opposed to vegetarianism have tried to learn about the vegetarian way of life or understand that vegetarian diets can be healthful. A small percentage of health professionals are vegetarians, and those who are not are often skeptical about how healthy a vegetarian diet can be. The possibility that something so different from the customary diet can be nourishing to the body may be rejected out of hand.

An objective look at vegetarianism reveals that both appropriately planned vegetarian diets and the lifestyle often followed by vegetarians can be very good for health.[3] Vegetarians tend to be health conscious; they often avoid using alcohol, tobacco, and illicit drugs, and engage in regular physical activity.[3] Diet is usually one of several characteristics shared by people practicing particular types of vegetarianism.

Key Nutrition Concepts

Key nutrition concepts that underlie this content on vegetarian diets include:

• Poor nutrition can result from both inadequate and excessive levels of nutrient intake.

• Some groups of people are at higher risk of becoming inadequately nourished than others.

• Poor nutrition can influence the development of certain chronic diseases.

• Adequacy, variety, and balance are key characteristics of healthful diets.

Reasons for Vegetarianism

Table 16.1 Reasons for vegetarianism[5,6]

• Lack of availability or affordability of animal products
• Desire not to cause harm to animals
• Religious beliefs
• Desire to "eat low on the food chain"
• Desire to preserve the world's food supply
• Health
• Desire to omit hormones, antibiotics, and possible contaminants in meats

Worldwide, vegetarians number in the hundreds of millions.[4] Much of the world's population subsists on vegetarian diets because meat and other animal products are scarce or too expensive (Table 16.1). In other societies, people have the luxury of choosing a healthy assortment of food from an abundant and affordable food supply that includes a wide variety of items acceptable to vegetarians (Illustration 16.1). When food availability is not an issue, people tend to adopt vegetarian diets because of a desire to cause no harm to animals, a personal commitment to preserve the environment and the world's food supply by "eating low on the food chain," or a belief that animal products are unhealthful or unsafe. They may avoid animal products as part of a value or religious belief system. Others follow vegetarian diets to keep their weight down or to lower the risk of developing specific diseases such as diabetes or heart disease.[6]

Photo Disc

Vegetarian Diets Come in Many Types There is no one vegetarian diet. People who consider themselves to be vegetarians range from those who eat all foods except red meat (mainly beef) to those who exclude all foods from animal sources, including honey. According to the American Vegetarian Society, a person is a vegetarian only if she or he eats *no* meat.

Vegetarian Diet Options The types of foods included in common vegetarian diets are summarized in Table 16.2. The least restrictive form of vegetarian diet has been unofficially labeled "far vegetarian" because it excludes only red meats. This diet is very much like that consumed by omnivores, or meat and plant eaters. Quasi-vegetarian diets (also called semivegetarian and flexitarian diets) vary somewhat but generally exclude beef, pork, and poultry while including fish, eggs, dairy products, and plant foods. The lacto-ovo vegetarian

Illustration 16.1   The growing selection of vegetarian foods in supermarkets.

Richard Anderson

diet includes only dairy foods, eggs, and plant foods. Individuals practicing this type of diet exclude all meats. The lacto-vegetarian diet, as the name implies, includes only milk and milk products and plant foods. Vegan (pronounced *vee-gun*) and raw food diets are a more restrictive type of vegetarian diet. Vegans and raw food dieters eat only plant foods; in addition, vegans may avoid honey and clothes made from wool, leather, or silk.

Macrobiotic Diets   Macrobiotic diets fall somewhere between quasi-vegetarian and vegan diets. The formulation of macrobiotic diets has changed dramatically over the past few decades. In addition to brown rice, other grains, and vegetables, these diets now include fish, dried beans, spices, fruits, and many other types of foods. No specific foods are prohibited, and locally grown and whole foods are emphasized.

Persons adhering to the macrobiotic philosophy place value on consuming organic foods and balancing the intake of "yin" and "yang" foods. Foods are classed as yin or yang based on beliefs about the food's relationship to the emotions and the physical condition of the body. Yin foods such as corn, seeds, nuts, fruits, and leafy vegetables are considered negative, dark, cold, and feminine. Yang foods represent opposing positive forces of light, warmth, and masculinity. Poultry, fish, eggs, and cereal grains such as buckwheat are yang foods.[7]

Photo Disc

Table 16.2 **Vegetarian diets come in many types**

Type of diet	Beef lamb, pork ("red meat")	Poultry	Fish	Eggs	Milk and milk products	Plant foods
"Far" vegetarian		X	X	X	X	X
Quasi-vegetarian[a]		X	X	X		X
Lacto-ovo vegetarian[b]				X	X	X
Lacto-vegetarian					X	X
Macrobiotic						X
Vegan						X
Raw food						X

[a]Quasi-vegetarian diets (also called semivegetarian) may include poultry; they tend to vary in the type of animal products consumed.
[b]Macrobiotic diets may include fish and other animal products.

Larry has been a vegetarian for the last seven years and wants to eat a hamburger again to see if it tastes as good as he remembers. He's a bit nervous about doing it because he thinks meat might make him feel sick.

Who gets the thumbs-up?

Answers on page 16-5

Susan: Larry should eat the hamburger if he wants to. It's a food, not an indigestion time bomb.

Doug: Larry's stomach isn't used to the heaviness of meat. It will make him nauseated.

Raw Food Diets Diets consisting primarily of raw foods have come in and out of fashion for hundreds of years. Raw food diets currently in vogue center on the health benefits of plant-based diets and the belief that raw foods provide enzymes helpful for digestion. Enzymes in raw foods do not promote normal digestion, however. Enzymes in food are inactivated by stomach acid and by digestive system enzymes that break down proteins.[8] Although there is no formal definition, raw food diets are usually described as an uncooked vegan diet. The diet consists of vegetables, fruits, nuts, seeds, sprouted grains, and beans. Uncooked foods usually make up between 50 and 100% of the diet.[8]

Extremely restrictive raw food diets are associated with impaired growth in children. Raw food diets, in general, lower beneficial HDL cholesterol levels and may lead to vitamin B_{12} deficiency and loss of bone mineral density. On the positive side, they are related to low body weights for height and healthy blood levels of triglycerides and total cholesterol.[9,10]

Other Vegetarian Diets For a number of other vegetarian regimes, the spiritual or emotional importance assigned to certain foods supersedes consideration of their contribution to an adequate diet.[12] Vegetarians adhering to the "living foods diet" consume uncooked and fermented plant foods only. This diet is inadequate in a number of nutrients, including vitamin B_{12}.[11] **Fruitarians** consume only fruit and olive oil. People adopting this type of diet rarely stick with it for long—it does not sustain health.

fruitarian A form of vegetarian diet in which fruits are the major ingredient. Such diets provide inadequate amounts of a variety of nutrients.

Vegetarian Diets and Health

Long-standing vegetarian dietary regimes that promote health in India and China have been tested by time (if not by science) and found to support health and life.[13] In fact, vegetarian diets are generally healthier than those of nonvegetarians.[14] On the other hand, highly restrictive vegetarian diets—such as fruitarian regimes, the raw food diet, and other popular vegetarian diets that stray from established regimes—can be harmful to health.[13] Vegans, who do not consume any animal products, are at higher risk of inadequate intakes of the omega-3 fatty acids EPA and/or DHA, iron, zinc, iodine, calcium, vitamin D, and vitamin B_{12} than vegetarians who consume some types of animal products.[12] In general, the more restrictive the diet, the more likely it is to be inadequate and lead to health problems. This is especially true for pregnant women, children, and persons who are ill, all of whom have relatively high needs for nutrients. Reports of caloric and nutrient deficiencies in young children of vegan parents are more common than is the case with children of parents with less restrictive vegetarian diets.[13]

The availability of a large assortment of vegan foods in the United States and other developed countries is making it easier to find healthful vegetarian food options. Vegetarians in economically developed countries are not at greater risk for iron deficiency than nonvegetarians. They generally obtain sufficient iron from plants and have ample intake of vitamin C, which enhances absorption of iron from plants.[15] In contrast, the diets of vegetarians in developing countries often include too few iron-rich plants, making iron deficiency common. Both the quality and the quantity of protein in vegetarian diets can also be a source of concern. However, vegetarians in developed countries generally have adequate protein intakes.[3] Vitamin B_{12} has consistently been shown to be the most likely nutrient to be lacking in the diets of individuals who do not consume animal products.[3] Vegans appear to be at greater risk for inadequate calcium intake and bone fractures than vegetarians who consume these foods.[16] Vegan diets, like most vegetarian diets, tend to contain above-average amounts of beneficial phytochemicals from plant foods.[15] Increasingly, vegetarian diets are being recognized for their beneficial effects on health and disease prevention.

What Are the Health Benefits of Vegetarian Diets?

Compared to the usual American diet and lifestyle, vegetarianism in general is associated with lower rates of obesity, lower blood pressure, better blood lipid levels, lower blood pressure, and a decreased risk of heart disease, type 2 diabetes, and certain types of cancer.[1] For children and adults at risk of early heart disease, the low-fat and high-fiber vegetable and fruit content of vegetarian diets may help reduce blood cholesterol levels and the risk of heart disease.[18]

Dietary Recommendations for Vegetarians

Because vegetarian diets exclude one or more types of foods, it's important that the foods included provide sufficient calories and the assortment and quantity of nutrients needed for health. No matter what the motivation underlying the assortment of foods included, vegetarian diets that fail to provide all the nutrients humans need in required amounts will not sustain health.

Well-Planned Vegetarian Diets There are a number of types of vegetarian diets, so there is no one set of dietary recommendations that is appropriate for all of them. Dietary guidelines for vegan diets generally recommend a variety of foods that includes grains, legumes, vegetables, fruits, fats and oils, nuts and seeds, and sweets (Illustration 16.2). Additional guidance for vegetarian diets is available from ChooseMyPlate.gov in the form of 10 healthy eating tips (Table 16.3). USDA's current food guide highlights the role of plants foods in the diet and emphasizes that vegetables, fruit, and grains should make up about three-fourths of a plate of food. It is generally recommended that Americans consume more plants foods and fewer animal products.[24]

Vegetarians who consume fish, dairy products, or eggs can reduce their reliance on legumes, nuts, and grains as protein and nutrient sources. Animal products are a major source of **complete proteins**, vitamin B_{12}, vitamin D, calcium, EPA (eicosapentaenoic acid), and DHA (docosahexaenoic acid). EPA and DHA are two important fatty acids primarily found in fish and seafood. Vitamin B_{12}, vitamin D, calcium, EPA, and DHA are the nutrients most likely to be lacking in the diets of vegetarians overall.[3]

complete proteins Proteins that contain all of the nine essential amino acids in amounts sufficient to support protein tissue construction by the body.

ANSWERS TO REALITY CHECK
Reintroducing Meat

The thought that meat could make one sick may be enough to trigger indigestion. Many self-defined vegetarians consume meat on rare occasion without reported ill effects.[17]

Susan:

Doug:

Illustration 16.2 Foundation foods for well-planned vegetarian diets.[21,25]

VEGAN DIETS

GRAINS, 6 or more servings[a]
- includes whole grain, enriched, and fortified grain products

LEGUMES AND NUTS, 5 or more servings[a]
- dry beans and peas, soy-based meat and fortified soy-based beverage vegetarian foods; nuts, seeds

VEGETABLES, 4 or more servings
- all types

FRUITS, 2 or more servings
- all types

FATS AND OILS, 2 or more servings[a]
- vegetable oils

SWEETS, 1–2 servings

[a]Number of serving depends on calorie need.

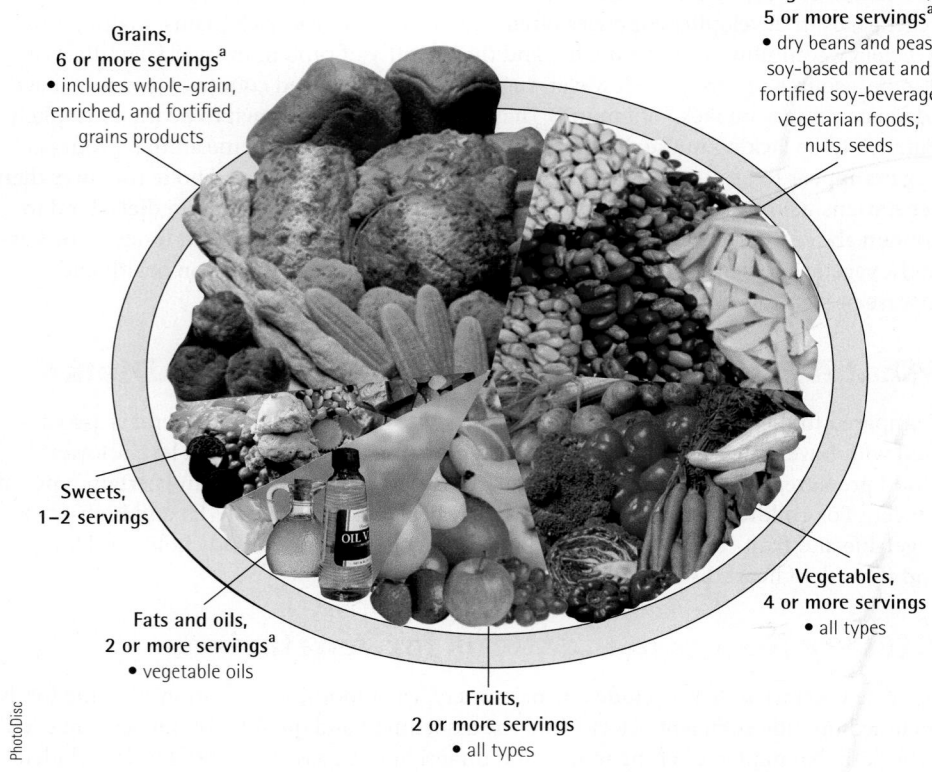

Grains, 6 or more servings[a]
- includes whole-grain, enriched, and fortified grains products

Legumes and nuts, 5 or more servings[a]
- dry beans and peas, soy-based meat and fortified soy-beverage vegetarian foods; nuts, seeds

Sweets, 1–2 servings

Fats and oils, 2 or more servings[a]
- vegetable oils

Fruits, 2 or more servings
- all types

Vegetables, 4 or more servings
- all types

PhotoDisc

A specific example of the food and nutrient composition of one day of a vegetarian's diet are shown in Table 16.4. The diet provides low amounts of cholesterol and saturated fat, is high in fiber, and provides adequate amounts of protein and most vitamins and minerals. These are fairly typical results for a vegetarian diet.[20] Of the four key nutrients most likely to be missing in the diet of vegetarians, this day's diet provides the recommended amount of vitamin D, but not of vitamin B_{12}, calcium, or EPA/DHA. Vitamin D intake reaches 6 mcg (240 IU) and B_{12} levels 1.9 mcg in this diet due to the inclusion of fortified soy milk.

Vegetarians can meet their needs for vitamin B_{12}, vitamin D, calcium, and EPA/DHA by consuming fortified foods (such as those shown in Table 16.5) or by use of B_{12}-fortified yeast, DHA algae supplements, or fortified plant foods (such as DHA-fortified soy or rice milk). EPA can be obtained, to some extent, from sources of DHA. EPA and DHA are closely related chemicals, and the body can convert one to the other to some degree.[21] Vitamin D is becoming easier to obtain by vegetarians. The FDA has approved the addition of vitamin D to soy-based foods such as soy beverages, tofu, burgers, and desserts. The change is reflected on food labels of soy vegetarian products.[24] Direct exposure of the face, arms, and hands to direct sunlight for about 5 to 15 minutes a day is one of the best ways to obtain vitamin D.

Vitamin and mineral supplements can be, and often are, used by vegetarians to meet specific nutrient needs. If needed, supplementary intake levels of vitamins B_{12} and D, calcium, and EPA/DHA should approximate those shown in Table 16.6. The best way to know if supplemental nutrients are needed is to carefully perform a dietary assessment that covers several typical days of food intake. (The Cengage/Wadsworth Diet Analysis Plus Program or the USDA's dietary analysis program available at ChooseMyPlate.gov can be used for the dietary assessment.)

Complementary Protein Foods Most well-planned vegetarian diets provide adequate amounts of protein, but the quality of the protein varies depending on which plant foods are consumed.

Table 16.3 ChooseMyPlate.gov's recommendations for healthy vegetarian diets

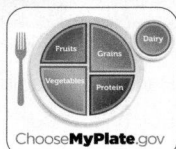

10 tips
Nutrition Education Series

healthy eating for vegetarians

10 tips for vegetarians

ChooseMyPlate.gov

A vegetarian eating pattern can be a healthy option. The key is to consume a variety of foods and the right amount of foods to meet your calorie and nutrient needs.

1 think about protein
Your protein needs can easily be met by eating a variety of plant foods. Sources of protein for vegetarians include beans and peas, nuts, and soy products (such as tofu, tempeh). Lacto-ovo vegetarians also get protein from eggs and dairy foods.

2 bone up on sources of calcium
Calcium is used for building bones and teeth. Some vegetarians consume dairy products, which are excellent sources of calcium. Other sources of calcium for vegetarians include calcium-fortified soymilk (soy beverage), tofu made with calcium sulfate, calcium-fortified breakfast cereals and orange juice, and some dark-green leafy vegetables (collard, turnip, and mustard greens; and bok choy).

3 make simple changes
Many popular main dishes are or can be vegetarian—such as pasta primavera, pasta with marinara or pesto sauce, veggie pizza, vegetable lasagna, tofu-vegetable stir-fry, and bean burritos.

4 enjoy a cookout
For barbecues, try veggie or soy burgers, soy hot dogs, marinated tofu or tempeh, and fruit kabobs. Grilled veggies are great, too!

5 include beans and peas
Because of their high nutrient content, consuming beans and peas is recommended for everyone, vegetarians and non-vegetarians alike. Enjoy some vegetarian chili, three bean salad, or split pea soup. Make a hummus-filled pita sandwich.

6 try different veggie versions
A variety of vegetarian products look—and may taste—like their non-vegetarian counterparts but are usually lower in saturated fat and contain no cholesterol. For breakfast, try soy-based sausage patties or links. For dinner, rather than hamburgers, try bean burgers or falafel (chickpea patties).

7 make some small changes at restaurants
Most restaurants can make vegetarian modifications to menu items by substituting meatless sauces or non-meat items, such as tofu and beans for meat, and adding vegetables or pasta in place of meat. Ask about available vegetarian options.

8 nuts make great snacks
Choose unsalted nuts as a snack and use them in salads or main dishes. Add almonds, walnuts, or pecans instead of cheese or meat to a green salad.

9 get your vitamin B$_{12}$
Vitamin B$_{12}$ is naturally found only in animal products. Vegetarians should choose fortified foods such as cereals or soy products, or take a vitamin B$_{12}$ supplement if they do not consume any animal products. Check the Nutrition Facts label for vitamin B$_{12}$ in fortified products.

10 find a vegetarian pattern for you
Go to www.dietaryguidelines.gov and check appendices 8 and 9 of the *Dietary Guidelines for Americans, 2010* for vegetarian adaptations of the USDA food patterns at 12 calorie levels.

USDA United States Department of Agriculture Center for Nutrition Policy and Promotion

DG TipSheet No. 8
June 2011
USDA is an equal opportunity provider and employer.

Go to www.ChooseMyPlate.gov for more information.

Animal products such as meat, eggs, and milk provide all of the nine essential amino acids in sufficient quantity to qualify as complete sources of protein. In addition, tests have shown soy proteins to be complete protein sources for children and adults.[23] The diet must include complete proteins because the body needs a sufficient supply of each essential amino acid if it is to build and replace protein substances such as red blood cells and enzymes. If any of the essential amino acids are missing in the diet, protein tissue construction stops, and the available amino acids will be used for energy instead. Essential amino acids consumed in foods are not stored for long, so the body needs a fresh supply each day or so.

Vegetarians who don't consume animal products can meet their need for essential amino acids by combining plant foods to yield complete proteins. This is done by consuming plant foods that *together* provide all the essential amino acids, although each individual food is missing some of these essential nutrients. The goal is to "complement" plant sources of essential amino acids, or to consume **complementary protein sources** from plant foods regularly.

complementary protein source Plant sources of protein that together provide sufficient quantities of the nine essential amino acids.

Table 16.4 Example of food intake and nutrient content in one day for the vegetarian diet of a 31-year-old, 130-pound female

One day's diet:	
Oatmeal, 1 cup	Salad dressing, 3 Tablespoons
Banana, 1 medium	Soy milk, 1 cup
Soy milk, 1 cup	Veggie burger, 1 patty
Brown sugar, 1 Tablespoon	Bun, 1
Almonds, 1 ounce (22 almonds)	French fries, 1 cup
Black beans, 2 cups	Herbal tea, 1 cup
Brown rice, 1 cup	Hummus, 1/2 cup
Lettuce salad with mixed vegetables, 1 cup	Baby carrots, 10

Selected nutrient analysis results: (calories: 2,170)

Nutrients	Amount consumed	Recommended intake
Protein, g	87	46
Total fiber, g	53	25
Total fat, g	100	53–92
Saturated fat, g	20	<26
Cholesterol, mg	27	<300
EPA + DHA, mg[a]	50	300–600
Vitamin A, mcg	723	700
Vitamin C, mg	54	75
Vitamin E, mg	14	15
Vitamin B_6, mg	3.2	1.3
Vitamin B_{12}, mcg	1.9	2.4
Thiamin, mg	3.1	1.1
Riboflavin, mg	1.6	1.1
Niacin, mg	19	14
Folate, mcg	561	400
Vitamin D, mcg[a]	6	5
Calcium, mg	491	1000
Magnesium, mg	614	320
Iron, mg	18.5	18
Zinc, mg	12.2	8
Selenium, mg	37	55

[a]EPA and DHA availability are estimated based on alpha-linolenic acid intake of 2.8 g; vitamin D intake is estimated based on fortified soymilk intake. Analysis is from mypyramidtracker.gov, September 2006.

Many different combinations of plant foods yield complete proteins. Basically, complete sources of protein can be obtained by combining grains such as rice, bulgur (whole wheat), millet, or barley with dried beans, tofu, or green peas; or corn with lima beans or dried beans; or seeds with dried beans. Some examples of complementary sources of plant proteins are shown in Illustration 16.3. Milk, meat, and eggs contain complete proteins and will complement the essential amino acids profile of any plant source of protein.

Where to Go for More Information on Vegetarian Diets

You can find information about vegetarian diets in a variety of sources. Web resources from credible organizations, as well as cookbooks that offer information about vegetar-

Table 16.5 Key nutrient levels in fortified and other vegetarian foods[a]

	Key Nutrients			
	Vitamin B$_{12}$	Vitamin D	Calcium	DHA
RDA for adults:	2.4 mcg	5 mcg	1,000 mg	250–500 mg[b]
Rice Dream, 1 cup	1.5 mcg	5 mcg	1,000 mg	—
Soy milk, 1 cup	2.6 mcg	3 mcg	300 mg	—
Meat analogs, 1 serv.	0–1.4 mcg	—	—	—
Tofu w/ calcium sulfate, 1/2 cup	—	—	130 mg	—
Fully fortified breakfast cereal, 1 cup	6 mcg	2.5 mcg	1,000 mg	—
Other breakfast cereals, 3/4 cup	1.5 mcg	1 mcg	0–100 mg	—
Nutritional yeast, 2 Tabl.	7.8 mcg	—	—	—
Algae, 1 capsule	—	—	—	180 mg
DHA-fortified fruit juice, 1 cup	—	—	—	32 mg
Collard greens, 1 cup	—	—	357	—
Kale, 1 cup	—	—	180	—
Turnip greens, 1 cup	—	—	107	—

[a]Check out product labels for product-specific information on nutrient content.
[b]DRIs for DHA are yet to be established. The figure is from Harris et al., 2009.[21]

Table 16.6 Amounts of key nutrients that may be needed by vegetarians daily[a]

Vitamin B$_{12}$	2–6 mcg
Vitamin D	200–400 IU (5-10 mcg)
Calcium	500 mg
EPA and DHA	300 mg

[a]Fortified foods may also supply these nutrients.

ian nutrition and delicious recipes are pictured in Illustration 16.4. Unfortunately, some of the information available on vegetarianism is wrong or misconstrued. Some vegetarian organizations are more committed to selling particular beliefs, memberships, or books and magazines than to promoting healthy vegetarian diets. Beware of vegetarian diets that claim to cure cancer, AIDS, or other serious illnesses or promise you'll experience inner peace or spiritual renewal. Appropriately planned vegetarian diets are health promoting, but they are not magic bullets that will cure all the ills of body and soul.

Illustration 16.3   Some combinations of plant foods that provide complete protein:

- *Rice and dried beans*
- *Rice and green peas*
- *Bulgur (wheat) and dried beans*
- *Barley and dried beans*
- *Corn and dried beans*
- *Corn and green peas*
- *Corn and lima beans (succotash)*
- *Soybeans and seeds*
- *Peanuts, rice, and dried beans*
- *Seeds and green peas*

Rice and black beans

Hummus and bread

Corn and black-eyed peas

Bulgur (whole wheat) and lentils

Tofu and rice

Corn and lima beans (succotash)

Tortilla with refried beans (e.g., a bean burrito)

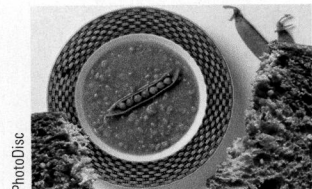
Pea soup and bread

Illustration 16.4 Many excellent vegetarian cookbooks are available; most provide information on diet as well as recipes. The books pictured here are *Moosewood Restaurant Low-Fat Favorites: Flavorful Recipes for Healthful Meals*, edited by Pam Krauss; *Vegetarian Times Complete Cookbook*, by the editors of *Vegetarian Times; Vegetarian Cooking for Everyone*, by Deborah Madison; *The Whole Soy Cookbook*, by Patricia Greenberg; and *Vegetable Heaven*, by Mollie Katzen.

© MIB Pictures/Getty Images/
UpperCut Images

NUTRITION
up close

Vegetarian Main Dish Options

Focal Point: Serving up vegetarian alternatives to meat dishes.

Assume you are a member of the "vegetarian option" planning committee for your college. You are asked to identify three meatless main dishes that could be served in the dining halls for lunch and another three that could be served for dinner. The one stipulation is that they should be main dishes you would enjoy eating.

What meatless dishes would you identify as options for lunch and for dinner?

Lunch Dishes	Dinner Dishes
1.	1.
2.	2.
3.	3.

Feedback to the Nutrition Up Close is located in Appendix G.

REVIEW QUESTIONS

- **Understand the health benefits and limitations of various types of vegetarian diets.**

1. The emotional and spiritual importance of certain foods in some vegetarian regimes supersedes any contribution the foods may make to an adequate diet.   **True/False**

2. Raw food diets supply ample amounts of all vitamins and required minerals and improve indigestion in people with digestive problems.   **True/False**

3. Compared to omnivore diets, vegetarian diets are associated with lower rates of obesity, heart disease, hypertension, and type 2 diabetes.   **True/False**

4. Vitamin B_{12} and EPA/DHA are largely found in animal foods.   **True/False**

5. The four nutrients most likely to be lacking in the diets of vegetarians are vitamin B_{12}, EPA/DHA, magnesium, and vitamin E.   **True/False**

6. Well-planned vegetarian diets provide adequate amounts of iron and protein.   **True/False**

7. When forming a new protein tissue, the body will respond to a deficiency of an essential amino acid by replacing it with another essential amino acid that is available.   **True/False**

8. Corn combined with cooked, dried beans is an example of a complementary protein source.   **True/False**

The next five questions refer to this situation:

James and Manda have both gained weight since they met and decide it's time to lose the weight. Manda was a vegetarian for 14 years, liked that type of diet, and had switched to eating meat only a year ago because she missed the taste of steak. James has always been a meat eater, but he needs to lose weight, too, and wants to lose it. They both decide to go on a vegan diet and eat only plant foods.

9. ____ Which food could they include in their vegan diet that would provide vitamin D?
   a. dried beans
   b. rice
   c. spinach
   d. fully fortified breakfast cereal

10 ____ Which food could they consume to help meet their need for vitamin B_{12}?
   a. nutritional yeast
   b. sweet potatoes
   c. cabbage
   d. hummus

11. ____ While on the vegan diet, intake of which of the following nutrients will likely increase?
   a. fiber
   b. calcium
   c. protein
   d. iron

12. ____ Which of the following represents a complementary protein source Manda and James could consume to get a complete source of protein?

    a. navy and pinto beans
    b. carrots and corn
    c. hummus and black beans
    d. tofu and rice

13. ____ Which of the following sets of foods could be used to provide good sources of calcium in the vegan diet?

    a. soy milk and bread
    b. veggie burgers and squash
    c. collard greens and soy milk
    d. tomato juice and rice

Answers to these questions can be found in Appendix G.

NUTRITION SCOREBOARD ANSWERS

1. Early humans developed on an omnivorous diet (meat and plant diet). However, health can be fostered on an omnivorous or a vegetarian diet. **False**

2. Persons who consume appropriately planned vegetarian diets do not have more health problems than other people. Furthermore, such diets may reduce the risk of developing several chronic diseases.[1] **False**

3. Macrobiotic diets have not been found to cure cancer. **False**

4. Vegetarians do need to plan their diets to ensure they obtain enough high-quality protein, but they need to consume complete sources of protein daily, not at every meal. Appropriately planned vegetarian diets provide sufficient amounts of high-quality protein. **False**

UNIT

17

Food Allergies and Intolerances

NUTRITION SCOREBOARD

1 About one in every three Americans is allergic to at least one food.   True/False

2 Food ingredient labels must indicate the presence of major food allergens in food products.   True/False

3 Skin prick tests are an accurate way to diagnose specific food allergies.   True/False

4 Food intolerances cause less severe reactions than food allergies do.   True/False

Answers can be found at the end of the unit.

After completing Unit 17
and its interactive learning
features, you will be able to:

- Describe the cause, effects,
diagnosis, treatment, and
prevention of true food
allergy.

- Separate true food allergy
from food intolerance, and
identify specific approaches
to the management
and prevention of food
intolerance.

Food Allergy

- **Describe the cause, effects, diagnosis, treatment, and prevention of true food allergy.**

In many circles, food allergy is a topic of heated debate and misconceptions. People use the term *food allergy* to refer to virtually any type of problem they have with food. At one extreme are people who believe allergies to milk, wheat, and sugar are to blame for hyperactivity and a host of other behavioral problems in children. At the other extreme are some health professionals who think people who complain of food allergies need to have their heads examined. In the middle is the fact that food allergies are increasing and are an important health concern in the United States and other Western countries.[4]

Food allergies are real and can be very serious. At the minimum, true food allergies can cause a rash or an upset stomach. At the maximum, they can result in death. Unreal food allergies can cause problems, too. They can lead people to eliminate nutritious foods from their diet unnecessarily, resulting in inadequate diets and eventually in health problems. One of the most intriguing aspects of food allergies is the frequency and ease with which foods are falsely blamed for a variety of mental and physical health problems.

Key Nutrition Concepts

This unit's content on food allergies and intolerance is directly relevant to two key nutrition concepts:

1. Health problems related to nutrition originate within cells.

2. Malnutrition can result from poor diets and from disease states, genetic factors, or combinations of these factors.

Prevalence of Food Allergy

The incidence of food allergies in adults is estimated to be to 4% of adults and 6% in children.[2] However, around 20–30% of the general public believe they are allergic to one or more foods.[1] The vast majority of complaints of food allergy fail to be confirmed by testing.[2] Here are two, real-life examples of self-diagnosed food allergies that went awry:

1. Isaiah, a lover of blueberries and blueberry pie, hasn't touched a blueberry since 2005. That year, Isaiah had a piece of blueberry pie in a restaurant and later became violently sick to his stomach. Bingo! Isaiah decided he must be allergic to blueberries.

   Actually, Isaiah gave up blueberries for no good reason. He was coming down with the flu when he ate the pie. Nevertheless, to this day when he thinks of blueberries, he gets a bit queasy.

2. For 11 years, Emilia rigidly avoided even small amounts of cow's milk. She was convinced that just a few drops of cow's milk would cause pressure in her head, blurred vision, dizziness, cramps, and nausea.

   Finally, Emilia's presumed allergy to cow's milk was put to the test. She was given liquid through a dark tube inserted into her stomach from her mouth. When she was told the fluid was cow's milk—even though it was actually water—the familiar symptoms appeared within 10 minutes. When she was given milk but was told it was water, no symptoms appeared.[5]

   Emilia was shocked. For 11 years she had scrutinized nearly everything she ate and had wasted hundreds of dollars on special food products and supplements. Finishing a frosted brownie with a glass of milk, Emilia contemplated the practical realities of the power of suggestion.

Adverse Reactions to Foods

People experience adverse reactions to food for three primary reasons. One is food poisoning. The other two involve the body's reaction to substances in food that are normally harmless. With **food allergies**, the body's **immune system** reacts to a substance in food

food allergy Adverse reaction to a normally harmless substance in food that involves the body's immune system. (Also called *food hypersensitivity*.)

immune system Body tissues that provide protection against bacteria, viruses, and other substances identified by cells as harmful.

food intolerance Adverse reaction to a normally harmless substance in food that does not involve the body's immune system.

food allergen A substance in food (almost always a protein) that is identified as harmful by the body and elicits an allergic reaction from the immune system.

antibodies In the case of allergies, proteins the body makes to combat allergens.

histamine (hiss-tah-mean) A substance released in allergic reactions. It causes dilation of blood vessels, itching, hives, and a drop in blood pressure and stimulates the release of stomach acids and other fluids. Antihistamines neutralize the effects of histamine and are used in the treatment of some cases of allergies.

Illustration 17.1 What do these foods have in common? Each food shown, plus many more not shown, can cause adverse reactions in some individuals.

(almost always a protein) that it identifies as harmful (Illustration 17.1).[6] The immune system is not involved in **food intolerances**, which encompass other adverse reactions to normally harmless substances in food.

Food Allergies and the Immune System It may seem odd that a system that helps the body conquer bacteria and viruses is involved in protecting us from normal constituents of food. In people with allergies, however, the cells recognize some food components as harmful—just as they recognize bacteria and viruses. Components of food that trigger the immune system are called **food allergens**.

For genetically susceptible people, exposure to trace amounts of an allergen triggers an allergic reaction. Highly sensitive people can develop an allergic reaction simply by inhaling vapors from allergenic food that is being prepared or cooked, or by touching the food.[6] Allergic reactions to peanuts have been triggered by kissing someone who has recently consumed peanuts or even by inhaling fumes in a confined space with people who are eating them.[7]

In response to the allergen, the immune system forms **antibodies** (Illustration 17.2). The antibodies attach to cells located in the nose, throat, lungs, skin, eyes, and other areas of the body. When the allergen appears again, the body is ready. The previously formed antibodies recognize the allergen, attach to it, and signal the body to secrete **histamine** and other substances that cause the physical signs of allergic reactions. Most commonly, allergic reactions cause a rash (Illustration 17.3), diarrhea, congestion, or wheezing (Table 17.1), but the symptoms may be much more serious.

Table 17.1 The three most common types of symptoms caused by food allergies

Some people experience more than one type of reaction.[7]

Symptom	Percentage of people with food allergies who develop the symptom
• Skin eruptions: rash, hives	84%
• Gastrointestinal upsets: diarrhea, vomiting, cramps, nausea	52
• Respiratory and other problems: congestion, swelling of the tongue and throat, runny nose, cough, wheezing, asthma	32

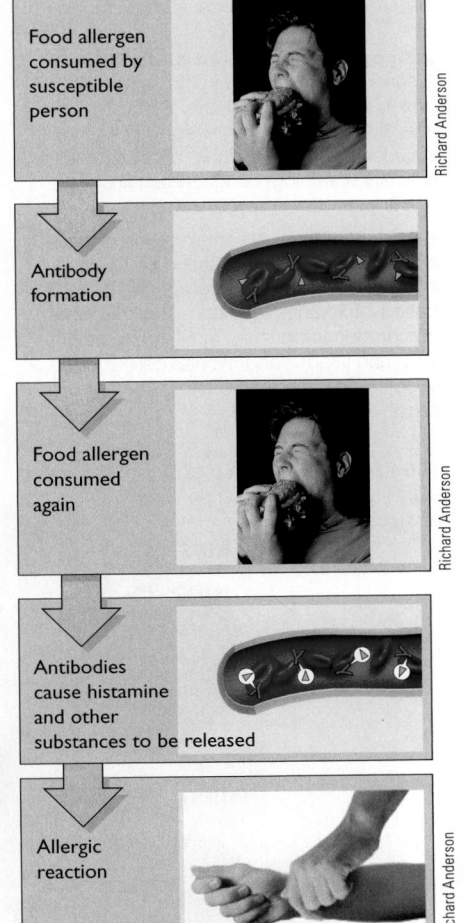

Food allergen consumed by susceptible person

Antibody formation

Food allergen consumed again

Antibodies cause histamine and other substances to be released

Allergic reaction

Illustration 17.2 The development of allergic reactions to foods.

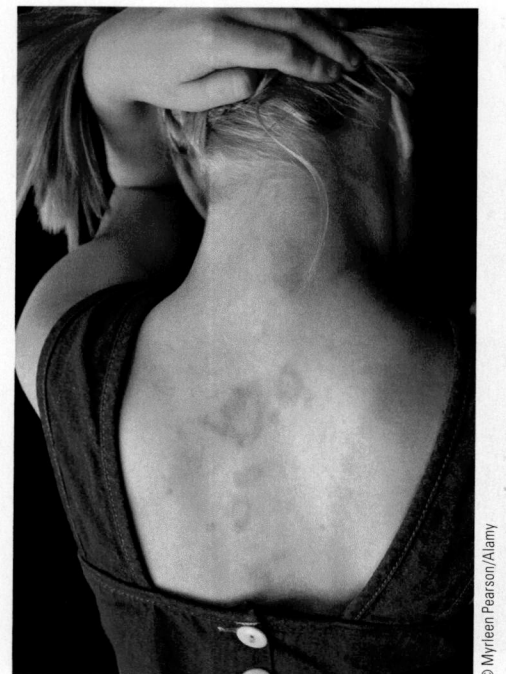

Illustration 17.3 What a common reaction to a food allergen can look like. (Most food allergies cause milder symptoms.)

anaphylactic shock (an-ah-fa-lac-tic) A serious allergic reaction that is rapid in onset and can cause death. Symptoms of anaphylactic shock (or "anaphylaxis") may include abdominal cramps, chest tightness, difficulty breathing, cough, hives, flushing, swelling, and/or itching.

celiac disease An autoimmune disease characterized by inflammation of the small intestine lining resulting from a genetically based intolerance to gluten. The inflammation produces diarrhea, fatty stools, weight loss, and vitamin and mineral deficiencies. (Also called *celiac sprue* and *gluten-sensitive enteropathy*.)

Exposure to trace amounts of an allergen in nuts, peanuts (peanuts are actually a legume, not a nut), fish, and shellfish can cause **anaphylactic shock**.

This massive reaction of the immune system can result in death, caused by the cutoff of the blood supply to tissues throughout the body. People experiencing anaphylactic shock can be revived by an injection of epinephrine.[6] Follow-up care should be provided because symptoms of anaphylactic shock may reoccur a number of hours later.[8]

Foods That Are Most Likely to Cause Allergic Reactions Many foods may cause allergic reactions in genetically susceptible individuals (Illustration 17.4). Nevertheless, approximately 90% of all food allergies are caused by eight foods: nuts, eggs, wheat, milk, peanuts, soy, shellfish, and fish.[9] Passage of the Food Allergen Labeling and Consumer Protection Act of 2004 is helping consumers with food allergies find offending foods and ingredients. Food ingredient labels are now required to state clearly whether the food contains one or more of the big eight allergenic foods.[9]

Wheat Allergy (Celiac Disease) Minnie is 46 years old and has three sisters and two brothers, all of whom are taller than she is. (She considers herself the "runt of the litter.") Since childhood, Minnie has had problems with diarrhea and cramps. Over the past two years, these problems have become worse, and she has been chronically tired. Assuming that she was lactose intolerant, Minnie stopped eating all dairy products. Her problems persisted, however. When she almost lost her job as a receptionist due to her frequent bathroom breaks and her obvious fatigue, she decided to see a doctor again.

Diagnosing anemia, the doctor sent Minnie to a registered dietitian. Hearing Minnie's story, the dietitian suspected celiac disease caused by an allergy to a component of wheat and other grains. After two weeks on an allergen-free diet, Minnie started to regain her strength and spent much less time in the bathroom. Diagnostic tests subsequently confirmed that Minnie had celiac disease, a condition that likely had existed since childhood.

An allergy to wheat, or more specifically to a component of gluten found in wheat and other grains such as barley, rye, and triticale has been underdiagnosed in the past. Called **celiac disease**, it affects approximately 1% of people worldwide.[10] It differs somewhat from other allergies in that an immune system reaction to gluten occurs that is localized in the lining of the small intestine.[10] Celiac disease can be difficult to diagnose because the symptoms (diarrhea, weight loss, cramps, anemia) are similar to those of other diseases and because symptoms are silent in some individuals. With new awareness about the disorder, doctors are ordering diagnostic tests more frequently, and the rate of confirmed cases of celiac disease is increasing.[11] The popularity of self-diagnosed celiac

Illustration 17.4 The "big eight" foods that can cause allergies.

NUTS EGGS PEANUTS SOY WHEAT MILK SEAFOOD FISH

disease appears to be increasing, too.[12] The gold standard, diagnostic test for celiac disease is small bowel biopsy and examination of cells for signs of damage due to the disease. The test has to be undertaken while individuals are consuming their normal diet, and not a gluten-free one. Removal of gluten from the diet corrects intestinal cell damage and may lead to an incorrect diagnosis.[10]

Celiac disease is treated with a lifelong, but rewarding, gluten-free diet. Because gluten may be included in food additives, processed and prepared foods, as well as grain products, it generally takes the help of an experienced, registered dietitian to design and monitor gluten-free diets.[13] Table 17.2 provides examples of some foods that contain gluten and some that don't.

The term *gluten-free* is standardized and allowed on food labels for qualifying foods. This labeling regulation is voluntary but is expected to be widely employed because the demand for gluten-free foods is increasing. Since the regulation is voluntary, checking for a "gluten-free" label to identify foods that can be safely consumed may not always work.[14] Ingredient labels listing wheat or products made from wheat, rye, barley, or triticale should not be consumed, nor should foods labeled as containing, or potentially containing, these ingredients.

Diagnosis: Is It a Food Allergy?

A variety of tests are used to diagnose food allergies, but there is only one "gold standard." That test is the **double-blind**, **placebo-controlled food challenge**.[16]

Suppose you suspect you are allergic to grapes because you got an annoying rash twice after eating them. Your doctor arranges an appointment for you at an allergy clinic. At the clinic, you will be given grapes, either in a concentrated pill form or blended into a liquid, and placebo pills (or a liquid) without being told which are the grapes and which is the placebo. After an appropriate amount of time, your reaction to the grapes and to the placebo will be noted. If the rash appears after the grapes but not after the placebo, you have an adverse reaction to grapes that may be an allergy.

double-blind, placebo-controlled food challenge   A test used to determine the presence of a food allergy or other adverse reaction to a food. In this test, neither the patient nor the care provider knows whether a suspected offending food or a placebo is being tested.

Table 17.2 Which foods contain gluten? Listed here are examples of foods that do or do not contain gluten[10,15]

Gluten-containing foods	Gluten-free foods[a]
Beer, ale	Fruit
Barley	Vegetables
Broth, bouillon powder/cubes	Dried beans
Brown rice syrup	Buckwheat
Bulgur	Cassava
Commercial soups	Grits
Bread, other wheat/rye flour products	Cornmeal
Imitation seafood	Nuts
Cakes, pies, cookies	Quinoa
Processed meats	Fresh meats, fish
Soy sauce	Soy flour, cereals
Wheat starch	Wild rice
Pizza	Eggs
Macaroni and cheese	Nuts, seeds
Seasonings	Cheese (not processed)
Marinades	Popcorn
Rye-containing products	Milk
Vegetarian meat substitutes	Chips (100% corn, potato)

[a]Assumes foods have not been contaminated with gluten during processing and are free of gluten-containing ingredients.

You and your coworkers decide to go to a Chinese restaurant for lunch. Trevor wants to go but hesitates because, as he said, "I must be allergic to Chinese food. Every time I eat it I feel dizzy, sweat, and get this ringing in my ears."

Who's got the better idea?

Answers on page 17-7

Haley: Trevor, that's not an allergy. It's all in your head.

Darrel: Have you ever asked them to leave out the MSG?

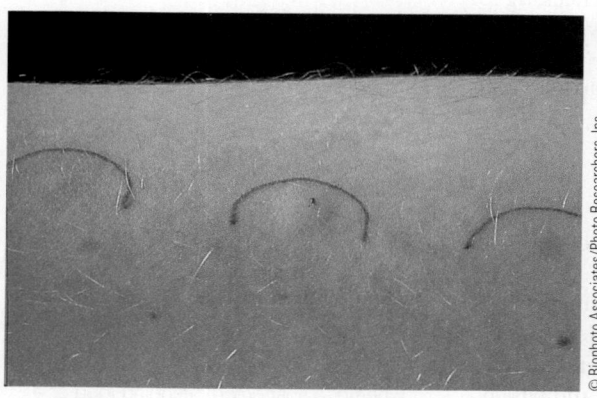

Illustration 17.5 A positive reaction to a skin prick test for a food allergy usually looks like this. The result indicates that an allergy to the food tested may be present but doesn't guarantee it.

Food challenges are best undertaken under medical supervision. True food allergies can cause serious reactions, and immediate help may be needed. Foods suspected of causing anaphylactic shock in the past should not be given in food challenge tests. In fact, they should be totally omitted from the diet.[6]

Other tests for food allergy are reliable to varying degrees. None of them is always accurate, and the results of bogus tests are not to be trusted at all.

Immunoglobulin E Tests Immunoglobulin E (IgE) is a protein produced by the immune system in response to an allergen. Identification of specific types of IgE in a person's blood may identify the presence of a food allergy, but often the results do not correctly identify a specific food allergy.[3]

Skin Prick Tests Skin prick tests (also called prick skin tests) for food allergy are commonly employed and are useful for identifying *the absence* of a food allergy. A positive test result isn't proof positive that an allergy exists, however, because positive test results are inaccurate most of the time.[3]

For this test, a few drops of food extract are placed on the skin, and the skin is then pricked with a needle. At approximately the same time, another area of skin is pricked using only water. If the area around the food skin prick becomes redder and more swollen than the area pricked with water (Illustration 17.5), it is concluded that the person *might* be allergic to the food tested.[3]

Bogus Tests A number of companies offer food allergy tests through the mail. In one study, researchers sent five of these companies the required samples, nine from adults who were allergic to fish and nine from adults who had no allergies. All five companies missed the fish allergy. When duplicate samples from the same adults were sent separately to each company, reports came back with different results for each of the pairs of samples. In addition, the companies reported their laboratories had identified various food allergies that none of the adults actually had.[5]

Treatment of Food Allergies

After a food allergy is confirmed, the food is eliminated from the diet. This is the *only* treatment currently available for food allergies. Allergy shots and other treatments for food are not yet available but may be in the future.[16] If the eliminated food is an important source of nutrients or is found in many food products, consultation with a registered dietitian is recommended.[17] Individuals with a history of anaphylaxis should be well informed about its signs and symptoms, and of the correct use of epinephrine self-injectors. Medical ID bracelets or necklaces are recommended.[6]

Some food allergies don't last forever. Many infants and young children outgrow allergies to cow's milk (64% by age 12), wheat (66% by age 12), and eggs (37% by age 10).[6] Food allergies that develop in adults, however, tend to persist.[17] Children and adults with a severe allergy to peanuts, nuts, fish, or shellfish may have to eliminate the food from their diet for a lifetime.[6]

Prevention of Food Allergies

Rather than prompting the development of allergies, delayed introduction of solid foods known to be allergenic may actually increase the onset of certain allergies.[25] There appears to be no rationale for eliminating foods during pregnancy, breastfeeding, or childhood in an attempt to prevent allergies in children. It is recommended that exclusive breast or formula feeding be undertaken in the first four to six months of life for health benefits other than allergy prevention.[16,17]

Food Intolerances

- **Separate true food allergy from food intolerance, and identify specific approaches to the management and prevention of food intolerance.**

Food intolerances produce some of the same reactions as food allergies, but the reactions develop by different mechanisms. Food intolerance reactions do not involve the immune system; they are due to a missing or abnormal enzyme, or other cause.[23]

It is clear that some people are intolerant of lactose, sulfite, histamine (a component of red wine and aged cheese), and other foods (Table 17.3). It is also clear that not all food intolerances are real. Just as with food allergies, the best way to separate the real from the unreal is the double-blind, placebo-controlled food challenge. True food intolerances produce predictable reactions. Problems such as headache, diarrhea, swelling, or stomach pain will occur every time a person consumes a sufficient amount of the suspected food.

Lactose Maldigestion and Intolerance

Lactose maldigestion, or the inability to break down lactose in dairy products due to the lack of the enzyme lactase, results in the condition known as **lactose intolerance**. Symptoms of lactose intolerance, such as flatulence, bloating, abdominal pain, diarrhea, and "rumbling in the bowel," occur in lactose maldigesters within several hours of consuming more lactose than can be broken down by the available lactase. These symptoms are due to the breakdown of undigested lactose by bacteria in the lower intestines (which produce gas as a by-product of lactose ingestion) and by fluid accumulation.[19] Lactose maldigestion is a common disorder. It occurs in about 25% of the human population.[20]

Reduced intake of lactose-containing dairy products (for example, milk, ice cream, and cottage cheese) may prevent lactose intolerance symptoms. Foods such as hard cheese, low- or no-lactose milk, buttermilk, and yogurt without added milk solids contain low amounts of lactose and can generally be consumed. Small amounts of milk or other lactose-containing dairy products are generally well tolerated by people with lactose maldigestion.[19] Care should be taken to ensure adequate intake of calcium and vitamin D if milk and dairy product intake is restricted.

Sulfite Sensitivity

Sulfite is a food additive used by food manufacturers in the past to keep vegetables and fruits looking fresh and to prevent mold growth. It is also added to some beers, wines,

Table 17.3 Foods and substances in them linked to food intolerance reactions

- Aged cheese
- Anchovies
- Beer
- Catsup
- Chocolate
- Dried beans
- Food coloring
- Lactose
- Mushrooms
- Pineapple
- Red wine
- Sausage (hard, cured)
- Soy sauce
- Spinach
- Tomatoes
- Yeast

lactose maldigestion   A disorder characterized by reduced digestion of lactose due to the low availability of the enzyme lactase.

lactose intolerance   The term for gastrointestinal symptoms (flatulence, bloating, abdominal pain, diarrhea, and "rumbling in the bowel") resulting from the consumption of more lactose than can be digested with available lactase.

ANSWERS TO **REALITY** CHECK

I think I'm allergic to . . .

Some people, like Trevor may be sensitive to MSG. It doesn't elicit an immune system reaction, so it's not a food allergy.

Haley:

Darrel:

Table 17.4 Some foods that may contain sulfite

- Wine
- Beer
- Hard cider
- Tea
- Fruit juices
- Vegetable juices
- Guacamole
- Dried fruit
- Potato products
- Canned vegetables
- Baked goods
- Spices
- Gravy
- Soup mixes
- Jam
- Trail mix
- Fish and seafood

processed foods, and medications as a preservative (see Table 17.4). Very small amounts of sulfite can cause anaphylactic shock and bring on an asthma attack in sensitive people.[21] Because many foods have added sulfites, people sensitive to sulfite should read food ingredient labels carefully. The FDA requires that processed foods containing sulfites list them on the ingredients label. It also prohibits the use of sulfite on fresh vegetables and fruits.

Red Wine, Aged Cheese, and Migraines

Some people develop migraine headaches when they drink red wine. The presence of histamine in wine has been blamed for the headaches. People with this intolerance are unable to break down histamine during digestion, so it accumulates in the blood and causes headaches.[22] Histamine is also found in beer, sardines, anchovies, hard cured sausage, pickled cabbage, spinach, and catsup. Tyramine, a compound closely related to histamine and found in aged cheese, soy sauce, and other fermented products, also causes migraine headaches in sensitive people.

MSG and the "Chinese Restaurant Syndrome"

Some people appear to be sensitive to MSG (monosodium glutamate), a flavor enhancer used on meats and in soups, stews, and many Chinese food items.[24] Because symptoms are reported to occur shortly after sensitive individuals eat Chinese food, sensitivity to MSG has been dubbed the "Chinese restaurant syndrome." Symptoms associated with MSG sensitivity include dizziness, sweating, flushing, a rapid heartbeat, and a ringing sound in the ears.[23]

Food Allergy and Intolerance Precautions

People with food allergies or intolerances have to be very careful about what they eat and should have a plan of action ready in case they develop a serious reaction. Becoming highly knowledgeable about which foods and food products contain ingredients that cause an adverse reaction is essential. When eating out, people with allergies and intolerances should ask a lot of questions to make sure they know what is being served. They have to become students of food ingredient labels.

Planning ahead is crucial for individuals with a history of anaphylactic shock. Although preparedness is the key factor related to quick treatment of anaphylaxis, individuals at risk are often not adequately prepared. Physicians may not have prescribed EpiPens (injectable epinephrine) to their patients with a history of anaphylactic shock or have instructed patients on their use. It is highly recommended that individuals with a history of anaphylactic shock have a plan for handling an emergency situation.[6] This topic is the subject of the unit's Take Action feature.

One final thought about food allergies and intolerances—they are never caused by studying about them.

take action   To Prevent Anaphylaxis

Are you at risk for anaphylactic shock due to a food allergy and unsure how to manage it? If yes, medical experts have advice to share with you about how to be prepared.[6] The advice can be translated into action by completing the following activities.

1. Scrupulously learn about and avoid eating foods that may trigger anaphylaxis. Make an appointment with a dietitian if needed.
2. Talk with your doctor about the symptoms of anaphylaxis and what you should do if it happens.
3. Get a prescription for two self-injection epinephrine pens (EpiPens) and any other medications that may be needed from your doctor.

4. Ask your doctor to teach you how to use injectable epinephrine and other medications needed when anaphylaxis develops.
5. Obtain these medications and carry the EpiPen and other prescribed, emergency medications with you at all times.
6. Get and wear a medical ID bracelet or necklace that contains information on the allergy.

AP Photo/Stephen Morton

Creatas/Fotosearch

NUTRITION
up close

Focal Point: How to serve gluten-free meals and snacks.

You're having a ball game–viewing party and want to offer snacks and a meal that can also be enjoyed by your best friend, who was diagnosed recently with celiac disease. What should you serve?

Place a check mark in front of the foods that are naturally gluten free and could be served. Assume you prepare the foods without adding gluten-containing ingredients. (*Hint*: Refer to Table 17.2.)

____ fresh meats, fish
____ tomatoes
____ pizza
____ macaroni and cheese
____ eggs
____ corn grits
____ corn chips
____ sourdough rolls
____ baked apples with sugar
____ hamburgers
____ grapes

____ tossed salad with oil and vinegar dressing
____ apple pie
____ black beans and rice
____ canned soup
____ sausage
____ popcorn
____ cheddar cheese
____ veggie burgers
____ lemonade
____ broccoli

Feedback to the Nutrition Up Close is located in Appendix G.

REVIEW QUESTIONS

• **Describe the cause, effects, diagnosis, treatment, and prevention of true food allergies.**

1. Peanuts, shellfish, cow's milk, and eggs are among the "big eight" foods that can cause food allergies.   **True/False**

2. The prevalence of food allergies is increasing in the United States.   **True/False**

3. The Food Allergen Labeling and Consumer Protection Act requires that manufacturers remove all potentially allergic ingredients from their foods.   **True/False**

4. Skin prick tests definitively identify allergies to specific foods.   **True/False**

5. Symptoms of food allergy can be triggered by exposure to a very small amount of a food allergen.   **True/False**

6. Food allergies can be treated with allergy shots.   **True/False**

7. Allergies to peanuts, nuts, fish, and shellfish are likely to last a lifetime.   **True/False**

8. ____ Food allergy is initiated by the ingestion of _____.
   a. antibodies
   b. histamine
   c. an antigen
   d. sulfite

9. ____ Food allergies can be prevented by:
   a. consuming only small amounts of allergenic foods
   b. keeping potentially allergenic foods out of the diets of infants
   c. allergy shots
   d. avoiding allergen-containing foods

• **Separate true food allergy from food intolerance, and identify specific approaches to the management and prevention of food intolerance.**

10. Adverse reactions to sulfite in foods are prompted by immune system–mediated processes.   **True/False**

11. Use of the term *gluten-free* on labels of qualifying food is allowed.   **True/False**

12. Recent research proves that the "Chinese Restaurant Syndrome" does *not* actually exist.   **True/False**

13. _____ Food intolerance is defined as:
    a. an adverse reaction to a normally harmless substance in food that involves the body's immune system.
    b. a condition that causes the regurgitation of food.
    c. a limited ability to ingest large amounts of specific types of food.
    d. an adverse reaction to a normally harmless substance in food that does *not* involve the body's immune system.

14. _____ Which of the following phrases about food intolerance is true?
    a. Food intolerance is *not* initiated by the immune system.
    b. Symptoms of food intolerance are different from those of food allergy.
    c. Food intolerance reactions are "hit and miss"; they sometimes are experienced when a specific food is consumed and sometimes not.
    d. Food intolerance does not cause anaphylactic shock.

Answers to these questions can be found in Appendix G.

NUTRITION SCOREBOARD ANSWERS

1. About one in three Americans *believes* he or she is allergic to at least one food.[1] Probably 3 to 4 in 100 adults and 6 in 100 infants and young children actually are.[2]   **False**

2. Listing potential, relatively common food allergens on food ingredient labels is required for packaged foods.   **True**

3. Skin prick test results assist in the diagnosis of food allergies but generally do not identify the food allergen that causes the allergic reaction.[3]   **False**

4. Food allergies and food intolerances have different causes, but both can produce life-threatening reactions.   **False**

Fats and Cholesterol in Health

NUTRITION SCOREBOARD

1 It is currently recommended that adults consume diets providing less than 30% of calories from fat.
True/False

2 The types of fat consumed are more important to health than total fat intake. **True/False**

3 Most all of the cholesterol in our diets comes from animal products. **True/False**

4 Saturated fat intake has a stronger influence on blood cholesterol levels than cholesterol intake. **True/False**

Answers can be found at the end of the unit.

After completing Unit 18 and its interactive learning features, you will be able to:

- Summarize the major functions, types, food sources, recommended intake, and health effects of various fats.

- Identify foods low in saturated and *trans* fat, and high in unsaturated fats.

Changing Views about Fat Intake and Health

- **Summarize the major functions, types, food sources, recommended intake, and health effects of various fats.**

- **Identify foods low in saturated and *trans* fat, and high in unsaturated fats.**

Scientific evidence and opinions related to the effects of fat on health have changed substantially in recent years—and so have recommendations about fat intake. In the past, it was recommended that Americans aim for diets providing less than 30% of total calories from fat. However, recent evidence indicates that the type of fat consumed is more important to health than total fat intake.[1] The watchwords for thinking about fat have become "Not all fats are created equal: Some are better for you than others." American adults are being urged to select food sources of "healthy" fats while keeping fat intake within the range of 20–35% of total caloric intake. Concerns that high-fat diets encourage the development of obesity have been eased by studies demonstrating that excessive caloric intakes—and not diets high in fat—are related to weight gain.[1]

New recommendations regarding fat intake do not encourage increased fat consumption. Rather, they focus on the consumption of certain types of fat. Diets providing as low as 20% of calories from fat, and those providing 30–35%, can be healthy—depending on the types of fat consumed and the quality of the rest of the diet.[1] This unit provides facts about fats, explains the reasons behind recent changes in recommendations for fat intake, and addresses the practical meaning of it all.

Key Nutrition Concepts

Three key nutrition concepts are particularly relevant to the content covered in this unit on fats and cholesterol. They are the following:

1. Poor nutrition can result from both inadequate and excessive levels of nutrient intake.

2. Poor nutrition can influence the development of certain chronic diseases.

3. Adequacy, variety, and balance are key characteristics of healthful diets.

Facts about Fats

Fats are a group of substances found in food. They have one major property in common: They are not soluble (or, in other words, will not dissolve) in water. If you have ever tried to mix vinegar and oil when making salad dressing, you have observed the principle of water and fat insolubility firsthand.

Fats are actually a subcategory of the fat-soluble substances known as **lipids**. Lipids include fats, oils, and cholesterol. Dietary fats such as butter, margarine, and shortening are often distinguished from oils by their property of being solid at room temperature. This physical difference between fats and oils is due to their chemical structures.

Functions of Fats Just like carbohydrates, fat molecules consist of carbon, hydrogen, and oxygen. The arrangement and number of carbon, hydrogen, and oxygen atoms determine the fat's structure and how it functions in the body.[1] Table 18.1 summarizes the major functions of fats.

Fats in Foods Supply Energy and Fat-Soluble Nutrients Dietary fats are a concentrated source of energy. Each gram of fat consumed supplies the body with nine calories worth of energy. That's enough energy for a 160-pound person to walk casually for a little over two minutes or to jog at a slow pace for about a minute. Fats in food supply

lipids Compounds that are insoluble in water and soluble in fat. Triglicerides, saturated and unsaturated fats, and essential fatty acids are examples of lipids, or "fats."

the **essential fatty acids** (linoleic acid and alpha-linolenic acid) and provide the fat-soluble vitamins D, E, K, and A (the "deka" vitamins). So, part of the reason we need fats in our diets is to obtain a supply of fat-soluble essential nutrients they "carry" in foods (Table 18.1). Diets containing little fat (less than 20% of total calories) often fall short on delivering adequate amounts of essential fatty acids and fat-soluble vitamins.[1]

Fat Contributes to the Body's Energy Stores Fat consumed as part of a dietary intake that exceeds calorie need is converted to triglycerides and stored in fat cells (Illustration 18.1). A pound of body fat can provide approximately 3,500 calories of energy to the body when needed.

Body fat is not just skin deep. Fat is also located around organs such as the kidneys and heart. It's there to cushion and protect the organs and keep them insulated. Cold-water swimmers can attest to the effectiveness of fat as an insulation material. They purposefully build up body fat stores because they need the extra layer of insulation (Illustration 18.2).

Fats Increase the Flavor and Palatability of Foods Although "pure" fats by them-selves tend to be tasteless, they absorb and retain the flavor of substances that surround them. Fats in meats, for example, pick up flavors from the meat and extend the meaty flavor to the fat. This characteristic of fat is why butter, if placed close enough to garlic in the refrigerator, tastes like garlic.

Fats Contribute to the Sensation of Feeling Full As they should, at nine calories per gram! Fats tend to stay in the stomach longer than carbohydrates or proteins and are absorbed over a longer period of time. Their presence in the stomach and small intestine trigger satiety signals and a feeling of fullness.[3]

Fats are a Component of Cell Membranes, Vitamin D, and Sex Hormones Some types of fats give cell membranes flexibility and help regulate the transfer of nutrients into and out of cells.[4] Others serve as precursors to vitamin D and sex hormones, such as estrogen and testosterone.[1]

Table 18.1 Functions of fats[1,2]

- Provide a concentrated source of energy
- Contribute to the body's energy reserves (fat stores)
- Carry the essential fatty acids, the fat-soluble vitamins, and certain phytochemicals
- Increase the flavor and palatability of foods
- Provide relief from hunger
- Serve as a structural component of cell membranes
- Serve as precursors to vitamin D, estrogen, testosterone
- Modify gene expression and the types of protein produced in cells
- Participate in the development of fetal and infant vision and central nervous system

essential fatty acids Components of fats (linoleic acid, pronounced *lynn-oh-lay-ick*, and alpha-linolenic acid, pronounced *lynn-oh-len-ick*) required in the diet.

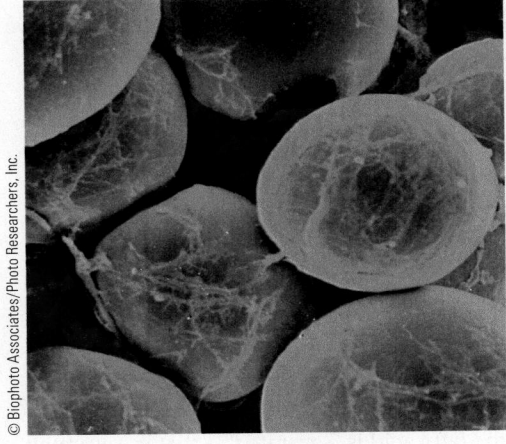

Illustration 18.1 A close look at fat cells (color-enhanced microphotograph).

Illustration 18.2 Although their body fat stores don't fit the image of the superb athlete, cold-water swimmers need the fat to help stay warm. Pictured here is the English swimmer Mike Read, who swam the English Channel 20 times by age 39. The narrowest width of the English Channel is 22 miles, or 35 km.

Table 18.2 Basic facts about the types of fat

Fats can be:
• Monoglycerides
• Diglycerides
• Triglycerides

Fats can be:
• Saturated
• Monounsaturated
• Polyunsaturated

Unsaturated fats come in:
• "*Cis*" forms
• "*Trans*" forms

Table 18.3 A glossary of fats

Triglycerides: Fats in which the glycerol molecule has three fatty acids attached to it; also called triacylglycerol. Triglycerides are the most common type of fat in foods and in body fat stores.

Saturated fats: Molecules of fat in which adjacent carbons within fatty acids are linked only by single bonds. The carbons are "saturated" with hydrogens; that is, they are attached to the maximum possible number of hydrogens. Saturated fats tend to be solid at room temperature. Animal products and palm and coconut oil are sources of saturated fats.

Unsaturated fats: Molecules of fat in which adjacent carbons are linked by one or more double bonds. The carbons are not saturated with hydrogens; that is, they are attached to fewer than the maximum possible number of hydrogens. Unsaturated fats tend to be liquid at room temperature and are found in plants, vegetable oils, meats, and dairy products.

Glycerol: A syrupy, colorless liquid component of fats that is soluble in water. It is similar to glucose in chemical structure.

Cholesterol: A fat soluble, colorless liquid primarily found in animals. Cholesterol is produced by the liver and is used by the body to form hormones such as testosterone and estrogen. It is a component of cell membranes. Cholesterol is present in plant cell membranes but the quantity is small and plants are not considered to be a significant dietary source of cholesterol.

Diglyceride: A fat in which the glycerol molecule has two fatty acids attached to it; also called diacylglycerol.

Monoglyceride: A fat in which the glycerol molecule has one fatty acid attached to it; also called monoacylglycerol.

Monounsaturated fats: Fats that contain a fatty acid in which one carbon-carbon bond is not saturated with hydrogen.

Polyunsaturated fats: Fats that contain a fatty acid in which two or more carbon–carbon bonds are not saturated with hydrogen.

***Trans* fats:** Fats containing fatty acids in the *trans* form. Also called *trans* fatty acids.

Glycerol + 3 fatty acids = Triglyceride

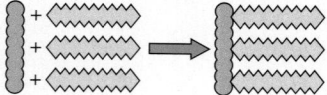

Illustration 18.3 A triglyceride.

The Different Types of Fat

There are many types of fat in food and in our bodies (Table 18.2). Of primary importance are *triglycerides* (or "triacylglycerols"), *saturated and unsaturated fats, cholesterol*, and *trans* fats (for definitions, see Table 18.3). The different types of fats have different effects on health.

Triglycerides, which consist of one *glycerol* unit (a glucoselike substance) and three fatty acids (Illustration 18.3), make up 98% of our dietary fat intake and the vast majority of our body's fat stores. Triglycerides are transported in blood attached to protein carriers and are used by cells for energy formation and tissue maintenance. A minority of fats have the form of *diglycerides* (glycerol plus two fatty acids) and *monoglycerides* (glycerol and one fatty acid). Diglycerides are present in some oils and small amounts are used in food products as emulsifiers—or to increase the blending of fat- and water-soluble substances. Monoglycerides are present in small amounts in some oils; we don't consume very much of them in foods.

As far as health is concerned, the glycerol component of fat is relatively unimportant. It's the fatty acids that influence what the body does with the fat we eat; and they are responsible, in part, for how fat affects health. Many different types of fatty acids are found in triglycerides. You've heard of the major ones: those that make fat "saturated" or "unsaturated."

Saturated and Unsaturated Fats Fatty acids found in fats consist primarily of hydrogen atoms attached to carbon atoms (Illustration 18.4). When the carbons are attached to as many hydrogens as possible, the fatty acid is "saturated"—that is, saturated with hydrogen. Saturated fats tend to be solid at room temperature. Except for palm and coconut oil, only animal products are rich in saturated fats (Illustration 18.5). Fatty acids

Two hydrogens are missing from each of these carbon–carbon links, making the fatty acid polyunsaturated. With fewer hydrogens to attach to, these carbons are doubly bonded to each other. Monounsaturated fatty acids have only one carbon–carbon bond that is "unsaturated" with hydrogen atoms.

that contain fewer hydrogens than the maximum are "unsaturated." They tend to be liquid at room temperature. By and large, plant foods are the best sources of unsaturated fats.

Unsaturated fats are classified by their degree of unsaturation. If only one carbon–carbon bond in the fatty acid is unsaturated, the fat is called "monounsaturated." If two or more carbon–carbon bonds are unsaturated with hydrogen, the fat qualifies as "polyunsaturated."

The Omega-6 and Omega-3 Fatty Acids

The essential fatty acids linoleic acid and alpha-linolenic acid are members of the fatty acid families of omega-6 (also called n-6 fatty acids) and omega-3 fatty acids (also known as n-3 fatty acids), respectively. Both are polyunsaturated, can be used as a source of energy, and are stored in fat tissue. Because they are essential, both linoleic and alpha-linolenic acid are required in the diet.

Linoleic acid is required for growth, maintenance of healthy skin, and normal functioning of the reproductive system. It is a component of all cell membranes and is found in particularly high amounts in nerves and the brain. A number of biologically active compounds produced in the body that participate in regulation of blood pressure and blood clotting are derived from linoleic acid. The major food sources of linoleic acid are sunflower, safflower, corn, and soybean oils.

Alpha-linolenic acid is a structural component of all cell membranes and is found in high amounts in the brain and other nervous system tissues. It also forms biologically active compounds used in the regulation of blood pressure and blood clotting, but these compounds have the opposite effects on blood pressure and blood clotting as do some of the derivatives of linoleic acid.[1] Omega-3 fatty acids are found in walnuts; dark, leafy green vegetables; and flaxseed, canola, and soybean oils in the form of alpha-linolenic acid.

Other, biologically important omega-3 fatty acids exist, and the two primary ones are EPA (eicosapentaenoic acid, pronounced *e-co-sah-pent-tah-no-ick*) and DHA (docosa-hexaenoic acid, pronounced *dough-cos-ah-hex-ah-no-ick*). These omega-3 fatty acids can be produced from alpha-linolenic acid, but the conversion process is slow and results in the availability of relatively small amounts of EPA and DHA.[1]

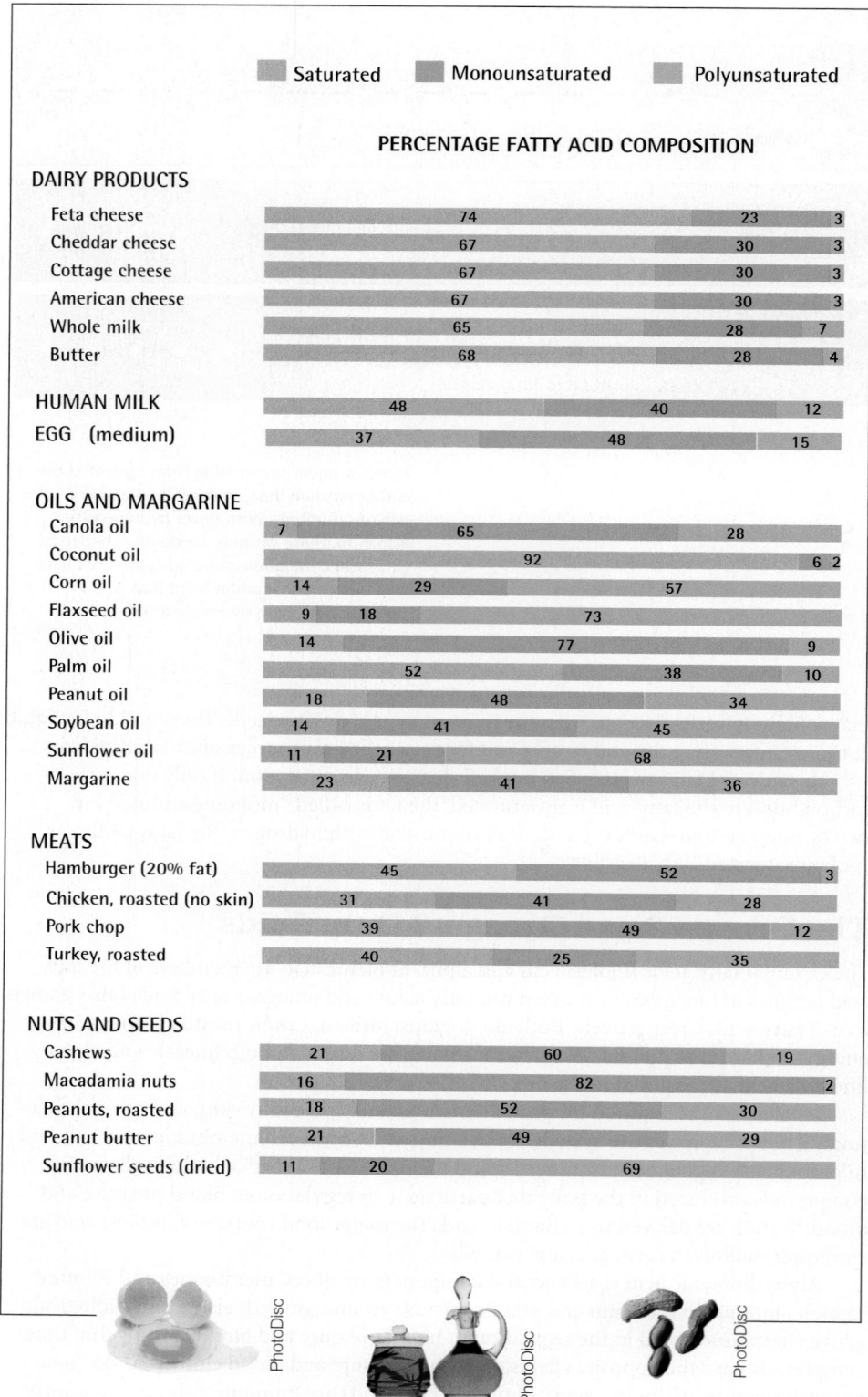

PERCENTAGE FATTY ACID COMPOSITION

Legend: ■ Saturated   ■ Monounsaturated   ■ Polyunsaturated

DAIRY PRODUCTS

Food	Saturated	Monounsaturated	Polyunsaturated
Feta cheese	74	23	3
Cheddar cheese	67	30	3
Cottage cheese	67	30	3
American cheese	67	30	3
Whole milk	65	28	7
Butter	68	28	4

	Saturated	Monounsaturated	Polyunsaturated
HUMAN MILK	48	40	12
EGG (medium)	37	48	15

OILS AND MARGARINE

Food	Saturated	Monounsaturated	Polyunsaturated
Canola oil	7	65	28
Coconut oil	92	6	2
Corn oil	14	29	57
Flaxseed oil	9	18	73
Olive oil	14	77	9
Palm oil	52	38	10
Peanut oil	18	48	34
Soybean oil	14	41	45
Sunflower oil	11	21	68
Margarine	23	41	36

MEATS

Food	Saturated	Monounsaturated	Polyunsaturated
Hamburger (20% fat)	45	52	3
Chicken, roasted (no skin)	31	41	28
Pork chop	39	49	12
Turkey, roasted	40	25	35

NUTS AND SEEDS

Food	Saturated	Monounsaturated	Polyunsaturated
Cashews	21	60	19
Macadamia nuts	16	82	2
Peanuts, roasted	18	52	30
Peanut butter	21	49	29
Sunflower seeds (dried)	11	20	69

PhotoDisc   PhotoDisc   PhotoDisc

EPA and DHA EPA and DHA are primarily ingredients of fish oils and perform a number of important functions in the body. DHA is a structural component of the brain and is found in high amounts in the retina of the eye. During the last three months of pregnancy and during infancy, DHA accumulates in these tissues and promotes optimal intellectual and visual development.[5]

EPA serves as a precursor of a number of biologically active compounds involved in blood pressure regulation, blood clotting, and anti-inflammatory reactions. Inflammation is a central component of many chronic diseases, including heart disease, type 2 diabetes, osteoporosis, cancer, Alzheimer's disease, and rheumatoid arthritis. It is part of the body's response to the presence of infectious agents or irritants. A by-product of the body's inflammatory processes, however, are oxidation reactions that can harm cells and tissues. Derivatives of EPA limit the harmful effects of inflammatory and oxidation reactions.[5]

Adequate intake of EPA and DHA for adults is considered to be 250–500 mg per day.[6] These levels of intake can often be achieved by consuming 8 ounces of fatty fish weekly. Deep-fried fish have been found to be a poorer source of EPA and DHA than baked or broiled fish.[7] Part of the reason for the difference appears to be the types of fish used in the commercial preparation of fried fish, and part to the effect of cooking method on the unsaturated fatty acids in fish. [22]

Fish and shellfish content of EPA and DHA is listed in Table 18.4. The table includes fish and shellfish that contain relatively low amounts of mercury. Mercury can cause nerve tissue damage and learning problems if consumed in excessive amounts. Pregnant and breast-feeding women can safely consume 12 ounces of low mercury fish and seafood weekly. Fish that are high in mercury, such as tilefish, swordfish, shark, and king mackerel, should be avoided.[8]

Consumption of two fish meals a week reduces the risk of heart disease, heart attack, sudden death, and type 2 diabetes; and improves fetal and infant development.[6,10] Higher amounts of EPA and DHA (1 to 3 grams) provided by fish or fish oils, are being used clinically to lower blood triglyceride levels, reduce the risk of heart disease and stroke, and reduce the need for anti-inflammatory drugs in people with rheumatoid arthritis.[11-13] The Food and Drug Administration recommends that consumers ingest no more than a total of three grams of EPA and DHA daily and limit supplementary intakes to two grams a day. Fish liver oils should be used with caution because they contain relatively high amounts of vitamins A and D. Fish oils, made from the body of the fish, do not.[14]

Increasing Omega-3 Fatty Acid Intake In the past it was thought that consuming high amounts of omega-6 fatty acids compared to omega-3 fatty acids could interfere with the availability of omega-3 fatty acids, particularly EPA and DHA. Although still somewhat controversial, it appears that rather high intakes of omega-6 fatty acids do not interfere with the availability of EPA and DHA.[2] What interferes most with the availability of EPA and DHA for body functions is inadequate intake. On average, adults in the United States and Canada consume around 100 mg EPA plus DHA daily, far short of an adequate intake of 250–500 mg daily.[16]

EPA and DHA Fortified Foods Fatty fish are clearly the richest sources of EPA and DHA. But what if you (like many other people) don't like fish?[21] In that case you can turn to seafoods like shrimp or EPA- and DHA-fortified foods.[17] (For more information on specific EPA- and DHA-fortified foods see this unit's Take Action feature.) Purified fish oils with no fishy taste are increasingly being added to products from fruit juice to yogurt and to animal feeds. The EPA and DHA in feed are incorporated into the animal's tissues. Consequently, beef, pork, eggs, milk and milk products consumed from animals "fortified" with EPA and DHA are fortified, too. Some animal feeds contain DHA from algae and provide only DHA in their food products.

To make sure you're choosing foods with EPA and DHA, or DHA, confirm that the label specifies that these fatty acids are contained in the product. Just because a product announces Omega-3 on the label doesn't mean it contains EPA and DHA. It may contain flax seed, flax seed oil, walnut oil, alpha-linolenic acid, or other sources of omega-3 fatty acids that are not equivalent to EPA or DHA.[6]

Hydrogenated Fats

Unsaturated fats aren't as stable as saturated fats. They are more likely to turn rancid with time and exposure to air (oxygen) and heat than saturated fats. Additionally, solid fats are preferable to oils for some cooking applications. These problems with unsaturated fats have a solution: It's called hydrogenation.

Table 18.4 EPA and DHA content of fish and seafoods containing, on average, less than 0.2 ppm mercury in a 3-ounce serving[15]

	EPA + DHA, mg
Fish oil (1 tsp.)	2,796
Shad	2,046
Salmon, farmed	1,825
Anchovies	1,747
Herring	1,712
Salmon, wild	1,564
Whitefish	1,370
Mackerel	1,023
Sardines	840
Whiting	440
Flounder	426
Trout, fresh water	420
Oysters	375
Snapper	273
Shrimp	268
Clams	241
Haddock	202
Catfish, wild	201
Crawfish	187
Sheepshead	162
Tuna, light, canned in oil	109
Lobster	71

take action To Consume Enough EPA and DHA from Foods Other Than Fish

If you would like to increase your EPA and DHA intake, here are some of your nonfish food choices. Indicate with a check mark foods you would be likely to consume. At the bottom of the list, answer this question: Based on your choices, about how much EPA and DHA would you be adding to your daily diet by consuming the foods you checked?

EPA and DHA/DHA content, mg

____ DHA fortified eggs, 1	150
____ Healthy Heart Omega-3 orange juice, 8 oz.	50
____ Omega Farm nonfat yogurt, 8 oz.	75
____ Egg Creations Liquid, 1/4 cup	260
____ Smart Balance Omega Buttery Spread 1 Tbsp.	32
____ Italica Omega-3 Olive Oil, 1 Tbsp.	120
____ Omega Farms Low-Fat Milk, 1 cup	75
____ Omega Farms Mild Cheddar, 1 oz.	75
____ Smart Balance Lactose-Free Milk with Omega-3, 1 cup	32
____ Minute Maid 100% Fruit Juice Blend with Omega 3/DHA, 1 cup	50
____ Shrimp, 3 oz.	268
____ Clams, 3 oz.	241
____ Crab, 3 oz.	375
____ Scallops, 3 oz	161

The approximate amount of EPA and DHA I would be adding to my diet daily based on my food choices is: ____ mg. If not between 250 to 500 mg, you would be getting closer.

What's Hydrogenation? **Hydrogenation** is a process that adds hydrogen to liquid unsaturated fats, thereby making them more saturated and solid. The shelf life, cooking properties, and sometimes the taste of vegetable oils are improved in the process. Hydrogenation has two drawbacks, however. Hydrogenated vegetable oils contain more saturated fat than the original oil. Corn oil, for example, contains only 6% saturated fats; but corn oil margarine has 17%. The other negative is that hydrogenation causes a change in the structure of the unsaturated fatty acids. Specifically, hydrogenation converts some unsaturated fats into *trans* **fats**.

***Trans* Fatty Acids** The bulk of *trans* fats in our diets comes from hydrogenated vegetable oils. Hydrogenation causes some of the unsaturated fatty acids to be converted from their naturally occurring *cis* to the *trans* form. Ruminant animals like cows, goats, and sheep form a small amount of *trans* fats in their stomachs. Consequently, milk and milk products from these animals will contain *trans* fats.[1]

Repositioned hydrogen molecules in *trans* fats change the way the body uses the fat. *Trans* fats raise blood cholesterol levels more than any other type of fat.[1] They increase the risk of heart disease, stroke, sudden death from heart disease, and type 2 diabetes, and they promote inflammation. Intake of *trans* fats as low as 1% of total calories per day strongly increases the risk of heart disease. An intake of 2.2 grams of *trans* fat daily would place a person consuming 2,000 calories a day at increased risk of heart disease.[18]

It is recommended that Americans consume as little *trans* fats as possible,[1] and nutrition information labeling requirements are making that easier to accomplish. Nutrition facts panels must include the *trans* fat content of food products (Illustration 18.6). The percent of daily value column (%DV) is not used for *trans* fats because there is no recommended level of intake. Products labeled "*trans* fat-free" (Illustration 18.7) must contain less than 0.5 gram of both *trans* and saturated fats. The requirement to label the *trans*

hydrogenation The addition of hydrogen to unsaturated fatty acids.

trans fats Unsaturated fatty acids in fats that contain atoms of hydrogen attached to opposite sides of carbons joined by a double bond:

```
    H
 —C=C—              H H
    H              —C=C—
Trans fatty acid   Cis fatty acid
```

Fats containing fatty acids in the *trans* form are generally referred to as *trans* fats. *Cis* fatty acids are the most common, naturally occurring form of unsaturated fatty acids. They contain hydrogens located on the same side of doubly bonded carbons.

Table 18.5 Where are the *trans* fats now?

Values may change as companies lower the *trans* fat content of foods.

Food	*Trans* fatty acids (grams)
Frozen apple pie, 1 slice	4.5
Chicken and biscuits, 2 small	3
Frozen biscuits, 1	2.5 (regular) 3.5 (grand)
Sausage, egg, cheese biscuit, 1	2.5
Frozen coconut cream pie, 1 slice	2
Canned frosting, 2 Tbsp.	1.5
Frozen dinner, chicken and pasta, 1 cup	0.5
Frozen dinner, cheese and manicotti, 7 oz.	0.5

Source: *Trans* fat content of foods is based on a "*trans* fat search" conducted by the author in supermarkets, January 2012.

Nutrition Facts
Serving Size 1 Entree
Serving Per Container 1

Amount Per Serving

Calories 380 Calories from Fat 170

	%Daily Value
Total Fat 19g	**29%**
Saturated Fat 10g	**50%**
Trans Fat 2g	
Cholesterol 85mg	**28%**
Sodium 810mg	**34%**
Total Carbohydrate 33g	**11%**
Dietary Fiber 3g	**12%**
Sugars 5g	
Protein 20g	

Vitamin A 10%	Vitamin C 0%
Calcium 10%	Iron 15%

Percent Daily Values are based on a 2000 calorie diet. Your daily values may be higher or lower depending on your calorie needs:

		Calories	2000	2500
Total Fat	Less Than		65g	80g
Sat Fat	Less Than		20g	25g
Cholesterol	Less Than		300mg	300mg
Sodium	Less Than		2400mg	2400mg
Total Carbohydrate			300g	375g
Dietary Fiber			25g	30g

Illustration 18.6 *Trans* fat: the newest addition to nutrition facts panels.

---

fat content of food products, regulations by various states and cities that ban their use in restaurant foods, and increased consumer awareness of the adverse effects of *trans* fat have encouraged food producers to take *trans* fats out of many prepared, processed, and fast foods.[1,19] Many major brands of potato and tortilla chips, bakery products, and frozen meals and desserts have taken the *trans* fats out of their products. Some products still contain them, however. The only way to know for sure if prepared, packaged foods contain *trans* fats is to look at the nutrition information label. A sampling of some of the remaining food products that contain *trans* fats is listed in Table 18.5.

Checking Out Cholesterol

Cholesterol is a lipid found primarily in animal products. It is tasteless and odorless and contained in both the lean and fat parts of animal products. Table 18.6 lists some sources of cholesterol. Cholesterol is present in plant cell membranes, but the quantity is small and plants are not considered to be a significant dietary source of cholesterol.

Sources of Cholesterol The cholesterol used by the body comes from two sources. Most (about two-thirds) of the cholesterol available to the body is produced by the liver and the brain (which makes its own supply of cholesterol). The rest comes from the diet (Illustration 18.8). Because the liver produces cholesterol from other substances in our diet, it does not qualify as an essential nutrient.

Richard Anderson

Illustration 18.7 Products that feature "no *trans* fats" and "*trans* fat-free" labels.

The Contributions of Cholesterol Would you be surprised to learn the following about cholesterol?

- It is found in every cell in your body.

- It serves as the building block for estrogen, testosterone, and the vitamin D.

- It is produced in your skin on exposure to sunlight.

- It is a major component of nerves and the brain.

- It cannot be used for energy (so it provides no calories).

As the above list indicates, the body has many uses for cholesterol—it doesn't just accumulate in arteries. The major functions of cholesterol are listed in Table 18.7.

The Fat Content of Food

Not all of the fat in food is visible. To avoid being fooled, it helps to use a reference on the fat composition of foods. Table 18.8 lists the fat content of common food sources of fat, including candy. Vegetables and fruits (except avocado and coconut) and grains are not listed because they contain relatively little fat. Other references can also be used, such

Table 18.6 Food sources of cholesterol

Note that all the leading food sources are animal products. Cholesterol in foods is a clear, oily liquid found in the fat and lean portions of many animal products.

Animal product	Amount	Cholesterol (milligrams)
Brain	3 oz	1,746
Liver	3 oz	470
Egg	1	186
Veal	3 oz	128
Shrimp	3 oz	107
Prime rib	3 oz	80
Chicken (no skin)	3 oz	75
Turkey (no skin)	3 oz	65
Hamburger, regular	3 oz	64
Pork chop, lean	3 oz	60
Fish, baked (haddock, flounder)	3 oz	58
Ice cream	1 cup	56
Sausage	3 oz	55
Hamburger, lean	3 oz	50
Milk, whole	1 cup	34
Crab, boiled	3 oz	33
Lobster	3 oz	29
Cheese (cheddar)	1 oz	26
Milk, 2%	1 cup	22
Yogurt, low-fat	1 cup	17
Milk, 1%	1 cup	14
Butter	1 tsp	10
Milk, skim	1 cup	7

Illustration 18.8 Food sources of cholesterol in the U.S. diet.[15] Percentages indicate the proportion of cholesterol each type of food contributes to the diet.

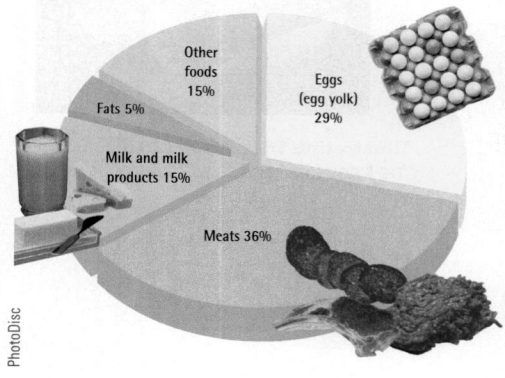

Other foods 15%
Eggs (egg yolk) 29%
Fats 5%
Milk and milk products 15%
Meats 36%

PhotoDisc

as the food composition tables in Appendix A, the Diet Analysis Plus Program software, and the nutrition labels on food products.

Knowledge of the caloric and fat content of a food can be used to calculate the percentage of calories provided by fat. For example, suppose that a slice of cherry pie provides 350 calories and 15 grams of fat. To calculate the percentage of fat calories, multiply 15 grams by 9 (the number of calories in each gram of fat), divide the result by 350 calories, and then multiply this result by 100:

$$15 \text{ grams fat} \times 9 \text{ calories/gram} = 135 \text{ calories}$$
$$135 \text{ calories} \div 350 = 0.39$$
$$0.39 \times 100 = 39\% \text{ of total calories from fat}$$

Fat Labeling Nutrition labeling regulations for fat require that food manufacturers adhere to standard definitions of "low-fat," "fat-free," and related terms on food labels. Similarly, claims made about the cholesterol content of food products must comply with standard definitions (Table 18.9). If a claim is made about the fat content of a food, the nutrition facts panel must specify the food's fat, saturated fat, *trans* fat, and cholesterol content. If a claim is made about cholesterol content (and claims can be made only for products that normally contain a meaningful amount of cholesterol), the nutrition panel must also reveal the product's fat and saturated fat content. To prevent the use of unrealistically small serving sizes as a way to appear to cut down on a product's fat content, standard serving sizes must also be used on food labels.

Table 18.7 How the body uses cholesterol

- Cholesterol is a component of all cell membranes, the brain, and nerves.
- Cholesterol is needed to produce estrogen, testosterone, vitamin D, and bile.

Table 18.8 The fat content of some foods

Food	Amount	Grams	Percentage of total calories from fat
Fats and oils			
Butter	1 tsp	4.0	100%
Margarine	1 tsp	4.0	100
Oil	1 tsp	4.7	100
Mayonnaise	1 Tbsp	11.0	99
Heavy cream	1 Tbsp	5.5	93
Salad dressing	1 Tbsp	6.0	83
Meats and fast foods			
Hot dog, 1	2 oz	17.0	83
Bologna	1 oz	8.0	80
Sausage	4 links	18.0	77
Bacon	3 pieces	9.0	74
Salami	2 oz	11.0	68
Hamburger, regular (20% fat)	3 oz	16.5	62
Chicken, fried with skin	3 oz	14.0	53
Big Mac	6.6 oz	31.4	52
Quarter Pounder with cheese	6.8 oz	28.6	50
Whopper	8.9 oz	32.0	48
Steak (rib eye)	3 oz	9.9	47
Veggie pita	1	17.0	38
Chicken, baked without skin	3 oz	4.0	25
Flounder, baked	3 oz	1.0	13
Shrimp, boiled	3 oz	1.0	10
Milk and milk products			
Cheddar cheese	1 oz	9.5	74
American cheese	1 oz	6.0	66
Milk, whole	1 cup	8.5	49
Cottage cheese, regular	1/2 cup	5.1	39
Milk, 2%	1 cup	5.0	32
Milk, 1%	1 cup	2.7	24
Cottage cheese, 1% fat	1/2 cup	1.2	13
Milk, skim	1 cup	0.4	4
Yogurt, frozen	3/4 cup	0.0–6.6	0–3
Other			
Olives	4 medium	1.5	90
Avocado	1/2	15.0	84
Almonds	1 oz	15.0	80
Sunflower seeds	1/4 cup	17.0	77
Peanuts	1/4 cup	17.5	75
Cashews	1 oz	13.2	73
Flax seed	1/4 cup	17.7	71
Egg	1	6.0	61
Potato chips	1 oz (13 chips)	11.0	61
French fries	14 fries	9.5	46
Taco chips	1 oz (10 chips)	6.2	41
Candy			
Peanut butter cups, 2 regular	1.6 oz	15.0	54
Milk chocolate	1.6 oz	14.0	53
Almond Joy	1.8 oz	14.0	50
Kit Kat	1.5 oz	12.0	47
M & M's, peanut	1.7 oz	13.0	47

Table 18.9 What claims about the cholesterol content of foods that normally contain cholesterol must mean

- No cholesterol or cholesterol-free: Contains less than three milligrams of cholesterol per serving

- Low cholesterol: Contains 20 milligrams or less of cholesterol per serving

- Reduced cholesterol: Contains at least 75% less cholesterol than normal

- Less cholesterol: Contains at least 25% less cholesterol than normal; the percentage less must be stated on the label

Illustration 18.9   A look at the cuisine of the Mediterranean diet.

PhotoDisc

Adherence to certain types of diets that are relatively high in total calories from fats, such as the Mediterranean diet highlighted in Illustration 18.9, reduce the risk of heart disease, stroke, obesity, or a number of other diseases.[1] Evidence established on the healthful effects of the Mediterranean diet and other examples of dietary fat intake and health in different populations prompted the development of new recommendations based primarily on the healthfulness of various types of fats.[1]

"Healthy" Fats, "Unhealthy" Fats   Fats come in many types in foods, and with few exceptions, they serve as a source of energy and provide a number of essential functions in the body. With regard to raising or lowering the risk of heart disease and stroke, however, fats differ. Those that elevate LDL-cholesterol levels are regarded as "bad" or unhealthy fats.

REALITY CHECK
Good fats, bad fats

What foods provide "healthy" fats?

Who gets thumbs up?

Answers on page 18-13

PhotoDisc

Kristen: How can I be wrong? Low-fat food products are best for healthy fat because they contain almost no fat!

PhotoDisc

Butch: I'm thinking foods like fish, peanut butter, and *trans* fat–free margarine contain healthy fats.

Table 18.10 Healthy and unhealthy fats and examples of food sources

Healthy fats	Unhealthy fats
DHA, EPA (omega-3 fatty acids) Fish and seafood	***Trans* fats** Snack and fried foods, bakery goods
Monounsaturated fats Olive and peanut oil, nuts, avocados	**Saturated fats** Animal fats
Polyunsaturated fats Vegetable oils	**Cholesterol** Eggs, seafood, and meat
Alpha-linolenic acid Soybeans, walnuts, flaxseed	

Those that lower LDL-cholesterol or raise blood levels of HDL-cholesterol (the one that helps the body get rid of cholesterol in the blood) are considered "good" or "healthy fats."[20]

The list of unhealthy fats includes *trans* fats, saturated fats, and cholesterol. Fats labeled "bad" are generally solid at room temperature and are included in foods such as high-fat meats and dairy products, hard margarines, shortening, and crispy snack foods. Monounsaturated fats, polyunsaturated fats, alpha-linolenic acid, DHA, and EPA are considered healthy fats and are present in food in the form of oils (Table 18.10).[1,9]

Recommendations for Fat and Cholesterol Intake

Current recommendations for adults call for consumption of 20–35% of total calories from fat. Average fat consumption in the United States is at the upper end of this range (34% of total calories).[16] The Adequate Intakes (AIs) for the essential fatty acid linoleic acid is set at 17 grams a day for men and 12 grams for women. AIs for the other essential fatty acid, alpha-linolenic acid, are 1.6 grams per day for men and 1.1 grams for women. It is recommended that intake of *trans* fats and saturated fats be as low as possible while consuming a nutritionally adequate diet. Only a small proportion of Americans consume too little linoleic acid, but intakes of alpha-linolenic acid tend to be low. Americans are being encouraged to increase consumption of EPA and DHA by eating fish more often. In addition, saturated fat intake averages 11% of calories, an amount that can increase the risk of heart disease in some individuals.[1]

There is no recommended level of cholesterol intake because there is no evidence to indicate that cholesterol is required in the diet. The body is able to produce enough cholesterol, and people do not develop a cholesterol deficiency disease if it is not consumed. Blood cholesterol levels tend to increase somewhat as consumption of cholesterol increases. Saturated and *trans* fats have a much stronger influence on raising blood cholesterol concentration than cholesterol does, however.[1] It is recommended that cholesterol intake be kept below 300 mg per day. Cholesterol intake averages 276 mg per day in the United States.[16]

Recent recommendations for fat intake represent an unusually large but necessary change in dietary intake guidance. Much remains to be understood about the effects of dietary fats on health, and how other components of diet, lifestyle, and genetic traits modify relationships between fat intake and health.

ANSWERS TO **REALITY** CHECK
Good fats, bad fats

Low-fat foods contain less fat than the regular version of the foods. But that doesn't mean the products contain no fat or only good fats, or are low in calories. Food sources of fish oils, unsaturated fat, and *trans* fat–free products provide the healthy fats. As always, healthy diets aren't based on individual foods, they are based on overall diets. You can emphasize foods providing healthy fats without feeling bad about occasionally eating foods branded with the "bad-fat" label.

Kristen:

Butch:

© James And James/Getty Images/FoodPix

NUTRITION
up close

The Healthy Fats in Your Diet

Focal Point: Identify your healthy fat food choices.

Are the fats in your diet the healthy type? Check it out by answering these questions.

How Often Do You Eat:

1. Sausage, hot dogs, ribs, and luncheon meats?
2. Heavily marbled steaks or roasts and chicken with the skin?
3. Soybean products such as tofu or soy nuts?
4. Nuts or seeds?
5. Whole milk, cheese, or ice cream?
6. Soft margarine or olive oil?
7. French fries, snack crackers, or commercial bakery products?
8. Rich sauces and gravies?
9. Fish or seafood?
10. Peanut butter?

Feedback for the Nutrition Up Close is located in Appendix G.

REVIEW QUESTIONS

- **Summarize the major functions, types, food sources, recommended intake, and health effects of various fats.**

Identify foods low in saturated and *trans* fat, and high in unsaturated fats.

1. Animal products are by far the leading source of saturated fats in the American diet.   **True/False**

2. Flaxseed oil and dark, leafy green vegetables are excellent sources of the essential fatty acid alpha-linolenic acid.   **True/False**

3. The best food sources of EPA and DHA are fish and shellfish.   **True/False**

4. Cholesterol is not a required nutrient because it serves no essential, or life-sustaining, function in the body.   **True/False**

5. Thanks in part to nutrition labeling requirements, there are lower amounts of *trans* fats in food products now than in the recent past.   **True/False**

6. *Trans* fats raise blood cholesterol levels more than any other type of fat.   **True/False**

7. If a claim is made on a food package about the product's fat content, the nutrition facts panel must list the food's fat, saturated fat, *trans* fat, and cholesterol content.   **True/False**

8. The type of fat consumed is more important to health than is total fat consumption.   **True/False**

9. Unhealthful or "bad" fats include *trans* fats and saturated fats.   **True/False**

10. It is recommended that adults consume 20–25% of total calories from fat.   **True/False**

The following five questions refer to this scenario:

You're sitting in the dentist's chair, your mouth is filled with a drill, fingers, a suction hose, and cotton swabs. The dental hygienist asks you what courses you're taking this term. You're able to get the word *nutrition* out of your mouth. She gets excited and tells you about her recipe for spinach salad that's rich in omega-3 fatty acids. The recipe consists of:

1 cup chopped spinach
1/4 cup chopped broccoli
6 baby carrots
1 hard-boiled egg
1/4 cup of flax seeds
2 tablespoons of Italian dressing

You have time to think about your response because you can't talk yet.

11. _____ Approximately how many milligrams (mg) of EPA+DHA is in the recipe?
    a. 0 mg
    b. 25 mg
    c. 78 mg
    d. 204 mg

12. _____ Which set of foods would contain the most polyunsaturated fats?
    a. spinach and broccoli
    b. carrots and spinach
    c. broccoli and egg
    d. flax seeds and Italian dressing

13. _____ Approximately how much DHA+EPA would be added to the recipe if the hygienist substituted the flax seeds for 3 ounces of shrimp?
    a. 126 mg
    b. 268 mg
    c. 89 mg
    d. 0 mg

14. _____ Approximately how many grams of fat are contained in the recipe?
    a. 78 g
    b. 24 g
    c. 30 g
    d. 36 g

15. _____ If the recipe provides 422 calories and 1.6 grams of saturated fat, approximately what percent of the total calories in the salad would come from saturated fat?
    a. 3%
    b. 6%
    c. 10%
    d. 15%

Answers to these questions can be found in Appendix G.

NUTRITION SCOREBOARD ANSWERS

1. The acceptable range of fat intake is 20–35% of total calories. **False**

2. The types of fat consumed are more important to health than the total amount of fat.[1] **True**

3. See Table 18.6. **True**

4. Saturated fat, which is found primarily in animal products, tends raise blood cholesterol levels to a greater extent than dietary cholesterol. **True**

photoshut/Shutterstock.com

Nutrition and Heart Disease

NUTRITION SCOREBOARD

1 Heart disease is the leading worldwide cause of death. **True/False**

2 Low-fat diets are recommended for the prevention and treatment of heart disease. **True/False**

3 Plaque that causes hardening of the arteries develops on the inside surface of artery walls. **True/False**

4 Chronic inflammation is a major contributor to the development and progression of heart disease. **True/False**

5 The major risk factors for heart disease are the same for women as for men. **True/False**

Answers can be found at the end of the unit.

After completing Unit 19 and its interactive learning features, you will be able to:

- List dietary, lifestyle, and biological factors that contribute to heart disease development.

- Evaluate components of your diet and lifestyle that help prevent the development of heart disease.

The Diet–Heart Disease Connection

- **List dietary, lifestyle, and biological factors that contribute to heart disease development.**

Suspicions that dietary fat may be related to heart disease were first raised over 200 years ago. During the late 18th century, physicians noted that people who died of heart attacks had fatty streaks and deposits in the arteries that led to the heart. Later it was discovered that people with heart disease tended to have high blood levels of cholesterol and high intakes of saturated fats and cholesterol. It was concluded that high intakes of these fats increased the risk of heart disease by raising blood cholesterol levels, and that diets low in saturated and total fat may lower blood cholesterol and help prevent and treat heart disease.[1]

Results of a number of large studies have demonstrated that this approach to heart disease prevention and treatment is not effective.[6] It is now known that low-fat diets alone do not decrease the risk of heart disease. It is clear that multiple, important dietary and other risk factors for heart disease exist, and that they all need to be addressed. Blood cholesterol level is one of them.[7] Revised recommendations for heart disease prevention and management feature the consumption of overall healthy diets and lifestyles.[2] State-of-the-science dietary and other lifestyle recommendations for the prevention and treatment of heart disease are presented in this unit.

iStockphoto.com/Creativeye99

Key Nutrition Concepts

Content in this unit on nutrition and heart disease directly relates to a number of key nutrition concepts, including the following:

- Health problems related to nutrition originate within cells.

- Poor nutrition can result from both inadequate and excessive levels of nutrient intake.

- Poor nutrition can influence the development of certain chronic diseases.

- Adequacy, variety, and balance are key characteristics of healthful diets.

A Primer on Heart Disease

There is no bigger health problem in the United States and other developed countries than heart disease. Heart disease is the leading cause of death worldwide, accounting for one out of every four deaths.[8] It is an "equal opportunity" disease, striking as many women as men, although women on average die 10 years later from heart disease than do men.[9] Deaths from heart disease are distributed unequally across the United States (Illustration 19.1), and disproportionately affect African and Hispanic Americans.[10]

The incidence and death rate from heart disease is declining in the United States. The decease in incidence is related to risk factor reduction and the decrease in death rate primarily to improved treatment.[10] Gains in life expectancy, quality of life, and reduced health care expenditures would be generated by further reduction of risk factors for heart disease.[13]

What Is Heart Disease? **Heart disease**, or more correctly "coronary" heart disease, is a term referring to several disorders that result from inadequate blood circulation to parts of the heart. Heart disease is almost always due to a narrowing of the arteries leading to the heart. Arteries become narrow due to a buildup of **plaque** (Illustration 19.2). People with narrowed arteries have **atherosclerosis**, or "hardening of the arteries" as it is often called. Heart disease develops silently over time (usually decades).[3]

When arteries are narrowed by 50% or more, the shortage of blood to the heart can produce chest pain (called "angina"). A heart attack occurs when an artery leading to the

heart disease One of a number of disorders that result when circulation of blood to parts of the heart is inadequate. Also called *coronary heart disease*. (*Coronary* refers to the blood vessels at the top of the heart. They look somewhat like a crown.)

plaque Deposits of cholesterol, other fats, white blood cells, calcium, and cell materials in the lining of the inner wall of arteries.

atherosclerosis "Hardening of the arteries" due to a buildup of plaque. Atherosclerosis is characterized by chronic inflammation of the arteries and develops over decades.

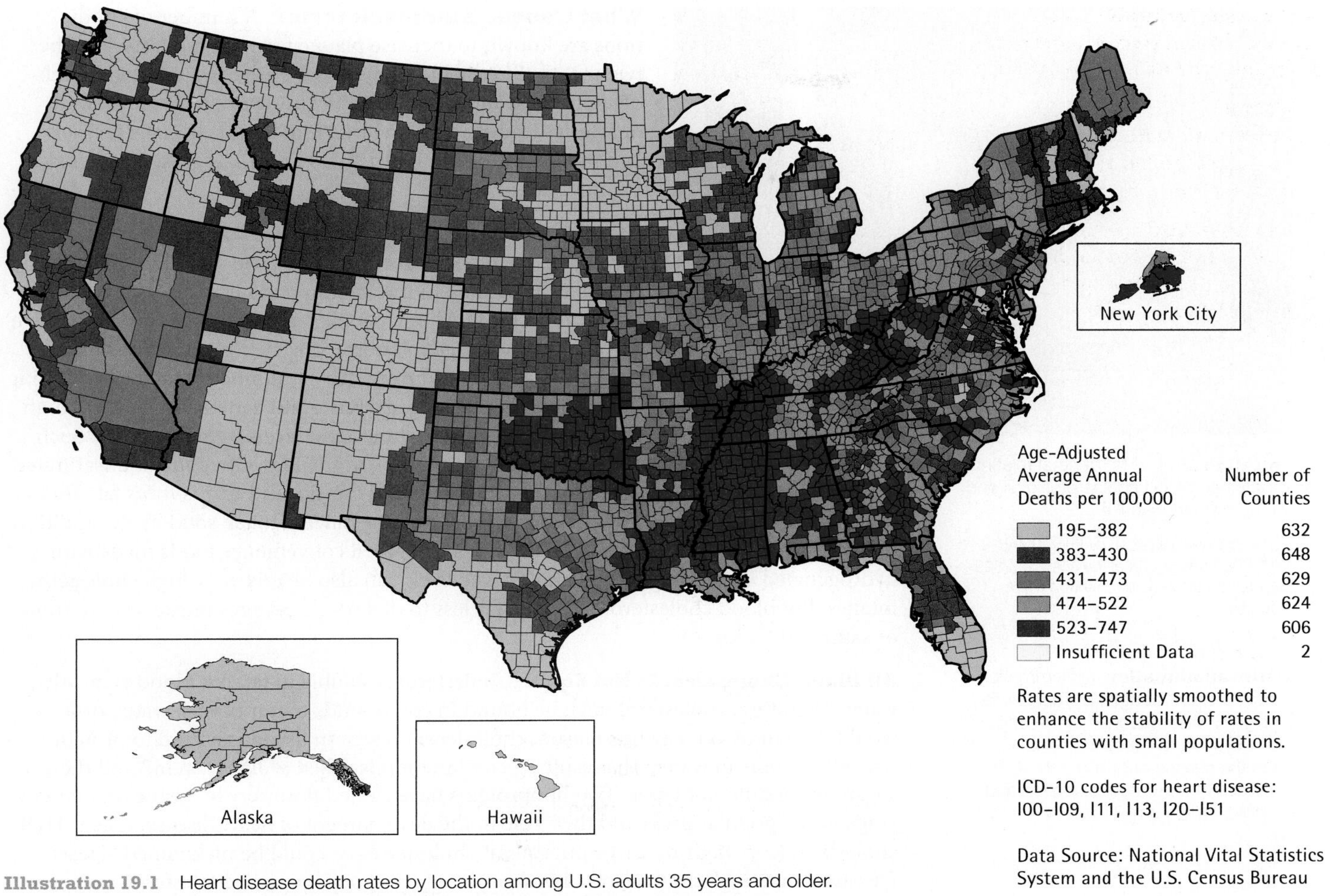

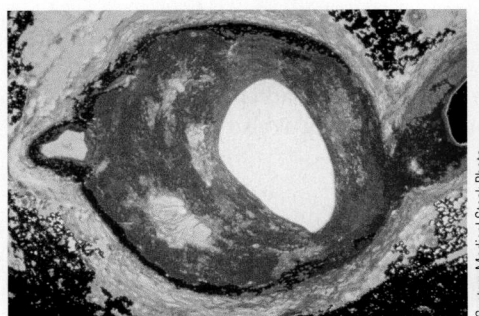

Illustration 19.1 Heart disease death rates by location among U.S. adults 35 years and older.

Alaska

Hawaii

New York City

Age-Adjusted
Average Annual
Deaths per 100,000

	Number of Counties
195–382	632
383–430	648
431–473	629
474–522	624
523–747	606
Insufficient Data	2

Rates are spatially smoothed to enhance the stability of rates in counties with small populations.

ICD-10 codes for heart disease: I00–I09, I11, I13, I20–I51

Data Source: National Vital Statistics System and the U.S. Census Bureau

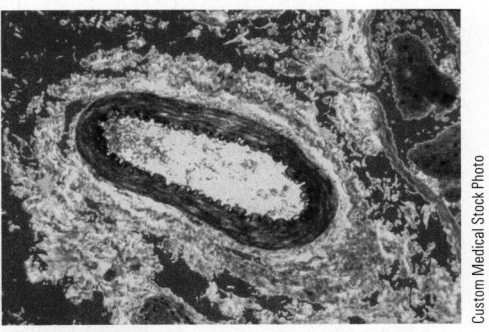

Custom Medical Stock Photo

Illustration 19.2 The progression of atherosclerosis. As plaque builds up, arteries narrow, reducing or stopping the supply of blood to the heart, brain, muscle, or other affected parts of the body.

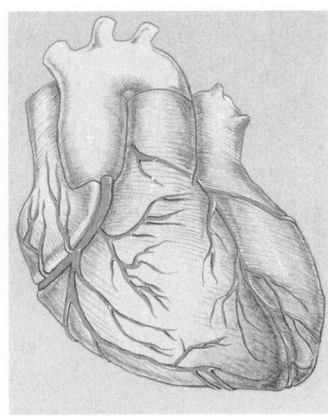

Illustration 19.3 The heart after a heart attack. The dark portion at the base of the heart is affected by the blockage in blood flow.

heart becomes clogged by a piece of plaque released by a ruptured portion of the artery wall or by a blood clot (Illustration 19.3).[3] Although heart disease primarily affects individuals over the age of 55, it's a progressive disease that may begin in childhood.[11]

Arteries leading to the heart aren't the only ones affected by atherosclerosis. Plaque can also build up in the arteries in the legs, neck, brain, and other body parts. If the blood supply to the legs is reduced, pain and muscle cramps may result after brief periods of exercise. Plaque buildup in arteries of the brain contributes to stroke—an event that occurs when the blood supply to a part of the brain is inadequate.[3] Health problems due to atherosclerosis in arteries of the heart, brain, neck, and legs are collectively referred to as **cardiovascular disease**.

cardiovascular disease Disorders related to plaque buildup in arteries of the heart, brain, and other organs and tissues.

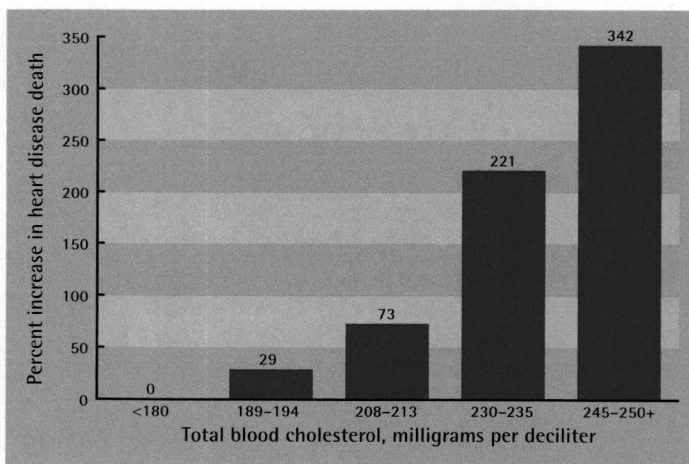

Illustration 19.4 The general relationship between blood cholesterol level and death from heart disease.

Sources: Shekelle RB et al. (*N Engl J Med* 304:65–70); Stamler J et al. (*JAMA* 256:2823–2828); Neaton JD et al. (*Arch Intern Med* 152:56–64).

chronic inflammation Inflammation that lasts weeks, months, or years. Inflammation is the first response of the body's immune system to infection or irritation. It triggers the release of biologically active substances that promote oxidation and other potentially harmful reactions in the body.

What Causes Atherosclerosis? A number of conditions are known to increase plaque formation in arteries. The two most influential identified so far are elevated blood levels of cholesterol and **chronic inflammation** in the inner walls of arteries leading to the heart. High blood cholesterol levels and inflammation are interrelated and act together to increase atherosclerosis.[3]

Blood Cholesterol Levels and Heart Disease In general, the higher the blood cholesterol level, the more likely it is that plaque will build up in the arteries and heart disease will occur (Illustration 19.4). (Some people with atherosclerosis don't have high blood levels of cholesterol, however.)[11] A person's blood cholesterol level is determined by a number of factors, including dietary intake, smoking, exercise, and genetic traits. Diets high in saturated fat elevate cholesterol levels in most people. Such diets are characterized by the regular consumption of animal fat. One type of unsaturated fat raises blood cholesterol levels more than saturated fats do, and that is *trans* fat. This type of fat is produced when vegetable oils are hydrogenated—made solid by the addition of hydrogen. *Trans* fats are found primarily in prepared convenience foods made with hydrogenated shortening. Blood cholesterol levels can also be raised by high cholesterol intakes. But blood cholesterol responds far less to dietary cholesterol intake than to *trans* or saturated fat intake.[12]

All Blood Cholesterol Is Not Equal Cholesterol is soluble in fat, but blood is mostly water. Therefore, cholesterol must be bound to compounds that mix with water, or it would float in blood. For this reason, cholesterol present in blood is bound to protein, which is soluble in water. The resulting combination is called a "lipoprotein," and there are a number of different types. Two lipoproteins have gained notoriety by virtue of their coverage in the popular press and their role in the development of heart disease. One is HDL cholesterol (for "high-density lipoprotein" cholesterol—it could be nicknamed "Heart-Disease-Lowering" cholesterol). It is considered the "good" type of cholesterol, and you want high levels of it in your blood. LDL cholesterol (low-density lipoprotein cholesterol) can be the villain. The composition and roles of both types of lipoproteins are summarized in Illustration 19.5.

Understanding HDL and LDL HDL gets its reputation as the good cholesterol because it helps remove cholesterol from the blood. An avid cholesterol acceptor, HDL escorts cholesterol to the liver for its eventual excretion from the body. It also reduces the forma-

Illustration 19.5 Composition and functions of LDL cholesterol and HDL cholesterol.

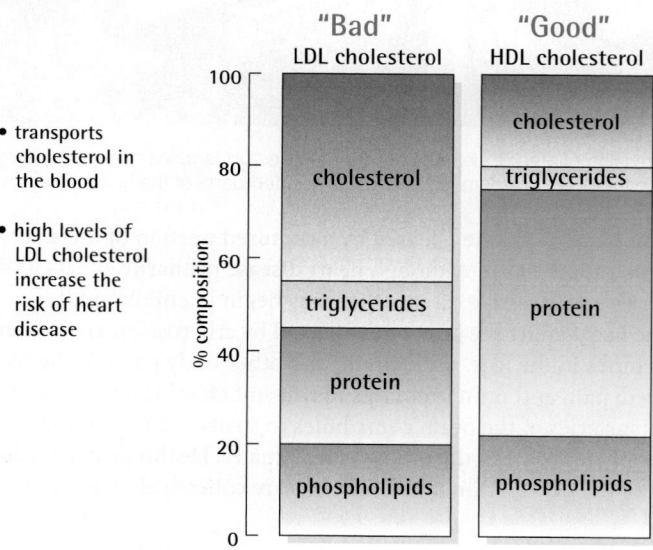

tion of plaque, speeds repair of the endothelial lining, and helps prevent oxidation and inflammation in plaque.[7] High HDL cholesterol levels (over 40 milligrams per deciliter in men and over 50 in women) are protective against heart disease.

LDL cholesterol carries more cholesterol than does HDL. It can be oxidized into reactive LDL particles that make it more likely it will enter plaque and contribute to plaque buildup. The higher the LDL cholesterol level, the greater the chances that atherosclerosis will develop and progress into heart disease.[11]

LDL and HDL cholesterol are components of total blood cholesterol measures. High levels of beneficial HDL cholesterol may elevate total cholesterol values and make it appear that a person is at risk of heart disease. High levels of HDL cholesterol (particularly those over 60 mg/dL) are strongly protective against heart disease.[2] Risk of heart disease is more accurately predicted based on individual measures of LDL and HDL cholesterol than by total blood cholesterol results.

Dietary and Lifestyle Factors That Affect HDL and LDL Diet and lifestyle factors, as well as genetic tendencies, affect HDL and LDL levels. The dietary and lifestyle changes that increase HDL cholesterol levels include:

- Physical activity
- Weight loss (if needed)
- Moderate alcohol consumption
- Increased EPA+DHA intake from fish and seafoods
- Not smoking

    Reduced HDL cholesterol levels are associated with:

- High intake of refined carbohydrates and added sugar
- Smoking
- *Trans* fat intake
- Obesity[7,12,14,15]

    Dietary and lifestyle changes that decrease LDL cholesterol levels include:

- Weight loss (if needed)
- Substitution of saturated fats for polyunsaturated fats
- Physical activity
- Frequent intake of whole grain, high-fiber foods
- Routinely high vegetable and fruit intake
- Intake of **plant stanols and sterols** from enriched food products
- Low intake of saturated fats

plant stanols and sterols Substances in corn, wheat, oats, rye, olives, wood, and some other plants that are similar in structure to cholesterol but that are not absorbed by the body. They bind cholesterol and decrease its absorption.

REALITY CHECK
Is Shrimp Off the Menu for People with High Blood Cholesterol Levels?

Who gets thumbs up?

Answers on page 19-6

Lupe: Shrimp has some cholesterol, but it also has good fats like DHA. You can still eat it on a cholesterol-lowering diet.

Sharon: Save your money and bring your cholesterol down. Forget about eating shrimp.

Is Shrimp Off the Menu for People with High Blood Cholesterol Levels?

Shrimp contains cholesterol—about 127 mg in three ounces (around five large shrimp). But that doesn't mean you should not eat it if you're watching your cholesterol intake. This amount of shrimp also contains approximately 268 mg DHA+EP, which are heart healthy, and only a trace of saturated fat. Shrimp intake increases HDL cholesterol more than it increases LDL cholesterol levels, and decreases triglycerides in most people.[45]

Lupe:

Sharon:

Increased LDL cholesterol levels are associated with:

- *Trans* fat intake

- High saturated fat intake

- High cholesterol intake[12,16]

Triglycerides and Heart Disease Risk Triglycerides are transported in blood attached to VLDL (very-low-density lipoprotein) cholesterol. Until recently, the risk of heart disease posed by elevated blood levels of triglycerides had taken a back seat to cholesterol. Research results confirm that high blood levels of triglycerides increase heart disease risk and that efforts to prevent and treat heart disease should include a focus on blood triglyceride levels.[16] In addition to increasing the risk of heart disease, elevated triglyceride levels may signal the development or presence of **metabolic syndrome** or type 2 diabetes. These conditions are characterized by elevated blood levels of triglycerides and glucose, **insulin resistance**, abdominal obesity, and low levels of HDL cholesterol. People with metabolic syndrome are at particularly high risk for heart disease.[14,17]

metabolic syndrome A constellation of metabolic abnormalities that increase the risk of heart disease and type 2 diabetes. It is characterized by insulin resistance, abdominal (central) obesity, high blood pressure and triglyceride levels, low levels of HDL cholesterol, and impaired glucose tolerance.

insulin resistance A condition in which cell membranes have reduced sensitivity to insulin so that more insulin is needed to transport glucose into cells. It is characterized by elevated levels of serum insulin, glucose, triglycerides, and increased blood pressure.

Diet and Lifestyle Factors That Affect Triglyceride Levels Dietary and lifestyle changes that decrease triglyceride level include:

- Weight loss (if needed)

- Increased physical activity

- Intake of fish and seafood; fish oils (EPA+DHA)

- Increased vegetable and fruit intake

- Increased intake of whole grains and high fiber, whole grain products

In most people, plasma triglyceride levels are increased by:

- Weight gain and obesity

- High intake of carbohydrates, particularly added sugars and refined carbohydrates

- Physical inactivity[12,16]

Genetic Effects on Blood Cholesterol Levels Genetic traits play important roles in how diet and exercise affect blood lipid levels. For example, some people are born with one or more genetic traits that reduce cholesterol absorption and therefore blood cholesterol levels. People who absorb cholesterol poorly do not experience the usual increase in blood cholesterol levels when they consume eggs or other high-cholesterol foods.[14] Other individuals have genetic traits that lower HDL cholesterol levels, raise LDL cholesterol levels substantially, or raise triglyceride levels.[18,20] Nutritional deprivation early in life can permanently modify the function of certain genes. Females born to mothers exposed to famine during pregnancy, for instance, have been linked to changes in gene function that promote the development of elevated LDL cholesterol and triglyceride levels later in life.[21]

Hundreds of gene-related risk factors that interact with environmental exposures such as dietary intake and exercise have been identified. Of these, 30 are considered "major" because they are present in 5% or more of the population.[22] Knowledge of individual genetic traits is being used increasingly to guide decisions about dietary and physical activity approaches to the prevention and treatment of heart disease.

Chronic Inflammation and Heart Disease   The link between chronic inflammation and heart disease appears to be related to oxidation reactions and prolonged inflammatory processes.[3] Oxidation and inflammation occur due to the presence of irritants stemming from high blood pressure, high LDL and low HDL cholesterol levels, smoking, the lack of antioxidants, and other factors. This environment prompts the development of oxidized LDL particles that are reactive. It also weakens the endothelium lining allowing the LDL cholesterol particles, white blood cells, and other substances to cross the endothelial lining, combine, and form plaque. Inflammatory reactions take place within the **endothelium**—the layer of cells that line the inside of artery walls (Illustration 19.6). As shown in Illustration 19.7, plaque accumulates within artery walls and causes a thickening, hardening, and narrowing of the arteries.[3] Removal of irritants, and increased availability of antioxidants and anti-inflammatory compounds from foods help reduce oxidation, inflammation, and plaque formation.[3,23]

Nutrients and other components of foods that increase oxidation and decrease chronic inflammation are summarized in Table 19.1. Antioxidants are principally found in plants foods such as fruits, vegetables, and whole grains. Some of the strongest natural antioxidants are phytochemicals that add color to fruits and vegetables. Nutrients in foods that function as antioxidants include vitamins C and E, selenium, and beta-carotene.[2,24] Foods rich in saturated fats such as butter, bacon, and fatty meats; foods with added sugars such as desserts and soft drinks with high added sugar; and refined grain products such as white rice, pasta, and white bread tend to promote inflammation. Fish and seafood, vegetable oils, nuts, whole grain products, and other high-fiber sources of carbohydrates do not promote inflammation and tend to reduce it.[12,25,26]

Who's at Risk for Heart Disease?

Frankly, you may be. Take a look at the Health Action feature and see. Most Americans are at risk for heart disease due to unhealthful diets, high LDL cholesterol or triglyceride level, low HDL cholesterol level, hypertension; or because they smoke, are physically inactive, obese, or have diabetes or high blood pressure (Illustration 19.8).[27] Elevated blood levels

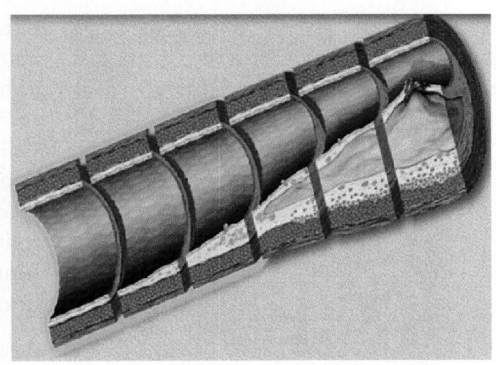

Illustration 19.6   The endothelium consists of cells that form the inside surface of arteries.

Illustration 19.7   Plaque (shown in yellow) forms under the endothelium and thickens, hardens, and narrows the artery.

Table 19.1 Foods that reduce oxidation and chronic inflammation[2,17,23,24]

- Fish and seafood
- Nuts
- Tea, coffee
- Most fruits
- Most vegetables
- Wine
- Whole grains, high-fiber foods
- Vegetable oils

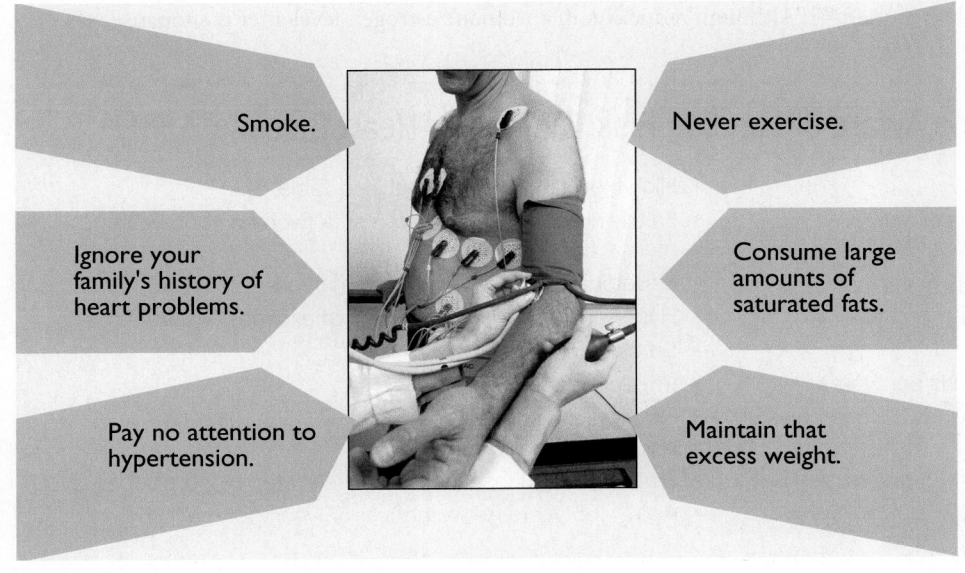

Smoke.

Never exercise.

Ignore your family's history of heart problems.

Consume large amounts of saturated fats.

Pay no attention to hypertension.

Maintain that excess weight.

PhotoDisc

Illustration 19.8   How to have a heart attack.

endothelium (n-dough-theil-e-um) The layer of cells lining the inside of blood vessels.

Table 19.2 Risk categories for total, LDL, and HDL cholesterol levels and triglyceride levels in adults[44]

	Total cholesterol (mg/dL)	LDL cholesterol (mg/dL)	HDL cholesterol (mg/dL)	Triglyceride (mg/dL)
Desirable/optimal	<200	<100	≥50 in women ≥40 in men	<150
Near optimal	—	100–129	—	—
Borderline high	200–239	130–159	(low) <50 in women, <40 in men	150–199
High	240+	160–189	—	200–499
Very high	—	190+	—	500+

of LDL cholesterol, low levels of HDL cholesterol, and high triglyceride levels account for about 50% of the risk for heart disease.[22] Blood levels of these lipids that increase the risk of heart disease have been established (Table 19.2). (Abnormal blood lipid levels by themselves do not cause symptoms so it's important to get them checked every five years.)[28] Unhealthful diets are a major risk factor due to their relationship with blood lipid levels, body fat, type 2 diabetes, and blood pressure.[2,13,22]

Researchers continue to look for additional risk factors because existing ones do not fully account for heart disease development. Some people identified to be at low risk for heart disease due to low LDL cholesterol or high HDL cholesterol, for example, develop heart disease.[2,29]

Are the Risks the Same for Women as for Men? It is generally assumed that risk factors for heart disease identified in men apply to women as well. This assumption, along with the notion that heart disease represents a major health problem only for men, led to the exclusion of women from research studies. One of these assumptions has been shown in studies of women to be largely correct and one incorrect. Major risk factors for heart disease identified for men apply to women as well.[4] Heart disease, however, is a major health problem of women. Indeed, it is the leading cause of death among women.[9]

Several sex-related differences in levels of risk factors have been noted, as have differences in the symptoms of heart attack and the recommended use of low-dose aspirin. Higher levels of HDL cholesterol are more protective against heart disease in women than is the case for men. High triglyceride levels are a stronger predictor of heart disease risk in women. LDL cholesterol levels and blood pressure increase more with age in women than in men.[13,30] Although reduction in a woman's estrogen level after menopause was

health action Become Aware of Leading Risk Factors for Heart Disease[2,3,12,13,16,17]

- Unhealthful diets:
  - High saturated fat intake (≥ 10% of total calories)
  - High *trans* fat intake (≥ 2 grams daily)
  - High added sugar intake (≥ 10% of total calories)
  - High-refined grain products intake (over half of all grain products consumed)
  - Low fish/seafood consumption (< 2 meals per week)
  - Low vegetable and fruit intake (< 3 servings of vegetables, < 2 servings fruit daily)
  - Low fiber intake (< 25 grams women, < 38 grams men daily)
- High LDL cholesterol level (see Table 19.2)
- Low HDL cholesterol level (see Table 19.2)

- High blood triglyceride level (see Table 19.2)
- Family history of early heart attack (women < 65 years, men < 55 years)
- Elevated levels of inflammation markers
- Hypertension (blood pressure that exceeds 140/90 mm Hg)
- Smoking
- Physical inactivity
- Obesity, especially central fat (waist circumference > 35" in women, > 40" in men)
- Diabetes or glucose intolerance (elevated blood glucose levels)
- Age (> 65 years for women, > 55 years for men)

once thought to increase the risk of heart disease, this does not appear to be the case.[31] Symptoms of heart attack tend to differ to an extent between men and women. Both sexes report feeling chest-related pain, but women are more likely to perceive pain in the neck, jaw, and throat than men.[32] Low-dose aspirin is recommended for men at risk of heart disease, but it is not recommended for women.[13]

Diet and Lifestyle in the Prevention and Management of Heart Disease

• **Evaluate components of your diet and lifestyle that help prevent the development of heart disease.**

Many factors are involved in the development of heart disease, so approaches to prevention and treatment need to be broad. Approaches begin and continue with dietary and lifestyle modifications. They include increased physical activity, reduction of high blood pressure and body weight (if needed), drugs in some cases, and smoking cessation if applicable. The goals of heart disease treatment are improved overall health, body weight, blood lipid levels, and inflammation status.[7] The benefits of such changes are clear. Exercising for 150 minutes a week (about 21 minutes daily), for example, can lower triglycerides 20–30%. A 5–10% loss of excess body weight lowers triglycerides an average of 20%, LDL cholesterol around 15%, and increases HDL cholesterol by 8–10%.[13]

Healthful Diets for Heart Disease Prevention and Management

Low-fat diets are no longer recommended for the prevention or treatment of heart disease because they do not reduce heart disease risk nor improve blood lipids levels.[6] Recommendations continue to call for a reduction in saturated fat intake, but saturated fat should be replaced primarily by polyunsaturated fats and not refined carbohydrates and added sugars. Replacement of saturated fats with refined carbohydrate foods and added sugars tends to increase blood triglyceride and lower HDL cholesterol levels.[6,16] The best defense against the risk of heart disease is consumption of an overall healthful diet.[2]

The healthy diet recommended for the prevention and treatment of heart disease centers on the consumption of the following:

• Plant foods.

Diets that consist primarily of vegetables, fruits, whole grains and whole grain products, dried beans, nuts, and vegetable oils are recommended. These foods tend to be good sources of antioxidants, fiber, vitamins, minerals, polyunsaturated fats, and plant stanols and sterols. Because plant stanols and sterols reduce cholesterol absorption and lower LDL cholesterol levels, they are being added to foods from margarine to orange juice, and yogurt drinks to granola bars (Illustration 19.9).[34]

• Limited amounts of animal fats.

Foods such as low-fat dairy products, poultry, fish, and seafood are recommended. Compared to fatty meats such as hamburgers, hot dogs, ribs, bacon, full-fat dairy products, and processed meats, these foods are low in saturated fats and cholesterol. They provide good levels of an array of vitamins and minerals, and some provide anti-inflammatory substances.[13]

• Limited amounts of refined, processed carbohydrates and added sugars.

Foods containing unrefined carbohydrates such as oats, oatmeal, whole grain cereals and bread, fruits, peas, carrots, orange juice, brown rice, barley, and couscous are recommended. These foods are often good sources of fiber and antioxidants, and tend to be digested slowly so they do not quickly release high amounts of glucose or fructose

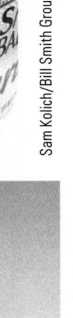

Illustration 19.9 Examples of foods fortified with LDL cholesterol–lowering plant stanols and sterols.

Illustration 19.10
Heart-healthy foods

PhotoDisc

into the blood stream after eating. Refined carbohydrates like white rice, instant mashed potatoes, non–whole grain breakfast cereals, and white bread; and added sugars in sugar-sweetened beverages, candy, cakes, cookies, and pies, include small carbohydrate particles that are quickly digested and ready glucose and fructose for rapid absorption into the blood stream. High blood glucose and fructose levels may increase the body's production of triglycerides and lower HDL cholesterol, especially in people with excess abdominal fat and insulin resistance.[6,17] It is recommended by the American Heart Association's Nutrition Committee that added sugar intake be limited to less than 10% of total calories, and that fructose consumption by individuals with elevated triglycerides be limited to fewer than 400 calories a day.[16] (*Note*: One 12-ounce serving of a soft drink sweetened with high-fructose corn syrup contains about 70 calories of fructose.)

- Fewer saturated fats and more polyunsaturated fats.

Substitution of foods high in saturated fats, such as butter, sausages, full-fat dairy products, and burgers with good sources of polyunsaturated fats such as canola, soybean, or flaxseed oil. Ingestion of salad dressings, soft margarine, fish, seafood, nuts, and lean poultry helps reduce LDL cholesterol.[2,12]

- Moderate amounts of alcohol (if any).

Alcohol consumption of one or two drinks a day helps raise HDL cholesterol levels and has been found to reduce the risk of heart disease. Red wine offers special benefits in that it also provides phytochemicals that act as antioxidants.[12]

Healthy diets recommended for the prevention and management of heart disease also reduce inflammation and the risk of type 2 diabetes, hypertension, and obesity.[2,12] Food choices that are compatible with current recommendations for the dietary treatment of heart disease are summarized in Illustration 19.10 and listed in Table 19.3. When implemented, dietary recommendations for the treatment of heart disease lower plaque formation and improve health and physical fitness levels in a large majority of people.[33]

In the recent past it was thought that supplemental doses of folic acid and vitamins B_{12}, C, and E reduce plaque buildup and decrease inflammation and oxidation. Their roles in the prevention of heart disease are now in doubt. It is not recommended that supplemental doses of these vitamins be used to specifically help prevent or treat heart disease.[13]

Cholesterol-lowering drugs and more aggressive dietary changes are indicated for the treatment of heart disease if blood lipid changes achieved by diet and lifestyle improvements are insufficient.[5]

Statins and the Portfolio Diet Statins have been hailed as wonder drugs for treating heart disease. They are the best-selling, prescription drugs sold in the United States.[35] Known by the names Lipitor, Vytorin, Zetia, and Crestor, statins markedly reduce cho-

Table 19.3 Food choices that promote health and lower heart disease risk[13,16]

- Oils: canola, peanut, olive, soy, safflower, flaxseed
- Fish and shellfish
- Nuts
- Vegetables
- Fruits
- Whole grain products and other high fiber foods
- Lean, low-fat, and unprocessed meats
- Low-fat dairy products
- Dried beans
- Spreads containing plant stanols or sterols

lesterol production in the liver; reduce LDL cholesterol 30–40%, raise HDL cholesterol concentrations, and decrease inflammation. Their use is related to a 30–35% reduction in heart disease and stroke in both women and men.[7,36] Statins are not a substitute for diet and lifestyle changes. They improve blood lipid levels more when combined with dietary and lifestyle changes than when used alone.[11]

Statins are used widely but are expensive. Their use is associated with side effects such as muscle pain and weakness, liver disease, kidney failure; and they may increase the risk of type 2 diabetes.[37,38] The cost and side effects of statins, and the fact that stains do not always achieve desired reductions in heart disease risk factors, have prompted researchers to take a close look at alternatives—like extreme cholesterol-lowering diets.[39] The portfolio diet is one of them. The portfolio diet is vegetarian and based on soy milk and soy-based foods, fiber, oat bran, nuts, plant sterols and stanols, and vegetables and fruits (Table 19.4). Different in a number of ways from many people's usual food choices, the diet may be a challenge to follow. The diet does, however, reduce LDL cholesterol levels substantially.[39,42] This and other therapeutic diets are increasingly viewed as one way to dramatically lower LDL cholesterol levels when standard dietary modifications and statins are not enough.[39]

Looking Toward the Future

Approaches to the prevention and treatment of heart disease have changed rather dramatically over recent years and will continue to evolve. Concerns about the cost of cholesterol-lowering drugs and their side effects, and the availability of low-cost preventive and treatment approaches, will factor into these changes. Approaches that use diet and lifestyle modification, changes in the quality of the food supply in stores and restaurants, and increased consumer involvement in risk reduction may lead the way to higher rates of decline in heart disease.[13]

An end to escalating rates of obesity and physical inactivity would serve our "collective heart" especially well. It is anticipated that escalating rates of child and adolescent obesity in the United States could lead to a 5–16% increase in the prevalence of heart disease by 2035.[43]

Table 19.4 Types of food included in the portfolio diet for substantial reduction in LDL cholesterol[40,41]

	LDL cholesterol lowering
Breakfast	Oat bran cereal with added psyllium fiber
	Soy milk
	Oat bran toast
	Plant-sterol enriched margarine
	Jam, strawberries
Lunch	Bean soup with rice
	Oat bran bread
	Plant-sterol enriched margarine
	Soy milk
Dinner	Spicy sautéed tofu
	Ratatouille
	Cooked barley
	Broccoli, cauliflower
	Plant-sterol enriched margarine
	Soy milk
	Almonds
Snacks	Nuts, soy bar
	Psyllium in juice
	Fruit and raw vegetables

photoshut/Shutterstock.com

NUTRITION
up close

Evaluate Your Dietary and Lifestyle Strengths for Heart Disease Prevention

Focal Point: Assess the positive aspects of your diet and lifestyle on the risk of heart disease.

A number of dietary characteristics and lifestyle behaviors decrease heart disease risk. Give yourself credit by checking those characteristics you call your own.

_____ I exercise at least 21 minutes daily.
_____ I eat three or more servings of colorful vegetables a day.
_____ I eat a serving of fruit at least twice a day.
_____ I don't smoke.
_____ I am normal weight or on my way to becoming that way.
_____ I eat more than two servings of whole grains or whole grain foods daily.
_____ I consume low- or no-fat milk when I drink milk.
_____ I eat fish or seafood twice a week or more.
_____ I eat nuts for snacks or otherwise consume them regularly.
_____ I choose lean meats if I eat meat.
_____ I don't drink more than one sugar-sweetened beverage a day.

Feedback to the Nutrition Up Close is located in Appendix G.

REVIEW QUESTIONS

• **List dietary, lifestyle, and biological factors that contribute to heart disease development.**

1. The term *cardiovascular disease* refers to disorders related to plaque buildup in arteries of the heart, brain, and other organs and tissues.   True/False

2. High levels of LDL cholesterol increase the risk of heart disease just as low levels of HDL cholesterol do. True/False

3. High levels of total cholesterol in the blood always represent a major risk factor for heart disease.   True/False

4. High levels of HDL cholesterol are more protective against heart disease in women than in men.   True/False

5. The major drawback to heart-healthy diets is that they help prevent heart disease only.   True/False

6. Healthful diets include ample vegetables and fruits, whole grain products, fish, and lean, unprocessed meats. True/False

7. Consumption of a low-fat diet is the most important element of diets that reduce the risk of heart disease. True/False

8. Metabolic syndrome is associated with an increased risk of heart disease.   True/False

9. Nutritional deprivation during pregnancy can modify the function of genes that influence blood lipid levels in offspring.   True/False

10. _____ Food sources of antioxidants include:
    a. colorful vegetables and lean meats
    b. whole grains and fruits
    c. food high in added sugars and fish
    d. regular soft drinks and tea

11. _____ Foods that tend to decrease chronic inflammation include:
    a. bacon and fatty meats
    b. white rice and pasta
    c. foods high in added sugars and fish
    d. nuts and vegetable oils

12. _____ Healthful diets are characterized by the regular consumption of foods such as:
    a. olive oil and vinegar
    b. potatoes and flaxseeds
    c. whole grain breakfast cereal and low-fat milk
    d. deep-fried fish and seafood

• **Evaluate components of your diet and lifestyle that help prevent the development of heart disease.**

13. _____ Diet and lifestyle factors associated with increases in blood HDL cholesterol level include:
    a. regular intake of high-carbohydrate and high-protein foods
    b. regular physical activity and fish consumption
    c. weight gain and low levels of physical activity
    d. increased intake of unsaturated fats and cholesterol

14. _____ Blood LDL cholesterol levels are decreased by:
    a. limiting intake of saturated fats and refined carbohydrates
    b. physical inactivity and fish intake
    c. high-protein, low-fat diets
    d. high saturated and *trans* fat intake

15. _____ Blood triglyceride levels can be reduced by:
    a. weight gain and obesity
    b. increased intake of carbohydrates and cholesterol
    c. moderate alcohol consumption and physical inactivity
    d. increased intake of fish and seafood, increased vegetable and fruit intake

The following five questions refer to this case scenario:

Shine is a 34-year-old junior marketing executive. Since she took this high-pressure job a year ago, Shine gained 15 pounds and became overweight. She recently decided to take advantage of the health screening her company offered and found out that her blood LDL cholesterol is 115 mg/dL, her HDL cholesterol is 42 mg/dL, and her triglyceride level is 168 mg/dL.

16. _____ Shine's blood level of LDL cholesterol of 115 mg/dL is:
    a. desirable/optimal
    b. near optimal
    c. borderline high
    d. very high

17. _____ Shine's blood level of HDL cholesterol of 42 mg/dL is:
    a. desirable/optimal
    b. near optimal
    c. low
    d. very low

18. ____ Shine's blood level of triglycerides of 168 mg/dL is:

   a. desirable/optimal
   b. near optimal
   c. borderline high
   d. very high

19. ____ Which of the following lifestyle changes would best help raise Shine's HDL cholesterol level?

   a. consumption of less saturated fat and refined grain products
   b. consumption of less fat and more high carbohydrate foods
   c. increased physical activity and weight loss
   d. use of vitamin C and vitamin E supplements

20. ____ Shine decides to work on lowering her triglyceride level. Which action would likely help her do that?

   a. decreased cholesterol intake
   b. increased intake of fish and seafood
   c. increased intake of low-fat dairy products
   d. decreased intake of unsaturated fat

Answers to these questions can be found in Appendix G.

NUTRITION SCOREBOARD ANSWERS

1. About one in four deaths of men and women across the globe are related to heart disease.[1]? **True**

2. Low-fats diets are no longer recommended for the prevention or treatment of heart disease.[2] **False**

3. Plaque develops underneath the inside surface of artery walls. **True**

4. Chronic inflammation in artery walls plays a major role in the development and progression of heart disease. Diet is a major factor influencing inflammation.[3] **True**

5. Major risk factors for heart disease do not differ between women and men.[4] **True**

UNIT 20

Vitamins and Your Health

NUTRITION SCOREBOARD

1 The only documented benefit of consuming sufficient amounts of vitamins is protection against deficiency diseases.   True/False

2 Vitamins provide energy.   True/False

3 Vitamin C is found only in citrus fruits.   True/False

4 Nearly all cases of illness due to excessive intake of vitamins result from the overuse of vitamin supplements.   True/False

Answers can be found at the end of the unit.

After completing Unit 20 and its interactive learning features, you will be able to:

- Identify the vitamins, their functions and food sources, and their effects on health.

- Assess the adequacy of your diet for vitamins that function as antioxidants.

vitamins Components of food required by the body in small amounts for growth and health maintenance.

Vitamins: They're on Center Stage

- **Identify the vitamins, their functions and food sources, and their effects on health.**

- **Assess the adequacy of your diet for vitamins that function as antioxidants.**

These are exciting times for people interested in **vitamins**. Long thought of as needed to prevent deficiency diseases, vitamins are now being viewed from a different vantage point. These essential nutrients clearly do more than the important job of protecting us from vitamin deficiency diseases. Vitamins also play roles in protecting the body from a host of ill effects, ranging from certain birth defects to osteoporosis.[1] This unit addresses the functions of vitamins in the body, their food sources, and health consequences related to the consumption of too little or too much of individual vitamins.

Key Nutrition Concepts

A number of key nutrition concepts underlie the content presented on vitamins, including:

- Foods provide energy (calories), nutrients, and other substances needed for growth and health.

- Poor nutrition can result from both inadequate and excessive levels of nutrient intake.

- Humans have adaptive mechanisms for managing fluctuations in nutrient intake.

Vitamin Facts

Vitamins are chemical substances that perform specific functions in the body. They are essential nutrients because, in general, the body cannot produce them or produce sufficient amounts of them. If we fail to consume enough of any of the vitamins, specific deficiency diseases develop.

Fourteen vitamins have been discovered so far, and they are listed in Table 20.1. It is possible that a few more substances will be added to the list of vitamins in years to come.

Water- and Fat-Soluble Vitamins Vitamins come in two basic types—those soluble in water (the B-complex vitamins and vitamin C) and those that dissolve in fat (vitamins D, E, K, and A, or the "deka" vitamins). Their key features are summarized in Table 20.2. With the exception of vitamin B_{12}, the water-soluble vitamins can be stored in the body only in small amounts. Consequently, deficiency symptoms generally develop within a few weeks to several months after the diet becomes deficient in water-soluble vitamins. Vitamin B_{12} is unique in that the body can build up stores that may last for a year or more after intake of the vitamin stops. Of the water-soluble vitamins, niacin, vitamin B_6, choline, and vitamin C are known to produce ill effects if consumed in excessive amounts.

Table 20.1 Fourteen vitamins are known to be essential for health. Ten are water-soluble and four are fat-soluble.

The water-soluble vitamins	The fat-soluble vitamins
B-complex vitamins	Vitamin A (retinol) (provitamin is beta-carotene)
Thiamin (B_1)	Vitamin D (1,25 dihydroxy-cholecalciferol)
Riboflavin (B_2)	Vitamin E (tocopherol)
Niacin (B_3)	Vitamin K (phylloquinone, menaquinone)
Vitamin B_6 (pyridoxine)	
Folate (folacin, folic acid)	
Vitamin B_{12} (cyanocobalamin)	
Biotin	
Pantothenic acid (pantothenate)	
Choline	
Vitamin C (ascorbic acid)	

The fat-soluble vitamins are primarily stored in body fat and the liver. Because the body is better able to store these vitamins, deficiencies of fat-soluble vitamins generally take longer to develop than deficiencies of water-soluble vitamins when intake from food is too low.

Bogus Vitamins Some substances are called "vitamins" even though they are not actually vitamins. A number of them are listed in Table 20.3. These bogus vitamins are called vitamins by some manufacturers of supplements, weight-loss products, and cosmetics. Although the label may help the product to sell, the ingredients aren't essential and therefore cannot be considered vitamins. People do not develop deficiency diseases when they consume too little of the bogus vitamins.

What Do Vitamins Do?

For starters, vitamins *don't* provide energy or, with the exception of choline, serve as components of body tissues such as muscle and bone. A number of vitamins do play critical roles as **coenzymes** in the conversion of proteins, carbohydrates, and fats into energy. Coenzymes are also involved in reactions that build and maintain body tissues such as bone, muscle, and red blood cells. Thiamin, for example, is needed for reactions that convert glucose into energy. People who are thiamin deficient tire easily and feel weak (among other things). Folate, another B-complex vitamin, is required for reactions that build body proteins. Without enough folate, proteins such as those found in red blood cells form abnormally and function poorly. Vitamin A is needed for reactions that generate new cells to replace worn-out cells lining the mouth, esophagus, intestines, and eyes. Without enough vitamin A, old cells aren't replaced, and the affected tissues are damaged. And vitamin C is required for reactions that build and maintain collagen, a protein found in skin, bones, blood vessels, gums, ligaments, and cartilage. Approximately 30% of the total amount of protein in the body is collagen. With vitamin C deficiency, collagen becomes weak, causing tissues that contain collagen to weaken and bleed easily.

These examples all relate to the physical effects of vitamins. Vitamins participate in reactions that affect behaviors, too. Alterations in behaviors such as reduced attention span, poor appetite, irritability, depression, or paranoia often precede the physical signs of vitamin deficiency.[2] Vitamins are truly "vital" for health.

coenzymes Chemical substances, including many vitamins, that activate specific enzymes. Activated enzymes increase the rate at which reactions take place in the body, such as the breakdown of fats or carbohydrates in the small intestine and the conversion of glucose and fatty acids into energy within cells.

Vitamins and the Prevention and Treatment of Disorders

Current research on vitamins centers around their effects on disease prevention and treatment. It is a very active area of research and, no doubt, new and important advances in knowledge about the vitamins will be gained. Here are a few examples of developments in research on vitamins and disease prevention and treatment.

Folate and Neural Tube Defects Folate plays key roles during pregnancy in the synthesis of proteins needed for the normal development of fetal tissues, including the spinal cord and brain. When folate status of women early in pregnancy is poor, the neural tube (that develops into the spinal cord and brain) may form abnormally and incompletely. Illustration 20.1 shows a photograph of spina bifida, an example of a potential outcome from inadequate folate status early in pregnancy. Daily consumption of 400 micrograms (mcg) of folic acid (the synthetic form of folate added to refined grain products) before and early in pregnancy significantly reduces the incidence of neural tube defects.[3,4]

In 1998 manufacturers started fortifying refined grain products such as bread, pasta, and rice with folic acid. The addition of folic acid to these foods has produced substantial gains in people's folate status. Blood levels of folate have nearly doubled, the prevalence of poor folate status has declined by 84%, and the incidence of neural tube defects has been reduced by 46%.[3,5] The folic acid intake of women of childbearing age, however, is often low, averaging 166 mcg per day compared to the recommended intake of 400 mcg.

Food sources of folate and other vitamins are listed in Table 20.5 that begins on page 20-15.

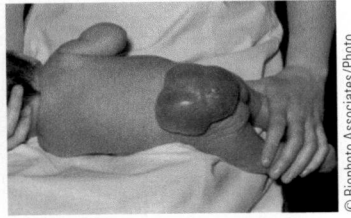

© Biophoto Associates/Photo Researchers, Inc.

Illustration 20.1 A baby with spina bifida, a form of neural tube defect associated with poor folate status early in pregnancy.

Table 20.2 An intensive course on vitamins

	The water-soluble vitamins	
	Primary functions	**Consequences of deficiency**
Thiamin (vitamin B$_1$) AI^a women: 1.1 mg men: 1.2 mg	• Helps body release energy from carbohydrates ingested • Facilitates growth and maintenance of nerve and muscle tissues • Promotes normal appetite	A look at beriberi, a thiamin-deficiency disease.	• Fatigue, weakness • Nerve disorders, mental confusion, apathy • Impaired growth • Swelling • Heart irregularity and failure
Riboflavin (vitamin B$_2$) AI women: 1.1 mg men: 1.3 mg	• Helps body capture and use energy released from carbohydrates, proteins, and fats • Aids in cell division • Promotes growth and tissue repair • Promotes normal vision		• Reddened lips, cracks at both corners of the mouth • Fatigue
Niacin (vitamin B$_3$) RDA women: 14 mg men: 16 mg UL: 35 mg (from supplements and fortified foods)	• Helps body capture and use energy released from carbohydrates, proteins, and fats • Assists in the manufacture of body fats • Helps maintain normal nervous system functions	Pellagra: the niacin-deficiency disease.	• Skin disorders • Nervous and mental disorders • Diarrhea, indigestion • Fatigue
Vitamin B$_6$ (pyridoxine) AI women: 1.3 mg men: 1.3 mg UL: 100 mg	• Needed for reactions that build proteins and protein tissues • Assists in the conversion of tryptophan to niacin • Needed for normal red blood cell formation • Promotes normal functioning of the nervous system	• Irritability, depression • Convulsions, twitching • Muscular weakness • Dermatitis near the eyes • Anemia • Kidney stones	

^{a}(Adequate Intakes) and RDAs (Recommended Dietary Allowances) are for 19–30-year-olds; UL (Upper Limits) are for 19–70-year-olds.

The water-soluble vitamins		
Consequences of overdose	**Primary food sources**	**Highlights and comments**
• High intakes of thiamin are rapidly excreted by the kidneys. Oral doses of 500 mg/day or less are considered safe.	• Grains and grain products (cereals, rice, pasta, bread) • Pork • Nuts	• Need increases with carbohydrate intake • There is no "e" on the end of thiamin! • Deficiency rare in the United States; may occur in people with alcoholism • Enriched grains and cereals prevent thiamin deficiency
• None known. High doses are rapidly excreted by the kidneys.	• Milk, yogurt, cheese • Grains and grain products (cereals, rice, pasta, bread) • Liver, fish, beef • Eggs	• Destroyed by exposure to light (that's why milk comes in opaque containers)
• Flushing, headache, cramps, rapid heart-beat, nausea, diarrhea, decreased liver function with doses above 0.5 g per day	• Meats (all types), fish • Grains and grain products (cereals, rice, pasta, bread) • Nuts	• Niacin has a precursor–tryptophan. Tryptophan, an amino acid, is converted to niacin by the body. Much of our niacin intake comes from tryptophan. • High doses raise HDL-cholesterol levels, lower LDL-cholesterol and triglycerides.
• Bone pain, loss of feeling in fingers and toes, muscular weakness, numbness, loss of balance (mimicking multiple sclerosis)	• Meats (all types) • Breakfast cereals • Bananas, avocados • Potatoes, brussels sprouts, sweet peppers	• Vitamins go from B_3 to B_6 because B_4 and B_5 were found to be duplicates of vitamins already identified.
		(continued)

VITAMINS: THEY'RE ON CENTER STAGE **20-5**

Table 20.2 **(continued)**

The water-soluble vitamins		
	Primary functions	**Consequences of deficiency**
Folate (folacin, folic acid) RDA. women: 400 mcg men: 400 mcg UL: 1,000 mcg (from supplements and fortified foods)	• Needed for reactions that utilize amino acids (the building blocks of protein) for protein tissue formation • Promotes the normal formation of red blood cells	• Megaloblastic anemia • Diarrhea • Red, sore tongue Normal red blood cells. • Increased rise of neural tube defects, and other malformations, preterm delivery • Elevated blood levels of homocysteine Red blood cells in megaloblastic anemia.
Vitamin B$_{12}$ (cyanocobalamin) AI women: 2.4 mcg men: 2.4 mcg	• Helps maintain nerve tissues • Aids in reactions that build up protein tissues • Needed for normal red blood cell development	• Neurological disorders (nervousness, tingling sensations, brain degeneration) • Pernicious anemia • Fatigue • Elevated blood level of homocysteine
Biotin AI women: 30 mcg men: 30 mcg	• Needed for the body's manufacture of fats, proteins, and glycogen	• Seizures • Vision problems • Muscular weakness • Hearing loss
Pantothenic acid (pantothenate) AI women: 5 mg men: 5 mg	• Needed for the release of energy from fat and carbohydrates	• Fatigue, sleep disturbances, impaired coordination • Vomiting, nausea
Vitamin C (ascorbic acid) RDA women: 75 mg men: 90 mg UL: 2,000 mg	• Needed for the manufacture of collagen • Helps the body fight infections, repair wounds • Acts as an antioxidant • Enhances iron absorption	 Gums that are swollen and bleed easily are signs of scurvy, the vitamin C-deficiency disease. • Bleeding and bruising easily due to weakened blood vessels, cartilage, and other tissues containing collagen • Slow recovery from infections and poor wound healing • Fatigue, depression
Choline AI women: 425 mg men: 550 mg UL: 3.5 g	• Serves as a structural and signaling component of cell membranes • Required for the normal development of memory and attention processes during early life • Required for the transport and metabolism of fat and cholesterol	• Fatty liver • Infertility • Hypertension

The water-soluble vitamins		
Consequences of overdose	Primary food sources	Highlights and comments
• May mask signs of vitamin B_{12} deficiency (pernicious anemia)	• Fortified, refined grain products (cereals, bread, pasta) • Dark green vegetables (spinach, collards, romaine) • Dried beans	• *Folate* means "foliage." It was first discovered in leafy green vegetables. • This vitamin is easily destroyed by heat. • Synthetic form (folic acid) added to fortified grain products is better absorbed than naturally occurring folates. • Some people have genetic traits that increase their need for folates. • May protect against colon cancer in early stages of development.
• None known. Excess vitamin B_{12} is rapidly excreted by the kidneys or is not absorbed into the bloodstream. • Vitamin B_{12} injections may cause a temporary feeling of heightened energy.	• Fish, seafood • Meat • Milk and cheese • Ready-to-eat cereals	• Older people, those who have had stomach surgery, and vegans are at risk for vitamin B_{12} deficiency. • Some people become vitamin B_{12} deficient because they are unable to absorb it. • Vitamin B_{12} is found in animal products and microorganisms only.
• None known. Excesses are rapidly excreted.	• Grain and cereal products • Meats, dried beans, cooked eggs • Vegetables	• Deficiency is extremely rare. May be induced by the overconsumption of raw eggs.
• None known. Excesses are rapidly excreted.	• Many foods, including meats, grains, vegetables, fruits, and milk	• Deficiency is very rare.
• Intakes of 1 g or more per day can cause nausea, cramps, and diarrhea and may increase the risk of kidney stones.	• Fruits: guava, oranges, lemons, limes, strawberries, cantaloupe, grapefruit, kiwi fruit • Vegetables: broccoli, green and red peppers, collards, tomato, potatoes • Ready-to-eat cereals	• Need increases among smokers (to 110–125 mg per day). • Is fragile; easily destroyed by heat and exposure to air. • Supplements may decrease duration and symptoms of colds in some people. • Deficiency may develop within three weeks of very low intake.
• Low blood pressure • Sweating, diarrhea • Fishy body odor • Liver damage	• Meat (all types) • Eggs • Dried beans • Milk	• Most of the choline we consume from foods comes from its location in cell membranes. • Lecithin, an additive commonly found in processed foods, is a rich source of choline. • Choline is primarily found in animal products. • It is considered a B-complex vitamin.

(continued)

Table 20.2 (*continued*)

	The fat-soluble vitamins		
	Primary functions		**Consequences of deficiency**

Vitamin A
I. Retinol
RDA
   women: 700 mcg
   men: 900 mcg
   UL: 3,000 mcg
	• Needed for the formation and maintenance of mucous membranes, skin, bone
	• Needed for vision in dim light
		
		Xerophthalmia. *Vitamin A deficiency is the leading cause of blindness in developing countries.*
			• Increased incidence and severity of infectious diseases
			• Impaired vision, blindness
			• Inability to see in dim light
			• Blindness

2. Beta-carotene (a vitamin A precursor or "provitamin") No RDA; suggested intake: 6 mg	• Acts as an antioxidant; prevents damage to cell membranes and the contents of cells by repairing damage caused by free radicals		• Deficiency disease related only to lack of vitamin A

Vitamin E
(alpha-tocopherol)
RDA women: 15 mg
   men: 15 mg
   UL: 1,000 mg
	• Acts as an antioxidant, prevents damage to cell membranes in blood cells, lungs, and other tissues by repairing damage caused by free radicals
	• Reduces the ability of LDL cholesterol (the "bad" cholesterol) to form plaque in arteries
	• Participates in the regulation of gene expression
			• Muscle loss, nerve damage
			• Anemia
			• Weakness

Vitamin D
(vitamin D$_2$ =
ergocalciferol,
vitamin D$_3$ =
cholecalciferol)
RDA
   women: 15 mcg
     (600 IU)
   men: 15 mcg
     (600 IU)
   UL: 100 mcg
     (4,000 IU)
	• Needed for the absorption of calcium and phosphorus, and for their utilization in bone formation, nerve and muscle activity.
	• Inhibits inflammation
	• Involved in insulin secretion and blood glucose level maintenance
		
		The vitamin D-deficiency disease: rickets
			• Weak, deformed bones (children)
			• Loss of calcium from bones (adults), osteoporosis
			• Increased risk of chronic inflammation

The fat-soluble vitamins		
Consequences of overdose	**Primary food sources**	**Highlights and comments**
• Vitamin A toxicity (hypervitaminosis A) with acute doses of 500,000 IU, or long-term intake of 50,000 IU per day. • Nausea, irritability, blurred vision, weakness, headache • Increased pressure in the skull, hip fracture • Liver damage • Hair loss, dry skin • Birth defects	• Vitamin A is found in animal products only. • Liver, clams, low-fat milk, eggs • Ready-to-eat cereals	• Symptoms of vitamin A toxicity may mimic those of brain tumors and liver disease. Vitamin A toxicity is sometimes misdiagnosed because of the similarities in symptoms. • 1 mcg vitamin A = 3.33 IU • Forms of retinol are used in the treatment of acne and skin wrinkles due to overexposure to the sun. Brittle hair and dry, rough, scaly, and cracked skin from vitamin A overdose. From: *American Journal of Clinical Nutrition,* Vol. 71, No. 4, 878–884. April 2000, Robert M. Russell.
• High intakes from supplements may increase lung damage in smokers. • With high intakes and supplemental doses (over 12 mg/day for months), skin may turn yellow-orange.	• Deep orange, and dark green vegetables. • Carrots, sweet potatoes, pumpkin, spinach, collards, cantaloupe, apricots, vegetable juice	• The body converts beta-carotene to vitamin A. Other carotenes are also present in food, and some are converted to vitamin A. Beta-carotene and vitamin A perform different roles in the body. • May decrease sunburn in certain individuals.
• Intakes of up to 800 IU per day are unrelated to toxic side effects; over 800 IU per day may increase bleeding (blood-clotting time). • Avoid supplement use if aspirin, anti-coagulants, or fish oil supplements are taken regularly.	• Nuts and seeds • Vegetable oils • Salad dressings, mayonnaise, • Whole grains, wheat germ • Leafy, green vegetables, asparagus	• Vitamin E is destroyed by exposure to oxygen and heat. • Oils naturally contain vitamin E. It's there to protect the fat from breakdown due to free radicals. • Eight forms of vitamin E exist, and each has different antioxidant strength. • 1 mg vitamin E = 1.49 IU • Intakes in the United States tend to be low.
• Mental retardation in young children • Abnormal bone growth and formation • Nausea, diarrhea, irritability, weight loss • Deposition of calcium in organs such as the kidneys, liver, and heart • Toxicity possible with long-term use of 10,000 IU daily	• Vitamin D-fortified milk, cereals, and other foods • Fish	• Vitamin D_3, the most active form of the vitamin, is manufactured from a form of cholesterol in skin cells upon exposure to ultraviolet rays from the sun. • Inadequate vitamin D status is common. • Breast-fed infants with little sun exposure benefit from vitamin D supplements. • 1 mcg vitamin D = 40 IU.

(continued)

Table 20.2 (continued)

The fat-soluble vitamins		
	Primary functions	**Consequences of deficiency**
Vitamin K (phylloquinone, menaquinone) AI women: 90 mcg men: 120 mcg	• Is an essential component of mechanisms that cause blood to clot when bleeding occurs • Aids in the incorporation of calcium into bones	© Photo Researchers, Inc The long-term use of antibiotics can cause vitamin K deficiency. People with vitamin K deficiency bruise easily.	• Bleeding, bruises • Decreased calcium in bones • Deficiency is rare. May be induced by the long-term use (months or more) of antibiotics.

Table 20.3 Nonvitamins
The "real" vitamins are listed in Table 20.1. These are some of the more popular nonvitamins.

Bioflavonoids (vitamin P)

Coenzyme Q_{10}

Gerovital H-3

Hesperidin

Inositol

Laetrile (vitamin B_{17})

Lecithin

Lipoic acid

Nucleic acids

Pangamic acid (vitamin B_{15})

Para-amino benzoic acid (PABA)

Provitamin B_5 complex

Rutin

chronic inflammation Low-grade inflammation that lasts weeks, months, or years. Inflammation is the first response of the body's immune system to infection or irritation. Inflammation triggers the release of biologically active substances that promote oxidation and other potentially harmful reactions in the body.

Vitamin A, Infectious Diseases, Acne, and Wrinkles Studies undertaken in both developing countries and in the United States indicate that adequate vitamin A status helps prevent and decreases the severity of measles and other infectious diseases. It has been known for decades that adequate vitamin A intake also prevents blindness, an all-too-common consequence of vitamin A deficiency in developing nations.[6] Vitamin A is needed for the synthesis of substances that keep the outer layer of the eye moist and resistant to infection. Without sufficient vitamin A, eyes dry out, become susceptible to infection, and cloud over.

Forms of vitamin A are used in the treatment of acne and skin wrinkles and blotches due to overexposure to the sun.[7,8] Very high intakes of vitamin A as retinol early in pregnancy (but not high intakes of the vitamin A-precursor beta-carotene) are related to the development of specific birth defects. Use of vitamin A-derived medications by women who may become or are pregnant should not occur.[9]

Vitamin D: From Osteoporosis to Chronic Inflammation Vitamin D is best known as the sunshine vitamin that helps build strong bones by facilitating the absorption and utilization of calcium. Vitamin D does much more than that, however. It plays key roles as a hormone in combating **chronic inflammation**. Low-grade, chronic inflammation is at the core of the development of disorders such as type 2 and type 1 diabetes, cardiovascular disease, multiple sclerosis, certain cancers, and rheumatoid arthritis.[10] Vitamin D reduces inflammation by entering cells and turning genes that produce inflammatory substances "off," and those that produce substances that reduce inflammation "on."[11] It also functions in the regulation of insulin secretion and blood glucose level regulation.[12]

Recommended Intake of Vitamin D Based on research results related to optimal doses of vitamin D for bone formation and maintenance, the recommended intake of vitamin D was increased in 2011 from 5 mcg (200 IU) to 15 mcg (600 IU) daily for adult women and men.[13] The recommended levels of intake of vitamin D reflect amounts needed from the diet and does not include vitamin D manufactured in the skin from exposure to the ultraviolet rays from the sun (Illustration 20.2). Most people do not consume the recommended amount of vitamin D from their diet, even if fortified foods are consumed, and get the majority of their vitamin D from exposure of their skin to direct sunlight.[14] Vitamin D is produced in the skin when the energy from ultraviolet rays from the sun is absorbed in skin cells. The energy absorbed initiates the conversion of a derivative of cholesterol in cells to an active form of vitamin D.[15]

The fat-soluble vitamins		
Consequences of overdose	**Primary food sources**	**Highlights and comments**
• Toxicity is only a problem when synthetic forms of vitamin K are taken in excessive amounts. That may cause liver disease.	• Leafy, green vegetables • Grain products	• Vitamin K is produced by bacteria in the gut. Part of our vitamin K supply comes from these bacteria. • Newborns are given vitamin K because they have "sterile" guts and consequently no vitamin K–producing bacteria. • Deficiency is rare.

Meeting the Need for Vitamin D through Foods and the Sun With the exception of fish, few foods are naturally rich in vitamin D. Consequently, most of the vitamin D in our diets comes from vitamin D–fortified foods, such as those shown in Illustration 20.3.[14,16] The availability of food products fortified with vitamin D is steadily increasing, and the increased availability of these foods will help increase vitamin D intake.[1] Because most types of yogurt, cheese, cottage cheese, ice creams, and dairy products other than milk are not fortified with vitamin D, it's important to look at product labels. Vitamin D–fortified foods can be identified from the nutrition information label on food packages. A good source of vitamin D would provide 10% or more of the % Daily Value in a serving. There are several ways individuals can increase their intakes of vitamin D and production of vitamin D in the skin. The Take Action feature in this unit provides a list of options for doing that.

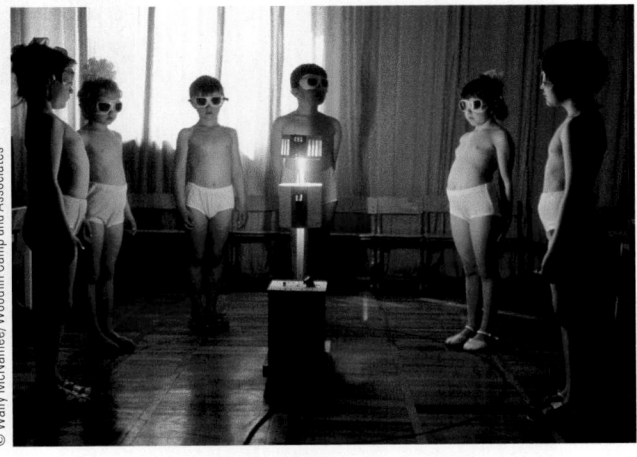

Illustration 20.2 Russian children are exposed to a quartz UV lamp to prevent vitamin D deficiency during the long winter.

© Wally McNamee/Woodfin Camp and Associates

The amount of vitamin D produced in the skin from exposure to direct sunlight varies depending on the lightness or darkness of skin color and the intensity of the sun's ultraviolet rays that reach the skin. Individuals with darker skin produce less vitamin D given the same circumstances than people with lighter skin because ultraviolet rays are less able to penetrate darker skin. Production of vitamin D in the skin is very low to zero in parts of the world during winter when sunlight is indirect, and higher in warmer parts of the world closer to the equator. The use of broad-spectrum suntan lotions with SPF factors of 15 and higher successfully block vitamin D production by absorbing the energy from the ultraviolet rays. Energy from ultraviolet rays does not pass through glass, windows, and plastic.[14,15]

Scientists have estimated that exposing the whole body to direct sunlight for 10 to 15 minutes generates around 500 mcg (20,000 IU) of vitamin D.[17] For most people, exposing the arms and legs to sunlight for 5 to 10 minutes between 10:00 a.m. and 5:00 p.m. two to three times a week is recommended as a sensible approach to getting enough vitamin D from the sun.[15] Maximum production of vitamin D in the skin is achieved before changes in skin color occur. You can get a healthy dose of vitamin D without risking a sunburn and skin damage.[18] And you cannot get too much vitamin D from the sun. Production of the vitamin stops when adequate amounts have been produced for use and for storage in body fat.[17] Vitamin D–rich food and supplements are recommended for individuals whose skin

take action To Improve Your Vitamin D Status

How do you become a person that stands out from the average in terms of vitamin D adequacy? If you're not that person already, here are some tips that will get you on your way.

Choose and check at least two of the following options for getting more vitamin D that appeal to you.
I would:

_____ substitute a cup of skim milk for a sweetened beverage at one meal or snack a day.

_____ eat salmon once a week at dinner.

_____ select a vitamin D–fortified orange juice when I buy orange juice.

_____ buy or select and consume vitamin D–fortified breakfast cereals.

_____ take a vitamin D supplement (400–600 IU) daily until I am able to get enough vitamin D in my diet or by brief exposure to direct sunshine.

_____ exercise or walk in sunshine for 10 minutes three times a week when the weather is warm while wearing shorts and a top.

Illustration 20.3 More foods are now fortified with Vitamin D than in the past. Check the labels on these foods for vitamin D fortification levels.

is sensitive to even short durations of sun exposure, and for individuals who, for other reasons, do not expose their skin to direct sunlight.[19]

The Antioxidant Vitamins

Vitamin C and the Common Cold A review of research studies confirm that vitamin C, the popular cold remedy, reduces the symptoms and duration of the common cold modestly but does *not* affect how often colds occur.[20]

Beta-carotene (a **precursor** to vitamin A), vitamin E, and vitamin C function as **antioxidants**. This means they prevent or repair damage to components of cells caused by exposure to **free radicals**. A free radical primarily results when an atom of oxygen loses an electron. Without the electron, there is an imbalance between the atom's positive and negative charges. This makes the atom reactive—it needs to steal an electron from a nearby atom or molecule to reestablish a balance between its positive and negative charges. Free radicals play a number of roles in the body, so they are always present. They are produced during energy formation, by breathing, and by the immune system to help destroy bacteria and viruses that enter the body. They can also be formed when the body is exposed to alcohol, radiation emitted by the sun, smoke, ozone, smog, and other environmental pollutants.

Atoms and molecules that have lost electrons to free radicals are said to be "oxidized." Oxidized substances can damage lipids, cell membranes, DNA, and other cell compo-

precursor In nutrition, a nutrient that can be converted into another nutrient (also called provitamin). Beta-carotene is a precursor of vitamin A.

antioxidants Chemical substances that prevent or repair damage to cells caused by exposure to free radicals. Beta-carotene, vitamin E, and vitamin C function as antioxidants.

free radicals Chemical substances (usually oxygen) that are missing an electron. The absence of the electron makes the chemical substances reactive and prone to oxidizing nearby atoms or molecules by stealing an electron from them.

nents. Antioxidants such as beta-carotene, vitamin E, and vitamin C donate electrons to stabilize oxidized molecules or repair them in other ways.[21] Illustration 20.4 shows visible effects of an oxidation and an antioxidation reaction.

Consumption of foods rich in beta-carotene, vitamin C, and vitamin E decrease the risk of heart disease, stroke, and cancer. Intake of these antioxidant vitamins in supplements, however, does not have the same protective effects.[1,22,23] Fruits, vegetables, whole grains, and other plant foods contain a variety of naturally occurring antioxidants that work together with the antioxidant vitamins and other nutrients in disease prevention.[22]

Controlling oxidation reactions is a very important process that does not solely rely on our intake of antioxidant vitamins. The body's sources of antioxidants also include colorful pigments in vegetables and fruits, and enzymes produced by the body that function as antioxidants.[21]

Illustration 20.4 Apple slices exposed to air turn brown due to oxidation. Coating the slices with lemon juice, a rich source of vitamin C, reduces the oxidation.

Preserving the Vitamin Content of Foods

The vitamin content of foods can be affected by food preparation and storage methods (Table 20.4). Food storage and preparation methods that involve heat lead to higher losses of heat-sensitive vitamins such as vitamin C and folate, and low or no loss of vitamin B_{12} and choline, for example, which are much less sensitive to heat. Vitamins in foods boiled in water can be lost down the drain if the water is thrown out. (See the nearby Reality Check for an example of this.) Vitamins in foods dissolve or are released into cooking water to some extent. In general, boiling or steaming foods using a small amount of water, using the cooking water in soups, stews, or sauces, or stir-frying lead to superior vitamin retention.[24]

Table 20.4 Percent of original vitamin content lost in fruits by food storage method and in dried beans by cooking method

	Fruits			Dried beans boiled 2–2.5 hours	
	Canned (%)	Frozen (%)	Dried (%)	Water drained (%)	Water used (%)
Vitamin C	50%	30%	80%	35%	30%
Thiamin	15	10	10	60	55
Riboflavin	5	5	5	25	20
Niacin	10	5	5	45	40
Vitamin B_6	10	10	5	50	45
Folate	35	25	15	70	65
Choline	0	0	0	0	0
Vitamin B_{12}	0	0	0	0	0
Vitamin A	5	5	10	15	10

Source: Table prepared by author from data presented in USDA's Nutrient retention in foods tables.

REALITY CHECK
To Peel or Not to Peel?

Jolene is peeling potatoes for dinner when she gets a tap on her shoulder from her mother. "Thanks you for helping with dinner, Sweetie, but quit peeling the potatoes! That's where the vitamins are!"

Who gets the thumbs-up?

Answer appears on page 20-14

Jolene: "Are you sure, mom? The peel is just fiber."

Mom: " Of course I'm sure, honey."

To Peel or Not to Peel?

Keeping the peel on potatoes does not appear to make a meaningful difference in the vitamin content of potatoes. On an equal weight basis (5.4 ounces each), the vitamin content of peeled and boiled versus nonpeeled and boiled potatoes is equivalent for thiamin, riboflavin, niacin, vitamin B_6, and folate. The content of vitamin C is approximately 60% higher in the nonpeeled and boiled potatoes (18 mg) versus the peeled and boiled potatoes (11 mg). This difference is likely due to vitamin C loss in water in the peeled and boiled potatoes. There is little difference in the fiber content of the potatoes.

Jolene:

Mom:

Vitamins: Getting Enough Without Getting Too Much

Table 20.5 lists good food sources of vitamins. Adequate amounts of vitamins can be obtained from diets that include a variety of basic foods including whole grain products, low-fat dairy products, vegetables, fruits, fish, and other foods recommended in MyPlate. gov food guidance materials. Most fruits and vegetables are good sources of vitamins, and eating five or more servings a day is one way for individuals to get an assortment of vitamins. Fortified foods such as ready-to-eat cereals, fruit juices, and dairy products contribute to adequate intakes of vitamins.[1,3]

Recommended Intake Levels of Vitamins Updated recommendations for vitamin intakes associated with the prevention of deficiency and chronic diseases are represented by standards called Dietary Reference Intakes (DRIs). DRIs include the Recommended Dietary Allowances (RDAs) for vitamins, which convincing scientific data establishes for intake standards. Adequate Intakes (AIs) are assigned to vitamins for which scientific information about levels of intake associated with chronic disease prevention is less convincing. Tolerable Upper Levels of Intake (ULs) are also assigned to vitamins and indicate levels of vitamin intake from foods, fortified foods, and supplements that should *not* be exceeded. The RDAs or AIs and the ULs for the vitamins are given in Table 20.2.

Although people can get all the vitamins they need from supplements, it makes more sense to get them from foods. Foods offer fiber, minerals, beneficial photochemicals, and other nutrients that don't come in supplements. Nutrients in foods interact to produce greater, positive effects on health maintenance than do vitamin supplements in general.[23,25]

Table 20.5 Food sources of vitamins

Thiamin			
Food		**Serving Size**	**Thiamin (mg)**
Meats:			
Ham		3 oz	0.6
Pork		3 oz	0.5
Beef		3 oz	0.4
Liver		3 oz	0.2
Nuts and seeds:			
Pistachios		1/4 cup	0.3
Macadamia nuts		1/4 cup	0.2
Peanuts, dry roasted		1/4 cup	0.2
Grains:			
Breakfast cereals		1 cup	0.3–1.4
Flour tortilla		1	0.2
Macaroni		1/2 cup	0.2
Rice		1/2 cup	0.2
Bread		1 slice	0.1
Vegetables:			
Peas		1/2 cup	0.2
Lima beans		1/2 cup	0.2
Corn		1/2 cup	0.2
Fruits:			
Orange juice		1 cup	0.2
Orange		1	0.1
Avocado		1/2	0.1

Riboflavin			
Food		**Serving size**	**Riboflavin (mg)**
Milk and milk products:			
Milk		1 cup	0.5
2% milk		1 cup	0.5
Yogurt, low-fat		1 cup	0.5
Skim milk		1 cup	0.4
Yogurt		1 cup	0.4
American cheese		1 oz	0.1
Cheddar cheese		1 oz	0.1
Meats:			
Liver		3 oz	3.6
Pork chop		3 oz	0.3
Beef		3 oz	0.2
Tuna		3 oz	0.1
Vegetables:			
Collard greens		1/2 cup	0.3
Spinach, cooked		1/2 cup	0.2
Broccoli		1/2 cup	0.1
Eggs:		1	0.2
Egg			

(continued)

Table 20.5 (continued)

Grains:

Food		Serving size	
Breakfast cereals		1 cup	0.1–1.7
Macaroni		½ cup	0.1
Bread		1 slice	0.1

Niacin

Food		Serving size	Niacin (mg)
Meats:			
Liver		3 oz	14.0
Tuna		3 oz	7.0
Turkey		3 oz	4.0
Chicken		3 oz	11.0
Salmon		3 oz	6.9
Veal		3 oz	6.4
Beef (round steak)		3 oz	4.0
Pork		3 oz	4.0
Haddock		3 oz	3.9
Shrimp		3 oz	2.2
Nuts and seeds:			
Peanuts, dry roasted		¼ cup	4.9
Almonds		¼ cup	1.3
Vegetables:			
Asparagus		½ cup	1.2
Corn		½ cup	1.2
Green beans		½ cup	1.2
Grains:			
Breakfast cereals		1 cup	5.0–20.0
Brown rice		½ cup	1.5
Noodles, enriched		½ cup	1.0
Rice, white, enriched		½ cup	1.2
Bread, enriched		1 slice	1.1

Gts/Shutterstock.com

Vitamin B$_6$

Food		Serving size	Vitamin B6 (mg)
Meats:			
Liver		3 oz	0.8
Other fish		3 oz	0.3–0.6
Chicken		3 oz	0.4
Ham		3 oz	0.4
Hamburger		3 oz	0.4
Veal		3 oz	0.4
Pork		3 oz	0.3
Beef		3 oz	0.2
Grains:			
Breakfast cereals		1 cup	0.5–7.0
Fruits:			
Banana		1	0.4
Avocado		½	0.3
Watermelon		1 cup	0.3
Vegetables:			
Brussels sprouts		½ cup	0.2
Potato		½ cup	0.4
Sweet potato		½ cup	0.3
Carrots		½ cup	0.2
Sweet peppers		½ cup	0.2

Maks Narodenko/Shutterstock.com

Folate Food		Serving size	Folate (mcg)
Vegetables:			
Garbanzo beans		1/2 cup	141
Spinach, cooked		1/2 cup	131
Navy beans		1/2 cup	128
Asparagus		1/2 cup	120
Lima beans		1/2 cup	76
Collard greens, cooked		1/2 cup	65
Romaine lettuce		1 cup	65
Peas		1/2 cup	47
Grains:[a]			
Ready-to-eat cereals		1 cup/1 oz	100–400
Rice		1/2 cup	77
Noodles		1/2 cup	45
Wheat germ		2 Tbsp	40

iStockphoto.com/Liv Friis-Larsen

Maks Narodenko/Shutterstock.com

Vitamin B$_{12}$ Food		Serving size	Vitamin B12 (mcg)
Fish and seafood:			
Oysters		3 oz	13.8
Scallops		3 oz	3.0
Salmon		3 oz	2.3
Clams		3 oz	2.0
Crab		3 oz	1.8
Tuna		3 oz	1.8
Meats:			
Liver		3 oz	6.8
Beef		3 oz	2.2
Veal		3 oz	1.7
Milk and milk products:			
Skim milk		1 cup	1.0
Milk		1 cup	0.9
Yogurt		1 cup	0.8
Cottage cheese		1/2 cup	0.7
American cheese		1 oz	0.2
Cheddar cheese		1 oz	0.2
Grains:			
Breakfast cereals		1 cup	0.6–12.0
Eggs:			
Egg		1	0.6

Robyn Mackenzie/Shutterstock.com

[a]Fortified, refined grain products such as bread, rice, pasta, and crackers provide approximately 60 micrograms of folic acid per standard serving.

(continued)

Table 20.5 (continued)

Vitamin C Food		Serving size	Vitamin C (mg)
Fruits:			
Guava		1/2 cup	180
Orange juice, vitamin C-fortified		1 cup	108
Kiwi fruit		1	108
Grapefruit juice, fresh		1 cup	94
Cranberry juice cocktail		1 cup	90
Orange		1	85
Strawberries, fresh		1 cup	84
Cantaloupe		1/4 whole	63
Grapefruit		1 medium	51
Raspberries, fresh		1 cup	31
Watermelon		1 cup	15
Vegetables:			
Sweet red peppers		1/2 cup	142
Cauliflower, raw		1/2 cup	75
Broccoli		1/2 cup	70
Brussels sprouts		1/2 cup	65
Green peppers		1/2 cup	60
Collard greens		1/2 cup	48
Vegetable (V-8) juice		3/4 cup	45
Tomato juice		3/4 cup	33
Cauliflower, cooked		1/2 cup	30
Potato		1 medium	29
Tomato		1 medium	23

iStockphoto.com/Joe Biafore

iStockphoto.com/NoDerog

Choline Food		Serving size	Choline (mg)
Meats:			
Beef		3 oz	111
Pork chop		3 oz	94
Lamb		3 oz	89
Ham		3 oz	87
Beef		3 oz	85
Turkey		3 oz	70
Salmon		3 oz	56
Eggs:			
Egg		1 large	126
Vegetables:			
Baked beans		1/2 cup	50
Navy beans, boiled		1/2 cup	41
Collards, cooked		1/2 cup	39
Black-eyed-peas (Cowpeas)		1/2 cup	39
Chickpeas (garbanzo beans)		1/2 cup	35
Brussels sprouts		1/2 cup	32
Broccoli		1/2 cup	32
Collard greens		1/2 cup	30
Refried beans		1/2 cup	29
Milk and milk products			
Milk, 2%		1 cup	40
Cottage cheese, low-fat		1/2 cup	37
Yogurt, low-fat		1 cup	35

Vitamin A			
Food		**Serving Size**	**Vitamin A Retinol (mcg)**
Meats:			
Liver		3 oz	9,124
Clams		3 oz	145
Fortified breakfast cereals		¾ cup	150
Milk and milk products:			
American cheese		1 oz	114
Fat-free/low-fat milk		1 cup	100
Whole milk		1 cup	58
Egg		1	84

Beta-Carotene			
Food		**Serving Size**	**Beta-carotene mcg retinol equivalents, RE**
Vegetables:			
Sweet potatoes		½ cup	961
Pumpkin, canned		½ cup	953
Carrots, raw		½ cup	665
Spinach, cooked		½ cup	524
Collard greens, cooked		½ cup	489
Kale, cooked		½ cup	478
Turnip greens, cooked		½ cup	441
Beet greens, cooked		½ cup	276
Swiss chard, cooked		½ cup	268
Winter squash, cooked		½ cup	268
Vegetable juice		1 cup	200
Romaine lettuce		1 cup	162
Fruit:			
Cantaloupe		½ cup	135
Apricots, fresh		4	134

Barbara Delgado/Shutterstock.com

Vitamin E			
Food		**Serving Size**	**Vitamin E (mg)**
Nuts and seeds:			
Sunflower seeds		1 oz	7.4
Almonds		1 oz	7.3
Hazelnuts (filberts)		1 oz	4.3
Mixed nuts		1 oz	3.1
Pine nuts		1 oz	2.6
Peanut butter		2 Tbsp	2.5
Peanuts		1 oz	2.2
Vegetable oil:			
Sunflower oil		1 Tbsp	5.6
Safflower oil		1 Tbsp	5.6
Canola oil		1 Tbsp	2.4
Peanut oil		1 Tbsp	2.1
Corn oil		1 Tbsp	1.9
Olive oil		1 Tbsp	1.9
Salad dressing		2 Tbsp	1.5
Fish and seafood:			
Crab		3 oz	4.5
Shrimp		3 oz	3.7
Fish		3 oz	2.4

iStockphoto.com/FotografiaBasica

(continued)

Table 20.5 (continued)

Grains:

Wheat germ		2 Tbsp	4.2
Whole wheat bread		1 slice	2.5

Vegetables:

Spinach, cooked		¹/₂ cup	3.4
Yellow bell pepper		1	2.8
Turnip greens, cooked		¹/₂ cup	2.2
Swiss chard, cooked		¹/₂ cup	1.7
Asparagus		¹/₂ cup	1.5
Sweet potato		¹/₂ cup	1.5

Vitamin D			Vitamin D	
Food		Serving size	(mcg)	IU

Fish and seafoods:

Ray Kachatorian/PhotoDisc/Jupiterimages

		Serving size	(mcg)	IU
Swordfish		3 oz	16	640
Trout		3 oz	16	480
Salmon		3 oz	12	228
Tuna, light, canned in oil		3 oz	5.7	196
Halibut		3 oz	4.9	
Tuna, light, canned in water		3 oz	3.8	152
Tuna, white, canned in water		3 oz	1.7	68

Vitamin D-fortified breakfast cereals:

		Serving size	(mcg)	IU
Whole grain Total		1 cup	3.3	132
Total Raisin Bran		1 cup	2.6	104
Corn Pops, Kellogg's		1 cup	1.2	48
Crispix, Kellogg's		1 cup	1.2	48

Other vitamin D-fortified foods:

		Serving size	(mcg)	IU
Orange juice		1 cup	2.5	100
Rice milk		1 cup	2.5	100
Soy milk		1 cup	2.5	100
Yogurt		1 cup	2.0	80
Margarine		2 tsp	1.2	48

Milk:

		Serving size	(mcg)	IU
Milk, whole		1 cup	3.2	128
Milk, 2%		1 cup	2.9	116
Milk, 1%		1 cup	2.9	116
Milk, skim (nonfat)		1 cup	2.9	116

PhotoDisc

NUTRITION
up close

Focal Point: Determine if you eat enough antioxidant-rich foods.

Vitamin C, beta-carotene, and vitamin E, the antioxidant vitamins, help to maintain cellular integrity in the body. Good food sources of these antioxidants reduce the risk of heart disease, certain cancers, and other ailments. Check below to find out how frequently you consume foods containing these important, health-promoting nutrients from foods.

How often do you eat:	Seldom or never	1–2 times per week	3–5 times per week	Almost daily
Vitamin C food sources:				
1. Grapefruit, lemons, oranges, or pineapple?				
2. Strawberries, kiwi, or honeydew melon?				
3. Orange juice, cranberry juice cocktail, or tomato juice?				
4. Green, red, or chili peppers?				
5. Broccoli, Chinese cabbage, or cauliflower?				
6. Asparagus, tomatoes, or potatoes?				
Beta-carotene food sources:				
7. Carrots, sweet potatoes, or winter squash?				
8. Spinach, collard greens, or swiss chard?				
9. Cantaloupe, papayas, or mangoes?				
10. Nectarines, peaches, or apricots?				
Vitamin E food sources:				
11. Whole grain breads, whole grain cereals, or wheat germ?				
12. Crab, shrimp, or fish?				
13. Peanuts, almonds, or sunflower seeds?				
14. Oils, margarine, butter, mayonnaise, or salad dressing?				

Feedback to the Nutrition Up Close is located in Appendix G.

REVIEW QUESTIONS

- **Identify the vitamins, their functions and food sources, and their effects on health.**

Assess the adequacy of your diet for vitamins that function as antioxidants.

1. Vitamins are essential. Specific deficiency diseases develop if we fail to consume enough of them.   **True/False**

2. Vitamin D, E, C, and B_6 are fat soluble.   **True/False**

3. The niacin deficiency disease is called pellagra.
   True/False

4. Some people are deficient in vitamin B_{12} because they are genetically unable to absorb it.   **True/False**

5. Vitamin A toxicity causes brain tumors.   **True/False**

6. Three good sources of vitamin D are sunshine, milk, and seafood.   **True/False**

7. Vitamin E, vitamin C, and beta-carotene supplements decrease the risk of heart disease.   **True/False**

8. Vitamin D acts as a hormone.   **True/False**

9. Adequate vitamin D status reduces chronic inflammation.   **True/False**

10. Very high intakes of each of the vitamins has been found to cause toxicity disease.   **True/False**

11. With the exception of vitamin B_{12}, the body is able to store high amounts of the water-soluble vitamins.   **True/False**

12. A number of vitamins act as coenzymes by activating specific enzymes.   **True/False**

13. Vitamin C functions in the replacement of cells that line the esophagus and eyes.   **True/False**

14. The sun is the primary source of vitamin D for people in general.   **True/False**

15. It is recommended that women who may become, or who are pregnant, consume 400 mcg folic acid daily to reduce the risk of fetal development of vision problems.
    True/False

16. _____ Which of the following is not considered a vitamin?

    a. pantothenic acid
    b. coenzyme Q_{10}
    c. biotin
    d. choline

17. _____ A person who never eats fish, seafood, nuts, seeds, or vegetable oil is at risk of developing a deficiency of

    _____.

    a. vitamin A
    b. vitamin K
    c. vitamin D
    d. vitamin E

18. _____ Which of the following is a consequence of vitamin A deficiency?

    a. rough, dry skin
    b. tingling sensation in the finger tips
    c. headache
    d. night blindness

19. _____ Assume you ate the following foods as snacks yesterday: an orange, 1/4 cup sunflower seeds, 1 cup yogurt, and a carrot. Which of these foods would provide the most vitamin E?

    a. orange
    b. sunflower seeds
    c. yogurt
    d. carrot

20. _____ Which of the following foods is not a good source of vitamin A (retinol)?

    a. milk
    b. eggs
    c. sweet peppers
    d. fortified breakfast cereal

21. _____ Which of the following vitamins is needed for blood to clot when bleeding occurs?

    a. vitamin K
    b. thiamin
    c. vitamin B_6
    d. vitamin B_{12}

Answers to these questions can be found in Appendix G.

NUTRITION SCOREBOARD ANSWERS

1. For some vitamins, intake levels above those known to prevent deficiency diseases help protect humans from certain cancers, heart disease, osteoporosis, depression, and other disorders.   **False**

2. Nope–only carbohydrates, proteins, and fats provide energy to the body. Vitamins are needed, however, to convert the energy in food into energy the body can use.   **False**

3. Citrus fruits are good sources of vitamin C, but so are red sweet peppers, strawberries, and other noncitrus fruits and vegetables.   **False**

4. True. Nearly all cases of illness due to vitamin overdoses result from the excessive intake of vitamin supplements.[1]   **True**

UNIT
21

Phytochemicals

NUTRITION SCOREBOARD

1 Phytochemicals are found only in plants. **True/False**

2 Phytochemicals are also called "phytonutrients" because they are biologically active in the body and have beneficial effects on health. **True/False**

3 Brightly colored vegetables and fruits are the richest sources of phytochemicals. **True/False**

4 Some chemical substances that occur naturally in food or result from food preparation may be harmful to health. **True/False**

Answers can be found at the end of the unit.

After completing Unit 21 and its interactive learning features, you will be able to

- Understand the functions and food sources of key phytochemicals and assess your intake of foods rich in phytochemicals.

Phytochemicals: The "What Else" Is in Your Food

- **Understand the functions and food sources of key phytochemicals and assess your intake of foods rich in phytochemicals.**

As recently as 25 years ago, the science of nutrition focused on the study of the functions and health effects of protein, fats, carbohydrates, vitamins, minerals, and water. These classes of essential nutrients have been extensively studied and a good deal is known about their effects on growth, reproduction, and health. However, essential nutrients do not account for all the benefits associated with healthy diets. There are other components in food that influence health.

We have ample evidence that diets rich in vegetables, fruits, whole grains, and other plant foods support health and reduce the risk of developing a number of diseases. It was largely assumed that the health benefits came from the vitamin and mineral content of fruits and vegetables. That conclusion turned out to be incorrect because supplementation with specific vitamins and minerals failed to yield the same health benefits as did diets rich in fruits and vegetables. In addition, use of individual vitamin and mineral supplements was found to increase health risks in some studies.[1] Now nutrition and other scientists are investigating the effects of thousands of other substances in food on health and their interactions with essential nutrients.[2] Illustration 21.1 shows some of the hundreds of chemical substances found in two plant foods.

The subjects of many current studies are plant chemicals, known as **phytochemicals** or *phytonutrients*. Phytochemicals are not considered essential nutrients because deficiency diseases do not develop when we fail to consumer them. They are considered to be nutrients, however, because they are biologically active and perform health-promoting functions in the body. Meats, eggs, dairy products, and other foods of animal origin also contain biologically active substances that affect body processes. Much less is known about these **zoochemicals**, and their effects on health are not yet as clear as those of some of the phytochemicals. Most bioactive food constituents are derived from plants.[2]

This unit presents information on the functions and health benefits of the most extensively studied phytochemicals and identifies their major food sources. It also highlights substances in foods considered to be natural toxins because they can be harmful to health if consumed in excess.

phytochemicals (phyto = *plant*) Biologically active, or "bioactive," substances in plants that have positive effects on health. Also called *phytonutrients*.

zoochemicals Chemical substances in animal foods, some of which may be biologically active in the body.

Illustration 21.1 A sampling of the chemical substances in two foods.

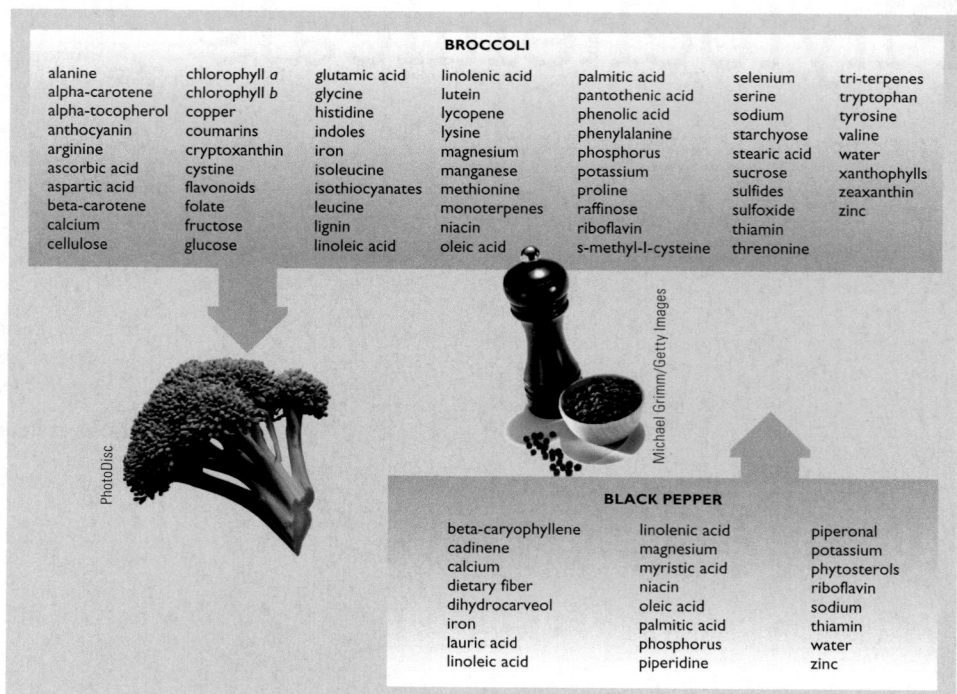

BROCCOLI

alanine	chlorophyll *a*	glutamic acid	linolenic acid	palmitic acid	selenium	tri-terpenes
alpha-carotene	chlorophyll *b*	glycine	lutein	pantothenic acid	serine	tryptophan
alpha-tocopherol	copper	histidine	lycopene	phenolic acid	sodium	tyrosine
anthocyanin	coumarins	indoles	lysine	phenylalanine	starchyose	valine
arginine	cryptoxanthin	iron	magnesium	phosphorus	stearic acid	water
ascorbic acid	cystine	isoleucine	manganese	potassium	sucrose	xanthophylls
aspartic acid	flavonoids	isothiocyanates	methionine	proline	sulfides	zeaxanthin
beta-carotene	folate	leucine	monoterpenes	raffinose	sulfoxide	zinc
calcium	fructose	lignin	niacin	riboflavin	thiamin	
cellulose	glucose	linoleic acid	oleic acid	s-methyl-l-cysteine	threnonine	

PhotoDisc

Michael Grimm/Getty Images

BLACK PEPPER

beta-caryophyllene	linolenic acid	piperonal
cadinene	magnesium	potassium
calcium	myristic acid	phytosterols
dietary fiber	niacin	riboflavin
dihydrocarveol	oleic acid	sodium
iron	palmitic acid	thiamin
lauric acid	phosphorus	water
linoleic acid	piperidine	zinc

Key Nutrition Concepts

The key nutrition concepts underlying content presented on phytochemicals relate to the importance of food as a source of nutrients and of healthy diets. These concepts are as follows:

- Foods provide energy (calories), nutrients, and other substances needed for growth and health.

- Adequacy, variety, and balance are key characteristics of healthful diets.

Characteristics of Phytochemicals

Phytochemicals play a variety of roles in plants. They provide protection against bacterial, viral, and fungal infection; ward off insects; and prevent tissue damage due to oxidation. Some act as plant hormones or participate in the regulation of gene function, while others provide plants with flavor and color.[3–5] More than 2,000 types of phytochemicals that act as pigments have been identified and give plants with a wide variety of colors (Illustration 21.2). Specific types of phytochemicals, their color, and their top food sources are listed in Table 21.1.

Sometimes the color of phytochemical contained in plants is obscured. Dark green vegetables, for example, are often good sources of orange and yellow carotenes, but the green chlorophyll obscures these colors. Many phytochemicals are colorless. You cannot identify rich sources of phytochemicals only by their color. The Reality Check feature for this unit emphasizes this point using the example of white vegetables and fruits.

Although thousands of biologically active substances in plants have been identified, knowledge of their effects on human health is currently known for only a small number of them. Rather than acting alone, phytochemicals appear to exert their beneficial effects primarily through complementary mechanisms of action and combined effects with other phytochemicals and nutrients in

Keith Weller/ARL/USDA

Illustration 21.2 The colors in plant foods are created by phytochemicals.

Table 21.1 The color and top food sources of specific phytochemicals[3]

Phytochemical	Color	Top food sources
Beta-carotene	orange	carrots
Lycopene (like-oh-pene)	red	tomatoes
Anthocyanins (an-tho-sigh-ah-nins)	blue to purple	blueberries, grapes
Allicin (all-is-on)	white	garlic
Lutein (loo-te-in) and zeaxanthin (ze-ah-zan-thin)	yellow-green	spinach

REALITY CHECK
Color and the Phytonutrients Content of Vegetables and Fruits

Are white vegetables and fruits good sources of phytonutrients?

Who gets the thumbs-up?

Answer appears on page 21-4

Hyde: I avoid eating white foods like onions, bananas, potatoes, and cauliflower. If you want phytonutrients, you have to eat the really colorful vegetables and fruits.

Ejay: I used to think that, too. But I just learned "white" doesn't mean that a white vegetable or fruit is a poor source of phytonutrients.

ANSWERS TO REALITY CHECK
Color and the Phytonutrients Content of Vegetables and Fruits

Ejay's got it right. White vegetables and fruits provide phytonutrients such as flavanols and allicin. You can't identify rich sources of phytonutrients based only on the color of vegetables and fruits. Not all phytonutrients that benefit health add bright colors to vegetables and fruits.

Hyde:

Ejay:

---

antioxidants Chemical substances that prevent or repair damage to cells caused by oxidizing agents such as pollutants, ozone, smoke, and reactive oxygen. Oxidation reations are a normal part of cellular processes. Vitamins C and E and certain phytochemicals function as antioxidants.

age-related macular degeneration (AMD) Eye damage caused by oxidation of the macula, the central portion of the eye that allows you to see details clearly.

Table 21.2 Top food sources of antioxidants[3]

Pomegranate
Red cabbage
Blackberries
Pecans
Walnuts
Cloves, ground
Peanuts
Sunflower seeds
Blueberries
Strawberries
Chocolate, dark
Raspberries
Cranberry juice
Kale
Artichokes
Wine, red
Grape juice
Cranberries
Pineapple juice
Green tea
Guava nectar
Coffee
Mango nectar

foods. The specific functions of phytochemicals may be diminished or expanded based on these interactions.[2]

Functions of Phytochemicals

Phytochemicals perform a variety of functions, including these roles:

- **antioxidants**
- inhibitors of inflammation
- preventers of infectious disease[6–8]

Although phytochemicals perform a number of different functions, the majority of them act as antioxidants. This is an important role because oxidation reactions contribute to the development of common diseases such as certain cancers, heart diseases, type 2 diabetes, and hypertension.[9,10]

Antioxidant Roles of Phytochemicals Oxidation reactions occur in the body all the time, primarily due to the use of oxygen for energy formation within cells and as part of the body's infection-fighting mechanisms. Reactive oxygen molecules are normally deactivated by antioxidants present in the body and supplied by phytochemicals and vitamins in foods that act as antioxidants. Antioxidants scavenge oxidized particles and neutralize their reactivity. A number of oxidation-related diseases result when the supply of antioxidants is inadequate, and an adequate supply from plant foods helps to prevent the diseases.[12] Table 21.2 provides a list of the top, commonly consumed plant sources of antioxidants.

Antioxidant Roles of Lutein, Zeaxanthin, and Beta-carotene Many pigments in plants act as antioxidants and participate in disease prevention. Lutein (pronounced loo-te-in) and zeaxanthin (pronounced ze-ah-zan-thin) are found together in plants and play key roles in the prevention and treatment of **age-related macular degeneration (AMD)**. AMD is the leading cause of blindness in people over the age of 50 years. AMD produces a loss of central vision due to degeneration of the macula (Illustration 21.3). The macula, located in the center of the eye, is yellow due to its content of lutein and zeaxanthin. When these two phytochemicals become depleted, vision declines. Increased intake of lutein and zeaxanthin helps prevent deterioration of the macula and improves the maintenance of vision.[10,11]

Regular intake of carrots and other vegetables high in beta-carotene are also related to a decrease in markers of oxidation and a reduced risk of cardiovascular disease and cancer. Beta-carotene in foods acts as a powerful antioxidant.[13]

Antioxidant Roles of Flavanols The good news about flavanols has made chocolate a health food. Cocoa, the main ingredient in chocolate, is a rich source of flavanols. Flavanols act as antioxidants. Regular intake of flavanols (such as daily consumption of a cup of hot chocolate made with cocoa powder) is related to improved blood flow, reduced blood pressure, and decreased risk of heart disease and stroke.[13,14] The content of flavanols in chocolate products increases with the amount of cocoa in the products. Chocolate products that contain 40% or more cocoa by weight have substantially higher amounts of flavanols than other chocolate products.[2] Milk chocolate made in the United States tends to have less than 30% cocoa whereas darker chocolate products usually have more than 40%. Flavanols are also found in good amounts in foods such as berries, wine, and tea.[2]

Due to their importance to health, this unit's Take Action feature attempts to remind you about specific vegetables, fruits, and nuts rich in antioxidant phytochemicals that may have escaped your attention.

Anti-inflammation Roles of Phytochemicals Quercetin (pronounced queer-sah-tin), a phytochemical present in apples, onions, and grapes, influences inflammation by modifying the expression (or the turning on or off) of genes that promote the production of pro-inflammatory compounds. Its presence helps decrease chronic inflammation due to excess body fat, improves insulin utilization, and may decrease bone loss.[7, 10] Similarly, phytochemicals in the **cruciferous family** of vegetables including broccoli and cauliflower (Illustration 21.4) appear to reduce the risk of certain types of cancer, particularly in people with specific gene types.[16,17]

Resveratrol (rez-ver-ah-trol) in the skin of red grapes also plays a role in the prevention of inflammation.[6] This phytochemical is also found in peanuts, tea, blueberries, and cranberries.[3]

Anti-infection Roles of Proanthocyanidins and Resveratrol Cranberries contain proanthocyanidins (pro-an-tho-sigh-an-ah-dins) that prevent E. coli and other types of bacteria from sticking to the walls of the urinary tract and limiting the spread of urinary tract infections. Cranberry juice by itself usually does not fully treat established urinary tract infections. Regular intake of cranberry juice (a cup daily) reduces the symptoms of urinary tract infection, and helps prevent them from developing and reoccurring.[18,19] Resveratrol also helps prevent the spread of bacteria that cause infection.[6]

Caffeine: A Phytochemical with Multiple Functions Caffeine is an example of a phytochemical that is not easily classified by function because it has antioxidant, anti-inflammatory, chemical-reaction–inhibiting, and gene-regulating effects on body processes.[20,21] Most of the caffeine in diets comes from coffee, one of the most commonly consumed beverages in the world.[22] Caffeine sources also include tea, some soft drinks,

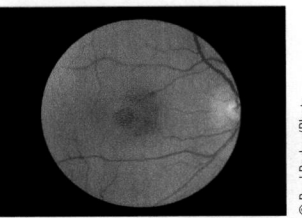

Illustration 21.3 Photograph of damage to the central part of the eye due to macular degeneration.

Illustration 21.4 Examples of cruciferous vegetables.

cruciferous vegetables Sulfur-containing vegetables whose outer leaves form a cross (or crucifix). Vegetables in this family include broccoli, cabbage, cauliflower, Brussels sprouts, mustard and collard greens, kale, bok choy, kohlrabi, rutabaga, turnips, broccoflower, and watercress.

take action   To Select and Consume Antioxidant-Rich Vegetables and Fruits

Sometimes we simply forget about some vegetables, fruits, and nuts we love to eat. Read through this list of the best sources of antioxidants and place a check in front of the ones you forgot you love, or that you would like to try.

_____ Pomegranate

_____ Blackberries

_____ Walnuts

_____ Blueberries

_____ Strawberries

_____ Raspberries

_____ Artichokes

_____ Cranberries

_____ Spinach

_____ Oranges

_____ Red cabbage

_____ Pecans

_____ Cloves, ground

_____ Grape juice

_____ Chocolate, dark

_____ Cranberry juice

_____ Cherries, sour

_____ Pineapple juice

_____ Guava nectar

_____ Mango nectar

and energy drinks (Table 21.3). Both positive and negative health effects are attributed to caffeine. On the positive side, caffeine or regular coffee intake has been found to:

© Peter Donaldson/Alamy

- Decrease the risk of type 2 diabetes by lowering blood glucose levels, improving insulin sensitivity and insulin production, and decreasing glucose output by the liver.[23]

- Decrease the risk of estrogen-sensitive cancers, potentially by blocking the effects of estrogen and insulin on cancer development.[24]

- Decrease the risk of Parkinson's disease and Alzheimer's disease.[21,25]

- Increase mental alertness and energy by blocking the activity of a chemical messenger that decreases heart rate and increases drowsiness.[26]

- Improve mood; decrease depression.[27]

Intakes of greater than three cups of coffee a day does not appear to be related to the risk of hypertension, malformation in newborns, miscarriage, or growth retardation.[28,29] It is being discovered that coffee, like chocolate, is more of a health food than a health hazard. High intakes of coffee and particularly of caffeine, however, can have important negative effects on health.

Negative Effects of Excess Caffeine Some people do not tolerate caffeine well and may experience increased blood pressure if coffee or caffeine is consumed regularly. Excess caffeine intake can lead to anxiety, nervousness, sleep problems, and a feeling of unusually strong heart beats (heart palpitations). Tolerance to caffeine occurs over time in most people and abrupt withdrawal can lead to headache, fatigue, and irritability within 12 to 24 hours.[30]

Medical visits for children and adults for symptoms of excess caffeine intake have sharply increased in the United States in recent years. The increase is largely due to the availability and popularity of caffeinated energy drinks and alcohol plus caffeine beverages.[31] The mix of alcohol and caffeine gives drinkers the feeling of being wide awake and not very drunk. Caffeine, however, does not counteract the effects of alcohol on impaired judgments and reduced reaction time.[32] The FDA recently banned the sale of beverages containing both caffeine and alcohol, and both Canada and the United States have set limits on the amount of caffeine that can be added to energy drinks.[32,33]

Table 21.3 Caffeine content of foods, beverages, and some drugs

Source	Caffeine (mg)
Coffee (1 cup)	
Drip	115–175
Decaffeinated (ground or instant)	0.5–4.0
Instant	61–70
Percolated	97–140
Espresso (2 oz)	100
Tea (1 cup)	
Black, brewed 5 minutes, U.S. brands	32–144
Black, brewed 5 minutes, imported brands	40–176
Green, brewed 5 minutes	25
Instant	40–80
Soft drinks	
Coca-Cola (12 oz)	47
Cherry Coke (12 oz)	47
Diet Coke (12 oz)	47
Dr Pepper (12 oz.)	40
Ginger ale (12 oz)	0
Mountain Dew (12 oz)	54
Pepsi-Cola (12 oz)	38
Diet Pepsi (12 oz)	37
7-Up (12 oz)	0
Energy drinks	
Ripped force (8 oz)	120
Energy shot (1 oz)	100–350
Red Bull (8 oz)	80
Full Throttle (8 oz)	72
Jolt (12 oz)	72
Kick (12 oz)	56
Surge (8 oz)	35
Chocolate	
Cocoa, chocolate milk (1 cup)	10–17
Milk chocolate candy (1 oz)	1–15
Chocolate syrup, one ounce (2 Tbsp)	4
Nonprescription drugs, two tablets	
Nodoz	200
Vivarin	200
Excedrin	130
Weight-control pills	150

Food Sources of Phytochemicals

The amount and type of phytochemicals present in plants vary a good deal, depending on a number of factors. The content varies based on growing condition, genetic strains used, storage, and processing and preparation methods. Some plant foods, such as spinach and carrots, which are considered rich sources of vitamins, contain a variety of phytochemicals. Celery, tea, and onions, foods determined by nutrient composition tables to be relative "vitamin weaklings," are a good source of a number of phytochemicals. Plant foods rich in phytochemicals may or may not be good sources of vitamins and minerals as well. Intake of phytochemicals, and the beneficial effects of vegetable and fruit intake on health increase, however, as vegetable and fruit consumption increases.[2] Established health benefits related to plant food intake are reflected in the dietary recommendations developed for the MyPlate.gov of "fill half your plate with vegetables and fruits." Similarly, the Dietary Guidelines recommends that people consume ample plant foods because of their benefits to health promotion and disease prevention.[34] The top five sources of phytochemicals and other leading sources of phytochemicals in the American diet are listed in Table 21.4.

Phytochemical Supplements Health benefits associated with the intake of specific phytochemicals, and the fact that some people eat few vegetables, have prompted supplement manufacturers to produce pills and extracts that contain them (Illustration 21.5). Phytochemical supplements take the form of dehydrated and powdered whole vegetables and fruits, such as blueberries and broccoli, and extracts that contain one or more specific phytochemical.

Do supplements and extracts work? Do they offer the same health benefits as their plants sources do? The answer is that some phytochemical supplements benefit health whereas others do not. Supplements containing lutein and zeaxanthin appear to improve macula content of these two phytochemicals and delay vision loss.[12] Cranberry and blueberry powders, and supplements containing resveratrol from grape skins have also been found to benefit indicators of disease risk.[35,37] Quercetin supplements have been found to have negligible effects on disease risks that are reduced by consumption of quercetin in plants.[38] Supplements made from powdered, whole broccoli have been found not to work because the manufacturing process de-actives an enzyme in broccoli needed to make the target phytochemical available for absorption.[16]

Naturally Occurring Toxins in Food

Some foods contain biologically active substances that can harm health if consumed in excess. These substances are considered naturally occurring toxins. Some naturally occurring toxins form from components of food during food processing and preparation.[39]

Spinach, collard greens, rhubarb, and other dark green, leafy vegetables contain oxalic acid. Eating too much of these foods can make your teeth feel as though they are covered with sand and give you a stomachache. Have you ever seen a potato that was partly colored green? (If not, take a look at Illustration 21.6) The green area contains solanine, a bitter-tasting, insect-repelling phytochemical that is normally found only in the leaves and stalks of potato plants. Small amounts of solanine are harmless, but large quantities (an ounce or so) can interfere with the transmission of nerve impulses.

Phytate is present in whole grains, seeds, dried beans, and nuts. It tightly binds zinc, iron, calcium, magnesium, and copper, and reduces their absorption. Diets high in phytate have been found to produce mineral deficiency diseases.[39] Cassava, a root consumed daily in many parts of tropical Africa, can be very toxic if not prepared properly because it contains cyanide. Soaking cassava roots in water for three nights will get rid of the cyanide, but soaking for shorter periods of time does not. When the soaking time is cut to one or two nights, as sometimes happens during periods of food shortage, enough of the toxin remains in the root to cause konzo, a disease caused by the cyanide overdose.[15] Konzo is characterized by permanent, spastic paralysis.

Ackee fruit is another potential hazard to health (Illustration 21.7). If you're from Jamaica, chances are excellent that you love the taste of the core of ackee—and also know the fruit can be deadly. The national fruit of Jamaica, the yellow fleshy part around the seeds tastes like butter and looks like scrambled eggs. The rest of the

Table 21.4 The top five and other leading sources of phytochemicals[10]

Top five sources
1. tomatoes
2. carrots
3. oranges
4. orange juice
5. strawberries

Other leading sources
• Coffee
• Tea
• Spinach
• Corn
• Lettuce
• Collards
• Watermelon
• Grapes
• Blueberries
• Strawberries
• Bananas
• Onions
• Apples
• Raspberries

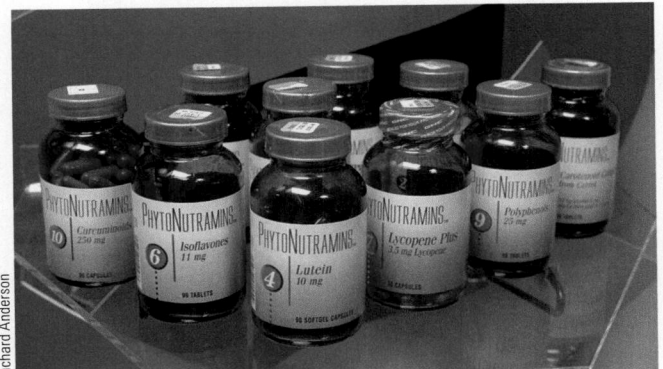

Richard Anderson

Illustration 21.5 Many phytochemicals supplements are available. Some are beneficial and some are not.

© Scott Goodwin Photography

Illustration 21.6 Potatoes grown partly above ground develop a green color in the part exposed to the sun. The green section contains solanine, a naturally occurring, potentially toxic phytochemical.

Illustration 21.7 Ackee fruit and seeds.

fruit, however, is not edible. The fruit of unopened, unripe ackee contains high concentrations of phytochemicals that cause severe vomiting and a drastic drop in blood glucose levels. Ingestion of the fruit has caused hundreds of deaths in Jamaica. Its sale was banned in the United States until 2000, and imported ackee is routinely analyzed by the FDA for ripeness.[39]

Acrylamide is an example of a toxin that develops from components of food due to food processing or preparation. Acrylamide is produced when starchy foods are fried, grilled, broiled, or roasted at high temperatures due to a reaction between the starch and an amino acid. High intake levels of acrylamide promote oxidation and inflammation, causing nerve and tissue damage. Exposure to acrylamide can be reduced by soaking potatoes in water before cooking; marinating meats in lemon juice, vinegar, or other acidic liquid; moist cooking, stewing, cooking at lower temperatures.[39,40]

© Scott Goodwin Photography

NUTRITION
up close

Focal Point: Consuming good sources of the "other" beneficial components of food.

Good food sources of beneficial phytochemicals are listed below. Indicate foods you consumed at least twice last week and those you have never tried eating.

	Foods Eaten	Foods Never Tried	Twice Last Week
Broccoli			
Cabbage			
Brussels sprouts			
Cauliflower			
Carrots			
Celery			
Collard greens			
Turnip greens			
Kale			
Swiss chard			
Spinach			
Tomatoes			
Peanuts			
Walnuts			
Apple/apple juice			
Orange/orange juice			
Grapefruit/grapefruit juice			
Grapes/grape juice			
Strawberries			
Blueberries			
Papaya			
Banana			
Pear			
Peaches			

Feedback to the Nutrition Up Close is located in Appendix G.

REVIEW QUESTIONS

- **Understand the functions and food sources of key phytochemicals and assess your intake of foods rich in phytochemicals.**

1. Phytochemicals that benefit health are found in high amounts in fish and organ meats.   **True/False**

2. Some plant foods that contain relatively low amounts of vitamins and minerals are rich sources of beneficial phytochemicals.   **True/False**

3. The best way to achieve the health benefits of phytochemicals is through the consumption of five or more servings of vegetables and fruits daily.   **True/False**

4. If a chemical substance occurs naturally in a plant food, it can be considered harmless to health.   **True/False**

5. Functions of phytochemicals that naturally occur in plants are different than their functions in the body.
   True/False

6. MyPlate.gov food guidance recommends that half your plate consist of vegetables and fruits.   **True/False**

7. Age-related macular degeneration has been related to a depletion of quercetin in the macula of the eye.
   True/False

8. Caffeine's functions are limited to roles in combating inflammation.   **True/False**

9. The combination of caffeine and alcohol prevents the intoxicating effects of alcohol from occurring.
   True/False

10. Most plant foods contain some level of phytochemicals, but onions and celery do not.   **True/False**

11. Orange juice is not a leading source of phytochemicals.   **True/False**

12. ____ Which of the following statements related to the effects of photochemical supplements on disease risk is true?

    a. Some phytochemical supplements have been found to reduce disease risk.
    b. None of phytochemical supplements available on the market have been found to reduce disease risk.
    c. Only supplements of lutein and zeaxanthin have been found to reduce disease risk.
    d. All phytochemical supplements tested so far have been found to reduce disease risk.

The next three questions refer the scenario below:

Assume you eat a salad that consists of spinach, strawberries, and walnuts with a salad dressing.

13. ____ Which of these plant foods would be the best source of lutein and zeaxanthin?

    a. salad dressing
    b. spinach
    c. strawberries
    d. walnuts

14. ____ The strawberries in the salad are considered to be a good source of:

    a. lutein
    b. zeaxanthin
    c. resveratrol
    d. lycopene

15. ____ The walnuts in the salad would contribute most to your intake of phytochemicals that play a role in:

    a. anti-inflammation processes
    b. anti-infection processes
    c. gene regulation processes
    d. antioxidation processes

16. ____ Which of the following plant foods do *not* belong to the cruciferous vegetable family?

    a. cabbage
    b. broccoli
    c. potatoes
    d. cauliflower

17. ____ Which of the following biologically active substances is *not* considered a potential "natural toxin"?

    a. phytates
    b. resveratrol
    c. oxalic acid
    d. acrylamide

Answers to these questions can be found in Appendix G.

NUTRITION SCOREBOARD ANSWERS

1. The "phyto" in phytochemicals means plants.   **True**

2. That's true.   **True**

3. You cannot judge a plant's overall content of phytochemicals by looking at its color.   **False**

4. Some foods contain naturally occurring toxins that can be harmful if consumed in excess.   **True**

© Digital Vision/Getty Images

Diet and Cancer

NUTRITION SCOREBOARD

1 Some types of cancer are contagious. **True/False**

2 Cancer is primarily an inherited disease. **True/False**

3 People who regularly consume a variety of fruits and vegetables are less likely to develop cancer than people who don't. **True/False**

4 High levels of body fat contribute to the development of some types of cancer. **True/False**

5 Six out of ten Americans never get cancer. **True/False**

Answers can be found at the end of the unit.

- Understand the processes involved in the development and progression of cancer.

- Identify dietary and lifestyle factors that affect cancer development and that decrease the risk of cancer.

What Is Cancer?

- **Understand the processes involved in the development and progression of cancer.**

Cancer, the second leading cause of death in the United States, is a group of conditions that result from the uncontrolled growth of abnormal cells. Although these cells can begin to grow in any tissue in the body, the lungs, colon, **prostate**, and breasts are the most common sites for cancer development (Illustration 22.1). Some forms of cancer are highly curable.

Key Nutrition Concepts

The key nutrition concepts presented in Unit 1 that underlie the complex relationships among diet and cancer are:

- Health problems related to nutrition originate within cells.

- Poor nutrition can influence the development of certain chronic diseases.

- Adequacy, variety, and balance are key characteristics of healthful diets.

cancer A group of diseases in which abnormal cells grow out of control and can spread throughout the body. Cancer is not contagious and has many causes.

prostate A gland located below the bladder in males. The prostate secretes a fluid that surrounds sperm.

initiation phase The start of the cancer process. It begins with damage to DNA.

promotion phase The period in cancer development when the number of cells with altered DNA increases.

Illustration 22.1 Percentage of cancer deaths by selected sites and sex.[2] About 23% of deaths in the United States, and 30% in Canada, are due to cancer.[7–9]

How Does Cancer Develop?

Cancer develops by complex processes that are not yet fully understood.[5] Adding to the complexity is the fact that cancer development often does not proceed in a straight line—cancer can progress two steps forward and then take a step or two back.

Illustration 22.2 summarizes the processes involved in the development of cancer, as we currently understand them. The risk of cancer development begins when DNA, the genetic material in cells that controls the body's production of proteins that regulate cell functions, becomes damaged. DNA is easily damaged by reactive oxygen molecules, radiation, toxins, and other reactive substances within cells. The structure of DNA is altered by the damage and that changes the accuracy of DNA codes for protein synthesis. Cells have multiple and redundant systems that continuously work to repair DNA to maintain its proper structure. Most of the time, DNA is successfully repaired. Problems arise when the repair processes fail to keep up with the damage. When this happens, errors in DNA codes for protein synthesis disrupt protein synthesis and normal cell functions in ways that can lead to the development of abnormal cell functions and structures. This situation can lead to the **initiation phase** of cancer cell development if not reversed by DNA repair processes. Cancer development enters the **promotion phase** if damaged DNA is allowed to accumulate over the course of years (10 to 30 years in most cases).[6] Ultimately, the growth of abnormal cells becomes rampant and can spread throughout the body. Although there are many factors that lead to DNA damage, all cancers share the hallmark of damaged DNA.[5]

What Causes DNA Damage?

About 80 to 90% of common forms of cancers are related to environmental factors that modify the structure and function of DNA.[2] Many environmental factors play a role in DNA damage development, and most are modifiable. The top nine modifiable risk factors for cancer development

	Percent				Percent	
Lung	31			26	Lung	
Colon & rectum	8			15	Breast	
Prostate	8			10	Colon & rectum	
Lymphoma & Leukemia	8			7	Lymphoma & Leukemia	
Urinary	6			6	Pancreas	
Pancreas	6			6	Ovary	
Esophagus	4			3	Urinary	
Liver	4			3	Cervix	
Oral	3			2	Uterus	
Skin	5			1	Oral	
Stomach	2			3	Skin	
All Other	17			18	All Other	

PhotoDisc

Illustration 22.2   Steps in the development of cancer.[5]

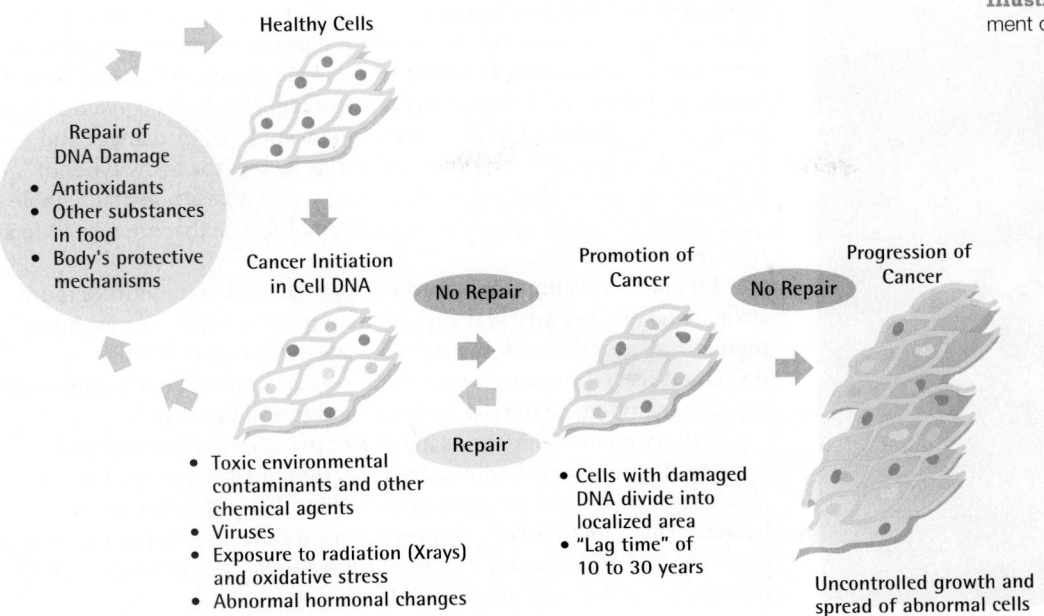

Healthy Cells

Repair of DNA Damage
- Antioxidants
- Other substances in food
- Body's protective mechanisms

Cancer Initiation in Cell DNA

No Repair

Promotion of Cancer

No Repair

Progression of Cancer

Repair

- Toxic environmental contaminants and other chemical agents
- Viruses
- Exposure to radiation (Xrays) and oxidative stress
- Abnormal hormonal changes

- Cells with damaged DNA divide into localized area
- "Lag time" of 10 to 30 years

Uncontrolled growth and spread of abnormal cells

worldwide are listed in Table 22.1. Poor diets, excess alcohol intake, and obesity are three important environmental factors, accounting for 30 to 40% of cancer risk.[13]

Some of the most convincing evidence of the robust relationship between environmental factors and cancer comes from studies of cancer rates in people migrating to other countries.[2] Rates of breast cancer, for example, are low in rural Asia. When individuals from rural parts of Asia immigrate to the United States, rates of breast cancer become the same as or higher than the U.S. rate by the third generation. Rates of prostate cancer similarly increase as people move from countries with low rates to countries with high rates. Rates of breast cancer in Japanese and Alaskan Eskimo women have increased substantially as they have adopted Westernized diets and lifestyles.[14]

Some people have genetic predispositions toward cancer, which means they have a tendency to develop cancer if regularly exposed to certain substances in the diet or environment. A genetically based susceptibility to cancer can develop in a fetus during pregnancy and in infancy. Exposure to calorie deficits, certain viruses, and specific other substances during these periods of rapid growth and development can modify the function of genes that help protect people from developing cancer.[15] Genetic factors appear to account for 42% of the risk for prostate cancer, 5 to 27% of the risk for breast cancer, and 36% of the risk for pancreatic cancer, for example. Endometrial cancer (cancer of the lining of the uterus), oral, thyroid, and bone cancer do not appear to be related to genetic factors.[6]

Given the high percentage of cancers related to diet and other environmental factors, cancer is considered a largely preventable disease. Increasing rates of new cases of cancer took a turn for the better after 1992 and correspond to declines in rates of tobacco use. Death rates from cancer in the United States are continuing to decline due to improved treatments, and the incidence of cancer is decreasing slightly overall.[16]

Fighting Cancer with a Fork

- **Identify dietary and lifestyle factors that affect cancer development and that decrease the risk of cancer.**

Specific characteristics of diets that have been linked to the development of cancer include low vegetable and fruits intake and a lack of variety of vegetables and fruits. Excess alcohol intake, or more than one drink a day by women, and two drinks a day by men, is

Table 22.1 The nine leading environmental factors related to cancer development worldwide[10]

1. Obesity
2. Low vegetable and fruit intake
3. Physical inactivity
4. Smoking
5. Excess alcohol intake
6. Unsafe sex
7. Air pollution
8. Indoor use of solid fuels
9. Hepatitis B or C viral infection

Illustration 22.3 The charred and black, oily coating on grilled or broiled meats is the part you shouldn't eat.

associated with the development of a number of cancers of the digestive system. Diets routinely low in whole grain products and fiber appear to promote the development of colorectal cancer. Regular intake of "charred meats," or the black, charred outer parts of high-fat meats cooked at high temperatures (Illustration 22.3) may also promote DNA damage and cancer development.[2,3,13,17] Other major risk factors for many types of cancer include smoking, physical inactivity, and excess body fat.[13] Table 22.2 summarizes characteristics of diets and lifestyle that are related to a reduced risk of cancer.

Frequent consumption of certain types of food is sometimes more strongly related to particular cancers than to other types. For example, regular consumption of tomato products is related in particular to decreased risk of prostate cancer,[18] and regular intake of black and green tea appears to reduce the risk of breast and ovarian cancer.[19]

Dietary recommendations for cancer prevention have changed recently. They no longer emphasize reduction of fat, saturated fat, and meat intake and center on properties of diets that promote overall health.[12] Diets and lifestyles that best prevent cancer are represented by a set of characteristics and not by hard rules about specific foods, dietary restrictions, or types of physical activities.

Table 22.2 Dietary patterns and lifestyles related to reduced risk of cancer[11,12]

1. Rely on foods for nutrient needs.
2. Consume a plant-based diet.
   - 5+ servings of a variety of vegetables and fruits daily, including those that are dark green, orange, and red
   - 3+ whole grains/products daily
3. Regularly consume dried beans, nuts, and seeds.
4. Include fish and seafood, lean beef, chicken, pork, and other meats.
5. Exclude charred or nitrate-preserved meats.
6. Exclude smoking.
7. Exclude excess alcohol consumption.
8. Include 30 minutes 5+ days a week of physical activity.
9. Maintain normal weight.

REALITY CHECK
Kumquats and Cancer

"KUMQUATS PREVENT CANCER" reads the headline in the newspaper. Researchers speculate that "tartaphil," the substance that makes kumquats tangy, may be the reason. Should you eat kumquats?

Who gets a thumbs up?

Answers appear on page 22-5.

Sebastian: Kumquats are too sour. I'd take a tartaphil supplement though.

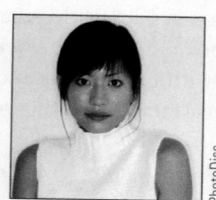

Flora: I already eat about three fruits every day. I'll include kumquats.

Table 22.3 Colorful vegetable and fruit antioxidant sources

Red: tomatoes, red raspberries, watermelon, strawberries, red peppers, cherries

Dark green: broccoli, brussels sprouts, kale, spinach, watercress, turnip greens, collard greens

Orange: carrots, mangos, papayas, apricots, sweet potatoes, pumpkins, oranges, tangerines, peaches, cantaloupe

PhotoDisc

How Do Good Diets Help Prevent Cancer?

Foods contain a variety of vitamins and minerals, as well as fiber and phytochemicals that help prevent DNA or assist in its repair. These substances in food, particularly plant foods, appear to work together in ways that provide the protection. Attempts to prevent cancer by giving large groups of people vitamin supplements or phytochemical extracts thought to account for the plant's beneficial effects on cancer development have not been successful.[3,20,21] In fact, a number of studies have noted that more harm than good results from the use of high amounts of individual supplements such as vitamin C, beta-carotene, and vitamin E.[24] Particular types of food clearly provide greater levels of protection against cancer than supplements. (The Reality Check feature for this unit addresses the issue of "magical" foods for cancer prevention. Take a look. It's nearby.) Some of the mechanisms underlying relationships between intake of specific foods and cancer are fairly well described, whereas others are yet to be elucidated.

A major role plants foods play in reducing cancer risk appears to be related to the antioxidant function of certain vitamin phytochemicals. These antioxidants in foods neutralize reactive oxygen and other molecules, preventing them from damaging DNA. They also participate in the repair of DNA.[25] Many brightly colored vegetables and fruits contain phytochemicals that act as antioxidants, and their consumption is being encouraged. A list of some of the best sources of the antioxidant phytochemicals that make vegetables and fruits colorful is given in Table 22.3.

ANSWERS TO **REALITY** CHECK
Kumquats and Cancer

No individual food or specific phytochemical has been proved to prevent cancer. Diets that help prevent cancer are not based on one or even a few foods or supplements. They are based on day-to-day intake of a healthy array of foods.

P.S. The author made up the headline about kumquats.

Sebastian:

PhotoDisc

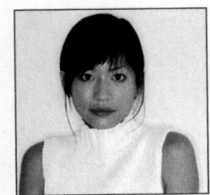
Flora:

PhotoDisc

Vegetables from the cruciferous family (e.g., broccoli, brussels sprouts, cabbage, and cauliflower) appear to "turn off" genes that help produce proteins that increase the ability of cancer cells to grow blood vessels that support the continued spread of cancer.[22,23] Substances in foods that reduce inflammation may also decrease cancer risk by reducing the amount of oxidized particles in cells that can damage DNA.[26]

Bogus Cancer Treatments

Unorthodox, purported cancer cures such as macrobiotic diets; hydrogen peroxide ingestion; laetrile tablets; vitamin, mineral, and herbal supplements; and animal gland therapy have not been shown to be effective treatments for cancer. Such remedies have been promoted since the early 1900s. They still exist because, although not proven to work, they offer some cancer patients a last ray of hope. They should not be used as a substitute for conventional cancer treatments.[27]

NUTRITION
up close

A Cancer Risk Checkup

Focal Point: Reducing cancer risk.

A number of behaviors that help protect people from developing cancer are listed below. Check those that apply to you.

	Yes	No	Don't know
1. I eat a dark green, orange, or red vegetable or fruit at least daily.			
2. I consume whole grain products daily.			
3. I eat broccoli, cauliflower, cabbage, or another vegetable of the cruciferous family twice a week.			
4. I avoid eating charred meats.			
5. I do not take high amounts of vitamin supplements.			
6. I do not smoke or chew tobacco.			
7. I exercise regularly.			
8. I do not have too much body fat.			

Feedback to the Nutrition Up Close is located in Appendix G.

REVIEW QUESTIONS

- **Understand the processes involved in the development and progression of cancer.**

- **Identify dietary and lifestyle factors that affect cancer development and that decrease the risk of cancer.**

1. Cancer development is related to unrepaired damaged to DNA. **True/False**

2. Little can be done to prevent cancer. **True/False**

3. Diets providing five or more servings per day of vegetables and fruits reduce the risk of cancer. **True/False**

4. Cancer has many causes. **True/False**

5. Some of the strongest evidence of the lack of a relationship between environmental factors and cancer development come from studies demonstrating a lack of change in cancer rates among groups of people migrating from one country to another. **True/False**

6. Some people have genetic predispositions that increase their risk of developing cancer if regularly exposed to certain substances in the diet or the environment. **True/False**

7. Diets high in fat, saturated fat, and meat represent a major risk factor for the development of a number of types of cancer. **True/False**

8. A number of studies have shown that supplements of vitamin C, beta-carotene, and vitamin E reduce the risk of cancer. **True/False**

9. A major role of plants foods in reducing cancer risk appears to be related to the antioxidant function of certain phytochemicals. **True/False**

10. Although the reasons for the effect are not yet understood, it appears that hydrogen peroxide ingestion and laetrile tablets reverse DNA damage and prevent cancer from developing. **True/False**

11. _____ Assume you get a call from your sister notifying you that your great uncle has been diagnosed with early stage cancer. Your sister tells you that your great uncle's daughter believes her father can get better if he drinks carrot juice often. What would be the evidence-based reaction to the daughter's belief about the effectiveness of carrot juice?
    a. Foods high in beta-carotene have been found to reverse the development of cancer.
    b. Carrots, but not carrot juice, reverse cancer development.
    c. Beta-carotene supplements, but not food high in beta-carotene, have been found to limit cancer progression.
    d. The daughter may believe that carrot juice is effective against cancer but that has not been shown by scientific studies to be true.

12. ____ After hearing about your great uncle's cancer, you decide to look at your diet and make a change to lower your risk of cancer development. Which of the following would be reasonable changes to make?

a. Reduce your intake of dairy products and spices.
b. Eat the dark green, orange, and red vegetables and fruits you like often.
c. Start taking a multivitamin and mineral supplement to enhance your intake of vitamins from food.
d. Eat more food because a few extra pounds of body weight helps prevent cancer.

Answers to these questions can be found in Appendix G.

NUTRITION SCOREBOARD ANSWERS

1. Cancer doesn't spread from person to person.   **False**

2. Cancer development is primarily related to environmental factors including diet, smoking, and exposure to radio-active particles and toxins. Genetic traits play a role by placing some individuals at increased risk for developing cancer.[1,2]   **False**

3. It's true.[3]   **True**

4. High levels of body fat are related to the development of some types of cancer.[4]   **True**

5. Most people don't development cancer.[2]   **True**

UNIT

23

Good Things to Know about Minerals

NUTRITION SCOREBOARD

1 The sole function of minerals is to serve as a component of body structures such as bones, teeth, and hair.   **True/False**

2 Bones continue to grow and mineralize through the first 30 years of life.   **True/False**

3 Ounce for ounce, spinach provides more iron than beef.   **True/False**

4 Worldwide, the most common nutritional deficiency is iron deficiency.   **True/False**

5 More than one in four American adults has hypertension.   **True/False**

Answers can be found at the end of the unit.

After completing Unit 23 and its interactive learning features, you will be able to:

• Identify key functions and food sources of five essential minerals.

Mineral Facts

• Identify key functions and food sources of five essential minerals.

What substances are neither animal nor vegetable in origin, cannot be created or destroyed by living organisms (or by any other ordinary means), and provide the raw materials from which all things on earth are made? The answer is the **mineral** elements, and they are displayed in full in the periodic table presented in Illustration 23.1. Minerals considered "essential," or required in the diet, are highlighted.

The body contains 40 or more minerals. Only 15 are an essential part of our diets; we obtain the others through the air we breathe or from other essential nutrients in the diet such as protein and vitamins.

Minerals are unlike the other essential nutrients in that they consist of single atoms. A single atom of a mineral typically does not have an equal number of protons (particles that carry a positive charge) and electrons (particles that carry a negative charge), and it therefore carries a charge. The charge makes minerals reactive. Many of the functions of minerals in the body are related to this property.

Key Nutrition Concepts

Material covered in this unit on minerals relates to the following key nutrition concepts:

• Foods provide energy (calories), nutrients, and other substances needed for growth and health.

• Humans have adaptive mechanisms for managing fluctuations in nutrient intake.

• Poor nutrition can influence the development of certain chronic diseases.

Getting a Charge Out of Minerals

The charge carried by minerals allows them to combine with other minerals of the opposite charge and form fairly stable compounds that become part of bones, teeth, cartilage, and other tissues. In body fluids, charged minerals serve as a source of electrical power that stimulates muscles to contract and nerves to react. The electrical current generated by charged minerals when performing these functions can be recorded by an electrocardiogram (abbreviated EKG or ECG) or an electroencephalogram (EEG). Abnormalities in the pattern of electrical activity in EKGs signal pending or past problems in the heart muscle. An EKG recording is shown in Illustration 23.2. Electroencephalograms similarly record electrical activity in the brain.

The charge minerals carry is related to many other functions. It helps maintain an adequate amount of water in the body and assists in neutralizing body fluids when they become too acidic or basic. Minerals that perform the roles of **cofactors** are components of proteins and enzymes, and they provide the "spark" that initiates enzyme activity.

Charge Problems Because minerals tend to be reactive, they may combine with other substances in food and form highly stable compounds that are not easily absorbed. Absorption of zinc from foods, for example, can vary from 0 to 100%, depending on what is attached to it. Zinc in whole grain products is very poorly absorbed because it is bound tightly to a substance called phytate. In contrast, zinc in meats is readily available because it is bound to protein. People whose sole source of zinc is whole grains have developed zinc deficiency, even though their intake of zinc is adequate.[3] The absorption of iron from foods in a meal decreases by as much as 50% if tea is consumed with the meal. In the intestines, iron binds with tannic acid in tea and forms a compound that cannot be broken down.[4] The calcium present in spinach and collard greens is poorly absorbed because it is firmly bound to oxalic acid. Many more examples could be given. The point is that you don't always get what you consume; the availability of minerals in food can vary a great deal.

minerals In the context of nutrition, minerals are specific, single atoms that perform particular functions in the body. There are 15 essential minerals—or minerals required in the diet.

cofactors Individual minerals required for the activity of certain proteins. For example:
• Iron is needed for hemoglobin's function in oxygen and carbon dioxide transport.
• Zinc is needed to activate or is a structural component of more than 200 enzymes.
• Magnesium activates over 300 enzymes involved in the formation of energy and proteins.

Groups

1 ← Atomic number
H
1s¹ ← Valence-electron configuration

$1s^1$

Period	1 1A	2 2A	3 3B	4 4B	5 5B	6 6B	7 7B	8	9 8B	10	11 1B	12 2B	13 3A	14 4A	15 5A	16 6A	17 7A	18 8A
1	1 H $1s^1$																	2 He $1s^2$
2	3 Li $2s^1$	4 Be $2s^2$											5 B $2s^2 2p^1$	6 C $2s^2 2p^2$	7 N $2s^2 2p^3$	8 O $2s^2 2p^4$	9 F $2s^2 2p^5$	10 Ne $2s^2 2p^6$
3	11 Na $3s^1$	12 Mg $3s^2$											13 Al $3s^2 3p^1$	14 Si $3s^2 3p^2$	15 P $3s^2 3p^3$	16 S $3s^2 3p^4$	17 Cl $3s^2 3p^5$	18 Ar $3s^2 3p^6$
4	19 K $4s^1$	20 Ca $4s^2$	21 Sc $4s^2 3d^1$	22 Ti $4s^2 3d^2$	23 V $4s^2 3d^3$	24 Cr $4s^1 3d^5$	25 Mn $4s^2 3d^5$	26 Fe $4s^2 3d^6$	27 Co $4s^2 3d^7$	28 Ni $4s^2 3d^8$	29 Cu $4s^1 3d^{10}$	30 Zn $4s^2 3d^{10}$	31 Ga $4s^2 4p^1$	32 Ge $4s^2 4p^2$	33 As $4s^2 4p^3$	34 Se $4s^2 4p^4$	35 Br $4s^2 4p^5$	36 Kr $4s^2 4p^6$
5	37 Rb $5s^1$	38 Sr $5s^2$	39 Y $5s^2 4d^1$	40 Zr $5s^2 4d^2$	41 Nb $5s^1 4d^4$	42 Mo $5s^1 4d^5$	43 Tc $5s^2 4d^5$	44 Ru $5s^1 4d^7$	45 Rh $5s^1 4d^8$	46 Pd $4d^{10}$	47 Ag $5s^1 4d^{10}$	48 Cd $5s^2 4d^{10}$	49 In $5s^2 5p^1$	50 Sn $5s^2 5p^2$	51 Sb $5s^2 5p^3$	52 Te $5s^2 5p^4$	53 I $5s^2 5p^5$	54 Xe $5s^2 5p^6$
6	55 Cs $6s^1$	56 Ba $6s^2$	57* La $6s^2 5d^1$	72 Hf $6s^2 5d^2$	73 Ta $6s^2 5d^3$	74 W $6s^2 5d^4$	75 Re $6s^2 5d^5$	76 Os $6s^2 5d^6$	77 Ir $6s^2 5d^7$	78 Pt $6s^1 5d^9$	79 Au $6s^1 5d^{10}$	80 Hg $6s^2 5d^{10}$	81 Tl $6s^2 6p^1$	82 Pb $6s^2 6p^2$	83 Bi $6s^2 6p^3$	84 Po $6s^2 6p^4$	85 At $6s^2 6p^5$	86 Rn $6s^2 6p^6$
7	87 Fr $7s^1$	88 Ra $7s^2$	89† Ac $7s^2 6d^1$	104 Unq $7s^2 6d^2$	105 Unp $7s^2 6d^3$	106 Unh $7s^2 6d^4$	107 Ns $7s^2 6d^5$	108 Hs $7s^2 6d^6$	109 Mt $7s^2 6d^7$									

Transition Metals

Metals ← → Nonmetals

Lanthanides

58 Ce $6s^2 4f^1 5d^1$	59 Pr $6s^2 4f^3$	60 Nd $6s^2 4f^4$	61 Pm $6s^2 4f^5$	62 Sm $6s^2 4f^6$	63 Eu $6s^2 4f^7$	64 Gd $6s^2 4f^7 5d^1$	65 Tb $6s^2 4f^9$	66 Dy $6s^2 4f^{10}$	67 Ho $6s^2 4f^{11}$	68 Er $6s^2 4f^{12}$	69 Tm $6s^2 4f^{13}$	70 Yb $6s^2 4f^{14}$	71 Lu $6s^2 4f^{14} 5d$

Actinides

90 Th $7s^2 6d^2$	91 Pa $7s^2 5f^2 6d^1$	92 U $7s^2 5f^3 6d^1$	93 Np $7s^2 5f^4 6d^1$	94 Pu $7s^2 5f^6$	95 Am $7s^2 5f^7$	96 Cm $7s^2 5f^7 6d^1$	97 Bk $7s^2 5f^9$	98 Cf $7s^2 5f^{10}$	99 Es $7s^2 5f^{11}$	100 Fm $7s^2 5f^{12}$	101 Md $7s^2 5f^{13}$	102 No $7s^2 5f^{14}$	103 Lr $7s^2 5f^{14} 6d$

Periods (vertical label on left)

Illustration 23.1 The periodic table lists all known minerals. The highlighted minerals are required in the human diet.

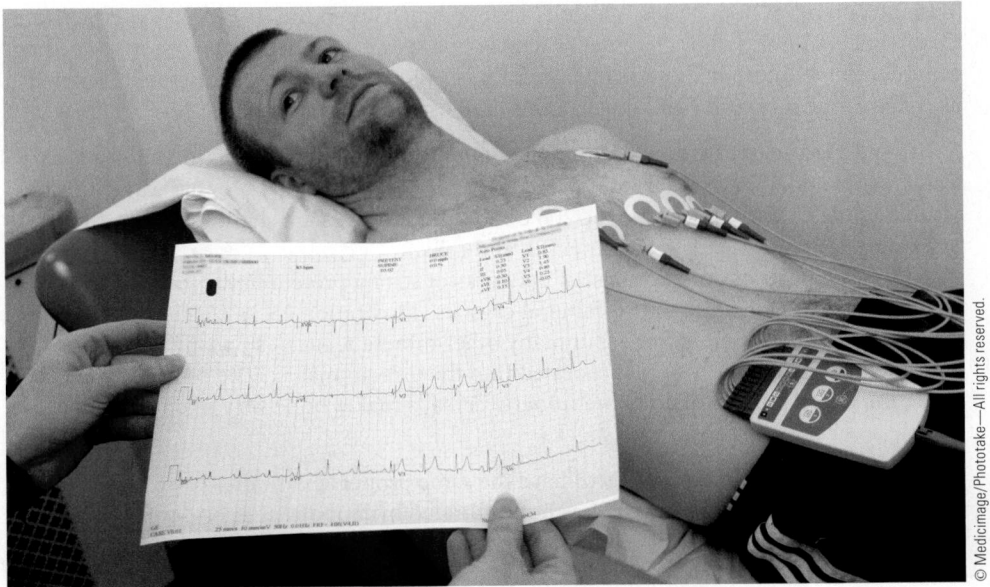

Illustration 23.2 The electrical current measured by an EKG results from the movement of charged minerals across membranes of the muscle cells in the heart.

Table 23.1 Percent of original mineral content lost in fruits by food storage method and in dried beans by cooking method[5]

	Fruits			Dried beans boiled 2–2.5 hours	
	Canned (%)	Frozen (%)	Dried (%)	Water drained (%)	Water used (%)
Calcium	5	5	0	35	30
Iron	0	0	0	25	20
Magnesium	0	0	0	30	25
Potassium	10	10	0	35	30
Zinc	0	0	0	15	10
Copper	10	10	0	45	40

Illustration 23.3 (top) Electron micrograph of healthy bone. (bottom) Electron micrograph of bone affected by osteoporosis.

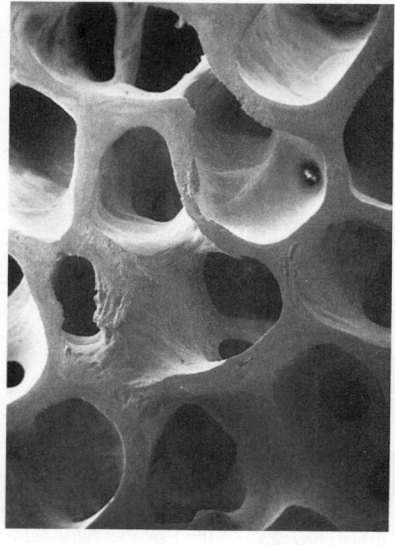

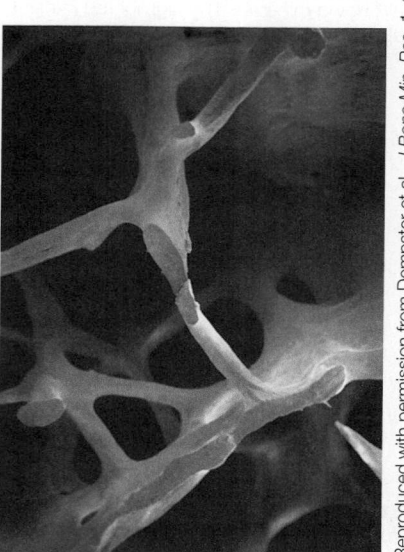

Reproduced with permission from Dempster et al., *J Bone Min. Res.* 1, 15–21, 1986.

Preserving the Mineral Content of Food Minerals in foods can be lost during food storage and preparation primarily due to the leaching out of minerals in cooking water and the drippings from the meats. Table 23.1 gives two examples of the percentage of minerals lost by fruit storage method and the method used to prepare dried beans. Dried foods retain minerals well, and those lost in cooking fluids can be recovered if the cooking water is minimized and consumed.

The Boundaries of This Unit All of the minerals could be the subject of fascinating stories, but in this unit we will concentrate on just three. A summary of the main features of all the essential minerals is provided in Table 23.2. This table lists the recommended intake level, functions, consequences of deficiency and overdose, and food sources of each mineral. Table 23.11 at the end of the unit lists food sources for many of the essential minerals.

The minerals highlighted are calcium, iron, and sodium. They have been selected primarily because they play important roles in the development of osteoporosis, iron-deficiency anemia, and hypertension, respectively. These disorders are widespread in the United States and in many other countries, and improved diets offer a key to their prevention and treatment.

Selected Minerals: Calcium

What you've heard about calcium is true: It's good for bones and teeth. About 99% of the 3 pounds of calcium in the body is located in bones and teeth. The remaining 1% is found in blood and other body fluids. We don't hear so much about this 1%, but it's very active. Every time a muscle contracts, a nerve sends out a signal, or blood clots to stop a bleeding wound, calcium in body fluids is involved. Calcium's most publicized function, however, is its role in bone formation and the prevention of osteoporosis.

A Short Primer on Bones Most of the bones we see or study are hard and dead. As a result, people often have the impression that bones in living bodies are that way. Nothing could be further from the truth. The 206 bones in our bodies are slightly flexible, living tissues infiltrated by blood vessels, nerves, and cells.

The solid parts of bones consist of networks of strong protein fibers (called the "protein matrix") embedded with mineral crystals (Illustration 23.3). Calcium is the most abundant mineral found in bone, but many other minerals, such as phosphorus, magnesium, and carbon, are also incorporated into the protein matrix. The combination of water, the tough protein matrix, and mineral crystals makes bone very strong yet slightly flexible and capable of absorbing shocks.

Teeth Are a Type of Bone Teeth have the same properties as other bone plus a hard outer covering called enamel, which is not infiltrated by blood vessels or nerves. Enamel serves to protect the teeth from destruction by bacteria and mechanical wear and tear.

Remodeling Your Bones Bones slowly and continually go through a repair and replacement process known, appropriately enough, as **remodeling**. During remodeling, the old protein matrix is replaced and remineralized. If insufficient calcium is available to complete the remineralization, or if other conditions such as vitamin D inadequacy prevent calcium from being incorporated into the protein matrix, **osteoporosis** results.[6]

Osteoporosis

If you are female, you have a one in four chance of developing osteoporosis in your lifetime. If you are a Caucasian, Hispanic, or Asian female, you have a higher risk of developing osteoporosis than if you are an African American woman or a male. If you are male, your chance of developing osteoporosis is one in eight. Approximately 8 million adults in the United States have osteoporosis, and 1.5 million suffer a broken wrist or hip or crushed spinal vertebrae each year because of the disease.[7] Far more adults, 34 million in all, are at risk of fractures due to osteopenia, a condition characterized by borderline low bone mineral density.[8]

Osteoporosis is a disabling disease that reduces quality of life and dramatically increases the need for health care.[9] Because the incidence of osteoporosis increases with age, its importance as a personal and public health problem is intensifying as the U.S. population ages. It currently appears, however, that a large percentage of the cases of osteoporosis can be prevented. The key to prevention is to build dense bones during childhood and the early adult years and then keep bones dense as you age.[7]

The Timing of Bone Formation Bones develop and mineralize throughout the first three decades of life. Even after the growth spurt occurs during adolescence and people think they are as tall as they will ever be, bones continue to increase in width and mineral content for 10 to 15 more years. Peak bone density, or the maximal level of mineral content in bones, is reached somewhere between the ages of 30 and 40 years. After that, bone mineral content no longer increases. The higher the peak bone mass, the less likely it is that osteoporosis will develop. People with higher peak bone mass simply have more calcium to lose before bones become weak and fracture easily. That also is the reason males experience osteoporosis less often than females do: They have more bone mass to lose.[8]

Bone size and density often remain fairly stable from age 30 to the mid-40s, but then bones tend to demineralize with increasing age. By the time women are 70, for example, their bones are 30–40% less dense on average than they once were.[10] A woman may lose an inch or more in height with age and develop the "dowager's hump" that is characteristic of osteoporosis in the spine (Illustration 23.4).

Role of Calcium and Vitamin D in Osteoporosis Since the late 1980s, a large number of studies have examined the relationships among dietary calcium intake, vitamin D status, and bone density. Results have been quite consistent: Bone mass before the age of 30 is increased by adequate calcium intakes and vitamin D status. One of the primary results of the studies indicates that both calcium and vitamin D are important in the prevention and treatment of osteoporosis.[11,12] Vitamin D is required for the absorption of calcium and the incorporation of calcium into bone. In adults over the age of 50, bone density tends to be preserved with calcium intakes of 1,000 to 1,200 milligrams of calcium and 10 to 20 mcg (400 to 800 IU) of vitamin D per day.[12,14] Low bone mineral density can be increased somewhat by switching from a low to an adequate intake of calcium and improved vitamin D status. The earlier in life this occurs, the more improvement in bone density is noted. High levels of bone density once achieved do not last forever; bone must be continually renewed by an adequate supply of calcium and vitamin D.[13]

The RDA (Recommended Dietary Allowance) for calcium intake by adults aged 19 to 50 years is 1,000 mg per day. The amount increases to 1,200 mg per day for women above the age of 50 and for men over the age of 70 years. The RDA for vitamin D is set at 15 mcg per day (600 IU) for males and females from the age of 1 through 70 years. These levels of intake correspond to amounts of calcium and vitamin D that help build bone density and contribute to the prevention and treatment of osteoporosis.[14]

Illustration 23.4 This woman's stooped appearance is due to osteoporosis.

remodeling The breakdown and buildup of bone tissue.

osteoporosis (*osteo* = bones; *poro* = porous, *osis* = abnormal condition) A condition characterized by porous bones; it is due to the loss of minerals from the bones.

Table 23.2 An intensive course on essential minerals

		Primary functions	Consequences of deficiency
  Calcium supplements are large because our daily need for calcium is high (1000 milligrams/day). Four pills provide 800 milligrams of calcium, the amount in 2 ²/₃ cups of milk. An aspirin is shown for comparison. 	**Calcium** Al[a] women: 1,000 mg men: 1,000 mg UL: 2,500 mg	• Component of bones and teeth • Needed for muscle and nerve activity, blood clotting	• Poorly mineralized, weak bones (osteoporosis) • Rickets in children • Osteomalacia (rickets in adults) • Stunted growth in children • Convulsions, muscle spasms
 	Phosphorus RDA women: 700 mg men: 700 mg UL: 4,000 mg	• Component of bones and teeth • Component of certain enzymes and other substances involved in energy formation • Needed to maintain the right acid–base balance of body fluids	• Loss of appetite • Nausea, vomiting • Weakness • Confusion • Loss of calcium from bones
	Magnesium RDA women: 310 mg men: 400 mg UL: 350 mg (from supplements only)	• Component of bones and teeth • Needed for nerve activity • Activates hundreds of enzymes involved in energy and protein formation and other body processes	• Stunted growth in children • Weakness • Muscle spasms • Personality changes

[a]AI (Adequate Intakes) ad RDAs (Recommended Dietary Allowances are for 19–30-year-olds: UL (Upper Limits) are for 19–70-year-olds, 1997–2004.

Consequences of overdose	Primary food sources	Highlights and comments
• Drowsiness • Calcium deposits in kidneys, liver, and other tissues • Suppression of bone remodeling • Decreased zinc absorption	• Milk and milk products (cheese, yogurt) • Calcium-fortified foods (some juices, breakfast cereals, soy milk)	• The average intake of calcium among U.S. women is approximately 60% of the DRI. • One in four women and one in eight men in the United States develop osteoporosis. • Adequate calcium and vitamin D status must be maintained to prevent bone loss.
• Muscle spasms	• Milk and milk products (cheese, yogurt) • Meats • Seeds, nuts • Phosphates added to foods Tony Freeman/PhotoEdit Phosphates are a common food additive.	• Deficiency is generally related to disease processes.
• Diarrhea • Dehydration • Impaired nerve activity due to disrupted utilization of calcium	• Plant foods (dried beans, nuts, potatoes, green vegetables) • Ready-to-eat cereals	• Magnesium is primarily found in plant foods where it is attached to chlorophyll. • Average intake among U.S. adults is below the RDA. PhotoDisc Magnesium is chlorophyll.

(continued)

Table 23.2 (continued)

		Primary functions	Consequences of deficiency
 Food and Drink Photos/age fotostock *Daniel Padavona/Shutterstock.com*	**Iron** RDA women: 18 mg men: 8 mg UL: 45 mg *Visuals Unlimited, Inc./Dr. Richard Kessel/Getty Images* *Visuals Unlimited, Inc./Dr. Gladden Willis/Getty Images* Iron-deficiency anemia is characterized by microcytic anemia and small, pale red blood cells (bottom photo). Normal red blood cells are shown in the top photo.	• Transports oxygen as a component of hemoglobin in red blood cells • Component of myoglobin (a muscle protein) • Needed for certain reactions involving energy formation	• Iron deficiency • Iron-deficiency anemia • Weakness, fatigue • Pale appearance • Reduced attention span and resistance to infection • Hair loss • Mental retardation, developmental delay in children
 sbarabu/Shutterstock.com	**Zinc** RDA women: 8 mg men: 11 mg UL: 40 mg	• Required for the activation of many enzymes involved in the reproduction of proteins • Component of insulin, many enzymes	• Growth failure • Delayed sexual maturation • Slow wound healing • Loss of taste and appetite • In pregnancy, low-birth-weight infants and preterm delivery
 Marie C Fields/Shutterstock.com	**Fluoride** AI women: 3 mg men: 4 mg UL: 10 mg	• Component of bones and teeth (enamel) • Helps rebuild enamel that is beginning to decay	• Tooth decay and other dental diseases
 D.A. Weinstein/Custom Medical Stock Photo/Newscom	**Iodine** RDA women: 150 mcg men: 150 mcg UL: 1,100 mcg	• Component of thyroid hormones that help regulate energy production and growth • Required for normal brain development	• Goiter, thyroid disease • Cretinism (mental retardation, hearing loss, growth failure) *© 2009 National Medical Slide Bank/Custom Medical Stock Photo* Goiter is a highly visible sign of deficiency.

Consequences of overdose	Primary food sources	Highlights and comments
• Hemochromatosis ("iron poisoning") • Vomiting, abdominal pain • Blue coloration of skin • Liver and heart damage, diabetes • Decreased zinc absorption • Atherosclerosis (plaque buildup) in older adults	• Liver, beef, pork • Dried beans • Iron-fortified cereals • Prunes, apricots, raisins • Spinach • Bread	• Cooking foods in iron and stainless steel pans increases the iron content of the foods. • Vitamin C, meat, and alcohol increase iron absorption. • Iron deficiency is the most common nutritional deficiency in the world. • Average iron intake of young children and women in the United States is low.
• Over 25 mg/day is associated with nausea, vomiting, weakness, fatigue, susceptibility to infection, copper deficiency, and metallic taste in mouth. • Increased blood lipids	• Meats (all kinds) • Dried beans • Grains • Nuts • Ready-to-eat cereals	• Like iron, zinc is better absorbed from meats than from plants. • Marginal zinc deficiency may be common, especially in children. • Zinc supplements taken within 24 hours of onset may decrease duration and severity of the common cold.
• Fluorosis • Brittle bones • Mottled teeth • Nerve abnormalities ©Dr. P. Marazzi/Photo Researchers, Inc. "Mottled teeth" result from excessive fluoride.	• Fluoridated water and foods and beverages made with it • White grape juice • Raisins • Wine	• Toothpastes, mouth rinses, and other dental care products may provide fluoride. • Fluoride overdose has been caused by ingestion of fluoridated toothpaste. • Fluoridated water is not related to cancer.
• Over 1 mg/day may produce pimples, goiter, decreased thyroid function, and thyroid disease. Medical-on-Line/Alamy Iodine deficiency during pregnancy produces cretinism in the offspring.	• Iodized salt • Milk and milk products • Seaweed, seafoods • Bread from commercial bakeries	• Iodine deficiency was a major problem in the United States in the 1920s and 1930s. Deficiency remains a major health problem in some developing countries. • Amount of iodine in plants depends on iodine content of soil. • Most of the iodine in our diet comes from the incidental addition of iodine to foods from cleaning compounds used by food manufacturers.

(continued)

Table 23.2 (continued)

		Primary functions	Consequences of deficiency
Dani Vincek/Shutterstock.com	**Selenium** RDA women: 55 mcg men: 55 mcg UL: 400 mcg	• Acts as an antioxidant in conjunction with vitamin E (protects cells from damage due to exposure to oxygen) • Needed for thyroid hormone production	• Anemia • Muscle pain and tenderness • Keshan disease (heart failure), Kashin-Beck disease (joint disease)
	Copper RDA women: 900 mcg men: 900 mcg UL: 10,000 mcg	• Component of enzymes involved in the body's utilization of iron and oxygen • Functions in growth, immunity, cholesterol and glucose utilization, brain development	• Anemia • Seizures • Nerve and bone abnormalities in children • Growth retardation
	Manganese AI women: 2.3 mg men: 1.8 mg	• Needed for the formation of body fat and bone	• Weight loss • Rash • Nausea and vomiting
	Chromium AI women: 35 mcg men: 25 mcg	• Required for the normal utilization of glucose and fat	• Elevated blood glucose and triglyceride levels • Weight loss
	Molybdenum RDA women: 45 mcg men: 45 mcg UL: 2,000 mcg	• Component of enzymes involved in the transfer of oxygen from one molecule to another	• Rapid heartbeat and breathing • Nausea, vomiting • Coma
Jill Battaglia/Shutterstock.com	**Sodium** AI women: 1,500 mg men: 1,500 mg UL: 2,300 mg	• Needed to maintain the right acid–base balance in body fluids • Helps maintain an appropriate amount of water in blood and body tissues • Needed for muscle and nerve activity	• Weakness • Apathy • Poor appetite • Muscle cramps • Headache • Swelling
DUSAN ZIDAR/Shutterstock.com Valentyn Volkov/Shutterstock.com	**Potassium** AI women: 4,700 mg men: 4,700 mg UL: Not determined	• Same as for sodium	• Weakness • Irritability, mental confusion • Irregular heartbeat • Paralysis
	Chloride AI women: 2,300 mg men: 2,300 mg UL: 3,600 mg	• Component of hydrochloric acid secreted by the stomach (used in digestion) • Needed to maintain the right acid–base balance of body fluids • Helps maintain an appropriate water balance in the body	• Muscle cramps • Apathy • Poor appetite • Long-term mental retardation in infants

Consequences of overdose	Primary food sources	Highlights and comments
• "Selenosis." Symptoms of selenosis are hair and fingernail loss, weakness, liver damage, irritability, and "garlic" or "metallic" breath.	• Fish • Eggs	• Content of foods depends on amount of selenium in soil, water, and animal feeds. • May play a role in the prevention of some types of cancer.
• Wilson's disease (excessive accumulation of copper in the liver and kidneys) • Vomiting, diarrhea • Tremors • Liver disease	• Potatoes • Grains • Dried beans • Nuts and seeds • Seafood • Ready-to-eat cereals	• Toxicity can result from copper pipes and cooking pans. • Average intake in the United States is below the RDA.
• Infertility in men • Disruptions in the nervous system, learning impairment • Muscle spasms	• Whole grains • Coffee, tea • Dried beans • Nuts	• Toxicity is related to overexposure to manganese dust in miners or contaminated ground water.
• Kidney and skin damage	• Whole grains • Wheat germ • Liver, meat • Beer, wine • Oysters	• Toxicity usually results from exposure in chrome-making industries or overuse of supplements. • Supplements do not build muscle mass or increase endurance.
• Loss of copper from the body • Joint pain • Growth failure • Anemia • Gout	• Dried beans • Grains • Dark green vegetables • Liver • Milk and milk products	• Deficiency is extraordinarily rare.
• High blood pressure in susceptible people • Kidney disease • Heart problems	• Foods processed with salt • Cured foods (corned beef, ham, bacon, pickles, sauerkraut) • Table and sea salt • Bread • Milk, cheese • Salad dressing	• Very few foods naturally contain much sodium. • Processed foods are the leading source of dietary sodium. • High-sodium diets are associated with the development of hypertension in "salt-sensitive" people.
• Irregular heartbeat, heart attack	• Plant foods (potatoes, squash, lima beans, tomatoes, plantains, bananas, oranges, avocados) • Meats • Milk and milk products • Coffee	• Content of vegetables is often reduced in processed foods. • Diuretics (water pills) and other antihypertension drugs may deplete potassium. • Salt substitutes often contain potassium.
• Vomiting	• Same as for sodium. (Most of the chloride in our diets comes from salt.)	• Excessive vomiting and diarrhea may cause chloride deficiency. • Legislation regulating the composition of infant formulas was enacted in response to formula-related chloride deficiency and subsequent mental retardation in infants.

Since many nutrients are involved in bone formation and maintenance, it is recommended that high-quality diets provide the recommended amounts of calcium and vitamin D. Foods such as milk and other dairy products provide the full array of nutrients needed by bone, including calcium, vitamin D, phosphorous, magnesium, and high-quality protein.[15] Exposure of the skin to direct sunlight for 10 to 15 minutes daily during warm parts of the year can boost vitamin D status.

Calcium and Vitamin D Supplements Calcium (1,000 to 1,200 mg per day) and vitamin D supplements (400 to 800 IU per day or higher if blood levels of vitamin D are very low) are often recommended components of the treatment of osteoporosis.[21] However, too much supplemental calcium and vitamin D can cause health problems. Supplemental doses of calcium above 2,000 mg per day along with vitamin D supplementation is associated with excessive absorption of calcium from food.[16] That can lead to the deposition of calcium in blood vessels and soft tissues, and an increased risk of kidney stone development.[17] Long-term intake of vitamin D from supplements in excess of 125 mcg (5,000 IU) per day has been associated with a greater risk of cancer at some sites like the pancreas, a greater risk of heart disease, and more falls and fractures among the elderly.[18] The Tolerable Upper Intake Level (UL) for vitamin D in adults is 100 mcg (4,000 IU) per day.

Food Sources of Calcium Most women over the age of 50 years and men over the age of 70 years consume less than the RDA for calcium.[19] Consequently, it is recommended that diets include good sources of calcium that would make up for deficits in intake.[2]

Over half of the calcium supplied by the diets of Americans comes from milk and milk products. Milk (including chocolate milk), cheese, and yogurt are all good sources of calcium (see Table 23.3). Some plants such as kale, broccoli, and bok choy provide appreciable amounts of calcium, too. On average, 32% of the calcium content of milk and milk products, calcium-fortified orange juice, and calcium supplements is absorbed, compared to approximately 5–60% of the calcium from different plants.[48]

Many foods rich in calcium aren't loaded with calories or fat. As Table 23.3 shows, low-fat yogurt, skim and soymilk, kale, and broccoli, for example, provide good amounts of calcium at a low cost in calories. Calcium is also appearing in unexpected places such as in candy, snack bars, waffles, and bread (Illustration 23.5). The array of calcium-fortified foods has increased in the United States in response to publicity about our need for more of it. Health claims relating to the benefits of calcium in reducing the risk of osteoporosis may appear on the labels of food products that qualify as good sources of calcium. Look for health claims on foods such as milk, yogurt, and calcium-fortified foods such as orange juice, breakfast cereals, and grain products (Illustration 23.6).

Diet, Lifestyles, and Osteoporosis Calcium and vitamin D status are important but are not the only factors related to the development and maintenance of bone, and the

Table 23.3 Caloric content, calcium level, and percentage of available calcium from different foods[48]

Food	Amount	Calories	Calcium (mg)	Calcium absorbed (%)	Available calcium (mg)
Yogurt, low fat	1 cup	143	13	32	132
Skim milk	1 cup	85	301	32	96
Soy milk (fortified)	1 cup	79	300	31	93
1% milk	1 cup	163	274	32	88
Tofu	1 cup	188	260	31	81
Cheese	1 oz	114	204	32	65
Kale, cooked	1 cup	42	94	49	46
Broccoli, cooked	1 cup	44	72	61	44
Bok choy, raw	1 cup	9	73	54	39
Dried beans	1 cup	209	120	24	29
Spinach, cooked	1 cup	42	244	5	12

Illustration 23.5 The number of calcium-fortified foods is increasing.

"Regular exercise and a healthy diet with enough calcium help maintain good bone health and may reduce the risk of osteoporosis later in life."

Nutrition Facts

Serving Size: 1 cup (240ml)
Servings per Container: 16

Amount per Serving

Calories 110   Calories from Fat 20

	% Daily Value*
Total Fat 2.5g	**4%**
Saturated Fat 1.5g	**8%**
Trans Fat 0g	
Cholesterol 15mg	**4%**
Sodium 135mg	**6%**
Total Carbohydrate 13g	**4%**
Dietary Fiber 0g	**0%**
Sugars 12g	
Protein 8g	

Vitamin A 10% • Vitamin C 4%

Calcium 30% Iron 0% Vitamin D 25%

Phosphorus 10%

* Percent Daily Values are based on a 2,000 calorie diet.

Illustration 23.6 Food products that are good sources of calcium can be labeled with a health claim.

---

prevention and treatment of osteoporosis. Other factors, such as age, physical activity, genetic traits, diet quality, and body size influence bone health (Table 23.4). If you were to find all of the various risk factors combined in one person, that individual would be a thin woman with light skin who has consumed too little calcium, has poor vitamin D status, is physically inactive, an excessive alcohol drinker, a smoker, and genetically "small-boned." Additionally, she would have had her ovaries surgically removed for medical reasons before the age of 45.[9] Few people meet every aspect of that description, but people who have several of these characteristics are more likely to develop osteoporosis than those who don't.

Overall healthful diets that mimic the food assortment described in MyPlate.gov food guidance materials is recommended for the prevention of osteoporosis. This type of diet is based around plant foods including vegetables, fruits, whole grains and whole grain products, low-fat dairy foods, dried beans, nuts, seeds, and lean meats and fish. Regular physical activity that improves strength, balance, and agility is an important component of lifestyles that reduce the risk of osteoporosis. The risk of osteoporosis is also lower in people who do not smoke, drink excessively, or maintain a very thin body shape.[20,21]

At one time, researchers thought that a relatively high intake of protein might intensify the effects of low intakes of calcium on the development of osteoporosis. Studies in humans, however, have failed to demonstrate harmful effects of high protein intake on the development of osteoporosis given adequate calcium intake.[20]

How Is Osteoporosis Treated? Calcium supplements or adequate calcium intakes (around 1,000 to 1,200 milligrams per day), vitamin D supplements (800 IU or more) or sunshine, five to nine servings of fruits and vegetables daily, and weight-bearing exercise like walking and tennis decrease the progression of osteoporosis in many people.[23]

Selected Minerals: Iron

Most of the body's iron supply is found in **hemoglobin**. Small amounts are present in **myoglobin**, and free iron is involved in processes that capture energy released during the breakdown of proteins and fats.

Table 23.4 Risk factors for osteoporosis[20]

- Female
- Menopause
- Poor overall diet
- Deficient calcium intake
- Caucasian or Asian heritage
- Thinness ("small bones")
- Cigarette smoking
- Excessive alcohol intake
- Ovarectomy (ovaries removed) before age 45
- Physical inactivity
- Deficient vitamin D status
- Genetic factors

hemoglobin The iron-containing protein in red blood cells.

myoglobin The iron-containing protein in muscle cells.

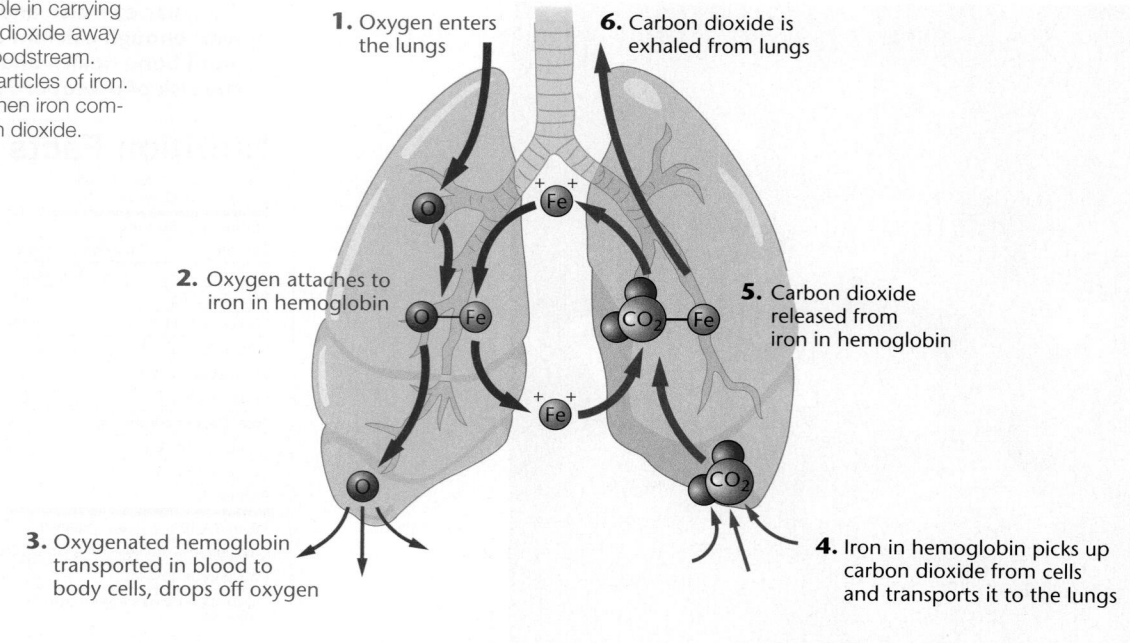

1. Oxygen enters the lungs

6. Carbon dioxide is exhaled from lungs

2. Oxygen attaches to iron in hemoglobin

5. Carbon dioxide released from iron in hemoglobin

3. Oxygenated hemoglobin transported in blood to body cells, drops off oxygen

4. Iron in hemoglobin picks up carbon dioxide from cells and transports it to the lungs

© Cengage Learning

The Role of Iron in Hemoglobin and Myoglobin What happens to a car when its paint gets scratched? After a while, the exposed metal rusts. The iron in the metal combines with oxygen in the air, and the result is iron oxide, or "rust." Iron readily combines with oxygen, and that property is put to good use in the body. From its location in hemoglobin in red blood cells, iron loosely attaches to oxygen when blood passes near the inner surface of the lungs. The bright red, oxygenated blood is then delivered to cells throughout the body. When the oxygenated blood passes near cells that need oxygen for energy formation or for other reasons, oxygen is released from the iron and diffuses into cells (Illustration 23.7). The free iron in hemoglobin then picks up carbon dioxide, a waste product of energy formation. When carbon dioxide attaches to iron, blood turns from bright red to dark bluish red. Blood then circulates back to the lungs, where carbon dioxide is released from the iron and exhaled into the air. The free iron attaches again to oxygen that enters the lungs, and the cycle continues.

Iron in myoglobin traps oxygen delivered by hemoglobin, stores it, and releases it as needed for energy formation for muscle activity. In effect, myoglobin boosts the supply of oxygen available to muscles.

The functions of iron just described operate smoothly when the body's supply of iron is sufficient. Unfortunately, that is often not the case.

Iron Deficiency Is a Big Problem The most widespread nutritional deficiency in both developing and developed countries is **iron deficiency** (Table 23.5). It is estimated that one out of every four people in the world is iron deficient.[1] For the most part, iron deficiency affects very young children and women of childbearing age, who have a high need for iron and frequently consume too little of it.[24] Iron deficiency may develop in people who have lost blood due to injury, surgery, or ulcers. Donating blood more than three times a year can also precipitate iron deficiency.[25]

iron deficiency A disorder that results from a depletion of iron stores in the body. It is characterized by weakness, fatigue, short attention span, poor appetite, increased susceptibility to infection, and irritability.

iron-deficiency anemia A condition that results when the content of hemoglobin in red blood cells is reduced due to a lack of iron. It is characterized by the signs of iron deficiency plus paleness, exhaustion, and a rapid heart rate.

Consequences of Iron Deficiency Many body processes sputter without sufficient oxygen. People with iron deficiency usually feel weak and tired. They have a shortened attention span and a poor appetite, are susceptible to infection, and become irritable easily. If the deficiency is serious enough, **iron-deficiency anemia** develops, and additional symptoms occur. People with iron-deficiency anemia look pale, are easily exhausted, and have rapid heart rates. Iron-deficiency anemia is a particular problem for infants and young children because it is related to lasting retardation in mental development.[26]

Table 23.5 Incidence of iron deficiency[1,25]

	Population with iron deficiency (%)
Worldwide (children under five years):	
Developing countries	51
Developed countries	12
United States:	
Children, 1–3 years	8
Pregnant women	12
Females, 20–49 years	7
Males, 12–49 years	1 or less

Food Sources of Iron "Enough" iron, according to the RDA, is 8 milligrams for men and 18 milligrams per day for women aged 19 to 50 years. Consuming that much iron can be difficult for women. On average, 1,000 calories' worth of food provides about 6 milligrams of iron. Women would have to consume around 2,500 calories per day to obtain even 15 milligrams of iron on average. Selection of good sources of iron has to be done on a better-than-average basis if women are to get enough.

Iron is found in small amounts in many foods, but only a few foods such as liver, beef, and prune juice are rich sources. (A list of food sources of iron is given in Table 23.11.) Foods cooked in iron and stainless steel pans can be a significant source of iron because some of the iron in the pan leaches out during cooking. On average, approximately 1 milligram of iron is added to each 3-ounce serving of food cooked in these pans.[27]

Most of the iron in plants and eggs is tightly bound to substances such as phytates or oxalic acid that limit iron absorption, making these foods relatively poor sources of iron even though they contain a fair amount of it. (Differences in the proportions of iron absorbed from various food sources are shown in Illustration 23.8.) A 3-ounce hamburger and a cup of asparagus both contain approximately 3 milligrams of iron, for example. But 20 times more iron can be absorbed from the hamburger than from the asparagus.[28] Absorption of iron from plants is increased substantially if foods containing vitamin C are included in the same meal.[4]

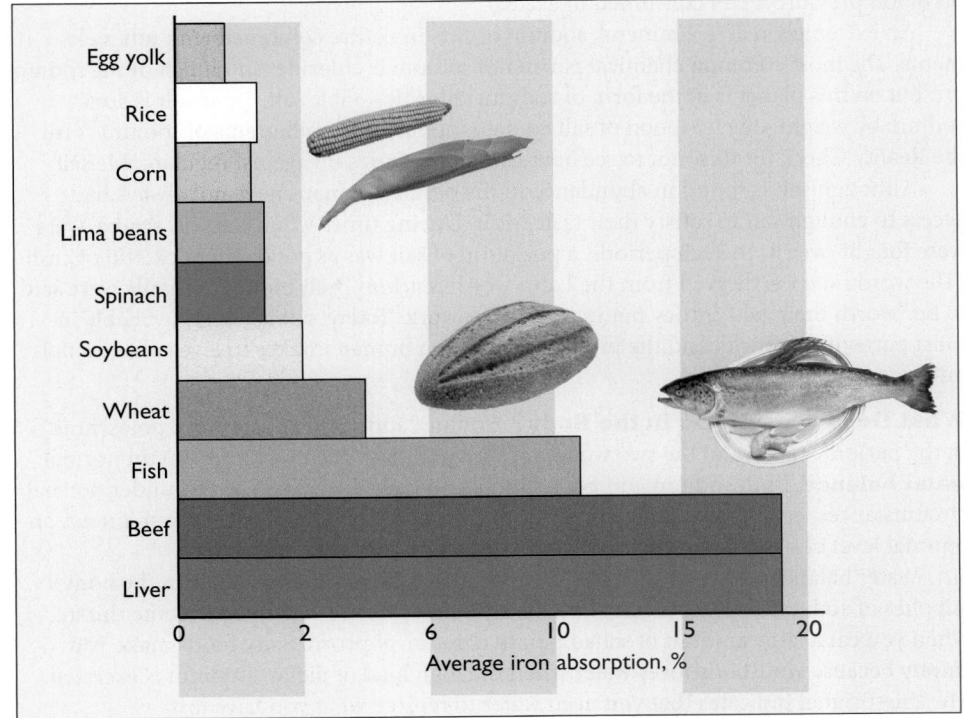

Illustration 23.8 Average percentage of iron absorbed from selected foods by healthy adults.[28]

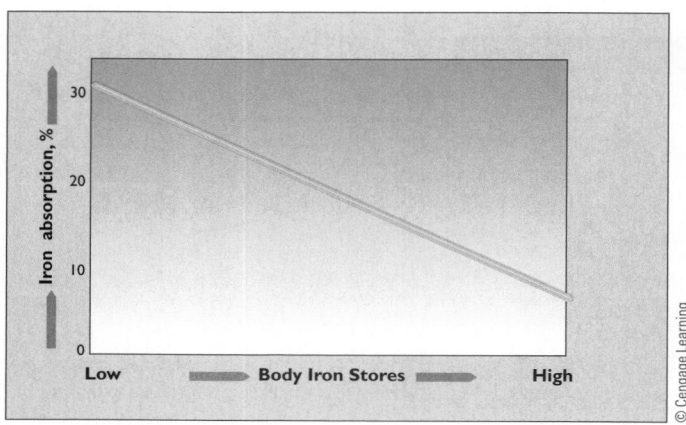

Illustration 23.9 People absorb more iron from foods and supplements when body stores of iron are low than when stores of iron are high.

Iron absorption is also increased by low levels of iron stores in the body (Illustration 23.9). In other words, if you're in need of iron, your body sets off mechanisms that allow more of it to be absorbed from foods or supplements. When iron stores are high, less iron is absorbed.

The body's ability to regulate iron absorption provides considerable protection against iron deficiency and overdose. The protection is not complete, however, as evidenced by widespread iron deficiency and the occurrence of iron toxicity.

Iron Toxicity Excess iron absorbed into the body cannot be easily excreted. Consequently, it is deposited in various tissues such as the liver, pancreas, and heart. There, the iron reacts with cells, causing damage that can result in liver disease, diabetes, and heart failure. One in two hundred people in the United States has an inherited tendency to absorb too much iron (a disorder called hemochromatosis and pronounced *hem-oh-chrom-ah-toe-sis*). Other people develop iron toxicity from consuming large amounts of iron with alcohol (alcohol increases iron absorption) or from very high iron intakes, usually due to overdoses of iron supplements.[25]

Table 23.6 Accidental overdoses of chemical substances by children up to age four in the United States[29]

Substance	Cases per 100,000 children
Aspirin and substitutes	12.7
Solvents/petroleum products	11.2
Tranquilizers	9.9
Iron supplements	8.7
Corrosives and caustics	7.4

Each year in the United States, more than 10,000 people accidentally overdose on iron supplements.[29] Victims are often young children who mistakenly think iron pills are candy (Table 23.6 and Illustration 23.10). The lethal dose of iron for a two-year-old child is about 3 grams,[30] the amount of iron present in 25 pills containing 120 milligrams of iron each. Iron supplements that are not being used should be thrown away or stored in a place where toddlers cannot get to them.

Selected Minerals: Sodium

On the one hand, sodium is a gift from the sea; on the other, it is a hazard to health. Sodium's life- and health-sustaining functions are frequently overshadowed by its effects on blood pressure when consumed in excess.

An extremely reactive mineral, sodium occurs in nature combined with other elements. The most common chemical partner of sodium is chloride, and much of the sodium present on this planet is in the form of sodium chloride—table salt. Table salt is 40% sodium by weight; one teaspoon of salt contains about 2,300 milligrams of sodium. Visit the Reality Check for this unit to see how sea salt measures up against regular table salt.

Although salt is found in abundance in the oceans, humans have not always had access to enough salt to satisfy their taste for it. During times when salt was scarce, wars were fought over it. In such periods, a pocketful of salt was as good as a pocketful of cash. (The word *salary* is derived from the Latin word *salarium*, "salt money.") People were said to be "worth their salt" if they put in a full day's work. Today, salt is widely available in most parts of the world, and the problem is limiting human intakes to levels that do not interfere with health.

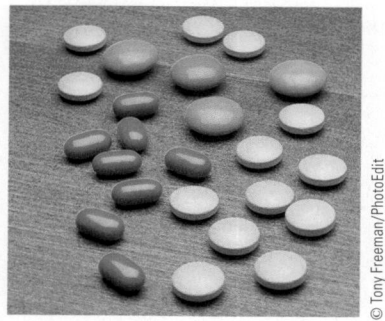

Illustration 23.10 If you were three years old, could you tell which "pills" are candy and which are the iron supplements? Overdoses of iron supplements are a leading cause of accidental poisoning in young children. (The iron supplements are the lightest green in color.)

What Does Sodium Do in the Body? Sodium appears directly above potassium in the periodic table, and the two work closely together in the body to maintain normal **water balance**. Both sodium and potassium chemically attract water, and under normal circumstances, each draws sufficient water to the outside or inside of cells to maintain an optimal level of water in both places.[33]

Water balance and cell functions are upset when there's an imbalance in the body's supplies of sodium and potassium. You have probably noticed that you become thirsty when you eat a large amount of salted potato chips or popcorn. Salty foods make you thirsty because your body loses water when the high load of dietary sodium is excreted. The thirst signal indicates that you need water to replace what you have lost.

water balance The ratio of the amount of water outside cells to the amount inside cells; a proper balance is needed for normal cell functioning.

The loss of body water that accompanies ingestion of large amounts of salt explains why seawater neither quenches thirst nor satisfies the body's need for water. When a person drinks seawater, its high concentration of sodium causes the body to excrete more water than it retains. Rather than increasing the body's supply of water, the ingestion of seawater increases the need for water.

In healthy people, the body's adaptive mechanisms provide a buffer against upsets in water balance due to high sodium intakes. It appears that many people are overwhelming the body's ability to cope with high sodium loads, however. High dietary intakes of sodium appear to play an important role in the development of **hypertension** in many people.[34]

A Bit About Blood Pressure To circulate through the body, blood must exist under pressure in the blood vessels. The amount of pressure exerted on the walls of blood vessels is greatest when pulses of blood are passing through them (that's when "systolic" blood pressure is measured) and least between pulses (that's when "diastolic" blood pressure is taken). Blood pressure measurements note the highest and lowest pressure in blood vessels (Illustration 23.11).

Blood pressure levels less than 120/80 millimeters of mercury (mm Hg) are considered normal, whereas levels between 120/80 through 139/89 mm Hg are classified as "**prehypertension.**" It is estimated that a third of U.S. adults qualify as having prehypertension.[3] Values of 140/90 mm Hg and higher qualify as hypertension (Table 23.7), and approximately one-third of American adults have this disorder.[2] Several blood pressure measurements, taken while a person is relaxed, are needed to obtain an accurate measure of blood pressure. Even going to a clinic or doctor's office to have blood pressure measured can raise it for some people and lead to a false diagnosis of hypertension. This condition is referred to as "white coat hypertension."[35]

Although not considered a disease by itself, the presence of hypertension substantially increases the risk that a person will develop heart disease or kidney failure or will experience a heart attack or stroke.[2]

What Causes Hypertension? Approximately 5–10% of all cases of hypertension can be directly linked to a cause. People who have hypertension with no identifiable cause (90-95% of all cases) are said to have **essential hypertension.**[36]

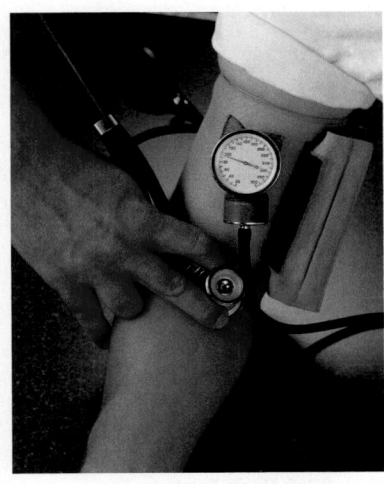

Illustration 23.11 A blood pressure test. Blood pressure is expressed as two numbers: Systolic pressure measures the force of the blood when the heart contracts. Diastolic pressure measures the force of the blood when the heart is at rest.

hypertension High blood pressure. It is defined as blood pressure exerted inside blood vessel walls that typically exceeds 140/90 millimeters of mercury. Hypertension in children and adolescents is defined as systolic or diastolic blood pressure equal to or greater than the 95th blood pressure percentile of sex-, age- and height-specific blood pressure percentiles.

prehypertension In adults, typical blood pressure levels of 120/80 mm Hg through 139/89 mm Hg. In children and adolescents, prehypertension is defined as systolic or diastolic blood pressure equal to or greater than the 90th percentile but less than the 95th percentile of sex-, age- and height-specific blood pressure percentiles.

essential hypertension Hypertension of no known cause; also called primary or idiopathic hypertension, it accounts for 90-95% of all cases of hypertension.

Table 23.7 **Hypertension categories**[41]

Category	Systolic (mm Hg)[a]	Diastolic (mm Hg)
Normal	<120	80
Prehypertension	120–139	80–89
Hypertension		
Stage 1	140–159	90–99
Stage 2	160+	100+

[a]mmHg = millimeters of mercury.

REALITY CHECK
Is Sea Salt a Better Choice Than Table Salt?

Matty and Miriam were shopping for snacks to serve when their friends came over for the game. They are now standing in front of the nut section at the store and can't decide if they should get nuts salted with sea salt or the regular salted nuts.

Who gets thumbs up?

Answer appears on page 23-18.

Matty: Sea salt is lower in sodium. Why don't we get the sea-salted nuts?

Miriam: I agree. It's amazing, though. I've had them before and they still taste salty enough.

A survey by the American Heart Association of 1,000 adults discovered that 6 out of 10 thought sea salt was a low-sodium alternative to table salt. Equal amounts of sea salt and table salt contain the same amount of sodium (40%). Unlike table salt, sea salt (as well as Kosher salt) is generally not fortified with iodine. Iodized table salt is a leading source of iodine in U.S. diets.[32]

Matty:

Miriam:

Table 23.8 Risk factors for hypertension[41]

- Age
- Family history
- High-sodium, low-potassium diet
- Obesity
- Physical inactivity
- Excessive alcohol consumption
- Smoking
- Frequent stress, anxiety

A number of dietary and other risk factors for hypertension have been identified (Table 23.8). Foremost among the evidence linking diet to hypertension are population studies that show a relationship between salt intake and hypertension. As salt intake rises, so do rates of hypertension in populations. When people with hypertension reduce their salt intake, their blood pressure tends to drop somewhat.[2] Further scrutiny of the data on salt and blood pressure, however, reveals that not everyone is equally susceptible to high-salt diets.[37] Some people are genetically susceptible to salt or have subtle forms of kidney disorders that raise their blood pressure more in response to high-sodium diets than is the case for other people.[33] This condition is known as **salt sensitivity**.

Salt Sensitivity   Approximately 51% of people with hypertension and 26% of people with normal blood pressure are salt sensitive.[39,40] Salt sensitivity is more common in African Americans than in other population groups. It has been identified in 73% of African Americans with hypertension.[38] Reduction in salt intake by people who are salt sensitive, along with weight loss if overweight, substantially improves blood pressure in most cases.[2] Because it is currently difficult to identify who in the population is salt sensitive and who is not; because most Americans consume much more sodium than needed; and because high-sodium diets tend to increase blood pressure somewhat in the population in general, official advice for adults is to limit their sodium intake to 1,500 milligrams per day (the equivalent of approximately two-thirds level teaspoon of salt from all sources).[2]

Other Risk Factors for Hypertension   Obesity is a major risk factor for hypertension. For obese people with hypertension, the most effective treatment is weight loss. Excessive alcohol intake can prompt the development of hypertension, and moderation of intake (to two or fewer alcoholic drinks per day) can bring blood pressure back down. Physically inactive lifestyles foster the development of hypertension. Low intake of foods rich in potassium also contribute to the development high blood pressure.[2]

Diets low in potassium are a well-established risk factor for hypertension, and most Americans consume too little of it from foods. On average, U.S. women consume half of the RDA for potassium of 4,700 mg daily and men consume a third less than the RDA.[19] Diets containing adequate amounts of potassium from foods help lower blood pressure and appear to counteract the effects of high sodium intake on blood pressure.[2] The Take Action feature in this unit attempts to increase your awareness of good sources of potassium. If you are like most Americans, you are consuming too little of it.

Reduction of Sodium Intake   Most children and adults in the United States consume over twice the recommended amount of sodium.[19] The leading sources of salt (and therefore of sodium) in the U.S. diet are processed foods (Table 23.9). Restricting the use of foods to which salt has been added during processing is the most effective way to lower salt intake. These foods account for around 75% of the total sodium intake by Americans.[44,47] Approximately 10% of total sodium intake comes from salt added at the table.[44]

High-salt processed foods include frozen meals, salad dressings, canned soups, ham, sausages, and biscuits. Only a small proportion of our total sodium intake enters our diet

salt sensitivity   A genetically influenced condition in which a person's blood pressure rises when high amounts of salt or sodium are consumed. Such individuals are sometimes identified by blood pressure increases of 5 or 10% or more when switched from a low-salt to a high-salt diet.

take action   To Increase Foods Rich in Potassium in Your Diet

The Dietary Guidelines for Americans has designated potassium as a nutrient likely to be underconsumed by Americans in general. Foods listed below are good sources of potassium, as well as many other nutrients. Read the list and check the foods you consume at least twice a week. If you can't check four or more of the foods, consider taking action to include more good sources of potassium in your diet.

____ bran buds/flakes
____ avocado
____ dried beans
____ orange juice
____ lima beans
____ banana
____ baked potato

____ sweet potato
____ winter squash
____ spinach
____ tomato juice
____ orange juice
____ yogurt
____ fish

Table 23.9 Major sources of sodium in the American diet[44,47]

Sodium source	Contribution to sodium intake (%)
Processed foods	75
Fresh foods	12
Salt added at the table	10
Salt added during cooking and food preparation	3

from fresh foods. Very few foods naturally contain much sodium—at least not until they are processed (Illustration 23.12).

How Is Hypertension Treated? The recommended approach to treatment of all cases of hypertension consists of dietary and lifestyle changes and the use of medications if necessary.[41] Weight loss and smoking cessation (if needed), a reduced-sodium diet, regular exercise, moderate alcohol consumption (if any), and the DASH diet are basic components of the approach to treatment (Table 23.10). The DASH diet (Illustration 23.13) is based on vegetables, fruits, low-fat dairy products, whole grains, and poultry and fish. Its composition is similar to that of diets recommended for heart disease and cancer prevention. People who adhere to the DASH diet often bring their blood pressure levels back into the normal range.[42] If blood pressure remains elevated after dietary and lifestyle changes have been implemented, or if blood pressure is quite high when diagnosed, anti-hypertension drugs are usually prescribed.[41]

Table 23.10 Approaches to the treatment of hypertension[36,41]

- Weight loss (if needed)
- Salt intake <1,500 mg of sodium per day
- Moderate alcohol consumption (if any)
- Regular physical activity (30 minutes per day)
- The DASH diet
- Antihypertension drugs (if hypertension not controlled by above measures)
- Meditation, yoga

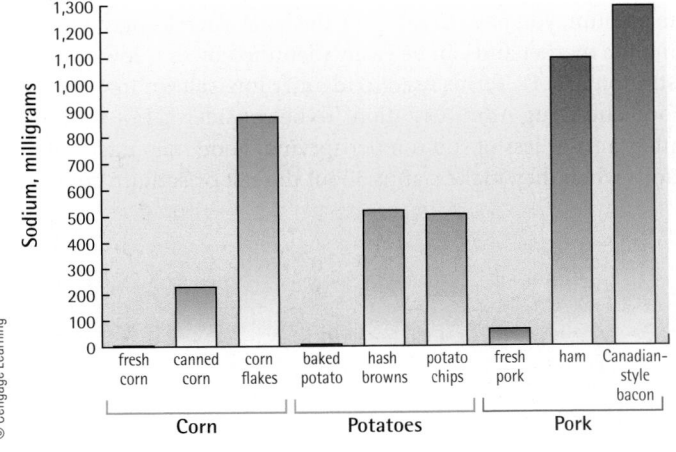

© Cengage Learning

Illustration 23.12   Examples of how processing increases the sodium content of foods. Sodium values are for a three-ounce serving of each food shown.

The Dash Eating Plan

In the mid-1990s a revolutionary approach to the control of mild and moderate hypertension was tested, and the results have changed health professionals' thinking about high blood pressure prevention and management. Called the DASH (Dietary Approaches to Stop Hypertension) Eating Plan, it didn't focus on salt restriction and was related to significant reductions in blood pressure within two weeks in most people tested. In some people, reductions in blood pressure were sufficient to erase the need for antihypertension medications, and for others the diet reduced the amount or variety of medications needed. A subsequent study showed that a low-sodium diet boosts the blood pressure–lowering effects of the DASH diet, especially in African Americans.[45] This dietary pattern is also associated with reduced risk of heart disease and stroke,[2] and is recommended by MyPlate.gov and the 2010 Dietary Guidelines as a health-promoting diet for people in general.

The DASH diet consists of eating patterns made up of the following food groups:

	Daily Servings
Vegetables	4–5
Fruits	4–5
Grain products, mostly whole grain	7–8
Low-fat milk and dairy products	2–3
Lean meats, fish, poultry	6 ounces
Nuts, seeds, dried beans	1

Although it does not work for all individuals with hypertension, the DASH diet is a mainstay in clinical efforts to reverse prehypertension and to manage hypertension.

A very useful Web site is available that describes refinements in the diet based on research studies and that gives practical tips on how to implement and follow the DASH eating pattern (www.nhlbi.nih.gov/health/public/heart/hbp/dash/new_dash.pdf). You can find the address quickly by searching the key term "DASH diet."

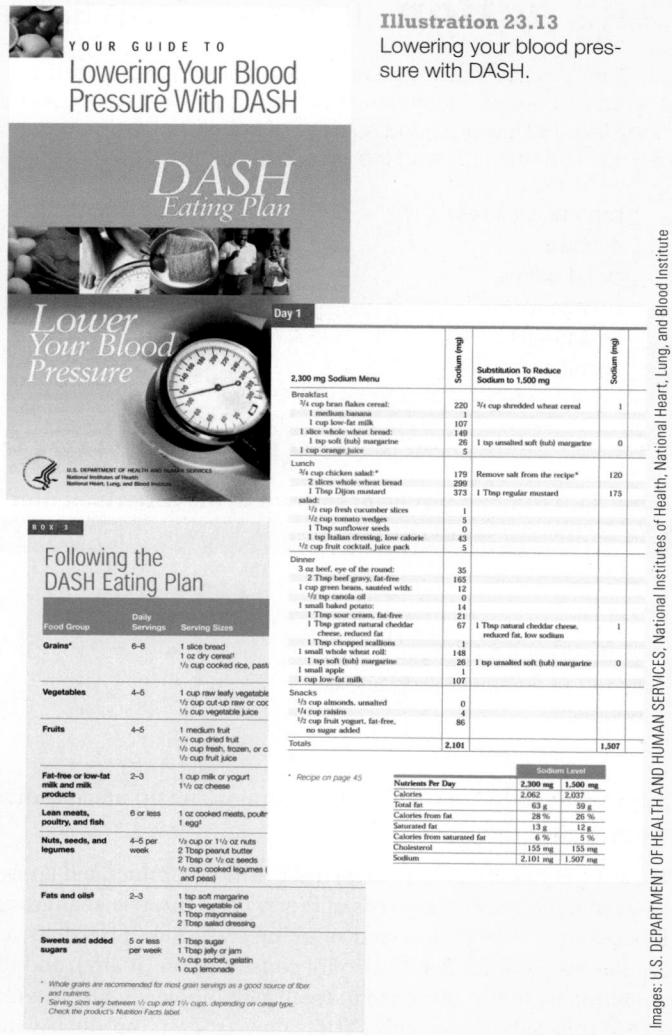

Images: U.S. DEPARTMENT OF HEALTH AND HUMAN SERVICES, National Institutes of Health, National Heart, Lung, and Blood Institute

Illustration 23.13 Lowering your blood pressure with DASH.

Using spices and lemon juice to flavor foods, consuming fresh vegetables (rather than processed ones) and fresh fruits, and checking out the sodium content of foods and then selecting low-sodium ones can also help reduce sodium intake.

Label Watch Not all processed foods with added sodium taste salty. To find out which processed foods are high in sodium, you have to examine the label. Increasingly, low-salt processed foods are entering the market and can be easily identified by the "low-salt" message on the label (Illustration 23.14). Terms used to identify low-salt (or low-sodium) foods are defined by the Food and Drug Administration. To be considered low sodium, foods must contain 140 milligrams or less of sodium per serving. Food manufacturers must adhere to the definitions when they make claims about the salt or sodium content of a food on the label.

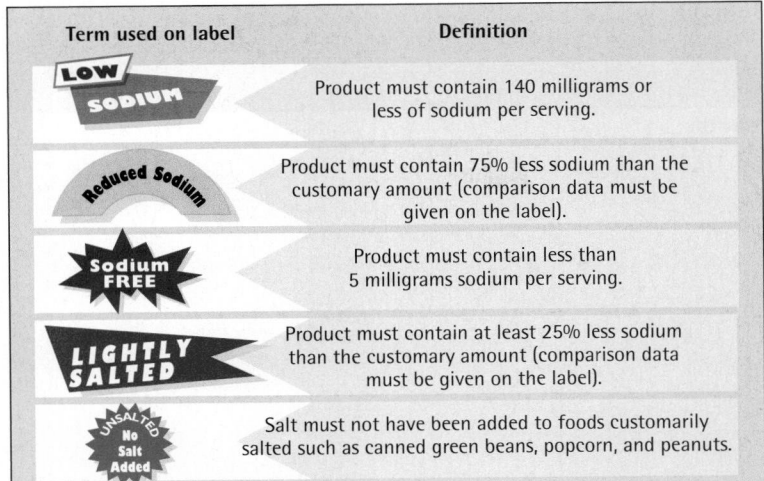

Term used on label	Definition
LOW SODIUM	Product must contain 140 milligrams or less of sodium per serving.
Reduced Sodium	Product must contain 75% less sodium than the customary amount (comparison data must be given on the label).
Sodium FREE	Product must contain less than 5 milligrams sodium per serving.
LIGHTLY SALTED	Product must contain at least 25% less sodium than the customary amount (comparison data must be given on the label).
UNSALTED No Salt Added	Salt must not have been added to foods customarily salted such as canned green beans, popcorn, and peanuts.

Illustration 23.14 When a label makes a claim about the sodium content of a food product, the label must list the amount of sodium in a serving and adhere to these definitions.

© Cengage Learning

Table 23.11 Food sources of minerals

Magnesium		
Food	**Amount**	**Magnesium (mg)**
Legumes:		
Lentils, cooked	1/2 cup	134
Split peas, cooked	1/2 cup	134
Tofu	1/2 cup	130
Nuts:		
Peanuts	1/4 cup	247
Cashews	1/4 cup	93
Almonds	1/4 cup	80
Grains:		
Bran buds	1 cup	240
Wild rice, cooked	1/2 cup	119
Breakfast cereal, fortified	1 cup	85
Wheat germ	2 Tbsp	45
Vegetables:		
Bean sprouts	1/2 cup	98
Black-eyed peas	1/2 cup	58
Spinach, cooked	1/2 cup	48
Lima beans	1/2 cup	32
Milk and milk products:		
Milk	1 cup	30
Cheddar cheese	1 oz	8
American cheese	1 oz	6
Meats:		
Chicken	3 oz	25
Beef	3 oz	20
Pork	3 oz	20

(continued)

Table 23.11 (continued)

Calcium[a]		
Food	**Amount**	**Calcium (mg)**
Milk and milk products:		
Yogurt, low-fat	1 cup	413
Milk shake (low-fat frozen yogurt)	1¼ cup	352
Yogurt with fruit, low-fat	1 cup	315
Skim milk	1 cup	301
1% milk	1 cup	300
2% milk	1 cup	298
3.25% milk (whole)	1 cup	288
Swiss cheese	1 oz	270
Milk shake (whole milk)	1¼ cup	250
Frozen yogurt, low-fat	1 cup	248
Frappuccino	1 cup	220
Cheddar cheese	1 oz	204
Frozen yogurt	1 cup	200
Cream soup	1 cup	186
Pudding	½ cup	185
Ice cream	1 cup	180
Ice milk	1 cup	180
American cheese	1 oz	175
Custard	½ cup	150
Cottage cheese	½ cup	70
Cottage cheese, low-fat	½ cup	69
Vegetables:		
Spinach, cooked	½ cup	122
Kale	½ cup	47
Broccoli	½ cup	36
Legumes:		
Tofu	½ cup	260
Dried beans, cooked	½ cup	60
Foods fortified with calcium:		
Orange juice	1 cup	350
Frozen waffles	2	300
Soy milk	1 cup	200–400
Breakfast cereals	1 cup	150–1000

[a]Actually, the richest source of calcium is alligator meat; 3 1/2 ounces contain about 1,231 milligrams of calcium, but just try to find it on your grocer's shelf!

Selenium		
Food	**Amount**	**Selenium (mg)**
Seafood:		
Lobster	3 oz	66
Tuna	3 oz	60
Shrimp	3 oz	54
Oysters	3 oz	48
Fish	3 oz	40
Meats/Eggs:		
Liver	3 oz	56
Egg	1 medium	37
Ham	3 oz	29
Beef	3 oz	22
Bacon	3 oz	21
Chicken	3 oz	18
Lamb	3 oz	14
Veal	3 oz	10

Zinc

Food	Amount	Zinc (mg)
Meats:		
Liver	3 oz	4.6
Beef	3 oz	4.0
Crab	½ cup	3.5
Lamb	3 oz	3.5
Turkey ham	3 oz	2.5
Pork	3 oz	2.4
Chicken	3 oz	2.0
Legumes:		
Dried beans, cooked	½ cup	1.0
Split peas, cooked	½ cup	0.9
Grains:		
Breakfast cereal, fortified	1 cup	1.5–4.0
Wheat germ	2 Tbsp	2.4
Oatmeal, cooked	1 cup	1.2
Bran flakes	1 cup	1.0
Brown rice, cooked	½ cup	0.6
White rice	½ cup	0.4
Nuts and seeds:		
Pecans	¼ cup	2.0
Cashews	¼ cup	1.8
Sunflower seeds	¼ cup	1.7
Peanut butter	2 Tbsp	0.9
Milk and milk products:		
Cheddar cheese	1 oz	1.1
Whole milk	1 cup	0.9
American cheese	1 oz	0.8

Sodium

Food	Amount	Sodium (mg)
Miscellaneous:		
Salt	1 Tbsp	2,132
Dill pickle	1 (4½ oz)	1,930
Sea salt	1 Tbsp	1,716
Ravioli, canned	1 cup	1,065
Spaghetti with sauce, canned	1 cup	955
Baking soda	1 tsp	821
Beef broth	1 cup	810
Chicken broth	1 cup	770
Gravy	¼ cup	720
Italian dressing	2 Tbsp	720
Pretzels	5 (1 oz)	500
Green olives	5	465
Pizza with cheese	1 wedge	455
Soy sauce	1 tsp	444
Cheese twists	1 cup	329
Bacon	3 slices	303
French dressing	2 Tbsp	220
Potato chips	1 oz (10 pieces)	200
Catsup	1 Tbsp	155
Meats:		
Corned beef	3 oz	808
Ham	3 oz	800
Fish, canned	3 oz	735
Meat loaf	3 oz	555
Sausage	3 oz	483
Hot dog	1	477
Fish, smoked	3 oz	444
Bologna	1 oz	370

(continued)

Table 23.11 **(continued)**

Sodium		
Food	**Amount**	**Iron (mg)**
Milk and milk products:		
Cream soup	1 cup	1070
Cottage cheese	1/2 cup	455
American cheese	1 oz	405
Cheese spread	1 oz	274
Parmesan cheese	1 oz	247
Gouda cheese	1 oz	232
Cheddar cheese	1 oz	175
Skim milk	1 cup	125
Whole milk	1 cup	120
Grains:		
Bran flakes	1 cup	363
Cornflakes	1 cup	325
Croissant	1 medium	270
Bagel	1	260
English muffin	1	203
White bread	1 slice	130
Whole wheat bread	1 slice	130
Saltine crackers	4 squares	125

Iron		
Food	**Amount**	**Iron (mg)**
Meat and meat alternates:		
Liver	3 oz	7.5
Round steak	3 oz	3.0
Hamburger, lean	3 oz	3.0
Baked beans	1/2 cup	3.0
Pork	3 oz	2.7
White beans	1/2 cup	2.7
Soybeans	1/2 cup	2.5
Pork and beans	1/2 cup	2.3
Fish	3 oz	1.0
Chicken	3 oz	1.0
Grains:		
Breakfast cereal, iron-fortified	1 cup	8.0 (4–18)
Oatmeal, fortified, cooked	1 cup	8.0
Bagel	1	1.7
English muffin	1	1.6
Rye bread	1 slice	1.0
Whole wheat bread	1 slice	0.8
White bread	1 slice	0.6
Fruits:		
Prune juice	1 cup	9.0
Apricots, dried	1/2 cup	2.5
Prunes	5 medium	2.0
Raisins	1/4 cup	1.3
Plums	3 medium	1.1
Vegetables:		
Spinach, cooked	1/2 cup	2.3
Lima beans	1/2 cup	2.2
Black-eyed peas	1/2 cup	1.7
Peas	1/2 cup	1.6
Asparagus	1/2 cup	1.5

Phosphorous

Food	Amount	Phosphorous (mg)
Milk and milk products:		
Yogurt	1 cup	327
Skim milk	1 cup	250
Whole milk	1 cup	250
Cottage cheese	1/2 cup	150
American cheese	1 oz	130
Meats:		
Pork	3 oz	275
Hamburger	3 oz	165
Tuna	3 oz	162
Lobster	3 oz	125
Chicken	3 oz	120
Nuts and seeds:		
Sunflower seeds	1/4 cup	319
Peanuts	1/4 cup	141
Pine nuts	1/4 cup	106
Peanut butter	1 Tbsp	61
Grains:		
Bran flakes	1 cup	180
Shredded wheat	2 large biscuits	81
Whole wheat bread	1 slice	52
Noodles, cooked	1/2 cup	47
Rice, cooked	1/2 cup	29
White bread	1 slice	24
Vegetables:		
Potato	1 medium	101
Corn	1/2 cup	73
Peas	1/2 cup	70
French fries	1/2 cup	61
Broccoli	1/2 cup	54
Other:		
Milk chocolate	1 oz	66
Cola	12 oz	51
Diet cola	12 oz	45

Potassium

Food	Amount	Potassium (mg)
Vegetables:		
Potato	1 medium	780
Winter squash	1/2 cup	327
Tomato	1 medium	300
Celery	1 stalk	270
Carrots	1 medium	245
Broccoli	1/2 cup	205
Fruits:		
Avocado	1/2 medium	680
Orange juice	1 cup	469
Banana	1 medium	440
Raisins	1/4 cup	370
Prunes	4 large	300
Watermelon	1 cup	158
Meats:		
Fish	3 oz	500
Hamburger	3 oz	480
Lamb	3 oz	382
Pork	3 oz	335
Chicken	3 oz	208

(continued)

Table 23.11 (continued)

Potassium		
Food	**Amount**	**Potassium (mg)**
Grains:		
Bran buds	1 cup	1080
Bran flakes	1 cup	248
Raisin bran	1 cup	242
Wheat flakes	1 cup	96
Milk and milk products:		
Yogurt	1 cup	531
Skim milk	1 cup	400
Whole milk	1 cup	370
Other:		
Salt substitutes	1 tsp	1,300–2,378

Fluoride		
Food	**Amount**	**Fluoride (mg)**
Grape juice, white	6 oz	350
Instant tea	1 cup	335
Raisins	3½ oz	234
Wine, white	3½ oz	202
Wine, red	3½ oz	105
French fries, McDonald's	1 medium	130
Dannon's Fluoride to Go	8 oz	178
Tap water, U.S.	8 oz	59
Municipal water, U.S.	8 oz	186
Bottled water, store brand	8 oz	37

Iodine		
Food	**Amount**	**Iodine (mg)**
Iodized salt	1 tsp	400
Haddock	3 oz	125
Cod	3 oz	87
Shrimp	3 oz	30
Bread	1 oz (1 slice)	35–142
Cottage cheese	½ cup	50
Egg	1	22
Cheddar cheese	1 oz	17

© mediablitzimages (uk) Limited/Alamy

NUTRITION
up close

The Salt of Your Diet

Focal Point: Are you consuming too much sodium from processed foods?

Below is a list of commonly consumed processed foods and their content of sodium. On the blank line in front of the foods write down how many times a day you eat the food, or another food that is very similar to it. (For example, if you eat sausage pizza, check cheese pizza.) Then answer this question: Are you most likely to be consuming more than 1,500 mg of sodium daily just from processed foods?

	Sodium Content (mg)
Biscuits, bread, crackers	
_____biscuit, refrigerated, 1	340–604
_____bread, 1 slice	215–380
_____cheese crackers , 30 (1 oz)	250
_____wheat crackers, 7 (1 oz)	180
_____thin wheat crackers, 16 (1 oz)	260
Cheese	
_____American cheese, 1 slice	260
_____cheddar cheese, 1 oz	180
_____cottage cheese, 1/2 cup	390–420
Frozen foods	
_____fried chicken dinner with	
_____mashed potatoes, corn, 10 oz	790–1,300
_____macaroni and cheese, 10 oz	630–1,100
_____chicken nuggets, 5 pieces	490–650
_____cheese pizza, 2 slices	530–1,090
_____pot pie, 1 pie	770–1,040
_____taquitos, 5	370–40

	Sodium Content (mg)
Meats	
_____bacon, 1 slice	210–330
_____ham, 2 oz	460–620
_____hot dog, 1	450–740
_____breakfast sausage, 3 links	450–550
_____turkey, deli, 2 oz	250–620
Other foods	
_____salad dressing, 2 Tbsp	170–1,067
_____catsup, 1 Tbsp	150–190
_____salsa, 2 Tbsp	100–800
_____spaghetti sauce, 1/2 cup	330–650
_____steak sauce, 1 Tbsp	170–300
_____soy sauce, 1 Tbsp	920–1,160
_____potato chips, 1 oz (20 chips)	120–211
_____soup, canned, 1 cup	340–1,100
Fast foods	
_____bacon, egg, and cheese biscuit, 1	1,230
_____sausage biscuit, 1	1,160–1,360
_____sausage and egg biscuit, 1	1,200
_____french fries, 1 medium	270–960
_____cheeseburger, 1 regular	750–970
_____hamburger, 1 regular	490–560

Feedback to the Nutrition Up Close is located in Appendix G.

REVIEW QUESTIONS

- **Identify key functions and food sources of five essential minerals.**

1. There are 40 essential minerals that perform specific functions in the body. Each must be obtained from the diet. **True/False**

2. Essential minerals are a structural component of teeth; they serve as a source of electrical power that stimulate muscles to contract and help maintain an appropriate balance of fluids in the body. **True/False**

3. The amount of certain minerals absorbed from foods varies a good deal based on the food source of the minerals. **True/False**

4. Calcium is the most abundant mineral found in bones, but other minerals such as magnesium, phosphorus, and carbon are also components of bone. **True/False**

5. Bones fully develop and mineralize during the first 18 years of life. **True/False**

6. Iodine deficiency and overdose are related to goiter and thyroid disease.   **True/False**

7. Wilson's disease is related to magnesium deficiency.   **True/False**

8. Vitamin D is required for the absorption and utilization of calcium by the body.   **True/False**

9. People with iron deficiency generally feel weak and tired, have a poor appetite, and are susceptible to infection.   **True/False**

10. Although iron deficiency is a major health problem in America, iron overdose is not.   **True/False**

11. Blood pressure tends to increase as salt (or sodium) intake increases.   **True/False**

12. The major source of salt (or sodium) in our diet comes from salt added to food at the table.   **True/False**

Pablo decides to lose 5 pounds in a week by going on a high-protein diet. The foods he'll eat consist of canned tuna, lean beef, skinless chicken, low-fat cottage cheese, and skim milk.

13. _____ Of the minerals listed below, which one is most likely to be lacking in his diet?

   a. iron
   b. calcium
   c. magnesium
   d. phosphorus

The next four questions refer to this scenario:

Dorthea was just told by her adult nurse practitioner that she has prehypertension and recommends that she meet with a registered dietitian for consultation on the DASH eating plan.

The appointment was made and Dorthea is speaking with the registered dietitian now.

14. _____ During the consultation, Dorthea will learn that the DASH eating plan emphasizes:

   a. foods very low in sodium and high in fiber
   b. elimination of fast foods
   c. low-calorie, low-fat foods
   d. vegetables, fruits, and grain products

15. _____ Dorthea asks the dietitian how many vegetables a day are included in the DASH eating plan. The likely response from the dietitian is:

   a. 1–2 servings
   b. 2–3 servings
   c. 3–4 servings
   d. 4–5 servings

16. _____ How many servings of fruits would Dorthea be urged to include in her diet under the DASH eating plan?

   a. 1–2 servings
   b. 2–3 servings
   c. 3–4 servings
   d. 4–5 servings

17. _____ What other types of food would Dorthea be encouraged to include in her DASH eating plan?

   a. lean meats, fish, and poultry
   b. milk products of all types
   c. coffee and tea
   d. chocolate and molasses

Answers to these questions can be found in Appendix G.

NUTRITION SCOREBOARD ANSWERS

1. Minerals serve as structural components of the body, but they also play important roles in stimulating muscle and nerve activity and in other functions.   **False**

2. Bones continue to grow and mineralize well after we reach adult height. Bone growth and development continues to about age 30.   **True**

3. Spinach is a nutritious food, providing 0.6 mg iron per ounce. It contains less iron than beef (1 mg iron per ounce). The iron in spinach is poorly absorbed, but the iron in meat is well absorbed by the body.   **False**

4. Iron deficiency is the most common nutritional deficiency in both developed and developing countries. Approximately one-fourth of the world's population is iron deficient.[1]   **True**

5. Approximately one-third of American adults have hypertension (high blood pressure).[2]   **True**

Dietary Supplements and Functional Foods

NUTRITION SCOREBOARD

1 Products classified as "dietary supplements" consist of herbs and vitamin and mineral supplements only. **True/False**

2 Dietary supplements must be tested for safety and effectiveness before they can be sold. **True/False**

3 Herbal remedies have been used for over 100 years in Germany, so they must be safe and effective. **True/False**

4 "Probiotics" are "friendly bacteria" that may benefit health. **True/False**

Answers can be found at the end of the unit.

After completing Unit 24 and its interactive learning features, you will be able to:

- Make evidence-based decisions about the probable safety and effectiveness of specific dietary supplements.

- Describe the strengths and limitations of regulatory standards for the production, promotion, and sale of dietary supplements.

dietary supplements Any products intended to supplement the diet, including vitamin and mineral supplements; proteins, enzymes, and amino acids; fish oils and fatty acids; hormones and hormone precursors; and herbs and other plant extracts. Such products must be labeled "Dietary Supplement."

Dietary Supplements

- **Make evidence-based decisions about the probable safety and effectiveness of specific dietary supplements.**

- **Describe the strengths and limitations of regulatory standards for the production, promotion, and sale of dietary supplements.**

What do vitamin E supplements, amino acid pills, and herbal remedies all have in common? They are members of the increasingly popular group of products called **dietary supplements**—and they are all discussed in this unit. Types of dietary supplements available to consumers are presented in Table 24.1. About half of U.S. adults use one or more of these products.[1] Dietary supplements are supposed to "supplement the diet." They are not intended to prevent, treat, or cure disease. Many that do supplement the diet have beneficial health effects for some people, but others neither supplement the diet nor provide health benefits. These disparate products are grouped together because they are regulated as a subcategory of food by a common set of rules.[2]

You will learn from this unit that taking some of the dietary supplements available on the market can be a benefit but others represent a gamble with one's heath. This "buyer beware" situation exists because of the loose rules that govern dietary supplements. You will also learn about "functional foods" and some exciting, new developments related to intestinal fertilizers and friendly bacteria (no kidding).

Key Nutrition Concepts

Foundation knowledge about nutrition that applies to the topic of dietary supplements is represented by the following three key nutrition concepts:

1. Foods provide energy (calories), nutrients, and other substances needed for growth and health.

2. Poor nutrition can result from both inadequate and excessive levels of nutrient intake.

3. Adequacy, variety, and balance are key characteristics of healthful diets

Regulation of Dietary Supplements

In 1994 Congress passed the Dietary Supplement Health and Education Act, which started the explosion in the availability of dietary supplements. Under the act, dietary supplements are minimally regulated by the Food and Drug Administration (FDA); they do not have to be tested prior to marketing or shown to be safe or effective.[2] Although often advertised to relieve certain ailments, they are not considered to be drugs. Dietary supplements are not subjected to vigorous testing to prove safety and effectiveness, as drugs must be. Responsibility for evaluating the safety of dietary supplements lies with manufacturers and not the FDA. Supplements are deemed unsafe when the FDA has proof they are harmful. Since few dietary supplements have been adequately tested, and because results of studies showing negative effects may never see the light of day,[3] it is difficult to prove them to be unsafe. The FDA largely relies on reports of ill effects from manufacturers, health professionals, and consumers to assess supplement safety. Since 1994, the FDA has received thousands of reports of adverse effects of supplements (primarily for herbs), including several hundred deaths.[4] In the past few years, the FDA has taken action against about 300 products because they contained prescription drugs or ingredients harmful to health. The actions were taken primarily against products labeled for use in sexual enhancement, body building, and weight loss.[4]

According to FDA regulations (Table 24.2), dietary supplements must be labeled with a Supplemental Facts panel that lists serving size, ingredients, and percent Daily Value (% DV) of essential nutrient ingredients. Products can be labeled with a health claim, such as "high in calcium" or "low fat," if the product qualifies according to the nutrition labeling regulations. Supplements can also be labeled with "structure/function" claims. These claims cannot refer to disease prevention or treatment effects. Claims such as "improves circulation," "pre-

Table 24.1 Types of dietary supplements

Type	Example
1. Vitamins and minerals	Vitamins C and E, selenium
2. Herbs (botanicals)	Ginkgo, ginseng, St. John's wort
3. Proteins and amino acids	Shark cartilage, chondroitin sulfate, creatine
4. Hormones, hormone precursors	DHEA, vitamin D
5. Fats	Fish oils, DHA, lecithin
6. Other plant extracts	Garlic capsules, fiber, cranberry concentrate, echinacea

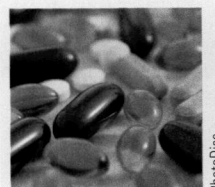
Vitamin, mineral, protein, and amino acid supplements.

Botanical supplements, such as dong quai, often come in gel caps.

Plant extracts can also be taken in liquid form, as tinctures.

Herbal teas are a familiar source of dietary supplements.

vents wrinkles," "supports the immune system," and "helps maintain mental health" can be used, whereas "prevents heart disease" or "cures depression" cannot be. If a function claim is made on the label or package inserts, the label or insert must include the FDA disclaimer that states the FDA does not support the claim. (This is done to reduce the FDA's liability for problems that may be caused by supplements.) Nonetheless, many people believe health claims made for supplements. Table 24.3 provides a summary of common consumer beliefs about dietary supplements and then lists the realities of how they are regulated.

The Federal Trade Commission (FTC) regulates claims for dietary supplements made in print and broadcast advertisements, including direct marketing, websites, infomercials, and mass e-mails. Claims made for dietary supplements in advertisements are supposed to be truthful, but often are not. Although some companies have been prosecuted for making false and misleading claims, neither the FDA nor the FTC has sufficient resources to fully monitor products and enforce laws related to dietary supplements. Rules and regulations related to dietary supplements are expected to become more stringent, and enforcement actions are being expanded in the United States and Canada.[4] The FDA has developed an "Adverse Events Reporting System" website (www.fda.gov/Safety/MedWatch/HowTo Report/ucm053074.htm) that simplifies recording and tracking of adverse effects of dietary supplements.

Table 24.2 FDA regulations for dietary supplement labeling

1. Product must be labeled "Dietary Supplement."

2. Product must have a Supplemental Facts label that includes serving size, amount of the product per serving, % Daily Value of essential nutrients, a list of other ingredients, and the manufacturer's name and address.

3. Nutrient claims (such as "low in sodium" and "high in fiber") can be made on labels of products that qualify based on nutrition labeling regulations.

4. Structure/function claims about how the product affects normal body structures (such as "helps maintain strong bones") or functions ("enhances normal bowel function") can be made on product labels. If a structure/function claim is made, this FDA disclaimer must appear:

This statement has not been evaluated by the FDA. This product is not intended to diagnose, treat, cure, or prevent any disease.

Nutrition Facts
Serving size 1 Tablet

Amount Per Serving	% DV
Melatonin 3 mg	*

*Daily Value (DV) not established

Other Ingredients: Dicalcium Phosphate, Cellulose (Plant Origin), Vegetable Stearic Acid, Vegetable Magnesium Stearate, Silica, Croscarmellose.

GUARANTEED FREE OF: wheat, yeast, soy, corn, sugar, starch, milk, eggs. No artificial colors, flavors. No chemical additives. No preservatives. No animal derivatives.

KEEP OUT OF REACH OF CHILDREN

Directions: As a dietary supplement for adults, take one (1) tablet, under the direction of a physician, only at bedtime as Melatonin may produce drowsiness. **DO NOT EXCEED 3 MG IN A 24 HOUR PERIOD.**

Warning: For Adults. Use only at bedtime. This product is not to be taken by pregnant or lactating women. If you are taking medication or have a medical condition such as an autoimmune condition or a depressive disorder, consult your physician before using this product. **NOT FOR USE BY CHILDREN 16 YEARS OF AGE OR YOUNGER.** Do not take this product when driving a motor vehicle, operating machinery or consuming alcoholic beverages.

In case of accidental overdose, seek professional assistance or contact a Poison Control Center immediately.

Table 24.3 Consumer beliefs about dietary supplements versus reality

Common consumer beliefs[5-8]
Consumers tend to believe dietary supplements:

- Are not drugs.
- Have fewer side effects than prescription drugs.
- Have accurate health claims.
- Are approved for use by the FDA.
- Will improve health and help maintain health.
- Are safe, high quality, and effective.
- May replace conventional medicines.

Dietary supplement realities[2,9-11]

- FDA does not approve, test, or regulate the manufacture or sale of dietary supplements.
- The FDA has limited power to keep potentially harmful dietary supplements off the market.
- Dietary supplements may not have been tested for safety or effectiveness before they are sold.
- Dietary supplements often do not list side effects, warnings, or drug or food interactions on product labels.
- Ingredients listed on dietary supplement labels may not include all active ingredients.
- Dietary supplements may not relieve problems or promote health and performance as advertised.

bioavailability   The amount of a nutrient consumed that is available for absorption and use by the body.

Table 24.4 Who may benefit from vitamin and mineral supplements? Here are some examples:[14,16,17]

- People with diagnosed vitamin and/or mineral deficiency diseases
- Newborns (vitamin K)
- People living in areas without a fluoridated water supply (fluoride)
- Vegans (vitamins B_{12} and D)
- Pregnant women (iron and folate)
- People experiencing blood loss (iron)
- Elderly persons on limited diets or who have low food intake (multiple vitamins and minerals)
- People on restricted diets (multiple vitamins and minerals)
- People at risk for osteoporosis due to low calcium intake and poor vitamin D status (calcium, vitamin D)
- People with alcoholism (multivitamins and minerals)
- Elderly people diagnosed with vitamin B_{12}, vitamin D, and/or folate deficiency

Some dietary supplements are safe and effective, and can help people maintain their own health. Unfortunately, the for-profit-only manufacturer and sale of dietary supplements taint good products with suspicion.

Vitamin and Mineral Supplements

Multivitamin and mineral supplements—such as vitamins E or C, calcium, or magnesium—are among the wide variety of vitamin and mineral supplements used by consumers. They represent the most popular type of dietary supplement, and 33% of Americans report using them mostly on a daily basis.[1] Intake levels of vitamins and minerals below the "Tolerable Upper Limits" of the Dietary Reference Intake values (given on the inside covers of this book) are safe for the vast majority of people, although lower amounts are related to their optimal functioning in the body.[12] Certain concerns related to vitamin and mineral supplements exist. The **bioavailability** of nutrients contained in some supplements remains uncertain. In general, the wider the assortment of vitamins and minerals in a supplement, the lower the absorption of some of them. Minerals are particularly prone to forming unabsorbable complexes with each other, reducing the bioavailability of multiple minerals in supplements.[13]

Vitamin and Mineral Supplements: Who Benefits? Supplements can have positive effects on health, as shown by the examples given in Table 24.4. This table lists examples of conditions for which specific vitamin and/or mineral supplements have been shown to be effective.

Using Vitamin and Mineral Supplements for the Wrong Reason People often take supplements as a sort of insurance policy against problems caused by poor diets. Although multivitamin and mineral supplements can help fill in some of the nutrient gaps caused by poor food habits or low food intake, they can't turn a poor diet into a healthful one. Whether a diet is healthful or unhealthful is determined by more than its vitamin and mineral content. The healthfulness of a diet also depends on its content of essential fatty acids, protein, fiber, water, and other nutrients and phytochemicals from food.

One of the most serious consequences of supplements results when they are used as a remedy for health problems that can be treated, but not by vitamins or minerals. Vitamin and mineral supplements have not been found to prevent or treat heart disease, cancer, diabetes, hypertension, premature death, behavioral problems, sexual dysfunction, hair loss, autism, chronic fatigue syndrome, obesity, cataracts, or stress, for example.[6,14,15] Some vitamin supplements, such as vitamin E, vitamin C, and beta-carotene may be harmful to certain groups of people.[15]

The Rational Use of Vitamin and Mineral Supplements Like all medications, vitamin and mineral supplements should be taken only if there is a need for them. If they are taken, dosages should not be excessive. Guidelines for the selection and use of vitamin and mineral supplements can be found in the accompanying Health Action.

Herbal Remedies

Herbal remedies (also called "botanicals") have a long history, are used worldwide, and are moving from alternative to mainstream health care in North America. Approximately 21% of U.S. adults use herbal dietary supplements each year.[18] The herb pharmacopoeia includes over 550 primary herbs known by at least 1,800 names. Many modern drugs are derived from plants,[19] and there's no doubt that many additional plants and plant ingredients benefit health. Plant products known to treat disease are considered drugs, however. Those that have not passed the scientific tests needed to demonstrate safety and effectiveness in disease treatment are often considered herbs. Yet, the truth is that some products sold as herbs in the United States have druglike effects on body functions. An example of an herbal dietary supplement that acts like a drug is given in Illustration 24.1.

Effects of Herbal Remedies Likely effects and identified adverse effects of some herbal and similar remedies are listed in Table 24.5. Herbal remedies, like drugs, have biologically active ingredients that can have positive, negative, and neutral effects on body processes. Basically, an herbal remedy (or a drug) is considered valuable if it has beneficial effects on body processes and if the benefits are not outweighed by the risks. Knowledge of the risks and benefits of many herbal supplements remains incomplete. However, available evidence suggests that some herbal remedies are safe and effective, while others appear to be neither.

Which herbal remedies are likely ineffective or unsafe? Human studies with various herbal remedies have helped identify herbs and other dietary supplements that lack beneficial effects or have adverse side effects. Table 24.6 lists some of these products. The extent to which the herbs included in the table pose a risk to health depends on the amount

Illustration 24.1 Herb or drug?

"Alternative medicine becomes standard medicine when it is proven true."

—PAUL OKUNIEFF, MD

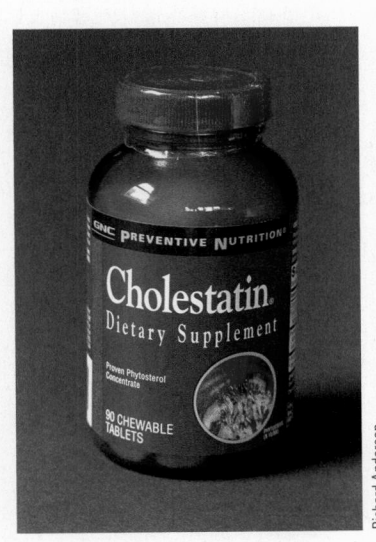

Richard Anderson

Statins are a popular type of prescription drugs that effectively lower blood levels of LDL cholesterol and triglycerides.[21] It turns out that a biologically active statin is also available in dietary supplements made from rice fermented with red yeast. The supplement was originally made available under the name "Cholestin." When banned by the FDA as containing a drug in 2000, manufacturers reformulated the statin in the red yeast product and changed the name of "Cholestatin."

Statin-containing red yeast products are banned from being sold in the United States but can be purchased from other countries over the Internet. Some red yeast products have been found to contain high amounts of statin, whereas others contain none; and some contain toxins introduced by low-quality manufacturing processes. Like statins, red yeast dietary supplements may cause muscle pain and damage and liver problems.[22]

High-quality red yeast products are being prescribed by some physicians for use by patients who cannot tolerate statin medications but can tolerate red yeast rice.[21]

Table 24.5 Likely effects and adverse effects of herbal and similar remedies[20]

Herb/other remedy	Likely effects	Likely adverse effects
Glucosamine Sulfate	Slows progression of osteoarthritis and its pain within 4–8 weeks	Gastrointestinal upset, fatigue, headache
Ginseng	Reduces blood glucose, may decrease colds and flu	Insomnia, nervousness, increased blood pressure, headache, diarrhea. Not for use in pregnancy.
Garlic	Lowers blood pressure	Heartburn, gas, prolonged bleeding, bad breath
Echinacea	May reduce cold symptoms	Allergies to plant components
DHEA	Improved mental health, skin appearance in the elderly, bone density, and symptoms of lupus	Acne, hair loss, menstrual changes, facial hair growth in women. Not for use by pregnant or breastfeeding women. Limit use to 50–100 mg/day for ≤ 2 months. Banned by the National Collegiate Athlete Association.
Creatine	Increased performance in short, high intensity athletic events; slows progression of Parkinson's disease, increases muscle strength and endurance in people with muscle disease.	Not for use by people with kidney disease, may cause stomach pain, nausea, diarrhea, muscle cramping.
Senna	Relieves constipation in short term (≤ 2 weeks)	Long-term use related to laxative dependency and liver damage.
Ginkgo	Increases learning skills, delays progression of Alzheimer's disease, increases blood flow, improves vertigo (lack of balance) and symptoms of PMS	Allergic reaction in some people, slow blood clotting, decreased fertility. Ginkgo plant seeds are unsafe. Not for use by pregnant and breastfeeding women, individuals with seizure disorders, diabetes, bleeding disorders, or with blood-thinning drugs.
Red Yeast	Lowers LDL-cholesterol and triglycerides	Muscle pain and damage, liver damage. Use should be supervised by a health care professional. Should not be taken by pregnant or breastfeeding women.
Saint John's wort	Relieves mild and moderate depression	Not to be used by people with major depression, bipolar disorder, schizophrenia, Alzheimer's disease, by people undergoing surgery; or by pregnant and breastfeeding women, women attempting pregnancy or on birth control pills. Does not work better than antidepressants. Banned in France based on medication interactions.
Coenzyme Q10	Remedy for coenzyme Q10 deficiency (rare), energy formation disorders	Nausea, diarrhea, rash, decreased blood pressure. Interacts with some drugs.

Table 24.6 Examples of dietary supplements that may not be effective or safe[20,27,28]

Apricot pits (laetrile)	Ephedra	Pennyroyal
Androstenedione (Andro)	Eyebright	Pokeroot
Aristolochic acid	Ginkgo seed	Hoodia
Sassafras	Saw palmetto	Shark cartilage
Belladonna	Kava	Skullcap
Black cohosh	Blue cohosh	Licorice root
Star anise	Bitter orange	Liferoot
Vinca	Borage	Lily of the valley
Wild yam	Broom	Lobelia
Wormwood	Chaparral	Yohimbe
Chinese yew	Mandrake	
Comfrey	Mistletoe	
Dong quai	Organ/glandular extracts	

taken and the duration of use, the age and health status of the user, and other factors. Not everyone reacts the same way to different herbal supplements.[20]

A Measure of Quality Assurance for Dietary Supplements Not all dietary supplements contain the amounts of herbal and other ingredients declared on the label, and some contain contaminants such as bacteria, mold, and lead. Analyses of the composition of 25 ginseng products, for example, found that concentrations of ginseng compounds in the supplements were up to 36 times different than labeled amounts.[23] Similar studies of echinacea products found that 10% of samples contained no echinacea, and half contained the labeled amount.[24] Male enhancement supplements are among the most common products recalled from the market and are taken for reasons such as content of unlisted prescription drugs or contamination.[41] None of the "herbal" or "all natural" sexual enhancement products work. No herbal or "all natural" substance has been shown to cure impotence.[42] Dietary supplements labeled "pure," "natural," or "quality assured" may or may not fit the description. You generally can't tell by the label.

There is no government body that monitors the contents of herbal supplements. Private groups, such as the U.S. Pharmacopeia (USP) and Consumer Laboratories (CL) offer testing services to ensure that dietary supplements meet standards for disintegration, purity, potency, and labeling (Illustration 24.2).[36,37] Products that pass these tests can display a "USP" logo or the CL symbol on product labels. These symbols represent quality ingredients and labeling but do not address product safety or effectiveness.

Due to the lack of studies and potential dangers, the FDA has advised dietary supplement manufacturers not to make claims related to pregnancy for herbs and other products, and to label products truthfully based on scientific evidence.[25,26]

Considerations for the use of herbal supplements are summarized in the nearby Health Action.

Functional Foods

Also known as "neutraceuticals," **functional foods** include a variety of foods and products that have in reality or theoretically been modified to enhance their contribution to a healthy diet (Table 24.7).[29] All foods are functional in that they provide nutrients. Foods considered "functional," however, are generally specifically formulated to supply one or more dietary ingredients that may improve health, or they are foods containing high amounts of substances that tend to prevent certain diseases.[30] Since there is no statutory definition for what constitutes functional foods, there are no specific regulations that apply to them. Health claims can be made for functional foods given approval by the FDA.[29] Functional foods containing food additives not on the Generally Accepted As Safe (GRAS) list must be tested and approved by the FDA before being sold.

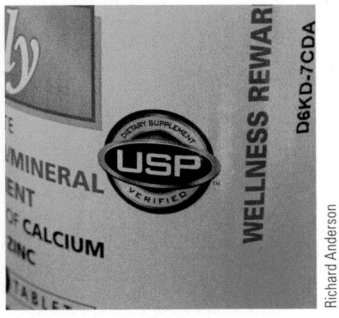

Richard Anderson

Illustration 24.2 These symbols on the labels of dietary supplements certify quality ingredients and accurate labeling but do not address product safety or effectiveness.

functional foods Generally taken to mean food or food ingredients that may provide a health benefit beyond the effects of traditional nutrients it contains.

---

health action Guidelines for Choosing and Using Vitamin and Mineral Supplements

1. Purchase supplements labeled "USP." They are tested for purity, ingredients, and dose.[11]
   - Terms such as *release assured*, *laboratory tested*, *quality tested*, and *scientifically blended* on supplement labels guarantee nothing.
2. Check the expiration date on supplements. Use unexpired supplements.
3. Choose supplements containing 100% of the Daily Value or less.

   Note: 1 mcg (or μg) vitamin A = 3.33 IU, 1 mcg vitamin D = 40 IU, and 1 mg of vitamin E = 1.49 IU.

4. Take supplements with meals.

5. If you have a diagnosed need for a specific vitamin or mineral, take that individual vitamin or mineral and not a multiple supplement.
   - Avoid calcium supplements made from oyster shells, bone, or coral calcium. They may contain lead or aluminum.
6. Store supplements where small children cannot get at them.

Remember! Tell your health care provider about the dietary supplements you take. Consult your health care provider about health problems before you start taking supplements to try to treat health problems with herbal supplements.

Table 24.7 How are foods made to be "functional"?

Foods are made to be "functional" by:
1. Taking out potentially harmful components (e.g., cholesterol in egg yolk and lactose in milk)
2. Increasing the amount of nutrients and beneficial non-nutrients (e.g., fiber-fortified liquid meals, calcium-, and vitamin C-fortified orange juice)
3. Using beneficial substances in food production or products (e.g., using "friendly" bacteria in fermented milk and soy products)

Table 24.8 Example of functional foods with apparent health benefits[30,34]

Functional food	Benefit
Stanol and sterol fortified margarine, psyllium fiber, whole oat products	Reduced blood levels of LDL cholesterol
Omega-3 fatty acids	Reduced blood triglycerides
Cranberry juice, extracts	Decreased urinary tract infections
Folic acid–fortified breads and cereals	Decreased risk of neural tube defects
Probiotics	Decreased risk of infection, lactose intolerance, diarrhea

Examples of functional foods that appear to benefit health are listed in Table 24.8. Increasingly, however, the list of functional foods is becoming infiltrated with sports bars, soups, beverages, and cereals spiked with vitamins, minerals, and herbs. Some of these products carry labels with unsubstantiated health claims and may be of no benefit.[30] For these products, the label *functional food* may be a marketing term.

Prebiotics and Probiotics

The terms *prebiotics* and *probiotics* were derived from "antibiotics" due to their probable effects on increasing resistance to various diseases. They are in a class of functional foods by themselves. **Prebiotics** are fiberlike, nondigestible carbohydrates that are broken down by bacteria in the colon. The breakdown products foster the growth of beneficial bacteria. For this reason they are considered "intestinal fertilizer." **Probiotics** is the term for live, beneficial—or "friendly"—bacteria that enter food through fermentation and aging processes.[31] Table 24.9 lists food and other sources of pre- and probiotics. Availability of foods and other products containing prebiotics and probiotics is much more common in Japan and European countries than in Canada or the United States.[27] However, availability of such products is increasing as research results shed light on their safety and effectiveness.

prebiotics Nondigestible food ingredients that beneficially affect a person by selectively stimulating the growth or activity of one or a limited number of bacteria in the colon. Inulin, an extract from chicory root, is a common prebiotic. It is also called "intestinal fertilizer."

probiotics Live bacteria that are taken orally to restore beneficial bacteria to the gastrointestinal tract. Strains of *lactobacillus* (lac-toe-bah-sil-us) and *bifidobacteria* (bif-id-dough bacteria) are the best known probiotics. Also called "friendly bacteria."

Table 24.9 Food and other sources of prebiotics and probiotics[31]

Prebiotics	Probiotics
Chicory	Fermented or aged milk and milk products:
Jerusalem artichokes	• Yogurt with live culture
Wheat	• Buttermilk
Barley	• Kefir
Rye	• Cottage cheese
Onions	• Dairy spreads with added inulin
Garlic	Other fermented products
Leeks	• Soy sauce
Prebiotics tablets and powders and nutritional beverages	• Tempeh
	• Fresh sauerkraut
	• Miso
	Breast milk

The digestive tract, particularly the colon, is home to over 500 species of microorganisms representing 100 trillion bacteria (and billions of viruses and fungi, too). Some species of bacteria such as E. coli may cause disease, whereas others such as lactobacillus and bifidobacteria (Illustration 24.3) prevent various diseases.[40]

Pre- and probiotics have been credited with important health effects (Table 24.10); the right combination and dose of each fosters the proliferation of healthful bacteria in the colon, nose, and some other internal canals of the body. The concept of the combined benefits of pre- and probiotics has been termed "symbiotics."[31] Because it is difficult to recolonize gut bacteria, the benefits of pre- and probiotics last only as long as dietary intake.[31]

Prebiotics appear to be safe in general; however, probiotics may be harmful to some individuals who develop blood infections related to their use.[35] The primary side effects associated with prebiotic and probiotic use are flatulence, bloating, and constipation.[40]

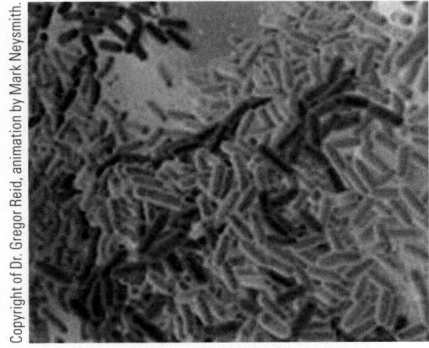

Illustration 24.3 A lactobacillus species (blue) taking over harmful E. coli bacteria (red).

Source: Bob Beale, Probiotics: Their tiny worlds . . . *The Scientist*, July 22, 2002, pp. 20–22.

Final Thoughts

From dietary supplements to bacteria: The universe of substances considered dietary ingredients is expanding. Knowledge about potential benefits of pre- and probiotics is charging ahead, and advances are catching the attention of consumers and health care professionals. Perhaps you never thought that "intestinal fertilizer" or "friendly bacteria" would ever intentionally pass through your lips. But that may well be the nature of dietary ingredients to come. If you didn't know before, you do now.

REALITY CHECK
Herbals on the Web

Can you trust information on herbal products you see on the web?

Who gets the thumbs-up?

Answer appears on page 24-10.

Sarah: When I'm sick, the first place I go is to the web to find an herb that will make me feel better.

Pablo: Herbal products I see advertised on the web look like they'll work for my problem. But I'm conflicted about buying them, because I'm not sure I can trust the information.

ANSWERS TO REALITY CHECK
Herbals on the Web

Many people use the web as a source of information about illness remedies. Unfortunately, many of the sites selling herbal products illegally claim the products prevent or cure specific diseases as if they were really drugs and do not include the required FDA disclaimer and ingredient list.[4,39]

Take the worry out of decisions about herbal supplements. Check them out using scientifically reliable websites.

Sarah: **Pablo:**

Table 24.10 **Apparent and potential benefits of prebiotics and probiotics**[32–34]

Prebiotics	Probiotics
• Prevention and treatment of diarrhea and constipation	• Treatment for antibiotic-related diarrhea, chronic diarrhea, traveler's diarrhea, and irritable bowel syndrome
• Prevention of colon cancer	• Treatment of lactose intolerance and some food allergies
• Increased mineral absorption	• Prevention of infections in urinary tract, gastrointestinal tract, ear canals
• Decreased blood triglyceride, glucose, and insulin levels	• Decreased dermatitis in infants
	• Increased bacterial production of some B vitamins
	• Decreased blood cholesterol levels
	• Decreased risk of certain types of cancer
	• Treatment of irritable bowel syndrome
	• Decreased dental caries
	• Decreased high blood pressure
	• Prevention of wound infections

PhotoDisc

NUTRITION
up close

Supplement Use and Misuse

Focal Point: Decide if a dietary supplement is warranted in these situations.

People take dietary supplements for many reasons, but is their use justified? Apply the information from this chapter to determine if you agree with the decisions made in each of the following scenarios.

1. Martha works part time and takes a full load of classes. Like many college students, she is always on the go, often grabbing something quick to eat at fast-food restaurants or skipping meals altogether. Nevertheless, Martha feels confident her health will not suffer, because she takes a daily vitamin and mineral supplement.

   Is a supplement warranted in this case? Why or why not?

   _____

2. Sylvia is a 23-year-old student diagnosed with iron-deficiency anemia. She has learned in her nutrition class that it is preferable to get vitamins and minerals from food instead of supplements. Therefore, instead of taking the iron pills her doctor has prescribed, Sylvia has decided to counteract the anemia by increasing her consumption of iron-rich foods.

   Is a supplement warranted in this case? Why or why not?

   _____

3. John is a 21-year-old physical education major involved in collegiate sports. He is very aware that nutrition plays an important role in the way he feels, so he is careful to eat well-balanced meals. In addition, John takes megadoses of vitamins and minerals daily. He is convinced they enhance his physical performance.

   Is a supplement warranted in this case? Why or why not?

   _____

4. Roberto, a native Californian, is backpacking through Europe when he is slowed down by constipation. He visits a pharmacy where English is spoken and is given senna by the pharmacist. Roberto has never taken an herb before and is not sure how his body will react to it, or if it will work.

   Should Roberto try the senna, ask for a nonherbal drug, or take another action? (Assume they cost the same.) What's the rationale for this decision?

   _____

5. While shopping at the mall, Yuen notices a kiosk selling "Hypermetabolite," a weight-loss product that guarantees you'll lose 5 pounds a week without dieting. Having gained 10 pounds since she started working full time, Yuen decides to try it. Her examination of the product's label reveals that an ephedra-derivative and Asian ginseng are major ingredients.

   Should Yuen take Hypermetabolite for weight loss? Why or why not?

   _____

Feedback to the Nutrition Up Close is located in Appendix G.

REVIEW QUESTIONS

- **Make evidence-based decisions about the probable safety and effectiveness of specific dietary supplements.**
- **Describe the strengths and limitations of regulatory standards for the production, promotion, and sale of dietary supplements**

1. Dietary supplements include vitamins, minerals, herbs, proteins, amino acids, and fish oil.   **True/False**

2. The FDA must approve dietary supplements before they are allowed to enter the market.   **True/False**

3. Dietary supplement labels must include a Supplement Facts panel and may include qualifying nutrient claims and structure/function claims.   **True/False**

4. Say you read a dietary supplement label that claims the product "helps enhance muscle tone or size." Because the statement appears on the label, it must be true. **True/False**

5. The Recommended Dietary Allowance (RDA) for vitamin A for breastfeeding women is 1,300 mcg per day. A supplement containing 3,300 IU vitamin A would supply less than the RDA amount.   **True/False**

6. Various vitamin and mineral supplements appear to benefit pregnant women, the elderly, and some adults at risk of osteoporosis.   **True/False**

7. Vitamin and mineral supplements available over-the-counter can be safely consumed at any dose levels. **True/False**

8. A USP label displayed on dietary supplements indicated the product has been tested for safety and effectiveness.   **True/False**

9. Prebiotics support the growth of beneficial bacteria in the colon.   **True/False**

10. Prebiotices have been demonstrated to benefit health under certain circumstances, but probiotics have not. **True/False**

The following four questions refer to the following scenario.

While reading a men's magazine you notice a full-page ad for a new dietary supplement that promises to "ignite sexual well-being." Containing a natural aphrodisiac, the supplement is specifically designed by scientists to encourage healthy, meaningful, and long-term relationships. The supplements have been tested on movie stars and college students, and you have heard from social media contacts that they work. Ingredients include vitamins, taurine, caffeine, dong quai, ginkgo, and ginseng.

11. _____ How many of the herbal ingredients listed have been shown to act as an aphrodisiac?
    a. 0
    b. 1
    c. 2
    d. 3

12. _____ Which herbal ingredient contained in the product does not appear to be effective for any health condition or be safe to consume?
    a. dong quia
    b. ginkgo
    c. ginseng
    d. taurine

13. _____ This product has been:
    a. tested for purity, safety, and effectiveness by the FDA before it became available.
    b. has probably not been tested for purity, safety, and effectiveness by the FDA before it became available.
    c. will be examined by the manufacturer for safety and effectiveness while on the market.
    d. will improve people's chances of finding a long-term relationship.

14. ___ Assume your friend bought and took the product and thinks it really works. What is the most likely reason for that?
    a. The vitamins in the product worked.
    b. Ginseng improves personality and the potential for long-term relationships.
    c. The caffeine provides extra energy for interpersonal relationships.
    d. Your friend believed the product would be effective (placebo effect).

Answers to these questions can be found in Appendix G.

NUTRITION SCOREBOARD ANSWERS

1. Dietary supplements include herbs, vitamin and mineral supplements, protein powders, amino acid and enzyme pills, fish oils and fatty acids, hormone extracts, and other products.   **False**

2. Dietary supplements can be sold without proof of their safety or effectiveness.   **False**

3. Results of clinical trials, and not historical use, are the gold standard for determining the safety and effectiveness of herbal remedies and other dietary supplements.   **False**

4. Yes! There are healthful bacteria.   **True**

Water Is an Essential Nutrient

NUTRITION SCOREBOARD

1 Bottled water is better for health than municipal water from your faucet.   **True/False**

2 Drinking lots of water helps flush toxins out of your body.   **True/False**

3 Few bottled water brands are fluoridated.   **True/False**

4 You can't drink too much water.   **True/False**

Answers can be found at the end of the unit.

After completing Unit 25 and its interactive learning features, you will be able to:

- List the main functions of water in the body and the consequences of water deficiency and toxicity.

- Describe factors that affect the availability of safe water supplies and the potential consequences to health of inadequate and unsanitary water supplies.

Photo Disc

Water: Where Would We Be Without It?

- **List the main functions of water in the body and the consequences of water deficiency and toxicity.**

Ask any three people you know to name as many essential nutrients as they can. If they mention water, give them a prize. Our need for water is so obvious that it is often taken for granted. Well, that's not going to happen in this text!

Water differs from other essential nutrients in that it is liquid, and our need for it is measured in cups rather than in grams or milligrams. It is required by every cell, tissue, and organ in the body. Without it, our days are limited to about six. The largest single component of our diet and body, water is a basic requirement of all living things.[3] Now, how could something this important be forgotten?

Key Nutrition Concepts

Three key nutrition concepts are directly related to the essential nutrient water and its functions in the body:

1. Poor nutrition can result from both inadequate and excessive levels of nutrient intake.

2. Humans have adaptive mechanisms for managing fluctuations in nutrient intake.

3. Health problems related to nutrition originate within cells.

Water's Role as an Essential Nutrient

Water qualifies in all respects as an essential nutrient. It is a required part of our diet; it performs specific functions in the body; and deficiency and toxicity signs develop when too little or too much is consumed.

Water is our body's main source of fluoride, an essential mineral needed for the formation and maintenance of enamel and resistance to tooth decay.[5] Fluoride is naturally present in some well waters and is added to 70% of all municipal water supplies in the United States.[6]

Water is the medium by which many chemical reactions take place within our body. Water plays key roles in digestion and energy formation. It also provides the medium by which digestive enzymes access and break down carbohydrates, proteins, and fats in food. Water is produced as an end product of energy formation from carbohydrates, proteins, and fats. We continue to produce and excrete water even if we quit drinking it for a while, because energy production and respiration are ongoing processes. Water is needed to "carry" nutrients to cells and waste products away from them. Additionally, water acts as the body's cooling system. When our internal temperature gets too high, water transfers heat to the skin and releases it in perspiration. When we're too cool, less water—and heat—are released through the skin. Water's functions are summarized in Table 25.1.

Water has been given credit for other functions in the body, but undeservedly. Drinking more water than is normally needed does not prevent dry, wrinkled skin, lead to weight loss, or flush toxins out of the body. Nor will it cure chronic fatigue, arthritis, migraines, or hypertension.[7,8] The nearby Reality Check addresses another misconception about water.

Water, Water, Everywhere The body of a 160-pound person contains about 12 gallons of water (Illustration 25.1). Adults are approximately 60 to 65% water by weight.[4] Water is distributed in the body in blood, in the spaces in-between cells, and in all cells. The proportion of water in body tissues varies: blood is 83% water, muscle 75%, and bone 22%. Even fat cells are 10% water.[9]

Most Foods Contain Lots of Water, Too Most beverages are more than 85% water, and fruits and vegetables are 75–90% (Illustration 25.2). Meats, depending on their type and

Table 25.1 Key functions of water in the body[3,4]

- Needed to maintain normal internal body temperature.
- Serves as the medium for chemical reactions that take place within the body.
- Serves as the medium for the transport of nutrients throughout the body in blood.
- Serves as the medium for the action of digestive enzymes.
- Required for the normal elimination of waste products in urine and stools.
- Participates in energy formation.

how well done they are, contain 50–70% water. Consequently, the water content of foods makes an important contribution to our daily intake of water. Approximately half of our water intake comes from plain water and other beverages, and half from foods.[10]

Meeting Our Need for Water

In general, individuals require enough water each day to replace water lost in urine, perspiration, stools, and exhaled air. Adequate Intakes (AIs) of water from fluids and food are set at 11 cups per day for women and 15 cups per day for men.[4] No-calorie water is the preferred source for meeting water needs.[8]

Built-in mechanisms that trigger thirst generally protect people from consuming too much or too little water. There is no evidence that water intake among Americans in general is either excessive or inadequate.[8] People who do strenuous work, athletic or otherwise, in hot and humid weather need to consume enough water to replace the amount that is lost in sweat, urine, respiration, and evaporation of water from the skin's surface. How much extra water this takes varies from person to person, but it is often within the ballpark of a 50% increase. You know you are drinking enough water if you haven't lost weight from before to after physical activity and if your urine is pale yellow and produced in normal volume.[3]

Exposure to both cold weather and high altitudes increases water need. Cold air holds little moisture, so you lose more water when you breathe in cold, dry air than warm, moist air. People exposed to cold weather tend to dress in layers and lose body water in the form of sweat that pools next to the skin. Exposure to cold environmental conditions at high altitudes can increase urine production and water loss. High levels of physical activity in cold, high altitude climates further increase water need (Illustration 25.3).[23]

Illustration 25.1 The body of a 160-pound person contains approximately 12 gallons of water. That's 96 pounds of water!

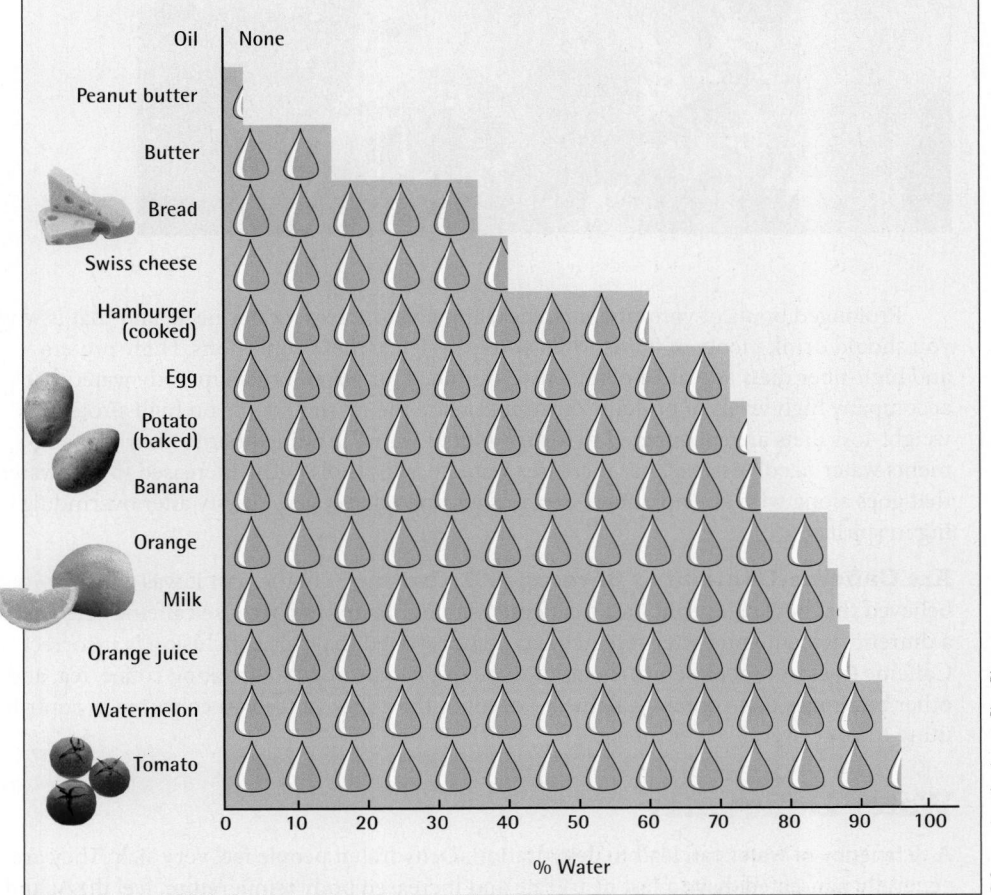

Illustration 25.2 The water content of some foods.

Illustration 25.3 Cold, high altitude conditions increase water need.

© Picimpact/Corbis

Prolonged bouts of vomiting, diarrhea, and fever increase water need, and that is why you should drink plenty of fluids when you experience these conditions. High-protein and high-fiber diets and alcohol increase your need for water. Losses in body water that accompany high levels of protein consumption are the reason people on high-protein weight-loss diets are encouraged to "drink a lot of water." Adding fiber to your diet augments water need because fiber increases water loss in stools.[10] The increased loss of water that goes along with alcohol intake explains why people get very thirsty after overindulging in spirits.

Are Caffeine-Containing Beverages Hydrating? In the past it was widely believed that beverages containing caffeine were not hydrating because caffeine acted as a diuretic. Recent and better research has demonstrated that this conclusion is incorrect. Caffeine does not increase urine output in people accustomed to drinking coffee, tea, and other beverages that contain caffeine.[24] So, count the coffee or tea you consume as contributing to your overall water intake.

Water Deficiency

A deficiency of water can lead to dehydration. Dehydrated people feel very sick. They are generally nauseated, have a fast heart rate and increased body temperature, feel dizzy, and may find it hard to move. The ingestion of fluids produces quick recovery in all but the most serious cases of dehydration. If it is not resolved, however, dehydration can lead to kidney failure and death.[25]

Water Toxicity

It is possible for people to overdose on water if they drink too much of it.[26] High intake of water can lead to a condition known as hyponatremia—or low blood sodium level and excessive water accumulation in the brain and lungs. The consequences can be devastating and include confusion, severe headache, nausea, vomiting, and even seizure, coma, and death.[2]

Water intoxication is rare, but it has occurred in marathon runners who consumed too much water during an event, in infants given too much water or overdiluted formula (Illustration 25.4), and psychotic patients taking medications that produce cravings for water. The drive for water created by antipsychotic medications can be so strong that access to water (even when showering) has to be limited.[27]

The Nature of Our Water Supply

- **Describe factors that affect the availability of safe water supplies and the potential consequences to health of inadequate and unsanitary water supplies.**

Water covers about three-quarters of the earth's surface, yet very little of it is drinkable. Nearly 97% of the total supply is salt water, and only 3% is fresh.[11] Of the freshwater supply, only a fourth of the total amount is available for use. The rest is located in polar and glacier ice. Although fresh water is readily available in most locations in the United States, this is not the case in a number of other countries.

Water is so scarce in parts of Russia that drinking-water dispensers are coin operated. Fresh water is so highly prized in sections of the arid Middle East that having a decorative fountain in one's home is considered a sign of affluence. Drinking water is sampled, judged, and celebrated in Middle Eastern countries as ceremoniously as fine wines are in France. Although water is not a subject of envy or a cause for celebration in most parts of America, it is being increasingly viewed as a precious resource that must be protected.[12]

Water scarcity has become a primary concern of many nations in the 21st century (Illustration 25.5). Demand for water worldwide is increasing due to population growth and expanding use of water-intensive energy production and farming techniques. Wasteful use of water, climate change, groundwater depletion, pollution, and leaky public water systems are contributing to water shortages and safety concerns (Illustration 25.6).[12]

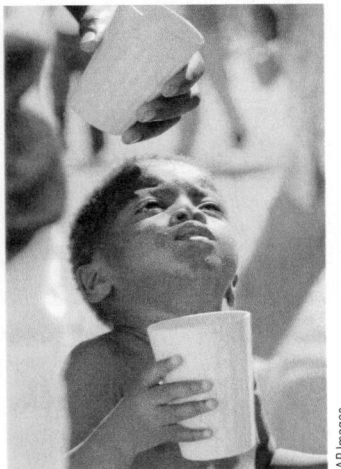

Rise in water intoxication found in babies on formula

Chicago, Ill.
Water intoxication, a potentially fatal condition, has increased dramatically among infants in part because poor parents give babies too much water when formula runs out, a study indicates.

Illustration 25.4 This headline is accurate—overdilution of infant formula can lead to water intoxication in infants.

Illustration 25.5 A child waits for water at a refugee camp.

REALITY CHECK
To Drink, or Not to Drink Water with Meals

According to a health and wellness website, people should not drink water with a meal because it dilutes digestive juices and interferes with digestion. Really?

Who gets the thumbs-up?

Answer appears on page 25-6

Barbara Ann: What website made that statement? Why would that happen? Do you know how watery food gets when it's in your stomach?

Carole: I get a bloated feeling when I drink water with food. I think the water is slowing down the digestion of food in my stomach.

Water intake during a meal does not interfere with digestion. On the contrary, it increases the exposure of digestive enzymes to food and facilitates the absorption of nutrients.[18]

Barbara Ann:

Carole:

Illustration 25.6 Contaminated water supplies are a threat to the public's health.

Without more effective protection of the world's water, the quantity of water available for agriculture and clean water for consumption may be insufficient, limiting food production and increasing the incidence of water-borne infectious diseases.[11,13]

The Environmental Protection Agency (EPA) in the United States is responsible for the safety of U.S. public water supplies and has set maximum allowable levels of contaminants. Water quality is monitored by local water utilities, and the results are reported to state and federal officials. Problems identified are remedied, and attempts are made to prevent future contamination in order to provide safe public water supplies.[14]

Does the Earth Supply "Gourmet" Water?

About 40 years ago, the French made drinking "fine waters" very stylish. In Paris, boulevardiers crowded sidewalk cafés for hours sipping chilled Perrier served with thinly sliced lemon. The popularity of bottled waters skyrocketed in many Western countries and it continues to grow.[15] In the United States, "mineral," "spring," and "seltzer" waters are now best sellers (Illustration 25.7). These waters have a strong, positive image. But what *are* these waters? How do they differ from the water we get from the faucet?

True *mineral water* is taken from protected underground reservoirs that are lodged between layers of rock. The water dissolves some of the minerals found in the rocks, and as a result, it contains a higher amount of minerals than most sources of surface water. Actually, most water contains some minerals and could be legitimately considered "mineral water." *Spring water* is taken from freshwater springs that form pools or streams on earth's surface. True "seltzers" (not the sweetened kind you often find for sale) are *sparkling waters* that are naturally carbonated. Most seltzers, however, become bubbly by the commercial addition of pressurized carbon dioxide.[16]

Bottled waters have their advantages. They are calorie free, are generally sodium free or low in sodium, and quench one's thirst better than their major competitor, soft drinks. But they are no "purer" or better for you than tap water.[8] As a matter of fact, 45% of single-serve bottles of water sold in the United States contain tap water.[16] Increasing rates of bottled water use on campuses are prompting students at a number of colleges to ask for changes. Their concerns are based primarily on the energy cost of making plastic bottles and environmental concerns about plastic waste. Colleges, including the University of Vermont, Dartmouth, and Harvard, are instituting partial or complete bans on the sale of bottled water on campus. Some colleges are retrofitting water fountains to serve as bottle-refilling stations.[17]

Is Bottled Water Safe? Every now and then, a story about contaminated bottled water makes the news, raising concerns about the safety of bottled water and the way the industry is monitored. The Food and Drug Administration (FDA) regulates and monitors the bottled water industry.[14] Domestic bottlers must conform to specified standards of water safety (such as allowable levels of chemical contaminants) and labeling requirements. "Mineral" or "spring" water must be just that under FDA regulations. Bottled tap water must be labeled as such. The FDA supplements its monitoring role by requiring that bot-

Richard Anderson

© Jerryb8/Dreamstime.com

Illustration 25.7 Is bottled water better for you than tap water? Consumers often perceive that bottled waters have fewer impurities and are better for health than tap water. Such beliefs are unfounded.

tlers periodically test their water. Foreign bottling plants are not under the FDA's jurisdiction, but imported waters may be tested when they enter the United States.

Safety concern about bottled water extends to the bottles. Certain types of plastic used in water bottles may contain bisphenol A (BPA) that acts like the hormone estrogen.[19] Current levels of BPA exposure may be related to changes in behavior and the functioning of the brain and prostate in fetuses and young children. Transfer of BPA from plastic to fluids increases with heat.[20]

Some manufacturers of plastic bottles (such as Nalgene) have taken BPA out of consumer plastic bottles and others are in the process of doing so.[21] The FDA banned BHA in baby bottles and sippy cups in mid-2012. If concerned, do not heat liquids in clear plastic bottles, look for plastic bottles labeled "without BPA" or "Bisphenol-A free," or use glass bottles.[20]

Water Gimmicks New waves of bottled water products are finding their way to grocery shelves throughout America. Tapping into consumer interest in vitamins, minerals, herbs, and fitness, producers are marketing "enhanced" bottled water, fortified with everything from caffeine to oxygen (Illustration 25.8). Although mineral waters provide

Illustration 25.8 "Specialty" bottled water products are of uncertain value to health or fitness.

Sam Kolich/Bill Smith Group/Cengage Learning

Illustration 25.9 Want to help protect your teeth from cavities? Look for the word *fluoride* on the label of bottled water products.

Robin Sterling/Sterling Research

absorbable calcium and magnesium that contribute to nutrient status and potentially to health,[22] the rationale for many of the products (other than their commercial appeal) is elusive. It makes little sense to fortify water with nutrients better obtained from foods. Safety of some of the combinations of herbs put into bottled water is unclear. Adding oxygen to beverages is total nonsense. Water is rich in oxygen (it consists of oxygen and hydrogen molecules), and the oxygen our bodies use for energy formation and metabolic processes comes from the air we breathe.

Fluoridated Bottled Water Bottled waters have been criticized for their lack of fluoride compared to fluoridated tap water. People of all ages need this mineral for bone health and the prevention of tooth decay. Children who regularly consume bottled water and infants who receive formula mixed with bottled water develop more tooth decay at an early age than do young children who receive fluoridated water.[1] The industry caught on to this shortcoming and responded by offering fluoridated waters (Illustration 25.9). Bottled waters with added fluoride must be labeled as "fluoridated."[28] A quick look at the label can tell you if you're getting fluoride along with your water.

Fotofeeling/Getty Images

NUTRITION
up close

Foods as a Source of Water

Focal Point: Water is a primary component of many foods.

About half of our requirement for water is met by fluids and solid foods. But how much water is in foods? You may be surprised.

Circle the food in each set that you think contains the highest percentage of water by weight.

1. avocado; potato, boiled; corn, cooked
2. egg; almonds; ripe (black) olives
3. watermelon; celery; pineapple, fresh
4. 2% milk; Coke; cranberry juice, low calorie
5. cheddar cheese; banana; refried beans
6. hot dog; pork sausage, cooked; ham, extra lean
7. apple; mushrooms, raw; orange
8. onions, cooked; lettuce; okra, cooked
9. peanut butter; butter; mayonnaise
10. cake with frosting; bagel; Italian bread

Feedback to the Nutrition Up Close is located in Appendix G.

REVIEW QUESTIONS

• **List the main functions of water in the body and the consequences of water deficiency and toxicity.**

1. Water plays key roles in the prevention of acne and dry skin and in the dilution of toxins formed during digestion. **True/False**

2. Water is our major source of fluoride. **True/False**

3. The functions of water include serving as a medium for chemical reactions and for nutrient transport in blood. **True/False**

4. Adult females living in moderate climates require an average of 8 cups of water a day, and men require twice that amount. **True/False**

5. Most of a person's body weight consists of water. **True/False**

6. Beverages containing caffeine are not hydrating. **True/False**

7. _____ Water need is increased by three of the four following conditions. Which condition is *not* related to an increased need for water?
   a. dry skin
   b. hot, humid environment
   c. vomiting
   d. alcohol intake

8. _____ Dehydration is *not* characterized by:
   a. accumulation of water in the brain and lungs
   b. nausea and vomiting
   c. dizziness
   d. severe headache

- **Describe factors that affect the availability of safe water supplies and the potential consequences to health of inadequate and unsanitary water supplies.**

9. The lack of adequate availability of safe water is a global health concern.

10. The Food and Drug Administration (FDA) requires that all bottled water be fluoridated.

11. Increased incidence of infectious disease is one consequence of shortages of clean water.

12. Approximately 20% of bottled waters contain tap water.

13. _____ Which of the following does *not* appear to contribute to water shortages?
   a. climate change
   b. groundwater depletion
   c. water-intensive farming techniques
   d. bottled water use

Answers to these questions can be found in Appendix G.

NUTRITION SCOREBOARD ANSWERS

1. Bottled waters have not been shown to be better for health or to be "purer" than tap water obtained from city water supplies.   **False**

2. Drinking lots of water flushes toxic wastes from our bodies.   **False**

3. Only a minority of bottled waters are fluoridated; these are labeled *fluoridated*.[1]   **True**

4. Yes, you can drink too much water, and it results in water intoxication.[2]   **False**

living the dream/Flickr/Getty Images

Nutrient–Gene Interactions in Health and Disease

NUTRITION SCOREBOARD

1 Most chronic diseases are genetically caused. **True/False**

2 Differences in genetic traits are responsible for large differences in essential nutrient needs between individuals. **True/False**

3 Heart disease, cancer, obesity, and hypertension primarily result from interactions among environmental and genetic factors. **True/False**

Answers can be found at the end of the unit.

After completing Unit 26 and its interactive learning features, you will be able to:

- Describe two ways that genes can affect nutrient needs and two ways that nutrients affect gene function.

- Identify specific ways in which gene variants interact with nutrient exposures to increase or decrease disease risk.

Nutrition and Genomics

- **Describe two ways that genes can affect nutrient needs and two ways that nutrients affect gene function.**

- **Identify specific ways in which gene variants interact with nutrient exposures to increase or decrease disease risk.**

In the history of the relatively young science of nutrition, at least two breakthroughs have defined the future. The first resulted from experiments in the late 1800s demonstrating that some constituents of foods are essential for life. The second is represented by the mapping of the human *genome* and development of the field of **nutrigenomics**. (Find definitions and explanations for many of the genetic terms used in this unit in Table 26.1.) That breakthrough marked the beginning of a new era in the discovery of underlying causes of a variety of common diseases and the specific roles of components of diets in disease prevention and treatment. It is being made possible, in part, by the identification of genetic codes for protein production and nutrients that affect, or are affected by, gene activity. Enzymes and other proteins produced as a result of genetic codes are central to life and health because they determine which chemical changes will take place within the body. These changes affect every body process, including growth, food digestion, nutrient absorption, nutrient utilization, disease resistance, healing, and blood pressure control.[4]

Key Nutrition Concepts

Key nutrition concepts are directly related to this unit's content on nutrient-gene interactions:

- Health problems related to nutrition originate within cells.

- Malnutrition can result from poor diets and from disease states, genetic factors, or combinations of these factors.

Genetic Secrets Unfolded

Our bodies may be new, but our genes have been around for over 40,000 years. Genes replicate themselves exactly over generations, and lasting modifications in them almost never occur. This means that changes in genetic traits do not account for increases or decreases in the incidence of disease. It is why genetic traits cannot be given as the cause of recent epidemics of obesity and diabetes, nor can they be given credit for major declines in heart disease and stroke rates.

© Cengage Learning; Photos (clockwise)—Broccoli: Valentyn Volkov/Shutterstock.com; butter: Stargazer/Shutterstock.com; cabbage: monticello/Shutterstock.com; muesli: Picsfive/Shutterstock.com; branch with olives and bottle of olive oil: Volosina/Shutterstock.com; salt: Photodisc; asparagus: Franco Deriu/Shutterstock.com; bok choi: Muellek Josef/Shutterstock.com

Humans are united as a species because we all share 99.9% of the same DNA.[5] We are, in part, unique as individuals due to the 0.1% difference. The human genome is estimated to contain 20,000–25,000 genes. The DNA that makes up each gene has four different chemical building blocks. These are called bases and abbreviated A, T, C, and G. All together, our genes contain about 3 billion bases attached to **chromosomes** with the nucleus of cells. To get an idea of the size of the human genome present in each of our cells, consider the following analogy: If the DNA sequence of the human genome were compiled in books, the equivalent of 200 volumes the size of a Manhattan telephone book (at 1,000 pages each) would be needed to hold it all.[4]

Not all segments of DNA code for protein formation, and evidence of individual uniqueness in genetic makeup, is located in these nonprotein coding segments of DNA. These segments contain instructions for the regulation of gene activity, such as when a specific gene should turn on or off, or when a gene should increase in activity or be silenced.[5] These segments appear to vary in composition based on environmental factors such as nutrient availability and exposure to infectious agents, diesel fuel exhaust, toxins, and other substances. The composition, pattern, and length

Table 26.1 Definitions and explanations of genetic terms

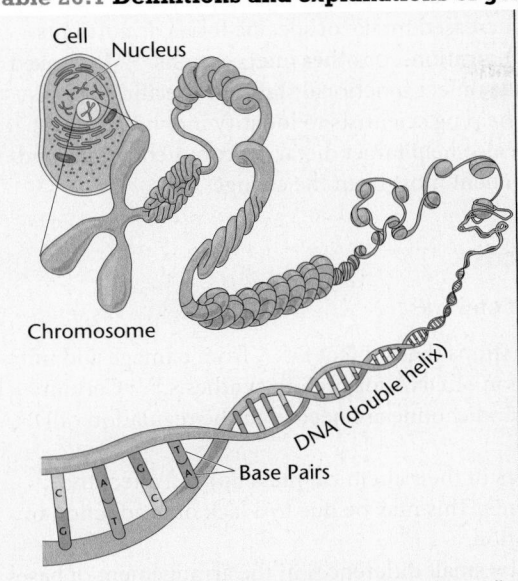

DNA is placed in chromosomes located in cell nuclei

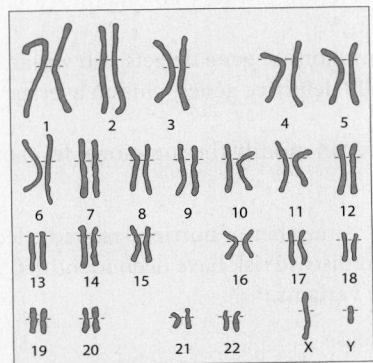

Human chromosomes

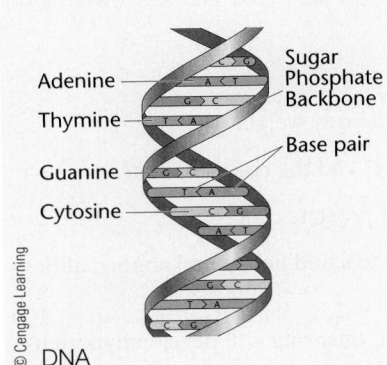

DNA

- **DNA (deoxyribonucleic acid, pronounced d-oxy-rye-bow-new-clay-ic)** Segments of genes that provide instructions for the synthesis, or production, of specific enzymes and other proteins needed by cells to carry out life- and health-sustaining processes. Enlarged, DNA looks something like an immensely long ladder twisted into a coil. The sides of the ladder structure of DNA are formed by a backbone of sugar-phosphate molecules, and the "rungs" consist of base pairs (abbreviated A, T, C, and G) joined by weak chemical bonds. Bases are the "letters" that spell out the genetic code, and there are over 3 billion of them in human DNA. Some sections of DNA do not code for protein synthesis. They signal genes to turn on or off, or to increase or decrease activity. Characteristics of these nonprotein coding segments of DNA vary among individuals, making it possible to identify individuals based on "DNA fingerprinting."

- **Genome** Combined term for "genes" and "chromosomes." It represents all the genes and DNA contained in an organism. It is estimated that the human genome consists of 20,000 to 25,000 genes.

- **Chromosomes** Structures in the nuclei of cells that contain genes. Humans have 23 pairs of chromosomes (shown on the left); half of each pair comes from the mother and half from the father.

- **Genes** The basic units of heredity that occupy specific places (loci) on chromosomes. Genes consist of large DNA molecules, each of which contains the code for the manufacture of a specific enzyme or other protein.

- **Genotype** The specific genetic makeup of an individual as coded by DNA.

- **Genomics** The study of the functions and interactions of all genes in the genome. Unlike genetics, it includes the study of genes related to common conditions and their interaction with environmental, or "epigenetic" factors.

- **Epigenetic** Alterations in gene activity that do not change the structure of DNA. Gene activity can be shut off or turned on, or slowed or sped up by epigenetic mechanisms.

- **Gene variant** An alteration in the normal sequence of a gene. The different forms of the same genes are considered "alleles."

- **Allele** A different version of the same gene. Alleles have a different arrangement of bases than the usual version of the gene.

- **Nutrigenomics** The study of nutrient-related functions and interactions with genes and their effects on health and disease.

of the chemical sequences in these segments is different for every individual (except is the case of identical twins early in life).[6,7]

Genetic Traits and Susceptibility to Disease The genes individuals inherit at birth from their mothers and fathers affect disease susceptibility.[9] In a minority of people, inborn genetic traits provide protection against diseases that may override some lifestyle factors. In other people, genetic traits increase the risk for disease when combined with certain lifestyle factors. Disease development among people with these types of genetic traits can be lowered by increased intake of specific foods or nutrients, weight loss, increased physical activity, medications, or other interventions.[8-10] Increased understanding of **epigenetic** processes that affect functional status of specific genes as well as person's inherited genetic traits is helping scientists to identify individuals most susceptible to diet-related diseases. It will also help target dietary recommendations and interventions to individuals who would benefit most from the changes.[9]

Nutrient-Gene Interactions

Nutrients interact with genes in a number of ways:

1. Nutrients are needed in chemical reactions that protect DNA from damage and initiate processes involved in the replication of DNA for protein synthesis.[10,11] Certain vitamins, such as folate, vitamin B_{12}, and choline are needed in the regulation of DNA activity.[12]
2. Single defects in gene structure (errors in their chemical makeup) can affect the body's ability to utilize certain nutrients. This may be due to a lack of production of enzymes required for nutrient utilization.
3. Nutrient utilization can be modified by small differences in the arrangement of bases in specific genes. Called **gene variants**, these different forms of the same gene can modify the characteristics of proteins produced by affected genes. Disease risk and nutrient requirements, utilization, storage, and excretion can vary among individuals depending on the characteristics and gene variants.[9,10]

    The Human Genome Project has identified millions of gene defects and variants in human DNA. The average person has 250 to 300 defective genes, plus an average of 75 gene variants associated with disease.[13]
4. Nutrient availability, particularly during early life, can modify the functional level of specific genes thereby and influence disease risk.

Gene Variants, Nutrients, and Disease Risk   A number of nutrient-related effects of gene variants on the body's metabolic processes and disease risk have been identified. Note the following effects in people with specific gene variants.[14-19]

*   High vitamin E intake deceases the risk of heart disease.

*   High folate intake decreases the risk of cancer.

*   High polyunsaturated fat, low dietary cholesterol, or low-saturated-fat diets lower blood cholesterol levels.

*   High sodium intake increases the risk of hypertension.

*   High fat intake increases blood triglyceride levels and body weight.

*   High-sugar diets increase fat accumulation in the liver and the risk of diabetes.

*   Regular consumption of green tea reduces the risk of prostate cancer.

*   High alcohol intake during pregnancy produces physical and behavioral abnormalities in the fetus.

*   Malnutrition during pregnancy increases the risk that offspring will develop hypertension and heart disease later in life.

Many questions about the effects of interactions among gene variants and environmental exposures on health remain to be answered. Currently, scientists know the most about disorders related to **single-gene defects**.

single-gene defects   Disorders resulting from one abnormal gene. Also called "inborn errors of metabolism." Over 6,000 single-gene disorders have been cataloged, and most are rare.

Single-Gene Defects

Thousands of rare diseases related to a defect in a single gene have been identified, and many of these affect nutrient needs. Such defects can alter the absorption or utilization

Table 26.2 Examples of single-gene disorders that affect nutrient need or utilization[20-23]

PKU (phenylketonuria, pronounced phen-eel-key-tone-uria)	A rare disorder caused by the lack of the enzyme phenylalanine hydroxylase. Lack of this enzyme causes phenylalanine, an essential amino acid, to build up in the blood. High blood levels of phenylalanine during growth lead to mental retardation, poor growth, and other problems. PKU is treated by low-phenylalanine diets.
Galactosemia (pronounced gal-lac-tos-see-me-ah)	Galactosemia is a single-gene-defect disorder that interferes with the body's utilization of the sugar galactose found in lactose ("milk sugar"). The signs and symptoms of galactosemia result from an inability to use galactose to produce energy. If infants with classic galactose-mia are not treated promptly with a low-galactose diet, life-threatening complications appear within a few days after birth. People with this condition must avoid all milk, milk-containing products (including dry milk), and other foods that contain galactose for life. It occurs in approximately 1 in 30,000 to 60,000 newborns.
Hemochromatosis (pronounced heme-oh-chrome-ah-toe-sis)	A single-gene defect disorder affecting 1 in 300 Caucasians. This fairly common single-gene defect that occurs most commonly in Caucasians. It occurs due to a defect in a gene that produces a protein that controls how much iron is absorbed from food. Individuals with hemochromatosis absorb more iron than normal and have excessive levels of body iron. High levels of body iron have toxic effects on tissues such as the liver and heart. Hemochromosto-sis is treated with medications and a low-iron and vitamin C diet. A high intake of vitamin C can make hemochromatosis worse because vitamin C increases the absorption of iron.

of nutrients such as amino acids, iron, zinc, and the vitamins B_{12}, B_6, or folate. Phenylketonuria (PKU), galacostemia, and hemochromatosis are three examples of single-gene defects that substantially affect nutrient needs or utilization (Table 26.2). Due to its potentially severe affects on mental development if not treated with a low phenylalanine diet starting after birth, all babies born in the United States and many other countries are tested for PKU after birth (Illustration 26.1).

Most diseases related to genetic traits are not as well defined as are single-gene defects. They are more likely to represent an interwoven mesh of gene variants and environmental factors.

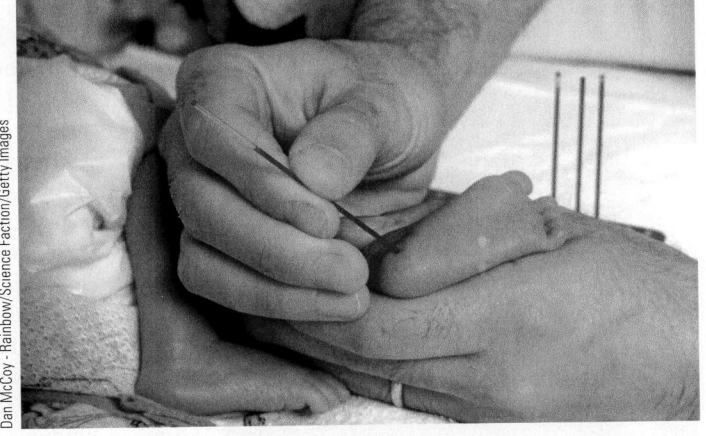

Illustration 26.1 A blood test for PKU is a routine part of newborn screening in most Western countries.[20]

Diseases Resulting from Multiple Gene Variants

Major health problems of our day, including heart disease, cancer, hypertension, obesity, and diabetes, are very rarely due to single-gene defects but commonly result from interactions of multiple gene variants with dietary components and other environmental factors.[1, 24] We take a closer look at the effects of nutrient-gene variant interactions on the development of cancer, hypertension, and obesity next.

Cancer Most types of cancer are primarily related to environmental exposures such as high fat and alcohol intakes, low vegetable and fruit diets, high levels of body fat, smoking, and toxins in the environment.[25] People who regularly consume vegetables from the cruciferous family (e.g., broccoli, cabbage, cauliflower) have a reduced risk of certain types of cancer *if* they have a particular gene.[26] The variant gene, on the other hand, codes for the excretion of the sulfur compound in cruciferous vegetables that helps prevent the development of cancer.

Studies of the roles played by gene variants and environmental factors, to the extent they are currently understood, show differences by cancer site (Table 26.3). As can be seen by the contributions of environmental factors to cancer development, the hope for cancer prevention primarily lies in our ability to remove harmful environmental exposures.

Hypertension The primary causes of most cases of hypertension are unknown, but factors such as high body fat, high alcohol intake, physical inactivity, and gene variants play a role.[27] High salt (sodium, really) intakes are also related, but not in all studies. Differences in results between studies may be related to the proportion of people in study samples who are genetically predisposed to the effects of sodium intake on blood pressure. Such

Table 26.3 Estimates of the contribution of environmental and genetic factors to development of some types of cancer[41,42]

Cancer Site	Environmental Factors (%)	Genetic Factors (%)
Endometrial uterine wall	100%	0%
Ovary	78	22
Lung	74	26
Breast	73–90	10–27
Stomach	72	28
Colon	65–95	5–35
Pancreas	64	36
Prostate	58	42

salt sensitivity A genetically determined condition in which a person's blood pressure rises when high amounts of salt or sodium are consumed. Such individuals are sometimes identified by blood pressure increases of 5-10% or more when switched from a lower-salt to a higher-salt diet.

people exhibit **salt sensitivity**; their blood pressure increases when they consume high amounts of sodium.[28] Approximately 51% of people with hypertension and 26% of people with normal blood pressure are salt sensitive.[29]

Obesity The current obesity epidemic appears to be driven by a mismatch between multiple components of our 40,000-year-old genetic endowment and current food and physical activity environments (Illustration 26.2). Genetic traits that helped our early ancestors survive times of famine and that encouraged food intake rather than discourage it, and those that set up metabolic systems around unrefined and unprocessed basic foods are at odds with much of today's food supply and physical activity requirements.[30]

Over 40 gene variants have been related to obesity development in people exposed to Western-type diets and low levels of physical activity.[30] Many of the traits prompt excess food intake in response to high levels of availability of palatable, energy-dense foods. Gene variants have been found to influence body mechanisms that increase appetite and decrease satiety, for example.[18] Gene functions that promote obesity can develop in the fetus and during other periods of rapid growth and development in response to energy and nutrient availability.[31]

The effect of gene variants on the development of obesity explain a relatively small percent of the risk for obesity.[1] Additional research is needed to identify why some individuals are at risk of developing obesity while others are not, and why some people are more genetically susceptible to obesogenic environments than others.[31,32]

Reversing the worldwide trend in obesity development will largely depend on decreasing our exposure to environmental triggers that encourage excess food intake and physical inactivity. Introducing safe, walkable areas into neighborhoods, expanding access

Illustration 26.2 Obesity is related to complex interactions among nutritional, other environmental, and genetic factors.

REALITY CHECK
Changing Family History

Elena and Alfredo both come from families with a number of relatives who have died from heart disease.

Who gets the thumbs-up?

Answer appears on page 26-7.

Elena: Heart disease is in my genes. There's nothing I can do about it.

Alfredo: I already know my LDL-cholesterol is high and my HDL-cholesterol is low. I'm eating and exercising to beat the odds.

to healthy foods, and increasing the availability of recreational facilities are examples of environmental changes that appear to help.[43,44]

Genetics of Taste

Food preferences are partly influenced by genetic traits. Genetically influenced food preferences apply to dogs, cats, and other animals as well humans (Illustration 26.3). There are more than 80 genes that help people taste bitter foods, for example, and some people get the set of genes that make them highly sensitive to bitter-tasting foods (mainly vegetables). People born with a high sensitivity to bitter tastes tend to dislike cooked cabbage, collard greens, spinach, Brussels sprouts, or other vegetables that taste bitter to them. People who tend to like these vegetables generally don't perceive them to be bitter tasting. A genetic tendency to reject these vegetables is likely to limit intake and therefore may be linked to diseases associated with low vegetable intake.[33]

Genetic traits also affect a person's perception of sweet and savory (umami) tastes. People who prefer sweet and savory tastes tend to have higher intake of sugar and fats, and higher body mass index than people who don't.[34]

Illustration 26.3 Dogs and many other animals have a taste for sweet, but cats don't. Somewhere during genetic evolution the gene in cats that enabled them to recognize sweetness was turned off.[40]

Nutrition Tomorrow

The promise of advances in knowledge of the genetic bases of disease is longer, healthier lives. No doubt drugs that counter ill effects of genetic traits will continue to be developed, and attempts to fix abnormal genes by gene therapy will broaden. Some of the most meaningful breakthroughs in upcoming decades will be in the area of disease prevention and treatment through nutritional changes.[35] Knowledge of genetic traits will be increasingly used to identify individual nutrient needs and responsiveness to dietary and other changes.[1]

Although used to some extent now, personalized modifications of dietary intake based on genotypes will become standard practice in clinical dietetics and medicine (Illustration 26.4).[36] It is already clear that lifestyle modifications, such as weight loss of overweight, increased physical activity and generous intake of vegetables and fruits, for example, will be components of health improvement efforts stemming from knowledge of genetic risk factors.[1]

The sequencing of a human's genome, or an individual's DNA profile, cost $350,000 in 2008 but can now be done for $3,000. In the next few years the price is expected to decline to $1,000.[37] Whole genome maps of DNA are becoming available before scientists and clinicians even know what the results mean and how to effectively apply the results broadly to disease prevention and treatment.[38] The availability of relatively inexpensive DNA tests will allow health care practitioners to focus on specific genetic traits for which enough is known to offer recommendations for beneficial therapies. For these and other reasons, whole genome DNA tests are not recommended for the pubic.[39]

Although tremendous advances in knowledge have been made, understanding the functions and interactions of the entire human genome is an extraordinarily complex undertaking and will evolve over time.[1] When understood, it will represent the next revolution in breakthrough knowledge about nutrition and health.

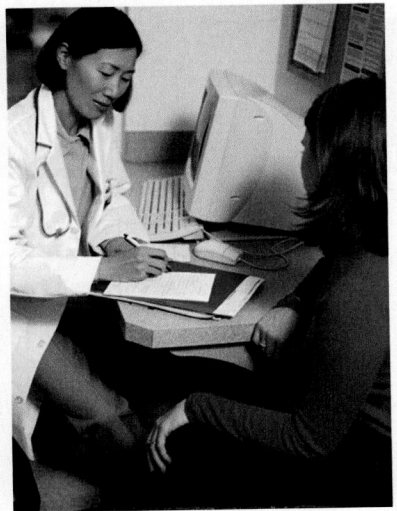

Illustration 26.4 In the future, advice given by registered dietitians may be tailored to the specific genetic traits of individuals.

ANSWERS TO REALITY CHECK
Changing Family History

Alfredo recognizes that a family history of heart disease doesn't mean he is destined to die from it. A minority of diseases are due solely to genetic factors; many are primarily due to interactions between genetic traits and environmental factors. Disease-promoting traits can be diminished by the right environmental changes.[1]

Elena:

Alfredo:

NUTRITION
up close

Nature and Nurture

Focal Point: Gaining insight into your family's and your own dietary health history.

Families often share dietary factors that interact with genetic traits to influence chronic disease development. Check out your dietary health history and that of your family by completing the following family food tree activity. If you don't know all of your relatives included in the activity or their dietary behaviors, fill out the parts based on what you do know.

Check each dietary behavior that applies:

Relative	Excess calorie consumption		Low vegetable and/or fruit intake		Low whole grains intake		Low fiber intake		Low fish intake		High alcohol intake	
	yes	no	yes	no	yes	no	yes	no	yes	no	yes	no
A. On your mother's side:												
Grandmother												
Grandfather												
B. On your father's side:												
Grandmother												
Grandfather												
C. Your mother												
D. Your father												
E. Yourself												

Feedback to the Nutrition Up Close is located in Appendix G.

REVIEW QUESTIONS

- Describe two ways that genes can affect nutrient needs and two ways that nutrients affect gene function.

- Identify specific ways in which gene variants interact with nutrient exposures to increase or decrease disease risk.

1. Genomics is the study of the functions and interactions of genes. True/False

2. Nutrigenomics is the study of the interaction of body weight and genes. True/False

3. Genes provide the codes for the production of enzymes and other proteins in cells. True/False

4. Unfavorable changes in genetic traits are the primary cause of the current, global epidemics of obesity and diabetes. True/False

5. Many chronic diseases result from single-gene defects. True/False

6. Many chronic diseases result from interactions among multiple genetic traits and environmental factors. True/False

7. How genes function, such as when they are prompted to turn on or off, can be programmed early in life by dietary exposures. True/False

8. Some of the most meaningful breakthroughs in the prevention and management of chronic diseases in the future will be due to advances in nutrigenomics. True/False

9. Galactosemia is a single-gene defect that interferes with the digestion of lactose. True/False

10. Hemochromotosis is primarily caused by a defect in a gene that affects iron absorption. True/False

11. Gene variants represent different genes that code for the synthesis of the same specific protein. True/False

12. Availability of certain vitamins can affect the functional level of specific genes. True/False

The following three questions refer to this scenario:

You're standing in a line to order Chinese takeout when you can't help but overhear the conversation about weight loss the gentleman behind you is having with his friend. The conversation goes like this. "Ya know George, no matter what I do, I can't lose weight. I eat like a bird and I still gain weight. I swear I've inherited the obesity gene."

13. _____ Which of the following statements most likely represents the truth about the gentleman's problem with losing weight?
   a. He has probably inherited the gene known to cause obesity from his mother or father.
   b. He may have inherited genetic traits that increase the likelihood he will overeat in an environment where plenty of highly palatable foods are available.
   c. He has developed a single-gene defect that causes obesity.
   d. He has gene variants that decrease his calorie need substantially so he will not be able to lose weight.

14. _____ If you could have turned around and talked with the gentleman about the likely reasons for obesity, what would you say that would represent state-of-the-science knowledge?
   a. Genetic traits do not influence a person's chances of becoming obese.
   b. Genetic defects in energy metabolism affect obesity development, not calorie consumption.
   c. It has become clear recently that multiple genetic traits make it impossible for some people to lose weight.
   d. Multiple genes may influence the development of obesity to some degree, but if you consume fewer calories than you use you will lose weight.

15. _____ The gentleman decides to have his genetic makeup tested to confirm that his genes produced his obesity. Which of the following statements about the results of the test is correct?
   a. The test will identify multiple genes that account for his obese state.
   b. The test will provide information needed to recommend a particular nutrient supplement that will enable him to lose weight.
   c. The test will be inconclusive because genes, environment, and behavior influence the development of obesity and not genes alone.
   d. His DNA will likely indicate that he is obese because he is missing a gene that increases a person's level of physical activity.

Answers to these questions can be found in Appendix G.

NUTRITION SCOREBOARD ANSWERS

1. That's incorrect. Only a very small proportion of diseases are directly caused by a person's genes.[1] **False**

2. Individual genetic traits alter essential nutrient needs to a small extent in many people, and to a large extent in relatively few.[2] **False**

3. Correct. Common diseases result from interactions between multiple genetic traits and environmental factors such as nutrient intake.[3] **True**

Nutrition and Physical Fitness for Everyone

NUTRITION SCOREBOARD

1 Physical fitness means being very muscular.   **True/False**

2 Overweight people can be physically fit.   **True/False**

3 Exercise is more effective than diet in preventing heart disease.   **True/False**

4 Physical fitness can only be achieved by exercising intensively for at least an hour every day.   **True/False**

Answers can be found at the end of the unit.

UNIT **27**

After completing Unit 27 and its interactive learning features, you will be able to:

- Describe the individual components of physical fitness and the health advantages of being physically fit.

- Be able to calculate how to measure heart rate and perform calculations related to estimating maximal heart rate (MHR) for age and percent MHR targets for aerobic activity.

Physical Activity: It Offers Something for Everyone

- **Describe the individual components of physical fitness and the health advantages of being physically fit.**
- **Be able to calculate how to measure heart rate and perform calculations related to estimating maximal heart rate (MHR) for age and percent MHR targets for aerobic activity.**

Physical activity has much to offer the athlete and nonathlete alike. It provides recreation, which is good for both the body and soul, it doesn't have to cost anything, and it benefits almost everyone. At its best, physical activity results from play and has positive consequences for health and well-being. As long as there are physical activities people enjoy doing, there's a physically fit "athlete" in everyone.

Key Nutrition Concepts

Physical activity interacts with healthy diets in a number of ways to promote overall health and well-being. Key nutrition concepts that relate to the "good diet" portion of these relationships include:

- Poor nutrition can influence the development of certain chronic diseases.

- Adequacy, variety, and balance are key characteristics of healthful diets.

The "Happy Consequences" of Physical Activity

Ask people who are in good physical condition what they get from physical activity and you're likely to hear a variety of responses. Someone who is 20 years old may say he wants to stay in shape and improve his stamina. A 40-year-old individual may say that exercise helps keep her blood triglyceride levels and weight from increasing. People who are 80 might tell you they exercise so that they won't have to use a walker and will be able to maintain their independence. Ask children why they exercise, however, and they may not understand the question. Children engage in active play; they don't exercise. For them, fitness is truly an unintended consequence of play.

Regular physical activity benefits both physical and psychological health in people of all ages (Table 27.1). By improving a person's physical health, regular physical activity helps ward off heart disease, obesity, some types of cancer, hypertension and stroke, osteoporosis, and diabetes. It tends to increase a person's feeling of well-being and helps relieve depression, anxiety, and stress.[1,4,5]

The Bonus Pack: Physical Activity Plus a Good Diet Physical activity benefits health most when combined with a good diet and other healthy behaviors.[3,7] Regular physical activity helps build bone mass and reduce the risk of bone fractures.[8] But the risk is lowered to a greater extent if physical activity is combined with a diet that supplies adequate amounts of calcium and vitamin D.[9] Some other benefits of exercise, such as a reduced chance of developing colon and breast cancer, may be related to the effect of exercise on reducing the ill effects of high levels of body fat on health. Regular physical activity is one of the few factors we can identify that helps people reduce weight gain with age.[10]

Exercise and Body Weight Combined with a moderate decrease in usual caloric intake (on the order of 200 calories per day), exercise helps people lose body fat, build muscle mass, and become physically fit.[7] Exercise alone is generally ineffective, or less effective as a weight reduction measure, than is exercise plus a reduced-calorie intake.[11,12] Because the body needs more calories to maintain muscle than fat, exercise that results in an increase in muscle mass leads to an increase in caloric requirements. For many people, this increase in calorie requirement makes it easier to maintain weight during adulthood and to keep lost weight off.[10,11]

Table 27.2 lists the calorie cost of various types of exercise. The values are calculated based on body weight because an individual's weight affects the energy requirement for

physical activity Body movements produced by muscles that require energy expenditure. Exercise is a subcategory of physical activity. It is generally planned, structured, and repetitive. The terms *physical activity* and *exercise* are often used interchangeably.

Table 27.1 Benefits of regular physical activity[1,4–6]

Reduced risk of certain diseases and disorders	Improved sense of well-being
• Heart disease	• Increases feeling of well-being
• Colon and breast cancer	• Decreases depression and anxiety
• Hypertension	• Helps relieve stress
• Stroke	• Decreases risk of dementia
• Osteoporosis	
• Back and other injuries	
• Obesity, excess abdominal fat	
• Type 2 diabetes	
• Bone and joint diseases	

Paul Bradbury/OJO Images/Jupiterimages

Table 27.2 Average calorie output per pound of body weight for selected types of exercise

Exercise	Intensity	Calories (pound/hour)	Exercise	Intensity	Calories (pound/hour)
Walking	3 mph (20 min/mi)	1.6	Weight lifting		2.9
	3½ mph (17 min/mi)	1.8	Wrestling		6.2
	4 mph (15 min/mi)	2.7	Handball	Moderate	4.8
	4½ mph (13 min/mi)	2.9		Vigorous	6.2
Jogging	5 mph (12 min/mi)	4.1	Swimming	Resting strokes	1.4
	5½ mph (11 min/mi)	4.5		20 yd/min (mod.)	2.9
	6 mph (10 min/mi)	4.9		40 yd/min (vig.)	4.8
	6½ mph (9 min/mi)	5.2	Rowing (sculling or machine)		4.8
	7 mph (8½ min/mi)	5.6			
	7½ mph (8 min/mi)	6.0	Downhill skiing		3.8
Running	8 mph (7½ min/mi)	6.3	Cross-country skiing (level)	4 mph (15 min/mi)	4.3
	8½ mph (7 min/mi)	6.7		6 mph (10 min/mi)	5.7
	9 mph (6⅔ min/mi)	7.1		8 mph (7½ min/mi)	6.7
	9½ mph (6⅓ min/mi)	7.4		10 mph (6 min/mi)	7.8
	10 mph (6 min/mi)	7.6	Aerobic dancing	Moderate	3.4
	11 mph (5½ min/mi)	8.5		Vigorous	4.3
	12 mph (5 min/mi)	9.5	Rebound trampoline	50–60 steps/min	4.1
Cycling (stationary)	Mild effort	2.9	Racquetball/squash	Moderate	4.3
	Moderate effort	3.4		Vigorous	4.8
	Vigorous effort	4.3	Tennis	Moderate	3.4
				Vigorous	4.3
Cycling (level)	6 mph (10 min/mi)	1.5	Volleyball	Moderate	3.4
	8 mph (7½ min/mi)	1.8		Vigorous	3.8
	10 mph (6 min/mi)	2.0	Basketball	Moderate	3.8
	12 mph (5 min/mi)	2.8		Vigorous	4.8
	15 mph (4 min/mi)	3.9	Football	Moderate	3.8
	20 mph (3 min/mi)	5.7		Vigorous	4.3
Skating		2.9	Baseball/golf/woodcutting/ horseback riding/badminton/ canoeing		2.4
Calisthenics	Moderate	2.4	Soccer/hill climbing/fencing/ judo/snowshoeing		5.3
	Vigorous	2.9			
			Bowling/archery/pool		1.2
Rope skipping	Moderate	4.8			
Bench stepping	12" high, 24 steps/min	3.2			

Source: Values calculated from *Guidelines for Graded Exercise Testing and Exercise Prescription*. 2nd ed. Philadelphia: American College of Sports Medicine, Lea and Febiger, 1984.

Illustration 27.1 An example of a strength-building exercise.

specific physical activities. If a 140-pound person cycles on a stationary bicycle at a moderate level of effort for an hour, and the calorie cost of the exercise is 3.4 calories per pound per hour, the person would use about 476 calories per hour for moderate cycling (3.4 calories per pound × 140 pounds = 476 calories). A person who weighs 212 pounds would use approximately 720 calories (3.4 calories per pound × 212 pounds). If the exercise is continued for 10 minutes, the calorie expenditure number would be divided by 6, or if done for 15 minutes, it would be divided by 4.

What Is Physical Fitness?

Many of the benefits of physical activity are related to the **physical fitness** it promotes. Physical fitness is not defined by bulging muscles, thin waistlines, or amount of physical activity. Overweight as well as thin people can be physically fit or not. Physical fitness is a state of health measured by **muscular strength** and **endurance**, **cardiorespiratory fitness**, and **flexibility**.[13]

The strength component of physical fitness relates to the level of maximum force that muscles can produce. Muscular endurance refers to the length of time muscles can perform physical activities, and flexibility refers to a person's range of motion. Cardiorespiratory fitness relates to the functioning of the circulatory system, heart, and the lungs. Physical fitness exists when all four are present at health-promoting levels.

Muscle Strength Muscle strength depends on the ability of a muscle or groups of muscles to lift, pull, push, or otherwise exert force against a weight or opposing force. Muscle-building activities are referred to as **resistance exercise**. Such activities require muscles to work harder than usual and over time increase muscular strength. Strength-building exercise includes lifting weights, doing pull-ups and push-ups, and the use of stretch bands (Illustration 27.1).

Muscle Endurance and Cardiorespiratory Fitness Muscular endurance is a measure of the ability of a muscle or group of muscles to sustain repeated muscular contractions against a weight or force over time. The length of time muscles can work against a weight or force largely depends on cardiorespiratory fitness and is measured in terms of the amount of oxygen an individual is able to deliver to muscles. (Cardiorespiratory fitness is sometimes referred to as **aerobic fitness**.) Levels of cardiorespiratory fitness build up with time as individuals increase their level of **aerobic exercise**.

Aerobic exercises use oxygen for energy formation by muscles. Performing these types of activities enhance the functioning of the heart and lungs and increase their ability to deliver oxygen to muscles. Fat-burning, aerobic exercises increase cardiovascular fitness in three major ways:

1. They strengthen and expand the capacity of the lungs to deliver oxygen.
2. They increase the ability of the circulatory systems to deliver blood and oxygen to muscles and other tissues throughout the body.
3. They strengthen the ability of the heart to move an increased volume of blood through the body.

Aerobic activities include jogging, basketball, swimming, soccer, zumba, aerobic dance, and other low- and moderate-intensity activities (Illustration 27.2). They give most of the body a workout.

How Is Cardiorespiratory Fitness Determined? Cardiorespiratory fitness is classically assessed by measuring **maximal oxygen consumption** (abbreviated as VO_2 max) in a specially equipped laboratory (Illustration 27.3). In the lab, individuals are exercised at increasingly higher intensities, for example, by elevating the grade of a treadmill or increasing the resistance on a stationary cycle. The individual performing the exercise breathes through a tube that delivers air. Monitoring equipment measures the amount of oxygen from air that is used during the exercise. The maximal amount of oxygen a person delivers to working muscles is the amount used when the intensity of exercise can no longer be increased. The higher the level of oxygen used at the peak level of activity (or 100%

physical fitness The health of the body as measured by muscular strength and endurance, cardiorespiratory fitness, and flexibility. Body composition, agility, and balance are sometimes included as components of physical fitness.

muscular strength The ability of a muscle or muscle group to exert force. It is assessed by the maximal amount of resistance or force that can be sustained in a single effort.

muscular endurance The ability of a muscle or group of muscles to sustain repeated contractions against a resistance for an extended period of time. Also called *muscular fitness*.

cardiorespiratory fitness The ability of the circulatory and respiratory systems to supply fuel during sustained physical activity and to eliminate fatigue products after supplying fuel. Cardiorespiratory fitness is also called *aerobic fitness and cardiorespiratory endurance*.

flexibility Range of motion of joints.

resistance exercise Physical activities that involve the use of muscles against a weight or force. Resistance exercise increases muscle strength, mass, power, and endurance. Also called resistance training, strength training, and weight training.

aerobic fitness A state of respiratory and circulatory health as measured by the ability to deliver oxygen to muscles and the capacity of muscles to use the oxygen for physical activity.

aerobic exercise Physical activity in which the body's large muscles move in a rhythmic manner for a sustained period of time. Aerobic exercise involves metabolic pathways that require oxygen for energy production and improve functioning of the cardiovascular and respiratory systems. Aerobic activities include walking, jogging, running, swimming, skiing, vacuuming, house cleaning, lawn mowing, raking, cycling, and aerobic dance.

VO$_2$ max), the higher the level of aerobic fitness and the longer physical activity can be performed.

People can perform physical activity at 100% of VO$_2$ max for only a few minutes. Consequently, aerobic fitness goals are set below that level. In general, it is recommended that beginners start a cardiorespiratory fitness program with a goal of exercising at 40 to 60% of VO$_2$ max and working up to a higher level.[13] Aerobically fit people generally train at 70 to 85% of VO$_2$ max. VO$_2$ max can be increased by exercising regularly at intensities that raise a person's heart rate.[23,26]

Maximum Heart Rate The amount of blood circulated to muscles is roughly equivalent to the number of times the heart sends out a pulse of blood (or "beats") each minute. The more oxygen muscles need to perform an activity, the more oxygen the lungs have to deliver and the faster the heart has to beat (that is, up to the point where the lungs and heart can no longer send an additional supply of oxygen to muscles). The rate at which the highest level of oxygen delivery to working muscles occurs is considered "maximum heart rate (MHR)" or 100% MHR. Moderate intensity activities are considered those undertaken between 64–76% MHR, and intense activities fall between 77–95% MHR.[23]

The formula for estimating maximal heart rate among adults is straightforward:

$$MHR = 208 - (age \times 0.7)^{28}$$

An individual's MHR calculated using this formula may over- or under-estimate true rate by about 10 bpm. To obtain a target heart rate, or heart rate range for exercise, multiply your MHR by the desired percent of MHR (Table 27.3 shows an example). The result is your "target heart rate" for aerobic exercise. Table 27.4 lists maximal heart rate, 70% of the maximum, and 85% of the maximum rate for adults between the ages of 20 and 50 years.

Illustration 27.2   An example of aerobic exercise.

maximal oxygen consumption   The highest amount of oxygen that can be delivered to, and used by, muscles for physical activity. Also called VO$_2$ max and maximal volume of oxygen.

Table 27.3 **Applying the 100% MHR formula to a 22-year-old who will exercise at a level of 60% MHR**

```
100% MHR = 208 – (age X 0.7)
         = 208 – (22 x 0.7)
         = 208 – 15 (rounded down) = 193 bpm

100% MHR = 193 bpm

60% MHR = 193 x 0.6 = 116 bpm
```

Table 27.4 **Target heart rates estimated at 100%, 70% (moderate intensity exercise) and 85% (vigorous intensity exercise) MHR for adults ages 20–50 years[23,26,28]**

Age (Years)	100% MHR	70% MHR	85% MHR
20	194	136	165
25	191	133	162
30	187	131	159
35	184	128	156
40	180	126	153
45	177	124	150
50	173	121	147

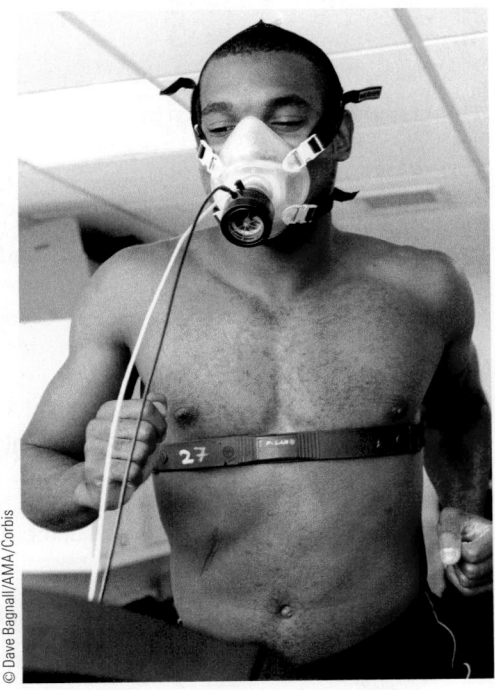

Illustration 27.3   Determining VO$_2$ max. VO$_2$ max is measured when a person's nose is plugged so she or he must breathe through a tube that delivers air. The tube is attached to an apparatus that measures the total amount of air breathed in and out, and calculates the difference between the amount of oxygen inhaled and exhaled. Every few minutes the resistance of the cycle is increased, making the exercise more intense. The amount of oxygen a person uses at the point when exercise intensity can go no higher is considered "maximal oxygen consumption," or 100% VO$_2$ max.

How Do You Know Your Heart Rate? During aerobic exercise, take a break, press a fingertip gently on the artery in your wrist directly below and in line with your thumb (Illustration 27.4). Count the number of pulses you feel in a 15-second period and multiply that number by 4. That's your heart rate measured in beats per minute (bpm).

Have you been thinking about increasing your level of aerobic exercise? If you have, refer to the first of two Take Action features in this unit to make a plan for doing that.

Flexibility

Flexibility refers to the range of motion of your muscles and connective tissues around your joints, or how "tight" or "loose" your body movements are. It affects your ability to stretch, react, bend, and maintain balance and agility.[14] It is important for day-to-day activities such as walking, reaching, moving smoothly and quickly out of danger, and balance. Flexibility is maintained and increased by many types of physical activities. Stretching exercises, in particular, increase flexibility (Illustration 27.5).[14]

Stretching helps improve flexibility by lengthening muscle and tendons. It involves extending large muscle groups, while breathing, to the point you feel muscle tightness rather than discomfort. This position is held for 10 to 30 seconds, released, and then repeated once or twice. You can stretch your shoulder muscles, for example, by extending as arm across your chest and placing a hand on your elbow and gently pulling the arm towards your chest. Or, by grabbing the bottom of one leg above the ankle and pulling the heel back toward the buttocks while pushing your hips out. Yoga, pilates, and tai chi movements are good examples of exercises that improve flexibility.

Stretching before or after exercising does not appear to protect against muscle soreness or reduce muscle injury, and it is not yet clear whether it increases performance.[14,15] It does, however, improve flexibility and muscle stiffness when it is done before exercise begins.[14]

Delayed Onset Muscle Soreness (DOMS) Have you ever exercised really hard one day when you were out of shape and felt it the next day? You are certainly not alone. People who begin an exercise program and overdo it, or perform physical activities they don't usually do, often hear back from the muscles within the next day or two. The muscle

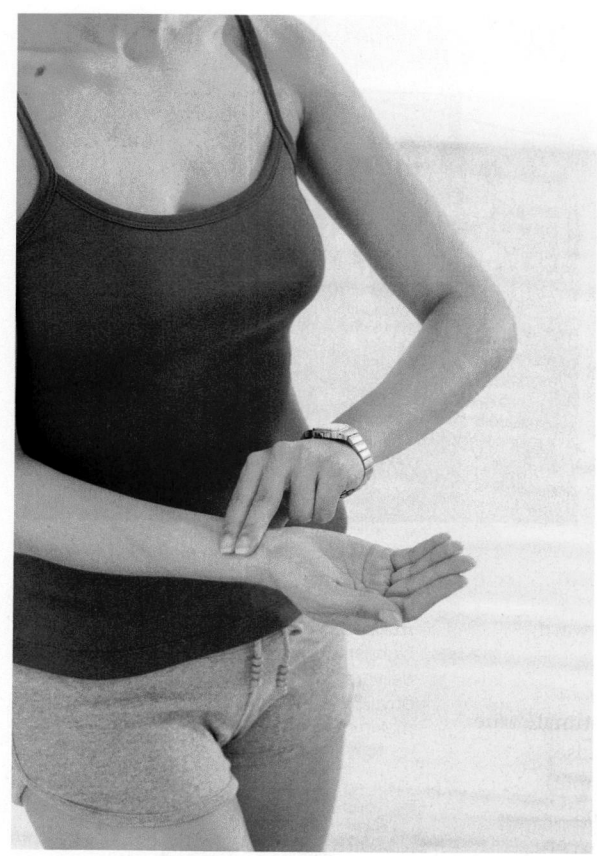

Illustration 27.4 Finger location for measuring pulse on your wrist.

© dk/Alamy

---

take action Increase Your Cardiorespiratory Fitness

Increase your level of cardiorespiratory fitness by choosing aerobic activities you enjoy. First, check the aerobic activities you like to do and can do. Then, list how long and on how many days per week you will engage in the activity.

Aerobic activity	Duration of activity	Which days?		Aerobic activity	Duration	
____ Swimming	____ minutes	_____		____ Basketball	____ minutes	_____
____ Walking	____ minutes	_____		____ Cycling	____ minutes	_____
____ Jogging	____ minutes	_____		____ Aerobic dance	____ minutes	_____
____ Running	____ minutes	_____		____ Zumba	____ minutes	_____
____ House-cleaning	____ minutes	_____		____ Ballroom dancing	____ minutes	_____
____ Table tennis	____ minutes	_____		____ Other: _____	____ minutes	_____
____ Treadmill	____ minutes	_____				

Refer to these goals periodically to see how you're doing or modify your activities into a more workable plan. You can obtain forms for developing goals and a plan for aerobic and strengthening exercise, and for tracking your progress in meeting the goals from the Physical Activity Guidelines for Americans site: www.health.gov/paguidelines/adultguide/keepingtrack.pdf.

groups that were used most feel stiff and sore, and you feel every move that's made using them.

This condition is called **delayed onset muscle soreness (DOMS)** and here's why it happens. Overused muscles develop microscopic tearing in muscle fibers that cause stiffness and soreness for a day or two. Within a few days your muscles recover and actually end up in a better state to be strengthened. Muscles rebuild and become ready for the next bout of use.[16]

DOMS is most likely to occur when you:

- Use muscles for vigorous physical activity you don't generally use.

- Engage in physical activity when you are out of shape.

- Dramatically increase exercise intensity or duration of physical activity.

- Run downstairs or downhill, or engage in other downward motions.[16]

iStockphoto.com/Mark Bowden

Illustration 27.5 Stretching increases flexibility and reduces muscle stiffness when it is done before beginning to exercise.

Fueling Physical Activity

Muscles use fat, glucose, and amino acids for energy. The proportion and amount of each that is used depends on the intensity of activity (Illustration 27.6). When we're inactive, fat supplies between 85 and 90% of the total amount of energy needed by muscles. The rest is provided by glucose (about 10%) and amino acids (5% at most).[17] Fat is also the primary source of fuel for activities of low-to-moderate intensity such as jogging and swimming. Because oxygen is required to convert fat into energy, low-to-moderate intensity activities are "aerobic"—or "oxygen requiring." This unit's Reality Check addresses the topic of

Illustration 27.6 Schematic representation showing the proportionate use of sources of energy for physical activities of various intensities. Most activities are fueled by both fat and glucose.

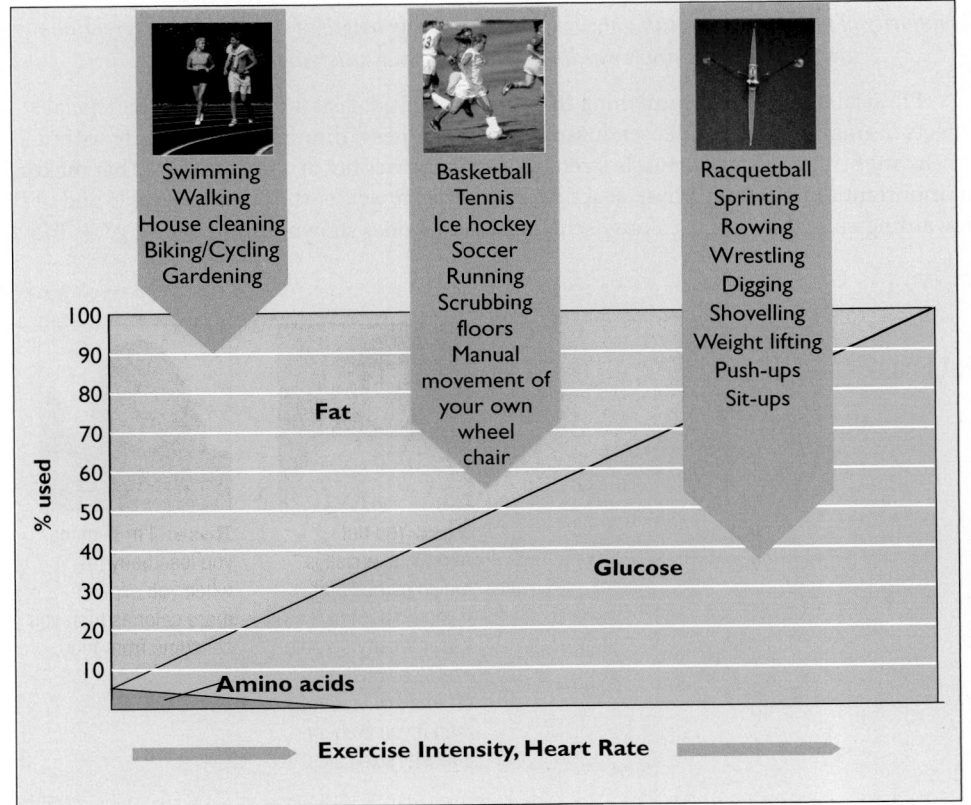

Swimming
Walking
House cleaning
Biking/Cycling
Gardening

Basketball
Tennis
Ice hockey
Soccer
Running
Scrubbing
floors
Manual
movement of
your own
wheel
chair

Racquetball
Sprinting
Rowing
Wrestling
Digging
Shovelling
Weight lifting
Push-ups
Sit-ups

100
90
80
70
60
50
40
30
20
10

% used

Fat

Glucose

Amino acids

Exercise Intensity, Heart Rate

PhotoEdit, Tony Freeman; © Cengage Learning

delayed onset muscle soreness (DOMS) Muscle pain, soreness, and stiffness that occurs a day or two after exercise of muscles that are not used to high levels of activity.

aerobic exercise and its relationship to body fat. If you are curious about that, take a look at the Reality Check.

High-intensity, short-duration activities like making a tackle, throwing a fastball, or sprinting down the block to catch a bus, are fueled primarily by glucose. Glucose is the primary fuel for these activities because metabolic processes exist within muscle cells that convert glucose to energy quickly. Our supply of glucose for intense activities comes principally from glycogen, the storage form of glucose. Glycogen is stored in muscles and the liver and can be rapidly converted to glucose when needed by working muscles. High-intensity, short-duration activities represent **anaerobic exercise** because the conversion of glucose to energy does not require oxygen (it's *an*aerobic). People can undertake very intense activity only as long as their stores of glycogen last.[18]

Activities such as basketball, hockey, tennis, football, and soccer that involve walking, running, and high-intensity, quick moves use both fat and glucose for energy.

A Reminder About Water Physical activity increases the body's need for water, and if the climate is hot and humid, this need increases even more. In general, people should drink in response to thirst and, overall, consume enough water to replace the amount lost in sweat, respiration, and urine during exercise.[20] (The amount of water lost during exercise is equivalent to the amount of weight that is lost during the exercise.) You are consuming the right amount of water if your urine is pale yellow and normal in volume. For exercise that lasts over an hour, consumption of a sports drink appears to improve hydration.[20] It's important to keep up fluid intake when exercising in cold weather. Cold air holds less water vapor than hot air, so you lose more water through breathing in cold weather.

Achieving Physical Fitness

Commenting on physical activity for the long run, Sir Roger Bannister, a noted physician and the first person to run a mile in less than four minutes, said:

> *The best advice to those who are unfit is to take exercise unobtrusively. I am not an enthusiast for fitness and exercise schemes that are boring because they tend to fizzle out. . . . The vast majority of people, I think, can only be attracted for any length of time towards recreational activities if these are rewarding, enjoyable, and satisfying in themselves.*

Physical fitness is not something that, once achieved, lasts a lifetime. The beneficial effects of training on muscular endurance (aerobic fitness) diminish dramatically within 2 weeks after training stops. Muscle strength also decreases but at a lower rate.[21] That makes it important to undertake physical activities that "wear well"—those are enjoyable and rewarding enough to fit into a busy schedule like the ones shown in Illustration 27.7. Too

anaerobic exercise Short-duration, intense activities. Anaerobic exercise requires glucose for energy production. Energy formation from glucose does not require oxygen.

REALITY CHECK
Does Aerobic Exercise Increase Body Fat Loss?

Who gets the thumbs-up?

Answer appears on page 27-9.

Tex: You bet. Aerobic exercise is fueled by body fat. If you do a lot of low-intensity aerobic exercise you'll burn off more of your fat than if you do high-intensity exercises.

Rose: I'm thinking you lose body fat when you expend more calories than you consume from food.

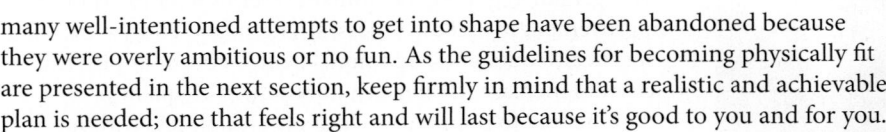

Illustration 27.7 Physical activity is often fun. (Even walkers and joggers smile sometimes.)

many well-intentioned attempts to get into shape have been abandoned because they were overly ambitious or no fun. As the guidelines for becoming physically fit are presented in the next section, keep firmly in mind that a realistic and achievable plan is needed; one that feels right and will last because it's good to you and for you.

Physical Activity Recommendations One out of three U.S. adults are physically inactive and only about 34% engage in regular physical activity (Illustration 27.8).[22] In response to the nationwide epidemic of inactivity, the Department of Health and Human Service developed recommendations aimed at improving the physical activity and fitness levels of Americans. Entitled "Physical Activity Guidelines for Americans,"[13] the recommendations urge Americans to undertake the following:

- At least 2 hours and 30 minutes (150 minutes) per week of moderate intensity aerobic activity. The activity can be performed in bouts of 10 minutes each if desired. Moderate intensity activities are those that increase heart and breathing rates. They include brisk walking and jogging, scrubbing floors, operating a wheelchair manually or playing a sport from a wheelchair, running around with children, fast dancing, tennis (doubles), and playing basketball.

- At least two sessions of strengthening activities per week. Strengthening activities should include activities like push-ups, sit-ups, weight lifting, or other resistance exercises. Exercises should work legs, hips, back, chest, stomach, shoulders, and arms. Each exercise should be repeated 8 to 12 times per session. Vigorous-intensity activities should be added to exercise programs a little at a time in order to reduce the risk of injuries.

ANSWERS TO REALITY CHECK
Does Aerobic Exercise Increase Body Fat Loss?

It's a myth that aerobic exercise is better for body fat loss than other types of exercise. Body fat levels decline when a person's calorie expenditure is greater than calorie intake.

Tex:

Rose:

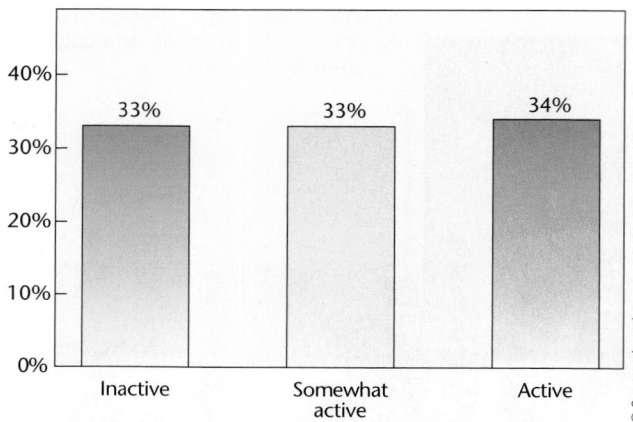

Illustration 27.8 Percent of U.S. adults who are inactive, somewhat active, and active.[22]

People who have a chronic disease should consult a physician before starting an exercise program. It may well turn out that exercise will be just what the doctor orders. Moderate exercise programs are beneficial and safe for almost everyone.[24]

Physical Activity Recommendations for Schoolchildren Regular physical activity is an important component of children's health and development. Yet, only one in four children exercise at the recommended level of an hour or more each day at a moderate to vigorous level.[13] Only one in three students are offered daily physical activity classes.[13]

Schools are being encouraged to include physical activity as a daily part of the curriculum for all children and adolescents. Physical activity should focus on meeting fitness goals rather than on competitive or sports performance goals.[28] Tests of physical fitness, such as the muscle strength and endurance assessment shown in Illustration 27.9, should be performed periodically on all children and the results used to modify instruction.

Illustration 27.9 An example of a flexibility exercise assessment in a child.

Some Exercise Is Better Than None Physical activity levels that are lower than the recommended levels still benefit health.[1] Sedentary people who take up walking, dancing, gardening, biking, golfing, or similar exercises tend to experience improvement in aerobic fitness, strength, and energy level. For most health outcomes, additional benefits occur as physical activities increase in intensity, duration, or frequency.

One way to improve levels of physical fitness without spending more time on exercise is to increase the intensity of physical activities. In general, 15 minutes of vigorous activity provides the same benefits as 30 minutes of moderate activity.[13] Vigorous activity burns more calories per minute and more effectively reduces body fat (including abdominal fat) than lower-intensity activities.[25] The second Take Action feature for this unit provides some choices for increasing exercise intensity

take action Ramping Up Exercise Intensity

Here are some options for increasing exercise intensity. Do you see an option or two that would work for you?

**I can
do that**   **No way**

_____   _____   1. Play "Beat your time." Trim seconds off the amount of time it takes you to walk a specific distance each time out for a month. Then maintain that level of walking intensity.

_____   _____   2. Gradually increase your cycling speed until you're safely traveling at least 1 mile in 6 minutes (10 mph).

_____   _____   3. Take the stairs two steps at a time. (Hold the handrail! Start with one floor of steps.)

_____   _____   4. Jog or run, rather than walk, whenever possible.

_____   _____   5. Carry light free weights with you when you walk or jog.

Average calorie output per pound of body weight for selected types of exercise.

Source: Values calculated from *Guidelines for Graded Exercise Testing and Exercise Prescription*. 2nd ed. Philadelphia: American College of Sports Medicine, Lea and Febiger; 1984.

Michael Blann/Getty Images/Riser

NUTRITION
up close

Exercise: Your Options

Focal Point: Assess your level of physical activity.

About a third of U.S. adults are physically inactive. How much physical activity are you getting?

Physical Activity

1. Check the category that best describes your usual overall daily activity level. Then answer two questions about exercise.

Overall Daily Activity Level

- *Inactive*: Sitting most of the day, with less than 2 hours moving about slowly or standing.
- *Average*: Sitting most of the day, walking or standing 2 to 4 hours each day, but not engaging in strenuous activity.
- *Active*: Physically active 4 or more hours each day. Little sitting or standing, engaging in some physically strenuous activities.

Exercise

2. Do you exercise at moderate intensity a total of 150 minutes per week?
3. Do you perform strength-building exercises twice a week?

Feedback to the Nutrition Up Close is located in Appendix G.

REVIEW QUESTIONS

- **Describe the individual components of physical fitness, and health advantages of being physically fit.**

- **Be able to calculate how to measure heart rate and perform calculations related to estimating maximal heart rate (MHR) for age and percent MHR targets for aerobic activity.**

1. Regular physical activity benefits psychological and physical health.   **True/False**

2. The primary components of physical fitness are body fat composition, strength, and heart rate.   **True/False**

3. Aerobic fitness is related to a person's ability to deliver oxygen to muscles and the capacity of muscles to use the oxygen for physical activity.   **True/False**

4. Endurance is related to aerobic fitness.   **True/False**

5. Maximal oxygen consumption is related to endurance.   **True/False**

6. A 44-year-old individual who wants to exercise at 70% of maximal heart rate would aim for a target heart rate of around 124 beats per minute.   **True/False**

7. Glycogen supplies the fuel to meet most of our energy needs while we are at rest.   **True/False**

8. Fat is the major fuel for low-to-moderate intensity physical activities.   **True/False**

9. Resistance training primarily benefits endurance.
True/False

10. To build and maintain a state of physical fitness it is recommended that adults engage in vigorous physical activity for 30 minutes daily.   **True/False**

11. Increased physical activity plus reduced calorie intake is more effective for reducing body weight and for preventing obesity than is increased exercise alone.   **True/False**

The next three questions refer to the following scenario:

Tacchi, now 45 years old, is well into an exercise program. He is exercising three times a week at 80% of maximum heart rate (MHR) for 30 minutes. During his exercise Tacchi takes a break to measure his heart rate. Yesterday he calculated his heart rate to be 155 beats per minute.

12. _____ Tacchi is:
   a. exercising at a level below 80% of his MHR.
   b. exercising at a level that is above 80% of his MHR.
   c. exercising at a level that is close to 80% of his MHR.
   d. exercising too short a time to accurately assess heart rate.

13. _____ Tacchi wants to expand his exercise program to activities that would increase his flexibility. Which of the following types of physical activities should be choose?
   a. weight lifting
   b. aerobic dance
   c. roller-blading
   d. stretching

14. ____ Assume Tacchi adds strength training to his exercise program. On the first day of strength training, Tacchi decides to go for it in a big way and spends 45 minutes running up and down hills and doing sit-ups. He felt fine all day after that, but the next day his muscles felt really sore and stiff. What probably happened to Tacchi's muscles?

a. He probably seriously injured a muscle.

b. He's probably coming down with the flu.

c. He developed delayed onset muscle soreness. His muscles will feel better in a few days.

d. He caused serious damage to his muscles and will be weaker for months because of the damaged caused.

Answers to these questions can be found in Appendix G.

NUTRITION SCOREBOARD ANSWERS

1. Muscle strength is one of four components of physical fitness. **False**

2. You can be fit and overweight. You don't have to be thin to be physically fit.[1] **True**

3. Regular exercise does help reduce the risk of heart disease—but not as much as a combined program of exercise, weight loss, and consumption of a healthy diet.[2,3] **False**

4. You don't have to exercise to that extent to become physically fit.[1] **False**

Nutrition and Physical Performance

NUTRITION SCOREBOARD

1 The diet an athlete consumes affects energy substrate availability to muscles during exercise. **True/False**

2 Carbohydrate loading is a waste of time. **True/False**

3 Supplementation with protein and amino acids increases muscle strength to about the same extent as does resistance training. **True/False**

4 Losing your period is normal when you are a female athlete. **True/False**

Answers can be found at the end of the unit.

After completing Unit 28
and its interactive learning
features, you will be able to:

- Describe how carbohydrates,
  proteins, and fats are utilized
  by muscles.

- Describe the recommended
  method for maintaining
  hydration during prolonged
  exercise, and recognize the
  signs of dehydration.

- Assess the safety and
  effectiveness of ergogenic
  aids offered to athletes.

Sports Nutrition

- **Describe how carbohydrates, proteins, and fats are utilized by muscles.**

> *BAM! Nobody heard it, but Lou felt it. She had "hit the wall." She was ahead
> of her planned pace, but now her legs felt like lead. She would have to finish
> the last two miles of the marathon at the slow pace her legs would allow.*

From her carefully crafted and scrupulously followed training program to her refined shaping of mental attitude, Lou thought she had done everything right. She had left one thing out, however, and that may have cost her the race. Lou failed to pay attention to her diet while training and ran out of **glycogen** too soon (Illustration 28.1).[3]

Three major factors affect **physical performance**: genetics, training, and nutrition.[3] The first gives some people an innate edge in sprinting or endurance, and nothing can be done about it. The second is acknowledged as a basic truth. Most athletes know a good bit about proper training, and the trick is to follow the right plan. The third is often ignored or, when taken seriously, misunderstood.

Nutrition has important effects on physical performance, but the legitimate role of nutrition is often poorly understood by athletes and coaches alike.[5] Incorrect information about nutrition and physical performance can be heard in locker rooms, coaches' offices, and health food stores. Misinformation about nutrition and physical performance can be found on dietary supplement labels, in online blogs, and in magazine articles. Buying into myths about nutrition and physical performance can cost you money or time, jeopardize health, or decrease performance. Common myths about nutrition and physical performance are listed in Table 28.1. Did you believe some of them? (Some coaches do.)[5] Knowledge of the science supporting connections between nutrition and physical performance will help you make solid decisions about food, nutrients, dietary supplements, and performance.

Illustration 28.1   Hitting the wall. This runner ran out of muscle glycogen before the finish line.

Key Nutrition Concepts

The roles played by carbohydrate, protein, fat, sodium, water, and other nutrients on physical performance relate to the key nutrition concepts of:

- Foods provide energy (calories), nutrients, and other substances needed for growth and health.

- Adequacy, variety, and balance are key characteristics of healthful diets.

glycogen   The storage form of glucose. Glycogen is stored in muscles and the liver.

physical performance   The ability to perform a physical task or sport at a desired or particular level.

ergogenic aids (ergo = work; genic = producing)   Substances that increase the capacity for muscular work.

ATP, ADP   Adenosine triphosphate (ah-den-o-scene tri-phos-fate) and adenosine diphosphate. Molecules containing a form of phosphorous that can trap energy obtained from the macronutrients. ADP becomes ATP when it traps energy and returns to being ADP when it releases energy for muscular and other work.

Basic Components of Energy Formation During Exercise

Understanding the role of nutrition and performance, and the potential effects of some **ergogenic aids** can be fostered by basic knowledge of how energy is formed within muscle cells. Illustration 28.2 summarizes the processes by which energy for muscle movement is formed.

There are two main substrates for energy formation in muscles: glucose from muscle and liver glycogen stores and fatty acids released from fat stores. How much of each is used depends on the intensity and duration of the exercise, as well as the body's ability to deliver each along with oxygen to muscle cells (Illustration 28.3). Each substrate is used to form **ATP** from **ADP**. ATP serves as the source of energy for muscle contraction.

Table 28.1 Eight common myths about nutrition and physical performance[2,6-9]

Myth	Reality
1. Very physically active people can eat all the fat and sugar they want.	Very active children and adults may stay thin no matter what they eat. However, that does not mean a poor diet won't affect their dental and heart health, or protect them from consuming low amounts of vitamins, minerals, and beneficial phytochemicals in plant foods. Calorie needs for athletes are best met through diets based on MyPlate.gov food groups.
2. Protein and amino acid supplements improve strength and endurance.	Resistance exercise is the key ingredient in strength building. Athletes can get all the protein and other nutrients they need for building and strengthening muscles from exercise and high-quality proteins from food. Excess protein adds calories and increases the workload of the kidneys because it has to excrete the excess nitrogen that results from high levels of protein breakdown.
3. Eating a pasta dinner the night before an endurance event improves glycogen stores.	A single high-carbohydrate meal the night before a competition will not build glycogen stores. Building up glycogen stores requires reducing the stores and then rebuilding them.
4. Athletes benefit from consuming a vitamin and mineral supplement.	Vitamins and minerals participate in energy formation but do not, by themselves, increase your ability to produce energy. Vitamin and mineral supplements benefit individuals with diagnosed deficiency disease, but do not improve performance in well-nourished individuals including athletes.
5. Chromium picolinate supplements improve muscle mass, strength, and body composition.	Recent research indicates that chromium picolinate does none of these things and it may be harmful to health. Use of chromium supplements is not recommended.
6. DHEA and related androstenedione supplements are natural boosters of analolic steroids and promote muscle formation.	There is virtually no evidence that these compounds affect training adaptations in younger men with normal hormone levels. In fact, most studies indicate that they do not affect testosterone and they may actually increase estrogen levels and reduce HDL-cholesterol. DHEA and related products may have clinical applications in older adults.
7. Drinking water during exercise decreases performance.	Drinking water before, during, and after exercise keeps athletes hydrated and prevents dehydration. Sports drinks do the same for endurance events lasting over an hour.
8. Body weight is more important than body composition to athletes' performance.	Body composition can be more important than weight for some types of sports.

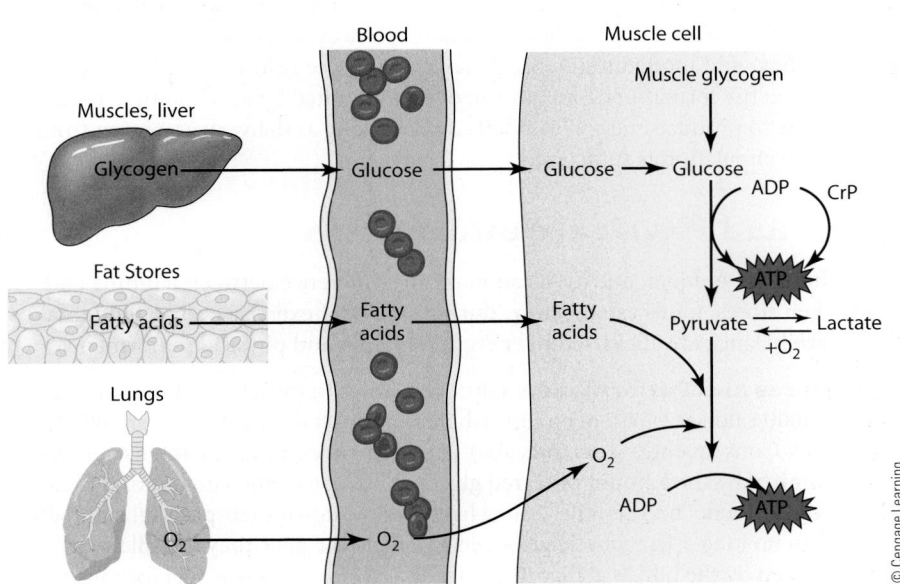

Illustration 28.2 Schematic representation of how ATP is formed for muscular movement.

© Cengage Learning

Activities primarily fueled by fat.

Activities primarily fueled by glycogen.

Illustration 28.3 Fat is the main source of energy for low- and moderate-intensity activities, whereas glycogen is the primary fuel for high-intensity activities.

Anaerobic Energy Formation Glucose from the liver and muscle glycogen (which is converted to glucose) form ATP without oxygen. This route of energy formation is "anaerobic," or "without oxygen," and it generates most of the energy used for intense muscular work (70% VO_2 max or higher).[1] Creatine phosphate (abbreviated CrP in Illustration 28.2), an amino acid containing a high-energy phosphate molecule in muscles, converts ADP to ATP to some extent. Creatine phosphate stores are limited and decrease rapidly during intensive exercise.

Glucose is converted to "pyruvate" during anaerobic energy formation. In the absence of oxygen, pyruvate is converted to lactate. Lactate can build up in muscles and blood if not reconverted to pyruvate by the addition of oxygen. Pyruvate yields additional energy when it enters "aerobic" energy formation pathways along with fatty acids from fat stores.

Aerobic Energy Formation The conversion of pyruvate and fatty acids to ATP requires oxygen. Much more ATP is delivered by the breakdown of fatty acids than glucose (fats provide 9 calories per gram; glucose only 4). The rate of energy formation from fatty acids is four times slower than that from glucose, however. It's the reason fatty acids are used to fuel low- and moderate-intensity exercise, or those below 60% VO_2 max.[10] Unlike glucose, energy formation from fatty acids is not limited by availability. Muscle cells can continue to produce energy from fatty acids as long as delivery of oxygen from the lungs and the circulation is sufficient.[11]

Nutrition and Physical Performance

When everything else is equal, nutrition can make the difference between winning and losing.[3] Glycogen stores, foods eaten before, during, and after exercise events, fluid intake, and **electrolyte** balance are all related to energy formation and physical performance.[7]

electrolytes Minerals such as sodium and potassium that carry a charge when in solution. Many electrolytes help the body maintain an appropriate amount of fluid.

Glycogen Stores and Performance Glycogen stores in muscles and the liver can deliver about 2,000 calories worth of energy, whereas adults have access to over 100,000 calories from fat. Consequently, a person's ability to perform continuous, intense physical activity is limited by the amount of stored glycogen.[12] People who run out of muscle glycogen during an event "hit the wall"—they have to slow down their pace substantially because they can no longer use muscle glycogen as a fuel. The pace they are able to maintain will be dictated by the body's ability to use fat and liver glycogen as fuel for muscular

Table 28.2 Effects of diet on muscle glycogen stores during training[45]

Carbohydrate intake level for previous three days	Muscle glycogen (grams per 100 grams of muscle)	Time to exhaustion at 75% Vo$_2$ max (in minutes)
Low	0.6 grams	60 minutes
Average	1.8	126
High (65–70% of total calories)	3.5	189

work. If athletes keep pushing themselves after muscle glycogen runs out, they may end up "bonking"—using up the liver's supply of glycogen. That's worse than hitting the wall because hypoglycemia (low blood sugar) can develop, and the person becomes dizzy and shaky and may pass out.[13] Obviously, endurance athletes do not want to exhaust their glycogen stores too soon.

Athletes consuming a typical U.S. diet normally have enough glycogen stores to fuel continuous, intense exercise for about an hour or two.[3] Both glycogen stores and endurance can be increased, however, by loading up muscles with glycogen prior to an event.

Carbohydrate Loading Carbohydrate loading increases muscle glycogen stores and benefits performance in athletes who participate in endurance events (Table 28.2).[1] Such events include activities like running, cross-country skiing, swimming, or cycling and last over 60 minutes. It does not improve performance in high-intensity, short-duration events such as sprints, short races, weight lifting, high jump, and baseball. These activities don't benefit from large glycogen stores.[8]

The currently recommended method for carbohydrate loading is much less stringent than in the past and effectively increases muscle glycogen stores. It calls for increasing carbohydrate intake to 60–70% of calories for one to four days prior to an endurance event. Glycogen stores accumulated are maintained by the tapering-off of exercise levels during the period of carbohydrate loading.[8,14]

Table 28.3 shows how many grams of carbohydrate correspond to 60 or 70% of total calories for people with different caloric needs. Average diets provide about 50% of calories from carbohydrates. See Illustration 28.4 for an example of one day's diet providing 68% of total calories from carbohydrate.

Carbohydrate-Loading Caveats There is a limit on glycogen storage in muscle. Each gram of glycogen binds with 2 grams of water, limiting the total amount that can "fit" into muscle tissue. Once the capacity of muscles to store glycogen is reached, any additional carbohydrate is largely converted to fat and stored.[31]

Not all athletes tolerate the relatively high-carbohydrate, low-fat, carbohydrate-loading diet equally well. Some athletes dislike the stiff, heavy feeling that may occur when muscles become "loaded" with glycogen.[15]

Table 28.3 Grams carbohydrate in diets providing 60 or 70% of total calories from carbohydrates

| Usual caloric intake | Grams of carbohydrate | |
	60% carb diet	70% carb diet
2,200 calories	330 grams	385 grams
2,600	390	455
2,800	420	490
3,200	480	560
3,600	540	630
4,000 [a]	600	700

[a]For each additional 200 calories of usual intake, add 30 grams carbohydrate for a 60% carbohydrate diet and 35 grams for a 70% carbohydrate diet.

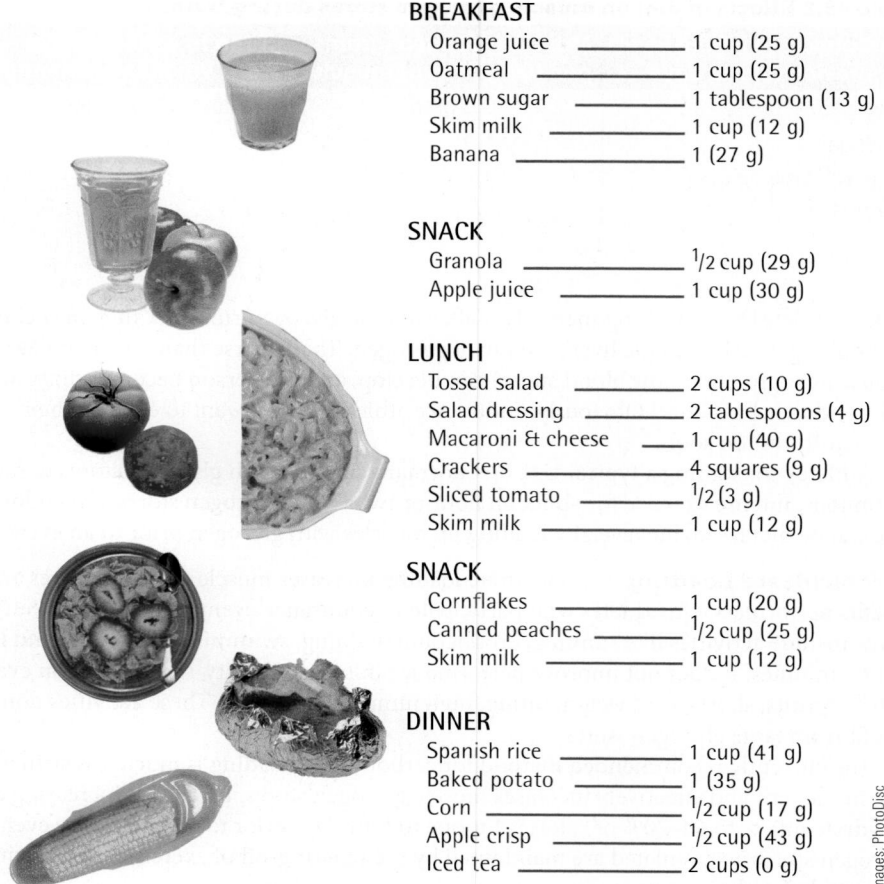

BREAKFAST

Orange juice	1 cup (25 g)
Oatmeal	1 cup (25 g)
Brown sugar	1 tablespoon (13 g)
Skim milk	1 cup (12 g)
Banana	1 (27 g)

SNACK

| Granola | 1/2 cup (29 g) |
| Apple juice | 1 cup (30 g) |

LUNCH

Tossed salad	2 cups (10 g)
Salad dressing	2 tablespoons (4 g)
Macaroni & cheese	1 cup (40 g)
Crackers	4 squares (9 g)
Sliced tomato	1/2 (3 g)
Skim milk	1 cup (12 g)

SNACK

Cornflakes	1 cup (20 g)
Canned peaches	1/2 cup (25 g)
Skim milk	1 cup (12 g)

DINNER

Spanish rice	1 cup (41 g)
Baked potato	1 (35 g)
Corn	1/2 cup (17 g)
Apple crisp	1/2 cup (43 g)
Iced tea	2 cups (0 g)

Images: PhotoDisc

Protein and Performance Many athletes require no more than their RDA of protein. Individuals undertaking strength or endurance training, however, may need 20–40 grams of additional protein daily to support muscle cell growth or repair.[1] For most athletes, this additional amount of protein will already be included in the diet. On average, adult females in the United States consume 67 grams of protein daily (the RDA is 48 grams), and males 98 grams (versus the RDA of 56 grams).[16] Protein intake generally increases with calorie intake, so athletes tend to have higher protein intakes than nonathletes due to their higher calorie intake levels. Diets providing up to 35% of total calories from protein are compatible with health, but may provide too little carbohydrate and lead to early fatigue in athletes.[3]

The Protein-Muscle Connection Strength training increases muscle mass and strength, and prepares muscles to continue increasing in mass and strength as training continues.[17,18] Muscular strength has been shown to increase 25–35%, and muscle mass by 9% after a 12-week, three session per week resistance training program, for example.[17] The question has been whether additional protein or amino acids would increase gains in muscle mass and strength achieved by resistance exercise.

Muscle is high in protein and it has been assumed for a long time that high protein intake helps build muscle mass and strength. Although high protein intake by itself does not build muscle mass or strength, it appear that intake of **high-quality protein** before and, in particular, after training does.[19–21]

Muscle fibers develop microscopic tears during training and protein is needed to help limit further muscle tissue breakdown, and to repair and rebuild the muscle. Intake of high-quality protein along with continued strength training provide a stimulus for muscles to get bigger and stronger.[9,18] Protein consumed before and 30–120 minutes after resistance training decreases muscle breakdown and increases muscle rebuilding.[18,22,23]

high-quality protein Proteins that contain all of the essential amino acids in amounts needed to support growth and tissue maintenance. Examples of high-quality proteins include eggs, soymilk, milk, meat, and beans and rice. Also referred to as "complete proteins."

Muscles use the amino acids in proteins for repair and rebuilding. Since amino acids are not stored in the body, excess amino acids from protein intake are broken down within an hour or two after consumption. That means amino acid incorporation into muscles takes place within an hour or two after exercise. The maximum amount of high-quality protein that can be used to repair and rebuild muscles is approximately 20 grams.[19,21] Twenty grams of high-quality protein can be supplied by the sandwich shown in Illustration 28.5, or by the following:

- 3 ounces of canned tuna

- 3 ounces lean beef or chicken

- 3 cups of soymilk

- 3 egg whites

- 1 cup white or brown rice plus 1 cup cooked dried beans

- 2$\frac{1}{2}$ cups of skim milk

Illustration 28.5 This sandwich of turkey breast on whole wheat bread supplies 20 grams of high-quality protein.

Protein consumed above the 20-gram amount is largely converted to fat or broken down for use in energy formation.[19] Some studies have found that consuming carbohydrate along with the protein enhances these effects, but other studies found the effects came only from protein.[18,23]

Protein Powders and Amino Acid Supplements Protein powders providing high-quality protein, and amino acid supplements that contain leucine and other essential amino acids, can increase muscle mass and strength post-resistance exercise.[9,18,22,24] Protein powders and amino acid supplements do not work any better than high-quality protein from foods, however.[18,23,25]

Pre-Event, Event, and Recovery Foods Recommendations about what athletes should eat and drink before, during, and after events in the past were largely based on untested assumptions. As more research has been done, previous recommendations are being replaced by those backed by research results on nutrition and performance. The evidence shows that consuming the right types and amounts of foods and fluids before, during, and after an endurance event benefits performance. Illustration 28.6 gives examples of foods and fluids that are good choices for pre-event, event and post-event recovery meals and snacks. Athletes are encouraged to experiment with foods before using them during an event.[1]

Pre-Event Foods and Fluids Foods and fluids consumed 2 to 4 hours prior to an endurance event should be high in carbohydrates to help fuel the upcoming exercise, low in fat and fiber to prevent digestive upsets, moderate in protein, and provide about 2 cups of water.[1] Food sources of carbohydrate that release glucose slowly into the bloodstream (or low glycemic index foods) are generally preferred over high glycemic foods.[26] Honey, oatmeal, corn, and bananas are examples of low glycemic index foods, while white bread, pretzels, cornflakes, and Gatorade have high glycemic index values.

Illustration 28.6 Examples of pre-event, event, and recovery foods and fluids.

Pre-event	Event	Post-Event	
Oatmeal	Bananas, other fruits	Non-fat milk	Bagel
Honey or brown sugar	Energy bars	Non-fat chocolate milk	Protein bars
Muffin	Sports gels	Water	Meat sandwich
Unsweetened fruit juice	Sports drinks	Fruit juice	Crackers and low-fat
Non-fat milk, coffee	Water	Sports drink	cheese
2 cups water		Eggs	Soup with pasta
		Soy products	or rice, vegetables,
		Low-fat yogurt	meat or beans
		Bananas	

Event Foods and Fluids Endurance athletes should consume enough fluids during the event to replace what is lost in sweat. (Additional information on hydration is presented later in this unit.) Adequate fluid and electrolyte consumption during events lasting over an hour prevents dehydration, reduces muscle fatigue, and helps prevent muscle cramps.[7] Sports drinks containing 4–8% carbohydrate and sodium should be consumed to replace some of the fluid loss.[1] Such drinks can lead to faster and more sustained rehydration than water alone,[27] and to increased endurance of 7.4% versus the 2% related to water.[28]

Carbohydrate should be consumed at the rate of 30–80 grams per hour after the first hour of the event.[28] It can be obtained from solid foods or beverages, depending on the preferences of the athlete. Bananas and other fruits can also be used as supplemental sources of carbohydrate.

Protein needs can be met by consuming 20 grams of high-quality protein from low-fat dairy products, lean meats, soy products, and food combinations such as grains and legumes.[19]

Post-Event Foods and Fluids Food sources of carbohydrate and high-quality protein, as well as fluids should be consumed within 30–60 minutes after completion of an endurance event. Carbohydrates help rebuild muscle glycogen stores, and protein increases muscle protein synthesis and repair in muscle cells. Enough water and other fluids should be consumed to bring body weight back up to its pre-event level.[7] The Reality Check feature in this unit addresses the issue of appetite post-exercise. The content may ring a bell if you feel really hungry, or else not hungry at all after exercise.

Hydration

- **Describe the recommended method for maintaining hydration during prolonged exercise, and recognize the signs of dehydration.**

hydration status The state of the adequacy of fluid in the body tissues. Dehydration indicates the presence of inadequate fluid in body tissues, and hyperhydration means there is too much.

hyponatremia A deficiency of sodium in the blood (135 mmol/L sodium or less).

Hydration status is a major factor affecting physical performance and health. Adequate hydration during training and competition enhances performance, prevents excessive body temperatures, delays fatigue, and helps prevent injuries. Dehydration has the opposite effects on physical performance and health. Hyperhydration, caused by overdrinking water before, during, and after endurance events, can lead to a loss of sodium from blood and body tissues and cause **hyponatremia**.[7]

Hydration status during exercise is primarily affected by how much a person sweats—or by how much water he or she loses through the skin while exercising. Muscular activity produces heat that must be eliminated to prevent the body from becoming overheated. To keep the body cool internally, the heat produced by muscles is collected in the blood and then released both through blood that circulates near the surface of the skin and in sweat. Sweating cools the body because heat is released when water evaporates on the skin. People sweat more during physical activity in hot, humid weather because it is harder to

REALITY CHECK

Does Strenuous Physical Activity Increase Appetite and Food Intake?

Who gets the thumbs-up?

Answer appears on page 28-9.

Junior: It must. You read about all these long-distance runners who chow down on a 3,000-calorie meal after the race is over.

Lakisha: I can't even think about eating after I've been working out on a hot day. I can't believe people would want to eat after that, either.

Table 28.4 Carbohydrate (based on weight) and sodium content of fluids

One-cup serving	Carbohydrate (%)	Sodium (in milligrams)	Potassium (in milligrams)
Gatorade, 1 cup	5%	108 mg	30 mg
PowerAde	6	55	30
Cola	11	15	7
Fruit punch	12	94	62
Koolaid	10	37	50
Apple juice	12	7	150

add moisture and heat to warm, moist air than to dry, cool air. To stay cool during exercise in hot, humid conditions, the body must release more heat and water than when exercise is undertaken in drier, cooler conditions.[29]

Maintaining Hydration Status During Exercise It is recommended that athletes engaged in events that last an hour or less drink about 2 cups of water one to two hours before an event and continue to consume water at regular intervals during the event. Athletes undertaking long events (over one hour in duration) should consume beverages that provide sodium (about 100 mg sodium and 30 mg potassium per 8 ounces), and carbohydrate (4–8% by weight) during the exercise.[30] The sodium is needed to replace what is lost in sweat and to prevent hyponatremia; the potassium is needed to maintain normal fluid balance; and the carbohydrate helps maintain blood glucose levels.[7] Hydration status is maintained when athletes do not lose weight during an event and when their urine remains pale yellow and is normal in volume.[1]

Fluids That Don't Hydrate Not all fluids help maintain hydration status. Fluids containing over 8% sugar don't quench a thirst and should not be consumed for fluid replacement. Their high-sugar content may draw fluid from the blood into the intestines, thereby increasing the risk of dehydration, nausea, and bloating.[28] Alcohol-containing beverages such as beer, wine, and gin and tonics are not hydrating, nor is water when consumed by a sodium-depleted person.[1]

Table 28.4 lists the carbohydrate, sodium, and potassium content of various beverages.

Estimating Fluid Needs: Sweat Rate The amount of fluid athletes need during an event can be estimated by calculating an hourly **sweat rate** for a specific activity (Table 28.5). A person preparing for a marathon who loses a pound in an hour of training and who drinks nothing during the hour would have a sweat rate of 1 pound (16 ounces). If that athlete drank 8 ounces of fluid in the hour and lost a pound—or 16 ounces—his or her sweat rate would be 24 ounces (16 ounces + 8 ounces). The sweat rate amount of fluid should be consumed per hour of the marathon event. Athletes who gain weight during an event have consumed too much water.[31]

Table 28.5 Calculating sweat rate: An example

1. Determine your body weight one hour before and one hour after exercise.
2. Subtract your postexercise weight from your pre-exercise weight.
3. Convert the number of pounds lost to ounces. (One pound equals 16 ounces.)
4. Add the number of ounces lost or gained to the number of ounces of fluid you consumed during the hour of exercise. The result is your sweat rate, and that's an approximation of the amount of fluid you need to consume during one hour of that exercise.

Example

1. Terrell weighed 172 pounds an hour before an hour-long bout of exercise and 171 pounds an hour after the exercise:
172 pounds – 171 pounds = 1 pound lost
1 pound × 16 ounce per pound = 16 ounces
2. He drank 16 ounces of fluid during the hour of exercise.
16 ounces + 16 ounces = 32 ounces, or "sweat rate."

Terrell would need to consume about 32 ounces of fluid per hour to remain hydrated.

sweat rate Fluid loss per hour of exercise. It equals the sum of body weight loss plus fluid intake.

ANSWERS TO REALITY CHECK
Does Strenuous Physical Activity Increase Appetite and Food Intake?

Two thumbs-up for these responses because exercise increases appetite and food intake in some athletes and it doesn't in others. Part of the difference seems to be due to the temperature. People who exercise in the heat, in particular, tend to feel less like eating after exercise.[44]

Junior:

Lakisha:

Illustration 28.7 Effects of dehydration. Even a low level of dehydration impairs physical performance.

AP Photo/The Journal, Doug Lindley

Dehydration: The Consequences Loss of more than 2% of body weight (2 to 4 pounds generally) during an event indicates that the body is becoming dehydrated. Effects of dehydration range from mild to severe, depending on how much body water is lost (Illustration 28.7). Any amount of dehydration impairs physical performance. At the extreme, dehydration can lead to heat exhaustion or heat stroke (Table 28.6). People who over-exercise in hot weather when they are out of condition are most likely to suffer heat exhaustion or heat stroke, but these conditions occasionally occur among seasoned athletes, too. Heat exhaustion can be remedied by fluids and electrolytes, but heat stroke requires emergency medical care.[32]

The Health Action feature for this unit highlights dehydration. The information may help you recognize and avoid it.

Hyponatremia and Excess Water Hyponatremia is a serious cause of death and life-threatening illness in marathon runners. In a recent Boston marathon, 13% of runners developed blood sodium levels that qualified as hyponatremia, and 0.6% became very ill due to extremely low levels of sodium.[33] The condition needs to be treated promptly. The major signs of hyponatremia are difficulty breathing, bloating, confusion, nausea, vomiting, and swelling around the brain.[34]

Hyponatremia is most likely to occur in athletes who gain weight during an event, consume more than 3 liters (12 cups) of low- or no-sodium fluids during the race, and engage in running for four or more hours. Individuals with a low body mass index (under 20 kg/m²) and women are at higher risk of developing hyponatremia than others.[25] Regular consumption of a beverage containing about 100 mg sodium per 8 ounces during a race or eating salty foods decreases the risk of hyponatremia.[1,30]

Body Fat and Weight: Heavy Issues for Athletes As a group, athletes tend to be very concerned about their weight and body shape, and they are leaner than the U.S. population in general (Table 28.7).[35] But the advantages of low body weight and body fat disappear when they become too low. Body fat levels of less than 5% in men and less than 12% in women can seriously interfere with health. These percentages represent the obliga-

Table 28.6 A primer on heat exhaustion and heat stroke[32]

Heat exhaustion: A condition caused by low body water and sodium content due to excessive loss of water through sweat in hot weather. Symptoms include intense thirst, weakness, paleness, dizziness, nausea, fainting, and confusion. Fluids with electrolytes and a cool place are the remedy. Also called "heat prostration" and "heat collapse."

Heat stroke: A condition requiring emergency medical care. It is characterized by hot, dry skin, labored and rapid breathing, a rapid pulse, nausea, blurred vision, irrational behavior, and, often, coma. Internal body temperature exceeds 105°F due to a breakdown of the mechanisms for regulating body temperature. Heat stroke is caused by prolonged exposure to environmental heat or strenuous physical activity. The person affected by heat stroke should be kept cool by any means possible, such as removing clothing and soaking the person in ice-cold water. If conscious, the person should be given fluids. Also called "sunstroke."

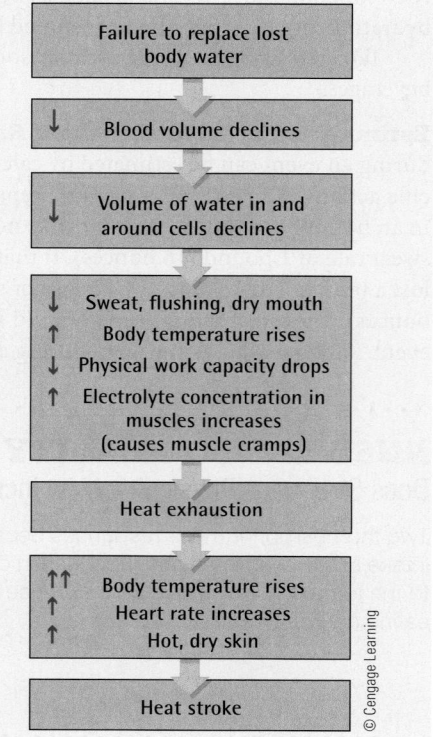

© Cengage Learning

health action Be Aware of the Signs and Symptoms of Dehydration!

The signs and symptoms of dehydration can include:

- Thirst
- Lightheadedness or fatigue
- Headache
- Inability to concentrate
- Feeing cold or having chills when it is hot out and you are sweating
- Collapse due to heat exhaustion or stroke
- Low amount of urine and dark yellow color

---

tory body fat content that people need for the functions of fat unrelated to energy production.[13] Everybody needs some fat to use for hormone production, to maintain normal body temperature in cold weather, and to cushion internal organs.

Exercise, Body Fat, and Health in Women It is not uncommon for female athletes to experience irregular or absent menstrual periods or the late onset of periods during adolescence (Table 28.8). Female athletes engaged in "leanness" sports such as gymnastics, diving, and figure skating are particularly at risk for inadequate calorie intake.[36] These aberrations in menstrual periods appear to be related to specific effects on hormone production of high levels of physical activity and deficits of caloric intake.[37] Female athletes with missing and irregular periods are at risk for developing low bone density, osteoporosis, and bone fractures. Women at particular risk of bone fractures are those with the "female-athlete triad" of disordered eating, amenorrhea (pronounced *a-men-or-re-ah* and meaning "no menstrual periods"), and osteoporosis. Abnormal menstrual cycles should not be dismissed as a normal part of training. Normal menstrual periods should be reinstated through increased caloric intake.[1] Suppressed testosterone levels or other metabolic changes due to caloric deficits in male athletes may be the parallel to menstrual irregularities in female athletes.[38]

Wrestling: The Sport of Weight Cycling "Making weight" is a common and recurring practice among wrestlers (Illustration 28.8). Most wrestlers will "cut" 1 to 20 pounds over a period of days between 50 and 100 times during a high school or college career.[39] That makes wrestling more than a test of strength and agility. It makes it a contest of rapid weight loss.

For competitive reasons, wrestlers often want to stay in the lowest weight class possible and may go to great lengths to achieve it. They may fast, "sweat the weight off" in saunas or rubber suits, or vomit after eating to lose weight before the weigh-in. These practices can be dangerous, even life threatening, if taken too far. After the match is over, the wrestlers may binge and regain the weight they lost.

Table 28.7 Average body fat content of various athletes[37,38]

	Body fat content (as percentage of body weight)	
	Women	Men
Bodybuilders	10%	6%
Long-distance runners	17	6–13
Baseball players	—	12–14
Basketball players	21–27	7–10
Football players	—	9–19
Gymnasts	10	5
Soccer players	—	10
Tennis players	20	16
Wrestlers	—	9

Table 28.8 Incidence of irregular or absent menstrual cycles in female athletes and sedentary women[37,46]

Joggers (5 to 30 miles per week)	23%
Runners (over 30 miles per week)	34
Long-distance runners (over 70 miles per week)	43
Competitive bodybuilders	86
Noncompetitive bodybuilders	30
Volleyball players	48
Ballet dancers	44
Sedentary women	13

©Charles Guptil/Stock Boston, Inc.

Illustration 28.8 Dropping weight quickly in the days before a match can wreck a wrestler's chances and harm his health.

Wrestlers, like other athletes involved in intense exercise, perform better if they have a good supply of glycogen and a normal amount of body water. Fasting before a weigh-in dramatically reduces glycogen stores, and withholding fluids or losing water by sweating puts the wrestler at risk of becoming dehydrated. Trying to stay within a particular weight class too long may also stunt or delay a young wrestler's growth.

The American Medical Association and the Association for Sports Medicine recommend that wrestling weight be determined after six weeks of training and normal eating. In addition, a minimum of 7% body fat should be used as a qualifier for assigning wrestlers to a particular weight class. In addition, weight classes are now based on normal weight for height and age, and weigh-ins are scheduled close to event times.[40]

Iron Status of Athletes Iron status is an important topic in sports nutrition because iron deficiency (or low iron stores) and iron-deficiency anemia (or low blood hemoglobin level) decrease endurance. Iron is a component of hemoglobin, a protein in blood that carries oxygen to cells throughout the body, and it works with enzymes involved in energy production. When iron stores or hemoglobin levels are low, less oxygen is delivered to cells, and less energy is produced than normal.[1]

Female athletes are at higher risk of iron-deficiency anemia than other females. One study of aerobically fit athletes found that 36% of women and 6% of males were iron deficient.[41] Consequently, it is recommended that female athletes especially pay attention to the amount of iron consumed.[1]

Table 28.9 Faster, higher, stronger, longer: A brief history of performance-enhancing substances used by athletes

B.C.	Large quantities of beef consumed by athletes in Greece to obtain "the strength of 10 men." Deer liver and lion heart consumed for stamina.
1880s	Morphine used to increase performance in (painful) endurance events.
1910s	Strychnine consumed for the same reason as morphine.
1930s	Amphetamines used to increase energy levels and endurance. Testosterone taken to increase muscle mass.
1980s	Blood doping, EPO used to increase endurance. Ephedra to increase energy.
2009	HGH; gene doping to increase strength and endurance.

Ergogenic Aids: The Athlete's Dilemma

- **Assess the safety and effectiveness of ergogenic aids offered to athletes.**

The quest for ownership of a competitive edge has drawn athletes to ergogenic aids throughout much of history. (See Table 28.9 for a historical review of ergogenic aids use.) Relatively few of the hundreds of available products work, and most are sold as "dietary supplements" so they do not have to be tested for safety. Those found to increase muscle mass, strength, or endurance are usually banned for use by competitive athletes. Several of the aids clearly pose serious risks to health. Some contain banned substances, and others simply represent misguided hopes and misspent money.[42] Table 28.10 lists dietary supplements and other erogenic aids that appear to enhance strength, endurance, or body composition; those about which too little is known to make a conclusion; and those that are banned by the FDA for sale to the public or by the U.S. Anti-Doping Agency. Products that do not increase strength or endurance to a greater extent than high-quality protein from foods or other normal components of diets are considered ineffective.

The Path to Improved Performance Genetics, training, and nutrition: These are the real keys to physical performance. Although other aids will be sought, those that exceed the boundaries of what is considered fair and safe will not be approved for use by athletes. After all, athletic competition is not a test of drugs or performance aids. It's a test of an individual's ability to excel. Anything less wouldn't be sporting.

Table 28.10 The status of dietary supplements and ergogenic aids on strength, endurance, or body composition[9,19,43]

A. Effective	B. Not shown to be effective	C. Banned substances[a]
Water and sports drinks	Glutamine	DHEA, androstenedione
High carbohydrate diets	Isoflavones	Ephedra/ephedrine, Ma Haung
Caffeine (moderate amounts)	Sulfo-Polysaccharides (Myostatin Inhibitors)	Growth hormone
Postexercise high-quality protein intake	Boron	Insulin
	Calcium Pyruvate	Sibutramine
	Chromium	Testosterone
	Creatine	
	Gamma Oryzanol (Ferulic Acid)	
	Herbal Diuretics	
	Tribulus terrestris	
	Vanadyl Sulfate (Vanadium)	
	Chitosan	
	Garcinia Cambogia (HCA)	
	L-Carnitine	
	Phosphates	
	Glycerol	
	Ribose	
	Inosine	
	Sodium Bicarbonate	
	β-HMB	
	Conjugated linoleic acid	
	Branched chain amino acids (BCAA)	
	Medium -chain triglycerides (MCT)	
	Omega-3 fatty acids	

[a]Note: You can get a list of substances banned for use by athletes by from Global Drug Reference Online at www.GlobalDRO.com.

Bill Milne/StockFood Creative/ Getty Images

NUTRITION
up close

Testing Performance Aids

Focal Point: The critical examination of studies on performance aids.

Read the following summary of a fictitious study of the effects of a phosphorous supplement on strength and then answer the critical thinking questions.

- *Purpose*: To assess the effect of a phosphorous supplement on strength.

- *Methods*: Twenty volunteers from the crew team were given the phosphorous supplement for a week. Strength, assessed as the maximum number of push-ups a study participant could do in one session, was assessed before and after supplementation. Participants recorded any supplement side effects.

- *Results*: The number of push-ups increased by an average of 5% after supplementation. Diarrhea was the only side effect consistently noted by participants.

Critical Thinking Questions

1. What is a major limitation in the basic design of the study? _____

2. Does the study demonstrate that the supplement is safe?

Yes _____ No _____ Give reasons for your answer. _____

3. Does the study demonstrate that the supplement increases strength?

Yes _____ No _____ Give reasons for your answer._____

Feedback to the Nutrition Up Close is located in Appendix G.

REVIEW QUESTIONS

- **Describe how carbohydrates, proteins, and fats are utilized by muscles.**

- **Describe the recommended method for maintaining hydration during prolonged exercise, and the signs of dehydration.**

- **Assess the safety and effectiveness of erogenic aids offered to athletes.**

1. The two main substrates for energy formation in muscles are amino acids and fatty acids.   **True/False**

2. Glucose is converted to energy by muscles anaerobically.   **True/False**

3. Much more ATP is produced from the breakdown of fatty acids than from amino acids or glucose.   **True/False**

4. Hypoglycemia results when athletes use up the liver's supply of glycogen.   **True/False**

5. Glucose is the major source of energy for low- and moderate-intensity activities.   **True/False**

6. Carbohydrate loading improves performance in short-duration, high-intensity events.   **True/False**

7. Post-event recovery meals should include rich sources of fat because fats facilitate muscle repair.   **True/False**

8. Hyponatremia is primarily caused by drinking excessive amounts of water during prolonged exercise.   **True/False**

9. An athlete who runs for an hour, loses two pounds in that hour, and drinks 12 ounces of fluid during the hour has a sweat rate of 60 ounces.   **True/False**

10. Signs of dehydration include thirst and lightheadedness or fatigue.   **True/False**

11. A few ergogenic aids improve performance somewhat, but most do not improve performance.   **True/False**

12. Athletes attempting to build muscle mass and strength should take protein and amino acid supplements.
True/False

The next three questions refer to this scenario:

Assume you attend a sports competition and are offered a free sample of MuscleMaxX, a new protein powder that promises to help you build muscle after resistance exercise. You take a look at the ingredient label and this is what it contains:

> triple-filtered water
> starch
> casein
> glucose
> lean lipids
> growth peptides
> micellar proteins
> lactalbumins

13. _____Which ingredient in this product would most likely help rebuild muscles after resistance exercise?
    a. triple-filtered water
    b. starch
    c. casein
    d. lean lipids

14. _____ Why do you think the ingredients are titled with unfamiliar terms?
    a. To ensure users know they are getting unique ingredients for muscle recovery.
    b. To confuse people about what's really in the product.
    c. To assure users they are getting ingredients for which scientific studies have shown effectiveness.
    d. The ingredients are highly technical and the accurate name for each is being used.

15. _____ Is it likely this product would work better for muscle re-building after resistance exercise than consuming $2^1/_2$ cups of skim milk after resistance exercise training?
    a. Yes, because it also contains growth peptides.
    b. Yes, because it provides starch.
    c. No, because high-quality protein such as casein from food also works.
    d. No, because skim milk also contain lactalbumins.

Answers to these questions can be found in Appendix G.

NUTRITION SCOREBOARD ANSWERS

1. Carbohydrate and fat availability to muscles during exercise varies based on usual diet. **True**

2. Carbohydrate loading increases endurance in trained athletes.[1] **False**

3. Resistance exercise increases muscle strength. Protein and amino acids supplements have not been shown to improve strength in healthy athletes.[1] **False**

4. It is not normal and for health reasons should be prevented.[2] **False**

UNIT **29**

Good Nutrition for Life: Pregnancy, Breast-Feeding, and Infancy

NUTRITION SCOREBOARD

1 The fetus is able to obtain needed nutrients from the mother regardless of the mother's diet or nutrient stores. **True/False**

2 The United States has the lowest rate of infant mortality in the world. **True/False**

3 Normal health, growth, and development of infants are defined by the experience of breast-fed infants. **True/False**

4 Drinking alcoholic beverages is safe during breast-feeding because the alcohol doesn't pass into the milk. **True/False**

Answers can be found at the end of the unit.

After completing Unit 29 and its interactive learning features, you will be able to:

- Define critical periods of growth and development and identify potential consequences of inadequate nutrient availability during these periods on future health status.

- Develop and evaluate a one-day diet for a pregnant or breast-feeding woman based on guidance from the ChooseMyPlate food groups.

- Identify three unique health advantages for infants associated with breast-feeding.

- Identify three examples of parental behaviors around eating that negatively influence, and three that positively influence infants' learning about food and eating.

PhotoDisc

infant mortality rate Deaths that occur within the first year of life per 1,000 live births.

low-birth-weight infants Infants weighing less than 2,500 grams (5.5 pounds) at birth.

preterm infants Infants born at or before 37 weeks of gestation (pregnancy).

Nutrition and a Healthy Start in Life

The day has finally arrived, the one Crystal and Raymond have anticipated for a year. Crystal's pregnancy test is positive!

This is a planned pregnancy, and both Crystal and Raymond have done all they could to prepare. Crystal's diet has been the picture of perfection, she has kept up her regular exercise schedule, and she has quit drinking alcohol entirely. Now they are ready to have the baby of their dreams. They are convinced that not only will this baby be healthy and strong, but it will be above average in every respect.

All parents want to give their children every advantage in life that they can. Although no one can guarantee that a baby will be born healthy and strong—no matter what the parents do—there are steps parents can take to make the healthiest baby possible.

Unfortunately, many infants born in the United States do not have the advantage of good health at birth. Among countries of the world, the United States ranks 34th in terms of **infant mortality rate** (Table 29.1).[3,7] The relatively high rate of infant deaths in the United States is due primarily to the proportion of infants born **low birth weight** or **preterm** (Illustration 29.1). One in 12 U.S. infants is born too small, and more than one in eight is born too early. Rates of low birth weight and preterm differ substantially among population groups, with the highest rates occurring in African Americans.[8] These infants are at particular risk for requiring intensive and continuing care and of dying within the first year of life. In contrast, infants weighing between 7 pounds, 11 ounces and 8 pounds, 13 ounces (or 3,500 to 4,000 grams) are least likely to die within the first year. The average weight of U.S. infants is about 7 pounds, 7 ounces (3,350 grams).[9]

Key Nutrition Concepts

Six of the 10 key nutrition concepts presented in Unit 1 are directly relevant to advances in knowledge about nutrition during growth and development and health. Four are highlighted here:

1. Some groups of people are at higher risk of becoming inadequately nourished than others.

2. Foods provide energy (calories), nutrients, and other substances needed for growth and health.

3. Poor nutrition can result from both inadequate and excessive levels of nutrient intake.

4. Adequacy, variety, and balance are key characteristics of healthful diets

Improving the Health of U.S. Infants

A high proportion of poor infant outcomes in the United States is attributed to a combination of factors including poverty, poor nutrition, limited access to health care, and a maternal lifestyle that includes the use of illicit drugs, cigarettes, and excessive amounts of alcohol. Many infant deaths, preterm infants, and low-birth-weight infants can be prevented, however, by optimizing maternal health and behaviors that influence health prior to pregnancy. Of the various behaviors that affect maternal health, none offers potentially greater advantage to both pregnant women and their infants than good nutrition.[5,10]

Nutrition and Pregnancy

- **Define critical periods of growth and development and identify potential consequences of inadequate nutrient availability during these periods on future health status.**

- **Develop and evaluate a one-day diet for a pregnant woman based on guidance from the ChooseMyPlate food groups.**

Nutrition is important during pregnancy because the **fetus** depends on it. In most all instances, the fetus is not protected from inadequate or excessive nutrient intake of mothers. Whether or not fetal needs for nutrients are met primarily depends on the availability of nutrients from the mother's diet and nutrient stores.[11] This situation makes sense in terms of survival of the species. By meeting the mother's needs first, nature protects the reproducer.[1,10,11]

Calcium is the exception to the rule that nutrients must be supplied by the mother's diet. Bones contain large amounts of calcium that are drawn upon all the time to maintain a certain blood level of calcium. Deviations in normal blood calcium levels can have serious health consequences, so the body has mechanisms in place that protect against fluctuating levels. The fetus has access to calcium from bone due to the maintenance of blood calcium levels.[12,13]

Rather striking evidence that maternal health is not placed in jeopardy by fetal nutrient utilization comes from studies showing that deficiencies of vitamin B_{12}, thiamin, iodine, folate, zinc, and other nutrients occur in newborns but not in their mothers. Similarly, infants born to women who consume excessive levels of vitamin or mineral supplements during pregnancy are more likely to display signs of nutrient overdose than are the mothers.[1,14,15]

Fetal growth and development can be compromised by insufficient and excessive availability of nutrients. The nature and extent of impairments primarily depends on when they occur during pregnancy and the extent of the deficiency or excess. The timing during pregnancy is important because organs and tissues growing most rapidly when the nutritional deficiencies or overdoses occur are affected most.[1,16]

Critical Periods of Growth and Development

Fetal **growth** and **development** proceed in a series of critical periods. A **critical period** of growth and development is an interval of time during which cells of a tissue or organ

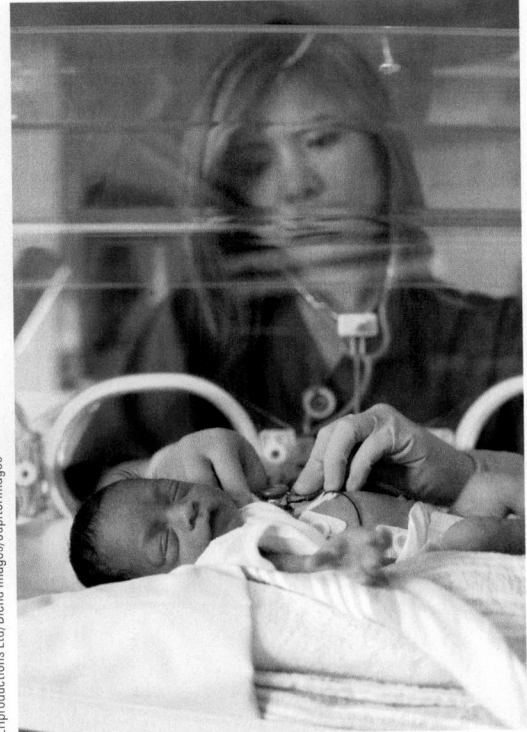

Illustration 29.1 In 2009, 8.2% of infants born in the United States had a low birth weight, and 12.2% were born preterm.[2]

Table 29.1 Examples of the infant mortality rate (IMR, deaths per 1,000 live births) in 2009 in various countries and the United States[2,3]

Country	IMR
Hong Kong	1.7
Singapore	1.9
Iceland	2.1
Japan	2.4
Sweden	2.5
Finland	2.6
Czech Republic	2.9
Norway, Greece, Denmark	3.1
Republic of Korea, Ireland	3.2
Spain	3.3
Italy	3.7
Israel	3.8
France	3.9
United Kingdom	4.7
Cuba	4.8
Canada	5.1
Croatia	5.3
United States	6.4

fetus A baby in the womb from the eighth week of pregnancy until birth. (Before then, it is referred to as an embryo.)

growth A process characterized by increases in cell number and size.

development Processes involved in enhancing functional capabilities. For example, the brain grows, but the ability to reason develops.

critical period A specific interval of time during which cells of a tissue or organ are genetically programmed to multiply.

Illustration 29.2 Baby's first picture. An ultrasound image of a rapidly growing and developing 16-week-old fetus.

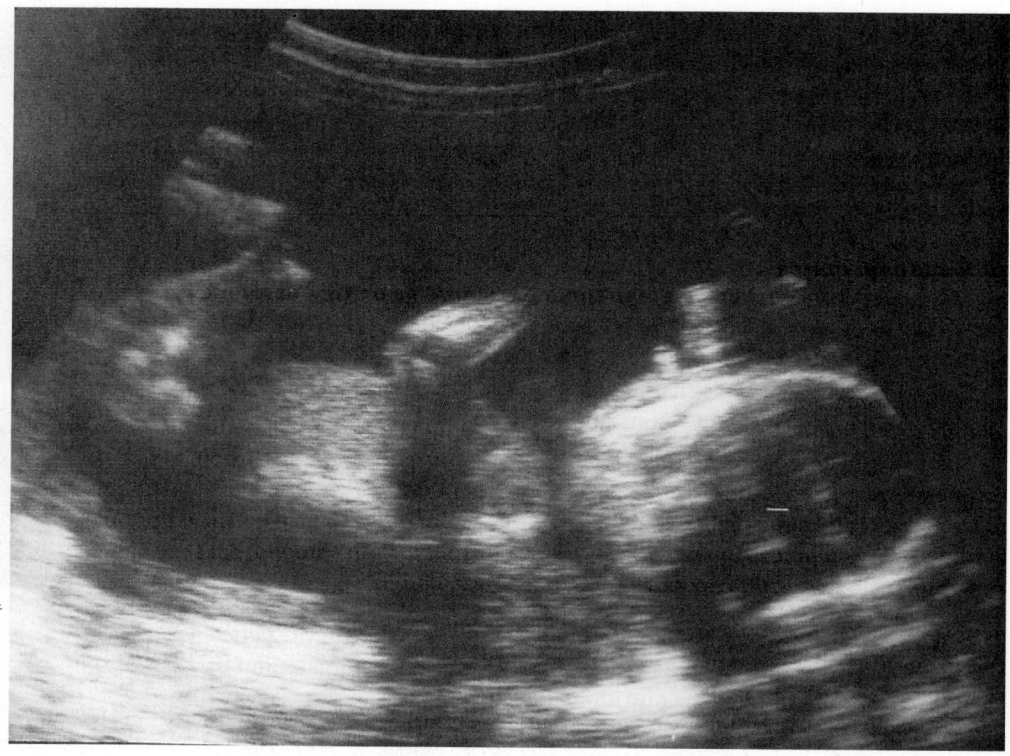

are genetically programmed to multiply. The period is considered critical because if the cells do not multiply as programmed during this set time interval, they cannot make up the deficiency later. The level of nutrients required for cell multiplication to occur normally must be available during this specific time interval. If the nutrients are not available, the developing tissue or organ will contain fewer cells than normal, will form abnormally, or will function less than optimally.[1]

The roof of the mouth (or the hard palate), for example, is formed early in the third month of pregnancy when two developing plates fuse together. This process can only occur early in the third month. If excessive amounts of vitamin A are present in fetal tissues during this period, the two plates may fail to combine, resulting in a cleft palate.[17] (The divided hard palate can be surgically corrected after birth.)

Critical periods of cell multiplication are most intensive in the first few months of pregnancy, when fetal tissues and organs are forming rapidly (Illustration 29.2)—hence the importance of the nutritional status of women at the time of conception and very early in pregnancy. For some organs and tissues, cell multiplication continues through the first two years after birth. Most of the growth that occurs in the fetus late in pregnancy and throughout the rest of the growing years, however, is due to increases in the size of cells within tissues and organs.[1]

The Fetal Origins Hypothesis of Later Disease Risk

One of the most striking advances in research on pregnancy concerns the potential effects of maternal nutrition on the baby's risk of developing certain chronic diseases later in life.[18] It appears that an important element of increased susceptibility to heart disease, stroke, diabetes, obesity, hypertension, and other disorders may be "programmed," or gene functions modified, by inadequate or excessive supplies of energy or nutrients during specific time periods during pregnancy.[19,20]

Given optimal conditions, fetal growth and development proceed according to the genetic blueprint established at conception. Organs and tissues are well developed and ready to function optimally after birth. If it exists in less than optimal growing conditions such as those introduced by maternal weight loss, poor nutrient intakes, or diseases in the mother that alter her ability to supply the fetus with energy or nutrients, fetal growth and development are modified. Fetal tissues undergoing critical phases of development at that time have to make adaptations to cope with the under- or oversupply of nutrients. These adaptations may produce long-term changes in the structure and function of tissues. Low maternal energy intake in the last few months of pregnancy, for example, may hinder the development of cells that produce insulin and limit the body's ability to control blood glucose levels later in life. This situation can increase the risk of developing diabetes later in life.[21,22]

A large body of evidence supports the fetal origins hypothesis. Nevertheless, much remains to be discovered about the effects of diet during pregnancy on the risk of later disease.[23]

Prepregnancy Weight Status, Pregnancy Weight Gain, and Pregnancy Outcomes

Women who enter pregnancy underweight or who fail to gain a certain minimum amount of weight during pregnancy are much more likely to deliver low-birth-weight and preterm infants than women who enter pregnancy at normal weight or above and gain an appropriate amount of weight.[24] The risk of low birth weight can be reduced by healthy diets that lead to a desired rate and amount of weight gain during pregnancy. Along with the duration of pregnancy and smoking, prepregnancy weight status and weight gain in pregnancy are the major factors known to influence an infant's birth weight (Table 29.2).

What's the Right Amount of Weight to Gain During Pregnancy? Weight-gain recommendations for pregnancy vary depending on whether a woman enters pregnancy underweight, normal weight, overweight, or obese. A woman who is obese when she becomes pregnant will need to gain less weight than a woman who enters pregnancy underweight. It is recommended that underweight women gain 28 to 40 pounds, normal-weight women gain 25 to 35 pounds, overweight women gain 15 to 25 pounds, and obese women gain 11 to 20 pounds during pregnancy. Normal-weight women carrying twins should gain 37 to 54 pounds during pregnancy.[31] Table 29.3 can be used to identify prepregnancy weight status by a woman's body mass index (BMI) category and the recommended weight gain during pregnancy. After the goal is identified, a woman may want to plot her prenatal weight gain on a chart like the one in Illustration 29.3. The weight gained should be the result of a high-quality diet that leads to gradual and consistent gains in weight throughout pregnancy.

Where Does the Weight Gain Go During Pregnancy? If healthy infants tend to weigh about 8 pounds at birth, why do women need to gain more than that? Where does the rest of the weight go? Many pregnant women ask themselves these questions, especially when they don't relish the thought of gaining weight. Illustration 29.4 shows where the weight gain goes.

Fetal growth is accompanied by marked increases in maternal blood volume, fat stores, and breast and uterus size, all of which contribute to weight gain. In addition, water accumulates in the amniotic fluid, which cushions and protects the fetus, and the volume of fluid that exists outside cells increases. The placenta (the tissue that transfers nutrients

Table 29.2 Major factors that directly influence birth weight[24,31]

- Duration of pregnancy
- Prenatal weight gain
- Prepregnancy weight status
- Smoking

Table 29.3 Prepregnancy weight status and recommended weight gain during pregnancy[26,31]

Prepregnancy weight status, body mass index	Recommended weight gain
Underweight, <18.5 kg/m²	28–40 pounds
Normal weight, 18.5–24.9 kg/m²	25–35 pounds
Overweight, 25–29.9 kg/m²	15–25 pounds
Obese, 30 kg/m² or higher	11–20 pounds
Twin pregnancy, normal weight status	37–54 pounds

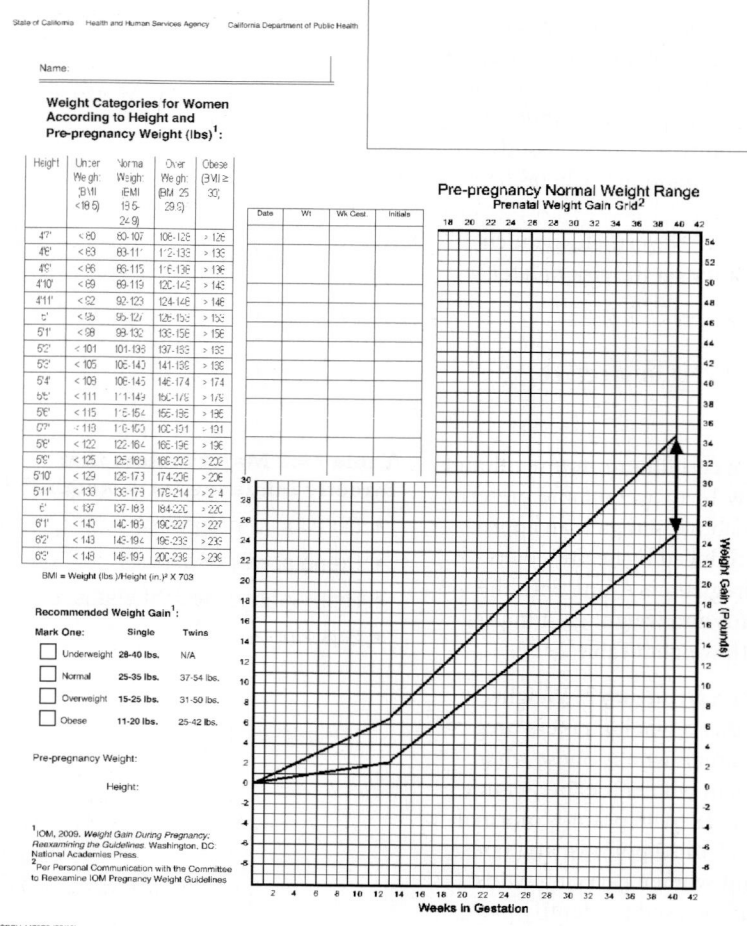

Illustration 29.3 An example of a prenatal weight-gain chart for women entering pregnancy normal weight from the California Department of Maternal and Child Health.

trimester One-third of the normal duration of pregnancy. The first trimester is 0 to 13 weeks, the second is 13 to 26 weeks, and the third is 26 to 40 weeks.

in the mother's blood supply to the fetus) also accounts for some of the weight that is gained during pregnancy.

Where Does the Weight Gain Go—After Pregnancy? Delivery is one of the greatest weight-loss plans known to humankind. On average, women lose 15 pounds within the first week after delivery, and weight loss generally continues for up to a year.[27] Women who gain more than the recommended amount of weight, however, and those who gain weight after delivery may have extra pounds to lose to get back to prepregnancy weight.[25] The Reality Check for this unit addresses a common issue related to weight gain during pregnancy among obese women.

The Need for Calories and Key Nutrients During Pregnancy

Pregnant women need more calories, protein, and essential nutrients than nonpregnant women. All of these are important; but calories, folate, vitamin B_6, vitamin A, calcium, vitamin D, iron, iodine, and EPA and DPH are of paramount concern. Although many pregnant women are concerned about protein, low protein intakes are rarely a problem in pregnancy. Women of child-bearing age consume an average of 69 grams of protein daily;[28] the recommended intake level for protein during pregnancy is 71 grams.

According to dietary intake recommendations, pregnant women need approximately 15% more calories and up to 50% more of various nutrients than do nonpregnant women (Illustration 29.5).[61] Relatively high nutrient requirements mean that pregnant women should increase their intake of nutrient-dense foods more than their consumption of calorie-rich foods.[6]

Calories On average, women need an additional 340 calories a day in the second **trimester**, and 450 calories in the third trimester of pregnancy.[61] Women entering pregnancy underweight will need more calories than this, and those entering overweight will need fewer. In addition, physically active pregnant women require higher caloric intakes than average. Rather than counting calories, however, it is generally more practical to monitor the adequacy of caloric intake by tracking weight gain.

REALITY CHECK
Should Obese Women Gain Weight During Pregnancy?

Elesha, who has been obese for five years, is 28 weeks pregnant. At her recent prenatal appointment she asked her health care provider about how much weight she should gain. Because she started pregnancy with a BMI indicating obesity, the health care provider tells her that she does not need to gain any weight during pregnancy. She returns home and asks her cousin Joanne if this is the right advice.

Who gets the thumbs up?

Answers appear on page 29-7.

Joanne: Makes sense to me. You have a good bit of stored energy for the baby to grow on!

Elesha: I thought all women were supposed to gain weight during pregnancy for the baby's sake.

Here's where the weight goes for a woman who gains 33 pounds

Breasts 1.1 lb

Fetus 8.3 lb

Placenta 1.6 lb

Amniotic fluid 2.0 lb

Uterus 2.4 lb

Fat stores 8.5 lb

Blood 3.0 lb

Other fluid 6.1 lb

TOTAL = 33 lb

Am I having a baby or a little elephant?!

© Cengage Learning

Illustration 29.4   Where does all the weight go?

Illustration 29.5 Percentage increases in the RDAs or AIs for pregnant and breast-feeding women compared to other women.

Key:
- Nonpregnant women
- Pregnant women
- Lactating women

Energy
Protein
Vitamin A
Vitamin D
Vitamin E
Vitamin K
Vitamin C
Folate
Niacin
Riboflavin
Thiamin
Vitamin B$_6$
Vitamin B$_{12}$
Calcium
Phosphorus
Magnesium
Iodine
Iron
Zinc
Selenium
Fluoride

0   50   100   150   200
Percent

The increased need for iron in pregnancy cannot be met by diet or by existing stores. Therefore, pregnant women need iron supplements.

© Cengage Learning

Folate   Folate is required for protein tissue construction and therefore is in high demand during pregnancy. It is considered an "at risk" nutrient because some pregnant women do not consume enough to meet their daily need for folate.[6]

Folate deficiency is related to the development of **neural tube defects** such as spina bifida (Illustration 29.6). In clinical trials, adequate folate status very early in pregnancy is associated with a 50–70% reduction in the occurrence of neural tube defects.[29] This finding, along with the knowledge that many women fail to consume enough folate, led to the fortification of refined grain products with folic acid in 1998. (Examples of fortified grain products are shown in Illustration 29.7.) Folic acid is a form of folate used in fortified foods and supplements. It is absorbed from foods about twice as completely as folate, the form of found in food. The vast majority of ready-to-eat breakfast cereals are fortified with at least 100 micrograms (0.1 milligram) of folic acid per serving.

Neural tube defects form before 30 days after conception, so it is important that women are consuming enough folate when they enter pregnancy. Enough is 400 mcg

neural tube defects   Malformations of the spinal cord and brain. They are among the most common and severe fetal malformations, occurring in approximately 1 in every 1,000 pregnancies. Neural tube defects include spina bifida (spinal cord fluid protrudes through a gap in the spinal cord; shown in Illustration 29.6), anencephaly (absence of the brain or spinal cord), and encephalocele (protrusion of the brain through the skull).

ANSWERS TO **REALITY** CHECK
Should Obese Women Gain Weight During Pregnancy?

Yes, they should. Obese women need to gain less (11 to 20 pounds) than women entering pregnancy normal weight (25 to 35 pounds) or overweight (15 to 25 pounds). Infant outcomes are generally better if women gain weight during pregnancy than if they lose it (weight gain during pregnancy).[26]

Joanne:

Elesha:

Vasiliy Koval/Shutterstock.com

iStockphoto.com/Jose Wilson Araujo

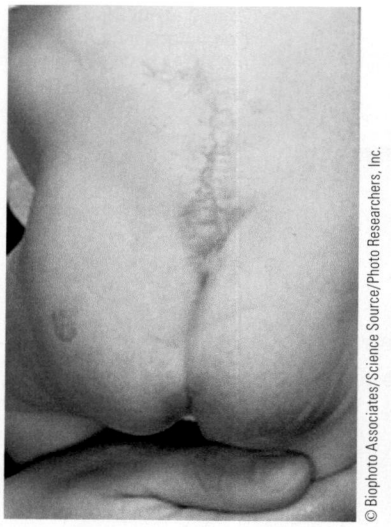

Illustration 29.6 Spina bifida: a neural tube defect. The interruption in the spinal cord results in paralysis below the injury. This photo shows the back after surgery to close the interruption.

Illustration 29.7 Foods fortified with folic acid.

(0.4 mg) folic acid before pregnancy from fortified foods or supplements, plus folate provided by a variety of vegetables. During pregnancy, it is recommended that women consume 600 mcg (0.6 mg) folate, including 400 mcg (0.4 mg) folic acid.[44] Women can generally obtain this level of folate by consuming two servings of fortified ready-to-eat breakfast cereal, six servings of grain products, and three servings of vegetables daily.[91] If the need for folate is not met by a diet that includes fortified foods, a 400-microgram folic acid supplement should be taken. Fortification of refined grain products with folic acid is improving the folate status in Americans and has contributed to reductions in the incidence of neural tube defects since 1998.[6,30]

Vitamin A Both low and high intakes of vitamin A may cause problems during pregnancy. Too little vitamin A is associated with poor fetal growth. Too much vitamin A in the form of retinol from supplements can cause fetal malformations.[31]

The effects of vitamin A overdoses during pregnancy came to public attention in the early 1990s, when women taking accutane or retinoic acid for acne were found to deliver far more than the expected number of infants with malformations. Use of these drugs before and very early in pregnancy increases the risk that babies will be born with malformations of facial features and the heart; the intake of more than 10,000 to 15,000 IU of retinol daily during the same period may have the same effect. As a precaution, the American College of Obstetrics and Gynecology recommends that women who are or could become pregnant should limit their intake of retinol to less than 5,000 IU per day and should not use medicines containing vitamin A.[32] Beta-carotene, a precursor to vitamin A, is not harmful, however.[33]

Calcium Uptake of calcium by the fetus is especially high during the third trimester, when the fetus's bones are mineralizing. Pregnant women who regularly consume low-calcium diets lose calcium from their bones during pregnancy but usually regain it after delivery. Because calcium losses are not from the teeth, the old saying, "for every baby a tooth" has no basis in fact.[1,34]

Vitamin D Vitamin D during pregnancy supports fetal growth and the addition of calcium to bone, and participates in the programming of genes in ways that may influence the development of chronic diseases such as rheumatoid arthritis and cancer later in life.[35,36] Insufficient vitamin D compromises fetal growth and bone development, and this appears to be happening during some pregnancies. About 42% of African American and 4% of Caucasian women have low blood levels of vitamin D.[37] Vegan women may also be at risk for poor vitamin D status because vitamin D is naturally present only in animal products.

An intake of 10 mcg (400 IU) of vitamin D daily is officially recommended for pregnancy. Intakes of vitamin D from foods and supplements should not exceed 100 mcg (4,000 IU) daily.[38]

Iron Iron deficiency is the most common nutrient deficiency in pregnant women in the United States.[39] It develops when a woman enters pregnancy with low iron stores and fails to consume enough iron during her pregnancy. Iron requirements increase during pregnancy due to increases in hemoglobin production and storage of iron by the fetus.

Because of the large increase in need, it is difficult for pregnant women to get enough iron from foods. Consequently, 30 milligrams of supplemental iron per day are recommended for the second and third trimesters of pregnancy.[39] Women who do not take supplemental iron are more likely to develop iron deficiency. They are also more likely to deliver infants who are small—and are at risk of developing iron deficiency in their first year—than are women who take supplements.[40]

Iodine Iodine is needed for normal thyroid function and plays important roles in protein tissue construction and maintenance. A lack of iodine during pregnancy can interfere with fetal development and, in extreme cases, cause mental and growth retardation, as well as malformations in children.[41]

About half of pregnant women in the United States consume less than the recommended 220 mcg of iodine daily, and 7% have moderate iodine deficiency.[42] The most reliable source of iodine is iodized salt. One teaspoon contains 400 mcg iodine. Salt with added iodine is clearly labeled as "iodized." Fish, shellfish, seaweed, and some types of tea also provide iodine. Women who consume iodized salt are not likely to need supplemental iodine.[43] Usual iodine intake should not exceed 1,100 mcg daily during pregnancy.[44]

EPA and DHA These two omega-3 fatty acids play important roles in fetal and infant development. EPA (eicosapentaenoic acid) and DHA (docosahexaenoic acid) are long-chain, highly unsaturated fatty acids. Women who have adequate EPA and DHA intake during pregnancy tend to deliver infants that develop a somewhat higher level of intelligence, better vision, and otherwise more mature central nervous system functioning than infants born to women with low intake of EPA and DHA. Rates of preterm delivery have been found to be lower among women with adequate EPA and DHA status.[6,45–47,49]

It is currently recommended that women consume 300 mg of EPA and DHA daily during pregnancy.[48] Most pregnant women in the United States consume less than half of this amount.[28] EPA and DHA are found together in fish, fish oils, and seafood. DHA is available in omega-3 fatty acid–fortified eggs, orange juice, soy products, margarines, and other products. Consumption of 12 ounces per week of low-mercury fish during pregnancy and breast-feeding is encouraged.[49] Safe fish that provide high levels of EPA and DHA include salmon, herring, anchovies, shrimp, and halibut (Table 29.4).

What's a Good Diet for Pregnancy?

Contrary to folklore, women do not instinctively select and consume a healthy diet during pregnancy. A good diet takes knowledge and planning. Regardless of a pregnant woman's age and whether or not she is a vegetarian, a healthy diet provides all the nutrients she needs with the possible exception of iron. Foods are recommended over supplements because foods provide fiber, antioxidants, and other beneficial substances that supplements do not. Pregnancy diets should include sufficient fluid and fiber and consist of regular meals and snacks. A pregnant woman should not consume alcohol.[6] Concerns about potential problems related to coffee and caffeine intake during pregnancy in the past have lessened due to recent study results. Intake caffeine from coffee does not appear to be related to negative effects on the course or outcome of pregnancy when consumed at levels of three cups or less of brewed coffee daily.[50,51]

Table 29.4 EPA and DHA content in a 3-ounce serving of fish and seafoods containing low levels of mercury[86]

	EPA + DHA (mg)
Shad	2,046 mg
Salmon	1,825
Anchovies	1,747
Herring	1,712
Salmon	1,564
Whitefish	1,370
Mackerel	1,023
Sardines	840
Tilefish	796
Whiting	440
Flounder	426
Trout	420
Oysters	375
Snapper	273
Shrimp	268
Clams	241
Haddock	202
Catfish	201
Crawfish	187
Sheepshead	162
Tuna, light, canned in oil	109
Lobster	71

Table 29.5 Dietary recommendations for pregnancy

- Consume sufficient calories for adequate weight gain.
- Eat a variety of foods from each ChooseMyPlate food group.
- Eat regular meals and snacks.
- Consume sufficient dietary fiber (about 28 grams per day).
- Consume 11 to 12 cups of water each day from fluids and foods.
- Use salt to taste (within reason).
- Do not drink alcoholic beverages.
- Eat foods you enjoy at pleasant meal times.

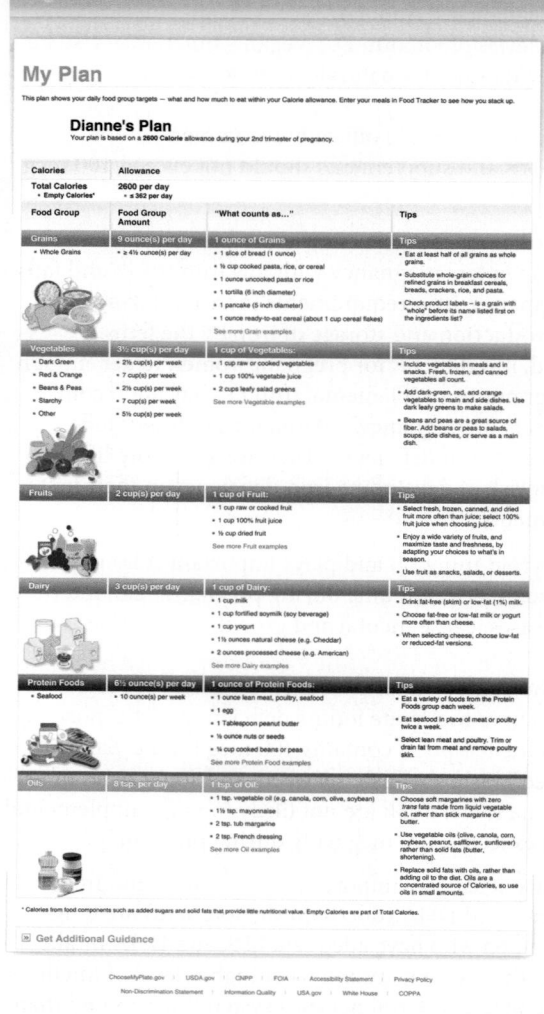

Illustration 29.8   An example of food group recommendations for pregnant and breast-feeding women based on ChooseMyPlate Plan for Moms.

www.choosemyplate.gov

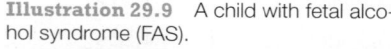

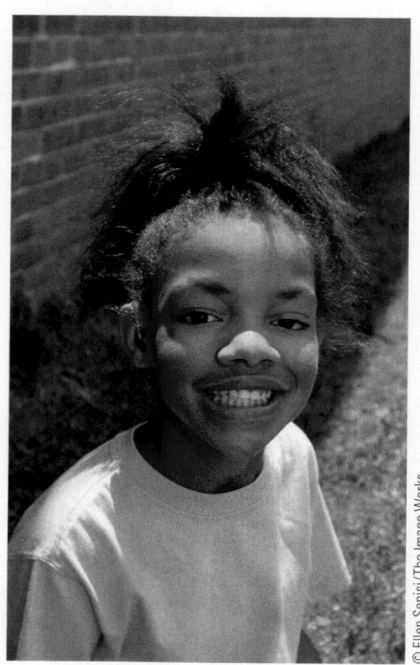

Illustration 29.9   A child with fetal alcohol syndrome (FAS).

© Ellen Senisi/The Image Works

Table 29.5 lists recommendations for diet during pregnancy. Planning a diet for pregnancy based on the basic food groups is the most straightforward approach to meeting nutrient needs. You can get a personalized plan for food intake during pregnancy from SuperTracker at ChooseMyPlate.gov. The "Daily Food Plan for Moms" generated shows the food types and amounts that meet nutrient needs in each trimester of pregnancy. The plan is personalized, based on age, gender, height, weight, and physical activity level. Illustration 29.8 shows an example of food group recommendations for Dianne, a 30-year-old women who is 5 foot, 8 inches tall, weighs 140 pounds, is in her second trimester of pregnancy, and is physically active for 30–60 minutes daily.[52]

Why Alcohol and Pregnancy Don't Mix   As early as the 1800s, maternal consumption of alcohol during pregnancy was said to cause the birth of "sickly" infants. The ill effects of alcohol on babies were not fully acknowledged, however, until the 1970s, when several research reports described a condition called fetal alcohol syndrome (Illustration 29.9). Women who drank heavily or frequently binged on alcohol during pregnancy were found to be at high risk of delivering infants with specific malformations and retarded physical and mental development. The effect of maternal alcohol consumption on the fetus worsens as intake increases. Heavy drinking in the first half of pregnancy is closely associated with the birth of malformed, small, mentally impaired infants. When excessive drinking occurs only in the second half of the pregnancy, infants are less likely to be malformed but are still likely to be small and to develop abnormal mental development.[53] These conditions are permanent; they cannot be fully corrected with special treatment, and the child does not outgrow them.

No amount of alcohol has been found to be absolutely safe during pregnancy. When only an occasional drink is consumed, however, the adverse effects of alcohol on fetal development appear to be small and rare.[54] To exclude the possibility of even small impairments in fetal growth and development, it is recommended that women not drink alcohol at all during pregnancy or when they are attempting to become pregnant.[6]

Vitamin and Mineral Supplements   Iron is the only supplement recommended for all pregnant women. About 83% of pregnant women take multiple vitamin and mineral supplements, however. Apparently, supplements are being prescribed as insurance against the possibility of poor diets. Women should be given supplements the same way they are given medications—when they are indicated. Under certain conditions, supplementation with nutrients besides iron is indicated. A multivitamin–mineral preparation is recommended for pregnant women who do not ordinarily consume an adequate diet and for those in high-risk categories, such as women carrying more than one fetus, heavy cigarette smokers, and alcohol and drug abusers.[5,31]

Teen Pregnancy

Approximately 4 of every 100 females between the ages of 15 and 19 years in the United States deliver babies each year.[55] Although teens who do not smoke, drink alcohol, or use drugs and who enter pregnancy in good physical and nutritional health tend to deliver healthy infants, many teens do not enter pregnancy in such good condition. Poor lifestyle habits are thought to be primarily responsible for the high rate of low-birth-weight and preterm infants born to teens. Few teens meet all of the food group recommendations for pregnancy. Individualized attention, nutrition intervention, and support all foster healthy outcomes of teen pregnancy.[56]

Breast-Feeding

- **Identify three unique health advantages for infants associated with breast-feeding.**

- **Develop and evaluate a one-day diet for a breast-feeding woman based on guidance from the ChooseMyPlate food groups.**

A woman's capacity to nourish a growing infant does not end at birth; it continues in the form of breast-feeding (Illustration 29.10). Breast milk from healthy, well-nourished women is ideally suited for infant nutrition and health.[57,58]

What's So Special About Breast Milk?

Breast milk is like a bonus pack. In addition to serving as a complete source of nutrition for infants for the first 4–6 months of life, breast milk contains substances that convey a significant degree of protection against a variety of illnesses (Table 29.6)—including infectious diseases such as polio and the flu and ear, respiratory tract, and gastrointestinal tract infections. Evidence indicates that breast milk may confer a degree of protection against the development of cancer of the lymph system (lymphoma), type 2 diabetes, and asthma during childhood.[58,59] In addition, breast milk contains essential and nonessential fats and other substances that appear to promote optimal growth and development of the nervous system and eyes. Intelligence, as measured by IQ, tends to be higher in babies receiving breast milk than in those given formula.[60] The disease-preventing and development-promoting components of human milk are lifesaving assets in many developing countries, where a safe water supply and medical care may be unavailable.[4] Although breast-feeding doesn't protect infants from all infectious diseases and food allergies, it's the ounce of prevention that's worth a pound of cure.

PhotoDisc

Illustration 29.10 Babies are born to be breast-fed.

Table 29.6 A sample of U.S. infants shows fewer significant episodes of illness among those who were breast-fed[87,88]

Illness	Percentage of infants fed:	
	Breast milk	**Formula/other**
Ear infections	3.7%	9.1%
Respiratory illness to age 7	17.0	32.0
Diarrhea, vomiting	3.5	6.9
Hospital admissions	1.0	3.0
Average number of episodes of illness per infant	8.2	21.1

Table 29.7 Fifteen reasons to breast-feed

1. The milk container is easy to clean.

2. Breast milk is a renewable resource.

3. There's no packaging to discard.

4. Breast milk comes in an attractive container.

5. The temperature of breast milk is always perfect right out of the container.

6. Breast milk tastes really good.

7. There are no leftovers.

8. You don't have to go to the kitchen in the middle of the night to get breast milk ready.

9. It takes just seconds to get a meal ready.

10. There's no bottle to repeatedly pick up off the floor.

11. The price is right.

12. The meal comes in a perfect serving size.

13. Meals and snacks are easy to bring along on a trip or outing.

14. Feeding units come in an assortment of beautiful colors and sizes.

15. One food makes a complete meal.

colostrum The milk produced during the first few days after delivery. It contains more antibodies, protein, and certain minerals than the mature milk that is produced later. It is thicker than mature milk and has a yellowish color.

Illustration 29.11 Percentage of breast-fed newborns in the United States.[63,89]

Breast-feeding also offers other benefits—ones that may mean a lot to parents. Table 29.7 presents some of them.

Is Breast-Feeding Best for All New Mothers and Infants?

Over 96% of women are biologically capable of breast-feeding, and the vast majority of infants thrive on breast milk.[62] In the United States, about 77% of new mothers breast-feed their infants to some extent (Illustration 29.11).[63] Successful breast-feeding involves more than biology; it is heavily influenced by environmental and psychological conditions.

The increase in women returning to work soon after delivery, a lack of health care provider and emotional support for breast-feeding, embarrassment, early hospital discharge, and inadequate knowledge about how to breast-feed all appear to be deterring U.S. women from breast-feeding. If more women are to have the opportunity to breast-feed, these and other barriers must be broken down.[64] A number of European countries that actively promote breast-feeding allow women to stay in the hospital after delivery until breast-feeding is going well. The usual practice in some African countries is to relieve a breast-feeding mother's workload so that she can devote nearly full time to feeding and caring for her young infant. A relative may move in with the family and take over household chores, or the mother and baby may live with her parents for a time.

Public health initiatives are in place to facilitate breast-feeding. To meet national health objectives for improving rates of breast-feeding initiation and continuation in the United States (see Table 29.8), it is recommended that:

• Health care workers encourage and facilitate breast-feeding

• "How-to" advice and problem-solving guidance be available for breast-feeding women from qualified health care staff

• Breast-feeding women returning to work have access to on-site child care facilities and private rooms for breast-feeding

Breast-feeding is at its best when both the mother and infant benefit from the experience. If mutual benefit is not possible, then formula feeding may be necessary. Infant growth and development are well supported by commercially available infant formulas.[57]

How Breast-Feeding Works

The mother's body prepares for breast-feeding during pregnancy. Fat is deposited in breast tissue, and networks of blood vessels and nerves infiltrate the breasts. Ducts that will channel milk from the milk-producing cells forward to the nipple—the milk collection ducts—also mature (Illustration 29.12). Hormonal changes that occur at delivery signal milk production to begin.

Breast milk produced during the first three days or so after delivery is different from the milk produced later. Called **colostrum**, this early milk contains higher levels of protein, minerals, and antibodies than "mature" milk.

While an infant is consuming one meal, she or he is "ordering" the next. The pressure produced inside the breast by the infant's sucking and the emptying of the breasts during a feeding cause a hormone to be released from specific cells in the brain. The hormone stimulates the production of milk so that more milk is produced for the next feeding. It generally takes about 2 hours for the milk-producing cells to manufacture enough milk for the next feeding. An important exception to this occurs when an infant enters a growth spurt and consumes more milk than usual. Then milk production takes longer, perhaps a day, to catch up with demand.

Table 29.8 Actual versus national goals for breast-feeding in the United States[63,89]

	Newborns	6 Months	12 Months
National goal	75%	50%	25%
Actual (all women)	77%	39%	20%

Only very rarely is a breast-feeding woman unable to produce enough milk. As long as an infant is allowed to satisfy her or his appetite by breast-feeding as often as desired, milk production will catch up with the baby's need.[57,58,65]

Nutrition and Breast-Feeding

A breast-feeding woman needs an adequate and balanced diet to replenish her body's nutrient stores, maintain her health, and produce sufficient milk for her baby. Increases in recommended dietary intakes for breast-feeding are generally higher than those for pregnancy. As during pregnancy, proportionately higher amounts of nutrients than calories are required, indicating the need for a nutrient-dense diet (see Illustration 29.5 on page 29-7).

Calorie and Nutrient Needs The RDA for calories is about 15% higher for breast-feeding women than for other women. Actually, a breast-feeding woman needs about 30% more calories than the RDA for women who are not breast-feeding, but she does not have to consume that level of calories from food. Energy supplied from fat stores that normally accumulate during pregnancy contributes to meeting these needs during breast-feeding, so not all of the calories must come from the mother's diet.[65]

A breast-feeding woman has a high requirement for DHA (docohexaenoic acid), an omega-3 fatty acid, because her infant used a good deal of it for central nervous system development. The DHA content of maternal milk directly reflects maternal intake. Adequate intake of DHA by women during breast-feeding is related to gains in intelligence and vision development in infants.[45]

What's a Good Diet for Breast-Feeding Women? The calories and nutrients needed by the breast-feeding woman can be obtained from a varied diet that includes foods from each of the basic food groups. The ChooseMyPlate for Moms recommended numbers of servings from each food group are shown in Illustration 29.8 (page 29-10); they are the same as for pregnant women. Dietary recommendations for breast-feeding women are summarized in Table 29.9.

Increases in hunger and food intake that accompany breast-feeding generally take care of meeting caloric needs. Failure to consume enough calories from food can decrease milk production, however. Low-calorie diets (those providing fewer than 1,500 calories per day) and weight loss that exceeds 1.5 to 2 pounds per week—even in women with a good supply of fat stores—can reduce the amount of milk women produce. Weight loss of about a pound a week starting a month after delivery appears to be safe and helpful in women's attempts to return to prepregnancy weight.[66]

Are Supplements Recommended for Breast-Feeding Women? Supplements are not recommended for breast-feeding women. Instead, breast-feeding women should try to get all of the nutrients they need from food.[65] Supplements, if needed, should be prescribed on an individual basis.

Mammary gland sagittal section

pectoralis major muscle

pectoral fascia

intercostal muscles

suspensory ligaments

lactiferous sinus

lactiferous duct

ribs

lung

gland lobules

fat

Encyclopaedia Britannica/UIG/Getty Images

Illustration 29.12 A view of the interior of the breast.

Table 29.9 Dietary recommendations for breast-feeding women[58,65]

- Diets should supply all of the nutrients breast-feeding women need. (The routine use of vitamin and mineral supplements is not recommended.)
- Fluids should be consumed to thirst (14 cups of fluid each day).
- Weight loss should not exceed 6 to 8 pounds per month after the first month after delivery; caloric intakes should not fall below 1,500.
- Alcohol intake should be avoided.

Dietary Cautions for Breast-Feeding Women Infants are much smaller than women, and it takes a smaller dose of caffeine, alcohol, drugs, or environmental contaminants to have an effect on them than on an adult. Whereas breast-feeding mothers may show no adverse effects from these substances, their infants may.

Almost anything a woman consumes may end up in her breast milk. Breast-fed infants of women who are heavy coffee drinkers (10 or more cups per day) may develop "caffeine jitters." Alcohol is transferred from a woman's body to breast milk. The development of the brain and nervous system of infants breast-fed by chronic, heavy drinkers appears to be retarded.[67] It is recommended that women do not drink alcohol during the first month of breast-feeding. Breast-feeding women are advised to avoid alcohol or limit it to one drink when breast-feeding will not be undertaken for three to four hours. It takes three to four hours to metabolize alcohol in one standard drink and clear it from the blood and breast milk.[6]

Infant Nutrition

- **Identify three examples of parental behaviors around eating that negatively influence, and three that positively influence infants' learning about foods and food intake.**

Illustration 29.13 Infants develop at a remarkable rate.

At no other time during life outside the womb do growth and development proceed at a faster pace than during infancy. Infants grow out of clothes long before they wear them out. Each day infants learn new behaviors, and their minds absorb large chunks of information that will serve them well in the future (Illustration 29.13). The rate at which growth and development proceed in the first year of life is truly amazing. Just as infants need security, love, and attention to flourish, so too do they need calories and essential nutrients.

Infant Growth

During the first week or two of life, it is not uncommon for infants to lose up to 10% of their birth weight while adjusting to the new surroundings.[68] After that, infants grow rapidly. Most infants double their birth weight by 4 months and triple it by 1 year. Length usually increases by 50% during the first year.[69] If this rate of growth were to continue, 10-year-old children would be about 10 stories high and weigh over 220 tons! After infancy, the growth rate declines and remains at a fairly low level until the adolescent growth spurt begins. Development proceeds in parallel with growth during the first year of life (Illustration 29.14).

Growth Charts for Infants The CDC recommends that the WHO growth standards be used to monitor growth of infants and children ages 0 to 2 years of age in the United States.[70]

Examples of WHO growth charts for weight for age for newborns to 2 year olds are shown in Illustration 29.15. The charts should be carefully plotted based on accurate, periodic measures of an infant's size. They are best used for screening growth problems; confirmation of underweight or obesity requires assessment of body fat by skinfold thickness and other measures.[70]

Body Composition Changes with Growth Humans grow up and out, and their body composition and proportions change as growth progresses (Illustration 29.16). Although generally measured by gains in pounds and inches, growth also reflects changes in bone mass, organ size, body proportions, and composition (the proportionate amounts of water, muscle, and fat). Infants are not shaped like little adults. Their heads are very large in relation to the rest of their bodies and their arms are short, for example. Brain growth takes precedence over trunk and limb growth early in life, and the body eventually grows to "fit" the head (Illustration 29.17). During the first ten years of life, a child who initially has the shape of a loaf of bread may normally come to resemble a string bean.

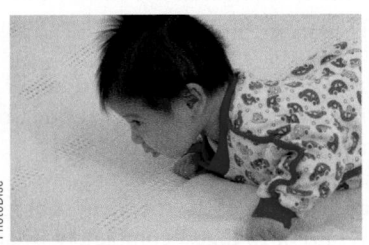

First Month
Generally weighs from 8 to 11 pounds; length is 20 to 23 inches. Head is relatively large and has soft spot on top. Startles and sneezes easily. Jaw may tremble. May hiccup and spit up. Eats every few hours.

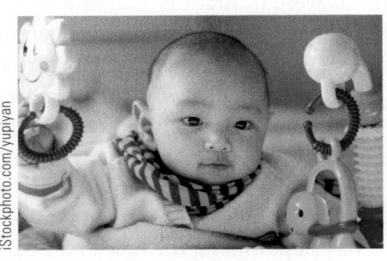

One to Three Months
Lifts head briefly when placed on stomach; smiles, coos, and gurgles. Whole body moves when infant is touched or lifted. Eats every three to four hours.

Four to Six Months
Weight nearly doubled. Has grown three to four inches. Follows objects with eyes. Reaches toward objects with both hands; puts fingers and objects into mouth. Turns over, sits unassisted. Awake longer at feeding time. Eats six to seven times per day. Sleeps six to seven hours at night.

Seven to Eleven Months
Gains in weight and height are less rapid, appetite has decreased. Stands up with help, hitches self along the floor. Reaches for, grasps, and examines objects with hands, eyes, and mouth. Has one or two teeth. Takes two naps a day.

Twelve Months
Usually has tripled birth weight and increased length by 50%. Grasps and releases objects with fingers. Holds spoon, but uses it poorly. Begins to walk unassisted.

Nutrition and Mental Development Malnutrition has the greatest impact on mental development when it is severe and occurs during the critical period for brain cell multiplication. For humans, this vulnerable period begins during pregnancy and ends after the first year of life. Impairment in mental development is less severe when malnutrition occurs only during pregnancy or only during infancy than if it occurs throughout both pregnancy and infancy.[71]

Children's mental development is also greatly influenced by the social and psychological environment in which they are raised. Because malnutrition is generally accompanied by both social and psychological deprivation, these factors often contribute jointly to poor mental development. For children in the United States, poverty, neglect, illnesses, and psychological problems appear to be the main cause of undernutrition and poor mental development.[72]

Weight-for-age BOYS
Birth to 2 years (percentiles)

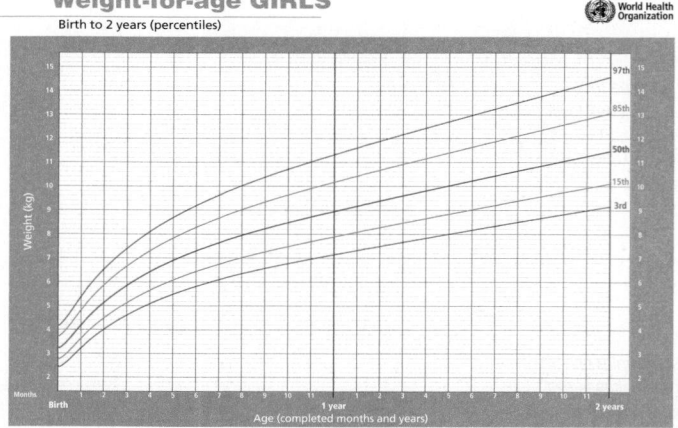

WHO Child Growth Standards

Weight-for-age GIRLS
Birth to 2 years (percentiles)

WHO Child Growth Standards

Illustration 29.15 Example of World Health Organization's growth charts for weight for age for males and females aged 0 to 2 years.[69]

Illustration 29.16 Infants' heads are proportionately large for the rest of the body. But, as you can see, there's much more growth to come.

Infant Feeding Recommendations

Current dietary recommendations for infants call, preferably, for exclusive breast-feeding for the first 6 months of life.[4,5,88] If breast-feeding is not an option, formula feeding for the first 12 months of life is recommended.[73] Infants should be fed "on demand," that is, when they indicate they are hungry, rather than on a rigid schedule set by the clock. In the first few months of life, most infants will get hungry every three or four hours. These and other recommendations for infant feeding are shown in Table 29.10.

Soft, solid foods should be added to an infant's diet of breast milk or infant formula around 6 months.[4] At one time, it was thought that infants should be offered solid foods within the first few months of life. Although such young infants are unable to swallow much of the food (most of it ends up on their face and bib) or to digest completely what they do swallow, solid foods were thought to help the baby grow and sleep through the night. This belief does not appear to be correct. Infants who receive cereal at bedtime before the age of 6 months are no more likely to grow normally or sleep through the night than are infants who start to receive solid foods around 6 months of age.[74] The age at which an infant begins to sleep for 6 or more hours during the night depends on other factors, including the infant's developmental level and how much he or she has slept during the day.[75] Neither the infant nor the infant's parents are likely to get a full night's sleep for at least 4 months.

Introducing Solid Foods It is recommended that iron-fortified rice cereal be given to breast-fed infants and to bottle-fed infants who are not receiving iron-fortified formula. The iron provided by iron-fortified rice cereal or formula helps to restock the infant's iron stores, which have been drawn on since birth. Generally, solid foods prepared for infants should consist of single, basic foods such as strained vegetables, fruits, or meats.

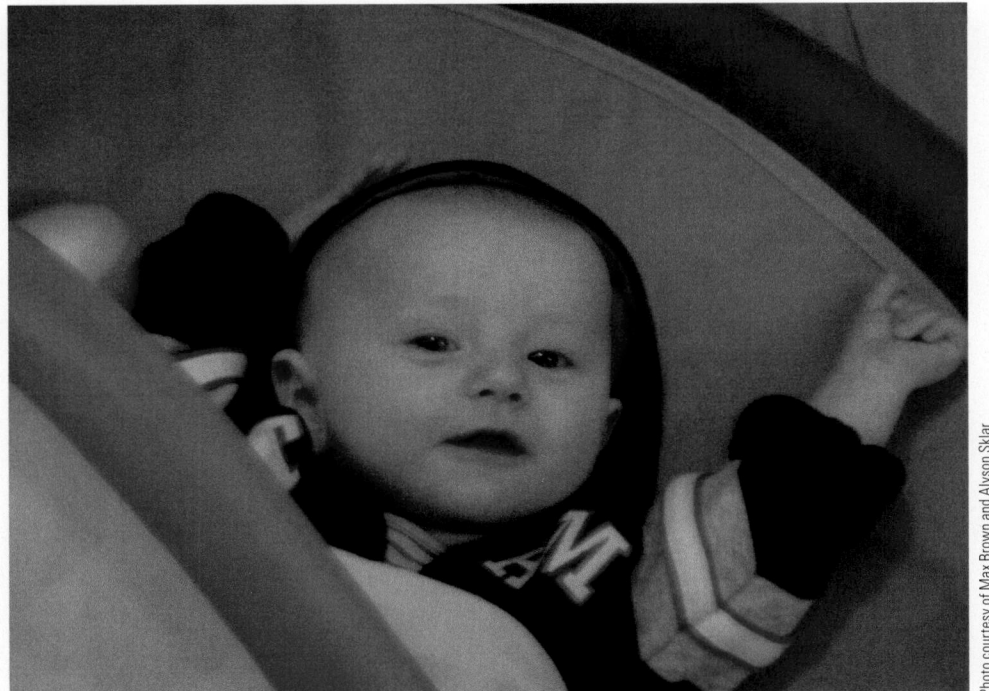

Illustration 29.17 When you hold your arm straight up, is it longer than your head? It's about the same for this 5-month-old.

Solid foods for infants can be purchased as commercial baby food or prepared at home. When baby foods are prepared at home, care should be taken to avoid contamination and to achieve the right consistency. Infants should be offered new foods one at a time. The new offerings should be separated by several days, so that any allergic reactions to a food can be identified. Variety is the key to achieving a healthy diet for infants in their second 6 months of life, and an assortment of basic, textured foods should be given.[76,77]

By 9 months of age, infants are ready for mashed foods and foods such as yogurt, applesauce, ripe banana pieces, and grits. Most infants have several teeth by this time, and they are able to bite into and chew soft foods.

Infants graduate to adult-type foods after the age of 12 months. Although most foods still need to be mashed or cut up into small pieces for them, 1-year-olds are able to eat the same types of food as the rest of the family. They can drink from a cup and nearly feed themselves with a spoon (Illustration 29.18). Infants have come a long way in 12 months.

Foods to Avoid Not all foods can be considered "baby foods." Some foods such as eggs, peanuts, tree nuts, and soy, are related to allergic reactions. These foods should be omitted from an infant's diet, especially in cases where the family has a history of allergic reaction to the foods.[78] Foods like hot dogs, peanut butter, and pieces of raw vegetables should not be given to infants because they are difficult for infants to chew into small pieces or swallow. Table 29.11 provides a list of foods that should not be offered to infants.

Illustration 29.18 Older infants develop self-feeding skills. It requires practice and parental patience.

Table 29.10 Overview of infant feeding recommendations[4,6,7,90]

- Exclusive breast-feeding for the first 6 months of life and continuation of breast-feeding through the first year of life are preferred. Iron-fortified infant formulas may be used as a secondary option to breast-feeding.

- Introduce solid foods around 6 months of age.

- An infant's first food should be iron-fortified, such as iron-fortified baby cereal.

- After 6 months, offer a variety of healthy foods appropriate for baby. Pay attention to the textures of baby's food, going from smooth to mashed to chopped to tiny pieces.

- Keep trying new foods—babies may need to try a new food ten times or more before they like it!

- Use a small baby toothbrush with a tiny dot of fluoride toothpaste when the teeth start to come in.

- Infants should be fed "on demand" and not by a set schedule. Feeding should stop when the baby loses interest in eating.

- Infants not receiving fluoridated water should be given fluoride drops (dose depends on age).

Table 29.11 **Foods to avoid in the first year of life**

Foods that may cause allergic reactions	Foods that may cause other problems	Foods that may lead to choking
Cow's milk	Soft drinks	Grapes
Egg white	Coffee	Frankfurter pieces
Fish, seafood	Corn	Hard candy
Nuts, peanuts	Fruit drinks	Meat chunks
Peanut butter	Honey (unpasteurized)	Nuts and seeds
Soy protein	Prune juice	Popcorn
Wheat products	Tea	Raw vegetables

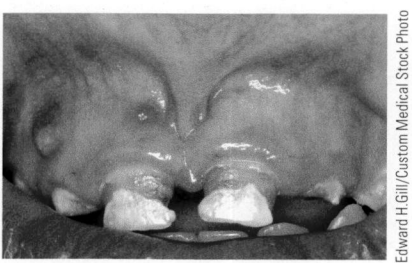

Edward H.Gill/Custom Medical Stock Photo

Illustration 29.19 Baby-bottle tooth decay.

Reduced-fat products are not recommended for infants. Infants need a relatively high-fat diet for brain and nervous system development. Unpasteurized honey is not recommended either, because it may cause botulism in infants due to their still maturing gastrointestinal defenses against bacteria. Beverages containing a lot of sugar, such as soft drinks and sweetened fruit drinks, should not be given to infants because they promote tooth decay and provide "empty" calories. To help prevent "baby-bottle tooth decay" shown in Illustration 29.19, infants should not be put to sleep with a bottle containing sweet fluids or formula.[73,76,77]

Do Infants Need Supplements? Two situations call for the use of supplements during infancy. Breast-fed infants and infants living in households that do not receive fluoridated water should be given a fluoride supplement (0.25 mg daily) after 6 months of age.[79] Most infants do not consume the recommended 400 IU of vitamin D daily and a vitamin D supplement is recommended to fill the vitamin D gap between intake and need.[80]

The Development of Healthy Eating Habits Begins in Infancy

Older infants and young children should be offered a wide variety of nutritious foods in a positive eating environment. They alone should make the decision about how much to eat any food offered. Food preferences are primarily learned and are unique to every individual. The learning process begins early in life and is affected by a variety of influences. "Instincts" that draw infants to select certain, nutritious foods are not one of the influences, however. Infants do not instinctively know what to eat. Rather, they are born with mechanisms that help regulate how much they eat. These basic characteristics of infant food intake were beautifully demonstrated by Clara Davis in experiments with 7- to 9-month-old infants in the 1920s and 1930s.[81] One of her most famous experiments is described in Illustration 29.20.

Subsequent research reinforced earlier conclusions that infants and young children should be offered a wide variety of nutritious foods and that decisions about how much to eat should be left up to the infant or child.[82] Beginning at birth, infants who are hungry eat enthusiastically. They stop eating when they are full. Coaxing, cajoling, and pleading by parents or caregivers can override infant's decisions about how much food to eat. That, unfortunately, may upset infants' built-in food intake regulating mechanisms and lead to over- or under-eating, and to "fussy" eaters.

Teaching Infants the Right Lessons About Food Recommendations for feeding infants are primarily based on energy and nutrient needs and the developmental readiness of infants for solid foods. But infant feeding recommendations also include a large educational component. Many of the lessons infants learn about food and eating make an impression that lasts a lifetime. Later food habits and preferences, appetite, and food intake regulation are all influenced by early learning experiences.[84, 85] Table 29.12 provides a lesson plan for teaching your infant the right lessons about food and eating.

Illustration 29.20   Infant food preferences.

PhotoDisc

Clara Davis conducted her most quoted feeding experiment with older infants (7 to 9 months old) living in an orphanage connected to a large hospital in Chicago. Three times a day a nurse would offer the infants a tray of various foods. The infants would pick up or grab for the foods they wanted, and the nurse would bring the food selected to the infant's hands or mouth.

Altogether 32 different foods were offered, consisting of fresh, unprocessed, unseasoned, and simply prepared basic foods. Foods included milk, beef, kidney, bone marrow, liver, brain, thymus, fish, whole wheat cereals, raw eggs, sea salt, 15 fresh fruits, and 10 fresh vegetables. Desserts, sugars, syrups, and other sweetened foods were not offered.

Infants quickly formed preferences and narrowed their choices to 14 of the 32 foods. Vegetables were the least liked, whereas milk, bone marrow, eggs, bananas, apple bits, oranges, and oatmeal were the best liked. They ate enough to grow normally and to remain in good health.

Dr. Davis concluded that appetite is the best guide to how much food an infant or young child needs, and that self-selection will have doubtful value only if the diet is chosen from inferior foods.[81,83]

Making Feeding Time Pleasurable T. Berry Brazelton, an expert on child development, says that feeding is an arena in which parents and baby work out the continuing struggle between dependence (being fed) and independence (feeding oneself). Independence must win. Pushing a child to eat is the surest way to create problems. Feeding has to be pleasurable. Dr. Brazelton points out that there is much more to the feeding of infants than meets the eye. Infants learn about communication, relationships, and independence during pleasurable eating experiences. If undertaken in positive and supportive circumstances, mealtimes for infants can provide lessons that will support their physical and psychological health well into the future.

Table 29.12 Tips for teaching infants the right lessons about food and eating

1. Infants learn to eat a variety of healthy foods by being offered an assortment of nutritious food choices. There are no inborn mechanisms that direct babies to select a nutritious diet.

2. Infants must be allowed to eat when they are hungry and to stop eating when they are full. Infants, not parents, know when they are hungry and have had enough to eat.

3. Food should be offered in a pleasant environment with positive adult attention.

4. Food should not be used as a reward, punishment, or pacifier.

5. Infants or children should never be coerced into eating anything.

6. Food preferences change throughout infancy. Because an infant rejects a food one time doesn't mean she or he will not accept the food if offered later. Offering a food on a number of occasions often improves acceptance of the food. Babies still may not like strong-flavored vegetables until they are older, however.

© Golden Pixels LLC/Alamy

NUTRITION
up close

You Be the Judge!

Focal Point: Apply your knowledge of nutrition to the diet of a pregnant woman.

Gloria's baby is due in 5 months. Before becoming pregnant, she never gave much thought to what she ate, but now she is trying to eat healthy foods for her baby and herself. Compare her choices with ChooseMyPlate recommendations in Illustration 29.8 to find out how well she is doing. Categorize Gloria's choices into the following food groups to determine if she ate the recommended servings from ChooseMyPlate. The grains group has been done for you.

Gloria's one-day diet:

Breakfast: Bran muffin (2 ounces), cold cereal (1 cup) with strawberries ($^1/_2$ cup), and a glass of milk (1 cup)
Lunch: Vegetable soup (1 cup vegetables), carrot sticks ($^1/_2$ cup), and a glass of orange juice (1 cup)
Afternoon snack: Yogurt (1 cup)
Dinner: Chicken (3 ounces), turnip greens ($^1/_2$ cup), a tossed salad (2 cups), and iced tea (2 cups)

Feedback to Nutrition Up Close is located in Appendix G.

	Grains
David Kay/Shutterstock.com	bran muffin, 2 ounces
	cold cereal, 1 cup
	Gloria's intake: 3 ounces
	Recommended intake: **8 ounces**

	Vegetables
	Gloria's intake:
Feng Yu/Shutterstock.com	Recommended intake: **3 cups**

	Fruits
Tatiana Popova/Shutterstock.com	
	Gloria's intake:
	Recommended intake: **2 cups**

	Milk and Milk Products
	Gloria's intake:
DenisNata/Shutterstock.com	Recommended intake: **3 cups**

	Meat and Beans
PHB.cz (Richard Semik)/Shutterstock.com	
	Gloria's intake:
	Recommended intake: **6 $^1/_2$ ounces**

	Oils
	Gloria's intake:
bonchan/Shutterstock.com	Recommended intake: **7 teaspoons**

REVIEW QUESTIONS

- **Define critical periods of growth and development and identify potential consequences of inadequate nutrient availability during these periods on future health status.**

1. Relatively high rates of infant mortality, low-birth-weight infants, and preterm delivery represent major public health problems in the United States.   True/False

2. Critical periods of growth and development are marked by large increases in the size of cells within developing organs.   True/False

3. Reduction in the size of organs related to the lack of nutrients during critical periods of growth and development can be corrected during the first year of life.   True/False

4. Inadequate or excessive levels of nutrient availability during pregnancy can modify the function of fetal genes in ways that increase the risk of chronic disease later in life.   True/False

- **Develop and evaluate a one-day diet for a pregnant or breast-feeding woman based on guidance from the ChooseMyPlate food groups.**

5. A women entering pregnancy with a BMI (body mass index) of 27 should gain 25 to 35 pounds during pregnancy.   True/False

6. The critical period of growth and development for formation of the neural tube in a fetus occurs during the second trimester of pregnancy.   True/False

7. Iodine, EPA and DHA, and vitamin D are considered "risk nutrients" during pregnancy because a sizeable portion of women fail to consume enough of them.   True/False

8. A daily multivitamin and mineral supplement is recommended for all breast-feeding women.   True/False

9. Women are advised not to eat any types of fish or seafood during pregnancy because these foods contain excessively high levels of mercury.   True/False

10. Moderate intakes of alcohol in the second and third trimester of pregnancy are safe for the mother and fetus, but intake of alcohol in the first trimester is not.   True/False

The next three questions refer to this situation:

Simonetta is 6 months pregnant. One night while visiting her aunt and uncle for dinner her aunt gives her advice on what and how much she should eat for a healthy pregnancy.

11. _____ Which of the following pieces of Simonetta's aunt's advice about a healthy diet for pregnant women is correct?
   a. "Simonetta, you're eating for two and you need to eat more than that."
   b. "Don't eat so many vegetables. They will take up space in your stomach that needs to be filled with the meat. It's full of protein, you know."
   c. "Glad you enjoyed the salmon. I heard this kind of fish is good for pregnant women so I made it especially for you."
   d. "Just eat enough food, honey. Your baby will get enough of the vitamins and minerals it needs from your body."

12. _____ Simonetta decides to have a cup of coffee after dinner. Both her aunt and uncle look at her while her uncle asks, "Are you sure you should be drinking coffee? Isn't it bad for the baby?" What would be a reasonable response that Simonetta could make to that question?
   a. "Decaffeinated coffee is okay for the baby if you have only a cup a day."
   b. " Maybe coffee used to be forbidden during pregnancy, but times have changed. A few cups of coffee a day won't harm the baby."
   c. "You're right! What was I thinking?"
   d. "Do you have any tea?"

13. _____ Assume Simonetta delivers a boy who weighs 5 kg (11 pounds) when he reaches 2 months of age. Referring to WHO growth charts in Illustration 29.15, the range of weight for age percentiles that would correspond to the baby's weight at 2 months of age would be in the:
   a. 3rd to the 15th percentiles.
   b. 15th to 50th percentiles.
   c. 50th to 85th percentiles.
   d. 85th to 97th percentiles.

- **Identify three unique health advantages for infants associated with breast-feeding.**

14. Breast-feeding promotes infant's health in part because breast milk contains substances that prevent infections and reduce the risk of development of certain diseases later in life.   True/False

15. Approximately one in four women is biologically unable to breast-feed successfully.   True/False

16. A woman's milk supply automatically increases as an infant consumes more breast milk.   True/False

- **Identify three examples of parental behaviors around eating that negatively influence, and three that positively influence infants' learning about foods and food intake.**

17. Infants should be given their first solid food around 6 months of age.   True/False

18. Infants are able to self-regulate the amount of food they consume. They will eat when they are hungry and stop eating when they have had enough food.   True/False

19. It has been demonstrated in multiple studies that infants will select and consume a well-balanced and adequate diet if given access to a wide variety of both "junk" and nutritious foods.   True/False

20. Body proportions and composition of infants change a good deal with age.   **True/False**

The next three questions refer to this scenario:

Your cousin and her husband have a 10-month-old girl who refuses to try any vegetable whatsoever. The parents find it totally frustrating and want their daughter to at least taste them.

21. _____ Which type of parental behavior would be *least* likely to convince their daughter to taste vegetables?

    a. Continuing to offer her different types of vegetables at meals during pleasant mealtimes.
    b. Letting the baby decide if she doesn't want to eat the vegetable offered but keep offering small portions of vegetables with a positive attitude at future meals.
    c. Showing your enjoyment of vegetables when you eat together.
    d. Stopping the meal as soon as the baby refuses to try the vegetable.

22. _____ Finally the baby tries sweet potatoes and likes them. What should be a parent's appropriate reaction?

    a. Give the baby a reward for eating her vegetables.
    b. Smile and continue eating your meal.
    c. Clap your hands and shout hurray!
    d. Offer her sweet potatoes at every meal.

23. _____ A month later the daughter tries chopped spinach, immediately makes a face, and spits the spinach out of her mouth. What's the most likely reason this happened?

    a. She didn't like the taste of spinach.
    b. She is exerting her independence and decides not to eat what her parents want her to eat.
    c. She was too full to eat anymore.
    d. She loves the look on her dad's face when she does something like that.

Answers to these questions can be found in Appendix G.

- -

NUTRITION SCOREBOARD ANSWERS

1. In most all instances, the fetus does not harm the mother for its own gain. If there is a shortage of essential nutrients, the mother generally gets access to them before the fetus does.[1]   **False**

2. More than 30 economically developed countries have lower infant mortality rates than the United States.[2,3]   **False**

3. That's the official recommendation.[4,5]   **True**

4. Alcohol does pass into breast milk.[6]   **False**

UNIT 30

Nutrition for the Growing Years: Childhood through Adolescence

NUTRITION SCOREBOARD

1 Gains in height and weight during childhood occur in spurts rather than gradually. **True/False**

2 Young children are able to associate the feeling of hunger with a need for food around the age of five. **True/False**

3 The incidence of overweight and obesity in U.S. children tripled between 1990 and 2010. **True/False**

4 Childhood and adolescence represent "grace periods" during which the diet consumed does not influence future health. **True/False**

Answers can be found at the end of the unit.

- Describe connections between growth and development, and eating behavior in children and adolescents.

- Demonstrate the use of the Centers for Disease Control and Prevention (CDC) charts for growth assessment in children and adolescents.

- Identify three positive and three negative characteristics of the average diet of children and adolescents.

- Identify three characteristics of diets of children and adolescents that increase disease risk later in life.

The Span of Growth and Development

- **Describe connections between growth and development, and eating behavior in children and adolescents.**

- **Demonstrate the use of the Centers for Disease Control and Prevention (CDC) charts for growth assessment in children and adolescents.**

Physical and mental development proceed at a high rate from infancy through adolescence (Illustration 30.1). Young children generally enter this phase of life able to take a step or two and to guide a spoon into their mouths sideways. They will likely leave adolescence with the ability to drive, work for pay, and solve complex problems. These are the formative years that lay the foundation for the rest of life.

Key Nutrition Concepts

Content presented in this unit on nutrition during childhood and adolescence relates most directly to these key nutrition concepts:

- Foods provide energy (calories), nutrients, and other substances needed for growth and health.

- Poor nutrition can result from both inadequate and excessive levels of nutrient intake.

- Some groups of people are at higher risk of becoming inadequately nourished than others.

- Poor nutrition can influence the development of certain chronic diseases.

- Adequacy, variety, and balance are key characteristics of healthful diets.

The Nutritional Foundation

Good nutrition takes on particular importance during the growing years for many reasons. Growth is an energy- and nutrient-requiring process. It will not proceed normally unless the diet supplies enough of both.

Children learn about food and its importance to health and well-being during these early years of life, and they also establish food preferences and physical activity patterns that may endure into the adult years. Food intake regulatory mechanisms are affected by the lessons children learn early in life. Given control over decisions about how much to eat, children generally become responsive to internal cues that signal when they should eat and when they should stop eating.[3] If the lessons go well, they learn to eat for the right reason.

Eating well and learning the right lessons about food and health have implications that transcend the growing years. Early diets may have long-term effects on the risk of developing a number of diseases later in life.[2]

Characteristics of Growth in Children

Between the ages of 2 and 10 years children normally gain somewhere around 5 pounds and 2–3 inches in height per year (Illustration 30.2). Gains in weight and height occur in "spurts" rather than continuously and gradually. Prior to a growth spurt, appetite and food intake increase (given an adequate food supply), and the child puts on a few pounds of fat stores (Illustration 30.3). During the growth spurt, these fat stores are normally used to supply energy needed for growth in height. When growth is not occurring, children are often disinterested in food and eat very little at times. They may go on "food jags," insisting on eating limited amounts of only a few favorite foods like peanut butter and jelly sandwiches or breakfast cereal.

The ups and downs of children's food intake can be unnerving for parents (see the Reality Check). But as long as growth continues normally and children remain in good health, there's little reason to worry about food jags and fluctuations in appetite and food intake.

PhotoDisc

One to Two Years
Gains in height and weight continue at a lower rate; appetite is less. Uses finger and thumb to pick up things. Soft spot grows smaller and then disappears. Baby teeth continue to appear. Usually takes one long nap a day. Drinks from a cup, attempts to feed self with a spoon. Likes to eat with hands. Pulls self up to standing position. Walks alone.

©Felicia Martinez/PhotoEdit

Two to Three Years
Slower and more irregular gains in height and weight. Has all 20 teeth. Runs and climbs, pushes, pulls, lugs, walks upstairs one step at a time. Feeds self using fingers, spoon, and cup; spills a lot. At times has one favorite food. By age three, associates the sensation of hunger with a need for food.

© Tony Freeman/PhotoEdit

Three to Four Years
Gains 4–6 pounds and grows about 2–3 inches. Feeds self and drinks from a cup quite neatly, carries things without spilling. May give up sleep at naptime, substituting quiet play.

Fuse./Jupiterimages

Four to Five Years
Gains in height and weight about same as previous year. Hops and skips, throws ball. Has increasingly good coordination, masters buttons and shoelaces. Can use knife and fork and is a good self-feeder.

© Mary Kate Denny/PhotoEdit

Five to Six Years
Growth continues at about the same rate. Legs lengthen. Six-year permanent molars usually appear (new teeth, not replacing baby teeth). Begins to lose front baby teeth. Prefers plain, bland, and unmixed foods.

(Continued)

Illustration 30.1 *(Continued)*

© Carl Purcell/Photo Researchers Inc.

Six to Nine Years
Slow gains in height and weight (2–3 inches and 4–6 pounds a year). Some additional permanent teeth appear. Likely to have the childhood communicable diseases. Sleeps 11 to 13 hours.

©David Young-Wolff/ PhotoEdit

Nine to Twelve Years
May be long legged and rangy, but health is generally sturdy. Permanent teeth continue to appear. Appetite good. Needs about 10 hours sleep. Growth spurt in girls usually begins. May get very irritable when hungry.

© David Young-Wolff/PhotoEdit

Twelve to Fourteen Years
Wide differences in height and weight in children of either sex of same age. Menstruation usually begins (sometimes earlier). Girls develop breasts, are usually taller and heavier than boys of the same age. Growth spurt of boys begins. Muscle growth rapid. Appetite increases. Likes to spend time with friends.

Ghislain & Marie David de Lossy/Riser/Getty Images

Fourteen to Sixteen Years
Boys are in period of growth spurt. Voice deepens. Girls have usually achieved maximum growth. Menstruation is established though may still be irregular. Enormous appetite. Pimples a common problem. Sleep reaches its adult pattern. All permanent teeth except wisdom teeth.

REALITY CHECK

The "Clean Your Plate" Club

Aiden and Zoe are eating lunch with a friend and his 4-year-old son, Jayden. During lunch, they notice Jayden has quit eating, becomes distracted, and wants to leave the table to play. Then their friend says, "Jayden, you can go out and play as soon as you finish all the food on your plate." Is this good advice?

Who gets the thumbs up?

Answers appear on page 30-5.

iStockphoto.com/drbimages

Aiden: I don't think it's good advice. Jayden isn't hungry anymore.

iStockphoto.com/drbimages

Zoe: It sounds like good advice to me. It's a shame to waste food.

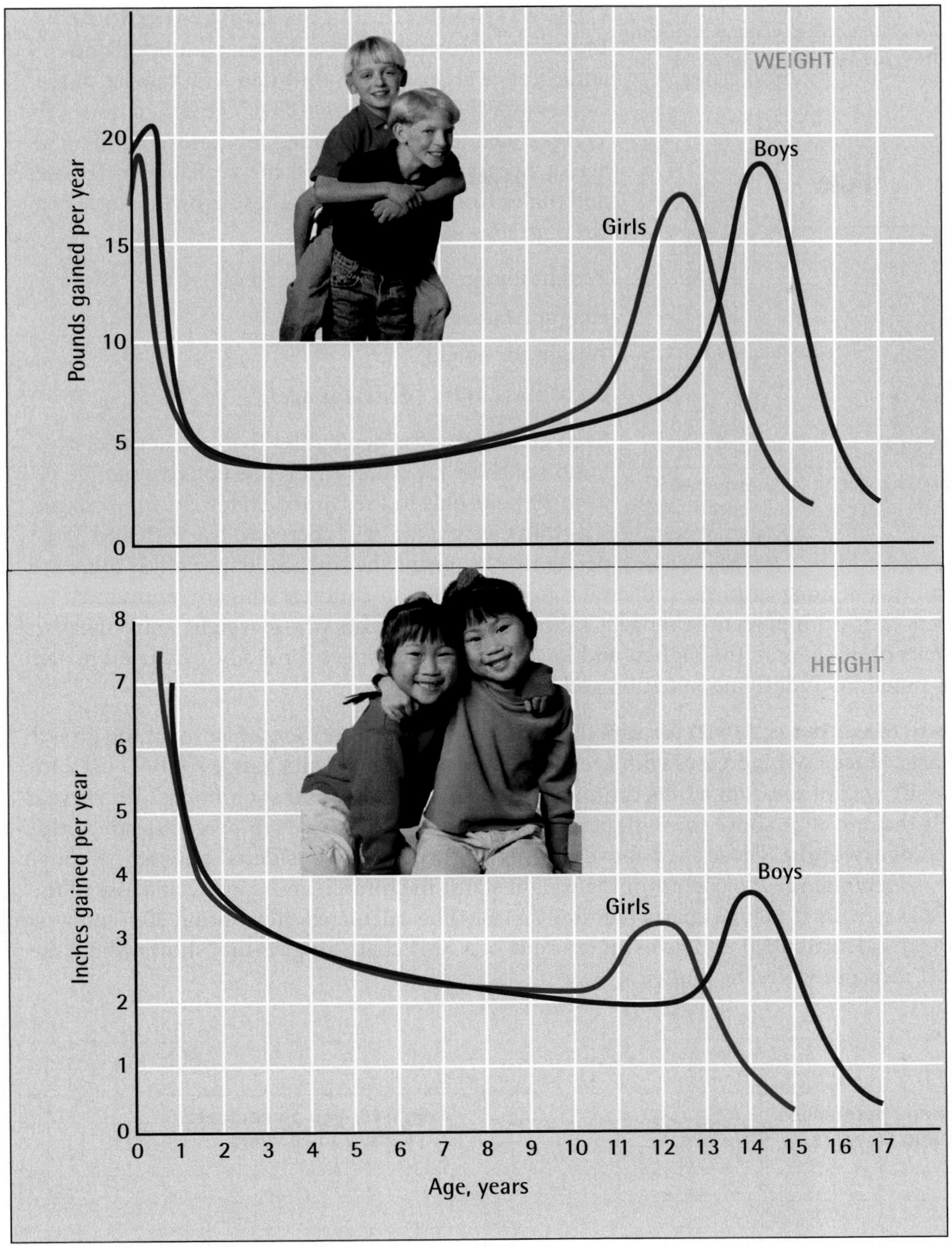
WEIGHT
Pounds gained per year
20
15
10
5
0
Girls
Boys
HEIGHT
Inches gained per year
8
7
6
5
4
3
2
1
0
Girls
Boys
0 1 2 3 4 5 6 7 8 9 10 11 12 13 14 15 16 17
Age, years

PhotoDisc; © Cengage Learning

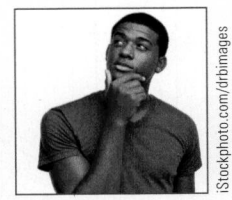

Illustration 30.2 Average yearly growth in weight and height during childhood and adolescence.

ANSWERS TO REALITY CHECK

The "Clean Your Plate" Club

Children should not be encouraged to eat more food than they want. They should be allowed to decide for themselves when they are full. Serving small portions, and then adding additional food if the child does not feel full can help reduce food waste.[30]

iStockphoto.com/drbimages

Aiden:

iStockphoto.com/drbimages

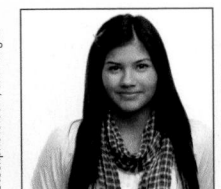

Zoe:

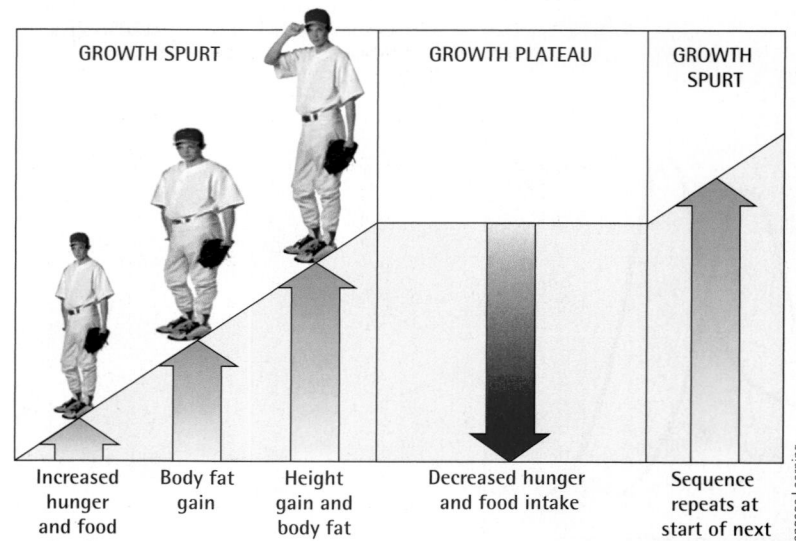

GROWTH SPURT | GROWTH PLATEAU | GROWTH SPURT

| Increased hunger and food intake | Body fat gain | Height gain and body fat loss | Decreased hunger and food intake | Sequence repeats at start of next growth spurt |

© Cengage Learning

Illustration 30.3 Growth occurs in a series of "spurts." Each spurt follows this sequence.

overweight In children and adolescents, overweight is defined as a BMI at or above the 85th percentile and lower than the 95th percentile for children of the same age and sex.

obese In children and adolescents, a BMI at or above the 95th percentile for children of the same age and sex.

CDC's Growth Charts for Children and Adolescents Growth progress during childhood and adolescence is generally monitored with the use of the Centers for Disease Control (CDC) growth charts. (The WHO charts are recommended for children under the age of 2 years.)[4] Growth charts are available for females and males from 2 to 20 years. These growth charts consist of graphs of:

Weight for age (see Illustration 30.4)

Height (stature) for age

Weight for height

Body mass index (BMI) for age

Each graph provides percentiles that reflect the distribution of these measures in a representative sample of 2- to 20-year-olds in the United States. So, for example, if a child's weight for age is between the 50th and 75th percentiles, it means that this child's weight is higher than that of most children and similar to that of 25% of children who are represented in the 50th to 75th percentile range. Children and adolescents whose weight status measurements place them in the highest and lowest percentile ranges should be evaluated further for potential underlying nutrition and health problems.[5]

Body Mass Index (BMI) for Age BMI charts by age and sex are a feature of the growth charts. Since BMI increases with age in children and adolescents, ranges of BMI used to classify weight status in adults cannot be used. BMIs for age and sex among 2- to 20-year-olds that are at or above the 85th percentile and lower than the 95th percentile are considered **overweight**. Those at or above the 95th percentile are considered **obese**.[6] Although now recommended for determining weight status in children and adolescents, use of the CDC's new BMI for age charts requires that BMI be calculated. Illustration 30.5 shows you how to calculate BMI and gives an example of a BMI chart for girls that shows where the BMI measure would be plotted.

Illustration 30.4 An example of the CDC's growth charts: Weight for age for boys and girls. The charts can be printed off the Internet using the address www.cdc.gov/growthcharts.

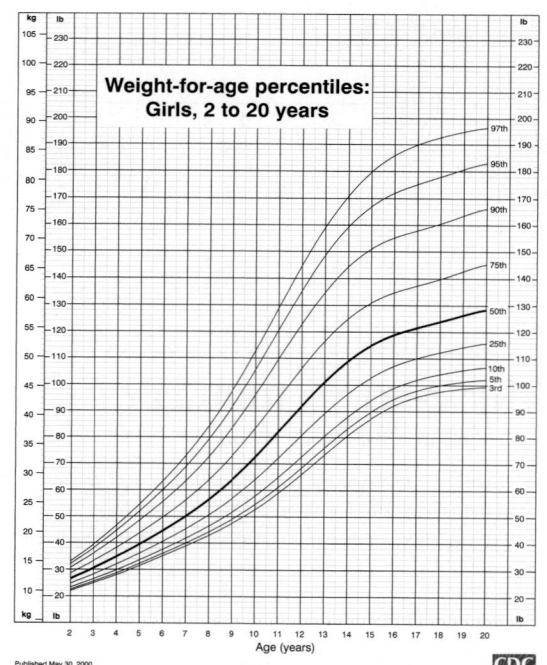

Weight-for-age percentiles: Boys, 2 to 20 years

Weight-for-age percentiles: Girls, 2 to 20 years

Published May 30, 2000.
SOURCE: Developed by the National Center for Health Statistics in collaboration with the National Center for Chronic Disease Prevention and Health Promotion (2000).

How to calculate BMI

Lucy just turned 7 years old and has had her weight and height carefully measured. She weighs 57 pounds (lb), and her height (stature) is 4 feet (ft), 2 inches (in.). What's her BMI?

1. Convert her height of 4 ft 2 in. to inches:

$$4 \text{ ft} \times 12 \text{ in. per foot} = 48 \text{ in.}$$
$$\underline{+2 \text{ in.}}$$
$$50 \text{ in.}$$

2. Square the inches:

$$50 \text{ in.} \times 50 \text{ in.} = 2500 \text{ in.}^2$$

3. Apply formula:

$$\text{BMI} = \frac{\text{kg}}{\text{m}^2} \quad \text{or} \quad \frac{\text{weight in. lbs} \times 700}{\text{height in in.}^2}$$

$$\text{BMI} = \frac{57 \text{ lbs} \times 700}{2500 \text{ in.}^2} = \frac{39{,}900}{2500 \text{ in.}^2} = 15.96 \text{ kg/m}^2$$

In this example, Lucy's BMI for age of 15.96 kg/m^2 falls between the 50th and 75th percentiles, or well within the normal range.

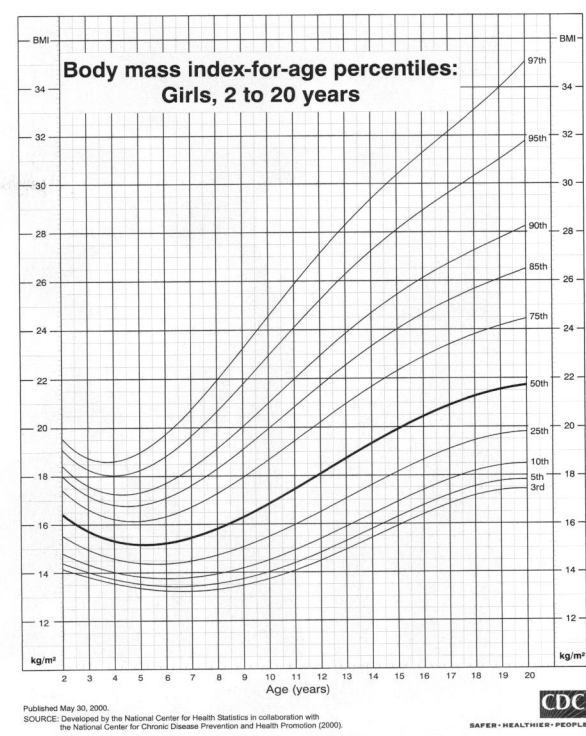

Body mass index-for-age percentiles: Girls, 2 to 20 years

Published May 30, 2000.
SOURCE: Developed by the National Center for Health Statistics in collaboration with the National Center for Chronic Disease Prevention and Health Promotion (2000).

CDC
SAFER · HEALTHIER · PEOPLE

http://www.cdc.gov/nchs/data/series/sr_11/sr11_246.pdf

Illustration 30.5 How to calculate BMI, and an example of plotting BMI on a CDC BMI-by-sex-and-age growth chart.

The Adolescent Growth Spurt The adolescent growth spurt usually occurs in girls between the ages of 9 and 12 years. For boys, this period of growth generally begins around the ages of 12 to 14. Actually, though, the age when adolescents start their growth spurt varies considerably. Pictured in Illustration 30.6 are three friends, all age 12. It's not hard to tell which one of them has experienced a growth spurt! The difference was also clear at the dinner table. By the time all three were 19 years old, Max (the one in the middle) had caught up with Ben (on the left), and David had grown taller but not as tall as Max or Ben.

During these years of growth, teenagers gain approximately 50% of their adult weight, 20–25% of adult height, and 45% of their total bone mass. In the year of peak growth, girls gain 18 pounds on average, and boys gain 20 pounds.[7]

Can You Predict or Influence Adult Height? Ultimate height is difficult to predict. On average, children tend to achieve adult heights that are between the heights of their biological parents.[8] There are many exceptions to this general finding, however, and that means that heredity is not the only influence on height. The dramatic increases in height of Japanese youth since World War II provide clear evidence that nongenetic factors have the strongest influence on height. Since the late 1940s, Japanese youth have grown an average of 2 inches taller each generation. (The increase in height has meant that everything from shoes to beds must be produced in larger sizes.) The increase in size of Japanese people is largely attributed to the availability of sufficient, nutritious foods.[9] Indeed, people in most economically developed countries continue to grow taller; a maximal genetically determined height has not yet been reached. If children are less well nourished than their parents, though, they tend to be shorter than their parents as adults.[10]

A healthy birthweight and diet during the growing years, and freedom from frequent bouts of illness, support growth in height.[11] There are no supplements, powders, or special diets that can be used to increase growth rate. Growth hormone injections can increase height somewhat, but the side effects of growth hormone are numerous and its use is limited.

Illustration 30.6   This photo was taken when these boys, born 2 months apart, were 12 years old. The photo (on the right) shows the same boys at age 19.

Nutrition for the Growing Years

- **Identify three positive and three negative characteristics of the average diet of children and adolescents.**

- **Identify three characteristics of diets of children and adolescents that increase disease risk later in life.**

Nutrient needs of adolescent children can be met by diets that regularly include grain products (including whole grains), dairy products, lean sources of protein, vegetables, and fruits. There is room in the diet for the occasional treat such as ice cream, cookies, or french fries. Illustration 30.7 provides an example of a ChooseMyPlate one-day diet appropriate for a 13-year-old girl who is moderately active 30–60 minutes a day.

Diets of children and adolescents rarely match the ChooseMyPlate recommendations. "Sometimes" foods such as chips, fries, candy, desserts, and soft drinks are often consumed as "regular" foods. Diets tend to be high in energy-dense foods rather than foods like vegetables, fruits, lean meats, and whole grain products that are nutrient dense. Regular intake of energy-dense foods is associated with the development of overweight and obesity in children and teens, and to the beginning of metabolic abnormalities that contribute to the risk of diseases such as type 2 diabetes, hypertension, and heart disease.[12,13]

The types of food most commonly consumed by children and teens affect nutrient intake and adequacy. The result of imbalanced diets on nutrient status is most pronounced during the adolescent years when the need for nutrients increases a good deal.

Nutrient Needs of Children and Adolescents

Recommended Dietary Allowances (RDAs) for most nutrients increase substantially as growth continues and body mass increases from childhood through adolescence. Some examples of the magnitude of these changes are shown in Table 30.1. This table compares the RDAs for children 4–8 years of age to those set for male and female adolescents ages 14–18.

The status of nutrient and total sugar intake in U.S. children and adolescents is shown in Tables 30.2 and 30.3. These tables compare nutrient and total sugar intake against recommended levels of intake. Those marked "adequate" are within approximately 10% of recommended amounts on average, and those in the "below recommended" and "above recommended" are below or above the recommended amounts by 10% or more. Average intake of fiber and vitamin D is low in both U.S children and adolescents. However, average nutrient intakes among children in the United States are more likely to meet recommended amounts than is the case for teenagers. Intakes of vitamin E, calcium, magnesium, iron, zinc, and potassium are below recommended amounts in teens but not in children. Both children and adolescents tend to consume too much sodium and total sugars.

Diet and Health Status of Children and Adolescents

Health status of children and teens, and health risks experienced during the adult years, are affected by dietary intake. Inadequate intake of calcium and vitamin D during the growing years is related to an increased risk of osteoporosis later in life. Inadequate fluoride intake from water and excess consumption of sugary foods places children and teens at

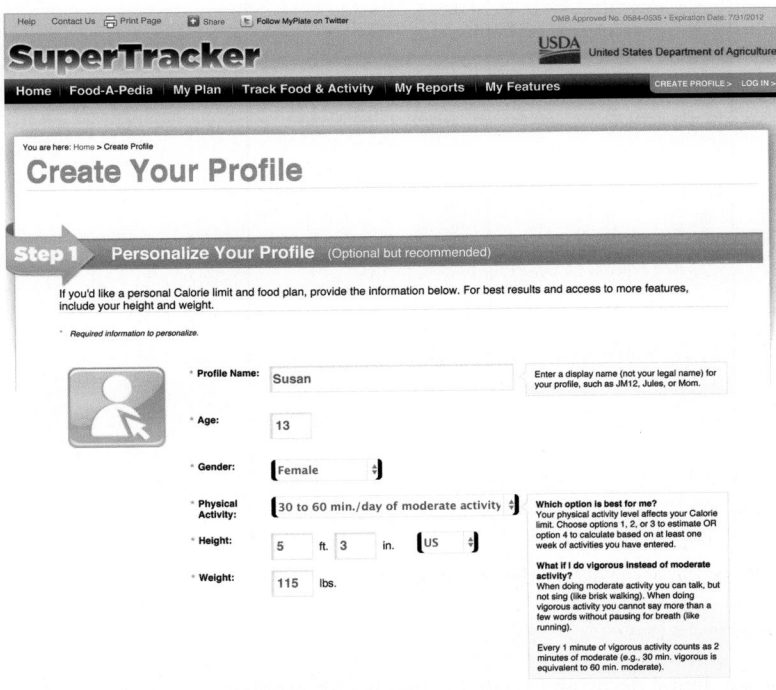

Illustration 30.7 ChooseMyPlate.gov one-day food plan for 13-year-old Susan who is physically active for 30–60 minutes a day, weighs 115 pounds, and is 5 foot, 3 inches tall.

Source: www.ChooseMyPlate.gov/SuperTracker/CreateProfile.aspx

Table 30.1 RDAs for selected nutrients for children (4–8 years) and adolescents (14–18 years)

| Nutrient | RDA | | |
| | Children | Adolescents | |
		Males	Females
Protein, g	19	52	46
Vitamin A, mcg	400	900	700
Vitamin B$_6$, mg	0.6	1.3	1.2
Folate, mcg	200	400	400
Vitamin C, mg	25	75	65
Vitamin D, mcg	15	15	15
Vitamin E, mg	7	15	15
Calcium, mg	1,000	1,300	1,300
Magnesium, mg	130	410	360
Iron, mg	10	11	15
Zinc, mg	5	11	9
Potassium, mg[a]	3,800	4,700	4,700
Sodium, mg[a]	1,200	1,500	1,500

[a]Adequate Intake (AI)

Table 30.2 Children: Average nutrient and total sugar intake compared to recommended levels of intake among children in the United States ages 2 to 12[2,18,19]

Food constituent	Near recommended	Below recommended	Above recommended
Protein			√
Fat	√		
Total sugar			√
Fiber		√	
Vitamin A			√
Vitamin B$_6$			√
Folate			√
Vitamin C			√
Vitamin D		√	
Vitamin E	√		
Calcium	√		
Magnesium			√
Iron			√
Zinc			√
Potassium		√	
Sodium			√

Data compiled by Judith Brown from national data sources.[2,18,19]

high risk of developing dental decay and related dental health problems.[14–17] Importantly, over-consumption of energy-dense and sugary foods and low levels of physical activity place children and adolescents at increased risk of becoming obese and developing disorders related to obesity later in life.[2,12]

Snacks are an important source of calories and nutrients in the diets of children and adolescents. Good choices for snack foods include the following:

Table 30.3 **Adolescents: Average nutrient and total sugar intake compared to recommended levels of intake among children in the United States aged 2 to 12[2,18,19]**

Food constituent	Near recommended	Below recommended	Above recommended
Protein			√
Fat	√		
Total sugar			√
Fiber		√	
Vitamin A			√
Vitamin B$_6$			√
Folate			√
Vitamin C	√		√
Vitamin D		√	
Vitamin E		√	
Calcium		√	
Magnesium		√	√
Iron		√	√
Zinc		√	√
Potassium		√	
Sodium			√

Data compiled by Judith Brown from national data sources.[2,18,19]

Yogurt	Bananas	Carrots
Cheese	Oranges	Cucumbers
Low-fat milk	Apples	Popcorn
Nuts, seeds	Dried fruit	Peanuts
Pears	Mangos	Cherry tomatoes
Melons	Grapes	Peanut butter

Obesity and Type 2 Diabetes in Children and Adolescents

Rates of obesity in U.S. children and adolescents have risen remarkably since 1960 (Illustration 30.8), and overweight- and obesity-related disorders have also increased. Conditions such as type 2 diabetes, bone and joint disorders, abnormal blood lipid levels, and elevated blood pressure that were very rarely observed in children and adolescents in the past are being diagnosed with increasing frequency.[1,20] Approximately 60% of overweight children have risk factors for cardiovascular disease.[21] The emergence of type 2 diabetes as a problem of the early years is of particular concern because diabetes generally worsens with time and causes long-term health impairments. It is estimated that 4% of children and adolescents have impaired glucose tolerance (a strong risk factor for type 2 diabetes), and that 6–17% of overweight and obese children and adolescents have type 2 diabetes.[22]

Causes of Overweight in Young People

Less than 10% of cases of overweight and obesity can be linked to a genetic or hormonal cause.[1] That means environmental factors are primarily responsible for the increased incidence of overweight and obesity. A number of "obesigenic" trends have developed over the past several decades that likely account for increased

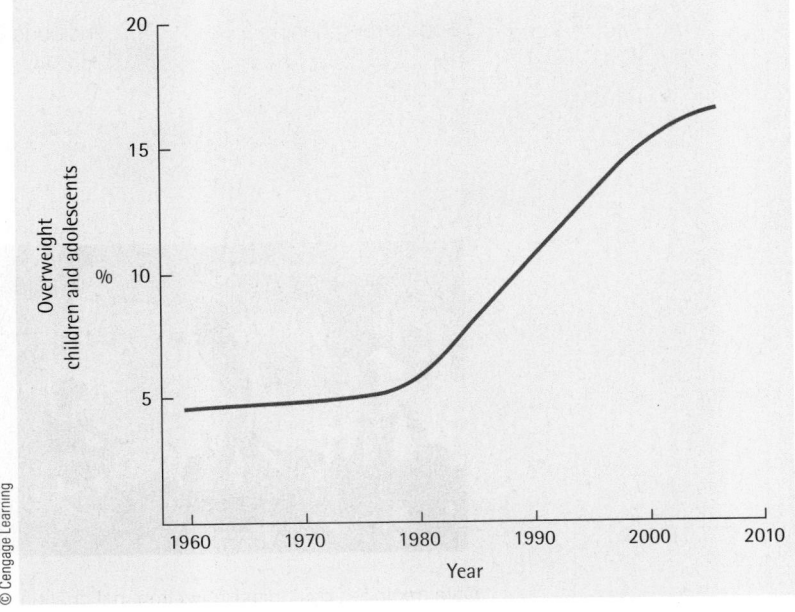

Illustration 30.8 Changes in the percent of obese children and adolescents in the United States.

Source: Developed by Judith Brown using data from Chartbook/Health, United States, 2008.

Illustration 30.9 An example of a part of the obesigentic environment. This fast-food beverage container holds over a quart of soft drink.

incidence (Illustration 30.9). Compared to 20 or so years ago, children and adolescents now have fewer opportunities for physical activity and are exposed to a generally plentiful supply of energy-dense and sugary foods.[23] Most schools offer too few physical activity classes and opportunities for exercise.[24] School vending machines in many districts are loaded with energy-dense, empty-calorie snack foods, and portion sizes of foods served to children in fast-food and other restaurants are often excessive.[25] Many communities lack bike paths or lanes, sidewalks, and safe playgrounds that invite physical activity.[26] Genetically based, inborn processes that regulate appetite and satiety are being overwhelmed by environmental conditions that influence food intake.

Prevention of Overweight If environmental factors play the predominant role in the obesity epidemic, then the path to prevention is paved with environmental changes. Children and adolescents need more opportunities to engage in physical activity. A wider array of healthy food options should be available at schools, and more nutritious and smaller portions in fast-food restaurants.[25] Children who regulate their food intake based on appetite and satiety rather than on environmental cues have a jump start on overweight prevention.

Physical Activity Guidelines for Children and Adolescents Regular exercise during childhood and adolescence helps prevent overweight and improves one's chances of having a healthy adulthood. How much is enough? Recommendations for healthy levels of physical activity for children and adolescents (Table 30.4) call for 60 minutes or more of moderate- to vigorous-intensity aerobic activity daily. The 60 minutes daily should include activities that build muscle, bone strength, and bone density.[27] Although exercises

Table 30.4 Physical activity guidelines for children and adolescents[27]

Activity	Duration	Examples
Moderate-to-vigorous aerobic exercise	60 minutes daily	Running, hopping, skipping, rope jumping, line dancing, cheerleading
Strength building exercise	Included in the 60 minutes daily • 3 days a week	Tree climbing, playing on playground equipment, hill climbing, weight lifting
Bone strengthening	Included in the 60 minutes daily • 3 days a week	Tennis, basketball, rope jumping, skiing, push-ups, soccer

Data compiled by Judith Brown from national data sources.[2,18,19]

that increase strength are recommended, power lifting and body building are not. These intense, high-impact exercises should not be undertaken before physical and skeletal maturity is reached.[28]

Helping Children Learn the Right Lessons about Food

Food preferences are a highly individual matter. It's hard to find two people who share the same food likes and dislikes, and many people are very picky about how food is prepared and served. How do food preferences form? Are they "shaped" by early learning experiences, or are they inborn and beyond anyone's control?

With the exception of genetically determined sensitivity to bitter-tasting foods and an inborn preference for sweet-tasting foods, there is no direct evidence that food preferences are inborn.[29] The belief that young children instinctively know what to eat is incorrect and can be hazardous to a child's health. Parents may take an overly relaxed attitude toward poor food habits and inadvertently contribute to the development of health problems later in life. Informed parents should decide what types of food their child should be offered, but the decision about how much to eat should be left to the child.[30]

Food likes and dislikes appear to be largely shaped by the environment in which children learn about food. Which foods are offered, the way they are offered, and how frequently particular foods are offered all influence whether a child will like a given food or not.[31]

Humans are born with a tendency to be cautious about accepting new things, including likes and dislikes appear to be largely shaped by the environment in which foods. They may need to get used to a new food before they trust it. Parents may have to offer a new food on five or more occasions before the child makes a decision.[31]

Sometimes children decide they like the food, and sometimes they really don't like it. Forcing a young child to eat foods she or he does not like, or totally restricting access to favorite foods, can have lifelong negative effects on food preferences and health. When attempts to get a child to eat a particular food turn the dinner table into a battleground for control, nobody wins. Foods should be offered in an objective, nonthreatening way so that the child has a fair chance to try the food and make a decision about it. Restricting access to, or prohibiting intake of children's favorite "junk" foods tends to strengthen their interest in the foods and consumption of those foods when they get a chance. Such prohibitions have the opposite effect of that intended because they make children want the foods even more.[32]

Healthy eating and activity levels during childhood and adolescence are clearly important to long-term health and quality of life. Improvements in both areas are needed. Families, schools, and communities—as well as children and adolescents themselves—can do many things to help improve eating and physical activity patterns.[33] The rewards of successful efforts would be for life.

Jose Luis Pelaez Inc/Getty Images/Blend Images

NUTRITION
up close

Obesity Prevention Close to Home

Focal Point: Tips for parents on preventing obesity in their children.

Debra and Dale have a 1-year-old daughter, and their second adopted infant will arrive in a few weeks. Both are acutely aware of the growing problem of obesity and type 2 diabetes in children and don't want their children to become obese.

Referring back to the information presented on page 30-9, list four actions Debra and Dale can take to help their children develop healthful food and activity habits:

1. _____

2. _____

3. _____

4. _____

Feedback to Nutrition Up Close is located in Appendix G.

REVIEW QUESTIONS

- **Describe connections between growth and development, and eating behavior in children and adolescents.**

1. Children generally become good "self-feeders" between the ages of four to five years.   **True/False**

2. As shown in Illustration 30.2, the rate of gain in height increases sharply between the ages of 10 to 12 years in girls, and between 13 to 14 years in boys.   **True/False**

3. Food likes and dislikes are strongly influenced by the environment in which children learn about food.   **True/False**

4. Young children instinctively know what to eat.   **True/False**

5. According to ChooseMyPlate food intake guidance, a 13-year-old girl should consume 3 cups of vegetables and 2.5 cups of fruits daily.   **True/False**

The next two questions refer to this scenario:

Deena and Britt are big football fans and want Greg, their 10-year-old, to grow up big and strong and play professional football. They figure that if they get him to eat enough he will grow tall and the extra calories will help put weight on his frame. At dinner, they encourage him to eat all the meat he can fit in his stomach.

6. _____ Will Greg grow taller if he eats a lot of meat?
   a. Maybe fatter, but overeating meat won't help him gain height.
   b. Yes, that's how adolescents grow taller and add muscle mass.
   c. If Greg eats more protein than he needs he will grow as tall as genetically possible.
   d. How tall Greg becomes depends on how many calories he consumes during the adolescent years only.

7. _____ What type of diet should Greg be offered to support his growth and physical development?
   a. A diet that will help him gain a pound week.
   b. A diet that includes the recommended assortment and amounts of foods recommended by ChooseMyPlate.gov food guidance materials for 10-year olds of his activity level.
   c. A diet that is adequate in calories.
   d. A diet that eliminates oils and other sources of fat.

- **Demonstrate the use of the Centers for Disease Control and Prevention (CDC) charts for growth assessment in children and adolescents.**

The next three questions refer to this scenario:

Ender just had his 10th birthday. He weighs 105 pounds (47.7 kg).

8. _____ According to the CDC weight-for-age chart (Illustration 30.4), Ender's weight-for-age would be located between the _____ and _____ percentiles.
   a. 95th–97th
   b. 50th–95th
   c. 50th–75th
   d. 25th–50th

9. _____ Ender's weight-for-age means that he may be:
   a. underweight for age
   b. normal weight for age
   c. overweight for age
   d. obese for age

10. _____ Ender is 4 feet tall. Using the formula given in Illustration 30.5, Ender's BMI is:
   a. 26.3 kg/m^2
   b. 21.6 kg/m^2
   c. 31.9 kg/m^2
   d. 35.3 kg/m^2

- **Identify three positive and three negative characteristics of the average diet of children and adolescents.**

11. Rates of overweight and obesity in children and adolescents have decreased in the United States since 2005.   **True/False**

12. Children and adolescents in the United States tend to consume too much sodium and sugar.   **True/False**

13. Diets of U.S. children and adolescents tend to provide adequate amounts of fiber and vitamin D.   **True/False**

14. Average protein intakes of U.S. children and adolescents are below recommended levels.   **True/False**

The next three questions refer to the following scenario and the follow-up scenario listed after question 15:

Your supervisor from work is having a cookout at his home in the country. You're in the kitchen with your supervisor and his two young children, who are having an early dinner. The two children are very active, running and playing most of the day, and appear to be thin. Your supervisor gives them their favorite dessert, which the children eat with relish, run out of the house, and go back to playing outside.

The parents make this dessert for their children often. Here is the recipe:

2 cups warm Rice Krispies marshmallow mix
4 fruity roll-ups
1 package gummy bears

The warm Rice Krispies mix is flattened out, lined with fruity rollups, stuffed with gummy bears, and rolled up.

15. _____ Is this dessert the type of food young children should eat often?

    a. Yes, if the children don't have a weight problem.
    b. Yes, because the Rice Krispies are probably fortified with vitamins.
    c. No, in part because sugary foods are related to dental decay in young children.
    d. No, because there aren't enough calories in the dessert to help the children gain weight.

Four months later you are chatting with your supervisor and he remarks that he's looking for a second job to pay for his children's dental care. He asks you if you know any way to help prevent cavities in children. The water supply where he lives is a well and the water contains no fluoride.

16. _____ What would *not* be a reasonable suggestion to make about preventing dental decay in his children?

    a. Cut down on the sugary foods in the children's diet.
    b. Investigate getting an appropriate fluoride supplement for the children.
    c. Stop offering the children dessert.
    d. Use bottled water fortified with fluoride for the children to drink.

17. _____ Which of the following options represent nutrient-dense and tasty dessert options for children?

    a. Caramel-covered apples
    b. Cupcakes from a natural foods market
    c. Vanilla cake with whipped cream
    d. Fruit and low-fat yogurt

- **Identify three characteristics of diets of children and adolescents that increase disease risk later in life.**

18. A number of chronic diseases such as heart disease, hypertension, and type 2 diabetes may begin to develop during childhood and adolescence. **True/False**

19. Inadequate intake of calcium and vitamin D during the growing years increases the risk of osteoporosis later in life. **True/False**

20. Overconsumption of energy-dense and sugary foods increases the risk of overweight and obesity during the adult years. **True/False**

Answers to these questions can be found in Appendix G.

NUTRITION SCOREBOARD ANSWERS

1. There really are "growth spurts." **True**

2. Most young children begin to equate the feeling of hunger with the need for food around age three. **False**

3. The incidence of overweight and obesity in U.S. children increased 50–60% between 1990 and 2010.[1] **False**

4. Childhood and adolescent diets can affect future health. Heart disease and diabetes, for example, can have "pediatric" origins that are related to dietary intake.[2] **False**

UNIT

31

Nutrition and Health Maintenance for Adults of All Ages

After completing Unit 31 and its interactive learning features, you will be able to:

- Identify three examples of interactions among diet, chronic disease development, and longevity.

- Explain why the need for calories and some nutrients changes with aging.

- Evaluate a one-day diet plan for a healthy adult based on ChooseMyPlate.gov food group intake guidance.

adult   Generally defined as people 18 to 65 years old.

older adult   Generally defined as individuals 65 or more years old.

life expectancy   The average length of life of people of a given age or from birth.

Nutrition for Adults of All Ages

- **Identify three examples of interactions among diet, chronic disease development, and longevity.**

Eating right during the **adult** and **older adult** years is a wise practice. Healthy diets, exercise, and normal weight status help adults feel healthy and vigorous as they age and improve overall sense of well-being. Adults who eat healthfully tend to develop heart disease, cancer, hypertension, and diabetes at older ages and may have more life in their years and years of life than adults who do not.[3,5]

Aging is a normal process, not a disease. Although the incidence of many diseases increases with age, the causes of the diseases can be unrelated to normal aging processes. They may fully or partially result from the cumulative effects of diets low in vegetables, fruits, dairy products, and whole grains; overeating and obesity, smoking, physical inactivity, excessive stress, or other habits that insidiously influence health on a day-to-day basis.[4,5] Aging cannot be prevented, but how healthy we are during aging can be influenced by what we do to our bodies.

Maintaining health as we age is becoming an increasingly important concern. There are more older adults in the United States than ever before, and the numbers are growing.[6]

Key Nutrition Concepts

Information about nutrition during the adult years presented in this unit most directly relates to the key nutrition concepts of:

- Foods provide energy (calories), nutrients, and other substances needed for growth and health

- Health problems related to nutrition originate within cells

- Poor nutrition can influence the development of certain chronic diseases

The Age Wave

The proportion of people in the United States aged 65 years and older is steadily increasing (Illustration 31.1). By the year 2050, approximately 20% of the U.S. population will be 65 years of age or older. Since 1900, **life expectancy** at birth has increased by over 30 years, from 47.3 to 78.7 years.[7]

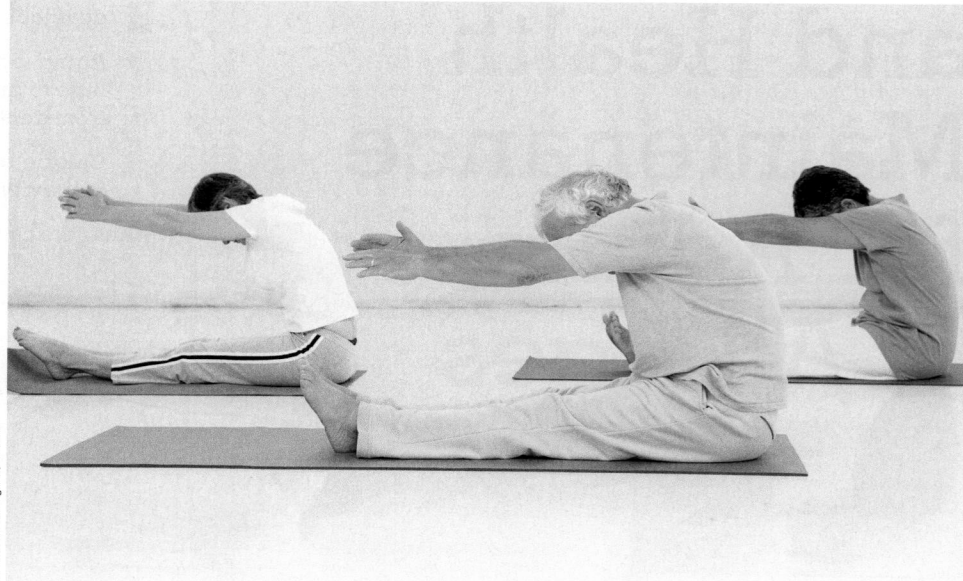

© Tomas Rodriguez/Corbis

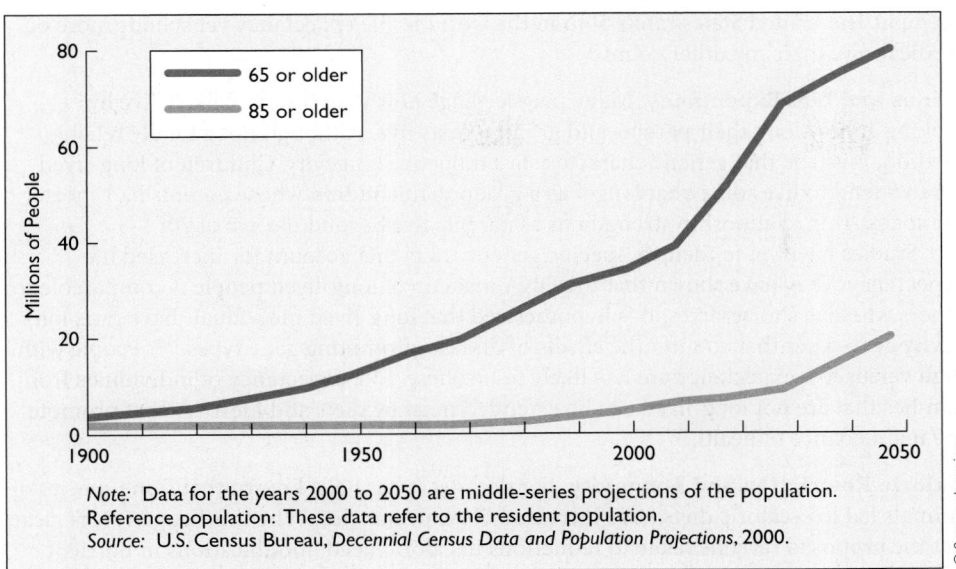

Illustration 31.1 Expected growth in the U.S. population of people aged 65 and 85 years and older.[38,39]

Note: Data for the years 2000 to 2050 are middle-series projections of the population.
Reference population: These data refer to the resident population.
Source: U.S. Census Bureau, *Decennial Census Data and Population Projections*, 2000.

Not all population groups in the United States have the same average life expectancy from birth. Life expectancy varies a good deal by sex and race. Table 31.1 shows overall life expectancy for Caucasian, African, and Hispanic American men and women over time. African American men have the shortest life expectancy, and Hispanic women have the longest. As in most other countries, life expectancy of females exceeds that of males.[8] Reasons for the increased average life expectancy in women compared to men is not yet fully understood. It appears that behaviors in men that accelerate disease progression (especially heart disease) are partly related to the 4- to 5-year shorter average life span in men. Female hormones do not appear to account for differences in life expectancy.[8,9]

Factors That Influence Life Expectancy Before World War II, most of the gains in life expectancy in the United States were due to public health advances like improved diet and nutrition, safe water and milk supplies, sanitation measures, better control of infectious diseases, and improved housing. After World War II, new drugs, and the development of effective medical devices and procedures accounted for most of the gains in life expectancy.[6] In 1900, 1 out of 10 newborns died in the first year of life but now less than 1 newborn in 100 does. Reductions in infant mortality increases a population's life expectancy.[10,11] Advances in medicine currently account for 80% of the increase in life expectancy of people aged 65 years and older.[6]

Due to high and rising costs of medical care, innovations in health and medical care that promote health and quality of life at lower cost than is the current case are urgently

Table 31.1 Average length of life for males and females by population group in the United States in 1900, 1960, and 2010[7,37]

	Life Expectancy (Years)		
	1900	1960	2010
Caucasian Americans			
Males	46.6	67.4	76.5
Females	48.7	74.1	81.3
African Americans			
Males	32.5	61.1	71.8
Females	33.5	63.3	78.0
Hispanic Americans			
Males	—	—	78.8
Females	—	—	83.8

needed.[6] The United States ranks 50th in the world in life expectancy yet spends more on medical care than any other country.[3,2]

Genes and Life Expectancy Many people gauge how long they will likely live by looking at how long their parents and grandparents live. Although not a totally reliable method, it is true that genetic characteristics influence longevity. Children of long-lived parents tend to live somewhat longer as a group than children whose parents had shorter life spans. This relationship strengthens as parents live beyond the age of 70.[13]

Studies have yet to identify specific genetic traits that account for increased life expectancy. They have shown that the DNA makeup of long-lived people is comparable to others who live shorter lives. It is hypothesized that long-lived individuals have rare, longevity gene types that counter the effects of disease-promoting gene types.[14,15] People with high versus low expectancy are less likely to be obese. Life expectancy of individuals from families that are not long-lived can be extended most by diets and lifestyles that promote the maintenance of health.[15]

Calorie Restriction and Longevity For decades it has been known that some species of animals fed low-calorie diets have increased life expectancies. Why this happens isn't clear, but the proposed reasons relate to reductions in calorie need, modifications in nutrient utilization, reduced metabolic rate, and decelerated aging processes. Human studies have not shown that calorie restriction increases longevity.[1,2]

Individuals who follow low-calorie diets during the adult years report being hungry most of the time, experience decreased libido, and feel cold. People who tend to live the longest have normal weights. Life expectancy decreases as body weight decreases below, or increases above normal.[16] (For a quick review of this topic, check out the Reality Check.)

Diet and Health Promotion for Adults of All Ages

- **Explain why the need for calories and some nutrients changes with aging.**

- **Evaluate a one-day diet plan for a healthy adult based on ChooseMyPlate.gov food group intake guidance.**

Researchers and others interested in nutrition and health during middle age and beyond have tended to focus their attention on the cumulative effects of diet on chronic disease. Nutrition exerts its effects on chronic disease development over time, and therefore diseases related to poor diets are most likely to express themselves during the adult years. The occurrence of diseases related to behavioral traits such as smoking and physical inactivity also increases among adults as they age.[17] In addition to the health effects of behaviors, people age biologically.[18]

REALITY CHECK
Do People Live Longer If They Are Underweight?

While golfing, old friends Evelyn and Dorothy discuss the news story about underweight and increased longevity in laboratory animals they had read about in the morning paper. As a result of the news story, Evelyn became convinced that if she went on a low-calorie diet, lost 30 pounds, and became and stayed underweight she would live longer. Dorothy wanted to live long, too, but didn't think research on laboratory animals was enough to convince her to go on a life-long diet.

Who gets the thumbs up?

Answers appear on page 31-5.

Evelyn: It's worth a try. It's probably true if it happens in animals.

Dorothy: Hmmm . . . I'm thinking results of animal studies are not enough to convince me it's true for humans.

Table 31.2 Examples of biological changes during aging and nutritional consequences[40–43]

Biological change	Nutritional consequences
• Lowered stomach acidity	• Decreased absorption of vitamin B_{12}
• Decreased lean muscle mass	• Reduced calorie need
• Reduced production of vitamin D in the skin	• Increased need for vitamin D
• Decreased sensation of thirst	• Dehydration risk

The combined effects of poor diets, other risky lifestyle behaviors, and biological aging increase the rates of serious illness during adulthood. How soon a disease develops largely depends on the intensity and duration of exposure to behavioral risks that contribute to disease development.

Nutrient Needs

Biological processes and lifestyle changes that generally accompany aging affect calorie and nutrient needs to some extent (Table 31.2). A person's need for calories generally declines with age as physical activity, muscle mass, and basal metabolic rate decrease. However, people who remain physically active into their older years maintain muscle mass, experience less muscle and bone pain, and gain less body fat than people who are inactive.[19] Reduction in calorie need with age increases the importance of nutrient-dense diets.

Nutrient needs of older adults can be affected by physiological changes that may develop with increasing age (Table 31.3). Decreased stomach acidity, for example, reduces absorption of vitamin B_{12} and the production of vitamin D in the skin upon exposure to sunlight is less than at younger ages. In various studies, 10–37% of older adults have been found to be deficient in vitamin D.[20]

Dietary Recommendations for Adults of All Ages

For the most part, the development of chronic disease in middle-aged and older adults can be viewed as a chain that represents the accumulation over time of problems that impair the functions of cells. Each link that is added to the chain, or each additional insult to cellular function, increases the risk that a chronic disease will develop. The presence of a disease indicates that the chain has gotten too long—that the accumulation of problems is sufficient to interfere noticeably with the normal functions of cells and tissues.[9,21]

Normal cell functions and health promotion are facilitated by healthful dietary and other behaviors. For example:

- Correcting obesity and stabilizing weight during the adult years tends to lengthen life expectancy.[3]

- Dietary intakes that correspond to the Dietary Guidelines for Americans are related to longer life expectancy.[5,9]

ANSWERS TO **REALITY** CHECK
Do People Live Longer If They Are Underweight?

Dorothy's thinking is on the right track. Research results in laboratory animals may not apply to humans. In addition, evidence shows that humans who are normal weight tend to live the longest.[12]

Purestock/Jupiterimages

iStockphoto.com/Jacob Wackerhausen

Evelyn:

Dorothy:

Table 31.3 RDAs for selected nutrients for adults 19–50 and over 70 years old

| Nutrient | RDA | | | |
| | Adults 19–50 | | Adults > 70 | |
	Male	Female	Male	Female
Protein, g	56	46	56	46
Vitamin A, mcg	900	700	900	700
Vitamin B$_6$, mg	1.3	1.3	1.7	1.5
Vitamin B$_{12}$, mcg	2.4	2.4	2.4	2.4
Folate, mcg	400	400	400	400
Vitamin C, mg	90	75	90	75
Vitamin D, mcg	15	15	20	20
Vitamin E, mg	15	15	15	15
Calcium, mg	1,000	1,000	1,200	1,200
Iron, mg	8	18	8	8
Zinc, mg	11	8	11	8
Potassium, mg[a]	4,700	4,700	4,700	4,700
Sodium, mg[a]	1,500	1,500	1,200	1,200

[a]Adequate Intake (AI)

- Maintaining adequate calcium, vitamin D, and protein intake; and engaging in regular physical activity during the adult years may prevent or postpone the development of osteoporosis, and help maintain muscle mass and strength.[22-25]

- Above average intake of vegetables, fruits, and whole grain foods may delay the development or help prevent a number of types of cancer, heart disease, hypertension, and cataracts.[5,26,27]

The health status of adults is not necessarily "fixed" by age; it can change for the better or the worse, or not much at all.[28]

ChooseMyPlate.gov Food Intake Guidance Diets that consist of food intake recommended by ChooseMyPlate are nutrient dense and low in saturated fat. They provide healthy amounts of calories, protein, vitamins, minerals, fiber, and beneficial phytonutrients. Illustration 31.2 shows an example of a one-day diet plan generated from the ChooseMyPlate.gov site for a 56-year-old male. Diet plans generated are based on age, sex, weight, height, and usual physical activity level.

Foods identified by the ChooseMyPlate plan can be used to guide food selection. Many different types of foods in the grain, dairy, vegetable, fruits, and protein groups can be selected to meet individual food preferences. SuperTracker, the ChooseMyPlate interactive feature that generates food guides, will develop dietary intake plans that help adults gradually lose weight if requested.

The Need for Water in Older Adults For reasons that aren't entirely clear, many older people don't get thirsty when their bodies are running low on water. The implication of this change is that older people may become dehydrated and need medical assistance. Approximately 1 million elderly men and women are admitted to hospitals each year due to dehydration.[29] The need for water need can be met through consumption of tap water, juices, teas, coffee, and other beverages. Women of all ages need about 11 cups of water a day from fluids and foods, and men need 15 cups.[30]

Does Taste Change with Age? Poor diets observed in some middle-aged and older adults have been ascribed to "declining taste" with age. Besides being a dreadful thought, it's not true that taste declines with age to the extent that it makes eating less pleasurable than before. Although taste sensitivity does diminish somewhat with age, chances are your favorite foods will taste as good to you in your older years as they did in your youth

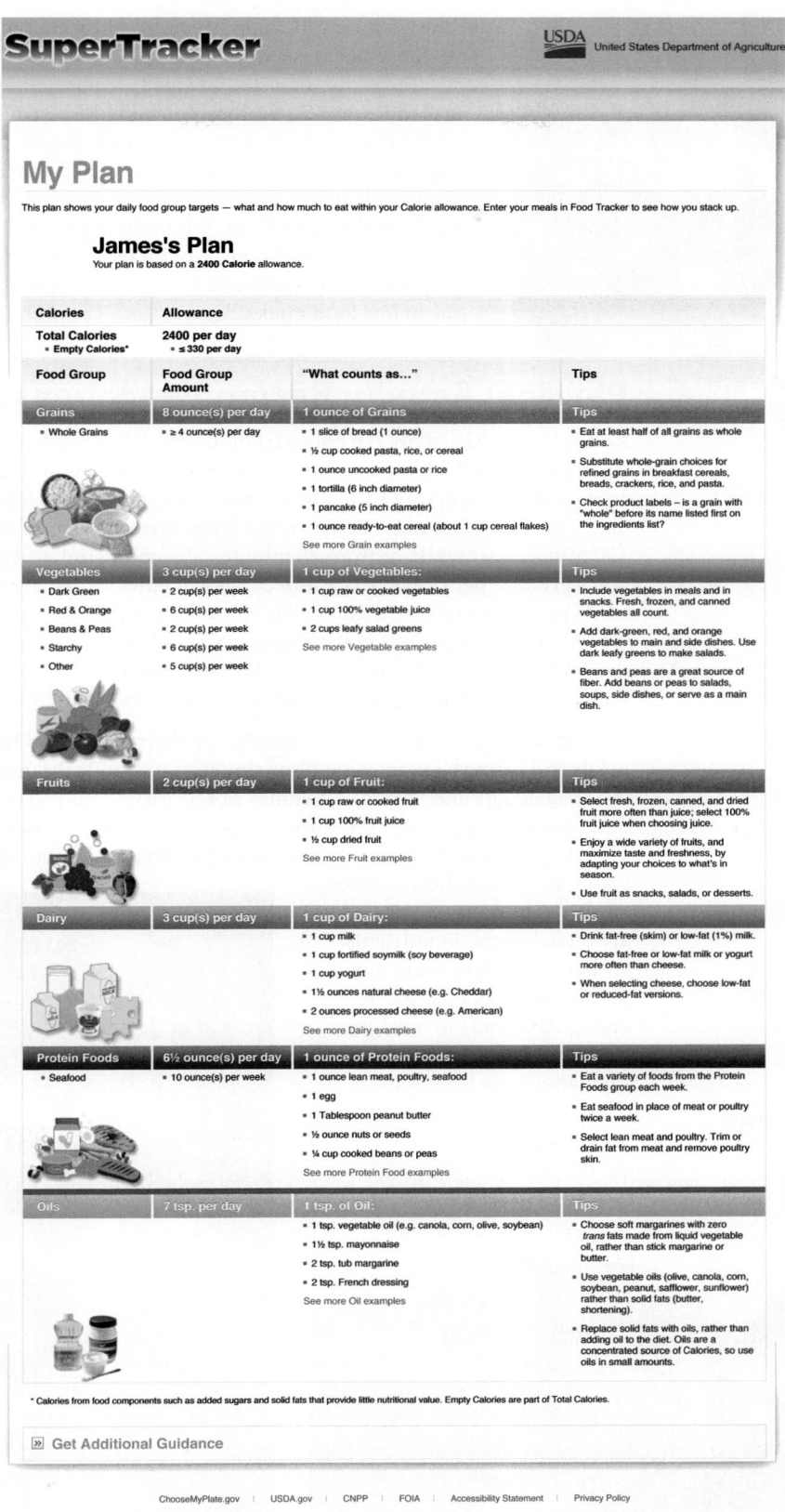

SuperTracker

USDA United States Department of Agriculture

My Plan

This plan shows your daily food group targets — what and how much to eat within your Calorie allowance. Enter your meals in Food Tracker to see how you stack up.

James's Plan
Your plan is based on a **2400 Calorie** allowance.

Calories	Allowance
Total Calories	**2400 per day**
• Empty Calories*	• ≤ 330 per day

Food Group	Food Group Amount	"What counts as…"	Tips
Grains	**8 ounce(s) per day**	**1 ounce of Grains**	**Tips**
• Whole Grains	• ≥ 4 ounce(s) per day	• 1 slice of bread (1 ounce) • ½ cup cooked pasta, rice, or cereal • 1 ounce uncooked pasta or rice • 1 tortilla (6 inch diameter) • 1 pancake (5 inch diameter) • 1 ounce ready-to-eat cereal (about 1 cup cereal flakes) See more Grain examples	• Eat at least half of all grains as whole grains. • Substitute whole-grain choices for refined grains in breakfast cereals, breads, crackers, rice, and pasta. • Check product labels – is a grain with "whole" before its name listed first on the ingredients list?
Vegetables	**3 cup(s) per day**	**1 cup of Vegetables:**	**Tips**
• Dark Green • Red & Orange • Beans & Peas • Starchy • Other	• 2 cup(s) per week • 6 cup(s) per week • 2 cup(s) per week • 6 cup(s) per week • 5 cup(s) per week	• 1 cup raw or cooked vegetables • 1 cup 100% vegetable juice • 2 cups leafy salad greens See more Vegetable examples	• Include vegetables in meals and in snacks. Fresh, frozen, and canned vegetables all count. • Add dark-green, red, and orange vegetables to main and side dishes. Use dark leafy greens to make salads. • Beans and peas are a great source of fiber. Add beans or peas to salads, soups, side dishes, or serve as a main dish.
Fruits	**2 cup(s) per day**	**1 cup of Fruit:**	**Tips**
		• 1 cup raw or cooked fruit • 1 cup 100% fruit juice • ½ cup dried fruit See more Fruit examples	• Select fresh, frozen, canned, and dried fruit more often than juice; select 100% fruit juice when choosing juice. • Enjoy a wide variety of fruits, and maximize taste and freshness, by adapting your choices to what's in season. • Use fruit as snacks, salads, or desserts.
Dairy	**3 cup(s) per day**	**1 cup of Dairy:**	**Tips**
		• 1 cup milk • 1 cup fortified soymilk (soy beverage) • 1 cup yogurt • 1½ ounces natural cheese (e.g. Cheddar) • 2 ounces processed cheese (e.g. American) See more Dairy examples	• Drink fat-free (skim) or low-fat (1%) milk. • Choose fat-free or low-fat milk or yogurt more often than cheese. • When selecting cheese, choose low-fat or reduced-fat versions.
Protein Foods	**6½ ounce(s) per day**	**1 ounce of Protein Foods:**	**Tips**
• Seafood	• 10 ounce(s) per week	• 1 ounce lean meat, poultry, seafood • 1 egg • 1 Tablespoon peanut butter • ½ ounce nuts or seeds • ¼ cup cooked beans or peas See more Protein Food examples	• Eat a variety of foods from the Protein Foods group each week. • Eat seafood in place of meat or poultry twice a week. • Select lean meat and poultry. Trim or drain fat from meat and remove poultry skin.
Oils	**7 tsp. per day**	**1 tsp. of Oil:**	**Tips**
		• 1 tsp. vegetable oil (e.g. canola, corn, olive, soybean) • 1½ tsp. mayonnaise • 2 tsp. tub margarine • 2 tsp. French dressing See more Oil examples	• Choose soft margarines with zero *trans* fats made from liquid vegetable oil, rather than stick margarine or butter. • Use vegetable oils (olive, canola, corn, soybean, peanut, safflower, sunflower) rather than solid fats (butter, shortening). • Replace solid fats with oils, rather than adding oil to the diet. Oils are a concentrated source of Calories, so use oils in small amounts.

* Calories from food components such as added sugars and solid fats that provide little nutritional value. Empty Calories are part of Total Calories.

» Get Additional Guidance

ChooseMyPlate.gov | USDA.gov | CNPP | FOIA | Accessibility Statement | Privacy Policy

Non-Discrimination Statement | Information Quality | USA.gov | White House | COPPA

Illustration 31.2 A one-day diet plan for James, who is 56 years old, weighs 160 pounds (normal weight), is 5 foot, 10 inches tall, and is physically active 30–60 minutes a day.

Source: www.ChooseMyPlate.gov/SuperTracker/CreateProfile.aspx

Illustration 31.3 Taste is less affected by age than are some other senses.

(Illustration 31.3). Sight, smell, and hearing senses usually decline with age more than taste does.[31]

The senses of taste and smell are affected by medications such as antibiotics, antihistamines, some lipid-lowering drugs, and cancer treatments; diseases including Alzheimer's disease, cancer, and allergies; and surgeries that affect parts of the brain and nasal passages. These treatments and conditions are more prevalent in older adults than in middle-aged populations and are primarily responsible for declines in the senses of taste and smell. Changes in these senses increase the likelihood that the person will experience food poisoning or consume an inadequate diet due to low food intake.[32]

Physical Activity Recommendations for Adults

Regular physical activity is a core part of healthy lifestyles that maintain health and delay aging processes.[19,33] It increases oxygen delivery to organs and tissues, decreases biologic aging, improves cognitive functions, raises HDL-cholesterol levels, and helps lowers body fat.[33,34] Substantial health benefits result from performing moderate- intensity exercise for 150 minutes per week, or from 75 minutes of vigorous-intensity exercise weekly. Resistance exercise performed twice a week further enhance the benefits of regular exercise while increasing strength, muscle mass, and balance.[19,35]

All adults, regardless of their age or physical condition, benefit from physical activity.[19] With proper training and spirit, physical fitness can be maintained during aging, or increased to the level of the "over 80" event winners pictured in Illustration 31.4.

Psychological and Social Aspects of Nutrition for Older Adults Preparing meals and eating right may not be as simple as it sounds for many older adults. Consuming an adequate and balanced diet may not be easy when you depend on someone else to take you shopping, when mealtimes involve little social life, or when you "don't feel up" to making a meal. Isolation, loneliness, depression, and poor health can be major contributors to poor diets in older adults.[36] The diets of older people are often lacking in nutrients, and because many older adults do not, or cannot, consume enough nutrients to meet their increased need for them, supplementation may be required.

Illustration 31.4 Athletes who are over 80.

PhotoDisc

NUTRITION
up close

Does He Who Laughs, Last?

Focal Point: Critically thinking about factors that influence longevity.

The following is an adaptation of a letter printed in the "Dear Abby" newspaper column:

> Dear Abby,
>   Since I've reached my 80s, my mail is full of ads for health products to help me live longer.
>   I once had many friends, all of whom were health vigilantes. They shook their heads knowingly as I avoided all health food fads and exercise. They made liquid out of good vegetables and spent fortunes buying all the latest supplements. They argued that "organic" was better and "natural" was best. I would tell them that snake venom, poison ivy and manure were "natural." But they wouldn't listen and they didn't laugh.
>   Now my friends are all dead and I have no one left to argue with.

© Cengage Learning

Critically think about the contents of this letter and identify three alternate explanations for why this individual outlived his friends.

Feedback to the Nutrition Up Close is located in Appendix G.

REVIEW QUESTIONS

- **Identify three examples of interactions among diet, chronic disease development, and longevity.**

1. One of the main reasons the United States ranks first among all countries in life expectancy is the high-quality diet consumed by adults in general.   **True/False**

2. Three characteristics of dietary intake related to increased longevity are ample consumption of fruits and vegetables, above-average consumption of whole grain products, and calorie restriction.   **True/False**

3. Obesity is associated with decreased life expectancy.   **True/False**

4. Adequate calcium, vitamin D, and protein intake during the adult years may prevent or postpone the development of osteoporosis and help build muscle mass and strength.   **True/False**

- **Explain why the need for calories and some nutrients change with aging.**

5. Elderly persons tend to have a lowered sensation of thirst.   **True/False**

6. The RDAs for older adults are the same as those for younger adults.   **True/False**

7. Reduction in calorie need with age increases the importance of nutrient-dense diets.   **True/False**

8. Decreased stomach acidity in older adults reduces absorption of vitamin B_{12}.   **True/False**

- **Evaluate a one-day diet plan for a healthy adult based on ChooseMyPlate.gov food group intake guidance.**

The next three questions refer to this scenario:

Assume your favorite uncle is visiting and explains to you that he is trying to eat a healthy diet now that he has turned 50. This is his new plan for his daily diet:

Breakfast: grits, eggs, English muffin with margarine, coffee
Lunch: Deli pastrami or corn beef sandwich on rye bread, coleslaw, soft drink
Dinner: Frozen dinner that includes meat, a vegetable, and potatoes; and coffee
Bedtime snack: cookies or peanuts

9. _____ How many ChooseMyPlate food groups are represented in your uncle's diet plan?

a. 1
b. 2
c. 3
d. 4

10. _____ Which ChooseMyPlate food groups are missing from his plan?

a. grains and fruit
b. protein and dairy
c. vegetables and fish
d. fruit and dairy

11. _____ Your uncle asks you if you have any suggestions for changes to his diet plan that would make it healthier. What would be the most appropriate changes to suggest:

a. substitute a blueberry muffin for the English muffin and leave off the margarine
b. have a whole grain cereal with low-fat milk and orange juice for breakfast
c. switch the cookies at bedtime to ice cream and top the ice cream with sliced fruit
d. Eat egg whites instead of whole eggs and replace the peanuts with a lower-fat snack like popcorn

Answers to these questions can be found in Appendix G.

NUTRITION SCOREBOARD ANSWERS

1. Healthy eating pays off. **True**

2. Aging is not a disease process! It's normal, and you can normally promote healthy aging through diet and lifestyles. **False**

3. Calorie restriction has not been shown to increase longevity in humans.[1,2] **False**

Daniel Pepper/Getty Images

The Multiple Dimensions of Food Safety

NUTRITION SCOREBOARD

1 Most cases of foodborne illness outbreaks can be traced back to eggs.   **True/False**

2 Freezing kills bacteria in food.   **True/False**

3 Alaska has the highest incidence of foodborne botulism in the United States.   **True/False**

4 You can tell if food has gone bad by smelling it.   **True/False**

Answers can be found at the end of the unit.

After completing Unit 32 and its interactive learning features, you will be able to:

- List the top causes of foodborne illness and the transmission and symptoms of each.

- Delineate five consumer practices that help prevent foodborne illness.

foodborne illnesses An illness related to consumption of foods or beverages containing disease-causing bacteria, viruses, parasites, toxic chemicals, or other harmful substances.

Table 32.1 Top 10 foods related to foodborne illness outbreaks[4]

1. Leafy green vegetables
2. Eggs
3. Tuna
4. Oysters
5. Potatoes
6. Cheese
7. Ice cream
8. Tomatoes
9. Sprouts
10. Berries

Threats to the Safety of the Food Supply

- **List the top causes of foodborne illness and the transmission and symptoms of each.**

Hamburgers contaminated with *E. coli*, cantaloupe tainted with Listeria, jalapeno peppers laced with *Salmonella*—it's enough to give you food fright. Actually, Americans have good reason to be concerned about the safety of some foods (Illustration 32.1). The Centers for Disease Control and Prevention (CDC) estimates that each year in the United States **foodborne illnesses** cause:

- Sickness in 48 million people

- 128,000 hospitalizations

- More than 3,000 deaths[3]

Foodborne illnesses can be caused by bacteria, viruses, algae, fungi, and toxins; as well as by chemical contaminants in foods or water. They are spread by a wide assortment of foods from contaminated eggs to hummus. The top 10 foods most commonly associated with foodborne illness outbreaks from all causes since 1990 are shown in Table 32.1 and begins with leafy green vegetables.

Key Nutrition Concepts

1. Health problems related to nutrition originate within cells.

2. Malnutrition can result from poor diets and from disease states, genetic factors, or combinations of these factors.

How Good Foods Go Bad

Bacteria and viruses are the most common causes of foodborne illnesses and largely enter the food supply during food processing, storage, or preparation. They are transferred to humans in foods through many different routes, a major one being the contamination of food with animal feces.

The lower intestines of many healthy farm animals are colonized by bacteria that may be harmful to humans. These bacteria contaminate food when unsanitary practices are used to prepare and process meats, and when vegetables and fruits are fertilized with animal manure. They can also be transferred to foods by humans who have colonies of harmful bacteria in their gastrointestinal tracts through the use of human sewage on crops, and by food handling by people carrying the bacteria on their hands (Illustration 32.2). Humans also transfer certain harmful microorganisms to food when fluids from

Illustration 32.1 The top three foodborne illness outbreaks in the United States in 2010 and 2011.[34]

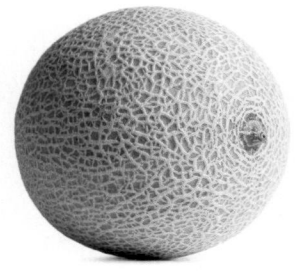
Cantaloupe *(Listeria)*

Ground turkey *(Salmonella)*

Eggs [in shell] *(Salmonella)*

Illustration 32.2 Can you find three potential opportunities for the spread of foodborne illness in this photograph?

© Dennis MacDonald/PhotoEdit

infected injuries or body secretions contact food.[5] These types of foodborne illnesses are mainly caused by bacteria that are "on" rather than "in" foods.

Although less common, bacteria can be present on the inside of foods. They can enter vegetables and fruits, for example, if the protective skin coatings are broken, allowing an entrance for bacteria. *Salmonella* bacteria can infect the ovaries of hens and cause them to lay normal-looking but infected eggs. Shellfish can concentrate microorganisms present in surrounding water, and although the bacteria may be harmless to the shellfish, they can provide large doses of harmful bacteria and viruses to humans.[6]

Cross-Contamination of Foods The source of many cases of foodborne illness is food that has come into contact with contaminated food. This situation, referred to as **cross-contamination**, increases the reach of foodborne illnesses.

Microorganisms, including bacteria, viruses, toxins, and other harmful substances, can contaminate safe foods during production, processing, shipping, preparation, or storage. Opportunities for cross-contamination of foods during processing, for example, are plentiful. The hamburger you eat may contain meat from hundreds of different cows, an omelet in a restaurant may contain the eggs of hundreds of different chickens, and the chicken you baked may have bathed with hundreds of others when they were all washed in the same vat of water at the meat-processing plant. Cross-contamination also occurs on cutting boards across America. The failure to routinely wash cutting boards in between the preparation of different raw foods is a major route to the spread of foodborne illness.[7]

cross-contamination The spread of bacteria, viruses, or other harmful substances from one surface to another.

Causes and Consequences of Foodborne Illness

More than 250 types of foodborne illnesses caused by infectious agents (bacteria, viruses, and parasites) and noninfectious agents (toxins and chemical contaminants) have been identified. Their impact on health ranges from a day or two of nausea and diarrhea to death within minutes. Effects of foodborne illnesses are generally most severe in people with weakened immune systems or certain chronic illnesses, pregnant women, young children, and older persons (Table 32.2). Symptoms of foodborne illness most commonly consist of nausea, vomiting, abdominal cramps, and diarrhea.[5] Many cases go unreported, making it difficult to identify the incidence and causes of foodborne illness. Estimates of

Table 32.2 **High-risk groups for severe effects of foodborne illness**[10]

- People with weakened immune systems due to HIV/AIDS and others with weakened defenses against infections
- People with certain chronic illnesses such as diabetes and cancer
- Pregnant women
- Young children
- Older persons

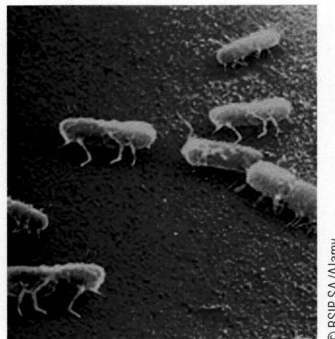

© BSIP SA/Alamy

Illustration 32.3 A magnified view of *Salmonella* bacteria.

the causes are usually based on cases reported by health care professionals to local health departments.[8] Table 32.3 summarizes facts about four types of foodborne illnesses that result from bacteria.

Salmonella It is estimated that over 1,027,000 cases of *Salmonella* infection occur in the United States each year.[9] A photograph of *Salmonella* bacteria is presented in Illustration 32.3. An outbreak of *Salmonella* infection that affected approximately 224,000 people was linked to the transport of an ice cream mixture in a tanker that had previously transported *Salmonella*-infected eggs. The infected ice cream was distributed in 41 states.

Campylobacter This bacterium is a common contaminant on chicken and causes an estimated 845,025 cases of foodborne illness each year.[3] The bacteria have been found in two-thirds of raw chickens, and some of the bacteria represent a strain of *Campylobacter* resistant to antibiotics.[22,37] In addition to poultry, *Campylobacter* infection has been traced to other foods, such as unpasteurized milk and contaminated water.[11]

E. Coli E. coli 0157:H7 is a relatively new strain of *E. coli* that has evolved into a potential killer. As few as 10 of these bacteria can lead to death by kidney failure in vulnerable people. Over 2,000 cases of *E. coli* 0157:H7 occur each year in the United States.[3] *E. coli* infections are particularly associated with consumption of undercooked ground beef, however, the bacteria can contaminate many types of food. A major outbreak of *E. coli* 0157:H7 infection occurred in the past when contaminated ground beef left over from one day's production was added to the next day's batch of beef. Use of the leftover, contaminated meat kept recontaminating subsequent batches.[12] Some 25 million pounds of ground beef had to be recalled as a result.

Noroviruses Noroviruses are an extremely common but underreported cause of foodborne illnesses. The viruses go underreported because laboratory tests required for diagnosis are not widely performed and people may not seek medical care. Noroviruses usually cause an acute bout of vomiting that resolves within two days. They are thought to be spread primarily by infected kitchen workers and by fishermen who have dumped sewage waste into waters above oyster beds. Over 5 million cases of illnesses related to noroviruses occur in the United States each year.[3]

Other Causes of Foodborne Illnesses

Of the hundreds of other causes of foodborne illness, seven have been selected for brief review here. They represent examples of foodborne illnesses ranging from mercury toxicity to mad cow disease.

Table 32.3 Description of selected bacteria-caused foodborne illness[36]

Bacteria	Onset	Illness Duration	Symptoms	Foods Most Commonly Affected	Usual Source of Contamination
Salmonella	1–3 days	4–7 days	Diarrhea, abdominal pain, chills, fever, vomiting, dehydration	Uncooked or undercooked eggs, unpasteurized milk, raw meat and poultry, vegetables and fruits	Infected animals, Human feces on food, contaminated water
Campylobacter	2–5 days	2–5 days	Diarrhea (may be bloody), abdominal cramps, fever, vomiting	Undercooked poultry, unpasteurized milk	Infected poultry and other animals
E. coli 0157:H7	1–8 days	5–10 days	Watery, bloody diarrhea, abdominal cramps, little or no fever	Raw or undercooked beef, unpasteurized milk, raw vegetables and fruits, contaminated water	Infected cattle
Noroviruses (Norwalk-like viruses)	1–2 days	1–3 days	Nausea, vomiting, diarrhea	Undercooked seafood	Human feces contamination of oysters and other shellfish beds

Mercury Contamination Seafoods have come under fire as a potential source of foodborne illnesses due to mercury contamination of waters and fish by fungicides, fossil fuel exhaust, smelting plants, pulp and paper mills, leather-tanning facilities, and chemical manufacturing plants. High levels of mercury are most likely to be present in large, long-lived fish such as shark and swordfish. Because mercury can interfere with fetal brain development, pregnant women should limit their consumption of fish with high mercury content. Women who are pregnant or may become pregnant are advised to not eat shark, swordfish, king mackerel, or tilefish. Consumption of "safe" fish by pregnant women and adults in general benefits health.[13]

Ciguatera Ciguatera poisoning from fish is caused by a neurotoxin (ciguatoxin) present in microorganisms called dinoflagellates that live in reefs (Illustration 32.4). The toxin is transferred through herbivorous reef fish to carnivorous fish and then to humans who eat these fish. Over 200 types of fish may cause ciguatera poisoning, the most common being grouper, red snapper, and barracuda. Primary areas affected by ciguatera poisoning include the Caribbean and South Pacific Islands. Symptoms develop within 1 to 30 hours after ingestion of poisoned fish and cause nausea, vomiting, abdominal cramps, watery diarrhea, and then numbness, shooting pains in the legs, and other symptoms. The toxin causing the poisoning is not destroyed by cooking, freezing, or digestive enzymes, and there is no effective treatment.[14]

Red Tide "Red tide" may occur between June and October on Pacific and Atlantic coasts. It is due to the accumulation of a microorganism in algae that produces a nerve toxin. Oysters and other shellfish that consume the microorganism become contaminated, as do humans who ingest the shellfish. Resistant to cooking, the toxin produced causes a burning or prickling sensation in the mouth from 5 to 30 minutes after contaminated shellfish are consumed. This symptom is followed by nausea, vomiting, muscle weakness, and a loss of feeling in the hands and feet. Recovery is usually complete, but it's easy to understand that the warning not to eat mussels, clams, oysters, scallops, or other shellfish from red-tide waters is for real.[15]

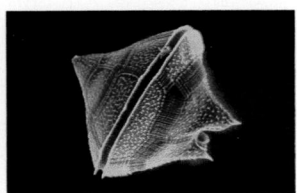
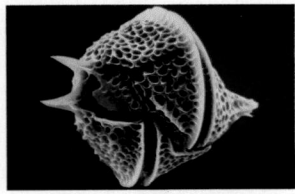
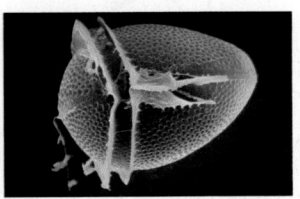

© Biophoto Associates/Photo Researchers, Inc.

Illustration 32.4 Magnified view of dinoflagellates that cause ciguatera poisoning.

Botulism *C. botulinum* bacteria produce a toxin that is one of the deadliest known. It can cause nerve damage and respiratory failure. (An antidote is available but must be given soon after the infection begins.) These bacteria are commonly present in soil and ocean and lake sediment, and they may contaminate crops, honey, animals, and seafood. In humans, foodborne botulism usually results from eating underheated, contaminated foods stored in airtight containers. The bacteria thrive without oxygen and produce gases as they grow. The gases expand the food container. For a real example of a can that exploded from the inside due to gas production by bacteria, see Illustration 32.5. Foods and fluids in cans, plastic bags and wraps, and other airtight containers that have bulged-out areas should *not* be consumed.

Alaska has the highest incidence of botulism in the United States.[1] Why does the 49th state rank first? The reason is related to traditional, Native Alaskan practices of placing uncooked or partially cooked salmon eggs, whale blubber, and other seafoods that harbor *C. botulinum* in plastic bags to ferment. The bags are squeezed to expel the air before sealing, and the no-oxygen environment can foster the growth of botulinum.[16]

Parasites Various parasitic worms, such as tapeworms, flatworms, and roundworms, can enter food and water through fecal material and soil. They are generally killed by freezing and always killed by high temperature. Once consumed, roundworms may attach to the lining of the intestine and feed on the person's blood. This can lead to anemia. One type of roundworm can bore a hole through the stomach within an hour after the worm's source—raw fish—is eaten. The severe pain that results sends people immediately to their doctors. Hundreds of Japanese people experience this foodborne illness every year.[17]

It's a benefit that most parasites are destroyed by freezing. One study in Seattle assessed the parasite content of raw fish used in sushi. About 40% of the raw fish samples contained roundworms, but since the fish had been deep frozen, all the worms were dead.[17]

Bisphenol A It is hard to drink from a bottle of water now without wondering if the plastic contains Bisphenol A (BPA). News about the potential effects of BPA on fetal and infant growth and development are popular topics for the media and a cause of concern

Richard Anderson

Illustration 32.5 Pressure caused by bacterial gases in this can of glucose drink made the can explode.

Toro Attila/Shutterstock.com

Illustration 32.6 Recycling codes are generally imprinted near or on the bottom of plastic containers. (Not all plastic containers have recycling codes.) A code of 3 or 7 indicates the plastic may contain BPA.[20]

Table 32.4 Avoiding BPA in plastic bottles and containers[20]

- Don't microwave food in plastic containers.
- Don't add hot liquids or foods to plastic containers.
- Check recycling codes on plastic bottles and containers (see Illustration 32.6). Plastic containers with the recycling code of 3 or 7 may contain BPA.
- Use glass bottles or containers.
- Consume clean, fresh, or frozen vegetables and fruits.

among parents and others.[18] Exposure to BPA is widespread: over 93% of individuals have detectable levels of BPA in their urine.[20] BPA is used in many plastic products like bottles and the inner coating of food cans. It helps makes plastic flexible.

Is BPA Safe? Few human studies have assessed the safety of BPA, but those that have indicate that high levels of exposure increase the risk of cardiovascular disease.[19] It has not been shown to adversely affect fetal or infant growth or development and has not been banned in the United States.[20] (BPA was banned in Canada in 2010.)[21] BPA is most likely to be released from plastic bottles and containers when plastic is exposed to high temperature. Until BPA is voluntarily taken out of plastic food and water containers, individuals concerned about BPA can reduce their exposure by following the tips given in Table 32.4.

Mad Cow Disease Technically called bovine spongiform encephalopathy (BSE), this rare disease in cattle is suspected of causing at least 150 human deaths in Europe. Only one case of the disease in humans—caused by consumption of affected beef—has been diagnosed in the United States, and it originated in England. The disease appears to have started when cows, which are herbivores by nature, were given sheep intestines and parts of the spinal cord in their feed. Some of the sheep harbored a protein called a prion that caused a deadly disease when consumed by cows. A prion is not a bacterium or other microorganism; it's a small protein that can transmit disease when consumed by another species. Only one other known foodborne illness is spread by the consumption of otherwise healthy body parts. That disease is kuru, and it is transmitted by cannibalism.[23,24]

Researchers concluded that mad cow disease was transferred to humans who ate the meat of prion-infected cows. In 1994, mad cow disease was discovered to be a new form of Creutzfeldt-Jakob disease (variant Creutzfeldt-Jakob disease, or VCJD) previously identified in humans. VCJD may take 5 to 50 years to develop.[24] It inevitably leads to death in humans after many years due to brain damage. Needless to say, it is no longer legal to feed cows animal parts that may transmit the disease. Although the risk of consuming beef from cattle with mad cow disease is extremely small, the possibility that animals may develop this disease exists. Stringent monitoring efforts are in place in the United States, Canada, and Europe, and rates of VCJD are declining.[24]

Antibiotics and Pesticides in Foods Other, potential foodborne illnesses of concern to consumers relate to antibiotic use in feeds and pesticides applied to crops. Chickens, cattle, and other farm-raised animals are commonly given antibiotics in feed to prevent infectious disease. In some cases, the antibiotics used are the same ones given to humans to treat infections. Microorganisms are very clever, however. They can transform themselves to become resistant to antibiotics used to kill them. People can become infected with these new strains of antibiotic-resistant microorganisms when they are consumed in foods.[25] These infections can be difficult to treat because the disease-causing microorganisms are not sensitive to the antibiotics used. Many common forms of foodborne illnesses today are represented by these new forms of bacteria.[26]

Pesticides containing organophosphates, mercury-containing fungicides, and DDT remain causes of foodborne illnesses.[27] Use of DDT on plants for insect control in the United States was phased out over 20 years ago due to links with cancer, but these long-lasting chemicals still contaminate some land, lakes, and streams. Pesticide levels in crops are periodically measured by the FDA (Illustration 32.7).[28] Fewer than 1% of the foods on the market contain an

Ed Kashi Photography/VII

Illustration 32.7 Gathering broccoli samples for pesticide testing. Pesticide residues do not appear to be a major cause of foodborne illnesses.

excessive level of pesticide.[25] Agriculture workers who fail to take safety precautions when applying pesticides are most likely to experience the illnesses related to pesticide exposure (Illustration 32.8). Health problems due to pesticides, although not extensively studied in humans, appear to be rare in other groups of people.[25]

Illustration 32.8 Protective gear—including a chemical-resistant suit, respirator, and goggles—is required for use with some of the most toxic pesticides but is not always worn.

Preventing Foodborne Illnesses

- **Delineate five consumer practices that help prevent foodborne illness.**

There are two major approaches to the prevention of foodborne illnesses. The first relies on food safety regulations that control food processing and handling practices, and the second involves consumer behaviors that lessen the risk of consuming contaminated foods. (Read what can happen when both of these principles are violated in the nearby Health Action.)

Food Safety Regulations

According to the Federal Food, Drug, and Cosmetic Act, it is illegal to produce or dispense foods that are contaminated with substances that cause illness in humans. Foods are considered "safe" if there is a reasonable certainty that no harm will result from repeated exposure to any substance added to foods. The act governs all substances that are added intentionally or accidentally to foods—except pesticides. According to legislation passed in 1996, pesticides are permitted if their consumption is associated with a "negligible risk" of cancer or other health problems.[29]

Irradiation of Foods Increasing rates of foodborne illness have led some health officials to recommend the expanded use of food irradiation. Food irradiation is a safe process when conducted under specified conditions. It destroys bacteria, parasites, and viruses present in or on foods. Availability of irradiated foods is growing but is still rather limited.[30]

Although the proposal has merits, food irradiation, as well as other food sterilization techniques, are not silver bullets that will prevent all cases of foodborne illness. Prions, toxins, pesticides, and mercury, for example, are resistant to radiation. Once irradiated or sterilized, foods can later become contaminated in such places as a packing plant, grocery store, restaurant, or home. The absence of all microorganisms in irradiated food may enable individual types of bacteria that contact the food after it is sterilized to grow at unusually high rates.[31]

health action Double Whammy

From the "true story" department of *Nutrition Now* comes the tale of a man who took a night off with his wife to dine at their favorite restaurant. The man, a physician, ordered his favorite item: the luscious chicken burritos. He was served more than he could eat, so he took the leftovers home and promptly put them in the refrigerator. About 8 hours later, he felt like his intestines were exploding; he thought he'd rather die than be so sick. Then the light bulb went off—he had probably somehow consumed *C. perfingens* toxin.

By the third day he began to feel better and noticed he was ravenously hungry. Then he remembered there was a leftover burrito in the refrigerator. . . . Happily, the second round of symptoms wasn't as bad as the first.

The message? It's difficult for consumers to prevent every case of foodborne illness, but we can prevent more cases than we do. Throw out all spoiled food.

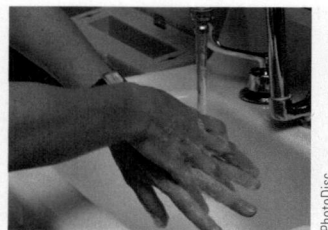

Illustration 32.9 There's a right way to wash your hands.

The Consumer's Role in Preventing Foodborne Illnesses

Due in large measure to the demands of individuals and consumer groups, many mechanisms are in place to help ensure the safety of the food supply. However, existing safeguards in food production and processing, and government regulations and enforcement efforts are insufficient to guarantee that all foods purchased by consumers will be free from contamination. This situation, along with the possibility that food can become contaminated in unanticipated ways, means that responsibility for food safety is shared by consumers.

Food Safety Basics he first rule of food safety is to wash your hands thoroughly with soap and water before and after handling food (Illustration 32.9). According to the CDC, this is the single most important means of preventing the spread of foodborne illness caused by bacteria.

There's a Right Way to Wash Your Hands Ever watch a TV dramatic series about doctors and hospitals and wondered why they show surgeons scrubbing their hands and upper arms before surgery for so long? TV has it right. It takes about 20 seconds (or the time it takes to repeat the word "Mississippi" 20 times) to sanitize your hands. First you need to lather up with soap and very warm water. The soap makes germs lose their grip on your skin. Next you have to make sure to scrub all the crevices between your fingers and under your fingernails. Rinse your hands thoroughly with really warm water and dry them with a paper towel. (If you use a dishcloth, you may reinfect your hands.)[32]

Keep Hot Foods Hot and Cold Foods Cold Good foods can go bad if stored improperly. One of the most effective ways to prevent foods from spoiling is to store them at, and heat them to, the right temperatures (see Table 32.5 and Illustrations 32.10 and 32.11. When holding prepared foods before serving, hot foods should be kept hot, and cold foods cold. Freezing foods halts the growth of all of the main types of bacteria that contaminate foods. Once the foods are thawed, however, bacteria growth may resume, and the foods may become contaminated with new bacteria while thawing.

Bacteria Bacteria grow best at temperatures between 40° and 135°F (see Illustration 32.10). The room temperature of homes is within this range, and that's why foods that spoil should not be left outside the refrigerator for more than an hour. If a food has been improperly stored or kept past the expiration date on the label, it should not be eaten even if it tastes, smells, and looks all right. Bacteria that most commonly cause foodborne illnesses usually do not change the taste, smell, or appearance of foods.[2] This general rule should apply: When in doubt, throw it out!

Don't Eat Raw Milk, Eggs, or Meats Unpasteurized milk and milk products, raw and partially cooked eggs, raw and undercooked meats and fish may be contaminated with microorganisms that cause foodborne illness. Certain raw oysters may be contaminated, too, and should be avoided.[33]

Food Handling and Storage Particular care needs to be taken when handling and storing foods. Dairy products such as milk and cheese may spoil in about a week or

REALITY CHECK
Is Moldy Cheese Safe to Eat?

Tino decides to make tacos for dinner and finds that the shredded cheese he wants to use has some mold on it. Is it safe to eat the cheese?

Who gets the thumbs up?

Answers appear on page 32-9.

Tino: I'm not sure if it's safe, but I think I'd better toss the cheese.

Nancy: Just take out the moldy pieces. The rest is okay to eat.

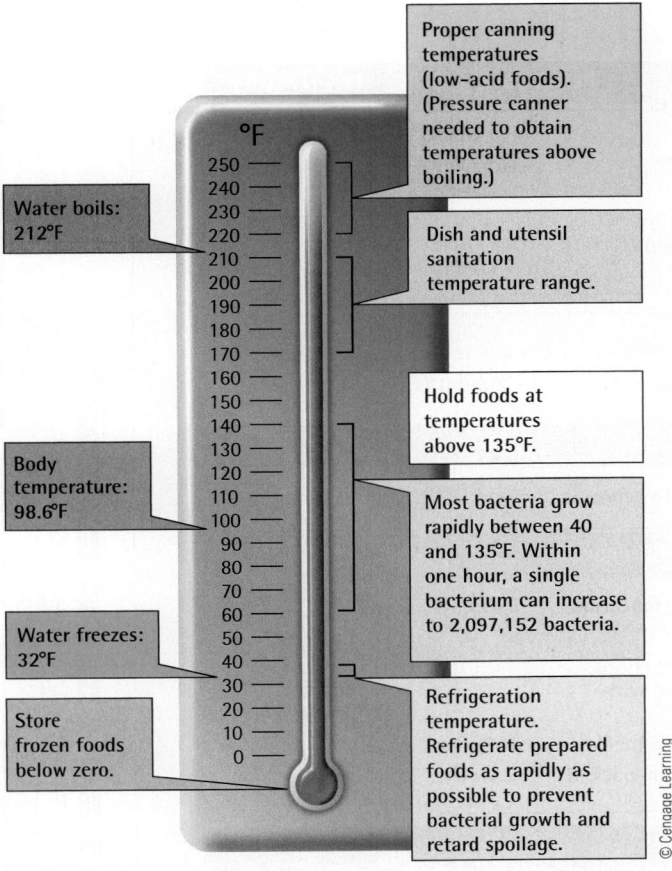

Water boils: 212°F

Body temperature: 98.6°F

Water freezes: 32°F

Store frozen foods below zero.

Proper canning temperatures (low-acid foods). (Pressure canner needed to obtain temperatures above boiling.)

Dish and utensil sanitation temperature range.

Hold foods at temperatures above 135°F.

Most bacteria grow rapidly between 40 and 135°F. Within one hour, a single bacterium can increase to 2,097,152 bacteria.

Refrigeration temperature. Refrigerate prepared foods as rapidly as possible to prevent bacterial growth and retard spoilage.

© Cengage Learning

Illustration 32.10 Temperature guide for safe handling of food in the home.

Source: Adapted from Minnesota Extension Service, University of Minnesota; updated in 2004, based on data from USDA.

Table 32.5 A guide for cooking foods to safe internal temperatures[40]

Food	Internal temperature
Ground meats	
Beef, veal, lamb, pork	160°F
Chicken, turkey	165°F
Roasts, steaks, chops	
Beef, veal, lamb	145°F
Chicken, turkey, duck	165°F
Stuffing	165°F
Pork	145°F
Ham, fresh	145°F
Ham, cooked, re-heated	140°F
Eggs and other foods	
Fried, poached	Yolk and white are firm
Casseroles, leftovers	165°F
Sauces, custards	160°F

Fish should be cooked until the flesh is milky white and flakes easily with a fork.

Shrimp, lobster, and crab until shells are red and the flesh is milky white, firm, and not raw looking.

Clams, oysters, and mussels should be cooked until the shells open, and not eaten if the shell is not open.

sooner if contaminated with bacteria from the air, surface areas, or hands. If you hold cheese by the wrapper when you take it out of the refrigerator to cut a slice off, it may last longer than if you touch it with dirty hands. (The Reality Check for this unit addresses the issue of the safety of cheese that has grown mold.) Handling foods with clean hands and using clean utensils on washed surface areas helps to reduce the number of bacteria that come in contact with the food. Raw meats should be separated from other foods, and utensils and surface areas used to prepare meats should be thoroughly cleaned after each use.[32]

The Safety of Canned Foods Commercially canned foods are heated to the point where all bacteria are killed so there is no bacterial growth. Consequently, the contents of canned foods are sterile and will not spoil if left on the shelf for years. Nevertheless,

ANSWERS TO REALITY CHECK
Is Moldy Cheese Safe to Eat?

Mold on soft cheeses such as cottage cheese, cream cheese, and brie; and sliced or shredded cheeses should be thrown out. Mold can easily penetrate throughout these cheeses. Mold generally cannot penetrate very far in blocks of hard cheeses, such as cheddar, Swiss, and Colby. The cheese that lies an inch below the mold is considered safe to eat.[39]

AISPIX by Image Source/ Shutterstock.com

Tino:

iStockphoto.com/ranplett

Nancy:

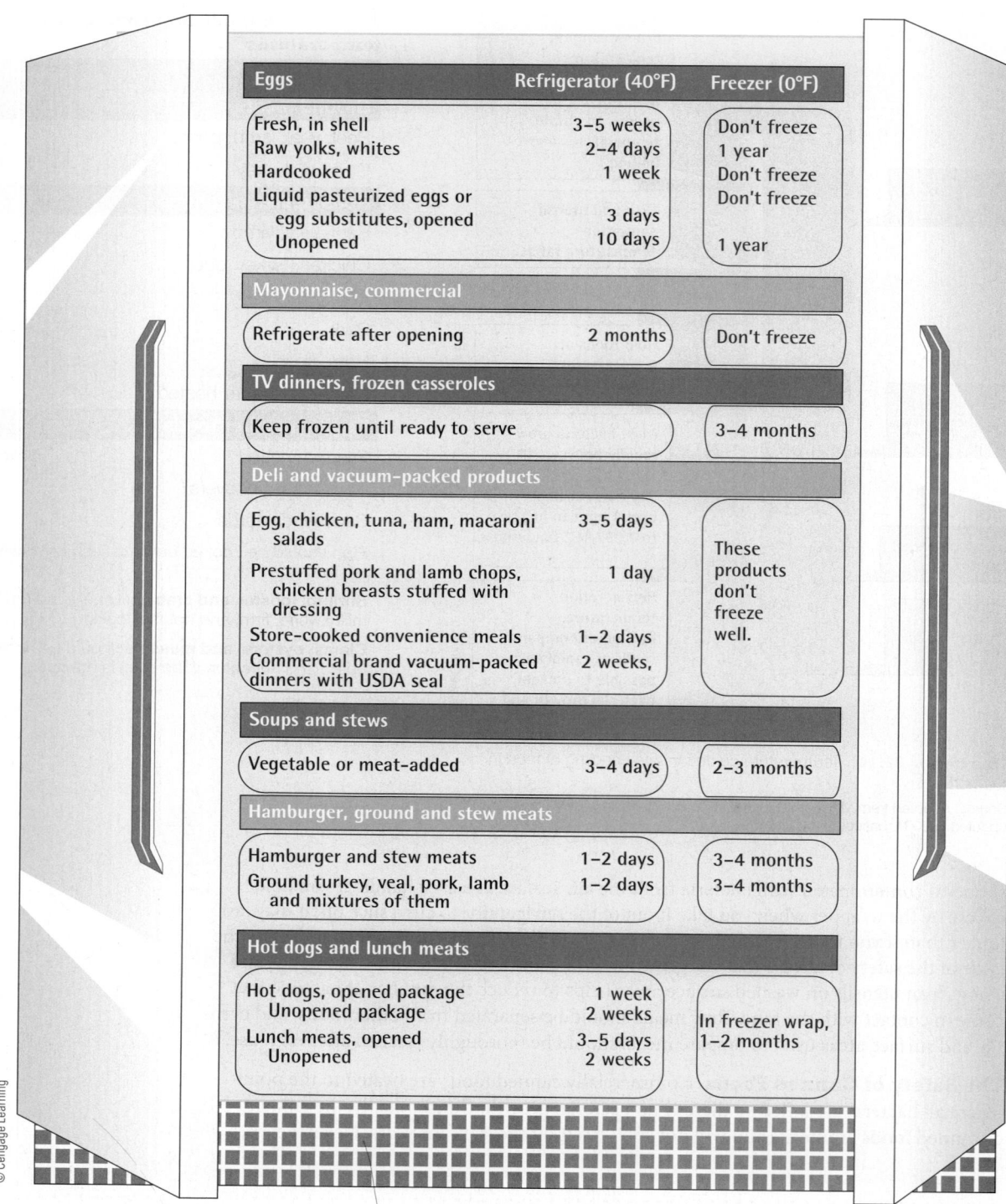

Eggs	Refrigerator (40°F)	Freezer (0°F)
Fresh, in shell	3–5 weeks	Don't freeze
Raw yolks, whites	2–4 days	1 year
Hardcooked	1 week	Don't freeze
Liquid pasteurized eggs or egg substitutes, opened	3 days	Don't freeze
Unopened	10 days	1 year
Mayonnaise, commercial		
Refrigerate after opening	2 months	Don't freeze
TV dinners, frozen casseroles		
Keep frozen until ready to serve		3–4 months
Deli and vacuum-packed products		
Egg, chicken, tuna, ham, macaroni salads	3–5 days	These products don't freeze well.
Prestuffed pork and lamb chops, chicken breasts stuffed with dressing	1 day	
Store-cooked convenience meals	1–2 days	
Commercial brand vacuum-packed dinners with USDA seal	2 weeks,	
Soups and stews		
Vegetable or meat-added	3–4 days	2–3 months
Hamburger, ground and stew meats		
Hamburger and stew meats	1–2 days	3–4 months
Ground turkey, veal, pork, lamb and mixtures of them	1–2 days	3–4 months
Hot dogs and lunch meats		
Hot dogs, opened package	1 week	In freezer wrap, 1–2 months
Unopened package	2 weeks	
Lunch meats, opened	3–5 days	
Unopened	2 weeks	

© Cengage Learning

Illustration 32.11   Cold storage.
These safe time limits will help keep refrigerated food from spoiling or becoming dangerous to eat. These time limits will keep frozen food at top quality.[35]

Source: U.S. Department of Agriculture, Food Safety and Inspection Service, September 1990 and 1997.

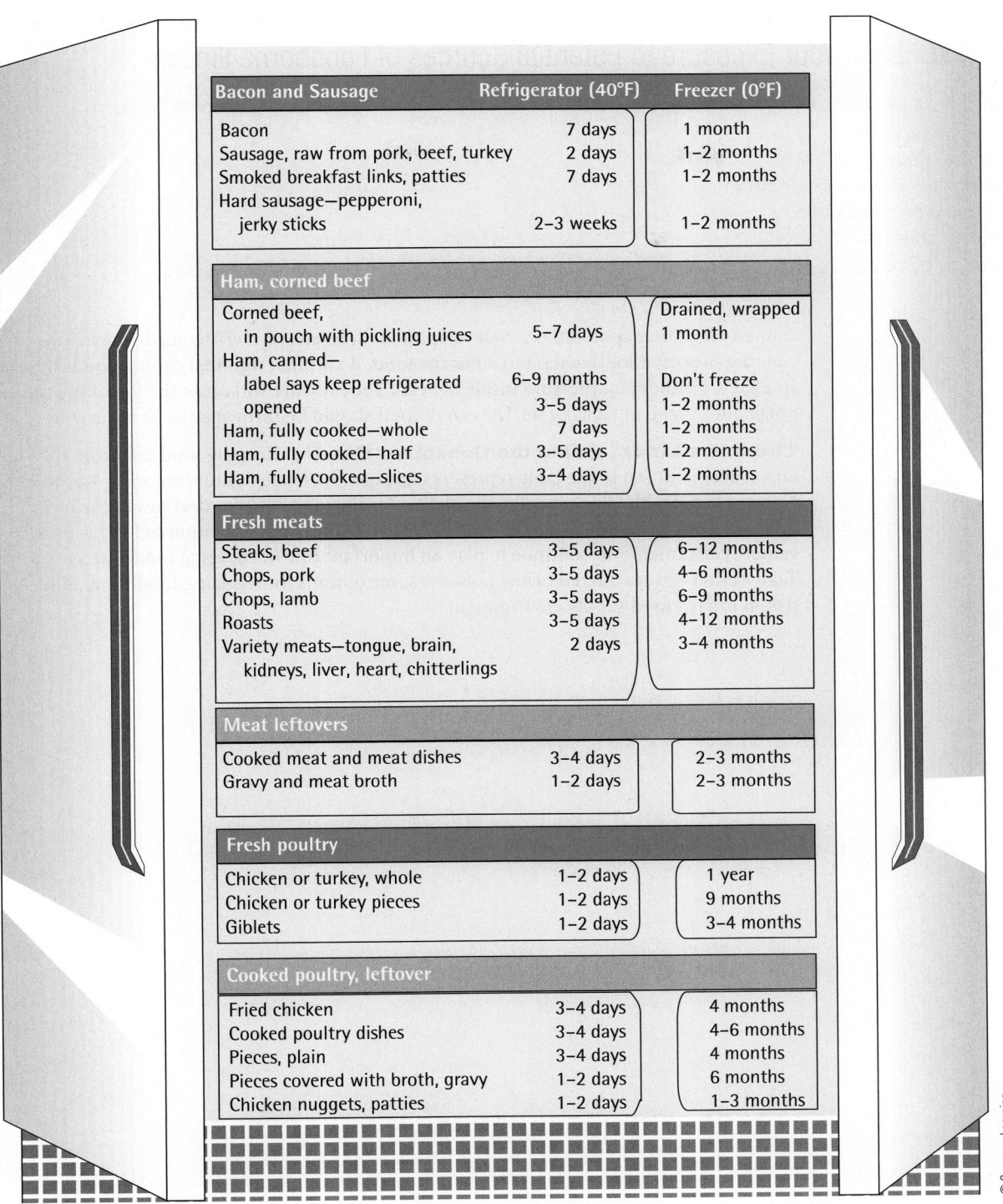

Bacon and Sausage	Refrigerator (40°F)	Freezer (0°F)
Bacon	7 days	1 month
Sausage, raw from pork, beef, turkey	2 days	1–2 months
Smoked breakfast links, patties	7 days	1–2 months
Hard sausage—pepperoni, jerky sticks	2–3 weeks	1–2 months

Ham, corned beef		
Corned beef, in pouch with pickling juices	5–7 days	Drained, wrapped 1 month
Ham, canned—label says keep refrigerated	6–9 months	Don't freeze
opened	3–5 days	1–2 months
Ham, fully cooked—whole	7 days	1–2 months
Ham, fully cooked—half	3–5 days	1–2 months
Ham, fully cooked—slices	3–4 days	1–2 months

Fresh meats		
Steaks, beef	3–5 days	6–12 months
Chops, pork	3–5 days	4–6 months
Chops, lamb	3–5 days	6–9 months
Roasts	3–5 days	4–12 months
Variety meats—tongue, brain, kidneys, liver, heart, chitterlings	2 days	3–4 months

Meat leftovers		
Cooked meat and meat dishes	3–4 days	2–3 months
Gravy and meat broth	1–2 days	2–3 months

Fresh poultry		
Chicken or turkey, whole	1–2 days	1 year
Chicken or turkey pieces	1–2 days	9 months
Giblets	1–2 days	3–4 months

Cooked poultry, leftover		
Fried chicken	3–4 days	4 months
Cooked poultry dishes	3–4 days	4–6 months
Pieces, plain	3–4 days	4 months
Pieces covered with broth, gravy	1–2 days	6 months
Chicken nuggets, patties	1–2 days	1–3 months

© Cengage Learning

Illustration 32.11 (*Continued*) Cold storage: safe time limits.

take **action** Limit Your Exposure to Potential Sources of Foodborne Illness

Pick two options from this list you have done, or believe you could do, to reduce your risk of developing foodborne illness.

_____ Buy locally grown produce
_____ Plant a vegetable garden
_____ Buy irradiated raw meats when available
_____ Buy pasteurized dairy products only
_____ Prevent cross-contamination

canned foods may spoil if the can develops a pinhole leak or if errors made during the canning process allow bacteria to enter the food. A cardinal sign that canned foods have spoiled is a buildup of pressure inside the can. The pressure will cause the top of the can to bulge out instead of curving in. The can of food should be thrown away or returned.

There Is a Limit to What the Consumer Can Do Keeping food safe from the farm, lake, or ocean to the table represents a major challenge to industry and government. None of the Health Objectives for the Nation for the year 2010 for declines in the incidence of foodborne diseases have been achieved.[38] Until contamination of food is prevented, consumers will continue to play an important role in ensuring food's safety. The Take Action feature offered below presents some options for reducing foodborne illness risk in today's food supply environment.

Daniel Pepper/Getty Images

NUTRITION
up close

Food Safety Detective

Focal Point: Investigating paths to foodborne illness.

Assume an outbreak of foodborne illness occurred among people who ate foods from the onion and lettuce platter shown being prepared in Illustration 32.2.

Identify three potential causes of the outbreak related to the preparation of the foods as shown in the illustration.

1. _____
2. _____
3. _____

Feedback to the Nutrition Up Close is located in Appendix G.

REVIEW QUESTIONS

- **List the top causes of foodborne illness and the transmission and symptoms of each.**

1. All foodborne illnesses are caused by bacteria.
   True/False

2. Foodborne illnesses often begin with the transfer of harmful bacteria or viruses from feces of an infected animal or person to food.   **True/False**

3. Cross-contamination can result from a failure to wash cutting boards after being used to cut raw meat but not raw vegetables.   **True/False**

4. Antibiotics given to livestock can result in the development of strains of bacteria that resist treatment with antibiotics.   **True/False**

5. Symptoms of foodborne illness most commonly consist of headache, blurred vision, and loss of balance.
   True/False

6. Infection with norovirus is an extremely common cause of foodborne illness. The illness is marked by episodes of vomiting that resolve with two days.   **True/False**

- **Delineate five consumer practices that help prevent foodborne illness.**

7. The first rule of food safety is to thoroughly wash your hands before and after food preparation.   **True/False**

8. Bacteria grow best at temperatures between 100° to 200°F.   **True/False**

9. Commercially canned vegetables, given the can is free from damage, can be safely consumed two years after purchase.   **True/False**

10. Ground chicken, beef, and turkey should be cooked until they reach a temperature of 160°F but ground beef is safe to eat after it reaches a temperature of 145°F.   **True/False**

The following three questions refer to this scenario:

Moshe comes home after an afternoon in the library and is really hungry. He hasn't shopped for a while so he has to eat what he can find in the fridge. This is what he finds:

- Half a package of deli turkey he bought 10 days ago
- Milk that expired five days ago
- Beef stew left over from his dinner a week ago
- Eggs he bought two weeks ago

11. _____ The milk smelled fine to Moshe. Should he drink it?
   a. If it smells fine, it is safe to drink.
   b. Moshe should look at it and taste it. If it looks and tastes okay, it would be safe to drink.
   c. It may be spoiled (contaminated). Moshe should throw it out.
   d. He could safely drink a small amount of it even if it's past the "use by" date.

12. _____ Which of the foods in Moshe's fridge is most likely to be safe to eat?
   a. deli turkey
   b. beef stew
   c. milk
   d. eggs

13. _____ Assume Moshe decides to eat the deli turkey on bread slices he had stored in the freezer. Two days later he develops diarrhea, abdominal cramps, fever, and vomiting. He has a foodborne illness. Which of the following causes of foodborne illness accounts best for his symptoms?
   a. *E. coli* 0157:H7   c. Botulinum
   b. Campylobacter   d. Ciguatera

Answers to these questions can be found in Appendix G.

NUTRITION SCOREBOARD ANSWERS

1. The answer is leafy green vegetables.   **False**

2. Freezing causes most bacteria to cease multiplying, but does not kill them. If a food is contaminated with bacteria before you freeze it, bacteria will be present when you thaw it.   **False**

3. Alaska has the highest incidence of botulism of all 50 states.[1]   **True**

4. Smell is not a foolproof indicator of contamination. Bacteria that most commonly cause foodborne illnesses may not change the smell, taste, or appearance of foods.[2]   **True**

UNIT

33

Aspects of Global Nutrition

NUTRITION SCOREBOARD

1 Approximately 30% of the world's population does not have access to a safe supply of water.   **True/False**

2 Many developing countries currently have rates of undernutrition and infectious disease that are similar to rates experienced in the United States and in many European countries 100 years ago.   **True/False**

3 Many developing countries now face the dual public health problems of increasing rates of obesity as well as high a prevalence of underweight.   **True/False**

Answers can be found at the end of the unit.

After completing Unit 33 and its interactive learning features, you will be able to:

- Identify effects of the nutrition transition on health status of people in developing countries.

- Identify five major causes of undernutrition in developing countries.

State of the World's Health

- **Identify effects of the nutrition transition on health status of people in developing countries.**

- **Identify five major causes of undernutrition in developing countries.**

If you represented the world's population with a group of 100 people, 60 would be Asian, 14 African, 12 European, 8 Latin American and Caribbean, 5 North American, and 1 person would be from Australia or New Zealand. One person would have a college education, 80 would live in poor housing, and 50 would be malnourished.[2]

The global community consists of countries grouped into the categories of "industrialized nations," "developing nations," and "least-developed nations." (Countries within each category are listed in Table 33.1.) Only 31 of 193 United Nations member countries are considered industrialized, most (113) are developing, and, of the 113 developing countries, 49 are least developed. All countries in the global community have interests and issues in common, such as the adequacy of the food supply, availability of health care and education, and safe water. But the countries of the world are also dissimilar in many ways. Differences in financial resources, population growth, and political and other systems translate into large disparities in health status and life expectancy. People in the least-developed and developing countries are more likely to have substantially shorter life expectancies, (Table 33.2), to die from infectious diseases, and to experience undernutrition than individuals living in industrialized countries.[3]

Key Nutrition Concepts

Two key nutrition concepts directly underlie this content on global nutrition:

1. Food is a basic need of humans.

2. Poor nutrition can result from both inadequate and excessive levels of nutrient intake.

The general state of health of populations in various countries is monitored by tracking key environmental, health, and behavioral characteristics. Health outcomes, such as the number of low birth-weight newborns, prevalence of child underweight, rates of breast-feeding, and access to safe drinking water are key indicators of the health status of a population.[4] Worldwide, 17% of infants are born with low birth weights and only 30% of people have access to safe water supplies.[5] As can be seen in Illustration 33.1, rates of underweight, a measure of undernutrition, vary substantially worldwide. When percentages of low birth weight in newborns and underweight in young children are high, and rates of breast-feeding and access to a safe water supply are low, one can rightfully assume that undernutrition, short stature (height for age), infection, and other health problems are common. When these key indicators show improvement, the health status of the population and longevity improve.[4,6]

The health and nutritional status of populations in developing countries is monitored by the World Health Organization (WHO), the Food and Agriculture Organization (FAO), and the United Nations International Children's Emergency Fund (UNICEF). Recent reports have identified leading problem areas related to undernutrition that must be addressed as part of global strategies aimed at improving health and well-being (Table 33.3). Important progress is being made, and rates of undernutrition, short stature, and infectious disease are decreasing slowly worldwide.[7,8] Much progress remains to be made, however.

Table 33.1 Countries classified by the U.N. International Children's Emergency Fund as industrialized, developing, and least developed[a]

Industrialized countries	Developing countries	Least developed countries
Andorra; Australia; Austria; Belgium; Canada; Denmark; Finland; France; Germany; Greece; Holy See; Iceland; Ireland; Israel; Italy; Japan; Liechtenstein; Luxembourg; Malta; Monaco; Netherlands; New Zealand; Norway; Portugal; San Marino; Slovenia; Spain; Sweden; Switzerland; United Kingdom; United States 	Afghanistan; Algeria; Angola; Antigua and Barbuda; Argentina; Armenia; Azerbaijan; Bahamas; Bahrain; Bangladesh; Barbados; Belize; Benin; Bhutan; Bolivia; Botswana; Brazil; Brunei Darussalam; Burkina Faso; Burundi; Cambodia; Cameroon; Cape Verde; Central African Rep.; Chad; Chile; China; Colombia; Comoros; Congo; Congo, Dem. Rep.; Cook Islands; Costa Rica; Cote d'Ivoire; Cuba; Cyprus; Djibouti; Dominica; Dominican Rep.; Ecuador; Egypt; El Salvador; Equatorial Guinea; Eritrea; Ethiopia; Fiji; Gabon; Gambia; Georgia; Ghana; Grenada; Guatemala; Guinea; Guinea-Bissau; Guyana; Haiti; Honduras; India; Indonesia; Iran; Iraq; Jamaica; Jordan; Kazakhstan; Kenya; Kiribati; Korea, Dem. People's Rep.; Korea, Rep. of; Kuwait; Kyrgyzstan; Lao People's Dem. Rep.; Lebanon; Lesotho; Liberia; Libya; Madagascar; Malawi; Malaysia; Maldives; Mali; Marshall Islands; Mauritania; Mauritius; Mexico; Micronesia, Fed. States of; Mongolia; Morocco; Mozambique; Myanmar; Namibia; Nauru; Nepal; Nicaragua; Niger; Nigeria; Niue; Oman; Pakistan; Palau; Panama; Papua New Guinea; Paraguay; Peru; Philippines; Qatar; Rwanda; Saint Kitts and Nevis; Saint Lucia; Saint Vincent/ Grenadines; Samoa; Sao Tome and Principe; Saudi Arabia; Senegal; Seychelles; Sierra Leone; Singapore; Solomon Islands; Somalia; South Africa; Sri Lanka; Sudan; Suriname; Swaziland; Syria; Tajikistan; Tanzania; Thailand; Togo; Tonga; Trinidad and Tobago; Tunisia; Turkey; Turkmenistan; Tuvalu; Uganda; United Arab Emirates; Uruguay; Uzbekistan; Vanuatu; Venezuela; Viet Nam; Yemen; Zambia; Zimbabwe	Afghanistan; Angola; Bangladesh; Benin; Bhutan; Burkina Faso; Burundi; Cambodia; Cape Verde; Central African Rep.; Chad; Comoros; Congo, Dem. Rep.; Djibouti; Equatorial Guinea; Eritrea; Ethiopia; Gambia; Guinea; Guinea-Bissau; Haiti; Kiribati; Lao People's Dem. Rep.; Lesotho; Liberia; Madagascar; Malawi; Maldives; Mali; Mauritania; Mozambique; Myanmar; Nepal; Niger; Rwanda; Samoa; Sao Tome and Principe; Sierra Leone; Solomon Islands; Somalia; Sudan; Tanzania; Togo; Tuvalu; Uganda; Vanuatu; Yemen; Zambia 

[a]Data for some eastern European countries, former Soviet republics, and the Baltic states are not available. Least-developed countries represent a subgroup of the developing countries.

Table 33.2 Life expectancy in selected countries from highest to lowest, 2012 estimates[3]

	Life expectancy (in years)
Japan	83.9
Singapore	83.8
Hong Kong	82.1
Italy	81.9
Canada	81.5
France	81.5
Spain	81.3
Sweden	81.2
Israel	81.1
Iceland	81.0
Ireland	80.3
Germany	80.2
United Kingdom	80.2
Panama	78.0
Costa Rica	77.9
Mexico	76.7
China	74.8
Brazil	73.0
Vietnam	69.2
Korea, North	69.2
Russia	66.5
Rwanda	58.4
Somalia	50.8
South Africa	49.4
Chad	48.7

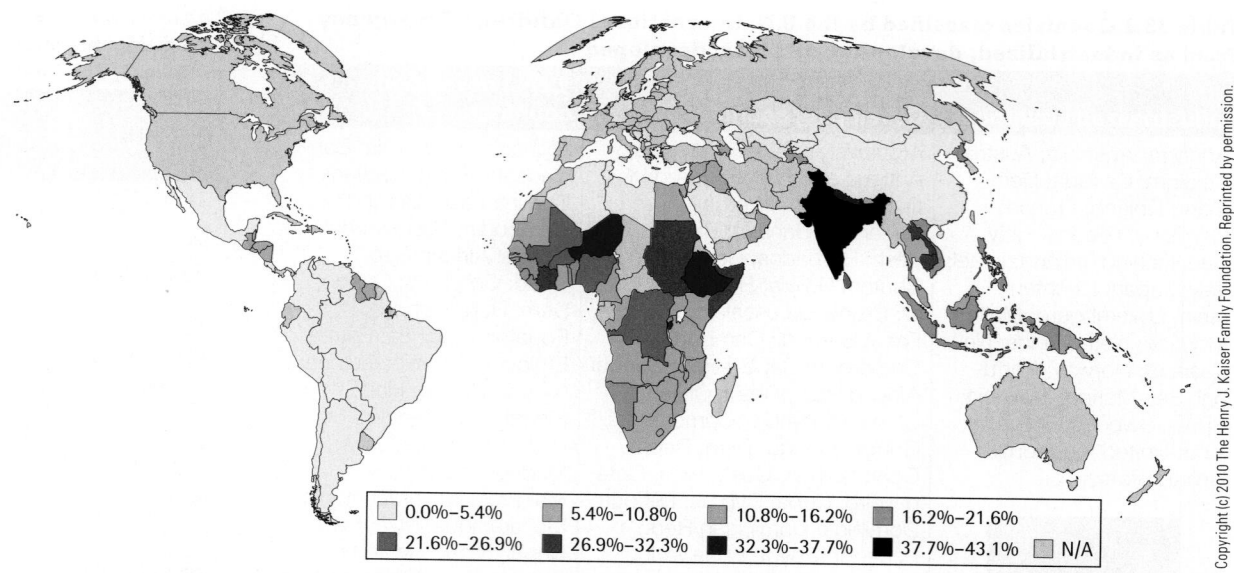

0.0%–5.4%	5.4%–10.8%	10.8%–16.2%	16.2%–21.6%	
21.6%–26.9%	26.9%–32.3%	32.3%–37.7%	37.7%–43.1%	N/A

Illustration 33.1   Worldwide prevalence of child malnutrition assessed as a percent of underweight among children under 5 years of age, 2000–2009.[15]

Source: www.globalhealthfacts.org/data/topic/map.aspx?ind=48.

Table 33.3 Priority problem areas related to malnutrition in developing countries[9–11]

1. Childhood protein-calorie malnutrition	7. Insufficient physical activity
2. Vitamin deficiencies: vitamin A, folate	8. Low vegetable and fruit intake
3. Mineral deficiencies: iodine, iron, zinc	9. Malnutrition and increased complications from HIV/AIDS
4. Lack of breast-feeding	10. Poor nutritional status of women of child-bearing age
5. Alcohol abuse	
6. Overweight and obesity	

The "Nutrition Transition"

Many developing countries are currently undergoing the **nutrition transition,** or changes in dietary intake and other lifestyle behaviors toward foods and lifestyles common in Western countries like the United States. The change is occurring due to economic development and modifications in the food supply. Children and adults in developing countries who have traditionally consumed foods such as cassava, millet, rice, chicken, beans, and bread or tortillas are switching to burgers, fries, soft drinks, and ice cream when they can. There is also a general trend of reduced levels of physical activity.[12] In populations undergoing the nutrition transition such as India and China, rates of underweight decline but rates of obesity, diabetes, hypertension, heart disease, dental disease, and cancer increase.[4] Illustration 33.2 shows WHO's projected increases in the incidence of chronic, noncommunicable diseases within countries by income level.

Countries undergoing the nutrition transition often experience rising rates of obesity on top of existing high rates of underweight and undernutrition.[10] In 1980, 8% of adults worldwide were obese and the figure increased to 14% by 2008.[1] Children born to poorly nourished women may be biologically programmed in the womb to conserve energy. Adaptations are made by the fetus to adjust growth to correspond to the available energy and nutrient supplies, and these adaptations persist throughout life. When exposed to a generous food supply after birth, they may gain fat more readily than height.[9,14] Excess accumulation of fat places children at increased risk for hypertension, diabetes, heart

nutrition transition   A change in a population's diet from traditional foods such as grains, roots, beans, and rice to calorie-dense foods high in sugar, fat, sodium, and other components that characterize Western diets.

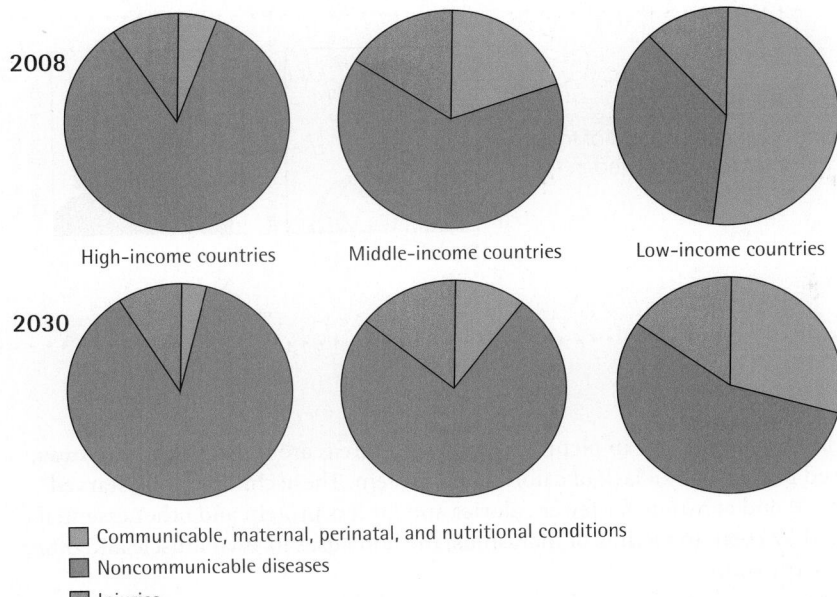

Illustration 33.2 Worldwide rates of chronic, noncommunicable disease, 2008, and projections for 2030.[31]

Source: www.nia.nig.gov/research/publication/global-health-and-aging/new-disease-patterns.

2008

High-income countries   Middle-income countries   Low-income countries

2030

☐ Communicable, maternal, perinatal, and nutritional conditions
☐ Noncommunicable diseases
☐ Injuries

disease, and a number of other "diseases of civilization" later in life. Underweight, short stature, undernutrition, and infections such as malaria and tuberculosis will continue to be top health problems in developing countries for many years to come while diseases of "Western civilization" will continue to gain ground.[1]

Food and Nutrition: The Global Challenge

We have all heard about starvation in Somalia, Ethiopia, Sudan, and Bangladesh. The pictures of starving children with desperation in their big eyes and heads too large for their frail bodies burn an image in our minds and makes many people wonder why it happens. (The Reality Check for this unit provides a follow-up discussion on this topic.)

What we don't hear about in the news, however, is the extent of starvation and malnutrition in the world. These are not just problems of a few isolated areas experiencing civil war or crop failures. In developing regions of the world, undernutrition is an ongoing problem for up to 45% of children under the age of 5, and short stature affects an average of 24%.[15,16] Approximately half of young children and pregnant women in the world are iron deficient, and 33% of young children are deficient in vitamin A. Vitamin A deficiency is related to frequent and severe infectious disease, decreased growth, and blindness. Deficiencies of iodine, zinc, and vitamin B_{12} are common in some developing countries.[1,11]

REALITY CHECK
Can the World Produce Enough Food?

The image of starving children upsets both Kai and Kari after they had seen the image posted on YouTube today. Kai says to Kari that he thinks the major reason why children starve is that the world cannot produce enough food for everyone. Kari believes that's not the reason. She thinks starvation is mostly related to causes such as civil unrest, discrimination, and abject poverty.

Who gets the thumbs up?

Answers appear on page 33-6.

szefei/Shutterstock.com

Kai: The planet is overpopulated, and we can't produce enough food for everyone.

Kharidehal Abhirama Ashwi/Shutterstock.com

Kari: There doesn't have to be starvation. Enough food can be produced.

The world can and does produce enough food for everyone. Children in need of food may not have access to it for the reasons Kari gave and for other reasons listed in Table 33.3.

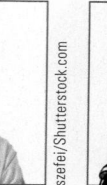

Kai:

Kari: 👍

The children we usually see in pictures from famine areas are victims of **marasmus**, a disease caused primarily by a lack of calories and protein. These children look starved (Illustration 33.3) and consume far fewer calories and far less protein and other essential nutrients than they need. In victims of marasmus, the body uses its own muscle and other tissues as an energy source.

Protein-energy malnutrition produces a disease called **kwashiorkor** in some young children. It is similar to marasmus in that it is primarily due to a chronic lack of protein and calories. Kwashiorkor is unlike marasmus, however, in some ways. Young children with kwashiorkor appear to be genetically susceptible to developing faulty protein utilization mechanisms when chronically exposed to protein and calorie shortages.[17] These children may appear fat due to massive swelling related to abnormal protein utilization that occurs with the disease (Illustration 33.4). Most victims of kwashiorkor show some of the symptoms of marasmus as well. Children with either type of severe malnutrition are generally deficient in multiple vitamins and minerals and at high risk of dying from infectious diseases due to weakened immune systems.[18, 19]

marasmus A severe form of malnutrition primarily due to a chronic lack of calories and protein. Also called protein-energy malnutrition.

kwashiorkor A severe form of protein-energy malnutrition in young children. It is characterized by swelling, fatty liver, susceptibility to infection, profound apathy, and poor appetite. The cause of kwashiorkor is unclear.

Survivors of Malnutrition Malnutrition in the form of undernutrition that occurs before a child's brain has completely grown has lasting effects on development.[20] Brain growth occurs primarily during pregnancy through the age of 2 years. If children suffer severe malnutrition during these years, they will experience permanent reductions in growth and mental development. The severity of the growth and mental development impairments depends on the timing and duration of the malnutrition. The longer it exists, the harder it becomes for young children to achieve a brighter future.[16] In addition, the state of hunger and starvation leads people to become self-centered and to lose any sense of well-being. The need for food for survival may prompt unethical behaviors such as stealing and injuring others to obtain food. The devastating psychological effects of starvation may follow children into adulthood and have a lasting effect on behavior throughout life. Persistent malnutrition saps the physical and mental energy of people and, ultimately, compromises the economic and social progress of nations.[8,16]

Illustration 33.3 The look of marasmus. Children with marasmus look like "skin and bones."

Illustration 33.4 The look of kwashiorkor. This child has the characteristic "moon face" (edema), swollen belly, and patchy dermatitis often seen with kwashiorkor.

Malnutrition and Infection There is a very close relationship between malnutrition and infection: Malnutrition weakens the immune system and increases the likelihood and severity of infection. Simultaneously, repeated infection and the bouts of diarrhea that often accompany it can produce undernutrition. Poor sanitary conditions, contaminated water supplies, and lack of refrigeration contribute to the spread of infectious diseases and thus to the malnutrition that diseases promote. Most deaths of children under the age of five years in developing countries are related to both malnutrition and infection.[10,19]

Low rates of breast-feeding promote the spread of infection, especially in countries where contaminated water is used to prepare formula. In addition to providing optimal nutrition, breast milk contains a number of substances that protect babies from infection.[21]

Vitamin A deficiency in children plays important roles in the spread and severity of infections. Children who are deficient in vitamin A are much more likely to die from a measles infection, for example, than are children whose vitamin A status is sufficient.[22] HIV infection progresses more rapidly into AIDS in malnourished individuals and shortens the time parents can work and support their families. Life expectancy in Sudan, for example, has decreased from 53.0 years in 1996 to 51.4 years in 2009.[5] Over 33 million people in the world have HIV infection.[23]

Why Do Starvation and Malnutrition Happen? Malnutrition and starvation don't have to happen anywhere. The world's agricultural systems have the capacity to produce enough food to feed everyone on earth.[13] People become malnourished or starve primarily because of poverty. Human-made disasters, including discrimination against women, the HIV/AIDS epidemic, racism, use of agricultural land for biofuel crops, corrupt governance, and other factors (Table 33.4) heavily contribute to malnutrition. In some instances, starvation and malnutrition are due to natural disasters that lead to crop failures or the inability to distribute food to those in need.[8]

The famine in Somalia that led to the death of over 1.5 million people was due to the collapse of the government, fighting among rival clans, and general lawlessness. Farmers and their families were driven off their land by bandits who stole their crops and possessions. Many had nowhere to go for help.[24] Mass starvation in Ethiopia and Sudan was similarly initiated by violent conflict within the country.[25] Seizing the opposing side's food and preventing them from obtaining food aid are among the primary weapons used to fight these civil wars. In Bangladesh, where poor people have no choice but to live on floodplains, cyclones wipe out crops and regularly result in the death of thousands due to floods, starvation, and infectious disease.[9] None of these cases of mass famine was due primarily to the ravages of nature.

Women and female children are at particular risk for malnutrition in some societies because cultural practices call for food to be allocated to men and boys first (Illustration 33.5). If too little food is available, what remains after meals for women and daughters may be too little to support health and growth.[11] In many developing countries, discrimination against women in education and employment, sanctions against the use of birth control, and violence toward women place women at high risk of developing malnutrition and having a low quality of life.[20]

Ending Malnutrition The long-term solution to malnutrition will depend on the ability of humans to work together to achieve educational and economic development, peace, population growth control, improved sanitation, social equity for women and children, and environmentally sound and productive agricultural policies and practices in developing countries (Illustration 33.6). Bottom-up approaches to problem solving work better than top-down approaches. The people most affected by malnutrition must be active participants in the planning and implementation of improvement programs. Efforts to improve the nutritional status of populations can and have paid off.

Table 33.4 Root causes of malnutrition in developing countries[6,20]

- Poverty
- Low levels of education
- Discrimination against females
- Inequitable distribution of the food supply
- HIV/AIDS epidemic
- Lack of economic opportunities
- Racism, ethnocentrism
- Low agricultural productivity
- Poor and corrupt governance
- Unsafe water due primarily to the ravages of nature.

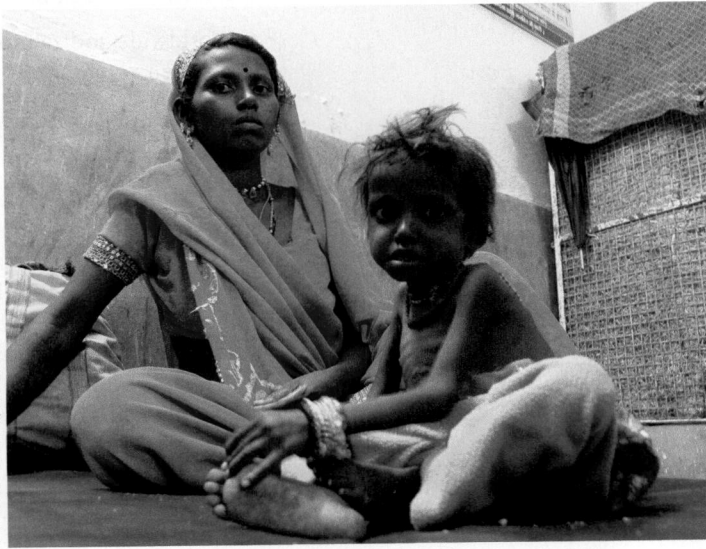

Illustration 33.5 Females in some developing countries are particularly vulnerable to malnutrition because they, and their nutritional needs, may be considered less important than those of males.

© SANJEEV GUPTA/epa/Corbis

©Porterfield/Chickering/Photo Researchers, Inc.

Success Stories Examples of strikingly successful efforts to reduce malnutrition in various countries around the world are available for telling. A repeated theme is the reduction in malnutrition as economic, educational, nutritional, and sanitary conditions improve. Improvements in these same areas were responsible for the dramatic declines in malnutrition and infectious diseases and the increases in life expectancy experienced in the United States, Canada, and many European countries around 100 years ago.[6,26] In other instances, specific nutrient deficiencies have been greatly reduced or eliminated by food fortification and nutrition education programs. Here are several specific examples:

- Vitamin A supplements and education on vitamin A-rich foods are related to a dramatic decrease in cases of severe and moderate vitamin A deficiency and infection in children in many countries.[22]

- Food supplements given to groups of infants in Russia, Brazil, South Africa, and China were associated with higher IQ scores at age 8.

- Iodization of salt has eliminated iodine deficiency in Bolivia and Ecuador.

- Worldwide, iodization of salt is associated with a drop in the number of iodine-deficient children.

- Fortification of flour with iron has led to a decrease in iron-deficiency anemia in the Philippines.[27,28]

In Thailand, a country with a serious iodine-deficiency problem, a highly creative approach to increasing iodine intake was implemented by the king in the 1990s. To celebrate his birthday, salt producers gave the king 10,000 tons of iodized salt. The king had the salt packaged in small plastic bags and, with the help of the Red Cross and the army, delivered a bag to every household in Thailand. A message from the king about the importance of using iodized salt was attached to each bag.[28]

Deaths of young children from malnutrition and its related diseases have dropped substantially in countries experiencing a resurgence of breast-feeding (Illustration 33.7).[21] Breast milk protects infants and young children from a variety of infectious diseases, supports the growth and health of infants, and protects them from the hazards of formulas reconstituted with contaminated water. Worldwide health initiatives have led to the development of the International Code of Marketing Breast Milk Substitutes and to Baby Friendly Hospitals. The international marketing code calls for the prohibition of free formula samples as well as the promotion of infant formulas by health care professionals, and staff training on breast-feeding support. More than 12,700 hospitals have adopted Baby Friendly Hospital policies that effectively promote and facilitate breast-feeding.[21]

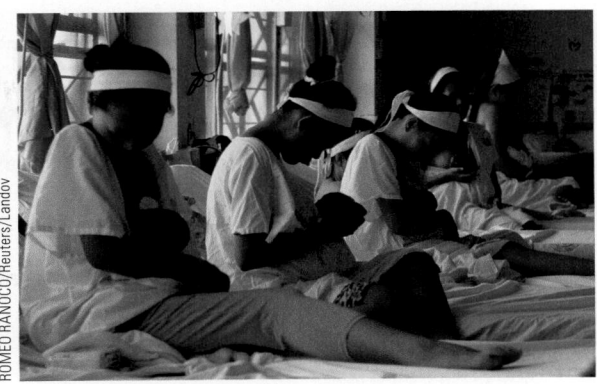

The World Food Summit The 1996 World Food Summit ended with a pledge that by 2015, countries would reduce by half the number of people in the world who are hungry and undernourished.[29] Progress is being made toward meeting the goal. Access to food and rates of undernutrition are improving, but at a slower pace than envisioned.[9] Attention is now being placed on expanding agriculture and food production, and decreasing food costs in developing countries. It is anticipated that it will take until the year 2030 to cut rates of hunger and undernutrition by half.[30]

Illustration 33.7 After a rooming-in policy that enabled mothers to breast-feed their infants on demand was introduced in one hospital in the Philippines, the incidence of early infant deaths due to infection dropped by 95%.[32]

Source: First Call for Children, A UNICEF Quarterly 1992 (January/ March):1.

The Future

Benefits derived from improving the nutritional status of the world's population are immense. Adequate nutrition and basic public assurances such as adequate supplies of affordable nutritious foods, clean water, vaccinations, and education are the bedrock on which present generations secure a future for themselves and the next one. People who are well nourished, grow normally, and are free from infectious disease are more likely to be productive members of society, earn enough money to buy material objects, and make sure their children are educated. In this type of environment, parents tend to have fewer children and are able to better take care of the children they have.[26] Members of families are more productive, happier, require less medical care, commit fewer crimes, and are more likely to be self-sufficient than undernourished people. When the undernutrition problem is solved, the world will likely wonder why it didn't happen sooner.[20]

© EyesWideOpen/Getty Images

NUTRITION

up close

Ethnic Foods Treasure Hunt

Focal Point: Values placed on foods are culturally specific.

The following questions deal with ethnic food preferences. Answer any three of the questions. Resources that will help you find the answers include knowledgeable individuals, the Internet, travel and anthropology books, and encyclopedias.

1. Name three foods commonly served at celebrations by the Igbo in Nigeria (Igbos were formerly referred to as Biafrans).
   a. _____
   b. _____
   c. _____

2. List two special foods given to Hmong women during pregnancy.
   a. _____
   b. _____

3. List the five foods that make up the traditional Costa Rican breakfast.
   a. _____
   b. _____
   c. _____
   d. _____
   e. _____

4. List three foods commonly sold at soccer matches in Italy.
   a. _____
   b. _____
   c. _____

5. Name two traditional foods of the Zulu in South Africa.
   a. _____
   b. _____

6. Name one of the staple foods of people in Nentsi (it's near Siberia).
   a. _____

7. List two foods considered to be "yin" and two considered to be "yang" in Chinese culture.

   Yin   Yang
   a. _____   a. _____
   b. _____   b. _____

Feedback to the Nutrition Up Close is located in Appendix G.

REVIEW QUESTIONS

- **Identify effects of the nutrition transition on health status of people in developing countries.**

1. Due to the expanding adoption of Western lifestyles and diet, some developing countries are experiencing increasing rates of overweight and chronic diseases.   True/False

2. The term *nutrition transition* is defined as improvements in a population's nutritional status and health that result from a country's economic development.   True/False

3. Children who were undernourished during their mother's pregnancy and early in their own lives may develop a tendency to gain fat more readily than height given access to an abundant food supply.   True/False

4. Over the past decade the leading cause of death in developing countries has changed from infectious diseases to chronic diseases such as diabetes and heart disease.   True/False

- **Identify five major causes of undernutrition in developing countries.**

5. Malnutrition and infectious diseases are common health problems in countries with low life expectancy. True/False

6. Rates of malnutrition are increasing worldwide. True/False

7. Deficiencies of iodine, iron, and vitamin A remain important public health problems in many developing countries.   True/False

8. The need for food for survival may prompt behaviors such as stealing and injuring others to obtain food. True/False

9. In many developing countries, discrimination places women at high risk of malnutrition.   True/False

10. Death rates for young children decrease in developing countries where rates of breast-feeding increase. True/False

The following three questions refer to this scenario:

In 2012, North Korea lowered the minimum height requirement for service in the military from 4 feet, 9 inches (145 cm) to 4 feet 8 inches (142 cm). The country is experiencing a shortage of taller men due to a famine in 1990 and continued undernutrition after that. The average height of the population is decreasing.[33]

11. _____ What is the most likely cause of the decreased stature of males?
    a. Their parents are short.
    b. Mothers lacked adequate health care during pregnancy.
    c. They experienced undernutrition during the growing years.
    d. They grew up lacking a safe and adequate supply of water.

12. _____ What other conditions are likely present in the North Korean population that are related to diminished growth in height?
    a. Increased agility due to low stature.
    b. Increased calorie need per pound of body weight.
    c. Decreased need for health care.
    d. Reduced life expectancy.

13. _____ Assume 30 years from now North Korea pulls back on military expenditures and the country grows economically. Peace prevails and people have jobs and money to spend. How would that likely change the health status or behaviors of the North Korean population?
    a. Individuals in future generations would gain height.
    b. Individuals would remain short because height is genetically determined.
    c. Heart disease would replace diabetes as the major cause of death.
    d. Couples would decide to have additional children because they can afford them.

Answers to these questions can be found in Appendix G.

NUTRITION SCOREBOARD ANSWERS

1. Contaminated water supplies are a major source of infection in many developing countries. **True**

2. Many developing countries are evolving just like the United States and Europe began to do a century ago. **True**

3. Rates of overweight are increasing in many developing countries.[1] **True**

Appendix

Appendix A
Table of Food Composition

Appendix B
Reliable Sources of Nutrition Information

Appendix C
The U.S. Food Exchange System

Appendix D
Table of Intentional Food Additives

Appendix E
Cells

Appendix F
Canadian Nutrition and Physical Activity Guidelines

Appendix G
Feedback and Answers to Nutrition Up Close and Review Questions

Appendix A

Table of Food Composition

This edition of the table of food composition includes a wide variety of foods from all food groups. It is updated with each edition to reflect nutrient changes for current foods, remove outdated foods, and add foods that are new to the marketplace.*

The nutrient database for this appendix is compiled from a variety of sources, including the USDA Standard Reference database (Release 16), literature sources, and manufacturers' data. The USDA database provides data for a wider variety of foods and nutrients than other sources. Because laboratory analysis for each nutrient can be quite costly, manufacturers tend to provide data only for those nutrients mandated on food labels. Consequently, data for their foods are often incomplete; any missing information is designated in this table as a blank space. Keep in mind that a blank space means only that the information is unknown and should not be interpreted as a zero.

Whenever using nutrient data, remember that many factors influence the nutrient contents of foods, including the mineral content of the soil, the diet of the animal or the fertilizer of the plant, the season of harvest, the method of processing, the length and method of storage, the method of cooking, the method of analysis, and the moisture content of the sample analyzed. With so many factors involved, users must view nutrient data as a close approximation of the actual amount.

For updates, corrections, and a list of 6000 foods and codes found in the diet analysis software that accompanies this text, visit **www.wadsworth.com/nutrition** and click on *Diet Analysis* in the right-hand column.

- *Fats* Total fats, as well as the breakdown of total fats to saturated, monounsaturated, and polyunsaturated fats, are listed in the table. The fatty acids seldom add up to the total due to rounding and to other fatty acid components that are not included in these basic categories, such as *trans*-fatty acids and glycerol. *Trans*-fatty acids can comprise a large share of the total fat in margarine and shortening (hydrogenated oils) and in any foods that include them as ingredients.

- *Vitamin A and Vitamin E* In keeping with the 2001 RDA for vitamin A, which established a new measure of vitamin A activity—retinol activity equivalents (RAE)—this appendix presents data for vitamin A in micrograms (μg) RAE. Similarly, because the 2000 RDA for vitamin E is based only on the alpha-tocopherol form of vitamin E, this appendix reports vitamin E data in milligrams (mg) alpha-tocopherol (listed in the table as mg α).

- *Bioavailability* Keep in mind that the availability of nutrients from foods depends not only on the quantity provided by a food, but also on the amount absorbed and used by the body—the bioavailability. The bioavailability of folate from fortified foods, for example, is greater than from naturally occurring sources. Similarly, the body can make niacin from the amino acid tryptophan, but niacin values in this table (and most databases) report preformed niacin only. Chapter 10 provides conversion factors and additional details.

- *Using the Table* The items in this table have been organized into several categories, which are listed at the head of each right-hand page. Page numbers have been provided, and each group has been color-coded to make it easier to find individual items.

*This food composition table has been prepared for Wadsworth Publishing Company and is copyrighted by ESHA Research in Salem, Oregon—the developer and publisher of the Food Processor and Genesis nutritional software programs. The nutritional data are supported by over 1300 references. Because the list of sources is so extensive, it is not provided here, but is available from the publisher.

In an effort to conserve space, the following abbreviations have been used in the food descriptions and nutrient breakdowns:

- diam = diameter
- ea = each
- enr = enriched
- f/ = from
- frzn = frozen
- g = grams
- liq = liquid
- pce = piece
- pkg = package
- w/ = with
- w/o = without
- t = trace
- 0 = zero (no nutrient value)
- blank space = information not available

Index to Appendix A

DA+ Code	Food Description	Quantity	Measure	Wt (g)	H₂O (g)	Ener (kcal)	Prot (g)	Carb (g)	Fiber (g)	Fat (g)	Sat	Mono	Poly	Trans
												Fat Breakdown (g)		
	BREADS, BAKED GOODS, CAKES, COOKIES, CRACKERS, CHIPS, PIES													
	Bagels													
8534	Cinnamon and raisin	1	item(s)	71	22.7	194	7.0	39.2	1.6	1.2	0.2	0.1	0.5	—
14395	Multi-grain	1	item(s)	61	—	170	6.0	35.0	1.0	1.5	0.5	0.1	0.4	—
8538	Oat bran	1	item(s)	71	23.4	181	7.6	37.8	2.6	0.9	0.1	0.2	0.3	—
4910	Plain, enriched	1	item(s)	71	25.8	182	7.1	35.9	1.6	1.2	0.3	0.4	0.5	0
4911	Plain, enriched, toasted	1	item(s)	66	18.7	190	7.4	37.7	1.7	1.1	0.2	0.3	0.6	0
	Biscuits													
25008	Biscuits	1	item(s)	41	15.8	121	2.6	16.4	0.5	4.9	1.4	1.4	1.8	—
16729	Scone	1	item(s)	42	11.5	148	3.8	19.1	0.6	6.2	2.0	2.5	1.3	—
25166	Wheat biscuits	1	item(s)	55	21.0	162	3.6	21.9	1.4	6.7	1.9	1.9	2.5	—
	Bread													
325	Boston brown, canned	1	slice(s)	45	21.2	88	2.3	19.5	2.1	0.7	0.1	0.1	0.3	—
8716	Bread sticks, plain	4	item(s)	24	1.5	99	2.9	16.4	0.7	2.3	0.3	0.9	0.9	—
25176	Cornbread	1	piece(s)	55	25.9	141	4.7	18.3	0.9	5.4	2.1	1.4	1.5	0
327	Cracked wheat	1	slice(s)	25	9.0	65	2.2	12.4	1.4	1.0	0.2	0.5	0.2	—
9079	Croutons, plain	¼	cup(s)	8	0.4	31	0.9	5.5	0.4	0.5	0.1	0.2	0.1	—
8582	Egg	1	slice(s)	40	13.9	113	3.8	19.1	0.9	2.4	0.6	0.9	0.4	—
8585	Egg, toasted	1	slice(s)	37	10.5	117	3.9	19.5	0.9	2.4	0.6	1.1	0.4	—
329	French	1	slice(s)	32	8.9	92	3.8	18.1	0.8	0.6	0.2	0.1	0.3	—
8591	French, toasted	1	slice(s)	23	4.7	73	3.0	14.2	0.7	0.5	0.1	0.1	0.2	—
42096	Indian fry, made with lard (Navajo)	3	ounce(s)	85	26.9	281	5.7	41.0	—	10.4	3.9	3.8	0.9	—
332	Italian	1	slice(s)	30	10.7	81	2.6	15.0	0.8	1.1	0.3	0.2	0.4	—
1393	Mixed grain	1	slice(s)	26	9.6	69	3.5	11.3	1.9	1.1	0.2	0.2	0.5	0
8604	Mixed grain, toasted	1	slice(s)	24	7.6	69	3.5	11.3	1.9	1.1	0.2	0.2	0.5	0
8605	Oat bran	1	slice(s)	30	13.2	71	3.1	11.9	1.4	1.3	0.2	0.5	0.5	—
8608	Oat bran, toasted	1	slice(s)	27	10.4	70	3.1	11.8	1.3	1.3	0.2	0.5	0.5	—
8609	Oatmeal	1	slice(s)	27	9.9	73	2.3	13.1	1.1	1.2	0.2	0.4	0.5	—
8613	Oatmeal, toasted	1	slice(s)	25	7.8	73	2.3	13.2	1.1	1.2	0.2	0.4	0.5	—
1409	Pita	1	item(s)	60	19.3	165	5.5	33.4	1.3	0.7	0.1	0.1	0.3	—
7905	Pita, whole wheat	1	item(s)	64	19.6	170	6.3	35.2	4.7	1.7	0.3	0.2	0.7	—
338	Pumpernickel	1	slice(s)	32	12.1	80	2.8	15.2	2.1	1.0	0.1	0.3	0.4	—
334	Raisin, enriched	1	slice(s)	26	8.7	71	2.1	13.6	1.1	1.1	0.3	0.6	0.2	—
8625	Raisin, toasted	1	slice(s)	24	6.7	71	2.1	13.7	1.1	1.2	0.3	0.6	0.2	—
10168	Rice, white, gluten free, wheat free	1	slice(s)	38	—	130	1.0	18.0	0.5	6.0	0	—	—	0
8653	Rye	1	slice(s)	32	11.9	83	2.7	15.5	1.9	1.1	0.2	0.4	0.3	—
8654	Rye, toasted	1	slice(s)	29	9.0	82	2.7	15.4	1.9	1.0	0.2	0.4	0.3	—
336	Rye, light	1	slice(s)	25	9.3	65	2.0	12.0	1.6	1.0	0.2	0.3	0.3	—
8588	Sourdough	1	slice(s)	25	7.0	72	2.9	14.1	0.6	0.5	0.1	0.1	0.2	—
8592	Sourdough, toasted	1	slice(s)	23	4.7	73	3.0	14.2	0.7	0.5	0.1	0.1	0.2	—
491	Submarine or hoagie roll	1	item(s)	135	40.6	400	11.0	72.0	3.8	8.0	1.8	3.0	2.2	—
8596	Vienna, toasted	1	slice(s)	23	4.7	73	3.0	14.2	0.7	0.5	0.1	0.1	0.2	—
8670	Wheat	1	slice(s)	25	8.9	67	2.7	11.9	0.9	0.9	0.2	0.2	0.4	—
8671	Wheat, toasted	1	slice(s)	23	5.6	72	3.0	12.8	1.1	1.0	0.2	0.2	0.4	—
340	White	1	slice(s)	25	9.1	67	1.9	12.7	0.6	0.8	0.2	0.2	0.3	—
1395	Whole wheat	1	slice(s)	46	15.0	128	3.9	23.6	2.8	2.5	0.4	0.5	1.4	—
	Cakes													
386	Angel food, prepared from mix	1	piece(s)	50	16.5	129	3.1	29.4	0.1	0.2	0	0	0.1	—
8772	Butter pound, ready to eat, commercially prepared	1	slice(s)	75	18.5	291	4.1	36.6	0.4	14.9	8.7	4.4	0.8	—
28517	Carrot	1	slice(s)	131	56.6	339	4.8	56.5	1.9	11.1	1.0	5.7	3.8	—
4931	Chocolate with chocolate icing, commercially prepared	1	slice(s)	64	14.7	235	2.6	34.9	1.8	10.5	3.1	5.6	1.2	—
8756	Chocolate, prepared from mix	1	slice(s)	95	23.2	352	5.0	50.7	1.5	14.3	5.2	5.7	2.6	—
393	Devil's food cupcake with chocolate frosting	1	item(s)	35	8.4	120	2.0	20.0	0.7	4.0	1.8	1.6	0.6	—
8757	Fruitcake, ready to eat, commercially prepared	1	piece(s)	43	10.9	139	1.2	26.5	1.6	3.9	0.5	1.8	1.4	—
1397	Pineapple upside down, prepared from mix	1	slice(s)	115	37.1	367	4.0	58.1	0.9	13.9	3.4	6.0	3.8	—
411	Sponge, prepared from mix	1	slice(s)	63	18.5	187	4.6	36.4	0.3	2.7	0.8	1.0	0.4	—
8817	White with coconut frosting, prepared from mix	1	slice(s)	112	23.2	399	4.9	70.8	1.1	11.5	4.4	4.1	2.4	—
8819	Yellow with chocolate frosting, ready to eat, commercially prepared	1	slice(s)	64	14.0	243	2.4	35.5	1.2	11.1	3.0	6.1	1.4	—
8822	Yellow with vanilla frosting, ready to eat, commercially prepared	1	slice(s)	64	14.1	239	2.2	37.6	0.2	9.3	1.5	3.9	3.3	—
	Snack cakes													
8791	Chocolate snack cake, creme filled, with frosting	1	item(s)	50	9.3	200	1.8	30.2	1.6	8.0	2.4	4.3	0.9	—
25010	Cinnamon coffee cake	1	piece(s)	72	22.6	231	3.6	35.8	0.7	8.3	2.2	2.6	3.0	—
16777	Funnel cake	1	item(s)	90	37.6	276	7.3	29.1	0.9	14.4	2.7	4.7	6.1	—
8794	Sponge snack cake, creme filled	1	item(s)	43	8.6	155	1.3	27.2	0.2	4.8	1.1	1.7	1.4	—
	Snacks, chips, pretzels													
29428	Bagel chips, plain	3	item(s)	29	—	130	3.0	19.0	1.0	4.5	0.5	—	—	—
29429	Bagel chips, toasted onion	3	item(s)	29	—	130	4.0	20.0	1.0	4.5	0.5	—	—	—

BREADS, BAKED GOODS, CAKES, COOKIES, CRACKERS, CHIPS, PIES —Continued

Chol (mg)	Calc (mg)	Iron (mg)	Magn (mg)	Pota (mg)	Sodi (mg)	Zinc (mg)	Vit A (µg)	Thia (mg)	Vit E (mg α)	Ribo (mg)	Niac (mg)	Vit B$_6$ (mg)	Fola (µg)	Vit C (mg)	Vit B$_{12}$ (µg)	Sele (µg)
0	13	2.69	19.9	105.1	228.6	0.80	14.9	0.27	0.22	0.19	2.18	0.04	78.8	0.5	0	22.0
0	60	1.08	—	—	310.0	—	0	—	—	—	—	—	—	0	0	—
0	9	2.18	22.0	81.7	360.0	0.63	0.7	0.23	0.23	0.24	2.10	0.03	69.6	0.1	0	24.3
0	63	4.29	15.6	53.3	318.1	1.34	0	0.42	0.07	0.18	2.82	0.04	103.0	0.7	0	16.2
0	65	2.97	15.8	56.1	316.8	0.85	0	0.39	0.07	0.17	2.88	0.04	86.5	0	0	16.6
0	38	0.94	6.0	47.4	206.0	0.20	—	0.16	0.01	0.12	1.20	0.01	31.7	0.1	0.1	7.1
49	79	1.35	7.1	48.7	277.2	0.29	64.7	0.14	0.42	0.15	1.19	0.02	32.3	0	0.1	10.9
0	57	1.21	16.1	81.0	321.1	0.42	—	0.19	0.01	0.14	1.65	0.03	35.3	0.1	0.1	0
0	32	0.94	28.4	143.1	284.0	0.22	11.3	0.01	0.14	0.05	0.50	0.03	5.0	0	0	9.9
0	5	1.02	7.7	29.8	157.7	0.21	0	0.14	0.24	0.13	1.26	0.01	38.9	0	0	9.0
21	94	0.91	10.5	71.5	209.8	0.48	—	0.14	0.32	0.15	1.03	0.04	34.6	1.7	0.2	6.2
0	11	0.70	13.0	44.3	134.5	0.31	0	0.09	—	0.06	0.92	0.08	15.3	0	0	6.3
0	6	0.30	2.3	9.3	52.4	0.06	0	0.04	—	0.02	0.40	0.00	9.9	0	0	2.8
20	37	1.21	7.6	46.0	196.8	0.31	25.2	0.17	0.10	0.17	1.93	0.02	42.0	0	0	12.0
21	38	1.23	7.8	46.6	199.8	0.31	25.5	0.14	0.10	0.16	1.77	0.02	36.3	0	0	12.2
0	14	1.16	9.0	41.0	208.0	0.29	0	0.13	0.05	0.09	1.52	0.03	47.4	0.1	0	8.7
0	11	0.89	7.1	32.2	165.6	0.24	0	0.10	0.04	0.09	1.24	0.02	32.2	0	0	6.8
6	48	3.43	15.3	65.5	279.8	0.29	0	0.36	0.00	0.18	3.91	0.03	103.8	—	0	15.8
0	23	0.88	8.1	33.0	175.2	0.25	0	0.14	0.08	0.08	1.31	0.01	57.3	0	0	8.2
0	27	0.65	20.3	59.8	109.2	0.44	0	0.07	0.09	0.03	1.05	0.06	19.5	0	0	8.6
0	27	0.65	20.4	60.0	109.7	0.44	0	0.06	0.10	0.03	1.05	0.07	16.8	0	0	8.6
0	20	0.93	10.5	44.1	122.1	0.26	0.6	0.15	0.13	0.10	1.44	0.02	24.3	0	0	9.0
0	19	0.92	9.2	33.2	121.0	0.28	0.5	0.12	0.13	0.09	1.29	0.01	18.6	0	0	8.9
0	18	0.72	10.0	38.3	161.7	0.27	1.4	0.10	0.13	0.06	0.84	0.01	16.7	0	0	6.6
0	18	0.74	10.3	38.5	162.8	0.28	1.3	0.09	0.13	0.06	0.77	0.02	13.3	0.1	0	6.7
0	52	1.57	15.6	72.0	321.6	0.50	0	0.35	0.18	0.19	2.77	0.02	64.2	0	0	16.3
0	10	1.95	44.2	108.8	340.5	0.97	0	0.21	0.39	0.05	1.81	0.17	22.4	0	0	28.2
0	22	0.91	17.3	66.6	214.7	0.47	0	0.10	0.13	0.09	0.98	0.04	29.8	0	0	7.8
0	17	0.75	6.8	59.0	101.4	0.18	0	0.08	0.07	0.10	0.90	0.01	27.6	0	0	5.2
0	17	0.76	6.7	59.0	101.8	0.19	0	0.07	0.07	0.09	0.81	0.02	23.5	0.1	0	5.2
0	100	1.08	—	—	140	—	—	0.15	—	0.10	1.20	—	32.0	0	—	—
0	23	0.90	12.8	53.1	211.2	0.36	0	0.13	0.10	0.10	1.21	0.02	35.2	0.1	0	9.9
0	23	0.89	12.5	53.1	210.3	0.36	0	0.11	0.10	0.09	1.09	0.02	29.9	0.1	0	9.9
0	20	0.70	3.9	51.0	175.0	0.18	0	0.10	—	0.08	0.80	0.01	5.3	0	0	8.0
0	11	0.91	7.0	32.0	162.5	0.23	0	0.11	0.05	0.07	1.19	0.03	37.0	0.1	0	6.8
0	11	0.89	7.1	32.2	165.6	0.24	0	0.10	0.04	0.09	1.24	0.02	32.2	0	0	6.8
0	100	3.80	—	128.0	683.0	—	0	0.54	—	0.33	4.50	0.04	—	0	—	42.0
0	11	0.89	7.1	32.2	165.6	0.24	0	0.10	0.04	0.09	1.24	0.02	32.2	0	0	6.8
0	36	0.87	12.0	46.0	130.3	0.30	0	0.09	0.05	0.08	1.30	0.03	21.3	0.1	0	7.2
0	38	0.94	13.6	51.3	140.5	0.34	0	0.10	0.06	0.09	1.44	0.04	19.8	0	0	7.7
0	38	0.94	5.8	25.0	170.3	0.19	0	0.11	0.06	0.08	1.10	0.02	27.8	0	0	4.3
0	15	1.42	37.3	144.4	159.2	0.69	0	0.13	0.35	0.10	1.83	0.09	29.9	0	0	17.8
0	42	0.11	4.0	67.5	254.5	0.06	0	0.04	0.01	0.10	0.08	0.00	9.5	0	0	7.7
166	26	1.03	8.3	89.3	298.5	0.34	111.8	0.10	—	0.17	0.98	0.03	30.8	0	0.2	6.6
0	65	2.18	23.0	279.6	367.7	0.44	—	0.25	0.01	0.19	1.73	0.10	43.4	4.6	0	14.7
27	28	1.40	21.8	128.0	213.8	0.44	16.6	0.01	0.62	0.08	0.36	0.02	10.9	0.1	0.1	2.1
55	57	1.53	30.4	133.0	299.3	0.65	38.0	0.13	—	0.20	1.08	0.03	25.7	0.2	0.2	11.3
19	21	0.70	—	46.0	92.0	—	—	0.04	—	0.05	0.30	—	2.1	0	—	2.0
2	14	0.89	6.9	65.8	116.1	0.11	3.0	0.02	0.38	0.04	0.34	0.02	8.6	0.2	0	0.9
25	138	1.70	15.0	128.8	366.9	0.35	71.3	0.17	—	0.17	1.36	0.03	29.9	1.4	0.1	10.8
107	26	0.99	5.7	88.8	143.6	0.37	48.5	0.10	—	0.19	0.75	0.03	24.6	0	0.2	11.7
1	101	1.29	13.4	110.9	318.1	0.37	13.4	0.14	0.13	0.21	1.19	0.03	34.7	0.1	0.1	12.0
35	24	1.33	19.2	113.9	215.7	0.39	21.1	0.07	—	0.10	0.79	0.02	14.1	0	0.1	2.2
35	40	0.68	3.8	33.9	220.2	0.16	12.2	0.06	—	0.04	0.32	0.01	17.3	0	0.1	3.5
0	58	1.80	18.0	88.0	194.5	0.52	0.5	0.01	0.54	0.03	0.46	0.07	13.0	1.0	0	1.7
26	55	1.36	9.9	91.9	277.6	0.30	—	0.17	0.23	0.16	1.29	0.02	36.1	0.3	0.1	9.6
62	126	1.90	16.2	152.1	269.1	0.65	49.5	0.23	1.54	0.32	1.86	0.04	50.4	0	0.3	17.7
7	19	0.54	3.4	37.0	155.1	0.12	2.1	0.06	0.50	0.05	0.52	0.01	17.0	0	0	1.3
0	0	0.72	—	45.0	70.0	—	0	—	—	—	—	—	—	0	0	—
0	0	0.72	—	50.0	300.0	—	0	—	—	—	—	—	—	0	0	—

Food Composition (Computer code is for Cengage Diet Analysis program) (For purposes of calculations, use "0" for t, <1, <.1, <.01, etc.)

DA+ Code	Food Description	Quantity	Measure	Wt (g)	H₂O (g)	Ener (kcal)	Prot (g)	Carb (g)	Fiber (g)	Fat (g)	Fat Breakdown (g) Sat	Mono	Poly	Trans
38192	Chex traditional snack mix	1	cup(s)	45	—	197	3.0	33.3	1.5	6.1	0.8	—	—	—
654	Potato chips, salted	1	ounce(s)	28	0.6	155	1.9	14.1	1.2	10.6	3.1	2.8	3.5	—
8816	Potato chips, unsalted	1	ounce(s)	28	0.5	152	2.0	15.0	1.4	9.8	3.1	2.8	3.5	—
5096	Pretzels, plain, hard, twists	5	item(s)	30	1.0	114	2.7	23.8	1.0	1.1	0.2	0.4	0.4	—
4632	Pretzels, whole wheat	1	ounce(s)	28	1.1	103	3.1	23.0	2.2	0.7	0.2	0.3	0.2	—
4641	Tortilla chips, plain	6	item(s)	11	0.2	53	0.8	7.1	0.6	2.5	0.3	0.8	0.5	0.3
	Cookies													
8859	Animal crackers	12	item(s)	30	1.2	134	2.1	22.2	0.3	4.1	1.0	2.3	0.6	—
8876	Brownie, prepared from mix	1	item(s)	24	3.0	112	1.5	12.0	0.5	7.0	1.8	2.6	2.3	—
25207	Chocolate chip cookies	1	item(s)	30	3.7	140	2.0	16.2	0.6	7.9	2.1	3.3	2.1	—
8915	Chocolate sandwich cookie with extra creme filling	1	item(s)	13	0.2	65	0.6	8.9	0.4	3.2	0.7	2.1	0.3	1.1
14145	Fig Newtons cookies	1	item(s)	16	—	55	0.5	11.0	0.5	1.3	0	—	—	0
8920	Fortune cookie	1	item(s)	8	0.6	30	0.3	6.7	0.1	0.2	0.1	0.1	0	—
25208	Oatmeal cookies	1	item(s)	69	12.3	234	5.7	45.1	3.1	4.2	0.7	1.3	1.8	—
25213	Peanut butter cookies	1	item(s)	35	4.1	163	4.2	16.9	0.9	9.2	1.7	4.7	2.3	—
33095	Sugar cookies	1	item(s)	16	4.1	61	1.1	7.4	0.1	3.0	0.6	1.3	0.9	—
9002	Vanilla sandwich cookie with creme filling	1	item(s)	10	0.2	48	0.5	7.2	0.2	2.0	0.3	0.8	0.8	—
	Crackers													
9012	Cheese cracker sandwich with peanut butter	4	item(s)	28	0.9	139	3.5	15.9	1.0	7.0	1.2	3.6	1.4	—
9008	Cheese crackers (mini)	30	item(s)	30	0.9	151	3.0	17.5	0.7	7.6	2.8	3.6	0.7	—
33362	Cheese crackers, low sodium	1	serving(s)	30	0.9	151	3.0	17.5	0.7	7.6	2.9	3.6	0.7	—
8928	Honey graham crackers	4	item(s)	28	1.2	118	1.9	21.5	0.8	2.8	0.4	1.1	1.1	—
9016	Matzo crackers, plain	1	item(s)	28	1.2	112	2.8	23.8	0.9	0.4	0.1	0	0.2	—
9024	Melba toast	3	item(s)	15	0.8	59	1.8	11.5	0.9	0.5	0.1	0.1	0.2	—
9028	Melba toast, rye	3	item(s)	15	0.7	58	1.7	11.6	1.2	0.5	0.1	0.1	0.2	—
14189	Ritz crackers	5	item(s)	16	0.5	80	1.0	10.0	0	4.0	1.0	—	—	0
9014	Rye crispbread crackers	1	item(s)	10	0.6	37	0.8	8.2	1.7	0.1	0	0	0.1	—
9040	Rye wafer	1	item(s)	11	0.6	37	1.1	8.8	2.5	0.1	0	0	0	—
432	Saltine crackers	5	item(s)	15	0.8	64	1.4	10.6	0.5	1.7	0.2	1.1	0.2	0.5
9046	Saltine crackers, low salt	5	item(s)	15	0.6	65	1.4	10.7	0.5	1.8	0.4	1.0	0.3	—
9052	Snack cracker sandwich with cheese filling	4	item(s)	28	1.1	134	2.6	17.3	0.5	5.9	1.7	3.2	0.7	—
9054	Snack cracker sandwich with peanut butter filling	4	item(s)	28	0.8	138	3.2	16.3	0.6	6.9	1.4	3.9	1.3	—
9048	Snack crackers, round	10	item(s)	30	1.1	151	2.2	18.3	0.5	7.6	1.1	3.2	2.9	—
9050	Snack crackers, round, low salt	10	item(s)	30	1.1	151	2.2	18.3	0.5	7.6	1.1	3.2	2.9	—
9044	Soda crackers	5	tem(s)	15	0.8	64	1.4	10.6	0.5	1.7	0.2	1.1	0.2	0.5
9059	Wheat cracker sandwich with cheese filling	4	item(s)	28	0.9	139	2.7	16.3	0.9	7.0	1.2	2.9	2.6	—
9061	Wheat cracker sandwich with peanut butter filling	4	item(s)	28	1.0	139	3.8	15.1	1.2	7.5	1.3	3.3	2.5	—
9055	Wheat crackers	10	item(s)	30	0.9	142	2.6	19.5	1.4	6.2	1.6	3.4	0.8	—
9057	Wheat crackers, low salt	10	item(s)	30	0.9	142	2.6	19.5	1.4	6.2	1.6	3.4	0.8	—
9022	Whole wheat crackers	7	item(s)	28	0.8	124	2.5	19.2	2.9	4.8	1.0	1.6	1.8	—
	Pastry													
16754	Apple fritter	1	item(s)	17	6.4	61	1.0	5.5	0.2	3.9	0.9	1.7	1.1	—
41565	Cinnamon rolls with icing, refrigerated dough	1	serving(s)	44	12.3	145	2.0	23.0	0.5	5.0	1.5	—	—	2.0
4945	Croissant, butter	1	item(s)	57	13.2	231	4.7	26.1	1.5	12.0	6.6	3.1	0.6	—
9096	Danish, nut	1	item(s)	65	13.3	280	4.6	29.7	1.3	16.4	3.8	8.9	2.8	—
9115	Doughnut with creme filling	1	item(s)	85	32.5	307	5.4	25.5	0.7	20.8	4.6	10.3	2.6	—
9117	Doughnut with jelly filling	1	item(s)	85	30.3	289	5.0	33.2	0.8	15.9	4.1	8.7	2.0	—
4947	Doughnut, cake	1	item(s)	47	9.8	198	2.4	23.4	0.7	10.8	1.7	4.4	3.7	—
9105	Doughnut, cake, chocolate glazed	1	item(s)	42	6.8	175	1.9	24.1	0.9	8.4	2.2	4.7	1.0	—
437	Doughnut, glazed	1	item(s)	60	15.2	242	3.8	26.6	0.7	13.7	3.5	7.7	1.7	—
10617	Toaster pastry, brown sugar cinnamon	1	item(s)	50	5.3	210	3.0	35.0	1.0	6.0	1.0	4.0	1.0	—
30928	Toaster pastry, cream cheese	1	item(s)	54	—	200	3.0	23.0	0	11.0	4.5	—	—	1.5
	Muffins													
25015	Blueberry	1	item(s)	63	29.7	160	3.4	23.0	0.8	6.0	0.9	1.5	3.3	—
9189	Corn, ready to eat	1	item(s)	57	18.6	174	3.4	29.0	1.9	4.8	0.8	1.2	1.8	—
9121	English muffin, plain, enriched	1	item(s)	57	24.0	134	4.4	26.2	1.5	1.0	0.1	0.2	0.5	—
29582	English muffin, toasted	1	item(s)	50	18.6	128	4.2	25.0	1.5	1.0	0.1	0.2	0.5	—
9145	English muffin, wheat	1	item(s)	57	24.1	127	5.0	25.5	2.6	1.1	0.2	0.2	0.5	—
8894	Oat bran	1	item(s)	57	20.0	154	4.0	27.5	2.6	4.2	0.6	1.0	2.4	—
	Granola bars													
38161	Kudos milk chocolate granola bars w/fruit & nuts	1	item(s)	28	—	90	2.0	15.0	1.0	3.0	1.0	—	—	—
38196	Nature Valley banana nut crunchy granola bars	2	item(s)	42	—	190	4.0	28.0	2.0	7.0	1.0	—	—	—
38187	Nature Valley fruit 'n' nut trail mix bar	1	item(s)	35	—	140	3.0	25.0	2.0	4.0	0.5	—	—	—
1383	Plain, hard	1	item(s)	25	1.0	115	2.5	15.8	1.3	4.9	0.6	1.1	3.0	—
4606	Plain, soft	1	item(s)	28	1.8	126	2.1	19.1	1.3	4.9	2.1	1.1	1.5	—
	Pies													
454	Apple pie, prepared from home recipe	1	slice(s)	155	73.3	411	3.7	57.5	2.3	19.4	4.7	8.4	5.2	—
470	Pecan pie, prepared from home recipe	1	slice(s)	122	23.8	503	6.0	63.7	—	27.1	4.9	13.6	7.0	—

BREADS, BAKED GOODS, CAKES, COOKIES, CRACKERS, CHIPS, PIES —Continued

Chol (mg)	Calc (mg)	Iron (mg)	Magn (mg)	Pota (mg)	Sodi (mg)	Zinc (mg)	Vit A (µg)	Thia (mg)	Vit E (mg α)	Ribo (mg)	Niac (mg)	Vit B$_6$ (mg)	Fola (µg)	Vit C (mg)	Vit B$_{12}$ (µg)	Sele (µg)
0	0	0.55	—	75.8	621.2	—	0	0.09	—	0.05	1.21	—	12.1	0	—	
0	7	0.45	19.8	465.5	148.8	0.67	0	0.01	1.91	0.06	1.18	0.20	21.3	5.3	0	2.3
0	7	0.46	19.0	361.5	2.3	0.30	0	0.04	2.58	0.05	1.08	0.18	12.8	8.8	0	2.3
0	11	1.29	10.5	43.8	514.5	0.25	0	0.13	0.10	0.18	1.57	0.03	51.3	0	0	1.7
0	8	0.76	8.5	121.9	57.6	0.17	0	0.12	—	0.08	1.85	0.07	15.3	0.3	0	—
0	19	0.25	15.8	23.2	45.5	0.26	0	0.00	0.46	0.01	0.13	0.02	2.2	0	0	0.7
0	13	0.82	5.4	30.0	117.9	0.19	0	0.10	0.03	0.09	1.04	0.01	30.9	0	0	2.1
18	14	0.44	12.7	42.2	82.3	0.23	42.2	0.03	—	0.05	0.24	0.02	7.0	0.1	0	2.8
13	11	0.69	12.4	62.1	108.8	0.24	—	0.08	0.54	0.06	0.87	0.01	17.8	0	0	4.1
0	2	1.01	4.7	17.8	45.6	0.10	0	0.02	0.25	0.02	0.25	0.00	6.0	0	0	1.1
0	10	0.36	—	—	57.5	—	0	—	—	—	—	—	—	0	—	—
0	1	0.12	0.6	3.3	21.9	0.01	0.1	0.01	0.00	0.01	0.15	0.00	5.3	0	0	0.2
0	26	1.93	48.8	176.7	311.1	1.42	—	0.26	0.23	0.13	1.35	0.09	34.9	0.3	0	17.4
13	27	0.65	21.1	112.8	154.1	0.46	—	0.08	0.73	0.09	1.85	0.05	23.5	0.1	0.1	4.8
18	5	0.30	1.7	12.2	49.4	0.08	—	0.04	0.28	0.05	0.31	0.01	9.5	0	0	3.1
0	3	0.22	1.4	9.1	34.9	0.04	0	0.02	0.16	0.02	0.27	0.00	5.0	0	0	0.3
0	14	0.76	15.7	61.0	198.8	0.29	0.3	0.15	0.66	0.08	1.63	0.04	26.3	0	0.1	2.3
4	45	1.43	10.8	43.5	298.5	0.33	8.7	0.17	0.01	0.12	1.40	0.16	45.6	0	0.1	2.6
4	45	1.43	10.8	31.8	137.4	0.33	5.1	0.17	0.09	0.12	1.40	0.16	26.7	0	0.1	2.6
0	7	1.04	8.4	37.8	169.4	0.22	0	0.06	0.09	0.08	1.15	0.01	12.9	0	0	2.9
0	4	0.89	7.1	31.8	0.6	0.19	0	0.11	0.01	0.08	1.10	0.03	4.8	0	0	10.5
0	14	0.55	8.9	30.3	124.4	0.30	0	0.06	0.06	0.04	0.61	0.01	18.6	0	0	5.2
0	12	0.55	5.9	29.0	134.9	0.20	0	0.07	—	0.04	0.70	0.01	12.8	0	0	5.8
0	20	0.72	—	10.0	135.0	—	—	—	—	—	—	—	—	0	—	—
0	3	0.24	7.8	31.9	26.4	0.23	0	0.02	0.08	0.01	0.10	0.02	4.7	0	0	3.7
0	4	0.65	13.3	54.5	87.3	0.30	0	0.04	0.08	0.03	0.17	0.03	5.0	0	0	2.6
0	10	0.84	3.3	23.1	160.8	0.12	0	0.01	0.14	0.06	0.78	0.01	20.9	0	0	1.5
0	18	0.81	4.1	108.6	95.4	0.11	0	0.08	0.01	0.06	0.78	0.01	18.6	0	0	2.9
1	72	0.66	10.1	120.1	392.3	0.17	4.8	0.12	0.06	0.19	1.05	0.01	28.0	0	0	6.0
0	23	0.77	15.4	60.2	201.0	0.31	0.3	0.13	0.57	0.07	1.71	0.04	24.1	0	0	3.0
0	36	1.08	8.1	39.9	254.1	0.20	0	0.12	0.60	0.10	1.21	0.01	27.0	0	0	2.0
0	36	1.08	8.1	106.5	111.9	0.20	0	0.12	0.60	0.10	1.21	0.01	27.0	0	0	2.0
0	10	0.84	3.3	23.1	160.8	0.12	0	0.01	0.14	0.06	0.78	0.01	20.9	0	0	1.5
2	57	0.73	15.1	85.7	255.6	0.24	4.8	0.10	—	0.12	0.89	0.07	17.9	0.4	0	6.8
0	48	0.74	10.6	83.2	226.0	0.23	0	0.10	—	0.08	1.64	0.03	19.6	0	0	6.1
0	15	1.32	18.6	54.9	238.5	0.48	0	0.15	0.15	0.09	1.48	0.04	35.1	0	0	1.9
0	15	1.32	18.6	60.9	84.9	0.48	0	0.15	0.15	0.09	1.48	0.04	15.0	0	0	10.1
0	14	0.86	27.7	83.2	184.5	0.60	0	0.05	0.24	0.02	1.26	0.05	7.8	0	0	4.1
14	9	0.26	2.2	22.4	6.8	0.09	7.1	0.03	0.07	0.04	0.23	0.01	6.3	0.2	0.1	2.6
0	—	0.72	—	—	340.1	—	0	—	—	—	—	—	—	—	—	—
38	21	1.15	9.1	67.3	424.1	0.42	117.4	0.22	0.47	0.13	1.24	0.03	50.2	0.1	0.1	12.9
30	61	1.17	20.8	61.8	236.0	0.56	5.9	0.14	0.53	0.15	1.49	0.06	54.0	1.1	0.1	9.2
20	21	1.55	17.0	68.0	262.7	0.68	9.4	0.28	0.24	0.12	1.90	0.05	59.5	0	0.2	9.2
22	21	1.49	17.0	67.2	249.1	0.63	14.5	0.26	0.36	0.12	1.81	0.08	57.8	0	0.2	10.6
17	21	0.91	9.4	59.7	256.6	0.25	17.9	0.10	0.90	0.11	0.87	0.02	24.4	0.1	0.1	4.4
24	89	0.95	14.3	44.5	142.8	0.23	5.0	0.01	0.08	0.02	0.19	0.01	18.9	0	0	1.7
4	26	0.36	13.2	64.8	205.2	0.46	2.4	0.53	—	0.04	0.39	0.03	13.2	0.1	0.1	5.0
0	0	1.80	—	70.0	190.0	—	—	0.15	—	0.17	2.00	0.20	40.0	0	0	—
10	100	1.80	—	—	220.0	—	—	0.15	—	0.17	2.00	—	40.0	0	0.6	—
20	56	1.02	7.8	70.2	289.4	0.28	—	0.17	0.75	0.15	1.25	0.02	34.0	0.4	0.1	8.8
15	42	1.60	18.2	39.3	297.0	0.30	29.6	0.15	0.45	0.18	1.16	0.04	45.6	0	0.1	8.7
0	30	1.42	12.0	74.7	264.5	0.39	0	0.25	—	0.16	2.21	0.02	42.2	0	0	—
0	95	1.36	11.0	71.5	252.0	0.38	0	0.19	0.16	0.14	1.90	0.02	43.5	0.1	0	13.5
0	101	1.63	21.1	106.0	217.7	0.61	0	0.24	0.25	0.16	1.91	0.05	36.5	0	0	16.6
0	36	2.39	89.5	289.0	224.0	1.04	0	0.14	0.37	0.05	0.23	0.09	50.7	0	0	6.3
0	200	0.36	—	—	60.0	—	0	—	—	—	—	—	—	0	0	—
0	20	1.08	—	120.0	160.0	—	0	—	—	—	—	—	—	0	—	—
0	0	0.00	—	—	95.0	—	0	—	—	—	—	—	—	0	—	—
0	15	0.72	23.8	82.3	72.0	0.50	0	0.06	—	0.03	0.39	0.02	5.6	0.2	0	4.0
0	30	0.72	21.0	92.3	79.0	0.42	0	0.08	—	0.04	0.14	0.02	6.8	0	0.1	4.6
0	11	1.73	10.9	122.5	327.1	0.29	17.1	0.22	—	0.16	1.90	0.05	37.2	2.6	0	12.1
106	39	1.80	31.7	162.3	319.6	1.24	100.0	0.22	—	0.22	1.03	0.07	31.7	0.2	0.2	14.6

DA+ Code	Food Description	Quantity	Measure	Wt (g)	H₂O (g)	Ener (kcal)	Prot (g)	Carb (g)	Fiber (g)	Fat (g)	Fat Breakdown (g) Sat	Mono	Poly	Trans
33356	Pie crust mix, prepared, baked	1	slice(s)	20	2.1	100	1.3	10.1	0.4	6.1	1.5	3.5	0.8	—
9007	Pie crust, ready to bake, frozen, enriched, baked	1	slice(s)	16	1.8	82	0.7	7.9	0.2	5.2	1.7	2.5	0.6	—
472	Pumpkin pie, prepared from home recipe	1	slice(s)	155	90.7	316	7.0	40.9	—	14.4	4.9	5.7	2.8	—
	Rolls													
8555	Crescent dinner roll	1	item(s)	28	9.7	78	2.7	13.8	0.6	1.2	0.3	0.3	0.6	—
489	Hamburger roll or bun, plain	1	item(s)	43	14.9	120	4.1	21.3	0.9	1.9	0.5	0.5	0.8	—
490	Hard roll	1	item(s)	57	17.7	167	5.6	30.0	1.3	2.5	0.3	0.6	1.0	—
5127	Kaiser roll	1	item(s)	57	17.7	167	5.6	30.0	1.3	2.5	0.3	0.6	1.0	—
5130	Whole wheat roll or bun	1	item(s)	28	9.4	75	2.5	14.5	2.1	1.3	0.2	0.3	0.6	—
	Sport bars													
37026	Balance original chocolate bar	1	item(s)	50	—	200	14.0	22.0	0.5	6.0	3.5	—	—	—
37024	Balance original peanut butter bar	1	item(s)	50	—	200	14.0	22.0	1.0	6.0	2.5	—	—	—
36580	Clif Bar chocolate brownie energy bar	1	item(s)	68	—	240	10.0	45.0	5.0	4.5	1.5	—	—	0
36583	Clif Bar crunchy peanut butter energy bar	1	item(s)	68	—	250	12.0	40.0	5.0	6.0	1.5	—	—	0
36589	Clif Luna Nutz over Chocolate energy bar	1	item(s)	48	—	180	10.0	25.0	3.0	4.5	2.5	—	—	0
12005	PowerBar apple cinnamon	1	item(s)	65	—	230	9.0	45.0	3.0	2.5	0.5	1.5	0.5	0
16078	PowerBar banana	1	item(s)	65	—	230	9.0	45.0	3.0	2.5	0.5	1.0	0.5	0
16080	PowerBar chocolate	1	item(s)	65	6.4	230	10.0	45.0	3.0	2.0	0.5	0.5	1.0	0
29092	PowerBar peanut butter	1	item(s)	65	—	240	10.0	45.0	3.0	3.5	0.5	—	—	0
	Tortillas													
1391	Corn tortillas, soft	1	item(s)	26	11.9	57	1.5	11.6	1.6	0.7	0.1	0.2	0.4	—
1669	Flour tortilla	1	item(s)	32	9.7	100	2.7	16.4	1.0	2.5	0.6	1.2	0.5	—
	Pancakes, waffles													
8926	Pancakes, blueberry, prepared from recipe	3	item(s)	114	60.6	253	7.0	33.1	0.8	10.5	2.3	2.6	4.7	—
5037	Pancakes, prepared from mix with egg and milk	3	item(s)	114	60.3	249	8.9	32.9	2.1	8.8	2.3	2.4	3.3	—
1390	Taco shells, hard	1	item(s)	13	1.0	62	0.9	8.3	0.6	2.8	0.6	1.6	0.5	0.6
30311	Waffle, 100% whole grain	1	item(s)	75	32.3	200	6.9	25.0	1.9	8.4	2.3	3.3	2.1	—
9219	Waffle, plain, frozen, toasted	2	item(s)	66	20.2	206	4.7	32.5	1.6	6.3	1.1	3.2	1.5	—
500	Waffle, plain, prepared from recipe	1	item(s)	75	31.5	218	5.9	24.7	1.7	10.6	2.1	2.6	5.1	—
	CEREAL, FLOUR, GRAIN, PASTA, NOODLES, POPCORN													
	Grain													
2861	Amaranth, dry	½	cup(s)	98	9.6	365	14.1	64.5	9.1	6.3	1.6	1.4	2.8	—
1953	Barley, pearled, cooked	½	cup(s)	79	54.0	97	1.8	22.2	3.0	0.3	0.1	0	0.2	—
1956	Buckwheat groats, cooked, roasted	½	cup(s)	84	63.5	77	2.8	16.8	2.3	0.5	0.1	0.2	0.2	—
1957	Bulgur, cooked	½	cup(s)	91	70.8	76	2.8	16.9	4.1	0.2	0	0	0.1	—
1963	Couscous, cooked	½	cup(s)	79	57.0	88	3.0	18.2	1.1	0.1	0	0	0.1	—
1967	Millet, cooked	½	cup(s)	120	85.7	143	4.2	28.4	1.6	1.2	0.2	0.2	0.6	—
1969	Oat bran, dry	½	cup(s)	47	3.1	116	8.1	31.1	7.2	3.3	0.6	1.1	1.3	—
1972	Quinoa, dry	½	cup(s)	85	11.3	313	12.0	54.5	5.9	5.2	0.6	1.4	2.8	—
	Rice													
129	Brown, long grain, cooked	½	cup(s)	98	71.3	108	2.5	22.4	1.8	0.9	0.2	0.3	0.3	—
2863	Brown, medium grain, cooked	½	cup(s)	98	71.1	109	2.3	22.9	1.8	0.8	0.2	0.3	0.3	—
37488	Jasmine, saffroned, cooked	½	cup(s)	280	—	340	8.0	78.0	0	0	0	0	0	0
30280	Pilaf, cooked	½	cup(s)	103	74.0	129	2.1	22.2	0.6	3.3	0.6	1.5	1.0	—
28066	Spanish, cooked	½	cup(s)	244	184.2	241	5.7	50.2	3.3	1.9	0.4	0.6	0.7	0
2867	White glutinous, cooked	½	cup(s)	87	66.7	84	1.8	18.3	0.9	0.2	0	0.1	0.1	—
484	White, long grain, boiled	½	cup(s)	79	54.1	103	2.1	22.3	0.3	0.2	0.1	0.1	0.1	—
482	White, long grain, enriched, instant, boiled	½	cup(s)	83	59.4	97	1.8	20.7	0.5	0.4	0	0.1	0	—
486	White, long grain, enriched, parboiled, cooked	½	cup(s)	79	55.6	97	2.3	20.6	0.7	0.3	0.1	0.1	0.1	—
1194	Wild brown, cooked	½	cup(s)	82	60.6	83	3.3	17.5	1.5	0.3	0	0	0.2	—
	Flour & grain fractions													
505	All purpose flour, self rising, enriched	½	cup(s)	63	6.6	221	6.2	46.4	1.7	0.6	0.1	0	0.2	—
503	All purpose flour, white, bleached, enriched	½	cup(s)	63	7.4	228	6.4	47.7	1.7	0.6	0.1	0	0.2	—
1643	Barley flour	½	cup(s)	56	5.5	198	4.2	44.7	2.1	0.8	0.2	0.1	0.4	—
383	Buckwheat flour, whole groat	½	cup(s)	60	6.7	201	7.6	42.3	6.0	1.9	0.4	0.6	0.6	—
504	Cake wheat flour, enriched	½	cup(s)	69	8.6	248	5.6	53.5	1.2	0.6	0.1	0.1	0.3	—
426	Cornmeal, degermed, enriched	½	cup(s)	69	7.8	255	5.0	54.6	2.8	1.2	0.1	0.2	0.5	0
424	Cornmeal, yellow whole grain	½	cup(s)	61	6.2	221	4.9	46.9	4.4	2.2	0.3	0.6	1.0	—
1978	Dark rye flour	½	cup(s)	64	7.1	207	9.0	44.0	14.5	1.7	0.2	0.2	0.8	—
1644	Masa corn flour, enriched	½	cup(s)	57	5.1	208	5.3	43.5	5.5	2.1	0.3	0.6	1.0	—
1976	Rice flour, brown	½	cup(s)	79	9.4	287	5.7	60.4	3.6	2.2	0.4	0.8	0.8	—
1645	Rice flour, white	½	cup(s)	79	9.4	289	4.7	63.3	1.9	1.1	0.3	0.3	0.3	—
1980	Semolina, enriched	½	cup(s)	84	10.6	301	10.6	60.8	3.2	0.9	0.1	0.1	0.4	—
2827	Soy flour, raw	½	cup(s)	42	2.2	185	14.7	14.9	4.1	8.8	1.3	1.9	4.9	—
1990	Wheat germ, crude	2	tablespoon(s)	14	1.6	52	3.3	7.4	1.9	1.4	0.2	0.2	0.9	—
506	Whole wheat flour	½	cup(s)	60	6.2	203	8.2	43.5	7.3	1.1	0.2	0.1	0.5	—
	Breakfast bars													
39230	Atkins Morning Start apple crisp breakfast bar	1	item(s)	37	—	170	11.0	12.0	6.0	9.0	4.0	—	—	—
10571	Nutri-Grain apple cinnamon cereal bar	1	item(s)	37	—	140	2.0	27.0	1.0	3.0	0.5	2.0	0.5	—

CEREAL, FLOUR, GRAIN, PASTA, NOODLES, POPCORN—Continued

Chol (mg)	Calc (mg)	Iron (mg)	Magn (mg)	Pota (mg)	Sodi (mg)	Zinc (mg)	Vit A (µg)	Thia (mg)	Vit E (mg α)	Ribo (mg)	Niac (mg)	Vit B$_6$ (mg)	Fola (µg)	Vit C (mg)	Vit B$_{12}$ (µg)	Sele (µg)
0	12	0.43	3.0	12.4	145.8	0.07	0	0.06	—	0.03	0.47	0.01	14.0	0	0	4.4
0	3	0.36	2.9	17.6	103.5	0.05	0	0.04	0.42	0.06	0.39	0.01	8.8	0	0	0.5
65	146	1.96	29.5	288.3	348.8	0.71	660.3	0.14	—	0.31	1.21	0.07	32.6	2.6	0.1	11.0
0	39	0.93	5.9	26.3	134.1	0.18	0	0.11	0.02	0.09	1.16	0.02	31.1	0	0.1	5.5
0	59	1.42	9.0	40.4	206.0	0.28	0	0.17	0.03	0.13	1.78	0.03	47.7	0	0.1	8.4
0	54	1.87	15.4	61.6	310.1	0.53	0	0.27	0.23	0.19	2.41	0.02	54.2	0	0	22.3
0	54	1.86	15.4	61.6	310.1	0.53	0	0.27	0.23	0.19	2.41	0.01	54.2	0	0	22.3
0	30	0.69	24.1	77.1	135.5	0.57	0	0.07	0.26	0.04	1.04	0.06	8.5	0	0	14.0
3	100	4.50	40.0	160.0	180.0	3.75	—	0.37	—	0.42	5.00	0.50	100.0	60.0	1.5	17.5
3	100	4.50	40.0	130.0	230.0	3.75	—	0.37	—	0.42	5.00	0.50	100.0	60.0	1.5	17.5
0	250	4.50	100.0	370.0	150.0	3.00	—	0.37	—	0.25	3.00	0.40	80.0	60.0	0.9	14.0
0	250	4.50	100.0	230.0	250.0	3.00	—	0.37	—	0.25	3.00	0.40	80.0	60.0	0.9	14.0
0	350	5.40	80.0	190.0	190.0	5.25	—	1.20	—	1.36	16.00	2.00	400.0	60.0	6.0	24.5
0	300	6.30	140.0	125.0	100.0	5.25	—	1.50	—	1.70	20.00	2.00	400.0	60.0	6.0	—
0	300	6.30	140.0	190.0	100.0	5.25	0	1.50	—	1.70	20.00	2.00	400.0	60.0	6.0	—
0	300	6.30	140.0	200.0	95.0	5.25	0	1.50	—	1.70	20.00	2.00	400.0	60.0	6.0	5.1
0	300	6.30	140.0	130.0	120.0	5.25	0	1.50	—	1.70	20.00	2.00	400.0	60.0	6.0	—
0	21	0.32	18.7	48.4	11.7	0.34	0	0.02	0.07	0.02	0.39	0.06	1.3	0	0	1.6
0	41	1.06	7.0	49.6	203.5	0.17	0	0.17	0.06	0.08	1.14	0.01	33.3	0	0	7.1
64	235	1.96	18.2	157.3	469.7	0.61	57.0	0.22	—	0.31	1.73	0.05	41.0	2.5	0.2	16.0
81	245	1.48	25.1	226.9	575.7	0.85	82.1	0.22	—	0.35	1.40	0.12	104.6	0.7	0.4	—
0	13	0.25	11.3	29.7	51.7	0.21	0.1	0.03	0.09	0.01	0.25	0.03	9.2	0	0	0.6
71	194	1.60	28.5	171.0	371.3	0.87	48.8	0.15	0.32	0.25	1.47	0.08	28.5	0	0.4	20.0
10	203	4.56	15.8	95.0	481.8	0.35	262.7	0.34	0.64	0.46	5.86	0.68	49.5	0	1.9	8.3
52	191	1.73	14.3	119.3	383.3	0.51	48.8	0.19	—	0.26	1.55	0.04	34.5	0.3	0.2	34.7
0	149	7.40	259.3	356.8	20.5	3.10	0	0.06	—	0.20	1.24	0.20	47.8	4.1	0	—
0	9	1.04	17.3	73.0	2.4	0.64	0	0.06	0.01	0.04	1.61	0.09	12.6	0	0	6.8
0	6	0.67	42.8	73.9	3.4	0.51	0	0.03	0.07	0.03	0.79	0.06	11.8	0	0	1.8
0	9	0.87	29.1	61.9	4.6	0.51	0	0.05	0.01	0.02	0.91	0.07	16.4	0	0	0.5
0	6	0.30	6.3	45.5	3.9	0.20	0	0.05	0.10	0.02	0.77	0.04	11.8	0	0	21.6
0	4	0.75	52.8	74.4	2.4	1.09	0	0.12	0.02	0.09	1.59	0.13	22.8	0	0	1.1
0	27	2.54	110.5	266.0	1.9	1.46	0	0.55	0.47	0.10	0.43	0.07	24.4	0	0	21.2
0	40	3.88	167.4	478.5	4.2	2.62	0.8	0.30	2.06	0.26	1.28	0.40	156.4	0	0	7.2
0	10	0.41	41.9	41.9	4.9	0.61	0	0.09	0.02	0.02	1.49	0.14	3.9	0	0	9.6
0	10	0.51	42.9	77.0	1.0	0.60	0	0.09	—	0.01	1.29	0.14	3.9	0	0	38.0
0	—	2.16	—	—	780.0	—	—	—	—	—	—	—	—	—	—	—
0	11	1.16	9.3	54.6	390.4	0.37	33.0	0.13	0.28	0.02	1.23	0.06	44.3	0.4	0	4.3
0	37	1.52	95.4	330.5	97.1	1.40	—	0.27	0.12	0.05	3.24	0.38	19.0	22.6	0	14.3
0	2	0.12	4.4	8.7	4.4	0.35	0	0.01	0.03	0.01	0.25	0.02	0.9	0	0	4.9
0	8	0.94	9.5	27.7	0.8	0.38	0	0.12	0.03	0.01	1.16	0.07	45.8	0	0	5.9
0	7	1.46	4.1	7.4	3.3	0.40	0	0.06	0.01	0.01	1.43	0.04	57.8	0	0	4.0
0	15	1.43	7.1	44.2	1.6	0.29	0	0.16	0.01	0.01	1.82	0.12	64.0	0	0	7.3
0	2	0.49	26.2	82.8	2.5	1.09	0	0.04	0.19	0.07	1.05	0.11	21.3	0	0	0.7
0	211	2.90	11.9	77.5	793.7	0.38	0	0.42	0.02	0.24	3.64	0.02	122.5	0	0	21.5
0	9	2.90	13.7	66.9	1.2	0.42	0	0.48	0.02	0.30	3.68	0.02	114.4	0	0	21.2
0	16	0.70	45.4	185.9	4.5	1.04	0	0.06	—	0.02	2.56	0.14	12.9	0	0	2.0
0	25	2.42	150.6	346.2	6.6	1.86	0	0.24	0.18	0.10	3.68	0.34	32.4	0	0	3.4
0	10	5.01	11.0	71.9	1.4	0.42	0	0.61	0.01	0.29	4.65	0.02	127.4	0	0	3.4
0	2	2.98	24.1	104.9	4.8	0.48	7.6	0.42	0.10	0.28	3.66	0.12	148.3	0	0	8.0
0	4	2.10	77.5	175.1	21.3	1.10	6.7	0.22	0.24	0.12	2.20	0.18	15.2	0	0	9.4
0	36	4.12	158.7	467.2	0.6	3.58	0.6	0.20	0.90	0.16	2.72	0.28	38.4	0	0	22.8
0	80	4.10	62.7	169.9	2.8	1.00	0	0.80	0.08	0.42	5.60	0.20	132.8	0	0	8.5
0	9	1.56	88.5	228.3	6.3	1.92	0	0.34	0.94	0.06	5.00	0.58	12.6	0	0	—
0	8	0.26	27.6	60.0	0	0.62	0	0.10	0.08	0.02	2.04	0.34	3.2	0	0	11.9
0	14	3.64	39.2	155.3	0.8	0.86	0	0.66	0.20	0.46	5.00	0.08	152.8	0	0	74.6
0	87	2.70	182.0	1067.0	5.5	1.65	2.5	0.24	0.82	0.48	1.83	0.18	146.4	0	0	3.2
0	6	0.90	34.4	128.2	1.7	1.76	0	0.27	—	0.07	0.97	0.18	40.4	0	0	11.4
0	20	2.32	82.8	243.0	3.0	1.74	0	0.26	0.48	0.12	3.82	0.20	26.4	0	0	42.4
0	200	—	—	90.0	70.0	—	—	0.22	—	0.25	3.00	—	—	9.0	—	—
0	200	1.80	8.0	75.0	110.0	1.50	—	0.37	—	0.42	5.00	0.50	40.0	0	—	—

DA+ Code	Food Description	Quantity	Measure	Wt (g)	H₂O (g)	Ener (kcal)	Prot (g)	Carb (g)	Fiber (g)	Fat (g)	Sat	Mono	Poly	Trans
10647	Nutri-Grain blueberry cereal bar	1	item(s)	37	5.4	140	2.0	27.0	1.0	3.0	0.5	2.0	0.5	—
10648	Nutri-Grain raspberry cereal bar	1	item(s)	37	5.4	140	2.0	27.0	1.0	3.0	0.5	2.0	0.5	—
10649	Nutri-Grain strawberry cereal bar	1	item(s)	37	5.4	140	2.0	27.0	1.0	3.0	0.5	2.0	0.5	—
	Breakfast cereals, hot													
1260	Cream of Wheat, instant, prepared	½	cup(s)	121	—	388	12.9	73.3	4.3	0	0	0	0	0
365	Farina, enriched, cooked with water and salt	½	cup(s)	117	102.4	56	1.7	12.2	0.3	0.1	0	0	0	—
363	Grits, white corn, regular and quick, enriched, cooked with water and salt	½	cup(s)	121	103.3	71	1.7	15.6	0.4	0.2	0	0.1	0.1	—
8636	Grits, yellow corn, regular and quick, enriched, cooked with salt	½	cup(s)	121	103.3	71	1.7	15.6	0.4	0.2	0	0.1	0.1	—
8657	Oatmeal, cooked with water	½	cup(s)	117	97.8	83	3.0	14.0	2.0	1.8	0.4	0.5	0.7	0
5500	Oatmeal, maple and brown sugar, instant, prepared	1	item(s)	198	150.2	200	4.8	40.4	2.4	2.2	0.4	0.7	0.8	—
5510	Oatmeal, ready to serve, packet, prepared	1	item(s)	186	158.7	112	4.1	19.8	2.7	2.0	0.4	0.7	0.8	—
	Breakfast cereals, ready to eat													
1197	All-Bran	1	cup(s)	62	1.3	160	8.1	46.0	18.2	2.0	0.4	0.4	1.3	0
1200	All-Bran Buds	1	cup(s)	91	2.7	212	6.4	72.7	39.1	1.9	0.4	0.5	1.2	0
1199	Apple Jacks	1	cup(s)	33	0.9	130	1.0	30.0	0.5	0.5	0	—	—	0
1204	Cap'n Crunch	1	cup(s)	36	0.9	147	1.3	30.7	1.3	2.0	0.5	0.4	0.3	—
1205	Cap'n Crunch Crunchberries	1	cup(s)	35	0.9	133	1.3	29.3	1.3	2.0	0.5	0.4	0.3	—
1206	Cheerios	1	cup(s)	30	1.0	110	3.0	22.0	3.0	2.0	0	0.5	0.5	—
3415	Cocoa Puffs	1	cup(s)	30	0.6	120	1.0	26.0	0.2	1.0	—	—	—	—
1207	Cocoa Rice Krispies	1	cup(s)	41	1.0	160	1.3	36.0	1.3	1.3	0.7	0	0	—
5522	Complete wheat bran flakes	1	cup(s)	39	1.4	120	4.0	30.7	6.7	0.7	—	—	—	0
1211	Corn Flakes	1	cup(s)	28	0.9	100	2.0	24.0	1.0	0	0	0	0	0
1247	Corn Pops	1	cup(s)	31	0.9	120	1.0	28.0	0.3	0	0	0	0	0
1937	Cracklin' Oat Bran	1	cup(s)	65	2.3	267	5.3	46.7	8.0	9.3	4.0	4.7	1.3	0
1220	Froot Loops	1	cup(s)	32	0.8	120	1.0	28.0	1.0	1.0	0.5	0	0	—
38214	Frosted Cheerios	1	cup(s)	37	—	149	2.5	31.1	1.2	1.2	—	—	—	—
372	Frosted Flakes	1	cup(s)	41	1.1	160	1.3	37.3	1.3	0	0	0	0	—
38215	Frosted Mini Chex	1	cup(s)	40	—	147	1.3	36.0	0	0	0	0	0	0
10268	Frosted Mini-Wheats	1	cup(s)	59	3.1	208	5.8	47.4	5.8	1.2	0	0	0.6	0
38216	Frosted Wheaties	1	cup(s)	40	—	147	1.3	36.0	0.3	0	0	0	0	0
1223	Granola, prepared	½	cup(s)	61	3.3	298	9.1	32.5	5.5	14.7	2.5	5.8	5.6	0
2415	Honey Bunches of Oats honey roasted	1	cup(s)	40	0.9	160	2.7	33.3	1.3	2.0	0.7	1.2	0.1	—
1227	Honey Nut Cheerios	1	cup(s)	37	0.9	149	3.7	29.9	2.5	1.9	0	0.6	0.6	—
2424	Honeycomb	1	cup(s)	22	0.3	83	1.5	19.5	0.8	0.4	0	—	—	—
10286	Kashi whole grain puffs	1	cup(s)	19	—	70	2.0	15.0	1.0	0.5	0	—	—	0
41142	Kellogg's Mueslix	1	cup(s)	83	7.2	298	7.6	60.8	6.1	4.6	0.7	2.4	1.5	0
1231	Kix	1	cup(s)	24	0.5	96	1.6	20.8	0.8	0.4	—	—	—	—
30569	Life	1	cup(s)	43	1.7	160	4.0	33.3	2.7	2.0	0.3	0.6	0.6	—
1233	Lucky Charms	1	cup(s)	24	0.6	96	1.6	20.0	0.8	0.8	0	—	—	—
38220	Multi Grain Cheerios	1	cup(s)	30	—	110	3.0	24.0	3.0	1.0	—	—	—	—
1201	Multi-Bran Chex	1	cup(s)	63	1.3	216	4.3	52.9	8.6	1.6	0	0	0.5	0
13633	Post Bran Flakes	1	cup(s)	40	1.5	133	4.0	32.0	6.7	0.7	0	—	—	—
1241	Product 19	1	cup(s)	30	1.0	100	2.0	25.0	1.0	0	0	0	0	0
32432	Puffed rice, fortified	1	cup(s)	14	0.4	56	0.9	12.6	0.2	0.1	0	—	—	—
32433	Puffed wheat, fortified	1	cup(s)	12	0.4	44	1.8	9.6	0.5	0.1	0	—	—	—
13334	Quaker 100% natural granola oats & honey	½	cup(s)	48	—	220	5.0	31.0	3.0	9.0	3.8	4.1	1.2	—
13335	Quaker 100% natural granola oats, honey & raisins	½	cup(s)	51	—	230	5.0	34.0	3.0	9.0	3.6	3.8	1.1	—
2420	Raisin Bran	1	cup(s)	59	5.0	190	4.0	46.0	8.0	1.0	0	0.1	0.4	—
1244	Rice Chex	1	cup(s)	31	0.8	120	2.0	27.0	0.3	0	0	0	0	0
1245	Rice Krispies	1	cup(s)	26	0.8	96	1.6	23.2	0	0	0	0	0	0
5593	Shredded Wheat	1	cup(s)	49	0.4	177	5.8	40.9	6.9	1.1	0.1	0	0.2	0
1248	Smacks	1	cup(s)	36	1.1	133	2.7	32.0	1.3	0.7	—	—	—	—
1246	Special K	1	cup(s)	31	0.9	110	7.0	22.0	0.5	0	0	0	0	0
3428	Total corn flakes	1	cup(s)	23	0.6	83	1.5	18.0	0.6	0	0	0	0	0
1253	Total whole grain	1	cup(s)	40	1.1	147	2.7	30.7	4.0	1.3	—	—	—	—
1254	Trix	1	cup(s)	30	0.6	120	1.0	27.0	1.0	1.0	—	—	—	—
382	Wheat germ, toasted	2	tablespoon(s)	14	0.8	54	4.1	7.0	2.1	1.5	0.3	0.2	0.9	—
1257	Wheaties	1	cup(s)	36	1.2	132	3.6	28.8	3.6	1.2	—	—	—	—
	Pasta, noodles													
449	Chinese chow mein noodles, cooked	½	cup(s)	23	0.2	119	1.9	12.9	0.9	6.9	1.0	1.7	3.9	—
1995	Corn pasta, cooked	½	cup(s)	70	47.8	88	1.8	19.5	3.4	0.5	0.1	0.1	0.2	—
448	Egg noodles, enriched, cooked	½	cup(s)	80	54.2	110	3.6	20.1	1.0	1.7	0.3	0.5	0.4	0
1563	Egg noodles, spinach, enriched, cooked	½	cup(s)	80	54.8	106	4.0	19.4	1.8	1.3	0.3	0.4	0.3	—
440	Macaroni, enriched, cooked	½	cup(s)	70	43.5	111	4.1	21.6	1.3	0.7	0.1	0.1	0.2	0
2000	Macaroni, tricolor vegetable, enriched, cooked	½	cup(s)	67	45.8	86	3.0	17.8	2.9	0.1	0	0	0	—
1996	Plain pasta, fresh-refrigerated, cooked	½	cup(s)	64	43.9	84	3.3	16.0	—	0.7	0.1	0.1	0.3	—

CEREAL, FLOUR, GRAIN, PASTA, NOODLES, POPCORN—Continued

Chol (mg)	Calc (mg)	Iron (mg)	Magn (mg)	Pota (mg)	Sodi (mg)	Zinc (mg)	Vit A (µg)	Thia (mg)	Vit E (mg α)	Ribo (mg)	Niac (mg)	Vit B$_6$ (mg)	Fola (µg)	Vit C (mg)	Vit B$_{12}$ (µg)	Sele (µg)
0	200	1.80	8.0	75.0	110.0	1.50	—	0.37	—	0.42	5.00	0.50	40.0	0	0	—
0	200	1.80	8.0	70.0	110.0	1.50	—	0.37	—	0.42	5.00	0.50	40.0	0	0	—
0	200	1.80	8.0	55.0	110.0	1.50	—	0.37	—	0.42	5.00	0.50	40.0	0	0	—
0	862	34.91	21.4	150.8	732.6	0.86	—	1.59	—	1.47	21.55	2.15	431.0	0	0	—
0	5	0.58	2.3	15.1	383.3	0.09	0	0.07	0.01	0.05	0.57	0.01	39.6	0	0	10.6
0	4	0.73	6.1	25.4	269.8	0.08	0	0.10	0.02	0.07	0.87	0.03	39.9	0	0	3.8
0	4	0.73	6.1	25.4	269.8	0.08	2.4	0.10	0.02	0.07	0.87	0.03	39.9			3.3
0	11	1.05	31.6	81.9	4.7	1.17	0	0.09	0.09	0.02	0.26	0.01	7.0	0	0	6.3
0	26	6.83	49.9	126.4	403.5	1.03	0	1.02	—	0.05	1.56	0.30	42.2	0	0	11.1
0	21	3.96	44.7	112.4	240.9	0.92	0	0.60	—	0.04	0.77	0.18	18.7	0	0	3.8
0	241	10.90	224.4	632.4	150.0	3.00	300.1	1.40	—	1.68	9.16	7.44	800.0	12.4	12.0	5.8
0	57	13.64	186.4	909.1	614.5	4.55	464.5	1.09	1.42	1.27	15.45	6.09	1222.7	18.2	18.2	26.3
0	0	4.50	8.0	30.0	130.0	1.50	150.2	0.37	—	0.42	5.00	0.50	100.0	15.0	1.5	2.4
0	5	6.80	20.0	73.3	266.7	5.00	2.5	0.51	—	0.57	6.68	0.67	133.5	0	0	6.7
0	7	6.53	18.7	73.3	240.0	5.13	2.4	0.51	—	0.57	6.68	0.67	133.7	0	0	6.7
0	100	8.10	40.0	95.0	280.0	3.75	150.3	0.37	—	0.42	5.00	0.50	200.0	6.0	1.5	11.3
0	100	4.50	8.0	50.0	170.0	3.75	0	0.37	—	0.42	5.00	0.50	100.0	6.0	1.5	2.0
0	53	6.00	10.7	66.7	253.3	2.00	200.1	0.49	—	0.56	6.67	0.67	133.3	20.0	2.0	5.8
0	0	24.00	53.3	226.7	280.0	20.00	300.1	2.00	—	2.27	26.67	2.67	533.3	80.0	8.0	4.1
0	0	8.10	3.4	25.0	200.0	0.16	149.8	0.37	—	0.42	5.00	0.50	100.0	6.0	1.5	1.4
0	0	1.80	2.5	25.0	120.0	1.50	150.0	0.37	—	0.42	5.00	0.50	100.0	6.0	1.5	2.0
0	27	2.40	80.0	293.3	200.0	2.00	299.9	0.49	—	0.56	6.67	0.67	133.3	20.0	2.0	14.4
0	0	4.50	8.0	35.0	150.0	1.50	150.1	0.37	—	0.42	5.00	0.50	100.0	15.0	1.5	2.3
0	124	5.60	19.9	68.4	261.3	4.67	—	0.46	—	0.52	6.22	0.62	124.4	7.5	1.9	—
0	0	6.00	3.7	26.7	200.0	0.20	200.1	0.49	—	0.56	6.67	0.67	133.3	8.0	2.0	1.8
0	133	12.00	—	33.3	266.7	4.00	—	0.49	—	0.56	6.67	0.67	266.7	8.0	2.0	—
0	0	16.66	69.4	196.7	5.8	1.74	0	0.43	—	0.49	5.78	0.58	115.7	0	1.7	2.4
0	133	10.80	0	46.7	266.7	10.00	—	1.00	—	1.13	13.33	1.33	533.3	8.0	4.0	—
0	48	2.58	106.8	329.4	15.3	2.45	0.6	0.44	6.77	0.17	1.30	0.17	50.0	0.7	0	17.0
0	0	10.80	21.3	0	253.3	0.40	—	0.49	—	0.56	6.67	0.67	133.0	0	2.0	—
0	124	5.60	39.8	112.0	336.0	4.67	—	0.46	—	0.52	6.22	0.62	248.9	7.5	1.9	8.8
0	0	2.03	6.0	26.3	165.4	1.13	—	0.28	—	0.32	3.74	0.37	75.0	0	1.1	—
0	0	0.36	—	60.0	0	0	0	0.03	—	0.03	0.80	0.00	—	0	—	—
0	48	6.83	74.2	363.3	257.5	5.67	136.7	0.67	6.00	0.67	8.33	3.08	615.0	0.3	9.2	14.4
0	120	6.48	6.4	28.0	216.0	3.00	120.2	0.30	—	0.34	4.00	0.40	160.0	4.8	1.2	4.8
0	149	11.87	41.3	120.0	213.3	5.33	0.9	0.53	—	0.60	7.12	0.71	142.4	0	0	10.7
0	80	3.60	12.8	48.0	168.0	3.00	—	0.30	—	0.34	4.00	0.40	160.0	4.8	1.2	4.8
0	100	18.00	24.0	85.0	200.0	15.00	—	1.50	—	1.70	20.00	2.00	400.0	15.0	6.0	—
0	108	17.50	64.8	237.7	410.6	4.05	171.1	0.40	—	0.45	5.40	0.54	432.2	6.5	1.6	4.9
0	0	10.80	80.0	266.7	280.0	2.00	—	0.49	—	0.56	6.67	0.67	133.3	0	2.0	—
0	0	18.00	16.0	50.0	210.0	15.00	225.3	1.50	—	1.70	20.00	2.00	400.0	60.0	6.0	3.6
0	1	4.43	3.5	15.8	0.4	0.14	0	0.36	—	0.25	4.94	0.01	2.7	0	0	1.5
0	3	3.80	17.4	41.8	0.5	0.28	0	0.31	—	0.21	4.23	0.02	3.8	0	0	14.8
0	61	1.20	51.0	220.0	20.0	1.05	0.5	0.13	—	0.12	0.82	0.07	15.0	0.2	0.1	8.3
0	59	1.20	49.0	250.0	20.0	0.99	0.5	0.13	—	0.12	0.80	0.08	14.1	0.4	0.1	8.8
0	20	10.80	80.0	360.0	360.0	2.25	—	0.37	—	0.42	5.00	0.50	100.0	0	2.1	—
0	100	9.00	9.3	35.0	290.0	3.75	—	0.37	—	0.42	5.00	0.50	200.0	6.0	1.5	1.2
0	0	1.44	12.8	32.0	256.0	0.48	120.1	0.30	—	0.34	4.80	0.40	80.0	4.8	1.2	4.1
0	18	2.90	60.3	179.3	1.1	1.37	0	0.14	—	0.12	3.47	0.18	21.1	0	0	2.0
0	0	0.48	10.7	53.3	66.7	0.40	200.2	0.49	—	0.56	6.67	0.67	133.3	8.0	2.0	17.5
0	0	8.10	16.0	60.0	220.0	0.90	225.1	0.52	—	0.59	7.00	2.00	400.0	21.0	6.0	7.0
0	752	13.53	0	22.6	157.9	11.28	112.8	1.13	22.56	1.28	15.04	1.50	300.8	45.1	4.5	1.2
0	1333	24.00	32.0	120.0	253.3	20.00	200.4	2.00	31.32	2.27	26.67	2.67	533.3	80.0	8.0	1.9
0	100	4.50	0	15.0	190.0	3.75	150.3	0.37	—	0.42	5.00	0.50	100.0	6.0	1.5	6.0
0	6	1.28	45.2	133.8	0.6	2.35	0.7	0.23	2.25	0.11	0.79	0.13	49.7	0.8	0	9.2
0	24	9.72	38.4	126.0	264.0	9.00	180.4	0.90	—	1.02	12.00	1.20	240.0	7.2	3.6	1.7
0	5	1.06	11.7	27.0	98.8	0.31	0	0.13	0.78	0.09	1.33	0.02	20.3	0	0	9.7
0	1	0.18	25.2	21.7	0	0.44	2.1	0.04	—	0.02	0.39	0.04	4.2	0	0	2.0
23	10	1.17	16.8	30.4	4.0	0.52	4.8	0.23	0.13	0.11	1.66	0.03	67.2	0	0.1	19.1
26	15	0.87	19.2	29.6	9.6	0.50	8.0	0.19	0.46	0.09	1.17	0.09	51.2	0	0.1	17.4
0	5	0.90	12.6	30.8	0.7	0.36	0	0.19	0.04	0.10	1.18	0.03	51.1	0	0	18.5
0	7	0.33	12.7	20.8	4.0	0.30	3.4	0.08	0.14	0.04	0.72	0.02	43.6	0	0	13.3
21	4	0.73	11.5	15.4	3.8	0.36	3.8	0.13	—	0.10	0.64	0.02	41.0	0	0.1	—

Food Composition   (Computer code is for Cengage Diet Analysis program)   (For purposes of calculations, use "0" for t, <1, <.1, <.01, etc.)

DA+ Code	Food Description	Quantity	Measure	Wt (g)	H₂O (g)	Ener (kcal)	Prot (g)	Carb (g)	Fiber (g)	Fat (g)	Fat Breakdown (g) Sat	Mono	Poly	Trans
1725	Ramen noodles, cooked	½	cup(s)	114	94.5	104	3.0	15.4	1.0	4.3	0.2	0.2	0.2	—
2878	Soba noodles, cooked	½	cup(s)	95	69.4	94	4.8	20.4	—	0.1	0	0	0	—
2879	Somen noodles, cooked	½	cup(s)	88	59.8	115	3.5	24.2	—	0.2	0	0	0.1	—
493	Spaghetti, al dente, cooked	½	cup(s)	65	41.6	95	3.5	19.5	1.0	0.5	0.1	0.1	0.2	—
2884	Spaghetti, whole wheat, cooked	½	cup(s)	70	47.0	87	3.7	18.6	3.2	0.4	0.1	0.1	0.1	—
	Popcorn													
476	Air popped	1	cup(s)	8	0.3	31	1.0	6.2	1.2	0.4	0	0.1	0.2	—
4619	Caramel	1	cup(s)	35	1.0	152	1.3	27.8	1.8	4.5	1.3	1.0	1.6	—
4620	Cheese flavored	1	cup(s)	36	0.9	188	3.3	18.4	3.5	11.8	2.3	3.5	5.5	—
477	Popped in oil	1	cup(s)	11	0.1	64	0.8	5.0	0.9	4.8	0.8	1.1	2.6	—
	FRUIT AND FRUIT JUICES													
	Apples													
952	Juice, prepared from frozen concentrate	½	cup(s)	120	105.0	56	0.2	13.8	0.1	0.1	0	0	0	—
225	Juice, unsweetened, canned	½	cup(s)	124	109.0	58	0.1	14.5	0.1	0.1	0	0	0	—
224	Slices	½	cup(s)	55	47.1	29	0.1	7.6	1.3	0.1	0	0	0	—
946	Slices without skin, boiled	½	cup(s)	86	73.1	45	0.2	11.7	2.1	0.3	0	0	0.1	—
223	Raw medium, with peel	1	item(s)	138	118.1	72	0.4	19.1	3.3	0.2	0	0	0.1	—
948	Dried, sulfured	¼	cup(s)	22	6.8	52	0.2	14.2	1.9	0.1	0	0	0	—
226	Applesauce, sweetened, canned	½	cup(s)	128	101.5	97	0.2	25.4	1.5	0.2	0	0	0.1	—
227	Applesauce, unsweetened, canned	½	cup(s)	122	107.8	52	0.2	13.8	1.5	0.1	0	0	0	—
38492	Crabapples	1	item(s)	35	27.6	27	0.1	7.0	0.9	0.1	0	0	0	—
	Apricot													
228	Fresh without pits	4	item(s)	140	120.9	67	2.0	15.6	2.8	0.5	0	0.2	0.1	—
229	Halves with skin, canned in heavy syrup	½	cup(s)	129	100.1	107	0.7	27.7	2.1	0.1	0	0	0	—
230	Halves, dried, sulfured	¼	cup(s)	33	10.1	79	1.1	20.6	2.4	0.2	0	0	0	—
	Avocado													
233	California, whole, without skin or pit	½	cup(s)	115	83.2	192	2.2	9.9	7.8	17.7	2.4	11.3	2.1	—
234	Florida, whole, without skin or pit	½	cup(s)	115	90.6	138	2.5	9.0	6.4	11.5	2.2	6.3	1.9	—
2998	Pureed	⅛	cup(s)	28	20.2	44	0.5	2.4	1.8	4.0	0.6	2.7	0.5	—
	Banana													
4580	Dried chips	¼	cup(s)	55	2.4	285	1.3	32.1	4.2	18.5	15.9	1.1	0.3	—
235	Fresh whole, without peel	1	item(s)	118	88.4	105	1.3	27.0	3.1	0.4	0.1	0	0.1	—
	Blackberries													
237	Raw	½	cup(s)	72	63.5	31	1.0	6.9	3.8	0.4	0	0	0.2	—
958	Unsweetened, frozen	½	cup(s)	76	62.1	48	0.9	11.8	3.8	0.3	0	0	0.2	—
	Blueberries													
959	Canned in Heavy Syrup	½	cup(s)	128	98.3	113	0.8	28.2	2.0	0.4	0	0.1	0.2	—
238	Raw	½	cup(s)	73	61.1	41	0.5	10.5	1.7	0.2	0	0	0.1	—
960	Unsweetened, frozen	½	cup(s)	78	67.1	40	0.3	9.4	2.1	0.5	0	0.1	0.2	—
	Boysenberries													
961	Canned in heavy syrup	½	cup(s)	128	97.6	113	1.3	28.6	3.3	0.2	0	0	0.1	—
962	Unsweetened, frozen	½	cup(s)	66	56.7	33	0.7	8.0	3.5	0.2	0	0	0.1	—
35576	**Breadfruit**	1	item(s)	384	271.3	396	4.1	104.1	18.8	0.9	0.2	0.1	0.3	—
	Cherries													
967	Sour red, canned in water	½	cup(s)	122	109.7	44	0.9	10.9	1.3	0.1	0	0	0	—
3000	Sour red, raw	½	cup(s)	78	66.8	39	0.8	9.4	1.2	0.2	0.1	0.1	0.1	—
3004	Sweet, canned in heavy syrup	½	cup(s)	127	98.2	105	0.8	26.9	1.9	0.2	0	0.1	0.1	—
969	Sweet, canned in water	½	cup(s)	124	107.9	57	1.0	14.6	1.9	0.2	0	0	0	—
240	Sweet, raw	½	cup(s)	73	59.6	46	0.8	11.6	1.5	0.1	0	0	0	—
	Cranberries													
3007	Chopped, raw	½	cup(s)	55	47.9	25	0.2	6.7	2.5	0.1	0	0	0	—
1717	Cranberry apple juice drink	½	cup(s)	123	102.6	77	0	19.4	0	0.1	0	0	0.1	—
1638	Cranberry juice cocktail	½	cup(s)	127	109.0	68	0	17.1	0	0.1	0	0	0.1	—
241	Cranberry juice cocktail, low calorie, with saccharin	½	cup(s)	119	112.8	23	0	5.5	0	0	0	0	0	—
242	Cranberry sauce, sweetened, canned	¼	cup(s)	69	42.0	105	0.1	26.9	0.7	0.1	0	0	0	—
	Dates													
244	Domestic, chopped	¼	cup(s)	45	9.1	125	1.1	33.4	3.6	0.2	0	0	0	—
243	Domestic, whole	¼	cup(s)	45	9.1	125	1.1	33.4	3.6	0.2	0	0	0	—
	Figs													
975	Canned in heavy syrup	½	cup(s)	130	98.8	114	0.5	29.7	2.8	0.1	0	0	0.1	—
974	Canned in water	½	cup(s)	124	105.7	66	0.5	17.3	2.7	0.1	0	0	0.1	—
973	Raw , medium	2	item(s)	100	79.1	74	0.7	19.2	2.9	0.3	0.1	0.1	0.1	—
	Fruit cocktail & salad													
245	Fruit cocktail, canned in heavy syrup	½	cup(s)	124	99.7	91	0.5	23.4	1.2	0.1	0	0	0	—
978	Fruit cocktail, canned in juice	½	cup(s)	119	103.6	55	0.5	14.1	1.2	0	0	0	0	—
977	Fruit cocktail, canned in water	½	cup(s)	119	107.6	38	0.5	10.1	1.2	0.1	0	0	0	—
979	Fruit salad, canned in water	½	cup(s)	123	112.1	37	0.4	9.6	1.2	0.1	0	0	0	—
	Gooseberries													
982	Canned in light syrup	½	cup(s)	126	100.9	92	0.8	23.6	3.0	0.3	0	0	0.1	—

FRUIT AND FRUIT JUICES —Continued

Chol (mg)	Calc (mg)	Iron (mg)	Magn (mg)	Pota (mg)	Sodi (mg)	Zinc (mg)	Vit A (µg)	Thia (mg)	Vit E (mg α)	Ribo (mg)	Niac (mg)	Vit B$_6$ (mg)	Fola (µg)	Vit C (mg)	Vit B$_{12}$ (µg)	Sele (µg)
18	9	0.89	8.5	34.5	414.5	0.30	—	0.08	—	0.04	0.71	0.03	4.0	0.1	0	—
0	4	0.45	8.5	33.2	57.0	0.11	0	0.09	—	0.02	0.48	0.03	6.6	0	0	—
0	7	0.45	1.8	25.5	141.7	0.19	0	0.01	—	0.03	0.08	0.01	1.8	0	0	—
0	7	1.00	12.4	51.5	0.5	0.35	0	0.12	0.04	0.07	0.90	0.04	7.8	0	0	40.0
0	11	0.74	21.0	30.8	2.1	0.57	0	0.08	0.21	0.03	0.50	0.06	3.5	0	0	18.1
0	1	0.25	11.5	26.3	0.6	0.25	0.8	0.01	0.02	0.01	0.18	0.01	2.5	0	0	0
2	15	0.61	12.3	38.4	72.5	0.20	0.7	0.02	0.42	0.02	0.77	0.01	1.8	0	0	1.3
4	40	0.79	32.5	93.2	317.4	0.71	13.6	0.04	—	0.08	0.52	0.08	3.9	0.2	0.2	4.3
0	0	0.22	8.7	20.0	116.4	0.34	0.9	0.01	0.27	0.00	0.13	0.01	2.8	0	0	0.2
0	7	0.31	6.0	150.6	8.4	0.05	0	0.00	0.01	0.02	0.05	0.04	0	0.7	0	0.1
0	9	0.46	3.7	147.6	3.7	0.04	0	0.03	0.01	0.02	0.12	0.04	0	1.1	0	0.1
0	3	0.06	2.7	58.8	0.5	0.02	1.6	0.01	0.10	0.01	0.05	0.02	1.6	2.5	0	0
0	4	0.16	2.6	75.2	0.9	0.03	1.7	0.01	0.04	0.01	0.08	0.04	0.9	0.2	0	0.3
0	8	0.16	6.9	147.7	1.4	0.05	4.1	0.02	0.24	0.03	0.12	0.05	4.1	6.3	0	0
0	3	0.30	3.4	96.8	18.7	0.04	0	0.00	0.11	0.03	0.20	0.03	0	0.8	0	0.3
0	5	0.44	3.8	77.8	3.8	0.05	1.3	0.01	0.26	0.03	0.24	0.03	1.3	2.2	0	0.4
0	4	0.14	3.7	91.5	2.4	0.03	1.2	0.01	0.25	0.03	0.22	0.03	1.2	1.5	0	0.4
0	6	0.12	2.5	67.9	0.4	—	0.7	0.01	0.20	0.01	0.03	—	2.0	2.8	0	—
0	18	0.54	14.0	362.6	1.4	0.28	134.4	0.04	1.24	0.05	0.84	0.07	12.6	14.0	0	0.1
0	12	0.38	9.0	180.6	5.2	0.14	80.0	0.02	0.77	0.02	0.48	0.07	2.6	4.0	0	0.1
0	18	0.87	10.5	381.5	3.3	0.12	59.1	0.00	1.42	0.02	0.85	0.05	3.3	0.3	0	0.7
0	15	0.66	33.3	583.0	9.2	0.78	8.0	0.08	2.23	0.16	2.19	0.31	102.3	10.1	0	0.4
0	12	0.19	27.6	403.6	2.3	0.45	8.0	0.02	3.03	0.04	0.76	0.08	40.3	20.0	0	—
0	3	0.14	8.0	134.1	1.9	0.17	1.9	0.01	0.57	0.03	0.48	0.07	22.4	2.8	0	0.1
0	10	0.69	41.8	294.8	3.3	0.40	2.2	0.04	0.13	0.01	0.39	0.14	7.7	3.5	0	0.8
0	6	0.30	31.9	422.4	1.2	0.17	3.5	0.03	0.11	0.08	0.78	0.43	23.6	10.3	0	1.2
0	21	0.45	14.4	116.6	0.7	0.38	7.9	0.01	0.84	0.02	0.47	0.02	18.0	15.1	0	0.3
0	22	0.60	16.6	105.7	0.8	0.19	4.5	0.02	0.88	0.03	0.91	0.05	25.7	2.3	0	0.3
0	6	0.42	5.1	51.2	3.8	0.09	2.6	0.04	0.49	0.07	0.14	0.05	2.6	1.4	0	0.1
0	4	0.20	4.4	55.8	0.7	0.12	2.2	0.03	0.41	0.03	0.30	0.04	4.4	7.0	0	0.1
0	6	0.14	3.9	41.9	0.8	0.05	1.6	0.03	0.37	0.03	0.40	0.05	5.4	1.9	0	0.1
0	23	0.55	14.1	115.2	3.8	0.24	2.6	0.03	—	0.04	0.29	0.05	43.5	7.9	0	0.5
0	18	0.56	10.6	91.7	0.7	0.15	2.0	0.04	0.57	0.02	0.51	0.04	41.6	2.0	0	0.1
0	65	2.07	96.0	1881.6	7.7	0.46	0	0.42	0.38	0.11	3.45	0.38	53.8	111.4	0	2.3
0	13	1.67	7.3	119.6	8.5	0.09	46.4	0.02	0.28	0.05	0.22	0.05	9.8	2.6	0	0
0	12	0.25	7.0	134.1	2.3	0.08	49.6	0.02	0.05	0.03	0.31	0.03	6.2	7.8	0	0
0	11	0.44	11.4	183.4	3.8	0.12	10.1	0.02	0.29	0.05	0.50	0.03	5.1	4.6	0	0
0	14	0.45	11.2	162.4	1.2	0.10	9.9	0.03	0.29	0.05	0.51	0.04	5.0	2.7	0	0
0	9	0.26	8.0	161.0	0	0.05	2.2	0.02	0.05	0.02	0.11	0.04	2.9	5.1	0	0
0	4	0.13	3.3	46.8	1.1	0.05	1.7	0.01	0.66	0.01	0.05	0.03	0.6	7.3	0	0.1
0	4	0.09	1.2	20.8	2.5	0.02	0	0.00	0.15	0.00	0.00	0.00	0	48.4	0	0
0	4	0.13	1.3	17.7	2.5	0.04	0	0.00	0.28	0.00	0.05	0.00	0	53.5	0	0.3
0	11	0.05	2.4	29.6	3.6	0.02	0	0.00	0.06	0.00	0.00	0.00	0	38.2	0	0
0	3	0.15	2.1	18.0	20.1	0.03	1.4	0.01	0.57	0.01	0.06	0.01	0.7	1.4	0	0.2
0	17	0.45	19.1	291.9	0.9	0.12	0	0.02	0.02	0.02	0.56	0.07	8.5	0.2	0	1.3
0	17	0.45	19.1	291.9	0.9	0.12	0	0.02	0.02	0.02	0.56	0.07	8.5	0.2	0	1.3
0	35	0.36	13.0	128.2	1.3	0.14	2.6	0.03	0.16	0.05	0.55	0.09	2.6	1.3	0	0.3
0	35	0.36	12.4	127.7	1.2	0.15	2.5	0.03	0.10	0.05	0.55	0.09	2.5	1.2	0	0.1
0	35	0.36	17.0	232.0	1.0	0.14	7.0	0.06	0.10	0.04	0.40	0.10	6.0	2.0	0	0.2
0	7	0.36	6.2	109.1	7.4	0.09	12.4	0.02	0.49	0.02	0.46	0.06	3.7	2.4	0	0.6
0	9	0.25	8.3	112.6	4.7	0.11	17.8	0.01	0.47	0.02	0.48	0.06	3.6	3.2	0	0.6
0	6	0.30	8.3	111.4	4.7	0.11	15.4	0.02	0.47	0.01	0.43	0.06	3.6	2.5	0	0.6
0	9	0.37	6.1	95.6	3.7	0.10	27.0	0.02	—	0.03	0.46	0.04	3.7	2.3	0	1.0
0	20	0.42	7.6	97.0	2.5	0.14	8.8	0.03	—	0.07	0.19	0.02	3.8	12.6	0	0.5

DA+ Code	Food Description	Quantity	Measure	Wt (g)	H₂O (g)	Ener (kcal)	Prot (g)	Carb (g)	Fiber (g)	Fat (g)	Fat Breakdown (g) Sat	Mono	Poly	Trans
981	Raw	½	cup(s)	75	65.9	33	0.7	7.6	3.2	0.4	0	0	0.2	—
	Grapefruit													
251	Juice, pink, sweetened, canned	½	cup(s)	125	109.1	57	0.7	13.9	0.1	0.1	0	0	0	—
249	Juice, white	½	cup(s)	124	111.2	48	0.6	11.4	0.1	0.1	0	0	0	—
3022	Pink or red, raw	½	cup(s)	114	100.8	48	0.9	12.2	1.8	0.2	0	0	0	—
248	Sections, canned in light syrup	½	cup(s)	127	106.2	76	0.7	19.6	0.5	0.1	0	0	0	—
983	Sections, canned in water	½	cup(s)	122	109.6	44	0.7	11.2	0.5	0.1	0	0	0	—
247	White, raw	½	cup(s)	115	104.0	38	0.8	9.7	1.3	0.1	0	0	0	—
	Grapes													
255	American, slip skin	½	cup(s)	46	37.4	31	0.3	7.9	0.4	0.2	0.1	0	0	—
256	European, red or green, adherent skin	½	cup(s)	76	60.8	52	0.5	13.7	0.7	0.1	0	0	0	—
3159	Grape Juice drink, canned	½	cup(s)	125	106.6	71	0	18.2	0.1	0	0	0	0	0
259	Grape Juice, sweetened, with added vitamin C, prepared from frozen concentrate	½	cup(s)	125	108.6	64	0.2	15.9	0.1	0.1	0	0	0	—
3060	Raisins, seeded, packed	¼	cup(s)	41	6.8	122	1.0	32.4	2.8	0.2	0.1	0	0.1	—
987	**Guava, raw**	1	item(s)	55	44.4	37	1.4	7.9	3.0	0.5	0.2	0	0.2	—
35593	**Guavas, strawberry**	1	item(s)	6	4.8	4	0	1.0	0.3	0	0	0	0	—
3027	**Jackfruit**	½	cup(s)	83	60.4	78	1.2	19.8	1.3	0.2	0	0	0.1	—
990	**Kiwi fruit or Chinese gooseberries**	1	item(s)	76	63.1	46	0.9	11.1	2.3	0.4	0	0	0.2	—
	Lemon													
262	Juice	1	tablespoon(s)	15	13.8	4	0.1	1.3	0.1	0	0	0	0	0
993	Peel	1	teaspoon(s)	2	1.6	1	0	0.3	0.2	0	0	0	0	—
992	Raw	1	item(s)	108	94.4	22	1.3	11.6	5.1	0.3	0	0	0.1	—
	Lime													
269	Juice	1	tablespoon(s)	15	14.0	4	0.1	1.3	0.1	0	0	0	0	—
994	Raw	1	item(s)	67	59.1	20	0.5	7.1	1.9	0.1	0	0	0	—
995	**Loganberries, frozen**	½	cup(s)	74	62.2	40	1.1	9.6	3.9	0.2	0	0	0.1	—
	Mandarin orange													
1038	Canned in juice	½	cup(s)	125	111.4	46	0.8	11.9	0.9	0	0	0	0	—
1039	Canned in light syrup	½	cup(s)	126	104.7	77	0.6	20.4	0.9	0.1	0	0	0	—
999	Mango	½	cup(s)	83	67.4	54	0.4	14.0	1.5	0.2	0.1	0.1	0	—
1005	**Nectarine, raw, sliced**	½	cup(s)	69	60.4	30	0.7	7.3	1.2	0.2	0	0.1	0.1	—
	Melons													
271	Cantaloupe	½	cup(s)	80	72.1	27	0.7	6.5	0.7	0.1	0	0	0.1	—
1000	Casaba melon	½	cup(s)	85	78.1	24	0.9	5.6	0.8	0.1	0	0	0	—
272	Honeydew	½	cup(s)	89	79.5	32	0.5	8.0	0.7	0.1	0	0	0	—
318	Watermelon	½	cup(s)	76	69.5	23	0.5	5.7	0.3	0.1	0	0	0	—
	Orange													
14412	Juice with calcium and vitamin D	½	cup(s)	120	—	55	1.0	13.0	0	0	0	0	0	0
29630	Juice, fresh squeezed	½	cup(s)	124	109.5	56	0.9	12.9	0.2	0.2	0	0	0	—
14411	Juice, not from concentrate	½	cup(s)	120	—	55	1.0	13.0	0	0	0	0	0	0
278	Juice, unsweetened, prepared from frozen concentrate	½	cup(s)	125	109.7	56	0.8	13.4	0.2	0.1	0	0	0	—
3040	Peel	1	teaspoon(s)	2	1.5	2	0	0.5	0.2	0	0	0	0	—
273	Raw	1	item(s)	131	113.6	62	1.2	15.4	3.1	0.2	0	0	0	—
274	Sections	½	cup(s)	90	78.1	42	0.8	10.6	2.2	0.1	0	0	0	—
	Papaya, Raw													
16830	Dried, strips	2	item(s)	46	12.0	119	1.9	29.9	5.5	0.4	0.1	0.1	0.1	—
282	Papaya, Raw	½	cup(s)	70	62.2	27	0.4	6.9	1.3	0.1	0	0	0	—
35640	**Passion fruit, purple**	1	item(s)	18	13.1	17	0.4	4.2	1.9	0.1	0	0	0.1	—
	Peach													
285	Halves, canned in heavy syrup	½	cup(s)	131	103.9	97	0.6	26.1	1.7	0.1	0	0	0.1	—
286	Halves, canned in water	½	cup(s)	122	113.6	29	0.5	7.5	1.6	0.1	0	0	0	—
290	Slices, sweetened, frozen	½	cup(s)	125	93.4	118	0.8	30.0	2.3	0.2	0	0.1	0.1	—
283	Raw, medium	1	item(s)	150	133.3	59	1.4	14.3	2.3	0.4	0	0.1	0.1	—
	Pear													
8672	Asian	1	item(s)	122	107.7	51	0.6	13.0	4.4	0.3	0	0.1	0.1	—
293	D'Anjou	1	item(s)	200	168.0	120	1.0	30.0	5.2	1.0	0	0.2	0.2	—
294	Halves, canned in heavy syrup	½	cup(s)	133	106.9	98	0.3	25.5	2.1	0.2	0	0	0	—
1012	Halves, canned in juice	½	cup(s)	124	107.2	62	0.4	16.0	2.0	0.1	0	0	0	—
291	Raw	1	item(s)	166	139.0	96	0.6	25.7	5.1	0.2	0	0	0	—
1017	**Persimmon**	1	item(s)	25	16.1	32	0.2	8.4	—	0.1	0	0	0	—
	Pineapple													
3053	Canned in extra heavy syrup	½	cup(s)	130	101.0	108	0.4	28.0	1.0	0.1	0	0	0	—
1019	Canned in juice	½	cup(s)	125	104.0	75	0.5	19.5	1.0	0.1	0	0	0	—
296	Canned in light syrup	½	cup(s)	126	108.0	66	0.5	16.9	1.0	0.2	0	0	0.1	—
1018	Canned in water	½	cup(s)	123	111.7	39	0.5	10.2	1.0	0.1	0	0	0	—
299	Juice, unsweetened, canned	½	cup(s)	125	108.0	66	0.5	16.1	0.3	0.2	0	0	0.1	—
295	Raw, diced	½	cup(s)	78	66.7	39	0.4	10.2	1.1	0.1	0	0	0	—
	FRUIT AND FRUIT JUICES —Continued													
1024	**Plantain, cooked**	½	cup(s)	77	51.8	89	0.6	24.0	1.8	0.1	0.1	0	0	—

Chol (mg)	Calc (mg)	Iron (mg)	Magn (mg)	Pota (mg)	Sodi (mg)	Zinc (mg)	Vit A (µg)	Thia (mg)	Vit E (mg α)	Ribo (mg)	Niac (mg)	Vit B$_6$ (mg)	Fola (µg)	Vit C (mg)	Vit B$_{12}$ (µg)	Sele (µg)
0	19	0.23	7.5	148.5	0.8	0.09	11.3	0.03	0.28	0.02	0.23	0.06	4.5	20.8	0	0.5
0	10	0.45	12.5	202.2	2.5	0.08	0	0.05	0.05	0.03	0.40	0.03	12.5	33.6	0	0.1
0	11	0.25	14.8	200.1	1.2	0.06	1.2	0.05	0.27	0.02	0.25	0.05	12.4	46.9	0	0.1
0	25	0.09	10.3	154.5	0	0.07	66.4	0.04	0.14	0.03	0.23	0.06	14.9	35.7	0	0.1
0	18	0.50	12.7	163.8	2.5	0.10	0	0.04	0.11	0.02	0.30	0.02	11.4	27.1	0	1.1
0	18	0.50	12.2	161.0	2.4	0.11	0	0.05	0.11	0.03	0.30	0.02	11.0	26.6	0	1.1
0	14	0.07	10.4	170.2	0	0.08	2.3	0.04	0.15	0.02	0.30	0.05	11.5	38.3	0	1.6
0	6	0.13	2.3	87.9	0.9	0.02	2.3	0.04	0.09	0.02	0.14	0.05	1.8	1.8	0	0
0	8	0.27	5.3	144.2	1.5	0.05	2.3	0.05	0.14	0.05	0.14	0.07	1.5	8.2	0	0.1
0	9	0.16	7.5	41.3	11.3	0.04	0	0.28	0.00	0.44	0.18	0.04	1.3	33.1	0	0.1
0	5	0.13	5.0	26.3	2.5	0.05	0	0.02	0.00	0.03	0.16	0.05	1.3	29.9	0	0.1
0	12	1.06	12.4	340.3	11.6	0.07	0	0.04	—	0.07	0.46	0.07	1.2	2.2	0	0.2
0	10	0.14	12.1	229.4	1.1	0.12	17.1	0.03	0.40	0.02	0.59	0.06	27.0	125.6	0	0.3
0	1	0.01	1.0	17.5	2.2	—	0.3	0.00	—	0.00	0.03	0.00	—	2.2	0	
0	28	0.49	30.5	250.0	2.5	0.35	12.4	0.02	—	0.09	0.33	0.09	11.5	5.5	0	0.5
0	26	0.23	12.9	237.1	2.3	0.10	3.0	0.02	1.11	0.01	0.25	0.04	19.0	70.5	0	0.2
0	1	0.00	0.9	18.9	0.2	0.01	0.2	0.00	0.02	0.00	0.02	0.01	2.0	7.0	0	0
0	3	0.01	0.3	3.2	0.1	0.01	0.1	0.00	0.01	0.00	0.01	0.00	0.3	2.6	0	0
0	66	0.75	13.0	156.6	3.2	0.10	2.2	0.05	—	0.04	0.21	0.11	—	83.2	0	1.0
0	2	0.02	1.2	18.0	0.3	0.01	0.3	0.00	0.03	0.00	0.02	0.01	1.5	4.6	0	0
0	22	0.40	4.0	68.3	1.3	0.07	1.3	0.02	0.14	0.01	0.13	0.02	5.4	19.5	0	0.3
0	19	0.47	15.4	106.6	0.7	0.25	1.5	0.04	0.64	0.03	0.62	0.05	19.1	11.2	0	0.1
0	14	0.34	13.7	165.6	6.2	0.64	53.5	0.10	0.12	0.04	0.55	0.05	6.2	42.6	0	0.5
0	9	0.47	10.1	98.3	7.6	0.30	52.9	0.07	0.13	0.06	0.56	0.05	6.3	24.9	0	0.5
0	8	0.10	7.4	128.7	1.7	0.03	31.4	0.05	0.92	0.04	0.48	0.11	11.6	22.8	0	0.5
0	4	0.19	6.2	138.7	0	0.12	11.7	0.02	0.53	0.02	0.78	0.02	3.5	3.7	0	0
0	7	0.17	9.6	213.6	12.8	0.14	135.2	0.03	0.04	0.01	0.59	0.05	16.8	29.4	0	0.3
0	9	0.29	9.4	154.7	7.7	0.06	0	0.01	0.04	0.03	0.20	0.14	6.8	18.5	0	0.3
0	5	0.15	8.8	201.8	15.9	0.07	2.7	0.03	0.01	0.01	0.37	0.07	16.8	15.9	0	0.6
0	5	0.18	7.6	85.1	0.8	0.07	21.3	0.02	0.04	0.01	0.13	0.03	2.3	6.2	0	0.3
0	175	0.00	12.0	225.0	0	—	0	0.08	—	0.03	0.40	0.06	30.0	36.0	0	—
0	14	0.25	13.6	248.0	1.2	0.06	12.4	0.11	0.05	0.04	0.50	0.05	37.2	62.0	0	0.1
0	10	0.00	12.5	225.0	0	0.06	0	0.08	—	0.03	0.40	0.06	30.0	36.0	0	0.1
0	11	0.12	12.5	236.6	1.2	0.06	6.2	0.10	0.25	0.02	0.25	0.06	54.8	48.4	0	0.1
0	3	0.01	0.4	4.2	0.1	0.01	0.4	0.00	0.01	0.00	0.01	0.00	0.6	2.7	0	0
0	52	0.13	13.1	237.1	0	0.09	14.4	0.11	0.23	0.05	0.36	0.07	39.3	69.7	0	0.7
0	36	0.09	9.0	162.9	0	0.06	9.9	0.07	0.16	0.03	0.25	0.05	27.0	47.9	0	0.4
0	73	0.30	30.4	782.9	9.2	0.21	83.7	0.06	2.22	0.08	0.93	0.05	58.0	37.7	0	1.8
0	17	0.07	7.0	179.9	2.1	0.05	38.5	0.02	0.51	0.02	0.24	0.01	26.6	43.3	0	0.4
0	2	0.28	5.2	62.6	5.0	0.01	11.5	0.00	0.00	0.02	0.27	0.01	2.5	5.4	0	0.1
0	4	0.35	6.6	120.5	7.9	0.11	22.3	0.01	0.64	0.03	0.80	0.02	3.9	3.7	0	0.4
0	2	0.39	6.1	120.8	3.7	0.11	32.9	0.01	0.59	0.02	0.63	0.02	3.7	3.5	0	0.4
0	4	0.46	6.3	162.5	7.5	0.06	17.5	0.01	0.77	0.04	0.81	0.02	3.8	117.8	0	0.5
0	9	0.37	13.5	285.0	0	0.25	24.0	0.03	1.09	0.04	1.20	0.03	6.0	9.9	0	0.2
0	5	0.00	9.8	147.6	0	0.02	0	0.01	0.14	0.01	0.26	0.02	9.8	4.6	0	0.1
0	22	0.50	12.0	250.0	0	0.24	—	0.04	1.00	0.08	0.20	0.03	14.6	8.0	0	1.0
0	7	0.29	5.3	86.5	6.7	0.10	0	0.01	0.10	0.02	0.32	0.01	1.3	1.5	0	0
0	11	0.36	8.7	119.0	5.0	0.11	0	0.01	0.10	0.01	0.25	0.02	1.2	2.0	0	0
0	15	0.28	11.6	197.5	1.7	0.16	1.7	0.02	0.19	0.04	0.26	0.04	11.6	7.0	0	0.2
0	7	0.62	—	77.5	0.3	—	0	—	—	—	—	—	—	16.5	0	
0	18	0.49	19.5	132.6	1.3	0.14	1.3	0.11	—	0.03	0.36	0.09	6.5	9.5	0	—
0	17	0.35	17.4	151.9	1.2	0.12	2.5	0.12	0.01	0.02	0.35	0.09	6.2	11.8	0	0.5
0	18	0.49	20.2	132.3	1.3	0.15	2.5	0.11	0.01	0.03	0.36	0.09	6.3	9.5	0	0.5
0	18	0.49	22.1	156.2	1.2	0.15	2.5	0.11	0.01	0.03	0.37	0.09	6.2	9.5	0	0.5
0	16	0.39	15.0	162.5	2.5	0.14	0	0.07	0.03	0.03	0.25	0.13	22.5	12.5	0	0.1
0	10	0.22	9.3	84.5	0.8	0.09	2.3	0.06	0.02	0.03	0.39	0.09	14.0	37.0	0	0.1
0	2	0.45	24.6	358.1	3.9	0.10	34.7	0.04	0.10	0.04	0.58	0.19	20.0	8.4	0	1.1

DA+ Code	Food Description	Quantity	Measure	Wt (g)	H₂O (g)	Ener (kcal)	Prot (g)	Carb (g)	Fiber (g)	Fat (g)	Sat	Mono	Poly	Trans
300	**Plum, raw, large**	1	item(s)	66	57.6	30	0.5	7.5	0.9	0.2	0	0.1	0	—
1027	**Pomegranate**	1	item(s)	154	124.7	105	1.5	26.4	0.9	0.5	0.1	0.1	0.1	—
	Prunes, Dried													
5644	Dried	2	item(s)	17	5.2	40	0.4	10.7	1.2	0.1	0	0	0	—
305	Dried, stewed	½	cup(s)	124	86.5	133	1.2	34.8	3.8	0.2	0	0.1	0	—
306	Juice, canned	1	cup(s)	256	208.0	182	1.6	44.7	2.6	0.1	0	0.1	0	—
	Raspberries													
309	Raw	½	cup(s)	62	52.7	32	0.7	7.3	4.0	0.4	0	0	0.2	—
310	Red, sweetened, frozen	½	cup(s)	125	90.9	129	0.9	32.7	5.5	0.2	0	0	0.1	—
311	**Rhubarb, cooked with sugar**	½	cup(s)	120	81.5	140	0.5	37.5	2.7	0.1	0	0	0.1	—
	Strawberries													
313	Raw	½	cup(s)	72	65.5	23	0.5	5.5	1.4	0.2	0	0	0.1	—
315	Sweetened, frozen, thawed	½	cup(s)	128	99.5	99	0.7	26.8	2.4	0.2	0	0	0.1	—
16828	**Tangelo**	1	item(s)	95	82.4	45	0.9	11.2	2.3	0.1	0	0	0	—
	Tangerine													
1040	Juice	½	cup(s)	124	109.8	53	0.6	12.5	0.2	0.2	0	0	0	—
316	Raw	1	item(s)	88	74.9	47	0.7	11.7	1.6	0.3	0	0.1	0.1	—
	VEGETABLES, LEGUMES													
	Amaranth													
1043	Leaves, boiled, drained	½	cup(s)	66	60.4	14	1.4	2.7	—	0.1	0	0	0.1	—
1042	Leaves, raw	1	cup(s)	28	25.7	6	0.7	1.1	—	0.1	0	0	0	—
8683	**Arugula leaves, raw**	1	cup(s)	20	18.3	5	0.5	0.7	0.3	0.1	0	0	0.1	—
	Artichoke													
1044	Boiled, drained	1	item(s)	120	100.9	64	3.5	14.3	10.3	0.4	0.1	0	0.2	—
2885	Hearts, boiled, drained	½	cup(s)	84	70.6	45	2.4	10.0	7.2	0.3	0.1	0	0.1	—
	Asparagus													
566	Boiled, drained	½	cup(s)	90	83.4	20	2.2	3.7	1.8	0.2	0	0	0.1	—
568	Canned, drained	½	cup(s)	121	113.7	23	2.6	3.0	1.9	0.8	0.2	0	0.3	—
565	Tips, frozen, boiled, drained	½	cup(s)	90	84.7	16	2.7	1.7	1.4	0.4	0.1	0	0.2	—
	Bamboo shoots													
1048	Boiled, drained	½	cup(s)	60	57.6	7	0.9	1.2	0.6	0.1	0	0	0.1	—
1049	Canned, drained	½	cup(s)	66	61.8	12	1.1	2.1	0.9	0.3	0.1	0	0.1	—
	Beans													
1801	Adzuki beans, boiled	½	cup(s)	115	76.2	147	8.6	28.5	8.4	0.1	0	—	—	—
511	Baked beans with franks, canned	½	cup(s)	130	89.8	184	8.7	19.9	8.9	8.5	3.0	3.7	1.1	—
513	Baked beans with pork in sweet sauce, canned	½	cup(s)	127	89.3	142	6.7	26.7	5.3	1.8	0.6	0.6	0.5	0
512	Baked beans with pork in tomato sauce, canned	½	cup(s)	127	93.0	119	6.5	23.6	5.1	1.2	0.5	0.7	0.3	—
1805	Black beans, boiled	½	cup(s)	86	56.5	114	7.6	20.4	7.5	0.5	0.1	0	0.2	—
14597	Chickpeas, garbanzo beans or bengal gram, boiled	½	cup(s)	82	49.4	134	7.3	22.5	6.2	2.1	0.2	0.5	0.9	—
569	Fordhook lima beans, frozen, boiled, drained	½	cup(s)	85	62.0	88	5.2	16.4	4.9	0.3	0.1	0	0.1	—
1806	French beans, boiled	½	cup(s)	89	58.9	114	6.2	21.3	8.3	0.7	0.1	0	0.4	—
2773	Great northern beans, boiled	½	cup(s)	89	61.1	104	7.4	18.7	6.2	0.4	0.1	0	0.2	—
2736	Hyacinth beans, boiled, drained	½	cup(s)	44	37.8	22	1.3	4.0	—	0.1	0.1	0.1	0	—
570	Lima beans, baby, frozen, boiled, drained	½	cup(s)	90	65.1	95	6.0	17.5	5.4	0.3	0.1	0	0.1	—
515	Lima beans, boiled, drained	½	cup(s)	85	57.1	105	5.8	20.1	4.5	0.3	0.1	0	0.1	—
579	Mung beans, sprouted, boiled, drained	½	cup(s)	62	57.9	13	1.3	2.6	0.5	0.1	0	0	0	—
510	Navy beans, boiled	½	cup(s)	91	58.1	127	7.5	23.7	9.6	0.6	0.1	0.1	0.4	0
32816	Pinto beans, boiled, drained, no salt added	½	cup(s)	85	53.8	122	7.7	22.4	7.7	0.6	0.1	0.1	0.2	—
1052	Pinto beans, frozen, boiled, drained	½	cup(s)	47	27.3	76	4.4	14.5	4.0	0.2	0	0	0.1	—
514	Red kidney beans, canned	½	cup(s)	128	99.0	108	6.7	19.9	6.9	0.5	0.1	0.2	0.2	—
1810	Refried beans, canned	½	cup(s)	127	96.1	119	6.9	19.6	6.7	1.6	0.6	0.7	0.2	—
1053	Shell beans, canned	½	cup(s)	123	111.1	37	2.2	7.6	4.2	0.2	0	0	0.1	—
1670	Soybeans, boiled	½	cup(s)	86	53.8	149	14.3	8.5	5.2	7.7	1.1	1.7	4.4	—
1108	Soybeans, green, boiled, drained	½	cup(s)	90	61.7	127	11.1	9.9	3.8	5.8	0.7	1.1	2.7	—
1807	White beans, small, boiled	½	cup(s)	90	56.6	127	8.0	23.1	9.3	0.6	0.1	0.1	0.2	—
575	Yellow snap, string or wax beans, boiled, drained	½	cup(s)	63	55.8	22	1.2	4.9	2.1	0.2	0	0	0.1	—
576	Yellow snap, string or wax beans, frozen, boiled, drained	½	cup(s)	68	61.7	19	1.0	4.4	2.0	0.1	0	0	0.1	—
	Beets													
584	Beet greens, boiled, drained	½	cup(s)	72	64.2	19	1.9	3.9	2.1	0.1	0	0	0.1	—
2730	Pickled, canned with liquid	½	cup(s)	114	92.9	74	0.9	18.5	3.0	0.1	0	0	0	—
581	Sliced, boiled, drained	½	cup(s)	85	74.0	37	1.4	8.5	1.7	0.2	0	0	0.1	—
583	Sliced, canned, drained	½	cup(s)	85	77.3	26	0.8	6.1	1.5	0.1	0	0	0	—
580	Whole, boiled, drained	2	item(s)	100	87.1	44	1.7	10.0	2.0	0.2	0	0	0.1	—
585	**Cowpeas or black-eyed peas, boiled, drained**	½	cup(s)	83	62.3	80	2.6	16.8	4.1	0.3	0.1	0	0.1	—
	Broccoli													
588	Chopped, boiled, drained	½	cup(s)	78	69.6	27	1.9	5.6	2.6	0.3	0.1	0	0.1	—
	VEGETABLES, LEGUMES—Continued													

Chol (mg)	Calc (mg)	Iron (mg)	Magn (mg)	Pota (mg)	Sodi (mg)	Zinc (mg)	Vit A (µg)	Thia (mg)	Vit E (mg α)	Ribo (mg)	Niac (mg)	Vit B$_6$ (mg)	Fola (µg)	Vit C (mg)	Vit B$_{12}$ (µg)	Sele (µg)
0	4	0.11	4.6	103.6	0	0.06	11.2	0.02	0.17	0.02	0.27	0.02	3.3	6.3	0	0
0	5	0.46	4.6	398.9	4.6	0.18	7.7	0.04	0.92	0.04	0.46	0.16	9.2	9.4	0	0.9
0	7	0.16	6.9	123.0	0.3	0.07	6.6	0.01	0.07	0.03	0.32	0.03	0.7	0.1	0	0
0	24	0.51	22.3	398.0	1.2	0.24	21.1	0.03	0.24	0.12	0.90	0.27	0	3.6	0	0.1
0	31	3.02	35.8	706.6	10.2	0.53	0	0.04	0.30	0.17	2.01	0.55	0	10.5	0	1.5
0	15	0.42	13.5	92.9	0.6	0.26	1.2	0.02	0.54	0.02	0.37	0.03	12.9	16.1	0	0.1
0	19	0.81	16.3	142.5	1.3	0.22	3.8	0.02	0.90	0.05	0.28	0.04	32.5	20.6	0	0.4
0	174	0.25	16.2	115.0	1.0	—	—	0.02	—	0.03	0.25	—	—	4.0	0	—
0	12	0.30	9.4	110.2	0.7	0.10	0.7	0.02	0.21	0.02	0.28	0.03	17.3	42.3	0	0.3
0	14	0.59	7.7	125.0	1.3	0.06	1.3	0.01	0.30	0.09	0.37	0.03	5.1	50.4	0	0.9
0	38	0.09	9.5	172.0	0	0.06	10.5	0.08	0.17	0.03	0.26	0.05	28.5	50.5	0	0.5
0	22	0.25	9.9	219.8	1.2	0.04	16.1	0.07	0.16	0.02	0.12	0.05	6.2	38.3	0	0.1
0	33	0.13	10.6	146.1	1.8	0.06	29.9	0.05	0.18	0.03	0.33	0.07	14.1	23.5	0	0.1
0	138	1.49	36.3	423.1	13.9	0.58	91.7	0.01	—	0.09	0.37	0.12	37.6	27.1	0	0.6
0	60	0.65	15.4	171.1	5.6	0.25	40.9	0.01	—	0.04	0.18	0.05	23.8	12.1	0	0.3
0	32	0.29	9.4	73.8	5.4	0.09	23.8	0.01	0.09	0.02	0.06	0.01	19.4	3.0	0	0.1
0	25	0.73	50.4	343.2	72.0	0.48	1.2	0.06	0.22	0.10	1.33	0.09	106.8	8.9	0	0.2
0	18	0.51	35.3	240.2	50.4	0.33	0.8	0.04	0.16	0.07	0.93	0.06	74.8	6.2	0	0.2
0	21	0.81	12.6	201.6	12.6	0.54	45.0	0.14	1.35	0.12	0.97	0.07	134.1	6.9	0	5.5
0	19	2.21	12.1	208.1	347.3	0.48	49.6	0.07	1.47	0.12	1.15	0.13	116.2	22.3	0	2.1
0	16	0.50	9.0	154.8	2.7	0.36	36.0	0.05	1.08	0.09	0.93	0.01	121.5	22.0	0	3.5
0	7	0.14	1.8	319.8	2.4	0.28	0	0.01	—	0.03	0.18	0.06	1.2	0	0	0.2
0	5	0.21	2.6	52.4	4.6	0.43	0.7	0.02	0.41	0.02	0.09	0.09	2.0	0.7	0	0.3
0	32	2.30	59.8	611.8	9.2	2.03	0	0.13	—	0.07	0.82	0.11	139.2	0	0	1.4
8	62	2.24	36.3	304.3	556.9	2.42	5.2	0.08	0.21	0.07	1.17	0.06	38.9	3.0	0.4	8.4
9	75	2.08	41.7	326.4	422.5	1.73	0	0.05	0.03	0.07	0.44	0.07	10.1	3.5	0	6.3
9	71	4.09	43.0	373.2	552.8	6.93	5.1	0.06	0.12	0.05	0.62	0.08	19.0	3.8	0	5.9
0	23	1.80	60.2	305.3	0.9	0.96	0	0.21	—	0.05	0.43	0.05	128.1	0	0	1.0
0	40	2.36	39.4	238.6	5.7	1.25	0.8	0.09	0.28	0.05	0.43	0.11	141.0	1.1	0	3.0
0	26	1.54	35.7	258.4	58.7	0.62	8.5	0.06	0.24	0.05	0.90	0.10	17.9	10.9	0	0.5
0	56	0.95	49.6	327.5	5.3	0.56	0	0.11	—	0.05	0.48	0.09	66.4	1.1	0	1.1
0	60	1.88	44.3	346.0	1.8	0.77	0	0.14	—	0.05	0.60	0.10	90.3	1.2	0	3.6
0	18	0.33	18.3	114.0	0.9	0.16	3.0	0.02	—	0.03	0.20	0.01	20.4	2.2	0	0.7
0	25	1.76	50.4	369.9	26.1	0.49	7.2	0.06	0.57	0.04	0.69	0.10	14.4	5.2	0	1.5
0	27	2.08	62.9	484.5	14.5	0.67	12.8	0.11	0.11	0.08	0.88	0.16	22.1	8.6	0	1.7
0	7	0.40	8.7	62.6	6.2	0.29	0.6	0.03	0.04	0.06	0.51	0.03	18.0	7.1	0	0.4
—	63	2.14	48.2	354.0	0	0.93	0	0.21	0.01	0.06	0.59	0.12	127.4	0.8	0	2.6
0	39	1.79	43	373	1.0	0.84	0	0.16	0.80	0.05	0.27	0.19	147.0	0.7	0	5.3
0	24	1.27	25.4	303.6	39.0	0.32	0	0.12	—	0.05	0.29	0.09	16.0	0.3	0	0.7
0	32	1.62	35.8	327.7	330.2	2.09	0	0.13	0.02	0.11	0.57	0.10	25.6	1.4	0	0.6
10	44	2.10	41.7	337.8	378.2	1.48	0	0.03	0.00	0.02	0.39	0.18	13.9	7.6	0	1.6
0	36	1.21	18.4	133.5	409.2	0.33	13.5	0.04	0.04	0.07	0.25	0.06	22.1	3.8	0	2.6
0	88	4.42	74.0	442.9	0.9	0.98	0	0.13	0.30	0.24	0.34	0.20	46.4	1.5	0	6.3
0	131	2.25	54.0	485.1	12.6	0.82	7.2	0.23	—	0.14	1.13	0.05	99.9	15.3	0	1.3
0	65	2.54	60.9	414.4	1.8	0.97	0	0.21	—	0.05	0.24	0.11	122.6	0	0	1.2
0	29	0.80	15.6	186.9	1.9	0.23	2.5	0.05	0.28	0.06	0.38	0.04	20.6	6.1	0	0.3
0	33	0.59	16.2	85.1	6.1	0.32	4.1	0.02	0.03	0.06	0.26	0.04	15.5	2.8	0	0.3
0	82	1.36	49.0	654.5	173.5	0.36	275.8	0.08	1.30	0.20	0.35	0.09	10.1	17.9	0	0.6
0	12	0.46	17.0	168.0	299.6	0.29	1.1	0.01	—	0.05	0.28	0.05	30.6	2.6	0	1.1
0	14	0.67	19.6	259.3	65.5	0.30	1.7	0.02	0.03	0.03	0.28	0.05	68.0	3.1	0	0.6
0	13	1.54	14.5	125.8	164.9	0.17	0.9	0.01	0.02	0.03	0.13	0.04	25.5	3.5	0	0.4
0	16	0.79	23.0	305.0	77.0	0.35	2.0	0.02	0.04	0.04	0.33	0.06	80.0	3.6	0	0.7
0	106	0.92	42.9	344.9	3.3	0.85	33.0	0.08	0.18	0.12	1.15	0.05	104.8	1.8	0	2.1
0	31	0.52	16.4	228.5	32.0	0.35	60.1	0.04	1.13	0.09	0.43	0.15	84.2	50.6	0	1.2

DA+ Code	Food Description	Quantity	Measure	Wt (g)	H₂O (g)	Ener (kcal)	Prot (g)	Carb (g)	Fiber (g)	Fat (g)	Fat Breakdown (g) Sat	Mono	Poly	Trans
590	Frozen, chopped, boiled, drained	½	cup(s)	92	83.5	26	2.9	4.9	2.8	0.1	0	0	0.1	—
587	Raw, chopped	½	cup(s)	46	40.6	15	1.3	3.0	1.2	0.2	0	0	0	—
16848	**Broccoflower, raw, chopped**	½	cup(s)	32	28.7	10	0.9	1.9	1.0	0.1	0	0	0	—
	Brussels sprouts													
591	Boiled, drained	½	cup(s)	78	69.3	28	2.0	5.5	2.0	0.4	0.1	0	0.2	—
592	Frozen, boiled, drained	½	cup(s)	78	67.2	33	2.8	6.4	3.2	0.3	0.1	0	0.2	—
	Cabbage													
595	Boiled, drained, no salt added	1	cup(s)	150	138.8	35	1.9	8.3	2.8	0.1	0	0	0	—
35611	Chinese (pak choi or bok choy), boiled with salt, drained	1	cup(s)	170	162.4	20	2.6	3.0	1.7	0.3	0	0	0.1	—
16869	Kim chee	1	cup(s)	150	137.5	32	2.5	6.1	1.8	0.3	0	0	0.2	—
594	Raw, shredded	1	cup(s)	70	64.5	17	0.9	4.1	1.7	0.1	0	0	0	—
596	Red, shredded, raw	1	cup(s)	70	63.3	22	1.0	5.2	1.5	0.1	0	0	0.1	—
597	Savoy, shredded, raw	1	cup(s)	70	63.7	19	1.4	4.3	2.2	0.1	0	0	0	—
35417	**Capers**	1	teaspoon(s)	4	—	2	0	0	0	0	0	0	0	0
	Carrots													
8691	Baby, raw	8	item(s)	80	72.3	28	0.5	6.6	2.3	0.1	0	0	0.1	—
601	Grated	½	cup(s)	55	48.6	23	0.5	5.3	1.5	0.1	0	0	0.1	0
1055	Juice, canned	½	cup(s)	118	104.9	47	1.1	11.0	0.9	0.2	0	0	0.1	—
600	Raw	½	cup(s)	61	53.9	25	0.6	5.8	1.7	0.1	0	0	0.1	0
602	Sliced, boiled, drained	½	cup(s)	78	70.3	27	0.6	6.4	2.3	0.1	0	0	0.1	—
32725	**Cassava or manioc**	½	cup(s)	103	61.5	165	1.4	39.2	1.9	0.3	0.1	0.1	0	—
	Cauliflower													
606	Boiled, drained	½	cup(s)	62	57.7	14	1.1	2.5	1.4	0.3	0	0	0.1	—
607	Frozen, boiled, drained	½	cup(s)	90	84.6	17	1.4	3.4	2.4	0.2	0	0	0.1	—
605	Raw, chopped	½	cup(s)	50	46.0	13	1.0	2.6	1.2	0	0	0	0	—
	Celery													
609	Diced	½	cup(s)	51	48.2	8	0.3	1.5	0.8	0.1	0	0	0	—
608	Stalk	2	item(s)	80	76.3	13	0.6	2.4	1.3	0.1	0	0	0.1	—
	Chard													
1057	Swiss chard, boiled, drained	½	cup(s)	88	81.1	18	1.6	3.6	1.8	0.1	0	0	0	—
1056	Swiss chard, raw	1	cup(s)	36	33.4	7	0.6	1.3	0.6	0.1	0	0	0	—
	Collard greens													
610	Boiled, drained	½	cup(s)	95	87.3	25	2.0	4.7	2.7	0.3	0	0	0.2	—
611	Frozen, chopped, boiled, drained	½	cup(s)	85	75.2	31	2.5	6.0	2.4	0.3	0.1	0	0.2	—
	Corn													
29614	Yellow corn, fresh, cooked	1	item(s)	100	69.2	107	3.3	25.0	2.8	1.3	0.2	0.4	0.6	—
615	Yellow creamed sweet corn, canned	½	cup(s)	128	100.8	92	2.2	23.2	1.5	0.5	0.1	0.2	0.3	—
612	Yellow sweet corn, boiled, drained	½	cup(s)	82	57.0	89	2.7	20.6	2.3	1.1	0.2	0.3	0.5	—
614	Yellow sweet corn, frozen, boiled, drained	½	cup(s)	82	63.2	66	2.1	15.8	2.0	0.5	0.1	0.2	0.3	—
618	**Cucumber**	¼	item(s)	75	71.7	11	0.5	2.7	0.4	0.1	0	0	0	—
16870	**Cucumber, kim chee**	½	cup(s)	75	68.1	16	0.8	3.6	1.1	0.1	0	0	0	—
	Dandelion greens													
620	Chopped, boiled, drained	½	cup(s)	53	47.1	17	1.1	3.4	1.5	0.3	0.1	0	0.1	—
2734	Raw	1	cup(s)	55	47.1	25	1.5	5.1	1.9	0.4	0.1	0	0.2	—
1066	**Eggplant, boiled, drained**	½	cup(s)	50	44.4	17	0.4	4.3	1.2	0.1	0	0	0	—
621	**Endive or escarole, chopped, raw**	1	cup(s)	50	46.9	8	0.6	1.7	1.5	0.1	0	0	0	—
8784	**Jicama or yambean**	½	cup(s)	65	116.5	49	0.9	11.4	6.3	0.1	0	0	0.1	—
	Kale													
623	Frozen, chopped, boiled, drained	½	cup(s)	65	58.8	20	1.8	3.4	1.3	0.3	0	0	0.2	—
29313	Raw	1	cup(s)	67	56.6	33	2.2	6.7	1.3	0.5	0.1	0	0.2	—
	Kohlrabi													
1072	Boiled, drained	½	cup(s)	83	74.5	24	1.5	5.5	0.9	0.1	0	0	0	—
1071	Raw	1	cup(s)	135	122.9	36	2.3	8.4	4.9	0.1	0	0	0.1	—
	Leeks													
1074	Boiled, drained	½	cup(s)	52	47.2	16	0.4	4.0	0.5	0.1	0	0	0	—
1073	Raw	1	cup(s)	89	73.9	54	1.3	12.6	1.6	0.3	0	0	0.1	—
	Lentils													
522	Boiled	¼	cup(s)	50	34.5	57	4.5	10.0	3.9	0.2	0	0	0.1	—
1075	Sprouted	1	cup(s)	77	51.9	82	6.9	17.0	—	0.4	0	0.1	0.2	—
	Lettuce													
625	Butterhead leaves	11	piece(s)	83	78.9	11	1.1	1.8	0.9	0.2	0	0	0.1	—
624	Butterhead, Boston or Bibb	1	cup(s)	55	52.6	7	0.7	1.2	0.6	0.1	0	0	0.1	—
626	Iceberg	1	cup(s)	55	52.6	8	0.5	1.6	0.7	0.1	0	0	0	—
628	Iceberg, chopped	1	cup(s)	55	52.6	8	0.5	1.6	0.7	0.1	0	0	0	—
629	Looseleaf	1	cup(s)	36	34.2	5	0.5	1.0	0.5	0.1	0	0	0	—
1665	Romaine, shredded	1	cup(s)	56	53.0	10	0.7	1.8	1.2	0.2	0	0	0.1	—
	Mushrooms													
15585	Crimini (about 6)	3	ounce(s)	85	—	28	3.7	2.8	1.9	0	0	0	0	0

VEGETABLES, LEGUMES—Continued

Chol (mg)	Calc (mg)	Iron (mg)	Magn (mg)	Pota (mg)	Sodi (mg)	Zinc (mg)	Vit A (µg)	Thia (mg)	Vit E (mg α)	Ribo (mg)	Niac (mg)	Vit B$_6$ (mg)	Fola (µg)	Vit C (mg)	Vit B$_{12}$ (µg)	Sele (µg)
0	30	0.56	12.0	130.6	10.1	0.25	46.9	0.05	1.21	0.07	0.42	0.12	51.5	36.9	0	0.6
0	21	0.33	9.6	143.8	15.0	0.19	14.1	0.03	0.36	0.05	0.29	0.08	28.7	40.6	0	1.1
0	11	0.23	6.4	96.0	7.4	0.20	2.6	0.02	0.01	0.03	0.23	0.07	18.2	28.2	0	0.2
0	28	0.93	15.6	247.3	16.4	0.25	30.4	0.08	0.33	0.06	0.47	0.13	46.8	48.4	0	1.2
0	20	0.37	14.0	224.8	11.6	0.18	35.7	0.08	0.39	0.08	0.41	0.22	78.3	35.4	0	0.5
0	72	0.24	22.5	294.0	12.0	0.30	6.0	0.08	0.20	0.04	0.36	0.16	45.0	56.2	0	0.9
0	158	1.76	18.7	630.7	459.0	0.28	360.4	0.04	0.14	0.10	0.72	0.28	69.7	44.2	0	0.7
0	144	1.26	27.0	379.5	996.0	0.36	288.0	0.06	0.36	0.10	0.80	0.32	88.5	79.6	0	1.5
0	28	0.33	8.4	119.0	12.6	0.12	3.5	0.04	0.10	0.02	0.16	0.08	30.1	25.6	0	0.2
0	31	0.56	11.2	170.1	18.9	0.15	39.2	0.04	0.07	0.05	0.29	0.14	12.6	39.9	0	0.4
0	24	0.28	19.6	161.0	19.6	0.18	35.0	0.05	0.11	0.02	0.21	0.13	56.0	21.7	0	0.6
0	0	0.00	—	—	140	—	0	—	—	—	—	—	—	0	—	—
0	26	0.71	8.0	189.6	62.4	0.13	552.0	0.02	—	0.02	0.44	0.08	21.6	2.1	0	0.7
0	18	0.16	6.6	176.0	37.9	0.13	459.2	0.03	0.36	0.03	0.54	0.07	10.4	3.2	0	0.1
0	28	0.54	16.5	344.6	34.2	0.21	1128.1	0.11	1.37	0.07	0.46	0.26	4.7	10.0	0	0.7
0	20	0.18	7.3	195.2	42.1	0.15	509.4	0.04	0.40	0.04	0.60	0.08	11.6	3.6	0	0.1
0	23	0.26	7.8	183.3	45.2	0.15	664.6	0.05	0.80	0.03	0.50	0.11	10.9	2.8	0	0.5
0	16	0.27	21.6	279.1	14.4	0.35	1.0	0.08	0.19	0.04	0.87	0.09	27.8	21.2	0	0.7
0	10	0.19	5.6	88.0	9.3	0.10	0.6	0.02	0.04	0.03	0.25	0.10	27.3	27.5	0	0.4
0	15	0.36	8.1	125.1	16.2	0.11	0	0.03	0.05	0.04	0.27	0.07	36.9	28.2	0	0.5
0	11	0.22	7.5	151.5	15.0	0.14	0.5	0.03	0.04	0.03	0.26	0.11	28.5	23.2	0	0.3
0	20	0.10	5.6	131.3	40.4	0.07	11.1	0.01	0.14	0.03	0.16	0.04	18.2	1.6	0	0.2
0	32	0.16	8.8	208.0	64.0	0.10	17.6	0.01	0.21	0.04	0.25	0.05	28.8	2.5	0	0.3
0	51	1.98	75.3	480.4	156.6	0.29	267.8	0.03	1.65	0.08	0.32	0.07	7.9	15.8	0	0.8
0	18	0.64	29.2	136.4	76.7	0.13	110.2	0.01	0.68	0.03	0.14	0.03	5.0	10.8	0	0.3
0	133	1.10	19.0	110.2	15.2	0.21	385.7	0.03	0.83	0.10	0.54	0.12	88.4	17.3	0	0.5
0	179	0.95	25.5	213.4	42.5	0.22	488.8	0.04	1.06	0.09	0.54	0.09	64.6	22.4	0	1.3
0	2	0.61	32.0	248.0	242.0	0.48	13.0	0.20	0.09	0.07	1.60	0.06	46.0	6.2	0	0.2
0	4	0.48	21.8	171.5	364.8	0.67	5.1	0.03	0.09	0.06	1.22	0.08	57.6	5.9	0	0.5
0	2	0.36	21.3	173.8	0	0.50	10.7	0.17	0.07	0.05	1.32	0.04	37.7	5.1	0	0.2
0	2	0.38	23.0	191.1	0.8	0.51	8.2	0.02	0.05	0.05	1.07	0.08	28.7	2.9	0	0.6
0	12	0.20	9.8	110.6	1.5	0.14	3.8	0.01	0.01	0.01	0.07	0.03	5.3	2.1	0	0.2
0	7	3.61	6.0	87.8	765.8	0.38	—	0.02	—	0.02	0.34	0.08	17.3	2.6	0	—
0	74	0.95	12.6	121.8	23.1	0.15	179.6	0.07	1.28	0.09	0.27	0.08	6.8	9.5	0	0.2
0	103	1.70	19.8	218.3	41.8	0.22	279.4	0.10	1.89	0.14	0.44	0.13	14.8	19.2	0	0.3
0	3	0.12	5.4	60.9	0.5	0.06	1.0	0.04	0.20	0.01	0.30	0.04	6.9	0.6	0	0
0	26	0.41	7.5	157.0	11.0	0.39	54.0	0.04	0.22	0.03	0.20	0.01	71.0	3.2	0	0.1
0	16	0.78	15.5	194.0	5.2	0.20	1.3	0.02	0.59	0.04	0.25	0.05	15.5	26.1	0	0.9
0	90	0.61	11.7	208.7	9.8	0.11	477.8	0.02	0.59	0.07	0.43	0.05	9.1	16.4	0	0.6
0	90	1.14	22.8	299.5	28.8	0.29	515.2	0.07	—	0.08	0.66	0.18	19.4	80.4	0	0.6
0	21	0.33	15.7	280.5	17.3	0.26	1.7	0.03	0.43	0.02	0.32	0.13	9.9	44.6	0	0.7
0	32	0.54	25.7	472.5	27.0	0.04	2.7	0.06	0.64	0.02	0.54	0.20	21.6	83.7	0	0.9
0	16	0.56	7.3	45.2	5.2	0.02	1.0	0.01	—	0.01	0.10	0.04	12.5	2.2	0	0.3
0	53	1.86	24.9	160.2	17.8	0.10	73.9	0.05	0.81	0.02	0.35	0.20	57.0	10.7	0	0.9
0	9	1.65	17.8	182.7	1.0	0.63	0	0.08	0.05	0.04	0.52	0.09	89.6	0.7	0	1.4
0	19	2.47	28.5	247.9	8.5	1.16	1.5	0.17	—	0.09	0.86	0.14	77.0	12.7	0	0.5
0	29	1.02	10.7	196.4	4.1	0.16	137.0	0.04	0.14	0.05	0.29	0.06	60.2	3.1	0	0.5
0	19	0.68	7.1	130.9	2.7	0.11	91.3	0.03	0.09	0.03	0.19	0.04	40.1	2.0	0	0.3
0	10	0.22	3.8	77.5	5.5	0.08	13.7	0.02	0.09	0.01	0.07	0.02	15.9	1.5	0	0.1
0	10	0.22	3.8	77.5	5.5	0.08	13.7	0.02	0.09	0.01	0.07	0.02	15.9	1.5	0	0.1
0	13	0.31	4.7	69.8	10.1	0.06	133.2	0.02	0.10	0.02	0.13	0.03	13.7	6.5	0	0.2
0	18	0.54	7.8	138.3	4.5	0.13	162.4	0.04	0.07	0.03	0.17	0.04	76.2	13.4	0	0.2
0	0	0.67	—	—	32.6	—	0	—	—	—	—	—	—	0	0	—

DA+ Code	Food Description	Quantity	Measure	Wt (g)	H₂O (g)	Ener (kcal)	Prot (g)	Carb (g)	Fiber (g)	Fat (g)	Sat	Mono	Poly	Trans
											fat breakdown			

DA+ Code	Food Description	Quantity	Measure	Wt (g)	H₂O (g)	Ener (kcal)	Prot (g)	Carb (g)	Fiber (g)	Fat (g)	Sat	Mono	Poly	Trans
8700	Enoki	30	item(s)	90	79.7	40	2.3	6.9	2.4	0.3	0	0	0.1	—
1079	Mushrooms, boiled, drained	½	cup(s)	78	71.0	22	1.7	4.1	1.7	0.4	0	0	0.1	—
1080	Mushrooms, canned, drained	½	cup(s)	78	71.0	20	1.5	4.0	1.9	0.2	0	0	0.1	—
630	Mushrooms, raw	½	cup(s)	48	44.4	11	1.5	1.6	0.5	0.2	0	0	0.1	—
15587	Portabella, raw	1	item(s)	84	—	30	3.0	3.9	3.0	0	0	0	0	0
2743	Shiitake, cooked	½	cup(s)	73	60.5	41	1.1	10.4	1.5	0.2	0	0.1	0	—
	Mustard greens													
2744	Frozen, boiled, drained	½	cup(s)	75	70.4	14	1.7	2.3	2.1	0.2	0	0.1	0	—
29319	Raw	1	cup(s)	56	50.8	15	1.5	2.7	1.8	0.1	0	0	0	—
	Okra													
16866	Batter coated, fried	11	piece(s)	83	55.6	156	2.1	12.7	2.0	11.2	1.5	3.7	5.5	—
32742	Frozen, boiled, drained, no salt added	½	cup(s)	92	83.8	26	1.9	5.3	2.6	0.3	0.1	0	0.1	—
632	Sliced, boiled, drained	½	cup(s)	80	74.1	18	1.5	3.6	2.0	0.2	0	0	0	—
	Onions													
635	Chopped, boiled, drained	½	cup(s)	105	92.2	46	1.4	10.7	1.5	0.2	0	0	0.1	—
2748	Frozen, boiled, drained	½	cup(s)	106	97.8	30	0.8	7.0	1.9	0.1	0	0	0	—
1081	Onion rings, breaded and pan fried, frozen, heated	10	piece(s)	71	20.2	289	3.8	27.1	0.9	19.0	6.1	7.7	3.6	—
633	Raw, chopped	½	cup(s)	80	71.3	32	0.9	7.5	1.4	0.1	0	0	0	—
16850	Red onions, sliced, raw	½	cup(s)	57	50.7	24	0.5	5.8	0.8	0	0	0	0	—
636	Scallions, green or spring onions	2	item(s)	30	26.9	10	0.5	2.2	0.8	0.1	0	0	0	—
16860	**Palm hearts, cooked**	½	cup(s)	73	50.7	84	2.0	18.7	1.1	0.1	0	0	0.1	—
637	**Parsley, chopped**	1	tablespoon(s)	4	3.3	1	0.1	0.2	0.1	0	0	0	0	—
638	**Parsnips, sliced, boiled, drained**	½	cup(s)	78	62.6	55	1.0	13.3	2.8	0.2	0	0.1	0	—
	Peas													
639	Green peas, canned, drained	½	cup(s)	85	69.4	59	3.8	10.7	3.5	0.3	0.1	0	0.1	—
641	Green peas, frozen, boiled, drained	½	cup(s)	80	63.6	62	4.1	11.4	4.4	0.2	0	0	0.1	—
35694	Pea pods, boiled with salt, drained	½	cup(s)	80	71.1	32	2.6	5.2	2.2	0.2	0	0	0.1	—
1082	Peas and carrots, canned with liquid	½	cup(s)	128	112.4	48	2.8	10.8	2.6	0.3	0.1	0	0.2	—
1083	Peas and carrots, frozen, boiled, drained	½	cup(s)	80	68.6	38	2.5	8.1	2.5	0.3	0.1	0	0.2	—
2750	Snow or sugar peas, frozen, boiled, drained	½	cup(s)	80	69.3	42	2.8	7.2	2.5	0.3	0.1	0	0.1	—
640	Snow or sugar peas, raw	½	cup(s)	32	28.0	13	0.9	2.4	0.8	0.1	0	0	0	—
29324	Split peas, sprouted	½	cup(s)	60	37.4	77	5.3	16.9	—	0.4	0.1	0	0.2	—
	Peppers													
644	Green bell or sweet, boiled, drained	½	cup(s)	68	62.5	19	0.6	4.6	0.8	0.1	0	0	0.1	—
643	Green bell or sweet, raw	½	cup(s)	75	69.9	15	0.6	3.5	1.3	0.1	0	0	0	—
1664	Green hot chili	1	item(s)	45	39.5	18	0.9	4.3	0.7	0.1	0	0	0	—
1663	Green hot chili, canned with liquid	½	cup(s)	68	62.9	14	0.6	3.5	0.9	0.1	0	0	0	—
1086	Jalapeno, canned with liquid	½	cup(s)	68	60.4	18	0.6	3.2	1.8	0.6	0.1	0	0.3	—
8703	Yellow bell or sweet	1	item(s)	186	171.2	50	1.9	11.8	1.7	0.4	0.1	0	0.2	—
1087	**Poi**	½	cup(s)	120	86.0	134	0.5	32.7	0.5	0.2	0	0	0.1	—
	Potatoes													
1090	Au gratin mix, prepared with water, whole milk and butter	½	cup(s)	124	97.7	115	2.8	15.9	1.1	5.1	3.2	1.5	0.2	—
1089	Au gratin, prepared with butter	½	cup(s)	123	90.7	162	6.2	13.8	2.2	9.3	5.8	2.6	0.3	—
5791	Baked, flesh and skin	1	item(s)	202	151.3	188	5.1	42.7	4.4	0.3	0.1	0	0.1	—
645	Baked, flesh only	½	cup(s)	61	46.0	57	1.2	13.1	0.9	0.1	0	0	0	—
1088	Baked, skin only	1	item(s)	58	27.4	115	2.5	26.7	4.6	0.1	0	0	0	—
5795	Boiled in skin, flesh only, drained	1	item(s)	136	104.7	118	2.5	27.4	2.1	0.1	0	0	0.1	—
5794	Boiled, drained, skin and flesh	1	item(s)	150	115.9	129	2.9	29.8	2.5	0.2	0	0	0.1	—
647	Boiled, flesh only	½	cup(s)	78	60.4	67	1.3	15.6	1.4	0.1	0	0	0	—
648	French fried, deep fried, prepared from raw	14	item(s)	70	32.8	187	2.7	23.5	2.9	9.5	1.9	4.2	3.0	—
649	French fried, frozen, heated	14	item(s)	70	43.7	94	1.9	19.4	2.0	3.7	0.7	2.3	0.2	—
1091	Hashed brown	½	cup(s)	78	36.9	207	2.3	27.4	2.5	9.8	1.5	4.1	3.7	—
652	Mashed with margarine and whole milk	½	cup(s)	105	79.0	119	2.1	17.7	1.6	4.4	1.0	2.0	1.2	0.7
653	Mashed, prepared from dehydrated granules with milk, water, and margarine	½	cup(s)	105	79.8	122	2.3	16.9	1.4	5.0	1.3	2.1	1.4	—
2759	Microwaved	1	item(s)	202	145.5	212	4.9	49.0	4.6	0.2	0.1	0	0.1	—
2760	Microwaved in skin, flesh only	½	cup(s)	78	57.1	78	1.6	18.1	1.2	0.1	0	0	0	—
5804	Microwaved, skin only	1	item(s)	58	36.8	77	2.5	17.2	4.2	0.1	0	0	0	—
1097	Potato puffs, frozen, heated	½	cup(s)	64	38.2	122	1.3	17.8	1.6	5.5	1.2	3.9	0.3	—
1094	Scalloped mix, prepared with water, whole milk and butter	½	cup(s)	124	98.4	116	2.6	15.9	1.4	5.3	3.3	1.5	0.2	—
1093	Scalloped, prepared with butter	½	cup(s)	123	99.2	108	3.5	13.2	2.3	4.5	2.8	1.3	0.2	—
	Pumpkin													
1773	Boiled, drained	½	cup(s)	123	114.8	25	0.9	6.0	1.3	0.1	0	0	0	—
656	Canned	½	cup(s)	123	110.2	42	1.3	9.9	3.6	0.3	0.2	0	0	—
	Radicchio													
8731	Leaves, raw	1	cup(s)	40	37.3	9	0.6	1.8	0.4	0.1	0	0	0	—
2498	Raw	1	cup(s)	40	37.3	9	0.6	1.8	0.4	0.1	0	0	0	—

VEGETABLES, LEGUMES—Continued

Chol (mg)	Calc (mg)	Iron (mg)	Magn (mg)	Pota (mg)	Sodi (mg)	Zinc (mg)	Vit A (µg)	Thia (mg)	Vit E (mg α)	Ribo (mg)	Niac (mg)	Vit B6 (mg)	Fola (µg)	Vit C (mg)	Vit B12 (µg)	Sele (µg)
0	1	0.98	14.4	331.2	2.7	0.54	0	0.16	0.01	0.14	5.31	0.07	46.8	0	0	2.0
0	5	1.35	9.4	277.7	1.6	0.67	0	0.05	0.01	0.23	3.47	0.07	14.0	3.1	0	9.3
0	9	0.61	11.7	100.6	331.5	0.56	0	0.06	0.01	0.01	1.24	0.04	9.4	0	0	3.2
0	1	0.24	4.3	152.6	2.4	0.25	0	0.04	0.01	0.19	1.73	0.05	7.7	1.0	0	4.5
0	39	0.35	—	—	9.9	—	0	—	—	—	—	—	—	0	0	—
0	2	0.31	10.2	84.8	2.9	0.96	0	0.02	0.02	0.12	1.08	0.11	15.2	0.2	0	18.0
0	76	0.84	9.8	104.3	18.8	0.15	265.5	0.03	1.01	0.04	0.19	0.08	52.5	10.4	0	0.5
0	58	0.81	17.9	198.2	14.0	0.11	294.0	0.04	1.12	0.06	0.45	0.10	104.7	39.2	0	0.5
2	54	1.13	32.2	170.8	109.7	0.44	14.0	0.16	1.50	0.12	1.29	0.11	39.6	9.2	0	3.6
0	88	0.61	46.9	215.3	2.8	0.57	15.6	0.09	0.29	0.11	0.72	0.04	134.3	11.2	0	0.6
0	62	0.22	28.8	108.0	4.8	0.34	11.2	0.10	0.21	0.04	0.69	0.15	36.8	13.0	0	0.3
0	23	0.24	11.5	174.3	3.1	0.21	0	0.03	0.02	0.02	0.17	0.12	15.7	5.5	0	0.6
0	17	0.32	6.4	114.5	12.7	0.06	0	0.02	0.01	0.02	0.14	0.06	13.8	2.8	0	0.4
0	22	1.20	13.5	91.6	266.3	0.29	7.8	0.19	—	0.09	2.56	0.05	46.9	1.0	0	2.5
0	18	0.16	8.0	116.8	3.2	0.13	0	0.03	0.01	0.02	0.09	0.09	15.2	5.9	0	0.4
0	13	0.10	5.7	82.4	1.7	0.09	0	0.02	0.01	0.01	0.04	0.08	10.9	3.7	0	0.3
0	22	0.44	6.0	82.8	4.8	0.11	15.0	0.01	0.16	0.02	0.15	0.01	19.2	5.6	0	0.2
0	13	1.23	7.3	1318.4	10.2	2.72	2.2	0.03	0.36	0.12	0.62	0.53	14.6	5.0	0	0.5
0	5	0.23	1.9	21.1	2.1	0.04	16.0	0.00	0.02	0.00	0.05	0.00	5.8	5.1	0	0
0	29	0.45	22.6	286.3	7.8	0.20	0	0.06	0.78	0.04	0.56	0.07	45.2	10.1	0	1.3
0	17	0.80	14.5	147.1	214.2	0.60	23.0	0.10	0.02	0.06	0.62	0.05	37.4	8.2	0	1.4
0	19	1.21	17.6	88.0	57.6	0.53	84.0	0.22	0.02	0.08	1.18	0.09	47.2	7.9	0	0.8
0	34	1.57	20.8	192.0	192.0	0.29	41.6	0.10	0.31	0.06	0.43	0.11	23.2	38.3	0	0.6
0	29	0.96	17.9	127.5	331.5	0.74	368.5	0.09	—	0.07	0.74	0.11	23.0	8.4	0	1.1
0	18	0.75	12.8	126.4	54.4	0.36	380.8	0.18	0.41	0.05	0.92	0.07	20.8	6.5	0	0.9
0	47	1.92	22.4	173.6	4.0	0.39	52.8	0.05	0.37	0.09	0.45	0.13	28.0	17.6	0	0.6
0	14	0.65	7.6	63.0	1.3	0.08	17.0	0.04	0.12	0.02	0.19	0.05	13.2	18.9	0	0.2
0	22	1.34	33.6	228.6	12.0	0.62	4.8	0.12	—	0.08	1.84	0.14	86.4	6.2	0	0.4
0	6	0.31	6.8	112.9	1.4	0.08	15.6	0.04	0.34	0.02	0.32	0.15	10.9	50.6	0	0.2
0	7	0.25	7.5	130.4	2.2	0.09	13.4	0.04	0.27	0.02	0.35	0.16	7.5	59.9	0	0
0	8	0.54	11.3	153.0	3.2	0.13	26.6	0.04	0.31	0.04	0.42	0.12	10.4	109.1	0	0.2
0	5	0.34	9.5	127.2	797.6	0.10	24.5	0.01	0.46	0.02	0.54	0.10	6.8	46.2	0	0.2
0	16	1.28	10.2	131.2	1136.3	0.23	57.8	0.03	0.47	0.03	0.27	0.13	9.5	6.8	0	0.3
0	20	0.85	22.3	394.3	3.7	0.31	18.6	0.05	—	0.04	1.65	0.31	48.4	341.3	0	0.6
0	19	1.06	28.8	219.6	14.4	0.26	3.6	0.16	2.76	0.05	1.32	0.33	25.2	4.8	0	0.8
19	103	0.39	18.6	271.0	543.3	0.29	64.4	0.02	—	0.10	1.16	0.05	8.7	3.8	0	3.3
28	146	0.78	24.5	485.1	530.4	0.85	78.4	0.08	—	0.14	1.22	0.21	13.5	12.1	0	3.3
0	30	2.18	56.6	1080.7	20.2	0.72	2.0	0.12	0.08	0.09	2.84	0.62	56.6	19.4	0	0.8
0	3	0.21	15.3	238.5	3.1	0.18	0	0.06	0.02	0.01	0.85	0.18	5.5	7.8	0	0.2
0	20	4.08	24.9	332.3	12.2	0.28	0.6	0.07	0.02	0.06	1.77	0.35	12.8	7.8	0	0.4
0	7	0.42	29.9	515.4	5.4	0.40	0	0.14	0.01	0.02	1.95	0.40	13.6	17.7	0	0.4
0	13	1.27	34.1	572.0	7.4	0.46	0	0.14	0.01	0.03	2.13	0.44	15.0	18.4	0	—
0	6	0.24	15.6	255.8	3.9	0.21	0	0.07	0.01	0.01	1.02	0.21	7.0	5.8	0	0.2
0	16	1.05	30.8	567.0	8.4	0.39	0	0.08	0.09	0.03	1.34	0.37	16.1	21.2	0	0.4
0	8	0.51	18.2	315.7	271.6	0.26	0	0.09	0.07	0.02	1.55	0.12	19.6	9.3	0	0.1
0	11	0.43	27.3	449.3	266.8	0.37	0	0.13	0.01	0.03	1.80	0.37	12.5	10.1	0	0.4
1	23	0.27	19.9	344.4	349.6	0.31	43.0	0.09	0.44	0.04	1.23	0.25	9.4	11.0	0.1	0.8
2	36	0.21	21.0	164.8	179.5	0.26	49.3	0.09	0.53	0.09	0.90	0.16	8.4	6.8	0.1	5.9
0	22	2.50	54.5	902.9	16.2	0.72	0	0.24	—	0.06	3.46	0.69	24.2	30.5	0	0.8
0	4	0.31	19.4	319.0	5.4	0.25	0	0.10	—	0.01	1.26	0.25	9.3	11.7	0	0.3
0	27	3.44	21.5	377.0	9.3	0.29	0	0.04	0.01	0.04	1.28	0.28	9.9	8.9	0	0.3
0	9	0.41	10.9	199.7	307.2	0.21	0	0.08	0.15	0.02	0.97	0.08	9.0	4.0	0	0.4
14	45	0.47	17.4	252.2	423.7	0.31	43.5	0.02	—	0.06	1.28	0.05	12.4	4.1	0	2.0
15	70	0.70	23.3	463.1	410.4	0.49	0	0.08	—	0.11	1.29	0.22	13.5	13.0	0	2.0
0	18	0.69	11.0	281.8	1.2	0.28	306.3	0.03	0.98	0.09	0.50	0.05	11.0	5.8	0	0.2
0	32	1.70	28.2	252.4	6.1	0.20	953.1	0.02	1.29	0.06	0.45	0.06	14.7	5.1	0	0.5
0	8	0.23	5.2	120.8	8.8	0.25	0.4	0.01	0.90	0.01	0.10	0.02	24.0	3.2	0	0.4
0	8	0.23	5.2	120.8	8.8	0.25	0.4	0.01	0.90	0.01	0.10	0.02	24.0	3.2	0	0.4

DA+ Code	Food Description	Quantity	Measure	Wt (g)	H₂O (g)	Ener (kcal)	Prot (g)	Carb (g)	Fiber (g)	Fat (g)	Fat Breakdown (g) Sat	Mono	Poly	Trans
657	**Radishes**	6	item(s)	27	25.7	4	0.2	0.9	0.4	0	0	0	0	—
1099	**Rutabaga, boiled, drained**	½	cup(s)	85	75.5	33	1.1	7.4	1.5	0.2	0	0	0.1	—
658	**Sauerkraut, canned**	½	cup(s)	118	109.2	22	1.1	5.1	3.4	0.2	0	0	0.1	—
	Seaweed													
1102	Kelp	½	cup(s)	40	32.6	17	0.6	3.8	0.5	0.2	0.1	0	0	—
1104	Spirulina, dried	½	cup(s)	8	0.4	22	4.3	1.8	0.3	0.6	0.2	0.1	0.2	—
1106	**Shallots**	3	tablespoon(s)	30	23.9	22	0.8	5.0	—	0	0	0	0	—
	Soybeans													
1670	Boiled	½	cup(s)	86	53.8	149	14.3	8.5	5.2	7.7	1.1	1.7	4.4	—
2825	Dry roasted	½	cup(s)	86	0.7	388	34.0	28.1	7.0	18.6	2.7	4.1	10.5	—
2824	Roasted, salted	½	cup(s)	86	1.7	405	30.3	28.9	15.2	21.8	3.2	4.8	12.3	—
8739	Sprouted, stir fried	½	cup(s)	63	42.3	79	8.2	5.9	0.5	4.5	0.6	1.0	2.5	0
	Soy products													
1813	Soy milk	1	cup(s)	240	211.3	130	7.8	15.1	1.4	4.2	0.5	1.0	2.3	0
2838	Tofu, dried, frozen (koyadofu)	3	ounce(s)	85	4.9	408	40.8	12.4	6.1	25.8	3.7	5.7	14.6	—
13844	Tofu, extra firm	3	ounce(s)	85	—	86	8.6	2.2	1.1	4.3	0.5	0.9	2.8	—
13843	Tofu, firm	3	ounce(s)	85	—	75	7.5	2.2	0.5	3.2	0	0.9	2.3	—
1816	Tofu, firm, with calcium sulfate and magnesium chloride (nigari)	3	ounce(s)	85	72.2	60	7.0	1.4	0.8	3.5	0.7	1.0	1.5	—
1817	Tofu, fried	3	ounce(s)	85	43.0	230	14.6	8.9	3.3	17.2	2.5	3.8	9.7	—
13841	Tofu, silken	3	ounce(s)	85	—	42	3.7	1.9	0	2.3	0.5	—	—	—
13842	Tofu, soft	3	ounce(s)	85	—	65	6.5	1.1	0.5	3.2	0.5	1.1	2.2	—
1671	Tofu, soft, with calcium sulfate and magnesium chloride (nigari)	3	ounce(s)	85	74.2	52	5.6	1.5	0.2	3.1	0.5	0.7	1.8	—
	Spinach													
663	Canned, drained	½	cup(s)	107	98.2	25	3.0	3.6	2.6	0.5	0.1	0	0.2	—
660	Chopped, boiled, drained	½	cup(s)	90	82.1	21	2.7	3.4	2.2	0.2	0	0	0.1	—
661	Chopped, frozen, boiled, drained	½	cup(s)	95	84.5	32	3.8	4.6	3.5	0.8	0.1	0	0.4	—
662	Leaf, frozen, boiled, drained	½	cup(s)	95	84.5	32	3.8	4.6	3.5	0.8	0.1	0	0.4	—
659	Raw, chopped	1	cup(s)	30	27.4	7	0.9	1.1	0.7	0.1	0	0	0	—
8470	Trimmed leaves	1	cup(s)	32	27.5	3	0.9	0	2.8	0.1	—	—	—	—
	Squash													
1662	Acorn winter, baked	½	cup(s)	103	85.0	57	1.1	14.9	4.5	0.1	0	0	0.1	—
29702	Acorn winter, boiled, mashed	½	cup(s)	123	109.9	42	0.8	10.8	3.2	0.1	0	0	0	—
29451	Butternut, frozen, boiled	½	cup(s)	122	106.9	47	1.5	12.2	1.8	0.1	0	0	0	—
1661	Butternut winter, baked	½	cup(s)	102	89.5	41	0.9	10.7	3.4	0.1	0	0	0	—
32773	Butternut winter, frozen, boiled, mashed, no salt added	½	cup(s)	121	106.4	47	1.5	12.2	—	0.1	0	0	0	—
29700	Crookneck and straightneck summer, boiled, drained	½	cup(s)	65	60.9	12	0.6	2.6	1.2	0.1	0	0	0.1	—
29703	Hubbard winter, baked	½	cup(s)	102	86.8	51	2.5	11.0	—	0.6	0.1	0	0.3	—
1660	Hubbard winter, boiled, mashed	½	cup(s)	118	107.5	35	1.7	7.6	3.4	0.4	0.1	0	0.2	—
29704	Spaghetti winter, boiled, drained, or baked	½	cup(s)	78	71.5	21	0.5	5.0	1.1	0.2	0	0	0.1	—
664	Summer, all varieties, sliced, boiled, drained	½	cup(s)	90	84.3	18	0.8	3.9	1.3	0.3	0.1	0	0.1	—
665	Winter, all varieties, baked, mashed	½	cup(s)	103	91.4	38	0.9	9.1	2.9	0.4	0.1	0	0.2	—
1112	Zucchini summer, boiled, drained	½	cup(s)	90	85.3	14	0.6	3.5	1.3	0	0	0	0	—
1113	Zucchini summer, frozen, boiled, drained	½	cup(s)	112	105.6	19	1.3	4.0	1.4	0.1	0	0	0.1	—
	Sweet potatoes													
666	Baked, peeled	½	cup(s)	100	75.8	90	2.0	20.7	3.3	0.2	0	0	0.1	—
667	Boiled, mashed	½	cup(s)	164	131.4	125	2.2	29.1	4.1	0.2	0.1	0	0.1	—
668	Candied, home recipe	½	cup(s)	91	61.1	132	0.8	25.4	2.2	3.0	1.2	0.6	0.1	—
670	Canned, vacuum pack	½	cup(s)	100	76.0	91	1.7	21.1	1.8	0.2	0	0	0.1	—
2765	Frozen, baked	½	cup(s)	88	64.5	88	1.5	20.5	1.6	0.1	0	0	0	—
1136	Yams, baked or boiled, drained	½	cup(s)	68	47.7	79	1.0	18.7	2.7	0.1	0	0	0	—
32785	**Taro shoots, cooked, no salt added**	½	cup(s)	70	66.7	10	0.5	2.2	—	0.1	0	0	0	—
	Tomatillo													
8774	Raw	2	item(s)	68	62.3	22	0.7	4.0	1.3	0.7	0.1	0.1	0.3	—
8777	Raw, chopped	½	cup(s)	66	60.5	21	0.6	3.9	1.3	0.7	0.1	0.1	0.3	—
	Tomato													
16846	Cherry, fresh	5	item(s)	85	80.3	15	0.7	3.3	1.0	0.2	0	0	0.1	—
671	Fresh, ripe, red	1	item(s)	123	116.2	22	1.1	4.8	1.5	0.2	0	0	0.1	—
675	Juice, canned	½	cup(s)	122	114.1	21	0.9	5.2	0.5	0.1	0	0	0	—
75	Juice, no salt added	½	cup(s)	122	114.1	21	0.9	5.2	0.5	0.1	0	0	0	—
1699	Paste, canned	2	tablespoon(s)	33	24.1	27	1.4	6.2	1.3	0.2	0	0	0.1	—
1700	Puree, canned	¼	cup(s)	63	54.9	24	1.0	5.6	1.2	0.1	0	0	0.1	—
1118	Red, boiled	½	cup(s)	120	113.2	22	1.1	4.8	0.8	0.1	0	0	0.1	—
3952	Red, diced	½	cup(s)	90	85.1	16	0.8	3.5	1.1	0.2	0	0	0.1	—
1120	Red, stewed, canned	½	cup(s)	128	116.7	33	1.2	7.9	1.3	0.2	0	0	0.1	—
1125	Sauce, canned	¼	cup(s)	61	55.6	15	0.8	3.3	0.9	0.1	0	0	0	—
	VEGETABLES, LEGUMES—Continued													
8778	Sun dried	½	cup(s)	27	3.9	70	3.8	15.1	3.3	0.8	0.1	0.1	0.3	—

Chol (mg)	Calc (mg)	Iron (mg)	Magn (mg)	Pota (mg)	Sodi (mg)	Zinc (mg)	Vit A (µg)	Thia (mg)	Vit E (mg α)	Ribo (mg)	Niac (mg)	Vit B_6 (mg)	Fola (µg)	Vit C (mg)	Vit B_{12} (µg)	Sele (µg)
0	7	0.09	2.7	62.9	10.5	0.07	0	0.00	0.00	0.01	0.06	0.01	6.8	4.0	0	0.2
0	41	0.45	19.6	277.1	17.0	0.30	0	0.07	0.27	0.04	0.61	0.09	12.8	16.0	0	0.6
0	35	1.73	15.3	200.6	780.0	0.22	1.2	0.03	0.17	0.03	0.17	0.15	28.3	17.3	0	0.7
0	67	1.12	48.4	35.6	93.2	0.48	2.4	0.02	0.32	0.04	0.16	0.00	72.0	1.2	0	0.3
0	9	2.14	14.6	102.2	78.6	0.15	2.2	0.18	0.38	0.28	0.96	0.03	7.1	0.8	0	0.5
0	11	0.36	6.3	100.2	3.6	0.12	18.0	0.02	—	0.01	0.06	0.09	10.2	2.4	—	0.4
0	88	4.42	74.0	442.9	0.9	0.98	0	0.13	0.30	0.24	0.34	0.20	46.4	1.5	0	6.3
0	120	3.39	196.1	1173.0	1.7	4.10	0	0.36	—	0.64	0.90	0.19	176.3	4.0	0	16.6
0	119	3.35	124.7	1264.2	140.2	2.70	8.6	0.08	0.78	0.12	1.21	0.17	181.5	1.9	0	16.4
0	52	0.25	60.4	356.6	8.8	1.32	0.6	0.26	—	0.12	0.69	0.10	79.9	7.5	0	0.4
0	60	1.53	60.0	283.2	122.4	0.28	0	0.14	0.26	0.16	1.23	0.18	43.2	0	0	11.5
0	310	8.27	50.2	17.0	5.1	4.16	22.1	0.42	—	0.27	1.01	0.24	78.2	0.6	0	46.2
0	65	1.16	84.1	—	0	—	0	—	—	—	—	—	—	0	0	—
0	108	1.16	56.1	—	0	—	0	—	—	—	—	—	—	0	0	—
0	171	1.36	31.5	125.9	10.2	0.70	0	0.05	0.01	0.05	0.08	0.06	16.2	0.2	0	8.4
0	316	4.14	51.0	124.2	13.6	1.69	0.9	0.14	0.03	0.04	0.08	0.08	23.0	0	0	24.2
0	56	0.34	33.1	—	4.7	—	0	—	—	—	—	—	—	0	1.7	—
0	108	1.16	35.5	—	0	—	0	—	—	—	—	—	—	0	1.9	—
0	94	0.94	23.0	102.1	6.8	0.54	0	0.04	0.01	0.03	0.45	0.04	37.4	0.2	0	7.6
0	136	2.45	81.3	370.2	28.9	0.48	524.3	0.02	2.08	0.14	0.41	0.11	104.8	15.3	0	1.5
0	122	3.21	78.3	419.4	63.0	0.68	471.6	0.08	1.87	0.21	0.44	0.21	131.4	8.8	0	1.4
0	145	1.86	77.9	286.9	92.2	0.46	572.9	0.07	3.36	0.16	0.41	0.12	115.0	2.1	0	5.2
0	145	1.86	77.9	286.9	92.2	0.46	572.9	0.07	3.36	0.16	0.41	0.12	115.0	2.1	0	5.2
0	30	0.81	23.7	167.4	23.7	0.16	140.7	0.02	0.61	0.06	0.22	0.06	58.2	8.4	0	0.3
0	25	2.13	25.5	134.1	38.0	0.18	—	0.03	—	0.05	0.18	0.07	0	7.5	0	—
0	45	0.95	44.1	447.9	4.1	0.17	21.5	0.17	—	0.01	0.90	0.19	19.5	11.1	0	0.7
0	32	0.68	31.9	322.2	3.7	0.13	50.2	0.12	—	0.01	0.65	0.14	13.5	8.0	0	0.5
0	23	0.70	10.9	161.9	2.4	0.14	203.3	0.06	0.14	0.05	0.56	0.08	19.5	4.3	0	0.6
0	42	0.61	29.6	289.6	4.1	0.13	569.1	0.07	1.31	0.01	0.99	0.12	19.4	15.4	0	0.5
0	23	0.70	10.9	161.2	2.4	0.14	202.4	0.06	—	0.05	0.56	0.08	19.4	4.2	0	0.6
0	14	0.31	13.6	137.1	1.3	0.19	5.2	0.03	—	0.02	0.29	0.07	14.9	5.4	0	0.1
0	17	0.48	22.4	365.1	8.2	0.15	308.0	0.07	—	0.04	0.57	0.17	16.3	9.7	0	0.6
0	12	0.33	15.3	252.5	5.9	0.11	236.0	0.05	0.14	0.03	0.39	0.12	11.8	7.7	0	0.4
0	16	0.26	8.5	90.7	14.0	0.15	4.7	0.02	0.09	0.01	0.62	0.07	6.2	2.7	0	0.2
0	24	0.32	21.6	172.8	0.9	0.35	9.9	0.04	0.12	0.03	0.46	0.05	18.0	5.0	0	0.2
0	23	0.45	13.3	247.0	1.0	0.23	267.5	0.02	0.12	0.07	0.51	0.17	20.5	9.8	0	0.4
0	12	0.32	19.8	227.7	2.7	0.16	50.4	0.04	0.11	0.04	0.39	0.07	15.3	4.1	0	0.2
0	19	0.54	14.5	216.3	2.2	0.22	10	0.05	0.13	0.04	0.43	0.05	8.9	4.1	0	0.2
0	38	0.69	27.0	475.0	36.0	0.32	961.0	0.10	0.71	0.10	1.48	0.28	6.0	19.6	0	0.2
0	44	1.18	29.5	377.2	44.3	0.33	1290.7	0.09	1.54	0.08	0.88	0.27	9.8	21.0	0	0.3
7	24	1.03	10.0	172.6	63.9	0.13	0	0.01	—	0.03	0.36	0.03	10.0	6.1	0	0.7
0	22	0.89	22.0	312.0	53.0	0.18	399.0	0.04	1.00	0.06	0.74	0.19	17.0	26.4	0	0.7
0	31	0.47	18.4	330.1	7.0	0.26	913.3	0.05	0.67	0.04	0.49	0.16	19.3	8.0	0	0.5
0	10	0.35	12.2	455.6	5.4	0.13	4.1	0.06	0.23	0.01	0.37	0.15	10.9	8.2	0	0.5
0	10	0.28	5.6	240.8	1.4	0.37	2.1	0.02	—	0.03	0.56	0.07	2.1	13.2	0	0.7
0	5	0.42	13.6	182.2	0.7	0.15	4.1	0.03	0.25	0.02	1.25	0.03	4.8	8.0	0	0.3
0	5	0.41	13.2	176.9	0.7	0.15	4.0	0.03	0.25	0.02	1.22	0.04	4.6	7.7	0	0.3
0	9	0.22	9.4	201.5	4.3	0.14	35.7	0.03	0.45	0.01	0.50	0.06	12.8	10.8	0	0
0	12	0.33	13.5	291.5	6.2	0.20	51.7	0.04	0.66	0.02	0.73	0.09	18.5	15.6	0	0
0	12	0.52	13.4	278.2	326.8	0.18	27.9	0.06	0.39	0.04	0.82	0.14	24.3	22.2	0	0.4
0	12	0.52	13.4	278.2	12.2	0.18	27.9	0.06	0.39	0.04	0.82	0.14	24.3	22.2	0	0.4
0	12	0.97	13.8	332.6	259.1	0.20	24.9	0.02	1.41	0.05	1.00	0.07	3.9	7.2	0	1.7
0	11	1.11	14.4	274.4	249.4	0.22	16.3	0.01	1.23	0.05	0.91	0.07	6.9	6.6	0	0.4
0	13	0.82	10.8	261.6	13.2	0.17	28.8	0.04	0.67	0.03	0.64	0.10	15.6	27.4	0	0.6
0	9	0.24	9.9	213.3	4.5	0.15	37.8	0.03	0.48	0.01	0.53	0.07	13.5	11.4	0	0
0	43	1.70	15.3	263.9	281.8	0.22	11.5	0.06	1.06	0.04	0.91	0.02	6.4	10.1	0	0.8
0	8	0.62	9.8	201.9	319.6	0.12	10.4	0.01	0.87	0.04	0.59	0.06	6.7	4.3	0	0.1
0	30	2.45	52.4	925.3	565.7	0.53	11.9	0.14	0.00	0.13	2.44	0.09	18.4	10.6	0	1.5

DA+ Code	Food Description	Quantity	Measure	Wt (g)	H₂O (g)	Ener (kcal)	Prot (g)	Carb (g)	Fiber (g)	Fat (g)	Sat	Mono	Poly	Trans
											Fat Breakdown (g)			
8783	Sun dried in oil, drained	¼	cup(s)	28	14.8	59	1.4	6.4	1.6	3.9	0.5	2.4	0.6	—
	Turnips													
678	Turnip greens, chopped, boiled, drained	½	cup(s)	72	67.1	14	0.8	3.1	2.5	0.2	0	0	0.1	—
679	Turnip greens, frozen, chopped, boiled, drained	½	cup(s)	82	74.1	24	2.7	4.1	2.8	0.3	0.1	0	0.1	—
677	Turnips, cubed, boiled, drained	½	cup(s)	78	73.0	17	0.6	3.9	1.6	0.1	0	0	0	—
	Vegetables, mixed													
1132	Canned, drained	½	cup(s)	82	70.9	40	2.1	7.5	2.4	0.2	0	0	0.1	—
680	Frozen, boiled, drained	½	cup(s)	91	75.7	59	2.6	11.9	4.0	0.1	0	0	0.1	—
7489	V8 100% vegetable juice	½	cup(s)	120	—	25	1.0	5.0	1.0	0	0	0	0	0
7490	V8 low sodium vegetable juice	½	cup(s)	120	—	25	0	6.5	1.0	0	0	0	0	0
7491	V8 spicy hot vegetable juice	½	cup(s)	120	—	25	1.0	5.0	0.5	0	0	0	0	0
	Water chestnuts													
31073	Sliced, drained	½	cup(s)	75	70.0	20	0	5.0	1.0	0	0	0	0	0
31087	Whole	½	cup(s)	75	70.0	20	0	5.0	1.0	0	0	0	0	0
1135	**Watercress**	1	cup(s)	34	32.3	4	0.8	0.4	0.2	0	0	0	0	0
	NUTS, SEEDS, AND PRODUCTS													
	Almonds													
32940	Almond butter with salt added	1	tablespoon(s)	16	0.2	101	2.4	3.4	0.6	9.5	0.9	6.1	2.0	—
1137	Almond butter, no salt added	1	tablespoon(s)	16	0.2	101	2.4	3.4	0.6	9.5	0.9	6.1	2.0	—
32886	Blanched	¼	cup(s)	36	1.6	211	8.0	7.2	3.8	18.3	1.4	11.7	4.4	—
32887	Dry roasted, no salt added	¼	cup(s)	35	0.9	206	7.6	6.7	4.1	18.2	1.4	11.6	4.4	—
29724	Dry roasted, salted	¼	cup(s)	35	0.9	206	7.6	6.7	4.1	18.2	1.4	11.6	4.4	—
29725	Oil roasted, salted	¼	cup(s)	39	1.1	238	8.3	6.9	4.1	21.7	1.7	13.7	5.3	—
508	Slivered	¼	cup(s)	27	1.3	155	5.7	5.9	3.3	13.3	1.0	8.3	3.3	0
1138	**Beechnuts, dried**	¼	cup(s)	57	3.8	328	3.5	19.1	5.3	28.5	3.3	12.5	11.4	—
517	**Brazil nuts, dried, unblanched**	¼	cup(s)	35	1.2	230	5.0	4.3	2.6	23.3	5.3	8.6	7.2	—
1166	**Breadfruit seeds, roasted**	¼	cup(s)	57	28.3	118	3.5	22.8	3.4	1.5	0.4	0.2	0.8	—
1139	**Butternuts, dried**	¼	cup(s)	30	1.0	184	7.5	3.6	1.4	17.1	0.4	3.1	12.8	—
	Cashews													
32931	Cashew butter with salt added	1	tablespoon(s)	16	0.5	94	2.8	4.4	0.3	7.9	1.6	4.7	1.3	—
32889	Cashew butter, no salt added	1	tablespoon(s)	16	0.5	94	2.8	4.4	0.3	7.9	1.6	4.7	1.3	—
1140	Dry roasted	¼	cup(s)	34	0.6	197	5.2	11.2	1.0	15.9	3.1	9.4	2.7	—
518	Oil roasted	¼	cup(s)	32	1.1	187	5.4	9.6	1.1	15.4	2.7	8.4	2.8	—
	Coconut, Shredded													
32896	Dried, not sweetened	¼	cup(s)	23	0.7	152	1.6	5.4	3.8	14.9	13.2	0.6	0.2	—
1153	Dried, shredded, sweetened	¼	cup(s)	23	2.9	116	0.7	11.1	1.0	8.3	7.3	0.4	0.1	—
520	Shredded	¼	cup(s)	20	9.4	71	0.7	3.0	1.8	6.7	5.9	0.3	0.1	—
	Chestnuts													
1152	Chinese, roasted	¼	cup(s)	36	14.6	87	1.6	19.0	—	0.4	0.1	0.2	0.1	—
32895	European, boiled and steamed	¼	cup(s)	46	31.3	60	0.9	12.8	—	0.6	0.1	0.2	0.2	—
32911	European, roasted	¼	cup(s)	36	14.5	88	1.1	18.9	1.8	0.8	0.1	0.3	0.3	—
32922	Japanese, boiled and steamed	¼	cup(s)	36	31.0	20	0.3	4.5	—	0.1	0	0	0	—
32923	Japanese, roasted	¼	cup(s)	36	18.1	73	1.1	16.4	—	0.3	0	0.1	0.1	—
4958	**Flax seeds or linseeds**	¼	cup(s)	43	3.3	225	8.4	12.3	11.9	17.7	1.7	3.2	12.6	0
32904	**Ginkgo nuts, dried**	¼	cup(s)	39	4.8	136	4.0	28.3	—	0.8	0.1	0.3	0.3	—
	Hazelnuts or filberts													
32901	Blanched	¼	cup(s)	30	1.7	189	4.1	5.1	3.3	18.3	1.4	14.5	1.7	—
32902	Dry roasted, no salt added	¼	cup(s)	30	0.8	194	4.5	5.3	2.8	18.7	1.3	14.0	2.5	—
1156	**Hickorynuts, dried**	¼	cup(s)	30	0.8	197	3.8	5.5	1.9	19.3	2.1	9.8	6.6	—
	Macadamias													
32905	Dry roasted, no salt added	¼	cup(s)	34	0.5	241	2.6	4.5	2.7	25.5	4.0	19.9	0.5	—
32932	Dry roasted, with salt added	¼	cup(s)	34	0.5	240	2.6	4.3	2.7	25.5	4.0	19.9	0.5	—
1157	Raw	¼	cup(s)	34	0.5	241	2.6	4.6	2.9	25.4	4.0	19.7	0.5	—
	Mixed nuts													
1159	With peanuts, dry roasted	¼	cup(s)	34	0.6	203	5.9	8.7	3.1	17.6	2.4	10.8	3.7	—
32933	With peanuts, dry roasted, with salt added	¼	cup(s)	34	0.6	203	5.9	8.7	3.1	17.6	2.4	10.8	3.7	—
32906	Without peanuts, oil roasted, no salt added	¼	cup(s)	36	1.1	221	5.6	8.0	2.0	20.2	3.3	11.9	4.1	—
	Peanuts													
2807	Dry roasted	¼	cup(s)	37	0.6	214	8.6	7.9	2.9	18.1	2.5	9.0	5.7	—
2806	Dry roasted, salted	¼	cup(s)	37	0.6	214	8.6	7.9	2.9	18.1	2.5	9.0	5.7	—
1763	Oil roasted, salted	¼	cup(s)	36	0.5	216	10.1	5.5	3.4	18.9	3.1	9.4	5.5	—
1884	Peanut butter, chunky	1	tablespoon(s)	16	0.2	94	3.8	3.5	1.3	8.0	1.3	3.9	2.4	—
30303	Peanut butter, low sodium	1	tablespoon(s)	16	0.2	95	4.0	3.1	0.9	8.2	1.8	3.9	2.2	—
30305	Peanut butter, reduced fat	1	tablespoon(s)	18	0.2	94	4.7	6.4	0.9	6.1	1.3	2.9	1.8	—
524	Peanut butter, smooth	1	tablespoon(s)	16	0.3	94	4.0	3.1	1.0	8.1	1.7	3.9	2.3	—
2804	Raw	¼	cup(s)	37	2.4	207	9.4	5.9	3.1	18.0	2.5	8.9	5.7	—
	Pecans													
32907	Dry roasted, no salt added	¼	cup(s)	28	0.3	198	2.6	3.8	2.6	20.7	1.8	12.3	5.7	—

NUTS, SEEDS, AND PRODUCTS —Continued

Chol (mg)	Calc (mg)	Iron (mg)	Magn (mg)	Pota (mg)	Sodi (mg)	Zinc (mg)	Vit A (µg)	Thia (mg)	Vit E (mg α)	Ribo (mg)	Niac (mg)	Vit B_6 (mg)	Fola (µg)	Vit C (mg)	Vit B_{12} (µg)	Sele (µg)
0	13	0.73	22.3	430.4	73.2	0.21	17.6	0.05	—	0.10	0.99	0.08	6.3	28.0	0	0.8
0	99	0.58	15.8	146.2	20.9	0.10	274.3	0.03	1.35	0.05	0.30	0.13	85.0	19.7	0	0.6
0	125	1.59	21.3	183.7	12.3	0.34	441.2	0.04	2.18	0.06	0.38	0.06	32.0	17.9	0	1.0
0	26	0.14	7.0	138.1	12.5	0.09	0	0.02	0.02	0.02	0.23	0.05	7.0	9.0	0	0.2
0	22	0.86	13.0	237.2	121.4	0.33	475.1	0.04	0.24	0.04	0.47	0.06	19.6	4.1	0	0.2
0	23	0.74	20.0	153.8	31.9	0.44	194.7	0.06	0.34	0.10	0.77	0.06	17.3	2.9	0	0.3
0	20	0.36	12.9	260.0	310.0	0.24	100.0	0.05	—	0.03	0.87	0.17	—	30.0	0	—
0	20	0.36	—	450.0	70.0	—	100.0	0.02	—	0.02	0.75	—	—	30.0	0	—
0	20	0.36	12.9	240.0	360.0	0.24	50.0	0.05	—	0.03	0.88	0.17	—	15.0	0	—
0	7	0.00	—	—	5.0	—	0	—	—	—	—	—	—	2.0	—	—
0	7	0.00	—	—	5.0	—	0	—	—	—	—	—	—	2.0	—	—
0	41	0.06	7.1	112.2	13.9	0.03	54.4	0.03	0.34	0.04	0.06	0.04	3.1	14.6	0	0.3
0	43	0.59	48.5	121.3	72.0	0.49	0	0.02	4.16	0.10	0.46	0.01	10.4	0.1	0	0.8
0	43	0.59	48.5	121.3	1.8	0.48	0	0.02	—	0.09	0.46	0.01	10.4	0.1	0	0.8
0	78	1.34	99.7	249.0	10.2	1.13	0	0.07	8.95	0.20	1.32	0.04	10.9	0	0	1.0
0	92	1.55	98.7	257.4	0.3	1.22	0	0.02	8.97	0.29	1.32	0.04	11.4	0	0	1.0
0	92	1.55	98.7	257.4	117.0	1.22	0	0.02	8.97	0.29	1.32	0.04	11.4	0	0	1.0
0	114	1.44	107.5	274.4	133.1	1.20	0	0.03	10.19	0.30	1.43	0.04	10.6	0	0	1.1
0	71	1.00	72.4	190.4	0.3	0.83	0	0.05	7.07	0.27	0.91	0.03	13.5	0	0	0.7
0	1	1.39	0	579.7	21.7	0.20	0	0.16	—	0.20	0.48	0.38	64.4	8.8	0	4.0
0	56	0.85	131.6	230.7	1.1	1.42	0	0.21	2.00	0.01	0.10	0.03	7.7	0.2	0	671.0
0	49	0.50	35.3	616.7	15.9	0.58	8.5	0.22	—	0.12	4.20	0.22	33.6	4.3	0	8.0
0	16	1.21	71.1	126.3	0.3	0.94	1.8	0.12	—	0.04	0.31	0.17	19.8	1.0	0	5.2
0	7	0.81	41.3	87.4	98.2	0.83	0	0.05	0.15	0.03	0.26	0.04	10.9	0	0	1.8
0	7	0.81	41.3	87.4	2.4	0.83	0	0.05	—	0.03	0.26	0.04	10.9	0	0	1.8
0	15	2.06	89.1	193.5	5.5	1.92	0	0.07	0.32	0.07	0.48	0.09	23.6	0	0	4.0
0	14	1.95	88.0	203.8	4.2	1.73	0	0.12	0.30	0.07	0.56	0.10	8.1	0.1	0	6.5
0	6	0.76	20.7	125.2	8.5	0.46	0	0.01	0.10	0.02	0.13	0.07	2.1	0.3	0	4.3
0	3	0.45	11.6	78.4	60.9	0.42	0	0.01	0.09	0.00	0.11	0.06	1.9	0.2	0	3.9
0	3	0.48	6.4	71.2	4.0	0.21	0	0.01	0.04	0.00	0.11	0.01	5.2	0.7	0	2.0
0	7	0.54	32.6	173.0	1.4	0.33	0	0.05	—	0.03	0.54	0.15	26.1	13.9	0	2.6
0	21	0.80	24.8	328.9	12.4	0.11	0.5	0.06	—	0.03	0.32	0.10	17.5	12.3	0	—
0	10	0.32	11.8	211.6	0.7	0.20	0.4	0.08	0.18	0.05	0.48	0.18	25.0	9.3	0	0.4
0	4	0.19	6.5	42.8	1.8	0.14	0.4	0.04	—	0.01	0.19	0.03	6.1	3.4	0	—
0	13	0.75	23.2	154.8	6.9	0.51	1.4	0.15	—	—	0.24	0.14	21.4	10.1	0	—
0	142	2.13	156.1	354.0	11.9	1.83	0	0.06	0.14	0.06	0.59	0.39	118.4	0.5	0	2.3
0	8	0.62	20.7	390.2	5.1	0.26	21.5	0.17	—	0.07	4.58	0.25	41.4	11.4	0	—
0	45	0.98	48.0	197.4	0	0.66	0.6	0.14	5.25	0.03	0.46	0.17	23.4	0.6	0	1.2
0	37	1.31	51.9	226.5	0	0.74	0.9	0.10	4.58	0.03	0.61	0.18	26.4	1.1	0	1.2
0	18	0.64	51.9	130.8	0.3	1.29	2.1	0.26	—	0.04	0.27	0.06	12.0	0.6	0	2.4
0	23	0.88	39.5	121.6	1.3	0.43	0	0.23	0.19	0.02	0.76	0.12	3.4	0.2	0	3.9
0	23	0.88	39.5	121.6	88.8	0.43	0	0.23	0.19	0.02	0.76	0.12	3.4	0.2	0	3.9
0	28	1.24	43.6	123.3	1.7	0.44	0	0.40	0.18	0.05	0.83	0.09	3.7	0.4	0	1.2
0	24	1.27	77.1	204.5	4.1	1.30	0.3	0.07	—	0.07	1.61	0.10	17.1	0.1	0	1.0
0	24	1.26	77.1	204.5	229.1	1.30	0	0.06	3.74	0.06	1.61	0.10	17.1	0.1	0	2.6
0	38	0.92	90.4	195.8	4.0	1.67	0.4	0.18	—	0.17	0.70	0.06	20.2	0.2	0	—
0	20	0.82	64.2	240.2	2.2	1.20	0	0.16	2.52	0.03	4.93	0.09	52.9	0	0	2.7
0	20	0.82	64.2	240.2	296.7	1.20	0	0.16	2.84	0.03	4.93	0.09	52.9	0	0	2.7
0	22	0.54	63.4	261.4	115.2	1.18	0	0.03	2.49	0.03	4.97	0.16	43.2	0.3	0	1.2
0	7	0.30	25.6	119.2	77.8	0.45	0	0.02	1.01	0.02	2.19	0.07	14.7	0	0	1.3
0	6	0.29	25.4	107.0	2.7	0.47	0	0.01	1.23	0.02	2.14	0.07	11.8	0	0	1.2
0	6	0.34	30.6	120.4	97.2	0.50	0	0.05	1.20	0.01	2.63	0.06	10.8	0	0	1.4
0	7	0.30	24.6	103.8	73.4	0.47	0	0.01	1.44	0.02	2.14	0.09	11.8	0	0	0.9
0	34	1.67	61.3	257.3	6.6	1.19	0	0.23	3.04	0.04	4.40	0.12	87.6	0	0	2.6
0	20	0.78	36.8	118.3	0.3	1.41	1.9	0.12	0.35	0.03	0.32	0.05	4.5	0.2	0	1.1

DA+ Code	Food Description	Quantity	Measure	Wt (g)	H₂O (g)	Ener (kcal)	Prot (g)	Carb (g)	Fiber (g)	Fat (g)	Fat Breakdown (g) Sat	Mono	Poly	Trans
32936	Dry roasted, with salt added	¼	cup(s)	27	0.3	192	2.6	3.7	2.5	20.0	1.7	11.9	5.6	—
1162	Oil roasted	¼	cup(s)	28	0.3	197	2.5	3.6	2.6	20.7	2.0	11.3	6.5	—
526	Raw	¼	cup(s)	27	1.0	188	2.5	3.8	2.6	19.6	1.7	11.1	5.9	—
12973	Pine nuts or pignolia, dried	1	tablespoon(s)	9	0.2	58	1.2	1.1	0.3	5.9	0.4	1.6	2.9	—
	Pistachios													
1164	Dry roasted	¼	cup(s)	31	0.6	176	6.6	8.5	3.2	14.1	1.7	7.4	4.3	—
32938	Dry roasted, with salt added	¼	cup(s)	32	0.6	182	6.8	8.6	3.3	14.7	1.8	7.7	4.4	—
1167	**Pumpkin or squash seeds, roasted**	¼	cup(s)	57	4.0	296	18.7	7.6	2.2	23.9	4.5	7.4	10.9	—
	Sesame													
32912	Sesame butter paste	1	tablespoon(s)	16	0.3	94	2.9	3.8	0.9	8.1	1.1	3.1	3.6	—
32941	Tahini or sesame butter	1	tablespoon(s)	15	0.5	89	2.6	3.2	0.7	8.0	1.1	3.0	3.5	—
1169	Whole, roasted, toasted	3	tablespoon(s)	10	0.3	54	1.6	2.4	1.3	4.6	0.6	1.7	2.0	—
	Soy nuts													
34173	Deep sea salted	¼	cup(s)	28	—	119	11.9	8.9	4.9	4.0	1.0	—	—	—
34174	Unsalted	¼	cup(s)	28	—	119	11.9	8.9	4.9	4.0	0	—	—	—
	Sunflower seeds													
528	Kernels, dried	1	tablespoon(s)	9	0.4	53	1.9	1.8	0.8	4.6	0.4	1.7	2.1	—
29721	Kernels, dry roasted, salted	1	tablespoon(s)	8	0.1	47	1.5	1.9	0.7	4.0	0.4	0.8	2.6	—
29723	Kernels, toasted, salted	1	tablespoon(s)	8	0.1	52	1.4	1.7	1.0	4.8	0.5	0.9	3.1	—
32928	Sunflower seed butter with salt added	1	tablespoon(s)	16	0.2	93	3.1	4.4	—	7.6	0.8	1.5	5.0	—
	Trail mix													
4646	Trail mix	¼	cup(s)	38	3.5	173	5.2	16.8	2.0	11.0	2.1	4.7	3.6	—
4647	Trail mix with chocolate chips	¼	cup(s)	38	2.5	182	5.3	16.8	—	12.0	2.3	5.1	4.2	—
4648	Tropical trail mix	¼	cup(s)	35	3.2	142	2.2	23.0	—	6.0	3.0	0.9	1.8	—
	Walnuts													
529	Dried black, chopped	¼	cup(s)	31	1.4	193	7.5	3.1	2.1	18.4	1.1	4.7	11.0	—
531	English or Persian	¼	cup(s)	29	1.2	191	4.5	4.0	2.0	19.1	1.8	2.6	13.8	—
	VEGETARIAN FOODS													
	Prepared													
34222	Brown rice & tofu stir-fry (vegan)	8	ounce(s)	227	244.4	302	16.5	18.0	3.2	21.0	1.7	4.7	13.4	0
34368	Cheese enchilada casserole (lacto)	8	ounce(s)	227	80.3	385	16.6	38.4	4.1	17.8	9.5	6.1	1.1	—
34247	Five bean casserole (vegan)	8	ounce(s)	227	175.8	178	5.9	26.6	6.0	5.8	1.1	2.5	1.9	0
34261	Lentil stew (vegan)	8	ounce(s)	227	227.9	188	11.5	35.9	11.0	0.7	0.1	0.1	0.3	0
34397	Macaroni and cheese (lacto)	8	ounce(s)	227	352.1	391	18.1	37.1	1.0	18.7	9.8	6.0	1.8	0
34238	Steamed rice and vegetables (vegan)	8	ounce(s)	227	222.9	587	11.2	87.9	5.8	23.1	4.1	8.7	9.1	0
34308	Tofu rice burgers (ovo-lacto)	1	piece(s)	218	77.6	435	22.4	68.6	5.6	8.4	1.7	2.4	3.5	—
34276	Vegan spinach enchiladas (vegan)	1	piece(s)	82	59.2	93	4.9	14.5	1.8	2.4	0.3	0.6	1.3	—
34243	Vegetable chow mein (vegan)	8	ounce(s)	227	163.3	166	6.5	22.1	2.0	6.4	0.7	2.7	2.5	0
34454	Vegetable lasagna (lacto)	8	ounce(s)	227	178.9	208	13.7	29.9	2.6	4.1	2.3	1.1	0.3	—
34339	Vegetable marinara (vegan)	8	ounce(s)	252	200.7	104	3.0	16.7	1.4	3.1	0.4	1.4	1.0	—
34356	Vegetable rice casserole (lacto)	8	ounce(s)	227	178.9	238	9.7	24.4	4.0	12.5	4.9	3.5	3.1	—
34311	Vegetable strudel (ovo-lacto)	8	ounce(s)	227	63.1	478	12.0	32.4	2.5	33.8	11.5	16.7	3.9	0
34371	Vegetable taco (lacto)	1	item(s)	85	46.5	117	4.2	13.6	2.9	5.6	2.1	1.9	1.3	—
34282	Vegetarian chili (vegan)	8	ounce(s)	227	191.4	115	5.6	21.4	7.1	1.5	0.2	0.3	0.7	0
34367	Vegetarian vegetable soup (vegan)	8	ounce(s)	227	257.9	111	3.2	16.0	3.2	5.0	1.0	2.1	1.6	0
	Boca burger													
32067	All American flamed grilled patty	1	item(s)	71	—	90	14.0	4.0	3.0	3.0	1.0	—	—	0
32074	Boca Chik n nuggets	4	item(s)	87	—	180	14.0	17.0	3.0	7.0	1.0	—	—	0
32075	Boca meatless ground burger	½	cup(s)	57	—	60	13.0	6.0	3.0	0.5	0	—	—	0
32072	Breakfast links	2	item(s)	45	—	70	8.0	5.0	2.0	3.0	0.5	—	—	0
32071	Breakfast patties	1	item(s)	38	—	60	7.0	5.0	2.0	2.5	0	—	—	0
35780	Cheeseburger meatless burger patty	1	item(s)	71	—	100	12.0	5.0	3.0	5.0	1.5	—	—	0
33958	Original meatless Chik 'n patties	1	item(s)	71	—	160	11.0	15.0	2.0	6.0	1.0	—	—	0
32066	Original patty	1	item(s)	71	—	70	13.0	6.0	4.0	0.5	0	—	—	0
32068	Roasted garlic patty	1	item(s)	71	—	70	12.0	6.0	4.0	1.5	0	—	—	0
37814	Roasted onion meatless burger patty	1	item(s)	71	—	70	11.0	7.0	4.0	1.0	0	—	—	0
	Gardenburger													
37810	BBQ chik'n with sauce	1	item(s)	142	—	250	14.0	30.0	5.0	8.0	1.0	—	—	0
39661	Black bean burger	1	item(s)	71	—	80	8.0	11.0	4.0	2.0	0	—	—	0
39666	Buffalo chick'n wing	3	item(s)	95	—	180	9.0	8.0	5.0	12.0	1.5	—	—	0
39665	Country fried chicken with creamy pepper gravy	1	item(s)	142	—	190	9.0	16.0	2.0	9.0	1.0	—	—	0
37808	Flamed grilled Chik'n	1	item(s)	71	—	100	13.0	5.0	3.0	2.5	0	—	—	0
37803	Garden vegan	1	item(s)	71	—	100	10.0	12.0	2.0	1.0	—	—	—	0
39663	Homestyle classic burger	1	item(s)	71	—	110	12.0	6.0	4.0	5.0	0.5	—	—	0
37807	Meatless breakfast sausage	1	item(s)	43	—	50	5.0	2.0	2.0	3.5	0.5	—	—	0
37809	Meatless meatballs	6	item(s)	85	—	110	12.0	8.0	4.0	4.5	1.0	—	—	0
37806	Meatless riblets with sauce	1	item(s)	142	—	160	17.0	11.0	4.0	5.0	0	—	—	0
29913	Original	1	item(s)	71	—	90	10.0	8.0	3.0	2.0	0.5	—	—	0
39662	Sun-dried tomato basil burger	1	item(s)	71	—	80	10.0	11.0	3.0	1.5	0.5	—	—	0

VEGETARIAN FOODS —Continued

Chol (mg)	Calc (mg)	Iron (mg)	Magn (mg)	Pota (mg)	Sodi (mg)	Zinc (mg)	Vit A (µg)	Thia (mg)	Vit E (mg α)	Ribo (mg)	Niac (mg)	Vit B$_6$ (mg)	Fola (µg)	Vit C (mg)	Vit B$_{12}$ (µg)	Sele (µg)
0	19	0.75	35.6	114.5	103.4	1.36	1.9	0.11	0.34	0.03	0.31	0.05	4.3	0.2	0	1.1
0	18	0.68	33.3	107.8	0.3	1.23	1.4	0.13	0.70	0.03	0.33	0.05	4.1	0.2	0	1.7
0	19	0.69	33.0	111.7	0	1.23	0.8	0.18	0.38	0.04	0.32	0.06	6.0	0.3	0	1.0
0	1	0.47	21.6	51.3	0.2	0.55	0.1	0.03	0.80	0.02	0.37	0.01	2.9	0.1	0	0.1
0	34	1.29	36.9	320.4	3.1	0.71	4.0	0.26	0.59	0.05	0.44	0.39	15.4	0.7	0	2.9
0	35	1.34	38.4	333.4	129.6	0.73	4.2	0.26	0.61	0.05	0.45	0.40	16.0	0.7	0	3.0
0	24	8.48	303.0	457.4	10.2	4.22	10.8	0.12	0.00	0.18	0.99	0.05	32.3	1.0	0	3.2
0	154	3.07	57.9	93.1	1.9	1.17	0.5	0.04	—	0.03	1.07	0.13	16.0	0	0	0.9
0	21	0.66	14.3	68.9	5.3	0.69	0.5	0.24	—	0.02	0.85	0.02	14.7	0.6	0	0.3
0	94	1.40	33.8	45.1	1.0	0.68	0	0.07	—	0.02	0.43	0.07	9.3	0	0	0.5
0	59	1.07	—	—	148.1	—	0	—	—	—	—	—	—	0	—	—
0	59	1.07	—	—	9.9	—	0	—	—	—	—	—	—	0	—	—
0	7	0.47	29.3	58.1	0.8	0.45	0.3	0.13	2.99	0.03	0.75	0.12	20.4	0.1	0	4.8
0	6	0.30	10.3	68.0	32.8	0.42	0	0.01	2.09	0.02	0.56	0.06	19.0	0.1	0	6.3
0	5	0.57	10.8	41.1	51.3	0.44	0	0.03	—	0.02	0.35	0.07	19.9	0.1	0	5.2
0	20	0.76	59.0	11.5	83.2	0.85	0.5	0.05	—	0.05	0.85	0.13	37.9	0.4	0	—
0	29	1.14	59.3	256.9	85.9	1.20	0.4	0.17	—	0.07	1.76	0.11	26.6	0.5	0	—
2	41	1.27	60.4	243.0	45.4	1.17	0.8	0.15	—	0.08	1.65	0.09	24.4	0.5	0	—
0	20	0.92	33.6	248.2	3.5	0.41	0.7	0.15	—	0.04	0.51	0.11	14.7	2.7	0	—
0	19	0.97	62.8	163.4	0.6	1.05	0.6	0.01	0.56	0.04	0.14	0.18	9.7	0.5	0	5.3
0	29	0.85	46.2	129.0	0.6	0.90	0.3	0.10	0.20	0.04	0.32	0.15	28.7	0.4	0	1.4
0	353	6.34	118.3	501.4	142.2	2.03	—	0.23	0.07	0.14	1.49	0.36	51.8	24.8	0	14.8
39	441	2.44	34.6	191.2	1139.7	1.84	—	0.31	0.05	0.35	2.23	0.11	87.1	20.4	0.4	20.0
0	48	1.78	40.8	364.1	613.6	0.61	—	0.10	0.52	0.07	0.93	0.11	39.4	8.3	0	3.3
0	34	3.23	50.0	548.8	436.5	1.42	—	0.24	0.14	0.16	2.31	0.29	91.4	26.4	0	12.1
43	415	1.71	45.4	267.8	1641.0	2.32	—	0.32	0.27	0.48	2.18	0.13	74.2	0.9	0.8	33.3
0	91	3.31	153.1	810.1	3117.8	2.04	—	0.37	3.03	0.21	6.16	0.64	61.7	35.2	0	18.8
52	467	9.01	89.7	455.6	2449.5	2.06	—	0.27	0.12	0.26	3.43	0.29	106.6	2.0	0.1	43.0
0	117	1.13	40.4	170.5	134.2	0.68	—	0.07	—	0.07	0.53	0.10	52.1	1.8	0	5.1
0	189	3.70	28.0	310.3	372.7	0.76	—	0.13	0.05	0.11	1.43	0.14	45.6	8.0	0	6.5
10	176	1.86	41.9	470.0	759.4	1.14	—	0.26	0.05	0.25	2.49	0.22	64.7	19.0	0.4	21.8
0	17	0.94	19.1	189.9	439.6	0.42	—	0.15	0.55	0.08	1.36	0.12	40.6	23.5	0	10.8
17	190	1.28	29.3	414.2	626.0	1.24	—	0.16	0.35	0.29	2.00	0.19	72.6	56.0	0.2	5.8
29	200	2.15	24.5	181.0	512.1	1.24	—	0.28	0.20	0.31	2.88	0.11	88.3	17.4	0.2	19.7
7	77	0.88	26.3	174.1	280.7	0.59	—	0.08	0.04	0.06	0.49	0.08	38.7	4.6	0	3.0
0	65	1.98	41.0	543.1	390.7	0.74	—	0.14	0.15	0.10	1.31	0.18	59.3	20.3	0	4.4
0	46	1.87	34.9	550.3	729.5	0.56	—	0.13	0.55	0.09	1.99	0.27	47.8	29.9	0	1.4
5	150	1.80	—	—	280.0	—	0	—	—	—	—	—	—	0	—	—
0	40	1.44	—	—	500.0	—	—	—	—	—	—	—	—	0	—	—
0	60	1.80	—	—	270.0	—	0	—	—	—	—	—	—	0	—	—
0	20	1.44	—	—	330.0	—	0	—	—	—	—	—	—	0	—	—
0	20	1.08	—	—	280.0	—	0	—	—	—	—	—	—	0	—	—
5	80	1.80	—	—	360.0	—	—	—	—	—	—	—	—	0	—	—
0	40	1.80	—	—	430.0	—	—	—	—	—	—	—	—	0	—	—
0	60	1.80	—	—	280.0	—	0	—	—	—	—	—	—	0	—	—
0	60	1.80	—	—	370.0	—	0	—	—	—	—	—	—	0	—	—
0	100	2.70	—	—	300.0	—	—	—	—	—	—	—	—	0	—	—
0	150	1.08	—	—	890.0	—	—	—	—	—	—	—	—	0	—	—
0	40	1.44	—	—	330.0	—	—	—	—	—	—	—	—	0	—	—
0	40	0.72	—	—	1000.0	—	—	—	—	—	—	—	—	0	—	—
5	40	1.44	—	—	550.0	—	—	—	—	—	—	—	—	0	—	—
0	60	3.60	—	—	360.0	—	—	—	—	—	—	—	—	0	—	—
0	40	4.50	—	—	230.0	—	—	—	—	—	—	—	—	0	—	—
0	80	1.44	—	—	380.0	—	—	—	—	—	—	—	—	0	—	—
0	20	0.72	—	—	120.0	—	—	—	—	—	—	—	—	0	—	—
0	60	1.80	—	—	400.0	—	—	—	—	—	—	—	—	0	—	—
0	60	1.80	—	—	720.0	—	—	—	—	—	—	—	—	3.6	—	—
0	80	1.08	30.4	193.4	490.0	0.89	—	0.10	—	0.15	1.08	0.08	10.1	1.2	0.1	7.0
5	60	1.44	—	—	260.0	—	—	—	—	—	—	—	—	3.6	—	—

DA+ Code	Food Description	Quantity	Measure	Wt (g)	H₂O (g)	Ener (kcal)	Prot (g)	Carb (g)	Fiber (g)	Fat (g)	Sat	Mono	Poly	Trans
												Fat Breakdown (g)		
29915	Veggie medley	1	item(s)	71	—	90	9.0	11.0	4.0	2.0	0	—	—	0
	Loma Linda													
9311	Big franks, canned	1	item(s)	51	—	110	11.0	3.0	2.0	6.0	1.0	1.5	3.5	0
9323	Fried Chik'n with gravy	2	piece(s)	80	45.9	150	12.0	5.0	2.0	10	1.5	2.5	5.0	0
9326	Linketts, canned	1	item(s)	35	21.0	70	7.0	1.0	1.0	4.0	0.5	1.0	2.5	0
9336	Redi-Burger patties, canned	1	slice(s)	85	50.5	120	18.0	7.0	4.0	2.5	0.5	0.5	1.5	0
9350	Swiss Stake pattie with gravy, frozen	1	piece(s)	92	65.7	130	9.0	9.0	3.0	6.0	1.0	1.5	3.5	0
9354	Tender Rounds meatball substitute, canned in gravy	6	piece(s)	80	53.9	120	13.0	6.0	1.0	4.5	0.5	1.5	2.5	0
	Morningstar Farms													
33707	America's Original Veggie Dog links	1	item(s)	57	—	80	11.0	6.0	1.0	0.5	0	—	—	0
9362	Better 'n Eggs egg substitute	¼	cup(s)	57	50.3	20	5.0	0	0	0	0	0	0	0
9371	Breakfast bacon strips	2	item(s)	16	6.8	60	2.0	2.0	0.5	4.5	0.5	1.0	3.0	0
9368	Breakfast sausage links	2	item(s)	45	26.8	80	9.0	3.0	2.0	3.0	0.5	1.5	1.0	0
33705	Chik 'n nuggets	4	piece(s)	86	—	190	12.0	18.0	2.0	7.0	1.0	2.0	4.0	0
11587	Chik patties	1	item(s)	71	36.3	150	9.0	16.0	2.0	6.0	1.0	1.5	2.5	0
2531	Garden veggie patties	1	item(s)	67	40.1	100	10.0	9.0	4.0	2.5	0.5	0.5	1.5	0
33702	Spicy black bean veggie burger	1	item(s)	78	—	140	12.0	15.0	3.0	4.0	0.5	1.0	2.5	0
9412	Vegetarian chili, canned	1	cup(s)	230	172.6	180	16.0	25.0	10.0	1.5	0.5	0.5	0.5	0
	Worthington													
9424	Chili, canned	1	cup(s)	230	167.0	280	24.0	25.0	8.0	10.0	1.5	1.5	7.0	0
9436	Diced Chik, canned	¼	cup(s)	55	42.7	50	9.0	2.0	1.0	0	0	0	0	0
9440	Dinner roast, frozen	1	slice(s)	85	53.2	180	14.0	6.0	3.0	11.0	1.5	4.5	5.0	0
9420	Meatless chicken slices, frozen	3	slice(s)	57	38.9	90	9.0	2.0	0.5	4.5	1.0	1.0	2.5	0
36702	Meatless chicken style roll, frozen	1	slice(s)	55	—	90	9.0	2.0	1.0	4.5	1.0	1.0	2.5	0
9428	Meatless corned beef, sliced, frozen	3	slice(s)	57	31.2	140	10.0	5.0	0	9.0	1.0	2.0	5.0	0
9470	Meatless salami, sliced, frozen	3	slice(s)	57	32.4	120	12.0	3.0	2.0	7.0	1.0	1.0	5.0	0
9480	Meatless smoked turkey, sliced	3	slice(s)	57	—	140	10.0	4.0	0	9.0	1.5	2.0	5.0	0
9462	Prosage links	2	item(s)	45	26.8	90	9.0	3.0	2.0	3.0	0.5	0.5	2.0	0
9484	Stakelets patty beef steak substitute, frozen	1	piece(s)	71	41.5	150	14.0	7.0	2.0	7.0	1.0	2.5	3.5	0
9486	Stripples bacon substitute	2	item(s)	16	6.8	60	2.0	2.0	0.5	4.5	0.5	1.0	3.0	0
9496	Vegetable Skallops meat substitute, canned	½	cup(s)	85	—	90	17.0	4.0	3.0	1.0	0	0	0.5	0
	DAIRY													
	Cheese													
1433	Blue, crumbled	1	ounce(s)	28	12.0	100	6.1	0.7	0	8.1	5.3	2.2	0.2	—
884	Brick	1	ounce(s)	28	11.7	105	6.6	0.8	0	8.4	5.3	2.4	0.2	—
885	Brie	1	ounce(s)	28	13.7	95	5.9	0.1	0	7.8	4.9	2.3	0.2	—
34821	Camembert	1	ounce(s)	28	14.7	85	5.6	0.1	0	6.9	4.3	2.0	0.2	—
5	Cheddar, shredded	¼	cup(s)	28	10.4	114	7.0	0.4	0	9.4	6.0	2.7	0.3	—
888	Cheddar or colby	1	ounce(s)	28	10.8	112	6.7	0.7	0	9.1	5.7	2.6	0.3	—
32096	Cheddar or colby, low fat	1	ounce(s)	28	17.9	49	6.9	0.5	0	2.0	1.2	0.6	0.1	—
889	Edam	1	ounce(s)	28	11.8	101	7.1	0.4	0	7.9	5.0	2.3	0.2	—
890	Feta	1	ounce(s)	28	15.7	75	4.0	1.2	0	6.0	4.2	1.3	0.2	—
891	Fontina	1	ounce(s)	28	10.8	110	7.3	0.4	0	8.8	5.4	2.5	0.5	—
8527	Goat cheese, soft	1	ounce(s)	28	17.2	76	5.3	0.3	0	6.0	4.1	1.4	0.1	—
893	Gouda	1	ounce(s)	28	11.8	101	7.1	0.6	0	7.8	5.0	2.2	0.2	—
894	Gruyere	1	ounce(s)	28	9.4	117	8.5	0.1	0	9.2	5.4	2.8	0.5	—
895	Limburger	1	ounce(s)	28	13.7	93	5.7	0.1	0	7.7	4.7	2.4	0.1	—
896	Monterey jack	1	ounce(s)	28	11.6	106	6.9	0.2	0	8.6	5.4	2.5	0.3	—
13	Mozzarella, part skim milk	1	ounce(s)	28	15.2	72	6.9	0.8	0	4.5	2.9	1.3	0.1	—
12	Mozzarella, whole milk	1	ounce(s)	28	14.2	85	6.3	0.6	0	6.3	3.7	1.9	0.2	—
897	Muenster	1	ounce(s)	28	11.8	104	6.6	0.3	0	8.5	5.4	2.5	0.2	—
898	Neufchatel	1	ounce(s)	28	17.6	74	2.8	0.8	0	6.6	4.2	1.9	0.2	—
14	Parmesan, grated	1	tablespoon(s)	5	1.0	22	1.9	0.2	0	1.4	0.9	0.4	0.1	—
17	Provolone	1	ounce(s)	28	11.6	100	7.3	0.6	0	7.5	4.8	2.1	0.2	—
19	Ricotta, part skim milk	¼	cup(s)	62	45.8	85	7.0	3.2	0	4.9	3.0	1.4	0.2	—
18	Ricotta, whole milk	¼	cup(s)	62	44.1	107	6.9	1.9	0	8.0	5.1	2.2	0.2	—
20	Romano	1	tablespoon(s)	5	1.5	19	1.6	0.2	0	1.3	0.9	0.4	0	—
900	Roquefort	1	ounce(s)	28	11.2	105	6.1	0.6	0	8.7	5.5	2.4	0.4	—
21	Swiss	1	ounce(s)	28	10.5	108	7.6	1.5	0	7.9	5.0	2.1	0.3	—
	Imitation cheese													
42245	Imitation American cheddar cheese	1	ounce(s)	28	15.1	68	4.7	3.3	0	4.0	2.5	1.2	0.1	—
53914	Imitation cheddar	1	ounce(s)	28	15.1	68	4.7	3.3	0	4.0	2.5	1.2	0.1	—
	Cottage cheese													
9	Low fat, 1% fat	½	cup(s)	113	93.2	81	14.0	3.1	0	1.2	0.7	0.3	0	—
8	Low fat, 2% fat	½	cup(s)	113	89.6	102	15.5	4.1	0	2.2	1.4	0.6	0.1	—
	Cream cheese													
11	Cream cheese	2	tablespoon(s)	29	15.6	101	2.2	0.8	0	10.1	6.4	2.9	0.4	—
17366	Fat free cream cheese	2	tablespoon(s)	30	22.7	29	4.3	1.7	0	0.4	0.3	0.1	0	—
10438	Tofutti Better than Cream Cheese	2	tablespoon(s)	30	—	80	1.0	1.0	0	8.0	2.0	—	6.0	—

DAIRY—Continued

Chol (mg)	Calc (mg)	Iron (mg)	Magn (mg)	Pota (mg)	Sodi (mg)	Zinc (mg)	Vit A (µg)	Thia (mg)	Vit E (mg α)	Ribo (mg)	Niac (mg)	Vit B$_6$ (mg)	Fola (µg)	Vit C (mg)	Vit B$_{12}$ (µg)	Sele (µg)
0	40	1.44	27.0	182.0	290.0	0.46	—	0.07	—	0.08	0.90	0.09	10.6	9.0	0	4.0
0	0	0.77	—	50.0	220.0	—	0	0.22	—	0.10	2.00	0.70	—	0	2.4	—
0	20	1.80	—	70.0	430.0	0.33	0	1.05	—	0.34	4.00	0.30	—	0	2.4	—
0	0	0.36	—	20.0	160.0	0.46	0	0.12	—	0.20	0.80	0.16	—	0	0.9	—
0	0	1.06	—	140.0	450.0	—	0	0.15	—	0.25	4.00	1.00	—	0	1.2	—
0	0	0.72	—	200.0	430.0	—	0	0.45	—	0.25	10.00	1.00	—	0	5.4	—
0	20	1.08	—	80.0	340.0	0.66	0	0.75	—	0.17	2.00	0.16	—	0	1.2	—
0	0	0.72	—	60.0	580.0	—	0	—	—	—	—	—	—	0	—	—
0	20	0.72	—	75.0	90.0	0.60	37.5	0.03	—	0.34	0.00	0.08	24.0	—	0.6	—
0	0	0.36	—	15.0	220.0	0.05	0	0.75	—	0.04	0.40	0.07	—	0	0.2	—
0	0	1.80	—	50.0	300.0	—	0	0.37	—	0.17	7.00	0.50	—	0	3.0	—
0	20	2.70	—	320.0	490.0	—	0	0.52	—	0.25	5.00	0.30	—	0	1.5	—
0	0	1.80	—	210.0	540.0	—	0	1.80	—	0.17	2.00	0.20	—	0	1.2	—
0	40	0.72	—	180.0	350.0	—	—	—	—	—	—	—	—	0	—	—
0	40	1.80	—	320.0	470.0	—	0	—	—	—	0.00	—	—	0	—	—
0	40	3.60	—	660.0	900.0	—	—	—	—	—	—	—	—	0	—	—
0	40	3.60	—	330.0	1130.0	—	0	0.30	—	0.13	2.00	0.70	—	0	1.5	—
0	0	1.08	—	100.0	220.0	0.24	0	0.06	—	0.10	4.00	0.08	—	0	0.2	—
0	20	1.80	—	120.0	580.0	0.64	0	1.80	—	0.25	6.00	0.60	—	0	1.5	—
0	250	1.80	—	250.0	250.0	0.26	0	0.37	—	0.13	4.00	0.30	—	0	1.8	—
0	100	1.08	—	240.0	240.0	—	0	0.37	—	0.13	4.00	0.30	—	0	1.8	—
0	0	1.80	—	130.0	460.0	0.26	0	0.45	—	0.17	5.00	0.30	—	0	1.8	—
0	0	1.08	—	95.0	800.0	0.30	0	0.75	—	0.17	4.00	0.20	—	0	0.6	—
0	60	2.70	—	60.0	450.0	0.23	0	1.80	—	0.17	6.00	0.40	—	0	3.0	—
0	0	1.44	—	50.0	320.0	0.36	0	1.80	—	0.17	2.00	0.30	—	0	3.0	—
0	40	1.08	—	130.0	480.0	0.50	0	1.20	—	0.13	3.00	0.30	—	0	1.5	—
0	0	0.36	—	15.0	220.0	0.05	0	0.75	—	0.03	0.40	0.08	—	0	0.2	—
0	0	0.36	—	10.0	390.0	0.67	0	0.03	—	0.03	0.00	0.01	—	0	0	—
21	150	0.08	6.5	72.6	395.5	0.75	56.1	0.01	0.07	0.10	0.28	0.04	10.2	0	0.3	4.1
27	191	0.12	6.8	38.6	158.8	0.73	82.8	0.00	0.07	0.10	0.03	0.01	5.7	0	0.4	4.1
28	52	0.14	5.7	43.1	178.3	0.67	49.3	0.02	0.06	0.14	0.10	0.06	18.4	0	0.5	4.1
20	110	0.09	5.7	53.0	238.7	0.67	68.3	0.01	0.06	0.14	0.18	0.06	17.6	0	0.4	4.1
30	204	0.19	7.9	27.7	175.4	0.87	74.9	0.01	0.08	0.10	0.02	0.02	5.1	0	0.2	3.9
27	194	0.21	7.4	36.0	171.2	0.87	74.8	0.00	0.07	0.10	0.02	0.02	5.1	0	0.2	4.1
6	118	0.11	4.5	18.7	173.5	0.51	17.0	0.00	0.01	0.06	0.01	0.01	3.1	0	0.1	4.1
25	207	0.12	8.5	53.3	273.6	1.06	68.9	0.01	0.06	0.11	0.02	0.02	4.5	0	0.4	4.1
25	140	0.18	5.4	17.6	316.4	0.81	35.4	0.04	0.05	0.23	0.28	0.12	9.1	0	0.5	4.3
33	156	0.06	4.0	18.1	226.8	0.99	74.0	0.01	0.07	0.05	0.04	0.02	1.7	0	0.5	4.1
13	40	0.53	4.5	7.4	104.3	0.26	81.6	0.02	0.05	0.10	0.12	0.07	3.4	0	0.1	0.8
32	198	0.06	8.2	34.3	232.2	1.10	46.8	0.01	0.06	0.09	0.01	0.02	6.0	0	0.4	4.1
31	287	0.04	10.2	23.0	95.3	1.10	76.8	0.01	0.07	0.07	0.03	0.02	2.8	0	0.5	4.1
26	141	0.03	6.0	36.3	226.8	0.59	96.4	0.02	0.06	0.14	0.04	0.02	16.4	0	0.3	4.1
25	211	0.20	7.7	23.0	152.0	0.85	56.1	0.00	0.07	0.11	0.02	0.02	5.1	0	0.2	4.1
18	222	0.06	6.5	23.8	175.5	0.78	36.0	0.01	0.04	0.08	0.03	0.02	2.6	0	0.2	4.1
22	143	0.12	5.7	21.5	177.8	0.82	50.7	0.01	0.05	0.08	0.02	0.01	2.0	0	0.6	4.8
27	203	0.11	7.7	38.0	178.0	0.79	84.5	0.00	0.07	0.09	0.02	0.01	3.4	0	0.4	4.1
22	21	0.07	2.3	32.3	113.1	0.14	84.5	0.00	—	0.05	0.03	0.01	3.1	0	0.1	0.9
4	55	0.04	1.9	6.3	76.5	0.19	6.0	0.00	0.01	0.02	0.01	0.00	0.5	0	0.1	0.9
20	214	0.14	7.9	39.1	248.3	0.91	66.9	0.01	0.06	0.09	0.04	0.02	2.8	0	0.4	4.1
19	167	0.27	9.2	76.9	76.9	0.82	65.8	0.01	0.04	0.11	0.04	0.01	8.0	0	0.2	10.3
31	127	0.23	6.8	64.6	51.7	0.71	73.8	0.01	0.06	0.12	0.06	0.02	7.4	0	0.2	8.9
5	53	0.03	2.1	4.3	60	0.12	4.8	0.00	0.01	0.01	0.00	0.00	0.4	0	0.1	0.7
26	188	0.15	8.5	25.8	512.9	0.59	83.3	0.01	—	0.16	0.20	0.03	13.9	0	0.2	4.1
26	224	0.05	10.8	21.8	54.4	1.23	62.4	0.01	0.10	0.08	0.02	0.02	1.7	0	0.9	5.2
10	159	0.08	8.2	68.6	381.3	0.73	32.3	0.01	0.07	0.12	0.03	0.03	2.0	0	0.1	4.3
10	159	0.09	8.2	68.6	381.3	0.73	32.3	0.01	0.07	0.12	0.04	0.03	2.0	0	0.1	4.3
5	69	0.15	5.7	97.2	458.8	0.42	12.4	0.02	0.01	0.18	0.14	0.07	13.6	0	0.7	10.2
9	78	0.18	6.8	108.5	458.8	0.47	23.7	0.02	0.02	0.20	0.16	0.08	14.7	0	0.8	11.5
32	23	0.34	1.7	34.5	85.8	0.15	106.1	0.01	0.08	0.05	0.02	0.01	3.8	0	0.1	0.7
2	56	0.05	4.2	48.9	163.5	0.26	83.7	0.01	0.00	0.05	0.04	0.01	11.1	0	0.2	1.5
0	0	0.00	—	—	135.0	—	0	—	—	—	—	—	—	0	—	—

DA+ Code	Food Description	Quantity	Measure	Wt (g)	H₂O (g)	Ener (kcal)	Prot (g)	Carb (g)	Fiber (g)	Fat (g)	Fat Breakdown (g) Sat	Mono	Poly	Trans
	Processed cheese													
24	American cheese food, processed	1	ounce(s)	28	12.3	94	5.2	2.2	0	7.1	4.2	2.0	0.3	—
25	American cheese spread, processed	1	ounce(s)	28	13.5	82	4.7	2.5	0	6.0	3.8	1.8	0.2	—
22	American cheese, processed	1	ounce(s)	28	11.1	106	6.3	0.5	0	8.9	5.6	2.5	0.3	—
9110	Kraft deluxe singles pasteurized process American cheese	1	ounce(s)	28	—	108	5.4	0	0	9.5	5.4	—	—	—
23	Swiss cheese, processed	1	ounce(s)	28	12.0	95	7.0	0.6	0	7.1	4.5	2.0	0.2	—
	Soy cheese													
10437	Galaxy Foods vegan grated parmesan cheese alternative	1	tablespoon(s)	8	—	23	3.0	1.5	0	0	0	0	0	0
10430	Nu Tofu cheddar flavored cheese alternative	1	ounce(s)	28	—	70	6.0	1.0	0	4.0	0.5	2.5	1.0	
	Cream													
26	Half and half cream	1	tablespoon(s)	15	12.1	20	0.4	0.6	0	1.7	1.1	0.5	0.1	—
32	Heavy whipping cream, liquid	1	tablespoon(s)	15	8.7	52	0.3	0.4	0	5.6	3.5	1.6	0.2	—
28	Light coffee or table cream, liquid	1	tablespoon(s)	15	11.1	29	0.4	0.5	0	2.9	1.8	0.8	0.1	—
30	Light whipping cream, liquid	1	tablespoon(s)	15	9.5	44	0.3	0.4	0	4.6	2.9	1.4	0.1	—
34	Whipped cream topping, pressurized	1	tablespoon(s)	3	1.8	8	0.1	0.4	0	0.7	0.4	0.2	0	—
	Sour cream													
30556	Fat free sour cream	2	tablespoon(s)	32	25.8	24	1.0	5.0	0	0	0	0	0	0
36	Sour cream	2	tablespoon(s)	24	17.0	51	0.8	1.0	0	5.0	3.1	1.5	0.2	—
	Imitation cream													
3659	Coffeemate non dairy creamer, liquid	1	tablespoon(s)	15	—	20	0	2.0	0	1.0	0	0.5	0	—
40	Cream substitute, powder	1	teaspoon(s)	2	0	11	0.1	1.1	0	0.7	0.7	0	0	—
904	Imitation sour cream	2	tablespoon(s)	29	20.5	60	0.7	1.9	0	5.6	5.1	0.2	0	—
35972	Non dairy coffee whitener, liquid, frozen	1	tablespoon(s)	15	11.7	21	0.2	1.7	0	1.5	0.3	1.1	0	—
35976	Non dairy dessert topping, frozen	1	tablespoon(s)	5	2.4	15	0.1	1.1	0	1.2	1.0	0.1	0	—
35975	Non dairy dessert topping, pressurized	1	tablespoon(s)	4	2.7	12	0	0.7	0	1.0	0.8	0.1	0	—
	Fluid milk													
60	Buttermilk, low fat	1	cup(s)	245	220.8	98	8.1	11.7	0	2.2	1.3	0.6	0.1	—
54	Lowfat, 1%	1	cup(s)	244	219.4	102	8.2	12.2	0	2.4	1.5	0.7	0.1	—
55	Lowfat, 1%, with nonfat milk solids	1	cup(s)	245	220.0	105	8.5	12.2	0	2.4	1.5	0.7	0.1	—
57	Nonfat, skim or fat free	1	cup(s)	245	222.6	83	8.3	12.2	0	0.2	0.1	0.1	0	—
58	Nonfat, skim or fat free with nonfat milk solids	1	cup(s)	245	221.4	91	8.7	12.3	0	0.6	0.4	0.2	0	—
51	Reduced fat, 2%	1	cup(s)	244	218.0	122	8.1	11.4	0	4.8	3.1	1.4	0.2	—
52	Reduced fat, 2%, with nonfat milk solids	1	cup(s)	245	217.7	125	8.5	12.2	0	4.7	2.9	1.4	0.2	—
50	Whole, 3.3%	1	cup(s)	244	215.5	146	7.9	11.0	0	7.9	4.6	2.0	0.5	—
	Canned milk													
62	Nonfat or skim evaporated	2	tablespoon(s)	32	25.3	25	2.4	3.6	0	0.1	0	0	0	—
63	Sweetened condensed	2	tablespoon(s)	38	10.4	123	3.0	20.8	0	3.3	2.1	0.9	0.1	—
61	Whole evaporated	2	tablespoon(s)	32	23.3	42	2.1	3.2	0	2.4	1.4	0.7	0.1	—
	Dried milk													
64	Buttermilk	¼	cup(s)	30	0.9	117	10.4	14.9	0	1.8	1.1	0.5	0.1	—
65	Instant nonfat with added vitamin A	¼	cup(s)	17	0.7	61	6.0	8.9	0	0.1	0.1	0	0	—
5234	Skim milk powder	¼	cup(s)	17	0.7	62	6.1	9.1	0	0.1	0.1	0	0	—
907	Whole dry milk	¼	cup(s)	32	0.8	159	8.4	12.3	0	8.5	5.4	2.5	0.2	—
909	**Goat milk**	1	cup(s)	244	212.4	168	8.7	10.9	0	10.1	6.5	2.7	0.4	—
	Chocolate milk													
33155	Chocolate syrup, prepared with milk	1	cup(s)	282	227.0	254	8.7	36.0	0.8	8.3	4.7	2.1	0.5	—
33184	Cocoa mix with aspartame, added sodium and vitamin A, no added calcium or phosphorus, prepared with water	1	cup(s)	192	177.4	56	2.3	10.8	1.2	0.4	0.3	0.1	0	—
908	Hot cocoa, prepared with milk	1	cup(s)	250	206.4	193	8.8	26.6	2.5	5.8	3.6	1.7	0.1	0.2
69	Low fat	1	cup(s)	250	211.3	158	8.1	26.1	1.3	2.5	1.5	0.8	0.1	—
68	Reduced fat	1	cup(s)	250	205.4	190	7.5	30.3	1.8	4.8	2.9	1.1	0.2	—
67	Whole	1	cup(s)	250	205.8	208	7.9	25.9	2.0	8.5	5.3	2.5	0.3	—
70	**Eggnog**	1	cup(s)	254	188.9	343	9.7	34.4	0	19.0	11.3	5.7	0.9	—
	Breakfast drinks													
10093	Carnation Instant Breakfast classic chocolate malt, prepared with skim milk, no sugar added	1	cup(s)	243	—	142	11.1	21.3	0.7	1.3	0.7	—	—	—
10092	Carnation Instant Breakfast classic French vanilla, prepared with skim milk, no sugar added	1	cup(s)	273	—	150	12.9	24.0	0	0.4	0.4	—	—	—
10094	Carnation Instant Breakfast stawberry sensation, prepared with skim milk, no sugar added	1	cup(s)	243	—	142	11.1	21.3	0	0.4	0.4	—	—	—
10091	Carnation Instant Breakfast strawberry sensation, prepared with skim milk	1	cup(s)	273	—	220	12.5	38.8	0	0.4	0.4	—	—	—
1417	Ovaltine rich chocolate flavor, prepared with skim milk	1	cup(s)	258	—	170	8.5	31.0	0	0	0	0	0	0
8539	**Malted milk, chocolate mix, fortified, prepared with milk**	1	cup(s)	265	215.8	223	8.9	28.9	1.1	8.6	5.0	2.2	0.5	—

DAIRY —Continued

Chol (mg)	Calc (mg)	Iron (mg)	Magn (mg)	Pota (mg)	Sodi (mg)	Zinc (mg)	Vit A (µg)	Thia (mg)	Vit E (mg α)	Ribo (mg)	Niac (mg)	Vit B$_6$ (mg)	Fola (µg)	Vit C (mg)	Vit B$_{12}$ (µg)	Sele (µg)
23	162	0.16	8.8	82.5	358.6	0.90	57.0	0.01	0.06	0.14	0.04	0.02	2.0	0	0.4	4.6
16	159	0.09	8.2	68.6	381.3	0.73	49.0	0.01	0.05	0.12	0.03	0.03	2.0	0	0.1	3.2
27	156	0.05	7.7	47.9	422.1	0.80	72.0	0.01	0.07	0.10	0.02	0.02	2.3	0	0.2	4.1
27	338	0.00	0	33.8	459.0	1.22	114.0	—	—	0.14	—	—	—	0	0.2	—
24	219	0.17	8.2	61.2	388.4	1.02	56.1	0.00	0.09	0.07	0.01	0.01	1.7	0	0.3	4.5
0	60	0.00	—	75.0	97.5	—	—	—	—	—	—	—	—	—	—	—
0	200	0.36	—	—	190.0	—	—	—	—	—	—	—	—	0	—	—
6	16	0.01	1.5	19.5	6.2	0.08	14.6	0.01	0.05	0.02	0.01	0.01	0.5	0.1	0	0.3
21	10	0.00	1.1	11.3	5.7	0.03	61.7	0.00	0.15	0.01	0.01	0.00	0.6	0.1	0	0.1
10	14	0.01	1.4	18.3	6.0	0.04	27.2	0.01	0.08	0.02	0.01	0.01	0.3	0.1	0	0.1
17	10	0.00	1.1	14.6	5.1	0.03	41.9	0.00	0.13	0.01	0.01	0.00	0.6	0.1	0	0.1
2	3	0.00	0.3	4.4	3.9	0.01	5.6	0.00	0.01	0.00	0.00	0.00	0.1	0	0	0
3	40	0.00	3.2	41.3	45.1	0.16	23.4	0.01	0.00	0.04	0.02	0.01	3.5	0	0.1	1.7
11	28	0.01	2.6	34.6	12.7	0.06	42.5	0.01	0.14	0.03	0.01	0.00	2.6	0.2	0.1	0.5
0	0	0.00	—	30.0	0	—	0	0.01	—	0.01	0.20	—	—	0	—	—
0	0	0.02	0.1	16.2	3.6	0.01	0	0.00	0.01	0.00	0.00	0.00	0	0	0	0
0	1	0.11	1.7	46.3	29.3	0.34	0	0.00	0.21	0.00	0.00	0.00	0	0	0	0.7
0	1	0.00	0	28.9	12.0	0.00	0.2	0.00	0.12	0.00	0.00	0.00	0	0	0	0.2
0	0	0.00	0.1	0.9	1.2	0.00	0.3	0.00	0.05	0.00	0.00	0.00	0	0	0	0.1
0	0	0.00	0	0.8	2.8	0.00	0.2	0.00	0.04	0.00	0.00	0.00	0	0	0	0.1
10	284	0.12	27.0	370.0	257.3	1.02	17.2	0.08	0.12	0.37	0.14	0.08	12.3	2.5	0.5	4.9
12	290	0.07	26.8	366.0	107.4	1.02	141.5	0.04	0.02	0.45	0.22	0.09	12.2	0	1.1	8.1
10	314	0.12	34.3	396.9	127.4	0.98	144.6	0.09	—	0.42	0.22	0.11	12.3	2.5	0.9	5.6
5	306	0.07	27.0	382.2	102.9	1.02	149.5	0.11	0.02	0.44	0.23	0.09	12.3	0	1.3	7.6
5	316	0.12	36.8	419.0	129.9	1.00	149.5	0.10	0.00	0.42	0.22	0.11	12.3	2.5	1.0	5.4
20	285	0.07	26.8	366.0	100.0	1.04	134.2	0.09	0.07	0.45	0.22	0.09	12.2	0.5	1.1	6.1
20	314	0.12	34.3	396.9	127.4	0.98	137.2	0.09	—	0.42	0.22	0.11	12.3	2.5	0.9	5.6
24	276	0.07	24.4	348.9	97.6	0.97	68.3	0.10	0.14	0.44	0.26	0.08	12.2	0	1.1	9.0
1	93	0.09	8.6	105.9	36.7	0.28	37.6	0.01	0.00	0.09	0.05	0.01	2.9	0.4	0.1	0.8
13	109	0.07	9.9	141.9	48.6	0.36	28.3	0.03	0.06	0.16	0.08	0.02	4.2	1.0	0.2	5.7
9	82	0.06	7.6	95.4	33.4	0.24	20.5	0.01	0.04	0.10	0.06	0.01	2.5	0.6	0.1	0.7
21	359	0.09	33.3	482.5	156.7	1.21	14.9	0.11	0.03	0.48	0.27	0.10	14.2	1.7	1.2	6.2
3	209	0.05	19.9	289.9	93.3	0.75	120.5	0.07	0.00	0.30	0.15	0.06	8.5	1.0	0.7	4.6
3	214	0.05	20.3	296.0	95.3	0.76	123.1	0.07	0.00	0.30	0.15	0.06	8.7	1.0	0.7	4.7
31	292	0.15	27.2	425.6	118.7	1.06	82.2	0.09	0.15	0.38	0.20	0.09	11.8	2.8	1.0	5.2
27	327	0.12	34.2	497.8	122.0	0.73	139.1	0.11	0.17	0.33	0.67	0.11	2.4	3.2	0.2	3.4
25	251	0.90	50.8	408.9	132.5	1.21	70.5	0.11	0.14	0.46	0.38	0.09	14.1	0	1.1	9.6
0	92	0.74	32.6	405.1	138.2	0.51	0	0.04	0.00	0.20	0.16	0.04	1.9	0	0.2	2.5
20	263	1.20	57.5	492.5	110.0	1.57	127.5	0.09	0.07	0.45	0.33	0.10	12.5	0.5	1.1	6.8
8	288	0.60	32.5	425.0	152.5	1.02	145.0	0.09	0.05	0.41	0.31	0.10	12.5	2.3	0.9	4.8
20	273	0.60	35.0	422.5	165.0	0.97	160.0	0.11	0.10	0.45	0.41	0.06	5.0	0	0.8	8.5
30	280	0.60	32.5	417.5	150.0	1.02	65.0	0.09	0.15	0.40	0.31	0.10	12.5	2.3	0.8	4.8
150	330	0.50	48.3	419.1	137.2	1.16	116.8	0.08	0.50	0.48	0.26	0.12	2.5	3.8	1.1	10.7
9	444	4.00	88.9	631.1	195.6	3.38	—	0.33	—	0.45	4.44	0.44	4.0	26.7	1.3	8.0
9	500	4.50	100.0	665.0	192.0	3.75	—	0.37	—	0.51	5.00	0.49	100.0	30.0	1.5	9.0
9	444	4.00	88.9	568.9	186.7	3.38	—	0.33	—	0.45	4.44	0.44	88.9	26.7	1.3	8.0
9	500	4.47	100.0	665.0	288.0	3.75	—	0.37	—	0.51	5.07	0.50	100.0	30.0	1.5	8.8
5	350	3.60	100.0	—	270.0	3.75	—	0.37	—	—	4.00	0.40	—	12.0	1.2	—
27	339	3.76	45.1	577.7	230.6	1.16	903.7	0.75	0.15	1.31	11.08	1.01	18.6	31.8	1.1	12.5

DA+ Code	Food Description	Quantity	Measure	Wt (g)	H₂O (g)	Ener (kcal)	Prot (g)	Carb (g)	Fiber (g)	Fat (g)	Fat Breakdown (g) Sat	Mono	Poly	Trans
	Milkshakes													
73	Chocolate	1	cup(s)	227	164.0	270	6.9	48.1	0.7	6.1	3.8	1.8	0.2	—
3163	Strawberry	1	cup(s)	226	167.8	256	7.7	42.8	0.9	6.3	3.9	—	—	—
74	Vanilla	1	cup(s)	227	169.2	254	8.8	40.3	0	6.9	4.3	2.0	0.3	—
	Ice cream													
4776	Chocolate	½	cup(s)	66	36.8	143	2.5	18.6	0.8	7.3	4.5	2.1	0.3	—
12137	Chocolate fudge, no sugar added	½	cup(s)	71	—	100	3.0	16.0	2.0	3.0	1.5	—	—	0
16514	Chocolate, soft serve	½	cup(s)	87	49.9	177	3.2	24.1	0.7	8.4	5.2	2.4	0.3	—
16523	Sherbet, all flavors	½	cup(s)	97	63.8	139	1.1	29.3	3.2	1.9	1.1	0.5	0.1	—
4778	Strawberry	½	cup(s)	66	39.6	127	2.1	18.2	0.6	5.5	3.4	—	—	—
76	Vanilla	½	cup(s)	72	43.9	145	2.5	17.0	0.5	7.9	4.9	2.1	0.3	—
12146	Vanilla chocolate swirl, fat free, no sugar added	½	cup(s)	71	—	100	3.0	14.0	2.0	3.0	2.0	—	—	0
82	Vanilla, light	½	cup(s)	76	48.3	125	3.6	19.6	0.2	3.7	2.2	1.0	0.2	—
78	Vanilla, light, soft serve	½	cup(s)	88	61.2	111	4.3	19.2	0	2.3	1.4	0.7	0.1	—
	Soy desserts													
10694	Tofutti low fat vanilla fudge non dairy frozen dessert	½	cup(s)	70	—	140	2.0	24.0	0	4.0	1.0	—	—	—
15721	Tofutti premium chocolate supreme non dairy frozen dessert	½	cup(s)	70	—	180	3.0	18.0	0	11.0	2.0	—	—	—
15720	Tofutti premium vanilla non dairy frozen dessert	½	cup(s)	70	—	190	2.0	20.0	0	11.0	2.0	—	—	—
	Ice milk													
16517	Chocolate	½	cup(s)	66	42.9	94	2.8	16.9	0.3	2.1	1.3	0.6	0.1	—
16516	Flavored, not chocolate	½	cup(s)	66	41.4	108	3.5	17.5	0.2	2.6	1.7	0.6	0.1	—
	Pudding													
25032	Chocolate	½	cup(s)	144	109.7	155	5.1	22.7	0.7	5.4	3.1	1.7	0.2	0
1923	Chocolate, sugar free, prepared with 2% milk	½	cup(s)	133	—	100	5.0	14.0	0.3	3.0	1.5	—	—	—
1722	Rice	½	cup(s)	113	75.6	151	4.1	29.9	0.5	1.9	1.1	0.5	0.1	—
4747	Tapioca, ready to eat	1	item(s)	142	102.0	185	2.8	30.8	0	5.5	1.4	3.6	0.1	—
25031	Vanilla	½	cup(s)	136	109.7	116	4.7	17.6	0	2.8	1.6	0.9	0.2	0
1924	Vanilla, sugar free, prepared with 2% milk	½	cup(s)	133	—	90	4.0	12.0	0.2	2.0	1.5	—	—	—
	Frozen yogurt													
4785	Chocolate, soft serve	½	cup(s)	72	45.9	115	2.9	17.9	1.6	4.3	2.6	1.3	0.2	—
1747	Fruit varieties	½	cup(s)	113	80.5	144	3.4	24.4	0	4.1	2.6	1.1	0.1	—
4786	Vanilla, soft serve	½	cup(s)	72	47.0	117	2.9	17.4	0	4.0	2.5	1.1	0.2	—
	Milk substitutes													
	Lactose free													
16081	Fat free, calcium fortified [milk]	1	cup(s)	240	—	80	8.0	13.0	0	0	0	0	0	0
36486	Low fat milk	1	cup(s)	240	—	110	8.0	13.0	0	2.5	1.5	—	—	—
36487	Reduced fat milk	1	cup(s)	240	—	130	8.0	12.0	0	5.0	3.0	—	—	—
36488	Whole milk	1	cup(s)	240	—	150	8.0	12.0	0	8.0	5.0	—	—	—
	Rice													
10083	Rice Dream carob rice beverage	1	cup(s)	240	—	150	1.0	32.0	0	2.5	0	—	—	—
17089	Rice Dream original rice beverage, enriched	1	cup(s)	240	—	120	1.0	25.0	0	2.0	0	—	—	—
10087	Rice Dream vanilla enriched rice beverage	1	cup(s)	240	—	130	1.0	28.0	0	2.0	0	—	—	—
	Soy													
34750	Soy Dream chocolate enriched soy beverage	1	cup(s)	240	—	210	7.0	37.0	1.0	3.5	0.5	—	—	—
34749	Soy Dream vanilla enriched soy beverage	1	cup(s)	240	—	150	7.0	22.0	0	4.0	0.5	—	—	—
13840	Vitasoy light chocolate soymilk	1	cup(s)	240	—	100	4.0	17.0	0	2.0	0.5	0.5	1.0	—
13839	Vitasoy light vanilla soymilk	1	cup(s)	240	—	70	4.0	10.0	0	2.0	0.5	0.5	1.0	—
13836	Vitasoy rich chocolate soymilk	1	cup(s)	240	—	160	7.0	24.0	1.0	4.0	0.5	1.0	2.5	—
13835	Vitasoy vanilla delite soymilk	1	cup(s)	240	—	120	7.0	13.0	1.0	4.0	0.5	1.0	2.5	—
	Yogurt													
3615	Custard style, fruit flavors	6	ounce(s)	170	127.1	190	7.0	32.0	0	3.5	2.0	—	—	—
3617	Custard style, vanilla	6	ounce(s)	170	134.1	190	7.0	32.0	0	3.5	2.0	0.9	0.1	—
32101	Fruit, low fat	1	cup(s)	245	184.5	243	9.8	45.7	0	2.8	1.8	0.8	0.1	—
29638	Fruit, non fat, sweetened with low calorie sweetener	1	cup(s)	241	208.3	123	10.6	19.4	1.2	0.4	0.2	0.1	0	—
93	Plain, low fat	1	cup(s)	245	208.4	154	12.9	17.2	0	3.8	2.5	1.0	0.1	—
94	Plain, nonfat	1	cup(s)	245	208.8	137	14.0	18.8	0	0.4	0.3	0.1	0	—
32100	Vanilla, low fat	1	cup(s)	245	193.6	208	12.1	33.8	0	3.1	2.0	0.8	0.1	—
5242	Yogurt beverage	1	cup(s)	245	199.8	172	6.2	32.8	0	2.2	1.4	0.6	0.1	—
38202	Yogurt smoothie, nonfat, all flavors	1	item(s)	325	—	290	10.0	60.0	6.0	0	0	0	0	0
	Soy yogurt													
34617	Stonyfield Farm O'Soy strawberry-peach pack organic cultured soy yogurt	1	item(s)	113	—	100	5.0	16.0	3.0	2.0	0	—	—	0
34616	Stonyfield Farm O'Soy vanilla organic cultured soy yogurt	1	item(s)	170	—	150	7.0	26.0	4.0	2.0	0	—	—	0
10453	White Wave plain silk cultured soy yogurt	8	ounce(s)	227	—	140	5.0	22.0	1.0	3.0	0.5	—	—	0
	EGGS													
	Eggs													
99	Fried	1	item(s)	46	31.8	90	6.3	0.4	0	7.0	2.0	2.9	1.2	—
	EGGS —Continued													

Chol (mg)	Calc (mg)	Iron (mg)	Magn (mg)	Pota (mg)	Sodi (mg)	Zinc (mg)	Vit A (µg)	Thia (mg)	Vit E (mg α)	Ribo (mg)	Niac (mg)	Vit B$_6$ (mg)	Fola (µg)	Vit C (mg)	Vit B$_{12}$ (µg)	Sele (µg)
25	300	0.70	36.4	508.9	252.2	1.09	40.9	0.10	0.11	0.50	0.28	0.05	11.4	0	0.7	4.3
25	256	0.24	29.4	412.0	187.9	0.81	58.9	0.10	—	0.44	0.39	0.10	6.8	1.8	0.7	4.8
27	332	0.22	27.3	415.8	215.8	0.88	56.8	0.06	0.11	0.44	0.33	0.09	15.9	0	1.2	5.2
22	72	0.61	19.1	164.3	50.2	0.38	77.9	0.02	0.19	0.12	0.14	0.03	10.6	0.5	0.2	1.7
10	100	0.36	—	—	65.0	—	—	—	—	—	—	—	—	0	—	—
22	103	0.32	19.0	192.0	43.3	0.45	66.6	0.03	0.22	0.13	0.11	0.03	4.3	0.5	0.3	2.5
0	52	0.13	7.7	92.6	44.4	0.46	9.7	0.02	0.02	0.08	0.07	0.02	6.8	5.6	0.1	1.3
19	79	0.13	9.2	124.1	39.6	0.22	63.4	0.03	—	0.16	0.11	0.03	7.9	5.1	0.2	1.3
32	92	0.06	10.1	143.3	57.6	0.49	85.0	0.03	0.21	0.17	0.08	0.03	3.6	0.4	0.3	1.3
10	100	0.00	—	—	65.0	—	—	—	—	—	—	—	—	0	—	—
21	122	0.14	10.6	158.1	56.2	0.55	97.3	0.04	0.09	0.19	0.10	0.03	4.6	0.9	0.4	1.5
11	138	0.05	12.3	194.5	61.6	0.46	25.5	0.04	0.05	0.17	0.10	0.04	4.4	0.8	0.4	3.2
0	0	0.00	—	8.0	90.0	—	0	—	—	—	—	—	—	0	—	—
0	0	0.00	—	7.0	180.0	—	0	—	—	—	—	—	—	0	—	—
0	0	0.00	—	2.0	210.0	—	0	—	—	—	—	—	—	0	—	—
6	94	0.15	13.1	155.2	40.6	0.36	15.7	0.03	0.05	0.11	0.08	0.02	3.9	0.5	0.3	2.2
16	76	0.05	9.2	136.2	48.5	0.47	90.4	0.02	0.05	0.11	0.06	0.01	3.3	0.1	0.2	1.3
35	149	0.46	31.3	226.7	137.0	0.71	—	0.05	0.00	0.22	0.15	0.06	8.3	1.2	0.5	4.9
10	150	0.72	—	330.0	310.0	—	—	0.06	—	0.26	—	—	—	0	—	—
7	113	0.28	15.8	201.4	66.4	0.52	41.6	0.03	0.05	0.17	0.34	0.06	4.5	0.2	0.2	4.8
1	101	0.15	8.5	130.6	205.9	0.31	0	0.03	0.21	0.13	0.09	0.03	4.3	0.4	0.3	0
35	146	0.17	17.2	188.9	136.4	0.52	—	0.04	0.00	0.22	0.10	0.05	8.0	1.2	0.5	4.6
10	150	0.00	—	190.0	380.0	—	—	0.03	—	0.17	—	—	—	0	—	—
4	106	0.90	19.4	187.9	70.6	0.35	31.7	0.02	—	0.15	0.22	0.05	7.9	0.2	0.2	1.7
15	113	0.52	11.3	176.3	71.2	0.31	55.4	0.04	0.10	0.20	0.07	0.04	4.5	0.8	0.1	2.1
1	103	0.21	10.1	151.9	62.6	0.30	42.5	0.02	0.07	0.16	0.20	0.05	4.3	0.6	0.2	2.4
3	500	0.00	—	—	125.0	—	100.0	—	—	—	—	—	—	0	0	—
10	300	0.00	—	—	125.0	—	100.0	—	—	—	—	—	—	0	—	—
20	300	0.00	—	—	125.0	—	98.2	—	—	—	—	—	—	0	—	—
35	300	0.00	—	—	125.0	—	58.1	—	—	—	—	—	—	0	—	—
0	20	0.72	—	82.5	100.0	—	—	—	—	—	—	—	—	1.2	—	—
0	300	0.00	13.3	60.0	90.0	0.24	—	0.06	—	0.00	0.84	0.07	—	0	1.5	—
0	300	0.00	—	53.0	90.0	—	—	—	—	—	—	—	—	0	1.5	—
0	300	1.80	60.0	350.0	160.0	0.60	33.3	0.15	—	0.06	0.80	0.12	60.0	0	3.0	—
0	300	1.80	40.0	260.0	140.0	0.60	33.3	0.15	—	0.06	0.80	0.12	60.0	0	3.0	—
0	300	0.72	24.0	200.0	140.0	0.90	—	0.09	—	0.34	—	—	24.0	0	0.9	—
0	300	0.72	24.0	200.0	120.0	0.90	—	0.09	—	0.34	—	—	24.0	0	0.9	—
0	300	1.08	40.0	320.0	150.0	0.90	—	0.15	—	0.34	—	—	60.0	0	0.9	—
0	40	0.72	—	320.0	115.0	—	0	—	—	—	—	—	—	0	—	—
15	300	0.00	16.0	310.0	100.0	—	—	—	—	0.25	—	—	—	0	—	—
15	300	0.00	16.0	310.0	100.0	—	—	—	—	0.25	—	—	—	0	—	—
12	338	0.14	31.9	433.7	129.9	1.64	27.0	0.08	0.04	0.39	0.21	0.09	22.1	1.5	1.1	6.9
5	369	0.62	41.0	549.5	139.8	1.83	4.8	0.10	0.16	0.44	0.49	0.10	31.3	26.5	1.1	7.0
15	448	0.19	41.7	573.3	171.5	2.18	34.3	0.10	0.07	0.52	0.27	0.12	27.0	2.0	1.4	8.1
5	488	0.22	46.6	624.8	188.7	2.37	4.9	0.11	0.00	0.57	0.30	0.13	29.4	2.2	1.5	8.8
12	419	0.17	39.2	536.6	161.7	2.03	29.4	0.10	0.04	0.49	0.26	0.11	27.0	2.0	1.3	12.0
13	260	0.22	39.2	399.4	98.0	1.10	14.7	0.11	0.00	0.51	0.30	0.14	29.4	2.1	1.5	—
5	300	2.70	100.0	580.0	290.0	2.25	—	0.37	—	0.42	5.00	0.50	100.0	15.0	1.5	—
0	100	1.08	24.0	5.0	20.0	—	0	0.22	—	0.10	—	0.04	—	0	0	—
0	150	1.44	40.0	15.0	40.0	—	—	0.30	—	0.13	—	0.08	—	0	0	—
0	400	1.44	—	0	30.0	—	0	—	—	—	—	—	—	0	—	—
210	27	0.91	6.0	67.6	93.8	0.55	91.1	0.03	0.56	0.23	0.03	0.07	23.5	0	0.6	15.7

DA+ Code	Food Description	Quantity	Measure	Wt (g)	H₂O (g)	Ener (kcal)	Prot (g)	Carb (g)	Fiber (g)	Fat (g)	Fat Breakdown (g)			
											Sat	Mono	Poly	Trans
100	Hard boiled	1	item(s)	50	37.3	78	6.3	0.6	0	5.3	1.6	2.0	0.7	—
101	Poached	1	item(s)	50	37.8	71	6.3	0.4	0	5.0	1.5	1.9	0.7	—
97	Raw, white	1	item(s)	33	28.9	16	3.6	0.2	0	0.1	0	0	0	—
96	Raw, whole	1	item(s)	50	37.9	72	6.3	0.4	0	5.0	1.5	1.9	0.7	—
98	Raw, yolk	1	item(s)	17	8.9	54	2.7	0.6	0	4.5	1.6	2.0	0.7	—
102	Scrambled, prepared with milk and butter	2	item(s)	122	89.2	204	13.5	2.7	0	14.9	4.5	5.8	2.6	0.7
	Egg substitute													
4028	Egg Beaters	¼	cup(s)	61	—	30	6.0	1.0	0	0	0	0	0	0
920	Frozen	¼	cup(s)	60	43.9	96	6.8	1.9	0	6.7	1.2	1.5	3.7	—
918	Liquid	¼	cup(s)	63	51.9	53	7.5	0.4	0	2.1	0.4	0.6	1.0	—
	SEAFOOD													
	Cod													
6040	Atlantic cod or scrod, baked or broiled	3	ounce(s)	85	64.6	89	19.4	0	0	0.7	0.1	0.1	0.2	—
1573	Atlantic cod, cooked, dry heat	3	ounce(s)	85	64.6	89	19.4	0	0	0.7	0.1	0.1	0.2	—
2905	**Eel, raw**	3	ounce(s)	85	58.0	156	15.7	0	0	9.9	2.0	6.1	0.8	—
	Fish fillets													
25079	Baked	3	ounce(s)	84	79.9	99	21.7	0	0	0.7	0.1	0.1	0.3	—
8615	Batter coated or breaded, fried	3	ounce(s)	85	45.6	197	12.5	14.4	0.4	10.5	2.4	2.2	5.3	—
25082	Broiled fish steaks	3	ounce(s)	85	68.1	128	24.2	0	0	2.6	0.4	0.9	0.8	—
25083	Poached fish steaks	3	ounce(s)	85	67.1	111	21.1	0	0	2.3	0.3	0.8	0.7	—
25084	Steamed	3	ounce(s)	85	72.2	79	17.2	0	0	0.6	0.1	0.1	0.2	—
25089	Flounder, baked	3	ounce(s)	85	64.4	113	14.8	0.4	0.1	5.5	1.1	2.2	1.4	—
1825	**Grouper, cooked, dry heat**	3	ounce(s)	85	62.4	100	21.1	0	0	1.1	0.3	0.2	0.3	—
	Haddock													
6049	Baked or broiled	3	ounce(s)	85	63.2	95	20.6	0	0	0.8	0.1	0.1	0.3	—
1578	Cooked, dry heat	3	ounce(s)	85	63.1	95	20.6	0	0	0.8	0.1	0.1	0.3	—
1886	**Halibut, Atlantic and Pacific, cooked, dry heat**	3	ounce(s)	85	61.0	119	22.7	0	0	2.5	0.4	0.8	0.8	—
1582	**Herring, Atlantic, pickled**	4	piece(s)	60	33.1	157	8.5	5.8	0	10.8	1.4	7.2	1.0	—
1587	**Jack mackerel, solids, canned, drained**	2	ounce(s)	57	39.2	88	13.1	0	0	3.6	1.1	1.3	0.9	—
8580	**Octopus, common, cooked, moist heat**	3	ounce(s)	85	51.5	139	25.4	3.7	0	1.8	0.4	0.4	0.4	—
1831	**Perch, mixed species, cooked, dry heat**	3	ounce(s)	85	62.3	100	21.1	0	0	1.0	0.2	0.2	0.4	—
1592	**Pacific rockfish, cooked, dry heat**	3	ounce(s)	85	62.4	103	20.4	0	0	1.7	0.4	0.4	0.5	—
	Salmon													
2938	Coho, farmed, raw	3	ounce(s)	85	59.9	136	18.1	0	0	6.5	1.5	2.8	1.6	—
1594	Broiled or baked with butter	3	ounce(s)	85	53.9	155	23.0	0	0	6.3	1.2	2.3	2.3	—
29727	Smoked chinook (lox)	2	ounce(s)	57	40.8	66	10.4	0	0	2.4	0.5	1.1	0.6	—
154	**Sardine, Atlantic with bones, canned in oil**	3	ounce(s)	85	50.7	177	20.9	0	0	9.7	1.3	3.3	4.4	—
	Scallops													
155	Mixed species, breaded, fried	3	item(s)	47	27.2	100	8.4	4.7	—	5.1	1.2	2.1	1.3	—
1599	Steamed	3	ounce(s)	85	64.8	90	13.8	2.0	0	2.6	0.4	1.0	0.8	—
1839	**Snapper, mixed species, cooked, dry heat**	3	ounce(s)	85	59.8	109	22.4	0	0	1.5	0.3	0.3	0.5	—
	Squid													
1868	Mixed species, fried	3	ounce(s)	85	54.9	149	15.3	6.6	0	6.4	1.6	2.3	1.8	—
16617	Steamed or boiled	3	ounce(s)	85	63.3	89	15.2	3.0	0	1.3	0.4	0.1	0.5	—
1570	**Striped bass, cooked, dry heat**	3	ounce(s)	85	62.4	105	19.3	0	0	2.5	0.6	0.7	0.9	—
1601	**Sturgeon, steamed**	3	ounce(s)	85	59.4	111	17.0	0	0	4.3	1.0	2.0	0.7	—
1840	**Surimi, formed**	3	ounce(s)	85	64.9	84	12.9	5.8	0	0.8	0.2	0.1	0.4	—
1842	**Swordfish, cooked, dry heat**	3	ounce(s)	85	58.5	132	21.6	0	0	4.4	1.2	1.7	1.0	—
1846	**Tuna, yellowfin or ahi, raw**	3	ounce(s)	85	60.4	92	19.9	0	0	0.8	0.2	0.1	0.2	—
	Tuna, canned													
159	Light, canned in oil, drained	2	ounce(s)	57	33.9	112	16.5	0	0	4.6	0.9	1.7	1.6	—
355	Light, canned in water, drained	2	ounce(s)	57	42.2	66	14.5	0	0	0.5	0.1	0.1	0.2	—
33211	Light, no salt, canned in oil, drained	2	ounce(s)	57	33.9	112	16.5	0	0	4.7	0.9	1.7	1.6	—
33212	Light, no salt, canned in water, drained	2	ounce(s)	57	42.6	66	14.5	0	0	0.5	0.1	0.1	0.2	—
2961	White, canned in oil, drained	2	ounce(s)	57	36.3	105	15.0	0	0	4.6	0.7	1.8	1.7	—
351	White, canned in water, drained	2	ounce(s)	57	41.5	73	13.4	0	0	1.7	0.4	0.4	0.6	—
33213	White, no salt, canned in oil, drained	2	ounce(s)	57	36.3	105	15.0	0	0	4.6	0.9	1.4	1.9	—
33214	White, no salt, canned in water, drained	2	ounce(s)	57	42.0	73	13.4	0	0	1.7	0.4	0.4	0.6	—
	Yellowtail													
8548	Mixed species, cooked, dry heat	3	ounce(s)	85	57.3	159	25.2	0	0	5.7	1.4	2.2	1.5	—
2970	Mixed species, raw	2	ounce(s)	57	42.2	83	13.1	0	0	3.0	0.7	1.1	0.8	—
	Shellfish, meat only													
1857	Abalone, mixed species, fried	3	ounce(s)	85	51.1	161	16.7	9.4	0	5.8	1.4	2.3	1.4	—
16618	Abalone, steamed or poached	3	ounce(s)	85	40.7	177	28.8	10.1	0	1.3	0.3	0.2	0.2	—
	Crab													
1851	Blue crab, canned	2	ounce(s)	57	43.2	56	11.6	0	0	0.7	0.1	0.1	0.2	—
1852	Blue crab, cooked, moist heat	3	ounce(s)	85	65.9	87	17.2	0	0	1.5	0.2	0.2	0.6	—
8562	Dungeness crab, cooked, moist heat	3	ounce(s)	85	62.3	94	19.0	0.8	0	1.1	0.1	0.2	0.3	—
1860	**Clams, cooked, moist heat**	3	ounce(s)	85	54.1	126	21.7	4.4	0	1.7	0.2	0.1	0.5	—
	SEAFOOD —Continued													

Chol (mg)	Calc (mg)	Iron (mg)	Magn (mg)	Pota (mg)	Sodi (mg)	Zinc (mg)	Vit A (µg)	Thia (mg)	Vit E (mg α)	Ribo (mg)	Niac (mg)	Vit B$_6$ (mg)	Fola (µg)	Vit C (mg)	Vit B$_{12}$ (µg)	Sele (µg)
212	25	0.59	5.0	63.0	62.0	0.52	84.5	0.03	0.51	0.25	0.03	0.06	22.0	0	0.6	15.4
211	27	0.91	6.0	66.5	147.0	0.55	69.5	0.02	0.48	0.20	0.03	0.06	17.5	0	0.6	15.8
0	2	0.02	3.6	53.8	54.8	0.01	0	0.00	0.00	0.14	0.03	0.00	1.3	0	0	6.6
212	27	0.91	6.0	67.0	70.0	0.55	70.0	0.03	0.48	0.23	0.03	0.07	23.5	0	0.6	15.9
210	22	0.46	0.9	18.5	8.2	0.39	64.8	0.03	0.43	0.09	0.00	0.06	24.8	0	0.3	9.5
429	87	1.46	14.6	168.4	341.6	1.22	174.5	0.06	1.33	0.53	0.09	0.14	36.6	0.2	0.9	27.5
0	20	1.08	4.0	85.0	115.0	0.60	112.5	0.15	—	0.85	0.20	0.08	60.0	0	1.2	—
1	44	1.18	9.0	127.8	119.4	0.58	6.6	0.07	0.95	0.23	0.08	0.08	9.6	0.3	0.2	24.8
1	33	1.32	5.6	207.1	111.1	0.82	11.3	0.07	0.17	0.19	0.07	0.00	9.4	0	0.2	15.6
47	12	0.41	35.7	207.5	66.3	0.49	11.9	0.07	0.68	0.06	2.13	0.24	6.8	0.8	0.9	32.0
47	12	0.41	35.7	207.5	66.3	0.49	11.9	0.07	0.68	0.06	2.13	0.24	6.8	0.9	0.9	32.0
107	17	0.42	17.0	231.3	43.4	1.37	887.0	0.13	3.40	0.03	2.97	0.05	12.8	1.5	2.6	5.5
44	8	0.31	29.1	489.0	86.1	0.48	—	0.02	—	0.05	2.47	0.46	8.1	3.0	1.0	44.3
29	15	1.79	20.4	272.2	452.5	0.37	9.4	0.09	—	0.09	1.78	0.08	14.5	0	0.9	7.7
37	55	0.97	96.7	524.3	62.9	0.49	—	0.05	—	0.08	6.47	0.36	12.6	0	1.2	42.5
32	48	0.85	84.0	455.6	54.7	0.42	—	0.05	—	0.07	5.92	0.33	11.5	0	1.1	37.0
41	12	0.29	24.7	319.3	41.7	0.34	—	0.06	—	0.06	1.89	0.21	6.1	0.8	0.8	32.0
44	19	0.34	47.3	224.7	280.2	0.20	—	0.06	0.40	0.07	2.02	0.18	7.4	2.8	1.6	33.5
40	18	0.96	31.5	404.0	45.1	0.43	42.5	0.06	—	0.01	0.32	0.29	8.5	0	0.6	39.8
63	36	1.15	42.5	339.4	74.0	0.40	16.2	0.03	0.42	0.03	3.94	0.29	6.8	0	1.2	34.4
63	36	1.14	42.5	339.3	74.0	0.40	16.2	0.03	—	0.03	3.93	0.29	11.1	0	1.2	34.4
35	51	0.91	91.0	489.9	58.7	0.45	45.9	0.05	—	0.07	6.05	0.33	11.9	0	1.2	39.8
8	46	0.73	4.8	41.4	522.0	0.31	154.8	0.02	1.02	0.08	1.98	0.10	1.2	0	2.6	35.1
45	137	1.15	21.0	110.0	214.9	0.57	73.7	0.02	0.58	0.12	3.50	0.11	2.8	0.5	3.9	21.4
82	90	8.11	51.0	535.8	391.2	2.85	76.5	0.04	1.02	0.06	3.21	0.55	20.4	6.8	30.6	76.2
98	87	0.98	32.3	292.6	67.2	1.21	8.5	0.06	—	0.10	1.61	0.11	5.1	1.4	1.9	13.7
37	10	0.45	28.9	442.3	65.5	0.45	60.4	0.03	1.32	0.07	3.33	0.22	8.5	0	1.0	39.8
43	10	0.29	26.4	382.7	40.0	0.36	47.6	0.08	—	0.09	5.79	0.56	11.1	0.9	2.3	10.7
40	15	1.02	26.9	376.6	98.6	0.56	—	0.13	1.14	0.05	8.33	0.18	4.2	1.8	2.3	41.0
13	6	0.48	10.2	99.2	1134.0	0.17	14.7	0.01	—	0.05	2.67	0.15	1.1	0	1.8	21.6
121	325	2.48	33.2	337.6	429.5	1.10	27.2	0.04	1.70	0.18	4.43	0.14	10.2	0	7.6	44.8
28	20	0.38	27.4	154.8	215.8	0.49	10.7	0.02	—	0.05	0.70	0.06	17.2	1.1	0.6	12.5
27	20	0.22	45.9	238.0	358.7	0.78	32.3	0.01	0.16	0.05	0.84	0.11	10.2	2.0	1.1	18.2
40	34	0.20	31.5	444.0	48.5	0.37	29.8	0.04	—	0.00	0.29	0.39	5.1	1.4	3.0	41.7
221	33	0.85	32.3	237.3	260.3	1.48	9.4	0.04	—	0.39	2.21	0.04	11.9	3.6	1.0	44.1
227	31	0.62	28.9	192.1	356.2	1.49	8.5	0.01	1.17	0.32	1.69	0.04	3.4	3.2	1.0	43.7
88	16	0.91	43.4	279.0	74.8	0.43	26.4	0.09	—	0.03	2.17	0.29	8.5	0	3.8	39.8
63	11	0.59	29.8	239.7	388.5	0.35	198.9	0.06	0.52	0.07	8.30	0.19	14.5	0	2.2	13.3
26	8	0.22	36.6	95.3	121.6	0.28	17.0	0.01	0.53	0.01	0.18	0.02	1.7	0	1.4	23.9
43	5	0.88	28.9	313.8	97.8	1.25	34.9	0.03	—	0.09	10.02	0.32	1.7	0.9	1.7	52.5
38	14	0.62	42.5	377.6	31.5	0.44	15.3	0.37	0.42	0.04	8.33	0.77	1.7	0.8	0.4	31.0
10	7	0.79	17.6	117.3	200.6	0.51	13.0	0.02	0.49	0.07	7.03	0.06	2.8	0	1.2	43.1
17	6	0.87	15.3	134.3	191.5	0.43	9.6	0.01	0.19	0.04	7.52	0.19	2.3	0	1.7	45.6
10	7	0.78	17.6	117.4	28.3	0.51	0	0.02	—	0.06	7.03	0.06	2.8	0	1.2	43.1
17	6	0.86	15.3	134.4	28.3	0.43	0	0.01	—	0.04	7.52	0.19	2.3	0	1.7	45.6
18	2	0.36	19.3	188.8	224.5	0.26	2.8	0.01	1.30	0.04	6.63	0.24	2.8	0	1.2	34.1
24	8	0.55	18.7	134.3	213.6	0.27	3.4	0.00	0.48	0.02	3.28	0.12	1.1	0	0.7	37.2
18	2	0.36	19.3	188.8	28.3	0.26	0	0.01	—	0.04	6.63	0.24	2.8	0	1.2	34.1
24	8	0.54	18.7	134.4	28.3	0.27	3.4	0.00	—	0.02	3.28	0.12	1.1	0	0.7	37.3
60	25	0.53	32.3	457.6	42.5	0.56	26.4	0.14	—	0.04	7.41	0.15	3.4	2.5	1.1	39.8
31	13	0.28	17.0	238.1	22.1	0.29	16.4	0.08	—	0.02	3.86	0.09	2.3	1.6	0.7	20.7
80	31	3.23	47.6	241.5	502.6	0.80	1.7	0.18	—	0.11	1.61	0.12	11.9	1.5	0.6	44.1
144	50	4.84	68.9	295.0	980.1	1.38	3.4	0.28	6.74	0.12	1.89	0.21	6.0	2.6	0.7	75.6
50	57	0.47	22.1	212.1	188.8	2.27	1.1	0.04	1.04	0.04	0.77	0.08	24.4	1.5	0.3	18.0
85	88	0.77	28.1	275.6	237.3	3.58	1.7	0.08	1.56	0.04	2.80	0.15	43.4	2.8	6.2	34.2
65	50	0.36	49.3	347.0	321.5	4.65	26.4	0.04	—	0.17	3.08	0.14	35.7	3.1	8.8	40.5
57	78	23.78	15.3	534.1	95.3	2.32	145.4	0.12	—	0.36	2.85	0.09	24.7	18.8	84.1	54.4

DA+ Code	Food Description	Quantity	Measure	Wt (g)	H₂O (g)	Ener (kcal)	Prot (g)	Carb (g)	Fiber (g)	Fat (g)	Sat	Mono	Poly	Trans
1853	**Crayfish, farmed, cooked, moist heat**	3	ounce(s)	85	68.7	74	14.9	0	0	1.1	0.2	0.2	0.4	—
	Oysters													
8720	Baked or broiled	3	ounce(s)	85	68.6	89	5.6	3.2	0	5.8	1.3	2.1	1.9	—
152	Eastern, farmed, raw	3	ounce(s)	85	73.3	50	4.4	4.7	0	1.3	0.4	0.1	0.5	—
8715	Eastern, wild, cooked, moist heat	3	ounce(s)	85	59.8	117	12.0	6.7	0	4.2	1.3	0.5	1.6	—
8584	Pacific, cooked, moist heat	3	ounce(s)	85	54.5	139	16.1	8.4	0	3.9	0.9	0.7	1.5	—
1865	Pacific, raw	3	ounce(s)	85	69.8	69	8.0	4.2	0	2.0	0.4	0.3	0.8	—
1854	**Lobster, northern, cooked, moist heat**	3	ounce(s)	85	64.7	83	17.4	1.1	0	0.5	0.1	0.1	0.1	—
1862	**Mussel, blue, cooked, moist heat**	3	ounce(s)	85	52.0	146	20.2	6.3	0	3.8	0.7	0.9	1.0	—
	Shrimp													
158	Mixed species, breaded, fried	3	ounce(s)	85	44.9	206	18.2	9.8	0.3	10.4	1.8	3.2	4.3	—
1855	Mixed species, cooked, moist heat	3	ounce(s)	85	65.7	84	17.8	0	0	0.9	0.2	0.2	0.4	—
	BEEF, LAMB, PORK													
	Beef													
4450	Breakfast strips, cooked	2	slice(s)	23	5.9	101	7.1	0.3	0	7.8	3.2	3.8	0.4	—
174	Corned beef, canned	3	ounce(s)	85	49.1	213	23.0	0	0	12.7	5.3	5.1	0.5	—
33147	Cured, thin siced	2	ounce(s)	57	32.9	100	15.9	3.2	0	2.2	0.9	1.0	0.1	—
4581	Jerky	1	ounce(s)	28	6.6	116	9.4	3.1	0.5	7.3	3.1	3.2	0.3	—
	Ground beef													
5898	Lean, broiled, medium	3	ounce(s)	85	50.4	202	21.6	0	0	12.2	4.8	5.3	0.4	—
5899	Lean, broiled, well done	3	ounce(s)	85	48.4	214	23.8	0	0	12.5	5.0	5.7	0.3	0.4
5914	Regular, broiled, medium	3	ounce(s)	85	46.1	246	20.5	0	0	17.6	6.9	7.7	0.6	—
5915	Regular, broiled, well done	3	ounce(s)	85	43.8	259	21.6	0	0	18.4	7.5	8.5	0.5	0.6
	Beef rib													
4241	Rib, small end, separable lean, 0" fat, broiled	3	ounce(s)	85	53.2	164	25.0	0	0	6.4	2.4	2.6	0.2	—
4183	Rib, whole, lean and fat, ¼" fat, roasted	3	ounce(s)	85	39.0	320	18.9	0	0	26.6	10.7	11.4	0.9	—
	Beef roast													
16981	Bottom round, choice, separable lean and fat, ⅛" fat, braised	3	ounce(s)	85	46.2	216	27.9	0	0	10.7	4.1	4.6	0.4	—
16979	Bottom round, separable lean and fat, ⅛" fat, roasted	3	ounce(s)	85	52.4	185	22.5	0	0	9.9	3.8	4.2	0.4	—
16924	Chuck, arm pot roast, separable lean and fat, ⅛" fat, braised	3	ounce(s)	85	42.9	257	25.6	0	0	16.3	6.5	7.0	0.6	—
16930	Chuck, blade roast, separable lean and fat, ⅛" fat, braised	3	ounce(s)	85	40.5	290	22.8	0	0	21.4	8.5	9.2	0.8	—
5853	Chuck, blade roast, separable lean, 0" trim, pot roasted	3	ounce(s)	85	47.4	202	26.4	0	0	9.9	3.9	4.3	0.3	—
4296	Eye of round, choice, separable lean, 0" fat, roasted	3	ounce(s)	85	56.5	138	24.4	0	0	3.7	1.3	1.5	0.1	—
16989	Eye of round, separable lean and fat, ⅛" fat, roasted	3	ounce(s)	85	52.2	180	24.2	0	0	8.5	3.2	3.6	0.3	—
	Beef steak													
4348	Short loin, t-bone steak, lean and fat, ¼" fat, broiled	3	ounce(s)	85	43.2	274	19.4	0	0	21.2	8.3	9.6	0.8	—
4349	Short loin, t-bone steak, lean, ¼" fat, broiled	3	ounce(s)	85	52.3	174	22.8	0	0	8.5	3.1	4.2	0.3	—
4360	Top loin, prime, lean and fat, ¼" fat, broiled	3	ounce(s)	85	42.7	275	21.6	0	0	20.3	8.2	8.6	0.7	—
	Beef variety													
188	Liver, pan fried	3	ounce(s)	85	52.7	149	22.6	4.4	0	4.0	1.3	0.5	0.5	0.2
4447	Tongue, simmered	3	ounce(s)	85	49.2	242	16.4	0	0	19.0	6.9	8.6	0.6	0.7
	Lamb chop													
3275	Loin, domestic, lean and fat, ¼" fat, broiled	3	ounce(s)	85	43.9	269	21.4	0	0	19.6	8.4	8.3	1.4	—
	Lamb leg													
3264	Domestic, lean and fat, ¼" fat, cooked	3	ounce(s)	85	45.7	250	20.9	0	0	17.8	7.5	7.5	1.3	—
	Lamb rib													
182	Domestic, lean and fat, ¼" fat, broiled	3	ounce(s)	85	40.0	307	18.8	0	0	25.2	10.8	10.3	2.0	—
183	Domestic, lean, ¼" fat, broiled	3	ounce(s)	85	50.0	200	23.6	0	0	11.0	4.0	4.4	1.0	—
	Lamb shoulder													
186	Shoulder, arm and blade, domestic, choice, lean and fat,¼" fat, roasted	3	ounce(s)	85	47.8	235	19.1	0	0	17.0	7.2	6.9	1.4	—
187	Shoulder, arm and blade, domestic, choice, lean, ¼" fat, roasted	3	ounce(s)	85	53.8	173	21.2	0	0	9.2	3.5	3.7	0.8	—
3287	Shoulder, arm, domestic, lean and fat, ¼" fat, braised	3	ounce(s)	85	37.6	294	25.8	0	0	20.4	8.4	8.7	1.5	—
3290	Shoulder, arm, domestic, lean, ¼" fat, braised	3	ounce(s)	85	41.9	237	30.2	0	0	12.0	4.3	5.2	0.8	—
	Lamb variety													
3375	Brain, pan fried	3	ounce(s)	85	51.6	232	14.4	0	0	18.9	4.8	3.4	1.9	—
3406	Tongue, braised	3	ounce(s)	85	49.2	234	18.3	0	0	17.2	6.7	8.5	1.1	—
	Pork, cured													
29229	Bacon, Canadian style, cured	2	ounce(s)	57	37.9	89	11.7	1.0	0	4.0	1.3	1.8	0.4	—
161	Bacon, cured, broiled, pan fried or roasted	2	slice(s)	16	2.0	87	5.9	0.2	0	6.7	2.2	3.0	0.7	0

BEEF, LAMB, PORK —Continued

Chol (mg)	Calc (mg)	Iron (mg)	Magn (mg)	Pota (mg)	Sodi (mg)	Zinc (mg)	Vit A (µg)	Thia (mg)	Vit E (mg α)	Ribo (mg)	Niac (mg)	Vit B_6 (mg)	Fola (µg)	Vit C (mg)	Vit B_{12} (µg)	Sele (µg)
117	43	0.94	28.1	202.4	82.5	1.25	12.8	0.03	—	0.06	1.41	0.11	9.4	0.4	2.6	29.1
43	36	5.30	37.4	125.0	403.8	72.22	60.4	0.07	0.98	0.06	1.04	0.04	7.7	2.8	14.7	50.7
21	37	4.91	28.1	105.4	151.3	32.23	6.8	0.08	—	0.05	1.07	0.05	15.3	4.0	13.8	54.1
89	77	10.19	80.8	239.0	358.9	154.45	45.9	0.16	—	0.15	2.11	0.10	11.9	5.1	29.8	60.9
85	14	7.82	37.4	256.8	180.3	28.27	124.2	0.10	0.72	0.37	3.07	0.07	12.8	10.9	24.5	131.0
43	7	4.34	18.7	142.9	90.1	14.13	68.9	0.05	—	0.20	1.70	0.04	8.5	6.8	13.6	65.5
61	52	0.33	29.8	299.4	323.2	2.48	22.1	0.01	0.85	0.05	0.91	0.06	9.4	0	2.6	36.3
48	28	5.71	31.5	227.9	313.8	2.27	77.4	0.25	—	0.35	2.55	0.08	64.6	11.6	20.4	76.2
150	57	1.07	34.0	191.3	292.4	1.17	0	0.11	—	0.11	2.60	0.08	15.3	1.3	1.6	35.4
166	33	2.62	28.9	154.8	190.5	1.32	57.8	0.02	1.17	0.02	2.20	0.10	3.4	1.9	1.3	33.7
27	2	0.71	6.1	93.1	509.2	1.44	0	0.02	0.06	0.05	1.46	0.07	1.8	0	0.8	6.1
73	10	1.76	11.9	115.7	855.6	3.03	0	0.01	0.12	0.12	2.06	0.11	7.7	0	1.4	36.5
23	6	1.53	10.8	243.2	815.9	2.25	0	0.04	0.00	0.10	2.98	0.19	6.2	0	1.5	16.0
14	6	1.53	14.5	169.2	627.4	2.29	0	0.04	0.13	0.04	0.49	0.05	38.0	0	0.3	3.0
58	6	2.00	17.9	266.2	59.5	4.63	0	0.05	—	0.23	4.21	0.23	7.6	0	1.8	16.0
69	12	2.21	18.4	250.0	62.4	5.86	0	0.08	—	0.23	5.10	0.16	9.4	0	1.7	19.0
62	9	2.07	17.0	248.3	70.6	4.40	0	0.02	—	0.16	4.90	0.23	7.6	0	2.5	16.2
71	12	2.30	18.5	242.4	72.4	5.18	0	0.08	—	0.23	4.93	0.17	8.5	0	1.6	18.0
65	16	1.59	21.3	319.8	51.9	4.64	0	0.06	0.34	0.12	7.15	0.53	8.5	0	1.4	29.2
72	9	1.96	16.2	251.7	53.6	4.45	0	0.06	—	0.14	2.85	0.19	6.0	0	2.1	18.7
68	6	2.29	17.9	223.7	35.7	4.59	0	0.05	0.41	0.15	5.05	0.36	8.5	0	1.7	29.3
64	5	1.83	14.5	182.0	29.8	3.76	0	0.05	0.34	0.12	3.92	0.29	6.8	0	1.3	23.0
67	14	2.15	17.0	205.8	42.5	5.93	0	0.05	0.45	0.15	3.63	0.25	7.7	0	1.9	24.1
88	11	2.66	16.2	198.2	55.3	7.15	0	0.06	0.17	0.20	2.06	0.22	4.3	0	1.9	20.9
73	11	3.12	19.6	223.7	60.4	8.73	0	0.06	—	0.23	2.27	0.24	5.1	0	2.1	22.7
49	5	2.16	16.2	200.7	32.3	4.28	0	0.05	0.30	0.15	4.69	0.34	8.5	0	1.4	28.0
54	5	1.98	15.3	193.1	31.5	3.95	0	0.05	0.34	0.13	4.37	0.31	7.7	0	1.5	25.2
58	7	2.56	17.9	233.9	57.8	3.56	0	0.07	0.18	0.17	3.29	0.27	6.0	0	1.8	10.0
50	5	3.11	22.1	278.1	65.5	4.34	0	0.09	0.11	0.21	3.93	0.33	6.8	0	1.9	8.5
67	8	1.88	19.6	294.3	53.6	3.85	0	0.06	—	0.15	3.96	0.31	6.0	0	1.6	19.5
324	5	5.24	18.7	298.5	65.5	4.44	6586.3	0.15	0.39	2.91	14.86	0.87	221.1	0.6	70.7	27.9
112	4	2.22	12.8	156.5	55.3	3.47	0	0.01	0.25	0.25	2.96	0.13	6.0	1.1	2.7	11.2
85	17	1.53	20.4	278.1	65.5	2.96	0	0.08	0.11	0.21	6.03	0.11	15.3	0	2.1	23.3
82	14	1.59	19.6	263.7	61.2	3.79	0	0.08	0.11	0.21	5.66	0.11	15.3	0	2.2	22.5
84	16	1.59	19.6	229.5	64.6	3.40	0	0.07	0.10	0.18	5.95	0.09	11.9	0	2.2	20.3
77	14	1.87	24.7	266.1	72.3	4.47	0	0.08	0.15	0.21	5.56	0.12	17.9	0	2.2	26.4
78	17	1.67	19.6	213.4	56.1	4.44	0	0.07	0.11	0.20	5.22	0.11	17.9	0	2.2	22.3
74	16	1.81	21.3	225.3	57.8	5.13	0	0.07	0.15	0.22	4.89	0.12	21.3	0	2.3	24.2
102	21	2.03	22.1	260.3	61.2	5.17	0	0.06	0.12	0.21	5.66	0.09	15.3	0	2.2	31.6
103	22	2.29	24.7	287.5	64.6	6.20	0	0.06	0.15	0.23	5.38	0.11	18.7	0	2.3	32.1
2130	18	1.73	18.7	304.5	133.5	1.70	0	0.14	—	0.31	3.87	0.19	6.0	19.6	20.5	10.2
161	9	2.23	13.6	134.4	57.0	2.54	0	0.06	—	0.35	3.13	0.14	2.6	6.0	5.4	23.8
28	5	0.38	9.6	195.0	798.9	0.78	0	0.42	0.11	0.09	3.53	0.22	2.3	0	0.4	14.2
18	2	0.22	5.3	90.4	369.6	0.56	1.8	0.06	0.04	0.04	1.76	0.04	0.3	0	0.2	9.9

DA+ Code	Food Description	Quantity	Measure	Wt (g)	H₂O (g)	Ener (kcal)	Prot (g)	Carb (g)	Fiber (g)	Fat (g)	Sat	Mono	Poly	Trans
35422	Breakfast strips, cured, cooked	3	slice(s)	34	9.2	156	9.8	0.4	0	12.5	4.3	5.6	1.9	—
189	Ham, cured, boneless, 11% fat, roasted	3	ounce(s)	85	54.9	151	19.2	0	0	7.7	2.7	3.8	1.2	—
29215	Ham, cured, extra lean, 4% fat, canned	2	2 ounce(s)	57	41.7	68	10.5	0	0	2.6	0.9	1.3	0.2	—
1316	Ham, cured, extra lean, 5% fat, roasted	3	ounce(s)	85	57.6	123	17.8	1.3	0	4.7	1.5	2.2	0.5	—
16561	Ham, smoked or cured, lean, cooked	1	slice(s)	42	27.6	66	10.5	0	0	2.3	0.8	1.1	0.3	—
	Pork chop													
32671	Loin, blade, chops, lean and fat, pan fried	3	ounce(s)	85	42.5	291	18.3	0	0	23.6	8.6	10	2.6	—
32672	Loin, center cut, chops, lean and fat, pan fried	3	ounce(s)	85	45.1	236	25.4	0	0	14.1	5.1	6.0	1.6	—
32682	Loin, center rib, chops, boneless, lean and fat, braised	3	ounce(s)	85	49.5	217	22.4	0	0	13.4	5.2	6.1	1.1	—
32603	Loin, center rib, chops, lean, broiled	3	ounce(s)	85	55.4	158	21.9	0	0	7.1	2.4	3.0	0.8	0.1
32478	Loin, whole, lean and fat, braised	3	ounce(s)	85	49.6	203	23.2	0	0	11.6	4.3	5.2	1.0	—
32481	Loin, whole, lean, braised	3	ounce(s)	85	52.2	174	24.3	0	0	7.8	2.9	3.5	0.6	—
	Pork leg or ham													
32471	Pork leg or ham, Rump portion, lean and fat, roasted	3	ounce(s)	85	48.3	214	24.6	0	0	12.1	4.5	5.4	1.2	—
32468	Pork leg or ham, Whole, lean and fat, roasted	3	ounce(s)	85	46.8	232	22.8	0	0	15.0	5.5	6.7	1.4	—
	Pork ribs													
32693	Loin, country style, lean and fat, roasted	3	ounce(s)	85	43.3	279	19.9	0	0	21.6	7.8	9.4	1.7	—
32696	Loin, country style, lean, roasted	3	ounce(s)	85	49.5	210	22.6	0	0	12.6	4.5	5.5	0.9	—
	Pork shoulder													
32626	Shoulder, arm picnic, lean and fat, roasted	3	ounce(s)	85	44.3	270	20.0	0	0	20.4	7.5	9.1	2.0	—
32629	Shoulder, arm picnic, lean, roasted	3	ounce(s)	85	51.3	194	22.7	0	0	10.7	3.7	5.1	1.0	—
	Rabbit													
3366	Domesticated, roasted	3	ounce(s)	85	51.5	168	24.7	0	0	6.8	2.0	1.8	1.3	—
3367	Domesticated, stewed	3	ounce(s)	85	50.0	175	25.8	0	0	7.2	2.1	1.9	1.4	—
	Veal													
3391	Liver, braised	3	ounce(s)	85	50.9	163	24.2	3.2	0	5.3	1.7	1.0	0.9	0.3
3319	Rib, lean only, roasted	3	ounce(s)	85	55.0	151	21.9	0	0	6.3	1.8	2.3	0.6	—
1732	Deer or venison, roasted	3	ounce(s)	85	55.5	134	25.7	0	0	2.7	1.1	0.7	0.5	—
	POULTRY													
	Chicken													
29562	Flaked, canned	2	ounce(s)	57	39.3	97	10.3	0.1	0	5.8	1.6	2.3	1.3	—
	Chicken, fried													
29632	Breast, meat only, breaded, baked or fried	3	ounce(s)	85	44.3	193	25.3	6.9	0.2	6.6	1.6	2.7	1.7	—
35327	Broiler breast, meat only, fried	3	ounce(s)	85	51.2	159	28.4	0.4	0	4.0	1.1	1.5	0.9	—
36413	Broiler breast, meat and skin, flour coated, fried	3	ounce(s)	85	48.1	189	27.1	1.4	0.1	7.5	2.1	3.0	1.7	—
36414	Broiler drumstick, meat and skin, flour coated, fried	3	ounce(s)	85	48.2	208	22.9	1.4	0.1	11.7	3.1	4.6	2.7	—
35389	Broiler drumstick, meat only, fried	3	ounce(s)	85	52.9	166	24.3	0	0	6.9	1.8	2.5	1.7	—
35406	Broiler leg, meat only, fried	3	ounce(s)	85	51.5	177	24.1	0.6	0	7.9	2.1	2.9	1.9	—
35484	Broiler wing, meat only, fried	3	ounce(s)	85	50.9	179	25.6	0	0	7.8	2.1	2.6	1.8	—
29580	Patty, fillet or tenders, breaded, cooked	3	ounce(s)	85	40.2	256	14.5	12.2	0	16.5	3.7	8.4	3.7	—
	Chicken, roasted, meat only													
35409	Broiler leg, meat only, roasted	3	ounce(s)	85	55.0	162	23.0	0	0	7.2	1.9	2.6	1.7	—
35486	Broiler wing, meat only, roasted	3	ounce(s)	85	53.4	173	25.9	0	0	6.9	1.9	2.2	1.5	—
35138	Roasting chicken, dark meat, meat only, roasted	3	ounce(s)	85	57.0	151	19.8	0	0	7.4	2.1	2.8	1.7	—
35136	Roasting chicken, light meat, meat only, roasted	3	ounce(s)	85	57.7	130	23.1	0	0	3.5	0.9	1.3	0.8	—
35132	Roasting chicken, meat only, roasted	3	ounce(s)	85	57.3	142	21.3	0	0	5.6	1.5	2.1	1.3	—
	Chicken, stewed													
1268	Gizzard, simmered	3	ounce(s)	85	57.8	124	25.8	0	0	2.3	0.6	0.4	0.3	0.1
1270	Liver, simmered	3	ounce(s)	85	56.8	142	20.8	0.7	0	5.5	1.8	1.2	1.7	0.1
3174	Meat only, stewed	3	ounce(s)	85	56.8	151	23.2	0	0	5.7	1.6	2.0	1.3	—
	Duck													
1286	Domesticated, meat and skin, roasted	3	ounce(s)	85	44.1	287	16.2	0	0	24.1	8.2	11.0	3.1	—
1287	Domesticated, meat only, roasted	3	ounce(s)	85	54.6	171	20	0	0	9.5	3.5	3.1	1.2	—
	Goose													
35507	Domesticated, meat and skin, roasted	3	ounce(s)	85	44.2	259	21.4	0	0	18.6	5.8	8.7	2.1	—
35524	Domesticated, meat only, roasted	3	ounce(s)	85	48.7	202	24.6	0	0	10.8	3.9	3.7	1.3	—
1297	Liver pate, smoked, canned	4	tablespoon(s)	52	19.3	240	5.9	2.4	0	22.8	7.5	13.3	0.4	—
	Turkey													
3256	Ground turkey, cooked	3	ounce(s)	85	50.5	200	23.3	0	0	11.2	2.9	4.2	2.7	—
3263	Patty, batter coated, breaded, fried	1	item(s)	94	46.7	266	13.2	14.8	0.5	16.9	4.4	7.0	4.4	—
219	Roasted, dark meat, meat only	3	ounce(s)	85	53.7	159	24.3	0	0	6.1	2.1	1.4	1.8	—
222	Roasted, fryer roaster breast, meat only	3	ounce(s)	85	58.2	115	25.6	0	0	0.6	0.2	0.1	0.2	—
220	Roasted, light meat, meat only	3	ounce(s)	85	56.4	134	25.4	0	0	2.7	0.9	0.5	0.7	—
1303	Turkey roll, light and dark meat	2	slice(s)	57	39.8	84	10.3	1.2	0	4.0	1.2	1.3	1.0	—
1302	Turkey roll, light meat	2	slice(s)	57	42.5	56	8.4	2.9	0	0.9	0.2	0.2	0.1	0
	PROCESSED MEATS													
	Beef													
1331	Corned beef loaf, jellied, sliced	2	slice(s)	57	39.2	87	13.0	0	0	3.5	1.5	1.5	0.2	—

PROCESSED MEATS —Continued

Chol (mg)	Calc (mg)	Iron (mg)	Magn (mg)	Pota (mg)	Sodi (mg)	Zinc (mg)	Vit A (µg)	Thia (mg)	Vit E (mg α)	Ribo (mg)	Niac (mg)	Vit B$_6$ (mg)	Fola (µg)	Vit C (mg)	Vit B$_{12}$ (µg)	Sele (µg)
36	5	0.67	8.8	158.4	713.7	1.25	0	0.25	0.08	0.12	2.58	0.11	1.4	0	0.6	8.4
50	7	1.13	18.7	347.7	1275.0	2.09	0	0.62	0.26	0.28	5.22	0.26	2.6	0	0.6	16.8
22	3	0.53	9.6	206.4	711.6	1.09	0	0.47	0.09	0.13	3.00	0.25	3.4	0	0.5	8.2
45	7	1.25	11.9	244.1	1023.1	2.44	0	0.64	0.21	0.17	3.42	0.34	2.6	0	0.6	16.6
23	3	0.39	9.2	132.7	557.3	1.07	0	0.28	0.10	0.10	2.10	0.19	1.7	0	0.3	10.7
72	26	0.74	17.9	282.4	57.0	2.71	1.7	0.52	0.17	0.25	3.35	0.28	3.4	0.5	0.7	29.7
78	23	0.77	24.7	361.5	68.0	1.96	1.7	0.96	0.21	0.25	4.76	0.39	5.1	0.9	0.6	33.2
62	4	0.78	14.5	329.1	34.0	1.76	1.7	0.44	—	0.20	3.66	0.26	3.4	0.3	0.4	28.4
56	22	0.57	21.3	291.7	48.5	1.91	0	0.48	0.08	0.18	6.68	0.57	0	0	0.4	38.6
68	18	0.91	16.2	318.1	40.8	2.02	1.7	0.53	0.20	0.21	3.75	0.31	2.6	0.5	0.5	38.5
67	15	0.96	17.0	329.1	42.5	2.10	1.7	0.56	0.17	0.22	3.90	0.32	3.4	0.5	0.5	41.0
82	10	0.89	23.0	318.1	52.7	2.39	2.6	0.63	0.18	0.28	3.95	0.26	2.6	0.2	0.6	39.8
80	12	0.85	18.7	299.4	51.0	2.51	2.6	0.54	0.18	0.26	3.89	0.34	8.5	0.3	0.6	38.5
78	21	0.90	19.6	292.6	44.2	2.00	2.6	0.75	—	0.29	3.67	0.37	4.3	0.3	0.7	31.6
79	25	1.09	20.4	296.8	24.7	3.24	1.7	0.48	—	0.29	3.96	0.37	4.3	0.3	0.7	36.0
80	16	1.00	14.5	276.4	59.5	2.93	1.7	0.44	—	0.25	3.33	0.29	3.4	0.2	0.6	28.6
81	8	1.20	17.0	298.5	68.0	3.46	1.7	0.49	—	0.30	3.66	0.34	4.3	0.3	0.7	32.7
70	16	1.93	17.9	325.7	40.0	1.93	0	0.07	—	0.17	7.17	0.40	9.4	0	7.1	32.7
73	17	2.01	17.0	255.1	31.5	2.01	0	0.05	0.37	0.14	6.09	0.28	7.7	0	5.5	32.7
435	5	4.34	17.0	279.8	66.3	9.55	17983.6	0.15	0.57	2.43	11.18	0.78	281.5	0.9	72.0	16.4
98	10	0.81	20.4	264.5	82.5	3.81	0	0.05	0.30	0.24	6.37	0.23	11.9	0	1.3	9.4
95	6	3.80	20.4	284.9	45.9	2.33	0	0.15	—	0.51	5.70	—	—	0	—	11.0
35	8	0.89	6.8	147.4	408.2	0.79	19.3	0.01	—	0.07	3.58	0.19	2.3	0	0.2	—
67	19	1.05	24.7	222.6	450.2	0.84	—	0.08	—	0.09	10.97	0.46	4.3	0	0.3	—
77	14	0.96	26.4	234.7	67.2	0.91	6.0	0.06	0.35	0.10	12.57	0.54	3.4	0	0.3	22.3
76	14	1.01	25.5	220.3	64.6	0.93	12.8	0.06	0.39	0.11	11.68	0.49	5.1	0	0.3	20.3
77	10	1.13	19.6	194.8	75.7	2.45	21.3	0.06	0.65	0.19	5.13	0.29	8.5	0	0.3	15.6
80	10	1.12	20.4	211.8	81.6	2.73	15.3	0.06	—	0.20	5.22	0.33	7.7	0	0.3	16.7
84	11	1.19	21.3	216.0	81.6	2.53	17.0	0.07	0.38	0.21	5.68	0.33	7.7	0	0.3	16.0
71	13	0.96	17.9	176.9	77.4	1.80	15.3	0.03	0.40	0.10	6.15	0.50	3.4	0	0.3	21.6
49	11	0.75	19.6	244.8	411.4	0.79	4.3	0.09	1.04	0.12	5.99	0.24	24.7	0	0.2	13.9
80	10	1.11	20.4	205.8	77.4	2.43	16.2	0.06	0.22	0.19	5.37	0.31	6.8	0	0.3	18.8
72	14	0.98	17.9	178.6	78.2	1.82	15.3	0.03	0.22	0.10	6.21	0.50	3.4	0	0.3	21.0
64	9	1.13	17.0	190.5	80.8	1.81	13.6	0.05	—	0.16	4.87	0.26	6.0	0	0.2	16.7
64	11	0.91	19.6	200.7	43.4	0.66	6.8	0.05	0.22	0.07	8.90	0.45	2.6	0	0.3	21.9
64	10	1.02	17.9	194.8	63.8	1.29	10.2	0.05	—	0.12	6.70	0.34	4.3	0	0.2	20.9
315	14	2.71	2.6	152.2	47.6	3.75	0	0.02	0.17	0.17	2.65	0.06	4.3	0	0.9	35.0
479	9	9.89	21.3	223.7	64.6	3.38	3385.8	0.24	0.69	1.69	9.39	0.64	491.6	23.7	14.3	70.1
71	12	0.99	17.9	153.1	59.5	1.69	12.8	0.04	0.22	0.13	5.20	0.22	5.1	0	0.2	17.8
71	9	2.29	13.6	173.5	50.2	1.58	53.6	0.14	0.59	0.22	4.10	0.15	5.1	0	0.3	17.0
76	10	2.29	17.0	214.3	55.3	2.21	19.6	0.22	0.59	0.39	4.33	0.21	8.5	0	0.3	19.1
77	11	2.40	18.7	279.8	59.5	2.22	17.9	0.06	1.47	0.27	3.54	0.31	1.7	0	0.3	18.5
82	12	2.44	21.3	330.0	64.6	2.69	10.2	0.07	—	0.33	3.47	0.39	10.2	0	0.4	21.7
78	36	2.86	6.8	71.8	362.4	0.47	520.5	0.04	—	0.15	1.30	0.03	31.2	0	4.9	22.9
87	21	1.64	20.4	229.6	91.0	2.43	0	0.04	0.28	0.14	4.09	0.33	6.0	0	0.3	31.6
71	13	2.06	14.1	258.5	752.0	1.35	9.4	0.09	0.87	0.17	2.16	0.18	38.5	0	0.2	20.8
72	27	1.98	20.4	246.6	67.2	3.79	0	0.05	0.54	0.21	3.10	0.30	7.7	0	0.3	34.8
71	10	1.30	24.7	248.3	44.2	1.48	0	0.03	0.07	0.11	6.37	0.47	5.1	0	0.3	27.3
59	16	1.14	23.8	259.4	54.4	1.73	0	0.05	0.07	0.11	5.81	0.45	5.1	0	0.3	27.3
31	18	0.76	10.2	153.1	332.3	1.13	0	0.05	0.19	0.16	2.72	0.15	2.8	0	0.1	16.6
19	4	0.21	10.8	242.1	590.8	0.50	0	0.01	0.07	0.08	4.05	0.23	2.3	0	0.2	7.4
27	6	1.15	6.2	57.3	540.4	2.31	0	0.00	—	0.06	0.99	0.06	4.5	0	0.7	9.8

Food Composition (Computer code is for Cengage Diet Analysis program) (For purposes of calculations, use "0" for t, <1, <.1, <.01, etc.)

DA+ Code	Food Description	Quantity	Measure	Wt (g)	H₂O (g)	Ener (kcal)	Prot (g)	Carb (g)	Fiber (g)	Fat (g)	Fat Breakdown (g) Sat	Mono	Poly	Trans
	Bologna													
13459	Beef	1	slice(s)	28	15.1	90	3.0	1.0	0	8.0	3.5	4.3	0.3	—
13461	Light, made with pork and chicken	1	slice(s)	28	18.2	60	3.0	2.0	0	4.0	1.0	2.0	0.4	—
13458	Made with chicken and pork	1	slice(s)	28	15.0	90	3.0	1.0	0	8.0	3.0	4.1	1.1	—
13565	Turkey bologna	1	slice(s)	28	19.0	50	3.0	1.0	0	4.0	1.0	1.1	1.0	0
	Chicken													
7125	Breast, smoked	1	slice(s)	10	—	10	1.8	0.3	0	0.2	0	—	—	—
	Ham													
7127	Deli-sliced, honey	1	slice(s)	10	—	10	1.7	0.3	0	0.3	0.1	—	—	—
7126	Deli-sliced, smoked	1	slice(s)	10	—	10	1.7	0.2	0	0.3	0.1	—	—	—
8614	**Beef & pork mortadella, sliced**	2	slice(s)	46	24.1	143	7.5	1.4	0	11.7	4.4	5.2	1.4	—
1323	**Pork olive loaf**	2	slice(s)	57	33.1	133	6.7	5.2	0	9.4	3.3	4.5	1.1	—
1324	**Pork pickle & pimento loaf**	2	slice(s)	57	34.2	128	6.4	4.8	0.9	9.1	3.0	4.0	1.6	—
	Sausages & frankfurters													
37296	Beerwurst beef, beer salami (bierwurst)	1	slice(s)	29	16.6	74	4.1	1.2	0	5.7	2.5	2.7	0.2	—
37257	Beerwurst pork beer salami	1	slice(s)	21	12.9	50	3.0	0.4	0	4.0	1.3	1.9	0.5	—
35338	Berliner, pork & beef	1	ounce(s)	28	17.3	65	4.3	0.7	0	4.9	1.7	2.3	0.4	—
37298	Bratwurst pork, cooked	1	piece(s)	74	42.3	181	10.4	1.9	0	14.3	5.1	6.7	1.5	—
37299	Braunschweiger pork liver sausage	1	slice(s)	15	8.2	51	2.0	0.3	0	4.5	1.5	2.1	0.5	—
1329	Cheesefurter or cheese smokie, beef & pork	1	item(s)	43	22.6	141	6.1	0.6	0	12.5	4.5	5.9	1.3	—
1330	Chorizo, beef & pork	2	ounce(s)	57	18.1	258	13.7	1.1	0	21.7	8.2	10.4	2.0	—
8600	Frankfurter, beef	1	item(s)	45	23.4	149	5.1	1.8	0	13.3	5.3	6.4	0.5	—
202	Frankfurter, beef & pork	1	item(s)	45	25.2	137	5.2	0.8	0	12.4	4.8	6.2	1.2	—
1293	Frankfurter, chicken	1	item(s)	45	28.1	100	7.0	1.2	0.2	7.3	1.7	2.7	1.7	0.1
3261	Frankfurter, turkey	1	item(s)	45	28.3	100	5.5	1.7	0	7.8	1.8	2.6	1.8	0.4
37275	Italian sausage, pork, cooked	1	item(s)	68	32.0	234	13.0	2.9	0.1	18.6	6.5	8.1	2.2	—
37307	Kielbasa, kolbassa, pork & beef	1	slice(s)	30	18.5	67	5.0	1.0	0	4.7	1.7	2.2	0.5	—
1333	Knockwurst or knackwurst, beef & pork	2	ounce(s)	57	31.4	174	6.3	1.8	0	15.7	5.8	7.3	1.7	—
37285	Pepperoni, beef & pork	1	slice(s)	11	3.4	51	2.2	0.4	0.2	4.4	1.8	2.1	0.3	—
37313	Polish sausage, pork	1	slice(s)	21	11.4	60	2.8	0.7	0	5.0	1.8	2.3	0.5	—
206	Salami, beef, cooked, sliced	2	slice(s)	52	31.2	136	6.5	1.0	0	11.5	5.1	5.5	0.5	—
37272	Salami, pork, dry or hard	1	slice(s)	13	4.6	52	2.9	0.2	0	4.3	1.5	2.0	0.5	—
40987	Sausage, turkey, cooked	2	ounce(s)	57	36.9	111	13.5	0	0	5.9	1.3	1.7	1.5	0.2
8620	Smoked sausage, beef & pork	2	ounce(s)	57	30.6	181	6.8	1.4	0	16.3	5.5	6.9	2.2	0
8619	Smoked sausage, pork	2	ounce(s)	57	32.0	178	6.8	1.2	0	16.0	5.3	6.4	2.1	0.1
37273	Smoked sausage, pork link	1	piece(s)	76	29.8	295	16.8	1.6	0	24.0	8.6	11.1	2.8	—
1336	Summer sausage, thuringer, or cervelat, beef & pork	2	ounce(s)	57	25.6	205	9.9	1.9	0	17.3	6.5	7.4	0.7	—
37294	Vienna sausage, cocktail, beef & pork, canned	1	piece(s)	16	10.4	37	1.7	0.4	0	3.1	1.1	1.5	0.2	—
	Spreads													
1318	Ham salad spread	¼	cup(s)	60	37.6	130	5.2	6.4	0	9.3	3.0	4.3	1.6	—
32419	Pork and beef sandwich spread	4	tablespoon(s)	60	36.2	141	4.6	7.2	0.1	10.4	3.6	4.6	1.5	—
	Turkey													
13604	Breast, fat free, oven roasted	1	slice(s)	28	—	25	4.0	1.0	0	0	0	0	0	0
13606	Breast, hickory smoked fat free	1	slice(s)	28	—	25	4.0	1.0	0	0	0	0	0	0
16049	Breast, hickory smoked slices	1	slice(s)	56	—	50	11.0	1.0	0	0	0	0	0	0
16047	Breast, honey roasted slices	1	slice(s)	56	—	60	11.0	3.0	0	0	0	0	0	0
16048	Breast, oven roasted slices	1	slice(s)	56	—	50	11.0	1.0	0	0	0	0	0	0
7124	Breast, oven-roasted	1	slice(s)	10	—	10	1.8	0.3	0	0.1	0	0	0	0
13567	Turkey ham, 10% water added	2	slice(s)	56	40.9	70	10.0	2.0	0	3.0	0	0.4	0.6	0
37270	Turkey pastrami	1	slice(s)	28	20.3	35	4.6	1.0	0	1.2	0.3	0.4	0.3	—
3262	Turkey salami	2	slice(s)	57	39.1	98	10.9	0.9	0.1	5.2	1.6	1.8	1.4	0
37318	Turkey salami, cooked	1	slice(s)	28	20.4	43	4.3	0.1	0	2.7	0.8	0.9	0.7	—
	BEVERAGES													
	Beer													
866	Ale, mild	12	fluid ounce(s)	360	332.3	148	1.1	13.3	0.4	0	0	0	0	0
686	Beer	12	fluid ounce(s)	356	327.7	153	1.6	12.7	0	0	0	0	0	0
16886	Beer, non alcoholic	12	fluid ounce(s)	360	328.1	133	0.8	29.0	0	0.4	0.1	0	0.2	0
31609	Bud Light beer	12	fluid ounce(s)	355	335.5	110	0.9	6.6	0	0	0	0	0	0
31608	Budweiser beer	12	fluid ounce(s)	355	327.7	145	1.3	10.6	0	0	0	0	0	0
869	Light beer	12	fluid ounce(s)	354	335.9	103	0.9	5.8	0	0	0	0	0	0
31613	Michelob Beer	12	fluid ounce(s)	355	323.4	155	1.3	13.3	0	0	0	0	0	0
31614	Michelob Light beer	12	fluid ounce(s)	355	329.8	134	1.1	11.7	0	0	0	0	0	0
	Gin, rum, vodka, whiskey													
857	Distilled alcohol, 100 proof	1	fluid ounce(s)	28	16.0	82	0	0	0	0	0	0	0	0
687	Distilled alcohol, 80 proof	1	fluid ounce(s)	28	18.5	64	0	0	0	0	0	0	0	0
688	Distilled alcohol, 86 proof	1	fluid ounce(s)	28	17.8	70	0	0	0	0	0	0	0	0
689	Distilled alcohol, 90 proof	1	fluid ounce(s)	28	17.3	73	0	0	0	0	0	0	0	0
856	Distilled alcohol, 94 proof	1	fluid ounce(s)	28	16.8	76	0	0	0	0	0	0	0	0

BEVERAGES —Continued

Chol (mg)	Calc (mg)	Iron (mg)	Magn (mg)	Pota (mg)	Sodi (mg)	Zinc (mg)	Vit A (µg)	Thia (mg)	Vit E (mg α)	Ribo (mg)	Niac (mg)	Vit B6 (mg)	Fola (µg)	Vit C (mg)	Vit B12 (µg)	Sele (µg)
20	0	0.36	3.9	47.0	310.0	0.56	0	0.01	—	0.03	0.67	0.04	3.6	0	0.4	—
20	40	0.36	5.6	45.6	300.0	0.45	0	—	—	—	—	—	—	0	—	—
30	20	0.36	5.9	43.1	300.0	0.39	0	—	—	—	—	—	—	0	—	—
20	40	0.36	6.2	42.6	270.0	0.51	0	—	—	—	—	—	—	0	—	—
4	0	0.00	—	—	100.0	—	0	—	—	—	—	—	—	0	—	—
4	0	0.12	—	—	100.0	—	0	—	—	—	—	—	—	0.6	—	—
4	0	0.12	—	—	103.3	—	0	—	—	—	—	—	—	0.6	—	—
26	8	0.64	5.1	75.0	573.2	0.96	0	0.05	0.10	0.07	1.23	0.06	1.4	0	0.7	10.4
22	62	0.30	10.8	168.7	842.9	0.78	34.1	0.16	0.14	0.14	1.04	0.13	1.1	0	0.7	9.3
33	62	0.75	19.3	210.7	740.7	0.95	44.3	0.22	0.22	0.06	1.41	0.23	21.0	4.4	0.3	4.5
18	3	0.44	3.5	66.5	264.9	0.71	0	0.02	0.05	0.03	0.98	0.04	0.9	0	0.6	4.7
12	2	0.15	2.7	53.3	261.0	0.36	0	0.11	0.03	0.04	0.68	0.07	0.6	0	0.2	4.4
13	3	0.32	4.3	80.2	367.7	0.70	0	0.10	—	0.06	0.88	0.05	1.4	0	0.8	4.0
44	33	0.95	11.1	156.9	412.2	1.70	0	0.37	0.01	0.13	2.36	0.15	1.5	0.7	0.7	15.7
24	1	1.42	1.7	27.5	131.5	0.42	641.0	0.03	0.05	0.23	1.27	0.05	6.7	0	3.1	8.8
29	25	0.46	5.6	88.6	465.3	0.96	20.2	0.10	0.10	0.06	1.24	0.05	1.3	0	0.7	6.8
50	5	0.90	10.2	225.7	700.2	1.93	0	0.35	0.12	0.17	2.90	0.30	1.1	0	1.1	12.0
24	6	0.67	6.3	70.2	513.0	1.10	0	0.01	0.09	0.06	1.06	0.04	2.3	0	0.8	3.7
23	5	0.51	4.5	75.2	504.0	0.82	8.1	0.09	0.11	0.05	1.18	0.05	1.8	0	0.6	6.2
43	33	0.52	9.0	90.9	379.8	0.50	0	0.02	0.09	0.11	2.10	0.14	3.2	0	0.2	10.4
35	67	0.66	6.3	176.4	485.1	0.82	0	0.01	0.27	0.08	1.65	0.06	4.1	0	0.4	6.8
39	14	0.97	12.2	206.7	820.8	1.62	6.8	0.42	0.17	0.15	2.83	0.22	3.4	0.1	0.9	15.0
20	13	0.44	4.9	84.4	283.0	0.61	0	0.06	0.06	0.06	0.87	0.05	1.5	0	0.5	5.4
34	6	0.37	6.2	112.8	527.3	0.94	0	0.19	0.32	0.07	1.55	0.09	1.1	0	0.7	7.7
13	2	0.15	2.0	34.7	196.7	0.30	0	0.05	0.00	0.02	0.59	0.04	0.7	0.1	0.2	2.4
15	2	0.29	2.9	37.3	199.3	0.40	0	0.10	0.04	0.03	0.71	0.03	0.4	0.2	0.2	3.7
37	3	1.14	6.8	97.8	592.8	0.92	0	0.04	0.08	0.08	1.68	0.08	1.0	0	1.6	7.6
10	2	0.16	2.8	48.4	289.3	0.53	0	0.11	0.02	0.04	0.71	0.07	0.3	0	0.4	3.3
52	12	0.84	11.9	169.0	377.1	2.19	7.4	0.04	0.10	0.14	3.24	0.18	3.4	0.4	0.7	0
33	7	0.42	7.4	101.5	516.5	0.71	7.4	0.10	0.07	0.06	1.66	0.09	1.1	0	0.3	0
35	6	0.33	6.2	273.9	468.9	0.74	0	0.12	0.14	0.10	1.59	0.10	0.6	0	0.4	10.4
52	23	0.87	14.4	254.6	1136.6	2.13	0	0.53	0.18	0.19	3.43	0.26	3.8	1.5	1.2	16.4
42	5	1.15	7.9	147.4	737.1	1.45	0	0.08	0.12	0.18	2.44	0.14	1.1	9.4	3.1	11.5
14	2	0.14	1.1	16.2	155.0	0.25	0	0.01	0.03	0.01	0.25	0.01	0.6	0	0.2	2.7
22	5	0.35	6.0	90.0	547.2	0.66	0	0.26	1.04	0.07	1.25	0.09	0.6	0	0.5	10.7
23	7	0.47	4.8	66.0	607.8	0.61	15.6	0.10	1.04	0.08	1.03	0.07	1.2	0	0.7	5.8
10	0	0.00	—	—	340.0	—	0	—	—	—	—	—	—	0	—	—
10	0	0.00	—	—	300.0	—	0	—	—	—	—	—	—	0	—	—
25	0	0.72	—	—	720.0	—	0	—	—	—	—	—	—	0	—	—
20	0	0.72	—	—	660.0	—	0	—	—	—	—	—	—	0	—	—
20	0	0.72	—	—	660.0	—	0	—	—	—	—	—	—	0	—	—
4	0	0.06	—	—	103.3	—	0	—	—	—	—	—	—	0	—	—
40	0	0.72	12.3	162.4	700.0	1.44	0	—	—	—	—	—	—	0	—	—
19	3	1.19	4.0	97.8	278.1	0.61	1.1	0.01	0.06	0.07	1.00	0.07	1.4	4.6	0.1	4.6
43	23	0.70	12.5	122.5	569.3	1.31	1.1	0.24	0.13	0.17	2.25	0.24	5.7	0	0.6	15.0
22	11	0.35	6.2	61.2	284.6	0.65	0.6	0.12	0.06	0.08	1.12	0.12	2.8	0	0.3	7.5
0	18	0.07	21.6	90.0	14.4	0.03	0	0.03	0.00	0.10	1.62	0.18	21.6	0	0.1	2.5
0	14	0.07	21.4	96.2	14.3	0.03	0	0.01	0.00	0.08	1.82	0.16	21.4	0	0.1	2.1
0	25	0.21	25.2	28.8	46.8	0.07	—	0.07	0.00	0.18	3.99	0.10	50.4	1.8	0.1	4.3
0	18	0.14	17.8	63.9	9.0	0.10	0	0.03	—	0.10	1.39	0.12	14.6	0	0	4.0
0	18	0.10	21.3	88.8	9.0	0.07	0	0.02	—	0.09	1.60	0.17	21.3	0	0.1	4.0
0	14	0.10	17.7	74.3	14.2	0.03	0	0.01	0.00	0.05	1.38	0.12	21.2	0	0.1	1.4
0	18	0.10	21.3	88.8	9.0	0.07	0	0.02	—	0.09	1.60	0.17	21.3	0	0.1	4.0
0	18	0.14	17.8	63.9	9.0	0.10	0	0.03	—	0.10	1.39	0.12	14.6	0	0	4.0
0	0	0.01	0	0.6	0.3	0.01	0	0.00	—	0.00	0.00	0.00	0	0	0	0
0	0	0.01	0	0.6	0.3	0.01	0	0.00	0.00	0.00	0.00	0.00	0	0	0	0
0	0	0.01	0	0.6	0.3	0.01	0	0.00	0.00	0.00	0.00	0.00	0	0	0	0
0	0	0.01	0	0.6	0.3	0.01	0	0.00	0.00	0.00	0.00	0.00	0	0	0	0
0	0	0.01	0	0.6	0.3	0.01	0	0.00	—	0.00	0.00	0.00	0	0	0	0

DA+ Code	Food Description	Quantity	Measure	Wt (g)	H₂O (g)	Ener (kcal)	Prot (g)	Carb (g)	Fiber (g)	Fat (g)	Sat	Mono	Poly	Trans
											Fat Breakdown (g)			

DA+ Code	Food Description	Quantity	Measure	Wt (g)	H₂O (g)	Ener (kcal)	Prot (g)	Carb (g)	Fiber (g)	Fat (g)	Sat	Mono	Poly	Trans
	Liqueurs													
33187	Coffee liqueur, 53 proof	1	fluid ounce(s)	35	10.8	113	0	16.3	0	0.1	0	0	0	—
3142	Coffee liqueur, 63 proof	1	fluid ounce(s)	35	14.4	107	0	11.2	0	0.1	0	0	0	—
736	Cordials, 54 proof	1	fluid ounce(s)	30	8.9	106	0	13.3	0	0.1	0	0	0	—
	Wine													
861	California red wine	5	fluid ounce(s)	150	133.4	125	0.3	3.7	0	0	0	0	0	0
858	Domestic champagne	5	fluid ounce(s)	150	—	105	0.3	3.8	0	0	0	0	0	0
690	Sweet dessert wine	5	fluid ounce(s)	147	103.7	235	0.3	20.1	0	0	0	0	0	0
1481	White wine	5	fluid ounce(s)	148	128.1	121	0.1	3.8	0	0	0	0	0	0
1811	Wine cooler	10	fluid ounce(s)	300	267.4	159	0.3	20.2	0	0.1	0	0	0	0
	Carbonated													
31898	7 Up	12	fluid ounce(s)	360	321.0	140	0	39.0	0	0	0	0	0	0
692	Club soda	12	fluid ounce(s)	355	354.8	0	0	0	0	0	0	0	0	0
12010	Coca-Cola Classic cola soda	12	fluid ounce(s)	360	319.4	146	0	40.5	0	0	0	0	0	0
693	Cola	12	fluid ounce(s)	368	332.7	136	0.3	35.2	0	0.1	0	0	0	—
2391	Cola or pepper-type soda, low calorie with saccharin	12	fluid ounce(s)	355	354.5	0	0	0.3	0	0	0	0	0	0
9522	Cola soda, decaffeinated	12	fluid ounce(s)	372	333.4	153	0	39.3	0	0	0	0	0	0
9524	Cola, decaffeinated, low calorie with aspartame	12	fluid ounce(s)	355	354.3	4	0.4	0.5	0	0	0	0	0	0
1415	Cola, low calorie with aspartame	12	fluid ounce(s)	355	353.6	7	0.4	1.0	0	0.1	0	0	0	—
1412	Cream soda	12	fluid ounce(s)	371	321.5	189	0	49.3	0	0	0	0	0	0
31899	Diet 7 Up	12	fluid ounce(s)	360	—	0	0	0	0	0	0	0	0	0
12031	Diet Coke cola soda	12	fluid ounce(s)	360	—	2	0	0.2	0	0	0	0	0	0
29392	Diet Mountain Dew soda	12	fluid ounce(s)	360	—	0	0	0	0	0	0	0	0	0
29389	Diet Pepsi cola soda	12	fluid ounce(s)	360	—	0	0	0	0	0	0	0	0	0
12034	Diet Sprite soda	12	fluid ounce(s)	360	—	4	0	0	0	0	0	0	0	0
695	Ginger ale	12	fluid ounce(s)	366	333.9	124	0	32.1	0	0	0	0	0	0
694	Grape soda	12	fluid ounce(s)	372	330.3	160	0	41.7	0	0	0	0	0	0
1876	Lemon lime soda	12	fluid ounce(s)	368	330.8	147	0.2	37.4	0	0.1	0	0	0	—
29391	Mountain Dew soda	12	fluid ounce(s)	360	314.0	170	0	46.0	0	0	0	0	0	0
3145	Orange soda	12	fluid ounce(s)	372	325.9	179	0	45.8	0	0	0	0	0	0
1414	Pepper-type soda	12	fluid ounce(s)	368	329.3	151	0	38.3	0	0.4	0.3	0	0	—
29388	Pepsi regular cola soda	12	fluid ounce(s)	360	318.9	150	0	41.0	0	0	0	0	0	0
696	Root beer	12	fluid ounce(s)	370	330.0	152	0	39.2	0	0	0	0	0	0
12044	Sprite soda	12	fluid ounce(s)	360	321.0	144	0	39.0	0	0	0	0	0	0
	Coffee													
731	Brewed	8	fluid ounce(s)	237	235.6	2	0.3	0	0	0	0	0	0	0
9520	Brewed, decaffeinated	8	fluid ounce(s)	237	234.3	5	0.3	1.0	0	0	0	0	0	0
16882	Cappuccino	8	fluid ounce(s)	240	224.8	79	4.1	5.8	0.2	4.9	2.3	1.0	0.2	—
16883	Cappuccino, decaffeinated	8	fluid ounce(s)	240	224.8	79	4.1	5.8	0.2	4.9	2.3	1.0	0.2	—
16880	Espresso	8	fluid ounce(s)	237	231.8	21	0	3.6	0	0.4	0.2	0	0.2	0
16881	Espresso, decaffeinated	8	fluid ounce(s)	237	231.8	21	0	3.6	0	0.4	0.2	0	0.2	0
732	Instant, prepared	8	fluid ounce(s)	239	236.5	5	0.2	0.8	0	0	0	0	0	0
	Fruit drinks													
29357	Crystal Light sugar free lemonade drink	8	fluid ounce(s)	240	—	5	0	0	0	0	0	0	0	0
6012	Fruit punch drink with added vitamin C, canned	8	fluid ounce(s)	248	218.2	117	0	29.7	0.5	0	0	0	0	0
31143	Gatorade Thirst Quencher, all flavors	8	fluid ounce(s)	240	—	50	0	14.0	0	0	0	0	0	0
260	Grape drink, canned	8	fluid ounce(s)	250	210.5	153	0	39.4	0	0	0	0	0	0
17372	Kool-Aid (lemonade/punch/fruit drink)	8	fluid ounce(s)	248	220.0	108	0.1	27.8	0.2	0	0	0	0	—
17225	Kool-Aid sugar free, low calorie tropical punch drink mix, prepared	8	fluid ounce(s)	240	—	5	0	0	0	0	0	0	0	0
266	Lemonade, prepared from frozen concentrate	8	fluid ounce(s)	248	221.6	99	0.2	25.8	0	0.1	0	0	0	—
268	Limeade, prepared from frozen concentrate	8	fluid ounce(s)	247	212.6	128	0	34.1	0	0	0	0	0	—
14266	Odwalla strawberry 'C' monster smoothie blend	8	fluid ounce(s)	240	—	160	2.0	38.0	0	0	0	0	0	0
10080	Odwalla strawberry lemonade quencher	8	fluid ounce(s)	240	—	110	0	28.0	0	0	0	0	0	0
10099	Snapple fruit punch fruit drink	8	fluid ounce(s)	240	—	110	0	29.0	0	0	0	0	0	0
10096	Snapple kiwi strawberry fruit drink	8	fluid ounce(s)	240	211.2	110	0	28.0	0	0	0	0	0	0
	Slim Fast ready-to-drink shake													
16054	French vanilla ready to drink shake	11	fluid ounce(s)	325	—	220	10.0	40.0	5.0	2.5	0.5	1.5	0.5	—
40447	Optima rich chocolate royal ready-to-drink shake	11	fluid ounce(s)	330	—	180	10.0	24.0	5.0	5.0	1.0	3.5	0.5	0
16055	Strawberries n cream ready to drink shake	11	fluid ounce(s)	325	—	220	10.0	40.0	5.0	2.5	0.5	1.5	0.5	—
	Tea													
33179	Decaffeinated, prepared	8	fluid ounce(s)	237	236.3	2	0	0.7	0	0	0	0	0	0
1877	Herbal, prepared	8	fluid ounce(s)	237	236.1	2	0	0.5	0	0	0	0	0	0
735	Instant tea mix, lemon flavored with sugar, prepared	8	fluid ounce(s)	259	236.2	91	0	22.3	0.3	0.2	0	0	0	—
734	Instant tea mix, unsweetened, prepared	8	fluid ounce(s)	237	236.1	2	0.1	0.4	0	0	0	0	0	0
733	Tea, prepared	8	fluid ounce(s)	237	236.3	2	0	0.7	0	0	0	0	0	0
	Water													
1413	Mineral water, carbonated	8	fluid ounce(s)	237	236.8	0	0	0	0	0	0	0	0	0

BEVERAGES —Continued

Chol (mg)	Calc (mg)	Iron (mg)	Magn (mg)	Pota (mg)	Sodi (mg)	Zinc (mg)	Vit A (µg)	Thia (mg)	Vit E (mg α)	Ribo (mg)	Niac (mg)	Vit B$_6$ (mg)	Fola (µg)	Vit C (mg)	Vit B$_{12}$ (µg)	Sele (µg)
0	0	0.02	1.0	10.4	2.8	0.01	0	0.00	0.00	0.00	0.05	0.00	0	0	0	0.1
0	0	0.02	1.0	10.4	2.8	0.01	0	0.00	—	0.00	0.05	0.00	0	0	0	0.1
0	0	0.02	0.6	4.5	2.1	0.01	0	0.00	0.00	0.00	0.02	0.00	0	0	0	0.1
0	12	1.43	16.2	170.6	15.0	0.14	0	0.01	0.00	0.04	0.11	0.05	1.5	0	0	—
0	—	—	—	—	—	—	—	—	—	—	—	—	—	—	0	—
0	12	0.34	13.2	135.4	13.2	0.10	0	0.01	0.00	0.01	0.30	0.00	0	0	0	0.7
0	13	0.39	14.8	104.7	7.4	0.18	0	0.01	0.00	0.01	0.15	0.06	1.5	0	0	0.1
0	18	0.75	15.0	129.0	24.0	0.18	—	0.01	0.03	0.03	0.13	0.03	3.0	5.4	0	0.6
0	—	—	—	0.6	75.0	—	—	—	—	—	—	—	—	—	—	—
0	18	0.03	3.5	7.1	74.6	0.35	0	0.00	0.00	0.00	0.00	0.00	0	0	0	0
0	—	—	—	0	49.5	—	0	—	—	—	—	—	—	0	—	—
0	7	0.41	0	7.4	14.7	0.06	0	0.00	0.00	0.00	0.00	0.00	0	0	0	0.4
0	14	0.06	3.5	14.2	56.8	0.11	0	0.00	0.00	0.00	0.00	0.00	0	0	0	0.3
0	7	0.08	0	11.2	14.9	0.03	0	0.00	0.00	0.00	0.00	0.00	0	0	0	0.4
0	11	0.06	0	24.9	14.2	0.03	0	0.02	0.00	0.08	0.00	0.00	0	0	0	0.3
0	11	0.39	3.5	28.4	28.4	0.03	0	0.02	0.00	0.08	0.00	0.00	0	0	0	0
0	19	0.18	3.7	3.7	44.5	0.26	0	0.00	0.00	0.00	0.00	0.00	0	0	0	0
0	—	—	—	77.0	45.0	—	—	—	—	—	—	—	—	—	—	—
0	—	—	—	18.0	42.0	—	0	—	—	—	—	—	—	0	—	—
0	—	—	—	70.0	35.0	—	—	—	—	—	—	—	—	—	—	—
0	—	—	—	30.0	35.0	—	—	—	—	—	—	—	—	—	—	—
0	—	—	—	109.5	36.0	—	0	—	—	—	—	—	—	0	—	—
0	11	0.65	3.7	3.7	25.6	0.18	0	0.00	0.00	0.00	0.00	0.00	0	0	0	0.4
0	11	0.29	3.7	3.7	55.8	0.26	0	0.00	—	0.00	0.00	0.00	0	0	0	0
0	7	0.41	3.7	3.7	33.2	0.14	0	0.00	0.00	0.00	0.05	0.00	0	0	0	0
0	—	—	—	0	70.0	—	—	—	—	—	—	—	—	—	—	—
0	19	0.21	3.7	7.4	44.6	0.36	0	0.00	—	0.00	0.00	0.00	0	0	0	0
0	11	0.14	0	3.7	36.8	0.14	0	0.00	—	0.00	0.00	0.00	0	0	0	0.4
0	—	—	—	0	35.0	—	—	—	—	—	—	—	—	—	—	—
0	18	0.18	3.7	3.7	48.0	0.26	0	0.00	0.00	0.00	0.00	0.00	0	0	0	0.4
0	—	—	—	0	70.5	—	0	—	—	—	—	—	—	0	—	—
0	5	0.02	7.1	116.1	4.7	0.04	0	0.03	0.02	0.18	0.45	0.00	4.7	0	0	0
0	7	0.14	11.8	108.9	4.7	0.00	0	0.00	0.00	0.03	0.66	0.00	0	0	0	0.5
12	144	0.19	14.4	232.8	50.4	0.50	33.6	0.04	0.09	0.27	0.13	0.04	7.2	0	0.4	4.6
12	144	0.19	14.4	232.8	50.4	0.50	33.6	0.04	0.09	0.27	0.13	0.04	7.2	0	0.4	4.6
0	5	0.30	189.6	272.6	33.2	0.11	0	0.00	0.04	0.42	12.34	0.00	2.4	0.5	0	0
0	5	0.30	189.6	272.6	33.2	0.11	0	0.00	0.04	0.42	12.34	0.00	2.4	0.5	0	0
0	10	0.09	9.5	71.6	9.5	0.01	0	0.00	0.00	0.00	0.56	0.00	0	0	0	0.2
0	0	0.00	—	160.0	40.0	—	0	—	—	—	—	—	—	0	—	—
0	20	0.22	7.4	62.0	94.2	0.02	5.0	0.05	0.04	0.05	0.05	0.02	9.9	89.3	0	0.5
0	0	0.00	—	30.0	110.0	—	0	—	—	—	—	—	—	0	—	—
0	130	0.17	2.5	30.0	40.0	0.30	0	0.00	0.00	0.01	0.02	0.01	0	78.5	0	0.3
0	14	0.45	5.0	49.6	31.0	0.19	—	0.03	—	0.05	0.04	0.01	4.3	41.6	0	1.0
0	0	0.00	—	10.1	10.1	—	0	—	—	—	—	—	—	6.0	—	—
0	10	0.39	5.0	37.2	9.9	0.05	0	0.01	0.02	0.05	0.04	0.01	2.5	9.7	0	0.2
0	5	0.00	4.9	24.7	7.4	0.02	0	0.01	0.00	0.01	0.02	0.01	2.5	7.7	0	0.2
0	20	0.72	—	0	20.0	—	0	—	—	—	—	—	—	600.0	0	—
0	0	0.00	—	70.0	10.0	—	0	—	—	—	—	—	—	54.0	0	—
0	0	0.00	—	20.0	10.0	—	0	—	—	—	—	—	—	0	0	—
0	0	0.00	—	40.0	10.0	—	0	—	—	—	—	—	—	0	0	—
5	400	2.70	140.0	600.0	220.0	2.25	—	0.52	—	0.59	7.00	0.70	120.0	60.0	2.1	17.5
5	1000	2.70	140.0	600.0	220.0	2.25	—	0.52	—	0.59	7.00	0.70	120.0	30.0	2.1	17.5
5	400	2.70	140.0	600.0	220.0	2.25	—	0.52	—	0.59	7.00	0.70	120.0	60.0	2.1	17.5
0	0	0.04	7.1	87.7	7.1	0.04	0	0.00	0.00	0.03	0.00	0.00	11.9	0	0	0
0	5	0.18	2.4	21.3	2.4	0.09	0	0.02	0.00	0.01	0.00	0.00	2.4	0	0	0
0	5	0.05	2.6	38.9	5.2	0.02	0	0.00	0.00	0.00	0.02	0.00	0	0	0	0.3
0	7	0.02	4.7	42.7	9.5	0.02	0	0.00	0.00	0.01	0.07	0.00	11.9	0	0	0
0	0	0.04	7.1	87.7	7.1	0.04	0	0.00	0.00	0.03	0.00	0.00	11.9	0	0	0
0	33	0.00	0	0	2.4	0.00	0	0.00	—	0.00	0.00	0.00	0	0	0	0

DA+ Code	Food Description	Quantity	Measure	Wt (g)	H₂O (g)	Ener (kcal)	Prot (g)	Carb (g)	Fiber (g)	Fat (g)	Fat Breakdown (g)			
											Sat	Mono	Poly	Trans
33183	Poland spring water, bottled	8	fluid ounce(s)	237	237.0	0	0	0	0	0	0	0	0	0
1821	Tap water	8	fluid ounce(s)	237	236.8	0	0	0	0	0	0	0	0	0
1879	Tonic water	8	fluid ounce(s)	244	222.3	83	0	21.5	0	0	0	0	0	0
	FATS AND OILS													
	Butter													
104	Butter	1	tablespoon(s)	14	2.3	102	0.1	0	0	11.5	7.3	3.0	0.4	—
2522	Butter Buds, dry butter substitute	1	teaspoon(s)	2	—	5	0	2.0	0	0	0	0	0	0
921	Unsalted	1	tablespoon(s)	14	2.5	102	0.1	0	0	11.5	7.3	3.0	0.4	—
107	Whipped	1	tablespoon(s)	9	1.5	67	0.1	0	0	7.6	4.7	2.2	0.3	—
944	Whipped, unsalted	1	tablespoon(s)	11	2.0	82	0.1	0	0	9.2	5.9	2.4	0.3	—
	Fats, cooking													
2671	Beef tallow, semisolid	1	tablespoon(s)	13	0	115	0	0	0	12.8	6.4	5.4	0.5	—
922	Chicken fat	1	tablespoon(s)	13	0	115	0	0	0	12.8	3.8	5.7	2.7	—
5454	Household shortening with vegetable oil	1	tablespoon(s)	13	0	115	0	0	0	13.0	3.4	5.5	2.7	2.2
111	Lard	1	tablespoon(s)	13	0	115	0	0	0	12.8	5.0	5.8	1.4	—
	Margarine													
114	Margarine	1	tablespoon(s)	14	2.3	101	0	0.1	0	11.4	2.1	5.5	3.4	2.1
5439	Soft	1	tablespoon(s)	14	2.3	103	0.1	0.1	0	11.6	1.7	4.4	2.1	3.0
32329	Soft, unsalted, with hydrogenated soybean and cottonseed oils	1	tablespoon(s)	14	2.5	101	0.1	0.1	0	11.3	2.0	5.4	3.5	—
928	Unsalted	1	tablespoon(s)	14	2.6	101	0.1	0.1	0	11.3	2.1	5.2	3.5	—
119	Whipped	1	tablespoon(s)	9	1.5	64	0.1	0.1	0	7.2	1.2	3.2	2.5	—
	Spreads													
54657	I Can't Believe It's Not Butter!, tub, soya oil (non-hydrogenated)	1	tablespoon(s)	14	2.3	103	0.1	0.1	0	11.6	2.8	2.0	5.1	0.1
2708	Mayonnaise with soybean and safflower oils	1	tablespoon(s)	14	2.1	99	0.2	0.4	0	11.0	1.2	1.8	7.6	—
16157	Promise vegetable oil spread, stick	1	tablespoon(s)	14	4.2	90	0	0	0	10.0	2.5	2.0	4.0	—
	Oils													
2681	Canola	1	tablespoon(s)	14	0	120	0	0	0	13.6	1.0	8.6	3.8	0.1
120	Corn	1	tablespoon(s)	14	0	120	0	0	0	13.6	1.8	3.8	7.4	0
122	Olive	1	tablespoon(s)	14	0	119	0	0	0	13.5	1.9	9.9	1.4	—
124	Peanut	1	tablespoon(s)	14	0	119	0	0	0	13.5	2.3	6.2	4.3	—
2693	Safflower	1	tablespoon(s)	14	0	120	0	0	0	13.6	0.8	10.2	2.0	—
923	Sesame	1	tablespoon(s)	14	0	120	0	0	0	13.6	1.9	5.4	5.7	—
128	Soybean, hydrogenated	1	tablespoon(s)	14	0	120	0	0	0	13.6	2.0	5.8	5.1	—
130	Soybean, with soybean and cottonseed oil	1	tablespoon(s)	14	0	120	0	0	0	13.6	2.4	4.0	6.5	—
2700	Sunflower	1	tablespoon(s)	14	0	120	0	0	0	13.6	1.8	6.3	5.0	—
357	**Pam original no stick cooking spray**	1	serving(s)	0	0.2	0	0	0	0	0	0	0	0	—
	Salad dressing													
132	Blue cheese	2	tablespoon(s)	30	9.7	151	1.4	2.2	0	15.7	3.0	3.7	8.3	—
133	Blue cheese, low calorie	2	tablespoon(s)	32	25.4	32	1.6	0.9	0	2.3	0.8	0.6	0.8	—
1764	Caesar	2	tablespoon(s)	30	10.3	158	0.4	0.9	0	17.3	2.6	4.1	9.9	—
29654	Creamy, reduced calorie, fat free, cholesterol free, sour cream and/or buttermilk and oil	2	tablespoon(s)	32	23.9	34	0.4	6.4	0	0.9	0.2	0.2	0.5	—
29617	Creamy, reduced calorie, sour cream and/or buttermilk and oil	2	tablespoon(s)	30	22.2	48	0.5	2.1	0	4.2	0.6	1.0	2.4	—
134	French	2	tablespoon(s)	32	11.7	146	0.2	5.0	0	14.3	1.8	2.7	6.7	—
135	French, low fat	2	tablespoon(s)	32	17.4	74	0.2	9.4	0.4	4.3	0.4	1.9	1.6	—
136	Italian	2	tablespoon(s)	29	16.6	86	0.1	3.1	0	8.3	1.3	1.9	3.8	—
137	Italian, diet	2	tablespoon(s)	30	25.4	23	0.1	1.4	0	1.9	0.1	0.7	0.5	—
139	Mayonnaise-type	2	tablespoon(s)	29	11.7	115	0.3	7.0	0	9.8	1.4	2.6	5.3	—
942	Oil and vinegar	2	tablespoon(s)	32	15.2	144	0	0.8	0	16.0	2.9	4.7	7.7	—
1765	Ranch	2	tablespoon(s)	30	11.6	146	0.1	1.6	0	15.8	2.3	5.2	7.6	—
3666	Ranch, reduced calorie	2	tablespoon(s)	30	20.5	62	0.1	2.2	0	6.1	1.1	1.8	2.9	—
940	Russian	2	tablespoon(s)	30	11.6	107	0.5	9.3	0.7	7.8	1.2	1.8	4.4	—
939	Russian, low calorie	2	tablespoon(s)	32	20.8	45	0.2	8.8	0.1	1.3	0.2	0.3	0.7	—
941	Sesame seed	2	tablespoon(s)	30	11.8	133	0.9	2.6	0.3	13.6	1.9	3.6	7.5	—
142	Thousand Island	2	tablespoon(s)	32	14.9	118	0.3	4.7	0.3	11.2	1.6	2.5	5.8	—
143	Thousand Island, low calorie	2	tablespoon(s)	30	18.2	61	0.3	6.7	0.4	3.9	0.2	1.9	0.8	—
	Sandwich spreads													
138	Mayonnaise with soybean oil	1	tablespoon(s)	14	2.1	99	0.1	0.4	0	11.0	1.6	2.7	5.8	0
140	Mayonnaise, low calorie	1	tablespoon(s)	16	10.0	37	0	2.6	0	3.1	0.5	0.7	1.7	—
141	Tartar sauce	2	tablespoon(s)	28	8.7	144	0.3	4.1	0.1	14.4	2.2	3.8	7.7	—
	SWEETS													
4799	**Butterscotch or caramel topping**	2	tablespoon(s)	41	13.1	103	0.6	27.0	0.4	0	0	0	0	—
	Candy													
1786	Almond Joy candy bar	1	item(s)	45	4.3	220	2.0	27.0	2.0	12.0	8.0	3.3	0.7	0
1785	Bit-O-Honey candy	6	item(s)	40	—	190	1.0	39.0	0	3.5	2.5	—	—	—
33375	Butterscotch candy	2	piece(s)	12	0.6	47	0	10.8	0	0.4	0.2	0.1	0	—
	SWEETS —Continued													

Chol (mg)	Calc (mg)	Iron (mg)	Magn (mg)	Pota (mg)	Sodi (mg)	Zinc (mg)	Vit A (µg)	Thia (mg)	Vit E (mg α)	Ribo (mg)	Niac (mg)	Vit B$_6$ (mg)	Fola (µg)	Vit C (mg)	Vit B$_{12}$ (µg)	Sele (µg)
0	2	0.02	2.4	0	2.4	0.00	0	0.00	—	0.00	0.00	0.00	0	0	0	0
0	7	0.00	2.4	2.4	7.1	0.00	0	0.00	0.00	0.00	0.00	0.00	0	0	0	0
0	2	0.02	0	0	29.3	0.24	0	0.00	0.00	0.00	0.00	0.00	0	0	0	0
31	3	0.00	0.3	3.4	81.8	0.01	97.1	0.00	0.32	0.01	0.01	0.00	0.4	0	0	0.1
0	0	0.00	0	1.6	120.0	0.00	0	0.00	0.00	0.00	0.00	0.00	0	0	0	—
31	3	0.00	0.3	3.4	1.6	0.01	97.1	0.00	0.32	0.01	0.01	0.00	0.4	0	0	0.1
21	2	0.01	0.2	2.4	77.7	0.01	64.3	0.00	0.21	0.00	0.00	0.00	0.3	0	0	0.1
25	3	0.00	0.2	2.7	1.3	0.01	78.0	0.00	0.26	0.00	0.00	0.00	0.3	0	0	0.1
14	0	0.00	0	0	0	0.00	0	0.00	0.34	0.00	0.00	0.00	0	0	0	0.1
11	0	0.00	0	0	0	0.00	0	0.00	0.34	0.00	0.00	0.00	0	0	0	0
0	0	0.00	0	0	0	0.00	0	0.00	—	0.00	0.00	0.00	0	0	0	—
12	0	0.00	0	0	0	0.01	0	0.00	0.07	0.00	0.00	0.00	0	0	0	0
0	4	0.01	0.4	5.9	133.0	0.00	115.5	0.00	1.26	0.01	0.00	0.00	0.1	0	0	0
0	4	0.00	0.3	5.5	155.4	0.00	142.7	0.00	1.00	0.00	0.00	0.00	0.1	0	0	0
0	4	0.00	0.3	5.4	3.9	0.00	103.1	0.00	0.98	0.00	0.00	0.00	0.1	0	0	0
0	2	0.00	0.3	3.5	0.3	0.00	115.5	0.00	1.80	0.00	0.00	0.00	0.1	0	0	0
0	2	0.00	0.2	3.4	97.1	0.00	73.7	0.00	0.45	0.00	0.00	0.00	0.1	0	0	0
0	4	0.00	0.3	5.5	155.3	0.00	142.6	0.00	0.72	0.00	0.00	0.00	0.1	0	0	0
8	2	0.06	0.1	4.7	78.4	0.01	11.6	0.00	3.03	0.00	0.00	0.08	1.1	0	0	0.2
0	10	0.18	—	8.7	90.0	—	—	0.00	—	0.00	0.00	—	—	0.6	—	
0	0	0.00	0	0	0	0.00	0	0.00	2.37	0.00	0.00	0.00	0	0	0	0
0	0	0.00	0	0	0	0.00	0	0.00	1.94	0.00	0.00	0.00	0	0	0	0
0	0	0.07	0	0.1	0.3	0.00	0	0.00	1.93	0.00	0.00	0.00	0	0	0	0
0	0	0.00	0	0	0	0.00	0	0.00	2.11	0.00	0.00	0.00	0	0	0	0
0	0	0.00	0	0	0	0.00	0	0.00	4.63	0.00	0.00	0.00	0	0	0	0
0	0	0.00	0	0	0	0.00	0	0.00	0.19	0.00	0.00	0.00	0	0	0	0
0	0	0.00	0	0	0	0.00	0	0.00	1.10	0.00	0.00	0.00	0	0	0	0
0	0	0.00	0	0	0	0.00	0	0.00	1.64	0.00	0.00	0.00	0	0	0	0
0	0	0.00	0	0	0	0.00	0	0.00	5.58	0.00	0.00	0.00	0	0	0	0
0	0	0.00	0	0.3	1.5	0.01	0.1	0.00	0.00	0.00	0.00	0.00	0	0	0	0
5	24	0.06	0	11.1	328.2	0.08	20.1	0.00	1.80	0.03	0.03	0.01	7.8	0.6	0.1	0.3
0	28	0.16	2.2	1.6	384.0	0.08	—	0.01	0.08	0.03	0.01	0.01	1.0	0.1	0.1	0.5
1	7	0.05	0.6	8.7	323.4	0.03	0.6	0.00	1.56	0.00	0.01	0.00	0.9	0	0	0.5
0	12	0.08	1.6	42.6	320.0	0.05	0.3	0.00	0.21	0.01	0.01	0.01	1.9	0	0	0.5
0	2	0.03	0.6	10.8	306.9	0.01	—	0.00	0.71	0.00	0.01	0.01	0	0.1	0	0.5
0	8	0.25	1.6	21.4	267.5	0.09	7.4	0.01	1.60	0.01	0.06	0.00	0	0	0	0
0	4	0.27	2.6	34.2	257.3	0.06	8.6	0.01	0.09	0.01	0.14	0.01	0.6	0	0	0.5
0	2	0.18	0.9	14.1	486.3	0.03	0.6	0.00	1.47	0.01	0.00	0.01	0	0	0	0.6
2	3	0.19	1.2	25.5	409.8	0.05	0.3	0.00	0.06	0.00	0.00	0.02	0	0	0	2.4
8	4	0.05	0.6	2.6	209.0	0.05	6.2	0.00	0.60	0.01	0.00	0.01	1.8	0	0.1	0.5
0	0	0.00	0	2.6	0.3	0.00	0	0.00	1.46	0.00	0.00	0.00	0	0	0	0.5
1	4	0.03	1.2	8.4	354.0	0.01	5.4	0.00	1.84	0.01	0.00	0.00	0.3	0.1	0	0.1
0	5	0.01	1.5	8.4	413.7	0.01	0.9	0.00	0.72	0.01	0.00	0.00	0.3	0.1	0	0.1
0	6	0.20	3.0	51.9	282.3	0.06	13.2	0.01	0.98	0.01	0.16	0.02	1.5	1.4	0	0.5
2	6	0.18	0	50.2	277.8	0.02	0.6	0.00	0.12	0.00	0.00	0.00	1.0	1.9	0	0.5
0	6	0.18	0	47.1	300.0	0.02	0.6	0.00	1.50	0.00	0.00	0.00	0	0	0	0.5
8	5	0.37	2.6	34.2	276.2	0.08	4.5	0.46	1.28	0.01	0.13	0.00	0	0	0	0.5
0	5	0.27	2.1	60.6	249.3	0.05	4.8	0.01	0.30	0.01	0.13	0.00	0	0	0	0
5	1	0.03	0.1	1.7	78.4	0.02	11.2	0.01	0.72	0.01	0.00	0.08	0.7	0	0	0.2
4	0	0.00	0	1.6	79.5	0.01	0	0.00	0.32	0.00	0.00	0.00	0	0	0	0.3
8	6	0.20	0.8	10.1	191.5	0.05	20.2	0.00	0.97	0.00	0.01	0.07	2.0	0.1	0.1	0.5
0	22	0.08	2.9	34.4	143.1	0.07	11.1	0.01	—	0.03	0.01	0.01	0.8	0.1	0	0
0	18	0.33	30.3	126.5	65.0	0.36	0	0.01	—	0.06	0.21	—	—	0	—	—
0	20	0.00	—	—	150.0	—	0	—	—	—	—	—	—	0	—	—
1	0	0.00	0	0.4	46.9	0.01	3.4	0.00	0.01	0.00	0.00	0.00	0	0	0	0.1

DA+ Code	Food Description	Quantity	Measure	Wt (g)	H₂O (g)	Ener (kcal)	Prot (g)	Carb (g)	Fiber (g)	Fat (g)	Sat	Mono	Poly	Trans
											Fat Breakdown (g)			
1701	Chewing gum, stick	1	item(s)	3	0.1	7	0	2.0	0.1	0	0	0	0	—
33378	Chocolate fudge with nuts, prepared	2	piece(s)	38	2.9	175	1.7	25.8	1.0	7.2	2.5	1.5	2.9	0.1
1787	Jelly beans	15	item(s)	43	2.7	159	0	39.8	0.1	0	0	0	0	—
1784	Kit Kat wafer bar	1	item(s)	42	0.8	210	3.0	27.0	0.5	11.0	7.0	3.5	0.3	0
4674	Krackel candy bar	1	item(s)	41	0.6	210	2.0	28.0	0.5	10.0	6.0	3.9	0.4	0
4934	Licorice	4	piece(s)	44	7.3	154	1.1	35.1	0	1.0	0	0.1	0	—
1780	Life Savers candy	1	item(s)	2	—	8	0	2.0	0	0	0	0	0	0
1790	Lollipop	1	item(s)	28	—	108	0	28.0	0	0	0	0	0	0
4679	M & Ms peanut chocolate candy, small bag	1	item(s)	49	0.9	250	5.0	30.0	2.0	13.0	5.0	5.4	2.1	—
1781	M & Ms plain chocolate candy, small bag	1	item(s)	48	0.8	240	2.0	34.0	1.0	10.0	6.0	3.3	0.3	—
4673	Milk chocolate bar, Symphony	1	item(s)	91	0.9	483	7.7	52.8	1.5	27.8	16.7	7.2	0.6	—
1783	Milky Way bar	1	item(s)	58	3.7	270	2.0	41.0	1.0	10.0	5.0	3.5	0.3	—
1788	Peanut brittle	1 ½	ounce(s)	43	0.3	207	3.2	30.3	1.1	8.1	1.8	3.4	1.9	—
1789	Reese's peanut butter cups	2	piece(s)	51	0.8	280	6.0	19.0	2.0	15.5	6.0	7.2	2.7	0
4689	Reese's pieces candy, small bag	1	item(s)	43	1.1	220	5.0	26.0	1.0	11.0	7.0	0.9	0.4	0
33399	Semisweet chocolate candy, made with butter	½	ounce(s)	14	0.1	68	0.6	9.0	0.8	4.2	2.5	1.4	0.1	—
1782	Snickers bar	1	item(s)	59	3.2	280	4.0	35.0	1.0	14.0	5.0	6.1	2.9	—
4694	Special Dark chocolate bar	1	item(s)	41	0.4	220	2.0	25.0	3.0	12.0	8.0	4.6	0.4	0
4695	Starburst fruit chews, original fruits	1	package(s)	59	3.9	240	0	48.0	0	5.0	1.0	2.1	1.8	—
4698	Taffy	3	piece(s)	45	2.2	179	0	41.2	0	1.5	0.9	0.4	0.1	0.1
4699	Three Musketeers bar	1	item(s)	60	3.5	260	2.0	46.0	1.0	8.0	4.5	2.6	0.3	—
4702	Twix caramel cookie bars	2	item(s)	58	2.4	280	3.0	37.0	1.0	14.0	5.0	7.7	0.5	—
4705	York peppermint pattie	1	item(s)	39	3.9	160	0.5	32.0	0.5	3.0	1.5	1.2	0.1	0
	Frosting, icing													
4760	Chocolate frosting, ready to eat	2	tablespoon(s)	31	5.2	122	0.3	19.4	0.3	5.4	1.7	2.8	0.6	—
4771	Creamy vanilla frosting, ready to eat	2	tablespoon(s)	28	4.2	117	0	19.0	0	4.5	0.8	1.4	2.2	0
17291	Dec-A-Cake variety pack candy decoration	1	teaspoon(s)	4		15	0	3.0	0	0.5	0	—	—	—
536	White icing	2	tablespoon(s)	40	3.6	162	0.1	31.8	0	4.2	0.8	2.0	1.2	—
	Gelatin													
13697	Gelatin snack, all flavors	1	item(s)	99	96.8	70	1.0	17.0	0	0	0	0	0	0
2616	Sugar free, low calorie mixed fruit gelatin mix, prepared	½	cup(s)	121	—	10	1.0	0	0	0	0	0	0	0
548	**Honey**	1	tablespoon(s)	21	3.6	64	0.1	17.3	0	0	0	0	0	0
	Jams, jellies													
550	Jam or preserves	1	tablespoon(s)	20	6.1	56	0.1	13.8	0.2	0	0	0	0	—
42199	Jams, preserves, dietetic, all flavors, w/ sodium saccarin	1	tablespoon(s)	14	6.4	18	0	7.5	0.4	0	0	0	0	—
552	Jelly	1	tablespoon(s)	21	6.3	56	0	14.7	0.2	0	0	0	0	—
545	**Marshmallows**	4	item(s)	29	4.7	92	0.5	23.4	0	0.1	0	0	0	—
4800	**Marshmallow cream topping**	2	tablespoon(s)	40	7.9	129	0.3	31.6	0	0.1	0	0	0	—
555	**Molasses**	1	tablespoon(s)	20	4.4	58	0	14.9	0	0	0	0	0	—
4780	Popsicle or ice pop	1	item(s)	59	47.5	47	0	11.3	0	0.1	0	0	0	—
	Sugar													
559	Brown sugar, packed	1	teaspoon(s)	5	0.1	17	0	4.5	0	0	0	0	0	0
563	Powdered sugar, sifted	⅓	cup(s)	33	0.1	130	0	33.2	0	0	0	0	0	—
561	White granulated sugar	1	teaspoon(s)	4	0	16	0	4.2	0	0	0	0	0	0
	Sugar substitute													
1760	Equal sweetener, packet size	1	item(s)	1	—	0	0	0.9	0	0	0	0	0	0
13029	Splenda granular no calorie sweetener	1	teaspoon(s)	1	—	0	0	0.5	0	0	0	0	0	0
1759	Sweet N Low sugar substitute, packet	1	item(s)	1	0.1	4	0	0.5	0	0	0	0	0	0
	Syrup													
3148	Chocolate syrup	2	tablespoon(s)	38	11.6	105	0.8	24.4	1.0	0.4	0.2	0.1	0	—
29676	Maple syrup	¼	cup(s)	80	25.7	209	0	53.7	0	0.2	0	0.1	0.1	—
4795	Pancake syrup	¼	cup(s)	80	30.4	187	0	49.2	0	0	0	0	0	0
	SPICES, CONDIMENTS, SAUCES													
	Spices													
807	Allspice, ground	1	teaspoon(s)	2	0.2	5	0.1	1.4	0.4	0.2	0	0	0	—
1171	Anise seeds	1	teaspoon(s)	2	0.2	7	0.4	1.1	0.3	0.3	0	0.2	0.1	—
729	Bakers' yeast, active	1	teaspoon(s)	4	0.3	12	1.5	1.5	0.8	0.2	0	0.1	0	—
683	Baking powder, double acting with phosphate	1	teaspoon(s)	5	0.2	2	0	1.1	0	0	0	0	0	—
1611	Baking soda	1	teaspoon(s)	5	0	0	0	0	0	0	0	0	0	0
8552	Basil	1	teaspoon(s)	1	0.8	0	0	0	0	0	0	0	0	—
34959	Basil, fresh	1	piece(s)	1	0.5	0	0	0	0	0	0	0	0	—
808	Basil, ground	1	teaspoon(s)	1	0.1	4	0.2	0.9	0.6	0.1	0	0	0	—
809	Bay leaf	1	teaspoon(s)	1	0	2	0	0.5	0.2	0.1	0	0	0	—
11720	Betel leaves	1	ounce(s)	28	—	17	1.8	2.4	0	0	—	—	—	—
730	Brewers' yeast	1	teaspoon(s)	3	0.1	8	1.0	1.0	0.8	0	0	0	0	—
11710	Capers	1	teaspoon(s)	5	0	0	0	0	0	0	0	0	0	—
1172	Caraway seeds	1	teaspoon(s)	2	0.2	7	0.4	1.0	0.8	0.3	0	0.2	0.1	—

SPICES, CONDIMENTS, SAUCES —Continued

Chol (mg)	Calc (mg)	Iron (mg)	Magn (mg)	Pota (mg)	Sodi (mg)	Zinc (mg)	Vit A (µg)	Thia (mg)	Vit E (mg α)	Ribo (mg)	Niac (mg)	Vit B$_6$ (mg)	Fola (µg)	Vit C (mg)	Vit B$_{12}$ (µg)	Sele (µg)
0	0	0.00	0	0.1	0	0.00	0	0.00	0.00	0.00	0.00	0.00	0	0	0	0
5	22	0.74	20.9	69.5	14.8	0.54	14.4	0.02	0.09	0.03	0.12	0.03	6.1	0.1	0	1.1
0	1	0.05	0.9	15.7	21.3	0.02	0	0.00	0.00	0.01	0.00	0.00	0	0	0	0.5
3	60	0.36	16.4	126.0	30.0	0.51	0	0.07	—	0.22	1.07	0.05	59.6	0	0.1	2.0
3	40	0.36	—	168.8	50.0		0	—	—	—	—	—	—	0	—	—
0	0	0.22	2.6	28.2	126.3	0.07	0	0.01	0.07	0.01	0.04	0.00	0	0	0	—
0	0	0.00	—	0	0	—	0	0.00	—	0.00	0.00	—	—	0	—	0
0	0	0.00	—	—	10.8	—	0	0.00	—	0.00	0.00	—	—	0	—	1.0
5	40	0.36	36.5	170.6	25.0	1.13	14.8	0.03	—	0.06	1.60	0.04	17.3	0.6	0.1	1.9
5	40	0.36	19.6	127.4	30.0	0.46	14.8	0.02	—	0.06	0.10	0.01	2.9	0.6	0.1	1.4
22	228	0.82	61.0	398.6	91.9	1.00	0	0.06	—	0.25	0.14	0.10	10.9	2.0	0.4	—
5	60	0.18	19.8	140.1	95.0	0.41	15.1	0.02	—	0.06	0.20	0.02	5.8	0.6	0.2	3.3
5	11	0.51	17.9	71.4	189.2	0.37	16.6	0.05	1.08	0.01	1.12	0.03	19.6	0	0	1.1
3	40	0.72	45.4	217.4	180.0	0.93	0	0.12	—	0.08	2.35	0.07	28.1	0	0.1	2.3
0	20	0.00	18.9	169.9	80.0	0.32	0	0.04	—	0.06	1.22	0.03	12.0	0	0.1	0.8
3	5	0.44	16.3	51.7	1.6	0.23	0.4	0.01	—	0.01	0.06	0.01	0.4	0	0	0.5
5	40	0.36	42.3	—	140.0	1.37	15.3	0.03	—	0.06	1.60	0.05	23.5	0.6	0.1	2.7
0	0	1.80	45.5	136.0	50.0	0.59	0	0.01	—	0.02	0.16	0.01	0.8	0	0	1.2
0	10	0.18	0.6	1.2	0	0.00	—	0.00	—	0.00	0.00	0.00	0	30	0	0.5
4	4	0.00	0	1.4	23.4	0.09	12.2	0.01	0.04	0.01	0.00	0.00	0	0	0	0.3
5	20	0.36	17.5	80.3	110.0	0.33	14.5	0.01	—	0.03	0.20	0.01	0	0.6	0.1	1.5
5	40	0.36	18.5	116.8	115.0	0.45	15.0	0.09	—	0.13	0.69	0.01	13.9	0.6	0.1	1.2
0	0	0.33	23.4	66.1	10.0	0.28	0	0.01	—	0.03	0.31	0.01	1.5	0	0	—
0	2	0.44	6.4	60.0	56.1	0.09	0	0.00	0.48	0.00	0.03	0.00	0.3	0	0	0.2
0	1	0.04	0.3	9.5	51.5	0.01	0	0.00	0.43	0.08	0.06	0.00	2.2	0	0	0
0	0	0.00	—	—	15.0	—	0	0.00	—	—	—	—	—	0	0	—
0	4	0.01	0.4	5.6	76.4	0.01	44.4	0.00	0.32	0.01	0.00	0.00	0	0	0	0.3
0	0	0.00	—	0	40.0	—	0	—	—	—	—	—	—	0	—	—
0	0	0.00	0	0	50.0	0.00	0	0.00	0.00	0.00	0.00	0.00	0	0	0	—
0	1	0.08	0.4	10.9	0.8	0.04	0	0.00	0.00	0.01	0.02	0.01	0.4	0.1	0	0.2
0	4	0.10	0.8	15.4	6.4	0.01	0	0.00	0.02	0.02	0.01	0.00	2.2	1.8	0	0.4
0	1	0.56	0.7	9.7	0	0.01	0	0.00	0.01	0.00	0.00	0.00	1.3	0	0	0.2
0	1	0.04	1.3	11.3	6.3	0.01	0	0.00	0.00	0.01	0.01	0.00	0.4	0.2	0	0.1
0	1	0.06	0.6	1.4	23.0	0.01	0	0.00	0.00	0.00	0.02	0.00	0.3	0	0	0.5
0	1	0.08	0.8	2.0	32.0	0.01	0	0.00	0.00	0.00	0.03	0.00	0.4	0	0	0.7
0	41	0.94	48.4	292.8	7.4	0.05	0	0.01	0.00	0.00	0.18	0.13	0	0	0	3.6
0	0	0.31	0.6	8.9	4.1	0.08	0	0.00	0.00	0.00	0.00	0.00	0	0.4	0	0.1
0	4	0.03	0.4	6.1	1.3	0.00	0	0.00	0.00	0.00	0.01	0.00	0	0	0	0.1
0	0	0.01	0	0.7	0.3	0.00	0	0.00	0.00	0.00	0.00	0.00	0	0	0	0.2
0	0	0.00	0	0.1	0	0.00	0	0.00	0.00	0.00	0.00	0.00	0	0	0	0
0	0	0.00	0	0	0	0.00	0	0.00	0.00	0.00	0.01	0.00	0	0	0	0
0	0	0.00	—	—	0	—	—	0.00	—	0.00	0.00	—	—	0	0	—
0	0	0.00	—	—	0	—	0	0.00	—	0.00	—	—	—	0	0	—
0	5	0.79	24.4	84.0	27.0	0.27	0	0.00	0.01	0.01	0.12	0.00	0.8	0.1	0	0.5
0	54	0.96	11.2	163.2	7.2	3.32	0	0.01	0.00	0.01	0.02	0.00	0	0	0	0.5
0	2	0.02	1.6	12.0	65.6	0.06	0	0.01	0.00	0.01	0.00	0.00	0	0	0	0
0	13	0.13	2.6	19.8	1.5	0.01	0.5	0.00	—	0.00	0.05	0.00	0.7	0.7	0	0.1
0	14	0.77	3.6	30.3	0.3	0.11	0.3	0.01	—	0.01	0.06	0.01	0.2	0.4	0	0.1
0	3	0.66	3.9	80.0	2.0	0.25	0	0.09	0.00	0.21	1.59	0.06	93.6	0	0	1.0
0	339	0.51	1.8	0.2	363.1	0.00	0	0.00	0.00	0.00	0.00	0.00	0	0	0	0
0	0	0.00	0	0	1258.6	0.00	0	0.00	0.00	0.00	0.00	0.00	0	0	0	0
0	2	0.02	0.6	2.6	0	0.01	2.3	0.00	0.01	0.00	0.01	0.00	0.6	0.2	0	0
0	1	0.01	0.4	2.3	0	0.00	1.3	0.00	—	0.00	0.00	0.00	0.3	0.1	0	0
0	30	0.58	5.9	48.1	0.5	0.08	6.6	0.00	0.10	0.00	0.09	0.03	3.8	0.9	0	0
0	5	0.25	0.7	3.2	0.1	0.02	1.9	0.00	—	0.00	0.01	0.01	1.1	0.3	0	0
0	110	2.29	—	155.9	2.0	—	—	0.04	—	0.07	0.19	—	—	0.9	0	—
0	6	0.46	6.1	50.7	3.3	0.21	0	0.41	—	0.11	1.00	0.06	104.3	0	0	0
0	—	—	—	—	105.0	—	—	—	—	—	—	—	—	0	—	—
0	14	0.34	5.4	28.4	0.4	0.11	0.4	0.01	0.05	0.01	0.07	0.01	0.2	0.4	0	0.3

TABLE A–1 Food Composition

(Computer code is for Cengage Diet Analysis program) (For purposes of calculations, use "0" for t, <1, <.1, <.01, etc.)

DA+ Code	Food Description	Quantity	Measure	Wt (g)	H₂O (g)	Ener (kcal)	Prot (g)	Carb (g)	Fiber (g)	Fat (g)	Sat	Mono	Poly	Trans
1173	Celery seeds	1	teaspoon(s)	2	0.1	8	0.4	0.8	0.2	0.5	0	0.3	0.1	—
1174	Chervil, dried	1	teaspoon(s)	1	0	1	0.1	0.3	0.1	0	0	0	0	—
810	Chili powder	1	teaspoon(s)	3	0.2	8	0.3	1.4	0.9	0.4	0.1	0.1	0.2	—
8553	Chives, chopped	1	teaspoon(s)	1	0.9	0	0	0	0	0	0	0	0	—
51420	Cilantro (coriander)	1	teaspoon(s)	0	0.3	0	0	0	0	0	0	0	0	—
811	Cinnamon, ground	1	teaspoon(s)	2	0.2	6	0.1	1.9	1.2	0	0	0	0	0
812	Cloves, ground	1	teaspoon(s)	2	0.1	7	0.1	1.3	0.7	0.4	0.1	0	0.1	—
1175	Coriander leaf, dried	1	teaspoon(s)	1	0	2	0.1	0.3	0.1	0	0	0	0	—
1176	Coriander seeds	1	teaspoon(s)	2	0.2	5	0.2	1.0	0.8	0.3	0	0.2	0	—
1706	Cornstarch	1	tablespoon(s)	8	0.7	30	0	7.3	0.1	0	0	0	0	—
1177	Cumin seeds	1	teaspoon(s)	2	0.2	8	0.4	0.9	0.2	0.5	0	0.3	0.1	—
11729	Cumin, ground	1	teaspoon(s)	5	—	11	0.4	0.8	0.8	0.4	—	—	—	—
1178	Curry powder	1	teaspoon(s)	2	0.2	7	0.3	1.2	0.7	0.3	0	0.1	0.1	—
1179	Dill seeds	1	teaspoon(s)	2	0.2	6	0.3	1.2	0.4	0.3	0	0.2	0	—
1180	Dill weed, dried	1	teaspoon(s)	1	0.1	3	0.2	0.6	0.1	0	0	0	0	—
34949	Dill weed, fresh	5	piece(s)	1	0.9	0	0	0.1	0	0	0	0	0	—
4949	Fennel leaves, fresh	1	teaspoon(s)	1	0.9	0	0	0.1	0	0	—	—	—	—
1181	Fennel seeds	1	teaspoon(s)	2	0.2	7	0.3	1.0	0.8	0.3	0	0.2	0	—
1182	Fenugreek seeds	1	teaspoon(s)	4	0.3	12	0.9	2.2	0.9	0.2	0.1	—	—	—
11733	Garam masala, powder	1	ounce(s)	28	—	107	4.4	12.8	0	4.3	—	—	—	—
1067	Garlic clove	1	item(s)	3	1.8	4	0.2	1.0	0.1	0	0	0	0	—
813	Garlic powder	1	teaspoon(s)	3	0.2	9	0.5	2.0	0.3	0	0	0	0	—
1068	Ginger root	2	teaspoon(s)	4	3.1	3	0.1	0.7	0.1	0	0	0	0	—
1183	Ginger, ground	1	teaspoon(s)	2	0.2	6	0.2	1.3	0.2	0.1	0	0	0	—
35497	Leeks, bulb and lower-leaf, freeze-dried	¼	cup(s)	1	0	3	0.1	0.6	0.1	0	0	0	0	—
1184	Mace, ground	1	teaspoon(s)	2	0.1	8	0.1	0.9	0.3	0.6	0.2	0.2	0.1	—
1185	Marjoram, dried	1	teaspoon(s)	1	0	2	0.1	0.4	0.2	0	0	0	0	—
1186	Mustard seeds, yellow	1	teaspoon(s)	3	0.2	15	0.8	1.2	0.5	0.9	0	0.7	0.2	—
814	Nutmeg, ground	1	teaspoon(s)	2	0.1	12	0.1	1.1	0.5	0.8	0.6	0.1	0	—
2747	Onion flakes, dehydrated	1	teaspoon(s)	2	0.1	6	0.1	1.4	0.2	0	0	0	0	—
1187	Onion powder	1	teaspoon(s)	2	0.1	7	0.2	1.7	0.1	0	0	0	0	—
815	Oregano, ground	1	teaspoon(s)	2	0.1	5	0.2	1.0	0.6	0.2	0	0	0.1	—
816	Paprika	1	teaspoon(s)	2	0.2	6	0.3	1.2	0.8	0.3	0	0	0.2	—
817	Parsley, dried	1	teaspoon(s)	0	0	1	0.1	0.2	0.1	0	0	0	0	—
818	Pepper, black	1	teaspoon(s)	2	0.2	5	0.2	1.4	0.6	0.1	0	0	0	—
819	Pepper, cayenne	1	teaspoon(s)	2	0.1	6	0.2	1.0	0.5	0.3	0.1	0	0.2	—
1188	Pepper, white	1	teaspoon(s)	2	0.3	7	0.3	1.6	0.6	0.1	0	0	0	—
1189	Poppy seeds	1	teaspoon(s)	3	0.2	15	0.5	0.7	0.3	1.3	0.1	0.2	0.9	—
1190	Poultry seasoning	1	teaspoon(s)	2	0.1	5	0.1	1.0	0.2	0.1	0	0	0	—
1191	Pumpkin pie spice, powder	1	teaspoon(s)	2	0.1	6	0.1	1.2	0.3	0.2	0.1	0	0	—
1192	Rosemary, dried	1	teaspoon(s)	1	0.1	4	0.1	0.8	0.5	0.2	0.1	0	0	—
11723	Rosemary, fresh	1	teaspoon(s)	1	0.5	1	0	0.1	0.1	0	0	0	0	—
2722	Saffron powder	1	teaspoon(s)	1	0.1	2	0.1	0.5	0	0	0	0	0	—
11724	Sage	1	teaspoon(s)	1	—	1	0	0.1	0	0	—	—	—	—
1193	Sage, ground	1	teaspoon(s)	1	0.1	2	0.1	0.4	0.3	0.1	0	0	0	—
30189	Salt substitute	¼	teaspoon(s)	1	—	0	0	0	0	0	0	0	0	0
30190	Salt substitute, seasoned	¼	teaspoon(s)	1	—	1	0	0.1	0	0	0			—
822	Salt, table	¼	teaspoon(s)	2	0	0	0	0	0	0	0	0	0	0
1194	Savory, ground	1	teaspoon(s)	1	0.1	4	0.1	1.0	0.6	0.1	0	—	—	—
820	Sesame seed kernels, toasted	1	teaspoon(s)	3	0.1	15	0.5	0.7	0.5	1.3	0.2	0.5	0.6	—
11725	Sorrel	1	teaspoon(s)	3		1	0.1	0.1	0	0	0	—	—	—
11721	Spearmint	1	teaspoon(s)	2	1.6	1	0.1	0.2	0.1	0	0	0	0	—
35498	Sweet green peppers, freeze-dried	¼	cup(s)	2	0	5	0.3	1.1	0.3	0	0	0	0	—
11726	Tamarind leaves	1	ounce(s)	28	—	33	1.6	5.2	0	0.6	—	—	—	—
11727	Tarragon	1	ounce(s)	28	—	14	1.0	1.8	0	0.3	—	—	—	—
1195	Tarragon, ground	1	teaspoon(s)	2	0.1	5	0.4	0.8	0.1	0.1	0	0	0.1	—
11728	Thyme, fresh	1	teaspoon(s)	1	0.5	1	0	0.2	0.1	0	0	0	0	—
821	Thyme, ground	1	teaspoon(s)	1	0.1	4	0.1	0.9	0.5	0.1	0	0	0	—
1196	Turmeric, ground	1	teaspoon(s)	2	0.3	8	0.2	1.4	0.5	0.2	0.1	0	0	—
11995	Wasabi	1	tablespoon(s)	14	10.7	10	0.7	2.3	0.2	0	—	—	—	—
Condiments														
674	Catsup or ketchup	1	tablespoon(s)	15	10.4	15	0.3	3.8	0	0	0	0	0	—
703	Dill pickle	1	ounce(s)	28	26.7	3	0.2	0.7	0.3	0	0	0	0	—
138	Mayonnaise with soybean oil	1	tablespoon(s)	14	2.1	99	0.1	0.4	0	11.0	1.6	2.7	5.8	0
140	Mayonnaise, low calorie	1	tablespoon(s)	16	10.0	37	0	2.6	0	3.1	0.5	0.7	1.7	—
1682	Mustard, brown	1	teaspoon(s)	5	4.1	5	0.3	0.3	0	0.3	—	—	—	—
700	Mustard, yellow	1	teaspoon(s)	5	4.1	3	0.2	0.3	0.2	0.2	0	0.1	0	0
706	Sweet pickle relish	1	tablespoon(s)	15	9.3	20	0.1	5.3	0.2	0.1	0	0	0	—
141	Tartar sauce	2	tablespoon(s)	28	8.7	144	0.3	4.1	0.1	14.4	2.2	3.8	7.7	—

SPICES, CONDIMENTS, SAUCES —Continued

Chol (mg)	Calc (mg)	Iron (mg)	Magn (mg)	Pota (mg)	Sodi (mg)	Zinc (mg)	Vit A (µg)	Thia (mg)	Vit E (mg α)	Ribo (mg)	Niac (mg)	Vit B_6 (mg)	Fola (µg)	Vit C (mg)	Vit B_{12} (µg)	Sele (µg)
0	35	0.89	8.8	28.0	3.2	0.13	0.1	0.01	0.02	0.01	0.06	0.01	0.2	0.3	0	0.2
0	8	0.19	0.8	28.4	0.5	0.05	1.8	0.00	—	0.00	0.03	0.01	1.6	0.3	0	0.2
0	7	0.37	4.4	49.8	26.3	0.07	38.6	0.01	0.75	0.02	0.20	0.09	2.6	1.7	0	0.2
0	1	0.01	0.4	3.0	0	0.01	2.2	0.00	0.00	0.00	0.01	0.00	1.1	0.6	0	0
0	0	0.01	0.1	1.7	0.2	0.00	1.1	0.00	0.01	0.00	0.00	0.00	0.2	0.1	0	0
0	23	0.19	1.4	9.9	0.2	0.04	0.3	0.00	0.05	0.00	0.03	0.00	0.1	0.1	0	0.1
0	14	0.18	5.5	23.1	5.1	0.02	0.6	0.00	0.17	0.01	0.03	0.01	2.0	1.7	0	0.1
0	7	0.25	4.2	26.8	1.3	0.02	1.8	0.01	0.01	0.01	0.06	0.00	1.6	3.4	0	0.2
0	13	0.29	5.9	22.8	0.6	0.08	0	0.00	—	0.01	0.03	—	0	0.4	0	0.5
0	0	0.03	0.2	0.2	0.7	0.01	0	0.00	0.00	0.00	0.00	0.00	0	0	0	0.2
0	20	1.39	7.7	37.5	3.5	0.10	1.3	0.01	0.07	0.01	0.09	0.01	0.2	0.2	0	0.1
0	20	—	—	43.6	4.8	—	—	—	—	—	—	—	—	—	—	—
0	10	0.59	5.1	30.9	1.0	0.08	1.0	0.01	0.44	0.01	0.06	0.02	3.1	0.2	0	0.3
0	32	0.34	5.4	24.9	0.4	0.10	0.1	0.01	—	0.01	0.05	0.01	0.2	0.4	0	0.3
0	18	0.48	4.5	33.1	2.1	0.03	2.9	0.00	—	0.00	0.02	0.01	1.5	0.5	0	—
0	2	0.06	0.6	7.4	0.6	0.01	3.9	0.00	0.01	0.00	0.01	0.00	1.5	0.9	0	—
0	1	0.02	—	4.0	0.1	—	—	0.00	—	0.00	0.01	0.00	—	0.3	0	—
0	24	0.37	7.7	33.9	1.8	0.07	0.1	0.01	—	0.01	0.12	0.01	—	0.4	0	—
0	7	1.24	7.1	28.5	2.5	0.09	0.1	0.01	—	0.01	0.06	0.02	2.1	0.1	0	0.2
0	215	9.24	93.6	411.1	27.5	1.07	—	0.09	—	0.09	0.70	—	0	0	0	—
0	5	0.05	0.8	12.0	0.5	0.03	0	0.01	0.00	0.00	0.02	0.03	0.1	0.9	0	0.4
0	2	0.07	1.6	30.8	0.7	0.07	0	0.01	0.01	0.00	0.01	0.08	0.1	0.5	0	1.1
0	1	0.02	1.7	16.6	0.5	0.01	0	0.00	0.01	0.00	0.02	0.01	0.4	0.2	0	0
0	2	0.20	3.3	24.2	0.6	0.08	0.1	0.00	0.32	0.00	0.09	0.01	0.7	0.1	0	0.7
0	3	0.06	1.3	19.2	0.3	0.01	0.1	0.01	—	0.00	0.02	0.01	2.9	0.9	0	0
0	4	0.23	2.8	7.9	1.4	0.03	0.7	0.01	—	0.01	0.02	0.00	1.3	0.4	0	0
0	12	0.49	2.1	9.1	0.5	0.02	2.4	0.00	0.01	0.00	0.02	0.01	1.6	0.3	0	0
0	17	0.32	9.8	22.5	0.2	0.18	0.1	0.01	0.09	0.01	0.26	0.01	2.5	0.1	0	4.4
0	4	0.06	4.0	7.7	0.4	0.04	0.1	0.01	0.00	0.00	0.02	0.00	1.7	0.1	0	0
0	4	0.02	1.5	27.1	0.4	0.03	0	0.01	0.00	0.00	0.01	0.02	2.8	1.3	0	0.1
0	8	0.05	2.6	19.8	1.1	0.04	0	0.01	0.01	0.00	0.01	0.02	3.5	0.3	0	0
0	24	0.66	4.1	25.0	0.2	0.06	5.2	0.01	0.28	0.01	0.09	0.01	4.1	0.8	0	0.1
0	4	0.49	3.9	49.2	0.7	0.08	55.4	0.01	0.62	0.03	0.32	0.08	2.2	1.5	0	0.1
0	4	0.29	0.7	11.4	1.4	0.01	1.5	0.00	0.02	0.00	0.02	0.00	0.5	0.4	0	0.1
0	9	0.60	4.1	26.4	0.9	0.03	0.3	0.00	0.01	0.01	0.02	0.01	0.2	0.4	0	0.1
0	3	0.14	2.7	36.3	0.5	0.04	37.5	0.01	0.53	0.01	0.15	0.04	1.9	1.4	0	0.2
0	6	0.34	2.2	1.8	0.1	0.02	0	0.00	—	0.00	0.01	0.00	0.2	0.5	0	0.1
0	41	0.26	9.3	19.6	0.6	0.28	0	0.02	0.03	0.01	0.02	0.01	1.6	0.1	0	0
0	15	0.53	3.4	10.3	0.4	0.04	2.0	0.00	0.02	0.00	0.04	0.02	2.1	0.2	0	0.1
0	12	0.33	2.3	11.3	0.9	0.04	0.2	0.00	0.01	0.00	0.03	0.01	0.9	0.4	0	0.2
0	15	0.35	2.6	11.5	0.6	0.03	1.9	0.01	—	0.01	0.01	0.02	3.7	0.7	0	0.1
0	2	0.04	0.6	4.7	0.2	0.01	1.0	0.00	—	0.00	0.01	0.00	0.8	0.2	0	—
0	1	0.07	1.8	12.1	1.0	0.01	0.2	0.00	—	0.00	0.01	0.01	0.7	0.6	0	0
0	4	—	1.1	2.7	0	0.01	—	0.00	—	—	—	—	—	—	0	—
0	12	0.19	3.0	7.5	0.1	0.03	2.1	0.01	0.05	0.00	0.04	0.01	1.9	0.2	0	0
0	7	0.00	0	603.6	0.1	—	0	—	—	—	—	—	—	0	0	—
0	0	0.00	—	476.3	0.1	—	0	—	—	—	—	—	—	0	0	—
0	0	0.01	0	0.1	581.4	0.00	0	0.00	0.00	0.00	0.00	0.00	0	0	0	0
0	30	0.53	5.3	14.7	0.3	0.06	3.6	0.01	—	—	0.05	0.02	—	0.7	0	0.1
0	3	0.21	9.2	10.8	1.0	0.27	0.1	0.03	0.01	0.01	0.15	0.00	2.6	0	0	0
0	—	—	—	—	0.1	—	—	—	—	—	—	—	—	—	0	—
0	4	0.22	1.2	8.7	0.6	0.02	3.9	0.00	—	0.00	0.01	0.00	2.0	0.3	0	—
0	2	0.16	3.0	50.7	3.1	0.03	4.5	0.01	0.06	0.01	0.11	0.03	3.7	30.4	0	0.1
0	85	1.48	20.2	—	—	—	—	0.06	—	0.02	1.16	—	—	0.9	0	—
0	48	—	14.5	128.1	2.6	0.17	—	0.04	—	—	—	—	—	0.6	0	—
0	18	0.51	5.6	48.3	1.0	0.06	3.4	0.00	—	0.02	0.14	0.03	4.4	0.8	0	0.1
0	3	0.14	1.3	4.9	0.1	0.01	1.9	0.00	—	0.00	0.01	0.00	0.4	1.3	0	—
0	26	1.73	3.1	11.4	0.8	0.08	2.7	0.01	0.10	0.01	0.06	0.01	3.8	0.7	0	0.1
0	4	0.91	4.2	55.6	0.8	0.09	0	0.00	0.06	0.01	0.11	0.04	0.9	0.6	0	0.1
0	13	0.11	—	—	—	—	—	0.02	—	0.01	0.07	—	—	11.2		
0	3	0.07	2.9	57.3	167.1	0.03	7.1	0.00	0.21	0.02	0.21	0.02	1.5	2.3	0	0
0	12	0.10	2.0	26.1	248.1	0.03	2.6	0.01	0.02	0.01	0.03	0.01	0.3	0.2	0	0
5	1	0.03	0.1	1.7	78.4	0.02	11.2	0.01	0.72	0.01	0.00	0.08	0.7	0	0	0.2
4	0	0.00	0	1.6	79.5	0.01	0	0.00	0.32	0.00	0.00	0.00	0	0	0	0.3
0	6	0.09	1.0	6.8	68.1	0.01	0	0.00	0.09	0.00	0.01	0.00	0.2	0.1	0	—
0	3	0.07	2.5	6.9	56.8	0.03	0.2	0.01	0.01	0.00	0.02	0.00	0.4	0.1	0	1.6
0	0	0.13	0.8	3.8	121.7	0.02	9.2	0.00	0.08	0.01	0.03	0.00	0.2	0.2	0	0
8	6	0.20	0.8	10.1	191.5	0.05	20.2	0.00	0.97	0.00	0.01	0.07	2.0	0.1	0.1	0.5

DA+ Code	Food Description	Quantity	Measure	Wt (g)	H₂O (g)	Ener (kcal)	Prot (g)	Carb (g)	Fiber (g)	Fat (g)	Fat Breakdown (g)			
											Sat	Mono	Poly	Trans
	Sauces													
685	Barbecue sauce	2	tablespoon(s)	31	18.9	47	0	11.3	0.2	0.1	0	0	0.1	0
834	Cheese sauce	¼	cup(s)	63	44.4	110	4.2	4.3	0.3	8.4	3.8	2.4	1.6	—
32123	Chili enchilada sauce, green	2	tablespoon(s)	57	53.0	15	0.6	3.1	0.7	0.3	0	0	0.1	0
32122	Chili enchilada sauce, red	2	tablespoon(s)	32	24.5	27	1.1	5.0	2.1	0.8	0.1	0	0.4	0
29688	Hoisin sauce	1	tablespoon(s)	16	7.1	35	0.5	7.1	0.4	0.5	0.1	0.2	0.3	—
1641	Horseradish sauce, prepared	1	teaspoon(s)	5	3.3	10	0.1	0.2	0	1.0	0.6	0.3	0	—
16670	Mole poblano sauce	½	cup(s)	133	102.7	156	5.3	11.4	2.7	11.3	2.6	5.1	3.0	—
29689	Oyster sauce	1	tablespoon(s)	16	12.8	8	0.2	1.7	0	0	0	0	0	—
1655	Pepper sauce or Tabasco	1	teaspoon(s)	5	4.8	1	0.1	0	0	0	0	0	0	—
347	Salsa	2	tablespoon(s)	32	28.8	9	0.5	2.0	0.5	0.1	0	0	0	—
52206	Soy sauce, tamari	1	tablespoon(s)	18	12.0	11	1.9	1.0	0.1	0	0	0	0	—
839	Sweet and sour sauce	2	tablespoon(s)	39	29.8	37	0.1	9.1	0.1	0	0	0	0	—
1613	Teriyaki sauce	1	tablespoon(s)	18	12.2	16	1.1	2.8	0	0	0	0	0	0
25294	Tomato sauce	½	cup(s)	150	132.8	63	2.2	11.9	2.6	1.8	0.2	0.4	0.9	0
728	White sauce, medium	¼	cup(s)	63	46.8	92	2.4	5.7	0.1	6.7	1.8	2.8	1.8	—
1654	Worcestershire sauce	1	teaspoon(s)	6	4.5	4	0	1.1	0	0	0	0	0	—
	Vinegar													
30853	Balsamic	1	tablespoon(s)	15	—	10	0	2.0	0	0	0	0	0	0
727	Cider	1	tablespoon(s)	15	14.0	3	0	0.1	0	0	0	0	0	0
1673	Distilled	1	tablespoon(s)	15	14.3	2	0	0.8	0	0	0	0	0	0
12948	Tarragon	1	tablespoon(s)	15	13.8	2	0	0.1	0	0	0	0	0	0
	MIXED FOODS, SOUPS, SANDWICHES													
	Mixed dishes													
16652	Almond chicken	1	cup(s)	242	186.8	281	21.8	15.8	3.4	14.7	1.8	6.3	5.6	—
25224	Barbecued chicken	1	serving(s)	177	99.3	327	27.1	15.7	0.5	17.1	4.8	6.8	3.8	0
25227	Bean burrito	1	item(s)	149	81.8	326	16.1	33.0	5.6	14.8	8.3	4.7	0.9	—
9516	Beef and vegetable fajita	1	item(s)	223	143.9	397	22.4	35.3	3.1	18.0	5.9	8.0	2.5	—
16796	Beef or pork egg roll	2	item(s)	128	85.2	225	9.9	18.4	1.4	12.4	2.9	6.0	2.6	—
177	Beef stew with vegetables, prepared	1	cup(s)	245	201.0	220	16.0	15.0	3.2	11.0	4.4	4.5	0.5	—
30233	Beef stroganoff with noodles	1	cup(s)	256	190.1	343	19.7	22.8	1.5	19.1	7.4	5.7	4.4	—
16651	Cashew chicken	1	cup(s)	242	186.8	281	21.8	15.8	3.4	14.7	1.8	6.3	5.6	—
30274	Cheese pizza with vegetables, thin crust	2	slice(s)	140	76.6	298	12.7	35.4	2.5	12.0	4.9	4.7	1.6	—
30330	Cheese quesadilla	1	item(s)	54	18.3	190	7.7	15.3	1.0	10.8	5.2	3.6	1.3	—
215	Chicken and noodles, prepared	1	cup(s)	240	170.0	365	22.0	26.0	1.3	18.0	5.1	7.1	3.9	—
30239	Chicken and vegetables with broccoli, onion, bamboo shoots in soy based sauce	1	cup(s)	162	125.5	180	15.8	9.3	1.8	8.6	1.7	3.0	3.1	—
25093	Chicken cacciatore	1	cup(s)	244	175.7	284	29.9	5.7	1.3	15.3	4.3	6.2	3.3	0
28020	Chicken fried turkey steak	3	ounce(s)	492	276.2	706	77.1	68.7	3.6	12.0	3.4	2.9	3.9	—
218	Chicken pot pie	1	cup(s)	252	154.6	542	22.6	41.4	3.5	31.3	9.8	12.5	7.1	—
30240	Chicken teriyaki	1	cup(s)	244	158.3	364	51.0	15.2	0.7	7.0	1.8	2.0	1.7	—
25119	Chicken waldorf salad	½	cup(s)	100	67.2	179	14.0	6.8	1.0	10.8	1.8	3.1	5.2	—
25099	Chili con carne	¾	cup(s)	215	174.4	198	13.7	21.4	7.5	6.9	2.5	2.8	0.5	0
1062	Coleslaw	¾	cup(s)	90	73.4	70	1.2	11.2	1.4	2.3	0.3	0.6	1.2	—
1574	Crab cakes, from blue crab	1	item(s)	60	42.6	93	12.1	0.3	0	4.5	0.9	1.7	1.4	—
32144	Enchiladas with green chili sauce (enchiladas verdes)	1	item(s)	144	103.8	207	9.3	17.6	2.6	11.7	6.4	3.6	1.0	0
2793	Falafel patty	3	item(s)	51	17.7	170	6.8	16.2	—	9.1	1.2	5.2	2.1	—
28546	Fettuccine alfredo	1	cup(s)	244	88.7	279	13.1	46.1	1.4	4.2	2.2	1.0	0.4	0
32146	Flautas	3	item(s)	162	78.0	438	24.9	36.3	4.1	21.6	8.2	8.8	2.3	—
29629	Fried rice with meat or poultry	1	cup(s)	198	128.5	333	12.3	41.8	1.4	12.3	2.2	3.5	5.7	—
16649	General Tso chicken	1	cup(s)	146	91.0	296	18.7	16.4	0.9	17.0	4.0	6.3	5.3	—
1826	Green salad	¾	cup(s)	104	98.9	17	1.3	3.3	2.2	0.1	0	0	0	—
1814	Hummus	½	cup(s)	123	79.8	218	6.0	24.7	4.9	10.6	1.4	6.0	2.6	—
16650	Kung pao chicken	1	cup(s)	162	87.2	434	28.8	11.7	2.3	30.6	5.2	13.9	9.7	—
16622	Lamb curry	1	cup(s)	236	187.9	257	28.2	3.7	0.9	13.8	3.9	4.9	3.3	—
25253	Lasagna with ground beef	1	cup(s)	237	158.4	284	16.9	22.3	2.4	14.5	7.5	4.9	0.8	—
442	Macaroni and cheese, prepared	1	cup(s)	200	122.3	390	14.9	40.6	1.6	18.6	7.9	6.4	2.9	—
29637	Meat filled ravioli with tomato or meat sauce, canned	1	cup(s)	251	198.7	208	7.8	36.5	1.3	3.7	1.5	1.4	0.3	—
25105	Meat loaf	1	slice(s)	115	84.5	245	17.0	6.6	0.4	16.0	6.1	6.9	0.9	0
16646	Moo shi pork	1	cup(s)	151	76.8	512	18.9	5.3	0.6	46.4	6.9	15.8	21.2	—
16788	Nachos with beef, beans, cheese, tomatoes and onions	1	serving(s)	551	253.5	1576	59.1	137.5	20.4	90.8	32.6	41.9	9.4	—
6116	Pepperoni pizza	2	slice(s)	142	66.1	362	20.2	39.7	2.9	13.9	4.5	6.3	2.3	—
29601	Pizza with meat and vegetables, thin crust	2	slice(s)	158	81.4	386	16.5	36.8	2.7	19.1	7.7	8.1	2.2	—
655	Potato salad	½	cup(s)	125	95.0	179	3.4	14.0	1.6	10.3	1.8	3.1	4.7	—
25109	Salisbury steaks with mushroom sauce	1	serving(s)	135	101.8	251	17.1	9.3	0.5	15.5	6.0	6.7	0.8	0
16637	Shrimp creole with rice	1	cup(s)	243	176.6	309	27.0	27.7	1.2	9.2	1.7	3.6	2.9	—
	MIXED FOODS, SOUPS, SANDWICHES —Continued													
497	Spaghetti and meat balls with tomato sauce, prepared	1	cup(s)	248	174.0	330	19.0	39.0	2.7	12.0	3.9	4.4	2.2	—

Chol (mg)	Calc (mg)	Iron (mg)	Magn (mg)	Pota (mg)	Sodi (mg)	Zinc (mg)	Vit A (µg)	Thia (mg)	Vit E (mg α)	Ribo (mg)	Niac (mg)	Vit B_6 (mg)	Fola (µg)	Vit C (mg)	Vit B_{12} (µg)	Sele (µg)
0	4	0.06	3.8	65.0	349.7	0.04	3.8	0.00	0.20	0.01	0.15	0.01	0.6	0.2	0	0.4
18	116	0.13	5.7	18.9	521.6	0.61	50.4	0.00	—	0.07	0.01	0.01	2.5	0.3	0.1	2.0
0	5	0.36	9.5	125.7	61.9	0.11	—	0.02	0.00	0.02	0.63	0.06	5.7	43.9	0	0
0	7	1.05	11.1	231.3	113.8	0.14	—	0.01	0.00	0.21	0.61	0.34	6.6	0.3	0	0.3
0	5	0.16	3.8	19.0	258.4	0.05	0	0.00	0.04	0.03	0.18	0.01	3.7	0.1	0	0.3
2	5	0.00	0.5	6.7	14.6	0.01	8.0	0.00	0.02	0.01	0.00	0.00	0.5	0.1	0	0.1
1	38	1.81	58.3	280.9	304.8	1.15	13.3	0.06	1.72	0.08	1.84	0.09	15.9	3.4	0.1	1.1
0	5	0.02	0.6	8.6	437.3	0.01	0	0.00	0.00	0.02	0.23	0.00	2.4	0	0.1	0.7
0	1	0.05	0.6	6.4	31.7	0.01	4.1	0.00	0.00	0.00	0.01	0.01	0.1	0.2	0	0
0	9	0.14	4.8	95.0	192.0	0.11	4.8	0.01	0.37	0.01	0.02	0.05	1.3	0.6	0	0.3
0	4	0.43	7.3	38.7	1018.9	0.08	0	0.01	—	0.03	0.72	0.04	3.3	0	0	0.1
0	5	0.20	1.2	8.2	97.5	0.01	0	0.00	—	0.01	0.11	0.03	0.2	0	0	—
0	5	0.30	11.0	40.5	689.9	0.01	0	0.01	0.00	0.01	0.22	0.01	1.4	0	0	0.2
0	23	1.24	28.9	536.8	268.6	0.36	—	0.08	0.52	0.08	1.64	0.20	23.2	32.0	0	1.0
4	74	0.20	8.8	97.5	221.3	0.25	—	0.04	—	0.11	0.25	0.02	3.1	0.5	0.2	—
0	6	0.30	0.7	45.4	55.6	0.01	0.3	0.00	0.00	0.01	0.03	0.00	0.5	0	0	0
0	0	0.00	—	—	0	—	0	—	—	—	—	—	—	0	—	—
0	1	0.03	0.7	10.9	0.7	0.01	0	0.00	0.00	0.00	0.00	0.00	0	0	0	0
0	1	0.09	0	2.3	0.1	0.00	0	0.00	0.00	0.00	0.00	0.00	0	0	0	5.0
0	0	0.07	—	2.3	0.7	—	—	0.07	—	0.07	0.07	—	—	0.3	0	—
41	68	1.86	58.1	539.7	510.6	1.50	31.5	0.07	4.11	0.22	9.57	0.43	26.6	5.1	0.3	13.6
120	26	1.70	32.4	419.7	500.9	2.67	—	0.09	0.01	0.24	6.87	0.40	15.0	7.9	0.3	19.5
38	333	3.01	52.5	447.6	510.6	1.98	—	0.28	0.01	0.30	1.92	0.19	122.9	8.2	0.3	15.9
45	85	3.65	37.9	475.0	756.0	3.52	17.8	0.38	0.80	0.29	5.33	0.39	69.1	23.4	2.1	28.3
74	31	1.68	20.5	248.3	547.8	0.89	25.6	0.32	1.28	0.24	2.55	0.18	38.4	4.0	0.3	17.5
71	29	2.90	—	613.0	292.0	—	—	0.15	0.51	0.17	4.70	—	—	17.0	0	15.0
74	69	3.25	35.8	391.7	816.6	3.63	69.1	0.21	1.25	0.30	3.80	0.21	48.6	1.3	1.8	27.9
41	68	1.86	58.1	539.7	510.6	1.50	31.5	0.07	4.11	0.22	9.57	0.43	26.6	5.1	0.3	13.6
17	249	2.78	28.0	294.0	739.2	1.42	47.6	0.29	1.05	0.33	2.84	0.14	61.6	15.3	0.4	18.6
23	190	1.04	13.5	75.6	469.3	0.86	58.3	0.11	0.43	0.15	0.89	0.02	21.6	2.4	0.1	9.2
103	26	2.20	—	149.0	600.0	—	—	0.05	—	0.17	4.30	—	—	0	—	29.0
42	28	1.19	22.7	299.7	620.5	1.32	81.0	0.07	1.11	0.14	5.28	0.36	16.2	22.5	0.2	12.0
109	47	1.97	40.0	489.3	492.1	2.13	—	0.11	0.00	0.20	9.81	0.57	16.1	14.0	0.3	22.6
156	423	8.79	110.3	1182.9	880.4	6.18	—	0.72	0.00	1.05	20.16	1.20	113.9	2.7	1.3	97.6
68	66	3.32	37.8	390.6	652.7	1.94	259.6	0.39	1.05	0.39	7.25	0.23	80.6	10.3	0.2	27.0
156	51	3.26	68.3	588.0	3208.6	3.75	31.7	0.15	0.58	0.36	16.68	0.88	24.4	2.0	0.5	36.1
42	20	0.82	23.9	202.5	246.5	1.13	—	0.05	0.62	0.09	4.06	0.25	15.8	2.5	0.2	10.7
27	42	2.83	50.6	636.8	864.8	2.36	—	0.15	0.01	0.22	3.18	0.19	58.1	10.3	0.6	7.3
7	41	0.53	9.0	162.9	20.7	0.18	47.7	0.06	—	0.05	0.24	0.11	24.3	29.4	0	0.6
90	63	0.64	19.8	194.4	198.0	2.45	34.2	0.05	—	0.04	1.74	0.10	31.8	1.7	3.6	24.4
27	266	1.07	38.5	251.4	276.3	1.26	—	0.07	0.02	0.16	1.27	0.17	44.6	59.3	0.2	6.0
0	28	1.74	41.8	298.4	149.9	0.76	0.5	0.07	—	0.08	0.53	0.06	47.4	0.8	0	0.5
9	218	1.83	38.4	163.9	472.9	1.24	—	0.41	0.00	0.35	2.85	0.09	96.2	1.6	0.4	38.4
73	146	2.66	61.3	222.9	885.7	3.43	0	0.10	0.10	0.16	3.00	0.26	95.7	0	1.2	36.7
103	38	2.77	33.7	196.0	833.6	1.34	41.6	0.33	1.60	0.18	4.17	0.27	97.0	3.4	0.3	22.0
66	26	1.46	23.4	248.2	849.7	1.40	29.2	0.10	1.62	0.18	6.28	0.28	23.4	12.0	0.2	19.9
0	13	0.65	11.4	178.0	26.9	0.21	59.0	0.03	—	0.05	0.56	0.08	38.3	24.0	0	0.4
0	60	1.91	35.7	212.8	297.7	1.34	0	0.10	0.92	0.06	0.49	0.49	72.6	9.7	0	3.0
65	50	1.96	63.2	427.7	907.2	1.50	38.9	0.15	4.32	0.14	13.22	0.58	42.1	7.5	0.3	23.0
90	38	2.95	40.1	493.2	495.6	6.60	—	0.08	1.29	0.28	8.03	0.21	28.3	1.4	2.9	30.4
68	233	2.22	40.1	420.1	433.6	2.70	—	0.21	0.21	0.29	3.06	0.22	50.4	15.0	0.8	21.4
34	310	2.06	40.0	258.0	784.0	2.06	180.0	0.27	0.72	0.43	2.18	0.08	64.0	0	0.5	30.6
15	35	2.10	20.1	283.6	1352.9	1.28	27.6	0.19	0.70	0.16	2.77	0.14	42.7	21.6	0.4	13.3
85	59	1.87	21.8	300.8	411.7	3.40	—	0.08	0.00	0.27	3.72	0.13	18.7	0.9	1.6	17.9
172	32	1.57	25.7	333.7	1052.5	1.82	49.8	0.49	5.39	0.36	2.88	0.31	21.1	8.0	0.8	30.0
154	948	7.32	242.4	1201.2	1862.4	10.68	259.0	0.29	7.71	0.81	6.39	1.09	148.8	16.0	2.6	44.1
28	129	1.87	17.0	305.3	533.9	1.03	105.1	0.26	—	0.46	6.09	0.11	73.8	3.3	0.4	26.1
36	258	3.14	31.6	352.3	971.7	2.02	49.0	0.37	1.13	0.37	3.77	0.19	64.8	15.6	0.6	22.8
85	24	0.81	18.8	317.5	661.3	0.38	40.0	0.09	—	0.07	1.11	0.17	8.8	12.5	0	5.1
60	74	1.94	23.8	314.5	360.5	3.45	—	0.10	0.00	0.27	3.95	0.13	20.8	0.7	1.6	17.4
180	102	4.68	63.2	413.1	330.5	1.72	94.8	0.29	2.06	0.11	4.75	0.21	75.3	12.9	1.2	49.3
89	124	3.70	—	665.0	1009.0	—	81.5	0.25	—	0.30	4.00	—	—	22.0	—	22.0

DA+ Code	Food Description	Quantity	Measure	Wt (g)	H₂O (g)	Ener (kcal)	Prot (g)	Carb (g)	Fiber (g)	Fat (g)	Sat	Mono	Poly	Trans
												Fat Breakdown (g)		
28585	Spicy thai noodles (pad thai)	8	ounce(s)	227	73.3	221	8.9	35.7	3.0	6.4	0.8	3.3	1.8	—
33073	Stir fried pork and vegetables with rice	1	cup(s)	235	173.6	348	15.4	33.5	1.9	16.3	5.6	6.9	2.6	0
28588	Stuffed shells	2 ½	item(s)	249	157.5	243	15.0	28.0	2.5	8.1	3.1	3.0	1.3	—
16821	Sushi with egg in seaweed	6	piece(s)	156	116.5	190	8.9	20.5	0.3	7.9	2.2	3.2	1.5	—
16819	Sushi with vegetables and fish	6	piece(s)	156	101.6	218	8.4	43.7	1.7	0.6	0.2	0.1	0.2	—
16820	Sushi with vegetables in seaweed	6	piece(s)	156	110.3	183	3.4	40.6	0.8	0.4	0.1	0.1	0.1	—
25266	Sweet and sour pork	¾	cup(s)	249	205.9	265	29.2	17.1	1.0	8.1	2.6	3.5	1.5	0
16824	Tabouli, tabbouleh or tabuli	1	cup(s)	160	123.7	198	2.6	15.9	3.7	14.9	2.0	10.9	1.6	—
25276	Three bean salad	½	cup(s)	99	82.2	95	1.9	9.7	2.6	5.9	0.8	1.4	3.5	0
160	Tuna salad	½	cup(s)	103	64.7	192	16.4	9.6	0	9.5	1.6	3.0	4.2	—
25241	Turkey and noodles	1	cup(s)	319	228.5	270	24.0	21.2	1.0	9.2	2.4	3.5	2.3	—
16794	Vegetable egg roll	2	item(s)	128	89.8	201	5.1	19.5	1.7	11.6	2.5	5.7	2.6	—
16818	Vegetable sushi, no fish	6	piece(s)	156	99.0	226	4.8	49.9	2.0	0.4	0.1	0.1	0.1	—
Sandwiches														
1744	Bacon, lettuce and tomato with mayonnaise	1	item(s)	164	97.2	341	11.6	34.2	2.3	17.6	3.8	5.5	6.7	—
30287	Bologna and cheese with margarine	1	item(s)	111	45.6	345	13.4	29.3	1.2	19.3	8.1	7.0	2.4	—
30286	Bologna with margarine	1	item(s)	83	33.6	251	8.1	27.3	1.2	12.1	3.7	5.0	2.1	—
16546	Cheese	1	item(s)	83	31.0	261	9.1	27.6	1.2	12.7	5.4	4.2	2.1	—
8789	Cheeseburger, large, plain	1	item(s)	185	78.9	564	32.0	38.5	2.6	31.5	12.5	10.2	1.0	1.8
8624	Cheeseburger, large, with bacon, vegetables and condiments	1	item(s)	195	91.4	550	30.8	36.8	2.5	30.9	11.9	10.6	1.3	1.5
1745	Club with bacon, chicken, tomato, lettuce and mayonnaise	1	item(s)	246	137.5	546	31.0	48.9	3.0	24.5	5.3	7.5	9.4	—
1908	Cold cut submarine with cheese and vegetables	1	item(s)	228	131.8	456	21.8	51.0	2.0	18.6	6.8	8.2	2.3	—
30247	Corned beef	1	item(s)	130	74.9	265	18.2	25.3	1.6	9.6	3.6	3.5	1.0	—
25283	Egg salad	1	item(s)	126	72.1	278	10.7	28.0	1.4	13.5	2.9	4.2	5.0	—
16686	Fried egg	1	item(s)	96	49.7	226	10.0	26.2	1.2	8.6	2.3	3.2	1.9	—
16547	Grilled cheese	1	item(s)	83	27.5	291	9.2	27.9	1.2	15.8	6.0	5.7	3.0	—
16659	Gyro with onion and tomato	1	item(s)	105	68.3	163	12.0	20.0	1.1	3.5	1.3	1.3	0.5	—
1906	Ham and cheese	1	item(s)	146	74.2	352	20.7	33.3	2.0	15.5	6.4	6.7	1.4	—
31890	Ham with mayonnaise	1	item(s)	112	56.3	271	13.0	27.9	1.9	11.6	2.8	4.0	4.0	—
756	Hamburger, double patty, large, with condiments and vegetables	1	item(s)	226	121.5	540	34.3	40.3	—	26.6	10.5	10.3	2.8	—
8793	Hamburger, large, plain	1	item(s)	137	57.7	426	22.6	31.7	1.5	22.9	8.4	9.9	2.1	—
8795	Hamburger, large, with vegetables and condiments	1	item(s)	218	121.4	512	25.8	40.0	3.1	27.4	10.4	11.4	2.2	—
25134	Hot chicken salad	1	item(s)	98	48.4	242	15.2	23.8	1.3	9.2	2.9	2.5	3.0	—
25133	Hot turkey salad	1	item(s)	98	50.1	224	15.6	23.8	1.3	6.9	2.3	1.6	2.5	—
1411	Hotdog with bun, plain	1	item(s)	98	52.9	242	10.4	18.0	1.6	14.5	5.1	6.9	1.7	—
30249	Pastrami	1	item(s)	134	71.2	328	13.4	27.8	1.6	17.7	6.1	8.3	1.2	—
16701	Peanut butter	1	item(s)	93	23.6	345	12.2	37.6	3.3	17.4	3.4	7.7	5.2	—
30306	Peanut butter and jelly	1	item(s)	93	24.2	330	10.3	41.9	2.9	14.7	2.9	6.5	4.4	—
1909	Roast beef submarine with mayonnaise and vegetables	1	item(s)	216	127.4	410	28.6	44.3	—	13.0	7.1	1.8	2.6	—
1910	Roast beef, plain	1	item(s)	139	67.6	346	21.5	33.4	1.2	13.8	3.6	6.8	1.7	—
1907	Steak with mayonnaise and vegetables	1	item(s)	204	104.2	459	30.3	52.0	2.3	14.1	3.8	5.3	3.3	—
25288	Tuna salad	1	item(s)	179	102.2	415	24.5	28.4	1.6	22.4	3.5	6.2	11.4	—
30283	Turkey submarine with cheese, lettuce, tomato and mayonnaise	1	item(s)	277	168.0	529	30.4	49.4	3.0	22.8	6.8	6.0	8.6	—
31891	Turkey with mayonnaise	1	item(s)	143	74.5	329	28.7	26.4	1.3	11.2	2.6	2.6	4.8	—
Soups														
25296	Bean	1	cup(s)	301	253.1	191	13.8	29.0	6.5	2.3	0.7	0.8	0.5	0
711	Bean with pork, condensed, prepared with water	1	cup(s)	253	215.9	159	7.3	21.0	7.3	5.5	1.4	2.0	1.7	—
713	Beef noodle, condensed, prepared with water	1	cup(s)	244	224.9	83	4.7	8.7	0.7	3.0	1.1	1.2	0.5	—
825	Cheese, condensed, prepared with milk	1	cup(s)	251	206.9	231	9.5	16.2	1.0	14.6	9.1	4.1	0.5	—
826	Chicken broth, condensed, prepared with water	1	cup(s)	244	234.1	39	4.9	0.9	0	1.4	0.4	0.6	0.3	—
25297	Chicken noodle soup	1	cup(s)	286	258.4	117	10.8	10.9	0.9	2.9	0.8	1.1	0.7	—
827	Chicken noodle, condensed, prepared with water	1	cup(s)	241	226.1	60	3.1	7.1	0.5	2.3	0.6	1.0	0.6	0
724	Chicken noodle, dehydrated, prepared with water	1	cup(s)	252	237.3	58	2.1	9.2	0.3	1.4	0.3	0.5	0.4	—
823	Cream of asparagus, condensed, prepared with milk	1	cup(s)	248	213.3	161	6.3	16.4	0.7	8.2	3.3	2.1	2.2	—
824	Cream of celery, condensed, prepared with milk	1	cup(s)	248	214.4	164	5.7	14.5	0.7	9.7	3.9	2.5	2.7	—
708	Cream of chicken, condensed, prepared with milk	1	cup(s)	248	210.4	191	7.5	15.0	0.2	11.5	4.6	4.5	1.6	—
715	Cream of chicken, condensed, prepared with water	1	cup(s)	244	221.1	117	3.4	9.3	0.2	7.4	2.1	3.3	1.5	—
709	Cream of mushroom, condensed, prepared with milk	1	cup(s)	248	215.0	166	6.2	14.0	0	9.6	3.3	2.0	1.8	0.1
716	Cream of mushroom, condensed, prepared with water	1	cup(s)	244	224.6	102	1.9	8.0	0	7.0	1.6	1.3	1.7	—
25298	Cream of vegetable	1	cup(s)	285	250.7	165	7.2	15.2	1.9	8.6	1.6	4.6	1.9	—
16689	Egg drop	1	cup(s)	244	228.9	73	7.5	1.1	0	3.8	1.1	1.5	0.6	—
MIXED FOODS, SOUPS, SANDWICHES —Continued														
25138	Golden squash	1	cup(s)	258	223.9	145	7.6	20.4	0.4	4.1	0.8	2.2	0.9	—
16663	Hot and sour	1	cup(s)	244	209.7	161	15.0	5.4	0.5	7.9	2.7	3.4	1.1	—

Chol (mg)	Calc (mg)	Iron (mg)	Magn (mg)	Pota (mg)	Sodi (mg)	Zinc (mg)	Vit A (µg)	Thia (mg)	Vit E (mg α)	Ribo (mg)	Niac (mg)	Vit B6 (mg)	Fola (µg)	Vit C (mg)	Vit B12 (µg)	Sele (µg)
37	31	1.56	49.4	181.3	591.6	1.05	—	0.18	0.35	0.13	1.82	0.17	44.6	22.6	0.1	3.2
46	38	2.71	33.0	396.9	569.5	2.08	—	0.51	0.38	0.20	5.07	0.30	103.0	18.8	0.4	22.8
30	188	2.26	49.4	403.0	471.5	1.41	—	0.26	0.00	0.26	3.83	0.24	80.0	17.9	0.2	28.9
214	45	1.84	18.7	135.7	463.3	0.98	106.1	0.13	0.67	0.28	1.35	0.13	62.4	1.9	0.7	20.3
11	23	2.15	25.0	202.8	340.1	0.78	45.2	0.26	0.24	0.07	2.76	0.14	76.4	3.6	0.3	13.9
0	20	1.54	18.7	96.7	152.9	0.68	25.0	0.19	0.12	0.03	1.86	0.13	73.3	2.3	0	9.8
74	40	1.76	35.6	619.9	621.8	2.53	—	0.81	0.20	0.37	6.69	0.66	14.7	11.9	0.7	49.6
0	30	1.21	35.2	249.6	796.8	0.48	54.4	0.07	2.43	0.04	1.11	0.11	30.4	26.1	0	0.5
0	26	0.96	15.5	144.8	224.2	0.30	—	0.02	0.88	0.04	0.26	0.04	32.2	10.0	0	2.7
13	17	1.02	19.5	182.5	412.1	0.57	24.6	0.03	—	0.07	6.86	0.08	8.2	2.3	1.2	42.2
77	69	2.56	33.2	400.8	577.1	2.51	—	0.23	0.28	0.30	6.41	0.29	61.4	1.4	1.1	33.4
60	29	1.65	17.9	193.3	549.1	0.48	25.6	0.15	1.28	0.20	1.59	0.09	46.1	5.5	0.2	11.3
0	23	2.38	21.8	157.6	369.7	0.82	48.4	0.28	0.15	0.05	2.44	0.12	85.8	3.7	0	8.1
21	79	2.36	27.9	351.0	944.6	1.08	44.3	0.32	1.16	0.24	4.36	0.21	73.8	9.7	0.2	27.1
40	258	2.38	24.4	215.3	941.3	1.88	102.1	0.31	0.55	0.35	2.97	0.14	59.9	0.2	0.8	20.4
17	100	2.24	16.6	138.6	579.3	1.02	44.8	0.29	0.49	0.21	2.92	0.12	58.1	0.2	0.5	16.0
22	233	2.04	19.9	127.0	733.7	1.24	97.1	0.25	0.47	0.30	2.25	0.05	58.1	0	0.3	13.1
104	309	4.47	44.4	401.5	986.1	5.75	0	0.35	—	0.77	8.26	0.49	109.2	0	2.8	38.9
98	267	4.03	44.9	464.1	1314.3	5.20	0	0.33	—	0.67	8.25	0.47	95.6	1.4	2.4	6.6
71	157	4.57	46.7	464.9	1087.3	1.82	41.8	0.54	1.52	0.40	12.82	0.61	118.1	6.4	0.4	42.3
36	189	2.50	68.4	394.4	1650.7	2.57	70.7	1.00	—	0.79	5.49	0.13	86.6	12.3	1.1	30.8
46	81	3.04	19.5	127.4	1206.4	2.26	2.6	0.23	0.20	0.24	3.42	0.11	59.8	0.3	0.9	31.2
219	85	2.25	18.8	159.0	423.1	0.87	—	0.27	0.12	0.43	2.06	0.15	73.6	0.9	0.6	29.9
206	104	2.79	17.3	117.1	438.7	0.92	89.3	0.26	0.66	0.40	2.26	0.10	79.7	0	0.6	24.2
22	235	2.05	19.9	128.7	763.6	1.26	129.5	0.19	0.72	0.28	2.05	0.05	38.2	0	0.2	13.2
28	47	1.77	22.1	218.4	235.2	2.33	9.5	0.23	0.26	0.20	3.12	0.13	44.1	3.2	0.9	18.3
58	130	3.24	16.1	290.5	770.9	1.37	96.4	0.30	0.29	0.48	2.68	0.20	75.9	2.8	0.5	23.1
34	91	2.47	23.5	210.6	1097.6	1.13	5.6	0.57	0.50	0.26	3.80	0.25	60.5	2.2	0.2	20.2
122	102	5.85	49.7	569.5	791.0	5.67	0	0.36	—	0.38	7.57	0.54	76.8	1.1	4.1	25.5
71	74	3.57	27.4	267.2	474.0	4.11	0	0.28	—	0.28	6.24	0.23	60.3	0	2.1	27.1
87	96	4.92	43.6	479.6	824.0	4.88	0	0.41	—	0.37	7.28	0.32	82.8	2.6	2.4	33.6
39	115	1.88	20.0	172.4	505.1	1.19	—	0.23	0.28	0.22	4.84	0.19	46.4	0.5	0.2	19.9
37	114	1.99	21.3	189.0	494.5	1.07	—	0.22	0.28	0.20	4.26	0.22	46.4	0.5	0.2	23.4
44	24	2.31	12.7	143.1	670.3	1.98	0	0.23	—	0.27	3.64	0.04	48.0	0.1	0.5	26.0
51	80	3.02	22.8	182.2	1364.1	2.70	2.7	0.28	0.26	0.26	4.97	0.14	60.3	0.3	1.0	14.3
0	110	2.92	66.0	226.0	580.3	1.32	0	0.31	2.39	0.23	6.72	0.18	92.1	0	0	13.2
0	94	2.50	56.7	198.1	492.9	1.12	—	0.26	2.01	0.20	5.66	0.15	78.1	0.1	0	11.2
73	41	2.80	67.0	330.5	844.6	4.38	30.2	0.41	—	0.41	5.96	0.32	71.3	5.6	1.8	25.7
51	54	4.22	30.6	315.5	792.3	3.39	11.1	0.37	—	0.30	5.86	0.26	57.0	2.1	1.2	29.2
73	92	5.16	49.0	524.3	797.6	4.52	0	0.40	—	0.36	7.30	0.36	89.8	5.5	1.6	42.0
59	78	2.97	35.9	316.3	724.7	1.02	—	0.27	0.34	0.26	12.07	0.46	60.8	1.8	2.4	76.9
64	307	4.59	49.9	534.6	1759.0	2.74	74.8	0.49	1.19	0.65	3.91	0.31	121.9	10.5	0.6	42.1
67	100	3.46	34.3	304.6	564.9	3.00	5.7	0.28	0.74	0.32	6.84	0.46	64.4	0	0.3	40
5	79	3.05	61.8	588.8	689.0	1.41	—	0.27	0.02	0.15	3.63	0.23	140.1	3.6	0.2	7.9
3	78	1.89	43.0	371.9	883.0	0.96	43.0	0.08	1.08	0.03	0.52	0.03	30.4	1.5	0	7.8
5	20	1.07	7.3	97.6	929.6	1.51	12.2	0.06	1.22	0.05	1.03	0.03	19.5	0.5	0.2	7.3
48	289	0.80	20.1	341.4	1019.1	0.67	358.9	0.06	—	0.33	0.50	0.07	10.0	1.3	0.4	7.0
0	10	0.51	2.4	209.8	775.9	0.24	0	0.01	0.04	0.07	3.34	0.02	4.9	0	0.2	0
24	25	1.38	16.4	340.1	774.5	0.77	—	0.15	0.02	0.16	5.57	0.13	37.2	1.8	0.3	10.2
12	14	1.59	9.6	53.0	638.7	0.38	26.5	0.13	0.07	0.10	1.30	0.04	19.3	0	0	11.6
10	5	0.50	7.6	32.8	577.1	0.20	2.5	0.20	0.12	0.07	1.08	0.02	17.6	0	0.1	9.6
22	174	0.86	19.8	359.6	1041.6	0.91	62.0	0.10	—	0.27	0.88	0.06	29.8	4.0	0.5	8.0
32	186	0.69	22.3	310.0	1009.4	0.19	114.1	0.07	—	0.24	0.43	0.06	7.4	1.5	0.5	4.7
27	181	0.67	17.4	272.8	1046.6	0.67	178.6	0.07	—	0.25	0.92	0.06	7.4	1.2	0.5	8.0
10	34	0.61	2.4	87.8	985.8	0.63	163.5	0.02	—	0.06	0.82	0.01	2.4	0.2	0.1	7.0
10	164	1.36	19.8	267.8	823.4	0.79	81.8	0.10	1.01	0.29	0.62	0.05	7.4	0.2	0.6	6.0
0	17	1.31	4.9	73.2	775.9	0.24	9.8	0.05	0.97	0.05	0.50	0.00	2.4	0	0	2.9
1	80	1.20	17.5	340.7	787.9	0.56	—	0.12	1.05	0.18	3.32	0.12	39.5	10.7	0.3	4.3
102	22	0.75	4.9	219.6	729.6	0.48	41.5	0.02	0.29	0.19	3.02	0.05	14.6	0	0.5	7.6
4	262	0.78	42.4	542.2	515.6	0.88	—	0.16	0.52	0.30	1.14	0.16	32.7	12.5	0.7	6.0
34	29	1.24	19.5	373.3	1561.6	1.43	—	0.26	0.12	0.24	4.96	0.20	14.6	0.5	0.4	19.3

Food Composition   (Computer code is for Cengage Diet Analysis program)   (For purposes of calculations, use "0" for t, <1, <.1, <.01, etc.)

DA+ Code	Food Description	Quantity	Measure	Wt (g)	H₂O (g)	Ener (kcal)	Prot (g)	Carb (g)	Fiber (g)	Fat (g)	Fat Breakdown (g)			
											Sat	Mono	Poly	Trans
28054	Lentil chowder	1	cup(s)	244	202.8	153	11.4	27.7	12.6	0.5	0.1	0.1	0.2	0
28560	Macaroni and bean	1	cup(s)	246	138.8	146	5.8	22.9	5.1	3.7	0.5	2.2	0.6	0
714	Manhattan clam chowder, condensed, prepared with water	1	cup(s)	244	225.1	73	2.1	11.6	1.5	2.1	0.4	0.4	1.2	—
28561	Minestrone	1	cup(s)	241	185.4	103	4.5	16.8	4.8	2.3	0.3	1.4	0.4	0
717	Minestrone, condensed, prepared with water	1	cup(s)	241	220.1	82	4.3	11.2	1.0	2.5	0.6	0.7	1.1	—
28038	Mushroom & wild rice	1	cup(s)	244	199.7	86	4.7	13.2	1.7	0.3	0	0	0.2	0
828	New England clam chowder, condensed, prepared with milk	1	cup(s)	248	212.2	151	8.0	18.4	0.7	5.0	2.1	0.7	0.6	0
28036	New England style clam chowder	1	cup(s)	244	227.5	61	3.8	8.8	1.8	0.2	0.1	0	0	0
28566	Old country pasta	1	cup(s)	252	183.3	146	6.5	18.3	3.6	4.5	2.0	2.4	0.9	0
725	Onion, dehydrated, prepared with water	1	cup(s)	246	235.7	30	0.8	6.8	0.7	0	0	0	0	0
16667	Shrimp gumbo	1	cup(s)	244	207.2	166	9.5	18.2	2.4	6.7	1.3	2.9	2.0	—
28037	Southwestern corn chowder	1	cup(s)	244	217.8	98	4.9	17.0	2.4	0.5	0.1	0.1	0.2	0
30282	Soybean (miso)	1	cup(s)	240	218.6	84	6.0	8.0	1.9	3.4	0.6	1.1	1.4	—
25140	Split pea	1	cup(s)	165	119.7	72	4.5	16.0	1.6	0.3	0.1	0	0.2	0
718	Split pea with ham, condensed, prepared with water	1	cup(s)	253	206.9	190	10.3	28.0	2.3	4.4	1.8	1.8	0.6	—
726	Tomato vegetable, dehydrated, prepared with water	1	cup(s)	253	238.4	56	2.0	10.2	0.8	0.9	0.4	0.3	0.1	0
710	Tomato, condensed, prepared with milk	1	cup(s)	248	213.4	136	6.2	22.0	1.5	3.2	1.8	0.9	0.3	—
719	Tomato, condensed, prepared with water	1	cup(s)	244	223.0	73	1.9	16.0	1.5	0.7	0.2	0.2	0.2	—
28595	Turkey noodle	1	cup(s)	244	216.9	114	8.1	15.1	1.9	2.4	0.3	1.1	0.7	—
28051	Turkey vegetable	1	cup(s)	244	220.8	96	12.2	6.6	2.0	1.1	0.3	0.2	0.3	0
25141	Vegetable	1	cup(s)	252	228.1	82	5.2	16.5	4.5	0.3	0	0	0.1	0
720	Vegetable beef, condensed, prepared with water	1	cup(s)	244	224.0	76	5.4	9.9	2.0	1.9	0.8	0.8	0.1	—
28598	Vegetable gumbo	1	cup(s)	252	184.5	170	4.4	28.9	3.6	4.7	0.7	3.2	0.5	0
721	Vegetarian vegetable, condensed, prepared with water	1	cup(s)	241	222.7	67	2.1	11.8	0.7	1.9	0.3	0.8	0.7	—
	FAST FOOD													
	Arby's													
36094	Au jus sauce	1	serving(s)	85	—	43	1.0	7.0	0	1.3	0.4	—	—	0.4
751	Beef 'n cheddar sandwich	1	item(s)	195	—	445	22.0	44.0	2.0	21.0	6.0	—	—	1.0
9279	Cheddar curly fries	1	serving(s)	198	—	631	8.0	73.0	7.0	37.4	6.8	—	—	5.7
34770	Chicken breast fillet sandwich, grilled	1	item(s)	233	—	414	32.0	36.0	3.0	17.0	3.0	—	—	0
36131	Chocolate shake, regular	1	serving(s)	397	—	507	13.0	83.0	0	13.0	8.0	—	—	0
36045	Curly fries, large size	1	serving(s)	198	—	631	8.0	73.0	7.0	37.0	7.0	—	—	6.0
36044	Curly fries, medium size	1	serving(s)	128	—	406	5.0	47.0	5.0	24.0	4.0	—	—	4.0
752	Ham 'n cheese sandwich	1	item(s)	167	—	304	23.0	35.0	1.0	7.0	2.0	—	—	0
36048	Homestyle fries, large size	1	serving(s)	213	—	566	6.0	82.0	6.0	37.0	7.0	—	—	5.0
36047	Homestyle fries, medium size	1	serving(s)	142	—	377	4.0	55.0	4.0	25.0	4.0	—	—	4.0
33465	Homestyle fries, small size	1	serving(s)	113	—	302	3.0	44.0	3.0	20.0	4.0	—	—	3.0
9249	Junior roast beef sandwich	1	item(s)	125	—	272	16.0	34.0	2.0	10.0	4.0	—	—	0
9251	Large roast beef sandwich	1	item(s)	281	—	547	42.0	41.0	3.0	28.0	12.0	—	—	2.0
39640	Market Fresh chicken salad with pecans sandwich	1	item(s)	322	—	769	30.0	79.0	9.0	39.0	10.0	—	—	0
39641	Market Fresh Martha's Vineyard salad, without dressing	1	serving(s)	330	—	277	26.0	24.0	5.0	8.0	4.0	—	—	0
34769	Market Fresh roast turkey & Swiss sandwich	1	serving(s)	359	—	725	45.0	75.0	5.0	30.0	8.0	—	—	1.0
9267	Market Fresh roast turkey ranch & bacon sandwich	1	serving(s)	382	—	834	49.0	75.0	5.0	38.0	11.0	—	—	1.0
39642	Market Fresh Santa Fe salad, without dressing	1	serving(s)	372	—	499	30.0	42.0	7.0	23.0	8.0	—	—	2.0
39650	Market Fresh Southwest chicken wrap	1	serving(s)	251	—	567	36.0	42.0	4.0	29.0	9.0	—	—	1.0
37021	Market Fresh Ultimate BLT sandwich	1	item(s)	294	—	779	23.0	75.0	6.0	45.0	11.0	—	—	1.0
750	Roast beef sandwich, regular	1	item(s)	154	—	320	21.0	34.0	2.0	14.0	5.0	—	—	1.0
36132	Strawberry shake, regular	1	serving(s)	397	—	498	13.0	81.0	0	13.0	8.0	—	—	0
2009	Super roast beef sandwich	1	item(s)	198	—	398	21.0	40.0	2.0	19.0	6.0	—	—	1.0
36130	Vanilla shake, regular	1	serving(s)	369	—	437	13.0	66.0	0	13.0	8.0	—	—	0
	Auntie Anne's													
35371	Cheese dipping sauce	1	serving(s)	35	—	100	3.0	4.0	0	8.0	4.0	—	—	0
35353	Cinnamon sugar soft pretzel	1	item(s)	120	—	350	9.0	74.0	2.0	2.0	0	—	—	0
35354	Cinnamon sugar soft pretzel with butter	1	item(s)	120	—	450	8.0	83.0	3.0	9.0	5.0	—	—	0
35372	Marinara dipping sauce	1	serving(s)	35	—	10	0	4.0	0	0	0	0	0	0
35357	Original soft pretzel	1	serving(s)	120	—	340	10.0	72.0	3.0	1.0	0	—	—	0
35358	Original soft pretzel with butter	1	item(s)	120	—	370	10.0	72.0	3.0	4.0	2.0	—	—	0
35359	Parmesan herb soft pretzel	1	item(s)	120	—	390	11.0	74.0	4.0	5.0	2.5	—	—	0
35360	Parmesan herb soft pretzel with butter	1	item(s)	120	—	440	10.0	72.0	9.0	13.0	7.0	—	—	0
35361	Sesame soft pretzel	1	item(s)	120	—	350	11.0	63.0	3.0	6.0	1.0	—	—	0
35362	Sesame soft pretzel with butter	1	item(s)	120	—	410	12.0	64.0	7.0	12.0	4.0	—	—	0
	FAST FOOD —Continued													
35364	Sour cream & onion soft pretzel	1	item(s)	120	—	310	9.0	66.0	2.0	1.0	0	—	—	0
35366	Sour cream & onion soft pretzel with butter	1	item(s)	120	—	340	9.0	66.0	2.0	5.0	3.0	—	—	0

Chol (mg)	Calc (mg)	Iron (mg)	Magn (mg)	Pota (mg)	Sodi (mg)	Zinc (mg)	Vit A (µg)	Thia (mg)	Vit E (mg α)	Ribo (mg)	Niac (mg)	Vit B_6 (mg)	Fola (µg)	Vit C (mg)	Vit B_{12} (µg)	Sele (µg)
0	48	4.38	59.3	626.2	26.7	1.57	—	0.24	0.06	0.12	1.87	0.32	176.3	16.1	0	3.5
0	59	1.90	32.4	275.9	531.0	0.51	—	0.16	0.37	0.12	1.44	0.10	58.2	9.1	0	8.8
2	27	1.56	9.8	180.6	551.4	0.87	48.8	0.02	1.22	0.03	0.77	0.09	9.8	3.9	3.9	9.0
0	62	1.70	29.9	287.5	442.7	0.42	—	0.09	0.23	0.09	0.70	0.07	47.3	13.3	0	3.5
2	34	0.91	7.2	313.3	911.0	0.74	118.1	0.05	—	0.04	0.94	0.09	36.2	1.2	0	8.0
0	30	1.38	27.3	376.9	283.8	1.00	—	0.06	0.07	0.23	3.21	0.14	18.1	3.7	0.1	4.7
17	169	3.00	29.8	456.3	887.8	0.99	91.8	0.20	0.54	0.43	1.96	0.17	22.3	5.2	11.9	10.9
3	89	1.30	29.2	503.7	256.8	0.52	—	0.06	0.02	0.10	1.30	0.15	24.2	11.8	3.0	3.8
5	57	2.43	51.9	500.4	355.7	0.78	—	0.21	0.01	0.14	2.64	0.20	80.0	22.0	0.1	10.3
0	22	0.12	9.8	76.3	851.2	0.12	0	0.03	0.02	0.03	0.15	0.06	0	0.2	0	0.5
51	105	2.85	48.8	461.2	441.6	0.90	80.5	0.18	1.90	0.12	2.52	0.19	85.4	17.6	0.3	14.9
1	83	1.03	26.3	434.4	217.4	0.57	—	0.08	0.09	0.13	1.81	0.21	32.1	39.6	0.2	1.7
0	65	1.87	36.0	362.4	988.8	0.86	232.8	0.06	0.96	0.16	2.61	0.15	57.6	4.6	0.2	1.0
0	28	1.26	29.8	328.1	602.4	0.52	—	0.10	0.00	0.07	1.50	0.16	49.5	8.1	0	0.7
8	23	2.27	48.1	399.7	1006.9	1.31	22.8	0.14	—	0.07	1.47	0.06	2.5	1.5	0.3	8.0
0	20	0.60	10.1	169.5	334.0	0.20	10.1	0.06	0.43	0.09	1.26	0.06	12.7	3.0	0.1	2.0
10	166	1.36	29.8	466.2	711.8	0.84	94.2	0.09	0.44	0.31	1.35	0.15	5.0	15.6	0.6	9.2
0	20	1.31	17.1	273.3	663.7	0.29	24.4	0.04	0.41	0.07	1.23	0.10	0	15.4	0	6.1
26	28	1.40	24.1	223.8	395.9	0.75	—	0.21	0.01	0.12	2.85	0.15	45.0	7.1	0.1	13.7
21	38	1.48	25.0	423.4	348.7	0.99	—	0.09	0.01	0.09	3.64	0.27	23.9	10.6	0.2	10.3
0	40	2.08	39.1	681.5	670.3	0.67	—	0.16	0.00	0.09	2.70	0.26	36.3	22.4	0	2.2
5	20	1.09	7.3	168.4	773.5	1.51	190.3	0.03	0.58	0.04	1.00	0.07	9.8	2.4	0.3	2.7
0	56	1.85	38.9	360.2	518.4	0.64	—	0.18	0.64	0.08	1.77	0.17	57.6	21.9	0	4.1
0	24	1.06	7.2	207.3	814.6	0.45	171.1	0.05	1.39	0.04	0.90	0.05	9.6	1.4	0	4.3
0	0	—	—	—	1510.0	—	—	—	—	—	—	—	—	—	—	—
51	80	3.96	—	—	1274.0	—	—	—	—	—	—	—	—	1.8	—	—
0	80	3.24	—	—	1476.0	—	—	—	—	—	—	—	—	9.6	—	—
9	90	3.06	—	—	913.0	—	—	—	—	—	—	—	—	10.8	—	—
34	510	0.54	—	—	357.0	—	—	—	—	—	—	—	—	5.4	—	—
0	80	3.24	—	—	1476.0	—	—	—	—	—	—	—	—	9.6	—	—
0	50	1.98	—	—	949.0	—	—	—	—	—	—	—	—	6.0	—	—
35	160	2.70	—	—	1420.0	—	—	—	—	—	—	—	—	1.2	—	—
0	50	1.62	—	—	1029.0	—	—	—	—	—	—	—	—	12.6	—	—
0	30	1.08	—	—	686.0	—	—	—	—	—	—	—	—	8.4	—	—
0	30	0.90	—	—	549.0	—	—	—	—	—	—	—	—	6.6	—	—
29	60	3.06	—	—	740.0	—	0	—	—	—	—	—	—	0	—	—
102	70	6.30	—	—	1869.0	—	0	—	—	—	—	—	—	0.6	—	—
74	180	4.32	—	—	1240.0	—	—	—	—	—	—	—	—	30.0	—	—
72	200	1.62	—	—	454.0	—	—	—	—	—	—	—	—	33.6	—	—
91	360	5.22	—	—	1788.0	—	—	—	—	—	—	—	—	10.2	—	—
109	330	5.40	—	—	2258.0	—	—	—	—	—	—	—	—	11.4	—	—
59	420	3.60	—	—	1231.0	—	—	—	—	—	—	—	—	36.6	—	—
88	240	4.50	—	—	1451.0	—	—	—	—	—	—	—	—	7.8	—	—
51	170	4.68	—	—	1571.0	—	—	—	—	—	—	—	—	16.8	—	—
44	60	3.60	—	—	953.0	—	0	—	—	—	—	—	—	0	—	—
34	510	0.72	—	—	363.0	—	—	—	—	—	—	—	—	6.6	—	—
44	70	3.78	—	—	1060.0	—	—	—	—	—	—	—	—	6.0	—	—
34	510	0.36	—	—	350.0	—	—	—	—	—	—	—	—	5.4	—	—
10	100	0.00	—	—	510.0	—	—	—	—	—	—	—	—	0	—	—
0	20	1.98	—	—	410.0	—	0	—	—	—	—	—	—	0	—	—
25	30	2.34	—	—	430.0	—	—	—	—	—	—	—	—	0	—	—
0	0	0.00	—	—	180.0	—	0	—	—	—	—	—	—	0	—	—
0	30	2.34	—	—	900.0	—	0	—	—	—	—	—	—	0	—	—
10	30	2.16	—	—	930.0	—	—	—	—	—	—	—	—	0	—	—
10	80	1.80	—	—	780.0	—	—	—	—	—	—	—	—	1.2	—	—
30	60	1.80	—	—	660.0	—	—	—	—	—	—	—	—	1.2	—	—
0	20	2.88	—	—	840.0	—	0	—	—	—	—	—	—	0	—	—
15	20	2.70	—	—	860.0	—	—	—	—	—	—	—	—	0	—	—
0	30	1.98	—	—	920.0	—	—	—	—	—	—	—	—	0	—	—
10	40	2.16	—	—	930.0	—	—	—	—	—	—	—	—	0	—	—

DA+ Code	Food Description	Quantity	Measure	Wt (g)	H₂O (g)	Ener (kcal)	Prot (g)	Carb (g)	Fiber (g)	Fat (g)	Fat Breakdown (g)			
											Sat	Mono	Poly	Trans
35373	Sweet mustard dipping sauce	1	serving(s)	35	—	60	0.5	8.0	0	1.5	1.0	—	—	0
35367	Whole wheat soft pretzel	1	item(s)	120	—	350	11.0	72.0	7.0	1.5	0	—	—	0
35368	Whole wheat soft pretzel with butter	1	item(s)	120	—	370	11.0	72.0	7.0	4.5	1.5	—	—	0
Boston Market														
34978	Butternut squash	¾	cup(s)	143	—	140	2.0	25.0	2.0	4.5	3.0	—	—	0
35006	Caesar side salad	1	serving(s)	71	—	40	3.0	3.0	1.0	20.0	2.0	—	—	1.5
35013	Chicken Carver sandwich with cheese and sauce	1	item(s)	321	—	700	44.0	68.0	3.0	29.0	7.0	—	—	0
34979	Chicken gravy	4	ounce(s)	113	—	15	1.0	4.0	0	0.5	0	—	—	0
35053	Chicken noodle soup	¾	cup(s)	283	—	180	13.0	16.0	1.0	7.0	2.0	—	—	0
34973	Chicken pot pie	1	item(s)	425	—	800	29.0	59.0	4.0	49.0	18.0	—	—	7.0
35054	Chicken tortilla soup with toppings	¾	cup(s)	227	—	340	12.0	24.0	1.0	22.0	7.0	—	—	0
35007	Cole slaw	¾	cup(s)	125	—	170	2.0	21.0	2.0	9.0	2.0	—	—	0
35057	Cornbread	1	item(s)	45	—	130	1.0	21.0	0	3.5	1.0	—	—	1.0
34980	Creamed spinach	¾	cup(s)	191	—	280	9.0	12.0	4.0	23.0	15.0	—	—	0
34998	Fresh vegetable stuffing	1	cup(s)	136	—	190	3.0	25.0	2.0	8.0	1.0	—	—	0
34991	Garlic dill new potatoes	¾	cup(s)	156	—	140	3.0	24.0	3.0	3.0	1.0	—	—	0
34983	Green bean casserole	¾	cup(s)	170	—	60	2.0	9.0	2.0	2.0	1.0	—	—	0
34982	Green beans	¾	cup(s)	91	—	60	2.0	7.0	3.0	3.5	1.5	—	—	0
34984	Homestyle mashed potatoes	¾	cup(s)	221	—	210	4.0	29.0	3.0	9.0	6.0	—	—	0
34985	Homestyle mashed potatoes and gravy	1	cup(s)	334	—	225	5.0	33.0	3.0	9.5	6.0	—	—	0
34988	Hot cinnamon apples	¾	cup(s)	145	—	210	0	47.0	3.0	3.0	0	—	—	0
34989	Macaroni and cheese	¾	cup(s)	221	—	330	14.0	39.0	1.0	12.0	7.0	—	—	0.5
51193	Market chopped salad with dressing	1	item(s)	563	—	580	11.0	31.0	9.0	48.0	9.0	—	—	1.0
34970	Meatloaf	1	serving(s)	218	—	480	29.0	23.0	2.0	33.0	13.0	—	—	0
39383	Nestle Toll House chocolate chip cookie	1	item(s)	78	—	370	4.0	49.0	2.0	19.0	9.0	—	—	0
34965	Quarter chicken, dark meat, no skin	1	item(s)	134	—	260	30.0	2.0	0	13.0	4.0	—	—	0
34966	Quarter chicken, dark meat, with skin	1	item(s)	149	—	280	31.0	3.0	0	15.0	4.5	—	—	0
34963	Quarter chicken, white meat, no skin or wing	1	item(s)	173	—	250	41.0	4.0	0	8.0	2.5	—	—	0
34964	Quarter chicken, white meat, with skin and wing	1	item(s)	110	—	330	50.0	3.0	0	12.0	4.0	—	—	0
34968	Roasted turkey breast	5	ounce(s)	142	—	180	38.0	0	0	3.0	1.0	—	—	0
35011	Seasonal fresh fruit salad	1	serving(s)	142	—	60	1.0	15.0	1.0	0	0	0	0	0
51192	Spinach with garlic butter sauce	1	serving(s)	170	—	130	5.0	9.0	5.0	9.0	6.0	—	—	0
34969	Spiral sliced holiday ham	8	ounce(s)	227	—	450	40.0	13.0	0	26.0	10.0	—	—	0
35003	Steamed vegetables	1	cup(s)	136	—	50	2.0	8.0	3.0	2.0	0	—	—	0
35005	Sweet corn	¾	cup(s)	176	—	170	6.0	37.0	2.0	4.0	1.0	—	—	0
35004	Sweet potato casserole	¾	cup(s)	198	—	460	4.0	77.0	3.0	17.0	6.0	—	—	0
Burger King														
29731	Biscuit with sausage, egg & cheese	1	item(s)	191	—	610	20.0	33.0	1.0	45.0	15.0	—	—	1.0
14249	Cheeseburger	1	item(s)	133	—	330	17.0	31.0	1.0	16.0	7.0	—	—	0.5
14251	Chicken sandwich	1	item(s)	219	—	660	24.0	52.0	4.0	40.0	8.0	—	—	2.5
3808	Chicken Tenders, 8 pieces	1	serving(s)	123	—	340	19.0	21.0	0.5	20.0	5.0	—	—	3.0
14259	Chocolate shake, small	1	item(s)	315	—	470	8.0	75.0	1.0	14.0	9.0	—	—	0
29732	Croissanwich with sausage & cheese	1	item(s)	106	37.2	370	14.0	23.0	0.5	25.0	9.0	12.7	3.3	2.0
14261	Croissanwich with sausage, egg & cheese	1	item(s)	159	71.4	470	19.0	26.0	0.5	32.0	11.0	15.8	6.1	2.5
3809	Double cheeseburger	1	item(s)	189	—	500	30.0	31.0	1.0	29.0	14.0	—	—	1.5
14244	Double Whopper sandwich	1	item(s)	373	—	900	47.0	51.0	3.0	57.0	19.0	—	—	2.0
14245	Double Whopper with cheese sandwich	1	item(s)	398	—	990	52.0	52.0	3.0	64.0	24.0	—	—	2.5
14250	Fish Filet sandwich	1	item(s)	250	—	630	24.0	67.0	4.0	30.0	6.0	—	—	2.5
14255	French fries, medium, salted	1	serving(s)	116	—	360	4.0	41.0	4.0	20.0	4.5	—	—	4.5
14262	French toast sticks (5)	1	serving(s)	112	37.6	390	6.0	46.0	2.0	20.0	4.5	10.6	2.9	4.5
14248	Hamburger	1	item(s)	121	—	290	15.0	30.0	1.0	12.0	4.5	—	—	0
14263	Hash brown rounds, small	1	serving(s)	75	27.1	230	2.0	23.0	2.0	15.0	4.0	—	—	5.0
14256	Onion rings, medium	1	serving(s)	91	—	320	4.0	40.0	3.0	16.0	4.0	—	—	3.5
39000	Tendercrisp chicken sandwich	1	item(s)	286	—	780	25.0	73.0	4.0	43.0	8.0	—	—	4.0
37514	TenderGrill chicken sandwich	1	item(s)	258	—	450	37.0	53.0	4.0	10.0	2.0	—	—	0
14258	Vanilla shake, small	1	item(s)	296	—	400	8.0	57.0	0	15.0	9.0	—	—	0
1736	Whopper sandwich	1	item(s)	290	—	670	28.0	51.0	3.0	39.0	11.0	—	—	1.5
14243	Whopper with cheese sandwich	1	item(s)	315	—	760	33.0	52.0	3.0	47.0	16.0	—	—	1.5
Carl's Jr														
33962	Carl's bacon Swiss crispy chicken sandwich	1	item(s)	268	—	750	31.0	91.0	—	28.0	28.0	—	—	—
10801	Carl's Catch Fish sandwich	1	item(s)	215	—	560	19.0	58.0	2.0	27.0	7.0	—	1.9	—
10862	Carl's Famous Star hamburger	1	item(s)	254	—	590	24.0	50.0	3.0	32.0	9.0	—	—	—
10785	Charbroiled chicken club sandwich	1	item(s)	270	—	550	42.0	43.0	4.0	23.0	7.0	—	2.9	—
10866	Charbroiled chicken salad	1	item(s)	437	—	330	34.0	17.0	5.0	7.0	4.0	—	1.0	—
10855	Charbroiled Santa Fe Chicken sandwich	1	item(s)	266	—	610	38.0	43.0	4.0	32.0	8.0	—	—	—
10790	Chicken stars (6 pieces)	6	item(s)	85	—	260	13.0	14.0	1.0	16.0	4.0	—	1.6	—
34864	Chocolate shake, small	1	serving(s)	595	—	540	15.0	98.0	0	11.0	7.0	—	—	—
FAST FOOD —Continued														
10797	Crisscut fries	1	serving(s)	139	—	410	5.0	43.0	4.0	24.0	5.0	—	—	—
10799	Double Western Bacon cheeseburger	1	item(s)	308	—	920	51.0	65.0	2.0	50	21.0	—	6.6	—

Chol (mg)	Calc (mg)	Iron (mg)	Magn (mg)	Pota (mg)	Sodi (mg)	Zinc (mg)	Vit A (µg)	Thia (mg)	Vit E (mg α)	Ribo (mg)	Niac (mg)	Vit B$_6$ (mg)	Fola (µg)	Vit C (mg)	Vit B$_{12}$ (µg)	Sele (µg)
40	0	0.00	—	—	120.0	—	0	—	—	—	—	—	—	0	—	—
0	30	1.98	—	—	1100.0	—	0	—	—	—	—	—	—	0	—	—
10	30	2.34	—	—	1120.0	—	—	—	—	—	—	—	—	0	—	—
10	59	0.80	—	—	35.0	—	—	—	—	—	—	—	—	22.2	—	—
0	60	0.43	—	—	75.0	—	—	—	—	—	—	—	—	5.4	—	—
90	211	2.85	—	—	1560.0	—	—	—	—	—	—	—	—	15.8	—	—
0	0	0.00	—	—	570.0	—	0	—	—	—	—	—	—	0	—	—
55	0	1.07	—	—	220.0	—	—	—	—	—	—	—	—	1.8	—	—
115	40	4.50	—	—	800.0	—	—	—	—	—	—	—	—	1.2	—	—
45	123	1.32	—	—	1310.0	—	—	—	—	—	—	—	—	18.4	—	—
10	41	0.48	—	—	270.0	—	—	—	—	—	—	—	—	24.5	—	—
5	0	0.71	—	—	220.0	—	0	—	—	—	—	—	—	0	—	—
70	264	2.84	—	—	580.0	—	—	—	—	—	—	—	—	9.5	—	—
0	41	1.48	—	—	580.0	—	—	—	—	—	—	—	—	2.5	—	—
0	0	0.85	—	—	120.0	—	0	—	—	—	—	—	—	14.3	—	—
5	20	0.72	—	—	620.0	—	—	—	—	—	—	—	—	2.4	—	—
0	43	0.38	—	—	180.0	—	—	—	—	—	—	—	—	5.1	—	—
25	51	0.46	—	—	660.0	—	—	—	—	—	—	—	—	19.2	—	—
25	100	0.59	—	—	1230.0	—	—	—	—	—	—	—	—	24.9	—	—
0	16	0.28	—	—	15.0	—	—	—	—	—	—	—	—	0	—	—
30	345	1.65	—	—	1290.0	—	—	—	—	—	—	—	—	0	—	—
10	—	—	—	—	2010.0	—	—	—	—	—	—	—	—	—	—	—
125	140	3.77	—	—	970.0	—	—	—	—	—	—	—	—	1.8	—	—
20	0	1.32	—	—	340.0	—	—	—	—	—	—	—	—	0	—	—
155	0	1.52	—	—	260.0	—	0	—	—	—	—	—	—	0	—	—
155	0	2.14	—	—	660.0	—	0	—	—	—	—	—	—	0	—	—
125	0	0.89	—	—	480.0	—	0	—	—	—	—	—	—	0	—	—
165	0	0.78	—	—	960.0	—	0	—	—	—	—	—	—	0	—	—
70	20	1.80	—	—	620.0	—	0	—	—	—	—	—	—	29.5	—	—
0	16	0.29	—	—	20.0	—	—	—	—	—	—	—	—	—	—	—
20	—	—	—	—	200.0	—	—	—	—	—	—	—	—	0	—	—
140	0	1.73	—	—	2230.0	—	0	—	—	—	—	—	—	24.0	—	—
0	53	0.46	—	—	45.0	—	—	—	—	—	—	—	—	5.8	—	—
0	0	0.43	—	—	95.0	—	—	—	—	—	—	—	—	9.8	—	—
20	44	1.18	—	—	210.0	—	—	—	—	—	—	—	—	—	—	—
210	250	2.70	—	—	1620.0	—	89.9	—	—	—	—	—	—	0	—	—
55	150	2.70	—	—	780.0	—	—	0.24	—	0.31	4.17	—	—	1.2	—	—
70	64	2.89	—	—	1440.0	—	—	0.50	—	0.32	10.29	—	—	0	—	—
55	20	0.72	—	—	960.0	—	—	0.14	—	0.11	10.93	—	—	0	—	—
55	333	0.79	—	—	350.0	—	—	0.11	—	0.61	0.26	—	—	2.7	0	—
50	99	1.78	20.1	217.3	810.0	1.51	—	0.34	1.03	0.33	4.33	—	—	0	0.6	22.2
180	146	2.63	28.6	313.2	1060.0	2.08	—	0.38	1.66	0.51	4.72	0.28	—	0	1.1	38.0
105	250	4.50	—	—	1030.0	—	—	0.26	—	0.44	6.37	—	—	1.2	—	—
175	150	8.07	—	—	1090.0	—	—	0.39	—	0.59	11.05	—	—	9.0	—	—
195	299	8.08	—	—	1520.0	—	—	0.39	—	0.66	11.03	—	—	9.0	—	—
60	101	3.62	—	—	1380.0	—	—	—	—	—	—	—	—	3.6	—	—
0	20	0.71	—	—	590.0	—	0	0.15	—	0.48	2.30	—	—	8.9	—	—
0	60	1.80	21.3	124.3	440.0	0.57	—	0.31	0.98	0.19	2.88	0.05	—	0	0	13.7
40	80	2.70	—	—	560.0	—	—	0.25	—	0.28	4.25	—	—	1.2	—	—
0	0	0.36	—	—	450.0	—	0	0.11	0.83	0.06	1.35	0.17	—	1.2	—	—
0	100	0.00	—	—	460.0	—	0	0.14	—	0.09	2.32	—	—	0	—	—
75	79	4.43	—	—	1730.0	—	—	—	—	—	—	—	—	8.9	—	—
75	57	6.82	—	—	1210.0	—	—	—	—	—	—	—	—	5.7	—	—
60	348	0.00	—	—	240.0	—	—	0.11	—	0.63	0.21	—	—	2.4	0	—
51	100	5.38	—	—	1020.0	—	—	0.38	—	0.43	7.30	—	—	9.0	—	—
115	249	5.38	—	—	1450.0	—	—	0.38	—	0.51	7.28	—	—	9.0	—	—
80	200	5.40	—	—	1900.0	—	—	—	—	—	—	—	—	2.4	—	—
80	150	2.70	—	—	990.0	—	60.0	—	—	—	—	—	—	2.4	—	—
70	100	4.50	—	—	910.0	—	—	—	—	—	—	—	—	6.0	—	—
95	200	3.60	—	—	1330.0	—	—	—	—	—	—	—	—	9.0	—	—
75	200	1.80	—	—	880.0	—	—	—	—	—	—	—	—	30.0	—	—
100	200	3.60	—	—	1440.0	—	—	—	—	—	—	—	—	9.0	—	—
35	19	1.02	—	—	470.0	—	0	—	—	—	—	—	—	0	—	—
45	600	1.08	—	—	360.0	—	0	—	—	—	—	—	—	0	—	—
0	20	1.80	—	—	950.0	—	0	—	—	—	—	—	—	12.0	—	—
155	300	7.20	—	—	1730.0	—	—	—	—	—	—	—	—	1.2	—	—

DA+ Code	Food Description	Quantity	Measure	Wt (g)	H₂O (g)	Ener (kcal)	Prot (g)	Carb (g)	Fiber (g)	Fat (g)	Sat	Mono	Poly	Trans
14238	French fries, small	1	serving(s)	92	—	290	5.0	37.0	3.0	14.0	3.0			
10798	French toast dips without syrup, 5 pieces	1	serving(s)	155	—	370	8.0	49.0	0	17.0	5.0	—	1.4	—
10802	Onion rings	1	serving(s)	128	—	440	7.0	53.0	3.0	22.0	5.0	—	0.8	—
34858	Spicy chicken sandwich	1	item(s)	198	—	480	14.0	48.0	2.0	26.0	5.0			
34867	Strawberry shake, small	1	serving(s)	595	—	520	14.0	93.0	0	11.0	7.0			
10865	Super Star hamburger	1	item(s)	348	—	790	41.0	52.0	3.0	47.0	14.0			
38925	The Six Dollar burger	1	item(s)	429	—	1010	40.0	60.0	3.0	66.0	26.0			
10818	Vanilla shake, small	1	item(s)	398	—	314	10.0	51.5	0	7.4	4.7	—	—	
10770	Western Bacon cheeseburger	1	item(s)	225	—	660	32.0	64.0	2.0	30.0	12.0	—	4.8	—
Chick Fil-A														
38746	Biscuit with bacon, egg & cheese	1	item(s)	163	—	470	18.0	39.0	1.0	26.0	9.0	—	—	3.0
38747	Biscuit with egg	1	item(s)	135	—	350	11.0	38.0	1.0	16.0	4.5	—	—	3.0
38748	Biscuit with egg & cheese	1	item(s)	149	—	400	14.0	38.0	1.0	21.0	7.0	—	—	3.0
38753	Biscuit with gravy	1	item(s)	192	—	330	5.0	43.0	1.0	15.0	4.0	—	—	4.0
38752	Biscuit with sausage, egg & cheese	1	item(s)	212	—	620	22.0	39.0	2.0	42.0	14.0	—	—	3.0
38771	Carrot & raisin salad	1	item(s)	113	—	170	1.0	28.0	2.0	6.0	1.0	—	—	0
38761	Chargrilled chicken Cool Wrap	1	item(s)	245	—	390	29.0	54.0	3.0	7.0	3.0	—	—	0
38766	Chargrilled chicken garden salad	1	item(s)	275	—	180	22.0	9.0	3.0	6.0	3.0	—	—	0
38758	Chargrilled chicken sandwich	1	item(s)	193	—	270	28.0	33.0	3.0	3.5	1.0	—	—	0
38742	Chicken biscuit	1	item(s)	145	—	420	18.0	44.0	2.0	19.0	4.5	—	—	3.0
38743	Chicken biscuit with cheese	1	item(s)	159	—	470	21.0	45.0	2.0	23.0	8.0	—	—	3.0
38762	Chicken Caesar Cool Wrap	1	item(s)	227	—	460	36.0	52.0	3.0	10.0	6.0	—	—	0
38757	Chicken deluxe sandwich	1	item(s)	208	—	420	28.0	39.0	2.0	16.0	3.5	—	—	0
38764	Chicken salad sandwich on wheat bun	1	item(s)	153	—	350	20.0	32.0	5.0	15.0	3.0	—	—	0
38756	Chicken sandwich	1	item(s)	170	—	410	28.0	38.0	1.0	16.0	3.5	—	—	0
38768	Chick-n-Strip salad	1	item(s)	327	—	400	34.0	21.0	4.0	20.0	6.0	—	—	0
38763	Chick-n-Strips	4	item(s)	127	—	300	28.0	14.0	1.0	15.0	2.5	—	—	0
38770	Cole slaw	1	item(s)	128	—	260	2.0	17.0	2.0	21.0	3.5	—	—	0
38776	Diet lemonade, small	1	cup(s)	255	—	25	0	5.0	0	0	0	0	0	0
38755	Hashbrowns	1	serving(s)	84	—	260	2.0	25.0	3.0	17.0	3.5	—	—	1.0
38765	Hearty breast of chicken soup	1	cup(s)	241	—	140	8.0	18.0	1.0	3.5	1.0	—	—	0
38741	Hot buttered biscuit	1	item(s)	79	—	270	4.0	38.0	1.0	12.0	3.0	—	—	3.0
38778	IceDream, small cone	1	item(s)	135	—	160	4.0	28.0	0	4.0	2.0	—	—	0
38774	IceDream, small cup	1	serving(s)	227	—	240	6.0	41.0	0	6.0	3.5	—	—	0
38775	Lemonade, small	1	cup(s)	255	—	170	0	41.0	0	0.5	0	—	—	0
38777	Nuggets	8	item(s)	113	—	260	26.0	12.0	0.5	12.0	2.5	—	—	0
38769	Side salad	1	item(s)	108	—	60	3.0	4.0	2.0	3.0	1.5	—	—	0
38767	Southwest chargrilled salad	1	item(s)	303	—	240	25.0	17.0	5.0	8.0	3.5	—	—	0
40481	Spicy chicken cool wrap	1	serving(s)	230	—	380	30.0	52.0	3.0	6.0	3.0	—	—	0
38772	Waffle potato fries, small, salted	1	serving(s)	85	—	270	3.0	34.0	4.0	13.0	3.0	—	—	1.5
Cinnabon														
39572	Caramellata Chill with whipped cream	16	fluid ounce(s)	480	—	406	10.0	61.0	0	14.0	8.0	—	—	0
39571	Cinnabon Bites	1	serving(s)	149	—	510	8.0	77.0	2.0	19.0	5.0	—	—	5.0
39570	Cinnabon Stix	5	item(s)	85	—	379	6.0	41.0	1.0	21.0	6.0	—	—	4.0
39567	Classic roll	1	item(s)	221	—	813	15.0	117.0	4.0	32.0	8.0	—	—	5.0
39568	Minibon	1	item(s)	92	—	339	6.0	49.0	2.0	13.0	3.0	—	—	2.0
39573	Mochalatta Chill with whipped cream	16	fluid ounce(s)	480	—	362	9.0	55.0	0	13.0	8.0	—	—	0
39569	Pecanbon	1	item(s)	272	—	1100	16.0	141.0	8.0	56.0	10.0	—	—	5.0
Dairy Queen														
1466	Banana split	1	item(s)	369	—	510	8.0	96.0	3.0	12.0	8.0	—	—	0
38552	Brownie Earthquake®	1	serving(s)	304	—	740	10.0	112.0	0	27.0	16.0	—	—	0.5
38561	Chocolate chip cookie dough blizzard®, small	1	item(s)	319	—	720	12.0	105.0	0	28.0	14.0	—	—	2.5
1464	Chocolate malt, small	1	item(s)	418	—	640	15.0	111.0	1.0	16.0	11.0	—	—	0.5
38541	Chocolate shake, small	1	item(s)	397	—	560	13.0	93.0	1.0	15.0	10.0	—	—	0.5
17257	Chocolate soft serve	½	cup(s)	94	—	150	4.0	22.0	0	5.0	3.5	—	—	0
1463	Chocolate sundae, small	1	item(s)	163	—	280	5.0	49.0	0	7.0	4.5	—	—	0
1462	Dipped cone, small	1	item(s)	156	—	340	6.0	42.0	1.0	17.0	9.0	4.0	3.0	1.0
38555	Oreo cookies blizzard, small	1	item(s)	283	—	570	11.0	83.0	0.5	21.0	10.0	—	—	2.5
38547	Royal Treats Peanut Buster® Parfait	1	item(s)	305	—	730	16.0	99.0	2.0	31.0	17.0	—	—	0
17256	Vanilla soft serve	½	cup(s)	94	—	140	3.0	22.0	0	4.5	3.0	—	—	0
Domino's														
31606	Barbeque buffalo wings	1	item(s)	25	—	50	6.0	2.0	0	2.5	0.5	—	—	—
31604	Breadsticks	1	item(s)	30	—	115	2.0	12.0	0	6.3	1.1	—	—	—
37551	Buffalo Chicken Kickers	1	item(s)	24	—	47	4.0	3.0	0	2.0	0.5	—	—	—
37548	CinnaStix	1	item(s)	30	—	123	2.0	15.0	1.0	6.1	1.1	—	—	—
37549	Dot, cinnamon	1	item(s)	28	7.6	99	1.9	14.9	0.7	3.7	0.7	—	—	—
31605	Double cheesy bread	1	item(s)	35	—	123	4.0	13.0	0	6.5	1.9	—	—	—
FAST FOOD —Continued														
31607	Hot buffalo wings	1	item(s)	25	—	45	5.0	1.0	0	2.5	0.5	—	—	—
Domino's Classic hand tossed pizza														

Chol (mg)	Calc (mg)	Iron (mg)	Magn (mg)	Pota (mg)	Sodi (mg)	Zinc (mg)	Vit A (µg)	Thia (mg)	Vit E (mg α)	Ribo (mg)	Niac (mg)	Vit B$_6$ (mg)	Fola (µg)	Vit C (mg)	Vit B$_{12}$ (µg)	Sele (µg)
0	0	1.08	—	—	170.0	—	0	—	—	—	—	—	—	21.0	—	—
3	0	0.00	—	—	470.0	—	0	0.25	—	0.23	2.00	—	—	0	—	—
0	20	0.72	—	—	700.0	—	0	—	—	—	—	—	—	3.6	—	—
40	100	3.60	—	—	1220.0	—	—	—	—	—	—	—	—	6.0	—	—
45	600	0.00	—	—	340.0	—	0	—	—	—	—	—	—	0	—	—
130	100	7.20	—	—	980.0	—	—	—	—	—	—	—	—	9.0	—	—
145	279	4.29	—	—	1960.0	—	—	—	—	—	—	—	—	16.7	—	—
30	401	0.00	—	—	234.0	—	0	—	—	—	—	—	—	0	—	—
85	200	5.40	—	—	1410.0	—	60.0	—	—	—	—	—	—	1.2	—	—
270	150	2.70	—	—	1190.0	—	—	—	—	—	—	—	—	0	—	—
240	80	2.70	—	—	740.0	—	—	—	—	—	—	—	—	0	—	—
255	150	2.70	—	—	970.0	—	—	—	—	—	—	—	—	0	—	—
5	60	1.80	—	—	930.0	—	0	—	—	—	—	—	—	0	—	—
300	200	3.60	—	—	1360.0	—	—	—	—	—	—	—	—	0	—	—
10	40	0.36	—	—	110.0	—	—	—	—	—	—	—	—	4.8	—	—
65	200	3.60	—	—	1020.0	—	—	—	—	—	—	—	—	6.0	—	—
65	150	0.72	—	—	620.0	—	—	—	—	—	—	—	—	30	—	—
65	80	2.70	—	—	940.0	—	—	—	—	—	—	—	—	6.0	—	—
35	60	2.70	—	—	1270.0	—	0	—	—	—	—	—	—	0	—	—
50	150	2.70	—	—	1500.0	—	—	—	—	—	—	—	—	0	—	—
80	500	3.60	—	—	1350.0	—	—	—	—	—	—	—	—	1.2	—	—
60	100	2.70	—	—	1300.0	—	—	—	—	—	—	—	—	2.4	—	—
65	150	1.80	—	—	880.0	—	—	—	—	—	—	—	—	0	—	—
60	100	2.70	—	—	1300.0	—	—	—	—	—	—	—	—	6.0	—	—
80	150	1.44	—	—	1070.0	—	—	—	—	—	—	—	—	0	—	—
65	40	1.44	—	—	940.0	—	—	—	—	—	—	—	—	36.0	—	—
25	60	0.36	—	—	220.0	—	—	—	—	—	—	—	—	15.0	—	—
0	0	0.36	—	—	5.0	—	0	—	—	—	—	—	—	0	—	—
5	20	0.72	—	—	380.0	—	—	—	—	—	—	—	—	0	—	—
25	40	1.08	—	—	900.0	—	—	—	—	—	—	—	—	0	—	—
0	60	1.80	—	—	660.0	—	0	—	—	—	—	—	—	0	—	—
15	100	0.36	—	—	80.0	—	—	—	—	—	—	—	—	0	—	—
25	200	0.36	—	—	105.0	—	—	—	—	—	—	—	—	15.0	—	—
0	0	0.36	—	—	10.0	—	0	—	—	—	—	—	—	0	—	—
70	40	1.08	—	—	1090.0	—	0	—	—	—	—	—	—	15.0	—	—
10	100	0.00	—	—	75.0	—	—	—	—	—	—	—	—	24.0	—	—
60	200	1.08	—	—	770.0	—	—	—	—	—	—	—	—	3.6	—	—
60	200	3.60	—	—	1090.0	—	—	—	—	—	—	—	—	1.2	—	—
0	20	1.08	—	—	115.0	—	0	—	—	—	—	—	—		—	—
46	—	—	—	—	187.0	—	—	—	—	—	—	—	—	—	—	—
35	—	—	—	—	530.0	—	—	—	—	—	—	—	—	—	—	—
16	—	—	—	—	413.0	—	—	—	—	—	—	—	—	—	—	—
67	—	—	—	—	801.0	—	—	—	—	—	—	—	—	—	—	—
27	—	—	—	—	337.0	—	—	—	—	—	—	—	—	—	—	—
46	—	—	—	—	252.0	—	—	—	—	—	—	—	—	—	—	—
63	—	—	—	—	600.0	—	—	—	—	—	—	—	—	—	—	—
30	250	1.80	—	—	180.0	—	—	—	—	—	—	—	—	15.0	—	—
50	250	1.80	—	—	350.0	—	—	—	—	—	—	—	—	0	—	—
50	350	2.70	—	—	370.0	—	—	—	—	—	—	—	—	1.2	—	—
55	450	1.80	—	—	340.0	—	—	—	—	—	—	—	—	2.4	—	—
50	450	1.44	—	—	280.0	—	—	—	—	—	—	—	—	2.4	—	—
15	100	0.72	—	—	75.0	—	—	—	—	—	—	—	—	0	—	—
20	200	1.08	—	—	140.0	—	—	—	—	—	—	—	—	0	—	—
20	200	1.08	—	—	130.0	—	—	—	—	—	—	—	—	1.2	—	—
40	350	2.70	—	—	430.0	—	—	—	—	—	—	—	—	1.2	—	—
35	300	1.80	—	—	400.0	—	—	—	—	—	—	—	—	1.2	—	—
15	150	0.72	—	—	70.0	—	—	—	—	—	—	—	—	0	—	—
26	10	0.36	—	—	175.5	—	—	—	—	—	—	—	—	0	—	—
0	0	0.72	—	—	122.1	—	—	—	—	—	—	—	—	0	—	—
9	0	0.00	—	—	162.5	—	—	—	—	—	—	—	—	0	—	—
0	0	0.72	—	—	111.4	—	—	—	—	—	—	—	—	0	—	—
0	6	0.59	—	—	85.7	—	—	—	—	—	—	—	—	0	—	—
6	40	0.72	—	—	162.3	—	—	—	—	—	—	—	—	0	—	—
26	10	0.36	—	—	254.5	—	—	—	—	—	—	—	—	1.2	—	—

DA+ Code	Food Description	Quantity	Measure	Wt (g)	H_2O (g)	Ener (kcal)	Prot (g)	Carb (g)	Fiber (g)	Fat (g)	Sat	Mono	Poly	Trans
31573	America's favorite feast, 12"	1	slice(s)	102	—	257	10.0	29.0	2.0	11.5	4.5	—	—	—
31574	America's favorite feast, 14"	1	slice(s)	141	—	353	14.0	39.0	2.0	16.0	6.0	—	—	—
37543	Bacon cheeseburger feast, 12"	1	slice(s)	99	—	273	12.0	28.0	2.0	13.0	5.5	—	—	—
37545	Bacon cheeseburger feast, 14"	1	slice(s)	137	—	379	17.0	38.0	2.0	18.0	8.0	—	—	—
37546	Barbeque feast, 12"	1	slice(s)	96	—	252	11.0	31.0	1.0	10.0	4.5	—	—	—
37547	Barbeque feast, 14"	1	slice(s)	131	—	344	14.0	43.0	2.0	13.5	6.0	—	—	—
31569	Cheese, 12"	1	slice(s)	55	—	160	6.0	28.0	1.0	3.0	1.0	—	—	0
31570	Cheese, 14"	1	slice(s)	75	—	220	8.0	38.0	2.0	4.0	1.0	—	—	0
37538	Deluxe feast, 12"	1	slice(s)	201	101.8	465	19.5	57.4	3.5	18.2	7.7	—	—	—
37540	Deluxe feast, 14"	1	slice(s)	273	138.4	627	26.4	78.3	4.7	24.1	10.2	—	—	—
31685	Deluxe, 12"	1	slice(s)	100	—	234	9.0	29.0	2.0	9.5	3.5	—	—	—
31694	Deluxe, 14"	1	slice(s)	136	—	316	13.0	39.0	2.0	12.5	5.0	—	—	—
31686	Extravaganzza, 12"	1	slice(s)	122	—	289	13.0	30.0	2.0	14.0	5.5	—	—	—
31695	Extravaganzza, 14"	1	slice(s)	165	—	388	17.0	40.0	3.0	18.5	7.5	—	—	—
31575	Hawaiian feast, 12"	1	slice(s)	102	—	223	10.0	30.0	2.0	8.0	3.5	—	—	—
31576	Hawaiian feast, 14"	1	slice(s)	141	—	309	14.0	41.0	2.0	11.0	4.5	—	—	—
31687	Meatzza, 12"	1	slice(s)	108	—	281	13.0	29.0	2.0	13.5	5.5	—	—	—
31696	Meatzza, 14"	1	slice(s)	146	—	378	17.0	39.0	2.0	18.0	7.5	—	—	—
31571	Pepperoni feast, extra pepperoni & cheese, 12"	1	slice(s)	98	—	265	11.0	28.0	2.0	12.5	5.0	—	—	—
31572	Pepperoni feast, extra pepperoni & cheese, 14"	1	slice(s)	135	—	363	16.0	39.0	2.0	17.0	7.0	—	—	—
31577	Vegi feast, 12"	1	slice(s)	102	—	218	9.0	29.0	2.0	8.0	3.5	—	—	—
31578	Vegi feast, 14"	1	slice(s)	139	—	300	13.0	40.0	3.0	11.0	4.5	—	—	—
	Domino's thin crust pizza													
31583	America's favorite, 12"	1	slice(s)	72	—	208	8.0	15.0	1.0	13.5	5.0	—	—	—
31584	America's favorite, 14"	1	slice(s)	100	—	285	11.0	20.0	2.0	18.5	7.0	—	—	—
31579	Cheese, 12"	1	slice(s)	49	—	137	5.0	14.0	1.0	7.0	2.5	—	—	—
31580	Cheese, 14"	1	slice(s)	68	27.0	214	8.8	19.0	1.4	11.4	4.6	2.9	2.5	—
31688	Deluxe, 12"	1	slice(s)	70	—	185	7.0	15.0	1.0	11.5	4.0	—	—	—
31697	Deluxe, 14"	1	slice(s)	94	—	248	10.0	20.0	2.0	15.0	5.5	—	—	—
31689	Extravaganzza, 12"	1	slice(s)	92	—	240	11.0	16.0	1.0	15.5	6.0	—	—	—
31698	Extravaganzza, 14"	1	slice(s)	123	—	320	14.0	21.0	2.0	20.5	8.0	—	—	—
31585	Hawaiian, 12"	1	slice(s)	71	—	174	8.0	16.0	1.0	9.5	3.5	—	—	—
31586	Hawaiian, 14"	1	slice(s)	100	—	240	11.0	21.0	2.0	13.0	5.0	—	—	—
31690	Meatzza, 12"	1	slice(s)	78	—	232	11.0	15.0	1.0	15.0	6.0	—	—	—
31699	Meatzza, 14"	1	slice(s)	104	—	310	14.0	20.0	2.0	20	8.0	—	—	—
31581	Pepperoni, extra pepperoni & cheese, 12"	1	slice(s)	68	—	216	9.0	14.0	1.0	14.0	5.5	—	—	—
31582	Pepperoni, extra pepperoni & cheese, 14"	1	slice(s)	93	—	295	13.0	20.0	1.0	19.0	7.5	—	—	—
31587	Vegi, 12"	1	slice(s)	71	—	168	7.0	15.0	1.0	9.5	3.5	—	—	—
31588	Vegi, 14"	1	slice(s)	97	—	231	10.0	21.0	2.0	13.5	5.0	—	—	—
	Domino's Ultimate deep dish pizza													
31596	America's favorite, 12"	1	slice(s)	115	—	309	12.0	29.0	2.0	17.0	6.0	—	—	—
31702	America's favorite, 14"	1	slice(s)	162	—	433	17.0	42.0	3.0	23.5	8.0	—	—	—
31590	Cheese, 12"	1	slice(s)	90	—	238	9.0	28.0	2.0	11.0	3.5	—	—	—
31591	Cheese, 14"	1	slice(s)	128	53.9	351	14.5	41.0	2.9	13.2	5.2	3.8	2.5	—
31589	Cheese, 6"	1	item(s)	215	—	598	22.9	68.4	3.9	27.6	9.9	—	—	—
31691	Deluxe, 12"	1	slice(s)	122	—	287	11.0	29.0	2.0	15.0	7.0	—	—	—
31700	Deluxe, 14"	1	slice(s)	156	—	396	15.0	42.0	3.0	20.0	7.0	—	—	—
31692	Extravaganzza, 12"	1	slice(s)	136	—	341	14.0	30.0	2.0	19.0	7.0	—	—	—
31701	Extravaganzza, 14"	1	slice(s)	186	—	468	20.0	43.0	3.0	25.5	9.5	—	—	—
31599	Hawaiian, 12"	1	slice(s)	114	—	275	12.0	30.0	2.0	13.0	5.0	—	—	—
31600	Hawaiian, 14"	1	slice(s)	162	—	389	17.0	43.0	3.0	18.0	6.5	—	—	—
31693	Meatzza, 12"	1	slice(s)	121	—	333	14.0	29.0	2.0	19.0	7.0	—	—	—
31703	Meatzza, 14"	1	slice(s)	167	—	458	19.0	42.0	3.0	25.0	9.5	—	—	—
31593	Pepperoni, extra pepperoni & cheese, 12"	1	slice(s)	110	—	317	13.0	29.0	2.0	17.5	6.5	—	—	—
31594	Pepperoni, extra pepperoni & cheese, 14"	1	slice(s)	155	—	443	18.0	42.0	3.0	24.0	9.0	—	—	—
31602	Vegi, 12"	1	slice(s)	114	—	270	11.0	30.0	2.0	13.5	5.0	—	—	—
31603	Vegi, 14"	1	slice(s)	159	—	380	15.0	43.0	3.0	18.0	6.5	—	—	—
31598	With ham and pineapple tidbits, 6"	1	item(s)	430	—	619	25.2	69.9	4.0	28.3	10.2	—	—	—
31595	With Italian sausage, 6"	1	item(s)	430	—	642	24.8	69.6	4.2	31.1	11.3	—	—	—
31592	With pepperoni, 6"	1	item(s)	430	—	647	25.1	68.5	3.9	32.0	11.7	—	—	—
31601	With vegetables, 6"	1	item(s)	430	—	619	23.4	70.8	4.6	28.7	10.1	—	—	—
	In-n-Out Burger													
34391	Cheeseburger with mustard & ketchup	1	serving(s)	268	—	400	22.0	41.0	3.0	18.0	9.0	—	—	0.5
34374	Cheeseburger	1	serving(s)	268	—	480	22.0	39.0	3.0	27.0	10.0	—	—	0.5
34390	Cheeseburger, lettuce leaves instead of buns	1	serving(s)	300	—	330	18.0	11.0	3.0	25.0	9.0	—	—	0
34377	Chocolate shake	1	serving(s)	425	—	690	9.0	83.0	0	36.0	24.0	—	—	1.0
34375	Double-Double cheeseburger	1	serving(s)	330	—	670	37.0	39.0	3.0	41.0	18.0	—	—	1.0
	FAST FOOD —Continued													
34393	Double-Double cheeseburger with mustard & ketchup	1	serving(s)	330	—	590	37.0	41.0	3.0	32.0	17.0	—	—	1.0

Chol (mg)	Calc (mg)	Iron (mg)	Magn (mg)	Pota (mg)	Sodi (mg)	Zinc (mg)	Vit A (μg)	Thia (mg)	Vit E (mg α)	Ribo (mg)	Niac (mg)	Vit B_6 (mg)	Fola (μg)	Vit C (mg)	Vit B_{12} (μg)	Sele (μg)
22	100	1.80	—	—	625.5	—	—	—	—	—	—	—	—	0.6	—	—
31	140	2.52	—	—	865.5	—	—	—	—	—	—	—	—	0.6	—	—
27	140	1.80	—	—	634.0	—	—	—	—	—	—	—	—	0	—	—
38	190	2.52	—	—	900.0	—	—	—	—	—	—	—	—	0	—	—
20	140	1.62	—	—	600.0	—	—	—	—	—	—	—	—	0.6	—	—
27	190	2.16	—	—	831.5	—	—	—	—	—	—	—	—	0.6	—	—
0	0	1.80	—	—	110.0	—	0	—	—	—	—	—	—	0	—	—
0	0	2.70	—	—	150.0	—	0	—	—	—	—	—	—	0	—	—
40	199	3.56	—	—	1063.1	—	—	—	—	—	—	—	—	1.4	—	—
53	276	4.84	—	—	1432.2	—	—	—	—	—	—	—	—	1.8	—	—
17	100	1.80	—	—	541.5	—	—	—	—	—	—	—	—	0.6	—	—
23	130	2.34	—	—	728.5	—	—	—	—	—	—	—	—	1.2	—	—
28	140	1.98	—	—	764.0	—	—	—	—	—	—	—	—	0.6	—	—
37	190	2.70	—	—	1014.0	—	—	—	—	—	—	—	—	1.2	—	—
16	130	1.62	—	—	546.5	—	—	—	—	—	—	—	—	1.2	—	—
23	180	2.34	—	—	765.0	—	—	—	—	—	—	—	—	1.2	—	—
28	130	1.80	—	—	739.5	—	—	—	—	—	—	—	—	0	—	—
37	190	2.52	—	—	983.5	—	—	—	—	—	—	—	—	0	—	—
24	130	1.62	—	—	670.0	—	70.9	—	—	—	—	—	—	0	—	—
33	180	2.34	—	—	920.0	—	104.7	—	—	—	—	—	—	0	—	—
13	130	1.62	—	—	489.0	—	—	—	—	—	—	—	—	0.6	—	—
18	180	2.34	—	—	678.0	—	—	—	—	—	—	—	—	0.6	—	—
23	100	0.90	—	—	533.0	—	—	—	—	—	—	—	—	2.4	—	—
32	140	1.26	—	—	736.5	—	—	—	—	—	—	—	—	3.0	—	—
10	90	0.54	—	—	292.5	—	60.0	—	—	—	—	—	—	1.8	—	—
14	151	0.48	17.7	125.1	338.0	0.02	64.6	0.05	1.01	0.07	0.69	—	—	2.4	0.5	24.1
19	100	0.90	—	—	449.0	—	—	—	—	—	—	—	—	2.4	—	—
24	130	1.08	—	—	601.0	—	—	—	—	—	—	—	—	3.6	—	—
29	140	1.08	—	—	671.5	—	—	—	—	—	—	—	—	2.4	—	—
38	190	1.44	—	—	886.5	—	—	—	—	—	—	—	—	3.6	—	—
17	130	0.72	—	—	454.0	—	—	—	—	—	—	—	—	3.0	—	—
24	180	0.90	—	—	637.5	—	—	—	—	—	—	—	—	3.6	—	—
29	140	0.90	—	—	647.0	—	—	—	—	—	—	—	—	1.8	—	—
38	190	1.26	—	—	865.5	—	—	—	—	—	—	—	—	2.4	—	—
26	130	0.72	—	—	577.0	—	80.0	—	—	—	—	—	—	1.8	—	—
35	80	1.08	—	—	792.5	—	105.8	—	—	—	—	—	—	2.4	—	—
14	130	0.72	—	—	396.5	—	—	—	—	—	—	—	—	2.4	—	—
19	180	1.08	—	—	550.5	—	—	—	—	—	—	—	—	3.0	—	—
25	120	2.34	—	—	796.5	—	—	—	—	—	—	—	—	0.6	—	—
34	170	3.24	—	—	1110.0	—	—	—	—	—	—	—	—	0.6	—	—
11	110	1.98	—	—	555.5	—	70.0	—	—	—	—	—	—	0	—	—
18	189	3.78	32.0	209.9	718.1	1.75	99.8	0.29	1.13	0.31	5.44	—	—	0	0.6	45.6
36	295	4.67	—	—	1341.4	—	174.0	—	—	—	—	—	—	0.5	—	—
20	120	2.16	—	—	712.0	—	—	—	—	—	—	—	—	1.2	—	—
26	170	3.06	—	—	974.5	—	—	—	—	—	—	—	—	1.2	—	—
31	160	2.52	—	—	934.5	—	—	—	—	—	—	—	—	1.2	—	—
40	220	3.42	—	—	1260.0	—	—	—	—	—	—	—	—	1.2	—	—
19	150	1.98	—	—	717.0	—	—	—	—	—	—	—	—	1.2	—	—
26	210	2.88	—	—	1011.0	—	—	—	—	—	—	—	—	1.8	—	—
31	160	2.34	—	—	910.5	—	—	—	—	—	—	—	—	0	—	—
40	220	3.24	—	—	1230.0	—	—	—	—	—	—	—	—	0.6	—	—
27	150	2.16	—	—	840.5	—	86.5	—	—	—	—	—	—	0	—	—
37	220	3.06	—	—	1166.0	—	115.4	—	—	—	—	—	—	0.6	—	—
15	150	2.16	—	—	659.5	—	—	—	—	—	—	—	—	0.6	—	—
21	220	3.06	—	—	924.0	—	—	—	—	—	—	—	—	1.2	—	—
43	298	4.84	—	—	1497.8	—	—	—	—	—	—	—	—	1.5	—	—
45	302	4.89	—	—	1478.1	—	—	—	—	—	—	—	—	0.6	—	—
47	299	4.81	—	—	1523.7	—	167.9	—	—	—	—	—	—	0.6	—	—
36	307	5.10	—	—	1472.5	—	—	—	—	—	—	—	—	4.7	—	—
60	200	3.60	—	—	1080.0	—	—	—	—	—	—	—	—	12.0	—	—
60	200	3.60	—	—	1000.0	—	—	—	—	—	—	—	—	9.0	—	—
60	200	2.70	—	—	720.0	—	—	—	—	—	—	—	—	12.0	—	—
95	300	0.72	—	—	350.0	—	—	—	—	—	—	—	—	0	—	—
120	350	5.40	—	—	1440.0	—	—	—	—	—	—	—	—	9.0	—	—
115	350	5.40	—	—	1520.0	—	—	—	—	—	—	—	—	12.0	—	—

											Fat Breakdown (g)			
DA+ Code	Food Description	Quantity	Measure	Wt (g)	H₂O (g)	Ener (kcal)	Prot (g)	Carb (g)	Fiber (g)	Fat (g)	Sat	Mono	Poly	Trans
	Double-Double cheeseburger, lettuce leaves instead													
34392	of buns	1	serving(s)	362	—	520	33.0	11.0	3.0	39.0	17.0	—	—	1.0
34376	French fries	1	serving(s)	125	—	400	7.0	54.0	2.0	18.0	5.0	—	—	0
34373	Hamburger	1	item(s)	243	—	390	16.0	39.0	3.0	19.0	5.0	—	—	0
34389	Hamburger with mustard & ketchup	1	serving(s)	243	—	310	16.0	41.0	3.0	10.0	4.0	—	—	0
34388	Hamburger, lettuce leaves instead of buns	1	serving(s)	275	—	240	13.0	11.0	3.0	17.0	4.0	—	—	0
34379	Strawberry shake	1	serving(s)	425	—	690	9.0	91.0	0	33.0	22.0	—	—	0.5
34378	Vanilla shake	1	serving(s)	425	—	680	9.0	78.0	0	37.0	25.0	—	—	1.0
	Jack in the Box													
30392	Bacon ultimate cheeseburger	1	item(s)	338	—	1090	46.0	53.0	2.0	77.0	30.0	—	—	3.0
1740	Breakfast Jack	1	item(s)	125	—	290	17.0	29.0	1.0	12.0	4.5	—	—	0
14074	Cheeseburger	1	item(s)	131	—	350	18.0	31.0	1.0	17.0	8.0	—	—	1.0
14106	Chicken breast strips, 4 piece	4	piece(s)	201	—	500	35.0	36.0	3.0	25.0	6.0	—	—	6.0
37241	Chicken club salad, plain, without salad dressing	1	serving(s)	431	—	300	27.0	13.0	4.0	15.0	6.0	—	—	0
14064	Chicken sandwich	1	item(s)	145	—	400	15.0	38.0	2.0	21.0	4.5	—	—	2.5
14111	Chocolate ice cream shake, small	1	serving(s)	414	—	880	14.0	107.0	1.0	45.0	31.0	—	—	2.0
14073	Hamburger	1	item(s)	118	—	310	16.0	30.0	1.0	14.0	6.0	—	—	1.0
14090	Hash browns	1	serving(s)	57	—	150	1.0	13.0	2.0	10.0	2.5	—	—	3.0
14072	Jack's Spicy Chicken sandwich	1	item(s)	270	—	620	25.0	61.0	4.0	31.0	6.0	—	—	3.0
1468	Jumbo Jack hamburger	1	item(s)	261	—	600	21.0	51.0	3.0	35.0	12.0	—	—	1.5
1469	Jumbo Jack hamburger with cheese	1	item(s)	286	—	690	25.0	54.0	3.0	42.0	16.0	—	—	1.5
14099	Natural cut french fries, large	1	serving(s)	196	—	530	8.0	69.0	5.0	25.0	6.0	—	—	7.0
14098	Natural cut french fries, medium	1	serving(s)	133	—	360	5.0	47.0	4.0	17.0	4.0	—	—	5.0
1470	Onion rings	1	serving(s)	119	—	500	6.0	51.0	3.0	30.0	6.0	—	—	10
33141	Sausage, egg & cheese biscuit	1	item(s)	234	—	740	27.0	35.0	2.0	55.0	17.0	—	—	6.0
14095	Seasoned curly fries, medium	1	serving(s)	125	—	400	6.0	45.0	5.0	23.0	5.0	—	—	7.0
14077	Sourdough Jack	1	item(s)	245	—	710	27.0	36.0	3.0	51.0	18.0	—	—	3.0
37249	Southwest chicken salad, plain, without salad dressing	1	serving(s)	488	—	300	24.0	29.0	7.0	11.0	5.0	—	—	0
14112	Strawberry ice cream shake, small	1	serving(s)	417	—	880	13.0	105.0	0	44.0	31.0	—	—	2.0
14078	Ultimate cheeseburger	1	item(s)	323	—	1010	40.0	53.0	2.0	71.0	28.0	—	—	3.0
14110	Vanilla ice cream shake, small	1	serving(s)	379	—	790	13.0	83.0	0	44.0	31.0	—	—	2.0
	Jamba Juice													
31645	Aloha Pineapple smoothie	24	fluid ounce(s)	730	—	500	8.0	117.0	4.0	1.5	1.0	—	—	—
31646	Banana Berry smoothie	24	fluid ounce(s)	719	—	480	5.0	112.0	4.0	1.0	0	—	—	—
31656	Berry Lime Sublime smoothie	24	fluid ounce(s)	728	—	460	3.0	106.0	5.0	2.0	1.0	—	—	—
31647	Carribean Passion smoothie	24	fluid ounce(s)	730	—	440	4.0	102.0	4.0	2.0	1.0	—	—	—
38422	Carrot juice	16	fluid ounce(s)	472	—	100	3.0	23.0	0	0.5	0	—	—	—
31648	Chocolate Moo'd smoothie	24	fluid ounce(s)	634	—	720	17.0	148.0	3.0	8.0	5.0	—	—	—
31649	Citrus Squeeze smoothie	24	fluid ounce(s)	727	—	470	5.0	110.0	4.0	2.0	1.0	—	—	—
31651	Coldbuster smoothie	24	fluid ounce(s)	724	—	430	5.0	100.0	5.0	2.5	1.0	—	—	—
31652	Cranberry Craze smoothie	24	fluid ounce(s)	793	—	460	6.0	104.0	4.0	0.5	0	—	—	—
31654	Jamba Powerboost smoothie	24	fluid ounce(s)	738	—	440	6.0	105.0	6.0	1.0	0	—	—	—
38423	Lemonade	16	fluid ounce(s)	483	—	300	1.0	75.0	0	0	0	0	0	0
31657	Mango-a-go-go smoothie	24	fluid ounce(s)	690	—	440	3.0	104.0	4.0	1.5	0.5	—	—	—
38424	Orange juice, freshly squeezed	16	fluid ounce(s)	496	—	220	3.0	52.0	0.5	1.0	0	—	—	—
38426	Orange/carrot juice	16	fluid ounce(s)	484	—	160	3.0	37.0	0	1.0	0	—	—	—
31660	Orange-a-peel smoothie	24	fluid ounce(s)	726	—	440	8.0	102.0	5.0	1.5	0	—	—	—
31662	Peach Pleasure smoothie	24	fluid ounce(s)	720	—	460	4.0	108.0	4.0	2.0	1.0	—	—	—
31665	Protein Berry Pizzazz smoothie	24	fluid ounce(s)	710	—	440	20.0	92.0	5.0	1.5	0	—	—	—
31668	Razzmatazz smoothie	24	fluid ounce(s)	730	—	480	3.0	112.0	4.0	2.0	1.0	—	—	—
31669	Strawberries Wild smoothie	24	fluid ounce(s)	725	—	450	6.0	105.0	4.0	0.5	0	—	—	—
38421	Strawberry Tsunami smoothie	24	fluid ounce(s)	740	—	530	4.0	128.0	4.0	2.0	1.0	—	—	—
38427	Vibrant C juice	16	fluid ounce(s)	448	—	210	2.0	50.0	1.0	0	0	0	0	—
38428	Wheatgrass juice, freshly squeezed	1	ounce(s)	28	—	5	0.5	1.0	0	0	0	0	0	0
	Kentucky Fried Chicken (KFC)													
31850	BBQ baked beans	1	serving(s)	136	—	220	8.0	45.0	7.0	1.0	0	—	—	0
31853	Biscuit	1	item(s)	57	—	220	4.0	24.0	1.0	11.0	2.5	—	—	3.5
51223	Boneless Fiery Buffalo Wings	6	item(s)	211	—	530	30.0	44.0	3.0	26.0	5.0	—	—	2.5
39386	Boneless Honey BBQ Wings	6	item(s)	213	—	570	30.0	54.0	5.0	26.0	5.0	—	—	2.5
51224	Boneless Sweet & Spicy Wings	6	item(s)	203	—	550	30.0	50.0	3.0	26.0	5.0	—	—	2.5
31851	Cole slaw	1	serving(s)	130	—	180	1.0	22.0	3.0	10.0	1.5	—	—	0
31842	Colonel's Crispy Strips	3	item(s)	151	—	370	28.0	17.0	1.0	20.0	4.0	—	—	2.5
31849	Corn on the cob	1	item(s)	162	—	150	5.0	26.0	7.0	3.0	1.0	—	—	0
51221	Double Crunch sandwich	1	item(s)	213	—	520	27.0	39.0	3.0	29.0	5.0	—	—	1.5
3761	Extra Crispy chicken, breast	1	item(s)	162	—	370	33.0	10.0	2.0	22.0	5.0	—	—	1.5
3762	Extra Crispy chicken, drumstick	1	item(s)	60	—	150	12.0	4.0	0	10.0	2.5	—	—	1.0
	FAST FOOD — Continued													
3763	Extra Crispy chicken, thigh	1	item(s)	114	—	290	17.0	16.0	1.0	18.0	4.0	—	—	1.5
3764	Extra Crispy chicken, whole wing	1	item(s)	52	—	150	11.0	11.0	1.0	7.0	1.5	—	—	0
51218	Famous Bowls mashed potatoes with gravy	1	serving(s)	531	—	720	26.0	79.0	6.0	34.0	9.0	—	—	3.5

Chol (mg)	Calc (mg)	Iron (mg)	Magn (mg)	Pota (mg)	Sodi (mg)	Zinc (mg)	Vit A (μg)	Thia (mg)	Vit E (mg α)	Ribo (mg)	Niac (mg)	Vit B$_6$ (mg)	Fola (μg)	Vit C (mg)	Vit B$_{12}$ (μg)	Sele (μg)
120	350	4.50	—	—	1160.0	—	—	—	—	—	—	—	—	12.0	—	—
0	20	1.80	—	—	245.0	—	0	—	—	—	—	—	—	0	—	—
40	40	3.60	—	—	650.0	—	—	—	—	—	—	—	—	9.0	—	—
35	40	3.60	—	—	730.0	—	—	—	—	—	—	—	—	12.0	—	—
40	40	2.70	—	—	370.0	—	—	—	—	—	—	—	—	12.0	—	—
85	300	0.00	—	—	280.0	—	—	—	—	—	—	—	—	0	—	—
90	300	0.00	—	—	390.0	—	—	—	—	—	—	—	—	0	—	—
140	308	7.38	—	540.0	2040.0	—	—	—	—	—	—	—	—	0.6	—	—
220	145	3.48	—	210.0	760.0	—	—	—	—	—	—	—	—	3.5	—	—
50	151	3.61	—	270.0	790.0	—	40.2	—	—	—	—	—	—	0	—	—
80	18	1.60	—	530.0	1260.0	—	—	—	—	—	—	—	—	1.1	—	—
65	280	3.35	—	560.0	880.0	—	—	—	—	—	—	—	—	50.4	—	—
35	100	2.70	—	240.0	730.0	—	—	—	—	—	—	—	—	4.8	—	—
135	460	0.47	—	840.0	330.0	—	—	—	—	—	—	—	—	0	—	—
40	100	3.60	—	250.0	600.0	—	0	—	—	—	—	—	—	0	—	—
0	10	0.18	—	190.0	230.0	—	0	—	—	—	—	—	—	0	—	—
50	150	1.80	—	450.0	1100.0	—	—	—	—	—	—	—	—	9.0	—	—
45	164	4.92	—	380.0	940.0	—	—	—	—	—	—	—	—	9.8	—	—
70	234	4.20	—	410.0	1310.0	—	—	—	—	—	—	—	—	8.4	—	—
0	20	1.42	—	1240.0	870.0	—	0	—	—	—	—	—	—	8.9	—	—
0	19	1.01	—	840.0	590.0	—	0	—	—	—	—	—	—	5.6	—	—
0	40	2.70	—	140.0	420.0	—	40.0	—	—	—	—	—	—	18.0	—	—
280	88	2.36	—	310.0	1430.0	—	—	—	—	—	—	—	—	0	—	—
0	40	1.80	—	580.0	890.0	—	—	—	—	—	—	—	—	0	—	—
75	200	4.50	—	430.0	1230.0	—	—	—	—	—	—	—	—	9.0	—	—
55	274	4.10	—	670.0	860.0	—	—	—	—	—	—	—	—	43.8	—	—
135	466	0.00	—	750.0	290.0	—	—	—	—	—	—	—	—	0	—	—
125	308	7.39	—	480.0	1580.0	—	—	—	—	—	—	—	—	0.6	—	—
135	532	0.00	—	750.0	280.0	—	—	—	—	—	—	—	—	0	—	—
5	200	1.80	60.0	1000.0	30.0	0.30	—	0.37	—	0.34	2.00	0.60	60.0	102.0	0	1.4
0	200	1.44	40.0	1010.0	115.0	0.60	—	0.09	—	0.25	0.80	0.70	24.0	15.0	0.2	1.4
5	200	1.80	16.0	510.0	35.0	0.30	—	0.06	—	0.25	6.00	0.70	140.0	54.0	0	1.4
5	100	1.80	24.0	810.0	60.0	0.30	—	0.09	—	0.25	5.00	0.50	100.0	78.0	0	1.4
0	150	2.70	80.0	1030.0	250.0	0.90	—	0.52	—	0.25	5.00	0.70	80.0	18.0	0	5.6
30	500	1.08	60.0	810.0	380.0	1.50	—	0.22	—	0.76	0.40	0.16	16.0	6.0	1.5	4.2
5	100	1.80	80.0	1170.0	35.0	0.30	—	0.37	—	0.34	1.90	0.60	100.0	180.0	0	1.4
5	100	1.08	60.0	1260.0	35.0	16.50	—	0.37	—	0.34	3.00	0.40	121.5	1302.0	0	1.4
0	250	1.44	16.0	500.0	50.0	0.30	—	0.03	—	0.25	5.00	0.60	120.0	54.0	0	1.4
0	1200	1.80	480.0	1070.0	45.0	16.50	—	5.55	—	6.12	68.00	7.40	640.0	288.0	10.8	77.0
0	20	0.00	8.0	200.0	10.0	0.00	—	0.03	—	0.17	14.00	1.80	320.0	36.0	0	0
5	100	1.08	24.0	780.0	50.0	0.30	—	0.15	—	0.25	5.00	0.70	120.0	72.0	0	1.4
0	60	1.08	60.0	990.0	0	0.30	—	0.45	—	0.13	2.00	0.20	160.0	246.0	0	0
0	100	1.80	60.0	1010.0	125.0	0.60	—	0.45	—	0.25	3.00	0.50	120.0	132.0	0	2.8
0	250	1.80	80.0	1380.0	160.0	0.90	—	0.45	—	0.42	2.00	0.50	140.0	240.0	0.6	1.4
5	100	0.72	32.0	740.0	60.0	0.30	—	0.06	—	0.25	4.00	0.60	80.0	18.0	0	1.4
0	1100	2.62	60.0	650.0	240.0	0.58	—	0.08	—	0.17	1.20	0.70	58.3	60.0	0	5.6
5	150	1.80	32.0	810.0	70.0	0.30	—	0.09	—	0.34	6.00	1.00	160.0	60.0	0	1.4
5	250	1.80	40.0	1050.0	180.0	0.90	—	0.12	—	0.34	0.80	0.40	40.0	60.0	0.6	1.4
5	100	1.08	24.0	480.0	10.0	0.30	—	0.06	—	0.34	14.00	1.80	320.0	90.0	0	1.4
0	20	1.08	40.0	720.0	0	0.30	—	0.30	—	0.10	1.60	0.40	80.0	678.0	0	0
0	0	1.80	8.0	80.0	0	0.00	0	0.03	—	0.03	0.40	0.04	16.0	3.6	0	2.8
0	100	2.70	—	—	730.0	—	—	—	—	—	—	—	—	1.2	—	—
0	40	1.80	—	—	640.0	—	—	—	—	—	—	—	—	0	—	—
65	40	1.80	—	—	2670.0	—	—	—	—	—	—	—	—	1.2	—	—
65	40	1.80	—	—	2210.0	—	—	—	—	—	—	—	—	1.2	—	—
65	60	1.80	—	—	2000.0	—	—	—	—	—	—	—	—	1.2	—	—
5	40	0.72	—	—	270.0	—	—	—	—	—	—	—	—	12.0	—	—
65	40	1.44	—	—	1220.0	—	0	—	—	—	—	—	—	1.2	—	—
0	60	1.08	—	—	10.0	—	—	—	—	—	—	—	—	6.0	—	—
55	100	2.70	—	—	1220.0	—	—	—	—	—	—	—	—	6.0	—	—
85	20	2.70	—	—	1020.0	—	—	—	—	—	—	—	—	1.2	—	—
55	0	1.44	—	—	300.0	—	0	—	—	—	—	—	—	0	—	—
95	20	2.70	—	—	700.0	—	—	—	—	—	—	—	—	—	—	—
45	20	1.08	—	—	340.0	—	—	—	—	—	—	—	—	0	—	—
35	200	5.40	—	—	2330.0	—	—	—	—	—	—	—	—	6.0	—	—

DA+ Code	Food Description	Quantity	Measure	Wt (g)	H₂O (g)	Ener (kcal)	Prot (g)	Carb (g)	Fiber (g)	Fat (g)	Sat	Mono	Poly	Trans
51219	Famous Bowls rice with gravy	1	serving(s)	384	—	610	25.0	67.0	5.0	27.0	8.0	—		2.5
31841	Honey BBQ chicken sandwich	1	item(s)	147	—	290	23.0	40.0	2.0	4.0	1.0	—		0
31833	Honey BBQ wing pieces	6	item(s)	157	—	460	27.0	26.0	3.0	27.0	6.0	—		2.0
10859	Hot wings pieces	6	piece(s)	134	—	450	26.0	19.0	2.0	30.0	7.0	—		2.0
42382	KFC Snacker sandwich	1	serving(s)	119	—	320	14.0	29.0	2.0	17.0	3.0	—		1.0
31848	Macaroni & cheese	1	serving(s)	136	—	180	8.0	18.0	0	8.0	3.5	—		1.0
31847	Mashed potatoes with gravy	1	serving(s)	151	—	140	2.0	20.0	1.0	5.0	1.0	—		0.5
10825	Original Recipe chicken, breast	1	item(s)	161	—	340	38.0	9.0	2.0	17.0	4.0	—		1.0
10826	Original Recipe chicken, drumstick	1	item(s)	59	—	140	13.0	3.0	0	8.0	2.0	—		0.5
10827	Original Recipe chicken, thigh	1	item(s)	126	—	350	19.0	7.0	1.0	27.0	7.0	—		1.0
10828	Original Recipe chicken, whole wing	1	item(s)	47	—	140	10.0	4.0	0	9.0	2.0	—		0.5
51222	Oven roasted Twister chicken wrap	1	item(s)	269	—	520	30.0	46.0	4.0	23.0	3.5	—		0
31844	Popcorn chicken, small or individual	1	item(s)	114	—	370	19.0	21.0	2.0	24.0	4.5	—		2.5
31852	Potato salad	1	serving(s)	128	—	180	2.0	22.0	2.0	9.0	1.5	—		0
10845	Potato wedges, small	1	serving(s)	102	—	250	4.0	32.0	3.0	12.0	2.0	—		1.5
31839	Tender Roast chicken sandwich with sauce	1	item(s)	236	—	430	37.0	29.0	2.0	18.0	3.5	—		0
	Long John Silver													
39392	Baked cod	1	serving(s)	101	—	120	22.0	1.0	0	4.5	1.0	—		0
3777	Batter dipped fish sandwich	1	item(s)	177	—	470	18.0	48.0	3.0	23.0	5.0	—		4.5
37568	Battered fish	1	item(s)	92	—	260	12.0	17.0	0.5	16.0	4.0	—		4.5
37569	Breaded clams	1	serving(s)	85	—	240	8.0	22.0	1.0	13.0	2.0	—		2.5
37566	Chicken plank	1	item(s)	52	—	140	8.0	9.0	0.5	8.0	2.0	—		2.5
39404	Clam chowder	1	item(s)	227	—	220	9.0	23.0	0	10.0	4.0	—		1.0
39398	Cocktail sauce	1	ounce(s)	28	—	25	0	6.0	0	0	0	0	0	0
3770	Coleslaw	1	serving(s)	113	—	200	1.0	15.0	3.0	15.0	2.5	1.8	4.1	0
39400	French fries, large	1	item(s)	142	—	390	4.0	56.0	5.0	17.0	4.0	—		5.0
3774	Fries, regular	1	serving(s)	85	—	230	3.0	34.0	3.0	10.0	2.5	—		3.0
3779	Hushpuppy	1	piece(s)	23	—	60	1.0	9.0	1.0	2.5	0.5	—		1.0
3781	Shrimp, batter-dipped, 1 piece	1	piece(s)	14	—	45	2.0	3.0	0	3.0	1.0	—		1.0
39399	Tartar sauce	1	ounce(s)	28	—	100	0	4.0	0	9.0	1.5	—		—
39395	Ultimate Fish sandwich	1	item(s)	199	—	530	21.0	49.0	3.0	28.0	8.0	—		5.0
	McDonald's													
50828	Asian salad with grilled chicken	1	item(s)	362	—	290	31.0	23.0	6.0	10.0	1.0	—	—	0
2247	Barbecue sauce	1	item(s)	28	—	45	0	11.0	0	0	0	0	0	0
737	Big Mac hamburger	1	item(s)	219	—	560	25.0	47.0	3.0	30.0	10.0	—		1.5
29777	Caesar salad dressing	1	package(s)	44	—	150	1.0	5.0	0	13.0	2.5	—		—
38391	Caesar salad with grilled chicken, no dressing	1	serving(s)	278	230.6	181	26.4	10.5	3.1	6.0	2.9	1.7	0.8	0.2
38393	Caesar salad without chicken, no dressing	1	serving(s)	190	170.4	84	6.0	8.1	3.0	3.9	2.2	0.9	0.3	0.1
738	Cheeseburger	1	item(s)	119	—	310	15.0	35.0	1.0	12.0	6.0	—		1.0
29775	Chicken McGrill sandwich	1	item(s)	213	—	400	27.0	38.0	3.0	16.0	3.0	—		0
1873	Chicken McNuggets, 6 piece	6	item(s)	96	—	250	15.0	15.0	0	15.0	3.0	—		1.5
3792	Chicken McNuggets, 4 piece	4	item(s)	64	—	170	10.0	10.0	0	10.0	2.0	—		1.0
29774	Crispy chicken sandwich	1	item(s)	232	121.8	500	27.0	63.0	3.0	16.0	3.0	5.7	7.4	1.5
743	Egg McMuffin	1	item(s)	139	76.8	300	17.0	30.0	2.0	12.0	4.5	3.8	2.5	0
742	Filet-O-Fish sandwich	1	item(s)	141	—	400	14.0	42.0	1.0	18.0	4.0	—		1.0
2257	French fries, large	1	serving(s)	170	—	570	6.0	70.0	7.0	30.0	6.0	—		8.0
1872	French fries, small	1	serving(s)	74	—	250	2.0	30.0	3.0	13.0	2.5	—		3.5
33822	Fruit 'n Yogurt Parfait	1	item(s)	149	111.2	160	4.0	31.0	1.0	2.0	1.0	0.2	0.1	0
739	Hamburger	1	item(s)	105	—	260	13.0	33.0	1.0	9.0	3.5	—		0.5
2003	Hash browns	1	item(s)	53	—	140	1.0	15.0	2.0	8.0	1.5	—		2.0
2249	Honey sauce	1	item(s)	14	—	50	0	12.0	0	0	0	0	0	0
38397	Newman's Own creamy caesar salad dressing	1	item(s)	59	32.5	190	2.0	4.0	0	18.0	3.5	4.6	9.6	0
38398	Newman's Own low fat balsamic vinaigrette salad dressing	1	item(s)	44	29.1	40	0	4.0	0	3.0	0	1.0	1.2	0
38399	Newman's Own ranch salad dressing	1	item(s)	59	30.1	170	1.0	9.0	0	15.0	2.5	9.0	3.7	0
1874	Plain Hotcakes with syrup and margarine	3	item(s)	221	—	600	9.0	102.0	2.0	17.0	4.0	—	—	4.0
740	Quarter Pounder hamburger	1	item(s)	171	—	420	24.0	40.0	3.0	18.0	7.0	—		1.0
741	Quarter Pounder hamburger with cheese	1	item(s)	199	—	510	29.0	43.0	2.0	25.0	12.0	—		1.5
2005	Sausage McMuffin with egg	1	item(s)	165	82.4	450	20.0	31.0	2.0	27.0	10.0	10.9	4.6	0.5
50831	Side salad	1	item(s)	87	—	20	1.0	4.0	1.0	0	0	0	0	0
	Pizza Hut													
39009	Hot chicken wings	2	item(s)	57	—	110	11.0	1.0	0	6.0	2.0	—		0.3
14025	Meat Lovers hand tossed pizza	1	slice(s)	118	—	300	15.0	29.0	2.0	13.0	6.0	—		0.5
14026	Meat Lovers pan pizza	1	slice(s)	123	—	340	15.0	29.0	2.0	19.0	7.0	—		0.5
31009	Meat Lovers stuffed crust pizza	1	slice(s)	169	—	450	21.0	43.0	3.0	21.0	10.0	—		1.0
14024	Meat Lovers thin 'n crispy pizza	1	slice(s)	98	—	270	13.0	21.0	2.0	14.0	6.0	—		0.5
	FAST FOOD —Continued													
14031	Pepperoni Lovers hand tossed pizza	1	slice(s)	113	—	300	15.0	30.0	2.0	13.0	7.0	—		0.5
14032	Pepperoni Lovers pan pizza	1	slice(s)	118	—	340	15.0	29.0	2.0	19.0	7.0	—		0.5
31011	Pepperoni Lovers stuffed crust pizza	1	slice(s)	163	—	420	21.0	43.0	3.0	19.0	10.0	—		1.0

Chol (mg)	Calc (mg)	Iron (mg)	Magn (mg)	Pota (mg)	Sodi (mg)	Zinc (mg)	Vit A (µg)	Thia (mg)	Vit E (mg α)	Ribo (mg)	Niac (mg)	Vit B_6 (mg)	Fola (µg)	Vit C (mg)	Vit B_{12} (µg)	Sele (µg)
35	200	4.50	—	—	2130.0	—	—	—	—	—	—	—	—	6.0	—	—
60	80	2.70	—	—	710.0	—	—	—	—	—	—	—	—	2.4	—	—
140	40	1.80	—	—	970.0	—	—	—	—	—	—	—	—	21.0	—	—
115	40	1.44	—	—	990.0	—	—	—	—	—	—	—	—	1.2	—	—
25	60	2.70	—	—	690.0	—	—	—	—	—	—	—	—	2.4	—	—
15	150	0.72	—	—	800.0	—	—	—	—	—	—	—	—	1.2	—	—
0	40	1.44	—	—	560.0	—	—	—	—	—	—	—	—	1.2	—	—
135	20	2.70	—	—	960.0	—	—	—	—	—	—	—	—	6.0	—	—
70	20	1.08	—	—	340.0	—	—	—	—	—	—	—	—	0	—	—
110	20	2.70	—	—	870.0	—	—	—	—	—	—	—	—	1.2	—	—
50	20	1.44	—	—	350.0	—	0	—	—	—	—	—	—	1.2	—	—
60	40	6.30	—	—	1380.0	—	—	—	—	—	—	—	—	15.0	—	—
25	40	1.80	—	—	1110.0	—	0	—	—	—	—	—	—	0	—	—
5	0	0.36	—	—	470.0	—	—	—	—	—	—	—	—	6.0	—	—
0	20	1.08	—	—	700.0	—	0	—	—	—	—	—	—	0	—	—
80	80	2.70	—	—	1180.0	—	—	—	—	—	—	—	—	9.0	—	—
90	20	0.72	—	—	240.0	—	—	—	—	—	—	—	—	0	—	—
45	60	2.70	—	—	1210.0	—	—	—	—	—	—	—	—	2.4	—	—
35	20	0.72	—	—	790.0	—	—	—	—	—	—	—	—	4.8	—	—
10	20	1.08	—	—	1110.0	—	0	—	—	—	—	—	—	0	—	—
20	0	0.72	—	—	480.0	—	0	—	—	—	—	—	—	2.4	—	—
25	150	0.72	—	—	810.0	—	—	—	—	—	—	—	—	0	—	—
0	0	0.00	—	—	250.0	—	—	—	—	—	—	—	—	0	—	—
20	40	0.36	—	222.7	340.0	0.70	—	0.07	—	0.08	2.34	—	—	18.0	—	—
0	0	0.00	—	—	580.0	—	0	—	—	—	—	—	—	24.0	—	—
0	0	0.00	—	370.0	350.0	0.30	0	0.09	—	0.01	1.60	—	—	15.0	—	—
0	20	0.36	—	—	200.0	—	0	—	—	—	—	—	—	0	—	—
15	0	0.00	—	—	160.0	—	0	—	—	—	—	—	—	1.2	—	—
15	0	0.00	—	—	250.0	—	0	—	—	—	—	—	—	0	—	—
60	150	2.70	—	—	1400.0	—	—	—	—	—	—	—	—	4.8	—	—
65	150	3.60	—	—	890.0	—	—	—	—	—	—	—	—	54.0	—	—
0	0	0.00	—	55.0	260.0	—	—	—	—	—	—	—	—	0	—	—
80	250	4.50	—	400.0	1010.0	—	—	—	—	—	—	—	—	1.2	—	—
10	40	0.18	—	30.0	400.0	—	—	—	—	—	—	—	—	0.6	—	—
67	178	1.77	—	708.9	767.3	—	—	0.15	—	0.19	10.62	—	127.9	29.2	0.2	—
10	163	1.15	17.1	410.4	157.7	—	—	0.08	—	0.07	0.40	—	102.6	26.8	0	0.4
40	200	2.70	—	240.0	740.0	—	60.0	—	—	—	—	—	—	1.2	—	—
70	150	2.70	—	510.0	1010.0	—	—	—	—	—	—	—	—	6.0	—	—
35	20	0.72	—	240.0	670.0	—	—	—	—	—	—	—	—	1.2	—	—
25	0	0.36	—	160.0	450.0	—	—	—	—	—	—	—	—	1.2	—	—
60	80	3.60	62.6	526.6	1380.0	1.53	41.8	0.46	2.27	0.39	12.85	—	104.4	6.0	0.4	—
230	300	2.70	26.4	218.2	860.0	1.59	—	0.36	0.82	0.51	4.31	0.20	109.8	1.2	0.9	—
40	150	1.80	—	250.0	640.0	—	36.2	—	—	—	—	—	—	0	—	—
0	20	1.80	—	—	330.0	—	0	—	—	—	—	—	—	9.0	—	—
0	20	0.72	—	—	140.0	—	0	—	—	—	—	—	—	3.6	—	—
5	150	0.67	20.9	248.8	85.0	0.53	0	0.06	—	0.17	0.35	—	19.4	9.0	0.3	—
30	150	2.70	—	210.0	530.0	—	5.0	—	—	—	—	—	—	1.2	—	—
0	0	0.36	—	210.0	290.0	—	0	—	—	—	—	—	—	1.2	—	—
0	0	0.00	—	0	0	—	0	—	—	—	—	—	—	0	—	—
20	61	0.00	3.0	16.0	500.0	0.20	—	0.01	15.43	0.02	0.01	0.64	2.4	0	0.1	0.1
0	4	0.00	1.3	8.8	730.0	0.01	—	0.00	0.00	0.00	0.00	0.00	0	2.4	0	0
0	40	0.00	1.8	70.4	530.0	0.03	0	0.01	—	0.08	0.01	0.02	0.6	0	0	0.2
20	150	2.70	—	280.0	620.0	—	—	—	—	—	—	—	—	0	—	—
70	150	4.50	—	390.0	730.0	—	10.0	—	—	—	—	—	—	1.2	—	—
95	300	4.50	—	440.0	1150.0	—	100.0	—	—	—	—	—	—	1.2	—	—
255	300	3.60	29.7	282.2	950.0	2.01	—	0.43	0.82	0.56	4.83	0.24	—	0	1.2	—
0	20	0.72	—	—	10.0	—	—	—	—	—	—	—	—	15.0	—	—
70	0	0.36	—	—	450.0	—	—	—	—	—	—	—	—	0	—	—
35	150	1.80	—	—	760.0	—	—	—	—	—	—	—	—	6.0	—	—
35	150	2.70	—	—	750.0	—	—	—	—	—	—	—	—	6.0	—	—
55	250	2.70	—	—	1250.0	—	—	—	—	—	—	—	—	9.0	—	—
35	150	1.44	—	—	740.0	—	—	—	—	—	—	—	—	6.0	—	—
40	200	1.80	—	—	710.0	—	57.7	—	—	—	—	—	—	2.4	—	—
40	200	2.70	—	—	700.0	—	57.7	—	—	—	—	—	—	2.4	—	—
55	300	2.70	—	—	1120.0	—	—	—	—	—	—	—	—	3.6	—	—

Food Composition   (Computer code is for Cengage Diet Analysis program)   (For purposes of calculations, use "0" for t, <1, <.1, <.01, etc.)

DA+ Code	Food Description	Quantity	Measure	Wt (g)	H₂O (g)	Ener (kcal)	Prot (g)	Carb (g)	Fiber (g)	Fat (g)	Fat Breakdown (g)			
											Sat	Mono	Poly	Trans
14030	Pepperoni Lovers thin 'n crispy pizza	1	slice(s)	92	—	260	13.0	21.0	2.0	14.0	7.0	—	—	0.5
10834	Personal Pan pepperoni pizza	1	slice(s)	61	—	170	7.0	18.0	0.5	8.0	3.0	—	—	1.0
10842	Personal Pan supreme pizza	1	slice(s)	77	—	190	8.0	19.0	1.0	9.0	3.5	—	—	1.0
39013	Personal Pan Veggie Lovers pizza	1	slice(s)	69	—	150	6.0	19.0	1.0	6.0	2.0	—	—	0.5
14028	Veggie Lovers hand tossed pizza	1	slice(s)	118	—	220	10.0	31.0	2.0	6.0	3.0	—	—	0.3
14029	Veggie Lovers pan pizza	1	slice(s)	119	—	260	10.0	30.0	2.0	12.0	4.0	—	—	0.3
31010	Veggie Lovers stuffed crust pizza	1	slice(s)	172	—	360	16.0	45.0	3.0	14.0	7.0	—	—	0.5
14027	Veggie Lovers thin 'n crispy pizza	1	slice(s)	101	—	180	8.0	23.0	2.0	7.0	3.0	—	—	0.5
39012	Wing blue cheese dipping sauce	1	item(s)	43	—	230	2.0	2.0	0	24.0	5.0	—	—	1.0
39011	Wing ranch dipping sauce	1	item(s)	43	—	210	0.5	4.0	0	22.0	3.5	—	—	0.5
	Starbucks													
38052	Cappuccino, tall	12	fluid ounce(s)	360	—	120	7.0	10.0	0	6.0	4.0	—	—	—
38053	Cappuccino, tall nonfat	12	fluid ounce(s)	360	—	80	7.0	11.0	0	0	0	0	0	0
38054	Cappuccino, tall soymilk	12	fluid ounce(s)	360	—	100	5.0	13.0	0.5	2.5	0	—	—	—
38059	Cinnamon spice mocha, tall nonfat w/o whipped cream	12	fluid ounce(s)	360	—	170	11.0	32.0	0	0.5	—	—	—	—
38057	Cinnamon spice mocha, tall w/ whipped cream	12	fluid ounce(s)	360	—	320	10.0	31.0	0	17.0	11.0	—	—	—
38051	Espresso, single shot	1	fluid ounce(s)	30	—	5	0	1.0	0	0	0	0	0	0
38088	Flavored syrup, 1 pump	1	serving(s)	10	—	20	0	5.0	0	0	0	0	0	0
32562	Frappuccino bottled coffee drink, mocha	9 ½	fluid ounce(s)	298	—	190	6.0	39.0	3.0	3.0	2.0	—	—	—
32561	Frappuccino coffee drink, all bottled flavors	9 ½	fluid ounce(s)	281	—	190	7.0	35.0	0	3.5	2.5	—	—	—
38073	Frappuccino, mocha	12	fluid ounce(s)	360	—	220	5.0	44.0	0	3.0	1.5	—	—	—
38067	Frappuccino, tall caramel w/o whipped cream	12	fluid ounce(s)	360	—	210	4.0	43.0	0	2.5	1.5	—	—	—
38070	Frappuccino, tall coffee	12	fluid ounce(s)	360	—	190	4.0	38.0	0	2.5	1.5	—	—	—
39894	Frappuccino, tall coffee, light blend	12	fluid ounce(s)	360	—	110	5.0	22.0	2.0	1.0	0	—	—	—
38071	Frappuccino, tall espresso	12	fluid ounce(s)	360	—	160	4.0	33.0	0	2.0	1.5	—	—	—
39897	Frappuccino, tall mocha, light blend	12	fluid ounce(s)	360	—	140	5.0	28.0	3.0	1.5	0	—	—	—
39887	Frappuccino, tall Strawberries & Creme, w/o whipped cream	12	fluid ounce(s)	360	—	330	10.0	65.0	0	3.5	1.0	—	—	—
38063	Frappuccino, tall Tazo chai creme w/o whipped cream	12	fluid ounce(s)	360	—	280	10.0	52.0	0	3.5	1.0	—	—	—
38066	Frappuccino, tall Tazoberry	12	fluid ounce(s)	360	—	140	0.5	36.0	0.5	0	0	0	0	0
38065	Frappuccino, tall Tazoberry Crème	12	fluid ounce(s)	360	—	240	4.0	54.0	0.5	1.0	0	—	—	—
38080	Frappuccino, tall vanilla w/o whipped cream	12	fluid ounce(s)	360	—	270	10.0	51.0	0	3.5	1.0	—	—	—
39898	Frappuccino, tall white chocolate mocha, light blend	12	fluid ounce(s)	360	—	160	6.0	32.0	2.0	2.0	1.0	—	—	—
38074	Frappuccino, tall white chocolate w/o whipped cream	12	fluid ounce(s)	360	—	240	5.0	48.0	0	3.5	2.5	—	—	—
39883	Java Chip Frappuccino, tall w/o whipped cream	12	fluid ounce(s)	360	—	270	5.0	51.0	1.0	7.0	4.5	—	—	—
33111	Latte, tall w/ nonfat milk	12	fluid ounce(s)	360	335.3	120	12.0	18.0	0	0	0	0	0	0
33112	Latte, tall w/ whole milk	12	fluid ounce(s)	360	—	200	11.0	16.0	0	11.0	7.0	—	—	—
33109	Macchiato, tall caramel w/ nonfat milk	12	fluid ounce(s)	360	—	170	11.0	30.0	0	1.0	0	—	—	—
33110	Macchiato, tall caramel w/ whole milk	12	fluid ounce(s)	360	—	240	10.0	28.0	0	10.0	6.0	—	—	—
33107	Mocha coffee drink, tall nonfat, w/o whip cream	12	fluid ounce(s)	360	—	170	11.0	33.0	1.0	1.5	0	—	—	—
38089	Mocha syrup	1	serving(s)	17	—	25	1.0	6.0	0	0.5	0	—	—	—
33108	Mocha, tall mocha w/ whole milk	12	fluid ounce(s)	360	—	310	10.0	32.0	1.0	17.0	10.0	—	—	—
38042	Steamed apple cider, tall	12	fluid ounce(s)	360	—	180	0	45.0	0	0	0	0	0	0
38087	Tazo chai black tea, soymilk, tall	12	fluid ounce(s)	360	—	190	4.0	39.0	0.5	2.0	0	—	—	—
38084	Tazo chai black tea, tall	12	fluid ounce(s)	360	—	210	6.0	36.0	0	5.0	3.5	—	—	—
38083	Tazo chai black tea, tall nonfat	12	fluid ounce(s)	360	—	170	6.0	37.0	0	0	0	—	—	—
38076	Tazo iced tea, tall	12	fluid ounce(s)	360	—	60	0	16.0	0	0	0	0	0	0
38077	Tazo tea, grande lemonade	16	fluid ounce(s)	480	—	120	0	31.0	0	0	0	0	0	0
38045	Vanilla creme steamed nonfat milk, tall w/whipped cream	12	fluid ounce(s)	360	—	260	11.0	33.0	0	8.0	5.0	—	—	—
38046	Vanilla crème steamed soymilk, tall w/whipped cream	12	fluid ounce(s)	360	—	300	8.0	37.0	1.0	12.0	6.0	—	—	—
38044	Vanilla creme steamed whole milk, tall w/whipped cream	12	fluid ounce(s)	360	—	330	10.0	31.0	0	18.0	11.0	—	—	—
38090	Whipped cream	1	serving(s)	27	—	100	0	2.0	0	9.0	6.0	—	—	—
38062	White chocolate mocha, tall nonfat w/o whipped cream	12	fluid ounce(s)	360	—	260	12.0	45.0	0	4.0	3.0	—	—	—
38061	White chocolate mocha, tall w/ whipped cream	12	fluid ounce(s)	360	—	410	11.0	44.0	0	20.0	13.0	—	—	—
38048	White hot chocolate, tall nonfat w/o whipped cream	12	fluid ounce(s)	360	—	300	15.0	51.0	0	4.5	3.5	—	—	—
38050	White hot chocolate, tall soymilk w/whipped cream	12	fluid ounce(s)	360	—	420	11.0	56.0	1.0	16.0	9.0	—	—	—
38047	White hot chocolate, tall w/whipped cream	12	fluid ounce(s)	360	—	460	13.0	50.0	0	22.0	15.0	—	—	—
	FAST FOOD —Continued													
	Subway													
15842	Cheese steak sandwich, 6", wheat bread	1	item(s)	250	—	360	24.0	47.0	5.0	10.0	4.5	—	—	0
40478	Chicken & bacon ranch sandwich, 6", white or wheat bread	1	serving(s)	297	—	540	36.0	47.0	5.0	25.0	10.0	—	—	0.5
38622	Chicken & bacon ranch wrap with cheese	1	item(s)	257	—	440	41.0	18.0	9.0	27.0	10.0	—	—	0.5
32045	Chocolate chip cookie	1	item(s)	45	—	210	2.0	30.0	1.0	10.0	6.0	—	—	0

Chol (mg)	Calc (mg)	Iron (mg)	Magn (mg)	Pota (mg)	Sodi (mg)	Zinc (mg)	Vit A (µg)	Thia (mg)	Vit E (mg α)	Ribo (mg)	Niac (mg)	Vit B_6 (mg)	Fola (µg)	Vit C (mg)	Vit B_{12} (µg)	Sele (µg)
40	200	1.44	—	—	690.0	—	58.0	—	—	—	—	—	—	2.4	—	—
15	80	1.44	—	—	340.0	—	38.5	—	—	—	—	—	—	1.4	—	—
20	80	1.86	—	—	420.0	—	—	—	—	—	—	—	—	3.6	—	—
10	80	1.80	—	—	280.0	—	—	—	—	—	—	—	—	3.6	—	—
15	150	1.80	—	—	490.0	—	—	—	—	—	—	—	—	9.0	—	—
15	150	2.70	—	—	470.0	—	—	—	—	—	—	—	—	9.0	—	—
35	250	2.70	—	—	980.0	—	—	—	—	—	—	—	—	9.0	—	—
15	150	1.44	—	—	480.0	—	—	—	—	—	—	—	—	9.0	—	—
25	20	0.00	—	—	550.0	—	0	—	—	—	—	—	—	0	—	—
10	0	0.00	—	—	340.0	—	0	—	—	—	—	—	—	0	—	—
25	250	0.00	—	—	95.0	—	—	—	—	—	—	—	—	1.2	0	—
3	200	0.00	—	—	100.0	—	—	—	—	—	—	—	—	0	0	—
0	250	0.72	—	—	75.0	—	—	—	—	—	—	—	—	0	0	—
5	300	0.72	—	—	150.0	—	—	—	—	—	—	—	—	0	0	—
70	350	1.08	—	—	140.0	—	—	—	—	—	—	—	—	2.4	0	—
0	0	0.00	—	—	0	—	0	—	—	—	—	—	—	0	0	—
0	0	0.00	—	—	0	—	0	—	—	—	—	—	—	0	0	—
12	219	1.08	—	530.0	110.0	—	—	—	—	—	—	—	—	0	—	—
15	250	0.36	—	510.0	105.0	—	—	—	—	—	—	—	—	0	—	—
10	150	0.72	—	—	180.0	—	—	—	—	—	—	—	—	0	0	—
10	150	0.00	—	—	180.0	—	—	—	—	—	—	—	—	0	0	—
10	150	0.00	—	—	180.0	—	—	—	—	—	—	—	—	0	0	—
0	150	0.00	—	—	220.0	—	—	—	—	—	—	—	—	0	—	—
10	100	0.00	—	—	160.0	—	—	—	—	—	—	—	—	0	0	—
0	150	0.72	—	—	220.0	—	—	—	—	—	—	—	—	0	—	—
3	350	0.00	—	—	270.0	—	—	—	—	—	—	—	—	21.0	—	—
3	350	0.00	—	—	270.0	—	—	—	—	—	—	—	—	3.6	0	—
0	0	0.00	—	—	30.0	—	0	—	—	—	—	—	—	0	0	—
0	150	0.00	—	—	125.0	—	0	—	—	—	—	—	—	1.2	0	—
3	350	0.00	—	—	370.0	—	—	—	—	—	—	—	—	3.6	0	—
3	150	0.00	—	—	250.0	—	—	—	—	—	—	—	—	0	—	—
10	150	0.00	—	—	210.0	—	—	—	—	—	—	—	—	0	0	—
10	150	1.44	—	—	220.0	—	—	—	—	—	—	—	—	0	—	—
5	350	0.00	39.8	—	170.0	1.35	—	0.12	—	0.47	0.36	0.13	17.5	0	1.3	—
45	400	0.00	46.6	—	160.0	1.28	—	0.12	—	0.54	0.34	0.14	16.8	2.4	1.2	—
5	300	0.00	—	—	160.0	—	—	—	—	—	—	—	—	1.2	—	—
30	300	0.00	—	—	135.0	—	—	—	—	—	—	—	—	2.4	—	—
5	300	2.70	—	—	135.0	—	—	—	—	—	—	—	—	0	—	—
0	0	0.72	—	—	0	—	0	—	—	—	—	—	—	0	0	—
55	300	2.70	—	—	115.0	—	—	—	—	—	—	—	—	0	—	—
0	0	1.08	—	—	15.0	—	0	—	—	—	—	—	—	0	0	—
0	200	0.72	—	—	70.0	—	—	—	—	—	—	—	—	0	0	—
20	200	0.36	—	—	85.0	—	—	—	—	—	—	—	—	1.2	0	—
5	200	0.36	—	—	95.0	—	—	—	—	—	—	—	—	0	0	—
Chol	0	0.00	—	—	0	—	0	—	—	—	—	—	—	0	0	—
0	0	0.00	—	—	15.0	—	0	—	—	—	—	—	—	4.8	0	—
35	350	0.00	—	—	170.0	—	—	—	—	—	—	—	—	0	0	—
30	400	1.44	—	—	130.0	—	—	—	—	—	—	—	—	0	0	—
65	350	0.00	—	—	140.0	—	—	—	—	—	—	—	—	0	0	—
40	0	0.00	—	—	10.0	—	—	—	—	—	—	—	—	0	0	—
5	400	0.00	—	—	210.0	—	—	—	—	—	—	—	—	0	0	—
70	400	0.00	—	—	210.0	—	—	—	—	—	—	—	—	2.4	0	—
10	450	0.00	—	—	250.0	—	—	—	—	—	—	—	—	0	0	—
35	500	1.44	—	—	210.0	—	—	—	—	—	—	—	—	0	0	—
75	500	0.00	—	—	250.0	—	—	—	—	—	—	—	—	3.6	0	—
35	150	8.10	—	—	1090.0	—	—	—	—	—	—	—	—	18.0	—	—
90	250	4.50	—	—	1400.0	—	—	—	—	—	—	—	—	21.0	—	—
90	300	2.70	—	—	1680.0	—	—	—	—	—	—	—	—	9.0	—	—
15	0	1.08	—	—	150.0	—	—	—	—	—	—	—	—	0	—	—

DA+ Code	Food Description	Quantity	Measure	Wt (g)	H₂O (g)	Ener (kcal)	Prot (g)	Carb (g)	Fiber (g)	Fat (g)	Sat	Mono	Poly	Trans
32048	Chocolate chip M&M cookie	1	item(s)	45	—	210	2.0	32.0	0.5	10.0	5.0	—	—	0
32049	Chocolate chunk cookie	1	item(s)	45	—	220	2.0	30.0	0.5	10.0	5.0	—	—	0
4024	Classic Italian B.M.T. sandwich, 6", white bread	1	item(s)	236	—	440	22.0	45.0	2.0	21.0	8.5	—	—	0
15838	Classic tuna sandwich, 6", wheat bread	1	item(s)	250	—	530	22.0	45.0	4.0	31.0	7.0	—	—	0.5
15837	Classic tuna sandwich, 6", white bread	1	item(s)	243	—	520	21.0	43.0	2.0	31.0	7.5	—	—	0.5
16397	Club salad, no dressing and croutons	1	item(s)	412	—	160	18.0	15.0	4.0	4.0	1.5	—	—	0
3422	Club sandwich, 6", white bread	1	item(s)	250	—	310	23.0	45.0	2.0	6.0	2.5	—	—	0
4030	Cold cut combo sandwich, 6", white bread	1	item(s)	242	—	400	20.0	45.0	2.0	17.0	7.5	—	—	0.5
34030	Ham & egg breakfast sandwich	1	item(s)	142	—	310	16.0	35.0	3.0	13.0	3.5	—	—	0
3885	Ham sandwich, 6", white bread	1	item(s)	238	—	310	17.0	52.0	2.0	5.0	2.0	—	—	0
3888	Meatball marinara sandwich, 6", wheat bread	1	item(s)	377	—	560	24.0	63.0	7.0	24.0	11.0	—	—	1.0
4651	Meatball sandwich, 6", white bread	1	item(s)	370	—	550	23.0	61.0	5.0	24.0	11.5	—	—	1.0
15839	Melt sandwich, 6", white bread	1	item(s)	260	—	410	25.0	47.0	4.0	15.0	5.0	—	—	—
32046	Oatmeal raisin cookie	1	item(s)	45	—	200	3.0	30.0	1.0	8.0	4.0	—	—	0
16379	Oven-roasted chicken breast sandwich, 6", wheat bread	1	item(s)	238	—	330	24.0	48.0	5.0	5.0	1.5	—	—	0
32047	Peanut butter cookie	1	item(s)	45	—	220	4.0	26.0	1.0	12.0	5.0	—	—	0
4655	Roast beef sandwich, 6", wheat bread	1	item(s)	224	—	290	19.0	45.0	4.0	5.0	2.0	—	—	0
3957	Roast beef sandwich, 6", white bread	1	item(s)	217	—	280	18.0	43.0	2.0	5.0	2.5	—	—	0
16378	Roasted chicken breast, 6", white bread	1	item(s)	231	—	320	23.0	46.0	3.0	5.0	2.0	—	—	0
34028	Southwest steak & cheese sandwich, 6", Italian bread	1	item(s)	271	—	450	24.0	48.0	6.0	20.0	6.0	—	—	0
4032	Spicy Italian sandwich, 6", white bread	1	item(s)	220	—	470	20.0	43.0	2.0	25.0	9.5	—	—	0
4031	Steak & cheese sandwich, 6", white bread	1	item(s)	243	—	350	23.0	45.0	3.0	10.0	5.0	—	—	0
32050	Sugar cookie	1	item(s)	45	—	220	2.0	28.0	0.5	12.0	6.0	—	—	0
40477	Sweet onion chicken teriyaki sandwich, 6", white or wheat bread	1	serving(s)	281	—	370	26.0	59.0	4.0	5.0	1.5	—	—	0
38623	Turkey breast & bacon melt wrap with chipotle sauce	1	item(s)	228	—	380	31.0	20.0	9.0	24.0	7.0	—	—	0
15834	Turkey breast & ham sandwich, 6", white bread	1	item(s)	227	—	280	19.0	45.0	2.0	5.0	2.0	—	—	0
16376	Turkey breast sandwich, 6", white bread	1	item(s)	217	—	270	17.0	44.0	2.0	4.5	2.0	—	—	0
15841	Veggie Delite sandwich, 6", wheat bread	1	item(s)	167	—	230	9.0	44.0	4.0	3.0	1.0	—	—	0
16375	Veggie Delite, 6", white bread	1	item(s)	160	—	220	8.0	42.0	2.0	3.0	1.5	—	—	0
32051	White chip macadamia nut cookie	1	item(s)	45	—	220	2.0	29.0	0.5	11.0	5.0	—	—	0
	Taco Bell													
29906	7-Layer burrito	1	item(s)	283	—	490	17.0	65.0	9.0	18.0	7.0	—	—	1.0
744	Bean burrito	1	item(s)	198	—	340	13.0	54.0	8.0	9.0	3.5	—	—	0.5
749	Beef burrito supreme	1	item(s)	248	—	410	17.0	51.0	7.0	17.0	8.0	—	—	1.0
33417	Beef Chalupa Supreme	1	item(s)	153	—	380	14.0	30.0	3.0	23.0	7.0	—	—	0.5
34474	Beef Gordita Baja	1	item(s)	153	—	340	13.0	29.0	4.0	19.0	5.0	—	—	0
29910	Beef Gordita Supreme	1	item(s)	153	—	310	14.0	29.0	3.0	16.0	6.0	—	—	0.5
2014	Beef soft taco	1	item(s)	99	—	200	10.0	21.0	3.0	9.0	4.0	—	—	0
10860	Beef soft taco supreme	1	item(s)	135	—	250	11.0	23.0	3.0	13.0	6.0	—	—	0.5
34472	Chicken burrito supreme	1	item(s)	248	—	390	20.0	49.0	6.0	13.0	6.0	—	—	0.5
33418	Chicken Chalupa Supreme	1	item(s)	153	—	360	17.0	29.0	2.0	20.0	5.0	—	—	0
34475	Chicken Gordita Baja	1	item(s)	153	—	320	17.0	28.0	3.0	16.0	3.5	—	—	0
29909	Chicken quesadilla	1	item(s)	184	—	520	28.0	40.0	3.0	28.0	12.0	—	—	0.5
29907	Chili cheese burrito	1	item(s)	156	—	390	16.0	40.0	3.0	18.0	9.0	—	—	1.5
10794	Cinnamon twists	1	serving(s)	35	—	170	1.0	26.0	1.0	7.0	0	—	—	0
29911	Grilled chicken Gordita Supreme	1	item(s)	153	—	290	17.0	28.0	2.0	12.0	5.0	—	—	0
14463	Grilled chicken soft taco	1	item(s)	99	—	190	14.0	19.0	1.0	6.0	2.5	—	—	—
29912	Grilled Steak Gordita Supreme	1	item(s)	153	—	290	15.0	28.0	2.0	13.0	5.0	—	—	0
29904	Grilled steak soft taco	1	item(s)	128	—	270	12.0	20.0	2.0	16.0	4.5	—	—	0
29905	Grilled steak soft taco supreme	1	item(s)	135	—	235	13.0	21.0	1.0	11.0	6.0	—	—	—
2021	Mexican pizza	1	serving(s)	216	—	530	20.0	42.0	7.0	30.0	8.0	—	—	1.0
29894	Mexican rice	1	serving(s)	131	—	170	6.0	23.0	1.0	11.0	3.0	—	—	0
10772	Meximelt	1	serving(s)	128	—	280	15.0	22.0	3.0	14.0	7.0	—	—	0.5
2011	Nachos	1	serving(s)	99	—	330	4.0	32.0	2.0	21.0	3.5	—	—	2.0
2012	Nachos Bellgrande	1	serving(s)	308	—	770	19.0	77.0	12.0	44.0	9.0	—	—	3.0
2023	Pintos 'n cheese	1	serving(s)	128	—	150	9.0	19.0	7.0	6.0	3.0	—	—	0.5
34473	Steak burrito supreme	1	item(s)	248	—	380	18.0	49.0	6.0	14.0	7.0	—	—	0.5
33419	Steak Chalupa Supreme	1	item(s)	153	—	360	15.0	28.0	2.0	21.0	6.0	—	—	0
747	Taco	1	item(s)	78	—	170	8.0	13.0	3.0	10.0	3.5	—	—	0
2015	Taco salad with salsa, with shell	1	serving(s)	548	—	840	30.0	80.0	15.0	45.0	11.0	—	—	1.5
	FAST FOOD —Continued													
14459	Taco supreme	1	item(s)	113	—	210	9.0	15.0	3.0	13.0	6.0	—	—	0
748	Tostada	1	item(s)	170	—	230	11.0	27.0	7.0	10.0	3.5	—	—	0.5
	CONVENIENCE MEALS													
	Banquet													
29961	Barbeque chicken meal	1	item(s)	281	—	330	16.0	37.0	2.0	13.0	3.0	—	—	—
14788	Boneless white fried chicken meal	1	item(s)	286	—	310	10.0	21.0	4.0	20.0	5.0	—	—	—
29960	Fish sticks meal	1	item(s)	207	—	470	13.0	58.0	1.0	20.0	3.5	—	—	—

Chol (mg)	Calc (mg)	Iron (mg)	Magn (mg)	Pota (mg)	Sodi (mg)	Zinc (mg)	Vit A (μg)	Thia (mg)	Vit E (mg α)	Ribo (mg)	Niac (mg)	Vit B$_6$ (mg)	Fola (μg)	Vit C (mg)	Vit B$_{12}$ (μg)	Sele (μg)
10	20	1.00	—	—	100.0	—	—	—	—	—	—	—	—	0	—	—
10	0	1.00	—	—	100.0	—	—	—	—	—	—	—	—	0	—	—
55	150	2.70	—	—	1770.0	—	—	—	—	—	—	—	—	16.8	—	—
45	100	5.40	—	—	1030.0	—	—	—	—	—	—	—	—	21.0	—	—
45	100	3.60	—	—	1010.0	—	—	—	—	—	—	—	—	16.8	—	—
35	60	3.60	—	—	880.0	—	—	—	—	—	—	—	—	30.0	—	—
35	60	3.60	—	—	1290.0	—	—	—	—	—	—	—	—	13.8	—	—
60	150	3.60	—	—	1530.0	—	—	—	—	—	—	—	—	16.8	—	—
190	80	4.50	—	—	720.0	—	66.7	—	—	—	—	—	—	3.6	—	—
25	60	2.70	—	—	1375.0	—	—	—	—	—	—	—	—	13.8	—	—
45	200	7.20	—	—	1610.0	—	—	—	—	—	—	—	—	36.0	—	—
45	200	5.40	—	—	1590.0	—	—	—	—	—	—	—	—	31.8	—	—
45	150	5.40	—	—	1720.0	—	—	—	—	—	—	—	—	24.0	—	—
15	20	1.08	—	—	170.0	—	—	—	—	—	—	—	—	0	—	—
45	60	4.50	—	—	1020.0	—	—	—	—	—	—	—	—	18.0	—	—
15	20	0.72	—	—	200.0	—	—	—	—	—	—	—	—	0	—	—
20	60	6.30	—	—	920.0	—	—	—	—	—	—	—	—	18.0	—	—
20	60	4.50	—	—	900.0	—	—	—	—	—	—	—	—	13.8	—	—
45	60	2.70	—	—	1000.0	—	—	—	—	—	—	—	—	13.8	—	—
45	150	8.10	—	—	1310.0	—	—	—	—	—	—	—	—	21.0	—	—
55	60	2.70	—	—	1650.0	—	—	—	—	—	—	—	—	16.8	—	—
35	150	6.30	—	—	1070.0	—	—	—	—	—	—	—	—	13.8	—	—
15	0	0.72	—	—	140.0	—	—	—	—	—	—	—	—	0	—	—
50	80	4.50	—	—	1220.0	—	—	—	—	—	—	—	—	24.0	—	—
50	200	2.70	—	—	1780.0	—	—	—	—	—	—	—	—	6.0	—	—
25	60	2.70	—	—	1210.0	—	—	—	—	—	—	—	—	13.8	—	—
20	60	2.70	—	—	1000.0	—	—	—	—	—	—	—	—	13.8	—	—
0	60	4.50	—	—	520.0	—	—	—	—	—	—	—	—	18.0	—	—
0	60	2.70	—	—	500.0	—	—	—	—	—	—	—	—	13.8	—	—
15	20	0.72	—	—	160.0	—	—	—	—	—	—	—	—	0	—	—
25	250	5.40	—	—	1350.0	—	—	—	—	—	—	—	—	15.0	—	—
5	200	4.50	—	—	1190.0	—	5.9	—	—	—	—	—	—	4.8	—	—
40	200	4.50	—	—	1340.0	—	9.9	—	—	—	—	—	—	6.0	—	—
40	150	2.70	—	—	620.0	—	—	—	—	—	—	—	—	3.6	—	—
35	100	2.70	—	—	780.0	—	—	—	—	—	—	—	—	2.4	—	—
40	150	2.70	—	—	620.0	—	—	—	—	—	—	—	—	3.6	—	—
25	100	1.80	—	—	630.0	—	—	—	—	—	—	—	—	1.2	—	—
40	150	2.70	—	—	650.0	—	—	—	—	—	—	—	—	3.6	—	—
45	200	4.50	—	—	1360.0	—	—	—	—	—	—	—	—	9.0	—	—
45	100	2.70	—	—	650.0	—	—	—	—	—	—	—	—	4.8	—	—
40	100	1.80	—	—	800.0	—	—	—	—	—	—	—	—	3.6	—	—
75	450	3.60	—	—	1420.0	—	—	—	—	—	—	—	—	1.2	—	—
40	300	1.80	—	—	1080.0	—	—	—	—	—	—	—	—	0	—	—
0	0	0.37	—	—	200.0	—	0	—	—	—	—	—	—	0	—	—
45	150	1.80	—	—	650.0	—	—	—	—	—	—	—	—	4.8	—	—
30	100	1.08	—	—	550.0	—	14.6	—	—	—	—	—	—	1.2	—	—
40	100	2.70	—	—	530.0	—	—	—	—	—	—	—	—	3.6	—	—
35	100	2.70	—	—	660.0	—	—	—	—	—	—	—	—	3.6	—	—
35	120	1.44	—	—	565.0	—	29.2	—	—	—	—	—	—	3.6	—	—
40	350	3.60	—	—	1000.0	—	—	—	—	—	—	—	—	4.8	—	—
15	100	1.44	—	—	790.0	—	—	—	—	—	—	—	—	3.6	—	—
40	250	2.70	—	—	880.0	—	—	—	—	—	—	—	—	2.4	—	—
3	80	0.71	—	—	530.0	—	0	—	—	—	—	—	—	0	—	—
35	200	3.60	—	—	1280.0	—	—	—	—	—	—	—	—	4.8	—	—
15	150	1.44	—	—	670.0	—	—	—	—	—	—	—	—	3.6	—	—
35	200	4.50	—	—	1250.0	—	9.9	—	—	—	—	—	—	9.0	—	—
40	100	2.70	—	—	530.0	—	—	—	—	—	—	—	—	3.6	—	—
25	80	1.08	—	—	350.0	—	—	—	—	—	—	—	—	1.2	—	—
65	450	7.20	—	—	1780.0	—	—	—	—	—	—	—	—	12.0	—	—
40	100	1.08	—	—	370.0	—	—	—	—	—	—	—	—	3.6	—	—
15	200	1.80	—	—	730.0	—	—	—	—	—	—	—	—	4.8	—	—
50	40	1.08	—	—	1210.0	—	0	—	—	—	—	—	—	4.8	—	—
45	80	1.44	—	—	1200.0	—	—	—	—	—	—	—	—	18.0	—	—
55	20	1.44	—	—	710.0	—	—	—	—	—	—	—	—	0	—	—

Food Composition (Computer code is for Cengage Diet Analysis program) (For purposes of calculations, use "0" for t, <1, <.1, <.01, etc.)

DA+ Code	Food Description	Quantity	Measure	Wt (g)	H₂O (g)	Ener (kcal)	Prot (g)	Carb (g)	Fiber (g)	Fat (g)	Fat Breakdown (g) Sat	Mono	Poly	Trans
29957	Lasagna with meat sauce meal	1	item(s)	312	—	320	15.0	46.0	7.0	9.0	4.0	—	—	—
14777	Macaroni and cheese meal	1	item(s)	340	—	420	15.0	57.0	5.0	14.0	8.0	—	—	—
1741	Meatloaf meal	1	item(s)	269	—	240	14.0	20.0	4.0	11.0	4.0	—	—	—
39418	Pepperoni pizza meal	1	item(s)	191	—	480	11.0	56.0	5.0	23.0	8.0	—	—	—
33759	Roasted white turkey meal	1	item(s)	255	—	230	14.0	30.0	5.0	6.0	2.0	—	—	—
1743	Salisbury steak meal	1	item(s)	269	196.9	380	12.0	28.0	3.0	24.0	12.0	—	—	—
	Budget Gourmet													
1914	Cheese manicotti with meat sauce entree	1	item(s)	284	194.0	420	18.0	38.0	4.0	22.0	11.0	6.0	1.3	—
1915	Chicken with fettucini entree	1	item(s)	284	—	380	20.0	33.0	3.0	19.0	10.0	—	—	—
3986	Light beef stroganoff entrée	1	item(s)	248	177.0	290	20.0	32.0	3.0	7.0	4.0	—	—	—
3996	Light sirloin of beef in herb sauce entrée	1	item(s)	269	214.0	260	19.0	30.0	5.0	7.0	4.0	2.3	0.3	—
3987	Light vegetable lasagna entrée	1	item(s)	298	227.0	290	15.0	36.0	4.8	9.0	1.8	0.9	0.6	—
	Healthy Choice													
9425	Cheese French bread pizza	1	item(s)	170	—	340	22.0	51.0	5.0	5.0	1.5	—	—	—
9306	Chicken enchilada suprema meal	1	item(s)	320	251.5	360	13.0	59.0	8.0	7.0	3.0	2.0	2.0	—
3821	Familiar Favorites lasagna bake with meat sauce entrée	1	item(s)	255	—	270	13.0	38.0	4.0	7.0	2.5	—	—	—
13744	Familiar Favorites sesame chicken with vegetables and rice entrée	1	item(s)	255	—	260	17.0	34.0	4.0	6.0	2.0	2.0	2.0	—
9316	Lemon pepper fish meal	1	item(s)	303	—	280	11.0	49.0	5.0	5.0	2.0	1.0	2.0	—
9322	Traditional salisbury steak meal	1	item(s)	354	250.3	360	23.0	45.0	5.0	9.0	3.5	4.0	1.0	—
9359	Traditional turkey breasts meal	1	item(s)	298	—	330	21.0	50.0	4.0	5.0	2.0	1.5	1.5	—
	Stouffers													
2313	Cheese French bread pizza	1	serving(s)	294	—	380	15.0	43.0	3.0	16.0	6.0	—	—	—
11138	Cheese manicotti with tomato sauce entrée	1	item(s)	255	—	360	18.0	41.0	2.0	14.0	6.0	—	—	—
2366	Chicken pot pie entrée	1	item(s)	284	—	740	23.0	56.0	4.0	47.0	18.0	12.4	10.5	—
11116	Homestyle baked chicken breast with mashed potatoes and gravy entrée	1	item(s)	252	—	270	21.0	21.0	2.0	11.0	3.5	—	—	—
11146	Homestyle beef pot roast and potatoes entrée	1	item(s)	252	—	260	16.0	24.0	3.0	11.0	4.0	—	—	—
11152	Homestyle roast turkey breast with stuffing and mashed potatoes entrée	1	item(s)	273	—	290	16.0	30.0	2.0	12.0	3.5	—	—	—
11043	Lean Cuisine Comfort Classics baked chicken and whipped potatoes and stuffing entrée	1	item(s)	245	—	240	15.0	34.0	3.0	4.5	1.0	2.0	1.0	0
11046	Lean Cuisine Comfort Classics honey mustard chicken with rice pilaf entrée	1	item(s)	227	—	250	17.0	37.0	1.0	4.0	1.0	1.0	1.0	0
9479	Lean Cuisine Deluxe French bread pizza	1	item(s)	174	—	310	16.0	44.0	3.0	9.0	3.5	0.5	0.5	0
360	Lean Cuisine One Dish Favorites chicken chow mein with rice	1	item(s)	255	—	190	13.0	29.0	2.0	2.5	0.5	1.0	0.5	0
11054	Lean Cuisine One Dish Favorites chicken enchilada Suiza with Mexican-style rice	1	serving(s)	255	—	270	10.0	47.0	3.0	4.5	2.0	1.5	1.0	0
9467	Lean Cuisine One Dish Favorites fettucini alfredo entree	1	item(s)	262	—	270	13.0	39.0	2.0	7.0	3.5	2.0	1.0	0
11055	Lean Cuisine One Dish Favorites lasagna with meat sauce entrée	1	item(s)	298	—	320	19.0	44.0	4.0	7.0	3.0	2.0	0.5	0
	Weight Watchers													
11164	Smart Ones chicken enchiladas suiza entrée	1	item(s)	255	—	340	12.0	38.0	3.0	10.0	4.5	—	—	—
39763	Smart Ones chicken oriental entrée	1	item(s)	255	—	230	15.0	34.0	3.0	4.5	1.0	—	—	—
11187	Smart Ones pepperoni pizza	1	item(s)	198	—	400	22.0	58.0	4.0	9.0	3.0	—	—	—
39765	Smart Ones spaghetti bolgnese entrée	1	item(s)	326	—	280	17.0	43.0	5.0	5.0	2.0	—	—	—
31512	Smart Ones spicy szechuan style vegetables & chicken	1	item(s)	255	—	220	11.0	34.0	4.0	5.0	1.0	—	—	—
	BABY FOODS													
787	Apple juice	4	fluid ounce(s)	127	111.6	60	0	14.8	0.1	0.1	0	0	0	—
778	Applesauce, strained	4	tablespoon(s)	64	56.7	26	0.1	6.9	1.1	0.1	0	0	0	—
779	Bananas with tapioca, strained	4	tablespoon(s)	60	50.4	34	0.2	9.2	1.0	0	0	0	0	—
604	Carrots, strained	4	tablespoon(s)	56	51.7	15	0.4	3.4	1.0	0.1	0	0	0	—
770	Chicken noodle dinner, strained	4	tablespoon(s)	64	54.8	42	1.7	5.8	1.3	1.3	0.4	0.5	0.3	—
801	Green beans, strained	4	tablespoon(s)	60	55.1	16	0.7	3.8	1.3	0.1	0	0	0	—
910	Human milk, mature	2	fluid ounce(s)	62	53.9	43	0.6	4.2	0	2.7	1.2	1.0	0.3	—
760	Mixed cereal, prepared with whole milk	4	ounce(s)	113	84.6	128	5.4	18.0	1.5	4.0	2.2	1.2	0.4	—
772	Mixed vegetable dinner, strained	2	ounce(s)	57	50.3	23	0.7	5.4	0.8	0	—	—	0	—
762	Rice cereal, prepared with whole milk	4	ounce(s)	113	84.6	130	4.4	18.9	0.1	4.1	2.6	1.0	0.2	—
758	Teething biscuits	1	item(s)	11	0.7	44	1.0	8.6	0.2	0.6	0.2	0.2	0.1	—

Chol (mg)	Calc (mg)	Iron (mg)	Magn (mg)	Pota (mg)	Sodi (mg)	Zinc (mg)	Vit A (µg)	Thia (mg)	Vit E (mg α)	Ribo (mg)	Niac (mg)	Vit B$_6$ (mg)	Fola (µg)	Vit C (mg)	Vit B$_{12}$ (µg)	Sele (µg)
20	100	2.70	—	—	1170.0	—	—	—	—	—	—	—	—	0	—	—
20	150	1.44	—	—	1330.0	—	0	—	—	—	—	—	—	0	—	—
30	0	1.80	—	—	1040.0	—	0	—	—	—	—	—	—	0	—	—
35	150	1.80	—	—	870.0	—	0	—	—	—	—	—	—	0	—	—
25	60	1.80	—	—	1070.0	—	—	—	—	—	—	—	—	3.6	—	—
60	40	1.44	—	—	1140.0	—	0	—	—	—	—	—	—	0	—	—
85	300	2.70	45.4	484.0	810.0	2.29	—	0.45	—	0.51	4.00	0.22	30.7	0	0.7	—
85	100	2.70	—	—	810.0	—	—	0.15	—	0.42	6.00	—	—	0	—	—
35	40	1.80	38.9	280.0	580.0	4.71	—	0.17	—	0.36	4.28	0.27	18.9	2.4	2.5	—
30	40	1.80	57.7	540.0	850.0	4.81	—	0.15	—	0.29	5.53	0.37	38.4	6.0	1.6	—
15	283	3.03	78.5	420.0	780.0	1.39	—	0.22	—	0.45	3.13	0.32	74.8	59.1	0.2	—
10	350	3.60	—	—	600.0	—	—	—	—	—	—	—	—	0	—	—
30	40	1.44	—	—	580.0	—	—	—	—	—	—	—	—	3.6	—	—
20	100	1.80	—	—	600.0	—	—	—	—	—	—	—	—	0	—	—
35	18	0.72	—	—	580.0	—	—	—	—	—	—	—	—	12.0	—	—
35	20	0.36	—	—	580.0	—	—	—	—	—	—	—	—	30.0	—	—
45	80	2.70	—	—	580.0	—	—	—	—	—	—	—	—	21.0	—	—
35	40	1.80	—	—	600.0	—	—	—	—	—	—	—	—	0	—	—
30	200	1.80	—	230.0	660.0	—	—	—	—	—	—	—	—	2.4	—	—
70	250	1.44	—	550.0	920.0	—	—	—	—	—	—	—	—	6.0	—	—
65	150	2.70	—	—	1170.0	—	—	—	—	—	—	—	—	2.4	—	—
55	20	0.72	—	490.0	770.0	—	0	—	—	—	—	—	—	0	—	—
35	20	1.80	—	800.0	960.0	—	—	—	—	—	—	—	—	6.0	—	—
45	40	1.08	—	490.0	970.0	—	—	—	—	—	—	—	—	3.6	—	—
25	40	1.16	—	500.0	650.0	—	—	—	—	—	—	—	—	3.6	—	—
30	64	0.38	—	370.0	650.0	—	—	—	—	—	—	—	—	0	—	—
20	150	2.70	—	300.0	700.0	—	—	—	—	—	—	—	—	15.0	—	—
25	40	0.72	—	380.0	650.0	—	—	—	—	—	—	—	—	2.4	—	—
20	150	0.72	—	350.0	510.0	—	—	—	—	—	—	—	—	2.4	—	—
15	200	0.72	—	290.0	690.0	—	0	—	—	—	—	—	—	0	—	—
30	250	1.47	—	610.0	690.0	—	—	—	—	—	—	—	—	2.4	—	—
40	200	0.72	—	—	800.0	—	—	—	—	—	—	—	—	2.4	—	—
35	40	0.72	—	—	790.0	—	—	—	—	—	—	—	—	6.0	—	—
15	200	1.08	—	401.0	700.0	—	69.1	—	—	—	—	—	—	4.8	—	—
15	150	3.60	—	—	670.0	—	—	—	—	—	—	—	—	9.0	—	—
10	40	1.44	—	—	890.0	—	—	—	—	—	—	—	—	0	—	—
0	5	0.72	3.8	115.4	3.8	0.03	1.3	0.01	0.76	0.02	0.10	0.03	0	73.4	0	0.1
0	3	0.12	1.9	45.4	1.3	0.01	0.6	0.01	0.36	0.02	0.04	0.02	1.3	24.5	0	0.2
0	3	0.12	6.0	52.8	5.4	0.04	1.2	0.01	0.36	0.02	0.08	0.04	3.6	10.0	0	0.4
0	12	0.20	5.0	109.8	20.7	0.08	320.9	0.01	0.29	0.02	0.25	0.04	8.4	3.2	0	0.1
10	17	0.40	9.0	89.0	14.7	0.32	70.4	0.03	0.12	0.04	0.44	0.04	7.0	0	0	2.4
0	23	0.40	12.0	87.6	3.0	0.12	10.8	0.02	0.04	0.04	0.20	0.02	14.4	0.2	0	0
9	20	0.02	1.8	31.4	10.5	0.10	37.6	0.01	0.04	0.02	0.10	0.01	3.1	3.1	0	1.1
12	249	11.82	30.6	225.7	53.3	0.80	28.4	0.49	—	0.65	6.54	0.07	12.5	1.4	0.3	
—	12	0.18	6.2	68.6	4.5	0.08	77.1	0.01	—	0.02	0.28	0.04	4.5	1.6	0	0.4
12	271	13.82	51.0	215.5	52.2	0.72	24.9	0.52	—	0.56	5.90	0.12	9.1	1.4	0.3	4.0
0	11	0.39	3.9	35.5	28.4	0.10	3.1	0.02	0.02	0.05	0.47	0.01	5.4	1.0	0	2.6

Appendix B

Reliable Sources of Nutrition Information

Many sources of nutrition information are available to consumers, but the quality of the information they provide varies widely. All of the sources listed here provide scientifically based information.

Expert Advice

Registered dietitians (hospitals and the yellow pages) Public health nutritionists (public health departments) College nutrition instructors/professors (colleges and universities) Extension Service home economists (state and county U.S. Department of Agriculture Extension Service offices) Consumer affairs staff of the Food and Drug Administration (national, regional, and state FDA offices)

You can find hundreds of toll-free telephone numbers for health information through the following Website: www.healthfinder.gov. After you are connected, search the term toll-free numbers.

U.S. Government

- **Federal Trade Commission (FTC)**
  Public Reference Branch
  (202) 326-2222
  www.ftc.gov

- **Food and Drug Administration (FDA)**
  Office of Consumer Affairs, HFE 1
  Room 16-85
  5600 Fishers Lane
  Rockville, MD 20857
  (301) 443-1544
  www.fda.gov

- **FDA Consumer Information Line**
  (301) 827-4420

- **FDA Office of Food Labeling, HFS 150**
  Washington, DC 20204
  (202) 205-4561; fax (202) 205-4564
  www.cfsan.fda.gov

- **FDA Office of Plant and Dairy Foods and Beverages**
  HFS 300
  200 C Street SW
  Washington, DC 20204
  (202) 205-4064; fax (202) 205-4422

- **FDA Office of Special Nutritionals,**
  HFS 450
  200 C Street SW
  Washington, DC 20204
  (202) 205-4168; fax (202) 205-5295

- **Food and Nutrition Information Center**
  National Agricultural Library,
  Room 304
  10301 Baltimore Avenue
  Beltsville, MD 20705-2351
  (301) 504-5719; fax (301) 504-6409
  www.nal.usda.gov/fnic

- **Food Research Action Center (FRAC)**
  1875 Connecticut Avenue NW,
  Suite 540
  Washington, DC 20009
  (202) 986-2200; fax (202) 986-2525

- **Superintendent of Documents**
  U.S. Government Printing Office
  Washington, DC 20402
  (202) 512-1071
  www.access.gpo.gov/su_docs

- **U.S. Department of Agriculture (USDA)**
  14th Street SW and Independence Avenue
  Washington, DC 20250
  (202) 720-2791
  www.usda.gov/fcs

- **USDA Center for Nutrition Policy and Promotion**
  1120 20th Street NW, Suite 200
  North Lobby
  Washington, DC 20036
  (202) 208-2417
  www.usda.gov/fcs/cnpp.htm

- **USDA Food Safety and Inspection Service**
  Food Safety Education Office,
  Room 1180-S
  Washington, DC 20250
  (202) 690-0351
  www.usda.gov/fsis

- **U.S. Department of Education (DOE)**
  Accreditation Agency Evaluation Branch
  7th and D Street SW
  ROB 3, Room 3915
  Washington, DC 20202-5244
  (202) 708-7417

- **U.S. Department of Health and Human Services**
  200 Independence Avenue SW
  Washington, DC 20201
  (202) 619-0257
  www.os.dhhs.gov

- **U.S. Environmental Protection Agency (EPA)**
  401 Main Street SW
  Washington, DC 20460
  (202) 260-2090
  www.epa.gov

- **U.S. Public Health Service**
  Assistant Secretary of Health
  Humphrey Building, Room 725-H
  200 Independence Avenue SW
  Washington, DC 20201
  (202) 690-7694

Health Canada

Headquarters

- **Health Canada**
  A.L. 0900C2
  Ottawa, Canada
  K1A 0K9
  Telephone: (613) 957-2991
  TTY: 1-800-267-1245
  http://www.hc-sc.gc.ca/

Regional Headquarters

- **British Columbia/Yukon**
  Suite 405, Winch Building
  757 West Hastings Street
  Vancouver, BC
  V6C 1A1
  Tel: (604) 666-2083
  Fax: (604) 666-2258

- **Alberta/NWT**
  Suite 710, Canada Place
  9700 Jasper Avenue
  Edmonton, AB
  T5J 4C3
  Tel: (780) 495-2651
  Fax: (780) 495-3285

- **Manitoba/Saskatchewan**
  391 York Avenue, Suite 425
  Winnipeg, MB
  R3C 0P4
  Tel: (204) 983-2508
  Fax: (204) 983-3972

- **Ontario/Nunavut**
  25 St. Clair Avenue East, 4th Floor
  Toronto, ON
  M4T 1M2
  Tel: (416) 973-4389
  Toll free: 1-866-999-7612
  Fax: (416) 973-1423

- **Quebec**
  Room 218, Complexe Guy-Favreau
  East Tower
  200 René Lévesque Blvd. West
  Montreal, QC
  H2Z 1X4
  Tel: (514) 283-2306
  Fax: (514) 283-6739

- **Atlantic**
  Suite 1525, 15th Floor,
  Maritime Centre
  1505 Barrington Street
  Halifax, NS B3J 3Y6
  Tel: (902) 426-2700
  Fax: (902) 426-9689

International Agencies

- **Food and Agriculture Organization of the United Nations (FAO)**
  Liaison Office for North America
  2175 K Street, Suite 300
  Washington, DC 20437
  (202) 653-2400
  www.fao.org

- **International Food Information Council Foundation**
  1100 Connecticut Avenue NW,
  Suite 430
  Washington, DC 20036
  (202) 296-6540
  ificinfo.health.org

- **UNICEF**
  3 United Nations Plaza
  New York, NY 10017
  (212) 326-7000
  www.unicef.com

- **World Health Organization (WHO)**
  Regional Office
  525 23rd Street NW
  Washington, DC 20037
  (202) 974-3000
  www.who.org

Professional Nutrition Organizations

- **Academy of Nutrition and Dietetics (formerly American Dietetics Association)**
  216 West Jackson Boulevard,
  Suite 800
  Chicago, IL 60606-6995
  (800) 877-1600; (312) 899-0040
  www.eatright.org

- **ADA, The Nutrition Hotline**
  (800) 366-1655

- **American Society for Clinical Nutrition**
  9650 Rockville Pike
  Bethesda, MD 20814-3998
  (301) 530-7110; fax (301) 571-1863
  www.faseb.org/ascn

- **Dietitians of Canada**
  480 University Avenue, Suite 604
  Toronto, Ontario M5G 1V2, Canada
  (416) 596-0857; fax (416) 596-0603
  www.dietitians.ca

- **Human Nutrition Institute (INACG)**
  1126 Sixteenth Street NW
  Washington, DC 20036
  (202) 659-0789
  www.ilsi.org

- **National Academy of Sciences/ National Research Council (NAS/NRC)**
  2101 Constitution Avenue, NW
  Washington, DC 20418
  (202) 334-2000
  www.nas.edu

- **National Institute of Nutrition**
  265 Carling Avenue, Suite 302
  Ottawa, Ontario K1S 2E1
  (613) 235-3355; fax (613) 235-7032
  www.nin.ca

- **Society for Nutrition Education**
  7101 Wisconsin Avenue, Suite 901
  Bethesda, MD 20814-4805
  (301) 656-4938

Aging

- **Administration on Aging**
  330 Independence Avenue SW
  Washington, DC 20201
  (202) 619-0724
  www.aoa.dhhs.gov

- **American Association of Retired Persons (AARP)**
  601 E Street NW
  Washington, DC 20049
  (202) 434-2277
  www.aarp.org

- **National Aging Information Center**
  330 Independence Avenue SW
  Washington, DC 20201
  (202) 619-7501
  www.aoa.dhhs.gov/naic

- **National Institute on Aging**
  Public Information Office
  31 Center Drive, MSC 2292
  Bethesda, MD 20892
  (301) 496-1752
  www.nih.gov/nia

Alcohol and Drug Abuse

- **Al-Anon Family Group Headquarters, Inc.**
  1600 Corporate Landing Parkway
  Virginia Beach, VA 23454-5617
  (800) 356-9996
  www.al-anon.alateen.org

- **Alateen**
  1600 Corporate Landing Parkway
  Virginia Beach, VA 23454-5617
  (800) 356-9996
  www.al-anon.alateen.org

- **Alcoholics Anonymous (AA)**
  General Service Office
  475 Riverside Drive
  New York, NY 10115
  (212) 870-3400
  www.aa.org

- **Narcotics Anonymous (NA)**
  P.O. Box 9999
  Van Nuys, CA 91409
  (818) 773-9999; fax (818) 700-0700
  www.wsoinc.com

- **National Clearinghouse for Alcohol and Drug Information (NCADI)**
  P.O. Box 2345
  Rockville, MD 20847-2345
  (800) 729-6686
  www.health.org

- **National Council on Alcoholism and Drug Dependence (NCADD)**
  12 West 21st Street
  New York, NY 10010
  (800) NCA-CALL or (800) 622-2255
  (212) 206-6770; fax (212) 645-1690
  www.ncadd.org

- **U.S. Center for Substance Abuse Prevention**
  1010 Wayne Avenue, Suite 850
  Silver Spring, MD 20910
  (301) 459-1591 ext. 244;
  fax (301) 495-2919
  www.covesoft.com/csap.html

Consumer Organizations

- **Center for Science in the Public Interest (CSPI)**
  1875 Connecticut Avenue NW,
  Suite 300
  Washington, DC 20009-5728
  (202) 332-9110; fax (202) 265-4954
  www.cspinet.org

- **Choice in Dying, Inc.**
  1035 30th Street NW
  Washington, DC 20007
  (202) 338-9790; fax (202) 338-0242
  www.choices.org

- **Consumer Information Center**
  Pueblo, CO 81009
  (888) 8 PUEBLO or (888) 878-3256
  www.pueblo.gsa.gov

- **Consumers Union of US Inc.**
  101 Truman Avenue
  Yonkers, NY 10703-1057
  (914) 378-2000
  www.consunion.org

- **National Council Against Health Fraud, Inc. (NCAHF)**
  P.O. Box 1276
  Loma Linda, CA 92354
  (909) 824-4690
  www.ncahf.org

Fitness

- **American College of Sports Medicine**
  P.O. Box 1440
  Indianapolis, IN 46206-1440
  (317) 637-9200
  _www.acsm.org/sportsmed

- **American Council on Exercise (ACE)**
  5820 Oberlin Drive, Suite 102
  San Diego, CA 92121
  (800) 529-8227
  www.acefitness.org

- **President's Council on Physical Fitness and Sports**
  Humphrey Building, Room 738
  200 Independence Avenue SW
  Washington, DC 20201
  (202) 690-9000; fax (202) 690-5211
  _www.indiana.edu/~preschal

- **Shape Up America!**
  6707 Democracy Boulevard,
  Suite 306
  Bethesda, MD 20817
  (301) 493-5368
  www.shapeup.org

- **Sport Medicine and Science Council of Canada**
  1600 James Naismith Drive, Suite 314
  Gloucester, Ontario K1B 5N4, Canada
  (613) 748-5671; fax (613) 748-5729
  www.smscc.ca

Food Safety

- **Alliance for Food & Fiber Food Safety Hotline**
  (800) 266-0200

- **FDA Center for Food Safety and Applied Nutrition**
  200 C Street SW
  Washington, DC 20204
  (800) FDA-4010 or (800) 332-4010
  vm.cfsan.fda.gov

- **National Lead Information Center**
  (800) LEAD-FYI or (800) 532-3394
  (800) 424-LEAD or (800) 424-5323

- **National Pesticide Telecommunications Network (NPTN)**
  Oregon State University
  333 Weniger Hall
  Corvallis, OR 97331-6502
  (541) 737-6091
  _www.ace.orst.edu/info/nptn

- **USDA Meat and Poultry Hotline**
  (800) 535-4555

- **U.S. EPA Safe Drinking Water Hotline**
  (800) 426-4791

Health and Disease

- **Alzheimer's Disease Education and Referral Center**
  P. O. Box 8250
  Silver Spring, MD 20907-8250
  (800) 438-4380
  www.alzheimers.org

- **Alzheimer's Disease Information and Referral Service**
  919 North Michigan Avenue,
  Suite 1000
  Chicago, IL 60611
  (800) 272-3900
  www.alz.org

- **American Academy of Allergy, Asthma, and Immunology**
  611 East Wells Street
  Milwaukee, WI 53202
  (414) 272-6071; fax (414) 276-3349
  www.aaaai.org

- **American Cancer Society National Home Office**
  1599 Clifton Road NE
  Atlanta, GA 30329-4251
  (800) ACS-2345 or (800) 227-2345
  www.cancer.org

- **American Council on Science and Health**
  1995 Broadway, 2nd Floor
  New York, NY 10023-5860
  (212) 362-7044; fax (212) 362-4919
  www.acsh.org

- **American Dental Association**
  211 East Chicago Avenue
  Chicago, IL 60611
  (312) 440-2800
  www.ada.org

- **American Diabetes Association**
  1660 Duke Street
  Alexandria, VA 22314
  (800) 232-3472 or (703) 549-1500
  www.diabetes.org

- **American Heart Association**
  Box BHG, National Center
  7320 Greenville Avenue
  Dallas, TX 75231
  (800) 275-0448 or (214) 373-6300
  www.amhrt.org

- **American Institute for Cancer Research**
  1759 R Street NW
  Washington, DC 20009
  (800) 843-8114 or (202) 328-7744;
  fax (202) 328-7226
  www.aicr.org

- **American Medical Association**
  515 North State Street
  Chicago, IL 60610
  (312) 464-5000
  www.ama-assn.org

- **American Public Health Association (APHA)**
  1015 Fifteenth Street NW, Suite 300
  Washington, DC 20005
  (202) 789-5600
  www.apha.org

- **American Red Cross**
  National Headquarters
  8111 Gatehouse Road
  Falls Church, VA 22042
  (703) 206-7180
  www.redcross.org

- **Canadian Diabetes Association**
  15 Toronto Street, Suite 800
  Toronto, ON M5C 2E3
  (800) BANTING or (800) 226-8464
  (416) 363-3373
  www.diabetes.ca

- **Canadian Public Health Association**
  400-1565 Carling Avenue
  Ottawa, Ontario K1Z 8R1
  (613) 725-3769; fax (613) 725-9826
  www.cpha.ca

- **Centers for Disease Control and Prevention (CDC)**
  1600 Clifton Road NE
  Atlanta, GA 30333
  (404) 639-3311
  www.cdc.gov

- **The Food Allergy Network**
  10400 Eaton Place, Suite 107
  Fairfax, VA 22030-2208
  (800) 929-4040 or (703) 691-3179
  www.foodallergy.org
- **Internet Health Resources**
  www.ihr.com
- **National AIDS Hotline
  (CDC)**
  (800) 342-AIDS (English)
  (800) 344-SIDA (Spanish)
  (800) 2437-TTY (Deaf)
  (900) 820-2437
- **National Cancer Institute**
  Office of Cancer Communications
  Building 31, Room 10824
  Bethesda, MD 20892
  (800) 4-CANCER or (800) 422-6237
  www.nci.nih.gov
- **National Diabetes Information
  Clearinghouse**
  1 Information Way
  Bethesda, MD 20892-3560
  (301) 654-3327
  www.niddk.nih.gov
- **National Digestive Disease Information
  Clearinghouse (NDDIC)**
  2 Information Way
  Bethesda, MD 20892-3570
  (301) 654-3810
  www.niddk.nih.gov
- **National Health Information Center
  (NHIC)**
  Office of Disease Prevention and Health
  Promotion
  (800) 336-4797
  nhic-nt.health.org
- **National Heart, Lung, and Blood
  Institute**
  Information Center
  P.O. Box 30105
  Bethesda, MD 20824-0105
  (301) 251-1222
  _www.nhlbi.nih.gov/nhlbi/nhlbi
  .htm
- **National Institute of
  Allergy and Infectious
  Diseases**
  Office of Communications
  Building 31, Room 7A50
  31 Center Drive, MSC2520
  Bethesda, MD 20892-2520
  (301) 496-5717
  www.niaid.nih.gov
- **National Institute of Dental Research
  (NIDR)**
  National Institute of Health
  Bethesda, MD 20892-2190
  (301) 496-4261
  www.nidr.nih.gov

- **National Institutes of Health (NIH)**
  9000 Rockville Pike
  Bethesda, MD 20892
  (301) 496-2433
  www.nih.gov
- **National Osteoporosis Foundation**
  1150 17th Street NW, Suite 500
  Washington, DC 20036
  (202) 223-2226
  www.nof.org
- **Office of Disease Prevention and Health
  Promotion**
  odphp.osophs.dhhs.gov
- **Office on Smoking and Health (OSH)**
  _www.americanheart.org/heart.org/
  Heart_and_stroke_A_Z_Guide/osh
  .html

Infancy and Childhood

- **American Academy of Pediatrics**
  141 Northwest Point Boulevard
  Elk Grove Village, IL 60007-1098
  (847) 228-5005
  www.aap.org
- **Association of Birth Defect Children, Inc.**
  930 Woodcock Road, Suite 225
  Orlando, FL 32803
  (407) 245-7035
  www.birthdefects.org
- **Canadian Paediatric Society**
  100-2204 Walkley Road
  Ottawa, ON K1G 4G8
  (613) 526-9397; fax (613) 526-3332
  www.cps.ca
- **National Center for Education in
  Maternal & Child Health**
  2000 15th Street North, Suite 701
  Arlington, VA 22201-2617
  (703) 524-7802
  www.ncemch.org

Pregnancy and Lactation

- **American College of Obstetricians and
  Gynecologists Resource Center**
  409 12th Street SW
  Washington, DC 20024-2188
  (202) 638-5577
  www.acog.org
- **La Leche International, Inc.**
  1400 N. Meacham Road
  Schaumburg, IL 60173
  (847) 519-7730
  www.lalecheleague.org
- **March of Dimes Birth Defects Foundation**
  1275 Mamaroneck Avenue
  White Plains, NY 10605
  (914) 428-7100
  www.sunkist.com

World Hunger

- **Bread for the World**
  1100 Wayne Avenue, Suite 1000
  Silver Spring, MD 20910
  (301) 608-2400
  www.bread.org
- **Center on Hunger, Poverty and Nutrition
  Policy**
  Tufts University School of Nutrition
  11 Curtis Avenue
  Medford, MA 02155
  (617) 627-3956
- **Freedom from Hunger**
  P.O. Box 2000
  1644 DaVinci Court
  Davis, CA 95617
  (530) 758-6200
  www.freefromhunger.org
- **Oxfam America**
  26 West Street
  Boston, MA 02111
  (617) 482-1211
  www.oxfamamerica.org
- **SEEDS Magazine**
  P.O. Box 6170
  Waco, TX 76706
  (254) 755-7745
  _www.helwys.com/seedhome.htm
- **Worldwatch Institute**
  1776 Massachusetts Avenue NW,
  Suite 800
  Washington, DC 20036
  (202) 452-1999
  www.worldwatch.org

Scientific Literature

Nutrition Journals

American Journal of Clinical Nutrition
British Journal of Nutrition
Human Nutrition, Applied Nutrition
Journal of the American College of Nutrition
Journal of the American Dietetic Association
Journal of the Canadian Dietetic Association
Journal of Food Composition and Analysis
Journal of Nutrition
Journal of Nutrition Education
Nutrition Abstracts and Reviews
Nutrition and Metabolism
Nutrition Reports International
Nutrition Research
Nutrition Reviews
Nutrition Today

Other Journals

American Journal of Epidemiology
American Journal of Nursing
American Journal of Public Health
Annals of Internal Medicine
Annals of Surgery
Canadian Journal of Public Health
Caries Research

Food Technology
Gastroenterology
International Journal of Obesity
Journal of the American Dental Association
Journal of the American Medical Association
Journal of Clinical Investigation
Journal of Food Science
Journal of Home Economics

Journal of Pediatrics
Lancet
New England Journal of Medicine
Pediatrics
Science
The Scientist

Appendix C

The U.S. Food Exchange System

The U.S. exchange system divides the foods suitable for use in planning a healthy diet into six lists—the starch/bread, meat/meat alternate, vegetable, fruit, milk, and fat lists.a These lists are shown in Tables C.1 through C.6. Following these lists are three other sets of foods: free foods, combination foods, and foods for occasional use (Tables C.7, C.8, and C.9).

aThe Exchange Lists are the basis of a meal planning system designed by a committee of the American Diabetes Association and The American Dietetic Association. While designed primarily for people with diabetes and others who must follow special diets, the Exchange Lists are based on principles of good nutrition that apply to everyone. © 2008 American Dietetic Association, The American Diabetes Association.

Table C.1

The U.S.. Exchange System: Starch/Bread List

15 g carbohydrate, 0-3 g protein, 0-15, 80 cal

Amount	Food	Amount	Food
Cereals/Grains/Pasta		**Bread**	
$1/2$ c	Bran cereals,	$1/4$ (1 oz)	Bagel, large (4 oz)
$1/2$ c	Bulgur, cooked	$1/2$	English muffins
$1/3$ c	Cooked cereals	$1/2$ (1 oz)	Frankfurter or hamburger buns
$1/2$ c	Grits, cooked	$1/2$ piece	Pita, 6" across
$3/4$ c	Other ready-to-eat unsweetened cereals	1 (1 oz)	Plain rolls, small
$1/3$ c	Pasta, cooked	1 slice (1 oz)	Raisin, unfrosted
$1^{1}/_{2}$ c	Puffed cereals	1 slice (1 oz)	Rye, pumpernickel
$1/3$ c	Rice, white or brown, cooked	1	Tortillas, 6" across
$1/2$ c	Shredded wheat	1 slice (1 oz)	White (including French, Italian)
3 tbs	Wheat germ	1 slice (1 oz)	Whole-wheat
Dried Beans/Peas/Lentils		2 slices ($1^{1}/_{2}$ oz)	Reduced-calorie
$1/3$ c	Baked beans	**Crackers/Snacks**	
$1/2$ c	Beans and peas, cooked, such as kidney, white, split, black-eyed	8	Animal crackers
$1/2$ c	Lentils, cooked	3	Graham crackers, $2^{1}/_{2}$" square
Starchy Vegetables		$3/4$ oz	Matzoh
$1/2$ c	Corn	4 pieces	Melba toast, 2" × 4" piece
$1/2$ cob (5 oz)	Corn on the cob, large	20	Oyster crackers
$1/2$ c	Lima beans	3 c	Popcorn, popped, no fat added
$1/2$ c	Peas, green, canned or frozen	$3/4$ oz	Pretzels
$1/3$ c	Plantains	6	Saltine-type crackers
1 small (3 oz)	Potatoes, baked	2 to 5 ($3/4$ oz)	Whole-wheat crackers, no fat added (crisp breads)
$1/2$ c	Potatoes, mashed	**Starch Foods Prepared with Fat**	
1 c	Squash, winter (acorn, butternut)	(Count as 1 starch/bread serving, plus 1 fat serving.)	
$1/2$ c	Yams, sweet potatoes, plain	1	Biscuits, $2^{1}/_{2}$" across
		1 ($1^{1}/_{2}$ oz)	Cornbread, $1^{3}/_{4}$" cube
		6	Crackers, round butter type
		1 cup (2 oz)	French fries, oven-baked
		1	Muffins, plain, small
		1	Pancakes, 4" across
		$1/3$ c	Stuffing, bread, prepared
		2	Taco shells, 5" across
		1	Waffles, 4" square
		2 to 5 ($3/4$ oz)	Whole-wheat crackers, fat added

3 grams or more dietary fiber per serving. Average fiber contents of whole-grain products is 2 grams per serving. For starch foods not on this list, the general rule is that ½ cup cereal, grain, or pasta is 1 serving: 1 ounce of a bread product is 1 serving.

Table C.2

U.S. Exchange System: Meat/Meat Alternate Lists

Lean meat = 7 g protein, 3 g fat, 55 cal; medium-fat meat = 7 g protein, 5 g fat, 75 cal; high-fat meat = 7 g protein, 8 g fat, 100 cal.

Category	Amount	Food
Lean Meat and Alternates		
Beef	1 oz	USDA Select or Choice grades of lean beef, trimmed of fat, such as round, sirloin, and flank steak; tenderloin;
Pork	1 oz	Lean pork, such as Canadian bacon 🖊, tenderloin, riborloin chop/roast, warm
Veal	1 oz	Chops, roasts
Poultry	1 oz	Chicken, turkey, Cornish hen (without skin)
Fish	1 oz	All fresh and frozen fish
	1 oz	Crab, lobster, scallops, shrimp, clams (fresh or canned in water 🖊)
	6 medium	Oysters
	1 oz	Tuna, canned in water or oil
	1 oz	Herring uncreamed or smoked
	2 medium	Sardines, canned
Wild game	1 oz	Venison, rabbit, buffalo, ostrich
Cheese	¼ c	Any cottage cheese
	1 oz	Cheeses with 3 grams of fat or less per oz
Other	1 oz	Lunch meats with 3 grams of fat or less per oz
	2 whites	Egg whites
	¼ c	Egg substitutes, plain
Medium-Fat Meat and Alternates		
Beef	1 oz	Most beef products fall into this category; examples: all ground beef, roasts (rib, chuck, rump), steak (cubed, porterhouse, T-bone), meatloaf
Pork	1 oz	Most pork products fall into this category; examples: cutlet, loin roast, shoulder roast
Lamb	1 oz	Most lamb products fall into this category; examples: chops, leg, roast, ground

Category	Amount	Food
Medium-Fat Meat and Alternates (*continued*)		
Veal	1 oz	Cutlet, ground or cubed, unbreaded
Poultry	1 oz	Chicken (with skin), wild duck or goose (well-drained of fat), ground turkey, fried chicken
Fish	1 oz	Any Fried Product
Cheese		Skim or part-skim milk cheeses, such as:
	¼ c	Ricotta
	1 oz	Mozzarella
	1 oz	Cheeses with 4-7 grams of fat per oz
Other	1	Eggs (high in cholesterol, limit to 3 per week)
	4 oz	Tofu, 2½" × 2¾" × 1"
	1 oz	Sausage with 4-7 grams of fat per oz
High-Fat Meat and Alternates[a]		
Pork	1 oz	Spareribs, ground pork, pork sausages (patties or links)
Cheese	1 oz	All regular cheeses, such as American, blue, Cheddar, Monterey, Swiss, brie
Other	1 oz	Lunch meats 🖊, such as bologna, salami, pastrami pimento loaf, with 8 grams of fat or more per oz
	1 oz	Sausage 🖊, such as Polish, Italian
	1 oz	Knockwurst, smoked 🖊
	1 oz	Bratwurst 🖊
	1 (10/lb)	Frankfurters 🖊 (turkey or chicken)
Count as 1 high-fat meat plus 1 fat exchange:		
1 frank	(10/lb)	Frankfurters 🖊 (beef, pork, or combination)

🖊 400 milligrams or more sodium per exchange. Meats contribute no fiber to the diet.

[a]These items are high in saturated fat, cholesterol, and calories and should be used no more than three times per week.

Table C.3

U.S. Exchange System: Vegetable List

5 g carbohydrate, 2 g protein, 25 cal
All portion sizes, except as otherwise noted, are $1/2$ c of any cooked vegetable or vegetable juice, 1 c of any raw vegetable.

Artichokes, $1/2$ medium	Cabbage, cooked (green, bok chog, Chinese)	Leeks	Spinach, cooked
Asparagus	Carrots	Mushrooms, fresh	Summer squash (crookneck)
Bean sprouts	Cauliflower	Okra	Tomatoes, 1 large
Beans (green, wax, Italian)	Eggplant	Onions	Tomato/vegetable juice 🖋
Beets	Peppers	Pea pods	Turnips
Broccoli	Greens (collard, mustard,	Rutabagas	Water chestnuts
Brussels sprouts	turnip)	Sauerkraut 🖋	
	Kohlrabi		

Starchy vegetables such as corn, peas, and potatoes are found on the Starch/Bread List.
For free vegetables, see the Free Food List (Table D.7).

🖋 400 milligrams or more sodium per serving. Most vegetable servings contain 2 to 3 grams dietary fiber.

Table C.4

U.S. Exchange System: Fruit List

15 g carbohydrate, 60 cal
All portion sizes, unless otherwise noted, are $1/2$ c fresh fruit or fruit juice, $1/4$ c dried fruit.

Amount	Food	Amount	Food
Fresh, Frozen, and Unsweetened Canned Fruit		$1/2$ c (2 halves)	Pears, canned
		$3/4$ c	Pineapple, raw
1	Apples, raw, 2" across	$1/2$ c	Pineapple, canned
$1/2$ c	Applesauce, unsweetened	2	Plums, raw, 2" across
4	Apricots, medium, raw	1 c	Raspberries, raw 🖋
$1/2$ c (4 halves)	Apricots, canned	$1 1/4$ c	Strawberries, raw, whole
1	Bananas, extra small	2	Tangerines, $2 1/2$" across
$3/4$ c	Blackberries, raw 🖋	$1 1/4$ c	Watermelon, cubes
$3/4$ c	Blueberries, raw 🖋	**Dried Fruit**	
$1/3$	Cantaloupe, 5" across		
1 c	Cantaloupe, cubes	4 rings	Apples 🖋
12	Cherries, large, raw	8 halves	Apricots 🖋
$1/2$ c	Cherries, canned	3 medium	Dates
2	Figs, raw, 2" across	$1 1/2$	Figs 🖋
$1/2$ c	Fruit cocktail, canned	3 medium	Prunes 🖋
$1/2$	Grapefruit, large	**Fruit Juice**	
$3/4$ c	Grapefruit, segments		
17	Grapes, small	$1/2$ c	Apple juice/cider
1 slice	Honeydew melon, medium	$1/3$ c	Cranberry juice cocktail
1 c	Honeydew melon, cubes	$1/3$ c	Grape juice
1	Kiwis, large	$1/2$ c	Grapefruit juice
$3/4$ c	Mandarin oranges	$1/2$ c	Orange juice
$1/2$	Mangoes, small	$1/2$ c	Pineapple juice
1	Nectarines, $1 1/2$" across 🖋	$1/3$ c	Prune juice
1	Oranges, $2 1/2$" across		
1 c	Papayas, cubes		
1 ($3/4$ c)	Peaches, $2 3/4$" across		
$1/2$ c (2 halves)	Peaches, canned		
$1/2$ large or 1 small	Pears		

🖋 3 grams or more dietary fiber per serving. Average fiber contents of fresh, frozen, and dry fruits: 2 grams per serving.

Table C.5

U.S. Exchange System: Milk List

Nonfat and very low-fat milk = 12 g carbohydrate, 8 g protein, trace fat, 100 cal; low-fat milk = 12 g carbohydrate, 8 g protein, = g fat, 120 cal; whole milk = 12 g carbohydrate, 8 g protein, 8 g fat, 160 cal.

Amount	Food	Amount	Food
Nonfat and Very Low-Fat Milk		**Low-Fat Milk**	
1 c	Nonfat milk	1 c fluid	2% milk
1 c	1% milk	6 oz	Plain low-fat yogurt, with added nonfat milk solids
½ c	Evaporated nonfat milk		
1 c	Low-fat buttermilk	**Whole Milk**	
6 oz	Plain nonfat yogurt	1 c	Whole milk, buttermilk, goat's milk
		½ c	Evaporated whole milk
		8 oz	Whole plain yogurt

Table C.6

U.S. Exchange System: Fat List

5 g fat, 45 cal

Amount	Food	Amount	Food
Unsaturated Fats		**Saturated Fats**	
2 tbs	Avocados (*trans* fat free)	1 slice	Bacon[a]
1 tsp	Margarine	1 tsp	Butter
1 tbs	Margarine, diet[a] (30%-50% vegetable oil, trans-fatfree)	½ oz (2 tbs)	Chitterlings
		2 tbs	Coconut, shredded
2 tsp	Mayonnaise	1 tbs	Cream (heavy, whipping)
1 tbs	Mayonnaise, reduced fat[a]	1½ tbs	Cream (light, coffee, table)
	Nuts and seeds:	2 tbs	Cream (sour)
6 whole	Almonds, dry roasted	1 tbs	Cream cheese
6 whole	Cashews, dry roasted	¼ oz	Salt pork[a]
10 whole	Peanuts		
2 whole	Pecans	Two tablespoons of low-calorie salad dressing is a free food.	
1 tbp	Pumpkin seeds		
1 tbs	Other nuts		
1 tbs	Seeds, pine nuts, sunflower seeds (without shells)		
2 whole	Walnuts		
1 tbp	Oil (canola, olive, peanut)		
8 large	Olives[a], black		
10 large	Olives, green		
1 tbs	Salad dressing, all varieties[a]		
1 tbp	Salad dressing, mayonnaise type		
1 tbs	Salad dressing, mayonnaise type, reduced calorie		
2 tbs	Salad dressing, reduced calorie		

Table C.7

U.S. Exchange System: Free Foods

A free food is any food or drink that contains less than 20 cal/serving. People with diabetes are advised to eat as much as they want of those items that have no serving size specified. They may eat two or three servings per day of those items that have a specific serving size. It is suggested that they spread the servings out through the day.

Amount	Food	Amount	Food
Drinks	Bouillon, low-sodium	**Condiments**	
	Bouillon 🔖 or broth without fat	1 tbs	Catsup
	Carbonated drinks, sugar-free		Horseradish
	Carbonated water		Mustard
	Club soda	1½ medium	Pickles 🔖, dill, unsweetened
1 tbs	Cocoa powder, unsweetened	1 tbs	Taco sauce
	Coffee/tea		Vinegar
	Drink mixes, sugar-free		
	Tonic water, sugar-free	**Seasonings**	Basil, fresh
			Celery seeds
Nonstick Pan Spray			Chili powder
			Chives
Fruit			Cinnamon
½ c	Cranberries, unsweetened		Curry
½ c	Rhubarb, unsweetened		Dill
Vegetables (raw, 1 c)	Cabbage		Flavoring extracts (almond, butter, lemon, peppermint, vanilla, walnut, etc.)
	Celery		
	Chinese cabbage ✱		
	Cucumbers		Garlic
	Green onions		Garlic powder
	Hot peppers		Herbs
	Mushrooms		Hot pepper sauce
	Radishes		Lemon
	Arugula ✱		Lemon juice
Salad Greens	Endive		Lemon pepper
	Escarole		Lime
	Lettuce		Lime juice
	Romaine		Mint
	Spinach		Onion powder
	Watercress		Oregano
			Paprika
Sweet Substitutes	Candy, hard, sugar-free		Pepper
	Gelatin, sugar-free		Pimento
	Gum, sugar-free		Spices
2 tsp	Jam/jelly, sugar-free	¼ c	Wine, used in cooking
2 tbs	Pancake syrup, sugar-free		Worcestershire sauce
	Sugar substitutes (saccharin, aspartame, acesulfame-K)		
2 tbs	Whipped topping, light or fat-free		

✱ 3 grams or more dietary fiber per serving.

🔖 400 milligrams or more sodium per serving.

Table C.8

U.S. Exchange Combination Foods

Much of the food we eat is mixed together in various combinations. These combination foods do not fit into only one exchange list. It can be quite hard to tell what is in a certain casserole dish or baked food item. This is a list of average values for some typical combination foods. This list will help you fit these foods into your meal plan. Ask your dietitian for information about any other foods you'd like to eat. The *American Diabetes Association/American Dietetic Association Family Cookbooks* and the *American Diabetes Associates Holiday Cookbook* have many recipes and further information about many foods, including combination foods. Check your library or local bookstore.

Food	Amount	Exchanges	Food	Amount	Exchanges
Casseroles, homemade	1 c (8 oz)	2 carbohydrates + 2 medium-fat meat	Spaghetti and meatballs 🖊, canned	1 c (8 oz)	2 carbohydrates + 2 medium-fat meats
Cheese pizza 🖊, thin crust	¼ of 12"	2 carbohydrates + 2 medium-fat meats	Sugar-free pudding, made with nonfat milk	½ c	1 carbohydrate
Chili with beans ⚶ 🖊, commercial	1 c (8 oz)	2 starch, 2 medium-fat meat, 2 fat			

If beans are used as a meat substitute:

Food	Amount	Exchanges
Chow mein ⚶ 🖊, with noodles and vegetables in source	1 c	2 carbohydrate + 1 fat	Dried beans ⚶, peas ⚶, lentils ⚶	1 c (cooked)	2 starch, 1 lean meat
Macaroni and cheese 🖊 meat, 2 fat	1 c (8 oz)	2 carbohydrates + 2 medium-fat meats			
Soups:					
Bean ⚶ 🖊	1 c (8 oz)	1 carbohydrate + 1 lean meat			
Cream 🖊 made with water	1 c (8 oz)	1 carbohydrate + 1 fat			
Vegetable 🖊 or broth 🖊	1 c (8 oz)	1 carbohydrate			

🖊 400 milligrams or more sodium per serving.

⚶ 3 grams or more dietary fiber per serving.

Table C.9

U.S. Exchange Foods for Occasional Use

The following list includes average exchange values for some foods high in sugar and fat. People are advised to use them only occasionally and in moderate amounts.

Food	Amount	Exchanges	Food	Amount	Exchanges
Angel food cake	$^1/_{12}$ cake	2 carbohydrates	Granola bars	1 small	$1^1/_2$ carbohydrate
Cake, no icing	$^1/_{12}$ cake or a 2" square	1 carbohydrate + 1 fat	Ice cream, any flavor, fat-free	$^1/_2$ c	$1^1/_2$ carbohydrate
Cookies	2 small, $2^1/_4$" across	1 carbohydrate + 2 fats	Sherbet, any flavor	$^1/_2$ c	2 carbohydrates
			Vanilla wafers	5 small	1 carbohydrate + 1 fat
Frozen fruit yogurt, fat free	$^1/_3$ c	1 carbohydrate			
Gingersnaps	3	1 carbohydrate			

🖊 If more than one serving is eaten, these foods have 400 milligrams or more sodium.

Appendix D

Table of Intentional Food Additives

Table D.1

A Guide to Intentional Food Additives

Abbreviation	Type of Additive	Uses
AA	Anticaking agents	Keeps dry powders and crystals from clumping together (e.g., salt, powdered sugar).
C	Colors	Synthetic (laboratory-made) vegetable and fruit concentrates and other substances used to color foods (e.g., soft drinks, frosting).
E	Emulsifiers	Used to make oil and water mix (e.g., salad dressings, sauces).
EX	Extenders	"Fillers" such as fruit pulp and texturized protein (e.g., fruit drinks, hamburger).
FA	Flavoring agents	Used to add particular flavors to food (e.g., pudding, rye bread).
N	Nutrients	Used to add vitamins or minerals to foods (e.g., breakfast cereals, skim milk).
P	Preservatives	Used to keep food from spoiling (e.g., breads, breakfast cereals).
S	Sweeteners	Used to sweeten foods. Some are "artificial," such as aspartame and saccharin, and some are extracted from plants, such as sugar cane and sugar beets (e.g., soft drinks, catsup).
T	Texturizers	Used to improve the texture of food by stabilizing moisture content, dryness, volume, tenderness, or hardness (e.g., cakes, breads).
TH	Thickeners	Used to improve the consistency of foods (e.g., low-fat salad dressings, low-calorie jams).

Common Food Additives and Their Primary Function

Alginates (T)
Alpha tocopherol (vitamin E) (N)
Alpha tocopheryl acetate (vitamin E) (N)
Ascorbic acid (vitamin C) (P)
Aspartame (NutraSweet™) (S)
Baking powder (T)
Beet juice (C)
Beet sugar (S)
Beta-carotene (N)
BHA (P)
BHT (P)
Calcium carbonate (calcium) (N)
Calcium pantothenate (pantothenic acid, a B vitamin) (N)
Calcium propionate (P)
Calcium silicate (AA)
Cane sugar (S)

Carotene (C)
Carrageenan (E, TH)
Cellulose gum (TH)
Chromium chloride (N)
Citric acid (FA, P)
Corn syrup (S)
Cupric oxide (copper) (N)
Cyanocobalamin (vitamin B12) (N)
Dextrin (TH)
Dextrose (S)
Dibasic calcium phosphate (calcium phosphorus) (N)
Diglycerides (E)
EDTA (P)
Extracts (FA)
FD&C blue no. 1 (C)
FD&C red no. 3 (use is being phased out) (C)

(continued)

Table D.1

A Guide to Intentional Food Additives

Common Food Additives and Their Primary Function

FD&C yellow no. 5 (C)	Salt (FA, P)
Ferrous fumarate (iron) (N)	Silicondioxide (AA)
Ferrous sulfate (iron) (N)	Sodium aluminum phosphate (T)
Fructose (S)	Sodium ascorbite (vitamin C) (N)
Fruit pulp (EX)	Sodium benzoate (P)
Gelatin (TH)	Sodium bicarbonate (baking soda) (T)
Glycerol (T)	Sodium bisulfite (P)
Glyceryl abietate (E, T)	Sodium chloride (P, FA)
Guar gum (T, TH)	Sodium citrate (P)
High-fructose corn syrup (S)	Sodium erythorbate (P)
Honey (S)	Sodium hexametaphosphate (P, T)
Hydrolyzed protein (EX)	Sodium metabisulfite (P)
Lecithin (E)	Sodium molydate (molybdenum) (N)
Magnesium oxide (magnesium) N	Sodium nitrate (P)
Maltodextrin (S, TH)	Sodium nitrite (P)
Manganese sulfate (manganese) (N)	Sodium propionate (P)
Modified food starch (TH)	Sodium saccharin (S)
Monoglycerides (E)	Sodium selenate (selenium) (N)
Monosodium glutamate (MSG) (FA)	Sodium stearolyn-2-lactylate (P, E)
Natural flavorings (FA)	Sodium sulfite (P)
Niacinamide (niacin, vitamin B3) (N)	Sorbitan monostearate (E)
Nitrates (P)	Sorbitol (S, T)
Nitrites (P)	Sorghum (S, P)
Paprika (C)	Starches (TH)
Pectin (TH)	Sucrose (S)
Phosphoric acid (P)	Sugar (P)
Phytonadione (vitamin K) (N)	Sulfur dioxide (P)
Polysorbates (P)	Sweeteners (FA)
Potassium benzoate (P)	Texturized protein (EX)
Potassium bicarbonate (T)	Thiamin hydrochloride (thiamin, B1) (N)
Potassium chloride (N)	Thiamin mononitrate (thiamin, B1) (N)
Potassium iodite (iodine) (N)	Tocopherols (P, N)
Potassium metabisulfite (P)	Turmeric (C)
Potassium sorbate (P)	Vitamin A palmitate (vitamin A) (N)
Propyl gallate (P)	Vitamin C (ascorbic acid) (N, P)
Propyleneglycol (T)	Vitamin E (N, P)
Pyridoxine hydrochloride (vitamin B6) (N)	Xanthan gum (T)
Reduced iron (iron) (N)	Xylitol (S)
Retinol (vitamin A) (N)	Yeast (T)
Riboflavin (vitamin B2) (N)	Zinc oxide (zinc) (N)
Saccharin (S)	

Appendix E

Cells

This appendix presents an overview of the basic structure and functions of cells in the human body. Cell structure and function are central to discussions of nutrition for it is within cells that nutrients are utilized to sustain life and health. The life-sustaining processes that take place within each of the more than one hundred trillion cells in the human body are maintained by the nutrients we consume in our diet. Chemical reactions within the cells produce energy from carbohydrates, proteins, and fats; cause proteins to be broken down or built up; and in thousands of other ways keep us in a state of health.

A cell is a basic unit of life. Any substance that does not consist of one or more cells cannot be alive. Cells are the "building blocks" of tissues (such as muscles and bones), organs (such as the kidneys and liver), and systems (the respiratory and digestive systems, for example). Normal cell health and functioning are maintained when a state of nutritional and environmental utopia exists within and around the cells. A disruption in the availability of nutrients or the presence of harmful substances in the cell's environment can initiate disorders that eventually affect our health or growth. Health problems in general begin with disruptions in the normal activity of cells.

The types and amounts of food and supplements people consume affect the cells' environment and their ability to function normally. Excessive or inadequate supplies of nutrients and other chemical substances disrupt cell functions and result in health problems. Humans remain in a state of health as long as their cells do.

Cells
The basic unit of life, of which all living things are composed. Every cell is surrounded by a membrane and contains cytoplasm, within which are organelles and a nucleus; the cell nucleus contains chromosomes.

Cell Structures and Functions

A generalized diagram of a human cell is shown in Illustration E.1. Although all cells have some structures and functions in common, the specific functions performed, and the structures that support those functions, can vary a good deal from cell to cell. Cells lining the esophagus, stomach, and intestines, for example, are specialized to produce and secrete mucus that helps food pass through the digestive tract. Red blood cells are specially formed to transport oxygen and carbon dioxide.

Every cell is surrounded by a cell membrane that helps move nutrients into and out of the cell. Inside the cell membrane lies the cytoplasm, a fluid material that contains many organelles, tubes, and particles. Among the organelles in the cytoplasm are ribosomes, mitochondria, and lysosomes. Each of these "little organs" is encased in a cell membrane and performs specific functions. Ribosomes assemble amino acids into proteins following the instructions of DNA and its messenger RNA. The mitochondria are made of intricately folded membranes that bear thousands of highly organized sets of enzymes on their surfaces. These enzymes are actively involved in the production of energy and are found in particularly dense quantities in muscle cells. The lysosomes are like packets of enzymes. The enzymes are used to break down old cell particles that are being recycled and to destroy substances that are harmful to the body.

Cytoplasm also contains a highly organized system of membranes called the endoplasmic reticulum. When these membranes are dotted with ribosomes, they are called "rough endoplasmic reticulum." When ribosomes are absent, the endoplasmic reticulum is referred to as "smooth." Some membranes within the cytoplasm form tubes that collect certain types of cellular material and transport it out of the cell. These membranous tubes are called "Golgi apparatus." The rough and smooth endoplasmic reticulum are continuous with the Golgi apparatus, so secretions produced throughout the cell can be collected and transported to its exterior.

Within each cell is a nucleus covered by a two-layer membrane. The nucleus contains chromosomes and the genetic material DNA. DNA encodes all of the instructions a cell needs to conduct protein synthesis and to replicate life.

All cells within the body are part of a complex communication system that uses hormones, electrical impulses, and other chemical messengers to link each cell to the others. No cell is an island that operates independently from the others.

Illustration E.1 Generalized structure of a human cell.

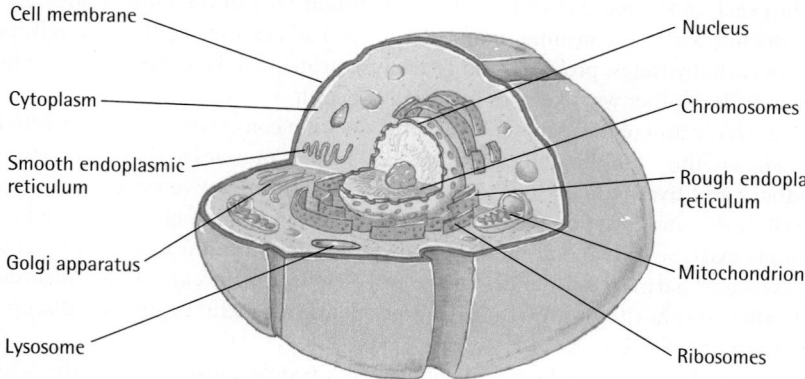

cell membrane
Cytoplasm
Smooth endoplasmic reticulum
Golgi apparatus
Lysosome
Nucleus
Chromosomes
Rough endopla reticulum
Mitochondrion
Ribosomes

cell membrane: the membrane that surrounds the cell and encloses its contents; made primarily of lipid and protein.

chromosomes: a set of structures within the nucleus of every cell that contain the cell's genetic material, DNA, associated with other materials (primarily proteins).

cytoplasm (SIGH-toe-plazm): the cell contents, except for the nucleus.
> *cyto* = cell
> *plasm* = a form

Golgi (GOAL-gee) **apparatus:** a set of membranes within the cell where secretory materials are packaged for export.

lysosomes: cellular organelles; membrane-enclosed sacs of degradative enzymes.
> *lysis* = dissolution

mitochondria (my-toe-KON-dree-uh; *singular* **mitochondrion**): the cellular organelles responsible for producing ATP aerobically; made of membranes (lipid and protein) with enzymes mounted on them.
> *mitos* = thread (referring to their slender shape)
> *chondros* = cartilage (referring to their external appearance)

nucleus: a major membrane-enclosed body within every cell, which contains the cell's genetic material, DNA, embedded in chromosomes.
> *nucleus* = a kernel

organelles: membrane-bound subcellular structures such as ribosomes, mitochondria, and lysosomes.
> *organelle* = little organ

ribosomes: protein-making organelles in cells; composed of RNA and protein.
> *ribo* = containing the sugar ribose
> *some* = body

rough endoplasmic reticulum (en-doh-PLAZ-mic reh-TIC-you-lum): intracellular membranes dotted with ribosomes, where protein synthesis takes place.
> *endo* = inside
> *plasm* = the cytoplasm

smooth endoplasmic reticulum: smooth intracellular membranes bearing no ribosomes.

Appendix F

Canadian Nutrition and Physical Activity Guidelines

Canada's Guidelines for Healthy Eating

Canada's Guidelines for Healthy Eating encourage healthy Canadians over 2 years of age to

- Enjoy a variety of foods.
- Emphasize cereals, breads, other grain products, vegetables, and fruits.
- Choose lower-fat dairy products, leaner meats, and foods prepared with little or no fat.
- Achieve and maintain a healthy body weight by enjoying regular physical activity and healthy eating.
- Limit salt, alcohol, and caffeine.

Canada's Recommended Nutrient Intakes

The Dietary Reference Intakes (DRI) reports have replaced both the 1989 RDA in the United States and the 1990 Recommended Nutrient Intakes (RNI) in Canada. The DRI are presented on Study Cards pages 27–28 at the back of this book.

Eating Well with Canada's Food Guide

Figure B-1 presents the 2007 *Eating Well with Canada's Food Guide*, which gives detailed information for selecting foods to meet *Canada's Guidelines for Healthy Eating*. The *Food Guide* was designed to meet the nutritional needs of all Canadians 2 years of age and older and takes a total diet approach. The *Food Guide* shows the four food groups with pictorial examples of foods in each group and shows the number of servings recommended for each food group.

For more detailed information and interactive tools using the *Food Guide*, go to www.healthcanada.gc.ca/foodguide. Additional publications are available from Health Canada's Office of Nutrition Policy and Promotion at www.hc-sc.gc.ca.

Health Canada   Santé Canada   Your health and safety… our priority.   Votre santé et votre sécurité… notre priorité.

Eating Well with

Canada's Food Guide

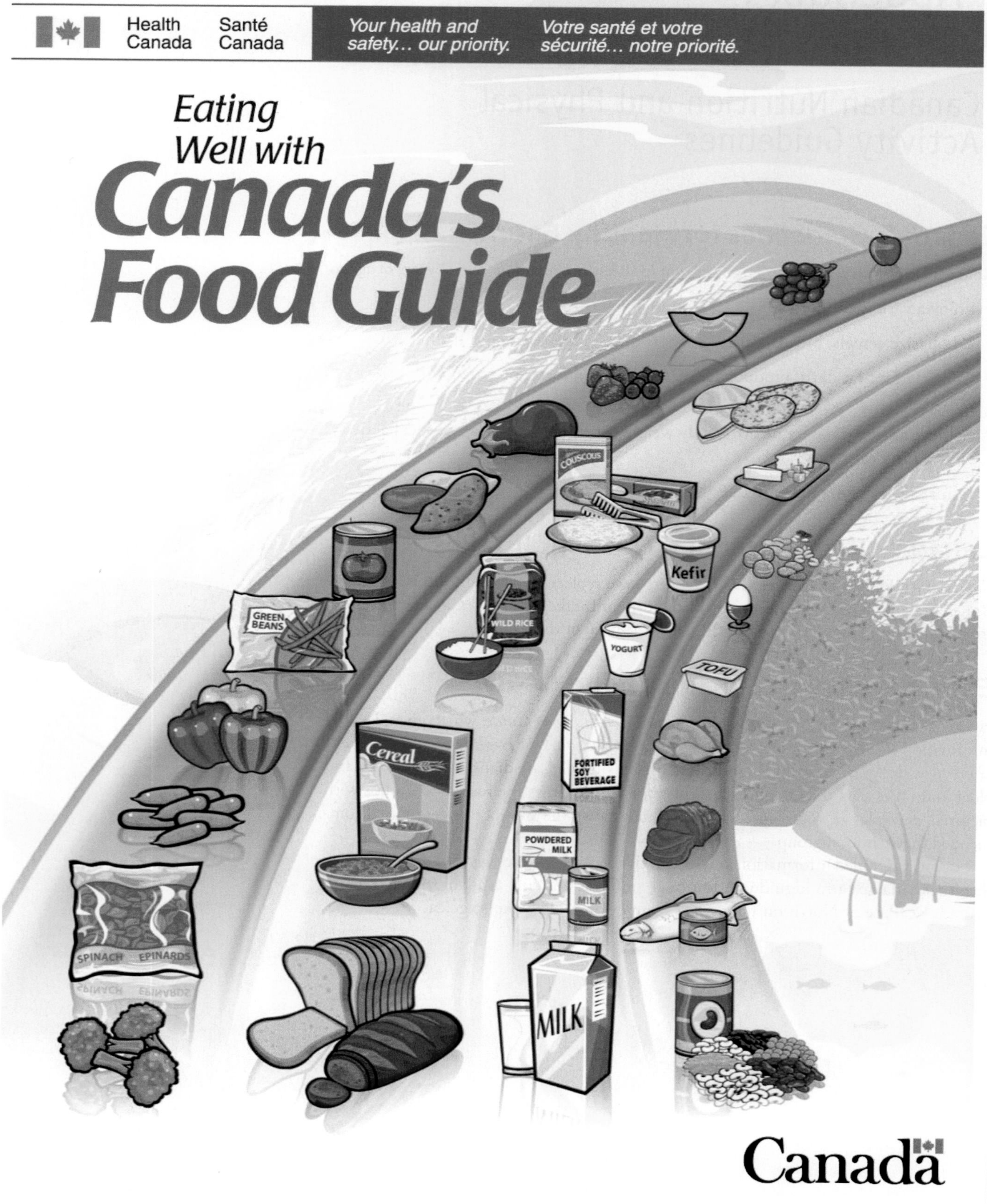

Canada

Eating Well with Canada's Food Guide—Continued

Recommended Number of *Food Guide Servings* per Day

Age in Years	Children			Teens		Adults			
	2-3	4-8	9-13	14-18		19-50		51+	
Sex	Girls and Boys			Females	Males	Females	Males	Females	Males
Vegetables and Fruit	4	5	6	7	8	7-8	8-10	7	7
Grain Products	3	4	6	6	7	6-7	8	6	7
Milk and Alternatives	2	2	3-4	3-4	3-4	2	2	3	3
Meat and Alternatives	1	1	1-2	2	3	2	3	2	3

The chart above shows how many Food Guide Servings you need from each of the four food groups every day.

Having the amount and type of food recommended and following the tips in *Canada's Food Guide* will help:

- Meet your needs for vitamins, minerals and other nutrients.
- Reduce your risk of obesity, type 2 diabetes, heart disease, certain types of cancer and osteoporosis.
- Contribute to your overall health and vitality.

FIGURE B-1

Eating Well with Canada's Food Guide—Continued

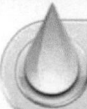

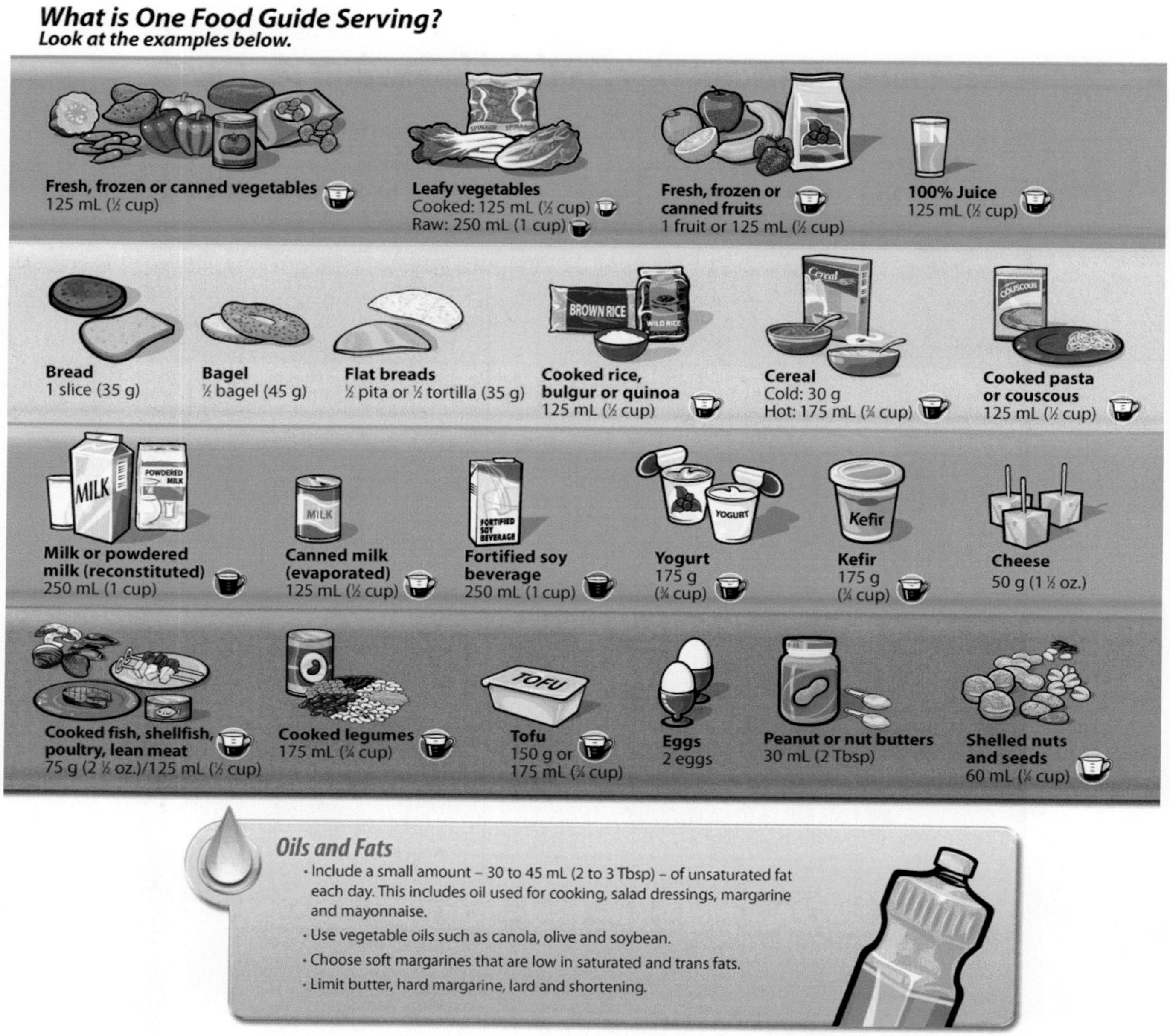
What is One Food Guide Serving?
Look at the examples below.
Fresh, frozen or canned vegetables
125 mL (½ cup)
Leafy vegetables
Cooked: 125 mL (½ cup)
Raw: 250 mL (1 cup)
Fresh, frozen or canned fruits
1 fruit or 125 mL (½ cup)
100% Juice
125 mL (½ cup)
Bread
1 slice (35 g)
Bagel
½ bagel (45 g)
Flat breads
½ pita or ½ tortilla (35 g)
BROWN RICE
WILD RICE
Cooked rice, bulgur or quinoa
125 mL (½ cup)
Cereal
Cereal
Cold: 30 g
Hot: 175 mL (¾ cup)
COUSCOUS
Cooked pasta or couscous
125 mL (½ cup)
MILK
POWDERED MILK
Milk or powdered milk (reconstituted)
250 mL (1 cup)
MILK
Canned milk (evaporated)
125 mL (½ cup)
FORTIFIED SOY BEVERAGE
Fortified soy beverage
250 mL (1 cup)
YOGURT
Yogurt
175 g
(¾ cup)
Kefir
Kefir
175 g
(¾ cup)
Cheese
50 g (1 ½ oz.)
Cooked fish, shellfish, poultry, lean meat
75 g (2 ½ oz.)/125 mL (½ cup)
Cooked legumes
175 mL (¾ cup)
TOFU
Tofu
150 g or
175 mL (¾ cup)
Eggs
2 eggs
Peanut or nut butters
30 mL (2 Tbsp)
Shelled nuts and seeds
60 mL (¼ cup)
Oils and Fats
• Include a small amount – 30 to 45 mL (2 to 3 Tbsp) – of unsaturated fat each day. This includes oil used for cooking, salad dressings, margarine and mayonnaise.
• Use vegetable oils such as canola, olive and soybean.
• Choose soft margarines that are low in saturated and trans fats.
• Limit butter, hard margarine, lard and shortening.

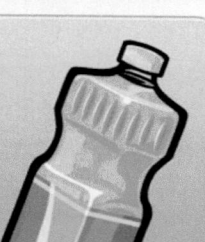

Eating Well with Canada's Food Guide—Continued

Make each Food Guide Serving count...
wherever you are – at home, at school, at work or when eating out!

▶ **Eat at least one dark green and one orange vegetable each day.**
- Go for dark green vegetables such as broccoli, romaine lettuce and spinach.
- Go for orange vegetables such as carrots, sweet potatoes and winter squash.

▶ **Choose vegetables and fruit prepared with little or no added fat, sugar or salt.**
- Enjoy vegetables steamed, baked or stir-fried instead of deep-fried.

▶ **Have vegetables and fruit more often than juice.**

▶ **Make at least half of your grain products whole grain each day.**
- Eat a variety of whole grains such as barley, brown rice, oats, quinoa and wild rice.
- Enjoy whole grain breads, oatmeal or whole wheat pasta.

▶ **Choose grain products that are lower in fat, sugar or salt.**
- Compare the Nutrition Facts table on labels to make wise choices.
- Enjoy the true taste of grain products. When adding sauces or spreads, use small amounts.

▶ **Drink skim, 1%, or 2% milk each day.**
- Have 500 mL (2 cups) of milk every day for adequate vitamin D.
- Drink fortified soy beverages if you do not drink milk.

▶ **Select lower fat milk alternatives.**
- Compare the Nutrition Facts table on yogurts or cheeses to make wise choices.

▶ **Have meat alternatives such as beans, lentils and tofu often.**

▶ **Eat at least two Food Guide Servings of fish each week.***
- Choose fish such as char, herring, mackerel, salmon, sardines and trout.

▶ **Select lean meat and alternatives prepared with little or no added fat or salt.**
- Trim the visible fat from meats. Remove the skin on poultry.
- Use cooking methods such as roasting, baking or poaching that require little or no added fat.
- If you eat luncheon meats, sausages or prepackaged meats, choose those lower in salt (sodium) and fat.

Enjoy a variety of foods from the four food groups.

Satisfy your thirst with water!

Drink water regularly. It's a calorie-free way to quench your thirst. Drink more water in hot weather or when you are very active.

* Health Canada provides advice for limiting exposure to mercury from certain types of fish. Refer to www.healthcanada.gc.ca for the latest information.

Advice for different ages and stages...

Children

Following *Canada's Food Guide* helps children grow and thrive.

Young children have small appetites and need calories for growth and development.

- Serve small nutritious meals and snacks each day.

- Do not restrict nutritious foods because of their fat content. Offer a variety of foods from the four food groups.

- Most of all... be a good role model.

Women of childbearing age

All women who could become pregnant and those who are pregnant or breastfeeding need a multivitamin containing **folic acid** every day. Pregnant women need to ensure that their multivitamin also contains **iron**. A health care professional can help you find the multivitamin that's right for you.

Pregnant and breastfeeding women need more calories. Include an extra 2 to 3 Food Guide Servings each day.

Here are two examples:
- Have fruit and yogurt for a snack, or
- Have an extra slice of toast at breakfast and an extra glass of milk at supper.

Men and women over 50

The need for **vitamin D** increases after the age of 50.

In addition to following *Canada's Food Guide*, everyone over the age of 50 should take a daily vitamin D supplement of 10 µg (400 IU).

How do I count Food Guide Servings in a meal?

Here is an example:

Vegetable and beef stir-fry with rice, a glass of milk and an apple for dessert		
250 mL (1 cup) mixed broccoli, carrot and sweet red pepper	=	2 **Vegetables and Fruit** Food Guide Servings
75 g (2 ½ oz.) lean beef	=	1 **Meat and Alternatives** Food Guide Serving
250 mL (1 cup) brown rice	=	2 **Grain Products** Food Guide Servings
5 mL (1 tsp) canola oil	=	part of your **Oils and Fats** intake for the day
250 mL (1 cup) 1% milk	=	1 **Milk and Alternatives** Food Guide Serving
1 apple	=	1 **Vegetables and Fruit** Food Guide Serving

Eat well and be active today and every day!

The benefits of eating well and being active include:

- Better overall health.
- Lower risk of disease.
- A healthy body weight.
- Feeling and looking better.
- More energy.
- Stronger muscles and bones.

Be active

To be active every day is a step towards better health and a healthy body weight.

Canada's Physical Activity Guide recommends building 30 to 60 minutes of moderate physical activity into daily life for adults and at least 90 minutes a day for children and youth. You don't have to do it all at once. Add it up in periods of at least 10 minutes at a time for adults and five minutes at a time for children and youth.

Start slowly and build up.

Eat well

Another important step towards better health and a healthy body weight is to follow *Canada's Food Guide* by:

- Eating the recommended amount and type of food each day.
- Limiting foods and beverages high in calories, fat, sugar or salt (sodium) such as cakes and pastries, chocolate and candies, cookies and granola bars, doughnuts and muffins, ice cream and frozen desserts, french fries, potato chips, nachos and other salty snacks, alcohol, fruit flavoured drinks, soft drinks, sports and energy drinks, and sweetened hot or cold drinks.

Read the label

- Compare the Nutrition Facts table on food labels to choose products that contain less fat, saturated fat, trans fat, sugar and sodium.
- Keep in mind that the calories and nutrients listed are for the amount of food found at the top of the Nutrition Facts table.

Nutrition Facts

Per 0 mL (0 g)

Amount		% Daily Value
Calories 0		
Fat 0 g		0 %
Saturates 0 g		0 %
+ Trans 0 g		
Cholesterol 0 mg		
Sodium 0 mg		0 %
Carbohydrate 0 g		0 %
Fibre 0 g		0 %
Sugars 0 g		
Protein 0 g		
Vitamin A 0 %	Vitamin C	0 %
Calcium 0 %	Iron	0 %

Limit trans fat

When a Nutrition Facts table is not available, ask for nutrition information to choose foods lower in trans and saturated fats.

Take a step today...

✓ Have breakfast every day. It may help control your hunger later in the day.

✓ Walk wherever you can – get off the bus early, use the stairs.

✓ Benefit from eating vegetables and fruit at all meals and as snacks.

✓ Spend less time being inactive such as watching TV or playing computer games.

✓ Request nutrition information about menu items when eating out to help you make healthier choices.

✓ Enjoy eating with family and friends!

✓ Take time to eat and savour every bite!

For more information, interactive tools, or additional copies visit Canada's Food Guide on-line at:
www.healthcanada.gc.ca/foodguide

or contact:

Publications
Health Canada
Ottawa, Ontario K1A 0K9
E-Mail: publications@hc-sc.gc.ca
Tel.: 1-866-225-0709
Fax: (613) 941-5366
TTY: 1-800-267-1245

Également disponible en français sous le titre :
Bien manger avec le Guide alimentaire canadien

This publication can be made available on request on diskette, large print, audio-cassette and braille.

Appendix G

Feedback and Answers to Nutrition Up Close and Review Questions

Unit 1

Nutrition Up Close

Nutrition Concepts Review
The nutrition concepts apply to the situations as follows:

1. 1
2. 2, 9
3. 10
4. 4
5. 3
6. 7
7. 8
8. 5
9. 6

Review Questions

1. True
2. c
3. False
4. False
5. False
6. True
7. False
8. False
9. True
10. d
11. b
12. plant chemicals/phytonutrients, antioxidants
13. Answers consist of pregnant women, breastfeeding women, infants, growing children, frail elderly, ill persons, and people recovering from illness.

14. g

15. a

16. e

17. b

18. f

19. d

20. h

21. c

Unit 2

Nutrition Up Close

Food Types for Healthful Diets

To reduce the selection of energy dense, nutrient-poor foods and excessive calorie intake levels, the MyPlate dietary pattern highlights foods in their basic form. That means, for example, food groups contains foods such as whole grain breads, broccoli, baked chicken, dried beans, and oranges.

Food types	Dietary pattern	
	ChooseMyPlate	Western
Pasta		X
Cold cuts (ham, bologna, salami)		X
Fish and seafood	X	
Whole grain breads	X	
Fruit jams and jellies		X
Potato, tortilla, and other snack chips		X
Dried beans	X	
Poultry (chicken, turkey)	X	
Fruit	X	
Vegetables	X	
Gravy		X
Ice cream		X
Soft drinks		X
Salad dressing, mayonnaise		X
Skim milk	X	
Cake		X

Review Questions

1. False

2. True

3. True

4. True

5. False

6. It is rare that modern humans have to engage in strenuous physical activity to obtain food. Feasts are rarely followed by famines.

7. c

8. a

9. True

10. True

11. False

12. b

13. d

14. a

15. c

Unit 3

Nutrition Up Close

Checking Out a Fat-Loss Product

The product advertised earns five "yes" responses, making it highly unlikely that the product works. One or two "yes" responses provide a strong clue that advertised products may not work.

Review Questions

1. True

2. True

3. False

4. True

5. False

6. False

7. True

8. c

9. a

10. b

11. False

12. True

13. True

14. c

15. e

16. a

17. d

18. b

Unit 4

Nutrition Up Close

Comparison Shopping

The **Health Granola Cereal**, providing 210 **calories** per serving, contributes 30 **calories from fat** (or 5% of the % **Daily Value**). Each serving also provides 120 mg of **sodium** (or 5% of the % **Daily Value**), 3 g of **dietary fiber** (or 12% of the % **Daily Value**), and 16 g of **sugars**. The **Whole-Grain Cereal** provides the better nutritional value because it is lower in calories, total fat, sodium, and sugar and higher in fiber than the **Health Granola Cereal.** Don't be fooled by an attractive-sounding product. The proof of nutritional value is in the label, not the name.

Review Questions

1. True
2. False
3. False
4. True
5. False
6. True
7. False
8. d
9. a
10. d
11. 760
12. 460
13. 6 grams, 30%
14. 3.5 grams, 17%
15. False
16. False
17. b
18. True
19. True
20. False

Unit 5

Nutrition Up Close

Improving Food Choices

There are no right or wrong answers to this unit's "Nutrition Up Close." Plans for dietary change that are specific, easy to accomplish, and highly acceptable are most likely to work in the long run.

Review Questions

1. False
2. True
3. False
4. True
5. True
6. d
7. a
8. False
9. True
10. True
11. True
12. b
13. c
14. False
15. False

Unit 6

Nutrition Up Close

Dietary Guidelines for Americans: What Are Healthful Types of Food Choices?

Types of foods	Increase consumption	Decrease consumption
Fish and seafood	X	
Cheese		X
Fruits	X	
High sodium		X
Beans and peas	X	
Nuts and seeds	X	
Oils	X	
Whole grains	X	
High sugar		X
Low-fat milk	X	
Refined grain products		X

Review Questions

1. False
2. True
3. d
4. c
5. False

6. True

7. True

8. False

9. True

10. c

11. a

12. True

13. d

14. b

15. b

16. a

Unit 7

Nutrition Up Close

Carbohydrate, Fat, and Protein Digestion

The primary enzymes involved in the digestion of the carbohydrate, fat, and protein in Nikki's meal, and the primary end products of carbohydrate, fat, and protein digestion that are absorbed are shown below.

	Primary enzymes	Primary end products of digestion
Carbohydrate	1. amylase	1. glucose
	2. sucrase	
	3. lactase	
	4. maltase	
Fat	1. lipase	1. fatty acids
		2. glycerol
Protein	1. pepsin	1. amino acids
	2. trypsin	

Review Questions

1. False

2. True

3. True

4. False

5. True

6. False

7. True

8. True

9. True

10. False

11. True

12. False

13. a

14. b

15. a

16. d

17. False

18. True

19. True

20. True

21. False

22. True

Unit 8

Nutrition Up Close

Food as a Source of Calories

Potato Chips
Serving size: 1 oz (about 20 chips)
Fat: 10g __9__ = __90__ calories from fat
Carbohydrate: 15g __4__ = __60__ calories from carbohydrate
Protein: 2g __4__ = __8__ calories from protein
__158__ Total calories
% of calories from fat: __90__ = __0.57__ × 100 = __57__%

Mini Pretzels
Serving size: 1 oz (about 17 pieces)
Fat: 0g __9__ = __0__ calories from fat
Carbohydrate: 24g __4__ = __96__ calories from carbohydrate
Protein: 3g __4__ = __12__ calories from protein
__108__ Total calories
% of calories from carbohydrate: __96__ = __0.89__ × 100= __89__%

The pretzels are lower in total calories and fat calories than the potato chips.

Review Questions

1. True

2. False

3. True

4. False

5. False

6. True

7. True

8. b

9. c

10. c

11. d

12. b

13. c

14. True

15. False

16. True

17. False

18. False

19. a

20. c

Unit 9

Nutrition Up Close

Are You an Apple?
If your waist circumference is over 35" (88 cm) and you're a female, and over 40" (120 cm) if male, you're an apple. (You are all a peach for doing this exercise.)

Review Questions

1. False

2. False

3. True

4. False

5. False

6. False

7. c

8. b

9. True

10. True

11. False

12. True

13. False

14. True

15. True

16. d

17. c

18. False

19. False

20. True

21. False

22. False

23. True

Unit 10

Nutrition Up Close

Setting Small Behavior Change Goals

The specific, small steps in behavioral change that work will be different for different individuals. They can be changed over time, and may end up being a rewarding part of an enhanced lifestype.

If you need more practice writing small behavioral change goals, take a look at a few more examples:

I'll use a low-calorie oil and vinegar dressing on my sandwich rather than mayonnaise at lunch three times a week.

I'll brighten up my plate and my diet with a serving of one of my favorite brightly colored vegetables three times a week.

I'll walk up the stairs to get to my office rather than take the elevator daily.

I'll lift a 10-pound weight for 15 minutes five times a week while I watch the nightly news.

Review Questions

1. False

2. False

3. False

4. True

5. False

6. False

7. c

8. b

9. True

10. True

11. True

12. True

13. b

14. a

15. d

Unit 11

Nutrition Up Close

Eating Attitudes Test
Never = 3
Rarely = 2
Sometimes = 1
Always, usually, and often = 0

    A total score under 20 points may indicate abnormal eating behavior. If you think you have an eating disorder, it is best to find out for sure. Careful evaluation by a qualified health professional is necessary to exclude any possible underlying medical reasons for your symptoms. Contacting a physician, nurse practitioner, dietitian, or the student health center is an important first step. You may wish to show your Eating Attitudes Test to the health professional.

Review Questions

1. True
2. True
3. True
4. False
5. False
6. True
7. True
8. False
9. False
10. True

Unit 12

Nutrition Up Close

Does Your Fiber Intake Measure Up?
The total number of points you scored approximates the **grams** of total fiber you typically consume daily.[a] Use this scale to find out if your fiber intake meets the recommended goal:

    0–10 grams: You consume less than the average American. Increase your fiber intake by including more fruits, vegetables, whole grains, and legumes in your diet overall.

    11–15 grams: Like other Americans, you consume too little fiber. Increase the number of servings of high-fiber foods you already enjoy, while substituting more high-fiber foods for refined food products. A quick way to add fiber to your diet is to consume more of the two fiber powerhouses: legumes and bran cereal.

[a]Because different foods within a food group contribute varying amounts of dietary fiber, the point values have been average. Make sure to check the Nutrition Facts panel on bran cereals because these cereals vary in the amount of dietary fiber they contain.

15–20 grams: You currently consume more fiber than the average American. Make sure you're including 5 or more servings of fruits and vegetables. Eat 6 to 11 servings of bread, cereal, rice, and pasta daily; choose whole-grain versions of these foods often.

20–40 grams: Congratulations! Your dietary fiber intake is in the vicinity of that recommended. Keep up the good work.

Review Questions

1. True
2. False
3. True
4. True
5. False
6. True
7. False
8. True
9. False
10. True
11. a
12. True
13. False
14. c
15. False
16. True
17. d
18. b

Unit 13

Nutrition Up Close

Calculating Glycemic Load

Food	Glycemic load 100
cola beverage	69.6
potato, baked, no skin	28.9
apple juice, bottled	11.6
milk chocolate, plain	7.3
hummus	1.5

Effects of carbohydrate-containing foods on blood glucose levels vary depending on the glycemic index of these foods and the amount of them we consume.

1. True
2. False
3. True
4. True
5. True
6. True
7. False
8. False
9. False
10. c
11. d
12. b
13. d
14. False
15. False

Unit 14

Nutrition Up Close

Effects of Alcohol Intake

A. Ligia's estimated % blood alcohol = 0.03%. Mark's estimated % blood alcohol = 0.02%.
B. Three side effects of these blood alcohol levels:
1. Loss of some control over muscle movements
2. Slowed reaction time
3. Impaired thought processes

Review Questions

1. True
2. True
3. False
4. False
5. True
6. False
7. True
8. False
9. False
10. False
11. d

12. a

13. b

14. a

Unit 15

Nutrition Up Close

My Protein Intake

Compare your subtotals to find out which protein source you prefer, plant or animal. Protein from animal products is often accompanied by fat. If you are concerned about calories and fat in your diet, choose plant protein sources more often. And, if you are similar to many Americans, your intake of protein will exceed the RDA by quite a bit.

Review Questions

1. False

2. False

3. True

4. True

5. True

6. False

7. True

8. False

9. True

10. False

11. True

12. True

13. False

14. b

15. c

16. c

Unit 16

Nutrition Up Close

Vegetarian Main Dish Options

Vegetarians definitely had the advantage in completing this exercise! For the meat eaters, were the main dishes you identified meat-free? Here are eight vegetarian options for lunch and dinner main dishes:

- Eggplant parmesan

- Falafel

- Red beans and rice
- Pasta with broccoli
- Melted cheese, tomato, and sprout sandwiches
- Vegetable lasagna
- Veggie burgers
- Vegetable pizza

Review Questions

1. True
2. False
3. True
4. True
5. False
6. True
7. False
8. True
9. d
10. a
11. a
12. d
13. c

Unit 17

Nutrition Up Close

Gluten-Free Cuisine
Checkmarks should be placed in front of these gluten-free foods:

✓ fresh meats, fish ✓ tossed salad with oil and vinegar dressing
✓ tomatoes ___ apple pie
___ pizza ✓ black beans and rice
___ macaroni and cheese ___ canned soup
✓ eggs ___ sausage
✓ corn grits ✓ popcorn
✓ corn chips ✓ cheddar cheese
___ sourdough rolls ___ veggie burgers
✓ baked apples with sugar ✓ lemonade
✓ hamburgers ✓ broccoli
✓ grapes

Additional information about gluten-free foods can be found on food package labels.

Review Questions

1. True
2. True
3. False
4. False
5. True
6. False
7. True
8. c
9. d
10. False
11. True
12. False
13. d
14. a

Unit 18

Nutrition Up Close

The Healthy Fats in Your Diet

Give yourself a point for each time you checked the "3–5 Times per Week" or "Almost Daily" columns for numbers 3, 4, 6, 9, and 10. These foods are sources of healthy unsaturated fats or DHA and EPA. Take a point away for each time your answer ended up in the same columns for foods listed in numbers 1, 2, 5, 7, and 8. These foods provide saturated or *trans* fats. If you have any points left, your selection of food sources of fat regularly include healthy fats.

Review Questions

1. True
2. True
3. True
4. False
5. True
6. True
7. True
8. True
9. True
10. False
11. c

12. d

13. b

14. d

15. a

Unit 19

Nutrition Up Close

Evaluate Your Dietary and Lifestyle Strengths for Heart Disease Prevention
How did you do? Did you find strengths in your diet and lifestyle? You are on your way.

There is no set number of right or wrong answers to this activity. The more heart-healthy choices you checked, the better for your health overall. Give yourself credit for that while becoming aware of the multiple influences of diet and lifestyle of health.

Review Questions

1. True

2. True

3. False

4. True

5. False

6. True

7. False

8. True

9. True

10. b

11. d

12. c

13. b

14. a

15. d

16. b

17. c

18. c

19. c

20. b

Unit 20

Nutrition Up Close

Antioxidant Vitamins: How Adequate Is Your Diet?
Several responses in the last two columns likely indicate adequate antioxidant vitamin consumption. If you need to boost your intake, increase the overall amount of fruits, vegetables, nuts, oils, and whole grains in your diet.

Review Questions

1. True
2. False
3. True
4. True
5. False
6. True
7. False
8. True
9. True
10. False
11. False
12. True
13. False
14. True
15. False
16. b
17. d
18. d
19. b
20. c
21. a

Unit 21

Nutrition Up Close

Have You Had Your Phytochemicals Today?
Did you eat five or more of the foods listed at least twice last week? If yes, listen carefully and you'll hear your cells say "thank you."

   If you didn't eat five or more twice last week, go for the foods you didn't eat but like. Be adventurous! Try some of the phytochemical-rich foods you have never eaten before.

Review Questions

1. False
2. True
3. True
4. False
5. False
6. True
7. False
8. False
9. False
10. False
11. False
12. a
13. b
14. d
15. d
15. c
17. b

Unit 22

Nutrition Up Close

A Cancer Risk Checkup

If you answered yes to 6 out of the 8 questions you are on the right track.

If you didn't know the answer to statement 8, can you pinch an inch of fat above your ribs while you are standing? If yes, you probably have too much body fat.

Review Questions

1. True
2. False
3. True
4. True
5. False
6. True
7. False
8. False
9. True
10. False
11. d
12. b

Unit 23

Nutrition Up Close

The Salt of Your Diet

Were you aware of how much sodium there may be in processed food? Now you know. Fresh foods naturally contain low amounts of sodium. For example, ham contains over 460 mg sodium in 2 ounces, but 2 ounces of roasted pork contains 23 mg sodium. A regular-sized hamburger on a bun you would make at home contains about 269 mg sodium, far less than the 490–560 mg sodium in fast food hamburgers. One slice of tomato has about 1-mg sodium versus the 150-190 mg sodium in a tablespoon of catsup. Most of the sodium in foods is added during processing. Eat fresh.

Review Questions

1. False
2. True
3. True
4. True
5. False
6. True
7. False
8. True
9. True
10. False
11. True
12. False
13. c
14. d
15. d
16. d
17. a

Unit 24

Nutrition Up Close

Supplement Use and Misuse

1. *Is a supplement warranted in this case?* No

   Why or why not? Martha may be getting the vitamins and minerals she needs by taking a supplement, but this cannot make up for her poor food habits. She needs to improve the overall quality of her diet. If Martha follows the ChooseMyPlate recommendations (MyPlate.gov), she should be able to get all the nutrients and beneficial phytochemicals in plant foods she needs from what she eats.

2. *Is a supplement warranted in this case?* Yes

    Why or why not? This is one of the times when a supplement is in order. Sylvia needs to follow her doctor's advice and take the prescribed iron preparation to increase her hemoglobin and replete her iron stores. However, she is correct in consuming more iron-rich foods, too, so that after the anemia has been treated, she will not experience a relapse.

3. *Is a supplement warranted in this case?* No

    Why or why not? There is no scientific evidence that high doses of vitamins and minerals enhance physical performance. In fact, John may be setting himself up for toxicity reactions with prolonged intake of supplements with high dosages.

4. *Should Roberto try the senna, ask for a nonherbal drug, or take another action?*

    If unsure, Roberto should get more information from the pharmacist or look up senna on the National Library of Medicine's Medlineplus.gov. site to get the information he needs to make a well-informed decision about senna use.

5. *Should Yuen take Hypermetabolite for weight loss?* No

    Why or why not? She shouldn't take the product because the ephedra derivative may have serious side effects and because quick weight-loss strategies don't work in the longer run. Individuals who have jobs that require drug testing have been fired after ephedra derivatives appeared in their urine. Plus, there's no guarantee that the product will work or that it isn't mislabeled.

Review Questions

1. True
2. False
3. True
4. False
5. True
6. True
7. False
8. False
9. True
10. False
11. a
12. a
13. b
14. d

Unit 25

Nutrition Up Close

Foods as a Source of Water

The percentage of water is listed after each food.

1. avocado (80%); potato, boiled (77%); corn, cooked (73%)
2. egg (75%); almonds (4%); ripe (black) olives (80%)

3. watermelon (91%); celery (95%); pineapple, fresh (86%)

4. 2% milk (89%); Coke (89%); cranberry juice, low calorie (95%)

5. cheddar cheese (37%); banana (74%); refried beans (72%)

6. hot dog (53%); pork sausage, cooked (45%); ham, extra lean (74%)

7. apple (84%); mushrooms, raw (92%); orange (87%)

8. onions, cooked (88%); lettuce (96%); okra, cooked (90%)

9. peanut butter (1%); butter (16%); mayonnaise (15%)

10. cake with frosting (22%); bagel (33%); Italian bread (36%)

Review Questions

1. False

2. True

3. True

4. False

5. True

6. False

7. a

8. a

9. True

10. False

11. True

12. False

13. d

Unit 26

Nutrition Up Close

Nature and Nurture

The more *yes* responses, the better the odds that some of your family's shared dietary traits may influence disease risk. The most important responses are those you gave yourself.

Review Questions

1. True

2. False

3. True

4. False

5. False

6. True

7. True

8. True

9. False

10. True

11. False

12. True

13. b

14. d

15. c

Unit 27

Nutrition Up Close

Exercise: Your Options

For those of you who answered "Active" to the first question and "Yes" to the second and third questions, congratulations! You qualify as a mover and shaker.

Review Questions

1. True

2. False

3. True

4. True

5. True

6. True

7. False

8. True

9. False

10. False

11. True

12. b

13. d

14. c

Unit 28

Nutrition Up Close

Testing Performance Aids

1. Major limitations of the design of the study: Any of these answers would be correct: no control group, no group given a placebo, use of self-reports. You may have thought of other limitations that are warranted for this study as well.

2. Answer: No. Side effects, such as blood pressure changes, diarrhea, abnormalities in kidney function, and weight loss, were not assessed. Safety of long-term use was not addressed.

3. Answer: No. A 5% increase in strength is small and may have been due to usual differences in the number of push-ups a person can do from time to time. Any increase in strength may have been due to increased training during the week the supplement was taken. There was no control group. All participants knew they were taking phosphorous, a supplement that some may believe increases strength. This belief may have changed performance on the push-up test.

Review Questions

1. True
2. True
3. False
4. False
5. False
6. False
7. False
8. True
9. False
10. True
11. False
12. False
13. c
14. b
15. c

Unit 29

Nutrition Up Close

You Be the Judge!

Gloria's choices:

- Grains 3 ounces (bran muffin, cereal)

- Vegetables 3 cups (vegetable soup, carrot sticks, turnip greens, tossed salad)

- Fruits $2^1/_2$ cups (strawberries, apple, orange juice)

- Milk and milk products 2 cups (milk, yogurt)

- Meat and beans 3 ounces (chicken)

- Oils 0 teaspoons

All of Gloria's choices are healthy ones, but she is missing some important foods and sufficient calories. While concentrating hard on eating enough fruits and vegetables, she has neglected to consume enough nutrient-dense foods from the remaining food groups.

She needs to include in her diet more dairy products, add more meat or protein alternates, and more grains, especially whole-grain products. With a few additions to her existing choices, the quality and quantity of her diet can be greatly improved. Here is a more balanced version of Gloria's menu for a day:

Breakfast: Bran muffin, cereal with strawberries, and a glass of milk.

Lunch: Vegetable soup, carrot sticks, an apple, **burrito,** and a glass of orange juice.

Afternoon snack: Yogurt, **cheese,** and **whole-grain crackers.**

Dinner: Chicken, grits, turnip greens, tossed salad, with **2 Tbsp ranch dressing, whole grain roll** and iced tea.

Review Questions

1. True
2. True
3. False
4. True
5. False
6. False
7. True
8. False
9. False
10. False
11. c
12. b
13. b
14. True
15. False
16. True
17. True
18. True
19. False
20. True
21. d
22. b
23. a

Unit 30

Nutrition Up Close

Obesity Prevention Close to Home

To help their children develop healthful food and activity habits, Debra and Dale can:

1. Provide lots of opportunities for fun physical activities

2. Provide a variety of healthful food choices

3. Limit meals at fast-food restaurants that promote energy-dense, low-nutrient foods to children

4. Work with other parents and school staff to increase physical activity opportunities at school

Other:

Work with other parents and school staff to increase availability of healthful food choices at school

Allow their children to eat when they are hungry and to stop eating when they are full

Never force their children to eat something or totally restrict access to their favorite foods

Offer foods in an objective, nonthreatening way

Review Questions

1. False

2. True

3. True

4. False

5. False

6. a

7. b

8. a

9. d

10. c

11. False

12. True

13. False

14. False

15. c

16. c

17. d

18. True

19. True

20. True

Unit 31

Nutrition Up Close

Does He Who Laughs, Last?
Alternate explanations:

1. The letter writer's friends may have switched to perceived health foods when they found out they were sick.

2. The letter writer may have consumed a lifelong healthy diet without choosing organic or "natural" foods.

3. The letter writer may come from a family whose members tend to have long lives.

   Other possible explanations:

   A sense of humor may help extend life.

   Many factors in addition to diet, exercise, and genetic background influence longevity.

   One individual's experience may not apply to the masses.

Review Questions

1. False
2. False
3. True
4. True
5. True
6. False
7. True
8. True
9. c
10. d
11. b

Unit 32

Nutrition Up Close

Food Safety Detective
Potential causes of the foodborne illness outbreak:

1. Disease spread from hands to foods.

2. Use of contaminated cutting board, knife, or food platter.

3. Cross-contamination between onions and lettuce or between other foods stored together.

Review Questions

1. False
2. True
3. False
4. True
5. False
6. True
7. True
8. False
9. True
10. False
11. c
12. d
13. b

Unit 33

Nutrition Up Close

Ethnic Foods Treasure Hunt

1. a. yams; b. kola nuts; c. breadfruit with corn and yams; d. cassava
2. a. chicken; b. rice; c. ripe mango; d. grapes; e. ginger f. milk; g. many more
3. "Gallo pinto" includes the following:

   a. rice and black beans; b. eggs; c. tortillas; d. coffee; e. fruit; f. sometimes steak
4. a. Italian ices; b. calzones; c. Italian sodas; d. pizza; e. pasta; f. hot dogs
5. a. beef; b. cassava; c. milk; d. yams; e. pumpkin; f. millet; g. corn; h. dried beans; i. beer; j. honey; k. yogurt; l. cow's blood
6. a. reindeer; b. polar bear; c. sour milk; d. caribou; e. tea
7. Yin: a. raw kelp (seaweed); b. boiled rice; c. raw vegetables; d. fruit; e. dairy products. Yang: a. cooked kelp; b. cooked vegetables; c. fried rice; d. fish; e. eggs; f. chicken

Review Questions

1. True
2. False
3. True
4. False
5. True

6. False

7. True

8. True

9. True

10. True

11. c

12. d

13. a

Glossary

Note to the reader: If you have a customized book, this glossary may include some entries that will not be found in your book.

absorption
The process by which nutrients and other substances are transferred from the digestive system into body fluids for transport throughout the body.

adequate diet
A diet consisting of foods that together supply sufficient protein, vitamins, and minerals and enough calories to meet a person's need for energy.

Adequate Intakes (AIs)
Provisional RDAs developed when there is insufficient evidence to support a specific level of intake.

adult
Generally defined as people 18 to 65 years old.

aerobic exercise
Physical activity in which the body's large muscles move in a rhythmic manner for a sustained period of time. Aerobic exercise involves metabolic pathways that require oxygen for energy production and improve functioning of the cardiovascular and respiratory systems. Aerobic activities include walking, jogging, running, swimming, skiing, vacuuming, house cleaning, lawn mowing, raking, cycling, and aerobic dance.

aerobic fitness
A state of respiratory and circulatory health as measured by the ability to deliver oxygen to muscles, and the capacity of muscles to use the oxygen for physical activity.

age-related macular degeneration
Eye damage caused by oxidation of the macula, the central portion of the eye that allows a person to see details clearly. It is the leading cause of blindness in U.S. adults over the age of 65. Antioxidants provided by the carotenoids in dark green, leafy vegetables such as kale, collard greens, spinach, and Swiss chard may help prevent macular degeneration.

alcohol sugars
Simple sugars containing an alcohol group in their molecular structure. The most common are xylitol, mannitol, and sorbitol.

alcoholism
An illness characterized by a dependence on alcohol and by a level of alcohol intake that interferes with health, family and social relations, and job performance.

allele
A different version of the same gene. Alleles have a different arrangement of bases than the usual version of the gene.

Alzheimer's disease
A brain disease that represents the most common form of dementia. It is characterized by memory loss for recent events that expands to more distant memories over the course of 5 to 10 years. It eventually produces profound intellectual decline characterized by dementia and personal helplessness.

amylophagia (am-e-low-phag-ah)
Laundry starch or cornstarch eating.

anaerobic exercise
Short-duration, intense activity requiring glucose for energy production. Energy formation from glucose does not require oxygen.

anaphylactic shock (an-ah-fa-lac-tic)
A serious allergic reaction that is rapid in onset and cause death. Symptoms of anaphylactic shock (or "anaphylaxis") may include abdominal cramps, chest tightness, difficulty breathing, cough, hives, flushing, swelling, and/or itching.

anorexia nervosa
An eating disorder characterized by extreme weight loss, poor body image, and irrational fears of weight gain and obesity.

antibodies
Blood proteins that help the body fight particular diseases. They help the body develop an immunity, or resistance, to many diseases.

antioxidants
Chemical substances that prevent or repair damage to cells caused by oxidizing agents such as pollutants, ozone, smoke, and reactive oxygen. Oxidation reactions are a normal part of cellular processes. Vitamins C and E and certain phytochemicals function as antioxidants.

appetite
The desire to eat; a pleasant sensation that is aroused by thoughts of the taste and enjoyment of food.

association
The finding that one condition is correlated with, or related to another condition, such as a disease or disorder. For example, diets low in vegetables are associated with breast cancer. Associations do not prove that one condition (such as a diet low in vegetables) causes an event (such as breast cancer). They indicate that a statistically significant relationship between a condition and an event exists.

atherosclerosis
"Hardening of the arteries" due to a buildup of plaque.

ATP, ADP
Adenosine triphosphate (ah-den-o-scene tri-phos-fate) and adenosine diphosphate. Molecules containing a form of phosphorous that can trap energy obtained from the macronutrients. ADP becomes ATP when it traps energy, and returns to being ADP when it releases energy for muscular and other work.

autoimmune disease
A disease related to the destruction of the body's own cells by substances produced by the immune system that mistakenly recognize certain cell components as harmful.

balanced diet
A diet that provides neither too much nor too little of nutrients and other components of food such as fat and fiber.

bariatrics
The field of medicine concerned with weight loss.

basal metabolism
Energy used to support body processes such as growth, health, tissue repair and maintenance, and other functions. Assessed while at rest, basal metabolism includes energy the body expends for breathing, the pumping of the heart, the maintenance of body temperature, and other life-sustaining, ongoing functions. Also called *resting metabolism*.

bile
A yellowish-brown or green fluid produced by the liver, stored in the gallbladder, and secreted into the small intestine. It acts like a detergent, breaking down globs of fat entering the small intestine to droplets, making the fats more accessible to the action of lipase.

binge eating
The consumption of a large amount of food in a small amount of time.

binge-eating disorder
An eating disorder characterized by periodic binge eating, which normally is not followed by vomiting or the use of laxatives. People must experience eating binges twice a week on

average over a period of 6 months to qualify for the diagnosis.

bioavailability
The amount of a nutrient consumed that is available for absorption and use by the body.

biotechnology
As applied to food products, the process of modifying the composition of foods by biologically altering their genetic makeup. Also called *genetic engineering* of foods. The food products produced are sometimes referred to as *GM* and *GMOs* (genetically modified organisms).

body mass index (BMI)
An indicator of the appropriateness of a person's weight for their height. It is calculated by dividing weight in kilograms by height in meters. It can also be calculated using the method shown in Unit 9.

bulimia nervosa
An eating disorder characterized by recurrent episodes of rapid, uncontrolled eating of large amounts of food in a short period of time. Episodes of binge eating are followed by compensatory behaviors such as such as self-induced vomiting, dieting, excessive exercise, or misuse of laxatives, to prevent weight gain.

calorie (calor = heat)
A unit of measure used to express the amount of energy produced by foods in the form of heat. The calorie used in nutrition is the large Calorie, or the kilocalorie (kcal). It equals the amount of energy needed to raise the temperature of 1 kilogram of water (about 4 cups) from 15° to 16°C (59° to 61°F). The term *kilocalorie*, or *calorie* as used in this text, is gradually being replaced by the *kilojoule* (kJ) in the United States; 1 kcal = 4.2 kJ.

cancer
A group of diseases in which abnormal cells grow out of control and can spread throughout the body. Cancer is not contagious and has many causes.

carbohydrates
Chemical substances in foods that consist of a simple sugar molecule or multiples of them in various forms.

cardiovascular disease
Disorders related to plaque buildup in arteries of the heart, brain, and other organs and tissues.

cardiorespiratory fitness
The ability of the circulatory and respiratory systems to supply fuel during sustained physical activity and to eliminate fatigue products after supplying fuel. Cardiorespiratory fitness is also called *aerobic fitness* and *cardiorespiratory endurance*.

cataracts
Complete or partial clouding over the lens of the eye.

cause and effect
A finding that demonstrates that a condition causes a particular event. For example, vitamin C deficiency causes the deficiency disease scurvy.

celiac disease
An autoimmune disease characterized by inflammation of the small intestine lining resulting from a genetically based intolerance to gluten. The inflammation produces diarrhea, fatty stools, weight loss, and vitamin and mineral deficiencies. (Also called celiac *sprue* and *gluten-sensitive enteropathy*.)

cholesterol
A fat-soluble, colorless liquid found primarily in animal products. Cholesterol is used by the body to form hormones such as testosterone and estrogen and is a component of cell membranes. Cholesterol is present in plant cell membranes, but the quantity is small and plants are not considered to be a significant dietary source of cholesterol.

chromosomes
Structures in the nuclei of cells that contain genes. Humans have 23 pairs of chromosomes; half of each pair comes from the mother and half from the father.

chronic diseases
Slow-developing, long-lasting diseases that are not contagious (e.g., heart disease, cancer, diabetes). They can be treated but not always cured.

chronic inflammation
Low-grade inflammation that lasts weeks, months, or years. Inflammation is the first response of the body's immune system to infectious agents, toxins, or irritants. It triggers the release of biologically active substances that promote oxidation and other reactions to counteract the infection, toxin, or irritant. A side effect of chronic inflammation is that it also damages lipids, cells, and tissues.

circulatory system
The heart, arteries, capillaries, and veins responsible for circulating blood throughout the body.

cirrhosis (sear-row-sis)
A disease of the liver characterized by widespread fibrous tissue buildup and disruption of normal liver structure and function. It can be caused by a number of chronic conditions that affect the liver such as alcoholism, diabetes, and obesity.

clinical trial
A study design in which one group of randomly assigned subjects (or subjects selected by the "luck of the draw") receives an active treatment and another group receives an inactive treatment, or sugar pill, called the *placebo*.

coenzymes
Chemical substances, including many vitamins, that activate specific enzymes. Activated enzymes increase the rate at which reactions take place in the body, such as the breakdown of fats or carbohydrates in the small intestine and the conversion of glucose and fatty acids into energy within cells.

cofactors
Individual minerals required for the activity of certain proteins. For example: iron is needed for hemoglobin's function in oxygen and carbon dioxide transport; zinc is needed to activate or is a structural component of over 200 enzymes; and magnesium activates more than 300 enzymes involved in the formation of energy and proteins.

colostrum
The milk produced during the first few days after delivery. It contains more antibodies, protein, and certain minerals than the mature milk that is produced later. It is thicker than mature milk and has a yellowish color.

complementary protein sources
Plant sources of protein that together provide sufficient quantities of the nine essential amino acids.

complete proteins
Proteins that contain all of the essential amino acids in amounts needed to support growth and tissue maintenance.

complex carbohydrates
The form of carbohydrate found in starchy vegetables, grains, and dried beans and in many types of dietary fiber. The most common form of starch is made of long chains of interconnected glucose units.

control group
Subjects in a study who do not receive the active treatment or who do not have the condition under investigation. Control periods, or times when subjects are not receiving the treatment, are sometimes used instead of a control group.

critical period
A specific interval of time during which cells of a tissue or organ are genetically programmed to multiply. If the supply of nutrients needed for cell multiplication is not available during the specific time interval, the growth and development of the tissue or organ are permanently impaired.

cross-contamination
The spread of bacteria, viruses, or other harmful substances from one surface to another.

cruciferous vegetables
Sulfur-containing vegetables whose outer leaves form a cross (or crucifix). Vegetables in this family include broccoli, cabbage, cauliflower, brussels sprouts, mustard and collard greens,

kale, bok choy, kohlrabi, rutabaga, turnips, broccoflower, and watercress.

Daily Values (DVs)
Scientifically agreed-upon daily dietary intake standards for fat, saturated fat, cholesterol, carbohydrate, dietary fiber, and protein intake compatible with health. DVs are intended for use on nutrition labels only and are listed in the Nutrition Facts panel. "% Daily Value" on Nutrition Facts panels is calculated as the percentage of each DV supplied by a serving of the labeled food.

development
Processes involved in enhancing functional capabilities. For example, the brain grows, but the ability to reason develops.

diabetes
A disease characterized by abnormal utilization of glucose by the body and elevated blood glucose levels. There are three main types of diabetes: type 1, type 2, and gestational diabetes. The word *diabetes* in this text refers to type 2 diabetes, by far the most common. Diabetes is short for the term *diabetes mellitus*.

diarrhea
The presence of three or more liquid stools in a 24-hour period.

dietary fiber
Naturally occurring, intact forms of nondigestible carbohydrates in plants and "woody" plant cell walls. Oat and wheat bran, and raffinose in dried beans, are examples of this type of fiber.

dietary supplements
Any products intended to supplement the diet, including vitamin and mineral supplements; proteins, enzymes, and amino acids; fish oils and fatty acids; hormones and hormone precursors; and herbs and other plant extracts. Such products must be labeled "dietary supplement."

dietary thermogenesis
Thermogenesis means "the production of heat." Dietary thermogenesis is the energy expended during food ingestion, the digestion of food, and the absorption and utilization of nutrients. Some of the energy escapes as heat. It accounts for approximately 10% of the body's total energy need. Also called *diet-induced thermogenesis* and *thermic effect* of foods or feeding.

digestion
The mechanical and chemical processes whereby ingested food is converted into substances that can be absorbed by the intestinal tract and utilized by the body.

disaccharides (di = two, saccharide = sugar)
Simple sugars consisting of two molecules of monosaccharides linked together. Sucrose, maltose, and lactose are disaccharides.

DNA (deoxyribonucleic acid) (d-oxy-rye-bow-new-clay-ic)
Segments of genes that provide instructions for the synthesis, or production, of specific enzymes and other proteins needed by cells to carry out life- and health-sustaining processes. Genetic material contained in cells that initiates and directs the production of proteins in the body.

double blind
A study in which neither the subjects participating in the research nor the scientists performing the research know which subjects are receiving the treatment and which are getting the placebo. Both subjects and investigators are "blind" to the treatment administered.

double-blind, placebo-controlled food challenge
A test used to determine the presence of a food allergy or other adverse reaction to a food. In this test, neither the patient nor the care provider knows whether a suspected offending food or a placebo is being tested.

duodenal (do-odd-en-all) and stomach ulcers
Open sores in the lining of the duodenum (the uppermost part of the small intestine) or the stomach.

edema
Swelling due to an accumulation of fluid in body tissues.

electrolytes
Minerals such as sodium and potassium that carry a charge when in solution. Many electrolytes help the body maintain an appropriate amount of fluid.

empty-calorie foods
Foods that provide an excess of calories in relation to nutrients. Soft drinks, candy, sugar, alcohol, and fats are considered empty-calorie foods.

endothelium (n-dough-theil-e-um)
The layer of cells lining the inside of blood vessels.

energy-dense foods
Food that provide relatively high levels of calories per unit weight of the food. Fried chicken, cheeseburgers, a biscuit, egg, and sausage sandwich, and potato chips are energy-dense foods.

energy density
The number of calories per gram of food. It is calculated by dividing the number of calories in a portion of food by the food's weight in grams. May also be calculated as calories per 100 grams of food.

enrichment
The replacement of thiamin, riboflavin, niacin, and iron lost when grains are refined.

environmental trigger
An environmental factor, such as inactivity, a high-fat diet, or a high sodium intake, that causes a genetic tendency toward a disorder to be expressed.

enzymes
Protein substances that speed up chemical reactions. Enzymes are found throughout the body but are present in particularly large amounts in the digestive system.

epidemiological studies
Research that seeks to identify conditions related to particular events within a population. This type of research does not identify cause-and-effect relationships. For example, much of the information known about diet and cancer is based on epidemiological studies that have found that diets low in vegetables and fruits are associated with the development of heart disease.

epigenetic
Alterations in gene activity that do not change the structure of DNA. Gene activity can be shut off or turned on, or slowed or sped up by epigenetic mechanisms.

epigenomics
Pertaining to changes in the regulation of the expression of gene activity without alteration of gene structure.

ergogenic aids (ergo = work; genic = producing)
Substances that increase the capacity for muscular work.

essential amino acids
Amino acids that cannot be synthesized in adequate amounts by humans and therefore must be obtained from the diet. They are sometimes referred to as "indispensable amino acids."

essential fatty acids
Components of fats (linoleic acid [pronounced lynn-oh-lay-ick and alpha-linolenic acid–lynn-oh-len-ick]) required in the diet.

essential hypertension
Hypertension of no known cause; also called *primary* or *idiopathic hypertension*, it accounts for 95% of all cases of hypertension.

essential nutrients
Nutrients required for normal growth and health that the body cannot generally produce, or produce in sufficient amounts; they must be obtained in the diet.

exercise
A subcategory of physical activity that is planned, structured, and repetitive. The terms *physical activity* and *exercise* are often used interchangeably.

experimental group
Subjects in a study who receive the treatment being tested or have the condition that is being investigated.

fat
Chemical substances that are insoluble in water and soluble in fat. Triglycerides, saturated and unsaturated fats, and essential fatty acids are examples of lipids, or "fats."

fatty liver disease
A reversible condition characterized by fat infiltration of the liver (10% or more by weight). If not corrected, fatty liver disease can produce liver damage and other disorders. The condition is primarily associated with obesity, diabetes, and excess alcohol consumption. The disease is called *steatohepatitis* when accompanied by inflammation.

fermentation
The process by which carbohydrates are converted to ethanol by the action of the enzymes in yeast.

fetus
A baby in the womb from the eighth week of pregnancy until birth. (Before then, it is referred to as an embryo.)

fiber
See dietary fiber, *functional fiber*.

flatulence (flat-u-lens)
Presence of excess gas in the stomach and intestines.

flexibility
Range of motion of joints.

food additives
Any substances added to food that become part of the food or affect the characteristics of the food. The term applies to substances added both intentionally and unintentionally to food.

food allergen
A substance in food (almost always a protein) that is identified as harmful by the body and elicits an allergic reaction from the immune system.

food allergy
Adverse reaction to a normally harmless substance in food that involves the body's immune system. (Also called *food hypersensitivity*.)

food insecurity
Limited or uncertain availability of safe, nutritious foods or the ability to acquire them in socially acceptable ways.

food intolerance
Adverse reaction to a normally harmless substance in food that does not involve the body's immune system.

food security
Access at all times to a sufficient supply of safe, nutritious foods.

foodborne illness
An illness related to consumption of foods or beverages containing disease-causing bacteria, viruses, parasites, toxic chemicals, or other harmful substances.

fortification
The addition of one or more vitamins and/or minerals to a food product.

free radicals
Chemical substances (often oxygen-based) that are missing electrons. The absence of electrons makes the chemical substance reactive and prone to oxidizing nearby molecules by stealing electrons from them. Free radicals can damage lipids, proteins, cells, DNA, and eventually tissues by altering their chemical structure and functions.

fruitarian
A form of vegetarian diet in which fruits are the major ingredient. Such diets provide inadequate amounts of a variety of nutrients.

functional fiber
Specific types of nondigestible carbohydrates that have beneficial effects on health. Two examples of functional fibers are psyllium and pectin.

functional foods
Generally taken to mean and food or food ingredient that may provide a health benefit beyond the effects of traditional nutrients they contain.

genes
The basic units of heredity that occupy specific places (loci) on chromosomes. Genes consist of large DNA molecules, each of which contains the code for the manufacture of a specific enzyme or other protein.

gene variants
An alteration in the normal sequence of a gene. The different forms of the same genes are considered "alleles."

genome
Combined term for *genes* and *chromosomes*. It represents all the genes and DNA contained in an organism. It is estimated that the human genome consists of 20,000 to 25,000 genes.

genomics
The study of the functions and interactions of all genes in the genome. Unlike genetics, it includes the study of genes related to common conditions and their interaction with environmental, or "epigenetic" factors.

genotype
The specific genetic makeup of an individual as coded by DNA.

geophagia (ge-oh-phag-ah)
Clay or dirt eating.

gestational diabetes
Diabetes first discovered during pregnancy.

glycemic index (GI)
A measure of the extent to which blood glucose levels are raised by consumption of an amount of food that contains 50 grams of carbohydrate compared to 50 grams of glucose. A portion of white bread containing 50 grams of carbohydrate is sometimes used for comparison instead of 50 grams of glucose.

glycemic load (GL)
A measure of the extent to which blood glucose level is raised by a given amount of a carbohydrate-containing food. GL is calculated by multiplying a food's GI by its carbohydrate content.

glycogen
The body's storage form of glucose. Glycogen is stored in the liver and muscles.

growth
A process characterized by increases in cell number and size.

health
The WHO defines health as state of complete physical, mental, and social well-being and not merely the absence of disease or infirmity.

heart disease
One of a number of disorders that results when circulation of blood to parts of the heart is inadequate. Also called *coronary heart disease*. (*Coronary* refers to the blood vessels at the top of the heart. They look somewhat like a crown.)

heartburn
A condition that results when acidic stomach contents are released into the esophagus, usually causing a burning sensation.

hemoglobin
The iron-containing protein in red blood cells.

hemorrhoids (hem-or-oids)
Swelling of veins in the anus or rectum.

high-quality protein
Proteins that contain all of the essential amino acids in amounts needed to support growth and tissue maintenance. Examples of high-quality proteins include eggs, milk, meat, and beans and rice. Also referred to as *complete proteins*.

histamine (hiss-tah-mean)
A substance released in allergic reactions. It causes dilation of blood vessels, itching, hives, and a drop in blood pressure and stimulates the release of stomach acids and other fluids. Antihistamines neutralize the effects of histamine and are used in the treatment of some cases of allergies.

homocysteine
A compound produced when the amino acid methionine is converted to another amino acid, cysteine. High blood levels of homocysteine increase the risk of hardening of the arteries, heart attack, and stroke.

hormone
A substance, usually a protein or steroid (a cholesterol-derived chemical), produced by one tissue and conveyed by the bloodstream to another. Hormones affect the body's metabolic processes such as glucose utilization and fat deposition.

hunger
Unpleasant physical and psychological sensations (weakness, stomach pains, irritability) that lead people to acquire and ingest food.

hydrogenation
The addition of hydrogen to unsaturated fatty acids.

hypertension
High blood pressure. It is defined as blood pressure exerted inside of blood vessel walls that typically exceeds 140/90 mm Hg (or millimeters of mercury). Hypertension in children and adolescents is defined as systolic or diastolic blood pressure equal to or greater than the 95th blood pressure percentile of sex-, age- and height-specific blood pressure percentiles.

hypoglycemia
A disorder resulting from abnormally low blood glucose levels. Symptoms of hypoglycemia include irritability, nervousness, weakness, sweating, and hunger. These symptoms are relieved by consuming glucose or foods that provide carbohydrate.

hypothesis
A statement made prior to initiating a study of the relationship sought to be tested by the research.

immunoproteins
Blood proteins such as antibodies that play a role in the functioning of the immune system (the body's disease defense system). Antibodies attack foreign proteins.

immune system
Body tissues that provide protection against bacteria, viruses, and other substances identified by cells as harmful.

incomplete proteins
Proteins that are deficient in one or more essential amino acids.

indirect calorimetry
A method of measuring energy expenditure from determination of the amount of oxygen utilized by the body during a specific unit of time.

infant mortality rate
Deaths that occur within the first year of life per 1,000 live births.

inflammation
Reactions of the body to the presence of infectious agents, toxins, or irritants. Inflammation triggers the release of biologically active substances that promote oxidation and other reactions that counteract the infectious agent, toxin, or irritant.

initiation
The start of the cancer process; it begins with the alteration of DNA within cells.

insulin resistance
A condition in which cell membranes have reduced sensitivity to insulin so that more insulin than normal is required to transport a given amount of glucose into cells. It is characterized by elevated levels of serum insulin, glucose, and triglycerides, and increased blood pressure.

iron deficiency
A disorder that results from a depletion of iron stores in the body. It is characterized by weakness, fatigue, short attention span, poor appetite, increased susceptibility to infection, and irritability.

iron-deficiency anemia
A condition that results when the content of hemoglobin in red blood cells is reduced due to a lack of iron. It is characterized by the signs of iron deficiency plus paleness, exhaustion, and a rapid heart rate.

irritable bowel syndrome (IBS)
A disorder of bowel function characterized by chronic or episodic gas, abdominal pain, diarrhea or constipation, or both.

kwashiorkor (kwa-she-or-kor)
A severe form of protein-energy malnutrition in young children. It is characterized by swelling, fatty liver, susceptibility to infection, profound apathy, and poor appetite.

lactose intolerance
The term for gastrointestinal symptoms (flatulence, bloating, abdominal pain, diarrhea, and "rumbling in the bowel") resulting from the consumption of more lactose than can be digested with available lactase.

lactose maldigestion
A disorder characterized by reduced digestion of lactose due to the low availability of the enzyme lactase.

life expectancy
The average length of life of people of a given age or from birth.

lipids
Compounds that are insoluble in water and soluble in fat. Triglycerides, saturated and unsaturated fats, and essential fatty acids are examples of lipids, or "fats."

low-birthweight infants
Infants weighing less than 2,500 grams (5.5 pounds) at birth.

lymphatic system
A network of vessels that absorb some of the products of digestion and transport them to the heart, where they are mixed with the substances contained in blood.

macronutrients
The group name for carbohydrate, protein, fat, and water. They are called *macronutrients* because we need relatively large amounts of them in our daily diet.

malnutrition
Poor nutrition resulting from an excess or lack of calories or nutrients.

marasmus
A severe form of malnutrition primarily due to a chronic lack of calories and protein. Also called *protein-energy malnutrition*.

maximal oxygen consumption
The highest amount of oxygen that can be delivered to, and utilized by, muscles for physical activity. Also called VO_2 max and *maximal volume of oxygen*.

meta-analysis
An analysis of data from multiple studies. Results are based on larger samples than the individual studies and are therefore more reliable. Differences in methods and subjects among the studies may bias the results of meta-analyses.

metabolic syndrome
A constellation of metabolic abnormalities that increase the risk of heart disease, hypertension, type 2 diabetes, and other disorders. Metabolic syndrome is characterized by insulin resistance, abdominal obesity, high blood pressure and triglycerides levels, low levels of HDL cholesterol, and impaired glucose tolerance. It is also called *syndrome X* and *insulin resistance syndrome*.

metabolism
The chemical changes that take place in the body. The conversion of glucose to energy or to body fat is an example of a metabolic process.

microbes
Microscopic organisms, including bacteria and fungi. Some microbes are beneficial, some pathogenic (harmful), and some have little effect on the body. Also called *microorganisms*.

minerals
In the context of nutrition, minerals are specific, single atoms that perform particular functions in the body. There are 15 essential minerals—or minerals required in the diet.

monosaccharides (mono = one, saccharide = sugar)
Simple sugars consisting of one sugar molecule. Glucose, fructose, and galactose are common monosaccharides.

muscular endurance
The ability of a muscle or group of muscles to sustain repeated contractions against a resistance for an extended period of time. Also called *muscular fitness*.

muscular strength
The ability of a muscle or muscle group to exert force. It is assessed by the maximal amount of resistance or force that can be sustained in a single effort.

myoglobin
The iron-containing protein in muscle cells.

neural tube defects
Malformations of the spinal cord and brain. They are among the most common and severe fetal malformations, occurring in approximately 1 in every 1,000 pregnancies. Neural tube defects include spina bifida (spinal cord fluid protrudes through a gap in the spinal cord; shown in Illustration 29.6), anencephaly (absence of the brain or spinal cord), and encephalocele (protrusion of the brain through the skull).

nitrogen balance
The difference between nitrogen intake and excretion. It is assessed as the difference between nitrogen intake and nitrogen excreted.

nonessential amino acids
Amino acids that can be readily produced by humans from components of the diet. Also referred to as "dispensable amino acids."

nonessential nutrients
Nutrients required for normal growth and health that the body can manufacture in sufficient quantities from other components of the diet. We do not require a dietary source of nonessential nutrients.

nutrigenomics
The study of diet- and nutrient-related functions and interactions of genes and their effects on health and disease.

nutrient-dense foods
Foods that contain relatively high amounts of nutrients compared to their calorie value. Broccoli, collards, bread, cantaloupe, and lean meats are examples of nutrient-dense foods.

nutrients
Chemical substances found in food that are used by the body for growth and health. The six categories of nutrients are carbohydrates, proteins, fats, vitamins, minerals, and water.

nutrition
The study of foods, their nutrients and other chemical constituents, and the effects of food constituents on health.

nutrition transition
A change in population's diet from traditional foods such as grains, roots, beans, and rice to calorie-dense foods high in sugar, fat, sodium and other components that characterize "Western" diets.

obese
In adults, a BMI (body mass index) of 30 kg/m² or higher. In children and adolescents, a BMI at or above the 95th percentile for children and adolescents of the same age and sex.

older adult
Generally defined as individuals 65 or more years old.

osteoporosis (osteo = bones; poro = porous, osis = abnormal condition)
A condition in which bones become fragile and susceptible to fracture due to a loss of calcium and other minerals.

overweight
In adults, a BMI (body mass index) of 25–29.9 kg/m². In children and adolescents, overweight is defined as a BMI at or above the 85th percentile and lower than the 95th percentile for children of the same age and sex.

oxidative stress
A condition that occurs when cells are exposed to more oxidizing molecules (such as free radicals) than to antioxidant molecules that neutralize them. Over time, oxidative stress causes damage to lipids, DNA, cells, and tissues. It increases the risk of heart disease, type 2 diabetes, cancer, and other diseases.

pagophagia (pa-go-phag-ah)
Ice eating.

peer review
Evaluation of the scientific merit of research or scientific reports by experts in the area under review. Studies published in scientific journals have gone through peer review prior to being accepted for publication.

% Daily Value (%DV)
Scientifically agreed-upon standards of daily intake of nutrients from the diet developed for use on nutrition labels. The "% Daily Values" listed in nutrition labels represent the percentages of the standards obtained from one serving of the food product.

phenylketonuria (feen-ol-key-tone-u-reah) (PKU)
A rare genetic disorder related to the lack of the enzyme phenylalanine hydroxylase. Lack of this enzyme causes the essential amino acid phenylalanine to build up in blood.

physical activity
Body movements produced by muscles that require energy expenditure.

physical fitness
The health of the body as measured by muscular strength and endurance, cardiorespiratory fitness, and flexibility. Body composition, agility, and balance are sometimes included as components of physical fitness.

phytochemicals (phyto = plant)
Biologically active, or "bioactive," substances in plants that have positive effects on health. Also called *phytonutrients*.

pica (pike-eh)
The regular consumption of nonfood substances such as clay or laundry starch.

placebo
A "sugar pill," an imitation treatment given to subjects in research.

placebo effect
Changes in health or perceived health that result from expectations that a "treatment" will produce an effect on health.

plant stanols or **sterols**
Substances in corn, wheat, oats, rye, olives, wood, and some other plants that are similar in structure to cholesterol but that are not absorbed by the body. They decrease cholesterol absorption.

plaque (dental)
A soft, sticky, white material on teeth; formed by bacteria.

plaque (arterial)
Deposits of cholesterol, other fats, calcium, and cell materials in the lining of the inner wall of arteries.

plumbism
Lead (primarily from old paint flakes) eating.

polysaccharides (poly = many, saccharide = sugar)
Carbohydrates containing many molecules of monosaccharides linked together. Starch, glycogen, and dietary fiber are the three major types of polysaccharides. Polysaccharides consisting of 3 to 10 monosaccharides can be referred to as *oligosaccharides*.

prebiotics
Nondigestible food ingredients that beneficially affect a person by selectively stimulating the growth or activity of one or a limited number of bacteria in the colon. Insulin, an extract from chicory root, is a common prebiotic. Also called *intestinal fertilizer*.

precursor
In nutrition, a nutrient that can be converted into another nutrient. (Also called *provitamin*.) Beta-carotene is a precursor of vitamin A.

prediabetes
A condition in which blood glucose levels are higher than normal but not high enough for the diagnosis of diabetes. It is characterized

by impaired glucose tolerance, or fasting blood glucose levels between 100 and 126 mg/dl.

prehypertension
In adults, typical blood pressure levels of 120/80 mm Hg through 139/89 mm Hg. In children and adolescents. prehypertension is defined as systolic or diastolic blood pressure equal to or greater than the 90th percentile but less than the 95th percentile of sex-, age- and height-specific blood pressure percentiles.

preterm
Infants born at or before 37 weeks of gestation (pregnancy).

probiotics
Live bacteria that are taken orally to restore beneficial bacteria to the gastrointestinal tract. Strains of *lactobacillus* (lac-toe-bah-sil-us) and *bifidobacteria* (bif-id-dough bacteria) are the best known probiotics. Also called *friendly bacteria*.

progression
The uncontrolled growth of abnormal cells.

promotion
The period in cancer development when the number of cells with altered DNA increases.

prostate
A gland located above the testicles in males. The prostate secretes a fluid that surrounds sperm.

protein
Chemical substance in foods made up of chains of amino acids.

purging
The use of self-induced vomiting, laxatives, or diuretics (water pills) to rid the body of food.

Recommended Dietary Allowances (RDAs)
Intake levels of essential nutrients that meet the nutritional needs of practically all healthy people while decreasing the risk of certain chronic diseases.

refined grains
Whole grains that have been processed to remove the bran covering the grain kernel and germ that stores nutrients for seedling growth. The endosperm, the portion of the grain kernel that stores complex carbohydrate and other nutrients, remains.

remodeling
The breakdown and buildup of bone tissue.

resistance exercise
Physical activities that involve the use of muscles against a weight or force. It increases muscle strength, mass, power, and endurance. Also called *resistance training, strength training,* and *weight training*.

restrained eating
The purposeful restriction of food intake below desired amounts in order to control body weight.

salt sensitivity
A genetically determined condition in which a person's blood pressure rises when high amounts of salt or sodium are consumed. Such individuals are sometimes identified by blood pressure increases of 5 to 10% or more when switched from a low-salt to a high-salt diet.

satiety
A feeling of fullness or of having had enough to eat.

saturated fats
The type of fat that tends to raise blood cholesterol levels and the risk for heart disease. They are solid at room temperature and are found primarily in animal products such as meat, butter, and cheese.

serotonin (sare-uh-tone-in)
A neurotransmitter, or chemical messenger, for nerve cell activities that excite or inhibit various behaviors and body functions. It plays a role in mood, appetite regulation, food intake, respiration, pain transmission, blood vessel constriction, and other body processes.

simple sugars
Carbohydrates that consist of a glucose, fructose, or galactose molecule; or a combination of glucose and either fructose or galactose. High-fructose corn syrup and alcohol sugars are also considered simple sugars. Simple sugars are often referred to as *sugars*.

single-gene defects
Disorders resulting from one abnormal gene. Also called *inborn errors of metabolism*. Over 6,000 single-gene defects have been cataloged, and most are very rare.

soluble fiber
Types of dietary fiber that dissolve in water, forming a gel. Good sources of soluble fiber include oats, oatmeal, oat bran, apples, pears, dried beans, carrots, and psyllium.

starch
Complex carbohydrates made up of complex chains of glucose molecules. Starch is the primary storage form of carbohydrate in plants. The vast majority of carbohydrate in our diet consists of starch, monosaccharides, and disaccharides.

statistically significant
Research findings that likely represent a true or actual result and not one due to chance.

steatohepatitis (ste-at-oh-hepah-tie-tis)
A disease characterized by inflammation of, and fat accumulation in the liver. It is associated with alcoholism and may occur in obesity and diabetes. Steatohepatitis may progress to cirrhosis.

stroke
The event that occurs when a blood vessel in the brain suddenly ruptures or becomes blocked, cutting off blood supply to a portion of the brain. Stroke is often associated with "hardening of the arteries" in the brain. (Also called a *cerebral vascular accident*.)

subcutaneous fat (sub-q-tain-e-ous)
Fat located under the skin.

tooth decay
The disintegration of teeth due to acids produced by bacteria in the mouth that feed on sugar. Also called *dental caries* or *cavities*.

total fiber
The sum of functional and dietary fiber.

trans fats
A type of unsaturated fat present in hydrogenated oils, margarine, shortening, pastries, and some cooking oils that increase the risk of heart disease. Fats containing fatty acids in the trans form are generally referred to as trans fats.

trimester
One-third of the normal duration of pregnancy. The first trimester is 0 to 13 weeks, the second is 13 to 26 weeks, and the third is 26 to 40 weeks.

tryptophan (trip-tuh-fan)
An essential amino acid that is used to form the chemical messenger serotonin (among other functions). Tryptophan is generally present in lower amounts in food protein than most other essential amino acids. It can be produced in the body from niacin, a B vitamin.

type 1 diabetes
A disease characterized by high blood glucose levels resulting from destruction of the insulin-producing cells of the pancreas. This type of diabetes was called *juvenile-onset diabetes* and *insulin-dependent diabetes* in the past, and its official medical name is *type 1 diabetes mellitus*.

type 2 diabetes
A disease characterized by high blood glucose levels due to the body's inability to use insulin normally, or to produce enough insulin. This type of diabetes was called *adult-onset diabetes* and *non-insulin-dependent diabetes* in the past, and its official medical name is *type 2 diabetes mellitus*.

umami (u-mam-e)
A Japanese word meaning "pleasant savory taste." The taste is described as brothy or meaty and is recognized as the fifth basic taste. Foods containing glutamate, an amino acid derivative, elicit the umami taste and include fish, meats, mushrooms, ripe tomatoes, and aged cheese.

underweight
Usually defined as a low weight-for-height. May also represent a deficit of body fat.

unsaturated fats
The type of fat that tends to lower blood cholesterol level and the risk of heart disease. They are liquid at room temperature and found in foods such as nuts, seeds, fish, shellfish, and vegetable oils.

urea
Nitrogen released from the breakdown of proteins for energy is largely excreted in the urine in the form of urea. It can be measured in urine, or in blood as "blood urea nitrogen" (BUN).

visceral fat (vis-sir-el)
Fat located under the skin and muscle of the abdomen.

vitamins
Components of food required by the body in small amounts for growth and health maintenance.

water balance
The ratio of the amount of water outside cells to the amount inside cells; this balance is needed for normal cell functioning.

whole grains
Cereal grains that consist of the intact, ground, cracked, or flaked kernel, which includes the bran, the germ, and the inner most part of the kernel (the endosperm).

zoochemicals
Chemical substances in animal foods, some of which may be biologically active in the body.

References

Unit 1

1. Anderson SA. Core indicators of nutritional state for difficult-to-sample populations. *J Nutr.* 1990;120:15–8.
2. Franklin B, et al. Exploring mediators of food insecurity and obesity: A review of recent literature, June 5, 2011. doi:10.1007/s10900-011-9420-4.
3. Coleman-Jensen A, et al. Household food security in the United States in 2010. *Economic Research Report No.* (ERR-125), September 2011, www.ers.usda.gov/Publications/ERR125.
4. Stang J et al. Position of the American Dietetic Association: Child and adolescent nutrition assistance programs. *J Am Diet Assoc.* 2010;110:791–99.
5. Household Food Insecurity In Canada in 2007–2008: Key Statistics and Graphics. www.hc-sc.gc.ca/fn-an/surveill/nutrition/commun/insecurit/key-stats-cles-2007-2008-eng.php. Accessed September 21, 2011.
6. Healthy People 2010. hp2010.nhlbihin.net/2010objs/19nutrition.html#_Toc471881947. Accessed September 21, 2011.
7. Knutsson R et al. Accidental and deliberate microbiological contamination in the feed and food chains—how biotraceability may improve the response to bioterrorism. *Int J Food Microbiol.* 2011;145 (suppl 1):S123-8.
8. Meinhardt PL. Water and bioterrorism: Preparing for the potential threat to U.S. water supplies and public health. *Ann Rev Public Health.* 2005;26:213–37.
9. Scalbert A et al. Databases on food phytochemicals and their health-promoting effects. *J Agric Food Chem.* 2011;1;59:4331–48.
10. Hayes JD et al. The cancer chemopreventive actions of phytochemicals derived from glucosinolates. *Eur J Nutr.* 2008:47 (suppl 2):73–88.
11. The Dietary Reference Intakes: The Essential Guide to Nutrient Requirements, National Academies Press: Washington, DC, 2006.
12. Herbert V. Chapter 26. Folic acid. In: Shils ME, et al., eds. *Modern Nutrition in Health and Disease,* 9th ed. Baltimore: Lippincott Williams Wilkins; 1999:433–58.
13. Fairbanks VF. Chapter 10. Iron in medicine and Nutrition. In: Shils ME, et al., eds. *Modern Nutrition in Health and Disease,* 9th ed. Baltimore: Lippincott Williams Wilkins; 1999:193–221.
14. Hayes DP. Adverse effects of nutritional inadequacy and excess: A hormetic model. *Am J Clin Nutr.* 2008;88(suppl):578S–81S.
15. King JC, et al. Zinc. In: Shils ME, et al., eds. *Modern Nutrition in Health and Disease,* 10th ed. Philadelphia: Lippincott, Williams & Wilkins; 2006;279–83.
16. Tanumihardjo SA. Vitamin A: biomarkers of nutrition for development. *Am J Clin Nutr.* 2011;94(suppl):658S-65S.
17. Goodman, AS. Vitamin A and retinols in health and disease. *N Engl J Med.* 1984;310:1023–31.
18. Mahoney DH Jr. Anemia in at-risk populations—what should be our focus? *Am J Clin Nutr.* 2008;88:1457–8.
19. Meksawan K, et al. Effect of low and high fat diets on nutrient intakes and selected cardiovascular risk factors in sedentary men and women. *J Am Coll Nutr.* 2004;23:131–40.
20. De Caterina R. Opportunities and challenges in nutrigenetics/nutrigenomics and health. *World Rev Nutr Diet.* 2010;101:1–7.
21. Reedy J, et al. Comparison of food-based recommendations and nutrient values of three food guides: USDA's MyPyramid, NHLBI's Dietary Approaches to stop Hypertension Eating Plan, and Harvard's Healthy Eating Pyramid. *J Am Diet Assoc.* 2008;108:522–8.
22. Discretionary Calories, www.health.gov/DIETARYGUIDELINES/dga2005/report/HTML/D3_DiscCalories.htm#tabled31. Accessed November 2008.
23. 2010 Dietary Guideline for Americans. Available at: www.dietaryguidelines.gov. Accessed February 2011.
24. Position of the American Dietetic Association: Total diet approach to communicating food and nutrition information. *J Am Diet Assoc.* 2007;107:1224–42.
25. Drewnowski A. Concept of a nutritious food: Toward a nutrient density score. *Am J Clin Nutr.* 2005;82:721–32.

Unit 2

1. Kochanek KD, et al. Deaths: Preliminary data for 2009. *National vital statistics reports;* vol. 59, no. 4. Hyattsville, MD: National Center for Health Statistics; 2011.
2. Gold, M. Haly's: Measuring lifestyle-related factors that contribute to premature death and disability. In: *Estimating the Contributions of Lifestyle-Related Factors to Preventable Death: A Workshop Summary.* Washington, DC: National Academies Press; 2005:1–54.
3. Krebs-Smith SM, et al. How does MyPyramid compare to other population-based recommendations for controlling chronic disease? *J Am Diet Assoc.* 2007;107:830–37.
4. Contributors to death among individuals under the age of 75 years in the United States. Available at: www.cdc.gov/nchs. Accessed June 2006.
5. CDC report finds people live longer if they practice one or more healthy lifestyle behaviors. August 18, 2011. Available at: www.cdc.gov/media/releases/2011/p0818_living_longer.html. Accessed September 22, 2011.
6. Kochanek KD, et al. Deaths: Preliminary data for 2009. *National vital statistics reports;* vol. 59, no. 4. Hyattsville, MD: National Center for Health Statistics; 2011.
7. Paez K. More Americans getting multiple chronic illnesses. Available at: www.medscape.com/view article/58636. Accessed January 2009.
8. World Cancer Research Fund/American Institute for Cancer Research. *Food, Nutrition, Physical Activity, and the Prevention of Cancer. A Global Perspective*, Washington, DC: AIRC; 2007.
9. Van Horn L et al. The evidence for dietary prevention and treatment of cardiovascular disease, *J Am Diet Assoc* 2008;108:287–331.
10. Utsugi MT, et al. Fruit and vegetable consumption and the risk of hypertension determined by self measurement of blood pressure at home: The Ohasama study. *Hyperten Res.* 2008;3:1435–43.
11. O'Keffe JH, et al. Dietary and lifestyle strategies for improving postprandial glucose, lipid profile, markers of inflammation, and cardiovascular health. *Am Coll Cardiol.* 2008;51:249–55.
12. Sofi F, et al. Effectiveness of the Mediterranean diet: Can it help delay or prevent Alzheimer's disease? *J Alzheimers Dis.* 2010;20:795–801.
13. Denova-Guterrez E, et al. Dietary patterns are associated with different indexes of adiposity and obesity in an urban Mexican population. *J Nutr.* 2011;141:921–27. Available at: medscape.com/viewarticle/530121. Accessed April 20, 2006.
14. Galland L. Diet and Inflammation. *Nutr Clin Pract.* 2010;25:634–40.
15. Valtuena S, et al. Food selection based on total antioxidant capacity can modify antioxidant intake, systemic inflammation, and liver function without altering markers of oxidative stress. *Am J Clin Nutr.* 2008;87:1290–97.
16. Shukla V, et al. Oxidative stress in neurodegeneration. *Adv Pharmacol Sci.* September 21, 2011. doi:10.1155/2011/572634.
17. Kalupahana NS, et al. (n-3) Fatty acids alleviate adipose tissue inflammation and insulin resistance: Mechanistic insights. *Adv Nutr.* 2011;2:304–16.
18. Nordmann AJ, et al. Meta-analysis comparing Mediterranean to low-fat diets for modification of cardiovascular risk factors. *Am J Med.* 2011;124:841–51.e2.

19. Hayes JD, et al. The cancer chemo-preventive actions of phytochemicals derived from glucosinolates. *Eur J Nutr*. 2008;47 (suppl 2):73–88.

20. Greene CM, et al. Maintenance of the LDL cholesterol: HDL cholesterol ratio in an elderly population given a dietary cholesterol challenge. *J Nutr*. 2005;135:2793–98.

21. Whalley LJ, et al. n-3 fatty acid erythrocyte membrane content, APOEe4, and cognitive variation: An observational follow-up study in late adulthood. *Am J Clin Nutr*. 2008;87:449–54.

22. Bachman JL, et al. Sources of food group intake among the US population. *J Am Diet Assoc*. 2008;108:804–14.

23. 2010 Dietary Guideline for Americans. Available at: www.dietaryguidelines.gov. Accessed February 2011.

24. Cordain L, et al. Origins and evolution of the Western diet: Health implications for the 21st century. *Am J Clin Nutr*. 2005:81:341–54.

25. Lindeberg S, et al. A Paleolithic diet improves glucose tolerance more than a Mediterranean-like diet in individuals with ischaemic heart disease. *Diabetologia*. 2007;50:1795–807.

26. U.S. Census Bureau, International Statistics. Available at: www.census.gov/compendia/statab/cats/international_statistics.html. Accessed September 2011.

27. Kato H, et al. Study of the epidemiology of health and dietary habits of Japanese and Japanese Americans. *Japan J Nutr*. 1989;47:121–30.

28. Kudo Y, et al. Evolution of meal patterns and food choices of Japanese-American females born in the United States. *Eur J Clin Nutr*. 2000;54:665–70.

29. Hu FB. Globalization of diabetes: The role of diet, lifestyle, and genes. *Diabetes Care*. 2011;34:1249–57.

30. Perez-Escamilla R. Acculturation, nutrition, and health disparities in Latinos. *Am J Clin Nutr*. 2011;93 (suppl):1163S–7S.

31. Ford ES, et al. Explaining the decrease in U.S. deaths from coronary disease, 1980–2000. *N Engl J Med*. 2007;356:2388–98.

32. Veronique LR, et al. Heart disease and stroke statistics—2011 update. *Circulation* 2011; 123: e18-e209, December 15, 2010, doi: 10.1161/CIR.0b013e3182009701.

33. *Healthy People 2020*. Available at: www.healthypeople.gov. Accessed September 23, 2011.

Unit 3

1. Who's minding the lab? Tufts University Health and Nutrition Letter May 1997:4.

2. Levine J, et al. Authors' financial relationships with the food and beverage industry and their published positions on the fat substitute Olestra. *Am J Pub Health*. 2003;93:664–69.

3. Fung-Berman A. Wyeth paid writers to promote hormone therapy. *PLoS Medicine*, September 7, 2010. dodi/10.1371/journal.pmed.1000335.

4. Braun JM, et al. Impact of early-life bisphenol: An exposure on behavior and executive function in children. *Pediatrics*, October 24, 2011. doi:10.1542/peds.2011-1335.

5. Bucher J, et al. Questions and answers about bisphenol A. September 8, 2011, www.niehs.nih.gov/news/sya/sya-bpa. Accessed October 26, 2011.

6. Steinbrook R. Searching for the right search—reaching the medical literature. *N Engl J Med*. 2006;354:4–7.

7. Al-Ubaydli M. Using search engines to find online medical information. *PLoS Med*. October 2, 2005:2(9):e228.

8. Busey JC, et al. Telehealth—opportunities and pitfalls. *J Am Diet Assoc*. 2008;108: 1296–1302.

9. *Pathways to excellence*. Chicago: The American Dietetic Association; 1997.

10. Alger SA, et al. Effect of a tricyclic antidepressant and opiate antagonist on binge-eating behavior in normal weight, bulimic, and obese, binge-eating subjects. *Am J Clin Nutr*. 1991;53:865–71.

11. Shapiro J. Hair loss in women. *N Engl J Med*. 2008;357:1620–30.

12. Wasko, CA et al. The sixty-second hair count for assessing hair loss in men. Arch Dermatol 2008;144:759–62.

Unit 4

1. Brecher SJ, et al. Status of nutrition labeling, health claims, and nutrient content claims for processed foods. *J Am Diet Assoc*. 2000;100:1057–62.

2. Post R, et al. A guide to federal food labeling requirements for meat and poultry products. Food Safety and Inspection Service, USDA, August, 2007. Available at: www.fsis.usda.gov/pdf/labeling_requirements_guide.pdf. *Accessed November 1, 2011*.

3. Pennington JA, Hubbard VS. Derivation of daily values used for nutrition labeling. J Am Diet Assoc. 1999;97:1407–12.

4. How to Understand Use the Nutrition Facts Label. Available at: www.cfsan.fda.gov/~dms/foodlab.html#nopercent. Accessed July 7, 2006.

5. Manufacturers and consumers lose faith in natural label claims. September 11, 2008. Available at: www. nutraingredients-usa.com. Accessed January 2009.

6. Health Claims Meeting Significant Scientific Agreement. Available at: http://www.fda.gov/food/labelingnutrition/labelclaims/healthclaimsmeetingsignificantscientificagreementssa/default.htm. Accessed October 29, 2011.

7. Marquart L, et al. Solid science and effective marketing for health claims. *Nutr Today*. 2001;36:107–14.

8. Food Allergen Labeling and Consumer Protection Act of 2004. Available at: www.cfsan.fda. gov/dms/alrgact.html. Accessed July 7, 2006.

9. Shea KM. Technical report: irradiation of food. *Pediatrics*. 2000;106:1505–9.

10. Glaberson H. E. coli crisis could prompt interest in irradiation for salads. Available at: www.foodproductiondaily.com/content/view/383143. Accessed June 28, 2011.

11. Ferrier P. Irradiation of produce imports: Small inroads, big obstacles. *Amber Waves*. June 2011. Available at: www.ers.usda.gov/AmberWaves/June11/Features/Irradiation. Accessed June 15, 2011.

12. Guidance for Industry, Substantiation for Dietary Supplement Claims Made Under Section 403(r) (6) of the Federal Food, Drug, and Cosmetic Act, December12/ 2008. Available at: www.cfsan.fda.gov/ guidance. html. Accessed January 1, 2009.

13. Fraud supplement makers fined $16 m, January 16, 2009. Available at: nutraingredients-usa.co/ content/view/233187. Accessed January 1, 2009.

14. Hitti M. New food labels show country of origin, January 10, 2008. Available at: www.medscape.com/ viewarticle/581365. Accessed January 1, 2009.

15. National organic program. Available at: http://www.ams.usda.gov/AMSv1.0/nop. Accessed November 2, 2011.

16. Dungour AD, et al. Nutrition-related health effects of organic foods: A systematic review. *Am J Clin Nutr*. 2010;92:203–10.

17. Tobin R, et al. The Irish organic food market: Shortfalls, opportunities, and the need for research. *J Sci Food Agriculture*. doi:10.1002/jsfa.4503, 2011.

18. Lupton JR, et al. The Smart Choices front-of-package nutrition labeling program: rational and development criteria. *Am J Clin Nutr*. 2010;91(suppl):1078S-89S.

19. Gerrior SA. Nutrient profiling systems: Are science and the consumer connected? *Am J Clin Nutr*. 2010;91(suppl):1116S-7S.

20. Hitti M. "Smart Choices" food labels are coming. Available at: www.medscape.com/viewarticle/582810. Accessed October 2008.

21. Nestle M. Health care reform in action—Calorie labeling goes national. *N Engl J Med*. 2010;362:2343–5.

22. Dumanovsky T, et al. Changes in energy content of lunchtime purchases from fast food restaurants after introduction of calorie labeling: Cross-sectional customer surveys. *BMJ*. July 26, 2011;343:d4464.

23. Ollberding NJ, et al. Food label use and its relation to dietary intake among U.S. adults. *J Am Diet Assoc*. 2010;111:S47–S51.

24. Fulgoni VL et al. Foods, fortificants, and supplements: Where do Americans get their nutrients? *J Nutr*. 2011;141;1847-54.

Unit 5

1. Keskitalo K, et al. Genetic and environmental contributions to food use patterns of young adult twins. *Physiol Behavior*. 2008;93:235–42.

2. Cooper SB, et al. Breakfast consumption and cognitive function in adolescent schoolchildren. *Physiol Behavior.* 2011;103:431–9.

3. Pyke M. *Man and food.* New York: McGraw-Hill; 1972.

4. Rozin P. The acquisition of food habits and preferences. In: Matarazzo JD, Weiss SM, Herd JA, Miller NE, Weiss SM, eds. *Behavioral health: A handbook of health enhancement and disease prevention.* New York: John Wiley, 1984:590–607.

5. Beauchamp GK, Mennella JA. Sensitive periods in the development of human flavor perception and preference. *Ann Nestle.* 1998;56:19–31.

6. The Determinants of Food Choice, European Food and Nutrition Council Review, April 2005. Available at: www.eufic.org/article /en/expid/review-food-choice. Accessed November 17, 2011.

7. Birch L. Development of food preferences. *Ann Rev Nutr.* 1999;19:41–62.

8. Story M, et al. Do young children instinctively know what to eat? The studies of Clara Davis revisited. *N Engl J Med.* 1987;316:103–6.

9. Birch LL, et al. Effects of a nonenergy fat substitute on children's energy and macronutrient intake. *Am J Clin Nutr.* 1993;58:326–33.

10. Steiner JE. The gustofacial response: Observation on normal and anencephalic newborn infants. In: Bosma JF, ed. *Fourth symposium on oral sensation and perception: Development in the fetus and infant.* Bethesda, MD: National Institute of Dental Research, DHEW Publication No. 73-546, US Dept HEW, NIH; 1973:254.

11. Packard V, *The status seekers.* New York: Random House; 1959, p. 146.

12. Van Oudenhove L, et al. Fatty acid-induced gut-brain signaling attenuates neural and behavioral effects of sad emotion in humans. *J Clin Invest.* 2011;121:3094–99.

13. Parraga IM. Determinants of food consumption. *J Am Diet Assoc.* 1990;90:661–3.

14. Stein K. Contemporary comfort foods: Bringing back old favorites. *J Am Diet Assoc.* 2008;108:412, 414.

15. Hare-Bruun H, et al. Adult food intake patterns are related to adult and childhood socioeconomic status. *J Nutr.* 2011;141:928–34.

16. Beerman KA. Variation in nutrient intake of college students: A comparison by students' residence. *J Am Diet Assoc.* 1991;91:343–4.

17. Cheadle A, et al. Community-level comparisons between the grocery store environment and individual dietary practices. *Prev Med.* 1991;20:250–6.

18. Hayes JE, et al. Allelic variation in TAS2R bitter receptor gene associates with variation in sensations from and ingestive behaviors toward common bitter beverages in adults. *Chem Senses.* 2011;36:311–31.

19. Krebs JR. The gourmet ape: Evolution and human food preferences. *Am J Clin Nutr.* 2009;90(suppl):705S–11S.

20. Diet Quality and Food Consumption: Dietary Trends from Food and Nutrient Availability Data. Available at: www.ers.usda.gov/ Briefing/DietQuality/Availability.htm, updated September 1, 2011. Accessed November 17, 2011.

21. ADA Nutrition and You: Trends, 2011, eatright.org. Accessed October 13. 2011.

22. ERS/USDA Charts of Note. Available at: www .ers.usda.gov/ChartsOfNote/Default .aspx?mode=detail&id=285. Accessed November 17, 2011.

23. Per-capita consumption. Available at: www .ers.usda.gov/data/foodconsumption/ spreadsheets/dyfluid.xls. Accessed November 17, 2011.

24. Nutrition and you: Trends 2008. American Dietetic Association. Available at; www .eatright.org. Accessed December 2008.

25. Chen X, et al. Americans with diet-related chronic diseases report higher diet quality than those without these diseases. *J Nutr.* 2011;141:1543–51.

26. Holdt CS, et al. Knowledge and behaviors of allied health professionals regarding meat (abs.). *J Am Diet Assoc.* 1991;91S, abs. A-13.

27. Handley B, et al. Using action plans to help primary care patients adopt healthy behaviors: A descriptive study. *J Am Board Fam Med.* 2006;19:224–31.

28. Hornick BA, et al. Menu modeling with MyPyramid food patterns: Incremental dietary changes lead to dramatic improvements in diet quality of menus. *J Am Diet Assoc.* 2008;108:2077–83.

29. Gould L. Consumers taking small steps toward big lifestyle changes. Results of Datmonitor's survey of diet trends in Europe and the U.S. Available at: nutraingredients.com/Europe, March 18, 2005.

30. Young SN. Nutrition 3. The fuzzy boundary between nutrition and psychopharmacology. *Can Med Assoc J.* Jan. 22, 2002;155.

31. Walker SP, et al. Early childhood stunting is associated with poor psychological functioning in late adolescence and effects are reduced by psychological stimulation. *J Nutr.* 2007;137:2464–9.

32. Stein AD, et al. Nutritional supplementation in early childhood, schooling, and intellectual functioning in adulthood: a prospective study in Guatemala. *Arch Pediatr Adolesc Med.* 2008;162:612–8.

33. D'Anci KE, et al. Low-carbohydrate weight-loss diets: Effect on cognition and mood. *Appetite.* 2009;52: 96–103.

34. McCann D, et al. Food additives and hyperactive behaviour in 3-year-old and 8/9-year-old children in the community: A randomised, double-blinded, placebo-controlled trial. *The Lancet.* 2007;370:1560–7.

35. Wright JP, et al. Association of prenatal and childhood blood lead concentrations with criminal arrests in early adulthood. *PLoS Med.* 2008;5(5):e101. Available at: www.medscape .com/ viewarticle/576717. Accessed January 2009.

36. Lozoff B. Iron deficiency and child development. *Food Nutr Bull.* 2007;28(4 suppl):S560–71.

37. Laus MF, et al. early postnatal protein-calorie malnutrition and cognition: A review of human and animal studies. *Int J Environ Res Public Health.* 2011;8:590–612.

38. Nordenberg T. The facts about aphrodisiacs. *FDA Consumer,* vol. 30, January–February 1996.

39. Giovannini M, et al. Symposium overview: Do we all eat breakfast and is it important? *Crit Rev Food Sci Nutr.* 2010;50:97–9.

40. Dietary Guidelines for Americans: Alcohol, 2010. Available at: *health.gov/dietary guidelines/dga2010/dietaryguidelines2010 .pdf. Accessed November 17, 2011.*

41. Kordas K. Iron, lead, and children's behavior and cognition. *Annual Review of Nutrition.* 2010;30:123–48.

42. Bellinger DC. Very low lead exposures and children's neurodevelopment. *Curr Opin Pediatr.* 2008;20:172–7.

43. Child Health USA, 1997. US Maternal and Child Health Bureau, PHS, Washington, DC, p. 29.

44. Lead. Available at: www.cdc.gov/nceh/lead. Accessed November 16, 2011.

45. Martí LF. Effectiveness of nutritional interventions on the functioning of children with ADHD and/or ASD. An updated review of research evidence. *Bol Asoc Med P R.* 2010;102:31–42.

46. Ballard W, et al. Do dietary interventions improve ADHD symptoms in children? *J Fam Practice.* 2010;59:234.

47. Brevard PB. Nutrition education improves the diet of college students. *J Am Diet Assoc.* 1991, abs. A-47.

48. Davis JN, et al. Is nutrition knowledge reflected in healthier dietary practices of college students? Experimental Biology 2003: Meeting Abstract 709.8, San Diego, April 2003.

49. Kolodinsky J, et al. Knowledge of current dietary guidelines and food choices by college students: Better eaters have higher knowledge of dietary guidance. *J Am Diet Assoc.* 2007;107:1409–13.

Unit 6

1. State Indicator Report on Fruits and Vegetables, 2009. Available at: http://www .fruitsandveggiesmatter.gov/health _professionals/data_behavioral.html. Accessed November 19, 2011.

2. Dietary Guidelines for Americans, 2010. www.dietaryguidelines.gov.

3. Dietary Intake Data: What We Eat in America, NHANES 2007–2008. Available at: www.ars .usda.gov/ba/bhnrc/fsrg. Accessed November 19, 2011.

4. Kimmons J, et al. Fruit and Vegetable Intake Among Adolescents and Adults in the United States: Percentage Meeting Individualized Recommendations. Posted: January 26, 2009; *Medscape J Med.* 2009;11(1):26.

5. Loss-Adjusted Food Availability. Available at: http://www.ers.usda.gov/Data/ FoodConsumption/FoodGuideIndex. htm#calories. Accessed November 19, 2011; http://www.medscape.com/ viewarticle/586492. Accessed November 19, 2011.

6. Drewnowski A. The naturally nutrient-rich food index. Presented at the Experimental Biology Annual Conference, San Diego, CA, April 2008.

7. Japanese Health and Nutrition Information. Available at: http://www.dietitian.or.jp/ english/news/dietary.html. Accessed November 25, 2011.

8. Dietary guidelines around the world. Available at: http://www.fao.org/ag/humannutrition/ nutritioneducation/fbdg/en. Accessed November 24, 2011.

9. Why are the *Dietary Guidelines* important? Available at: www.choosemyplate.gov/ guidelines/QandA.pdf. Accessed October 31, 2011.

10. Dietary Guidelines for Americans, Appendix A-1: The DASH Eating Plan at 1,600–3,200 calories. Available at: www.health.gov/ dietaryguidelines/dga2005/document/html/ appendixA.htm.

11. Your guide to lowering your blood pressure with *DASH*. Available at: www.nhlbi.nih.gov/ health/public/heart/hbp/dash/new_dash.pdf. Accessed November 23, 2011.

12. Fung TT, et al. The Mediterranean and Dietary Approaches to Stop Hypertension (DASH) diets and colorectal cancer. *Am J Clin Nutr* 2010;92:1429-35.

13. Chen ST, et al. The effect of dietary patterns on estimated coronary heart disease risk: Results from the Dietary Approaches to Stop Hypertension (DASH) trial. Circ Cardiovasc Qual Outcomes. 2010;3:484–9.

14. Krebs-Smith SM, et al. How does MyPyramid compare to other population-based recommendations for controlling chronic disease? *J Am Diet Assoc.* 2007;107:830–7.

15. Simopoulos AP. The Mediterranean diet: What is so special about the diet of Greece? The scientific evidence. *J Nutr.* 2001;131 (Suppl):S3065–73.

16. Sofi F, et al. Accruing evidence on benefits of adherence to the Mediterranean diet on health: an updated systematic review and meta-analysis. *Am J Clin Nutr.* 2010;92:1189–96.

17. Gardener H, et al. Mediterranean-style diet and risk of ischemic stroke, myocardial infarction, and vascular death: the Northern Manhattan Study. *Am J Clin Nutr.* 2011;94:1458–64.

18. Young YR. How to reduce portions: What to tell patients. Available at: www.medscape. com/viewarticle/750739?src=mp&spon=2. Accessed November 23, 2011.

19. Portion Distortion Slide Show, National Institutes of Health. Available at: http:// hp2010.nhlbihin.net/oei_ss/menu.htm#sl2. Accessed November 23, 2011.

20. Nutrient Data Laboratory, USDA. Available at: http://www.nal.usda.gov/fnic/foodcomp/ search. Accessed November 23, 2011.

21. Nestle M. Increasing portion sizes in American diets: More calories, more obesity. *J Am Diet Assoc.* 2003;103:39–40.

22. Duffey KJ, et al. Energy Density, Portion Size, and Eating Occasions: Contributions to Increased Energy Intake in the United States, 1977–2006, June 2011 Issue of *PLoS Medicine.* Available at: http://www .plosmedicine.org/article/info:doi/10.1371/ journal.pmed.1001050. Accessed November 26, 2011.

23. Fisher JO, et al. Portion size effects on daily energy intake in low-income Hispanic and African American children and their mothers. *Am J Clin Nutr.* December 2007;86(6):1709–16.

24. Larson N, et al. Young adults and eating away from home: associations with dietary intake patterns and weight status differ by choice of restaurant. *J Am Diet Assoc.* 2011;111:1696–703.

25. Kant AK, et al. Eating out in America, 1987–2000: Trends and nutritional correlates. *Prev Med.* 2004;38:243–49.

26. Todd, et al. *The impact of food away from home on adult diet quality.* ERR-90, U.S. Department of Agriculture, Econ. Res. Serv. February 2010.

27. Bridges A. Restaurants should shrink portions. Available at: cnn.netscape.com/news/story. Accessed June 3, 2006.

28. Hammons AJ, et al. Is frequency of shared family meals related to the nutritional health of children and adolescents? *Pediatrics.* 2011;127:e1565–e1574.

29. Young LR, Nestle M. Expanding portion sizes in the U.S. marketplace: Implications for nutrition counseling. *J Am Diet Assoc.* 2003;103:231–4.

30. Slow food movement finally picking up speed, August 24, 2008. Available at: www.msnbc .msn. com/id/26378691. Accessed February 2, 2009.

31. Slow Food USA, www.slowfoodusa.org, accessed 2/2/09.

32. Riley NS. The joys of cooking class. *Wall Street Journal.* March 9, 2006: D9.

33. Piernas C, et al. Increased portion sizes from energy-dense foods affect total energy intake at eating occasions in U.S. children and adolescents. *Am J Clin Nutr.* 2011;94:1324–32.

34. Rolls BJ, et al. The effect of large portion sizes on energy intake is sustained for 11 days. *Obesity.* 2007;15:1535–43.

35. Larson N, et al. Young adults and eating away from home: Associations with dietary intake patterns and weight status differ by choice of restaurant. *J Am Diet Assoc.* 2011;111:1696-1703.

36. USDA Center for Nutrition Policy and Promotion, August 2011.

37. Welsh JA, et al. Consumption of added sugars is decreasing in the United States. *Am J Clin Nutr.* 2011;94:726–34.

Unit 7

1. Hospitalization Statistics, 2009. Available at: www.ct.gov/dph/cwp/view .asp?a=3131&q=397512. Accessed December 3, 2011.

2. http://Health Canada, www.phac-aspc.gc.ca/ publicat/lcd-pcd97/table2-eng.php. Accessed December 4, 2011.

3. Peptic ulcers. Available at: www.mayoclinic .com/health/peptic-ulcer/DS00242. Updated January 2011. Accessed December 4, 2011.

4. Singh N, et al. Peptic Ulcer Disease: Topic Overview. Available at: www.webmd.com/ digestive-disorders/tc/peptic-ulcer-disease- topic-overview. Updated January 2011. Accessed December 4, 2011.

5. Pellettieri J, et al. Cell turnover and adult tissue homeostasis: From humans to planarians. *Ann Rev Genetics.* 2007;41:83-105.

6. Schneeman B. Nutrition and gastrointestinal function. *Nutr Today.* January/February 1993;20–24.

7. Liao TH, et al. Fat digestion by lingual lipase: mechanism of lipolysis in the stomach and upper small intestine. *Pediatric Research.* 1984;18:402-9.

8. Mandel AL, et al. Individual differences in AMY1 gene copy number, salivary α-amylase levels, and the perception of oral starch. *PLoS One.* October 13, 2010;5(10):e13352.

9. Salles C, et al. In-mouth mechanisms leading to flavor release and perception. Crit Rev *Food Sci Nutr.* January 2011;51(1):67-90.

10. Camilleri M, et al. Human gastric emptying and colonic filling of solids characterized by a new method. *Am J Physiol Gastrointest Liver Physiol.* 1989;284–90.

11. Bengmark S. Econutrition and health maintenance: A new concept to prevent GI inflammation, ulceration, and sepsis. *Clin Nutr.* 1996;15:1–10.

12. Levitt MD, et al. Use of measurements of ethanol absorption from stomach and intestine to assess human ethanol metabolism. *Am J Physiol Gastrointestinal Liver Physiol.* 1997;273(4):G951-57.

13. Stewart JE, et al. Marked differences in gustatory and gastrointestinal sensitivity to oleic acid between lean and obese men. *Am J Clin Nutr.* 2011;93:703-11.
14. Finger TE, et al. Matters of taste. *The Scientist.* November 12, 2011;34-45.
15. Hoerauf A. Microflora, helminthes, and the immune system: Who controls them? *N Engl J Med.* 2010;363:1476-8.
16. Hamann L, et al. Components of gut bacteria as immunomodulators. Int. J Food Microbio. 1998;41:141-54.
17. Parfrey LW, et al. Microbial eukaryotes in the human microbiome: ecology, evolution, and future directions. *Front Microbiol.* July 2011;2:153. doi: 10.3389/fmicb.2011.00153.
18. McNeil NI. The contribution of the large intestine to energy supplies in man *Am J Clin Nutr.* 1984;39:338-42.
19. Conly JM, et al. The production of menaquinones (vitamin K2) by intestinal bacteria and their role in maintaining coagulation homeostasis. *Prog Food Nutr Sci.* 1992;16:307-43.
20. Binbuam J, et al. Biosynthesis of biotin in microorganisms v. control of vitamen production. *J Bacteriol.* 1967;94:1846–53.
21. Digestive Diseases. Available at: www.cdc .gov/nchs/fastats/digestiv.htm. Accessed December 6, 2011.
22. Lashner BA. What is recommended for chronic constipation? Available at: www.medscape .com/viewarticle/744652?src=mp&spon=17. Accessed December 6, 2011.
23. Constipation. Available at: http://digestive .niddk.nih.gov/ddiseases/pubs/constipation. Accessed December 6, 2011.
24. Zelman KM, et al. Dietary Fiber: Insoluble vs. Soluble. Available at: www.webmd.com/diet /fiber-health-benefits-11/insoluble-soluble-fiber. Accessed December 6, 2011.
25. Mishori R, et al. The dangers of colon cleansing. J Family Practice. 2011;60:454-7.
26. Constipation. Available at: http://digestive .niddk.nih.gov/ddiseases/pubs/constipation. Accessed December 6, 2011.
27. Digestive Diseases Statistics for the United States. June 2010. Available at: http:// digestive.niddk.nih.gov/statistics/statistics .aspx#specific. Accessed December 4, 2011.
28. Suerbaum S, et al. Helicobacter pylori infection. *N Engl J Med.* 2002;347:1175–86.
29. Peptic Ulcer Disease: Topic Overview. Available at: www.webmd.com/digestive-disorders/tc/peptic-ulcer-disease-topic-overview. Updated January 2011. Accessed December 4, 2011.
30. Peptic ulcers. Available at: www.mayoclinic .com/health/peptic-ulcer/DS00242. Updated January 2011. Accessed December 4, 2011.
31. Heartburn, Gastroesophageal Reflux (GER), and Gastroesophageal Reflux Disease (GERD), May 2007. Available at: http://digestive .niddk.nih.gov/ddiseases/pubs/gerd. Accessed December 6, 2011.

32. Khan BA, et al. Effect of bed head elevation during sleep in symptomatic patients of nocturnal gastro esophageal reflux. *J Gastroenterol Hepatol.* November 18, 2011. doi: 10.1111/j.1440-1746.2011.06968.x.
33. El-Serag HB, et al. Dietary intake and the risk of gastro-oesophageal reflux disease: A cross-sectional study in volunteers. *Gut.* 2005;54:11-7.
34. Fock K, et al. Gastroesophageal reflux disease. *J Gastroenterol.* 2010;45:808-15.
35. Festi D, et al. Body weight, lifestyle, dietary habits and gastroesophageal reflux disease. *World J Gastroenterol.* 2009;15:1690-701.
36. Irritable Bowel Syndrome. Available at: http:// digestive.niddk.nih.gov/ddiseases/pubs/IBS. Accessed December 6, 2011.
37. Bijkerk CJ, et al. Soluble fiber may be effective for symptoms of IBS. BMJ. 2009;339:b3154. Available at: cme.medscape.com/ viewarticle/708569. Accessed December 6, 2011.
38. National Institute of Digestive Diseases Health Information. Available at: www.niddk.nih.gov. Accessed February 2009.
39. Kilgore PE, et al. Trends in diarrheal disease associated mortality in U.S. children, 1968 through 1991. *JAMA.* 1995;274:1143–8.
40. Meyers A. Oral rehydration therapy: What are we waiting for? *Am Fam Phys.* 1993;47:740–2.
41. Lima AAM, et al. Persistent diarrhea in children: Epidemiology, risk factors, pathophysiology, nutritional impact, and management. *Epidemiol Rev.* 1992;34:222-42.
42. Digestive disease statistics, National Digestive Disease Clearinghouse, digestive.niddk.nih .gov/statistics/ statistics.htm, accessed 2/09.
43. Picco MF. What makes my stomach growl when I'm hungry? Available at: www .mayoclinic.com/health/stomach-noise/ nu00189. Accessed December 6, 2011.
44. Muller-Lissner SA, et al. Dispelling myths about constipation. *Am J Gastroenterol.* 2005;100:232–42.
45. Granado F, et al. Bioavailability of carotenoids and tocopherols from broccoli: In vivo and in vitro assessment. *Exp Biol Med (Maywood).* 2006 ;231:1733-8.
46. Tydeman EA, et al. Effect of carrot (Daucus carota) microstructure on carotene bioaccessibility in the upper gastrointestinal tract. 2. In vivo digestions. *J Agric Food Chem.* 2010;58:9855-60.
47. Tydeman EA, et al. Effect of carrot (Daucus carota) microstructure on carotene bioaccessibility in the upper gastrointestinal tract. 1. In vitro simulations of carrot digestion. *J Agric Food Chem.* 2010;58:9847-54.

Unit 8
1. Poehlman ET, et al. Energy needs: assessment and requirement in humans. In: Shils ME, et al. Modern Nutrition in Health and Disease,

9e.. Philadelphia: Lippincott, Williams, and Wilkins; 1998: p. 96.
2. Bosy-Westphal A, et al. Contribution of individual organ mass loss to weight-loss associated decline in resting energy expenditure. *Am J Clin Nutr.* 2009;90:993–1001.
3. Slow metabolism: Is there any such thing? Available at: www.mayoclinic.com/health/ slow-metabolism/AN00618. Accessed July 7, 2006.
4. Flancbaum L, et al. Comparison of indirect calorimetry, the Fick method, and prediction equations in estimating the energy requirements of critically ill patients. *Am J Clin Nutr.* 1999;69:461–6.
5. Harris JA, et al. A biometric study of basal metabolism in men. Publication no. 279 of the Carnegie Institute of Washington; 1919.
6. Klausen B, et al. Age and sex effects on energy expenditure. *Am J Clin Nutr.* 1997;65:895–907.
7. Frankenfield DC, et al. The Harris-Benedict studies of human basal metabolism: History and limitations. *J Am Diet Assoc.* 1998;98:439–45.
8. Boothby WM, et al. Studies of the energy of metabolism of normal individuals: A standard for basal metabolism, with a nomogram for clinical application. *Am J Physiol.* 1936;116:468–84.
9. Sjodin AM, et al. The influence of physical activity on BMR. *Med Sci Sports Exercise.* 1996;28:85–91.
10. Irwin ML, et al. Estimation of Energy Expenditure from Physical Activity Measures: Determinants of accuracy. *Obesity Research.* 2001;9:517–25.
11. Loss-Adjusted Food Availability. Available at: http://www.ers.usda.gov/Data/ FoodConsumption/FoodGuideIndex .htm#calories. Accessed November 19, 2011.
12. O'Sullivan HL, et al. Effects of repeated exposure on liking for a reduced-energy-dense food. *Am J Clin Nutr.* 2010;91:1584–9.
13. Dietary Guidelines for Americans. Available at: www.dietaryguidelines.gov. Accessed December 12, 2011.
14. Wang J, et al. Dietary energy density predicts the risk of incident type 2 diabetes: The European Prospective Investigation of Cancer (EPIC)–Norfolk Study. *Diabetes Care.* 2008;31:2120–5.
15. Savage JS, et al. Dietary energy density predicts women's weight change over 6 y. *Am J Clin Nutr.* 2008;88:677–84.
16. Rolls BJ. The relationship between dietary energy density and energy intake. *Physiol Behav.* 2009;97:609–15.
17. Schroder H, et al. Low energy density diets are associated with favorable nutrient intake profile and adequacy in free-living elderly men and women. *J Nutr.* 2008;138:1476–81.

18. Jacobs DR, Jr. Fast food and sedentary lifestyle: A combination that leads to obesity. *Am J Clin Nutr.* 2006;83:189–90.
19. Zorrilla G. Hunger and satiety: Deceptively simple words for the complex mechanisms that tell us when to eat and when to stop. *J Am Diet Assoc.* 1998;98:1111.
20. Marmonier C, et al. Snacks consumed in a nonhungry state have poor satiating efficiency. *Am J Clin Nutr.* 2002;76: 518–28.
21. Ferrannini E. The theoretical bases of indirect calorimetry: A review. *Metabolism.*1988;37:287–301.
22. Melby CL, et al. Reported diet and exercise behaviors, beliefs and knowledge among university undergraduates. *Nutr Res.* 1986;6:799–808.

Unit 9

1. Machann J. et al. Fat distribution and metabolic alterations in obesity. *Int Soc Magnetic Resonance in Medicine*, 2011, abs 739, presented May 13, 2011. Available at: www.medscape.com.viewarticle/743193. Accessed December 21, 2011.
2. Germain N, et al. Constitutional thinness and lean anorexia display opposite concentrations of peptide YY, glucagon-like peptide 1, ghrelin, and leptin. *Am J Clin Nutr.* 2007;85:967–71.
3. Levine JA. Poverty and obesity in the U.S. *Diabetes.* 2011;60:2667–8.
4. Rossner S. Ideal body weight—for whom? (editorial). *Acta Medica Scand.* 1984;216:241–2.
5. Grogan S. Body image: Understanding body dissatisfaction in men, women, and children. In: *Culture and Body Image*, 2e. East Sussex: Routledge Publishing; 2008, pp. 9–40.
6. de Gonzalez AB, et al. Body mass index and mortality among 1.46 million white adults. *N Engl J Med.* 2010;362:2211–9.
7. Fernández JR, et al. Is percentage body fat differentially related to body mass index? *Am J Clin Nutr.* 2003;77:71–5.
8. Adult Obesity Prevalence in Canada and the United States. Available at: http://www.cdc.gov/nchs/data/databriefs/db56.htm. Accessed December 18, 2011.
9. Overweight and obesity. Available at: http://www.cdc.gov/nchs/fastats/overwt.htm. Accessed December 18, 2011.
10. Prevalence of obesity maps. Available at: www.cdc.gov/obesity/data/trends.html#State. Accessed December 16, 2011.
11. Stancliffe RA, et al. Dairy attenuates oxidative and inflammatory stress in metabolic syndrome. *Am J Clin Nutr.* 2011;94:422–30.
12. Bray GA. Physiology and consequences of obesity. Medical Education Collaborative, Diabetes and endocrinology clinical management. Available at: www.medscape.com/Medscape/ endocrinology/Clinical Mgmt/Cm.v03/ public/index.CM.v03.html. Accessed January 8, 2001.

13. Wildman RP, et al. The obese without cardiometabolic risk factor clustering and the normal weight with cardiometabolic risk factor clustering: prevalence and correlates of 2 phenotypes among the U.S. population: (NHANES 1999–2004). *Arch Intern Med.* 2008;168:1617–24.
14. Pi-Sunyer X, et al. Obesity associated inflammation, presented at the Experimental Biology annual meetings, Washington, DC, April 3, 2007.
15. Dalmas E, et al. Variations in circulating inflammatory factors are related to changes in calorie and carbohydrate intakes early in the course of surgery-induced weight reduction. *Am J Clin Nutr.* 2011;94:450–8.
16. WHO Global status report on noncommunicable diseases; 2010. Available at: www.who.int/nmh/publications/ncd_report_full_en.pdf. *Accessed December 18, 2011.*
17. Puhl M. The Obese Patient in the Healthcare Environment. Available at: www.medscape.com/viewarticle/749440_3. Accessed September 13, 2011.
18. Lewis S, et al. *"I don't eat a hamburger and large chips every day!"* A qualitative study of the impact of public health messages about obesity on obese adults. *BMC Public Health.* 2010;10:309.
19. Appendix H, Table H-1. Body measurement summary statistics. In: *Dietary Reference Intakes for Energy through Amino Acids*, Washington, DC: The National Academies Press; 1999.
20. Demerath EW, et al. Visceral adiposity and its anatomical distribution as predictors of the metabolic syndrome and cardiometabolic risk factor levels. *Am J Clin Nutr.* 2008;88:1263–71.
21. Forgarty AW, et al. A prospective study of weight change and systemic inflammation over 9 y. *Am J Clin.* 2008;87:30–5.
22. Pischon T, et al. General and abdominal adiposity and risk of death. *N Engl J Med.* 2008;359:2105–20.
23. Davidson LE, et al. Resistance plus aerobic exercise may be best for sedentary, abdominally obese older adults. *Arch Intern Med.* 2009;169:122–31.
24. Tucker AM, et al. Prevalence of cardiovascular disease risk factors among national football league players. *JAMA.* 2009;301:2111–19.
25. Farin HMK, et al. Body mass index and waist circumference both contribute to differences in insulin-mediated glucose disposal in nondiabetic adults. *Am J Clin Nutr.* 2006;83:47–51.
26. Japan legislates mandatory national waist circumference measurement. June 13, 2008. Available at: amicor.blogspot.com/2008/06/japan-legislates-mandatory-national. Html. Accessed February 2009.
27. Ahmadi N, et al. Effects of intense exercise and moderate caloric restriction on

cardiovascular risk factors and inflammation. *Am J Med.* 2011;124:978–82.
28. Steinberg BA, et al. Measuring waist circumference. *Medscape Cardiology.* 2006;10(2). Available at: www.medscape.com/ viewarticle/542635. Accessed October 2006.
29. Gibson RS. *Principles of Nutritional Assessment.* New York: Oxford University Press; 1990.
30. Romao I, et al. Genetic and environmental interactions in obesity and type 2 diabetes. *J Am Diet Assoc.* 2008;108:S24–S28.
31. Malone M, et al. Medication associated with weight gain may influence outcome in weight management program. *Ann pharmacotherapy.* 2005;39:1204–13.
32. den Hoed M, et al. Postprandial responses in hunger and satiety are associated with the rs9939609 single nucleotide polymorphism in FTO. *Am J Clin Nutr.* 2009;90:1426–32.
33. Cooper R, et al. Associations between parental and offspring adiposity up to midlife: The contribution of adult lifestyle factors in the 1968 British Birth Cohort Study. *Am J Clin Nutr.* 2010;92:946–53.
34. Li L, et al. Intergenerational influences on childhood body mass index: The effect of parental BMI trajectories. *Am J Clin Nutr.* 2009;89:551–7.
35. Kuzawa CW, et al. Birth weight, postnatal weight gain, and adult body composition in five low and middle-income countries. *Am J Hum Biol.* 2012;24:5–13.
36. Swinburn B, et al. Increased food energy supply is more than sufficient to explain the U.S. obesity epidemic. *Am J Clin Nutr.* 2009;90:1453–6.
37. Ledikwe JH, et al. Dietary energy density is associated with energy intake and weight status in U.S. adults. *Am J Clin Nutr.* 2006;83:1362–8.
38. Duerksen SC, et al. Family restaurant choices are associated with child and adult overweight status in Mexican-American families. *J Am Diet Assoc.* 2007;107:849–53.
39. Rolls BJ, et al. Larger portion sizes lead to sustained increases in energy intake over 2 days. *J Am Diet Assoc.* 2006;106:543–9.
40. Yanover T, et al. Eating beyond satiety and body mass index. *Eat Weight Disord.* 2008;13:119–28.
41. Levine JA, et al. The role of free-living daily walking in human weight gain and obesity. *Diabetes.* 2008;57:548–54.
42. Danner FW. A national longitudinal study of the association between hours of TV viewing and the trajectory of BMI growth among U.S. children. *J Pediatr Psychol.* 2008;33:1100–7.
43. Position of the American Dietetic Association: Individual-, family-, school-, and community-based interventions for pediatric overweight. *J Am Diet Assoc.* 2006;106:925–45.

44. Jeffery RW, et al, Prevalence and correlates of large weight gains and losses in adults. *Int J Obes.* 2002;26:969–72.

45. Mozaffarian D, et al. Changes in diet and lifestyle and long-term weight gain in women and men. *N Engl J Med.* 2011;364:2392–404.

46. Lombard C, et al. Preventing weight gain: The baseline weight related behaviors and delivery of a randomized controlled intervention in community based women. *BMC Public Health.* January 3, 2009;9:2.

47. Rolls BJ. W. O. Atwater Memorial Lecture, presented at the Experimental Biology annual meetings, Washington, DC; May 5, 2007.

48. Benedict C, et al. Acute sleep deprivation reduces energy expenditure in healthy men. *Am J Clin Nutr.* 2011;93:1229–36.

49. Leibel RL. Energy in, energy out, and the effects of obesity-related genes. *N Engl J Med.* 2008;359:2603–4.

50. Byers T, et al. Public health response to the obesity epidemic too soon or too late? *J Nutr.* 2007;137:488–92.

51. Kumanyika SK, et al. American Heart Association's Obesity Statement: Population-based prevention of obesity, Circulation 2008. Available at: http:// circ.ahajournals.org.

52. Appendix H, Table H-1. Body measurement summary statistics. In: *Dietary Reference Intakes for Energy through Amino Acids*, Washington, DC: The National Academies Press; 1999.

53. Jacquemont S, et al. Mirror extreme BMI phenotypes associated with gene dosage at the chromosome 16p11.2 locus. *Nature.* 2011. 78:97–102.

54. Sabia S, et al. BMI over the adult life course and cognition in late midlife: The Whitehall II Cohort Study. *Am J Clin Nutr.* 2009;89:601–7.

55. Fontana L. The scientific basis of caloric restriction leading to longer life. *Curr Opin Gastroenterol.* 2009;25:144–50.

56. Obesity and diabetes risk on the rise, CDC survey finds. Available at: www.medscape .com/viewarticle/723644. Accessed June 10, 2010.

57. UK panel calls for media to end pressure on girls to be thin. Report of the UK's cabinet's Body Image Summit. *Reuter Health.* June 22, 2000.

58. Chapman GE, et al. Canadian dietitians' approaches to counseling adult clients seeking weight-management advice. *J Am Diet Assoc.* 2005;105:1275–9.

59. Bacon L, et al. Size acceptance and intuitive eating improve health for obese, female chronic dieters. *J Am Diet Assoc.* 2005;105:929–36.

60. Ingall M. End of dieting. Available at: http:// www.fitwoman.com/end-of-dieting. Accessed December 21, 2011.

61. Defining overweight and obesity. Available at: http://www.cdc.gov/obesity/defining.html,. Accessed December 21, 2011.

62. Whitaker JA, et al. Predicting obesity in young adulthood from childhood parental obesity. *N Engl J Med.* 1997;337:869–73.

63. Meyer TE, et al. Long-term calorie restriction ameliorates the decline in diastolic functions in humans. *J Am Coll Cardio.* 2006:47:398–402.

64. Friedman J. Leptin, the most recent advances, presented at the Experimental Biology annual meetings, San Diego, CA, April 6, 2008. *Nutr.* 2008;87:30–5.

65. Puhl RM, et al. Parental perceptions of weight terminology that providers use with youth. *Pediatrics.* 2011;128:e786–93.

66. Hankinson AL, et al. Maintaining a high physical activity level over 20 years and weight gain. *JAMA.* 2010;304:2603–10.

67. Kanaka-Gantenbein C. Fetal origins of adult diabetes. *Ann N Y Acad Sci.* 2010;1205:99–105.

68. Corella D, et al. A high intake of saturated fatty acids strengthens the association between fat mass and obesity-associated gene and BMI. *J Nutr.* 2011;141:2219–25.

69. Carlson JJ, et al. Pre- and post-prandial appetite hormone levels in normal weight and severely obese women. *J Nutr.* 2011;141:2219–25

70. Despres J-P, et al. A focus on visceral adiposity and liver fat. 2nd International Congress on Abdominal Obesity, February 26, 2011. Available at: www.medscape.com/ viewarticle/738448,. Accessed March 19, 2011.

Unit 10

1. Franz MJ, et al. Weight-loss outcomes: A systemic review and meta-analysis of weight-loss clinical trials with a minimum of 1-year follow-up. *J Am Diet Assoc.* 2007;107:1755–67.

2. Hill JO. Can a small-changes approach help address the obesity epidemic? A report of the joint task force of the ASN, IFT, and the IFIC. *Am J Clin Nutr.* 2009;89:477–84.

3. , et al. Perceived weight status, overweight diagnosis, and weight control among U.S. adults: The NHANES 2003-2008 Study. *Int J Obes (Lond).* 2011;35:1063–70.

4. Ethics opinion: Weight-loss products and medications. *J Am Diet Assoc.* 2008;108:2109–13.

5. Goodrick GK, et al. Why treatments for obesity don't last. *J Am Diet Assoc.* 1991;91:234–47.

6. Ludwig DS, et al. Weight-loss maintenance-mind over matter? *N Engl J Med.* 2010;363:2159–61.

7. Yanovski SZ. Obesity treatment in primary care-are we there yet? *N Eng J Med.* 2011;365:2030–1.

8. Varnado-Sullivan PJ, et al. Opinions and acceptability of common weight-loss practices. *Eat Weight Disord.* 2010;15:e256–64.

9. Thomas SL, et al. "They all work . . . when you stick to them": A qualitative investigation of dieting, weight loss, and physical exercise, in obese individuals. *Nutr J.* 2008;7:34 [online] November 24, 2008. doi: 10.1186/1475-2891-7-34.

10. Ness-Abramof R, et al. Diet modification for treatment and prevention of obesity. *Endocrine.* 2006;29:5–9.

11. Blackburn GL. Weight loss advertising: An analysis of current trends. 2002. Available at: www.ftc.gov. Accessed December 26, 2011.

12. Wadden TA, et al. Very low calorie diets: Their efficacy, safety, and future. *Ann Intern Med.* 1983;99:675–84.

13. Federal Trade Commission reaches "new year's" resolutions with four major weight-control pill marketers. Available at: http:// www.casewatch.org/ftc/news/2007/diet_pills .shtml. Accessed December 26, 2011.

14. Greenwood MRC. Help not hype: Getting real about weight loss. *Obesity Management* 2007;3:11–21.

15. Current trends in weight-loss advertising. *J Am Diet Assoc.* 2003;103:150.

16. Beware of fraudulent weight-loss "dietary supplements." March 15, 2011. Available at: http://www.fda.gov/ForConsumers/ ConsumerUpdates/ucm246742.htm#2. Accessed December 17, 2011.

17. FTC cracks down of hoodia for weight loss. Available at: http://www.fda.gov/ ForConsumers/ConsumerUpdates/ucm246742 .htm#2. Accessed November 14, 2011.

18. HCG Diet Products Are Illegal. Available at: http://www.fda.gov/ForConsumers/ ConsumerUpdates/ucm281333.htm. Accessed 11, 2011.

19. ElBoghdady D. FTC continues push against fake-news Web sites peddling weight-loss products. Available at: http://www. washingtonpost.com/business/economy/ftc-continues-push-against-fake-news-web-sites-peddling-weight-loss-products/2011/12/01/ gIQATgnSIO_story.html. Accessed December 26, 2011.

20. Taubes G. Dietary approaches to weight control. What works? Experimental Biology Annual Meeting, San Diego, April 14, 2003.

21. Cunningham A, et al. Is it possible to burn calories by eating grapefruit or vinegar? *J Am Diet Assoc.* 2001;101:1198.

22. Elias MC, et al. Effect of 6-month nutritional intervention on non-alcoholic fatty liver disease. *Nutrition.* 2010;26:1094–9.

23. Christian JG, et al. A computer support program that helps clinicians provide patients with metabolic syndrome tailored counseling to promote weight loss. *J Am Diet Assoc.* 2011;111:75–83.

24. Dansinger MG, et al. Comparisons of the Atkins, Ornish, Weight Watchers, and Zone diets for weight loss and heart disease risk reduction: A randomized trial. *JAMA.* 2005;293:43–50.

25. King NA, et al. Dual-process action of exercise on appetite control: Increase in orexigenic drive but improvement in meal-induced satiety. *Am J Clin Nutr.* 2009;90:921–7.

26. Hill JO, et al. Weight maintenance: What's missing? *J Am Diet Assoc.* 2005;105:S63–S66.

27. Arterburn D, et al. Comparison of commercial weight loss diets. *JAMA.* 2006;292:101–3.

28. Gardner CD, et al. Micronutrient quality of weight-loss diets that focus on macronutrients: Results from the A TO Z study. *Am J Clin Nutr.* 2010;92:304–12.

29. Bray GA, et al. Effect of Dietary Protein Content on Weight Gain, Energy Expenditure, and Body Composition During Overeating: A Randomized Controlled Trial. JAMA 2012;307:47–55.

30. Food and Drug Administration. FDA expands warnings to consumers about tainted weight loss pills. January 8, 2008. Available at: www.fad.gov/bbs/topics/NEWS/2008/ NEW01933.html. Accessed February 2009.

31. The possible dangers of buying medicines over the Internet. Available at: www.fda.gov/downloads/forconsumers. Updated November 2010.

32. Jebb SA, et al. Primary care referral to a commercial provider for weight loss treatment versus standard care: A randomized controlled trial. *Lancet.* 2011;378:1485–92.

33. Rock CL, et al. Effect of a free prepared meal and incentivized weight loss program on weight loss and weight loss maintenance in obese and overweight women, a randomized controlled trial. *JAMA.* 2010;304:1803–10.

34. Rosenbaum M, et al. Long-term persistence of adaptive thermogenesis in subjects who have maintained a reduced body weight. *Am J Clin Nutr,* 2008;88:906–12.

35. Wadden TA, et al. Randomized trial of lifestyle modification and pharmacotherapy for obesity. *N Engl J Med.* 2005;353:211–20.

36. Laddu D. A review of evidence-based strategies to treat obesity in adults. *Nutr Clin Pract.* 2011;26:512-25.

37. Macdonald IA. In search of the basis of successful maintenance of weight loss. *Am J Clin Nutr* 2009;90:908–9.

38. Appel LJ, et al. Comparative effectiveness of weight-loss interventions in clinical practice. *N Engl J Med.* 2011;364:1959–68.

39. May AM, et al. Effect of change in physical activity on body fatness over a 10-y period in the Doetinchem Cohort Study. *Am J Clin Nutr.* 2010;92:491–9.

40. American College of Sports Medicine, updated guidelines for appropriate physical activity interventions for weight loss and prevention of weight regain. *Med Sci Sports Exer.* 2009;41:459–71.

41. Gould L. Consumers taking small steps toward big lifestyle changes. Datmonitor survey of diet trends and advertising in the United Kingdom and the United States.
NUTRAingredients. com/Europe. March 18, 2005.

42. Fenske W, et al. Pharmacotherapy for obesity—A field in crisis? *Expert Rev Endocrinol Metab.* 2011;6:563–77.

43. Pray WS. New nonprescription weight loss product. *U.S. Pharmacist.* 2007;32:10–5.

44. Krempf M,, et al. Weight reduction and long-term maintenance after 18 months treatment with orlistat for obesity. *Int J Obes Relat Metab Disord.* 2003;27:591–7.

45. Yanovski SZ, et al. Obesity. *N Engl J Med.* 2002;346:591–602.

46. Stokowski LA. Weight loss drugs: What works? Available at: www.medscape.com .viewarticle/733585. Accessed February 9, 2011.

47. O'Riodan M, et al. Lap-band device gets wider indications: FDA lowers BMI requirements for weight-loss surgery. Available at: www .medscape.con/niewarticle/737792. Accessed February 23, 2011.

48. Courcoulas, A, et al. Safety data favors less-invasive weight-loss surgery. *Archives of Surgery.* Available at: http://bit.ly/sdDPNc. Accessed November 21, 2011.

49. Dayyeh BKA, et al. Obesity and bariatrics for the endoscopist: New techniques. *Ther Adv Gastroenterol.* 2011;4:433–42.

50. Pontiroli AE, et al. Long-term prevention of mortality in morbid obesity through bariatric surgery: A systematic review and meta-analysis of trials performed with gastric banding and gastric bypass. *Ann Surg.* 2011;253:484–7.

51. Buchwald H. Surgical intervention of the treatment of morbid obesity. *Future Lipidol.* 2007;2:513–25.

52. Dorman RB, et al. Case-matched outcomes in bariatric surgery for treatment of type 2 diabetes in the morbidly obese patient. *Ann Surg.* October 4, 2011.

53. DeMaria EJ. Bariatric surgery for morbid obesity. *N Engl J Med.* 2007;356:2176–83.

54. Position of the American Dietetic Association: Weight management. *J Am Diet Assoc.* 2009;109:330–46.

55. Malinowski SS. Nutritional and metabolic complications of bariatric surgery. *Am J Med Sci.* April 2006;331(4):219–25.

56. Marcason W. What are the dietary guidelines following bariatric surgery? *J Am Diet Assoc.* 2004;104:487–8.

57. Awad W, et al. Ten years experience of banded gastric bypass: Does it make a difference? *Obes Surg.* December 8, 2011. [Epub ahead of print]. Accessed December 27, 2011.

58. Eisenberg D, et al. Update on obesity surgery. *World J Gastroenterol.* 2006;12:3196–3203.

59. The skinny on liposuction. Available at: http://www.fda.gov/forconsumers/consumerupdates/ucm049314.htm. Accessed December 27, 2011.

60. Liposuction comes of age, and new liposuction technique faster, less invasive. Available at: www.healthfinder.gov. Accessed September 2003.

61. Swanson E, at al. Does liposuction offer more than a cosmetic benefit? *Plastic Surgery.* 2011: American Society of Plastic Surgeons (ASPS) Annual Meeting. Presented September 25, 2011.

62. Wing RR, et al. Long-term weight maintenance. *Am J Clin Nutr.* 2005;82(suppl):222S–5S.

63. Field AE, et al. Weight-control behaviors and subsequent weight change among adolescents and young adult females. *Am J Clin Nutr.* 2010;91:147–53.

64. Weinsier RL, et al. Free-living energy expenditure in women successful and unsuccessful at maintaining a normal body weight. *Am J Clin Nutr.* 2002.

65. Schlundt DG, et al. The role of breakfast in the treatment of obesity: A randomized trial. *Am J Clin Nutr.* 1992;55:645–51.

66. Ello-Martin JA, et al. Dietary energy density in the treatment of obesity: A year-long trial comparing 2 weight-loss diets. *Am J Clin Nutr.* 2007;85:1465–77.

67. Cunningham E. How can I help my client who is experiencing a weight-loss plateau? *J Am Diet Assoc.* 2011;101:1966–7.

Unit 11

1. Makino M, et al. Prevalence of eating disorders: A comparison of Western and non-Western countries. *MedGenMed.* 2004;6:49.

2. Eating Disorders. Available at: www .dsm5.org/ProposedRevisions/Pages/EatingDisorders.aspx. Accessed December 29, 2011.

3. Konstantynowicz J, et al. Thigh circumference as a useful predictor of body fat in adolescent girls with anorexia nervosa. *Ann Nutr Metab.* 2011;58:181–7.

4. Position of the American Dietetic Association: Nutrition intervention in the treatment of anorexia nervosa, bulimia nervosa, and other eating disorders. *J Am Diet Assoc.* 2006;10:2073–82.

5. Benninghoven D, et al. [Body images of male patients with eating disorders]. *Psychother Psychosom Med Psychol.* 2007;57:120–7.

6. Mangweth B, et al. Body fat perception in eating-disordered men. *Int J Eat Disord.* 2004;35:102–8.

7. Misra M, et al. Percentage extremity fat, but not percentage trunk fat, is lower in adolescent boys with anorexia nervosa than in healthy adolescents. *Am J Clin Nutr.* 2008;88:1478–84.

8. Manore MM, et al. The female athlete triad: Components, nutrition issues, and health consequences. *J Sports Sci.* 2007;25 (suppl)1:S61–71.

9. Barrack MT, et al. Dietary restraint and low bone mass in female adolescent endurance runners. *Am J Clin Nutr.* 2008;87:36–43.

10. Palacio, LE, et al. Nutrition for the female athlete. October 31, 2008. Available at: http://emedicine.medscape.com/article/108994. Accessed September 28, 2011.

11. Beals KA, et al. Disorders of the female athlete triad among collegiate athletes. *Int J Sports Nutr Exer Metab.* 2002;12:281–93.

12. Currie A. Eating disorders in elite athletes. International Congress of the Royal College of Psychiatrists 2001, presented July 1, 2011. Available at: www.medscape.com.viewarticle/746021. Accessed July 20, 2011.

13. American Psychiatric Association. *Diagnostic and statistical manual of mental disorders: DMS-IV,* 4th ed., Text Revision. Washington, DC: 2000.

14. Omizo SA, et al. Anorexia nervosa: Psychological considerations for nutrition counseling. *J Am Diet Assoc.* 1988;88:49–51.

15. Field AE, et al. Factors related to bingeing and purging in girls and boys: The growing up today study. *Arch Pediatr Adolesc Med.* 2008;162:574–9.

16. Feldman W, et al. Culture versus biology: Children's attitudes toward thinness and fatness. *Pediatrics.* 1988;81:190–4.

17. Yager J, et al. Anorexia nervosa. *N Engl J Med.* 2005;353:1481–8.

18. Treating eating disorders. *Harvard Women's Health Watch.* May 4–5, 1996; and When eating goes awry: An update on eating disorders. *Food Insight.* January/February 1997:35.

19. Naessen, S et al. Women with bulimia nervosa exhibit attenuated secretion of glucagon-like peptide 1, pancreatic polypeptide, and insulin in response to a meal. *Am J Clin Nutr.* 2011;94:967–72.

20. Herzog DB, et al. Bulimia nervosa—psyche and satiety (editorial). *N Engl J Med.* 1988;319:716–8.

21. Hohlstein LA. Eating disorders program. American Dietetic Association annual meeting, Boston, October 27, 1997.

22. Branson R, et al. Binge eating as a major phenotype of melanocortin 4 receptor gene mutations. *N Engl J Med.* 2003;348:1096–103.

23. Hudson JI, et al. Longitudinal study of the diagnosis of components of the metabolic syndrome in individuals with binge-eating disorder. *Am J Clin Nutr.* 2010;91:1568–73.

24. Mathieu J. What should you know about mindful eating and intuitive eating? *J Am Diet Assoc.* 2009;109:1982–3.

25. Wanden-Berghe RG, et al. The application of mindfulness to eating disorders treatment: A systematic review. *Eat Disord.* 2011;19:34–48.

26. Fairburn CG, et al. Distinctions between binge eating disorder and bulimia nervosa. *Arch Gen Psychiatry.* 2000;57:659–65.

27. Rodgers R, et al. An exploration of the tripartite influence model of body dissatisfaction and disordered eating among Australian and French college women. *Body Image.* 2011;8:208–15.

28. Hogan MJ, et al. Body image, eating disorders, and the media. *Adolesc Med State Art Rev.* 2008;19:521–46.

29. Zuckerman DM, et al. The prevalence of bulimia among college students. *Am J Public Health.* 1986;76:1135–7.

30. Elgin J, et al. Gender differences in disordered eating and its correlates. *Eat Weight Disord.* 2006;11:e96–101.

31. Calado M, et al. The mass media exposure and disordered eating behaviours in Spanish secondary students. *Eur Eat Disord Rev.* 2010;18:417–27.

32. Passareillo C, et al. French law targets extreme thinness. *The Wall Street Journal,* April 16, 2008:A11.

33. Cooper M. Pica: *A Survey of the Historical Literature as Well as Reports from the Fields of Veterinary Medicine and Anthropology.* Springfield, IL: Charles C. Thomas; 1957.

34. Edwards CH, et al. Clay- and corn starch eating women. *J Am Diet Assoc.* 1959;35:810–15.

35. Woywodt A, et al. Geophagia: The history of earth-eating. *J R Soc Med.* 2002;95: 143–46.

36. Coltman CA Jr. Pagophagia and iron lack. *JAMA.* 1969;207:513–16.

37. Rainville AJ. Pica practices of pregnant women are associated with lower maternal hemoglobin level at delivery. *J Am Diet Assoc.* 1998;98:293–6.

38. Jones RL, et al. Blood lead levels of children in the United States. *Pediatrics.* 2009;123:e376–85.

39. Hildebrand MP. Lead toxicity in a newborn. *J Pediatr Health Care.* 2011;25:328–30.

Unit 12

1. Dietary intake data: What we eat in America. NHANES 2007–2008. Available at: *www.ars.usda.gov/ba/bhnrc/fsrg.* Accessed November 19, 2011.

2. Dietary Guidelines for Americans, 2010. Available at: www.dietaryguidelines.gov.

3. Grabitske HA, et al. Low-digestible carbohydrates in practice. *J Am Diet Assoc.* 2008;108:1677–81.

4. Stanhope KL, et al. Consuming fructose-sweetened, not glucose-sweetened, beverages increases visceral adiposity and lipids and decreases insulin sensitivity in overweight/obese humans. *J Clin Invest.* 2009;119:1322–34.

5. Institute of Medicine, Food and Nutrition Board. *Dietary Reference Intakes for Energy, Carbohydrates, Fiber, Fat, Fatty Acids, Cholesterol, Protein, and Amino Acids.* Washington, DC: National Academies Press; 2002.

6. Position of the American Dietetic Association: Use of nutritive and nonnutritive sweeteners. *J Am Diet Assoc.* 2004;104:255–58.

7. Johnson RK, et al. Dietary sugars intake and cardiovascular health: A scientific statement from the American Heart Association circulation. 2009;120:1011–20.

8. Food and Drug Administration (Department of Health and Human Services), Nutrition labeling, *Federal Register.* November 27, 1991.

9. de Koning L, et al. Sugar-sweetened and artificially sweetened beverage consumption and risk of type 2 diabetes in men. *Am J Clin Nutr.* 2011;93:1321–7.

10. Sievenpiper J, et al. Fructose does not cause weight gain in isocaloric diet. International Diabetes Federation (IDF) World Diabetes Congress 2011: Abstract O-0476. Presented December 6, 2011.

11. Welsh JA, et al. *Consumption of added sugars and indicators of cardiovascular disease risk among US adolescents. Circulation.* 2011;123:249–57.

12. Davis JN, et al. Effects of breastfeeding and low sugar-sweetened beverage consumption on obesity prevalence in Hispanic toddlers. *Am Clin Nutr.* 2012;95:3–8.

13. *Kristoffersen L,* et al. Pain reduction on insertion of a feeding tube in preterm infants: a randomized controlled trial. *Pediatrics.* 2011;127:e1449–54.

14. Nadimi H, et al. Are sugar-free confections really beneficial for dental health? *Br Dent J.* October 7, 2011;211(7):E15. [online] doi: 10.1038/sj.bdj.2011.823.

15. Mattes RD, et al. Nonnutritive sweetener consumption in humans: Effects on appetite and food intake and their putative mechanisms. *Am J Clin Nutr.* 2009;8:1–14.

16. Saccharin deemed safe. Community Nutrition Institute, June 2, 2000:8.

17. U.S. EPA says saccharin not a threat. Available at: www.medscape.com.viewarticle/734270. Accessed December 17, 2010.

18. Butchko HH, et al. Acceptable daily intake vs. actual intake: The aspartame example. *J Am Coll Nutr.* 1991;10:258–66.

19. Spiers PA, et al. Aspartame: Neuropsychologic and neurophysiologic evaluation and chronic effects. *Am J Clin Nutr.* 1998;68:531–7.

20. Magnuson B, et al. Safety of aspartame. *Crit Rev Toxicol.* October 2007. Available at: www.medscape.com/viewarticle/564923. Accessed March 2009.

21. Bray GA. Fructose—how worried should we be? *Medscape J Med.* 2008;10:159. Available at: www.medscape.com/viewarticle/575891. Accessed February 2009.

22. Park Y, et al. Dietary fiber intake and mortality in the NIH-AARP diet and health study. *Arch Intern Med.* 2011;171:1061–8.

23. Jonnalagadda SS, et al. Putting the whole grain puzzle together: Health benefits associated with whole grains—summary of

American Society for Nutrition 2010 Satellite Symposium. *J Nutr.* 2011;141:1011S–22S.

24. Du H, et al. Dietary fiber and subsequent changes in body weight and waist circumference in European men and women. *Am J Clin Nutr.* 2010;91:329–36.

25. What We Eat in America, NHANES, 2007–2008. Available at www.ars.usda.gov/Services/docs.htm?docid=13793. Accessed February 25, 2011.

26. Position of the American Dietetic Association: Health implications of dietary fiber. *J Am Diet Assoc.* 2008;108:1716–31.

27. Scott-Thomas C. Prepare for higher calorie count for fibre, say scientists. December 19, 2008. Available at: www.foodnavigator/europe.com. Accessed December 2008.

28. Turner ND, et al. Dietary fiber. *Adv Nutr.* 2011:2;151–52.

29. USDA. Food labeling: Health claims; soluble dietary fiber from certain foods and coronary heart disease. Final rule, 71. *Federal Register* 29248–29250, 2006.

30. Zelman KM, et al. Dietary fiber: Insoluble vs. soluble. Available at; http://www.webmd.com/diet/fiber-health-benefits-11/insoluble-soluble-fiber. Accessed December 6, 2011.

31. Hur IY, et al. Relationship between whole-grain intake, chronic disease risk indicators, and weight status among adolescents in the national health and nutrition examination survey, 1999–2004. *J Acad Nutr Diet.* 2012;112:46–55.

32. Williams CL. A summary of conference recommendations on dietary fiber in childhood. *Pediatrics.* 1995;96:1023–8.

33. Corwin RL, et al. Symposium overview—Food addiction: Fact or fiction? *J Nutr.* 2009;139:617–9.

34. Meule A. How prevalent is "food addiction"? *Front Psychiatry.* 2011;2:61.

35. Liu S, et al. Dietary glycemic load and type 2 diabetes: modeling the glucose-raising potential of carbohydrates for prevention. *Am J Clin Nutr.* 2010;92:675–7.

36. Malik VS, et al. Sugar-sweetened beverages and health: Where does the evidence stand? *Am J Clin Nutr.* 2011;94:1161–2.

37. Stanhope KL, et al. Endocrine and metabolic effects of consuming beverages sweetened with fructose, glucose, sucrose, or high-fructose corn syrup. *Am J Clin Nutr.* 2008;88(suppl):1733S–7S.

38. Cornero S, et al. Diet and nutrition of prehistoric populations at the alluvial banks of the Parana River. *Medicina.* 2000;60:109–14.

39. Parajas IL. Sugar content of commonly eaten snack foods of school children in relation to their dental health status. *J Phillipine Dent Assoc.* 1999;51:4–21.

40. Bar A. EFSA boosts tooth-friendly developing world dental campaign. Available at: www.nutringredients.com/content/view/389940. Accessed August 31, 2011.

41. Jensen ME. Diet and dental caries. *Clin North Am.* 1999;43:615–33.

42. Ismail A. Food cariogenicity in Americans aged from 9 to 29 years accessed in a national cross-sectional study, 1971–1974. *J Dental Res.* 1986;65:1435–40.

43. Edgar WM. Sugar substitutes, chewing gum, and dental caries—a review. *British Dental J.* 1998;184:29–32.

44. Schachtele CF, et al. Will the diets of the future be less cariogenic? *J Canadian Dental Assoc.* 1984;3:213–9.

45. Palmer CA, et al. Position of the American Dietetic Association: The impact of fluoride on dental health. *J Am Diet Assoc.* 2005;105:1620–28.

46. Bailey W, et al. Populations receiving optimally fluoridated public drinking water—United States, 1992–2006. *MMWR.* 2008;57:737–41.

47. Sampaio FC, et al. Systemic fluoride. *Monogr Oral Sci.* 2011;22:133-45.

48. National Maternal and Child Oral Health Resource Center. Trends in children's oral health. 1999. Available at: www.ncemch.org.

49. Dykibra J, et al. Lactose maldigestion revisited: Diagnosis, prevalence in ethnic minorities. *Am J Lifestyle Med.* 2009;3:212–18.

50. Krebs JR. The gourmet ape: Evolution and human food preferences. *Am J Clin Nutr.* 2009;90(suppl):707S–11S.

51. NIH Consensus Development Conference Statement on Lactose Intolerance and Health, vol. 27, no. 2, February 22–24, 2010.

52. Lactose intolerance: A self-fulfilling prophecy leading to osteoporosis. *Nutr Rev.* 2003;61:221–3.

53. Montalto M, et al, Management and treatment of lactose malabsorption. *Gastroenterol.* 2006;12:187–91.

54. Simmons FJ. Primary adult lactose intolerance and the milking habit: A problem in biological and cultural interrelationships. I. Review of the medical research. *Am J Digestion.* 1981;14:819.

55. Inman-Felton AE. Overview of lactose maldigestion (lactose nonpersistence). *J Am Diet Assoc.* 1999;99:481–9.

56. Johnson RJ, et al. Attention-deficit/hyperactivity disorder: Is it time to reappraise the role of sugar consumption? *Postgrad Med.* 2011;123:39–49.

Unit 13

1. Dietary Guidelines for Americans, 2010. Available at: www.dietaryguidelines.gov.

2. van't Riet, et al. Progression of glucose disturbances prior to onset of diabetes. European Association Study Diabetes 47th annual meeting, abstract 335. September 13, 2011.

3. Foster-Powell K, et al. International table of glycemic index and glycemic load values: 2002. *Am J Clin Nutr.* 2002; 76:5–56.

4. Type 2 affects 6% of adults worldwide. Available at: www.nlm.nih.gov/medlineplus/news/ fullstory_34711.html. Accessed July 2006.

5. 2011 National Diabetes Fact Sheet. Available at: http://www.cdc.gov/diabetes/pubs/estimates11.htm#8. Accessed January 14, 2012.

6. Bennett WL, et al. Comparative effectiveness and safety of medications for type 2 diabetes: An update including new drugs and 2-drug combinations. *Ann Int Med.* 2011;154:602–13.

7. Rother KI. Diabetes treatment—bridging the divide. *N Engl J Med.* 2007;356:1499–1501.

8. Alexander CM, et al. NCEP-defined metabolic syndrome, diabetes, and prevalence of coronary heart disease among NHANES III participants age 50 years and older. *Diabetes.* 2003;52:1210–14.

9. Hyperglycemia and Diabetes. Available at: http://diabetes.webmd.com/diabetes-hyperglycemia. Accessed January 15, 2012.

10. Hemmingsen B, et al. Intensive glycaemic control for patients with type 2 diabetes: Systematic review with meta-analysis and trial sequential analysis of randomised clinical trials. *BMJ.* Available at: http://www.medscape.com/viewarticle/754363?src=mp&spon=2. Accessed December 16, 2011.

11. Jones V. The "diabesity" epidemic: Let's rehabilitate America. *Medscape Gen Med.* 2006;8:34.

12. The CDC's Database Modeling Project: developing a new tool for chronic disease prevention and control. Available at: www.cdc.gov. Accessed August 2006.

13. Song SH, et al. Early-onset diabetes mellitus: an increasing phenomenon of elevated cardiovascular risk. *Expert Rev Cardiovasc Ther.* 2008;6:315–22.

14. Health, United States, 2005. Available at: www.cdc. gov. Accessed August 2006.

15. Marin C, et al. The Ala54Thr polymorphism of the fatty-acid binding protein 2 gene is associated with a change in insulin sensitivity after a change in the type of dietary fat. *Am J Clin Nutr.* 2005;82:196–200.

16. American Diabetes Association. Standards of medical care in diabetes—2011. *Diabetes Care.* 2011;34:S11–S61.

17. Cowie CC. CDC surveillance system. *MMWR.* 2003:52:833–7.

18. Hsueh WA, et al. Prediabetes: The importance of early identification and intervention. *Postgrad Med.* 2010;122:129–43.

19. Diabetes Prevention Program Research Group. Reduction in the incidence of type 2 diabetes with lifestyle intervention or metformin. *N Engl J Med.* 2002; 346:393–403.

20. The Hemoglobin A1c (HbA1c) Test for Diabetes. Available at: http://diabetes.webmd.com/guide/glycated-hemoglobin-test-hba1c. Accessed November 13, 2012.

21. Bugianesi E, et al. Insulin resistance in nonalcoholic fatty liver disease. *Curr Pharm Des.* 2010;16:1941–51.

22. Harmon RC, et al. Inflammation in nonalcoholic steatohepatitis. *Expert Rev Gastroenterol Hepatol.* 2011;5:189–200.

23. Basu PP, et al. Combination antioxidant therapy may help fatty liver. American College of Gastroenterology (ACG) 2011 annual scientific meeting and postgraduate course, abstract 38. Presented November 3, 2011. Available at: http://www.medscape.com/view article/753034?sssdmh=dm1.732339&src=nl dne. Accessed November 14, 2011.

24. *Nomura K,* et al. The role of fructose-enriched diets in mechanisms of nonalcoholic fatty liver disease. *J Nutr Biochem.* November 28, 2011. doi.org/10.1016/j.jnutbio.2011.09.006. Accessed February 15, 2011.

25. Coppell KJ, et al. Nutritional intervention in patients with type 2 diabetes who are hyperglycaemic despite optimised drug treatment—Lifestyle Over and Above Drugs in Diabetes (LOADD) study: Randomised controlled trial. *BMJ.* 2010;341:c3337.

26. Ezmaillzadeh A, et al. Dietary patterns, insulin resistance, and prevalence of metabolic syndrome in women. *Am J Clin Nutr.* 2007;85: 910–8.

27. Ehrmann DA. Polycystic ovary syndrome. *N Engl J Med.* 2005;352:1223–36.

28. Symptoms and Diagnosis of Metabolic Syndrome. Available at: http://www .heart.org/HEARTORG/Conditions/ More/MetabolicSyndrome/Symptoms-and-Diagnosis-of-Metabolic-Syndrome_UCM_301925_Article.jsp#.TxO0GSMZA7A. Accessed January 15, 2012.

29. Riediger ND, et al. Prevalence of metabolic syndrome in the Canadian adult population, *CMAJ.* 2011:doi:10.1503/cmaj.110070.

30. Pittas AG, et al. High levels of vitamin D, calcium may lower type 2 diabetes risk. *Diabetes Care.* 2006;29:650–56.

31. Akbaralay TM, et al. Overall diet history and reversibility of the metabolic syndrome over 5 years: The Whitehall II prospective cohort study. Diabetes Care. November 2010;33:2339–41.

32. Ripsin CM, et al. Review of blood glucose management for type 2 diabetes. Am Fam Physician. 2009;79:29–36.

33. Franz MJ et al. The evidence for medical nutrition therapy for type 1 and type 2 diabetes in adults. J Am Diet Assoc 2010;110:1852-89.

34. Peters AL. ADA's practice guidelines for 2011. Available at: www.medscape.com/index/ list_3376_0. Accessed March 16, 2011.

35. Pournaras DJ, et al. Effect of the definition of type II diabetes remission in the evaluation of bariatric surgery for metabolic disorders. *Brit J Surg.* [online]. October 21, 2011. doi:10.1002/ bjs.7704.

36. Kashyap SR, et al. Weight loss as a cure for type 2 diabetes. *Expert Rev Endocrinol Meta.* 2011;6:557–61.

37. Hayes C, et al. Role of physical activity in diabetes management. *J Am Diet Assoc.* 2008;108:S19–S23.

38. Riccardi G, et al. Role of glycemic index and glycemic load in the healthy state, in prediabetes, and in diabetes. *Am J Clin Nutr.* 2008;87(suppl):269S–74S.

39. Atkinson FS, et al. International Tables of Glycemic Index and Glycemic Load Values. *Diabetes Care.* 2008;31:12 2281–83.

40. Fabricatore AN, et al. *Continuous glucose monitoring to assess the ecologic validity of dietary glycemic index and glycemic load. Am J Clin Nutr.* 2011;94:1519–24.

41. Liu S, et al. Dietary glycemic load and type 2 diabetes: modeling the glucose-raising potential of carbohydrate for prevention. *Am J Clin Nutr.* 2010;92:675–7.

42. Wolever TMS, et al. The Canadian Trial of Carbohydrates in Diabetes, a 1-y controlled trial of low-glycemic index dietary carbohydrate in type 2 diabetes. *Am J Clin Nutr.* 2008;87:114–25.

43. Mitchell HL. The glycemic index concept in action.. Am J Clin Nutr. 2008;87:244S–246S.

44. Finley CE, et al. Glycemic index, glycemic load, and prevalence of the metabolic syndrome in the Cooper Center Longitudinal Study. *J Am Diet Assoc.* 2010;110:1820–9.

45. Bantle JP, et al. American Diabetes Association updated guidelines for Medical Nutrition Therapy to prevent diabetes, manage existing diabetes, and prevent or slow the rate of development of diabetes complications. *Diabetes Care.* 2008;31(suppl):S61–78.

46. Ding EL, et al. Convergence of obesity and high glycemic diet on compounding diabetes and cardiovascular risks in modernizing China: An emerging public health dilemma. *Global Health.* February 26, 2008;4:4.

47. Kochan AM, et al. Glycemic index predicts individual glucose responses after self-selected breakfasts in free-living, abdominally obese adults. *J Nutr.* 2012;142:27–32.

48. Hu FB. Are refined carbohydrates worse than saturated fat? *Am J Clin Nutr.* 2010;91:1541–2.

49. Davis JN, et al. Effects of breastfeeding and low sugar-sweetened beverage consumption on obesity prevalence in Hispanic toddlers. *Am Clin Nutr.* 2012;95:3–4.

50. De Koning L, et al. Sugar-sweetened and artificially sweetened beverage consumption and risk of type 2 diabetes in men. *Am J Clin Nutr.* 2011;93:1321–7.

51. Ludwig DS. Glycemic load comes of age. *J Nutr.* 2003;133:2728–32.

52. Franz MJ, et al. Evidence-based nutrition principles and recommendations for the treatment and prevention of diabetes and related complications. *Diabetes Care.* 2002;25:148–66.

53. McKeown N, et al., Whole-grain intake is favorably associated with metabolic risk factors for type 2 diabetes and cardiovascular disease. *Am J Clin Nutr.* 2002;76:390–8.

54. Reis JP, et al. Lifestyle factors and risk for new-onset diabetes: A population-based cohort study. *Ann Intern Med.* 2011;155:292–99.

55. Goto A, et al. Caffeine and type 2 diabetes on postmenopausal women. *Diabetes.* 2011;60:269–75.

56. Mitri J, et al., *et al.* Effects of vitamin D and calcium supplementation on pancreatic β cell function, insulin sensitivity, and glycemia in adults at high risk of diabetes: The calcium and vitamin D for diabetes mellitus randomized controlled trial. *Am J Clin Nutr.* 2011;94:486–94.

57. Schienkiewitz A, et al. Body mass index history and risk of type 2 diabetes: Results from the European Prospective Investigation into Cancer and Nutrition (EPIC)—Potsdam Study. *Am J Clin Nutr.* 2006;84:427–33.

58. National Agenda for Public Health Action: A National Public Health Initiative on Diabetes and Women's Health. Available at: http:// diabetes.webmd.com/tc/type-1-diabetes-cause. Accessed January 14, 2012.

59. Diabetes. Available at: http://diabetes.niddk .nih. gov/dm/pubs/statistics/index.htm#7. Accessed August 2006.

60. Type 1 diabetes-cause. Available at: http:// diabetes.webmd.com/tc/type-1-diabetes-cause. Accessed January 13, 2012.

61. Fillippi CM, et al. Viral triggers for type 1 diabetes: Pros and cons. *Diabetes.* 2008;57:2863–71.

62. Virtanen SM, et al. Infant feeding in Finnish children over the age of 4 years with newly diagnosed IDDM. *Diabetes Care.* 1991;14:415–7.

63. Knip M, et al. Infant feeding and the risk of type 1 diabetes. *Am J Clin Nutr.* 2010;91:1506S–13S.

64. Menato G, et al. Current management of gestational diabetes mellitus. *Expert Rev of Obstet Gynecol.* 2008;3:73–91.

65. Waknine Y. FDA approves insulin patch-pen for use with insulin. Available at: http://www .medscape.com/viewarticle/725976. Accessed January 15, 2012.

66. Zhang C, et al. Effect of dietary and lifestyle factors on the risk of gestational diabetes: Review of epidemiological evidence. *Am J Clin Nutr.* 2011;95(suppl):1975S–9S.

67. Couri CEB, et al. Stem cell transplantation for type 1 diabetes. JAMA 2009;301:1573–9.

68. Dabelea D, et al. Increasing prevalence of gestational diabetes mellitus (GDM) over time and by birth cohort: Kaiser Permanente of Colorado GDM Screening Program. *Diabetes Care.* 2005;28:579–84.

69. Hypoglycemia. Available at: http://diabetes.niddk.nih.gov/dm/pubs/hypoglycemia/hypoglycemia.pdf. Accessed November 18, 2011.

70. Wild S, et al. Global prevalence of diabetes: Estimates for the year 2000 and projections for 2030. *Diabetes Care.* 2004;27:1047–53.

71. Brand-Miller JC, et al. Glycemic index, postprandial glycemia, and the shape of the curve in healthy subjects: Analysis of a database or more than 1000 foods. *Am J Clin Nutr.* 2009;89:97–105.

72. Atkinson FS, Foster-Powell K, Brand-Miller JC. International Tables of Glycemic Index and Glycemic Load Values. *Diabetes Care.* 2008;31:2281–83.

73. International tables of glycemic index and glycemic load values: Update 2008. *Diabetes Care.* 2008;31:2281–3.

74. Gestational diabetes in the United States. http:// Available at: www.cdc.gov/diabetes/pubs/estimates11.htm - 8. Accessed January 14, 2012.

Unit 14

1. Heilig M. Triggering addiction. *The Scientist.* 2008;Dec:28–35.

2. Joosten MM, et al. Combined effect of alcohol consumption and lifestyle behaviors on risk of type 2 diabetes. *Am J Clin Nutr.* 2010;91:1777–83.

3. Neafsey EJ, et al. Moderate alcohol consumption and cognitive risk. *Neuropsychiatric Disease and Treatment.* 2011;7: 465–84.

4. Dietary Guidelines for Americans. 2010. Available at: www.dietaryguidelines.gov.

5. Hamed S, et al. Red wine consumption improves in vitro migration of endothelial progenitor cells in young, healthy individuals. *Am J Clin Nutr.* 2010;92:161–9.

6. Boggs DA, et al. Tea and coffee intake in relation to risk of breast cancer in the Black Women's Health Study. *Cancer Causes Control.* 2010;21:1941–8.

7. Britton A, et al. Alcohol consumption and cognitive function in the Whitehall II Study. *Am J Epidemiol.* 2004;160:240–7.

8. Ferreira MP, et al. Alcohol consumption by aging adults in the United States: Health benefits and detriments. *J Am Diet Assoc.* 2008;108:1668–76.

9. Sacanella E, et al. Down-regulation of adhesion molecules and other inflammatory biomarkers after moderate wine consumption in healthy women: A randomized trial. *Am J Clin Nutr.* 2007;86:1463–9.

10. Ovaskainen M-L, et al. Dietary intake and major food sources of polyphenols in Finnish adults. *J Nutr.* 2008;138:562–9.

11. Goldberg IJ. To drink or not to drink? *N Engl J Med.* 2003;348:163–64.

12. Fedirko V, et al. Alcohol drinking and colorectal cancer risk: an overall and dose–response meta-analysis of published studies. *Ann Oncol.* 2011;22:1958–72.

13. Parker RN. Alcohol and violence: Connections, evidence and possibilities for prevention. *J Psychoactive Drugs.* 2004;Suppl 2:157–63.

14. Markel H. Dying for a drink: Alcohol on campus. *Medscape Public Health and Prevention,* June 7, 2006;4.

15. Cirrhosis. Updated January 2012. Available at: www.nlm.nih.gov/medlineplus/cirrhosis.html. Accessed January 18, 2012.

16. Fetal Alcohol Spectrum Disorders. Available at: www.cdc.gov/Features/FASD. Accessed January 16, 2012.

17. Sood B, et al. Prenatal alcohol exposure and childhood behavior at age 6 to 7 years: I. Dose-response effect. *Pediatrics.* 2001;108(2).

18. Kesse E, et al., Do eating habits differ according to alcohol consumption? *Am J Clin Nutr.* 2001;74:322–7.

19. Frazier TH, et al. Treatment of alcoholic liver disease. *Therap Adv Gastroenterol.* 2011;4:63–81.

20. Martin PR, et al. The role of thiamine deficiency in alcoholic brain disease. *Alcohol Res Health.* 2003;27:134-42.

21. Lieber CS. The influence of alcohol on nutritional status. *Nutr Rev.* 1988;46:241–54.

22. Levitt MD, et al. Use of measurements of ethanol absorption from stomach and intestine to assess human ethanol metabolism. *Am J Physiol Gastrointestinal Liver Physiol.* 1997 vol. 273 no. 4 G951–G957.

23. Alcohol Metabolism: An Update, July, 2007. Available at: http://pubs.niaaa.nih.gov/publications/AA72/AA72.htm. Accessed January 17, 2012.

24. Baik I, et al. Genome-wide association studies identify genetic loci related to alcohol consumption in Korean men. *Am J Clin Nutr.* 2011;93:809–16.

25. Morey TE, et al. Measurement of ethanol in gaseous breath using a miniature gas chromatograph. *J Anal Toxicol.* 2011;35:134–42.

26. Alcohol and the Human Body. Available at: http://www.intox.com/t-Physiology.aspx. Accessed January 19, 2012.

27. Alcohol and your health. Weighing the pros and cons. Available at: www.mayoclinic.com/health/alcohol/SC00024. Accessed August 2006.

28. Franzen J. How do you determine the alcohol proof of a given liquid? *MadSci Network: Chemistry.* February 1998.

29. Distilled spirits labels. Available at: www.ttb.gov/pdf/brochures/p51902.pd*a. Accessed January 18, 2012.*

30. Arria AM, et al. The "high" risk of energy of energy drinks. *JAMA.* 2011;305:600–1.

31. Alcoholism: Getting the facts. National Institute on Alcohol Abuse and Alcoholism. Available at: www.niaa.nih.gov. Accessed June 2003.

32. Moss HB. NESARC findings changing understanding of alcoholism. Available at: www. medscape.com/viewarticle584983. Accessed December 2008.

33. Agrawal A, et al. Genome-wide association studies of alcohol intake—a promising cocktail? *Am J Clin Nutr.* 2011;93:681–3.

34. Alcohol Alert. Underage drinking. Available at: www. niaaa.nih.gov. Accessed January 2006.

35. Fact Sheets: Underage Drinking. Available at: www.cdc.gov/alcohol/fact-sheets/underage-drinking.htm. Updated July 2010. Accessed January 19, 2012.

36. Suter PM. Is alcohol consumption a risk factor for weight gain and obesity? *Crit Rev Clin Lab Sci.* 2005;42:197–227.

37. Women and alcohol. Available at: http://pubs.niaaa.nih.gov/publications/womensfact/womensfact.htm. February 2011. Accessed January 20, 2012.

Unit 15

1. Trappe TA, et al. Influence of concurrent exercise or nutrition countermeasures on thigh and calf muscle size and function during 60 days of bed rest in women. *Acta Physiol (Oxf).* 2007;191:147–59.

2. 2010 Dietary Guideline for Americans. Available at: www.dietaryguidelines.gov.

3. Dietary intake data: What We Eat in America, NHANES 2007–2008. Available at: www.ars.usda.gov/ba/bhnrc/fsrg. Accessed January 19, 2012.

4. Van den Akker CHP, et al. Human fetal albumin synthesis rates during different periods of gestation. *Am J Clin Nutr.* 2008;88:997–1003.

5. Dietary Reference Intakes. Energy, carbohydrate, fiber, fat, fatty acids, cholesterol, protein, and amino acids. Institute of Medicine, National Academy of Sciences, Washington, DC: National Academies Press; 2002.

6. Matthews DE. Proteins and amino acids. In: Shils ME, et al., eds. *Modern Nutrition in Health and Disease,* 9th ed. Philadelphia: Lippincott, Williams & Wilkins; 1998:11–48.

7. Michelfelder AJ. Soy: A complete source of protein. *Am Fam Physician.* 2009;79:43–7.

8. Renwick AG. Establishing the upper end of the range of adequate and safe intakes for amino acids: a toxicologist's viewpoint. *J. Nutr.* 2004;134:1617S–24S.

9. Garlick PJ. Toxicity of methionine in humans. *J Nutr.* 2006;136:1722S–25S.

10. Van de Poll MC, et al. Adequate range for sulfur-containing amino acids and biomarkers for their excess: lessons from enteral and parenteral nutrition. *J Nutr.* 2006;136(6 suppl):1694S–1700S.

11. Kay LK. Melatonin. *Today's Dietitian.* November 2004:54–6.

12. Pencharz PB, et al. An approach to defining the upper safe limits of amino acid intake. *J Nutr.* 2009;138:1995S–2002S.
13. Iglay HB, et al. Resistance training and dietary protein: effects on glucose tolerance and contents of skeletal insulin signaling proteins in older persons. *Am J Clin Nutr.* 2007;85:1005–13.
14. Moore DR, et al. Ingested protein dose response of muscle and albumin protein synthesis after resistance exercise in young men. *Am J Clin Nutr.* 2009; 8:161–8.
15. Pennings B, et al. Whey protein stimulates postprandial muscle protein accretion more effectively than do casein and casein hydrolysate in older men. *Am J Clin Nutr.* 2011;93:997–1005.
16. Pasiakos SM, et al. Leucine-enriched essential amino acid supplementation during moderate steady state exercise enhances postexercise muscle protein synthesis. *Am J Clin Nutr.* 2011 94:809–18.
17. Hayashi Y. Application of the concepts of risk assessment to the study of amino acid supplements. *J Nutr.* 2003;133(6 suppl 1):2021S–24S.
18. Thalacker-Mercer AE, et al. Inadequate protein intake affects skeletal muscle transcript profiles in older humans. *Am J Clin Nutr.* 2007;85:1344–52.
19. Hass VK, et al. Total body protein in healthy adolescent girls: validation of estimates derived from simpler measures with neutron activation analysis. *Am J Clin Nutr.* 2007;85:66–72.
20. Amadi B, et al. Reduced production of sulfated glycosaminoglycans occurs in Zambian children with kwashiorkor but not marasmus. *Am J Clin Nutr.* 2009; 89:592–600.
21. Badaloo AV, et al. Lipid kinetic differences between children with kwashiorkor and those with marasmus. *Am J Clin Nutr.* 2006;83:1283–8.
22. Young VR, et al. Long-term nitrogen balance studies and other criteria for protein requirement estimations. 1981. Available at: http://www.fao.org/DOCREP/MEETING/004/M3021E/M3021E00.HTM. Accessed January 21, 2012.

Unit 16

1. Fraser GE. Vegetarian diets: What do we know of their effects on common chronic diseases? *Am J Clin Nutr.* 2009;89(suppl):1607S–12S.
2. Lanham-New SA. Is "vegetarianism" a serious risk factor of osteoporotic fracture? *Am J Clin Nutr.* 2009;90:910–1.
3. 2010 Dietary Guideline for Americans. Available at: www.dietaryguidelines.gov.
4. Antony AC. Vegetarianism and vitamin B-12 (cobalamin) deficiency. *Am J Clin Nutr.* 2003;78:3–6.

5. Buffonge I. The vegetarian/vegan lifestyle. *West Indian Med J.* 2000;49:17–19.
6. Barr SL, et al. Perceptions and practices of self-defined current vegetarian, former vegetarian, and nonvegetarian women. *J Am Diet Assoc.* 2002;102:354–6.
7. Kushi LH, et al. The macrobiotic diet and cancer. *J Nutr.* 2001;131:3056S–64S.
8. Cunningham E. What is a raw food diet and are there any risks or benefits associated with it? *J Am Diet Assoc.* 2004;104:1623.
9. Koebnick C. Long-term consumption of a raw food diet is associated with favorable serum LDL cholesterol and triglycerides but also with elevated plasma homocysteine and low serum HDL cholesterol in humans. *J Nutr.* 2005;135:2372–8.
10. Fontana L, et al. Low bone mass in subjects with long-term raw vegetarian diet. *Arch Int Med.* 2005;165:684–9.
11. Mutch PB. Food guides for the vegetarian. *Am J Clin Nutr.* 1988;48:913–19.
12. Rauma A-L, et al. Vitamin B-12 status of long-term adherents of a strict, uncooked vegan diet ("living foods diet") is compromised. *J Nutr.* 1995;125:2511–15.
13. Sabate J. The contribution of vegetarian diets to health and disease: a paradigm shift? *Am J Clin Nutr.* 2003;78(suppl):501S–7S.
14. Haddad EH, and Tanzman JS. What do vegetarians in the United States eat? *Am J Clin Nutr.* 2003;78(suppl):626S–32S.
15. Craig WJ. Health effects of vegan diets. *Am J Clin Nutr.* 2009;89(suppl):1627S–33S.
16. Pham LT, et al. Effect of vegetarian diets on bone mineral density: A Bayesian meta-analysis. *Am J Clin Nutr.* 2009;90:943–50.
17. Position of the American Dietetic Association and Dietitians of Canada: Vegetarian diets. *J Am Diet Assoc.* 2003;103:748–65.
18. Dunham L, et al. Vegetarian eating for children and adolescents. *J Pediatr Health Care.* 2006;20:27–34.
19. Craig WJ, et al. Position of the American Dietetic Association: Vegetarian diets. *J Am Diet Assoc.* 2009;109:1266–82.
20. Aronson D. Vegetarian nutrition: What every dietitian should know. *Today's Dietitian.* March 2005;31–7.
21. Harris WS et al. Towards establishing dietary reference intakes for eicosapentaenoic and docosahexaenoic acids. *J Nutr.* 2009;139:804S–19S.
22. Heller L. FDA approves vitamin D fortification of soy foods. Available at: nutraingredientsusa.com. Accessed March 2009.
23. Michelfelder AJ. Soy: A complete source of protein. *Am Fam Physician.* 2009;79:43–7.
24. Thedford K, et al. A vegetarian diet for weight management. *J Am Diet Assoc.* 2011;111:816–28.
25. Messina V et al. A new food guide for North American vegetarian and vegan diets. *J Am Diet Assoc.* 2003;65:771–5.

Unit 17

1. Bahna S, et al. Food allergies over-estimated by public. WHO 2010 International Scientific Conference. Presented December 5, 2010. Available at: www.medscape.com/viewarticle/734305. Accessed January 24, 2012.
2. Somers LS, et al. Peanut allergy: Case of an 11-year old boy with a selective diet. *J Am Diet Assoc.* 2011;111:301–6.
3. Fleischer DM, et al. Oral food challenges in children with a diagnosis of food allergy. *J Pediatr.* 2011;158:578–83.e1.
4. Wahn HU. Strategies for atopy prevention. *J Nutr.* 2008;138:1770S–72S.
5. May CD. Food sensitivity: Facts and fancies. *Nutr Rev.* 1984;42:72–8.
6. Jarvinen KM. Food-induced anaphylaxis. *Curr Opin Allergy Clin Immunol.* 2011;11:255–61.
7. Formanek R. Food allergies: When food becomes the enemy. *FDA Consumer Magazine.* April 2004 update. Available at: www. fda.gov/FDAC/features/2001/401_food. html. Accessed March 2009.
8. Peak I, et al. Physician prescriptions for EpiPen among patients with a history of anaphylaxis. Paper presented at the American Academy of Asthma, Allergy, and Immunology Annual Meeting, March 15, 2009.
9. Bren L. Food labels identify allergens more clearly. *FDA Consumer Magazine,* March–April 2006.
10. Catassi C, et al. Celiac disease, *Curr Opin Gastroenterol.* 2008;24:687–91.
11. Telega G, et al. Asymptomatic patients with celiac disease. *Arch Pediatr Adolesc Med.* 2008;162:164–8.
12. Copelton DA, et al. "You don't need a prescription to go gluten-free": The scientific self-diagnosis of celiac disease. *Soc Sci Med.* 2009;69:623–31.
13. Niewinski MM. Advances in celiac disease and gluten-free diet. *J Am Diet Assoc.* 2008;108:661–72.
14. A Glimpse at "Gluten-Free" Food Labeling. Available at: http://www.fda.gov/ForConsumers/ConsumerUpdates/ucm265212. htm. Accessed January 24, 2012.
15. Burrowes JD. Helping adults with celiac disease to eat well. *Nutr Today.* 2008;43:250–6.
16. Burks AW, et al. NIAID-sponsored 2010 guidelines for managing food allergy: Applications in the pediatric population. *Pediatrics.* October 10, 2011. [online] doi: 10.1542/peds.2011-053
17. Fenton N, et al. Food allergy guidelines. Available at: www.medscape.com/viewarticle/739532?src=mp&spon=17. Accessed June 1, 2011.
18. Tromp II, et al. The introduction of allergenic foods and the development of reported wheezing and eczema in childhood: The

generation R study. *Arch Pediatr Adolesc Med.* 2011;165: 933–38.

19. NIH Consensus and State of the Science Statements. NIH Consensus Development Conference Statement on Lactose Intolerance and Health. February 22–24, 2010;27,

20. Krebs JR. The gourmet ape: Evolution and human food preferences. *Am J Clin Nutr.* 2009;90(supp):707S–11S.

21. Yang WH, et al. Adverse reactions to sulfites. *Can Med Assoc J.* 1985;133:865–7.

22. Jarisch R, et al. Wine and headache. *Int Arch Allergy Immunol.* 1996;110:7–12.

23. Definition of Chinese restaurant syndrome, Available at: www.medterms.com/script/main/art. asp?articlekey=15584. Accessed September 2006.

24. Yamamoto S, et al. Can dietary supplementation of monosodium glutamate improve the health of the elderly? *Am J Clin Nutr.* 2009;90(suppl):844S–9S.

25. Poole JA, et al. Timing of initial exposure to cereal grains and the risk of wheat allergy. *Pediatrics.* 2006;117:2175–82.

Unit 18

1. Dietary Guidelines for Americans, 2010. Part D, Section 3: Fatty Acids and Cholesterol. Available at: www.dietaryguidelines.gov.

2. Kuipers RS, et al. Maternal DHA equilibrium during pregnancy and lactation is reached at an erythrocyte DHA content of 8 g/100 g fatty acids. *J Nutr.* 2011;41:418–27.

3. Maljaars J, et al. Effect of fat saturation on satiety, hormone release, and food intakes. *Am J Clin Nutr.* 2009;89:1019–24.

4. Vos E, et al. n-3 fatty acids and cardiovascular events. *N Engl J Med.* 364:880-1.2011.

5. Swanson D, et al. Omega-3 fatty acids EPA and DHA: Health benefits throughout life. *Adv Nutr.* 2012;3:1–7.

6. Harris WS, et al. Towards establishing dietary reference intakes for eicosapentaenoic and docosahexaenoic acids. *J Nutr.* 2009;139:804S–19S.

7. Belin RJ, et al. Fish intake and the risk of incident heart failure: The Women's Health Initiative. *Circ Heart Fail.* 2011; 4:404–13.

8. FDA affirms position on mercury in fish. Available at: www.fda.gov. Issued June 7, 2006.

9. Position of the American Dietetic Association and Dietitians of Canada: Dietary fatty acids. *J Am Diet Assoc.* 2007;107:1599–1611.

10. Kromhout D, et al. Omega-3 fatty acids protect against arrhythmia and fatal MI. *Diabetes Care.* 2011;34:2515–20.

11. Skulas-Ray AC, et al. Dose-response effects of omega-3 fatty acids on triglycerides, inflammation, and endothelial function in healthy persons with moderate hypertriglyceridemia. *Am J Clin Nutr.* 2011;93:243–52.

12. Larsson SC, et al. Fish consumption and risk of stroke in Swedish women. *Am J Clin Nutr.* 2011;93:487–93. [online] doi: 10.3945/ajcn.110.003871

13. Deckelbaum RJ. n-3 fatty acids. Experimental biology pre-conference satellite presentation, April 8, 2011.

14. Oh R. Practical applications of fish oil (omega-3 fatty acids) in primary care. *J Am Board Fam Pract.* 2005;18:28–36.

15. Nutrient composition of foods, USDA Nutrient Database. Available at: www.ars.usda.gov/nutrientdata and www.ars.usda.gov/ba/bhnrc/ndl. Accessed September 2008.

16. What We Eat in America, NHANES, 2007–08. Available at: www.ars.usda.gov/Services/docs.htm?docid=13793. Accessed February 25, 2011.

17. Harris WS. International recommendations for consumption of long-chain omega-3 fatty acids. *J Cardiovasc Med.* 2007;8(suppl)1:S50–2.

18. Mozaffarian D, et al. Trans fatty acids and cardiovascular disease. *N Engl J Med.* 2006;345:1602–12.

19. Harnack L. Trans fat content of fast foods. 34th National Nutrient Database Conference, July 12–14, 2010.

20. Weighing in on dietary fats. Available at: http://newsinhealth.nih.gov/issue/Dec2011/Feature1. Accessed December 11, 2011.

21. Vannice GK. n-3s from fish and the risk of metabolic syndrome. *J Am Diet Assoc.* 2010;101:1014–7.

22. Behrman EJ, et al. Cholesterol and plants. *J Chem Ed.* 2005;82:1791–-3.

23. Ansorena D, et al. Effect of fish and oil nature on frying process and nutritional product quality. *J Food Sci.* 2010;75:H62–7.

Unit 19

1. Cardiovascular disease. Available at: http://www.who.int/mediacentre/factsheets/fs317/en/index.html, September 2011. Accessed January 31, 2012.

2. Mozaffarian D. The great fat debate: Taking the focus off saturated fat. *J Am Diet Assoc.* 2011;111;665–6.

3. Nabel EG, et al. A tale of coronary artery disease and myocardial infarction. *N Engl J Med.* 2012;366:54–63.

4. Executive Writing Committee. Effectiveness-based guidelines for the prevention of cardiovascular disease in women—2011 update. *J Am Coll Cardio.* 2011;57:1404–23.

5. Hu FB. Are refined carbohydrates worse than saturated fat? *Am J Clin Nutr.* 2010;91:1541–2.

6. WC. The WHI joins MRFIT: A revealing look beneath the covers. *Am J Clin Nutr.* 2010;91:829–30.

7. Mogadam M. HDL: Keep aiming high? Available at: www.medscape.com/viewarticle/745338. Accessed July 1, 2011.

8. Deaths, percent of total deaths, and death rates for the 15 leading causes of death. Available at: www.cdc.gov/nchs/nvss/mortality/lcwk9.htm. Accessed February 3, 2012.

9. Leading causes of death, 2010. Available at: http://www.cdc.gov/nchs/fastats/lcod.htm, Accessed January 31, 2012.

10. Prevalence of Coronary Heart Disease—United States, 2006—2010. *CDC Weekly.* October 14, 2011;60(40);1377–81. Available at: ww.cdc.gov/mmwr/preview/mmwrhtml/mm6040a1.htm. Accessed January 31, 2012.

11. Caballero E. Dyslipidemia and vascular function: The rationale for early and aggressive intervention. Available at: www.medscape.com. Accessed April 18, 2006.

12. Dietary Guidelines for Americans, 2010. Part D, Section 3: Fatty Acids and Cholesterol, Available at: www.dietaryguidelines.gov.

13. Mosca L, et al. Effectiveness-based guidelines for the prevention of cardiovascular disease in women. 2011 update. *Circulation* 2011;123:1243–62.

14. Daniels SR. Lipid concentrations in children and adolescents: It is not all about obesity. *Am J Clin Nutr.* 2011;94:699–700.

15. Stanhope KL, et al. Consumption of fructose and high fructose corn syrup increase postprandial triglycerides, LDL cholesterol, and apolipoprotein-B in young men and women. *J Clin Endoccrinol Metab.* 2011;96:E596–605.

16. Miller M, et al. Triglycerides and cardiovascular disease. *Circulation* 2011 [online] doi:10.1161/CIR.Ob013c3182160726. Accessed April 18, 2011.

17. Hu FB. Are refined carbohydrates worse than saturated fat? *Am J Clin Nutr.* 2010;91:1541–2.

18. Herron KL, et al. The ABCG5 polymorphism contributes to individual responses to dietary cholesterol and carotenoids in eggs. *J Nutr.* 2006;136:1161–5.

19. Ned RM, et al. CDC Expert Commentary Series on Medscape. Available at: www.medscape.com/viewarticle/728541?src=mp&spon=2&auc=61212SX. Accessed September 28, 2010.

20. Frikke-Schmidt R, et al. LCAT, HDL cholesterol, and risk of heart disease. *J Clin Endocrinol Metab.* November 16, 2011.

21. Lumey LH, et al. Prenatal famine and adult health. *Ann Rev Public Health.* 2011;32:237–62.

22. O'Donnell CJ, et al. Genomics of cardiovascular disease. *N Engl J Med.* 2011;365:2098–109.

23. Marin C, et al. Mediterranean diet reduces endothelial damage and improves the regenerative capacity of endothelium. *Am J Clin Nutr.* 2011;93:267–74.

24. Rautianen S, et al. Total antioxidant capacity of diet and risk of stroke: A population-based prospective cohort study of women. [online] doi:10.1162/STAROKEAHAAA.111.635557. Accessed December 1, 2012.

25. Kuipers RS, et al. Maternal DHA equilibrium during pregnancy and lactation is reached at an erythrocyte DHA content of 8 g/100 g fatty acids. *J Nutr.* 2011;41: 418–27

26. van Bussel BCT, et al. Fish consumption in healthy adults in associated with decreased circulating biomarkers of endothelial dysfunction and inflammation during a 6-year follow-up. *J Nutr.* 2011;141:1719–25.

27. Roger V, et al. Heart disease and stroke statistics—2012 update. A report from the American Heart Association. *Circulation.* 2012. Available at: http://circ.ahajournals.org. Accessed December 16, 2011. [online] doi: 10.1161/CIR.0b013e31823ac046.

28. Cholesterol Fact Sheet. Available at: http://www.cdc.gov/dhdsp/data_statistics/fact_sheets/fs_cholesterol.htm. Accessed January 30, 2012.

29. Khera AV, et al. Cholesterol efflux capacity, high-density lipoprotein function, and atherosclerosis. *N Engl J Med.* 2011;364:127–35.

30. Foody JM, et al. Primary prevention in women: the role of lipids. Available at: cme.medscape.com/viewaricle/70787. Accessed August 25, 2009.

31. Munir JA, et al. Relationships between calcified and noncalcified coronary, aortic, and carotid atherosclerosis and risk factors and hormone levels. *Menopause.* 2012;19:10–15.

32. Mackay MH, et al. Gender differences in symptoms of myocardial ischaemia. *Euro Heart J.* 2011;32:31071224.

33. Varady KA, et al. Combination diet and exercise interventions for the treatment of dyslipidemia: An effective preliminary strategy to lower cholesterol levels? *J Nutr.* 2005;135:1829–35.

34. Marinangeli CP, et al. Plant sterols, marine-derived omega-3 fatty acids and other functional ingredients: A new frontier for treating hyperlipidemia. *Nutr Metab (Lond).* 2010;28;7:76.

35. The 10 most prescribed drugs. Available at: www.medscape.com/viewarticle/741526. Accessed April 20, 2011.

36. Armitage J, et al. Study of the effectiveness of additional reductions in cholesterol and homocysteine. (SEARCH) Collaborative Group, *Lancet.* 2010;376:1658–69.

37. Waters DD, et al. Predictors of new-onset diabetes in patients treated with atorvastatin: Results from 3 large randomized clinical trials. *J Am Coll Cardiol.* 2011;57:1535–45.

38. Steinberg D. The statins in preventive cardiology. *N Engl J Med.* 2008;359:1426–8.

39. Shanes JG. A review of the rationale for additional therapeutic interventions to attain lower ldl-c when statin therapy is not enough. *Curr Atheroscler Rep.* 2012;14:33–40.

40. Jenkins DJA, et al. Assessment of the longer-term effects of a dietary portfolio of cholesterol-lowering foods in hypercholesterolemia. *Am J Clin Nutr.* 2006;83:582–91.

41. Jenkins DJ, et al. Effect of a dietary portfolio of cholesterol-lowering foods given at two levels of intensity of dietary advice on serum lipids in hyperlipidemia. *JAMA.* 2011;306:831–9.

42. Van Horn L, et al. The evidence for dietary prevention and treatment of cardiovascular disease. *J Am Diet Assoc.* 2008;108:287–331.

43. Detection, evaluation, and treatment of high blood cholesterol in adults. Adult Treatment Panel III. National Heart, Lung, and Blood Institute of the National Institutes of Health, 2001.

44. American Heart Association. Guidelines for the secondary prevention of cardiovascular disease. *Circulation.* posted online May 15, 2006. Available at: www. medscape.com/viewarticle/532327.

45. ER De Oliverira e Silva, et al. Effects of shrimp consumption on plasma lipoproteins. *Am J Clin Nutr.* 1996;64:712–7.

Unit 20

1. Dietary Guidelines for Americans, 2010. Available at: www.dietaryguidelines.gov.

2. Buzina R, et al. Workshop on functional significance of mild-to-moderate malnutrition. *Am J Clin Nutr.* 1989;50:172–6.

3. Shakur YA, et al. Folic acid fortification above mandated levels results in a low prevalence of folate inadequacy among Canadians. *Am J Clin Nutr.* 2010;92:818–25.

4. Shelke N, et al. Folic acid supplementation for women of childbearing age versus supplementation for the general population: A review of the known advantages and risks. *Int J Family Med.* May 23, 2011:173705. [online] doi: 10.1155/2011/173705.

5. Ganji V, et al. Trends in serum folate, RBC folate, and circulating total homocysteine concentrations in the US. *J Nutr.* 2006;136:153–8.

6. Mayo-Wilson E, et al. Vitamin A supplements for preventing mortality, illness, and blindness in children aged under 5: Systematic review and meta-analysis. *BMJ.* August 25, 2011;343:d5094. [online] doi: 10.1136/bmj.d5094.7.

7. Whitney KM, et al. Management strategies for acne vulgaris. *Clin Cosmet Investig Dermatol.* 2011;4: 41–53.

8. Darlenski R, et al. Topical retinoids in the management of photodamaged skin: From theory to evidence-based practical approach. *Br J Dermatol.* 2010;163:1157–65.

9. Panchaud A, et al. Pregnancy outcome following exposure to topical retinoids: A multicenter prospective study. *J Clin Pharmacol.* December 15, 2011. [online] doi: 10.1177/0091270011429566.

10. Amer M, et al. Relation between serum 25-hydroxyvitamin d and c-reactive protein in asymptomatic adults. *Am J Cardiol.* 2012;109:226–30.

11. Arnson Y, et al. Vitamin D and autoimmunity: New aetiological and therapeutic considerations. *Ann Rheum Dis.* 2007;66:1137–42.

12. Nikooyeh B. Daily consumption of vitamin D– or vitamin D + calcium–fortified yogurt drink improved glycemic control in patients with type 2 diabetes: A randomized clinical trial. *Am J Clin Nutr.* February 2, 2011. [online] doi: 10.3945.

13. Dietary reference intakes: Calcium and vitamin D. Food and Nutrition Board, National Academy of Sciences, Washington, DC: National Academies Press, 2011.

14. Godar DE, et al. Solar UV doses of young Americans and vitamin D_3 production. *Environ Health Perspect.* 2012;120:139–43.

15. Holick MF. Sunlight and vitamin D for bone health and prevention of autoimmune diseases, cancers, and cardiovascular disease. Am J Clin Nutr. 2004;80(6 suppl):1678S–88S.

16. Whiting SJ, et al. Symposium: Optimizing vitamin D intake for populations with special needs: barriers to effective food fortification and supplementation. *J Nutr.* 2006:136:1114–16. 17. Hollis BW. Circulating 25-hydroxyvitamin D levels indicative of vitamin D sufficiency: Implications for establishing a new effective dietary intake recommendations for vitamin D. *J Nutr.* 2005;135:31–22.

18. Gilchrest BA. Sun exposure and vitamin D deficiency. *Am J Clin Nutr.* 2008;88(suppl):570S–7S.

19. Davies JR, et al. The determinants of serum vitamin D levels in participants in a melanoma case-control study living in a temperate climate. *Cancer Causes Control.* 2011;22;1471–82.

20. Douglas RM, et al. Vitamin C for preventing and treating the common cold. *Cochrane Database Syst Rev.* October 18, 2004;(4):CD000980.

21. Talegawkar SA, et al. Total antioxidant performance is associated with diet and serum antioxidants in participants of the diet and physical activity substudy of the Jackson Heart Study. *J Nutr.* 2009;139:1964–71.

22. Bjelakovic G, et al. Mortality in randomized trials of antioxidant supplements for primary and secondary prevention: Systematic review and meta-analysis. Available at: www.medscape.com/viewarticle/553714?src=ptalk. Accessed February 9, 2012.

23. Crowe FL, et al. European prospective investigation into cancer and nutrition. Fruit and vegetable intake and mortality from ischaemic heart disease. *Eur Heart J.* 2011. [online] doi:10.1093/eurheartj/ehq465. Accessed January 21, 2011.

24. USDA Tables of Nutrient Retention in Foods, Release 6 (2007). Available at: www.ars.usda.gov/Services/docs.htm?docid=9448.

25. Nutrition's dynamic duos. *Harvard Health Newsletter.* June 2009. Available at: www.health.harvard.edu/newsletters/Harvard_Health_Letter/2009/July/Nutritions-dynamic-duos. Accessed February 14, 2012.

Unit 21

1. Bjelakovic G, et al. Mortality in randomized trials of antioxidant supplements for primary and secondary prevention: systematic review and meta-analysis. www.medscape.com/viewarticle/553714?src=ptalk. Accessed February 9, 2010.rlsen MH, et al. The total antioxidant content of more than 3,100 foods, beverages, spices, herbs and supplements used worldwide. Nutr J. January 22, 2010;9:3 [online] doi: 10.1186/1475-2891-9-3.
3. Murphy MM, et al. Phytonutrient intake by adults in the United States in relation to fruit and vegetable consumption. *J Acad Nutr Diet.* 2012;112:222–9.
4. Tsumbu CN, et al. Polyphenol content and modulatory activities of some tropical dietary plant extracts on the oxidant activities of neutrophils and myeloperoxidase. *Int J Mol Sci.* 2012;13: 628–50.
5. Chang CH, et al. Phytochemical characteristics, free radical scavenging activities, and neuroprotection of five medicinal plant extracts. *Evid Based Complement Alternat Med.* 2012:984295.
6. Hamel S, et al. Red wine consumption improves in vitro gene migration of progenitor cells in young, healthy individuals. *Am J Clin Nutr.* 2010; 92:161–9.
7. Larsson SC, et al. Coffee consumption and risk of stroke: A dose-response meta-analysis of prospective studies. *Am J Epidemiol.* 2011;174:993–1001.
8. Clarke JD, et al. Comparison of isothiocyanate metabolite levels and histone deacetylase activity in human subjects consuming broccoli sprouts or broccoli supplement. *Agric Food Chem.* September 19, 2011.
9. Martin C, et al. How can research on plants contribute to promoting human health? *Plant Cell.* 2011;23:1685–99.
10. Murphy MM, et al. Phytonutrient intake by adults in the United States in relation to fruit and vegetable consumption. *J Acad Nutr Diet.* 2012;112:222–9.
11. Chang CH, et al. Phytochemical characteristics, free radical scavenging activities, and neuroprotection of five medicinal plant extracts. *Evid Based Complement Alternat Med.* 2012; [online] doi:10.1155/2012/984295.
12. Ma L, et al. Lutein and zeaxanthin intake and the risk of age-related macular degeneration: A systematic review and meta-analysis. *Br J Nutr.* 2012;107:350–9.
13. Yang Y, et al. Effects of some common food constituents on cardiovascular disease. *ISRN Cardiol.* June 16, 2011:397136. [online] doi: 10.5402/2011/397136.

14. Buitrago-Lopez A, et al. Chocolate consumption and cardiometabolic disorders: Systematic review and meta-analysis. *BMJ.* August 29. 2011;343:d4488. [online] doi: 10.1136/bmj.d4488.
15. Tylleskar T, et al. Dietary determinants of a non-progressive spastic paparesis (konzo). *Int J Epidemiol.* 1995;24:949–56.
16. Clarke JD, et al. Comparison of isothiocyanate metabolite levels and histone deacetylase activity in human subjects consuming broccoli sprouts or broccoli supplement. *Agric Food Chem.* September 19, 2011. [online] doi:10.1021/jf202887c.
17. Higdon JV, et al. Cruciferous vegetables and human cancer risk: Epidemiologic evidence and mechanistic basis. *Pharmacol Res.* 2007;55:224–36.
18. Salo J, et al. Cranberry juice for the prevention of recurrences of urinary tract infections in children: A randomized placebo-controlled trial. *Clin Infect Dis.* 2012;54:340–46.
19. Sengupta K, et al. A randomized, double-blind, controlled, dose dependent clinical trial to evaluate the efficacy of a proanthocyanidin standardized whole cranberry (vaccinium macrocarpon) powder on infections of the urinary tract. *Current Bioactive Compounds.* 2011;7:39–46.
20. Kang NJ, et al. Coffee phenolic phytochemicals suppress colon cancer metastasis by targeting MEK and TOPK. *Carcinogenesis.* 2011;32:921–8.
21. Patil H, et al. Cuppa joe: friend or foe? Effects of chronic coffee consumption on cardiovascular and brain health. *Mo Med.* 2011;108:431–8.
22. Sartorelli DS, et al. Differential effects of coffee on the risk of type 2 diabetes according to meal consumption in a French cohort of women: the E3N/EPIC cohort study. *Am J Clin Nutr.* 2010 :91;1002–12.
23. Wedick NM, et al. Effects of caffeinated and decaffeinated coffee on biological risk factors for type 2 diabetes: A randomized controlled trial. *Nutr J.* 2011. September 13, 2011. [online] doi: 10.1186/1475-2891-10-93.
24. Youjin J, et al. A prospective cohort study of coffee consumption and risk of endometrial cancer over a 26-year follow-up. *Cancer Epidemiol Biomarkers Prev.* November 22, 2011. [online] doi: 10.1158/1055-9965. EPI-11-0766.
25. Hamza TH, et al Genome-wide gene-environment study identifies glutamate receptor gene *grin2a* as a Parkinson's disease modifier gene via interaction with coffee. August 11, 2011. Available at: www.plosgenetics.org/article/info%3Adoi%2F10.1371%2Fjournal.pgen.1002237.
26. Gorby HE, et al. Do specific dietary constituents and supplements affect mental energy? Review of the evidence. *Nutr Rev.* 2010;68:697–718.

27. Lucas M, et al. Coffee, caffeine, and risk of depression among women. *Arch Intern Med.* 2011;171:1571–78.
28. Brent RL, et al. Evaluation of the reproductive and developmental risks of caffeine. *Birth Defects Res B Dev Reprod Toxicol.* 2011;92:152–87.
29. Zhang Z, et al. Habitual coffee consumption and risk of hypertension: A systematic review and meta-analysis of prospective observational studies. *Am J Clin Nutr.* 2011;93:1212–9.
30. Merideth A, et al. The effect of daily caffeine use on cerebral blood flow: How much caffeine can we tolerate? *Hum Brain Mapp.* 2009;30:3102–14.
31. Harrison L. Surge reported in energy drink emergency department visits. Available at: http://www.medscape.com/viewarticle/754146?src=mp&spon=42. Accessed December 12, 2011.
32. Arria AM, et al. The "high" risk of energy drinks. *JAMA.* 2011;305:600–1.
33. Canada to limit caffeine in energy drinks (Reuters). October 7, 2011. Available at: www.medscape.com/viewarticle/751155?src=mp&spon=42OTTAWA.
34. Dietary Guidelines for Americans, 2010. Available at: www.dietaryguidelines.gov.
35. Stull AP, et al. Bioactives in blueberries improve insulin sensitivity in obese, insulin-resistant men and women. *J Nutr.* 2010;140:764–1768.
36. Ghanim H, et al. An antiinflammatory and reactive oxygen species suppressive effects of an extract of *polygonum cuspidatum* containing resveratrol. *J Clin Endocrinol Metab.* 2010;95: E1–E8. June 9, 2010. [online]. doi: 10.1210/jc.2010-0482.
37. Sengupta K, et al. A randomized, double blind, controlled, dose dependent clinical trial to evaluate the efficacy of a proanthocyanidin standardized whole cranberry (vaccinium macrocarpon) powder on infections of the urinary tract. *Current Bioactive Compounds.* 2011;7:39–46.
38. Knab AM, et al. Influence of quercetin supplementation on disease risk factors in community-dwelling adults. *J Am Diet Assoc.* 2011;111:542–9.
39. Dolan LC, et al. Naturally occurring food toxins. *Toxins* (Basel). 2010;2:2289–32.
40. Uribarri J, et al. Advanced glycation end products in foods and a practical guide to their reduction in the diet. *J Am Diet Assoc.* 2010;110:911–16.e12.

Unit 22

1. Shannon K, et al. Genetics, epigenetics, and leukemia. *N Engl J Med.* 2010;363:2460–3.
2. Christiani DC. Combating environmental causes of cancer. *N Engl J Med.* 2011;364;791–4.

3. Russo M, et al. Phytochemicals in cancer prevention and therapy: Truth or dare? *Toxins* (Basel). 2010;2:517–55.

4. Peto R. The fraction of cancer attributable to lifestyle and environmental factors in the UK in 2010. *British Journal of Cancer.* 2011;105:S1–S1.

5. Hoeijmakers JHJ. DNA damage, aging, and cancer. *N Engl J Med.* 2009;361:1475–84.

6. Lichtenstein P, et al. Environmental and heritable factors in the causation of cancer. *N Engl J Med.* 2000;343:78–85.

7. American Cancer Society. Cancer facts and figures 2008 Available at: www.cancer.org. Accessed April 3, 2009; Finn OJ. Cancer immunology. *N Engl J Med.* 2008;358:2704–15.

8. Wood S. Cancer deaths overtake heart disease in Canada. Available at: www.medscape.com/viewarticle/752829?src=rss. Accessed November 2011.

9. Kochanek KD, et al. Pediatrics. January 30, 2012. [online] doi: 10.1542/peds.2011-3435.

10. Danaei G, et al. One-third of cancer deaths may be attributable to nine modifiable risk factors. *Lancet.* 2005;366:1784–93.

11. World Cancer Research Fund/American Institute for Cancer Research. Food, nutrition, physical activity, and the prevention of cancer: A global perspective, Washington, D.C.: AIRC, 2007. Available at: www.medscape.com/viewarticle/565197. Accessed April 2009.

12. New dietary guidelines could reduce cancer risk. Available at: www.medscape.com/viewarticle/736888. Accessed February 2, 2011.

13. R Peto. The fraction of cancer attributable to lifestyle and environmental factors in the UK in 2010. *British Journal of Cancer.* 2011;105:S1–S1.

14. Hoover RN. Cancer-nature, nurture, or both. *N Eng J Med.* 2000;343:135–6.

15. Jackson AA. Integrating the ideas of life across cellular, individual, and population levels in cancer causation. *J Nutr.* 2006;135:2927S–33S.

16. Siegel R, et al. Cancer statistics. *Cancer J Clin.* January 4, 2012. [online]. doi. 10.3322/caac.20138.

17. Büchner FL, et al. Fruits and vegetables consumption and the risk of histological subtypes of lung cancer in the European Prospective Investigation into Cancer and Nutrition (EPIC). *Cancer Causes Control.* 2010;21:357–71.

18. Konijeti R, et al. Chemoprevention of prostate cancer with lycopene in the tramp model. *Prostate.* 2010;70:1547–54.

19. Larsson SC, et al. Tea consumption and the risk for ovarian cancer. *Arch Intern Med.* 2005;165:2683–6.

20. Byers T. Anticancer vitamins du jour—the ABCED's so far. *Am J Epidemiol.* 2010;172:1–3.

21. Kristal AR, et al. Diet, supplement use, and prostate cancer risk: Results from the prostate cancer prevention trial. *Am J Epidemiol.* 200;172:566–77.

22. Li F, et al. Human gut bacteria communities are altered by addition of cruciferous vegetables to a controlled fruit- and vegetable-free diet. *J Nutr.* 2009;139:1685–91.

23. Syed Alwi, SS. In vivo modulation of 4E binding protein 1 (4E-BP1) phosphorylation by watercress: A pilot study. *Br J Nutr.* 2010;104:1288–96.

24. Bouayed J, et al. Exogenous antioxidants—Double-edged swords in cellular redox state. *Oxid Med Cell Longev.* July–August 2010;3(4):228–37.

25. Lü J-M, et al. Chemical and molecular mechanisms of antioxidants: Experimental approaches and model systems. *J Cell Mol Med.* 2010;14:840–60.

26. Shukla V, et al. Oxidative stress in neurodegeneration. *Adv Pharmacol Sci.* September 21, 2011. [online] doi: 10.1155/2011/572634. Accessed September 29, 2011.

27. Straus SE. Complementary and alternative medicine and cancer: What you should know. Available at: www.peoplelivingwithcancer.org. Accessed October 2003.

Unit 23

1. Mahoney DH Jr. Anemia in at-risk populations—what should be our focus? *Am J Clin Nutr.* 2008;88:1457–8.

2. Dietary Guidelines for Americans, 2010. Available at: www.dietaryguidelines.gov.

3. Prasad AS. Discovery and importance of zinc in human nutrition. *Federation Proc.* 1984;43:2829–34.

4. Thankachan P, et al. Iron absorption in young Indian women: the interaction of iron status with the influence of tea and ascorbic acid. *Am J Clin Nutr.* 2008;87:881–6.

5. USDA Tables of Nutrient Retention in Foods, Release 6 (2007). Available at: www.ars.usda.gov/Services/docs.htm?docid=9448.

6. Nordin BEC. Calcium absorption revisited. *Am J Clin Nutr.* 2010;92:673–4.

7. Raisz LG. Screening for osteoporosis. *N Engl J Med.* 2005;353:164–5.

8. Bone Health and Osteoporosis: A Report of the Surgeon General. Chapter 4: The Frequency of Bone Disease. Available at: www.surgeongeneral.gov/library/bonehealth/chapter_4.html. Accessed February 6, 2012.

9. Prestwood KM, et al. Prevention and treatment of osteoporosis. *Clin Cornerstone.* 2002;4:31–41.

10. Maximizing peak bone mass: Calcium supplementation increases bone mineral density in children. *Nutr Rev.* 1992;50:335–7.

11. Chung M, et al. Vitamin D with or without calcium supplementation for prevention of cancer and fractures: An updated meta-analysis for the U.S. Preventive Services Task Force. *Ann Intern Med.* 2011;155:827–38.

12. Tice JA. Vitamin D for the prevention of osteoporotic fractures: A technology assessment. August 4, 2011. Available at: www.medscape.com/viewarticle/744333. Accessed March 7, 2012.

13. Bischoff-Ferrari HA, et al. Effect of calcium supplementation on fracture risk: A double-blind randomized controlled trial. *Am J Clin Nutr.* 2008;87:1945–52.

14. Clinche R. DRIs for calcium and vitamin D: IOM report. Webinar, December 16, 2010. [online] www.nutrition.org.

15. Heaney RP, et al. Phosphate and carbonate salts of calcium support robust bone building in osteoporosis. *Am J Clin Nutr.* 2010;92:101–5.

16. Moyad, MA. Vitamin D: A rapid review. *Dermatology Nursing.* 2009;21(1). Available at: www.medscape.com/viewarticle/589256. Accessed March 7, 2012.

17. Brannon PM, et al. Vitamin D in pregnancy and lactation in humans. *Ann. Rev. Nutr.* 2011. 31:89–115.

18. Dietary Supplement Fact Sheet: Vitamin D. Available at: ods.od.nih.gov/factsheets/VitaminD-HealthProfessional. Accessed March 7, 2012.

19. What We Eat in America, NHANES, 2007–2008. Available at: www.ars.usda.gov/Services/docs.htm?docid=13793. Accessed February 25, 2011.

20. Osteoporosis fact sheet. Available at: www.womenshealth.gov/publications/our-publications/fact-sheet/osteoporosis.cfm#. Accessed March 7, 2012.

21. NAMS. Management of osteoporosis in postmenopausal women: 2010 position statement of the North American Menopause Society, Menopause 2010;17:25–54.

22. Sahni S, et al. Protective effect of high protein and calcium intake on the risk of hip fracture in the Framingham offspring cohort. *J Bone Miner Res.* 2010;25:2770–76.

23. Sweet MG, et al. Review of osteoporosis management. *Am Fam Physician.* 2009;79:193–200.

24. Stofzfus RJ. Iron interventions for women and children in low-income countries. *J Nutr.* 2011;141:756S–62S.

25. Fairbanks VF. Iron in medicine and nutrition. In: Shils ME, et al, eds. *Modern Nutrition in Health and Disease,* 9th ed. Philadelphia: Lippincott Williams & Wilkins; 1999:193–221.

26. Lozoff B. Early iron deficiency has brain and behavior effects consistent with dopaminergic dysfunction. *J Nutr.* 2011;141:740S–6S.

27. Park J, et al. Increased iron content of food due to stainless steel cookware. *J Am Diet Assoc.* 1997;97:659–61.

28. Monsen ER, et al. Estimation of available dietary iron. *Am J Clin Nutr.* 1978;31:134–41.

29. Shannon M. Ingestion of toxic substances by children. *N Engl J Med.* 2000;342:186–89.

30. Spanierman CS, et al. Iron toxicity in emergency medicine. Available at: emedicine.medscape.com/article/815213-overview. Accessed March 7, 2012.

31. Fryhofer SA. Salt from the sea or the earth: Is one better? August 25, 2011. Available at: www.medscape.com/viewarticle/748320. Accessed March 8, 2012.

32. Zhang W, et al. How much iodine is too much? *Am J Clin Nutr.* Available at: http://bit.ly/zQWJF4. Accessed December 28, 2011.

33. McCarron DA, et al. Science trumps politics: Uriary sodium data challenge US dietary sodium guideline. *Am J Clin Nutr.* 2010;92:1005–6.

34. Todd AS, et al, Dietary salt loading impairs arterial vascular reactivity. *Am J Clin Nutr.* 2010;91:557–64.

35. de la Sierra A, et al. Clinical features of 8,295 patients with resistant hypertension classified on the basis of ambulatory blood pressure monitoring. *Hypertension.* March 28, 2011. [online] doi: 10.1161/HYPERTENSIONAHA.110.168948. Accessed March 8, 2012.

36. Dugdale DC. Hypertension. Update. June 10, 2011.Available at: www.nlm.nih.gov/medlineplus/ency/article/000468.htm. Accessed March 7, 2012.

37. Graudal N, et al. Cochrane review of sodium restriction and health. *Am J Hypertension.* 2011. Available at: http://bit.ly/sM39PG.

38. Salt sensitivity and hypertension in African Americans. 2007. Available at: www.medscape.com/viewarticle/407741_3. Accessed March 7, 2012.

39. Kawasaki T, et al. The effect of high-sodium and low-sodium intakes on blood pressure and other related variables in human subjects with idiopathic hypertension. *Am J Med.* 1978;64:193–98.

40. Rhee M-Y, et al. Novel genetic variations associated with salt sensitivity in the Korean population. *Hypertension Res.* January 13, 2011;34:606–611. [online] doi:10.1038/hr.2010.278.

41. NHLBI issues new high blood pressure clinical practice guidelines. NHLBI Communications Office, May 14, 2003. Available at: www.nhlbi.nih.gov/index.htm.

42. Fung T, et al. DASH-style diet and the risk of heart disease and stroke. American Heart Association 2007 Scientific Sessions: Abstract 2369. November 5, 2007. Available at: www.medscape.com/ viewarticle/566449.

43. Apple LJ. The science supporting the 2005 Dietary Guidelines for Americans: Sodium, potassium, and physical activity. Available at: www.healthierus.gov. Accessed September 2005.

44. Gunn JP, et al. Sodium intake among adults— United States, 2005–2006, MMWR June 25, 2010;59:746–9.

45. Conlin PR, et al. DASH diet can control stage 1 hypertension. *Am J Hypertens.* 2000;13:949–55.

46. Ervin RB, et al. Dietary intake of selected minerals for the United States population: 1999–2000. *Advanced Data,* no. 341, April 27, 2004.

47. Sodium in Canada. December 9, 2011. Available at: www.hc-sc.gc.ca/fn-an/nutrition/sodium/index-eng.php. Accessed March 7, 2012.

48. Weaver CM, et al. Choices for achieving adequate dietary calcium with a vegetarian diet. *Am J Clin Nutr.* 1999;70:543S–548S.

Unit 24

1. Bailey RL et al. Dietary supplement use in the United States, 2003–2006. *J Nutr.* 2011;141:261–6.

2. Denham BE. Dietary supplements—regulatory issues and implications for public health. JAMA July 5, 2011. [online] doi:10.1001/jama.2011.982.

3. De Smet P. Herbal remedies. *N Engl J Med.* 2002;347:2046–56.

4. Beware of Fraudulent Weight-Loss "Dietary Supplements," March 15, 2011. Available at: http://www.fda.gov/ForConsumers/ConsumerUpdates/ucm246742.htm#2. Accessed December 17, 2011.

5. Obasanjo O. Dietary supplements and the clinical encounter, November 14, 2011.

6. Park SY, et al. Multivitamin use and the risk of mortality and cancer incidence. *Am J Epidemiol.* March 2011. [online] doi:10.1093/aje/kwq447.

7. Council for Responsible Nutrition: Consumer Survey on Dietary Supplements. Available at: www.crnusa.org/CRNPR2009CRNConsumerSurvey_UsageConfidence.html, 2009.

8. Kapsak WR, et al. Functional foods: Consumer attitudes, perceptions, and behaviors in a growing market. *J Am Die Assoc.* 2011;111:804, 806–10.

9. Raynor DK, et al. Buyer beware? Does the information provided with herbal products available over the counter enable safe use? *BMC Med.* 2011;9:94. [online] doi:10.1186/1741-7015-9-94.

10. Cohen PA. American roulette—contaminated dietary supplements. *N Engl J Med.* 2009;361:1523–6.

11. Anastasi JK, et al. Herbal supplements: Talking with your patients. *J Nurse Pract.* 2011;7:29–35.

12. Fletcher RH, et al. Vitamins for chronic disease prevention in adults: Clinical applications. *JAMA.* 2002;287:3127–9.

13. Yetley EA. Multivitamin and multimineral dietary supplements: Definitions, characterization, bioavailability, and drug interactions. *Am J Clin Nutr.* 2007; 85:269S–76S.

14. Dietary Guidelines for Americans, 2010. Available at: www.dietaryguidelines.gov.

15. Mursu J, et al. Best evidence review of dietary supplements and mortality rates in older women. *Arch Intern Med.* 2011;171:1625–33.

16. Willett WC, et al. What vitamins should I be taking, doctor? *N Engl J Med.* 2001;345:1819–24.

17. Fairfield KM, et al. Vitamins for chronic disease prevention in adults: Scientific review. *JAMA.* 2002;9:288:1720–4.

18. Gardiner P, et al. Factors Associated with herb and dietary supplement use by young adults young adults in the United States. *BMC Complem Altern Med.* 2007. [online] 7:39 doi:10.1186/1472-6882-7-39.

19. Chin Y-W, et al. Drug discovery from natural sources. *AAPS Journal.* 2006;8:E239–E253. [online] doi:10.1208/aapsj080228.

20. Herbs and Supplements. Available at: www.nlm.nih.gov/medlineplus/druginfo/herb_All.html. Accessed February 2012.

21. Halbert SC, et al. Toxicity of red yeast rice (2,400 mg twice a day) versus pravastatin (20 mg twice a day) in patients with previous statin intolerance. *Am J Cardiol.* 2010;105:198–204.

22. Red yeast. Available at: www.nlm.nih.gov/medlineplus/druginfo/natural/925.html#Safety. Accessed March 10, 2012.

23. Harkey MR, et al. Variability in commercial ginseng products: An analysis of 25 preparations. *Am J Clin Nutr.* 2001;73:1101–60.

24. Gilroy CM, et al. Echinacea and truth in labeling. *Arch Intern Med.* 2003;163:699–704.

25. HHS Statement: FDA statement concerning structure/function rule and pregnancy claims. January 6, 2000. [online] fda.gov.

26. Substantiation for dietary supplement claims made under section 403 (r) (6) of the Federal Food, Drug, and Cosmetic Act, December 2008. Available at: www.fds/cfsan.gov. Accessed April 2009.

27. Use of herbal supplements in chronic kidney disease. Available at: www.kidney.org/atoz/content/herbalsupp.cfm. Accessed March 10, 2012.

28. About herbs, botanical, and other products. Available at: www.mskcc.org/mskcc/html/11570.cfm. Accessed October 2006.

29. Hyman P. Claims for functional foods under the current food regulatory scheme. *Nutr Today.* 2002;37:217–9.

30. Position of the American Dietetic Association: Functional foods. 2009;109:735–46.

31. Douglas DC, et al. Probiotics and prebiotics in dietetics practice. *J Am Diet Assoc.* 2008;108:510–21.

32. Szajewska H, et al. The Effects of *Lactobacillus rhamnosus* GG supplementation for the prevention of healthcare-associated diarrhoea in children. *Alimen Pharmacol Ther.* 2011;34:1079–87.

33. Kligler B, et al. Probiotics. *Am Fam Physician.* 2008;78:1073–8.

34. Get the facts: An introduction to probiotics. Available at: www.nccam.nih.gov. Accessed April 2009.

35. Boyle RJ, et al. Risks of probiotic treatment. *Am J Clin Nutr.* 2006;83:1254–64.

36. USP Dietary Supplement Standards. Available at: www.usp.org/dietary-supplements/ overview. Accessed February 2012.

37. USP–NF. Available at: www.usp.org/usp-nf. Accessed February 2012.

38. Morris, et al. Herbal remedy sellers. Available at: www.nlm.nih.gov/medline plus/news/ fullstory 14069.html.

39. Raynor DK, et al. Buyer beware? Does the information provided with herbal products available over the counter enable safe use? *BMC Med.* 2011;9:94. [online] doi:10.1186/1741-7015-9-94.

40. Wallace TC, et al. The safety of probiotics: Considerations following the 2011 U.S. Agency for Health Research and Quality Report. *J Nutr.* September 14, 2011. [online] doi:10.3945/jn.111.147629.

41. Tainted sexual enhancement products. Available at: www.fda.gov/Drugs/ ResourcesForYou/Consumers/ BuyingUsingMedicineSafely/ MedicationHealthFraud/ucm234539.htm. Accessed March 11, 2012.

42. Pray W. Dangers of sexual enhancement supplements. *US Pharmacist.* 2007;32:10–15.

Unit 25

1. Mendoza RA, et al. Bottled water a risk factor for early childhood caries. APHA 137th Annual Meeting: Abs. no. 204297, presented November 8, 2009.

2. Noakes TD. Too many fluids as bad as too few. *BMJ.* 2003;327:113–4.

3. Shmerling RH. How do you know if you're dehydrated? Available at: http://www.intelihealth.com/IH/ ihtPrint/WSIHW000/24479/29730. html?hide=t&k=base. Accessed February 22, 2011.

4. Dietary Reference Intakes for Water, Potassium, Sodium Chloride, and Sulfate. Food and Nutrition Board, National Academy of Science. http://books.nap. edu. Accessed May 2004.

5. Markowitz DL. Fluoride supplementation: The ongoing debate. Available at: www. medscape.com/viewarticle/753193?src=mp&s pon=17. Accessed February 18, 2012.

6. Fluoridation of U.S. municipal water supplies. *MMWR.* 2008;57:737–41.

7. Fiske H. Measuring water's benefits and optimal intake recommendations. Today's Dietitian. January 2003;22–4.

8. Dietary Guidelines for Americans, 2010. Available at: www.dietaryguidelines.gov.

9. Askew EW. Water. In: Ziegler EE, Filer LJ Jr., eds. Present knowledge in nutrition. Washington, DC: ILSI Press;1996:98–108.

10. Kant AK, et al. Intakes of plain water, moisture in foods and beverages, and total water in the adult US population—nutritional, meal pattern, and body weight correlates: National Health and Nutrition Examination Surveys 1999–2006. *Am J Clin Nutr.* 2009;90:655–63.

11. Cullen H. In: The future of water. *Discover.* December 2011;47–53.

12. Myer SS, et al. The coming health crisis. *The Scientist,* January 2011;32–37.

13. Cheng JJ, et al. An ecological quantification of the relationships between water, sanitation and infant, child, and maternal mortality. *Environ Health.* 2012;11:4. [online] doi:10.1186/1476-069X-11-4. Accessed February 16, 2012,

14. HHS and EPA announce new scientific assessments and actions on fluoride. January 7, 2011. Available at: www.hhs.gov/news/ press/2011pres/01/20110107a.html. Accessed March 13, 2012.

15. Current research on the bottled water manufacturing industry. Available at: www. anythingresearch.com/industry/Bottled-Water-Manufacturing.htm. Accessed March 13, 2012.

16. Boyles S. Bottled water: FAQ on safety and purity. November 7, 2008. Available at: www. webmd.com/food-recipes/news/20081107/ bottled-water-faq-on-safety and-purity. Accessed March 14, 2012.

17. Student movement seeks to ban sale of bottled water on campuses. Available at: www.bgnews.com/campus/student-movement-seeks-to-ban-sale-of-bottled-water-on/article_39209f2a-6cc9-11e1-846c-0019bb2963f4.html. Accessed March 13, 2012.

18. Picco M. Does drinking water during or after a meal disturb digestion? Available at: www. mayoclinic.com/health/digestion/ AN01776. Accessed April 29, 2009.

19. Le HH, et al. Bisphenol A is released from polycarbonate drinking bottles and mimics the neurotoxic actions of estrogen in developing cerebellar neurons. *Toxicol Lett.* 2008;30;176:149–56.

20. Cooper JE, et al. Assessment of bisphenol A released from reusable plastic, aluminum and stainless steel water bottles. *Chemosphere.* 2011;85:943–7.

21. Hitti M. Baby bottle makers ditch BPA. Available at: www.medscape.com/ viewarticle/589362. Accessed March 10, 2009.

22. Heaney RP. Absorbability and utility of calcium in mineral waters. *Am J Clin Nutr.* 2006;84:371–4.

23. Wingo JE. Isolated effects of elevated temperatures on sweating in humans.

Presented at Experimental Biology Annual Meetings, New Orleans, April 19, 2009.

24. Grandjean AC, et al. Hydration: Issues for the 21st century. *Nutr Rev.* 2003;61:261–71.

25. Haskins J. A third of the world population faces water scarcity today. Available at: www .eurekalert.com. Accessed August 21, 2006.

26. Horswill CA, et al. Hydration and health. *Lifestyle Med.* 2011;5:304–15.

27. Goldman MB, et al. Mechanisms of altered water metabolism in psychotic patients with polydipsia and hyponatremia. *N Engl J Med.* 1988;318:397–403.

28. Quock RL, et al. Fluoride content of bottled water and its implications for the general dentist. *Gen Dent.* 2009;57:29–33.

Unit 26

1. Temelkova-Kurktschiev T, et al. Lifestyle and genetics in obesity and type 2 diabetes. *Exp Clin Endocrinol Diabetes.* 2012;120:1–6.

2. Stover PJ. Influence of human genetic variation on nutritional requirements. *Am J Clin Nutr.* 2006;83(suppl):436S–42S.

3. Gohlke JM, et al. Genetic and environmental pathways to complex diseases. *BMC Systems Biology.* 2009;3:46. [online]. doi:10.1186/1752-0509-3-46.

4. Human Genome Project Information. September 19, 2011. Available at: www.ornl .gov/sci/techresources/Human_Genome/faq/ faqs1.shtml#genetics. Accessed March 18, 2012.

5. Lee C, et al. Structural genomic variation and personalized medicine. *N Engl J Med.* 2008;358:740–1.

6. Weinhold B. Epigenetics: The science of change. *Environ Health Perspect.* 2006;114:A160–A167.

7. History of Forensic DNA Analysis. Available at: www.dna.gov/basics/analysishistory. Accessed March 16, 2012.

8. Rajpathak SN, et al. Lifestyle factors of people with exceptional longevity. *J Am Geriatric Soc.* August 3, 2011. [online] doi: 10.1111/j.1532-5415.2011.03498.

9. Sebastiani P, et al. Whole genome sequences of a male and female supercentenarian, ages greater than 114 years. *Front Gene.* 2012;2:90. [online] doi: 10.3389/ fgene.2011.00090.

10. Moore JP, et al. Proteomics and systems biology: Current and future applications in the nutritional sciences. *Adv Nutr.* 2011;2:355–64.

11. Depeint F, et al. Mitochondrial function and toxicity: Role of B vitamins on the one-carbon transfer pathways. *Chem-Biol Interactions.* 2006;163:113–32.

12. Choi S-W, et al. Epigenetics: A new bridge between nutrition and health. *Adv Nutr.* 2010. Available at: advances.nutrition.org/ content/1/1/8.full. Accessed November 29, 2010.

13. Gene sequencing for all, even Neanderthals. *Science News*. January 1, 2011, pp. 28–30.

14. Blum S, et al. Vitamin E reduces cardiovascular disease in individuals with diabetes mellitus and the haptoglobin 2-2 genotype. *Pharmacogenomics*. 2010;11:65–84.

15. Davis JN, et al. Increased hepatic fat in overweight Hispanic youth influenced by interaction between genetic variation in PNPLA3 and high dietary carbohydrate and sugar consumption. *Am J Clin Nutr*. 2010;92;1522–27.

16. Scrimshaw NS. The relation between fetal malnutrition and chronic disease in later life. *BMJ*. 1997;315:825.

17. Stover PJ. One-carbon metabolism-genome interactions in folate-associated pathologies.

18. Sanchez-Moreno C, et al. APOA5 gene variant interacts with dietary fat intake to modulate obesity and circulating triglycerides in a Mediterranean population. *J Nutr*. 2011;141:380–5.

19. Cornelis MC, et al. GSTT1 genotype modifies the association between cruciferous vegetable intake and the risk of myocardial infarction. *Am J Clin Nutr*. 2007;86:752–8.

20. Williams RA, et al. Phenylketonuria: An inborn error of phenylalanine metabolism. *Clin Biochem Rev*. 2008;29:31–41.

21. Brissot P, et al. Molecular diagnosis of genetic iron-overload disorders. *Expert Rev Mol Diagn*. 2010;10:755–63.

22. Galactosemia. March 12, 2012. Available at: ghr.nlm.nih.gov/condition=galactosemia.

23. Hemochromatosis. April 2007. Available at: digestive.niddk.nih.gov/ddiseases/pubs/hemochromatosis. Accessed March 17, 2012.

24. O'Donnell, CJ et al. Genomics of cardiovascular disease. *N Engl J Med*. 2011;365:2098–2108.

25. Lichtenstein P, et al. Environmental and heritable factors in the causation of cancer—analyses of cohorts of twins from Sweden, Denmark, and Finland. *N Engl J Med*. 2000;343:78–85.

26. Brauer HA, et al. Cruciferous vegetable supplementation in a controlled diet study alters the serum peptidome in a GSTM1-genotype dependent manner. *Nutr J*. January 27, 2011. [online] doi: 10.1186/1475-2891-10-11.

27. Dugdale DC. Hypertension. June 10, 2011. Available at: www.nlm.nih.gov/medlineplus/ency/article/000468.htm. Accessed March 7, 2012.

28. Kawasaki T, et al. The effect of high-sodium and low-sodium intakes on blood pressure and other related variables in human subjects with idiopathic hypertension. *Am J Med*. 1978;64:193–98.

29. McCarron DA, et al. Science trumps politics: Uniary sodium data challenge U.S. dietary sodium guideline. *Am J Clin Nutr*. 2010;92:1005–6.

30. Lee YS. The role of genes in the current obesity epidemic. *Ann Acad Med Singapore*. 2009;38:45–7.

31. Herrera BM, et al Genetics and epigenetics of obesity. *Maturitas*. 2011;69:41–49.

32. Spruijt-Metz D et al. Etiology, treatment and prevention of obesity in childhood and adolescence: A decade in review. *J Res Adolesc*. 2011;21:129–52.

33. Drewnowski A et al. Genetic taste markers and preferences for vegetables and fruit of female breast care patients. *J Am Diet Assoc*. 2000;100:191–7.

34. Feeney E et al. Genetic variation in taste perception: Does it have a role in healthy eating? *Proc Nutr Soc*. 2011;70:135–43.

35. Afman L, et al. Nutrigenomics: From molecular nutrition to prevention of disease. *J Am Diet Assoc*. 2006;106: 569–78.

36. Russell R. Nutrition society president says eat less, move more. *Science News*. July 17, 2010, p. 32. Available at: www.sciencenew.org.

37. Winslow R, et al. Soon, $1,000 will map genes. *Wall St J*. January 10, 2012, p. A2.

38. McDermott U, et al. Genomics and the continuum of cancer care. *N Engl J Med*. 2011;364:340–50.

39. Khoury M. CDC commentary: Personal genomic information testing, Information for health care providers. Available at: www.medscape.com.viewarticle.724963?src=mp&spon=42&uac=61213X. Accessed March 18, 2012.

40. Xia Li, et al. Cats lack a sweet taste receptor. *J Nutr*. 2006;136:1932S–34S.

41. Lichtenstein P, et al. Environmental and heritable factors in the causation of cancer—analyses of cohorts of twins from Sweden, Denmark, and Finland. *N Engl J Med*. 2000;343:78–85.

42. Foulkes WD. Inherited susceptibility to common cancers. *N Engl J Med*. 2008;359:2143–54.

43. Sallis JF, et al. Physical activity and food environments: Solutions to the obesity epidemic. *Milbank Q*. 2009;87:123–54.

44. Why people become overweight. Available at: www.health.harvard.edu/newsweek/Why-people-become-overweight.htm. Accessed March 17, 2012.

Unit 27

1. Dietary Guidelines for Americans, 2010. Available at: www.dietaryguidelines.gov.

2. Witkowski S, et al. Enhancing treatment for cardiovascular disease. *Exerc Sports Sci Rev*. 2011;39:93–101.

3. Lightfoot JT. Current understanding of the genetic basis for physical activity. *J Nutr*. 2010;141:526–30.

4. Tsai J, et al. Physical activity and optimal self-rated health of adults with and without diabetes. *BMC Public Health*. 2010;10:365. [online] doi:10.1186/1471-2458-10-365.

5. Global Strategy on Diet, Physical Activity and Health. WHO, 2012. Physical Activity. Available at: www.who.int/dietphysicalactivity/pa/en/index.htm. Accessed March 21, 2012.

6. Sattelmair J, et al. Dose response between physical activity and risk of coronary heart disease: A meta-analysis. *Circulation* 2011. [online] doi: 10.1161/CIRCULATIONAHA.110.010710.

7. Villareal DT, et al. Weight loss, exercise, or both and physical function in obese older adults. *N Engl J Med*. 2011;362:1218–29.

8. Michaëlsson K, et al. Leisure physical activity and the risk of fracture in men. *PLoS Med*. June 4, 2007;(6):e199. [online] doi: 10.1371/journal.pmed.0040199.

9. Gass MD, et al. Preventing osteoporosis-related fractures: An overview. *Am J Med*. 2006;119:(4 Suppl 1):S3–S11.

10. Hankinson AL, et al. Maintaining a high physical activity level over 20 years and weight gain. *JAMA*. 2010;304:2603–10.

11. King NA, et al. Dual-process action of exercise on appetite control: Increase in orexigenic drive but improvement in meat induced satiety. *Am J Clin Nutr*. 2009;90:921–7.

12. Curioni CC, et al. Long-term weight loss after diet and exercise: A systematic review. *Int J Obes*. 2005;29:1168–74.

13. Physical Activity Guidelines for Americans, 2008, U.S. Department of Health and Human Service. Available at: http://www. health.gov/PAGuidelines. Accessed October 2008.

14. Page P. Current concepts in muscle stretching for exercise and rehabilitation. *Int J Sports Phys Ther*. 2012;7:109–19.

15. Herbert RD, et al. Effects of stretching before and after exercising on muscle soreness and risk of injury: Systematic review. *MJ*. August 31, 2002;325(7362):468.

16. Watson S. What causes muscle soreness after exercise? Available at: www.webmd.com/fitness-exercise/features/art-sore-muscles-joint-pain. Accessed March 22, 2012.

17. Horton ES. Metabolic fuels, utilization, and exercise. *Am J Clin Nutr*. 1989; 49(suppl):931–2.

18. Jones NL, et al. Exercise limitations in health and disease. *N Engl J Med*. 2000;343:632–41.

19. Nieman DC, et al. Nutrient intake of marathon runners. *J Am Diet Assoc*. 1989;89:1273–8.

20. Position of the American Dietetic Association, Dietitians of Canada, and the American College of Sports Medicine: Nutrition and Physical Performance. *J Am Diet Assoc*. 2009;109:509–27.

21. Detraining: How long does it take to get out of shape? Available at: www.nehc.med.navy.mil/hp/fitness/index.htm. Accessed March 23, 2012.

22. Summary of Health Statistics for U.S. Adults: National Health Interview Survey, 2009. Available at: www.cdc.gov/*nhs*/data/nhis/

physicalactivity/pa_tableB.htm. Accessed March 22, 2012.

23. Garber C, et al. Quantity and quality of exercise for developing and maintaining cardiorespiratory, musculoskeletal, and neuromotor fitness in apparently healthy adults: Guidance for prescribing exercise. *Med Sci Sports Exer.* 2011;43:1334–59.

24. Morey MC, et al. Medical assessment for health advocacy and practical strategies for exercise initiation. *Am J Preventive Med.* 2003;29:1168–74.

25. Irving BA, et al. Effect of exercise training intensity in abdominal visceral fat and body composition. *Med Sci Sports Exer.* 2008;49:1863–72.

26. Target Heart Rate and Estimated Maximum Heart Rate. Available at: www.cdc.gov/ physicalactivity/everyone/measuring/heartrate. html. Accessed March 22, 2012.

27. Physical Activity and the Health of Young People. Available at: www.cdc.gov/ healthyyouth/physicalactivity/facts.htm. Accessed March 24, 2012.

28. Tanaka H, et al. J Age-predicted maximal heart rate revisited. *Am Coll Cardiol.* 2001;37:153–6.

Unit 28

1. Position of the American Dietetic Association, Dietitians of Canada, and the American College of Sports Medicine: Nutrition and physical performance. *J Am Diet Assoc.* 2009;109:509–27.

2. Pantano KJ. Current knowledge, perceptions, and interventions used by collegiate coaches in the U.S. regarding the prevention and treatment of the female athlete triad. *N Am J Sports Phys Ther.* 2006;1:195–207.

3. Maughan R. The athlete's diet: Nutritional goals and dietary strategies. *Proc Nutr Soc.* 2002;61:87–96.

4. Rosenbloom C. Athletes (and coaches) say the darndest things. *Nutr Today.* 2009;44:77–80.

5. Smith-Rockwell M, et al. Nutrition knowledge, opinions, and practices of coaches and athletic trainers at a division 1 university. *Int J Sport Nutr Exerc Metab.* 2001;11:174–85.

6. Peeke P. Hits and myths about spot reducing, Available at: blogs.webmd.com/pamela-peeke-md/2012/03/hits-and-myths-about-spot-reducing.html. Accessed March 16, 2012.

7. Rodriquez NR, et al. American College of Sports Medicine position stand:. Nutrition and athletic performance. *Med Sci Sports Exerc.* 2009;41:709–31.

8. Carbohydrate-loading diet. Available at: www.mayoclinic.com/health/carbohydrate-loading/MY00223. Accessed November 14, 2011.

9. Kreider RB, et al. ISSN exercise and sport nutrition review: Research and recommendations. *J Int Soc Sports Nutr.* February 2, 2010;7:7. [online] doi: 10.1186/1550-2783-7-7.

10. Horton ES. Metabolic fuels, utilization, and exercise. *Am J Clin Nutr.* 1989;49(suppl):931–1.

11. Brown RC. Nutrition for optimal performance during exercise: Carbohydrate and fat. *Curr Sports Med.* 2002;1:222–9.

12. Jones NL, et al. Exercise limitations in health and disease. *N Engl J Med.* 2000;343:632–41.

13. Coleman E. *Eating for endurance.* Palo Alto, CA: Bull Publishing Co., 1988.

14. Hawley JA, et al. Carbohydrate-loading and exercise performance. An update. *Sports Med.* 1997;24:73–81.

15. Minehan M for the Australian Sports Commission. Carbohydrate Loading, 2004. Available at: www.ausport.gov.au/ais/ nutrition/factsheets/competition_and_ training2/carbohydrate_loading. Accessed May 16, 2009.

16. What We Eat in America, NHANES, 2007–08. Available at: www.ars.usda.gov/Services/ docs.htm?docid=13793. Accessed February 25, 2011.

17. Verdijk LB, et al. Protein supplementation before and after exercise does not further augment skeletal muscle hypertrophy after resistance training in elderly men. *Am J Clin Nutr.* 2009;89:608–16.

18. Schwarz N, et al. Effects of 28 days of resistance exercise while consuming commercially available pre- and post-workout supplements, NO-Shotgun and NO-Synthesize on body composition, muscle strength and mass, markers of protein synthesis, and clinical safety markers in males. *Nutr Metab.* 2011;8:78 [online] doi:10.1186/1743-7075-8-78. Accessed November 14, 2011.

19. Moore DR, et al. Ingested protein dose response of muscle and albumin protein synthesis after resistance exercise in young men. *Am J Clin Nutr.* 2009;8:161–8.

20. Watson P, et al. A comparison of the effects of milk and a carbohydrate-electrolyte drink on the restoration of fluid balance and exercise capacity in a hot, humid environment. *Eur J Appl Physiol.* 2008;104:633–42.

21. Pennings B, et al. Exercising before protein intake allows for greater use of dietary protein-derived amino acids for de novo muscle protein synthesis in both young and elderly men. *Am J Clin Nutr.* 2011;93:322–31.

22. Atherton PJ, et al. Muscle full effect after oral protein: time-dependent concordance and discordance between human muscle protein synthesis and mTORC1 signaling. *Am J Clin Nutr.* 2010;92:1080–8.

23. Cramer JT, et al. Effects of creatine supplementation and three days of resistance training on muscle strength, power output, and neuromuscular function. *J Strength Cond Res.* 2007;21:668–77.

24. Pasiakos SM, et al. Leucine-enriched essential amino acid supplementation during moderate steady state exercise enhances postexercise muscle protein synthesis. *Am J Clin Nutr.* 2011;94:809–18.

25. Bemben MG, et al. The effects of supplementation with creatine and protein on muscle strength following a traditional resistance training program in middle-aged and older men. *J Nutr Health Aging.* 2010;14:155–9.

26. Stevenson EM, et al. Influence of high-carbohydrate mixed meals with different glycemic indexes on substrate utilization during subsequent exercise in women. *Am J Clin Nutr.* 2006;84:354–60.

27. Wong SH, et al. Effect of a carbohydrate-electrolyte beverage, lemon tea, or water on rehydration during short-term recovery from exercise. *Int J Sport Nutr Exerc Metab.* 2011;21:300–10.

28. Temesi J, et al. Carbohydrate ingestion during endurance exercise improves performance in adults. *J Nutr.* 2011;141:890–7.

29. Schneid S. Thermoregulation, presented at the Experimental Biology Annual Meeting, New Orleans, April 19, 2009.

30. Zelman KM. The best beverages to help you stay hydrated. Available at: www.webmd .com/fitness-exercise/features/drink-up-sports-fitness. Accessed March 27, 2012.

31. Stout A. Fueling and weight management strategies in sports nutrition. *J Am Diet Assoc.* 2007;107:1475–9.

32. Casa DJ, et al. Cold water immersion: the gold standard for exertional heatstroke treatment. *Exerc Sport Sci Res.* 2007;35:141–9.

33. Almond CSD, et al. Hyponatremia in nonelite marathon runners. *N Engl J Med.* 2005;352:1516–8,1550–6.

34. Goh KP. Management of hyponatremia. *Am Fam Physician.* 2004;15;69:2387–94.

35. Wilmore JH. Body composition in sports and exercise: Directions for future research. *Med Sci Sports Exer.* 1983;15:21–31.

36. Palacio, LE et al. Nutrition for the female athlete. October 31, 2008. Available at: http:// emedicine.medscape.com/article/108994. Accessed September 28, 2011.

37. Lloyd T, et al. Interrelationships of diet, athletic activity, menstrual status, and bone density in collegiate women. *Am J Clin Nutr.* 1987;46:681–4.

38. Brownell KP, et al. Weight regulation practices in athletes: Analysis of metabolic and health effects. *Med Sci Sports Exer.* 1987;19:546–56.

39. Oppliger RA, et al. The Wisconsin wrestling minimum weight project: A model for weight control among high school wrestlers. *Med Sci Sports Exer.* 1995;27:1220–24.

40. Cunningham E. Where can I find resources to address weight practices for wrestlers? *J Am Diet Assoc.* 2006;106:761.

41. Sinclair LM, et al. Prevalence of iron deficiency with and without anemia in recreationally active men and women. *J Am Diet Assoc.* 2005;105:975–8.

42. A to Z of nutritional supplements: Conclusions—What can ahletes do? *Br J Sports Med.* 2011;45:752–4.

43. Global Drug Reference Online. Available at: www.GlobalDRO.com. Accessed March 26, 2012.

44. Shorten AL, et al. Acute effect of environmental temperature during exercise on subsequent energy intake in active men. *Am J Clin Nutr.* 2009;90:1215–21.

45. Foster C et al. Effects of pre-exercise feeding on endurance performance. *Med Sci Sports Exer.* 1979;11:1–5.

46. Ireland ML. *The Female Athlete.* Philadelphia: W.B. Saunders; 2003.

Unit 29

1. Rosso P. *Nutrition and Metabolism in Pregnancy.* New York: Oxford University Press; 1990.

2. Kochanek KD, et al. Annual summary of vital statistics: 2009. *Pediatrics.* January 30, 2012. [online] doi 10.1542/peds.2011-3435.

3. United Nations. Live births, deaths, and infant mortality rate. Available at: unstats. un.org/unsd/demographic/products/vitstats/serATab3.pdf. Accessed April 1, 2012.

4. Allen LH. Adequacy of family foods for complementary feeding. *Am J Clin Nutr.* 2012;95:785–6.

5. Position of the American Dietetic Association: Nutrition and lifestyle for a healthy pregnancy outcome. *J Am Diet Assoc.* 2008;108:553–60.

6. Dietary Guidelines for Americans, 2010. Available at: www.dietaryguidelines.gov.

7. World Factbook. Available at: www.cia.gov/library/publications/the-world-factbook/rankorder/2091rank.html. Accessed March 30, 2012.

8. Births and natality. Available at: www.cdc .gov/nchs/fastats/births.htm. Accessed March 30, 2012.

9. MacDorman MF, et al. Annual summary of vital statistics. *Pediatrics.* 2002;110:1037–50.

10. Mendez MA, et al. A comparative analysis of dietary intakes during pregnancy in Europe: A planned pooled analysis of birth cohort studies. *Am J Clin Nutr.* 2011;94(suppl):1993S–9S.

11. Zeisel SH. Is maternal diet supplementation beneficial? Optimal development of infant depends on mother's diet. *Am J Clin Nutr.* 2009;89: 685S–687S.

12. Bocheva G, et al. Epigenetic regulation of fetal bone development and placental transfer of nutrients: Progress for osteoporosis. *Interdiscip Toxicol.* 2011;4: 167–72.

13. Heringhausen J, et al. Continuing education module—Maternal calcium intake and metabolism during pregnancy and lactation. *J Perinat Educ.* 2005;14:52–57.

14. Ladipo OA. Nutrition in pregnancy: Mineral and vitamin supplements. *Am J Clin Nutr.* 2000;72(suppl):280S–90S.

15. Miller DR. Vitamin excess and toxicity. In: Hathcock JN, ed. *Nutritional Toxicology.* New York: Academic Press; 1982:81–131.

16. Langley-Evans SC, et al. Nutritional programming of disease: Unraveling the mechanism. *J Anat.* July 2009; 215(1):36–51.

17. Johansen AMW, et al. Maternal dietary intake of vitamin A and risk of orafacial clefts: A population-based case-control study in Norway. *Am J Epidemiol.* 2008;167:1164–70.

18. Annet FM, et al. Survival effects of prenatal famine exposures. *Am J Clin Nutr.* 2012;95:179–83.

19. Godfrey KM, et al. The long-term effects of prenatal development on growth and metabolism. *Semin Reprod Med.* 2011;29:257–65.

20. Langley-Evans SC, et al. Developmental origins of adult disease. *Med Princ Pract.* 2010;19:87–98.

21. Langley-Evans SC. Developmental programming of health and disease. *Proc Nutr Soc.* 2006;65:97–105.

22. Barker DJ. Developmental origins of chronic disease. *Public Health.* 2012;126:185–9.

23. Gabory A, et al. Developmental programming and epigenetics. *Am J Clin Nutr.* 2011;94(suppl):1943S–52S.

24. Frederick IO, et al. Pre-pregnancy body mass index, gestational weight gain, and other maternal characteristics in relation to infant birth weight. *Matern Child Health J.* 2008;12:557–67.

25. Siega-Riz AM. A systematic review of outcomes of maternal weight gain according to the Institute of Medicine recommendations: Birth weight, fetal growth, and postpartum weight retention. *Am J Obstet Gynecol.* 2009;201:339.e1–14.

26. Fox NS, et al. Weight gain in twin pregnancies and adverse outcomes: Examining the 2009 Institute of Medicine guidelines. *Obstet Gynecol.* 2010;116:100–6.

27. Ohlin A, et al. Maternal body weight development after pregnancy. *Int J Obesity.* 1990;14:159–73.

28. What We Eat in America, NHANES, 2007–2008. Available at: www.ars.usda.gov/Services/docs.htm?docid=13793. Accessed February 25, 2011.

29. Daly LE, et al. Folate levels and neural tube defects. *JAMA.* 1995;274:1698–1702.

30. Tamura T, et al. Fotale and human reproduction. *Am J Clin Nutr.* 2006;83:993–1016.

31. Rasmussen KM, et al. *Weight Gain During Pregnancy: Reexamining the Guidelines.* Institute of Medicine (US) and National Research Council (US) Committee to Reexamine IOM Pregnancy Weight Guidelines; Washington (DC): National Academies Press (US); 2009.

32. Dibley MJ, et al. Safety and toxicity of vitamin A supplements in pregnancy. *Food Nutr Bulletin.* 2001;22:248–66.

33. Meyers DG, et al. Safety of antioxidant vitamins. *Arch Intern Med.* 1996;156:925–35.

34. Steinburg BJ. Women's oral health issues. Presented at the California Dental Association Meetings, September 13, 2000. Available at: www.cda.org/Library/cda_member/pubs/journal/jour0900/women.html. Accessed April 1, 2012.

35. Hollis BW, et al. Vitamin D deficiency during pregnancy: An ongoing epidemic. *Am J Clin Nutr.* 2006;84:350–3.

36. Mulligan ML, et al. Implications of vitamin D deficiency in pregnancy and lactation. *Am J Obstet Gynecol.* 2009;202:429.e1-429.e9.

37. Nesby-O'Dell, et al. Hypovitaminosis D prevalence and determinants among African American and white women of reproductive age: Third National Health and Nutrition Examination Survey, 1998–1994. *Am J Clin Nutr.* 2007;76:187–92.

38. Calcium and Vitamin D: *Dietary Reference Intakes.* Food and Nutrition Board, National Academy of Sciences, Washington, DC: National Academies Press; 2011.

39. Mei Z, et al. Assessment of iron status in U.S. pregnant women from the National Health and Nutrition Examination Survey (NHANES), 1999–2006. *Am J Clin Nutr.* 2011;93:1312–20.

40. Preziosi P, et al. Effect of iron supplementation on the iron status of pregnant women: Consequences for newborns. *Am J Clin Nutr.* 1997;66:1178–82.

41. Zimmermann MB. Iodine deficiency in pregnancy and the effects of maternal iodine supplementation on the offspring: A review. *Am J Clin Nutr.* 2009;89:668S–72S.

42. Utiger RD. Iodine nutrition: More is better. *N Engl J Med.* 2006;354:2819–21.

43. Vermiglio F, et al. Thyroid function in pregnant women from a mildly iodine deficient area. *J Clin Endocrinol Metab.* 2008;93:2466–8, 2616–21.

44. Dietary Reference Intakes for Vitamin A through Zinc. Available at: www.nap.edu.

45. Carlson SE. DHA supplementation in pregnancy and lactation. *Am J Clin Nutr.* 2009;89:678S–84S.

46. Jordan RG. Prenatal omega-3 fatty acids: Review and recommendations. *J Midwifery Womens Health.* 2010;55:520–8.

47. Swanson D et al. Omega-3 fatty acids EPA and DHA: Health benefits throughout life. *Adv Nutr.* 2012;3:1–7.

48. Harris WS, et al. Towards establishing dietary reference intakes for eicosapentaenoic and docosahexaenoic acids. *J Nutr.* 2009;139:804S–19S.

49. Swanson D, et al. Omega-3 fatty acids EPA and DHA: Health benefits throughout life. *Adv Nutr.* 2012;3:1–7.

50. Brent RL, et al. Evaluation of the reproductive and developmental risks of caffeine. *Birth Defects Res B Dev Reprod Toxicol.* 2011;92:152–87.

51. Jarosz M, et al. Maternal caffeine intake and its effect on pregnancy outcomes. *Eur J Obstet Gynecol Reprod Biol.* 2012;160:156–60.

52. SuperTracker. Available at: www.choosemyplate.gov/SuperTracker/myplan.asp. Accessed April 1, 2012.

53. Mattson SN, et al. Heavy prenatal alcohol exposure with or without physical features of fetal alcohol syndrome leads to IQ deficits. *J Pediatr.* 1997;131:718–21.

54. Sokol RJ, et al. Fetal alcohol spectrum disorder. 2003;290:2996–99.

55. Kochanek KD, et al. Annual summary of vital statistics: 2009. *Pediatrics.* January 30, 2012. [online] doi: 10.1542/peds.2011-3435.

56. Lenders C, et al. Nutrition in adolescent pregnancy. *Current Opin Pediatrics.* 2000;12:291–6.

57. Committee on Nutrition, American Academy of Pediatrics. Breastfeeding and the use of human milk. *Pediatrics.* 2005;115:496–506.

58. James DC, et al. Position of the American Dietetic Association: Promoting and supporting breastfeeding. *J Am Diet Assoc.* 2009;109:1926–42.

59. Bartick M, et al. The burden of suboptimal breastfeeding in the United States: A pediatric cost analysis. *Pediatrics.* 2010;125:e1048–56.

60. Kramer MS, et al. Prolonged and exclusive breast-feeding and cognitive development in children age 6.5 y. *Arch Gen Psychiatr.* 2008;65:578–84.

61. Energy, Carbohydrate, Fiber, Fatty Acids, Cholesterol, Protein, and Amino Acids: *Dietary Reference Intakes.* Washington, DC: National Academies Press; 2002.

62. Simopoulos AP, et al. Factors associated with the choice and duration of infant-feeding practices. *Pediatrics.* 1984;74(suppl):603–14.

63. McDowell MM, et al. Breast-feeding in the United States: Findings from NHANES, 1999–2006, NCHS Data Brief, no. 5, April 2008.

64. Few hospitals take steps needed to encourage breast-feeding, MMWR, [online] August 2, 2011.

65. National Academy of Sciences (Institute of Medicine). *Nutrition During Lactation.* Washington, DC: National Academy Press; 1991.

66. Lovelady CA, et al. The effect of weight loss in overweight, lactating women on the growth of their infants. *N Engl J Med.* 2000;342:449–53.

67. Menella JA, et al. The transfer of alcohol to human milk. *N Engl J Med.* 1991;325:981–85.

68. Grossman X, et al. Neonatal weight loss at a US Baby-Friendly Hospital. *Am Acad Nutr Diet.* 2012;112:410–3.

69. The WHO Child Growth Standards. Available at: www.who.int/childgrowth/publications/en. Accessed April 2, 2012.

70. WHO growth standards are recommended for use in the U.S. for infants and children 0 to 2 years of age. Available at: http://www.cdc.gov/growthcharts/who_charts.htm.

71. Winick M. *Malnutrition and Brain Development.* New York: Oxford University Press; 1976.

72. Lloyd-Still JD, et al. Intellectual development after severe malnutrition in infancy. *Pediatrics.* 1974;54:306–11.

73. Rappo PD, et al. Pediatrician's responsibility for infant nutrition. *Pediatrics.* 1997;99:749–50.

74. Macknin ML, et al. Infant sleep and bedtime cereal. *Arch Periatr Adol Med.* 1989;143:1066–8.

75. Beal VA. Termination of night feeding in infancy. *J Pediatr.* 1969;75:690–2.

76. New infant feeding guidelines. Available at: www.wicworks.ca.gov/ education/nutrition/ Infant_Feedltn/ InfantFeeding.htm. Accessed October 2006

77. Fiocchi A, et al. Consensus document for introducing solid foods into an infant's diet to avoid food allergies. *Am Allergy Asthma Immunol.* 2006;97:10–21.

78. Tromp II, et al. The introduction of allergenic foods and the development of reported wheezing and eczema in childhood: The Generation R study. *Arch Pediatr Adolesc Med.* 2011;165:933–8.

79. Markowitz DL. Fluoride supplementation: The ongoing debate. Available at: www.medscape.com/viewarticle/753193?src=mp&spon=17. Accessed February 18, 2012.

80. Wagner CL, et al. Prevention of rickets and vitamin D deficiency in infants, children, and adolescents. *Pediatrics.* 2008;122:1142–52.

81. Davis C. Self-selection of diets by newly weaned infants: An experimental study. *Am J Dis Child.* 1928;36:651–79.

82. Birch LL, et al. Caloric compensation and sensory specific satiety: Evidence for self-regulation of food intake by young children. *Appetite.* 1986;7:323–31.

83. Story M, et al. Do young children instinctively know what to eat? The studies of Clara Davis revisited. *N Engl J Med.* 1987;316:103–6.

84. Skinner JD, et al. Do food-related experiences in the first 2 years of life predict variety in school-aged children? *J Nutr Ed Behav,* 2002;34:310–5.

85. Mennella JA, et al. The timing and duration of a sensitive periods in human flavor learning: A randomized trial. *Am J Clin Nutr.* 2011;93:1019–24.

86. Nutrient Composition of Foods, USDA Nutrient Database. Available at: www.ars.usda.gov/ nutrientdata and www.ars.usda.gov/ba/bhnrc/ndl. Accessed September 2008.

87. Cummings AS. Morbidity in breast-feed and artificially fed infants. *J Pediatr.* 1979;90:726–9.

88. James DC, et al. Position of the American Dietetic Association: Promoting and supporting breastfeeding. *J Am Diet Assoc.* 2009;109:1926–42.

89. Health, United States, 2003. Available at: www.cdc.gov/nchs/hus.htm. Accessed October 2006.

90. Briefel R, et al. Feeding infants and toddlers study: Improvements needed in meeting infant feeding recommendations. *J Am Diet Assoc.* 2004;101(suppl 1):S31–S37.

91. Brown JE, et al. Predictors of red cell folate level in women attempting pregnancy. *JAMA.* 1997;277:548–52.

Unit 30

1. Schwartz SM et al. Obesity in children. February 15, 2012. Available at: emedicine.medscape.com/article/985333. Accessed April 9, 2012.

2. Dietary Guidelines for Americans, 2010. Available at: www.dietaryguidelines.gov.

3. Birch LL, et al. Calorie compensation and sensory specific satiety: Evidence for self-regulation of food intake by young children. *Appetite.* 1986;7:323–31.

4. WHO growth standards are recommended for use in the U.S. for infants and children 0 to 2 years of age. Available at: http://www.cdc.gov/growthcharts/who_charts.htm. Accessed April 9, 2012.

5. Growth charts. Available at: cdc.gov/growthcharts. Accessed April 9, 2012.

6. Basics about childhood obesity. Available atwww.cdc.gov/obesity/childhood/basics.html. Accessed April 5, 2012.

7. Tanner JM. *Fetus into Man: Physical Growth from Conception to Maturity.* Cambridge, MA: Harvard University Press; 1978.

8. Smith GD. *Growth and Its Disorders.* Philadelphia: W. B. Saunders; 1979.

9. Ohyama S, et al. Some secular changes in body height and proportion of Japanese medical students. *Am J Phys Anthropol.* May 3, 2005. [online], doi: 10.1002/ajpa.1330730204.

10. Bronner F. Adaptation and nutritional needs (letter to the editor). *Am J Clin Nutr.* 1997;65:1570.

11. Gigante DP, et al. Early life factors are determinants of female height at age 19 years in a population-based birth cohort. *J Nutr.* 2006;136:473–8.

12. Vernarelli JA, et al. Dietary energy density is associated with body weight status and vegetable intake in U.S. children. *J Nutr.* 2011;141:2204–10.

13. Daniels SR. Lipid concentrations in children and adolescents: It is not all about obesity. *Am J Clin Nutr.* 2011;94:699–700.

14. Scaglioni S, et al. Determinants of children's eating behavior. *Am J Clin Nutr.* 2011;94(suppl):2006S–11S.

15. Carpenter TO, et al. Demographic, dietary, and biochemical determinants of vitamin D status in inner-city children. *Am J Clin Nutr.* 2012;95:37–46.

16. Lamb MM, et al. Association of body fat percentage with lipid concentrations in children and adolescents: United States, 1999–2004. *Am J Clin Nutr.* 2011;94:877–83.

17. Mendoza RA, et al. Bottled water a risk factor for early childhood caries. APHA 137th Annual

Meeting: Abs. no. 204297, November 8, 2009.

18. What We Eat in America, NHANES, 2007–2008. Available at: www.ars.usda.gov/Services/docs.htm?docid=13793. Accessed February 25, 2011.

19. DRI tables. [online] www.nap.edu.

20. Denney-Wilson E, et al. Measures of adiposity and metabolic risk factors in adolescents. *Arch Pediatr Adolesc Med.* 2008;162:566–73.

21. Dietz WH, et al. Overweight children and adolescents. *N Engl J Med.* 2005; 352:2100–9.

22. Goran MI. Impaired glucose tolerance in obese children and adolescents (letter to the editor). *N Engl J Med.* 2002;347:290.

23. Waters E, et al. Interventions for preventing obesity in children. Cochrane Heart Group. December 7, 2011. [online]. doi: 10.1002/14651858.CD001871. Accessed December 12, 2011.

24. Brewer JD, et al. Increasing student physical activity during the school day: Opportunities for the physical educator. *Strategies: A Journal for Physical and Sport Educators.* January 1 2009. Available at: www.articlearchives.com/medicine-health/diseases-disorders-cancer-breast/2310574-1.html. Accessed May 2009.

25. Story M. The Third School Nutrition Dietary Assessment Study: Findings and policy implications or improving the health of U.S. children. *J Am Diet Assoc.* 2009;109:S7–13.

26. Surrounding our kids with the opportunity for physical activity. American Public Health Association. Available at: www. medscape.com/viewarticle/527767. Accessed March 23, 2006.

27. Physical Activity Guidelines for Americans, 2008, U.S. Department of Health & Human Service. Available at: www. health.gov/PAGuidelines.

28. Small EW, et al. Guidelines on strength training for children. *Pediatrics.* 2008;121:835–40.

29. Story M, et al. Do young children instinctively know what to eat? The studies of Clara Davis revisited. *N Engl J Med.* 1987;316:103–6.

30. Satter E. The feeding relationship: Problems and interventions. *J Pediatr.* 1990;117:S181–S9.

31. Birch LL. The role of experience in children's food acceptance patterns. *J Am Diet Assoc.* 1987;87(suppl):S36–S40.

32. Fisher JO, et al. Restricting access to palatable foods affects children's behavioral response, food selection, and intake. *Am J Clin Nutr.* 1999;69:1264–72.

33. Birch LL. Presentation at the National Conference on Nutrition Education Research, Chicago, September 1986.

Unit 31

1. Dillin A. Aging processes and calorie restriction, presented at the Experimental Biology Annual Meeting, New Orleans, April 18, 2009.

2. Johnson MA, et al. Challenges and opportunities for clinical nutrition interventions in the aged. *J Nutr.* 2011;141:535–41.

3. Preston SH, et al. Contribution of obesity to international differences in life expectancy. *American Journal of Public Health.* 2011;101:2137–43.

4. Rivlin RS. Keeping the young-elderly healthy: Is it too late to improve our health through nutrition? *Am J Clin Nutr.* 2007;86(suppl):1572S–6S.

5. Heidemann C, et al. Dietary patterns and risk of mortality from cardiovascular disease, cancer, and all causes in a prospective cohort of women. *Circulation.* 2008;118:230–7.

6. Fuchs VR. New priorities for future biomedical innovations. *N Engl J Med.* 2010;363:704–6.

7. Life expectancy. Available at: www.cdc.gov/nchs/fastats/lifexpec.htm. Accessed April 11, 2012.

8. Crimmims EM, et al. Gender differences in health: results from SHARE, ELSA and HRS. *Eur J Public Health.* 2011;21:81–91.

9. Vaidya D, et al. Ageing, menopause, and ischaemic heart disease mortality in England, Whales, and the United States. *BMJ.* 2011;343. [online] doi:10.1136/bmj.d5170. Accessed April 10, 2012.

10. Births and Natality. Available at: www.cdc.gov/nchs/fastats/births.htm. Accessed March 30, 2012.

11. National Vital Statistics Report, Deaths: preliminary data for 2010. Jan. 11 2012, www.cdc.gov/nchs/fastats/lifexpec.htm, accessed 4/11/12.

12. Country comparison life expectancy at birth. *CIA World Factbook.* Available at: www.cia.gov/library/publications/the-world-factbook/rankorder/2102rank.htm. Accessed April 11, 2012.

13. Gudmundsson H, et al. Inheritance of human longevity in Iceland. *Euro J Hum Genetics.* 2000;8:743–9

14. Sebastiani P, et al. Whole genome sequences of a male and female supercentenarian, ages greater than 114 years. *Front. Gene.* 2012;2:90. [online] doi: 10.3389/fgene.2011.00090.

15. Crandall JP, et al. Genes versus healthy living to age 100. *J Am Geriatr Soc.* 2011. Available from: www.health.harvard.edu/blog/living-to-100-and-beyond-the-right-genes-plus-a-healthy-lifestyle-201201114092. Accessed April 12, 2012.

16. de Gonzalez A B, et al. Body mass index and mortality among 1.46 million white adults. *N Engl J Med.* 2010;362:2211–9.

17. National diabetes fact sheet. CDC's Division of Diabetes Translation news release, January 27, 2011.

18. Olshansky SJ, et al. What if humans were designed to last? *The Scientist.* March 2007, pp. 28–35.

19. Physical activity guidelines for Americans. U.S. Department of Health & Human Service, 2008. Available at: http:// www.health.gov/PAGuidelines.

20. Visser M, et al. Low serum concentrations of 25-hydroxyvitamin D in older persons and the risk of nursing home admission. *Am J Clin Nutr.* 2006;84:616–22.

21. Hoeijmakers JHJ. DNA damage, aging, and cancer. *N Engl J Med.* 2009;361:1475–85.

22. Dawson-Hughes B. Serum 25-hydroxyvitamin D and functional outcomes in the elderly. *Am J Clin Nutr.* 2008;88(suppl):537S–40S.

23. NAMS. Management of osteoporosis in postmenopausal women: 2010 position statement of the North American Menopause Society. *Menopause.* 2010:17:25–54.

24. Hamidi M, et al. Healthy diet may promote bone health in aging women. Available at: www.medscape.com/viewarticle/743350. Accessed June 30, 2011.

25. Zhu K, et al. A randomized controlled trial of the effects of vitamin D on muscle strength and mobility in older women with vitamin D insufficiency. *J Am Geriatr Soc.* 2010;58:2063–8.

26. Tighe P, et al. Effect of increased consumption of whole grain foods on blood pressure and other cardiovascular risk markers in healthy middle-aged persons: A randomized controlled trial. *Am J Clin Nutr.* 2010;92:733–40.

27. Tan AG, et al. Antioxidant nutrient intake and the long-term incidence of age-related cataract: the Blue Mountain Eye Study. *Am J Clin Nutr.* 2008;87:1899–905.

28. Rowe JW, Kahn RL. Human aging: usual and successful. *Clin Nutr.* 1990; 9:26–33.

29. Vogelzang JL, et al. Overview of fluid maintenance/prevention of dehydration. *J Am Diet Assoc.* 1999;99:605–9.

30. Dietary reference intakes for water, potassium, sodium, chloride, and sulfate. Food and Nutrition Board. 2004. Available at: books.nap.edu/openbook/0309091691/gifmid/74.gif.

31. Rolls BJ. Do chemosensory changes influence food intake in the elderly? *Physiol Behav.* 1999;66:193–7.

32. Schiffman SS. Taste and smell losses in normal aging and disease. *JAMA.* 1997;278:1357–62.

33. Shepard RJ, et al. Regular exercise through middle age delays biological aging. *Brit J Sports Med.* August 10, 2008. Available at: www.medscape.com/viewarticle/ 573636.

34. Erickson KI, et al. Physical activity predicts gray matter volume in late adulthood: The Cardiovascular Health Study. *Neurology.* 2010;75;1415–22.

35. Terre L, et al. Resistance for women's health. *J Lifestyle Med.* 2010;4:314–6.

36. Isenring EA, et al. Beyond malnutrition screening: appropriate methods to guide nutrition care for aged care residents. *J Acad Nutr Diet.* 2012;112:376–81.

37. Health, United States 2008. Available at: www .cdc.gov/ nchs/hus.htm. Accessed May 2009.

38. U.S. Census Bureau. Decennial Census Data Population Projections, 2000.

39. Administration on Aging, USDHHS, Minority aging. Available at: www.aoa.gov/AoRoot/ Aging_Statistics/Minority_Aging/index.asp. Accessed June 3, 2010.

40. Dietary Guidelines for Americans, 2010. Available at: www.dietaryguidelines.gov.

41. Garry PJ, et al. Aging and nutrition. In: Ziegler EE, Filer LJ eds. *Present Knowledge in Nutrition.* ISLI Press: Washington, DC; 1998:404–19.

42. Bossingham MJ, et al. Water balance, hydration status, and fat-free mass hydration in younger and older adults. *Am J Clin Nutr.* 2005;81:1342–50.

43. Francesco VD, et al. Unbalanced serum leptin and ghrelin dynamics prolong postprandial satiety and inhibit hunger in healthy elderly: Another reason for the "anorexia of aging." *Am J Clin Nutr.* 2006:83;1149–52.

Unit 32

1. Botulism Annual Summary, 2010. December 2011. Available at: www.cdc. gov/nationalsurveillance/PDFs/Botulism_ CSTE_2010.pdf. Accessed April 13, 2012.

2. Safe food handling: Myth busters. Available at: www.fightbac.org/content/view/151/2. Accessed October 2006.

3. CDC 2011 Estimates: Findings. February 2012. Available at: www.cdc.gov/ foodborneburden/2011-foodborne-estimates. html. Accessed April 13, 2012.

4. Klein S. The ten riskiest foods regulated by the U.S. FDA. October 6, 2009. Available at: www.medscape.com/viewarticle/710182. Accessed November 2, 2009.

5. Foodborne illness. Available at: www .cdc.gov/ncidod/dbmd/diseaseinfo/ foodborneinfections_g.htm. Accessed May 2009.

6. Foodborne illnesses. Available at: www.cdc. gov. Accessed May 2009.

7. Zhao P, et al. Development of a model for evaluation of microbial cross-contamination in the kitchen. *J Food Protection.* 1998;61:960–3.

8. Mahon B. CDC expert commentary: Foodborne illness—a handy review. Available at: www.medscape.com/viewarticle/735505 ?src=mp&spon=9. Posted January 24, 2011. Accessed April 13, 2012.

9. CDC 2011 Estimates: Findings. February 2012. Available at: www.cdc.gov/ foodborneburden/2011-foodborne-estimates .html. Accessed April 13, 2012.

10. Mead PS, et al. Food-related illness and death in the United States. Available at: www.cdc .gov. Accessed October 2003.

11. Wegener HC. The consequences for food safety of the use of fluoroquinolones in food animals. *N Engl J Med.* 1999;340:1581–2.

12. Welland D. *E. coli* ... salmonella ... Listeria ... unwelcome house guests. *Envir Nutr.* October 1997;20:1, 6.

13. What you need to know about mercury in fish and shellfish. Available at: water.epa.gov/ scitech/swguidance/fishshellfish/outreach/ advice_index.cfm. Accessed April 13, 2012.

14. Treatment of ciguatera poisoning with gabapentin. *N Engl J Med.* 2001;344:692–3.

15. Dolan LC, et al. Naturally occurring food toxins. *Toxins* (Basel). 2010;2:2289–32.

16. Botulism in Alaska 2005 update: A guide for physicians and health care providers. Available at: www.epi.hss.state.ak.us/pubs/botulism/ Botulism.pdf. Accessed April 12, 2012.

17. Weir E. Sushi, nemotodes, and allergies. *CMAJ.* 2005;1;172:329.

18. Braun JM, et al. Impact of early-life bisphenol A exposure on behavior and executive function in children. *Pediatrics.* October 24, 2011. [online] doi:10.1542/peds.2011-1335.

19. Melzer D, et al. Urinary bisphenol A: A concentration and risk of future coronary artery disease in healthy men and women. *Circulation.* 2012. [doi] 10.1161/ CIRCULATIONAHA.111.069.153.

20. Bucher J, et al. Questions and answers about bisphenol A. Available at: www.niehs.nih .gov/news/sya/sya-bpa, Accessed October 26, 2011.

21. Bisphenol A. October 13, 2010. Available at: www.cbc.ca/news/health/story/2009/01/28/f-health-bisphenol.htm. Accessed April 13, 2012.

22. Consumer reports: Two-thirds of chickens carry bacteria. November 2009. Available at: abcnews.go.com. Accessed April 13, 2012.

23. Mad cow disease found in California; no human threat seen. www.nlm.nih.gov/ medlineplus/news/fullstory_124454.html. Available April 25, 2012.

24. Khan ZZ, et al. Kuru. July 12, 2010. Available at: emedicine.medscape.com/article/220043. Accessed February 22, 2011.

25. Safer food for a healthier you. Available at: www.webmd.com/diet/features/safer-food-healthier-you. Accessed April 12, 2012.

26. Ross V. Hard living breeds superbugs. *Discover Science, Technology, and the Future.* September 2010:21.

27. Holtz S. Antibiotics and hormone growth promoters in Canadian agriculture. April 2009. Available at: www.cielap.org/pdf/ AHGPs.pdf. Accessed April 13, 2012.

28. U.S Food and Drug Administration—Total Diet Study, 2005, Available at: www .fda.gov/downloads/Food/FoodSafety/ FoodContaminantsAdulteration/ TotalDietStudy/UCM291686.pdf. Accessed April 13, 2012.

29. Madigan D. After Delaney: How will the new pesticide law work? *CNI Weekly Report.* October 1996;11:4–5.

30. Byrne J. USDA may allow beef carcass irradiation as "processing aid." Available at: www. nutraingredientsusa.com. Accessed October 2008.

31. Shea KM. Technical report: Irradiation of food. *Pediatrics.* 2000;106:1505–9.

32. Dietary Guidelines for Americans, 2010. Available at: www.dietaryguidelines.gov.

33. Available at: www.fightbac.org. Accessed October 2006, 5/09.

34. Wilson J. The decade's 10 biggest foodborne illness outbreaks (based on CDC surveillance data). November 3, 2011. Available at: www .cnn.com/2011/09/30/health/high-profile-food-borne-illnesses-gallery/index.html. Accessed April 14, 2012.

35. Keep it cool. Refrigerator/freezer storage chart. Available at: www.homefoodsafety.org/ pub/file.cfm?item_type=xm_file&id=1165. Accessed September 15, 2011.

36. What are the most common foodborne diseases? Available at: www.cdc.gov/ncidod/ dbmd/ diseaseinfo/foodborneinfections_g .htm#mostcommon. Accessed May 2009.

37. Kent AJ, et al. Antibiotic-resistant *Campylobacter.* An increasing problem. *Postgrad Med J.* 2008;84:106–8.

38. Preliminary FoodNet data on the incidence of infection with pathogens transmitted commonly through food—10 states. 2008. *MMWR.* 2009;51(3):333–7.

39. Zeratsky K. Moldy cheese: Is it unsafe to eat? www.mayoclinic.com/print/food-and-nutrition/AN01024. February 26, 2011.

40. Safe Minimum Cooking Temperatures. www .foodsafety.gov/keep/charts/mintemp.html. Accessed June 21, 2012.

Unit 33

1. Essential nutrition actions improving maternal-newborn-infant and young child health and nutrition draft, May 2011. Available at: www.who.int/nutrition/ EB128_18_backgroundpaper2_A_ reviewofhealthinterventionswithan effectonnutrition.pdf. Accessed April 16, 2012.

2. Meadows DH. The population map: State of the village report. Available at: www.odt.org/ pop.htm. Accessed June 2009.

3. CIA World Factbook. Available at: www.cia. gov/library/publications/the-world-factbook/ rankorder/2102rank.html. Accessed April 15, 2012.

4. May M. The challenge of changing health. *Scientific American.* 2010. Available at: www .sa-pathways.com/the-challenge-of-changing-health/the-challenge. Accessed April 12, 2012.

5. U.S. Census Bureau, international data base. Available at: www.census.gov/cgi-bin/ipc/ agggen. Accessed June 2009.

6. Branca F. WHO launches e-library of evidence for nutrition actions. Available at: www.who .int/elena/en/index.html. Accessed August 10, 2011.

7. Lauria J. Anti-poverty plan lags in funding. *Wall St J.* September 20, 2010;A17. Also

Global crisis slowed U.N antipoverty effort. *Wall St J*. September 21, 2010;A13.

8. The State of Food Insecurity in the World 2011, FAO report. October 10, 2011. Available at: www.medscape.com/viewarticle/751296?src=mp&spon=42. Accessed October 18, 2011.

9. Oddo VM, et al. Predictors of maternal; and child double burden of malnutrition in rural Indonesia and Bangladesh. *Am J Clin Nutr*. 2012;95:951–8.

10. Narayan KMV, et al. Global noncommunicable disease—where worlds meet. *N Engl J Med*. 2010;363:1196–98.

11. Ruel MT, et al. Developing simple measures of women's diet quality in developing countries: overview. *J Nutr*. September 29, 2010. [online] doi: 10.3945/jn.110.123695.

12. Global status report on noncommunicable diseases 2010. WHO 2011. September 14, 2011. Available at: www.who.int/nmh/publications/ncd_report2010/en. Accessed April 17, 2012.

13. 2012 World Hunger and Poverty Facts and Statistics. Available at: www.worldhunger.org/articles/Learn/world%20hunger%20facts%202002.htm#Does_the_world_produce_enough_food_to_feed_everyone. Accessed April 17, 2012.

14. Obesity in the midst of unyielding food insecurity in developing countries. Available at: www.ers.usda.gov/AmberWaves/September08/Features/Obesitycountries.htm. Accessed June 2009.

15. Worldwide prevalence of child malnutrition assessed as percent underweight among children under 5 years of age, 2000–2009. Available at: www.globalhealthfacts.org/data/topic/map.aspx?ind=48. Accessed April 12, 2012.

16. Crookstson BT, et al. Children who recover from early stunting and children who are not stunted demonstrate similar level of cognition. *J Nutr*. 2010;140:1996–2001.

17. Amadi B, et al. Reduced production of sulfated glycosaminoglycans occurs in Zambian children with kwashiorkor but not for marasmus. *Am J Clin Nutr*. 2009;89:592–600.

18. Neumann CG. Background, symposium: food-based approaches to combating micronutrient deficiencies in children of developing countries. *J Nutr*. 2007;137:1091–2.

19. Scrimshaw NS. Historical concepts of interactions, synergism and antagonism between nutrition and infection. *J Nutr*. 2003;133:316S–21S.

20. Position of American Dietetic Association. Addressing world hunger, malnutrition, and food insecurity. *J Am Diet Assoc*. 2003;103:1046–56.

21. DelliFraine J, et al. Baby friendly hospitals don't cost more than other facilities. Available at: www.medscape.com/viwearticle/739528. Accessed June 2, 2011.

22. Evidence shows vitamin A for children can save lives. *BMJ*. August 25, 2011. Available at: www.medscape.com/viewarticle/748608?src=mpnews&spon=9.

23. AIDS epidemic update. December 2007. Available at: data.unaids.org/pub/EPISlides/2007/2007_epiupdate_en.pdf. Accessed June 2009.

24. Moloney GM, et al. Notes from the field: Malnutrition and mortality—Southern Somalia. August 5, 2011. *MMWR*. 2011;60:1026–7.

25. Alarming malnutrition in Sudan conflict zone. Available at: www.medscape.com/viewarticle/756399?scr-mp&spon=42. Accessed January 23, 2012.

26. Rosling H. Making data dance. *The Economist Technology Quarterly*. December 11, 2010;25–26.

27. Decimosexta Conferencia Internacional de Nutricion, Montreal. *Nutr View*. Fall 1997;1–36.

28. Food fortification to end micronutrient malnutrition: State of the art. *Nutr View*. Special issue 1997;1–8.

29. Scrimshaw NS. Historical concepts of interactions, synergism and antagonism between nutrition and infection. *J Nutr*. 2003;133:316S–21S.

12. United Nations International Children's Emergency Fund. *The State of the World's Children*. New York: United Nations; 1998.

30. End Poverty 2015 Campaign. Available at: www.endpoverty2015.org/faofoodsummit. Accessed June 2009.

31. Global health and aging, new disease patterns. March 28 2012. Available at: www.nia.nih.gov/research/publication/global-health-and-aging/new-disease-patterns. Accessed April 16, 2012.

32. First call for children: A UNICEF quarterly report. January/March 1992.

33. North Korea lowers height requirements for military service, April 3, 2012. Available at: www.outsidethebeltway.com/north-korea-lowers-height-requirements-for-military-service. Accessed April 17, 2012.

Index

Note to the reader: If you have a customized book, this index may include some entries that will not be found in your book.

Page references are chapter number followed by individual page number. Page ranges of information are indicated by an & separating the first and last pages.

NAME: _____

Activity 1
Family Tree Health History

Fill in the table to the extent you can by placing a ✓ under any of your relatives (biological or adoptive) who have a history of the disease or disorder listed.

Disease or Disorder	Maternal Grandparents	Paternal Grandparents	Mother	Father	Brother or Sister
Hypertension					
Heart disease/heart attack					
Cancer					
Diabetes					
Osteoporosis					
Tooth decay					

In general, the greater the number of relatives with a specific disease or disorder, the greater the likelihood that other family members may develop the same disease or disorder. Remember that family history is only one of many indicators of disease risk among family members. Adopted persons with unknown family history may have genetic characteristics that increase or decrease disease risk in ways that cannot be estimated by this activity.

Use the back of the form to record the ages of your family members and to expand the information if you wish.

Ages of family members/General health status (Excellent/Good/Fair/Poor)

Maternal grandmother

Maternal grandfather

Paternal grandmother

Paternal grandfather

Mother

Father

Brother(s)

Sister(s)

Other family members you wish to include:

If you are interested in creating an in-depth family health history, you can do so at https://familyhistory.hhs.gov/fhh-web/home.action

NAME: _____

Activity 2
Creating Your Own Fraudulent Nutrition Product

1. Identify a common appearance, health, or vitality concern or problem that will be "fixed" by your fraudulent nutrition product.

2. Develop components of an advertisement for the product:
   a. Give the product a name, state what the product is made from (for example vitamins, fatty acids, herbs), and connect the product to a biological process in the body.

   b. Develop a scientific-sounding explanation for why the product works, and refer to a scientific study that proves it does.

   c. Dream up a few testimonials, or fake "expert" statements concerning the effectiveness of the product.

d. Create an advertisement headline for your product and a statement about a money-back guarantee.

e. List nutrition-related products you have recently seen advertised that employ advertising techniques similar to those you just developed.

NAME: _____

. . . .

Activity 3
Understanding Nutrition Labels

Instructions: For this activity, you will evaluate the nutrition information given on a box of breakfast cereal.

Nutrition Facts

Serving Size: 3/4 cup (28g)

Amount Per Serving		
Calories 110	Calories from Fat 14	
		% Daily Value*
Total Fat 1.5 g		2%
Saturated Fat 0 g		0%
Trans Fat 0 g		
Cholesterol 0 mg		0%
Sodium 190.12 mg		8%
Potassium 115.08 mg		3%
Total Carbohydrate 21.65 g		7%
Dietary Fiber 1.99 g		8%
Sugars 9 g		
Sugar Alcohols		
Protein 3 g		
Vitamin A 500.08 IU		10%
Vitamin C 5.99 mg		10%
Calcium 99.96 mg		10%
Iron 4.5 mg		25%

Scott Goodwin Photography

1. On what basis is the health claim "Can help lower cholesterol" made?

2. Does this food product contain added sugars?

   If yes, list them:

3. Is the cereal free of saturated fats?

4. If you consumed ³/₄ cup of the dry cereal as a snack, your snack could be considered high in which nutrients?

5. Does this cereal potentially contain one or more food allergen?

   If yes, list the potential, common allergens:

6. Is the cereal gluten free?

7. Which nutrient is added to the product as a preservative?

NAME: _____

. . . .

Activity 4
Cultural Influences on Food Preferences

Introduction: Food preferences are, in part, defined by an individual's experiences with the foods and the meanings attached to them. Foods served for celebrations, those given to children when they are ill, and foods served at holiday gatherings of relatives often have their roots in long-standing practices of specific cultural groups. Cultural practices surrounding food are often strong enough to last from generation to generation, and become part of a person's cultural identity.

1. Name a holiday that your family celebrates with a special meal: _____

2. Name two foods that are typically served at the meal.

    _____

3. What specific food or beverage were you given by your parent or guardian when you were young and sick with a cold, the flu, or other common childhood illness?

    _____

4. Name two foods that were typically served at birthday parties when you were growing up.

    _____

5. Do you still celebrate holidays and birthdays with your parents/grandparents?

    ☐ Yes   ☐ No

6. If your answer is no, do you celebrate these occasions in other ways—for example, with friends or your spouse and children?

    ☐ Yes   ☐ No

7. If you no longer celebrate the occasions with your birth family, do you maintain the food traditions you practiced in your home?

    ☐ Yes   ☐ No

8. Will you continue some or all of these food traditions in the future?

☐ **Yes**. If yes, state why:

☐ **No**. If no, state why:

☐ **Not sure**

9. Based on the families you know, give examples of differences and similarities in food traditions within their families and your family, the importance of cultural influences within other families, and the origins of the traditions.

NAME: _____

Worksheet 5
Portion Size and Serving Size Comparison

Instructions: For this activity you will measure the next five different foods or beverages you serve yourself at home from a food or beverage container that has a Nutrition Facts panel. Foods from the protein group such as a hamburger patty, a piece of fish, or a slice of cheese cut from a block of cheese that have serving sizes measured in ounces are excluded from this activity; water is also excluded from this activity.

1. Serve yourself a portion of food or beverage you intend to consume.

2. Measure the amount of food or beverage you served yourself using a measuring cup, measuring spoon, or the number of pieces (for crackers and fresh fruit, for example). Record the amount on the form below.

3. Enter the standard serving size stated on the Nutrition Facts panel for the food or beverage on the form below.

4. Answer the follow-up questions.

Food/Beverage	My Portion Size	Nutrition Facts Panel Serving Size
1.		
2.		
3.		

10

a. Did the results of the comparisons of your portion sizes to the labeled serving sizes differ from what you expected they would be?

If yes, note how the results differed from what you expected:

b. For which foods or beverages were the labeled serving sizes smaller than the amount you served yourself?

For which foods or beverages were they larger?

c. Do you think your experience with measuring foods and beverages you serve yourself at home will influence your estimate of portion sizes when you eat away from home? State why it will or will not influence your estimate.

NAME: _____

Worksheet 6
Behavioral Change Plan Activity

Instructions: Complete the assignment on this form. Do not attach pages. Steps in one approach to improving a component of dietary intake are given below. Identify a weakness in your diet identified by your dietary assessment results (low calcium intake, for example). Then think about a small, specific, and acceptable change you could make to your diet to improve your intake of that component of your diet.

1. Identify a weakness in your diet revealed by your dietary assessment results that you would like to improve.

2. Identify two foods you like that, if consumed, would strengthen the weakness identified.

3. Identify two foods you could replace in your diet to make room for the foods identified in activity 2.

4. State which food addition option identified in activity 2 would be easiest to incorporate into your diet, and which food option listed in activity 3 would be easiest to replace.

   Easiest food to add: Easiest food to replace:

5. Plan how to incorporate this change in your diet.

    a. When during a day would you consume the "easiest food to add"?

    b. How much of this food would you likely consume?

    c. How would you make sure this food would be available when you plan to consume it?

    d. When during a day would you normally consume the "easiest food to replace"?

    e. How much of this food would you likely replace with the food you'll be adding to your diet?

6. Track your success for the next seven days. Check below on which days you ate the healthier food choice.

Day	Yes	No
Monday		
Tuesday		
Wednesday		
Thursday		
Friday		
Saturday		
Sunday		

NAME: _____

Worksheet 7
Anthropometry Lab

A. Use the best estimate of your current height and weight to calculate your Body Mass Index (BMI); or, alternatively, calculate the BMI of the hypothetical adult in question 4.

Calculating BMI from Your Estimated Height and Weight

1. What is your height without shoes? Round the measurement to the nearest quarter inch (e.g. 0.25, 0.50, or 0.75 inches). _____ feet _____._____ inches

2. What is your weight without clothing? _____ pounds

3. Based on your estimated height and weight, what is your BMI? _____ kg/m^2

$$BMI = \frac{Weight\ in\ pounds}{(Height\ in\ inches) \times (Height\ in\ inches)} \times 703 \quad \text{(Show calculations.)}$$

or Calculating BMI for a Hypothetical Adult

4. Calculate BMI for a hypothetical adult who reports that his height is 6 feet, 3 inches, and his weight is 220 pounds. (Show calculations.)

BMI = _____ kg/m^2

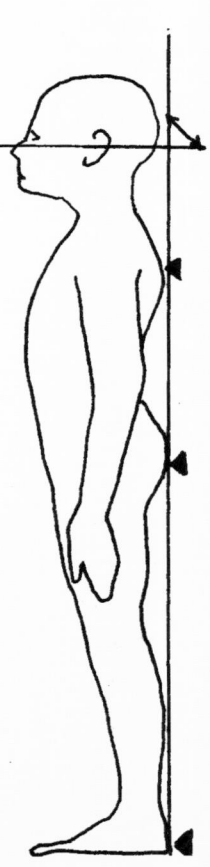

B. Measure your height for a more accurate calculation of BMI.

1. Height is measured without shoes (socks are okay) while a person is standing with his or her back at the center of the measuring tape. The person should be:

   a. Flat on his or her feet with heels almost together,

   b. Looking straight ahead, and

   c. Positioned so that shoulder blades, buttocks, and heels are touching the measurement surface (see the illustration).[a]

---

[a]CDC, government publication.

3. When the person is in this position:

    a. Place one flat surface of a right angle board on the top of the middle of the person's head, and the other flat surface directly against the measuring wall.

    b. Lightly press the angle board down on the person's head to flatten hair if needed.

    c. Draw a line that follows the base of the right angle board to the measuring tape and read the measurement.

    d. Round the measurement to the nearest quarter inch (e.g. 0.25, 0.50, or 0.75 inches).

4. What is your measured height? _____ feet _____._____ inches

5. Recalculate your BMI if your measured height is different than the height estimated for question A1. If the height figures are the same, complete question 6 below. (Show calculations.)

    _____ kg/m²   *or. . .*

6. Recalculate BMI for the hypothetical adult who is measured to be 6 feet, 1.5 inches tall but still weighs 220 pounds. (Show calculations.)

    _____ kg/m²

C. Waist Circumference

7. a. Measure your waist circumference while in light, indoor clothing if possible.

    b. Place the measuring tape around the smallest circumference below the last rib of the rib cage and above the navel. The tape should be flat against your waist and level.

    c. Take your measurement of waist circumference where the beginning of the tape meets the part of the tape you have placed around your waist.

NAME: _____

Worksheet 8
Physical Activity Assessment

Instructions:

1. On the form below, record the types of physical activities you perform in a usual day, including sleeping. Carefully estimate in minutes the amount of time you spent actively engaged in each activity. Total minutes of activity time in the day should add up to 1,440.

2. Access the Physical Activity Tool:

   a. Go to www.ChooseMyPlate.gov and select Physical Activity Tracker. From there:

   b. Login

   c. Complete Personal Profile and then select Physical Activity Tracker.

   e. Under Physical Activity Tracker enter each of the activities performed. After you enter the activity and press Go, a box will open with the direction "Enter the duration." Be sure to check the day on which you do the activity and press Add to enter the activity.

f. Print a copy of Your Physical Activity Results.

3. Staple a copy of Your Physical Activity Results printout to this form before you submit the assignment to your instructor.

Usual Day's Physical Activities

Type of Activity	Duration (minutes)

1. What is the total amount of time you spent doing moderate to vigorous physical activity?

2. What is the total amount of time you spent in sedentary activities?

3. Note here your reactions to the results of this tracking.

NAME: _____

Worksheet 9
Assessing Calcium Intake

Instructions: Enter the number of servings of foods listed that you usually consume in a day, and calculate the milligrams of calcium consumed from the foods (see example). Then add the results in the Total Calcium column to get an estimated total of your calcium intake. Subtract your total from the RDA of 1,000 mg per day to estimate how close your calcium intake is to the RDA for calcium.

Food	Serving Size	Number of Servings	Calcium/ Serving (mg)[a]	Calcium (mg)
Example: Orange juice, calcium fortified	1 cup	$1\frac{1}{2}$	× 300 =	450
			Total =	450
Yogurt	1 cup		× 400 =	
Pudding made with milk	1 cup		× 300 =	
Orange juice, calcium fortified	1 cup		× 300 =	
Milk (cow's, or calcium-fortified soy or rice)	1 cup		× 300 =	
Tofu, processed with calcium	$\frac{1}{2}$ cup		× 300 =	
Sesame seeds	1 ounce		× 300 =	
Cheese	1 ounce		× 200 =	
Ice cream	1 cup		× 200 =	
Collard greens	$\frac{1}{2}$ cup		× 150 =	
Cottage cheese	1 cup		× 150 =	
Green soybeans, cooked	$\frac{1}{2}$ cup		× 130 =	
Tahini	2 tablespoons		× 130 =	
Almonds	$\frac{1}{4}$ cup		× 100 =	
Dried beans, cooked	1 cup		× 100 =	
Turnip greens, cooked	$\frac{1}{2}$ cup		× 100 =	
Soy nuts	$\frac{1}{4}$ cup		× 100 =	
Bok choy, cooked	$\frac{1}{2}$ cup		× 80 =	
Tempeh	$\frac{1}{2}$ cup		× 80 =	
Kale, cooked	$\frac{1}{2}$ cup		× 80 =	
Okra, cooked	$\frac{1}{2}$ cup		× 70 =	
Broccoli, cooked	$\frac{1}{2}$ cup		× 50 =	
Swiss chard, cooked	$\frac{1}{2}$ cup		× 50 =	
Tortilla, 6-inch diameter	1	× 50 =		
		Total =		

[a]Values are rounded and in some cases average values for food types are given.

RDA for calcium	=	1,000 mg
Total	=	_____ mg
Difference (subtract total from RDA)	=	_____ mg

Worksheet 10
Should Herbs Be Regulated as Drugs?

Scenario: You are a consumer member of an FDA subcommittee that will provide the FDA with consumer views on whether herbal remedies should be regulated as drugs and not as food. Drugs have to pass tests related to their safety and effectiveness before they can be sold to prevent or treat disease. You are asked to prepare your responses to the following questions before the first meeting of the subcommittee takes place. What would your responses be?

1. Have you used herbal remedies to prevent or treat a disease?

☐ **Yes** ☐ **No**

If yes, did you consider the herbal remedy to be a drug or a food?

☐ **food** ☐ **drug**

2. State three reasons why herbal remedies should not be regulated as drugs.

a.

b.

c.

3. State three reasons why herbal remedies should be regulated as drugs.

a.

b.

c.

4. Should testing the safety and effectiveness of herbal products be required before they are sold to the public?

☐ **Yes**   ☐ **No**

Why or why not?

5. Would you feel the right to freedom of speech would be compromised if herbal remedies were regulated as drugs?

☐ **Yes**   ☐ **No**

Why or why not?

NAME: _____

Worksheet 11
Child Nutrition Dilemmas 1

Instructions: Consider the scenarios presented and answer the questions posed. Base your responses on the information presented in the section "Helping Children Learn the Right Lessons about Food" and other information presented in Unit 30 of the textbook.

Scenario 1: Lola is almost 3 years old and her parents want to introduce her to a variety of new foods. Tonight at dinner they decide to serve Lola beets for the first time. Lola eats the mashed potatoes and chicken that were on her plate, but she doesn't touch the beets. What should you do?

1. Develop three optional responses to the dilemma:

   a.

   b.

   c.

2. Identify the worst option and state why it is the worst one.

3. Identify the best option and state why it is the best one.

Scenario 2: Your 4-year-old son comes into the kitchen announcing he's hungry and asks for a cookie. Dinner is an hour away and you don't want him to spoil his appetite. What should you do?

1. Develop three optional responses to the dilemma:

    a.

    b.

    c.

2. Identify the worst option and state why it is the worst one.

3. Identify the best option and state why it is the best one.

NAME: _____

. . . .

Worksheet 12
Child Nutrition Dilemmas 2

Instructions: Consider the scenarios presented and answer the questions posed. Base your responses on the information presented in the section "Helping Children Learn the Right Lessons about Food" and other information presented in Unit 30 of the textbook.

Scenario 3: It's Friday night and Elesha and Raymond are excited. Their parents are taking them to the Country Cookin' Buffet for dinner. On the way to the restaurant their father tells the kids that they can choose any foods from the buffet they like, but that they should eat all the food they take. Elesha and Raymond take too much food and can't eat it all. If you were the dad, what should you do?

1. Develop three optional responses to the dilemma:

   a.

   b.

   c.

2. Identify the worst option and state why it is the worst one.

3. Identify the best option and state why it is the best one.

Scenario 4: It's Saturday afternoon and the proud parents of 6-month-old Elena are busy playing with her on the floor. After about 10 minutes of playing, Elena starts to fuss and then she cries. Her parents don't know what's wrong, but decide she must be hungry and feed her. What would you have done?

1. Develop three optional responses to the dilemma:

   a.

   b.

   c.

2. Identify the worst option and state why it is the worst one.

3. Identify the best option and state why it is the best one.

NAME: _____

Worksheet 13
Dietary Assessment Assignment

Instructions: Follow these instructions carefully when completing this assignment. You will be asked to hand in your completed Dietary Intake Recording Form, three Nutrient Intake Results printouts, and the completed Evaluation of Dietary Assessment Results form.

1. Record your dietary intake for one weekend day and two week days.

    a. Try to select days that represent your usual food intake.

    b. Carry the Dietary Intake Recording Form with you during the days you will be recording your food intake.

    c. Write down foods, beverages, and ingredients in mixed dishes, and the amount of each you consumed, on the form. Fully describe each food item.

    d. Try to record your food intake after each meal or snack.

    e. To increase the accuracy of estimates of food portion sizes, refer to Nutrition Facts panels on food labels and note the weight or measure of a standard amount of the food item. (For example, a slice of cheese may be labeled as weighing an ounce.) Serve yourself foods and beverages using a cup measure, or bowls, mugs, and glasses of known volume. Estimate diameters of round foods, such as pancakes, tortillas, and bagels using a piece of $8^{1}/_{2}"\times 11"$ notebook or tablet paper.

2. Review your dietary intake record for completeness.

3. To enter and analyze your dietary intake:

    a. Go to www.ChooseMyPlate.gov.

    (*Hint:* ChooseMyPlate Tracker can get busy and slow. *Try to avoid the peak period of usage, which is from 12 noon to 4 p.m. EST Monday through Thursday. It may take an hour to complete this analysis.*)

    b. Select SuperTracker.

    c. Log in, complete personal profile, and select Food Tracker. Note that after you complete your profile, you will receive an individualized food plan that you can refer to as you do this activity.

    d. Enter food items one at a time from your dietary intake record. After you find and select the food, enter the amount eaten and at which meal. When you have finished entering the foods, check to make sure you have entered all of them and made no entry mistakes.

    e. When all quantities have been entered and reviewed for accuracy, click the Print Page button and make a copy to hand in.

    f. Then click the Nutrient Intake button. Make a copy of this page to hand in.

    g. Repeat the above process to enter and analyze your dietary intake for the two other days.

4. Complete the Evaluation of Dietary Assessment Results form.

5. Prepare to hand in your assignment by stapling together your Dietary Intake Recording Forms, copies of the Food Record that show types and amounts of foods entered for the three days, and your three Nutrient Intakes printouts and the Evaluation of Dietary Assessment Results form.

. . . .

Evaluation of Dietary Assessment Results

Instructions: Use the results from the Nutrient Intakes printouts to complete the following activities.

1. Calculate your average intake of Calories (kcal) for the three days.

Example: Food Energy (kcal): Day 1 = 2310, Day 2 = 1990, Day 3 = 2250:
2310 + 1990 + 2250 = 6550 kcal
6550 kcal/3 = 2183 kcal per day

Your average, total energy (kcal) intake per day (show all calculations):

_____ kcal per day

2. Calculate the average percentage of Calories from carbohydrate for the three days.

a. Add daily results for Carbohydrate (g) intake. Multiply this total times 4 Calories per gram. Then, divide the total by 3 to get an average intake of Calories from carbohydrate for one day.

Example: carbohydrate (g) Day 1 = 191, Day 2 = 78, Day 3 = 245:
191 + 202 + 245 = 638 g
638 g x 4 kcal/g = 2552 kcal
2552 Calories /3 = 851 Calories (kcal) from carbohydrate per day

Your results (show all calculations):

_____ kcal from carbohydrate per day

b. Divide your average, daily intake of Calories from carbohydrate by your average, daily total energy (kcal) intake. Multiply the result by 100 to obtain the percentage.

Example: 851 /2183 kcal = 0.32:
0.32 x 100 = 39% total energy (kcal) from carbohydrate
Your results (show all calculations):

_____% total energy (kcal) from carbohydrate

3. Calculate the average percentage of Calories from protein for the three days.

    a. Add daily results for protein (g) intake. Multiply this total times 4 Calories per gram. Then, divide the total by 3 to get an average intake of Calories from protein for one day.

       Your results (show all calculations):

       _____ kcal from protein per day

    b. Divide your average daily intake of Calories from protein by your average daily total energy (kcal) intake. Multiply the result by 100 to obtain the percentage.

       Your results (show all calculations):

       _____% total energy (kcal) from protein

4. Calculate the average percentage of Calories from fat in your diet.

    a. Add daily results for total fat (g) intake from the three Nutrient Intakes printouts. Multiply this total times 9 Calories per gram. Then, divide the total by 3 to get an average intake of Calories from fat for one day.

       Your results (show all calculations):

       _____ kcal from total fat per day

    b. Divide your average daily intake of Calories from fat by your average daily total energy (kcal) intake. Multiply the result by 100 to obtain the percentage.

       Your results (show all calculations):

       _____% total energy (kcal) from fat

5. Add the average daily percent of Calories from carbohydrate, protein, and fat together.

    Your results (show the calculation):

    a. Does the result add up to 100%?   ☐ **Yes**   ☐ **No**

    b. If the sum is not 100%, and your calculation is off by more than 2%, recalculate the percentages of Calories from carbohydrate, protein, and fat. The percentages should be 100% or within a percent or two of 100%.

6. Calculate the average percentage of Calories from saturated fat in your diet.

    a. Add daily results for saturated fat (g) intake from the three Nutrient Intakes printouts. Multiply this total times 9 Calories per gram. Then, divide the total by 3 to get an average intake of Calories from saturated fat for one day.

    Your results (show all calculations):

    _____ kcal from saturated fat per day

    b. Divide your average daily intake of Calories from saturated fat by your average daily total energy (kcal) intake. Multiply the result by 100 to obtain the percentage.

    Your results (show all calculations):

    _____% of Calories (kcal) from saturated fat

7. Calculate the percent of Calories (kcal) from mono- and polyunsaturated fats in your average daily diet by subtracting your percent of Calories (kcals) from saturated fat from the percent of Calories (kcal) consumed from fat.

    Your results (show all calculations):

    _____% of Calories (kcal) from mono- and polyunsaturated fats

8. Calculate your average, daily intake of cholesterol.

    Your results (show all calculations):

    _____ mg cholesterol per day

9. Complete the following table:

    a. Record your average daily intake of food energy (kcal), and the percent of Calories from carbohydrate, protein, fat, saturated fat, and mono- and polyunsaturated fat determined previously in the table below.

    b. Record average daily cholesterol intake previously determined.

c. Record daily potassium and sodium intake in mg, then calculate and record average daily intake in mg.

d. Record the percent figures in the Percent of Recommended Intake column (the third column) on the Nutrient Intakes printout to fill in the blanks for vitamins and minerals. Calculate and record average daily percent of recommended intakes (% rec) for each vitamin and mineral listed in the table.

| | Percent of Recommended Intake | | | |
	Day 1	Day 2	Day 3	Average Daily Intake
Food energy (kcal)				
Protein (% kcal)				
Carbohydrate (% kcal)				
Dietary fiber (%)				
Total fat (% kcal)				
Saturated fat (% kcal)				
Mono- and polyunsaturated fat (% kcal)				
Cholesterol (mg)				
Vitamin A (% rec)				
Vitamin E (% rec)				
Vitamin C (% rec)				
Thiamin (% rec)				
Riboflavin (% rec)				
Niacin (% rec)				
Folate (% rec)				
Vitamin B6 (% rec)				
Vitamin B12 (% rec)				
Calcium (% rec)				
Iron (% rec)				
Magnesium (% rec)				
Phosphorus (% rec)				
Zinc (% rec)				
Potassium (mg)				
Sodium (mg)				

10. List two strong points of your diet based on the results of your dietary analysis.

   a. _____   b. _____

11. List two weak points of your diet based on the results of your dietary analysis.

   a. _____   b. _____

12. Answer the following questions about your average daily dietary intakes.

   a. Does your intake of protein exceed 100% of the recommended intake?

      ☐ Yes   ☐ No

   b. Does the percentage of Calories you consume from total fat fall within the recommended range of 20–35% of total Calories?

      ☐ Yes   ☐ No

   c. Is the percentage of Calories you consume from saturated fat less than the recommended limit of 7% of total Calories?

      ☐ Yes   ☐ No

   d. Do you consume over 300 mg of cholesterol daily?

      ☐ Yes   ☐ No

   e. Does your average daily intake of potassium fall within the range of 3,525–5,875 mg? (The DRI is 4,700 mg per day.)

      ☐ Yes   ☐ No

   f. Is your average daily intake of sodium less than 2,300 mg? (The DRI is 1,500 mg per day.)

      ☐ Yes   ☐ No

   g. List the vitamins consumed in amounts over twice the recommended intake level.

   h. List vitamins consumed in amounts under 75% of the recommended intake level.

   i. List minerals consumed in amounts over twice the recommended intake level.

   j. List minerals consumed in amounts under 75% of the recommended intake level.

   Prepare your forms for submission (see point 5 of the Dietary Assessment instructions).

Congratulations! You have completed your dietary assessment.

Dietary Intake Recording Form – Day _____

Food/Beverage	Description	Quantity